Common Fragment Ions

m/z	Ion	Origin	m/z	Ion	Origin
29	HCO^+	Aldehydes	85	(tetrahydropyran ion)	Tetrahydropyranyl ethers
30	$CH_2NH_2^+$	Amines			
31	$H_2C=\overset{+}{O}H$ CH_3O^+	Alcohols Methyl esters	88	$CH_3OCO-CH$ $\overset{+}{N}H_2$	Amino acid esters
36	$HCl^{+\bullet}$	Chloro compounds			
43	$C_3H_7^+$ CH_3CO^+	Propyl ion Acetyl groups	91	$C_7H_7^+$	Aromatic hydrocarbons with side chains
47	CH_3S^+	Sulfides			
49	CH_2Cl^+	Chloro compounds	92	$C_7H_8^{+\bullet}$	Benzyl compounds with a γ-hydrogen
55	$C_4H_7^+$	Alkyl groups	95	(furyl-CO ion)	Furyl-CO-X
57	$C_4H_9^+$ $C_2H_5CO^+$	Alkyl groups Acylium ion			
58	$[H_2C\overset{H}{CO}CH_3]^{+\bullet}$	Ketones with a γ-hydrogen	97	(thiophene ion)	Alkyl thiophenes
59	$[COOCH_3]^+$	Methyl esters	99	(dioxolane ion)	Ethylene ketals of cyclic compounds (steroids)
61	$CH_3\overset{OH}{C}=O^+H$	Esters of high molecular weight alcohols			
70	(pyrrolidine ion)	Pyrrolidines	104	$C_8H_8^{+\bullet}$	Alkyl aromatics
74	$[CH_2=\overset{OH}{C}-OCH_3]^{+\bullet}$	Methyl esters with a γ-hydrogen	105	$C_6H_5CO^+$ $C_8H_9^+$	Benzoyl compounds Aromatic hydrocarbons
77	$C_6H_5^+$	Aromatics	106	$CH_2=\overset{+}{N}H_2$	Amino benzyl
78	$C_5H_4N^+$	Pyridines and alkyl pyrroles	107	$C_7H_7O^+$	Phenolic hydrocarbons
80	(pyrrole ion)	Pyrroles	117	$C_9H_9^+$	Styrenes
			128	$HI^{+\bullet}$	Iodo compounds
80(82)	$HBr^{+\bullet}$	Bromo compounds	130	(indole ion)	Indoles
81	(furan ion)	Furans	131	(cinnamate CO ion)	Cinnamates
	(aliphatic chain ion)	Aliphatic chain with two double bonds	149	(phthalate ion)	Dialkyl phthalates (rearrangement)
83	$C_6H_{11}^+$ $CHCl_2^+$	Cyclohexanes or hexenes Chloro compounds			

"An investment in knowledge always pays the best interest."

~Benjamin Franklin

"Either write something worth reading or do something worth writing."

~Benjamin Franklin

INTRODUCTION TO MASS SPECTROMETRY

Instrumentation, Applications and Strategies for Data Interpretation

FOURTH EDITION

J. THROCK WATSON
Professor of Biochemistry
and of Chemistry
Michigan State University
East Lansing, Michigan

O. DAVID SPARKMAN
Adjunct Professor of Chemistry
College of the Pacific
University of the Pacific
Stockton, California

WILEY

John Wiley & Sons, Ltd

Other Wiley Editorial Offices
John Wiley & Sons Inc., 111 River Street, Hoboken, NJ 07030, USA
Jossey-Bass, 989 Market Street, San Francisco, CA 94103-1741, USA
Wiley-VCH Verlag GmbH, Boschstr. 12, D-69469 Weinheim, Germany
John Wiley & Sons Australia Ltd, 42 McDougall Street, Milton, Queensland 4064, Australia
John Wiley & Sons (Asia) Pte Ltd, 2 Clementi Loop #02-01, Jin Xing Distripark, Singapore 129809
John Wiley & Sons Ltd, 6045 Freemont Blvd, Mississauga, Ontario L5R 4J3, Canada
Wiley also publishes its books in a variety of electronic formats. Some content that appears in print may not be
available in electronic books.
Anniversary Logo Design: Richard J. Pacifico

Library of Congress Cataloging-in-Publication Data
Watson, J. Throck.
Introduction to mass spectrometry : instrumentation, applications, and
strategies for data interpretation / J. Throck Watson, O. David Sparkman.
-- 4th ed.
p. cm.
Includes index.
ISBN 978-0-470-51634-8 (cloth)
1. Mass spectrometry. 2. Biomolecules--Analysis. I. Sparkman, O. David
(Orrin David), 1942- II. Title.
QC454.M3W38 2007
543'.65--dc22
2007024030
British Library Cataloguing in Publication Data
A catalogue record for this book is available from the British Library
ISBN 9780470516348
Typeset by the authors
Printed and bound in Great Britain by Antony Rowe, Chippenham, Wiltshire
This book is printed on acid-free paper responsibly manufactured from sustainable forestry
in which at least two trees are planted for each one used for paper production.

Contents

Chapter 5 Strategies for Data Interpretation

PREFACE

This edition of *Introduction to Mass Spectrometry* is far more than a revision of the third edition, which appeared in 1997. Completely updated and more than 75% rewritten, it covers strategies for data interpretation, fundamental operating principles of instrumentation, and representative applications for all areas of organic, environmental, and biomedical mass spectrometry. A majority of the chapters have bibliographies containing several hundred references to research articles and reviews, mostly published since 2000. Most chapters, but especially the first two, provide a historical perspective on the development of mass spectrometry as well as commentary on the evolution of commercial developments of the instrumentation. Careful attention to nomenclature is provided throughout the book. In addition to serving as a general reference for the subject of mass spectrometry as it pertains to organic and biochemistry, this book is designed for use as a textbook for courses on mass spectrometry. The readily comprehensible approach to the topic, honed through the teamwork of the coauthors in teaching hundreds of classes on various aspects of mass spectrometry for nearly 30 years under the auspices of the American Chemical Society, will benefit the reader.

The physical instrument is dissected and described in Chapter 2 in a systematic manner from the ion source through ion guides to the *m/z* analyzer to the detection system with attention to the vacuum system. The fundamental physics for each type of *m/z* analyzer, as well as for common detectors and vacuum pumps, are provided together with a "common sense" description of the operating principles of each.

Chapter 3 describes the concept of MS/MS with emphasis on collisionally activated dissociation. *Tandem-in-space* is distinguished from *tandem-in-time*, and several qualitative and quantitative applications of both types of technology are presented in the context of environmental and biomedical fields. In addition, information on analyte identification from MS/MS is provided along with explanations and sources of spectral databases and how to use them.

Various means of transporting the sample into the low-pressure environment of the mass spectrometer are described in Chapter 4. The operating mechanics of "batch" inlet systems as well as continuous sampling systems are presented together with representative and/or illustrative examples. Descriptions of nonchromatographic continuous inlets include DART, DESI, DAPCI, SIFT, MIMS, CRIMS, pyrolysis, electrophoresis, laser ablation, continuous-flow FAB, and ICP. Continuous inlets in combination with chromatography include SFC and pyrolysis GC and are presented in Chapters 10 and 11, respectively.

A general strategy for interpretation of a mass spectrum, regardless of the type of ionization involved, is presented in Chapter 5. The *Nitrogen Rule* is introduced and used in a variety of situations. The importance of isotope peak-intensity ratios is introduced; several carefully detailed examples are described that show the relationship between isotope peak-intensity ratios and the elemental composition of the corresponding ion. The basis for recognizing peaks representing odd-electron vs even-electron ions is introduced; the importance of recognizing such ions is illustrated with appropriate examples of mass spectra resulting from a variety of ionization types, including EI, CI, and electrospray.

Chapter 6 is one of the highlights of the book, providing a solid introduction to the formation, appearance, and interpretation of EI mass spectra. Emphasis is placed on recognizing the most probable site of electron deficiency (site of the +•, the "plus/dot") in the molecular ion, which is the precursor of a majority of the ions represented by the fragmentation pattern in an EI mass spectrum. Four major pathways of fragmentation of a molecular ion (sigma-bond cleavage, homolytic cleavage, heterolytic cleavage, and hydrogen-shift rearrangements) are introduced in a clear manner, then supported systematically with nearly 100 fragmentation schemes to facilitate interpretation of dozens of representative mass spectra of various types of compounds. This chapter also includes detailed information on EI mass spectral databases and library search programs along with descriptions of their use.

The basis for chemical ionization is described in Chapter 7. Whereas positive-ion formation is emphasized, attention is also given to negative-ion formation, with careful distinction between negative-ion CI (NCI, the result of an ion/molecule reaction involving an anion) and electron capture negative ionization (ECNI), a resonant process involving capture of a thermal electron. Atmospheric pressure CI (APCI) is introduced, which serves as an important interface for LC/MS applications that are not amenable to electrospray ionization. The specialized technique of desorption CI (DCI) is also described. Descriptions of the various types of CI are supported with illustrative examples of application to environmental and biomedical problems.

The operating principles of electrospray ionization (ESI) are described in Chapter 8 together with some of the mechanical aspects of the interface that make it one of the most viable for LC/MS applications. The basis for automated computation of the mass of the analyte is illustrated in an example that dissects the peaks in an ESI mass spectrum and sets up simultaneous equations based on first principles relating to the *m/z* value of the mass spectral peaks. Although introduced in Chapter 4, the developing technique of DESI is covered. Many current applications of the ESI technology are reviewed, which results in more than 300 references in this chapter.

The operating principles of matrix-assisted laser desorption/ionization (MALDI) are described in Chapter 9, including some commentary on current theories of the mechanism of ionization. Attention is given to sample preparation, including descriptions of specialized sample probes to facilitate sample cleanup. Examples of typical MALDI spectra are described to illustrate the effect and use of delayed extraction and ion mirrors (reflectrons). The technology of atmospheric pressure MALDI (AP MALDI) is described. Many current applications of MALDI technology are reviewed, also making this chapter rich in citations (more than 500).

Chapter 10 describes the basis for trade-offs in individual operation of GC and MS that are necessary for successful operation of the combined technique. Introductory protocols for proper syringe/sample handling in the mature technology of GC/MS are presented. The important technology of selected ion monitoring (SIM) is described in the context of qualitative and quantitative applications in the biomedical and environmental fields. Strategies and procedures for data processing with mass chromatograms are described in the context of suspected overlapping data obtained from samples containing chromatographically unresolved components. Some current applications of the technology are reviewed along with explanations of software used for component deconvolution through processing complex data. This chapter has nearly 200 references.

Chapter 11 on LC/MS emphasizes how conventional protocols of HPLC operation must be modified to become compatible with MS operation for combined operation. Although electrospray is the dominant interface for LC/MS, the specialized ionization techniques of APCI and APPI (atmospheric pressure photoionization), which lend themselves to particular applications, are given serious consideration. Several current applications of LC/MS technology are reviewed resulting in almost 250 citations.

Methodology for proteomics is emphasized in Chapter 12, which also describes some basic approaches to the characterization of carbohydrates and nucleotides. The strategy and procedure for sequencing a peptide from CAD MS/MS data are described in detail as supported by results for a simple didactic example. The concept of peptide mass mapping is described, which is the basis, sometimes in combination with data from CAD MS/MS, for automated identification of proteins by software that is often purchased as part of a data system or that is used in conjunction with notable Web sites for such purposes. Methodology for identifying/characterizing a variety of post-translational modifications to proteins, including phosphorylation and disulfide-bond formation, is described in the context of several step-by-step examples. Hundreds of current applications are reviewed, bringing the number of references in this chapter to more than 800.

Because the book is designed for use as a textbook for courses on mass spectrometry, Power Point presentations, including figures from the book and animations developed by the authors, are available for downloading to site-registered instructors to support their teaching efforts. For the benefit of students, the authors will maintain a Web site (through and with the support of the publisher) that will contain exercises together with downloadable answer keys. These materials will be updated on a regular basis.

"Determination is often the first chapter in the book of excellence."

~Unknown

ACKNOWLEDGMENTS

Many of the realistic examples of mass spectral data and applications of mass spectrometry derive from experiments conducted in the Watson Laboratory by some 50 Ph.D. graduate students or postdoctoral fellows in the context of biomedical research applications. Recent contributors include graduate assistants Xue Li, Jose-Luis Gallegos-Perez, Nalini Sadagopan, Naxing Xu, Yingda Xu, Wei Wu, Jianfeng Qi, David Wagner, and professorial students Heidi Bonta, Brad Sauter, and Greg Boyd. Other illustrative applications of mass spectrometry derive from the Sparkman Laboratory, with the help of Teresa Vail and Matthew Curtis, in the context of environmental chemistry and computational approaches to preparing and interrogating standard libraries of mass spectra. Thanks to Leslie Behm and Susan Kendall in the MSU Library System for assistance and counsel to JTW in dealing with the vagaries of the EndNote algorithm.

The integrity of information and data interpretation contained herein has been bolstered by critiques from prominent colleagues in the field, including Professors Gavin Reid, John Allison, Jack Holland, Vernon Reinhold, Robert Brown, J.A. McCloskey, A. Daniel Jones, and Drs. Christian Rolando, J. Lemoine, Steven Pomerantz, Charles Ngowe, Chad Borges, Robin Hood, John Stults, and J. David Pinkston.

A special thanks to Professor Jean-Francois Gal, who generously provided lab/office space for JTW at the University of Nice for his sabbatical leave in 2002 during the formative stages of this project. The logical and systematic approach to presenting scientific/technical information that JTW learned from his mentor, Professor Klaus Biemann at MIT, was of continuing benefit during this project, and some of the critique/suggestions by Dr. Brian Sweetman and Professor John Oates at Vanderbilt University during preparation of the first edition of the book survive in this fourth edition. Thanks also to Patrick R. Jones, ODS's colleague at the University of the Pacific, for great discussions and reflections especially on instrumentation and physical chemistry. We appreciate the counsel of Frederick E. Klink, who has been our co-instructor in short courses for the last 10 years, and who has greatly added to the portions of this book involving HPLC and LC/MS. We also appreciate the cooperation of Harold G. Walsh, Director of the ACS Short Course Program, during our tenure from 1978 to 2006 with his program, and to the thousands of students who have participated in these short courses as well as the instrument manufacturers who provided equipment and other support for our "hands-on" courses.

Special thanks go to Stephen E. Stein at the Mass Spectrometry Data Center of the National Institute of Standards and Technology for permission to use many of the EI mass spectra, which come from the NIST05 NIST/EPA/NIH Mass Spectral Database. Unless otherwise designated, spectra were taken from the NIST Mass Spectral Database. Also, the NIST Mass Spectral Search Program proved invaluable in the preparation of many of the non-EI mass spectra contained in this book.

Both authors offer a special thanks to ODS's wife, Joan A. Sparkman, who spent a great number of hours implementing the suggestions of the Wiley contract-copyeditor and making sure that there was consistency throughout the book. Because of the sometimes orthogonally opposed styles of the authors, the inputs of the copyeditor and the proofreader, and further complications with the delivery of camera-ready copy, Joan has paraphrased the title of a song, saying that the style of this book can be considered "a little bit country, a little bit rock 'n roll". Thanks also goes to those two great canine mass spectrometrists, Maggie and Chili Sparkman, who endured the final edits with ODS and Joan.

"I feel sure that there are many problems in Chemistry which could be solved with far greater ease by the application of Positive Rays to chemical analysis than by any other method."

~Joseph John Thomson

Introduction to Mass Spectrometry

Instrumentation, Applications

and Strategies for Data Interpretation

<u>Fourth Edition</u>

Chapter 1   Introduction

Introduction to Mass Spectrometry, 4th Edition: Instrumentation, Applications, and Strategies for Data Interpretation; J.T. Watson and O.D. Sparkman, © 2007, John Wiley & Sons, Ltd

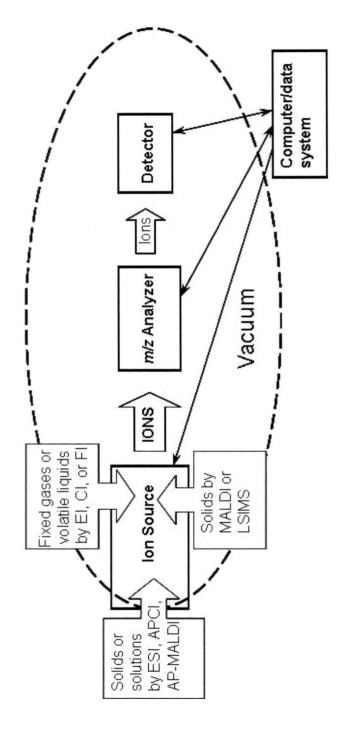

Figure 1-1. This conceptual illustration of the mass spectrometer shows the major components of the instrumentation, i.e., sample inlets (dependent on sample and ionization technique; ion source (origin of gas phase ions); m/z analyzer (portion of instrument responsible for separation of ions according to their individual m/z values); detector (generates the signals that are a recording of the m/z values and abundances of the ions); vacuum system (the components that remove molecules, thereby providing a collision-free path for the ions from the ion source to the detector); and the computer (coordinates the functions of the individual components and records and stores the data).

I. Introduction

Mass spectrometry is a microanalytical technique that can be used selectively to detect and determine the amount of a given analyte. Mass spectrometry is also used to determine the elemental composition and some aspects of the molecular structure of an analyte. These tasks are accomplished through the experimental measurement of the mass of gas-phase ions produced from molecules of an analyte. Unique features of mass spectrometry include its capacity for direct determination of the nominal mass (and in some cases, the molar mass) of an analyte, and to produce and detect fragments of the molecule that correspond to discrete groups of atoms of different elements that reveal structural features. Mass spectrometry has the capacity to generate more structural information per unit quantity of an analyte than can be determined by any other analytical technique.

Much of mass spectrometry concerns itself with the mass of the isotopes of the elements, not the *atomic mass*[1] of the elements. The *atomic mass* of an element is the weighted average of the naturally occurring stable isotopes that comprise the element. Mass spectrometry does not directly determine mass; it determines the *mass-to-charge ratio* (*m/z*) of ions. More detailed explanations of *atomic mass* and *mass-to-charge ratios* follow in this chapter.

It is a fundamental requirement of mass spectrometry that the ions be in the gas phase before they can be separated according to their individual *m/z* values and detected. Prior to 1970, only analytes having significant vapor pressure were amenable to mass spectrometry because gas-phase ions could only be produced from gas-phase molecules by the techniques of electron ionization (EI) or chemical ionization (CI). Nonvolatile and thermally labile molecules were not amenable to these otherwise still-valuable gas-phase ionization techniques. EI (Chapter 6) and CI (Chapter 7) continue to play very important roles in the combined techniques of gas chromatography/mass spectrometry (GC/MS, Chapter 10) and liquid chromatography/mass spectrometry (LC/MS, Chapter 11). After 1970, the capabilities of mass spectrometry were expanded by the development of desorption/ionization (D/I) techniques, the generic process of generating gas-phase ions directly from a sample in the condensed phase. The first viable and widely accepted technique[2] for D/I was fast atom bombardment (FAB), which required nanomoles of analyte to produce an interpretable mass spectrum. During the 1980s, electrospray ionization (ESI) and matrix-assisted laser desorption/ionization (MALDI) eclipsed FAB, in part because they required only picomoles of analyte for analysis. ESI and MALDI are mainly responsible for the dominant role of mass spectrometry in the biological sciences today because they are suitable for analysis of femtomole quantities of thermally labile and nonvolatile analytes; therefore, a chapter is devoted to each of these techniques (Chapters 8 and 9).

Mass spectrometry is not limited to analyses of organic molecules; it can be used for the detection of any element that can be ionized. For example, mass spectrometry can analyze silicon wafers to determine the presence of lead and iron, either of which can

---

[1] In the United States, the term *atomic weight* is used for the relative mass of the elements. In the rest of the world, which is based on the metric system, the term *atomic mass* is used. This book uses the term *atomic mass* instead of the more widely accepted term in the U.S., *atomic weight*.

[2] It should be mentioned that the techniques of ^{252}Cf (Ron MacFarlane) and Laser Microprobe Mass Analysis (LAMMA) (Franz Hillenkamp and Michael Karas) were less popular D/I techniques that were developed in the same temporal arena as FAB, but they were not commercially viable. More information on these two techniques can be found in Chapter 9.

cause failure of a semiconductor for microprocessors; similarly, drinking water can be analyzed for arsenic, which may have health ramifications. Mass spectrometry is extensively used in geology and material sciences. Each of these two disciplines has developed unique analytical capabilities for the mass spectrometer: isotope ratio mass spectrometry (IRMS) in geology and secondary ion mass spectrometry (SIMS) in material sciences. Both of these techniques, along with the analysis of inorganic ions, are beyond the scope of this present book, which concentrates on the mass spectrometry of organic substances.

1. The Tools and Data of Mass Spectrometry

The tools of mass spectrometry are *mass spectrometers*, and the data are *mass spectra*. Figure 1-1 is a conceptual representation of a mass spectrometer. Each of the individual components of the instrument will be covered at logical stages throughout this book. Figure 1-2 depicts the three ways of displaying the data recorded by the mass spectrometer. The acquired mass spectra can be displayed in many different ways, which allow the desired information about the analyte to be easily extracted. These various techniques for data display and their utility are covered later in this chapter.

2. The Concept of Mass Spectrometry

Ions are charged particles and, as such, their position in space can be manipulated with the use of electric and magnetic fields. When only individual ions are present, they can be grouped according to their unique properties (mass and the number of charges) and moved from one point to another. In order to have individual ions free from any other forms of matter, it is necessary to analyze them in a vacuum. This means that the ions must be in the gas phase. Mass spectrometry takes advantage of ions in the gas phase at low pressures to separate and detect them according to their mass-to-charge ratio (m/z) – the mass of the ion on the atomic scale divided by the number of charges that the ion possesses. This definition of the term m/z is important to an understanding of mass spectrometry. It should be noted that the m/z value is a dimensionless number. The m/z term is always used as an adjective; e.g., the ions with m/z 256, or the ion has an m/z value of 256. A recording of the number of ions (abundance) of a given m/z value as a function of the m/z value is a mass spectrum. Only ions are detected in mass spectrometry. Any particles that are not ionic (molecules or radicals[3]) are removed from the mass spectrometer by the continuous pumping that maintains the vacuum.

The mass component that makes up the dimensionless m/z unit is based on an atomic scale rather than the physical scale normally considered as mass. Whereas the mass physical scale is defined as one kilogram being the mass of one liter of water at a specific temperature and pressure, the atomic mass scale is defined based on a fraction of a specific isotope of carbon; i.e., 1 mass unit on an atomic scale is equal to 1/12 the mass of the most abundant naturally occurring stable isotope of carbon, ^{12}C. This definition of mass, as represented by the symbol u, which is synonymous with dalton (Da), will be used throughout this book [1].

A previous standard for the atomic mass unit was established in chemistry in 1905 (based on the earlier suggestion of the Belgium chemist, Jean Servais

---

[3] Both molecules and radicals are particles that have no charge. Molecules are characterized by an even number of electrons and radicals by an odd number of electrons.

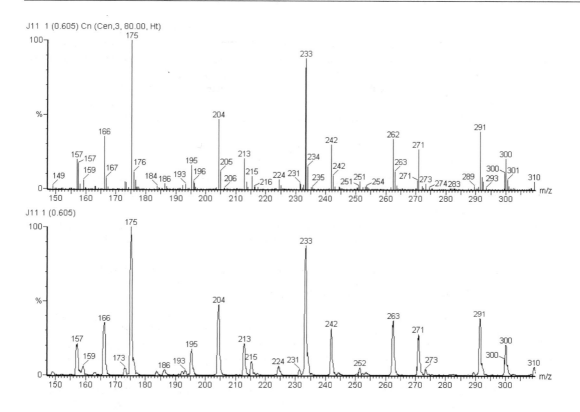

Figure 1-2. *The top part of this figure is a bar-graph presentation of a mass spectrum; this is the presentation most often used for data acquired by GC/MS. The middle display is the same mass spectrum presented in profile mode; this type of display is often used with LC/MS data because the mass spectral peaks represent ions of different m/z values that may not be well resolved by the mass spectrometer, such as is sometimes the case with multiple-charge data. The third way spectra are displayed is in a tabular format (not shown). The tabular format is a listing of pairs of m/z values and intensities. Often mass spectral peaks of significant intensity are observed in the tabular display, but not in a graphical display because of its limited resolution. The graphical displays provide the general mass spectral image of the analyte; the tabular display provides the mass spectral details.*

Stas, 1813–1891) when it was agreed that all masses would be relative to the atomic mass of oxygen. This later became known as the "chemistry mass scale". By setting the atomic mass of oxygen to an absolute value of 16, it was relatively easy to determine the atomic mass of new elements (in the form of their oxides) as they were discovered. Francis William Aston (British physicist and 1922 chemistry Nobel laureate for the development of the mass spectrograph and the measurement of the nuclides of the elements, 1877–1945) realized that the "chemistry mass scale" was not usable with his mass spectrograph (a device used to determine the existence of individual isotopes of the elements) because, *rather than dealing with the atomic mass of elements*, he *was measuring the mass of individual isotopes*, and oxygen had three naturally occurring stable isotopes, the most abundant of which accounted for only 99.76%. Therefore, ca. 1920, Aston established the "physics mass scale" by declaring the exact mass of the most abundant stable isotope of oxygen, ^{16}O, to be 16. This meant that there were now two different definitions for the *atomic mass unit* (amu). In one case, 1 amu was equal to 1/16 the mass of ^{16}O (the physics mass scale) and, in the second case, 1 amu was 1/16 the weighted average of the three naturally occurring isotopes of oxygen [1]. The amu on the physics mass scale was a factor of 1.000275 greater than that on the chemistry mass scale. This created confusion. Based on the 1957 independent recommendations of D. A. Ölander and A. O. Nier, the International Union of Physicists at Ottawa in 1960 and the International Union of Chemists at Montreal in 1961 adopted the *carbon-12 standard*, which, as stated above, establishes a single *unified atomic mass unit* (u) as 1/12 the most abundant naturally occurring stable isotope of carbon (^{12}C). At the same time, to keep from having three different values associated with the amu term, the symbol for the *unified atomic mass unit* was established as u [1]. Unfortunately, an atomic mass unit based on carbon-12 is incorrectly assigned the amu symbol in many textbooks with current copyright dates.

Another symbol used for the unified atomic mass unit is Da, dalton. Although not officially recognized by any standard governing boards, the dalton has become an accepted standard term for the unified atomic mass unit. This arbitrary scale for the atomic mass unit is closely related to that established ca. 1805 by John Dalton (1766–1844), which assigned a value of 1 for the lightest element, hydrogen [2]. In 1815, the Swedish scientist, Jöns Jacob Berzelius (1779–1848), set the atomic mass of oxygen to 100 in his table of atomic masses [3]; however, the Berzelius standard of mass was never adopted by others.

In the study of mass spectrometry, it is important to always keep in mind that the entity measured in the mass spectrometer is the mass-to-charge ratio of an ion, not the mass of the ion. In the case where there is only a single charge on the ion, the *m/z* value and the mass are the same. This statement is not true in the case where ions have multiple charges. It is inappropriate to use a unit of mass when describing the mass-to-charge ratio of an ion. Ions have both mass and an *m/z* value.

The mass spectrometer first must produce a collection of ions in the gas phase. These ions are separated according to their *m/z* values in a vacuum where the ions cannot collide with any other forms of matter during the separation process. The functionality of the mass spectrometer's vacuum system and its components are described in Chapter 2. Ions of individual *m/z* values are separated and detected in order to obtain the mass spectrum. Separation of ions in an evacuated environment is mandatory. If an ion collides with neutrals in an elastic collision during the ion separation process, the ion's direction of travel could be altered and the ion might not reach the detector. If an ion's collision with a neutral is inelastic, sufficient energy transfer may

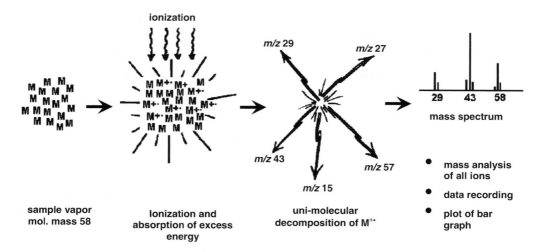

Figure 1-3. *Conceptual illustration of gas-phase ionization of analytes followed by ion separation according to the m/z value.*

cause it to decompose, meaning that the original ion will not be detected. Close encounters between ions of the same charge can cause deflection in the path of each. Contact between ions of opposite charge sign will result in neutralization.

Figure 1-3 is a conceptual illustration of the entire process of a mass spectral analysis by electron ionization (EI), culminating in a bar-graph mass spectrum that is often seen in published literature. In this illustration, M represents molecules of a pure compound in the gas phase. For ionization to occur in the gas phase, the sample must have a vapor pressure greater than 10 Pa because molecules of the sample must migrate by diffusion from the inlet system into the ionization chamber. For EI, samples may be introduced into the mass spectrometer using a direct probe or a batch inlet for pure solids or volatile liquids. Analytes purified by separation techniques (GC, LC, CE, etc.) can enter the mass spectrometer as the separation takes place in an on-line process. As the neutral molecules randomly diffuse throughout the ion source, only a few hundredths to a few thousandths of a percent of them are ionized.

The most common ionization process for gas-phase analysis, EI, transfers energy to the neutral molecule (a species characterized as having an even number of electrons) in the vapor state, giving it sufficient energy to eject one of its own electrons, thereby leaving a residual positive charge on the now ionic species. This process produces a molecular ion with a positive charge and odd number of electrons, as represented by the $M^{+\bullet}$ in Figure 1-3. This $M^{+\bullet}$ may have considerable excess energy that can be dissipated through fragmentation of certain chemical bonds. Cleavage of various chemical bonds leads to the production of positive-charge fragment ions whose mass is equal to the sum of the atomic masses of the constituent atoms. Not all of the molecular ions necessarily decompose into fragment ions. For compounds producing a relatively stable $M^{+\bullet}$, such as those stabilized through resonance, like aromatic compounds, an intense molecular-ion peak will be recorded because the $M^{+\bullet}$ tends to survive or resist fragmentation. For compounds that do not produce stable molecular ions, like aliphatic alcohols, nearly all of them decompose into fragment ions. In these cases, the mass spectrum contains only a small peak representing the $M^{+\bullet}$. Various combinations of the above-described processes are the basis of the chemical "fingerprint" in the form of a mass spectrum for a given compound.

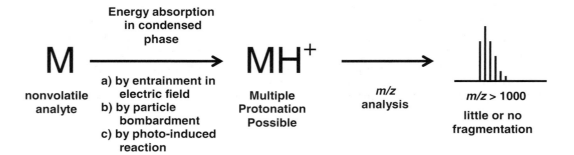

Figure 1-4. *Conceptual illustration of generic condensed-phase analysis (desorption/ionization) by mass spectrometry.*

Although not manipulable or directly detected in the mass spectrometer, radicals and molecules (neutral species) formed during the fragmentation of ions are represented in the mass spectrum. This is the *dark matter* of the mass spectrum. The difference between two *m/z* values (that of the precursor and that of the product) indicates that an ion of lesser mass was formed by the loss of a radical (e.g., a $^\bullet CH_3$ which has a mass of 15 Da) or the loss of a molecule (e.g., H_2O (mass 18 Da) or NH_3 (mass 17 Da)). The exact mass of the *dark matter* is as important as the exact *m/z* value of an ion. This exact mass allows for differentiation between the loss of a $H_3C^\bullet CH_2$ radical and a $H^\bullet C=O$ radical. In this way, the *dark matter* is an important part of the chemical fingerprint of an analyte.

For nonvolatile analytes, ions of the intact molecule are produced during entrainment of a solution into an electric field (see Chapter 8 on Electrospray) or through interaction with a photon-energized matrix compound (see Chapter 9 on Matrix-Assisted Laser Desorption/Ionization). The nonvolatile molecules are ionized by adduct formation usually involving a proton through a wide variety of processes, as summarized in an oversimplified fashion in Figure 1-4.

Mass spectrometry involves many different techniques for producing gas-phase ions. Ionization can take place in either the gas or condensed phase; however, the end result is ions in the gas phase. Many of these different ionization techniques are discussed in this book with respect to when to use them and the details of how they work. Today, ions are separated according to their *m/z* values by six specific types of *m/z* analyzers: magnetic sectors; transmission quadrupoles; quadrupole ion traps (both linear and three-dimensional); time-of-flight (TOF) analyzers (both linear and reflectrons); ion cyclotron resonance mass spectrometers (magnetic ion traps) that Fourier transform oscillating image currents to record the mass spectrum (FTICR); and most recently, by orbitraps, which store ions using electrostatic fields and, like the ICR, detect them through Fourier transformation of oscillating image currents. The operation of all these types of instruments is described in Chapter 2.

Mass spectrometers do not measure a physical property on an absolute scale such as is measured in various spectroscopies. For example, in infrared or ultraviolet spectroscopy, the absorbance or transmittance at a specific wavelength of electromagnetic energy is measured, and appropriate instrument parameters are calibrated and set during construction and final testing of the hardware. In mass spectrometry, ions are analyzed under control of the *m/z* analyzers that require day-to-day verification or calibration. The *m/z* value of each ion as a function of instrument settings can only be determined by calibrating the entire *m/z* scale of the instrument using substances that produce ions of known *m/z* values. The requirement of calibration and instrument suitability is more important in the mass spectrometer than in other spectra-producing instruments. Methods of calibrations of different types of *m/z* analyzers and different types of ion sources are detailed in Chapter 2. Descriptions of different methods of ion detection and various vacuum systems are found in Chapter 2. However, discussions of different methods of obtaining gas-phase ions are covered in several chapters.

II. History

Most chronicles of science are somewhat divided on whom to give credit for the development of the mass spectrometer. Many credit Sir John Joseph Thomson (British physicist and 1906 physics Nobel laureate for the discovery of the electron, 1856–1940) (Figure 1-5, upper right). Others credit Francis William Aston (British physicist and 1922 chemistry Nobel laureate for the development of the mass spectrograph and the measurement of the isotopes of the elements, 1877–1945) (Figure 1-5, upper left); however, Thomson's curiosity about the behavior of electrical discharges under reduced pressure has its origin in the work of the German physicist, Eugene Goldstein (1850–1930) (Figure 1-5, lower right). While at the Berlin Observatory, Goldstein reported that luminous rays in a discharge tube containing gases at low pressure traveled in straight lines from holes in a perforated metal disk used as a cathode. He called the rays *Kanalstrahlen* (canal rays) [4]. In his book, *Introduction to Mass Spectrometry Instrumentation and Techniques*, John Roboz [5] calls the Goldstein paper the first mass spectrometry publication, even though the terms *mass spectra* and *mass spectrometer* were coined much later by Aston and Josef Heinvich Elizabeth Mattauch (Austrian physicist and designer of mass spectrometers) (Figure 1-5, lower left) ca. 1920 and 1926, respectively [6].

Goldstein's work was expanded upon in 1895 by Jean-Baptiste Perrin's (French physicist, 1870–1942) report that the *Kanalstrahlen* were associated with a positive charge [7]. Perrin's suggestion was confirmed by Wilhelm Carl Werner Otto Fritz Franz Wien (German physicist and 1911 physics Nobel laureate for discoveries regarding the laws of heat radiation, 1862–1928) [8, 9]. Wien's publications demonstrated that these rays were deflected in a magnetic field and that their behavior could be studied by the combined effects of magnetic and electric fields. The device used by Wien in this study of Goldstein's *Kanalstrahlens*, the *Wien Filter*, has endured longer than any of the other devices used for ion separation developed in the same era. The Wien filter is an integral part of many ion sources on modern secondary ion mass spectrometers and is a significant component of the accelerator mass spectrometer used in ^{14}C dating and other isotope studies, as well as in other low-pressure analyzers. At the same time that Wien was exploring the works of Goldstein and Perrin, Thomson was developing a device that allowed for the determination of the difference in the *e/m* of an electron and a hydrogen atom (nucleus) [10]. It was a later refinement of this apparatus (the parabola machine) that Thomson used to observe two distinct signals when looking at the *positive rays* of neon, although it took Aston some 20 years later to realize that these data represented two of the three naturally occurring stable isotopes of neon.

Wien did not pursue the possibilities of the Wien filter. Thomson's changes to his original apparatus were reportedly based in part on Wien's work [11]. Thomson's student, Aston, refined the previous two apparatuses developed by Thomson and produced the *mass spectrograph*, which he used in *mass spectroscopy*. Aston's use of the term *mass spectroscopy* was in part due to the fact that his instrument used an

Figure 1-5.   *Five people who may be considered the founding fathers of*
mass spectrometry. Clockwise from upper left: Francis
William Aston (British physicist and 1922 chemistry Nobel
laureate for the development of the mass spectrograph and
the measurement of the isotopes of the elements, 1877–1945);
Sir John Joseph Thomson (British physicist and 1906 physics
Nobel laureate for the discovery of the electron, 1856–1940);
German physicist, Eugene Goldstein (1850–1930); Josef
Heinvich Elizabeth Mattauch (Austrian physicist and designer
of mass spectrometers); and (center) Canadian-American
physicist, Arthur Jeffery Dempster (1886–1950).

arrangement of the electric and magnetic fields for ion separation that was analogous to
that of an achromatic set of prisms without lenses, which produced a spectrum of lines
such as an optical spectrograph. The term *mass spectroscopy* grew to encompass many
different types of studies involving ions and, as such, was too broad and is no longer
recommended for what is currently referred to as *mass spectrometry*. The preferred term
for techniques involved with the measurement (electrically metered output) of ions
according to their *m/z* values and their abundances is mass spectrometry.

Aston's instrumentation was ideally suited for the accurate measurement of mass of an ion relative to the mass of a standard, ^{16}O. While Aston was perfecting the mass spectrograph using electric and magnetic fields to perform velocity focusing of an ion beam, the Canadian-American physicist, Arthur Jeffery Dempster (1886–1950) (Figure 1-5, center) independently developed a single magnetic-sector instrument that employed direction focusing of constant-energy ions at the University of Chicago. Dempster's instrument provided accurate ion abundances as opposed to accurate mass measurements. This distinction between velocity-focusing and direction-focusing was the difference between the mass spectrograph and the mass spectrometer. It is interesting to note that the term *mass spectrography* was coined by Aston along with the term used to describe the data recording of such instruments, the *mass spectrum*, ca. 1920; *mass spectrometer* was a term first used by two well-known early pioneers of mass spectrometry, William R. Smythe (U.S. scientist) and Josef Heinvich Elizabeth Mattauch (Austrian physicist) ca. 1926.

Thomson, Aston, and Dempster, and to a lesser extent Wien, are considered to be the founders of the field of mass spectrometry. Many others have followed in the succeeding years up to World War II. During this time, Thomson published two editions of his popular book *Rays of Positive Electricity* [12], and Aston published two editions each of two books: *Isotopes* and *Mass Spectra and Isotopes* [13]. Other than these six volumes, three French-language books published in 1937 and 1938 [14–16], and books on negative ions by Sir Harrie Massey [17, 18] and electrical discharge in gases by Leonard Loeb [19], no other books were published regarding the field until the 1950s. The most notable of these books was published by The Institute of Physics entitled *Modern Mass Spectrometry* by G. P. Barnard. Barnard's book, one by Henry E. Duckworth, a small monograph by A. J. B. Robertson (all in the UK), a book by Heinz Ewald and Heinrich Hintenberger (translated from German into English in 1965 by the U.S. Atomic Energy Commission), a Russian-language book by G. Rich published in 1953 with a German-language version published in 1956, and the 1954 publication by Mark G. Inghram and Richard J. Hayden published by the U.S. National Academy of Science–National Research Council, Committee on Nuclear Science, Division of Physical Science, Subcommittee on Instruments and Techniques were the only books published during the 1950s other than the proceedings of three different meetings held in the United Kingdom and one held in the United States. As the privy of mass spectrometry turned from physics to organic chemistry, the data became more complex; and the 1960s saw the publication of several books dealing with data as much as, if not more than, the operating principles of the instrumentation, some of which are still relevant in dealing with today's data (Biemann, Beynon, McLafferty, Budzikiewicz, etc.). A complete list of the books published between Thomson's 1902 book and 1970 can be found at the end of this chapter.

In 1949, two years after the American Chemical Society changed the name of the *Analytical Edition of Industrial and Engineering Chemistry*, then in its 19th year, to *Analytical Chemistry*, the first issue of the year carried the first *Analytical Chemistry* Review of mass spectrometry [20]. Then the youngest division in the ACS, the Analytical and Micro Chemistry Division, in its third year was aiming for a membership of 1000 by the end of the year. The five-page review by John A. Hipple and Martin Shepherd at the National Bureau of Standards in Washington DC began with a rather interesting statistic, "*Chemical Abstracts* reported 11 references to mass spectrometry in 1943, 15 in 1944, 17 in 1945, 26 in 1946, and 40 in 1947". This review had 176 citations. Even taking into account the effect World War II had on the number of publications, as Hipple and Shepherd pointed out, there was an unprecedented expanding interest in the field at that time; it continues at a similar rate today. When the last of these reviews appeared in *Analytical Chemistry* in 1998, it had grown to 70 pages in length. To some extent, the Review has been replaced by *CA Selects Plus: Mass Spectrometry*.

Mass spectrometry's primary role was in the study of small molecules and isotopes from Thomson's first parabola machine until just before World War II. The only instruments that were available during these formative years of mass spectrometry were those that were designed by individual researchers or those that were custom built to order by craftsman such as Mattauch and his German colleague, Richard Herzog. These instruments were employed by physicists who used them in the determination of isotopes and the study of ion formation. Thomson had talked of the potential of the mass spectrometer due to the fact that ions were not only formed during the initial absorption of external energy, but some secondary ions also were produced by decomposition of these initially formed ions. Physicists were annoyed by the presence of peaks in their mass spectra that could be attributed to the instrument's background. As the organic chemists looked at these peaks more closely, they realized that they represented ions of various hydrocarbon substances, and they predicted that the mass spectrometer would have broader applications.

Based on the possibility of using such an instrument in hydrocarbon analysis, in 1937, Herbert Hoover Jr, son of the 31st President of the United States (1929–1933) and the first scientific person (chemical engineer) to become a U.S. President, formed the Consolidated Engineering Company (CEC) as the engineering and manufacturing subsidiary of the United Geophysical Company. This company, which had close ties with the California Institute of Technology and the petroleum industry, was founded to develop instrumentation to locate petroleum deposits by detecting hydrocarbon gases emanating from the ground. However, due to the ubiquitous nature of methane, such a device was not possible. This business venture would have died at this point except for the growing interest in using mass spectrometry to increase the speed of analyzing aviation gasoline, which was becoming increasingly important because of the nearing possibility of World War II [21].

During World War II, mass spectrometry played a pivotal role in the preparation of weapons-grade plutonium [22]. Preparative mass spectrometers such as the Calcutron described by Yergy were used to produce weapons-grade fissionable materials. Another important consideration of mass spectrometry was its need in research for the development of synthetic rubber. Because of the Japanese occupation of Malaysia in 1941, there was no longer a supply of natural rubber for the U.S. [23]. The mass spectrometer saw an ever-increasing role. CEC's newly developed instrument, the CEC 21-101, looked like the answer, with a single exception. CEC mandated that purchasers of its instruments disclose to them all data and information made with purchased instruments so that the technology could be shared. This requirement did not sit well with the U.S. petroleum industry, which consisted of very competitive, and just as arrogant, companies. Another company, Westinghouse Electric, had also announced a commercial mass spectrometer, but had not started delivery. Westinghouse was pressured by the U.S. Government, acting on behalf of the petroleum industry, to produce instruments similar to the one being offered by CEC to aid the war effort.

After World War II, mass spectrometry began to have a broad number of applications in organic chemistry. The recovering economies of Germany, France, England, and Japan all saw developments of mass spectrometry instrumentation. There was also work on new instruments occurring in the then Soviet Union, but because of the closed nature of that country, not much is known of the details. By the very early 1950s, there were three companies building magnetic-sector mass spectrometers in the U.S. (CEC, General Electric, and Westinghouse). Soon, first Westinghouse and later GE left the market. A new technology was introduced by the aircraft manufacturer, the Bendix Corporation, with the publication of a seminal paper on time-of-flight mass spectrometry in December of 1955 by two of their researchers (William C. Wiley and Ian H. McLaren) working in Bendix Aviation Corporation Research Laboratories in Detroit, MI [24]. This instrument was the first incarnation of the time-of-flight mass spectrometer, which eluded earlier researchers such as Smythe and Mattauch in 1932 [25], W. E. Stephens [26, 27], while at the University of Pennsylvania (Philadelphia, PA) in 1946, and Henry S. Katzenstein and Stephen S. Friedland [28] in the Physics Department at the University of Connecticut in April 1955.

By the mid-1960s, there were significant offerings from England's Associated Electronics Industry (AEI), Germany's Mes und Analysen-Technik (MAT), Japan's Hitachi and JEOL, and, to a lesser extent, France's Thompson's Electronics, as well as those from the U.S. that had been known for pioneering developments in other fields of analytical instrumentation such as Varian, Beckman, and Perkin-Elmer. Another factor that had a profound effect on the development of mass spectrometry in the analysis of organic compounds was the revelation of gas chromatography (GC). The gas chromatograph separated mixtures of volatile nonthermally labile compounds into individual purified components and delivered them in the gas phase (a requirement at the time for mass spectrometry) to the mass spectrometer. GC did present one formidable problem for the mass spectrometer. Early gas chromatographs delivered the individual analytes to the mass spectrometer in a very dilute concentration in a large volume of the GC's mobile phase, helium, hydrogen, or, in rare cases, nitrogen. In order to detect the analyte, the mass spectrometer had to "digest" this large volume of superfluous carrier of the compound of interest. After this problem was overcome by the development of analyte-enrichment devices such as the jet, effusion, and membrane separators, invented, and later patented, by Einar Stenhagen (Swedish medical scientist) and perfected by Ragnar Ryhage [29], J. Throck Watson (while a PhD student of Klaus Biemann at the Massachusetts Institute of Technology in Cambridge, MA) [30, 31] and Duane Littlejohn and Peter Llewellyn at the Varian Research Center in Palo Alto, CA [32], respectively, gas chromatography/mass spectrometry (GC/MS) became the instrument that produced more information for amenable analytes from less sample than any other analytical technique. This made GC/MS an indispensable tool in the environmental, medical, and other biological sciences as well as in forensics, the food and flavor industry, and so forth (see Chapter 10).

During this era, analytes were converted to ions (the principal requirement of mass spectrometry) using the technique of electron ionization (EI). EI used a beam of high-energy electrons (50–70 eV) to produce molecular ions ($M^{+\bullet}$). Some of these $M^{+\bullet}$ then reproducibly fragment to produce a spectrum of ions that have various masses and usually a single-charge state; i.e., ions of various mass-to-charge ratios (*m/z*). These fragmentation patterns in the form of a mass spectrum are what allow for the unambiguous identification of a compound by GC/MS. Unfortunately, some compounds have such energetic $M^{+\bullet}$ produced by EI that almost none of the $M^{+\bullet}$ remains intact; therefore, it is not possible to determine the analyte's molecular mass. Without the nominal mass of an intact analyte, no matter how much information was available from the fragments, it is not possible to identify the analyte. This lack of molecular-ion current led to a way to have less energy imparted to the analyte during a different type of ionization process. The resulting technique, developed by Burnaby Munson and Frank Field, is known as chemical ionization (CI) [33–37]. CI is an ion/molecule reaction that usually produces protonated molecules (MH^+) of the analyte. These MH^+ are much lower in energy than the $M^{+\bullet}$ and are much more likely to remain intact for detection in the mass spectrometer. Together, EI and CI GC/MS have advanced many areas of science and have resulted in a much better quality of life through a better understanding of the chemistry of organic compounds.

GC/MS was taken to the next higher plane of advancement by the commercialization of the transmission quadrupole, invented by Wolfgang Paul and colleagues [38] at the University of Bonn (Bonn, Germany) in the early 1950s.[4] In the mid-1960s, Robert A. Finnigan and associates [39, 40] produced a quadrupole GC-MS at Electronic Associates Inc. (EAI) (Long Branch, NJ), an analog computer company. This instrument became the basis for the first instrument (the Finnigan 1015) produced by the subsequent company formed by Finnigan and T. Z. Chu (Finnigan Corporation, founded in Sunnyvale, CA, now known as Thermo Finnigan, a subsidiary of Thermo Electron). In addition to Finnigan Corporation's development, Hewlett-Packard of Palo Alto, CA, and, to a lesser extent, Extra Nuclear Corporation of Pittsburgh, PA, also contributed to this emerging technology. Today, the transmission quadrupole is the most ubiquitous of all mass spectrometers. Its contributions to GC/MS were due to the speed at which the analyzer could be scanned for fast data acquisition, the linear nature of the *m/z* scale, simple operation, its lower acceleration potential, and smaller size compared to that of the sector-based instruments, and the ease with which it could be operated and controlled by the then emerging minicomputer technology. As gas chromatography evolved into the use of capillary columns resulting in higher concentrations of analyte in the eluant and narrower peaks, the transmission quadrupole resulted in fast, easy, and reliable instruments for GC/MS. See Chapter 10 for GC/MS and Chapter 6 for strategies in dealing with EI data.

Another significant development in mass spectrometry came about with the introduction of resonance electron capture ionization (ECI) [41, 42]. Gas-phase analytes with high electron affinities are ionized by capturing thermal energy (0.1 eV) electrons resulting in the formation of negative-charge molecular ions ($M^{-\bullet}$). This technique, developed by George Stafford, while working on his doctorate with Don Hunt at the University of Virginia in the mid-1970s, allows for the analysis, at unprecedented detection levels (fg μL^{-1} injected into the instrument), of halogenated pesticides in very complex matrices such as those produced by extraction from the skins of fruits and vegetables. ECI also had a significant impact on analytical methods for drug metabolism. Using fluorinated reagents, derivatives of nonvolatile drugs and their metabolites could be formed through reactions with the polar sites on these drugs and metabolites, thus making them volatile as well as electrophilic. These electrophilic derivatives of drugs and metabolites extracted from blood and urine could be analyzed by GC/MS based on ECD without interference from endogenous substances. Before the LC/MS became a practical technique in the late 1990s, ECI GC/MS was a major technique in studies of drug metabolism.

---

[4] An important note about the transmission quadrupole might have disappeared from recorded history if Bob Finnigan had not published an interesting article in the A-pages of *Analytical Chemistry* on the history of transmission quadrupole; namely, while Paul pursued the quadrupole technology, Richard Post at the University of California Lawrence Berkeley Laboratory carried out independent research during the 1950s on the same technology. Post did not publish or apply for patents on his findings. Other than information contained in his personal notebooks, the only record of his work is a University of California Radiation Laboratory report (UCRL 2209) published in 1953.

El and Cl GC/MS continued to be the mainstay of mass spectrometry until the development of desorption/ionization (DI) techniques. These latter techniques allowed for the determination and fragmentation of nonvolatile thermally labile analytes such as peptides. DI techniques not only expanded the types of analytes that were amenable to mass spectrometry, but also opened the door to use of liquid chromatography as a technique to separate and purify components of a mixture. The first of these techniques to have a major impact on mass spectrometry was fast atom bombardment (FAB), a variation of secondary ion mass spectrometry (SIMS) carried out using a liquid matrix. SIMS is a process of producing ions from a solid surface by bombarding it with a beam of high-energy ions. SIMS is used in the characterization of organic and inorganic surfaces as well as metal or composite materials, such as those in the wings of aircraft. FAB, developed by Mickey (Michael) Barber [43] in the Department of Chemistry at the University of Manchester Institute of Science and Technology in the mid-1970s, employs a beam of high-energy (5–10 keV) atoms of a nonreactive element such as xenon to bombard the surface of a glycerol solution of analyte molecules; e.g., a peptide. In this way, FAB causes the desorption of protonated molecules from the condensed phase into the gas phase. FAB revolutionized the study of biopolymers such as DNA and proteins known for their nonvolatility and thermal lability.

The next phase of development in DI was ^{252}Cf desorption/ionization (Cf DI) (also known as plasma desorption, PD) as pioneered by Ronald D. MacFarlane at Texas A&M [44]. This technique remained a laboratory curiosity throughout its useful life. The dependence of PD on time-of-flight mass spectrometry was not sufficient to prevent the last U.S. manufacturer of these instruments (CVC Corporation) from discontinuing their manufacture in the late 1970s. There was a single commercial attempt at a PD instrument in the 1980s by the Uppsala, Sweden, company, Bio-Ion. The instrument was later marketed by Kratos Analytical, UK. Bio-Ion was later acquired by Applied Biosystems (now PE Biosystems). The primary reason for the lack of popularity of the ^{252}Cf-DI technique was the radioactive nature of the ionization source, which presented a significant safety and disposal problem for a number of laboratories. However, PD was what Frans Hillenkamp and Michael Karas credit with inspiring them to look at other possible desorption techniques as they developed matrix-assisted desorption/ionization (MALDI) [45], one of the two most significant DI tools (the other being John B. Fenn's electrospray ionization [46], ESI), which has continued as a primary factor in the development of proteomics and areas of biopolymer analysis.

Although the 2002 Nobel Prize in chemistry was not shared by Hillenkamp and Karas (the prize was awarded to John Fenn and Koichi Tanaka, who was the first to use laser desorption in a matrix (glycerol/metal filings in a one-time, never reproduced experiment for the mass spectrometry part and Kurt Wüthrich for the NMR part of "… methods in chemical analysis applied to biomacromolecules"), their contributions to the technique have spawned numerous commercial instrument designs, the resurgence of the time-of-flight mass spectrometer, and technology used in thousands of analyses performed each day in analyses of protein, DNA, and synthetic polymer samples. The use of MALDI involves mixing the analyte with a thousand-fold excess of a solid matrix of small organic molecules that absorb the energy of a laser to explosively discharge protonated molecules of the analyte in to the gas phase. When used with extended-range mass spectrometers such as the time-of-flight instrument, the mass of intact heavy (up to tens of thousands of daltons) proteins can be measured with an accuracy of 0.01% as compared to 1% for electrophoretic techniques (see Chapter 9 for details on MALDI).

Unlike MALDI, which primarily produces single-charge protonated molecules, ESI can produces multiple-charge ions, provided that the analyte molecule has multiple sites that can be protonated; e.g., a peptide with multiple basic amino acid residues. In ESI, ions are produced in solution through acid/base chemistry. The ions are then desorbed into the gas phase as the analyte-containing solution is sprayed from a charged needle into an electric field; this facilitates ion evaporation or coulombic ejection gas-phase ions. One significant advantage of ESI is that instruments such as the transmission quadrupole with a normal m/z range to ~1000–4000 can be used to analyze ions that have a mass of several tens of thousands of daltons because of the high charge state of the ions. This is the very reason that it is important to remember that the mass spectrometer detects ions based on their mass-to-charge ratio, not their mass alone. Another advantage of ESI is that the sample is analyzed as a solution, which can be the eluant from an HPLC, which allows mixtures of compounds in solution to be separated on-line while performing an analysis by LC/MS. The combination with MS can produce challenges such as having to rethink the use of traditional buffer systems for LC mobile phases (see Chapter 11).

LC/MS did not have as seamless a start as did GC/MS. Like GC/MS, soon after the development of high-performance liquid chromatography (HPLC), efforts began to connect this purification technique to the mass spectrometer. These efforts were frustrated by the fact that many of the analytes were thermally labile and/or nonvolatile. In the transformation of the HPLC mobile phase from liquid to gas, the mass spectrometer's vacuum system was often overwhelmed by the gas load. Several attempts were made to develop an interface, such as Patrick Arpino's direct inlet system [47] and Bill McFadden's moving belt [48] of the mid-1970s, both of which relied on conventional gas-phase EI and CI. Other attempts included Marvin Vestal's thermospray (a new method of ion formation) and Ross Willoughby's particle interface [49] of the 1980s (based on conventional gas-phase EI and CI), and, of course, the one enduring technique of the 1970s, Evan Horning's atmospheric pressure chemical ionization (APCI) [50], another new method of ionization. Another factor that frustrated the development of LC/MS was the mass spectrometry paradigm that dominated the 1960s, 1970s, and 1980s, as well as the early part of the 1990s, which was that for the technique to be useful, ions representing the intact molecule had to fragment so that structural information would be forthcoming. This paradigm is why APCI remained a research laboratory curiosity for many years. LC/MS came into an era of practicality when ESI was developed, and at the same time MS/MS became much more user friendly. For more on LC/MS, the history of development of its interfaces, and its applications, see Chapter 11.

The development of MALDI and ESI prompted improvements in the TOF mass spectrometer. One of the major factors involved with advancement of ESI and a companion technique that has become very popular in LC/MS, atmospheric pressure chemical ionization (APCI), was the development of the triple-quadrupole mass spectrometer by Rick Yost and Christie Enke at Michigan State University in the late 1970s [51]. This development made the technique of mass spectrometry/mass spectrometry (MS/MS) practical. MS/MS is a tandem process by which ions of a specific m/z value formed by an initial ionization are isolated in the first stage of mass spectrometry, caused fragment through some process such as inelastic collisions with an inert gas particle, and then the resulting fragment ions are separated according to their m/z values and abundances by a second stage of mass spectrometry. MS/MS is sometimes referred to as tandem mass spectrometry (one iteration of mass spectrometry (ionization and ion separation) followed by a second iteration of mass spectrometry). ESI and APCI primarily form ions containing the intact molecule; therefore, they yield little structure information such as that obtained through fragmentation of molecular ions in EI. The MS/MS process allows for the additional structural information following ionization by ESI or MALDI, and in many cases, a higher degree of specificity for target-compound analysis. Today, MS/MS is carried out in triple-quadrupole mass spectrometers and in hybrid *tandem-in-space* instruments such as a transmission quadrupole analyzer used for the first iteration of mass spectrometry followed by a time-of-flight analyzer used for the second iteration (as in a Q-TOF mass spectrometer). See Chapter 3 for more on the technique of MS/MS.

At the same time ESI was being advanced by tandem mass spectrometry and MALDI was finding increased success with new developments in TOF mass spectrometry, new *m/z* analyzers such as the three-dimensional quadrupole ion (3D QIT), the linear QIT, and, most recently, the orbitrap (only commercialized in mid-2005) were developed. Also, a more dated technique of separating ions according to their *m/z* values, the ion cyclotron resonance (ICR) mass spectrometer (developed in the mid-1960s), was given new life through the detection of ions via image currents with data processing by Fourier transformation [52, 53]. All of these mass spectrometers have the advantage that they are tandem-in-time instruments, meaning that the different iterations of MS/MS take place using a single piece of hardware with the essential actions (precursor-ion selection, precursor-ion dissociation, product-ion analysis) performed as a function of time.

The 3D QIT was also a result of Wolfgang Paul's research in the 1950s; it was commercialized by the same company that is primarily responsible for the introduction of the transmission quadrupole mass spectrometer, Finnigan Corp. The 3D QIT, again like the transmission quadrupole, began its life as a GC/MS system. Both of these analyzers have now seen even bigger roles in LC/MS. Finnigan Corp.'s successor, Thermo Fisher, is responsible for the development of the linear QIT as a standalone mass spectrometer (2003) and the orbitrap, which is used as the second stage of a tandem-in-space instrument (2005). Details of functionality and more on the history of these and other *m/z* analyzers can be found in Chapter 2.

The latest chapter in ionization is currently being written with the developments of techniques for the formation and desorption of ions on the surface of various substances. All mass spectral analyses (except maybe the early analyses of petroleum) of organic compounds have involved some type of sample preparation; i.e., extraction, concentration, derivatization to stabilize nonvolatile, thermally labile substances, separation though a GC or an LC (connected to the mass spectrometer or not), etc. This *sample-preparation/extraction* requirement is no longer an issue. In late 2004 through late 2005, several new ionization techniques were developed. These techniques allow for ionization and desorption of ions of the analyte directly into the gas phase from the surface containing the analyte; the sampling, analysis and detection of ions in accomplished in seconds. This means that pesticides on the skin of a piece of fruit can be detected without the need to remove the skin from the fruit, extract the pesticide, and then submit the concentrated extract to analysis by GC/MS or LC/MS. In cases of analyzing the urine of subjects suspected of driving while impaired for illicit drugs, there is no need to extract the urine and wait for a chromatographic process to complete the analysis. In forensic analyses such as determining whether a piece of porous concrete block might contain traces of a substance like VX nerve agent, there is no risk of having the analyte decompose as it might during analysis by GC/MS or LC/MS, etc. These techniques are DART (direct analysis in real time) [54], DESI (desorption electrospray ionization) [55], and ASAP (atmospheric pressure solid analysis probe) [56]. As of mid-2006, only DART is provided as a technique with a mass spectrometer. The DESI interface is an add-on for existing mass spectrometers and most reports have been associated with unit-resolution MS/MS instruments such as the 3D QIT mass spectrometers. At this time, the ASAP technique is still a research curiosity. ASAP involves modifying an existing APCI interface on a Q-TOF instrument, exposing the operator to a rather high voltage. DART is part of a JEOL AccuTOF atmospheric pressure ionization time-of-flight mass spectrometer with a resolving power of >7000. This high resolving power allows for accurate mass measurement, resulting in unambiguous elemental compositions for analytes of >500 Da, which is a major part of the reason that the DART technique has proven so successful. More about all three of these new-era ionization techniques can be found in Chapter 4.

Two of the real challenges in mass spectrometry from its beginnings through the later 1960s and early 1970s were involved with the vacuum system and the data. Vacuum became less of a problem with the development of the turbomolecular pump, and dealing with the huge quantities of data (especially those generated during GC/MS) went from being a nightmare to being reasonable and straightforward with the development of the minicomputer.

As can be imagined while viewing Figure 1-6, when Francis William Aston designed his instruments to determine the masses of various nuclides, implementation was a monumental challenge. Crude mechanical devices and water aspirators were

Figure 1-6. Francis William Aston in his workshop at the Cavendish Laboratory at Cambridge ca. 1921. In addition to his many other accomplishments, he was a very talented glass blower which was important in these early days of mass spectrometry.
From the American Institute of Physics archive, with permission.

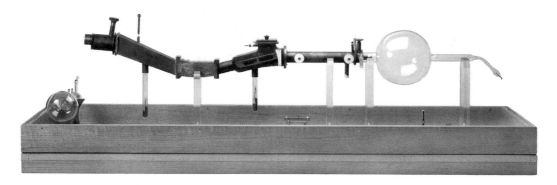

Figure 1-7. Replica of Aston's third design commissioned by the American Society for Mass Spectrometry after the restored instrument housed in the Thomson Museum at Cambridge.

used for the primary vacuum, but a clean high vacuum could only be achieved through the use of a mercury diffusion pump. Like a good portion of the mass spectrometer, except for the electromagnet (Figure 1-7), the mercury diffusion pump was constructed of glass similar to the device shown in Figure 1-8. The mercury diffusion pump was invented simultaneously by Wolfgang Max Paul Gaede (German physicist, 1878–1945) in Germany and Irving Langmuir (American chemist, 1881–1957) in the United States in 1915–1916. Operation of the mercury diffusion pump involves heating the liquid to 110 °C to force a stream of mercury vapor through the volume to be evacuated; collisions between the atoms of Hg and gas molecules force the fixed gas toward a fore pump as the Hg atoms condense back to the liquid state on the relatively cool walls of the chamber being evacuated. These pumps presented many challenges, including dealing with the toxic effects of breathing the mercury vapors. Oil was also used in diffusion pumps; although the background from these pumps interfered with the measurements of the various nuclides, these background spectra gave the organic chemist the idea of using the mass spectrometer for the identification of organic molecules. The oil and mercury were replaced first with silicon-based oils, but these led to the accumulation of silicon oxides on the slits of the instrument, causing a reduction in resolving power. The silicone oils have been largely replaced with synthetic organic polymers, usually polyphenyl ethers.

Recording data from these early mass spectrometers evolved from the hand-tracing of the phosphorescent patterns on willemite screens to the use of photographic plates. Because partial pressures of the analytes were relatively constant, mass spectra could

Figure 1-8. A modern glass mercury diffusion pump. *Permission of Yves-Marie Savoret, Chemistry Department, University of Guelph, Ontario, Canada.*

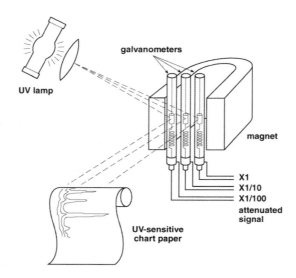

Figure 1-9. Schematic illustration of the light-beam oscillographic recorder. Mirrors attached to galvanometers having different torques reflect light beams to provide correspondingly different amplitudes of the instrument signal, thereby expanding the dynamic range of this analog recording device.
From McFadden WH, Ed. Techniques of Combined Gas Chromatography/Mass Spectrometry: Application in Organic Analysis, Wiley-Interscience, New York, 1973, with permission.

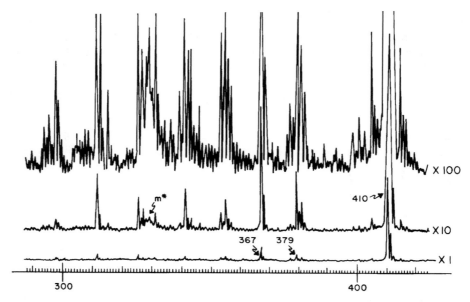

Figure 1-10. Artist's tracing of light-beam oscillographic data from photosensitive recording paper. The numeric m/z scale shown at the bottom was not present on the original data presentation.

be recorded over long periods of time, and these recording methods were satisfactory. In the early days of organic mass spectrometry, strip-chart recorders could be used to make a permanent record of the mass spectrum because there were no transient changes in sample concentration associated with chromatography. As the need developed for rapid acquisition of spectra as dictated by GC/MS, Polaroid® photographs were taken of the displays of mass spectra on oscilloscopic screens. The last device to be used to create an analog recording of the mass spectrum before the age of digitization and the use of the minicomputer was the light-beam oscillographic (LBO) recorder. This device reflected discrete light beams onto photosensitive paper from tiny mirrors attached to galvanometers as illustrated in Figure 1-9. These devices provided a good dynamic range for measuring the ion current through the use of several galvanometers, each having a different spring-loaded torque resistance, which produced different deflection amplifications as illustrated in Figure 1-9.

Such a device does allow for a rapid acquisition of a spectrum, but does not alleviate the cumbersome problem of dealing with huge numbers of analog mass spectral recordings. As seen in the illustration of the data obtained using the LBO recorder in Figure 1-10, this data record must be manually processed to produce a bar-graph mass spectrum that can be used for interpretation. Not only is there no *m/z* scale on the original data output, but the scale is not linear. Often the *m/z* scale must be created from the presence of peaks representing ions of known *m/z* values produced by substances that are present in the ion source with the analyte. Spectra in the published literature on mass spectrometry such as the one seen in Figure 1-11 are manual presentations prepared from LBO data. Another problem with the LBO output is that it is based on photosensitive paper, which when exposed to sunlight or fluorescent lighting will darken to the point that the multiple data profiles cannot be seen.

Spectra were acquired individually during GC/MS by pressing a button on the mass spectrometer that initiated a scan of the *m/z* range of the instrument. The analyst would initiate a scan based on observing the magnitude of the total ion current from the ionization source (a recording of this signal resembled a chromatogram). The operator would label the resulting spectrum to coordinate with a mark on the total ion current

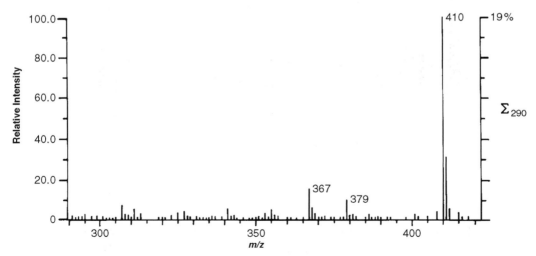

Figure 1-11. A bar-graph mass spectrum resulting from laborious manual processing of the analog data shown in Figure 1-10.

chromatogram so that during data analysis, the acquisition time of the spectrum could be correlated with the chromatogram. Usually no more than two or three spectra could be recorded for a single chromatographic peak. If a spectrum was to be used as a background spectrum, it was often acquired on the back side of the chromatographic peak because of the difficulty of anticipating when to acquire the spectrum on the front side of the peak. There was no real way to record mass chromatograms (defined later in this chapter); the concept of the mass chromatogram was introduced with computerized data systems.

As can be appreciated from this description of the manual effort involved in obtaining interpretable data, organic mass spectrometry in the years before the minicomputer was very challenging.

III. Some Important Terminology Used in Mass Spectrometry

1. Introduction

The definitions of the terms mass-to-charge ratio (m/z), dalton (Da), and unified atomic mass unit (u) and their significances have already been stated in the beginning of this chapter. Other important terms used in mass spectrometry (isotope, nominal mass, monoisotopic mass, and mass defect) are found under the heading "IV. Elemental Composition of an Ion and the Ratios of Its Isotope Peaks" in Chapter 5. In addition, there are some other very important terms that are necessary to the understanding of the literature and discussions of mass spectrometry. These definitions follow.

2. Ions

A **molecular ion** is a charged species that has an odd number of electrons and is formed from a molecule (an even-electron neutral species) through the addition or removal of an electron. A molecular ion is not a charged species that results from the addition of a charged species that has a significant mass such as a proton, sodium ion, chloride ion, etc. A molecular ion is not a species that represents an intact molecule that has had a proton or hydride (a proton with two electrons) removed from it or resulted from it. Under no circumstances should a molecular ion ever be called a *parent ion*. At one time, the precursor ion involved in collisional activation dissociation (CAD) analyses were inappropriately referred to as *parent ions*, prompting the discontinuances of the use of this anthropomorphic term as a synonym for molecular ion. The term parent ion has no place in mass spectrometry to describe a molecular ion, a precursor ion, or any other type of ion.

A **fragment ion** results from the decomposition of another ion. This term usually refers to an ion that is produced by the fragmentation of a molecular ion or a species that represents the intact molecule such as a protonated molecule, a deprotonated molecule, an ion produced by hydride abstraction, a sodiated molecule, etc. Fragment ions can be the results of the fragmentation of a fragment ion formed from a molecule ion: a secondary fragmentation. Fragment ions are formed through the breakage of chemical bonds. They do not result from the loss of nuclear matter from one of the atoms comprising the ion. No fragment ion can result from the loss of 12 Da from a molecular ion. Fragment ions always have a mass that is less than that of their precursor. This statement illustrates the importance of separating the terms related to mass and the mass-to-charge ratio of an ion. The fragment ion will have a mass less than the mass of its precursor, but it may have an m/z value greater than that of the precursor ion because the precursor ion has multiple charges and the fragment ion has fewer charges. In most forms of mass spectrometry, the products of fragmentation are an ion of lesser mass and the same charge sign as the percussor ion and a neutral species that is a radical (an odd-electron species) or molecule (an even-electron species). In some cases involving multiple-charge ions, the fragmentation involves the formation of two or more ions with the same charge sign, both of which can be detected in the mass spectrometer. In some very rare cases, mass spectrometry fragmentation results in an ion-pair formation (two ions of opposite charge sign).

3. Peaks

In mass spectrometry **peaks** appear in the recording of the mass spectrum and these peaks represent the **ions** that are formed in the mass spectrometer. Ions are not found on mass spectra, and peaks do not occur in mass spectrometers. Another important characteristic of peaks and ions is that *peaks* have **intensities** and *ions* have **abundances**. The peak with the highest intensity in the mass spectrum is called the **base peak**. The peak at *m/z* 175 in the two graphical presentations of the mass spectrum in Figure 1-2 is the base peak. The presentation of the mass spectrum can be absolute with the peak intensity representing the actual signal strength of the ion current for an ion with that *m/z* value, or the peaks can have a *relative* intensity which presents the data in such a way that the intensity of the base peak is 100%. This latter presentation has the disadvantage that the signal strength of the mass spectrum is not known. Unless the mass spectrum has something in the header of its presentation to indicate signal strength, the relevance of missing mass spectral peaks may remain a mystery. Commercial instruments vary in the way that they display graphical mass spectra. Some have an absolute intensity display and others have a relative intensity display. Those that display mass spectra with relative intensities usually have an indication of the base peak's signal strength in the header. All mass spectral databases (mass spectral libraries) use a relative intensity display.

When using the word "peak" in a discussion of GC/MS or LC/MS data it is important to make sure that there is a distinction between a *mass spectral peak* and a *chromatographic peak*. In today's modern mass spectrometry, analyses are carried out by acquiring one spectrum after another at a constant rate. These individual spectra can be displayed as a bar graph, in a profile graphic presentation, or in a tabular format as described in Figure 1-2. It may be convenient to display a group of spectra in a contiguous presentation of all the spectra acquired during a given time period. This type of display is often created by summing the ion current at each *m/z* value that exhibits any ion current in each spectrum. The sum of the ion current is then presented graphically as a plot of the total ion current for each spectrum vs the spectrum number. This plot is referred to as a **reconstructed total ion current (RTIC) profile**. The term "reconstructed" is appropriate because the profile displayed is created from the only data that are acquired, mass spectral data; this profile is not created directly by monitoring the total ion current as a function of time. When spectra are acquired of the eluate from a chromatographic process, the RTIC profile is called an **RTIC chromatogram**. The *RTIC profile/chromatogram* is a fourth way of displaying or viewing mass spectral data.

Another way of viewing mass spectra, where data are acquired over a range of *m/z* values, is to prepare a plot of the ion current at a single *m/z* value, a subset of the acquisition range (a range of *m/z* values), or the sum of the ion current of selected *m/z* values in the acquisition range. Such a display is called an **extracted ion current (EIC) profile** or, more appropriately for data acquired during a chromatographic analysis, a

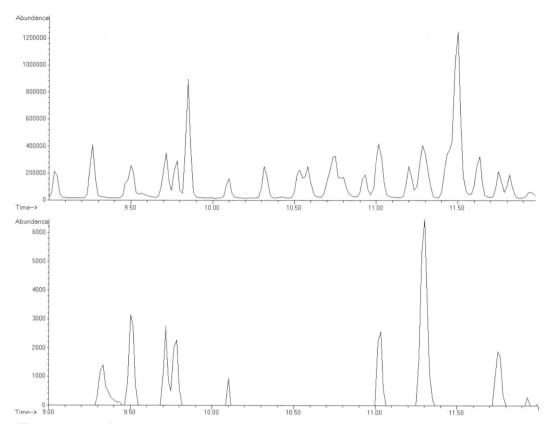

Figure 1-12. (Top) Reconstructed total ion current chromatogram of mass spectral data acquired at the rate of 1 spectrum sec^{-1} over an m/z range of 40–450. (Bottom) Mass chromatogram/extracted ion current chromatogram for m/z 232 from same data set.

mass chromatogram [57]. Because data are acquired during infusion or flow injection in LC/MS and when probes are used for sample inlets in GC/MS, the profile term is as appropriate as the chromatogram term. However, the chromatography terms (RTIC chromatogram or mass chromatograms) should be used when the data are a result of chromatography. If the data are a result of capillary electrophoresis, the term should be **RTIC pherogram** or **EIC pherogram**. These EIC profiles/mass chromatograms should not be confused with SIM profiles and SIM chromatograms, which result from data acquired during selected ion monitoring experiments. There can also be SRM profiles and SRM chromatograms, which are displays of data acquired during experiments using selected reaction monitoring. Figure 1-12 is an example of an RTIC chromatogram (top) and a mass (or an EIC) chromatogram (bottom).

The importance of employing a distinguishing adjective to differentiate between a reconstructed chromatographic peak and a mass spectral peak cannot be overemphasized. Both have height and width, which is significant for each. The peaks representing the chromatographic processes in chromatography/mass spectrometry are different from the chromatographic peaks observed when ultraviolet (UV) and refractive index (RI) detectors are used with LC or when flame ionization (FI) and electron capture

(EC) detectors are used with GC. These latter detection systems are analog devices and produce the chromatographic profile directly. In chromatography/mass spectrometry, the process of recreating (reconstructing) the chromatographic data profile from mass spectral data is a digital process.

4. Resolution and Resolving Power

There are two terms that are inappropriately and, more often than not, incorrectly used in an interchangeable way in mass spectrometry: **resolution** and **resolving power**. A venerable mass spectrometrist, Keith Jennings, said, "Resolution pertains to the data of mass spectrometry, whereas resolving power is a function of the mass spectrometer." The term *resolution* is always relevant to the separation of ions of two different *m/z* values. In mass spectrometers that separate ions using quadrupole fields (quadrupole ion traps and transmission quadrupoles), the resolution is constant throughout the *m/z* scale. That is to say, peaks representing two pairs of ions that differ by 1 *m/z* unit will have the same separation at *m/z* 100 and 101 as they do at *m/z* 2000 and 2001.

The term *resolving power* is defined as the difference in *m/z* values of ions that can be separated (according to some definition) from one another (Δm) divided into a specific *m/z* value (M) (i.e., R = M / Δm). This means that the resolution at this particular *m/z* value can be considered to be Δm [58]. Therefore, for an instrument that operates at constant unit resolution (Δm) throughout the *m/z* scale, the values for R would be 100 (R = 100 / 1) at *m/z* 100 and 2000 (R = 2000 / 1) at *m/z* 2000. There is no single resolving power for these types of instruments, and the value for R will be different at every *m/z* value. However, by saying that the instrument will function at a *constant resolution of X m/z units*, the differences in *m/z* values of ions that can be separated is clearly understood. Instruments that do not operate at a constant resolution throughout the *m/z* scale, such as the TOF and double-focusing mass spectrometers and instruments that use Fourier transforms for the detection of ions, do operate at constant resolving power. An instrument that has a resolving power (R) of 1000 will have a resolution of 1 at *m/z* 1000. This instrument will separate an ion with *m/z* 1001 from an ion with *m/z* 1000. This same instrument will produce data with a resolution of 0.1 at *m/z* 100; the mass spectrometer will be able to separate ions with *m/z* 100.1 from ions with *m/z* 100.0.

When the term *high-resolution mass spectrometry* is used, it usually means that the instrument is capable of *high resolving power*. Unfortunately, the terms *high resolution* and **accurate mass measurement** are also often used interchangeably. The term accurate mass measurement means mass measurement performed to a sufficient number of significant figures to allow for an unambiguous determination of an elemental composition. The number of decimal places needed will be a function of the ion's actual mass. For ions with a mass of less than 500 Da, an accuracy of 0.0025 Da should allow for an unambiguous assignment of an elemental composition. This type of mass accuracy can be obtained with any mass spectrometer provided that the ion is monoisotopic (there is a single elemental composition and all the atoms of all the elements present represent only the most abundant naturally occurring stable isotopes of those elements). The need for a high-resolving power mass spectrometer occurs when the *m/z* value represents two different ions (a **doublet** and a **multiplet**) as would be the case for a single-integer *m/z* value representing both an [M – OH]$^+$ and an [M – NH$_3$]$^+$ ion. A separation of these two ions at *m/z* 220 would require an R value of >7000 using a 10% valley definition to describe the overlap of the corresponding peaks.

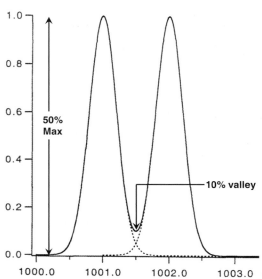

Figure 1-13.   *Illustration of 10% valley definition of resolving power.*

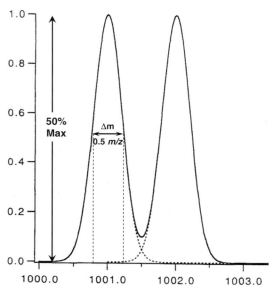

Figure 1-14.   *Illustration of FWHM definition of resolving power.*

When *resolving power* was defined above, the definition included a qualifier. This qualifier has to do with the determination of Δm. Resolving power qualifier definitions are **% valley** or **full width at half maximum (FWHM)**. Two mass spectral peaks of equal intensity will separate with an overlap sometimes referred to as the crosstalk between two ions of adjacent *m/z* values. The difference in these two *m/z* values is Δm. The distance from the baseline to the point of overlap (see Figure 1-13) expressed as a percentage of the mass spectral peak height is the percent valley. An acceptable separation (or percent valley) is 10–20%. Some instrument manufacturers (usually of double-focusing instruments) have used resolving power definitions of 50% valley. Such an instrument would be incapable of resolving ^{13}C-isotope peaks above *m/z* 150.

It is very difficult to obtain data resulting in two mass spectral peaks that have the same intensity and that are separated to the limit achievable by the particular instrument. When adjacent mass spectral peaks of equal intensity, overlapping designated extent are not available, the value for Δm is taken as the FWHM of a mass spectral peak. From an examination of the FWHM of the peak on the left side of Figure 1-14, it is clear that the value for Δm is half that obtained using the 10% valley definition. This means that values for R obtained using the FWHM definition are twice those obtained with the 10% definition.

The use of the full width at 5% maximum of a single mass spectral peak can be used to define the resolving power; however, accurate measurements at this position on the mass spectral peak often prove to be very difficult.

When using the FWHM definition, it is best to determine the number of *m/z* units represented by the physical distance between two non-overlapping mass spectral peaks. The difference in two mass spectra, both with the same resolving power but calculated using two different definitions, FWHM (left) and 10% valley (right), is illustrated in Figure 1-15.

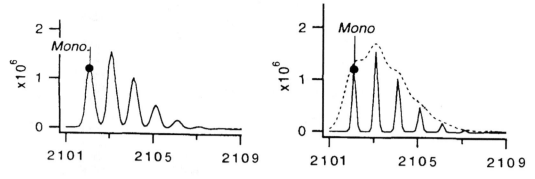

Figure 1-15. Data recorded at a resolving power value of 5000. R for the spectrum on the left was calculated using the FWHM definition, whereas R for the spectrum on the right was calculated using the 10% valley definition.

The resolving power of a time-of-flight mass spectrometer has a special significance. From the resolving power, the value of two *m/z* units that can be separated can be calculated. In the case of a linear TOF mass spectrometer (see Chapter 2), the resolving power (R) is 2000. This mass spectrometer can be used to determine accurately the mass-to-charge ratio of a single-charge ion that represents a protonated molecule of an intact protein with a mass of 200 kDa within a single *m/z* unit. However, at an *m/z* value of 200,000 and an R value of 2000, ions that differ by no less than 100 *m/z* units can be separated. The integer *m/z* unit determined for this ion represents the weighted average of the ions of all the different isotopic compositions within that window of 100 *m/z* units.

IV. Applications

From an instrumentation standpoint, mass spectrometry has had an exciting history of development of ionization techniques and in the way that ions have been separated according to their *m/z* value. In some cases, an analyzer type was developed and ways of applying that analyzer to specific analytical challenges came about. In other cases, an ionization method resulted from trying to find a way to determine the mass of a heavy compound such as an intact protein. Some instrument developments, like MALDI, were application driven. Some applications like the analysis of the skins of fruits and vegetables for pesticides were developed because someone designed an instrument that could ionize electrophilic compounds through resonance electron capture. DART and DESI were developed to form ions of analytes on surfaces and desorb these ions into the gas phase. The applications for such an instrumental technique are developing with each passing day. Refinements in these specific analyses will eventually result in routine analytical techniques such as those performed today by GC/MS and LC/MS. Some applications have resulted because certain kinds of instruments have been available for a required analysis. In other cases, instruments had to be developed for a required analysis. This has been the story of developments in mass spectrometry ever since its utility was first observed.

The following representative examples are designed to provide a brief exposure to some types of qualitative and quantitative applications of mass spectrometry. Some of the examples also provide an illustrative introduction to some of the operating principles or fundamental aspects of the technology and provide hints to the strategy of data interpretation.

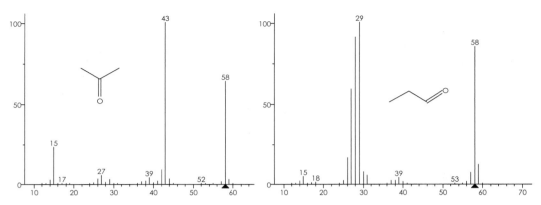

Figure 1-16.   EI mass spectrum of acetone (left) and propionaldehyde (right). Both compounds have an elemental composition of C_3H_6O and a nominal mass of 58.

1. Example 1-1: Interpretation of Fragmentation Patterns (Mass Spectra) to Distinguish Positional Isomers

One way of analyzing an organic compound by mass spectrometry involves conversion of the analyte molecule to a charged species (the molecular ion, $M^{+\bullet}$) by electron ionization (described in detail in Chapter 6). In the case of positional isomers such as acetone and propionaldehyde (both of which has the elemental composition C_3H_6O), a peak at m/z 58, which represents the $M^{+\bullet}$, does not distinguish the two compounds. However, in this case, the fragmentation pattern can be used to distinguish and identify the two compounds.

Compare the bar-graph spectra of acetone and propionaldehyde in Figure 1-16. The differences in the fragmentation patterns are related to structural differences between these two molecules. The base peak (the most intense peak in the spectrum) appears at m/z 43 in the mass spectrum of acetone (left side of Figure 1-16); in the mass spectrum of propionaldehyde (right side of Figure 1-16) there is no significant ion current at m/z 43. Conversely, the mass spectrum of propionaldehyde shows a substantial amount of ion current at m/z 29 and 28; the spectrum of acetone shows no appreciable ion current in this region. In addition, the propionaldehyde spectrum has a significant peak at m/z 57, the $[M-1]^+$ or $[M-H]^+$ peak.

Many of the driving forces that led to cleavage of certain chemical bonds in the $M^{+\bullet}$ are described in detail in Chapters 5 and 6. In this preliminary example, let it suffice that the bond on either side of the carbonyl carbon will break, as illustrated in the

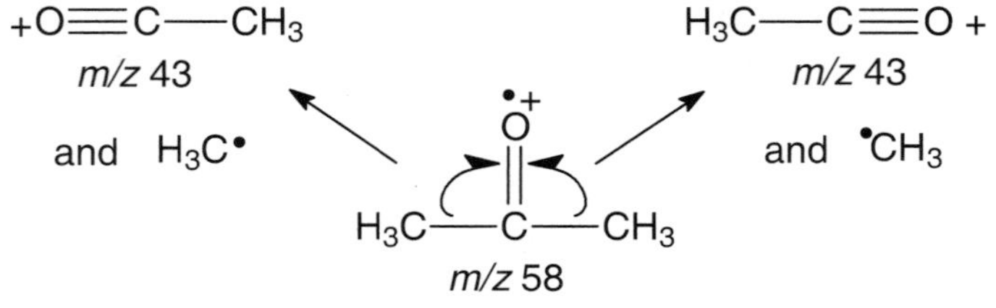

$$+O\equiv C—CH_3 \qquad\qquad H_3C—C\equiv O +$$
$$m/z\ 43 \qquad\qquad\qquad\qquad m/z\ 43$$

and $H_3C^{\bullet}$ and $^{\bullet}CH_3$

$$H_3C—\underset{\underset{m/z\ 58}{}}{C}—CH_3$$

Scheme 1-1

following schemes. Scheme 1-1 shows the $M^{+\bullet}$ of acetone with a positive charge and an odd electron on the oxygen atom. The carbon–carbon bond is cleaved as one of the electrons moves to "pair up" with the odd electron on the oxygen. Cleavage of either carbon–carbon bond therefore results in expulsion of a methyl radical (15 Da) from the $M^{+\bullet}$ of m/z 58 to yield a fragment ion of m/z 43. The ion of m/z 43 does not have an odd electron; i.e., all of its electrons have been paired. The high abundance of this stable acylium ion is reflected by the intense peak at m/z 43 (represented by the $[M-15]^+$ peak) in the left side of Figure 1-16.

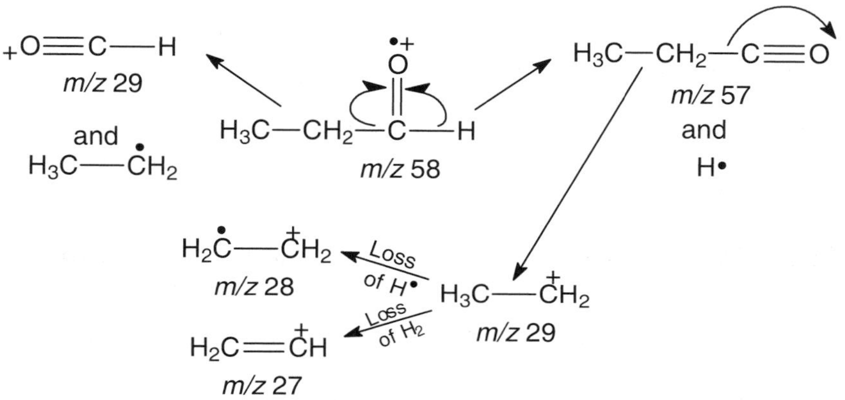

Scheme 1-2

Scheme 1-2 shows the $M^{+\bullet}$ of propionaldehyde, also with an odd electron on the oxygen atom. Again, the bonds adjacent to the carbonyl can transfer a single electron to pair with the odd electron on the oxygen. Cleavage of the carbon–carbon bond results in expulsion of an ethyl radical (29 Da) from the $M^{+\bullet}$, which has a nominal mass of 58 Da to yield a fragment ion of m/z 29. Again, this type of resonance-stabilized fragment ion (an acylium ion) is represented by an intense peak, but at m/z 29 in this case (see right side of Figure 1-16). Cleavage of the carbon–hydrogen bond (on the other side of the carbonyl group) results in expulsion of a hydrogen radical (1 Da) from the $M^{+\bullet}$ to produce the fragment ion of m/z 57; this ion is represented by the peak at m/z 57 (representing the $[M-1]^+$ ion) in the right side of Figure 1-16. The acylium ion with m/z 57 can undergo a fragmentation to lose a molecule of CO. The resulting ethyl ion can then fragment by either the loss of a hydrogen radical or a molecule of hydrogen to account for the peaks at m/z 28 and m/z 27, respectively.

Certain structural features in a molecule can manifest themselves in a particular mode of fragmentation. The phenomenon of structure-related fragmentation is the basis by which each compound that can be ionized by the EI process has its own unique and characteristic fragmentation pattern, a chemical fingerprint.

2. Example 1-2: Drug Overdose: Use of GC/MS to Identify a Drug Metabolite

In an effort to recommend rational therapy for a comatose drug-overdose victim, physicians and toxicologists prefer to analyze the patient's blood for suspected drugs. Figure 1-17 is a gas chromatogram showing separation of two major components from other constituents of a plasma extract of such a patient [59]. Peak 1 has the same retention time as that of authentic glutethimide (Doriden, a Schedule II hypnotic

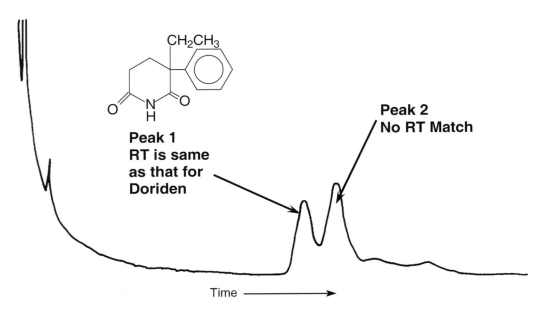

Peak 1
RT is same
as that for
Doriden

Peak 2
No RT Match

Time ⟶

Figure 1-17. *Gas chromatogram resulting from analysis of blood sample*
extract from a comatose patient following drug overdose.

sedative), which was an expected finding because an empty prescription bottle labeled
for that drug had been found near the victim. What did Peak 2 represent? Had the victim
ingested yet another drug?

Analysis of successive blood samples indicated that whereas Peak 1
(glutethimide) represented a decreasing amount with time, Peak 2 remained constant in
intensity. Because the patient's condition was deteriorating, the possibility of
hemodialysis was being considered, pending identification of the substance represented
by Peak 2 in the chromatogram (Figure 1-17). The retention time of Peak 2 under the
analytical conditions did not match that for any drug known to be available to the victim or
those for other drugs commonly abused.

Analysis of the plasma extract by GC/MS confirmed that Peak 1 does represent
glutethimide. The mass spectrum (Figure 1-18, top) obtained during the elution of Peak 1
was identical with that from authentic glutethimide [59]. The peak at m/z 217, labeled A
in the top panel of Figure 1-18, represents the M$^{+\bullet}$. The mass spectrum (Figure 1-18,
bottom) obtained during gas chromatographic elution that accounted for Peak 2 (Figure
1-17) indicates a M$^{+\bullet}$ peak at m/z 233 (also labeled A). Note that both M$^{+\bullet}$ peaks have an
odd m/z value (217 and 233), which means that each of the corresponding molecular
ions contains an odd number of nitrogen atoms (see Chapter 5 for explanation of the
Nitrogen Rule). Furthermore, the m/z values of the two M$^{+\bullet}$ peaks differ by 16, and
several other prominent peaks (B, C, D) in the mass spectra differ by 16 m/z units. This
shift of 16 m/z units suggests incorporation of oxygen into the molecule of glutethimide;
i.e., Peak 2 in the gas chromatogram (Figure 1-17) could represent a hydroxylated
metabolite of glutethimide. To confirm this possibility, the material represented by Peak 2
in the gas chromatogram was treated with acetic anhydride and pyridine, and was
reanalyzed by GC/MS. The mass spectrum of the derivatized (chemically modified)
material (data not shown) was identical with that of the acetate derivative

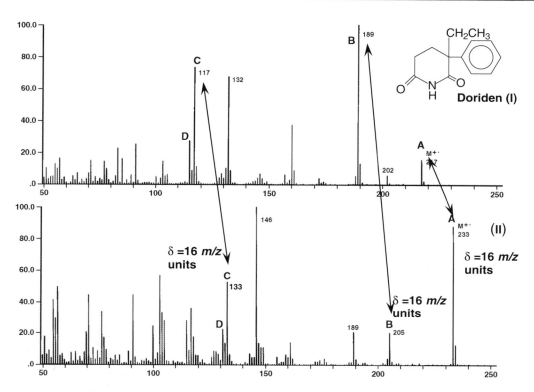

Figure 1-18. *(Top) EI mass spectrum obtained during elution of Peak 1 in Figure 1-17. This spectrum is identical to that of glutethimide (Doriden®). (Bottom) EI mass spectrum obtained during the elution of Peak 2 in Figure 1-17. Note the $M^{+\bullet}$ peak in both spectra is at an odd m/z value and that major peaks in the bottom spectrum occur 16 m/z units higher than the corresponding peaks in the top spectrum.*

of 4-hydroxy-2-ethyl-2-phenylglutarimide, an active metabolite of glutethimide that has demonstrated toxicity in animals [60].

Hemodialysis was then considered to be of reasonable risk because the 4-hydroxy metabolite was sufficiently polar to be removed by this technique. During hemodialysis, the plasma level of the 4-hydroxy metabolite dropped more rapidly than that of the less polar parent drug and the patient regained consciousness [59], possibly living happily ever after.

This example illustrates that several clues to the possible identity of the compound can be obtained from even cursory examination of a mass spectrum. In this case, the molecular ions of odd mass immediately suggested nitrogen-containing compounds (via the *Nitrogen Rule*), and the shift of 16 *m/z* units for major mass spectral peaks for the two compounds suggested that the heavier one (II) was an oxygenated variant of the other (I). A more detailed examination of the mass spectra than described here was required for complete verification. A similar result has been reported for the metabolism of aminoglutethimide [61].

3. Example 1-3: Verification that the Proper Derivative of the Compound of Interest Has Been Prepared

Many compounds have polar functional groups that do not permit direct analysis by the vapor-phase techniques of GC or GC/MS. Figure 1-19 shows a chemical scheme for modifying the structure of normetanephrine (NMN), a metabolite of norepinephrine via catechol-O-methyltransferase (COMT), which has free hydroxyl and amino groups that are unstable at temperatures of 150 to 250°C, temperatures usually employed in GC.

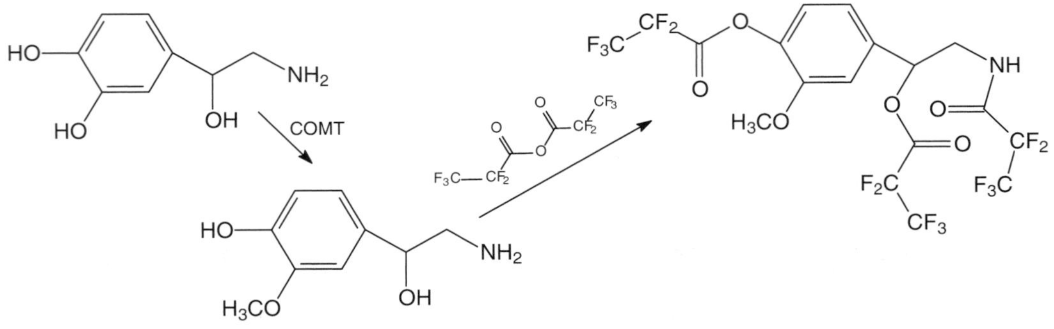

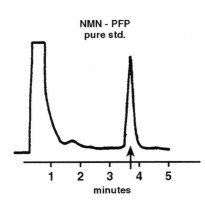

Figure 1-19.   *Norepinephrine is metabolized to normetanephrine by the enzyme catechol-O-methyltransferase (COMT). Normetanephrine is reacted with pentafluoropropionyl (PFP) anhydride to form a thermally stable and volatile derivative.*

The functional groups on these molecules tend to undergo hydrogen bonding, which is partially responsible for the low vapor pressures of these compounds. The chemical reaction at the bottom of Figure 1-19 shows the treatment of NMN with pentafluoropropionic (PFP) anhydride to produce a (*O,O',N*-tris-pentafluoropropionyl) derivative, which greatly increases the volatility and thermal stability of the molecule.

NMN - PFP pure std.

Figure 1-20. *Gas chromatogram resulting from injection NMN-PFP reaction mixture.*

The PFP derivatives of biogenic amines, such as NMN, have good vapor-phase properties. During the development of vapor-phase methods for such compounds it is good practice to verify that the proper derivative has been formed as anticipated under realistic analytical conditions, such as in low-level samples and in the presence of biological residue. During such a process of verification, analysis of an aliquot of the reaction mixture by GC produced the chromatogram illustrated in Figure 1-20. It is encouraging that Figure 1-20 indicates only a single GC peak, but does this chromatographic peak represent the structure at the bottom of Figure 1-19? That is, has the NMN molecule taken on three PFP groups, or has it taken on four? Four is a reasonable possibility because there is another replaceable hydrogen on the nitrogen atom of NMN, as seen in the structure at the bottom of Figure 1-19. Obtaining the mass spectrum should resolve the question.

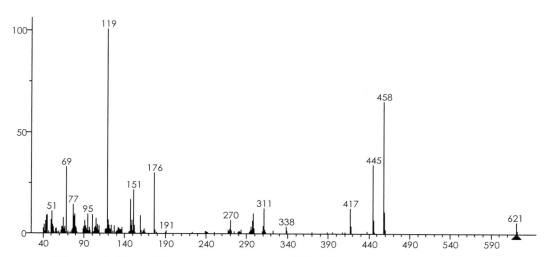

Figure 1-21. El mass spectrum of NMN as the PFP derivative.

The nominal mass of the derivative is substantially increased (a net increase of 146 Da for each PFP moiety (COC_2F_5)) over that of the underivatized compound. The mass spectrum (Figure 1-21) of the product of derivatization exhibits an apparent $M^{+\bullet}$ peak at m/z 621. A nominal mass of 621 daltons is consistent with the structure at the bottom of Figure 1-19. Had there been four PFP groups on the molecule, its nominal mass would have been 767 Da; however, no peak was detected at m/z 767 during analysis by GC/MS. Note that the $M^{+\bullet}$ peak at m/z 621 is consistent with the *Nitrogen Rule*, which states that a compound containing an odd number of nitrogen atoms will have an odd nominal mass.

The $M^{+\bullet}$ (an odd-electron species) is positively charged because it loses an electron during the process of ionization. Reasonable assumptions (described in detail in Chapters 5 and 6) concerning the site of electron deficiency in the $M^{+\bullet}$ can be proposed as an aid in rationalizing the formation of fragment ions. For example, formation of the ion of m/z 445 from NMN-PFP could be rationalized by the fragmentation shown in Scheme 1-3. Observing the peak at m/z 445 (an odd value) is consistent with a corollary to the *Nitrogen Rule*, which states that a fragment ion (as an even-electron species) having an even number of nitrogen atoms (including zero) will have an odd nominal mass.

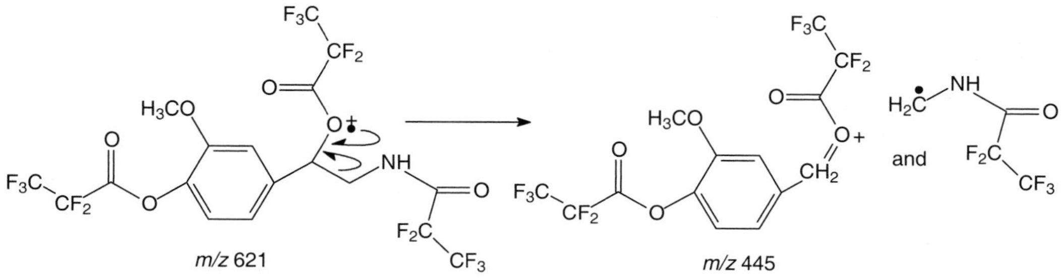

Scheme 1-3

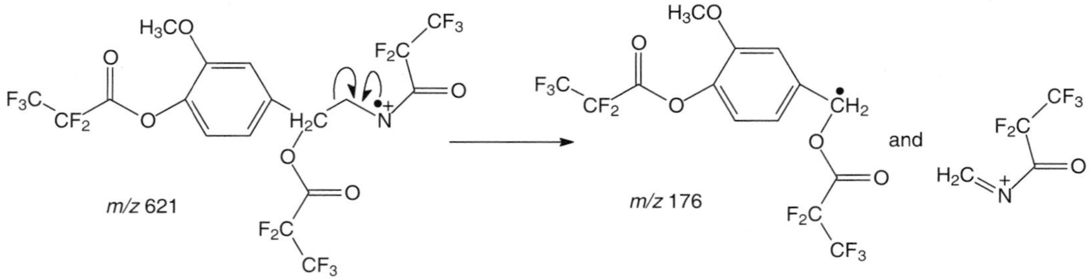

Scheme 1-4

The fragmentation pathway leading to formation of the ion of *m/z* 176 may involve a $M^{+\bullet}$ species that has an electron-deficient site on the amide nitrogen, as shown in Scheme 1-4. Observing the peak at *m/z* 176, an even value, is consistent with another corollary to the *Nitrogen Rule*, which states that a *fragment ion* that retains an odd number of nitrogen atoms will have an even nominal mass.

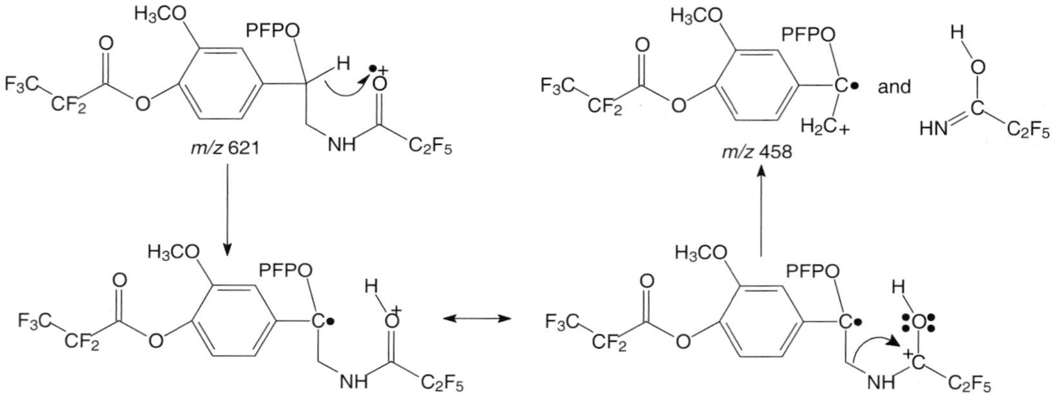

Scheme 1-5

The predominant ion of *m/z* 458 is apparently formed by a gamma-hydrogen rearrangement [62], which in this case also involves charge migration as indicated in Scheme 1-5. Observing the peak at *m/z* 458, an even value, the designated species in Scheme 1-5, is consistent with the same corollary to the *Nitrogen Rule* as described above for the peak at *m/z* 445. The fragment ion of *mass* 458 is an odd-electron species; the fragment ion of *mass* 445 is an even-electron species.

The ion of *mass* 417 is formed by expulsion of CO (28 daltons) from the fragment ion of *mass* 445 [63]. This process is supported by the observation of a "metastable" peak at *m/z* 391 (not shown in Figure 1-21), which is consistent with the transition *m/z* 445 → *m/z* 417 (417^2 / 445 = 390.8). An explanation of metastable peaks [64] and their utility in interpretation of mass spectra is rarely used these days because they are not revealed with the use of modern data systems; instead, the technique of MS/MS is commonplace, as explained in Chapter 5. Further details on rationalization of fragmentation can be found in the strategy for interpretation of EI mass spectra in Chapter 6.

The predominant ion of *m/z* 458 is apparently formed by a so-called gamma-hydrogen rearrangement [62] (explained in detail in Chapter 6), which in this case also involves charge migration as indicated in Scheme 1-5. Observing the peak at *m/z* 458, an even value, for the designated species in Scheme 1-5 is consistent with the same corollary to the *Nitrogen Rule* as described above for the peak at *m/z* 445. The fragment ion of *m/z* 458 is an odd-electron species; the fragment ion of *m/z* 445 is an even-electron species.

The ion of *m/z* 417 is formed by expulsion of CO (28 daltons) from the fragment ion of *m/z* 445 [63]. Further details on rationalization of fragmentation can be found in the strategy for interpretation of EI mass spectra in Chapter 6.

The major peaks in the mass spectrum (Figure 1-21) can be explained by reasonable cleavage processes for the proposed structure, thereby verifying the presence of such a structure. This means, the molecule whose structure is illustrated at the bottom of Figure 1-19 and is represented by the GC peak in Figure 1-20 is the product of the derivatization reaction.

4. Example 1-4: Use of a CI Mass Spectrum to Complement an EI Mass Spectrum

The barbituric acid derivative pentobarbital can be saponified by treatment with sodium hydroxide (Scheme 1-6). Because alkali can cleave one or more of the C–N bonds in the barbituric acid ring [65], it was of interest to verify that only C-2 had been removed from pentobarbital (III) to form the desired product, the malonamide (IV), which has a nominal mass of 200 Da. The purified reaction product was introduced into a mass spectrometer via the direct inlet probe to obtain the EI mass spectrum in Figure 1-21. No peak at *m/z* 200 was discernible in this mass spectrum or in others obtained at lower electron energies. Hence, there was considerable uncertainty concerning the sample and the EI mass spectra obtained from it. The peak at *m/z* 183 could represent a M$^{+\bullet}$, but not of the expected compound. Furthermore, if the peak at *m/z* 183 does, in fact, represent a M$^{+\bullet}$, the peaks at *m/z* 157 and *m/z* 130 would indicate losses of 26 and 53 Da, respectively, from an ion of *m/z* 183; these fragmentations are not easily rationalized. A preliminary explanation might be that the spectrum in Figure 1-22 is possibly the combined EI mass spectra of two or more impurities in the sample that distilled off the probe before the expected compound. However, increasing the temperature of the probe to temperatures much higher than required for barbituric acid derivatives produced no additional mass spectra. Furthermore, as the putative reaction product had been recrystallized, it seemed unlikely that the sample would be impure.

It is also possible that the malonamide of pentobarbital (IV) does not produce a stable M$^{+\bullet}$ under EI; such a M$^{+\bullet}$ could readily lose ammonia (17 daltons) to produce an ion of *m/z* 183. However, for purposes of identification, it would be desirable to observe an ion that consists of the intact molecule.

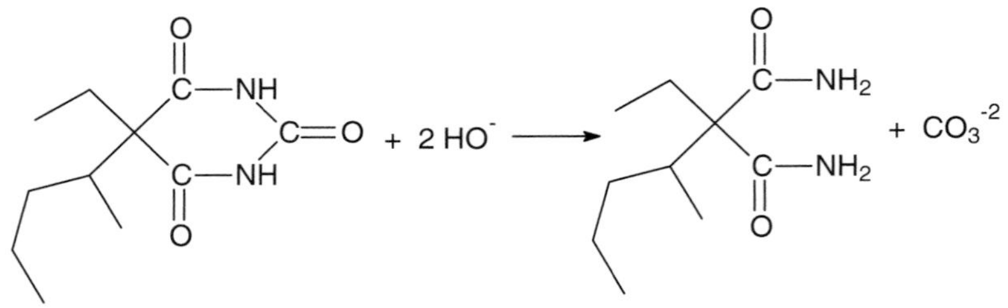

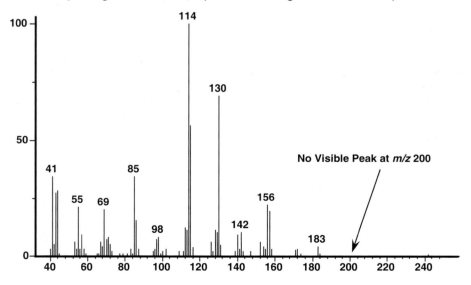

Pentobarbital (III)   Malonamide of Pentobarbital (IV)

Scheme 1-6

Chemical ionization (CI) using methane produced the mass spectrum in panel A of Figure 1-23. The series of peaks at *m/z* 201, 229, and 241 is consistent with the expected adduct ions of the series $[M + 1]^+$, $[M + 29]^+$, and $[M + 41]^+$, respectively, for CI by methane (see Chapter 7). Thus, the series of high-mass peaks appears to include a molecular species having a nominal mass of 200 Da. On the other hand, it is somewhat unusual to observe such intense low-mass peaks at *m/z* 156 and 184, although CI by methane can effect a significant amount of fragmentation of the intact molecule.

The ammonia CI spectrum shown in panel B of Figure 1-23 gives a more distinct indication of the intact molecule of the malonamide (IV). Because ammonia has a higher proton affinity than methane, its reagent ions effect protonation of the sample with less energy transfer; therefore, CI with ammonia results in less fragmentation of the ion representing the intact molecule than CI by methane (see Chapter 7). The predominant peak for the protonated molecule, MH^+, in panel B Figure 1-23 is a clear indication that the nominal mass of the analyte is 200 Da.

The CI spectra in panels A and B of Figure 1-23 confirm that the malonamide of pentobarbital is present and that an $[M + 1]^+$ of nominal mass 200 Da should be assumed when interpreting the EI mass spectrum in Figure 1-22. The peak at *m/z* 183 does

Figure 1-22.   *EI mass spectrum of the malonamide (IV) of pentobarbital.*

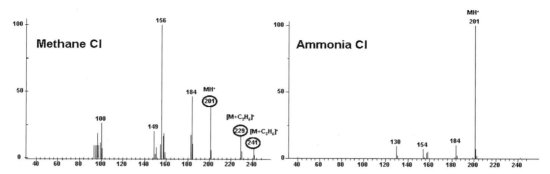

Figure 1-23. *CI spectra of the malonamide of pentobarbital using methane (panel A) or ammonia (panel B) as reagent gas.*

represent a fragment ion $[M - 17]^+$ formed by loss of ammonia from the $[M + 1]^+$, $([M - NH_3]^+)$. The ion of m/z 156 results from loss of one of the amide groups from the $[M + 1]^+$. The ion of m/z 130 is formed by elimination of pentene via a gamma-hydrogen rearrangement (explained in Chapter 6). Two possible routes of formation for an ion of m/z 114 could be seriously considered; the use of accurate mass data obtained from the EI of this malonamide would clarify which pathway is extant.

The important point illustrated by this example is the complementary information available from EI and CI. Structural information is available from the fragmentation pattern in EI, whereas CI provides nominal mass information of the intact molecule.

5. Example 1-5: Use of Exact Mass Measurements to Identify Analytes According to Elemental Composition

Several different types of modern mass spectrometers have sufficient resolving power and mass accuracy that two or more ionic species having the same nominal mass can be separated and analyzed accurately. For example, two ionic species of m/z 114.0555 and 114.0918 can be resolved easily into two separate peaks by a high-resolving power mass spectrometer, whereas a low-resolving power instrument (R = 500) would produce only a single mass spectral peak representing both species. A resolving power of 10,000 and mass accuracy of a few parts per million (ppm) is usually sufficient to define the elemental composition of an unknown ion having a nominal mass <500 Da. The availability of such data strengthens the basis for proper interpretation of a mass spectrum for purposes of structure identification [66, 67]. For example, measuring the mass of oligonucleotides to 0.01% restricts the possible compositions that must be considered for an unknown [68].

An appreciation of the value of accurate mass measurements can be gained by considering problems in the interpretation of a low-resolution unknown mass spectrum such as that in Figure 1-24. A reasonable start for interpreting such a mass spectrum might be to guess that the $M^{+\bullet}$ peak is at m/z 308. If the peak at m/z 308 is a $M^{+\bullet}$ peak, then the unknown does not contain an odd number of nitrogen atoms. It might even be reasonable to assume that the unknown does not contain any atoms of nitrogen because zero is an even number; further, the peak at m/z 279 is an odd number, also indicating that the corresponding ion does not contain an odd number of nitrogen atoms. The loss of 29 Da could represent the loss of an ethyl radical or the loss of an $HC^{\bullet}O$ radical from the $M^{+\bullet}$. The peak at m/z 238 might represent the loss of a C_5H_{10} (70 Da) molecule from the $M^{+\bullet}$ although this would be rather unusual. Unfortunately, as will become evident

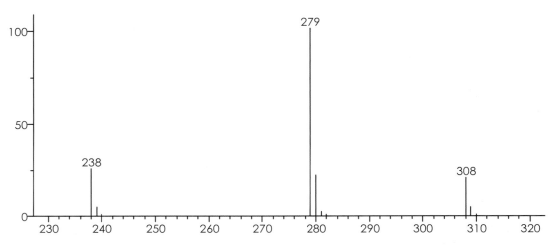

Figure 1-24.   An abbreviated EI mass spectrum of an unknown compound.

from data in Tables 1-1 and 1-2, some of these preliminary interpretations are off-base, even though they do not violate the *Nitrogen Rule* or involve illogical losses.

If Figure 1-24 represented the only information available for this unknown, several different compounds having a nominal of 308 Da, including those that contained an even number of nitrogen atoms, would have to be considered. In fact, there are over 200 different combinations of carbon, hydrogen, nitrogen, and oxygen that have a nominal mass of 308 Da. Table 1-1 contains some representative listings of possible elemental compositions that have a nominal molecular weight of 308 Da, but different exact masses [69]. Some of these possibilities are implausible, such as the third entry, which indicates that 18 carbon atoms are substituted with only 4 hydrogen, 4 nitrogen, and 2 oxygen atoms. However, the vast majority are reasonable elemental compositions that are worthy of serious consideration unless specifically eliminated by other data that may be available for the unknown compound.

Table 1-1. Abbreviated listing of C, H, N, and O combinations that have a nominal mass of 308.

Elemental Composition				
C	**H**	**N**	**O**	**Exact Mass**
18	20	4	1	308.1637
18	12	–	5	308.0685
18	4	4	2	308.0334
19	2	1	4	307.9984
19	20	2	2	308.1525
19	36	2	1	308.2827
20	12	4	–	308.1062
20	24	2	1	308.1888

Table 1-2 shows experimentally determined accurate mass measurements for the unknown compound obtained during analysis with a double-focusing instrument. The second column lists the elemental composition that most closely agrees with the experimentally determined mass. The third column shows the calculated exact mass of the elemental composition in column two; this value was calculated by summing the exact monoisotopic masses of the designated elements and is provided to show the accuracy of the experimental value. As a further indication of the high probability that the elemental composition corresponds to the measured accurate mass, the fourth and fifth columns show the deviation in exact mass and in ppm between the measured accurate mass and the calculated exact mass, respectively.

Table 1-2. **High-resolution mass spectral data on selected peaks of unknown mass spectrum.**

Accurate Mass Determination	Most likely Corresponding Elemental Composition	Calculated Exact Mass	Δ mmu	Δ ppm
308.1878	$C_{20}H_{24}N_2O$	308.1888	−1.0	3.2
279.1510	$C_{18}H_{19}N_2O$	279.1497	1.3	4.6
238.1249	$C_{16}H_{16}NO$	238.1232	1.7	7.1

The data in Table 1-2 indicate that there were some incorrect assumptions in the preliminary interpretation described above of the unknown low-resolution mass spectrum (Figure 1-24). For example, the exact mass of the molecular ion corresponds to an elemental composition of $C_{20}H_{24}N_2O$. There was no way to tell from the low-resolution data that the peak at *m/z* 308 represented an ion containing two atoms of nitrogen; a compound containing two atoms of nitrogen has an even nominal mass, just like compounds containing no nitrogen according to the *Nitrogen Rule*. The assumption that the $M^{+\bullet}$ lost an ethyl radical to give an $[M - 29]^+$ ion represented by the peak at *m/z* 279 was correct. The data in Table 1-2 indicate that peaks at *m/z* 308 and *m/z* 279 both represent ions containing two atoms of nitrogen and one of oxygen. The other incorrect assumption (besides the fact that the $M^{+\bullet}$ contained no atoms of nitrogen) was that the peak at *m/z* 238 (loss of 70 Da) represented the loss of C_5H_{10} from the $M^{+\bullet}$; in fact, it represents the loss of a C_4H_8N radical. Note that the last entry in column two of Table 1-2 indicates that the ion of nominal mass 238 contains only one atom of nitrogen and one atom of oxygen; therefore, one of the two nitrogen atoms in the $M^{+\bullet}$ is lost in forming this ion.

For many analyses, low-resolution mass spectra are adequate to solve the problem, because reference spectra or other data are available to help clarify any ambiguities that may arise from the data. In a few cases in which the sample is truly an unknown, and especially if no ancillary data are available, accurate mass measurements are essential to guide correct interpretation of the mass spectrum.

6. Example 1-6: Is This Protein Phosphorylated? If So, Where?

Determination of whether a given protein or peptide is phosphorylated involves analyzing the protein before and after treatment with phosphatase (an enzyme that cleaves a phosphate ester). If comparison of the data before and after treatment with phosphatase shows a shift in the molecular weight of the protein by multiples of 80 Da (the additional mass associated with attaching a phosphate group), then the original peptide or protein contains a number of phosphate groups equal to the multiples of 80 Da corresponding to the difference in molecular weights of the treated and untreated protein. Matrix-assisted laser desorption/ionization (MALDI) as described in detail in Chapter 9 can be used to analyze the peptide.

The MALDI mass spectrum of the phosphorylated model peptide amide is shown in the top panel of Figure 1-25; this spectrum has a peak at *m/z* 1424 corresponding to the protonated molecule. Treatment of the model peptide with alkaline phosphatase yields a peptide which, when analyzed by MALDI, presents the mass spectrum shown in panel B of Figure 1-25, which exhibits the appearance of a peak at *m/z* 1344, 80 *m/z* units lower than the peak at *m/z* 1424, corresponding to the removal of a phosphate group [70].

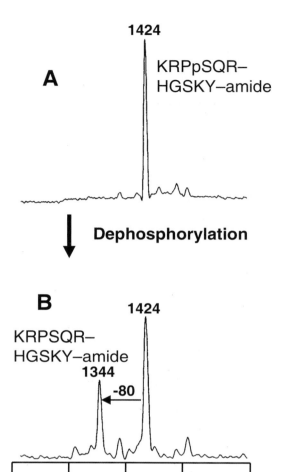

A

1424

KRPpSQR–
HGSKY–amide

Dephosphorylation

B

1424

KRPSQR–
HGSKY–amide

1344

−80

| 1200 | 1300 | 1400 | 1500 | 1600 |

m/z

Figure 1-25.   Matrix-assisted laser desorption/ionization (MALDI) mass spectrum of a phosphopeptide before (A) its treatment with phosphatase; (B) shift of 80 from m/z 1424 to m/z 1344 indicates loss of the phosphate group.

Because the location of the phosphate is of interest, all that is needed is to know the sequence of the peptide. If there were more than one potential site for phosphorylation (e.g., at serine, threonine, or tyrosine), the peptide could be treated enzymatically and then the mixture of degradation products would be analyzed by MALDI MS. The analytical process of peptide mass mapping to recognize the site of phosphorylation is demonstrated in the example illustrated in Figure 1-26 for a peptide having the following sequence: p-KRPSQRHGSKY-amide. Upon treatment of the peptide with trypsin, it would be expected that cleavage at the C-terminal side of arginine (R) would occur to yield two peptide fragments as represented in the middle panel of Figure 1-26. The peak at *m/z* 591 corresponds to HGSKY-amide, which computes to a protonated mass of 591 Da without the phosphate on either of the two possible phosphorylation sites at S9 or Y11. The peak at *m/z* 852 corresponds to KRPpSQR as a protonated molecule containing the phosphate group. Whereas the spectrum in panel B shows, by deduction, that the phosphate group must be on S4, treatment of the peptide mixture with alkaline phosphatase with subsequent analysis of that mixture by MALDI

produces the data in panel C to confirm this conclusion; i.e., the peak at *m/z* 591 remains in the MALDI mass spectrum in panel C, indicating that the peptide HGSKY-amide did not contain the phosphate as the mass of this peptide fragment did not shift. On the other hand, there is no peak at *m/z* 852 in panel C, whereas there is a peak at *m/z* 772, 80 *m/z* units lower than the peak at *m/z* 852. This gives direct evidence that treatment of this peptide fragment with alkaline phosphatase removed a phosphate group, which had to be at S4 as this is the only phosphorylatable site present in this peptide fragment [70].

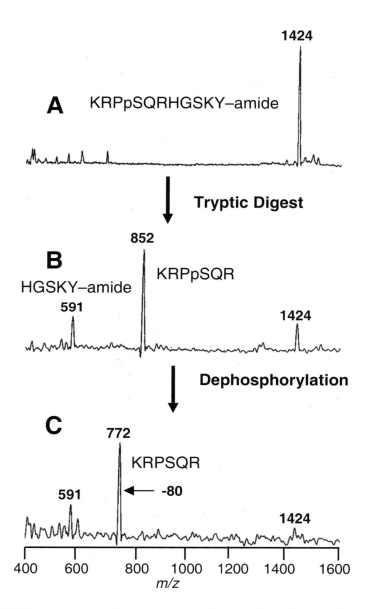

Figure 1-26. **MALDI mass spectrum of phosphopeptide before (A) and after (B) treatment with trypsin in an effort to map the location of the phosphate group. (C): MALDI spectrum of tryptic digest after treatment with phosphatase.** *Reprinted from Liao P-C, Leykam J, Andrews PV, Gage DA, and Allison J "An approach to locate phosphorylation sites in proteins by MALDI" Anal. Biochem. 1994, 219, 9-20, with permission from Academic Press.*

7. Example 1-7: Clinical Diagnostic Tests Based on Quantitation of Stable Isotopes by Mass Spectrometry in Lieu of Radioactivity

Many diagnostic methods for assessment of metabolic disorders are not available to a large segment of the population (pregnant or pediatric patients) because the tests involve oral or intravenous administration of radioactivity. However, with the increasing availability of stable isotope-labeled compounds, many existing methods can be converted to a protocol utilizing nonradioactive biochemical probes to obtain the same diagnostic information without the hazard of radiation [71–75].

A breath test for malabsorption is a noninvasive means of evaluating or recognizing this disorder. The absorption of lipid nutrients is essential for the growth and development of newborn infants [76]. Malabsorption in adults can be evaluated by administering $1\text{-}^{14}C$-fatty esters and measuring the rate of production of radioactive CO_2 in the breath as the radioactive lipids are absorbed by the gut and metabolized by the liver [77]. A similar method involving administration of a ^{13}C-labeled lipid and measurement of expired $^{13}CO_2$ by mass spectrometry can be used to assess malabsorption in children, because the test imposes no radiation hazard [78].

The breath test for malabsorption consists of administering a test meal containing a dose (10 mg/kg) of $1\text{-}^{13}C$-trioctanoin, a medium-chain triglyceride [78]. Octanoic acid, liberated in the small intestine by lipase, is absorbed and rapidly transported in portal blood to the liver, where it is quantitatively oxidized to CO_2. Expired air is collected by means of a face mask for a 5-min period every 30 min for 2 hr. The CO_2 is trapped by bubbling the expired air through 10 ml of carbonate-free sodium hydroxide. This solution can be stored or transported conveniently prior to analysis, at which time it is treated with acid to release quantitatively the carbon dioxide, which is transferred to an inlet reservoir on the mass spectrometer.

A typical plot of analytical results of the breath test is presented in Figure 1-27. At 1 hr after ingesting the dose of $1\text{-}^{13}C$-trioctanoin, the abundance of $^{13}CO_2$ in the respiratory carbon dioxide of normal children is approximately five times greater than that in the breath of children suffering from cystic fibrosis (see dashed line in Figure 1-27) [71]. This means, this breath test readily differentiates patients with normal fat absorption from those with significant fat malabsorption.

Other tests involving biochemical probes with appropriate ^{13}C- and ^2H-labeled biochemicals play an important role in the diagnosis of metabolic disorders, because the tests are simple, noninvasive, and sensitive and involve no radiation hazard. However, it must be emphasized that reliable quantitation of the slight changes in the relative abundances of ^2H and ^{13}C requires an isotope ratio mass spectrometer [79, 80].

V. The Need for Chromatography

In the evolution of mass spectrometry, there have been efforts to analyze samples that are complex mixtures without preliminary separation. In some cases, LC/MS instrumentation is used to analyze a sample by direct infusion without ever trying for a chromatographic separation. MS/MS has changed mass spectrometry. At one point, shortly after the development of the triple-quadrupole mass spectrometer, Graham Cooks, another venerable mass spectrometrist said, "There is no longer a need for GC/MS. All necessary separations of analytes of interest can be accomplished using MS/MS." Interestingly, some thirty years later, chromatography plays an even bigger role than it did at the time the triple quadrupole was being developed.

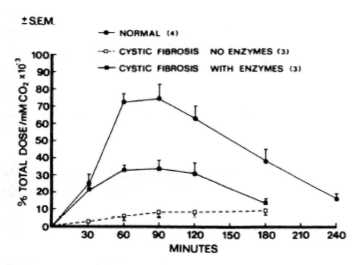

Figure 1-27. **Plot of excess $^{13}CO_2$ beyond natural abundance in the breath of normal vs cystic fibrosis patients following an oral dose (10 mg kg^{-1}) of 1-^{13}C-trioctanoin.** *Reprinted from Barr RG, Perman JA, Schoeller DA, and Watkins JB "Breath tests in pediatric gastrointestinal disorders: new diagnostic opportunities" Pediatrics 1978, 62(3), 393–401, with permission from the American Academy of Pediatrics.*

Mass spectrometry can provide useful qualifying information for a pure substance; however, mass spectral data obtained from a mixture are often not useful. This is especially true of EI mass spectra, which result from extensive fragmentation. In the case of analyzing a mixture, the fragmentation patterns of the various components overlap, and much of the interpretative value is lost by confounding. The mass spectrometer is more reliable as a separation device when soft ionization techniques like ESI, APCI, and APPI are employed. Even in these situations, analytes of different structures and elemental composition can form ions representing the intact molecule that have the same nominal *m/z* value. These analyses require chromatography for separation before they ever "see" the mass spectrometer. Simplified mass spectra, consisting principally of ions representing the intact molecules, can be acquired from simple mixtures of two to five components. The complexity of other mixtures may need to be simplified by some form of coarse separation prior to analysis by the above-referenced soft ionization techniques or MALDI.

Chromatography, with its capacity to separate similar compounds based on subtle structural features, can be used to isolate components so that they can be analyzed individually. In many cases, it is possible to use chromatography as an inlet system for MS to provide for on-line separations and acquisitions of mass spectra of individual components as described in detail in Chapters 10 (GC/MS) and 11 (LC/MS). Although it is easy to view a chromatography system as an inlet for a mass spectrometer, it is important to keep in mind that a chromatograph/mass spectrometer system is as different from either a chromatograph or a mass spectrometer as a chromatograph and a mass spectrometer are from one another.

VI. Closing Remarks

The field of mass spectrometry has grown beyond imagination since the beginning of commercial instruments in the 1940s. This is exemplified by the six current journals dedicated to mass spectrometry and the fact that two of these journals have been published since 1968: *Journal of Mass Spectrometry* (formed in 1995 by the merger of *Organic Mass Spectrometry* (1968) and *Biological Mass Spectrometry* (1974) by John Wiley and Sons, Inc.) and the *International Journal of Mass Spectrometry* (originated in 1968 as the *International Journal of Mass Spectrometry and Ion Physics* by Elsevier). The other mass spectrometry journals are *Mass Spectrometry Reviews* (John Wiley & Sons, Inc., 1982), *Rapid Communications in Mass Spectrometry* (John Wiley & Sons, Inc., 1987), the *Journal of the American Society for Mass Spectrometry* (American Society for Mass Spectrometry, 1990), and *Journal of European Mass Spectrometry* (IM Publications, 1995). In addition to these six dedicated publications, many articles dealing primarily with mass spectrometry appear in countless other journals such as the *Journal of Chromatography A* and *B*, *Analytical Chemistry*, *Journal of Chromatographic Science*, *Journal of Toxicology*, etc., to name just a few.

The American Society for Mass Spectrometry's Annual Meeting on Mass Spectrometry and Allied Topics, held for the past fifty-five years in various locations in North America, now attracks over six thousand people. The triennial International Mass Spectrometry Conference (held in various locations in Europe to this point), having its official beginning in 1958 but tracing its origins back to a meeting held in Manchester, England, April 20–21, 1950, and organized by The Institute of Petroleum, is seeing the same types of increases in attendance as being seen by ASMS. All of this is occurring while other meetings such as the annual Pittsburgh Conference, held in various cities in the United States, is seeing significant declines in attendance. As mass spectrometry went from the accurate mass determination of nuclides to the determination of the nominal mass and the elucidation of the structure of volatile organic compounds to the ability to provide mass and structure information of macromolecules in the condensed phase, it has proven to be the technique that provides more information from less sample than any other spectrometric method or other analytical technique, to paraphrase the statement at the end of the first paragraph of this chapter.

VII. Monographs on Mass Spectrometry Published Before 1970.

Pre-1920 (1)

1. Thomson, JJ *Rays of Positive Electricity and their Application to Chemical Analysis*, 1st ed., 1913.

1920–1930 (2)

1. Thomson, JJ *Rays of Positive Electricity and their Application to Chemical Analysis*, 2nd ed.; Longmans Green: London, 1921.

2. Aston, FW *Isotopes*, 2nd ed.; Edward Arnold: London, 1924; 1st ed., 1922.

1930–1940 (7)

1. Aston, FW *Mass Spectrometry and Isotopes*, 2nd ed.; Edward Arnold: London, 1st ed., 1933.

2. Mott, NF; Massey, HSW *The Theory of Atomic Collisions*, 1st ed., 1933; Oxford University Press: London.

3. Bauer, SH *A Mass Spectrograph: Products and Processes of Ionization in Methyl Chloride*; 1935; iv, 51 pp; illustrations.

4. Cartan, L *Spectrographie de Masse: Les Isotopes et Leurs Masses*; No. 550 in the series *Actualités Scientifiques et Industrielles*; No. VI in the subseries *Exposés de Physique Atomique Expérimentale*; Preface by Maurice de Broglie; Hermann & Cie: Paris, 1937; pp 3–90, pbk.

5. Massey, HSW *Negative Ions*, 3 rd ed. (741 pages); Cambridge at the University: London, 1976; 1950; 2nd ed. (136 pages); 1st ed. (105 pages), 1938.

6. Cartan, L *L'optique des Rayons Positifs et ses Applications à la Spectrographie de Masse*; Hermann & Cie: Paris, 1938; p 80.

7. Thibaud, J; Cartan, L; Comparat, P *Quelques Techniques Actuelles en Physique Nucléaire*, subtitle that appears to be the contents *Méthode de la Trochoïde: Électrons Positifs: Spectrographie de Masse: Isotopes: Compteurs de Particules à Amplification Linéaire Compteurs de Geiger et Müller*; Hermann & Cie: Paris, 1938; p 276.

8. Loeb, LB *Fundamental Processes of Electrical Discharge in Gases*; Wiley: New York, 1939.

1940–1950 (2)

1. Aston, FW *Mass Spectrometry and Isotopes*, 2nd ed.; Edward Arnold: London, 1942.

2. Mott, NF; Massey, HSW *The Theory of Atomic Collisions*, 2nd ed., 1949; Oxford University Press: London.

1950–1960 (17, 6 Proceedings)

1. Massey, HSW; Burhop EHS *Electronic and Ionic Impact Phenomena*; Oxford University Press: London, 1952; 2nd printing, 1956.

2. *Mass Spectrometry,* a report of a conference organized by The Mass Spectrometry Panel of The Institute of Petroleum, London, April 20, 21, 1950; The Institute of Petroleum: London, 1952. (a **Proceedings**)

3. Ewald, H; Hintenberger, H *Methoden und Anwendunyngen der Massenspektroskopie*; Verlag Chemie: Weinheim, Germany, 1952 (English translation by USAEC, Translation Series AEC-tr-5080; Office of Technical Service: Washington, DC, 1962).

4. Dowben, RM *Mass Spectrometry*; Joint Establishment for Nuclear Energy Research: Kjeller per Lillestrøm: Norway, 1952; 74 pp; diagrams.

5. Bainbridge, KT "Part V: Charged Particle Dynamics and Optics, Relative Isotopic Abundances of the Elements, Atomic Masses" in *Experimental Nuclear Physics*, Vol. 1; Sergè, E, Ed.; Wiley: New York, 1953; pp 559–767.

6. Barnard, GP *Modern Mass Spectrometry*; American Institute of Physics: London, 1953.

7. Robertson, AJB *Mass Spectrometry: Methuen's Monographs on Chemical Subjects*; Wiley: New York, 1954.

8. Hipple, JA; Aldrich, LT; Nier, AOC; Dibeler, VH; Mohler, FL; O'Dette, RE; Odishaw, H; Sommer, H (Mass Spectroscopy Committee) *Mass Spectrometry in Physics Research*, National Bureau of Standards Circular 522; United States Government Printing Office: Washington, DC, 1953. (a **Proceedings**)

9. Inghram, MG; Hayden, RJ *A Handbook on Mass Spectroscopy*, Nuclear Science Report No. 14; National Academy of Science, National Research Council Publication 311: Washington, DC, 1954 (manuscript completed in April 1952 and intended as a chapter in a proposed handbook on nuclear instruments and techniques for the Subcommittee on Nuclear Instruments and Techniques of the Committee on Nuclear Science of the National Research Council that was never completed.

10. Blears, J (Chairman, Mass Spectrometry Panel, Institute of Petroleum) *Applied Mass Spectrometry*, a report of a conference organized by The Mass Spectrometry Panel of The Institute of Petroleum, London, 29–31 October 1953; The Institute of Petroleum: London, 1954. (a **Proceedings**)

11. Loeb, LB *Basic Processes of Gaseous Electronics*; University of California Press: Berkeley, CA, 1955 (revised and reprinted as the 2nd edition in 1960 with Appendix I).

12. Smith, ML, Ed. *Electromagnetically Enriched Isotopes and Mass Spectrometry*, Proceedings of the Harwell Conference, Sept. 13–16, 1955; Butterworth: London, 1956. (a **Proceedings**)

13. Rieck, GR *Einführung in die Massenspektroskopie* (translated from Russian); VEB Deutscher Verlag der Wissebschaften: Berlin, 1956.

14. Hinteberger, H, Ed. *Nuclear Masses and their Determination*, Proceedings of the Conference "Max-Planck-Institute für Chemie" Mainz, 10–12 July 1955; Pergamon: London, 1957. (a **Proceedings**)

15. Duckworth, HE *Mass Spectroscopy*; Cambridge: London, 1958.

16. Loeb, LB *Static Electrification*; Springer-Verlag: Berlin, 1958.

17. Waldron, JD, Ed. *Advances in Mass Spectrometry*, Vol. 1; Pergamon: New York, 1959 (a **Proceedings**).

1960–1970 (56)

1. Beynon, JH *Mass Spectrometry and Its Applications to Organic Chemistry*; Elsevier: Amsterdam, 1960 (reprinted by ASMS, 1999).

2. Biemann, K *Mass Spectrometry: Organic Chemical Applications*; McGraw-Hill: New York, 1962 (reprinted by ASMS, 1998).

3. Reed, RI *Ion Production by Electron Impact*; Academic: London, 1962.

4. McLafferty, FW Mass Spectrometry. In *Determination of Organic Structures by Physical Methods*, Vol. II; Nachod, FC; Phillips, WD, Eds.; Academic: New York, 1962.

5. Beynon, JH; Williams, AE *Mass and Abundance Tables for Use in Mass Spectrometry*; Elsevier: New York, 1963.

6. Elliott, RM, Ed. *Advances in Mass Spectrometry*, Vol. 2; Pergamon: New York, 1963.

7. McDowell, CA, Ed. *Mass Spectrometry*; McGraw-Hill: New York, 1963 (reprinted by Robert E. Krieger: Huntington, NY, 1979).

8. McLafferty, FW, Ed. *Mass Spectra of Organic Ions*; Academic: New York, 1963.

9. Budzikiewicz, H; Djerassi, C; Williams, DH *Structure Elucidation of Natural Products by Mass Spectrometry*, Vol. I *Alkaloids*; Vol. II *Steroids, Terpenoids, Sugars, and Miscellaneous Natural Products*; Holden-Day: San Francisco, CA, 1964.

10. Budzikiewicz, H; Djerassi, C; Williams, DH *Interpretation of Mass Spectra of Organic Compounds*; Holden-Day: San Francisco, CA, 1964.

11. Lederberg, J *Computation of Molecular Formulas for Mass Spectrometry*; Holden-Day: San Francisco, CA, 1964.

12. Brunnee, C; Voshage, H *Massenspektrometrie*; K. Thiemig: Munich, 1964.

13. McDaniel, EW Collision Phenomena in Ionized Gases; Wiley: New York, 1964.

14. Kiser, RW *Introduction to Mass Spectrometry and Its Application*; Prentice-Hall: Englewood Cliffs, NJ, 1965.

15. Beynon, JH; Sanders, RA; Williams, AE *Table of Meta-stable Transitions*; Elsevier: New York, 1965.

16. Reed, RI, Ed. *Mass Spectrometry*, Proceedings of the 1st NATO Advanced Study Institute on Theory, Design, and Applications, Glasgow, Scotland, August 1964; Academic: London, 1965.

17. Quayle, A; Reed, RI Interpretation of Mass Spectra. In *Interpretation of Organic Spectra*; Mathieson, DW, Ed.; Academic: New York, 1965.

18. Thomson, GP *J. J. Thomson and the Cavendish Laboratory in His Day*; Doubleday: New York, 1965.

19. Reed, RI, Ed. *Mass Spectrometry*, Proceedings of the 1st NATO Advanced Study Institute of Mass Spectrometry on Theory, Design, and Applications; Academic: London, 1965.

20. Loeb, LB *Electric Coronas: Their Basic Physical Mechanisms*; University of California Press: Berkeley, CA, 1965.

21. Mott, NF; Massey, HSW *The Theory of Atomic Collisions*, 3rd ed.; Oxford University Press: London, 1965; 1st ed., 1933; 2nd ed., 1949.

22. Reed, RI *Applications of Mass Spectrometry to Organic Chemistry*; Academic: New York, 1966.

23. Spiteller, G *Massenspektrometrische Strukturanalyse Organischer Verbindungen*; Verlag Chemie: Weinheim, Germany, 1966.

24. Polyakova, AA; Khmel'nitskii, RA *Introduction to Mass Spectrometry of Organic Compounds*; Schmorak, J, Translator; Israel Program For ScientificTranslations: Jerusalem, Israel, 1968 (original Russian language edition, *Vvedenie V Mass Spektrometriyu Organicheskikh Soedinenii*; Izdatel'stvo "Khimya": Moskva-Leningrad, 1966).

25. Blauth, EW *Dynamic Mass Spectrometers* (translated from German); Elsevier: Amsterdam, 1966.

26. Jayaram, R *Mass Spectrometry: Theory and Applications*; Plenum: New York, 1966.

27. Mead, WL, Ed. *Advances in Mass Spectrometry*, Vol. 3; Pergamon: New York, 1966.

28. Thomson, GP *J. J. Thomson: Discoverer of the Electron*; Doubleday Anchor: Garden City, NY, 1966.

29. Ausloos, PJ, Ed. *Ion-Molecule Reactions in the Gas Phase*; ACS Advances in Chemistry Series 58; American Chemical Society: Washington, DC, 1966.

30. Budzikiewicz, H; Djerassi, C; Williams, DH *Mass Spectrometry of Organic Compounds*; Holden-Day: San Francisco, CA, 1967.

31. Kientiz, H; Aulinger, FG; Habfast, K; Spiteller, G *Mass Spectrometry*; Verlag Chemie: Weinheim, Germany, 1968.

32. Roboz, J *Introduction to Mass Spectrometry Instrumentation and Techniques*; Wiley: New York, 1968 (reprinted by ASMS, 2000).

33. Pierce, AE *Silylation of Organic Compounds*; Pierce Chemical: Rockford, IL, 1968.

34. White, FA *Mass Spectrometry in Science and Technology*; Wiley: New York, 1968.

35. Tatematsu, A; Tsuchiya T, Editors-in-Chief *Structure Indexed Literature of Organic Mass Spectra, 1966*; Organic Mass Spectral Data Division, Society of Mass Spectrometry of Japan; Academic Press of Japan: Tokyo, 1968.

36. Beynon, JH; Saunders, RA; Williams, AE *The Mass Spectra of Organic Molecules*; Elsevier: Amsterdam, 1968.

37. Horning EC; Brooks, CJW; Vanden Heuvel, WJA *Gas Phase Analytical Methods for the Study of Steroids*, Vol. 6; Academic: New York, 1968.

38. Reed, RI, Ed. *Modern Aspects of Mass Spectrometry*, Proceedings of the 2nd NATO Advanced Study Institute of Mass Spectrometry on Theory, Design, and Applications, July 1966, University of Glasgow, Glasgow, Scotland; Plenum: New York, 1968.

39. Moiseiwitsch, BL; Smith, SJ *Electron Impact Excitation of Atoms*; National Standard Reference Data Series; National Bureau of Standards; Astin, AV, Director; Smith, CR, Director; United States Department of Commerce: Washington, DC, August, 1968; (reprinted from *Rev. Mod. Phys.* **1968** [April], *40*[2]).

40. Kientiz, H *Massenspektrometrie*; Verlag Chemie: Weinheim, Germany, 1968.

41. Reed, RI, Ed. *Modern Aspects of Mass Spectrometry*, Proceedings of the 2nd NATO Advanced Study Institute of Mass Spectrometry on Theory, Design, and Applications, July 1966, University of Glasgow, Glasgow, Scotland; Plenum: New York, 1968.

42. Brymner, R; Penney, JR, Eds. *Mass Spectrometry*, Proceedings of the Symposium on Mass Spectrometry, Enfield College of Technology, July 5, 6, 1967; Chemical: New York, 1969.

43. Knewstubb, PF *Mass Spectrometry and Ion-molecule Reactions*; Cambridge University Press: Cambridge, UK, 1969.

44. Ettre, LS; McFadden, WH, Eds. *Ancillary Techniques of Gas Chromatography*, Wiley–Interscience, New York, 1969, ISBN:471246700.

The following is a listing of the proceedings published for the two major regularly held mass spectrometry meetings during the period of 1960 to 1970. These meetings continue to be held and should be continuing.

1. *Proceedings of the nth Annual ASMS Conference on Mass Spectrometry and Allied Topics* published annually; American Society for Mass Spectrometry: Santa Fe, NM (ASTM E14 Committee Meetings began in 1952; published proceedings annually from 1961 through 1969 by the ASTM E14 Committee as the *Proceedings of the nth Annual Conference on Mass Spectrometry and Allied Topics*; ASMS began in 1970 and published the Proceedings of the 18th Conference). Beginning with the 1999 meeting, the Proceedings are only available on CD-ROM.

2. *Advances in Mass Spectrometry,* Proceedings of the Triennial International Mass Spectrometry Conference. Vols. 1 (1958, UK), 2 (1961, UK), 3 (Paris, France, 1964), 4 (Berlin, Germany, 1967, were published during the 1960–1970 period. Over the past 50+ years, this conference has been held. Different publishers have produced different ones of the proceedings of these conferences.

References

1. Duckworth HE and Nier AO, Evolution of the unified scale of atomic mass , carbon 12 = 12u. *International Journal of Mass Spectrometry and Ion Processes* **86:** 1-19, 1988.
2. Dalton J, *A New System of Chemical Philosophy*, Manchester, 1808.
3. Berzelius JJ, *Jahresbericht* **8:** 73, 1828.
4. Goldstein E, Über eine noch nicht unterschte Strahlungsform an der Katode inducirter Entladungen. *Sitzungsberichte der Königlichen Akademie der Wissenschaften zu Berlin* **39:** 691-699, 1886.
5. Roboz J, *Introduction to Mass Spectrometry Instrumentation and Techniques*. Wiley, New York, 1968.
6. Kiser RW, *Introduction to Mass Spectrometry and Its Application*. Prentice-Hall, Englewood Cliffs, NJ, 1965.
7. Perrin J-B, Nouvelles Proriétés Des Rayons Cathodiques. *Comptes Rendus Hebdomadaires des Séances de l'Académie des Science* **121:** 1130, 1895.
8. Wien WC, Die elektrostatische und magnetische Ablenkung der Kanalstrahlem. *Berlin Physikalische Gesellschaft Verhandlungen* **17:** 10-12, 1898.
9. Wien WC, Untersuchuugen über die elektrische Entladung in verdünnten Gasen. *Ann. Phys. Leipzig* **17:** 224-266, 1902.
10. Thomson JJ, Cathode Rays. *The Electrician* **17:** 10-12, 1897.
11. Dahl PF, *Flash of the Cathode Rays – A History of JJ Thomson's Electron*. Institute of Physics, Philadelphia, PA, 1997.
12. Thomson JJ, *Rays of Positive Electricity and Their Application to Chemical Analyses*. Longmans Green and Co., London, 1913.
13. Aston FW, *Isotopes*, 2nd ed.; Edward Arnold and Co., London, 1924 and *Mass Spectra and Isotopes*, 2nd ed.; Edward Arnold and Co., London, 1942.
14. Cartan L, *L'optique des Rayons Positifs et ses Applications à la Spectrographie de Masse;*. Hermann & Cie, Paris, 1938.
15. Cartan L, *Spectrographie de Masse: Les Iotopes et Leurs Masses; No. 550 in the series Actualités Scientifiques et Industrielles; No. VI in the subseries Exposés de Physique Atomique Expérimentale*. Hermann & Cie, Paris, 1937.
16. Thibaud J, Cartan L and Comparat P, *Quelques Techniques Actuelles en Physique Nucléaire: Méthode de la Trochoïde: Électrons Positifs: Spectrographie de Masse: Isotopes: Compteurs de Particules à Amplification Linéaire Compteurs de Geiger et Müller*. Hermann & Cie: Paris, Paris, 1938.
17. Massey H, *Negative Ions*. Cambridge at the University, London, 1976.
18. Massey H, *Negative Ions*. Cambridge at the University, London, 1938.
19. Loeb L, *Fundamental Processes of Electrical Discharge in Gases*. Wiley, NYC, 1939.
20. Hipple J and Shepherd M, Mass Spectrometry. *Anal. Chem.* **21:** 32-36, 1949.
21. Judson C and Grayson M, A Workhorse Comes to CHF: Exxon Donates a CEC 21-103C Mass Spectrometer. *Chemical Heritage* **16:** 2-21, 1998.
22. Yergy J, Heller D, Hansen G, Cotter RJ and Fenselau C, Isotopic Distributions in MS of Large Molecules. *Anal Chem* **55:** 353-356, 1983.
23. Field FH, Letter to the Editor. *Mass Spectrom. Rev.* **13:** 99–101, 1994.
24. Wiley WC and McLaren IH, Time-of-Flight Mass Spectrometer with Improved Resolution. *Rev. Sci. Instr.* **26:** 1150-1157, 1955.
25. Smythe WR and Mattauch JHE, A New Mass Spectrometer. *Phys. Rev.* **40:** 429-433, 1932.
26. Stephens WE, A Pulsed Mass Spectrometer with Time Dispersion. *Phys. Rev.* **69:** 691, 1946.
27. Wolff MM and Stephens WE, A Pulsed Mass Spectrometer with Time Dispersion. *Rev. Sci. Instr.* **24:** 616-617, 1953.
28. Katzenstein H and Friedland S, New Time-of-Flight Mass Spectrometer. *Rev. Sci. Instr.* **26:** 324-327, 1955.
29. Ryhage R, MS as a Detector for GC. *Anal. Chem.* **36:** 759-764, 1964.
30. Watson JT and Biemann K, High-resolution MS with GC. *Anal. Chem.* **36:** 1135-1137, 1964.
31. Watson JT and Biemann K, High resolution MS of GC effluents. *Anal. Chem.* **37:** 844-851, 1965.
32. Llewellyn PM and Littlejohn DP, Membrane Separator for GCMS. *Proc. Pittsburgh Conference on Analytical Chemistry and Applied Spectroscopy*, 1966.
33. Field FH and Munson MSB, Chemical ionization mass spectrometry. V. Cycloparaffins. *Journal of the American Chemical Society* **89**(17)**:** 4272-80, 1967.
34. Field FH, Chemical ionization mass spectrometry. *Accounts of Chemical Research* **1**(2)**:** 42-9, 1968.
35. Munson B, Chemical ionization mass spectrometry. *Analytical Chemistry* **43**(13)**:** 28A-32A,34A,36A, 39A-40A,42A-43A, 1971.
36. Munson B, Development of chemical ionization mass spectrometry. *International Journal of Mass Spectrometry* **200**(1/3)**:** 243-251, 2000.
37. Munson MSB and Field FH, Chemical ionization mass spectrometry. IV. Aromatic hydrocarbons. *Journal of the American Chemical Society* **89**(5)**:** 1047-52, 1967.
38. Paul W and Steinwedel H, Ein neues Massenspektrometer ohne Magnetfeld. *Z. Naturforsch, A* **8:** 448-450, 1953.

39. Finnigan RE, Quadrupole Mass Spectrometers: From Development to Commercialization. *Anal. Chem.* **66:** 969A-975A, 1994.
40. Chu TZ and Finnigan RE, Thirty Years of a Mass Spectrometry Company. *Am Lab.* **October 1998:** 90s-109s, 1998.
41. Dougherty RC, Dalton J and Biros FJ, Negative Ionization of Chlorinated Insecticides. *Org. Mass Spectrom.* **6:** 1171-1181, 1972.
42. Hunt DF, Stafford GC, Jr., Crow FW and Russell JW, Pulsed Positive Negative CI. *Anal. Chem.* **48:** 2098–2105, 1976.
43. Barber M, Bordoli RS, Sedgwick RD and Tyler AN, F.A.B. *J. Chem. Soc. Chem. Commun.* **1981:** 325-327, 1981.
44. Torgerson DF, Skowronski RP and Macfarlane RD, New Approach to the Mass Spectroscopy of Non-volatile Compounds. *Biochem. Biophys. Res. Commun.* **60:** 616-621, 1974.
45. Karas M, Bachmann D, Bahr U and Hillenkamp F, Matrix-Assisted Laser Desorption Ionization Mass Spectrometry. *Int. J. Mass Spec. Ion Proc.* **78:** 53-68, 1987.
46. Fenn JB, Mann M, Meng CK, Wong SK and Whitehouse CM, ESI of Large Biomolecules. *Science* **246:** 64-71, 1989.
47. Arpino P, Baldwin MA and McLafferty FW, Liquid chromatography-mass spectrometry. II. Continuous monitoring. *Biomedical mass spectrometry* **1**(1): 80-2, 1974.
48. McFadden WH, Schwartz HL and Evans S, Direct Analysis of HPLC Effluents. *J. Chromatogr* **122:** 389-396, 1976.
49. Willoughby RC and Browner RF, Monodisperse Aerosol Generation Interface for LC-MS. *Anal. Chem.* **56:** 2626-2631, 1984.
50. Horning EC, Horning MG, Carroll DI, Dzidic I and Stillwell RN, New picogram detection system based on a mass spectrometer with an external ionization source at atmospheric pressure. *Analytical Chemistry* **45**(6): 936-43, 1973.
51. Yost RA and Enke CG, TQMS for Mixture Analysis. *Anal. Chem.* **51:** 1251A-1264A, 1979.
52. Marshall AG, Ion cyclotron resonance mass spectrometry: a brief history. *Actualite Chimique* (1): 18-22, 2001.
53. Marshall AG, Hendrickson CL and Jackson GS, Fourier transform ion cyclotron resonance mass spectrometry: a primer. *Mass Spectrometry Reviews* **17**(1): 1-35, 1998.
54. Cody RB, Laramee JA and Durst HD, Versatile new ion source for the analysis of materials in open air under ambient conditions. *Analytical Chemistry* **77**(8): 2297-2302, 2005.
55. Takats Z, Wiseman JM, Gologan B and Cooks RG, MS Sampling Under Ambient Conditions with DESI. *Science* **306**(5695): 471-473, 2004.
56. McEwen CN and McKay RG, A Combination Atmospheric Pressure LC/MS:GC/MS Ion Source: Advantages of Dual AP-LC/MS:GC/MS Instrumentation. *Journal of the American Society for Mass Spectrometry* **16**(11): 1730-1738, 2005.
57. Hites RA and Biemann K, Mass Spectrometry – Computer System Particularly Suited for Gas Chromatography of Complex Mixtures. *Anal. Chem.* **40:** 1736–1739, 1968.
58. McLafferty FW, Billionfold data increase from mass spectrometry instrumentation. *Journal of the American Society for Mass Spectrometry* **8**(1): 1-7, 1997.
59. Evans MA, Nies AS, Watson JT and Harbison RD, Identity of an Active Metabolite in Glutethimide Intoxication. *Journal* **1:** 229-232, 1977.
60. Hansen AR and Fischer LJ, Glutethimide and an Active Hydroxylated Metabolite in Tissues, Plasma, and Urine. *J Clin. Chem* **20:** 236-242, 1974.
61. Jarman M, Foster AB, Goss PE, Griggs LJ, Howe I and Coombes RC, Metabolism of Aminoglutethimide in Humans. *Journal* **10,** 1983.
62. Anggard E and Sedvall G, GC of Catecholamine Metabolites using ECD and MS. *Journal* **41:** 1250-1256, 1969.
63. King GS, Rearrangement of Pentafluoropropionic Esters under EI. *Journal* **9:** 1239-1241, 1974.
64. Cooks RG, Beynon JH, Caprioli RM and Lester GR, *Metastable Ions.* Elsevier, New York., 1973.
65. Aspelund H, Action of Alkali on some 1,3,5,5-Tetrasubstituted Barbituric Acid Acta Acad. Abo. %J Math. Phys. *Journal* **29:** 1-13 (Chem. Abstr., 73 45407m.), 1969.
66. Biemann K and Burlingame AL, High Resolution MS. Wiley Interscience, New York, 1970.
67. Biemann K, Fennessey PV, Pomerantz S, Kowalak J and McCloskey J, Progress in High Resolution MS: Determination of Oligonucleotide Composition from Mass Spectrometrically Measured Molecular Weight. *Chimia* **21:** 226-235, 1967.
68. Pomerantz SC, Kowalak JA and McCloskey JA, Determination of oligonucleotide composition from mass spectrometrically measured molecular weight. *Journal of the American Society for Mass Spectrometry* **4**(3): 204-9, 1993.
69. Beynon JH and Williams AE, *Mass and Abundance Tables for Use in Mass Spectrometry.* Elsevier, New York, 1963.
70. Liao PC, Leykam J, Andrews PC, Gage DA and Allison J, An Approach to locate Phosphorylation Sites in Proteins by MALDI. *Anal. Biochem.* **219:** 9-20, 1994.

71. Barr RG, Perman JA, Schoeller DA and Watkins JB, Breath tests in pediatric gastrointestinal disorders: new diagnostic opportunities. *Pediatrics* **62**(3): 393-401, 1978.
72. Schoeller DA, Schneider JF, Solomons NW, Watkins JB and Klein PD, Clinical diagnosis with the stable isotope 13C in CO2 breath tests: methodology and fundamental considerations. *The Journal of laboratory and clinical medicine* **90**(3): 412-21, 1977.
73. Schoeller DA, Klein PD, Watkins JB, Heim T and MacLean WC, Jr., Carbon-13 abundances of nutrients and the effect of variations in carbon-13 isotopic abundances of test meals formulated for 13CO2 breath tests. *American Journal of Clinical Nutrition* **33**(11): 2375-85, 1980.
74. Watkins JB, Schoeller DA, Klein PD, Ott DG, Newcomer AD and Hofmann AF, 13C-trioctanoin: a nonradioactive breath test to detect fat malabsorption. *The Journal of laboratory and clinical medicine* **90**(3): 422-30, 1977.
75. Watkins JB, Klein PD, Schoeller DA, Kirschner BS, Park R and Perman JA, Diagnosis and differentiation of fat malabsorption in children using 13C-labeled lipids: trioctanoin, triolein, and palmitic acid breath tests. *Gastroenterology* **82**(5 Pt 1): 911-7, 1982.
76. Watkins JB, Bile Acid Metabolism and Fat Absorption in Newborn Infants. *Pediatr. Clin. North Am.* **21**: 501-512, 1974.
77. Kaihara S and Wagner HN, Jr., Intestinal Fat Absorption with 14C-labeled Tracers. *J. Lab. Clin. Med.*: 400-411, 1968.
78. Watkins JB, Szczepanik P, Gould JB, Klein P and Lester R, Bile Salt Metabolism in the Human Premature Infant. *Gastroenterology* **69**: 706-713, 1975.
79. Schoeller DA and Hayes JM, Computer Controlled Ion Counting Isotope Ratio MS. *Anal. Chem.* **47**: 408-415, 1975.
80. Brenna JT, High-precision Gas Isotope Ratio MS: Instrumentation and Biomedical Applications. *Acc. Chem. Res.* **27**, 1994.

Chapter 2   The Mass Spectrometer

    1. Time-of-Flight *m/z* Analyzers
        A. Linear
            1) Resolving Power of the Linear TOF Instrument
            2) Time-Lag Focusing
            3) Beam Deflection
        B. Reflectron
        C. Orthogonal Acceleration
        D. Ion Detection in the TOF Analyzer
            1) Time-Slice Detection
            2) Time-Array Detection
            3) TAD with Transient Recorders
            4) TAD with an Integrating Transient Recorder
            5) Hadamard Transform TOF-MS
    2. Quadrupole Ion Traps
        A. 3D Quadrupole Ion Trap
        B. Linear Quadrupole Ion Trap (LIT)
        C. Performance Trade-Offs in the Ion Trap
    3. The Orbitrap
        A. Historical Aspects
        B. Operating Principles
            1) Role of the C Trap in Success of the Orbitrap
            2) Figures of Merit for the Orbitrap as an *m/z* Analyzer
    4. Transmission Quadrupoles
        A. QMF Equations of Motion
        B. The Stability Diagram
        C. Characteristics of Output
        D. Spectral Skewing
        E. Performance Limitations
    5. Magnetic-Sector Instruments
        A. Single-Focusing Instruments
            1) Operating Principles
            2) Magnetic Versus Scanning
            3) Performance Limitations
        B. Double-Focusing Instruments
    6. FTICR-MS
        A. Hardware Configuration
        B. Operational Considerations
        C. Representative Applications
    7. Ion Mobility Spectrometry (IMS)
        A. Operating Principles of IMS
        B. FAIMS
        C. Applications

Introduction to Mass Spectrometry, 4th Edition: Instrumentation, Applications, and Strategies for Data Interpretation; J.T. Watson and O.D. Sparkman, © 2007, John Wiley & Sons, Ltd

I. Introduction

No area of mass spectrometry has seen more changes in the last decade than the development of new *m/z* analyzers [1, 2]. When the last edition of this book was going to press, the time-of-flight (TOF) mass spectrometer was undergoing a renaissance as the *m/z* analyzer of choice for MALDI because of its pulsed mode of operation and its high *m/z* range [3, 4]. The TOF mass spectrometer gained even further popularity in later years because of the high resolving power that could be achieved through reflectron technology and its accurate mass capabilities as the secondary *m/z* analyzer in *tandem-in-space* Q-TOF MS/MS instruments [5]. The TOF has also become a popular single-stage analyzer for both GC/MS and LC/MS (high resolving power for accurate mass and nominal *m/z* value for rapid acquisition rates). During this time, the quadrupole ion trap [6, 7], especially attractive because of its MS/MS capabilities, was maturing as a GC/MS product and was just being introduced as an analyzer for LC/MS (ESI and APCI) because of the need for MS/MS to obtain structural information in conjunction with use of these soft ionization techniques.   The venerable double-focusing magnetic-sector technology [8–10] was beginning to wane because of increased emphasis on ease of use, the lack of its suitability for ESI and APCI, and the need for rapid data acquisition techniques in GC/MS due to use of narrower diameter columns.   The transmission quadrupole had matured with advances in solid state electronics [11, 12]; it had become the instrument of choice for obtaining specificity in quantitative LC/MS and MS/MS techniques.

These new developments were scrutinized during widespread use in the late 1990s, which revealed some imperfections. For example, limitations were found in the linear dynamic range of the TOF instrument, restricting its use as a quantitative tool. Limitations of the three-dimensional (3D) quadrupole ion traps due to constraints on the number of ions that could be stored and imperfections in the storage fields themselves resulted in a less-than-desired performance for quantitation and for some MS/MS experiments. Except for the specialized application of high-resolution GC/MS selected ion monitoring to quantitate dioxanes in the presence of other coeluting chlorinated substances, use of the double-focusing magnetic-sector *m/z* analyzer has almost disappeared. However, there are many pieces of information that still can only be obtained with these instruments. There has been a continued increase in the desire for accurate mass measurement and low detection limits; very good advancements have been seen in both of these areas through the use of TOF technology during the last ten years. Business considerations in the form of work-a-rounds for patented technology have resulted in superior devices for ion transmission, which translates into better detection limits. New microprocessor and other solid state technology also have led to advances in performance and ease-of-use technology.   Mass spectrometry has expanded into many fields of science where its use had never been considered. This expanded use has brought about better instruments from the core group of manufacturers as they continue to compete in the development of new instruments and techniques, trying to gain a share of this unending expansion of use.

The operating pressure of the *m/z* analyzer is an important parameter. In order to ensure that ions can efficiently traverse the region from the ion source to the ion detector without interference, the mass spectrometer is maintained in a matter-free state, a vacuum. The entire *m/z* analyzer is operated at a low pressure (high vacuum) of 10^{-4}–10^{-7} Pa, as illustrated conceptually in Figure 2-1. Once an ion is set in motion, it is important for it to avoid collision with molecules and other neutrals (radicals) or with other ions that could deflect its trajectory, result in collisional activation that could bring about

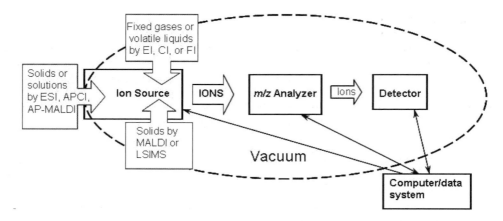

Figure 2-1. Conceptual illustration of the mass spectrometer.

fragmentation, or cause its disappearance due to charge neutralization. The high vacuum requirements of the mass spectrometer cannot be achieved with a single pump. Pumps that can maintain the pressure at the low levels required for operation of the mass spectrometer do not begin to function until the pressure has been reduced by fore pumps to approximately 10^{-1} Pa. Both the high-vacuum pump ($<10^{-2}$ Pa) and its fore pump (sometimes called the rough pump because the initial vacuum is not as fine as that obtained with the high-vacuum pump) are used to remove the continuous onslaught of gases associated with analyte introduction. The different types of pumps used with *m/z* analyzers are discussed in this chapter.

Another important part of the *m/z* analyzer discussed in this chapter is the ion detection device. The output signal results from amplification of the secondary emission of electrons or photons generated by ions striking a reactive surface. These devices have different geometries depending on the operating principles of the *m/z* analyzer.

In all mass spectrometers, ions are moved from one location to another as illustrated conceptually in Figure 2-1. In the ideal instrument, this ion transfer would have 100% efficiency. Unfortunately, this is not the case because the ions are moved as a beam, not as a discrete package; these ion beams have a tendency to diverge. Therefore, proper ion optics are necessary to minimize ion beam divergence [13]. The diverging ion beam can be controlled through the use of pre- and post-filters as well as special lenses at the entrance and exit of the *m/z* analyzers, respectively. These ion guides are as important as the analyzer itself, the detector, or any other single component of the mass spectrometer. A discussion of these ion pipes or ion guides precedes that of the *m/z* analyzer in this chapter.

II. Ion Guides

Ion losses between major components of a mass spectrometer are inevitable. Ion losses occur between the source and the analyzer, within the analyzer, and between the analyzer and the detector. Atmospheric pressure ion sources, such as encountered with electrospray ionization (ESI), atmospheric pressure chemical ionization (APCI), atmospheric pressure photoionization (APPI), etc., present a particular challenge with respect to efficient transfer of ions from the ion source into the *m/z* analyzer. Even in systems such as the electron ionization (EI) transmission quadrupole GC-MS, ion transfer from the ion source to the analyzer and from the analyzer to the detector is an important consideration. Various devices have been developed to reduce losses in ion current between major components in a mass spectrometer. The problem is more egregious in systems that have ion beams of low kinetic energy, like those in the transmission quadrupole and those involved in various versions of the quadrupole ion trap, but high kinetic energy systems such as those found in the TOF and double-focusing *m/z* analyzers also can experience ion transfer problems. Radio frequency (RF) *ion guides*, *lenses*, and *ion funnels* have been developed to facilitate efficient ion transfer, as described in the following sections.

As seen later in this chapter, a potential problem with the transmission quadrupole *m/z* analyzer is a phenomenon called *mass discrimination*, which refers to the apparent ability to transmit ions of lower *m/z* values more efficiently than those of higher *m/z* values. One of the reasons for mass discrimination in the transmission quadrupole is the existence of fringe fields at the entrance to the analyzer. The quadrupole field exists in the central space between the four electrodes because of the RF and DC potential applied to the two pairs of electrical poles. However, because both the DC and the RF fields extend beyond the termini of these electrical poles (fringe fields) the DC fringe fields can adversely affect the trajectory of ions at the entrance to the quadrupole, a region where there is no compensatory effect of the quadrupole field that exists only within the central space between the poles. These DC fringe fields can result in deflection of ions that enter the analyzer at less than a straight (180°) axial trajectory. Ions of higher *m/z* values (lower velocity) are more affected by the fringe fields than those of lower values.

To correct for this problem of fringe fields in the early transmission quadrupole GC/MS systems, either a Brubaker RF-only prefilter [14] or a Turner–Kruger lens [15] was used (Figure 2-2). Both of these devices are still used in modern GC/MS and LC/MS instruments.

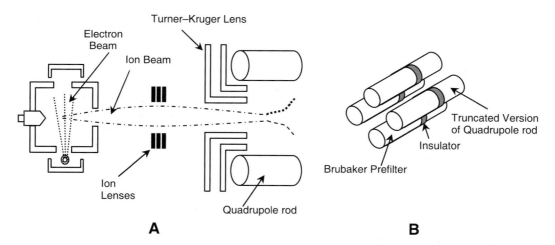

Figure 2-2. Devices used in transmission quadrupole mass spectrometers to compensate for DC fringe-field effect (A) the Turner–Kruger lens popular with the Agilent and Shimadzu GC/MS instruments and (B) the Brubaker prefilter popular with the Waters and Varian instruments.

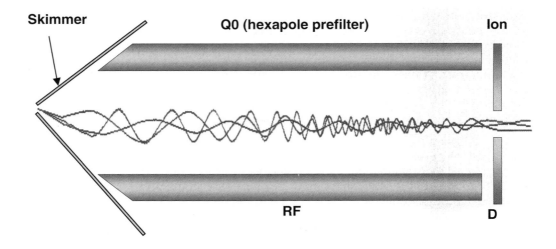

**Figure 2-3. *Illustration of effect of collisional cooling in an RF-only
quadrupole ion guide.*** *Courtesy of Varian, Inc.*

The RF ion guides used in the atmospheric pressure ionization (API)[1] LC/MS systems are often referred to as q_0. These ions guides can be quadrupoles, hexapoles, or octupoles [16].   As RF-only devices, these ion guides can be considered as wide-bandpass filters.   Hexapoles and octopoles have better characteristics as wide-bandpass filters than do quadrupoles. This is why these two devices are more often used as q_0 than the quadrupole.   The use of these ion guides does not eliminate the possible need for prefilters or entrance lenses to correct for fringe-field effects.

The q_0 RF-only prefilter can take advantage of the relatively high pressure ($\sim 10^{-1}$ Pa) in what is referred to as the "first pumped region" of the *m/z* analyzer to bring about *collisional focusing* or *collisional cooling* [17, 18].   As ions come out of an API source or even out of a reduced pressure ion source, they have a dispersion of initial kinetic energies, which are vector quantities expressed randomly in three-dimensional space; it is this dispersion of initial kinetic energy vectors that gives a macroscopic dimension to the ion beam and also causes it to diverge.   The collisional cooling dampens or diminishes the magnitude of the random vector quantities, thereby causing the ions to appear to coalesce along the *z*-axis as shown in Figure 2-3. The collision cell in *tandem-in-space* MS/MS instruments can also be used as a collisional cooling device to increase ion transfer efficiency between the cell and the second *m/z* analyzer.

Instead of an ion guide, an ion funnel can be used (see Figure 2-4).   These devices, developed at the Pacific Northwest National Laboratory (PNNL) by Richard D. Smith and his group [19, 20] to improve the transmission of ions from an API source into an FTICR mass spectrometer, have been employed in the Waters/Micromass and Bruker/Daltonics instrumentation as a way to circumvent the patent position of Applied Biosystems/MDS Sciex on the use of collisional cooling in an RF-only prefilter.   This device uses a series of ring electrodes with progressively smaller internal diameters that have RF and DC potentials simultaneously applied.   The combination of collisions of ions

---

[1]   API is a generic term applied to all ionization that operates at atmospheric pressure. These include electrospray ionization (ESI), atmospheric pressure chemical ionization (APCI), and atmospheric pressure photoionization (APPI).

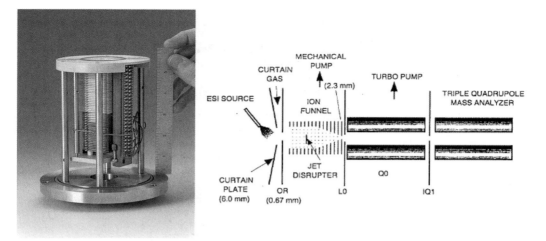

Figure 2-4. *The 100-electrode ion funnel; 55 of these electrodes have constant inside diameter (25.4 mm) ring electrodes at the funnel entrance. These are followed by a rear section that has 45 ring electrodes with decreasing inner diameters from 25.4 to 1.5 mm. The front section reduces certain gas dynamic effects that inhibit ion confinement and also serves to reduce the gas load downstream of the ion funnel.*

with neutral gas atoms or molecules and the combined RF and DC fields cause the ions to be focused along the *z*-axis, thereby aiding in ion transmission. This improved efficiency of ion transfer by use of an ion funnel [20–23] or an ion conveyer [24] significantly enhances the sensitivity of the mass spectrometer.

An einzel (equipotential) lens can be used to focus the ion beam without altering its overall kinetic energy [25]; this lens is often used in systems that do not provide collisional cooling by a neutral gas such as MALDI TOF instruments and EI GC/MS systems. The einzel lens is composed of three tubes of the same size with the first and third equally spaced from the second, as illustrated in the cross-sectional diagram in Figure 2-5. In this example, the first and third elements are grounded, while a potential of 600 V is applied to the center element. As a beam of 700 eV positive ions passes through the series of three tubes, it will first be defocused slightly by the diverging fringing field (see divergence of ion trajectory within and just after the first gap), then refocused by the converging fringe field within the second tube. Because the ions are decelerated in the first gap, the effect of the converging field within the second tube is greater than that of the diverging field of the first gap, giving an overall effect of focusing the beam toward the *z*-axis without changing the overall kinetic energy (i.e., the potential energy in the two gaps is equal, but of opposite sign). Einzel lenses are often found in TOF mass spectrometers in the first part of the field-free drift tube because the ion packets produced by such an ion source are somewhat divergent.

Ion lenses provide another means for collimating an ion beam as it enters or leaves an analyzer. As illustrated in Figure 2-6, ions of the same *m/z* value and similar, if not the same, kinetic energy can have different trajectories at the exit of a quadrupole analyzer due to the effects of the quadrupole field as they transverse the analyzer. This means that the ion beam will be divergent as it leaves the analyzer, resulting in ion loss

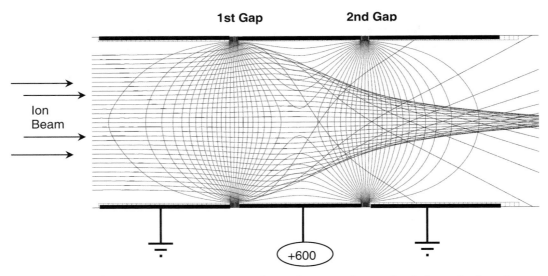

Figure 2-5. *Conceptual representation of operating principles of einzel lens; cross-sectional view of decelerating (1st Gap), then accelerating (2nd Gap) electric fields between cylindrical electrodes compressing the ion beam onto its longitudinal axis. Equipotential lines of fields and trajectory of ions modeled using SIMION 8 einzel lens example.* (www.simion.com)
Courtesy of David Manura at Scientific Instrument Services, Inc.

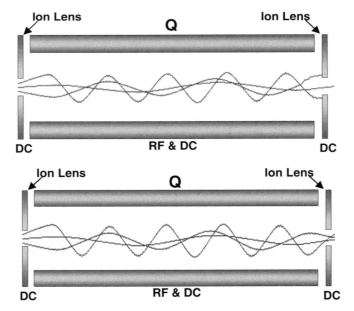

Figure 2-6. *The top system has a nodding voltage of 3 eV, and the bottom has a nodding voltage of 3.2 eV. Notice the difference in the angular dispersion of the ion beam exiting the two m/z analyzers.* Courtesy of Varian Inc.

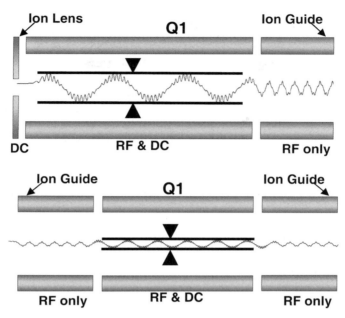

Figure 2-7. ***In both the top and bottom illustration, the ion energy is 3 eV. However, it is clear that the radial displacement of the ion beam is much less in the systems that have both a pre- and a post-RF-only filter.*** *Courtesy of Varian Inc.*

before the beam enters a collision cell or reaches the detector. The lenses at the front and at the end of the analyzer can be used to bring about small changes in the ion kinetic energy (e.g., increase from 3.0 to 3.2 eV). This change can have a large effect on the radial orientation of the ion trajectory at the exit of field. Application of a so-called nodding (because it affects the ion trajectory) potential can have a significant effect on collimating ions of a single *m/z* value, as seen in Figure 2-6. Finding the correct nodding potential requires a significant amount of manipulation for each *m/z* value.

The use of an entrance lens and an exit RF-only filter can compensate for the radial displacement of the ions of a single *m/z* value, as illustrated in the top part of Figure 2-7 [26]. However, best results are obtained when pre- and post-RF-only filters are both used, as seen in the bottom part of Figure 2-7. The reduction in the radial spread of the ion beam translates to better ion transmission through the device.

III. Types of *m/z* Analyzers

The *m/z* analyzer, or more commonly called the mass analyzer, is used to separate ions according to their mass-to-charge ratios based on their characteristic behavior in electric and/or magnetic fields [27, 28]. Mass spectrometry takes advantage of these different behaviors to separate the ions of different *m/z* values in space or time so that their individual abundances can be determined.

For many years, all mass spectrometry was accomplished using instruments that employed either magnetic fields [27, 28] or a serial combination of magnetic and electric fields (the double-focusing mass spectrometer [8–10]). In the 1950s, the time-of-flight mass spectrometer began to be used and studied [29–31]. In that same period, the use of the quadrupole field for the manipulation of ions was discovered; however, it was not until the late 1960s that technology from Wolfgang Paul's discoveries [32, 33] found their way into the analytical laboratory with Finnigan Corporation's (now Thermo Electron) and Extranuclear's (later Extrel) commercialization of the transmission quadrupole GC-MS [11, 34]. The transmission quadrupole mass filter (QMF) mass spectrometer is credited with the growth that has caused mass spectrometry to become so ubiquitous today. Over two decades passed before the other of Paul's inventions in quadrupole technology, the quadrupole ion trap (QIT) became a mass spectrometer. [7]. During those same two decades, the TOF mass spectrometer fell into almost total obscurity and the use of the Fourier transform for ion detection found a home with the ion cyclotron resonance (FTICR) mass spectrometer [35, 36].

Over the past decade, the TOF instrument has undergone a tremendous renaissance, the QIT technology has expanded via several iterations that have advanced mass spectrometry as much as the original transmission quadrupole did, and the double-focusing instrument has almost fallen into total disuse. The most recent addition to the family of *m/z* analyzers is the Thermo Electron Orbitrap, which represents the first new ion separation technology to be commercialized in over 20 years. What follows is a description of the *m/z* analyzers of today.

1. Time-of-Flight *m/z* Analyzers

The operating principles of the time-of-flight (TOF) mass spectrometer [37–39] involve measuring the time required for an ion to travel from an ion source to a detector usually located 1 to 2 m from the source (Figure 2-8). All the ions receive the same kinetic energy during instantaneous acceleration, but because the ions have different *m/z* values, they have correspondingly different velocities. As the ions traverse the "field-free" region between the ion source and detector, they separate into groups or packets according to velocity, which is a function of their *m/z* values. In principle, this method of separating ions of different *m/z* values has the advantage of no upper *m/z* limit. As will be seen, all other types of *m/z* analyzers have upper limits for *m/z* values that can be transmitted and separated based on the way the electric and/or magnetic fields are used to separate ions of different mass-to-charge ratios.

The earliest report of experimentation with the concept of time-of-flight for gas-phase ion separation was by William R. Smythe (U.S. scientist) and Josef Heinrich Elizabeth Mattauch (Austrian physicist) in 1932 [40]. According to an account by Robert W. Kiser in his book *Introduction to Mass Spectrometry and Its Applications*, the history of the early development of the time-of-flight mass spectrometer involved a number of people. During his tenure at the University of Pennsylvania (Philadelphia, PA), W. E. Stephens reported on the concept of the TOF-MS in an abstract appearing in the American Physical Society's program in 1946 [41]. Some seven years later, Stephens published a paper with M. M. Wolff with the same title as his 1946 abstract, "A Pulsed Mass Spectrometer with Time Dispersion", which described the instrument they had built based on the time-of-flight principle [42]. The seminal paper on time-of-flight mass spectrometry, "Time-of-Flight Mass Spectrometer with Improved Resolution", was published in 1955 by two researchers working for Bendix Aviation Corporation Research Laboratories in Detroit, MI [29]. This paper provides a detailed treatment of ion focusing in the TOF-MS. The Wiley–McLaren paper was considered sufficiently important to be republished in the *Journal of Mass Spectrometry* in 1997 [43].

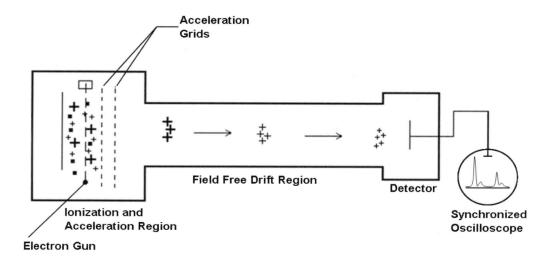

Figure 2-8. *Schematic representation of a linear time-of-flight (TOF) mass spectrometer.*

According to a citation in Robert Cotter's book *Time-of Flight Mass Spectrometry: Instrumentation and Applications in Biological Research* (ACS: Washington, DC, 1997), "…it was estimated that in 1962, one-third of the mass spectrometers in use in the United States were time-of-flight instruments" [44]. In his 1965 book, Kiser [45] states that the bulk of the U.S. market had been provided by CEC (approximately 700 instruments). Considering these two statements, it is possible that by the mid-1960s, only 10 years after the Wiley–McLaren paper, there may have been over 300 TOF mass spectrometers in use in the United States, most of them manufactured by the Bendix Corporation. In the mid-1960s, the future for the TOF-MS looked very bright. However, this was before the commercialization of the transmission quadrupole mass spectrometer.

These early TOF mass spectrometers had limited resolving power and relatively slow data recording systems in the form of oscillographic recorders or photography resulting in a 3 inch x 4 inch Polaroid photograph of the spectrum (or an expanded segment thereof). The distribution in the initial kinetic energy of ions resulted in broad spectral peaks and in resolving powers (R) of no more than 500. This meant that ions of *m/z* 450 could barely be separated from ions of *m/z* 451. The problem of resolving power was further exacerbated by the non-Gaussian mass spectral peak shape and the nonlinear (quadratic function) of the *m/z* scale (i.e., the peaks at high *m/z* were closer together than those at low *m/z*). The non-Gaussian peak shape required that the resolving power be calculated using the full width at half-maximum height (FWHM) method rather than the 10% valley method.

Another important consideration of the TOF mass spectrometer is its requirement for a high vacuum. In most models, the ions travel a distance of 1 to 2 meters from the ion source to the detector. This large distance of ion travel in the TOF instrument translates into a higher probability of ions encountering neutrals than would be the case in instrument types with significantly shorter ion paths. The ion beam also has a certain degree of divergence leading again to a higher probability of contact with neutrals than in the case of a collimated ion beam. In order to compensate for these increased possibilities of interaction with neutral particles through the large volume swept by the ion beam, the number of neutral particles must be reduced. This reduction in the number of neutral particles is achieved by reducing the pressure to a level that the mean-free path (the distance a particle travels without collision) increases to approximately 1 meter. For these reasons, a higher vacuum is required in the TOF mass spectrometer than in some other types of instruments.

An additional problem with the early TOF instruments was a drift in the *m/z* scale calibration caused by even the slightest changes in the instrument's ambient temperature. This important consideration is dealt with by using materials of compensating coefficients of thermal expansion in constructing modern instruments. Elaborate designs with selected materials that expand and contract with temperature variations now compensate for changes in the distances and angles associated with ion mirrors and flight tubes.

As the commercialization of the transmission quadrupole (which had better resolution throughout the *m/z* scale than the TOF instrument of that era, and a linear *m/z* scale that was better suited for data recording) progressed, interest in the TOF mass spectrometer declined to the point that they were no longer manufactured and marketed in the United States by the mid 1980s. A resurgence in the technology began with developments in the field of desorption/ionization, which culminated with the development of MALDI (matrix-assisted laser desorption/ionization) in the late 1980s [46]. This renewed interest in the TOF instrument coincided with the collapse of the Soviet Union, where extensive research had been taking place in TOF mass spectrometry and the use of the reflectron to obtain better resolving power [3, 4]. Advancements in electronics have resulted in instrumentation that is capable of acquiring a complete spectrum over a large range of *m/z* values, rapid spectral acquisition rates, and high resolving power. The high resolving power allows for accurate mass measurements, and the high acquisition rates result in better reconstructed chromatographic data and lower detection limits due to spectral summing. The TOF *m/z* analyzer is much better suited for chromatography/mass spectrometry because all the ions of all *m/z* values are detected during each pulsed cycle of the instrument; therefore, no matter how narrow the chromatographic peak, there is no spectral skewing. This total ion detection also leads to better sensitivity and higher quality spectra from low analyte concentrations. The TOF *m/z* analyzer is on its way to becoming the instrument of choice for chromatography applications. Because of the nearly unlimited upper *m/z* values capable of being detected, the TOF mass spectrometer is already the most practical instrument for MALDI [39].

A. Linear

As stated above, the principle of time-of-flight mass spectrometry involves packets of ions sequentially striking the detector in order of increasing *m/z* value, creating a time-based waveform, or simply an *m/z*-dependent transient.[2] Ions of lower *m/z* values reach the detector before those of higher *m/z* values because the latter have a lower velocity, as indicated schematically in Figure 2-8. The kinetic energy ($\frac{1}{2}mv^2$) gained by the ion is equal to the potential energy (zeV where V is the accelerating potential) available in the electric field between the accelerator plates:

$$zeV = \tfrac{1}{2}mv^2 \qquad\qquad\qquad \text{(Eqn. 2-1)}$$

---

[2] As used in this context, the word "transient" means a "continuum" of ion current consisting of the sequential arrival at the detector of individual ion packets representing ions of progressively increasing *m/z* value; i.e., it is the complete mass spectrum derived from a single extraction pulse of the ion source.

Solving Equation 2-1 for v gives:

$$v = \left(\frac{2zeV}{m}\right)^{1/2}$$

(Eqn. 2-2)

Equation 2-2 shows the inverse quadratic relationship between v and *m/z*. Because it is impractical to measure ion velocity directly, the time-of-flight (*tof*) from the source to the detector (separated by distance L) provides the useful experimental information:

$$tof = \frac{L}{v} = L\left(\frac{m}{2zeV}\right)^{1/2}$$

(Eqn. 2-3)

The *m/z* value of an ion is determined by its *tof* to the detector as calibrated by reference to the *tof* of at least two other ions of known *m/z* values. The transit time of a 3000-eV ion of *m/z* 800 is approximately 70 μsec through a 1-meter flight tube. This means that the cycle of ion production, acceleration, and detection is complete in a period of approximately 100 μsec for an *m/z* range to approximately 1000. A time-focusing equation [47] has been developed to improve *tof* assessments of the ions; a pseudo-random analysis of *tof* has also been reported [48].

As seen from the above explanation of the time-of-flight (*tof*) for an ion, there is no theoretical limit to the upper *m/z* value that can be detected; practically speaking, it is a matter of how long it takes for ions to arrive at the detector. This is one of the significant advantages of the TOF mass spectrometer when used with MALDI. An intact biopolymer with a mass of several hundred kilodaltons can be measured to within a few daltons, whereas earlier techniques such as electrophoresis resulted in average molecular mass (M_r) values having an error ±1%; i.e., an error of ±3000 Da for a 300-kDa protein!

1) Resolving Power of the Linear TOF Instrument

As pointed out in the introduction, the resolving power of the TOF *m/z* analyzer has always been problematic. The resolving power of the linear TOF instrument is a function of flight-tube length, accelerating voltage, and, most importantly, the spatial and velocity distributions of the initial ion packet [37, 49]. In theory, this ion packet should have an infinitesimal initial thickness and be monoenergetic; but in practice, because of various directions of travel, pre-acceleration velocities, and variations in thermal energies of nascent ions, the packet has a finite initial thickness that will ultimately limit the discreteness of detecting the *tof* of a packet of ions of a particular *m/z* value [47]. Broadening of a given ion packet caused by thermal-energy variation and spatial coordinates can be reduced by optimizing electrode positions and voltages in the ion optics [50, 51] .

At present, the resolving power (R) of commercially available linear conventional (i.e., no orthogonal acceleration) TOF instruments with a 2-meter flight tube is approximately 1000 FWHM. Three basic phenomena are responsible for this low resolving power. First, there is the spread in flight times due to differences in kinetic energies of the ions of the same *m/z* value. Second, there is the spread in flight times due to the different positions from which the ions start when they are pulsed out of the source; i.e., the spatial spread. The third problem is termed the "turnaround time" and can be visualized as the difference in flight times between two isomass ions starting in the same plane in the source and moving with the same initial velocities, but in opposite directions. Because these isomass ions will eventually leave the ion source at different times, but with the same energy, no degree of "space focusing" to correct for the initial spatial spread, or of "energy focusing" to correct for the initial energy spread, using conservative (time-independent) fields can reduce this turnaround time, which is increasingly significant at higher masses [52].

2) Time-Lag Focusing

Wiley and McLaren [29] addressed the spatial and energy focusing requirements and led the way to the first commercial instrumentation for TOF mass spectrometry. The spatial variations were corrected by a unique two-field extraction and acceleration source that enabled the temporal focus plane to be as remote as 1 to 2 meters from the source. This allowed ions of differing *m/z* value to separate easily in the extended field-free region, at the same time still accomplishing isomass ion focusing. The energy variations were corrected by a unique technique called *time-lag focusing*. This technique presents a time delay between the end of the ionization pulse and the start of the extraction pulse. During this time, the ions move in the direction and with the velocity dictated by their initial energy, displacing themselves from their initial positions in the source and, in essence, converting their energetic variation into a spatial variation. The subsequent application of the extraction pulse, because of the two-field process, corrects for this spatial variation, bringing the isomass ions into temporal focus at the detector.

Time-lag focusing was a novel and excellent approach, and it is being revisited today under different names, such as *delayed extraction* (DE), in some MALDI techniques as described in a later chapter. Unfortunately, time-lag focusing is *m/z* value dependent, and the delay must be slowly increased from just a few to several hundred nanoseconds as the *m/z* values to be focused are increased. This progressive change must be made during the acquisition of a single spectrum or the capture of a single transient from the pulsed ion source. This procedure was practiced in earlier days during use of the archaic "scanning" type operation of the TOF mass spectrometer with a box-car integrator. Today, the modern TOF instrument is operated in the time-array detection (TAD) mode by summing successive transients with an integrating transient recorder that is inherent in the computer interface of the data system. Because the complete spectrum is represented in each transient, a compromised delay time is chosen to provide optimal resolving power in the *m/z* range of greatest interest. The work of Wiley and McLaren was truly outstanding, so much so that it actually misled the scientific community into thinking of the TOF as a scanning instrument, ignoring its potential for TAD.

Several investigators have attempted improvements for energy correction in the ion optics of the TOF instrument. Studier [30] used a two-step extraction pulse with continuous ionization to achieve a resolving power of 400. Marable and Sanzone refined the time-dependent operation of the ion extraction pulse and formulated the impulse-field focusing theory [31]. Experimental confirmation of the impulse-field theory [53] was encouraging, with demonstration of a resolving power of 650 (50% valley) for an organic compound. Stein [52] pointed out that space focusing and velocity focusing are mutually exclusive applications of the linear, conservative-field ion sources in TOF mass spectrometry.

It should be noted that the term *time-lag focusing* was first used by Wiley and McLaren. This same technique has since been called *delayed extraction* [54–57] and *pulsed ion extraction* [58]. Manufacturers of instruments are always trying to differentiate their instrument from those of the competition. Which means, today, the technique of time-lag focusing is called delayed extraction (DE) by Applied Biosystems, time-lag focusing (TLF) by Waters/Micromass, and pulsed ion extraction (PIE) by Bruker Daltonics, three major suppliers of TOF mass spectrometers.

3) Beam Deflection

Pinkston *et al.* [59] addressed all three fundamental problems or nonideal attributes of the conventional TOF ion source by employing a continuous source to produce a steady stream of ions, thereby eliminating the turnaround-time problem, as all ions are moving in the same direction when the continuous beam is deflected past a set of slits to produce a time-resolved packet of ions. An electric sector reduced the energy spread of the ions admitted to the beam deflector. The ions were pulsed into the flight tube by using a beam-deflection technique similar to that described by Bakker [60]. The initial spatial spread is analogous to the "width-in-time" of the ion packet that is pulsed into the flight tube. This width-in-time can be controlled by varying the voltage pulse applied to the deflection-plate assembly. The deleterious effects of the initial spatial spread can therefore be minimized. Lubman and Jordan [61] used a pulsed molecular beam orthogonal with the ionizing beam and the ion beam to minimize energy aberrations of resolving power in TOF mass spectrometry. Other improvements in, and applications of, time-of-flight mass spectrometry have been reviewed by Price and Milnes [4].

As pointed out in the introduction to the TOF mass spectrometer, spectra are obtained by the acceleration of a packet of ions from the ion source to the detector through a field-free region. Each one of these pulses of ion packets results in a single spectrum. That spectrum by itself is of little value. The spectra that are recorded are the average of ten or more transients resulting from corresponding extraction pulses. This pulsed operation of the ion source to obtain a mass spectrum as opposed to sampling a continuous ion beam makes the TOF instrument ideally suited for the technique of matrix-assisted laser desorption/ionization (MALDI) as described in a later chapter. Another advantage of being able to detect all ions of all *m/z* values from a single pulse is the lack of spectral skewing that can occur when using scanning instruments in conjunction with chromatography, where there are transitory changes in analyte concentration, as described in great detail in later chapters on gas and liquid chromatography/mass spectrometry.

Today's linear TOF mass spectrometer is used primarily for MALDI. These instruments will have resolving power improvements that are a result of delayed extraction (time-lag focusing); however, they are still considered to be instruments that have low resolving power. From a practical standpoint, the effective resolving power achieved with a linear TOF instrument in MALDI applications can be limited by the thermodynamic and physicochemical properties of the sample as illustrated in Figure 2-9, which shows the MALDI TOF mass spectrum of ribonuclease A (molar mass = 13,690 Da) [62]. Within the 15 *m/z* unit center section of the peak for the protonated molecule, as delineated by the vertical dashed lines in the expanded format in the top panel, approximately 7 *m/z* units account for the multiplicity of isotopic variants, 2–4 *m/z* units are attributed to the residual distribution of initial velocities of the ions (beyond that compensated by delayed extraction), and 4 *m/z* units relate to the poor resolving power of the instrument. The broadening of the core peak at the base beyond the vertical dashed lines derives from metastable decomposition of the protonated molecule in the flight tube (leading edge) and from adduct formation with matrix and alkali metal ions (trailing edge) [62]. This observed diminution in resolving power due to physicochemical characteristics of the analyte is even more dramatic during analyses of dendrimers in which structural heterogeneity accrues with additional generations of synthesis [63].

A list of advantages and disadvantage of the linear TOF mass spectrometer appears in Table 2-1.

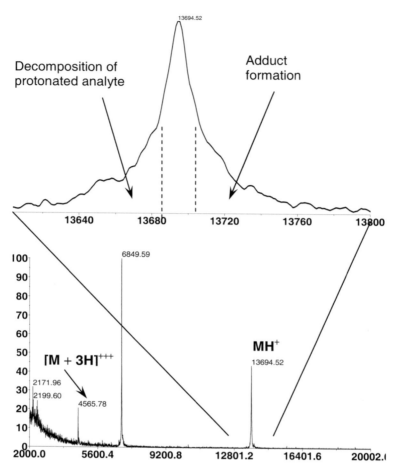

Figure 2-9.   *MALDI TOF mass spectrum of ribonuclease A (molar mass = 13,690 Da).*

Table 2-1.   *Advantages and disadvantages of linear TOF mass spectrometer.*

Advantages	Disadvantages
Pulsed mode of operation makes it suitable for use with MALDI	Low resolving power (R = 1000 to 2000)
Capacity to acquire spectra rapidly for averaging	Requires higher vacuum than most other types of mass spectrometers
No upper limit for the *m/z* scale Can detect neutrals in linear mode, which greatly increases signal from fragile compounds, e.g., carbohydrates	Operates at a constant resolving power (R); i.e., resolution changes with *m/z* value

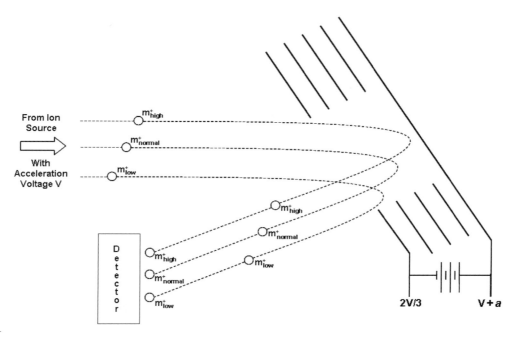

Figure 2-10. *Ions retain original velocity, but arrive at the detector at the same time because of the amount of time they spend in the reflectron.*

B. Reflectron

In order to improve the resolving power of the TOF mass spectrometer, an ion mirror can be used to reflect ions of the same *m/z* values that have different energies [39, 64–67]. The reflectron TOF (*re*TOF) mass spectrometer[3] has an ion mirror in the form of an electric field (as illustrated in Figure 2-10) that opposes and is of greater magnitude than the electric field in the ion acceleration region; the mirror is positioned at an angle less than 180° to avoid reflection of the ions directly back into the source. The operating principles of an ion mirror are represented conceptually in Figure 2-10, in which three ions of the same *m/z* value, but with high or low initial kinetic energy (KE) relative to that of a "normal" ion. The three ions of the same *m/z* value, but of slightly different kinetic energies (i.e., +/– a few volts out of several kilovolts), enter the opposing electric field with the ion (m^+_{high}) of highest kinetic energy entering first (because it has the highest velocity) and penetrating the deepest. The ions continue decelerating into the opposing electric field until their kinetic energy reaches zero. At this point, the ions begin to be accelerated by the electric field in the opposite direction. Ions of highest initial kinetic energy for a given *m/z* value, having penetrated deeper into the mirror, will acquire more kinetic energy during reacceleration than ions of lowest initial kinetic energy because the former traverse a longer segment of the electric field in the mirror. Therefore, the ions leave the mirror with the same distribution of kinetic energies and velocities as when they entered; however, the lengths of their flight paths now differ because of their differential penetration into the opposing electric field. Because the ions with the highest initial

---

[3] The reflectron TOF mass spectrometer is often designated as the *re*TOF to differentiate it from the linear TOF instrument.

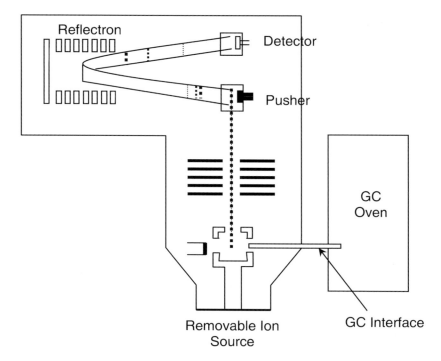

Figure 2-11. **Schematic representation of a GC/TOF-MS with orthogonal ion injection.** *Adapted from a Waters Corp. illustration of the GCT.*

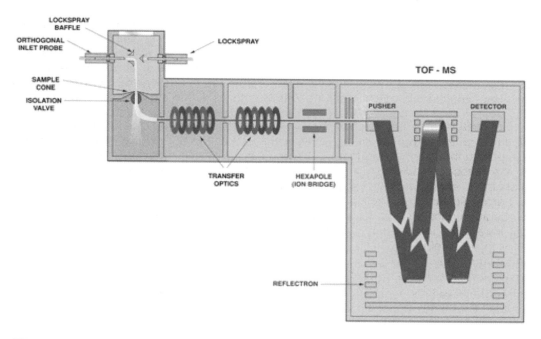

Figure 2-12. **Illustration of the Waters LC/TOF-MS (LCT) with orthogonal ion injection.** *Courtesy of Waters Corp.*

kinetic energy penetrate farther into the mirror, their flight path is longer, a feature that compensates for their having the highest velocity. Notice that the ion, m^+_{high}, is at the leading edge of the packet of three ions entering the mirror, but the ion, m^+_{low}, is at the leading edge of the packet of three ions as they leave the mirror enroute to the detector. Because of their higher velocities, m^+_{high} and m^+_{normal} will catch up with m^+_{low} just as all three impact the detector surface (see Figure 2-10). In this way, the ion mirror serves to focus ions of the same *m/z* value, but having a distribution of KE, at the detector with respect to their arrival time.

Schematic diagrams of selected commercially available *re*TOF mass spectrometers are shown in Figures 2-11 and 2-12. A V-shaped reflectron based on the early work of Mamyrin [68] is shown in Figure 2-11 in the context of a schematic diagram of an orthogonal-acceleration TOF instrument. Waters/Micromass has a unique variation on the reflectron TOF mass spectrometer shown in Figure 2-12; this ion optic geometry is referred to as the W optics. It involves an additional ion mirror, which doubles the instrument's resolving power. This is not without a price. When this instrument is operated in the W optics mode, the sensitivity is reduced by a factor of 3 compared with that obtained during operation in the traditional V optics mode, which offers a resolving power of 5000 (FWHM) compared to 10,000 for the W optics operation. This loss in sensitivity is due to ion loss in the dual reflectron mode.

One of the unique features of the *re*TOF mass spectrometer is that sensitivity is not a function of resolving power as is the case with double-focusing magnetic-sector instruments, which prior to the 1990s had been the only way to obtain accurate mass data. With the double-focusing magnetic mass spectrometer, it was necessary to choose whether the instrument was operated at high sensitivity or high resolving power. In general, the *re*TOF mass spectrometer gives *both* high resolving power and high sensitivity. As will become clear in the section on the double-focusing magnetic instrument described later in this chapter, they can be manipulated to give better resolving power than specified by their manufacturer, but only by sacrificing sensitivity. This is not the case with the *re*TOF mass spectrometer; the resolving power specified by the manufacturer is the only resolving power that can be achieved with a particular instrument and the sensitivity remains constant.

There are also in-line designs of the reflectron based on other work by Mamyrin [69], in which the ions are reflected at essentially 180°. In such an arrangement, it might be expected that the ions would be reflected directly back through the acceleration region into the detector. However, it must be remembered that most of the ions have some initial kinetic energy and that a component of this vector quantity is expressed radially; i.e., perpendicular to the longitudinal axis of the flight tube. As the ions enter the mirror, they begin to decelerate and eventually stop before being reaccelerated out of the mirror back toward the ion source; this process of deceleration and reacceleration causes the ions to have a relatively long residence time within the mirror, a time during which they can move radially. This radial movement of the ions causes the ion beam to bloom, such that as the ions leave the mirror in the direction of the ion source they have been displaced through 2π radians so that a circle of ion current will impact an annular detector (a hole in the annular detector allows ions to pass from the source into the mirror) located back toward the ion source, as illustrated in Figure 2-13.

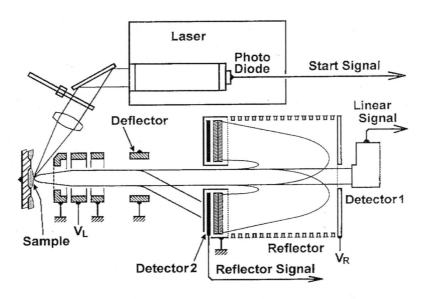

Figure 2-13.   *Illustration of reflectron TOF mass spectrometer with Detector 1 located in position for linear operation and Detector 2 in position for reflectron operation.*

Table 2-2.   *Advantages and disadvantages of the reTOF mass spectrometer.*

Advantages	Disadvantages
Pulsed mode of operation makes it suitable for use with MALDI	Limited *m/z* range compared to that of the linear TOF instrument
Capable of acquiring spectra rapidly for averaging and good definition of narrow chromatographic peaks	Requires higher vacuum than some other types of mass spectrometers
High resolving power results in accurate mass measurements to the nearest 0.1 millimass unit for determining elemental compositions for ions less than 500 Da	Need for carefully controlled ambient temperatures
Good sensitivity due to detection of all ions of all *m/z* values eliminating the need for SIM	Pulsed mode of operation is not necessarily ideal for use as chromatographic detector
No problems with spectral skewing in GC/MS	High price partially due to the requirement of a high vacuum

A significant advantage of the *re*TOF mass spectrometer as compared to the linear TOF instrument is the increase in resolving power. There are commercial instruments used for MALDI that have resolving powers as high as 20,000 (FWHM). The commercial GC/MS and LC/MS instruments that claim high resolving power have specifications of R = 7000–15,000 (FWHM).

Early versions of ion mirrors were often constructed with metal grids (somewhat like the household screen wire for windows) that would permit about 90% transmission of visible light [70]. However, such mirrors might have as little as 20% ion transmission. The ion loss in gridded mirrors was due to the fact that ions passing close to one of the wires in such a grid would be severely deflected so that it would not reach the detector [71]. A gridless ion mirror consists of an array of metallic disks, each with a circular hole cut in the middle through which ions can pass [71, 72]. The donut-like plates are arranged in a stack with individual plates spaced at regular intervals of 10–20 mm. The electrostatic potential along such an array can be regulated to compensate for an energy spread of as much as 10% among the ions [72].

Most ion mirrors employ a linear electric field; this design has been adopted by Bruker Daltonics for its *re*TOF instruments. Another variation of *re*TOF mass spectrometer incorporates Robert Cotter's *curved-field reflectron* (i.e., a parabolic electric field) [73], which is the design that has been incorporated into its MALDI TOF and hybrid TOF instruments commercialized by Shimadzu, the *Kratos AXMIA* series of instruments [74]. These instruments have many more grids than the conventional two-stage design to provide a nonlinear electric field to focus a wide range of *m/z* values at one time. Reflectrons with linear electric fields can focus ions only throughout a limited range of *m/z* values; therefore, to obtain data over the complete range of *m/z* values requires stitching together the results of several individual analyses. For this reason, use of the curved field reflectron has gained favor for peptide sequencing by the technique of post-source decay as described in Chapter 9 on MALDI [75].

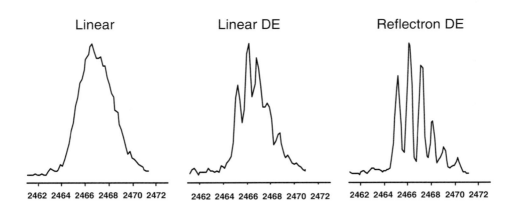

Figure 2-14. **Comparison of MALDI mass spectra obtained in linear mode, with delayed extraction in linear mode, and in reflectron mode.** *Data were obtained by N. Xu, Graduate Assistant, Department of Chemistry, Michigan State University, East Lansing, MI, USA.*

One important aspect of the *re*TOF mass spectrometer that cannot be ignored is the improved performance that comes about from lengthening the ion flight path. The lengthening of the flight path could also be accomplished by lengthening the flight tube; however, the advantage of a longer flight tube could be canceled by the loss of ions due to angular dispersion of ion trajectories from the ion source. The reflectron increases the ion flight path, but it also aids in reducing the angular dispersion of the ion trajectories.

The benefits of a reflectron, in combination with DE, on the resolving power of a TOF mass spectrometer are illustrated by the appearance of the abbreviated spectra shown in Figure 2-14. A third way to improve the resolving power of a TOF mass spectrometer is by orthogonal extraction of the ions into the flight tube, as described in the next section.

C. Orthogonal Acceleration

Instruments using *orthogonal acceleration* (*oa*) are sometimes referred to as *oa*TOF mass spectrometers. Most instruments that have orthogonal acceleration are also *re*TOF instruments; this is true for all commercial *oa*TOF instruments. The orthogonal injection involves the sampling of an ion beam traveling in a direction perpendicular to the axis of the flight path in the TOF instrument [5, 12, 76–79]. Because the original, continuous ion beam is collimated in the ion guide of the orthogonal interface, where some collisional cooling of the ions also takes place, there is very little radial energy dispersion in the ion beam, a direction that ultimately will be along the axis of the flight tube as illustrated in Figure 2-15. Importantly, having little or no radial dispersion of kinetic energy in the original ion beam is equivalent to having little or no dispersion of kinetic energy in the initial packet of ions extracted from a conventional TOF ion source, which would translate into improved resolving power. For this reason, the resolving power of an *oa*TOF instrument is vastly superior to that of a linear TOF instrument or even a conventional *re*TOF mass spectrometer. Stated differently, the *oa*TOF instrument is designed to minimize the energy dispersion in the ion packets, rather than to compensate for such an energy dispersion with a reflectron. Of course, the *oa*TOF source is not perfect, and some energy dispersion survives (though greatly attenuated) among the ions, and so combining an *oa*TOF source with a reflectron provides the best possible resolving power. The duty cycle of an *oa*TOF can be improved by nearly an order of magnitude with the coordinated use of a pair of linear ion traps for efficient injection of ion packets [80].

The packet of ions from an *oa*TOF source is wider (in the direction parallel to the velocity vector of the original, continuous ion beam) than that which would be pulsed or extracted from a conventional ion source into the TOF flight tube. This means that the surface area of the detector and the cross-sectional area of flight tube must be larger than those of a conventional TOF instrument. The first commercialized *oa*TOF instruments were used in LC/MS and in LC/MS *tandem-in-time* applications (see Chapter 3 on MS/MS). Recently, Perkin-Elmer has begun selling a MALDI *oa*TOF instrument that is manufactured by the Canadian company MDS Sciex. As the price of TOF instrumentation drops, there is a good chance that this *m/z* analyzer will become as ubiquitous as the transmission quadrupole *m/z* analyzer is today.

There are two substantial and related problems in ion detection and recording for the TOF *m/z* analyzer: (1) providing a detector with a sufficiently short response time and (2) providing a hardcopy of an analog trace that is commensurate with the electrical transient at the detector [81].

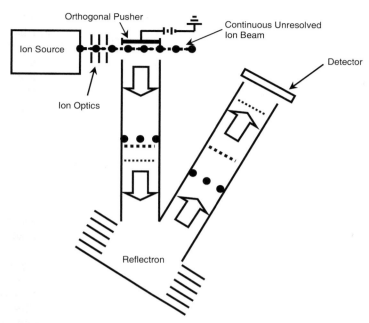

Figure 2-15.   Schematic diagram of an oaTOF mass spectrometer.

D. Ion Detection in the TOF Analyzer

Because the time interval between arrivals of packets of ions at the detector in a TOF instrument is on the order of ns, an amplifier with a bandwidth of at least 100 MHz is used. An oscilloscope has a sufficiently short time constant that it can be synchronized with the TOF cycle for facile presentation of the spectrum in real time for tuning the instrument or for examination of a particular region of the mass spectrum. Similarly, the time constant of an oscillographic recorder is sufficiently short so as to provide a hardcopy of an analog trace on photosensitive paper as used in early applications in conjunction with *analog detection via a variable-delay gating system* for ion detection, as described below. In modern TOF instruments, the detector is a microchannel plate, which is an array of thousands of micron-diameter channels, each one of which is effectively a continuous-dynode electron multiplier of sufficiently high response time (short time constant) to respond cleanly to the arrival of individual ion packets during the transient signal from each pulsed extraction from the ion source. The array of microchannels is fused into a ceramic plate about the size of a 3-inch pancake; this ceramic plate is mounted at the end of the flight tube with the plane of the plate perpendicular to the longitudinal axis of the flight tube. The output from the microchannel plate detector is digitized directly by the computer interface to the data system; no hardcopy is prepared via a variable delay gating system as was used in early TOF applications as described below.

Data acquisition from a TOF mass spectrometer originally used a gating system interface [82]. The gating system, in effect, is a special modification of or an attachment to a novel electron multiplier, which diverts the output to the anode of the gated channel for that brief time interval (e.g., 50 nsec or less) of a TOF cycle that corresponds to the arrival time of a packet of ions of a particular *m/z* value. The time delay of the gating pulse (which instantaneously diverts the electron multiplier output) is held constant for several TOF cycles, and the gated channel integrates the ion current at the particular *m/z* value. The channel can then be sampled with a meter, an analog recording device, or an analog-to-digital converter in a computer interface. In this way, a mass spectrum is acquired by incrementing the delay of the gating pulse over a specified range (corresponding to a range of *m/z* values) at a specified rate.

The spectrum-production rate from the TOF mass spectrometer is fundamentally limited by the transit time of the heaviest (slowest) ion in the mass spectrum. For a 2-meter flight tube and an ion energy of 2 keV, a single-charge ion of 800 Da requires approximately 90 µsec to travel from the source to the detector. This means, for an instrument with these specifications, it is possible to acquire a complete mass spectrum from *m/z* 1 to *m/z* 800 every 90 µsec – a repetition rate of approximately 11 kHz. This high rate should be interpreted only as an index of the fast response of this unique instrument and, of course, should not be taken as an indication of potential sample load for high-throughput analysis! In reality, using the *analog detection via a variable-delay gating system* ion detection system results in an acquisition rate of no more than 10 spectra per minute, a rate far too low for applications such as capillary GC/MS. The response time of the TOF mass spectrometer is ideally suited for monitoring selected ion currents during fast-reaction kinetic studies in the gas phase, such as in flames, explosions, etc., or in monitoring volatiles and fixed gases in breath-by-breath analyses for physiological studies, but the spectral acquisition rate over a range of several hundred *m/z* units by this archaic recording system limits its use as an instrument for GC/MS.

1) Time-Slice Detection

The original TOF mass spectrometers used as GC/MS instruments typically acquire spectra by sampling only a single arrival-time window (time slice or time bin) in each *transient* (the time-dependent signal following the pulsed extraction of a bundle of ions from the ion source) corresponding to only one *m/z* value position in the spectrum described in the previous section. By collecting other time slices (snap shots) at different time delays after successive source pulses, a complete mass spectrum can be reconstructed. To acquire the complete spectrum, the time delay of the gating pulse is progressively increased stepwise so that the ion current is acquired, stored, and recorded

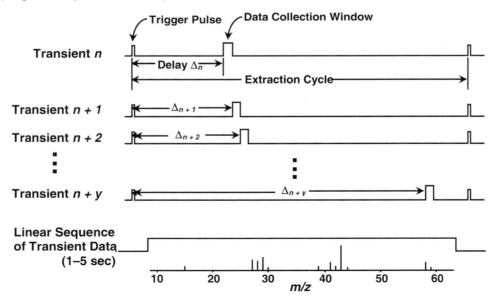

Figure 2-16.   Conceptual illustration of time-slice detection (TSD).

for consecutive *m/z* values. This inefficient process, called *time-slice detection* (TSD), of capturing only a small slice of the time-resolved ion current from each transient is illustrated in Figure 2-16. For simplicity, the figure shows these time slices as detection windows or time bins having a "width" of 0.1 μsec. The location of the opened time bin is scanned along the time axis, from low values to higher values of arrival time. In this way, only a 0.1-μsec-wide portion of this complete hypothetical spectrum presented to the detector would be collected for each extraction pulse of the ion source.

To maintain adequate resolution in the mass spectrum, the width of the time slice in functional instrumentation must be small, on the order of 5 nsec. To acquire a single 5-nsec slice from each source extraction pulse transient throughout the complete range of arrival times (up to 90 μsec) would require an overall or composite acquisition time of 1.8 sec corresponding to an *m/z* range from 50 to 500. An equally important consideration is the fact that while the signal for one time bin is being detected, signals at thousands of other time bins are being ignored or wasted. Furthermore, an acquisition time of nearly 2 sec is far too slow for the narrow chromatographic peaks (peak width of tens of msec) that result from modern capillary gas chromatographic columns. Another problem with this mode of detection is spectral skewing due to the fact that the partial pressure of the analyte is changing in the ion source due to the elution process from the GC column.

The representation of TSD in Figure 2-16 is greatly oversimplified. Rather than using one time bin to represent 1 mass unit, the *m/z* scale is actually divided into more than 10,000 time bins. This means that in TSD, when one time bin is sampled after each pulse of the ion source, only 0.0001 (0.01%) of the available mass-spectral data are collected. In reality, TSD makes TOF mass spectrometry a scanning technique completely ignoring its potential as an array (in time) detector, which can result in high spectral acquisition rates and can eliminate spectral skewing as described below for fast gas and liquid chromatographic analyses.

2) Time-Array Detection

One solution to the problems with TSD is *time-array detection* (TAD). When TAD is employed, all time bins are sampled throughout the transient following each extraction pulse of the ion source; therefore all the ions at all *m/z* values are detected during each TOF cycle (see bottom of Figure 2-17). In using TAD rather than TSD, 10,000 times more data will be available from which to derive a greatly improved signal-to-noise ratio (S/N). Maximal advantages in detection limits and S/N can be realized by summing successive transients in a time-slice registry, but they must be traded off against the frequency of producing a complete mass spectrum. In early applications, the usable spectral generation rate of the TOF mass spectrometer was limited by the methods of data collection and storage, not by fundamental principles of mass analysis. The concept of TAD can be used with a variety of ionization sources [83].

The data detected during TAD are analog data; therefore, it is necessary to sample an analog-to-digital converter (ADC). Recently, ADCs have become sufficiently fast to accommodate the data rates available from TOF instruments. Advancements in data collection from the TOF instrument have resulted in mass spectral generation rates much higher than can be achieved from a scanning mass spectrometer (e.g., quadrupole) by even the fastest ADC. The rate at which data are converted by the ADC limits the mass spectral generation rate, which in turn affects the observed resolution in a chromatogram reconstructed from consecutively recorded mass spectra.

In addition, the effective resolving power of the mass spectrometer is limited by the maximum digitization rate of the ADC. Figure 2-18 illustrates the effect of digitization

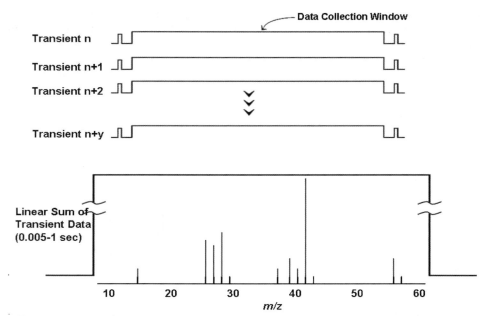

Figure 2-17.   Conceptual illustration of time-array detection (TAD).

rate on the effective resolving power achieved while scanning through the isotope peaks for the molecular ion of fullerene at *m/z* 720 [84].   A digitization rate of 1000 MHz provides a dwell time of 1 nsec at each data point along the *m/z* scale producing a resolving power of 6300.   When the digitization rate is only 100 MHz (dwell time of 10 nsec), the effective resolving power is only 1000 (see right panel in Figure 2-18). Most modern instruments use a 4-GHz 8-bit ADC, which converts the detector output to a value of 0–255, providing a dynamic range of little more than an order-of-magnitude when 10 to 100 spectra are summed.   As the ion optics in TOF improved, manufacturers went from using the ADC to a time-to-digital converter (TDC), which operates at much higher speeds than does the ADC, thereby allowing the possibility of even higher spectral acquisition rates.   However, the Achilles heel of the TDC is its dynamic range, which may be insufficient for quantitation or for representing isotope peak patterns that are useful in determining the elemental composition of an ion.   Recent advances in 8-bit ADC

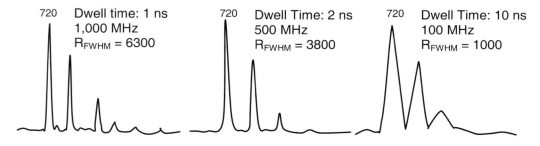

Figure 2-18.   Effect of three different digitization rates on mass spectral peak resolution. *Reprinted from Jurgen Gross, Mass Spectrometry: A Textbook, 2004, with permission from Springer-Verlag, Berlin.*

technology have led to the use of these devices in LC/MS instruments from JEOL and Agilent Technologies that exhibit dynamic ranges of at least four orders of magnitude after spectral summing. The Waters/Micromass TOF instruments still use the TDC; however, through the use of a technique that they call programmable dynamic range enhancement, their TOF instruments now exhibit dynamic ranges comparable to those seen in the ADC instruments, approximately four orders-of-magnitude.

3) TAD with Transient Recorders

Improvements over the simple TSD approach to collecting TOF mass spectra have been made. The principal problems with recording all information available in the transient signal at the detector following each extraction pulse of the ion source are the volume of data and the rate at which they are produced. Employment of high-speed transient recorders allow many time bins to be sampled for each individual ion-source pulse; an early illustrative approach uses a 100-MHz analog-to-digital converter with 16,000 time bins, allowing sequential spectra to be acquired without loss of information following a single source extraction pulse [83]. With these data systems, the accumulated data in the time bins are processed off-line after each sample has been analyzed.

4) TAD with an Integrating Transient Recorder

To achieve TAD without loss of information during GC/MS applications, the data system must acquire up to 10,000 full 100 μsec transients every second with a resolution of 5 nsec in a continuous mode [83]. If all the information is to be captured and used, the need for continuous data collection is mandatory because of the unpredictable dynamics of chromatography. This led to the development of the integrating transient recorder (ITR); this device collects and uses all of the information in each and every transient [83]. Two memory register banks are used alternately so the data collection can be continuous without loss of data. Because each transient contains information sufficient to create a mass spectrum, theoretically up to 10K spectra sec^{-1} could be produced. This rate is unnecessarily high, even for modern chromatographic applications. As a result, the ITR can take advantage of the benefits of integration by summing successive transients in a time-locked registry to produce a single mass spectrum. The number of transients summed to produce each spectrum varies with the application. Because significant advantages are gained by summation (S/N, dynamic range, good detection limits) the rule of thumb is to sum as much as possible while preserving the chromatographic information. Spectral (over an *m/z* range of 50–500) generation rates from 10 to 300 spectra sec^{-1} have been used with this device, and its potential goes as high as 1000 spectra sec^{-1}[83].

The mechanics of data acquisition from the TOF mass spectrometer eliminates the need for selected ion monitoring (SIM) to achieve increased sensitivity (lower limits-of-detection). Because the *m/z* value of a given ion is determined by its arrival time at the detector, and all arrival times are recorded in TAD, several ions from all regions of the mass spectrum can be monitored without affecting the optimum operating conditions of the instrument [85]. With the TAD approach, truly simultaneous monitoring of multiple ion currents with a single detector is achieved [83]. The patented TAD and Integrating Transit Recorder have resulted in two commercial implementations from LECO Corporation: the *Pegasus* EI-only GC-MS, a *re*TOF instrument that generates spectra of unit resolution at a rate of 500 per second over an *m/z* range of 1000, and the *Unique*, which is a linear *oa*TOF instrument (R = 2000 at *m/z* 609; accuracy to 5 ppm) having an *m/z* range to 6000 designed for LC/MS applications to small molecules. The rapid spectral acquisition rate of both the *Pegasus* and the *Unique* (up to 100 summed spectra per second) can accommodate fast chromatography allowing for quantitative analyses that require tens of minutes by conventional MS to be accomplished in tens of seconds [86, 87].

An important consideration for TOF instruments used in conjunction with chromatography is that ions of all *m/z* values reach the detector unless they are removed from the ion source before the transient is formed. This means that helium ions can reach the detector when using a GC/TOF mass spectrometer; in such an application, there is a very large flux of helium ions, especially when compared to the number of analyte ions. This is not a consideration with any other type of *m/z* analyzer because all other instrument types allow only ions of a specified *m/z* value or range to reach the detector. Such a large flux of helium ions would soon burn out the detector; this would occur because electron multipliers function on the principle of secondary electron emission resulting from a destructive interaction of the ions with the surface of the detector.

5) Hadamard Transform TOF-MS

In a Hadamard transform TOF-MS, up to 100% duty cycles can be achieved for maximum efficiency in utilizing the information available from a continuous ion source [36, 88]; i.e., the ion source and ion detection systems are modulated to sort out overlapping transients in such a way that all the ions are detected all the time [89].

The TAD system described above was designed to capture all the information available in a given extraction pulse from the ion source of a TOF mass spectrometer. For a pulsed source, as in a conventional MALDI source, such a TAD system is fine; however, for a continuous source, as in ESI or in EI, no information regarding the analyte is collected during the interval between extraction pulses from the ion source. For a conventional TOF ion source, the effective sampling time of the continuous ion beam is approximately 1 microsecond, which translates into a duty cycle of 1% or less (i.e., 99% of the continuously produced ion current is never captured). The situation is considerably better in an *oa*TOF ion source, in which an ion guide collimates the ion beam and effectively integrates as much as 10% of the continuously available ion current into the orthogonal extraction region during the interval between extraction pulses; however, a 10% duty cycle in an *oa*TOF source still means that 90% of the available ion current is not being utilized in the analysis.

The efficiency of analyzing a continuously available total ion current beam is an especially important issue in the combination of a TOF mass spectrometer with a chromatographic inlet because of the transitory nature of the analyte. For example, in capillary column GC/MS or in capillary electrophoresis LC/MS, the peak width representing the residence time of the analyte in the ion source might be as low as 10 msec. In such cases, it is imperative that the available ion current from the analyte during this interval be utilized as efficiently as possible. Furthermore, complete mass spectra must be captured in a very short timeframe to minimize distortion of the mass spectra due to the time-varying nature of the sample concentration in the ion source; this requirement also demands efficient utilization (high duty cycle) of the continuous ion source in order to provide good ion-counting statistics (good signal-to-noise ratio) in the mass spectral data.

A duty cycle approaching 100% in the TOF mass spectrometer can be achieved by modification of the operational mode of the instrument to permit use of the Hadamard transform as described by Zare and coworkers [89]. In conventional (data wasting) TOF mass spectrometry, the interval between extraction pulses of the ion source is long

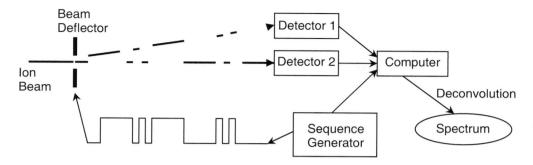

Figure 2-19. *Conceptual illustration of a Hadamard transform TOF-MS.*

enough to allow all ions, even the slowest (heaviest), to reach the detector before another pulse is released. To achieve a 100% duty cycle, multiple extraction pulses of the ion source during the flight time of the slowest ions are necessary to utilize all the ion current available from a continuous source. However, such a procedure would cause havoc in the detector/data system of a conventional TOF mass spectrometer because many bursts (pulses) of new batches of ions would be injected into the flight tube before all ions from the first pulse arrived at the detector; such a process would permit fast ions from a given extraction pulse to pass slow ions from a previous pulse, and so forth, as many velocity-dependent ion packets from different extraction pulses overlap one another *en route* to the detector. The resulting data record of the convolution of hundreds of conventional mass spectra, each shifted in time, would be nonsensical, were it not for an operational arrangement using beam modulation between two different detectors in a manner based on the Hadamard transform [88].

The beam deflector shown in Figure 2-19 effectively chops the ion stream into packets of different lengths. The ions in these packets travel at different velocities as they continue their journey, with faster ions passing slower ions from other packets before reaching the detector, a process that once again would produce a nonsensical output without employing the Hadamard transform. However, consistent with use of the Hadamard transform, the beam chopper or deflector is multiplexed at a variable, but known, rate; i.e., each chopped ion packet is coded (a particular column in a Hadamard mathematical array) for synchronized detection by a specified detector, as illustrated in Figure 2-19. During data processing, the transient at a specific detector (actually, a specified *x–y* region on a ceramic electron multiplier array (position-sensitive CEMA)) plate is deconvoluted into an intelligible mass spectrum according to an inverse mathematical function that coordinated the deflection of the ion beam to a designated detector [89]. In this way, multiple overlapping transients are analyzed in parallel to capture all information available from a continuous ion source [90]. This 100% duty cycle maximizes the S/N in the mass spectra, which translates to the lowest possible detection limits for a given level of ionization efficiency and allows the maximum spectral generation rate for chromatography/mass spectrometry.

Although this concept has been illustrated experimentally, as of the publication date of this book, there are no commercially produced TOF mass spectrometers that use the Hadamard transform technology.

2. Quadrupole Ion Traps

Commercially introduced in the mid-1980s, the quadrupole ion trap (now called the 3D quadrupole ion trap to differentiate it from the linear QIT and the orbitrap) mass spectrometer was an outgrowth of quadrupole technology, which includes the transmission quadrupole and monopole mass filters. The use of quadrupole electric fields to manipulate ions was first explored by Wolfgang Paul and colleagues at the University of Bonn (Bonn, Germany) in the early 1950s [33]. The initial paper on quadrupole technology was followed by a detailed description of the theory [32]. Paul shared half of the 1989 Nobel Prize in physics with the German-born American physicist, Hans Georg Dehmelt, the developer of the Penning (magnetic) trap, "for the developments of ion trap techniques". The seminal reference on the quadrupole technology is Peter Dawson's *Quadrupole Mass Spectrometry and Its Applications* (Elsevier: Amsterdam, 1976; reprinted by the American Institute of Physics: Woodbury, NY, 1995); a more recent book by Ray March and John Todd entitled *Quadrupole Ion Trap Mass Spectrometry*, 2nd ed. (Wiley-Interscience: New York, 2005) provides information on the newer technologies that are missing in Dawson's original treatise. The general fundamentals of trapping ions by a variety of devices including the QIT have been reviewed elsewhere [91].

Use of the word "quadrupole" as an adjective for an ion trap mass spectrometer is to differentiate this type of ion trap from the magnetic ion trap (ion cyclotron) mass spectrometer, which was well known by the time the QIT mass spectrometer was commercially introduced. The QIT mass spectrometer and its subsequent variants have had as great an impact on the general field of mass spectrometry as the *re*TOF and transmission quadrupole mass spectrometers, although for different reasons.

The name quadrupole is derived from the fact that an electric field is created between four opposing electrical poles. The shape of the electric field is a function of the geometric arrangement of these four surfaces and the magnitude of the AC and DC potentials. The primary component of the quadrupole electric field that surrounds the ions is based on radio frequency (RF) potentials applied to two pairs of opposing electrodes. In some cases, the effect of the RF is enhanced by the application of a DC potential. This is the case for the transmission quadrupole mass spectrometer. The transmission quadrupole mass spectrometer employs a two-dimensional electric field that pushes and pulls ions in the *x*- and *y*-directions as they travel along the *z*-axis (the axis parallel to the longitudinal axis of the quadrupole surfaces). This process filters out ions of all *m/z* values except for those of interest. In the QIT, the quadrupole field is three dimensional, which means that ions of all *m/z* values are stored in the device. Ions are affected in all directions so that they travel in discrete orbits within the field. The RF potential is usually applied at a fixed frequency in a transmission quadrupole filter, but with a variable amplitude; however, quadrupole ion traps are distinguished by the use of supplemental alternating potentials to the RF potential for various purposes.

The QIT differs from all other types of mass spectrometers in that it operates at a relatively high pressure of $\sim 10^{-1}$ Pa as opposed to 10^{-4} Pa for the transmission quadrupole and 10^{-7} Pa for *re*TOF mass spectrometers. This increased pressure allows for sufficient resolution to be achieved so that the device can be used as a mass spectrometer. The proper operating pressure is maintained by a continuous flow of helium or argon gas into the mass spectrometer. The use of this *buffer gas* collisionally cools the ions, reducing their rotational and vibrational energies so that the amplitude of their random displacement about the *z*-axis is diminished; this damping of the ion motions extends the *m/z* range of ions that can be trapped with good efficiency and resolution [92]. The choice of the buffer gas depends on the purpose of the trap. In most QIT mass spectrometers, helium is used; however, in the case of the Shimadzu ESI QIT-TOF hybrid *tandem-in-space* instrument, better resolving power is obtained when argon is used as the buffer gas because, in this case, the QIT is used for ion storage, precursor ion selection, and collisionally activated dissociation as opposed to serving as a mass spectrometer (a device that separates ions according to their *m/z* values).

The requirement for a buffer gas was serendipitously discovered by George Stafford and coworkers at Finnigan Corporation (San Jose, CA) in their attempts to produce a GC-MS based on the QIT mass spectrometer [92]. Interestingly, helium is a very good buffering gas, and of course helium is the carrier gas of choice for GC applications. Until that time, attempts to use the QIT for mass spectrometry had failed due to a lack of adequate resolution.

> The term used to describe the separation of ions of different *m/z* values when referring to the TOF mass spectrometer was *resolving power*. The term used for this same purpose when discussing the QIT and transmission quadrupole mass spectrometers is *resolution*. As was detailed in Chapter 1, *resolving power* is a number that determines the capacity of the instrument to separate ions at different points on the *m/z* scale. A numerical value for resolving power can be determined from the peak profiles distributed along the *m/z* scale for two ions differing by a known value of mass (Δm). The *resolving power*, R, is expressed as ($M/\Delta m$), where M is the *m/z* value for ions represented by one of the two mass spectral peaks. The TOF and double-focusing magnetic mass spectrometers operate at a constant value for R, which means that Δm increases with increasing *m/z* values; i.e., Δm is continually changing. In contrast, for instruments that separate ions using quadrupole fields, the *resolution*, Δm, is constant throughout the *m/z* scale; therefore, R has no meaning for these types of instruments.

Several methods have been employed for obtaining a mass spectrum from a QIT. The method that has been most widely used to obtain a mass spectrum is the mass-selective ejection method, although Varian has recently introduced a GC-MS and LC-MS 3D QIT that uses a technique referred to as triple resonance ejection. Other manufacturers of the 3D QIT use the mass-selective instability method to obtain a mass spectrum. Like the discovery of the necessity of a buffer gas for usable resolution, the mass-selective instability technique for operating the QIT as a mass spectrometer was developed by Stafford and coworkers [92] at what was then Finnigan Corp. (now Thermo Electron). Ray March, in the preface to *Practical Aspects of Ion Trap Mass Spectrometry*, Vol. 1, *Fundamentals of Ion Trap Mass Spectrometry* (CRC: Boca Raton, FL, 1995), states that Stafford's development of the mass-selective instability scan is as significant as Paul's work that resulted in the Nobel Prize.

A. 3D Quadrupole Ion Trap m/z Analyzers

Figure 2-20. **Representation of three-dimensional orbital of ions of a single m/z value in a 3D QIT.**

The 3D quadrupole ion trap (3D QIT) mass spectrometer uses a three-dimensional quadrupole electric field to store ions of multiple *m/z* values in concentric three-dimensional orbitals (as represented in Figure 2-20) as opposed to filtering out ions of all *m/z* values except those of a selected value from an ion beam [91, 93–96]. These concentric 3D orbitals can be visualized as layers of an onion with ions of the lowest stored *m/z* value composing the outer layer and ions of successively higher *m/z* values occupying orbitals of progressively shorter radii. The 3D QIT consists of two end-cap electrodes that are electrically isolated from either side of a ring electrode (Figure 2-21). This three-dimensional device has a hyperbolic cross-sectional surface, which is consistent with those used in quadrupole technology.

Stafford's initial breakthroughs of using a buffer gas and the mass-selective scan led Finnigan Corp. to announce the first commercial GC-MS based on a QIT—the ITD™ 700 at the 1983 Pittsburgh Conference on Analytical Chemistry and Applied Spectroscopy (Atlantic City, NJ). This early design of the instrument used the trap not only to store ions but also to provide a place for ions to be produced (internal ion production). The two end caps were held at ground at the same time as a fixed-frequency RF was applied to the ring electrode during an ionization period as electrons were accelerated into the trap at 70 eV to interact with analyte molecules. Ions of all *m/z* values above a certain lower limit were stored in consecutive orbitals. After the ionization period, the ions were cooled briefly. Then, ramping the amplitude of the RF applied to the ring electrode caused ions of successively higher *m/z* values to become unstable and fly towards the two end caps. One of the two end caps had an opening to allow ions to pass through to reach a detector. Improvements have been made in the 3D QIT that allow for *unidirectional ion ejection*; this makes all ions stored in the trap available for detection, not just those ejected in one of two opposing directions.

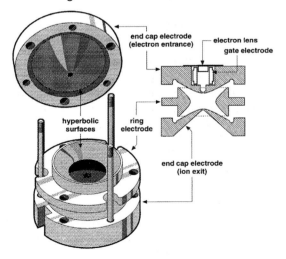

Figure 2-21. **Three electrodes that comprise the 3D QIT.**

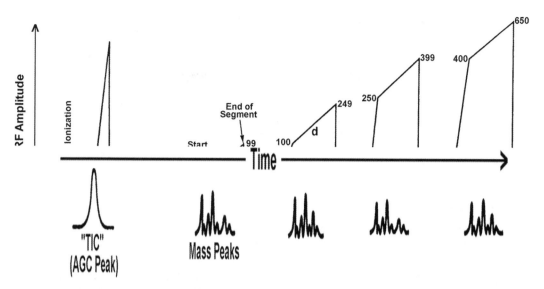

Figure 2-22. *Graphical representation of the scan function of a typical 3D QIT mass spectrometer.*

Some internal ionization mass spectrometers use what is called a *segmented scan function*. This process is described in detail later in this section. The Finnigan internal ionization QIT GC/MS instruments and the Varian Saturn series used a four-segment scan function. The new Varian QIT instruments have a six-segment scan function. The current Thermo Electron, Agilent, and Bruker Daltonics 3D QIT instruments (all having external ionization only) have only a single scan segment. The *m/z* range and the number of segments used in an acquisition is operator settable. As shown in the sequence represented in Figure 2-22, there is what is referred to as a "microscan" at the beginning of each acquisition (multiple or single segment). This microscan is used to determine the ionization time for each of the segments based on the amount of ionizable material in the trap.

In the QIT, the rate at which ions are sequentially scanned out of a QIT according to their individual *m/z* values is very high, 5K–10K *m/z* units per second. This high scan rate makes it possible to cover a large range of *m/z* values in a brief period. Of course, as with any mass spectrometer, a certain minimum of ions must be stored in the trap to yield a good-quality mass spectrum. On the other hand, storing too many ions in the trap can cause space-charge problems that lead to anomalies in the mass spectra. To guard against this problem an operational feature called automatic gain control (AGC) was developed to limit the maximum number of ions trapped, as explained in the next section of this chapter.

Following their introduction of a GC-MS based on the 3D QIT in late winter 1983, Finnigan had planned to begin shipping instruments by mid-1983; however, shipments did not begin until December of 1984. One of the problems that delayed shipment of the ITD 700 was the incorrect assignment of the *m/z* value for the molecular ion peak of some analytes; this problem was unrelated to space-charge effects or ion/molecule reactions (both related to pressure as explained below). This miss-mass assignment was compound (ion) specific. The final solution resulted in the "stretched" trap. This stretched trap solved the problem of mis-mass assignment even though it did not produce a homogeneous quadrupole electric field, which had been thought to be a requirement for proper ion separation [97]. Today, all 3D QIT mass spectrometers employ this *stretched geometry*.

Even after overcoming many of the initial problems in bringing the internal ionization QIT GC-MS to market through an intensive development cycle as described by John Syka [97], the instrument was met with mixed reactions regarding its usefulness as a GC-MS by those who received the first units. One of the limitations of QIT mass spectrometer is called the *space-charge effect*. From Figure 2-21, it is obvious that the trap has a finite volume. A calculated maximum number of ions can be stored in a given volume due to space-charge constraints. Once that maximum is reached, newly generated ions of different *m/z* values begin to displace those of other *m/z* values already occupying space in the trap. This results in a loss of resolution that produces unusable mass spectra. Because packed GC columns were predominant among GC users (especially in the U.S.) at the time of the ITD 700's introduction, the interface between the GC and the mass spectrometer was a device that split the GC eluate, with 10% going to the mass spectrometer and the remainder going to atmosphere as a discard (the open-split interface). This split arrangement often resulted in a higher-than-desired amount of analyte in the trap. Ionization took place in the trap for a fixed time period (settable by the operator), and when a larger number of ions was produced than could be stored, a space-charge mass spectrum would result with clusters of peaks on either side of the nominal *m/z* value peaks. There was no way to prevent this space-charge effect because the partial pressure of the analyte could not be controlled nor the ionization varied as a function of analyte partial pressure.

A further instrumental development, called *automatic gain control* (AGC), at Finnigan Corp. helped alleviate the consequences of the space-charge effect. AGC is a process by which the flux of ion current formed in the trap is determined during a pre-ionization run (the microscan). The ionization period is then adjusted to produce a prescribed number of ions from the determined ion flux in the trap. The ionization time does have a minimum, and if the maximum number of ions is exceeded during this minimum period, then the effects of the space charge will be still observed.

> Often manufacturers create names for processes that will differentiate their product from that of a competitor. This is the case with the name of the process used for varying the ionization time based on the partial pressure of the analyte. Thermo Electron and Varian, Inc. use the AGC term. Bruker Daltonics and Agilent Technologies use the term ion-current control (ICC) so that their instrument is not directly compared with the Thermo Electron technology. Neither term is particularly descriptive.

A subsequent development by Finnigan Corp. increased the number of ions that could be stored in the trap through the use of electric fields that further reduced the rotational and vibrational motion of the ions allowing for a greater storage capacity. This technique is called *axial modulation*, resulting from applying an RF potential to the end caps that is half the frequency applied to the ring electrode.

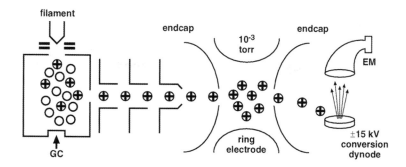

Figure 2-23. Schematic illustration of the external ion source 3D QIT.

The development of AGC and axial modulation can be considered the genesis of the second generation of the 3D QIT. The *internal* ionization 3D QIT technology was sold by Finnigan to Varian, which was able to promote the use of these instruments with their Saturn Series that began in 1987 and continues today as a very popular internal ionization QIT GC-MS. As a prelude to the LC-MS based on the 3D QIT, Finnigan developed an *external* ionization QIT GC-MS illustrated schematically in Figure 2-23. In this device ions formed external to the trap are transferred through a series of ion optics into the trap, where the ions are stored and separated into individual orbitals before they are sequentially scanned (ejected) from the trap one *m/z* value at a time. LC/MS instrumentation requires external ionization because of the high-pressure conditions associated with the various atmospheric pressure techniques used for liquid sample introduction into the mass spectrometer.

The third generation of the 3D QIT came about in 2004 when Bruker Daltonics and Agilent Technologies introduced a jointly developed 3D QIT LC-MS that used more complex electronic technology to increase the storage capacity. This development addressed the problem of sensitivity when ions representing the analyte were present in small numbers as compared to ions from background sources such as the sample matrix; this development also increased the dynamic range of the trap for quantitation. There is anecdotal evidence that this 3D QIT has a storage capacity that rivals that of the linear QIT. Varian has stated that in customer comparisons of the Thermo Fisher LTQ and the Varian 500 both instruments have shown comparable sensitivities.

As shown in the side bar, the problem of unwanted ion/molecule reactions can be avoided through proper selection of operational parameters. It is important to realize that the ion trap mass spectrometer is a pulsed technique; i.e., one in which the sample is not continuously ionized, but rather ionized for a discrete period of time followed by a mass-selective ion ejection ramp over designated segments of the *m/z* range to accumulate the mass spectrum. Even in those cases where an external ion source is used with an ion trap mass spectrometer, at designated times during the operational cycle, ions are either introduced into the trap or expelled from the trap; therefore, in all cases, the ion trap operation is in the pulsed mode.

SIDEBAR

A typical operational cycle of the Varian Saturn 2000 ion trap parameters to accomplish a complete single sweep of the mass axis (a microscan) is illustrated chronologically in Figure 2-22. Such a schedule is necessary for an instrument in which ion formation and ion analysis take place in the same volume. In order to maximize performance, the scan is performed in multiple segments. The scan function of the ion trap mass spectrometer is divided into five parts: the prescan (A) and four scan segments constituting a microscan (B,C,D,E). Dependent on the desired scan range, all or as few as one of the scan segments will be used in the acquisition of data for a spectrum.

During the prescan, (A in Figure 2-22), the electron beam is allowed to penetrate into the trap volume for a brief (1 microsecond) interval while the detector is turned off. During this period, all ions from m/z 20 to 650 are stored. The RF is pulsed to expel all ions below an operator-selectable background m/z value. Then the RF is quickly ramped to expel all ions from the background m/z value (operator selectable) to the upper limit (m/z 650 in this particular instrument). Depending on the magnitude of the signal from this short-burst ionization, the electron beam will be turned on automatically for a defined period of time ranging from 10 µsec up to 25,000 µsec during each scan segment of the microscan. This is the operational basis for the automatic gain control (AGC). If the magnitude of total ion current registered by the electron multiplier from the initial burst were 5.0 V, the period of ionization for analytical purposes might be 25 µsec. On the other hand, if during the initial assessment burst of ionization the total ion current were only 0.5 volts, the ionization period would be extended to 250 µsec by the AGC. As can be seen from the two scenarios described, the total number of ions accumulated in the trap during each of the two situations is the same; in the second case, the ionization was allowed to proceed for a ten-fold longer period because the concentration of the analyte and/or the background was apparently one-tenth that in the first. In this way, the AGC helps keep operation of the ion trap mass spectrometer under reasonably optimum conditions, or at least avoids situations in which ion accumulation to the point exceeding a space-charge limit can be avoided. The efficiency of the trapping is such that the ion current generated is a linear function of the ionization time, which is used to normalize the stored peak intensity values.

The second step in the scan function is the acquisition of data over the first scan segment for the microscan. Assuming that the starting point for the microscan is m/z 99, data will be acquired for the first scan segment. The following steps (a–e) are employed during each segment (B–E) in Figure 2-22:

a. The RF potential is turned on and set to an amplitude that will store all ions between m/z 20 and 650.

b. With the EM turned off, the electron gate allows electrons from the filament to flow into the ion trap for a period set by the AGC ionization period calculation. All ions formed during this period are stored.

c. At the end of the AGC ionization period, the RF amplitude is raised to the appropriate value to store all ions from the starting m/z value for the scan segment to m/z 650. The EM is turned on and allowed to stabilize during the brief period that the RF is pulsed to the proper value for the starting m/z.

d. The RF amplitude is scanned at a rate equivalent to 5600u per second until the end of the scan range or the end of the scan segment (m/z 99 for the first scan segment) is reached, whichever is greater. (The breaks in the default segment are defined by the m/z value of principal ions from PFTBA used in mass scale calibration.)

e.   The RF voltage and the EM are turned off, and all the remaining ions in the trap are pumped away.

If the scan range is greater than *m/z* 99, the five steps (a–e) of the first (B) scan segment are repeated for the second scan segment (C), except that during step (c) the RF is ramped to a starting value of *m/z* 100 and then scanned to an ending value of *m/z* 249.

If the scan range is greater than *m/z* 249, the five steps (a–e) of the first (B) scan segment are repeated for the third scan segment (D), which has a starting value of *m/z* 250 and an ending value of *m/z* 399.

If the scan range is greater than *m/z* 399, the five steps of the first (B) scan segment are repeated for the fourth scan segment (E), which has a starting value of *m/z* 400 and an ending value of *m/z* 650.

If the scan time specified in the analysis method is longer than the time required for one duty cycle of the instrument, then multiple scans (microscans) will be acquired and averaged to produce the analytical scan, which is stored as a mass spectrum.

Another problem with early versions of the 3D QIT mass spectrometer had to do with the potentially long interval between the formation and detection of ions. Unlike the beam-type mass spectrometers (sector, TOF, and transmission quadrupole), which detect ions a few to tens of microseconds after their formation, the 3D QIT stores ions for long periods (hundreds of msec or longer during the spectral acquisition process). In the case of the internal ionization process used with some GC/MS instruments, the ions are stored in a volume that also contains analyte molecules. Depending on the types of ions formed, products of a subsequent ion/molecule reaction might be represented by peaks having an *m/z* value higher than would be anticipated according to the molecular mass of the analyte. This phenomenon (which has been referred to as *self-CI*) is analyte dependent. A dramatic example of such ion/molecule reactions can be observed in the analysis of 2-octanone (nominal mass 128 Da). Figure 2-24 shows the mass spectrum of 2-octanone obtained using an internal ionization QIT as well as the spectrum that appears in the NIST05 Mass Spectral Database. In the mass spectrum of 2-octanone produced in the internal ionization 3D QIT, most of the ion current appears as a single peak at *m/z* 129, which represents the protonated molecule $[M + 1]^+$. The protonated molecule is produced by an ion/molecule reaction between the analyte molecule and its own fragment ions; such ion/molecule reactions can easily be avoided in modern QIT instruments. The ion/molecule reaction and space-charge problems are exacerbated by having too much sample in the trap. The QIT operates optimally with an order of magnitude less sample than is required for most transmission quadruple instruments.

The 3D QIT mass spectrometer was commercialized before the introduction of *re*TOF mass spectrometers in chromatography/mass spectrometry. The fact that as many as half of all the ions formed during a single cycle of the instrument were available for detection was responsible for the unprecedented sensitivity by the QIT mass spectrometer over the *m/z* ranges used in GC/MS as compared to that of either magnetic-sector or transmission quadrupole instruments. The QIT is rivaled in sensitivity by the *re*TOF because all the ions of all *m/z* values formed during a single acquisition cycle are detected, giving it an equal advantage in sensitivity. Another important issue in GC/MS addressed by the 3D QIT was the problem of spectral skewing. Because all ions of all *m/z* values were sampled during acquisition of a single spectrum, the effect of any change in partial pressure of the analyte was evenly distributed throughout the spectrum, thereby precluding spectral skewing. This is the same advantage that was later realized by the TOF mass spectrometer when used with time-array detection.

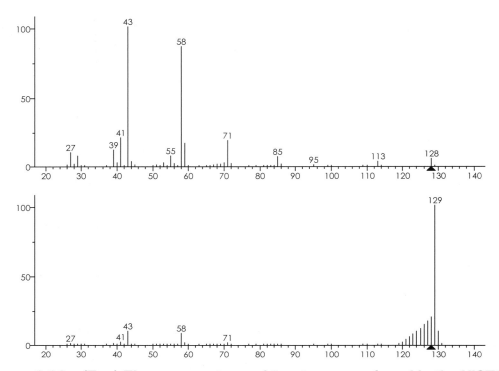

Figure 2-24.   (Top) EI mass spectrum of 2-octanone as found in the NIST05 NIST/EPA/NIH Mass Spectral Database.   (Bottom) EI mass spectrum at the apex of an RTIC chromatographic peak obtained by the injection of 1 µg into an internal ionization 2D QIT GC-MS.

Many of the problems of the original internal ionization 3D QIT used as a detector for a GC were overcome with the widespread use of capillary columns. This resulted in a reduced gas load for the instruments, with less sample being introduced, which meant less of an opportunity for the overproduction of ions that could lead to a space-charge problem.   Finnigan Corp., after licensing the patented QIT technology and other intellectual property to Varian, Inc., made an attempt to get around the *self-CI* problem by using an external conventional ion source (Figure 2-23).   From the continuous external ion source, there was an ion injection period during which the ions flowed from the source into the trap.   This flow was then interrupted and a mass spectrum was obtained from the stored batch of ions.   However, during the interval of no ion flow from the external ion source, it appears that the same ion/molecule reactions observed in the internal ionization QIT take place and the same spectral anomalies are observed. Because of the lack of reactive fragment ions in ESI and APCI systems, such ion/molecule reactions are not observed in LC/MS systems using 3D QIT mass spectrometers.

A potential problem with all ion traps used in GC/MS and LC/MS, beyond the *self-CI* issue, is the effect of storage time and environment in which ions are stored prior to being analyzed according to their *m/z* values.   Some ions are more reactive or unstable than others. This is more of a problem with aliphatic molecular ions in GC/MS than it is with protonated molecules in LC/MS, but none the less, it is an important consideration. In other types of instruments, ions are detected within less than 10 µsec after their formation. In the QIT, ions may be stored for as long as 25,000 μsec. This storage occurs in an inhospitable environment where the ions experience collisions with He atoms that are used to maintain the pressure required for ion orbital separation. The analyte ions can become collisionally activated and undergo dissociation.

Even with the reported difficulties, the 3D QIT mass spectrometer has some outstanding features [93]. For GC/MS, the internal ionization QIT has a very significant advantage in applications of chemical ionization. Chemical ionization (CI) results from an ion/molecule reaction as elaborated in Chapter 7. The analyte molecules of interest react with a proton-rich reagent ion to achieve protonation of the analyte. This occurs when the proton affinity of the analyte is greater than that of the precursor (conjugate base) of the reagent ion. The yield of an ion/molecule reaction, such as that used in CI, is dependent upon the number of collisions occurring during the experiment, which is dependent upon at least three interacting variables: (1) the partial pressure of the analyte, (2) the concentration of reagent ions, and (3) the duration of the ion/molecule interaction. In ion trap MS, the reagent ions can be trapped for a relatively long period (hundreds of μsec, even msec). This means that the concentration of reagent ions in the QIT can be much lower than that in a conventional CI experiment carried out in conventional ion sources to produce comparable mass spectra [98].

For the reasons described in the previous paragraph, a typical CI mass spectrum can be produced in an ion trap at operating pressures of the reagent gas approaching 10^{-3} Pa because the residence time of the analyte ions is on the order of one millisecond in the trap [94]. In a quadrupole mass filter, the ion source pressure must be ~10^2 Pa for the reagent gas because the residence time of reagent ions in the ion source is closer to one microsecond [14, 34]. The CI mass spectra produced by the two experiments described above are comparable because the number of collisions between analyte molecules and reagent ions (concentration of reagent ions times reaction time) is comparable. Negative CI in an ion trap has also been described [99]. However, it should be noted that resonance electron capture ionization (ECI) is not a possibility in the internal ionization QIT because of the requirement to store electrons; because the electrons have such a small mass (~1/2000 Da), they are almost immediately expelled from the trap by the same RF that is used to store the ions.

The problem with *self-CI* in an internal ionization QIT can be overcome; however, it resulted in early criticism of the ITD 700. On the other hand, the sensitivity of the instrument was unprecedented; users achieved detection limits lower than they had ever believed possible. Development of the AGC by Finnigan Corp. went a long way toward correcting the space-charge problem in early versions of the 3D QIT, a fix that could be accomplished with only a software modification; no additional hardware was required. Eventually, Finnigan Corp. entered into an OEM agreement with Perkin Elmer Corporation (Norwalk, CT) whereby PE sold the ITD 700 to customers who were not disclosed to Finnigan. Although Finnigan upgraded all of its customers to the next generation of the QIT software (ITD 800), unfortunately Perkin Elmer did not, thereby resulting in many dissatisfied users of the early ion trap instrumentation.

Another significant advantage of all QIT mass spectrometers, not just the 3D QIT, is the capability for MS/MS experiments. The trap can be used to isolate the precursor ion, cause collisional activation of the isolated precursor ion with the buffer gas, and then store and subsequently *m/z*-analyze the product ions that result from dissociation of the precursor ions. There are some limitations to the use of the 3D QIT for MS/MS as described in the MS/MS chapter, but the reality is that much can be accomplished with *tandem-in-time* mass spectrometry, which allows for an inexpensive solution to obtaining fragmentation information from protonated molecules.

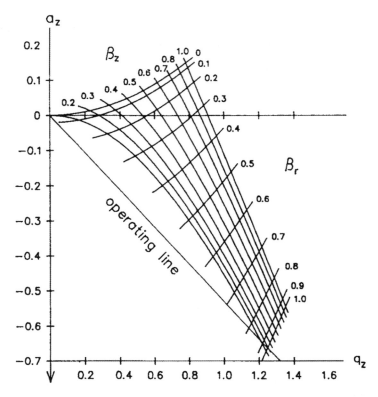

Figure 2-25.   Stability diagram for the 3D QIT showing the operating line.

SIDEBAR

Principles of Operation

Quadrupole ion traps confine ions in a small volume between a ring electrode and two end-cap electrodes by appropriately oscillating (radio frequency, RF) electric fields [93, 94]. The typical spatial configuration and "cutaway" views of an ion trap are given in Figure 2-21, which show the important parameters: r_0, the internal radius of the ring electrode, and z_0, the distance from the center of the trap to the apex of the end-cap electrode. Hyperbolic end-cap electrodes (electrically common) are arranged on opposite sides of (and in planes parallel to that of) a ring electrode also having a hyperbolic inner surface. The ion trap is operated by applying a sinusoidal potential (RF at a fixed frequency) to the ring electrode, while the end-cap electrodes may be grounded, biased at a constant potential (DC, often zero), or maintained at an oscillating (AC) potential. The variety of possible potentials applied to the end caps allow for trapping of all ions within a specified *m/z* range, trapping of ions above a specified *m/z* value, trapping of ions at only a selected *m/z* value, or ejection of ions of specified *m/z* values. The best references for understanding the physics of the QIT is *Quadrupole Storage Mass Spectrometry* by R. E. March and R. J. Hughes (John Wiley & Sons, Inc., New York, 1989) and the second edition of the book *Quadrupole Ion Trap Mass Spectrometry* by R. J. March and J. F. J Todd (John Wiley & Sons, Inc.: New York, 2005).

Ion motion within the ion trap, just as in the quadrupole mass filter, can be described by the Mathieu equations [93, 94]. Because DC potentials can be applied to the trap, stability diagrams based on a and q space (as defined for quadrupole mass filters in a later section) can be used to indicate which ions according to the *m/z* value will remain stable within the ion trap, and which will be ejected under a particular set of experimental parameters.

Field conditions for ion stability within an ion trap can be plotted in Mathieu a, q space as illustrated in Figure 2-25. The values of a and q from the Mathieu equations depend on the dimensions of the trap according to the following relationships:

$$a_z = -2a_r = -16\frac{e}{m}\frac{U}{r_0^2\omega_{RF}^2}$$

(Eqn. 2-4)

$$q_z = -2q_r = -8\frac{e}{m}\frac{V}{r_0^2\omega_{RF}^2}$$

(Eqn. 2-5)

The subscripts z and r represent axial and radial motion perpendicular to and between the end caps, respectively, U is the DC bias on the end-cap electrodes (if any), V is the RF potential applied to the ring electrode, r_0 is the radius of the ring electrode, and ω_{RF} is the RF angular frequency. Ions whose a and q values lie inside the cross-hatched envelope have stable trajectories inside the trap.

The region shown in the closed area in Figure 2-25 is of paramount importance as it represents those values that can be met with present technology to ensure that ions are stable in both the r and z directions simultaneously so that they can be confined in the trap.

The parameter β in Figure 2-25 helps convey conceptually some important behavioral aspects of ions in the trap; i.e., depending on the particular coordinates of a and q, the value of β at any given coordinate of a and q relates the extent to which the ion follows the imposed radio frequency field. In this regard, there are three categories of behavior of the ion as illustrated by the magnitude of β: (a) as the value of β approaches zero, the ion is apparently not affected by the oscillating fields and tends to drift out of the trap; (b) when the value of β equals unity, the ion oscillates in resonance with the RF field, absorbing kinetic energy to the point that its magnitude of oscillation increases so that the ion escapes the trap or collides with one of the end-cap surfaces; and (c) when β has a value between zero and unity, the ion is trapped by the oscillating fields and it oscillates in a periodic mode; its average kinetic energy can be approximately related to the value of β.

The 3D QIT has played a significant role in the advancement of LC/MS using ESI and APCI techniques [95]; a miniaturized version of the cylindrical ion trap also has been used in atmospheric ionization applications [100]. The early GC/MS instruments had upper *m/z* limits of 650. This was satisfactory for the mass-to-charge ratio of most ions that could be produced by electron ionization of analytes that had sufficient volatility to pass through the GC. However, the *m/z* range requirements for LC/MS were much greater. The modern 3D QIT is capable of upper *m/z* limits of ~6000. There are trade-offs involved in achieving this upper *m/z* scan limit, such as raising the *m/z* value at which the scan is started. These issues have not proven to be a particular problem when using these devices in LC/MS.

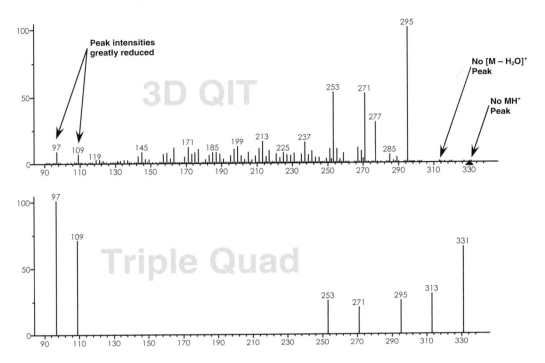

Figure 2-26. *(Top) Product-ion mass spectrum obtained in the 3D QIT of the protonated molecule of 17-α-hydroxyprogesterone. (Bottom) Product-ion mass spectrum obtained for the same precursor ion using a CAD cell in a tandem-in-space triple quadrupole mass spectrometer.*

The 3D QIT does offer an advantage in its ability to perform MS/MS in time as opposed to the use of multiple instruments in *tandem-in-space*. There is a problem with what is called the *low-mass limit* for product ions. Because of the way that the precursor ion is stored and collisionally excited to bring about dissociation, it is generally considered that product ions with *m/z* values of less than one-third that of the precursor ion are not efficiently stored. A good example of the adverse effects of this limitation is seen in a comparison of the two mass spectra in Figure 2-26. This low-mass limit presents a significant problem in the analysis of peptides because immonium ions fall into the *m/z* range below this cutoff. Recently, some manufacturers of 3D QIT traps have introduced the techniques of *electron transfer dissociation* (*ETD*) for the fragmentation of multiple-charge ions as described by McLuckey and coworkers [101]. Don Hunt at the University of Virginia has patented implementation of ETD in the linear quadrupole ion trap (LIT) [102]. Because there is no need simultaneously to store and excite the precursor ion when using ETD, the low-mass limit is completely eliminated. Implementation of ETD in the LIT offers a higher degree of control than can be obtained in a 3D QIT. This will be described in more detail in the LIT section of this chapter. There are other factors limiting the operation of the 3D QIT as discussed in more detail in the MS/MS chapter.

Table 2-3. **Advantages and disadvantages of the 3D QIT spectrometer.**

Advantages	Disadvantages
Availability of all ions of all *m/z* values for detection during a single instrument cycle (spectral acquisition)	Low resolving power, difficult to obtain better than *m/z* 0.3 resolution
No spectral skewing	Buffer gas can result in unwanted collisional activation of stored ions
Can perform low-pressure CI in the internal ionization GC/MS instruments	Long times between ion formation and ion detection
Increased sensitivity over scan-type instruments in full spectrum acquisition mode	Low-mass storage limit and inability to perform common neutral loss and common product-ion analysis in MS/MS
Can perform product ion analysis MS/MS within a time domain	Scan range is limited by upper *m/z* value of ~6000
Less collateral fragmentation of product ions than with collision cells in *tandem-in-space* instruments	Low-mass cutoff in MS/MS

Because of the increased number of ions of any given *m/z* value that are available for detection, the 3D QIT has a sensitivity advantage over the scan-type instruments such as the transmission quadrupole and the double-focusing magnetic-sector instruments when spectral data are acquired in what is called the *full spectrum mode* (a measurement of ion currents at successively increasing *m/z* values over a designated *m/z* range). However, for techniques such as selected ion monitoring (SIM) or selected reaction monitoring (SRM) as carried out using tandem mass spectral techniques, the QIT is significantly less sensitive than the scan-type instruments. SIM is not an option in the 3D QIT. SRM can be carried out on a single product–precursor ion pair with the same efficiency as by a scan-type instrument, but if more than one transition pair is to be monitored, a precursor-ion isolation, fragmentation, and product-ion storage and acquisition must be done for each transition pair.

With the development of the linear quadrupole ion trap and the orbitrap, the future of the 3D QIT may be in question; it may be the shortest lived of all the commercial mass spectrometer types. During its tenure, the 3D QIT has proven to be a very important tool and will have some, even if limited, future [95]. Not only has this type of mass spectrometer proved its value in LC/MS and GC/MS, it has also been a significant factor in MALDI because of its use in AP MALDI, which requires ion accumulation from multiple laser shots (see Chapter 10), and it has been used to advantage for analyses of permanent gases [103].

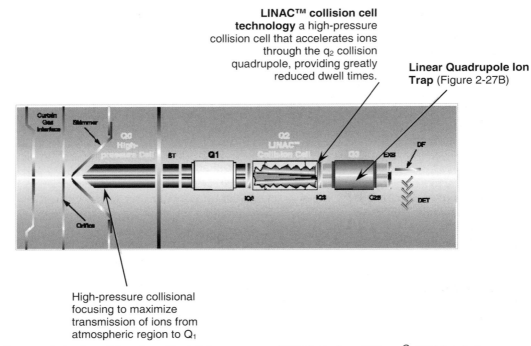

LINAC™ collision cell technology a high-pressure collision cell that accelerates ions through the q₂ collision quadrupole, providing greatly reduced dwell times.

Linear Quadrupole Ion Trap (Figure 2-27B)

High-pressure collisional focusing to maximize transmission of ions from atmospheric region to Q_1

Figure 2-27A. **The Applied Biosystems/MDS Sciex QTrap® 4000 triple quadrupole mass spectrometer with linear ion trap technology.** *Courtesy of Applied Biosystems, Foster City, CA, USA.*

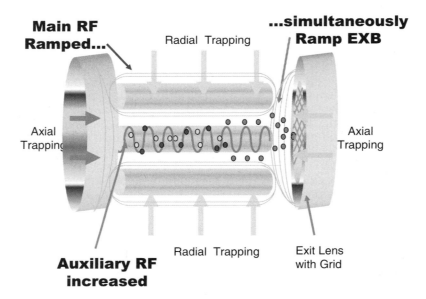

Main RF Ramped...

Radial Trapping

...simultaneously Ramp EXB

Axial Trapping

Axial Trapping

Auxiliary RF increased

Radial Trapping

Exit Lens with Grid

Figure 2-27B. **The Applied Biosystems/MDS Sciex linear quadrupole ion trap (LIT) uses axial ion ejection for the detection process and radial ejection in the ion isolation process in their QTrap 4000.** *Courtesy of Applied Biosystems, Foster City, CA, USA.*

B. Linear Quadrupole Ion Trap (LIT) *m/z* Analyzers

The linear ion trap (LIT), sometimes referred to as the two-dimensional quadrupole ion trap (2D QIT), is one of the more recent additions to the single *m/z* analyzer group [104–106] (with the orbitrap being the most recent). The LIT makes use of the basic structure of a transmission quadrupole; i.e., an array of four electrical surfaces; however, instead of being used to filter ions of all *m/z* values except for those of a desired value from an unresolved ion beam, they are used for trapping, manipulation of ion trajectories, and *m/z*-selective ion ejection [107]. The shape of the surfaces that compose the sides of the LIT apparently is not critical because one constructed from just four elongated planar electrodes [108], mounted in parallel and employing an RF potential for ion trapping in the radial and axial directions, still shows good figures of merit [109]. Ions in an LIT are confined radially (*x*- and *y*-directions) by a two-dimensional RF field, similar to that employed in a transmission quadrupole *m/z* analyzer, which in that instrument allow ions of only a single *m/z* value to reach the detector, and axially (*z*-direction) by potentials applied to the end caps (electrodes), which limit the flow of ions longitudinally [110]. A commercial single-stage *m/z* analyzer based on the LIT (the Thermo Scientific LTQ) uses radial ejection to obtain a mass spectrum. The LIT has been used for precursor-ion isolation and collisionally activated dissociation in a tandem instrument that uses an FTICR as the product-ion analyzer. The LIT has also been used as the third stage (Q3) of a triple quadrupole instrument [111]; this variation of the LIT employs axial ejection of ions [107]. A schematic representation of axial vs radial ejection is shown in Figure 2-27B, which is a diagram of the LIT produced by MDS Sciex.

As pointed out by Douglas [104], the LIT can store a larger number of ions than the 3D QIT. The volume of the LIT can be increased by increasing its length; however, the length cannot be increased indefinitely because the mass spectral peaks become too wide for unit resolution, and missed-mass assignments become a problem. The *exit barrier potential* (EXB) applied to the electrode on the end where ions are detected as ejected is used to aid in the containment of ions other than those of a specific *m/z* value that are being extracted.

The first commercially available linear quadrupole ion trap *m/z* analyzer was used as the third stage (MS2) of a multiple-*m/z* analyzer MS/MS (*tandem-in-space*) instrument, the QTrap® 2000, introduced (at PittCon 2002) by Applied Biosystems/MDS Sciex. In the QTrap 2000, MS2 (the second *m/z* analyzer) can function as a transmission quadrupole *m/z* filter or as a quadrupole ion trap. At the American Society for Mass Spectrometry (ASMS) Meeting in Montreal, Canada, in late May 2003, AB/MDS Sciex introduced the QTrap® 4000 as an optional configuration of the API 4000, which had also been introduced at PittCon 2002. A schematic representation of the QTrap 4000 is shown in Figure 2-27A.

Thermo Scientific introduced a linear quadrupole ion trap as the first stage to its tandem FTICR instrument at PittCon 2003, the LTQ FT-MS System [112]. This LIT is similar in design to the AB/MDS Sciex device except that ion ejection is radial as opposed to axial. Thermo Electron introduced this same LIT design as a standalone instrument at ASMS 2003; this standalone instrument has a maximum *m/z* range to 2000.

The LIT has higher injection efficiencies and higher storage capacities than the conventional 3D QIT (sometimes referred to as the Paul trap). Two-dimensional multipole fields are well known for their capacity to trap and manipulate ions [113]. A higher multipole field (the quadrupole field) is the basis of the transmission quadrupole

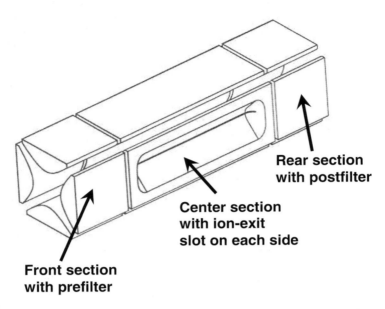

Rear section
with postfilter

Center section
with ion-exit
slot on each side

Front section
with prefilter

Figure 2-28. ***Schematic representation of the linear quadrupole ion trap***
mass spectrometer with radial ion ejection. Reprinted with permission
from Reference [107].

m/z analyzer [34, 114]. Multipole fields are often used as ion guides in conjunction with
collisional cooling (described earlier in this chapter) and to facilitate ion-beam collimation
in collision cells of *tandem-in-space* analyzers (described in the MS/MS chapter). The
enhanced storage capacity of the LIT over the newly introduced Varian 500 and the
high-capacity 3D QITs from Bruker Daltonics and Agilent Technologies may be
somewhat questionable because both of these instruments claim similar detection limit
performance to the LIT which would indicate similar stage capacities.

The linear quadrupole ion trap using radial ejection is schematically represented
in Figure 2-28 [110, 115]. The device has three discrete segments, all electrically
isolated; apparently this was not an absolute requirement because Thermo Scientific now
has a single detector, nonsegmented radial-ejection LIT available at a lower price than for
the segmented version. However, the only version of the LTQ that offers *electron
transfer dissociation* (*ETD*) is the segmented trap. Through computer control of
potentials on the three segments (as shown in Figure 2-28) it is possible to use the LTQ
to isolate precursor ions of a specific *m/z* value, store these ions in the first segment, and
then isolate reagent ions of a specific *m/z* value formed in an external EI/CI source prior
to the ETD step. This added degree of specificity makes the LTQ ideally suited for ETD.

There are slots in the two *y*-direction center sections to allow ions to be ejected in
the *x*-direction toward detectors that are placed on either side of the device, just beyond
the slots, to provide maximum sensitivity. The minimum number of ions that must be
stored to obtain a mass spectrum of a specific resolution and accuracy (the spectral
space-charge limit) was reported to be 2×10^4. The maximum number of ions that could
be stored without a space-charge effect (the space-charge ion capacity) was reported to
be 7×10^6. Both of these values were reported to be about ten times higher than
analogous values for the conventional 3D Paul trap [115]. At this time, there is no

reported comparison with the 3D QIT that has dipole and hexapole fields superimposed on the quadrupole field (the high-capacity QIT); however, according to the Quadrupole Ion Trap Product Manager at Varian, Inc., Varian compared the Varian 500 with the Thermo Scientific LTQ and was able to achieve similar limits of detection for small molecules. The radial-ejection LIT mass spectrometer, when compared with the Thermo Scientific LCQ (a conventional 3D Paul trap), shows a five-fold lower detection limit for the analysis of 500 fg of alprazolam ($C_{17}H_{13}ClN_4$) in an LC/MS/MS (m/z 309 → m/z 274) analysis [115].

After the commercial introduction of the standalone LIT mass spectrometer by Thermo Scientific, demand for their traditional 3D QIT instrument subsided quickly because the latter had lower performance in terms of sensitivity [116]. It is important to note that operational features such as the triple-resonance scan function and the increased capacity of the 3D QIT (due to its superimposed hexapole and dipole fields) has yet to be evaluated against those of the LIT. This is why the future of the 3D QIT is somewhat questionable.

Another interesting aspect of the LIT is its lack of dependence on a buffer gas. Apparently, based on the description of the MDS Sciex use of the LIT as either or both of the second and third stages of a *tandem-in-space* instrument, ion isolation and ion separation are possible at the low operating pressure of the third stage or of the second stage (collision cell). In their description of operating specifications for the radial-ejection LIT, Thermo Scientific states a pressure in the LIT housing of 2×10^{-3} Pa. Interestingly, in their original report of the radial ejection LIT, Thermo Scientific disputed earlier reports showing a dependence on pressure for ion isolation and trapping efficiency. There is no observable change in operating performance of the Thermo Scientific instrument over a pressure range of 5×10^{-2} to 10^{-3} Pa using He as the collision gas; the earlier report had been based on the use of nitrogen as the collision gas [117–119].

It is interesting to note that the MDS SCIEX instrument uses nitrogen as the collision gas. This is in contrast to the operational conditions for conventional triple quadrupole instruments, which often employ Ar as a collision gas.

Another important factor, all ready mentioned, regarding the LIT is its lack dependence on the phase and amplitude of the RF used for the trapping field. In the 3D Paul trap, the lack of a specific phase and amplitude for the RF voltage results in a trapping efficiency as low as 5 %. However, in the LIT the trapping efficiency can be as high as 100%.

In the book, *Quadrupole Ion Trap Mass Spectrometry*, 2nd ed., by Raymond E. March and John F. J. Todd (Wiley-Interscience: New York, 2005) a comparative example is cited of a product-ion analysis carried out on a conventional triple quadrupole mass spectrometer and on a triple quadrupole instrument using an LIT as the third stage of 100 pg μL^{-1}. The LIT instrument produces a spectrum identical to that from the conventional triple quadrupole m/z analyzer with respect to the m/z values of the product ions and their relative intensities; however, the spectrum from the low-pressure LIT MS/MS instrument exhibits a sixteen-fold increase in signal strength, better signal-to-noise background, and better resolution. This improved performance of the LIT MS/MS instrument is due to its capacity to accumulate ions in MS2 before a mass spectrum is acquired. Use of the LIT mode in both the collision cell (q2) and in MS2 also gives this type of triple-*m/z* analyzer MS/MS (*tandem-in-space)* instrument greater flexibility in its operation and application compared to that of the conventional triple quadruple analyzer. Refer to Chapter 3 on MS/MS regarding more about MS/MS, the LIT, and the differences in CAD in the collision cell and in MS2 of the MSD Sciex instrument.

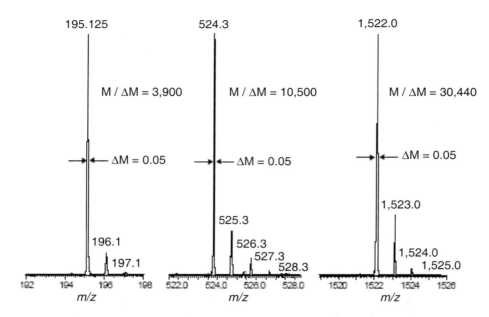

Figure 2-29. *Illustration of m/z 0.05 resolution across a full-scan mass spectrum of a calibration mixture containing caffeine (m/z 195), the peptide MRFA (m/z 524), and Ultramark 1621 acquired at 27 m/z units sec^{-1}. From Schwartz JC, Senko MW, Syka JEP "A Two-dimensional Quadrupole Ion Trap Mass Spectrometer" 50th ASMS Conference on Mass Spectrometry and Allied Topics, Orlando, Fl, 2002, AO21187.*

Like the 3D QIT *m/z* analyzer, the 2D LIT *m/z* analyzer is scanned by applying RF at a fixed frequency, typically 1 MHz to the trap; the amplitude of the RF; i.e., the magnitude of the applied potential (typically, up to 10 kV), determines the *m/z* value of the ions that are ejected from the trap. Normal operation of the LIT is at unit resolution, as accomplished by rapidly scanning ions out of the trap (usually at 11K *m/z* units sec^{-1}) using the *m/z*-instability mode described earlier for the 3D QIT. Resolution of 0.05 *m/z* units over an *m/z* range of 200–1522 has been reported for the Thermo Electron radial-ejection LIT when the scan rate was lowered to 27 *m/z* units sec^{-1}, as illustrated in the mass spectra in Figure 2-29. This same resolution enhancement can be obtained by reducing the scan rate when using the 3D QIT.

C. Performance Trade-Offs in the Ion Trap

The trade-offs in or interplay between resolution, scan speed, and sensitivity (detection limits) can be explained by the operating mechanics of the ion trap. Complete ejection of ions of a given *m/z* from the trap requires that the ion volume be exposed to approximately 200 cycles of RF. The amplitude of the RF can then be incremented to eject ions of the next higher desired increment in *m/z* value. If only unit resolution is desired along the *m/z* axis, each increment in RF amplitude (typically, a few volts) corresponds to one *m/z* unit, and an *m/z* range of 11,000 can be scanned in 1 second at the normal scan rate of 11K *m/z* units sec^{-1}. On the other hand, if higher resolution is desired, say 0.001 *m/z* unit, then the ion volume must be exposed to smaller incremental changes in RF amplitude, say millivolts. The scan rate under such conditions of high resolving power would be established during 1000 consecutive intervals (each allowing 200 cycles of exposure at a given increment in RF amplitude) of progressively increasing

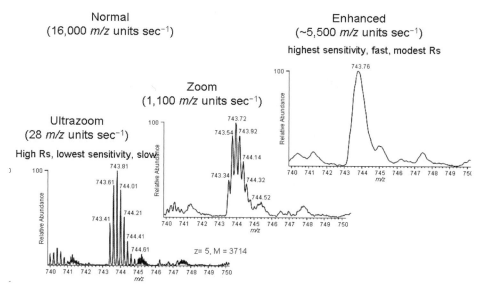

Figure 2-30. ***Effect of resonance ejection scan rate on resolution, sensitivity and data acquisition time.*** *Data courtesy of Professor Gavin Reid, Michigan State University, East Lansing, MI, USA.*

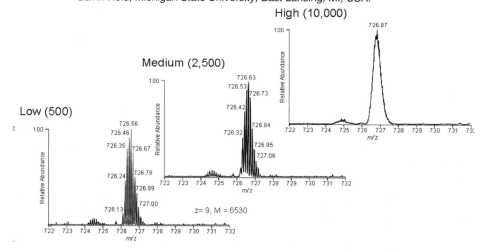

Figure 2-31. ***Effect of target number on resolution and sensitivity.*** *Data courtesy of Professor Gavin Reid, Michigan State University, E. Lansing, MI, USA.*

RF potential for each *m/z* unit along the *m/z* axis; clearly, this is a low-mass spectral scan rate, namely 11 *m/z* units sec^{-1} in this case. These features are illustrated in Figure 2-30 in the abbreviated mass spectra of an ion of mass 3714 Da containing five protons; i.e., *m/z* 743 (= 3714/5) obtained with a Thermo LTQ linear ion trap mass spectrometer [120]. Clearly, the resolution in the mass spectra deteriorates as the data are collected at unit resolution at 5500 *m/z* units sec^{-1} compared to those obtained in the ultrazoom mode at the much lower rate of 28 *m/z* units sec^{-1}.

Whereas the ion trap must be scanned slowly in order to capture high-resolution data, the resolving power of the instrument can also be influenced by space-charge effects in the trapping volume. This feature is illustrated in the abbreviated mass spectra in Figure 2-31 as collected at different charge densities (ion concentrations) in the trap for an ion of 6530 Da containing nine protons resulting in $m/z = 726.5$ ($= 6530/9$). The target value indicated next to each spectrum in Figure 2-31 is an arbitrary value that is proportional to the charge density in the trapping volume of the instrument. The automatic gain control (AGC) was adjusted to allow the largest number of ions to be trapped in the spectrum labeled High (target $=$ 10,000), and exhibits the poorest resolution. This results from adverse space-charge effects in the trapping volume; the space charge distorts the local interaction of some ions with the RF field such that inhomogeneous ejection occurs. Note that a trend of improving resolution is apparent in the mass spectrum labeled Medium (target $=$ 2500); namely, the resolution is better on the high-mass edge of the multiplet as the result of most of the other ions having been ejected from the trapping volume, thereby diminishing the space-charge effect as the scan proceeds from low to high m/z values. Clearly, the best resolution is achieved at a target value of 500, but this condition is associated with the lowest sensitivity because the lowest number of ions was collected before the resonance ejection scan began.

> The QIT should be viewed as a pulsed device, as is the TOF. Although the final output mass spectrum obtained with a QIT m/z analyzer is a result of scanning ions out of the trap, it should be realized that the bolus of ions analyzed during a duty cycle of the instrument was captured as a snapshot in time, a process that eliminates the problem of spectral skewing while acquiring data from samples introduced via on-line chromatography.

Table 2-3. *Advantages and disadvantages of the linear quadrupole ion trap mass spectrometer.*

Advantages	Disadvantages
High sensitivity (better than some 3D traps)	Normal operation provides unit resolution; a lower scan rate provides only 50 milli-m/z unit resolution
No problem with spectral skewing when used as a single-stage analyzer	Standalone analyzer limited to m/z 2000 maximum range
Buffer gas not essential, thereby differing from the 3D QIT	Standalone instrument is limited to the types of MS/MS experiments that can be performed. Not an issue when part of a *tandem-in-space* instrument
Low-cost MS/MS (MS^n) when used as a standalone instrument	Low-mass cutoff in MS/MS is an issue in the linear QIT as it is in the 3D trap
Can be used in combination with other analyzers for more flexibility	Cost is considerably higher than for a 3D trap
Better SRM sensitivity when used as third stage of a triple quadrupole	

The rectilinear ion trap [121], a simplified version of the linear trap, has been incorporated into a miniaturized mass spectrometer the size of a shoe box [122]. This miniature mass spectrometer has a power consumption of less than 70 W, similar to that of a laptop computer, yet has an *m/z* range of 500 together with tandem MS capabilities.

3. The Orbitrap

A. Historical Aspects

The orbitrap mass spectrometer is the latest development in trapping devices used as an *m/z* analyzer. This device was developed by Alexandor Makarov while at HD Technologies, Ltd in the United Kingdom [123, 124]. The orbitrap has now been commercialized as part of a *tandem-in-space* mass spectrometer using an upstream linear quadrupole ion trap developed by Thermo Electron (San Jose, CA) and introduced at the *2005 ASMS Meeting on Mass Spectrometry and Allied Topics* held in San Antonio, Texas, June 5–9.

The orbitrap has its origins in the device developed in 1922 by K. H. Kingdon at the Research Laboratory of General Electric Company in Schenectady, New York [125]. The Kingdon trap uses an electrostatic field for trapping ions; it does not employ either a magnetic or an RF (dynamic electric) field. A radial nonlinear (logarithmic) electric field is created between two end-cap electrodes by applying a DC potential between an outer cylindrical electrode and an electrically isolated thin wire that acts as a central electrode.

In 1981, R. D. Knight modified the shape of the outer electrode to add an axial electric field described by a quadrupole term in the equation of ion motion characterizing the original Kingdon trap [126]. This quadrupole potential "confines ions in the axial directions allowing them to undergo harmonic oscillations in the *z*-direction" [123, 127]. Knight split the outer electrode radially in the middle to allow ions to be injected into the trap. Both the axial and the radial ion-signal resonance were found to be much weaker than that observed in the 3D QIT [128]. No mass spectra were reported for the "Knight-style" Kingdon trap. The challenge in using this fundamental design of a Knight-style Kingdon trap for a practical *m/z* analyzer has been in manifesting some practical means for injecting ions into the device.

B. Operating Principles

The orbitrap is a new mass analyzer [124, 127, 129]; however, it is useful to consider the orbitrap as a modified Knight-style Kingdon trap with specially shaped inner axial (a spindle) and outer coaxial (a barrel) electrodes [123]. Unlike the quadrupole ion trap, which uses a dynamic electric field typically oscillating at ~1 MHz, the orbitrap uses a static electrostatic field to sustain ion trapping following the specialized dynamic injection pulse (*vide infra*).

The orbitrap consists of two electrodes in the form of coaxial axisymmetric electrodes, an outer barrel-shaped surface, and an inner spindle-shaped electrode oriented, as indicated in Figure 2-32B. A constant electric potential is

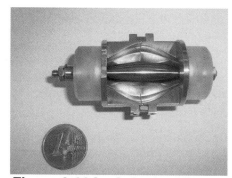

Figure 2-32A. Actual orbitrap showing its size relative to a one Euro coin.

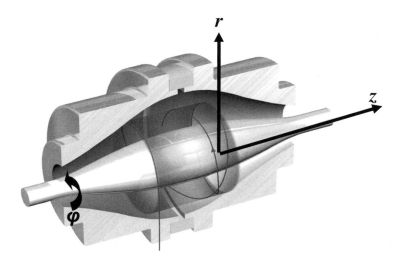

***Figure 2-32B.   Cutaway of the orbitrap mass analyzer. Ions are injected at
an off-axis point with a velocity perpendicular to the z-axis.***
Courtesy of Thermo Fisher Scientific.

imposed between these two axisymmetric electrodes (no magnetic field or oscillating electric potentials are involved). The opposing surfaces of the axisymmetric coaxial electrodes are nonparallel, as illustrated in the cross-sectional diagram in Figure 2-33. Thus, the electric field between the two surfaces varies as a function of position along the z-axis, the longitudinal axis of the two coaxial electrodes, reaching a minimum at the point of greatest separation of the electrode surfaces, namely the center of the orbitrap.

Ions are injected into the orbitrap at right angles to the z-axis at a nonzero position along the x- and y-axes at an optimal position between the surfaces, and displaced from the point of greatest separation of the two surfaces (center of the orbitrap) along the z-axis, as illustrated in Figure 2-33. Ions are injected from a C trap as a discrete bunch (see notes on the method of injection below) with a kinetic energy matching the opposing potential energy of the radial electric field between the two axisymmetric surfaces [130]. Properly injected ions follow a circular orbit above the outer surface of the inner spindle-like electrode but below the inner surface of the barrel-like outer electrode.   The radius of the circular orbit is established by balancing an electrodynamic centripetal force with the centrifugal forces acting on the ion related to its initial tangential velocity according to the relationship shown in Equation 2-6.

$$r = 2\,eV/eE \qquad\qquad \text{(Eqn. 2-6)}$$

Because there is no mass dependence of the ion in Equation 2-6, the radius of orbital migration (amplitude of the orbit) is the same for all ions regardless of mass. In this way, ions of all masses are trapped radially in the orbitrap.

Because the electric field between the surfaces of the coaxial electrodes is inhomogeneous in a symmetric manner as a function of position along the z-axis relative to the point of greatest separation between the two surfaces (center of the orbitrap), as shown in Figure 2-33, the ions have a natural tendency to oscillate axially (i.e., along the

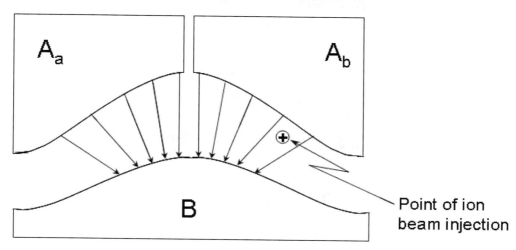

Figure 2-33. *Cross-sectional sketch of the upper annular chamber of the orbitrap showing the inhomogeneous electric field between the inner surface of the outer barrel-like electrode (A_a + A_b) and the outer surface of the inner spindle-like electrode (B,) as illustrated in Figure 2-32. The circle to the right in the gap indicates the point of entry of the ion packet from the C trap.*

z-axis) in the orbitrap. The electric field is inhomogeneous in two ways. First, the field strength is at a minimum in the center of the orbitrap; this is the position along the z-axis at which the separation of the surfaces is the greatest. The electric field strength increases uniformly in opposing directions along the z-axis from the center as the surfaces defining the annular space between the two coaxial electrodes become progressively closer in an axisymmetrical manner. The second aspect of inhomogeneity in the electric field is the more important; namely, the vectors of the electric field are nonparallel, with the angle of divergence from orthogonality increasing in opposing directions from the center point of the orbitrap along the z-axis. It is this second aspect of inhomogeneity of the electric field that causes the mass-dependent oscillation of the ions along the z-axis.

The electric field gradient from the point of ion entry (point of ion beam entering the plane of the page is indicated by the circled tail of an arrow in the right-hand side of the gap between electrodes in Figure 2-33; this symbol indicates that the injected beam of ions is oriented into the plane of the page at an angle of 90°) to the center of the orbitrap accelerates the ions toward the middle of the trap. The force that the ion feels toward the center of the trap is proportional to the projection of the electric field on to the z-axis. As can be appreciated from examining Figure 2-33, the greater the displacement from the center of the orbitrap, not only does the magnitude of the electric field between electrodes A (segmented into two parts so as to serve also as detectors of image currents as explained in next paragraph) and B increase, but so does the projection of the field to the z-axis [124]. The ions continue to migrate through the center of the trap (point of zero force) along the z-axis, but decelerate as they continue toward the opposite end of the orbitrap, expending the kinetic energy previously gained in traversing the electric field gradient from the entry point to the center of the orbitrap. Having spent their kinetic energy, the ions stop and then are accelerated back toward the center of the trap by the symmetric electric field along the z-axis. In this way, the ions oscillate naturally along the *z*-axis in a manner analogous to pulling back a pendulum bob and then releasing it to oscillate [123, 127].

The frequency of natural oscillation along the *z*-axis is mass dependent according to Equation 2-7 [123, 127]. It is this feature that allows the orbitrap to function as a mass spectrometer. These oscillations are detected as a time-domain signal using image current detection as sensed by the two electrically isolated components that constitute the barrel electrode (parts A_a and A_b in Figure 2-33). The frequencies of oscillating image current are transformed into mass spectra using a fast Fourier transform in a manner similar to that used in FTICRMS as developed by Marshall *et al.* [131–133].

$$\dot{\omega} = [\,(z/m)\,k\,]^{1/2} \qquad\qquad\qquad \text{(Eqn. 2-7)}$$

1) Role of the C Trap in the Success of the Orbitrap

Injection of an ion packet into the *static* electric field of the orbitrap would not result in trapping the ions; the packet of ions would essentially pass through the device [134]. Radial trapping of an injected packet requires use of a *dynamic* electric field between the electrodes that is carefully coordinated with the injection process; proper injection requires that the electric field between the two electrodes of the orbitrap be increased during the brief interval (>300 nsec) that the ion packet enters the orbitrap [123, 124, 135]. The missing link for efficient ion transfer in making the orbitrap a practical *m/z* analyzer was the C trap, which tightly focuses the injected ion packet in time and space [135, 136]. The C trap is a curved linear trap that electrodynamically squeezes the ion packet in time and space during a brief interval in which the RF is ramped down and the DC potential is ramped up to accelerate the ion packet out of the C trap and into the orbitrap. The acceleration process out of the C trap is carefully coordinated with proper dynamic ramping of the potential on the central electrode of the orbitrap to effect efficient transfer (approx. 30%) and trapping of the ion packet in the electric field between the two electrodes of the orbitrap [136].

An early configuration of the orbitrap arranged downstream from a linear ion trap and a C trap to realize a practical *m/z* analyzer is shown diagrammatically in Figure 2-34. The linear ion trap accumulates ions in the usual manner from an external ion source. A bolus of ions is then transferred to the C trap, which then electrodynamically squeezes the ion packet in time and space for proper injection into the orbitrap for *m/z* analysis. In

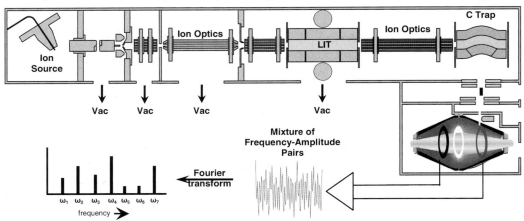

Figure 2-34.   Schematic of the LIT and C trap "front end" to the orbitrap.

principle, CAD can be accomplished in either of three regions: the linear trap, the C trap, or the orbitrap. Although the orbitrap can function as an efficient high-energy CAD chamber, the problem is with inefficient trapping of the product ions, which travel in unstable elliptical orbits because they have the same velocity as the precursor ions. Thus, the recommended protocol calls for CAD to be performed in either the linear ion trap or the C trap followed by *m/z* analysis in the orbitrap [134, 136]. Additional benefits to the orbitrap are likely to derive from mass-selective excitation or de-excitation resulting from application of a dipolar AC signal applied to the split outer electrodes (parts A_a and A_b in Figure 2-33) in phase or 180º out of phase, respectively, with ion motion; selective de-excitation and re-excitation can be performed with unit mass selection, leaving ion current associated with the adjacent ^{13}C-isotopic peak unaffected [67, 137, 138].

2) Figures of Merit for the Orbitrap as an *m/z* Analyzer

The orbitrap has *m/z* range and resolving power capabilities that make it a powerful *m/z* analyzer [124, 129, 130, 138–145]. Recent reports indicate a resolving power of 150,000; an *m/z* range of 6000; an *m/z* accuracy at high mass of 2–5 ppm; and a dynamic range of 10^3 [136, 146]. The axial frequency is independent of the energy and spatial spread of the ions in the trap. Because of the mass dependence and the energy independence of the axial frequency of ions, the ability to separate ions and accurately determine the *m/z* values is possible. The high resolving power of the orbitrap is possible because the *m/z* value of the ions is a function of frequency, and frequency is a parameter that can be measured with very high accuracy by existing technology. The orbitrap has a larger trapping volume compared to the 3D QIT or the FTICR-MS, and an increase in space-charge capacity. The orbitrap is compact, rugged, and essentially maintenance free.

Ions are accumulated, cooled, and pulsed into the trap at a pressure of about 10^{-8} Pa; even at this pressure, the ion packet can suffer a loss of phase coherence (undergo dephasing) through collisions with background molecules that results in a decrease in signal intensity because coupling of the ion-image current is highly dependent on the coherence of the ion packet. Further, mass accuracy and resolving power will deteriorate with dephasing of the ion packet. The orbitrap has the highest vacuum requirement of any *m/z* analyzer.

Table 2-4. Advantages and disadvantages of the orbitrap mass spectrometer.

Advantages	Disadvantages
Small, simple device	Requires low pressure because a mean-free path of ~100 km is required!
Resolving power of 70,000 +	High cost compared to that of the 3D and linear QIT
Fellget advantage as in FT-ICR	Inefficient trapping of the product ions
Comparable performance to FT-ICR without the need for cryogen	
CAD can be performed in the LTQ, C trap, or the orbitrap	

4. Transmission Quadrupole

The commercial development of the transmission quadrupole *m/z* analyzer, sometimes called the quadrupole mass filter (QMF), began in the mid-1960s, not long after the gas chromatograph (GC) was reaching widespread use in analytical laboratories. The QMF was ideally suited for use with the GC because it allowed for more rapid data acquisition than had been possible with the conventional magnetic mass spectrometers of the time. The QMF's operation was far simpler than that of magnetic instruments, and it could be manufactured at much lower cost due to the much shorter flight path of the ions and the lack of a need for a high-voltage power supply. Demand for the QMF rapidly eclipsed that for the magnetic-based instruments, and today it is the most prevalent of all mass spectrometers. The pioneering commercial developers of these early instruments changed the use of mass spectrometry forever by taking it out of the hands of skilled physicists and putting it at the disposal of the analytical chemistry technician.

This nonmagnetic *m/z* analyzer employs a combination of direct current (DC) and radio frequency (RF) electric fields as a mass "filter". Mechanically, the quadrupole consists of four parallel surfaces, ideally with a hyperbolic cross section. This arrangement can be accomplished with four longitudinally parallel round rods, as illustrated in Figure 2-35. This symmetrical arrangement allows nearly hyperbolic electric fields to be produced according to quadrupole theory [34, 147–149]. Opposing surfaces (i.e., those diagonally opposite) are connected together electrically, and to RF and DC power sources as indicated in Figure 2-35. This constitutes the same two-dimensional field used in the LIT described above. Extracted ions from an ion source are accelerated (5–15 V) into the central space that constitutes the quadrupole electric field along the longitudinal axis toward the detector. Miller and Denton [147] have described the concept of a quadrupole mass filter as the combination or overlap of a low-pass and a high-pass filter. At least for ions of relatively low *m/z* values (i.e., less than *m/z* 300), *m/z* separation is not affected by the structure of the ion [150].

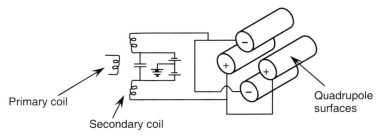

Primary coil

Secondary coil

Quadrupole surfaces

Figure 2-35. *Illustration of electrical means of superposing RF and DC potentials to opposing pairs of quadrupole surfaces.*

A mass spectrum is obtained using a QMF by increasing or decreasing the magnitude of the RF amplitude and DC potentials at a fixed ratio; i.e., the DC and RF voltages are ramped (scanned) at a constant ratio. The resolution of the instrument is established by the ratio of the RF to DC potentials. The RF frequency is held constant. As illustrated in Figure 2-36, the ion trajectories through the central space between the electrical poles are complicated, and for any given set of DC and RF potentials, only ions of a specific *m/z* value avoid collision with poles and successfully traverse the quadrupole filter along the *z*-axis to reach the detector; all other ions collide with the quadrupole surfaces at those values of RF and DC potential. The entire mass spectrum can be

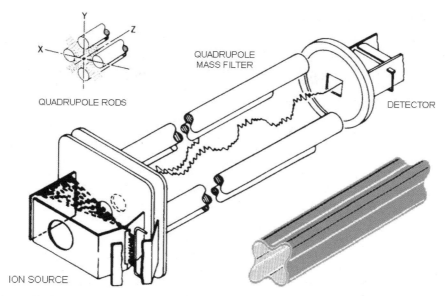

Figure 2-36. *Illustration of a transmission quadrupole filter, showing ion trajectories with ions of only one m/z value remaining stable enough to reach the detector. The quartz mandrel shown at the lower right can be used as a quadrupole mass filter.*

scanned as the potentials are swept from a pre-established minimum to a maximum value, but at a constant DC/RF ratio. It should be noted that spectra can be obtained by scanning from a high potential to a low potential as well.

> Illustrations often show the QMF as being constructed of round rods. A number of different shapes have been used including metal hyperbolic surfaces. Figure 2-36 shows a quartz mandrel fashioned in such a way as to create the two orthogonally opposed hyperbolic surfaces. The innermost surfaces are coated with a conducting material such as gold to create a device capable of generating the quadrupole electric field.

A. QMF Equations of Motion

A rigorous examination of the trajectory of ions through the QMF necessarily takes into account a large number of physical variables that affect the instantaneous electric fields experienced by the ions. For the classical hyperbolic mass filter, the potential distribution (Φ) at any time (*t*) is described by the following equation:

$$\Phi = \left[U + V Cos(\omega t) \right] \frac{x^2 - y^2}{2r_0^{\,2}} \qquad \text{(Eqn. 2-8)}$$

where x and y are the distances along the coordinate axes, r_0 is the distance from the *z*-axis in Figure 2-36 to either of the quadrupole surfaces, ω is the angular frequency ($2\pi f$) of the applied AC signal, V is the magnitude of the applied RF signal, and U is the magnitude of the DC potential applied to the surfaces.

The instantaneous electric field along any of the three axes (*x-y-z*) in Figure 2-36 can be computed by taking the partial derivative of the expression above (Equation 2-8) as a function of a particular axis. This exercise shows the useful result that the applied DC and AC (RF) potentials to the surfaces impress no acceleration on the ions in the *z* direction (i.e., along the axis between the ion source and the detector) because the general formula for the potential shows no dependence on the term *z*.

Definition of the a and q terms allows the Mathieu equations [34, 147, 148] to be somewhat simplified; a is related to DC and q is related to RF. From the standpoint of the QMF, a bounded solution to the Mathieu equation corresponds to a finite displacement of the ion along the *x*- or *y*-axis (i.e., the ion follows a trajectory that avoids collision with the surfaces and eventually reaches the detector). The other general class of solutions, the unbounded solution, describes the situation in which the radial displacement of the ion increases without bound, and therefore it collides with (is filtered out by) the surfaces to preclude transmission through the quadrupole to the detector:

$$a_x = -a_y = 4zeU / m^2 r_0^2 \qquad\qquad \text{(Eqn. 2-9)}$$

$$q_x = -q_y = 2zeV / m^2 r_0^2 \qquad\qquad \text{(Eqn. 2-10)}$$

The ions for which there is a bounded solution have a and q values that correspond to *stable trajectories* in the quadrupole mass filter. These stable trajectories allow the ion to be transmitted from the ion source to the detector without collision with the surfaces of the quadrupole filter. Stable coordinates of a and q space are those in the shaded area of the a-q diagram in Figure 2-37.

The mathematical transformations that lead to the formation of the a-q diagram serve a useful purpose in simplifying the complex array of parameters that affect motion of the ion in the quadrupole mass filter. In general, the stability diagram as shown in Figure 2-37 is a two-dimensional plot showing how the six-dimensional problem (involving e, ω, r_0, m, U, and V) can be reduced to a two-dimensional problem involving only the mathematically reduced a and q parameters. It is interesting to note that the stability region shown in Figure 2-37 is only a small area of the possible a and q coordinates for the ion trap shown in Figure 2-25.

B. The Stability Diagram

Two important points can be made while viewing the stability diagram in Figure 2-37. First, if the quadrupole is operated with a DC potential and an RF amplitude that define the operating line in the stability diagram [34, 147, 148], only those ions with *m/z* values defined by the (a-q) space coordinates at the tip of the diagram will have stable trajectories [151]. If the DC/RF ratio is adjusted to raise the operating line even higher, the resolving power of the instrument increases, but the sensitivity decreases, because fewer ions follow a stable trajectory to reach the detector. The second point relates to "RF-only" operation of a quadrupole. In the RF-only mode, there is no DC field; this means, $a = 0$ ($U = 0$ in Equation 2-8). Therefore, the operating line lies horizontally along the abscissa, indicating theoretically that ions of all *m/z* values have stable trajectories and are transmitted through the quadrupole filter.

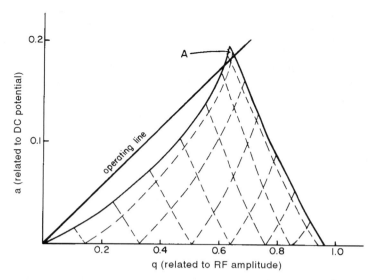

Figure 2-37.   *A stability diagram in *a-q* space for ions in a transmission quadrupole.*

As might be anticipated from the very large number of a and q values that constitute the shaded area in the stability diagram, a quadrupole mass filter could, in principle, be operated at any value in the a-q space corresponding to a point in the shaded region. However, in practice, the values of a and q space are limited by adjusting the RF and DC potentials to a fixed ratio. In fact, in the typical quadrupole MS, the DC potential is held at a fixed fraction of the RF potential. This fixed ratio leads to an operating mass scan line with a slope of $2U/V$, which is shown at the approximately 45° angle in Figure 2-37 of the stability diagram. By fixing the ratio of RF and DC potentials to establish the resolving power, the only values of a and q space available to ions in the instrument lie along the operating line within the shaded region; i.e., a circumstance exists in which ions with *m/z* 300 would be transmitted through the quadrupole, but ions with *m/z* 299 and lower and ions with *m/z* 301 and higher would not be transmitted. The complete mass spectrum is obtained by scanning the magnitude of RF (V) and DC (U) potentials through increasing or decreasing values. As explained in previous equations, the mass of an ion is inversely proportional to both a and q; therefore, by scanning the magnitude of RF and DC potentials from a low to a high value, ions of increasing mass will sequentially take on stable values of a and q space, and thereby be transmitted to the detector.

C. Characteristics of Output

The QMF instruments can be scanned in a few milliseconds to provide the convenience of real-time visualization of the mass spectrum on an oscilloscope while tuning the instrument. However, complete mass spectra are obtained from slower scans (several hundred milliseconds) for better ion counting statistics (better signal-to-noise ratio) from analytes. The capability for rapid adjustment of the potentials that control the *m/z* scale is also an important advantage of the quadrupole in applications of selected ion monitoring.

The QMF instrument produces mass spectra with a linear *m/z* scale; i.e., if the data are recorded by an oscillograph, the separation of the peaks at *m/z* 400 and 401 will be the same as that at *m/z* 30 and 31 (constant resolution (separation) as opposed to constant resolving power). These instruments are more tolerant of high pressures than the magnetic or TOF mass spectrometers; this may be an advantage in "hyphenated" techniques for chromatography or in CI applications. This tolerance to higher pressures accrues apparently because the QMF mass spectrometer has relatively large apertures and a short flight path from the source to the detector, and does not require tightly focused optics. The only harmful effect of higher pressures (10^{-2} Pa) is that once the mean free path of the gas is less than the flight path, molecular collisions result in loss of ions to the surfaces; this means that the sensitivity is inversely proportional to pressure in the analyzer.

The QMF is very well suited for selected ion monitoring (SIM) because ion currents from any region of the *m/z* range can be monitored sequentially without diminishing optimum conditions in the ion source or mass analyzer. Furthermore, the parameters (superimposed RF and DC fields) that control the mass scale can be changed rapidly, with good response throughout the *m/z* range of the instrument.

Mass discrimination (signals resulting from *X* number of ions with one *m/z* value do not equal signals from *X* number of ions with another *m/z* value) is a consideration in quadrupole mass spectrometry. The effect of high-mass discrimination as observed in the QMF can be easily visualized by imagining the mass spectrum of a hypothetical molecule that fragments to produce equal numbers of ions with *m/z* 30, 300, and 600. Its mass spectrum, as obtained with an ideal mass spectrometer, should show three equally intense peaks at *m/z* 30, 300, and 600, as illustrated in Figure 2-38. In contrast, an instrument exhibiting mass discrimination might produce peaks at *m/z* 300 and *m/z* 600 with significantly less intensity than that at *m/z* 30 (Figure 2-38). Severe mass discrimination is clearly undesirable as it may complicate, among other things, the use of reference libraries of standard mass spectra for automated search and identification.

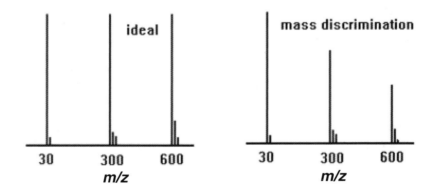

Figure 2-38. **Illustration of the effect of mass discrimination sometimes encountered in the transmission quadrupole mass filter.**

High-mass discrimination in quadrupole mass spectrometers has been minimized by improvements in instrument design. The problem is caused by fringe fields (DC) near the entrance to the central space of the quadrupole; these fields tend to divert the ions from efficient entry into the mass analyzer [14]. High-mass ions are affected more by these fringe fields than are low-mass ions, which have a greater velocity and therefore a shorter residence time in the fringe-field region. If an ion spends more than 3 RF cycles in this region, its velocity components and displacement are affected to the extent that its transmission into the quadrupole filter is significantly attenuated [26]. Several ion optical approaches have been used to overcome or minimize the undesired effects of mass discrimination [14, 15, 26]. The Brubaker filter [14] uses end-cap electrodes on the rods to produce an RF-only field to guide the ions through the DC fringe fields into the central space between the rods. The Turner–Kruger entrance lens [15] protrudes into the central space of the quadrupole assembly as shown schematically in Figure 2-2 to transmit ions more efficiently from the ion source through the DC fringe fields into the quadrupole filter. A "leaky" dielectric arrangement has also been used to facilitate ion transport through the adverse fringe fields [152].

Another factor that contributes to high-mass discrimination in the QMF is the relationship of detector response to the *m/z* value of an ion. Most mass spectrometers rely on a secondary emission from a physical impact of an ion. Detection based on these secondary emissions is described in more detail later in this chapter. This secondary emission is a function of the ion's momentum (mass times velocity). All quadrupole *m/z* analyzers accelerate ions at a low potential such as 5–15 eV. This means that the momentum is more a function of the ions *m/z* value rather than its velocity and the ions of higher *m/z* values will have lower momenta resulting in a smaller response when they strike the detector. The main component of momentum for ions associated with the TOF and double-focusing instruments is velocity; therefore, the degree of secondary emission does not vary significantly over wide ranges of *m/z* values.

To compensate for this lack of momentum, which causes mass discrimination, further acceleration of the ions is carried out just before they impact the detector. This post-acceleration (meaning after *m/z* separation of the ions) also improves the sensitivity of the instruments. The ions are accelerated by a potential difference of 10–30 kV. These post-analyzer acceleration dynodes can also be used to convert negative ions to positive ions in order to allow for negative ion detection using a conventional positive ion detector.

D. Spectral Skewing

Care must be taken not to confuse spectral skewing with mass discrimination. An example of spectral skewing is shown in Figure 10-19 in the context of GC/MS. Spectral skewing is the term used to describe the phenomenon of changes in relative intensities of mass spectral peaks due to the changes in concentration of the analyte in the ion source as the mass spectrum is scanned; this situation occurs routinely as chromatographic components elute into a continuous ion source.

Spectral skewing is not observed in ion trap (quadrupole or magnetic) or time-of-flight (TOF) mass spectrometers because potentially all ions formed in an operational cycle (a snapshot in time) of the instrument are available for detection.

All commercially manufactured GC/MS and LC/MS systems based on the transmission quadrupole *m/z* analyzer scan from low *m/z* to high *m/z* values except for those produced by Agilent Technologies. Those produced by Agilent scan from high *m/z* to low *m/z* values; the original reason for this high-to-low scan function was to avoid measurement of low-intensity signals following the measurement of high-intensity signals. This issue of low-intensity peaks after high-intensity peaks is seen among clusters of peaks throughout the mass spectrum as in the measurement of a monoisotopic peak intensity followed by the measurement peaks representing less abundant isotopes for a given ion. Hysteresis associated with the electronics in early instruments had a tendency to introduce a significant error in the lower signals. This is no longer an issue, but Agilent continues to use their original approach to scanning the mass spectrum.

Table 2-5.   Advantages and disadvantages of the transmission quadrupole mass spectrometer.

Advantages	Disadvantages
Simple to operate, rugged, and easy to maintain resulting in high sample capacity and throughput	Low resolution, 0.3 m/z units in spite of the fact that many data systems report to the nearest 0.05 or 0.01 m/z unit
Ideal for MS/MS (the triple quadrupole)	Limited to MS/MS – limited MS^3 possible with in-source collisionally activated dissociation (CAD) – MS^3 has been done with pentaquads (no longer commercially available)
Constant separation throughout the m/z scale	Full-scale acquisition rate limited to about 1 spectrum/second; however, increased data acquisition rates are possible because a narrow acquisition range is usually employed
Can monitor ions of wide differences in m/z values for selected ion monitoring (SIM) and selected reaction monitoring (SRM) – ideally suited for SIM	Maximum m/z range to 4000 – most commercial LC/MS instruments have a maximum obtainable m/z value of 1500 to 3000; range can be expanded by sacrificing resolution
A large number of manufacturers of commercial instruments	Unlike sector-based instruments, the m/z analyzers deteriorate in performance with time
Ready availability of third-party parts and service	Requires SIM for quantitation
Low initial cost for single-analyzer instruments	Spectral skewing is a potential problem

E. Performance Limitations

The quadrupole mass spectrometer has been at the focal point of developments in GC/MS instrumentation [11] primarily because of its capacity for rapid scanning; however, there is a practical limit to rapid scanning on this instrument. It must be remembered that in the operation of a quadrupole, the ions interact with an oscillating electric field in such a way that only ions of a specific *m/z* are allowed to pass to the detector. For maximum transmission of ions at that specific value of *m/z*, the RF and DC potentials should not change during the transit time of the corresponding ion through the filter.

An ion with an *m/z* value of 800 and an energy of 10 eV requires 129 μsec to pass through a 20-cm quadrupole filter. Using an intuitive approach, at a scan rate such that the *m/z* filter changes by no more than 0.1 u during the transit time of this ion, a maximum scan rate of 780 μsec can be reached before resolution, peak shape, and intensity are significantly sacrificed. Under these conditions, 90% of the ion current at *m/z* 800 is transmitted to the detector. Somewhat higher scanning rates have been achieved without sacrificing sensitivity by progressively increasing the ion energy concurrently with the mass scan ramp in order to reduce the transit time for heavier ions, thereby approaching conditions for a more nearly constant velocity for all the ions. Faster scanning can also be achieved if the *m/z*-selecting potentials are stepped rather than scanned; however, the resolution of the mass axis is severely degraded as fewer steps per mass unit are used. By a combination of these techniques, a useful upper scan repetition limit of approximately 10 Hz can be achieved with the QMF mass spectrometer [83].

5. Magnetic-Sector Instruments

As described in Chapter 1, it is somewhat debatable as to who qualifies as the father of mass spectrometry. Significant contributions were made by the German physicist and 1911 Nobel laureate, Wilhelm Wien (1862–1928) [153, 154] , Sir Joseph John Thomson, Francis William Aston (one of Thomson's collaborators at the Cavendish Laboratory), who was awarded the 1922 Nobel Prize in chemistry partly for the development of the mass spectrograph, and Arthur Jeffery Dempster (1886–1950), who in an independent effort developed a single magnetic-sector instrument that employed direction-focusing of constant-energy ions at the University of Chicago at the same time of Aston's contribution. Aston's instrument provided accurate *m/z* values, while Dempster's instrument provided accurate ion abundances. This distinction between the two developments could also be termed as a difference between velocity-focusing and direction-focusing of ions, which resulted in mass spectrographs and mass spectrometers, respectively. In the decade that followed (1911–1925), mass spectrometry became a distinct field mainly as a result of the efforts of these three pioneers. In the late 1920s, Dempster proposed the combination of direction-focusing and velocity-focusing with W. Bartky [155]. The first double-focusing mass spectrometers began to appear ca. 1934 with the Dempster, Mattauch–Herzog, and Bainbridge–Jordan geometries.[4] It was a single magnetic-sector instrument similar to the design developed by Alfred Otto Carl Nier (1911–1994) while studying with Kenneth T. Bainbridge at Harvard University. This design was similar to Dempster's 1918 180° instrument except that it had an ion source with both electric and magnetic fields. These early instruments used an electron ionization source to ionize gas-phase molecules that were present at very low pressures.

---

[4] The word "geometry" refers to the arrangement of the electric and magnetic sectors between the ion source and detector in the mass spectrometer.

This design became the basis for the Consolidated Engineering Company's CEC 21-104, which saw service at Exxon until 1997.

Single-focusing magnetic-sector mass spectrometers are no longer manufactured except for some esoteric areas of process control. These instruments are no longer in demand for the analytical chemistry laboratories, but their operating principles are included in all modern mass spectrometry texts because they constitute a practical application of the fundamental laws of electricity and magnetism in the separation of ions according to their individual mass-to-charge ratios.

A. Single-Focusing Instruments

1) Operating Principles

The mass analyzer in a single-focusing instrument uses a wedge-shaped (or sector) magnetic field to separate the total ion beam (i.e., ions of a variety of masses) from the ion source into discrete ion beams of individual *m/z* values through momentum dispersion and direction focusing [156, 157]. As indicated in Figure 2-39, the optical design of the instrument is symmetrical, the ion source and detector being equidistant from the magnet. This arrangement permits relatively free access to both the source and detector modules; it also minimizes any influence of the strong magnetic field on either the ion source or the detector.

The *m/z* value of any ion traversing the magnetic field at a fixed radius (fixed magnet geometry) can be related to H and V by Equation 2-11:

$$m/z = er^2 H^2 / 2V$$ (Eqn. 2-11)

The ions obtain their kinetic energy during acceleration through an electric field created by the magnitude of the accelerating potential, V (e.g., $V = 4000$ volts). Equation 2-12 shows this balance of energies, where z is the number of charges on the ion, e is the magnitude of the electronic charge, m is the mass of the ion, and v is its velocity. Further, a charged particle, in traversing a magnetic field at a right angle to the lines of magnetic flux, will be exposed to a force orthogonal to the field and to the velocity

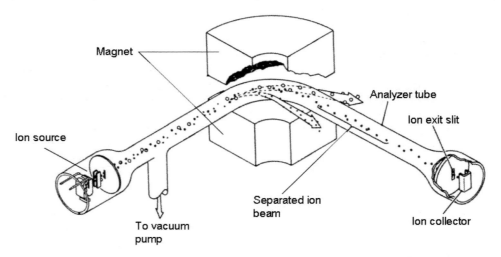

Figure 2-39. Schematic diagram of a magnetic-sector mass spectrometer.
Adapted from Interpretation of Mass Spectra, Fourth Edition; McLafferty FW and Tureček, F, University Science Books, Mill Valley, CA, 1993, with permission.

vector of the ion. This centripetal force, mv^2/r, is balanced (Equation 2-13) by a centrifugal force, $zevH$, causing the ions to traverse a circular path of radius r while in the magnetic field H. (The lines of magnetic flux in Figure 2-39 are perpendicular to the plane of the paper.)

The kinetic energy ($\frac{1}{2}mv^2$) gained by the ion is equal to the potential energy (zeV) available in the electric field between the accelerator plates.

$$zeV = \frac{1}{2}mv^2 \qquad \text{(Eqn. 2-12)}$$

Under conditions of energetic stability, the centrifugal and centripetal forces experienced by the ion in the magnetic field are equated:

$$Mv^2 / r = zevH \qquad \text{(Eqn. 2-13)}$$

If Equations 2-12 and 2-13 are solved for velocity, v, and thus equated, the expression in Equation 2-11 can be derived.

A previous important point of dispersion as a function of momentum is reinforced by solving Equation 2-13 for r:

$$r = mv / zeH \qquad \text{(Eqn. 2-14)}$$

Equation 2-14 indicates that the radius of curvature of an ion trajectory is proportional to its momentum (mv). Therefore, the total ion beam is dispersed into many individual ion beams, each having its characteristic radius of curvature while in the magnetic field. As illustrated in Figure 2-39, the imposed accelerating potential (V) and magnetic field (H) permit those ions of mass m_1 to follow a trajectory to the detector, whereas the heavier ions of mass m_2 follow a trajectory under these given conditions of V and H that leads to collision with the wall of the analyzer. Because the instrument has a fixed radius of curvature, either H or V must be scanned to collect the mass spectrum.

2) Magnetic Versus Voltage Scanning

Equation 2-11 indicates that the entire mass spectrum might be scanned by systematic and appropriate control of either the magnetic field or the accelerating potential. Either procedure will effectively move the ion current image of the source slit from one *m/z* value after another across the collector slit of the detector. However, because the quality of a mass spectrum in terms of ion counting statistics and ion focus is related to the magnitude of the accelerating potential, it can be varied only over a limited range in the process of scanning. In principle, the accelerating potential could be changed by a factor of 10 (e.g., from 4000 V to 400 V), producing a scan from *m/z* 60 to 600, for example, at a given setting of the magnetic field. In practice, the efficiency of ion transmission and the response of the detector drop in proportion to the reduction in V. Therefore, it is not desirable for instruments that were designed for optimum focus at high potential to be operated at low V. Another disadvantage of voltage scanning is the requirement for obtaining overlapping scans, each obtained at a different value of H, when the mass range is large or greater than one decade, as in the case of scanning from *m/z* 20 to *m/z* 600.

Magnetic scanning is advantageous in that the entire mass range may be scanned while the instrument maintains optimum V. A minor disadvantage is related to the quadratic dependence of *m/z* on the value of H and the fact that the resolution of this type of ion optics is constant. As the mass range is scanned, the peaks at the high-mass end appear closer together than those at the low-mass end. At high values of H, an incremental change in H will cause a much larger shift in the mass scale than that imposed at a low value of H, requiring a well-stabilized or controlled magnetic-current power supply for effective operation. Calibration of the mass scale is facilitated by use of a Hall probe, a transducer that produces a voltage proportional to the magnetic-field strength. The square of the Hall probe voltage can be converted to a digital display of the m/z value for an indication of the mass scale. The m/z scale can be calibrated during the analysis of a calibration compound such as perfluorokerosene.

3) Performance Limitations

Advances in chromatography continue to impose demands for faster scan rates on the MS in hyphenated applications; this is especially true in capillary GC/MS, as emphasized in Chapter 10. The scan rate of a magnetic instrument is limited by the reluctance of the electromagnet; reluctance is a kind of "inertia" of an electromagnet that limits the rate at which it can be forced to change field strength. Laminated magnets permit more rapid changes in the magnetic field such that the m/z range can be scanned at rates approaching 0.1 sec/decade (an m/z decade is 50–500 or 7–70 m/z units, for example). Problems with hysteresis (path from X to Y is not the same as from Y to X) of the magnet are minimized by discharging it completely between scans, adding a "settling time" interval between scan cycles, and only scanning uni-directionally for data collection, all of which improve the precision of the scan function.

Magnetic-sector mass spectrometers are constructed in such a way that each ion experiences a constant and homogeneous H. However, if scan speeds exceed a certain rate, H can change significantly during the ion transit period (the time required for an ion to traverse the H), causing ions to follow a distorted path toward the detector. Consequently, when scanning rapidly toward increasing field strength, the effective radius of curvature of the ion will be shorter than that corresponding to the constant-field value. When scanning toward the decreasing field strength, the opposite effect will be produced. In either case, losses in both resolution and sensitivity will result because the exit slit will not correspond to the effective (dynamic) focal point of the ion optical system. At scan rates of 0.1 sec/decade, the typical ion path variation will be of the order of ±2%, and, thus, this type of defocusing is not a critical limitation [83] although it does indicate the need for uni-directional scanning at rates comparable to those used in calibration.

B. Double-Focusing Instruments

Double-focusing instruments have an electric sector (E), separate from the ion accelerator, for selecting monoenergetic ions as well as a magnetic sector to analyze the momentum of the ions. These double-focusing instruments [27, 28] are designed to accomplish at least two major objectives: (a) to provide sufficient resolving power to separate various ionic species that share the same nominal mass so that each can be measured separately and (b) to focus the ion beams so carefully that the exact mass of the ions can be determined to six significant figures.

Exact mass measurements with an accuracy of a few parts per million (±1 milli mass unit for an ion of mass-to-charge ratio <500 Da) require a beam of well-collimated monoenergetic ions. In practice, however, the assemblage of ions accelerated out of the ion source has a distribution of initial kinetic energies. For example, if the initial kinetic energies of ions range from 0 to 3 V in a single-focusing magnetic instrument (Figure 2-39) and they are accelerated out of the ion source by an electric field of 3000 V, the final kinetic energy distribution of the ions will range from 2997 to 3003 V or 3000 ±3 V. This will mean that mass resolution of the ions cannot be achieved to better than 1 part per 1000 without further energy refinement. This energy dispersion is one of the limiting factors for performance of the single-focusing magnetic-sector mass spectrometer. The electric (E) sector (not to be confused with the electric field of the ion-accelerator (V) plates) in a double-focusing instrument can be used to select from the same initial ion beam only those ions having a kinetic energy of 3000 ± 0.03 V, which means mass resolution up to 1 part per 100,000 can be achieved on the transmitted ion beam, thereby allowing for accurate mass measurements of all ions (fragment ions as well as molecular ions). This is only possible if compensations are made for the small disparities in initial ion energies (smaller differences than in the foregoing example) and angular dispersions by energy and momentum focusing, respectively.

Double-focusing instruments achieve these two processes with tandem electric (E) and magnetic (H) sectors [10, 27, 28]. The electric sector eliminates those ions of greatly divergent energies, but focuses those having energies near that of the main beam. In energy focusing, each collimated component beam of isomass ions (same *m/z* value) of heterogeneous energy undergoes energy dispersion in traversing a radial electric field. The divergent isomass (same *m/z*) ion beams then pass through a homogeneous magnetic field in which isomass ions of different velocities (i.e., different energies) undergo different, but compensating deflections. The condition of double focusing requires that the energy dispersion imposed on isomass ions in the electrostatic field be counterbalanced by the direction-focusing effect of the magnetic field [10, 27, 28]. The principal function of the electric sector is in energy selection (the most deviant ions are screened out by an energy slit), but some energy focusing is accomplished for less deviant ions as described above.

There have been several different geometries used for the construction of double-focusing mass spectrometers. These geometries are designated by whether the electric sector precedes the magnetic sector (forward geometry), as illustrated in Figure 2-40 or the electric sector follows the magnetic sector (reverse geometry), as illustrated in Figure 2-41. In some cases, the geometry of the double-focusing mass spectrometer bears the name of its designer. One notable example of a forward geometry named after the developers is the Mattauch–Herzog instrument [8], which separates and focuses all the ions onto a focal plane such that they all fulfill the double-focusing conditions simultaneously (Figure 2-40). A photographic plate or a microchannel array

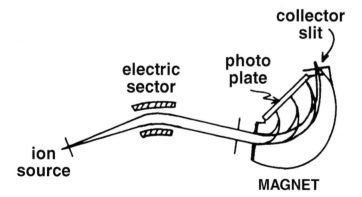

Figure 2-40. *Schematic diagram of a double-focusing mass spectrometer of the Mattauch–Herzog design; photoplate (array detector) is placed along the focal plane.*

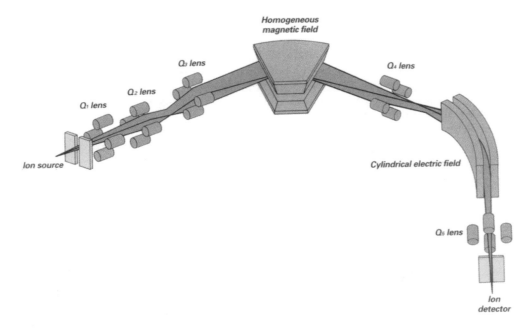

Figure 2-41.   Ion optic system of the JEOL GCmate® reverse geometry double-focusing mass spectrometer. *Courtesy of JEOL, USA.*

detector placed at this focal plane records all the ions all the time. An electron multiplier can be positioned at one end of the focal plane (beyond the collector slit in Figure 2-40) for use in "tuning" the instrument, or for electrical recording if a scan of the mass spectrum is required.

Another forward geometry double-focusing mass spectrometer whose geometry bares the name of the developer is the Nier–Johnson instrument [9, 157], in which a radial electric (E) field and sector magnetic (H) field are also used for double focusing (Figure 2-42), but in this case, the ions of interest are brought to a focal point, rather than a focal plane. The divergent ion beam from the ion source undergoes velocity dispersion in the electric sector (E). Ions with no initial kinetic energy are focused at B_1; those that have initial kinetic energy are focused toward B_2 (B_1 and B_2 define the energy spread of the total ion beam). The energy-compensated ion beam continues into H, which disperses the ions according to their momentum and also refocuses each isomass (same m/z value) ion beam. Directional focusing of all the ions occurs along the line b; velocity focusing occurs along the line v. Only ions of one m/z value are optimally focused at the intersection of lines v and b for any given value of V and H. The collector slit is positioned at this locus of optimum focus for electrical recording of the ion current.

Unlike the situation with the TOF mass spectrometer regarding high resolving power, with the double-focusing mass spectrometer the process of increasing the resolving power reduces the number of ions surviving for detection. When ions of a narrow energy spread are selected, this means that a large number of ions outside that energy spread are excluded, and thus not detected. The highest resolving power results in the poorest sensitivity. However, the resolving power used in an analysis is under the control of the operator. The double-focusing mass spectrometer can provide very good sensitivity when operated at a low resolving power.

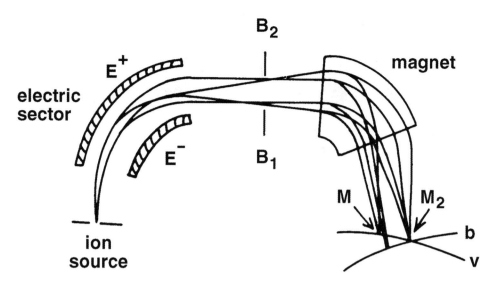

Figure 2-42.   Schematic diagram of a double-focusing mass spectrometer of the Nier–Johnson design; collector slit is placed at focal point (intersection of lines b and v).

Double-focusing mass spectrometers offer more flexibility than other types of mass spectrometers. Like all other types of mass spectrometers, these instruments have specifications set by the manufacturer. These specifications include a maximum *m/z* value obtainable, a maximum resolving power, and a maximum sensitivity. These three interplaying factors are strictly controlled by the design of the instrument except in the case of the double-focusing mass spectrometer. In the double-focusing instrument, a sensitivity is specified at a specific resolving power for a maximum *m/z* value. However, the operator can exceed that maximum *m/z* value by reducing the acceleration potential out of the ion source. This will result in poorer sensitivity, but will expand the *m/z* range. The *m/z* range of all other types of mass spectrometers is a strict function of instrument design and is not affected by operator control. The specified resolving power of 5000 (10% valley) for a given instrument may be insufficient to separate ions of two different *m/z* values. The operator can reduce the gap in the energy slits to less than the value required for a resolving power of 5000 and obtain a resolving power of 8000 at the sacrifice of sensitivity, but allowing for the separation of the two ions.

The technique of selected ion monitoring (SIM) enjoys some special benefits when performed on the double-focusing mass spectrometer. SIM on a single-focusing magnetic-sector mass spectrometer is done by holding the magnetic field constant and switching between various acceleration voltages. This range in acceleration voltage values that can be varied must be small to maintain good *m/z* scale calibration; typical allowable variations translate to a small *m/z* range (<50 *m/z* units) that can be examined during the SIM experiment. In the double-focusing instrument, jumping voltages in the electric sector will allow for wide variations in *m/z* values, while carrying out selected ion monitoring experiments.

Table 2-6. *Advantages and disadvantages of the double-focusing mass spectrometer.*

Advantages	Disadvantages
Capable of separating ions of very small differences in *m/z* values	Reduced sensitivity at high resolving power
Trade-offs: can increase *m/z* range or resolving power by sacrificing sensitivity	High accelerating voltages in the ion source
Very good sensitivity when operated at unit resolution	Requires *m/z* scale calibration points to be closer together than is required for quadrupole instruments
No problems with mass discrimination, as can be seen in transmission quadrupole instruments	Can be maintenance intensive and require highly skilled operators
Can perform all types of MS/MS in using a single analyzer through manipulation of the two sectors (see MS/MS chapter)	Resolution decreases with increasing *m/z* values because the instrument has a constant resolving power throughout the *m/z* scale

6.　FTICR-MS

Fourier transform mass spectrometry (FTMS) is based on the original technique [35, 131, 158–160] called ion cyclotron resonance (ICR) mass spectrometry. The operating principles of ICR are based on the observation that a charged particle will precess in a magnetic field at a frequency related to its *m/z* value [132, 161–164]. The name of the technology derives from the cyclotron frequency of a precessing ion in an orbit, the plane of which is perpendicular to the applied magnetic field, as illustrated in Figure 2-43, and from the fact that energy can be transferred to the oscillating ion provided that the energy is available at the cyclotron frequency (i.e., resonance condition). A didactic description of the operational aspects of an ICR cell in trapping ions has been reviewed elsewhere [91]. The resonance absorption of energy by precessing ions is involved in the detection process, as will be described in detail below.

The complete mass spectrum from an ICR mass spectrometer can be represented as an array of *m/z*-dependent frequencies. In the time domain, this array of frequencies appears as a complicated waveform that consists of an overlay of the individual frequencies and amplitudes (determined by the abundance of the ions with that frequency) of all the ions in the mass spectrometer. The Fourier transform expresses the array of frequencies that corresponds to the ICR mass spectrum in the time domain as an infinite series of sine and cosine functions [36]. The impact of Fourier transformation is illustrated in Figure 2-44 where a conceptual sketch shows that an almost incomprehensible waveform in the time domain has been transformed into discrete peaks in the frequency domain.

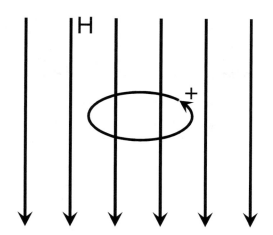

Figure 2-43.   Precessing ion in a magnetic field.

Because frequency is the physical parameter most easily and accurately measurable, FTMS or FTICRMS has the highest potential for mass accuracy determinations in mass spectrometry [165–167].   Achievable accuracy in experimental frequency determinations is also the basis for the extraordinarily good resolving power available in FTMS.   Unfortunately, the high resolution achievable in FTMS rapidly falls off with increasing *m/z* value, although developments in ICR instrumentation are improving this feature [131, 132, 168].

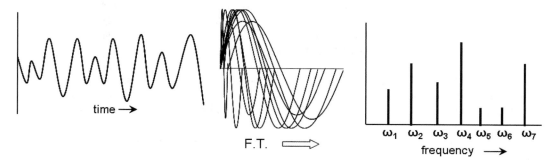

Figure 2-44.   Concept of Fourier transform (FT) in converting a complex signal from the time domain to the frequency domain.

A. Hardware Configuration

A generic ICR cell is illustrated in Figure 2-45 in which the magnetic field is oriented from the front panel through the back panel of the cubic ICR cell.   The orbits of the oscillating ions are illustrated in the center of the ICR cell in a vertical plane at right angles to the lines of magnetic field flux. In classical ICRMS operation, energy at a specific frequency (for example, that corresponding to the precession frequency of the

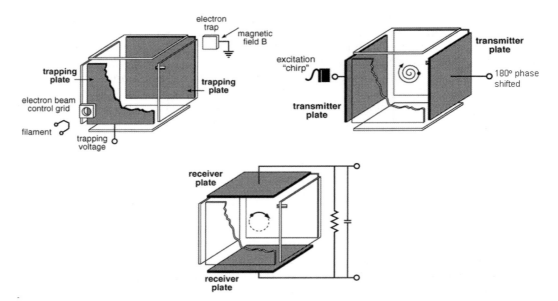

**Figure 2-45. Six-sided cell for ion cyclotron resonance. A broadband of
frequencies ("chirp") is provided to the transmitter plates to
excite the ions of various m/z values. Fourier transforms of
the time-dependent transients from the coherently excited
ions detected by the receiver plates provide the technique
now called Fourier transform mass spectrometry.**

ion of *m/z* 43) is transmitted into the ICR cell. If an ion of *m/z* 43 is present in the ICR
cell, it will absorb the energy because of the resonance condition and move to an orbit of
increasing radius while maintaining its characteristic precession frequency. Once the
radius of the precession orbit exceeds the internal dimensions of the ICR cell, the ions of
m/z 43 collide with the walls of the ICR cell, producing a measurable electrical signal, the
strength of which can be related to the abundance of ions of *m/z* 43. In this classical ICR
approach, only ions of a given *m/z* value can be detected under a given set of operational
conditions.

     In the presence of a constant and uniform magnetic field, H, all ions will move in
circular orbits with characteristic cyclotron frequencies, ω_c, that depend only on the ion's
m/z value:

$$\omega_c = zeH / 2\pi m \qquad\qquad \text{(Eqn. 2-15)}$$

In the FT mode of operation of the ICR cell illustrated in Figure 2-45, a wide variety of
energies (frequencies) is transmitted to the ICR cell in the form of an excitation "chirp".
This means, ions of many different *m/z* values can absorb energy at the same time.
When ions in the ICR cell are irradiated with a pulse of energy at a frequency that is
identical to their natural precession frequency (resonance), they absorb the energy and
not only thereby increase the radius of their orbit but they bunch up within their orbit,
becoming coherent or oscillating as a group in phase with the exciting field. In the case
of FT operation, irradiation is very brief so that the ions absorbing the energy will not
achieve a cyclotron orbit that exceeds the dimensions of the cell. Because the ions in the

ICR cell become coherent within their cyclotron orbit, they induce an oscillating charge (an image current) in the walls of the ICR cell as they precess. The frequency of the induced charge oscillation in the panels of the ICR cell can be detected and amplified, and related to the mass-to-charge ratio of the ions precessing in the ICR cell. Because the overall induced charge oscillation in the ICR cell walls consists of an overlay of all the component frequencies of different ions oscillating in the ICR cell, the FTMS approach allows ions of all *m/z* values to be determined simultaneously. Therefore, the FTMS approach is more comprehensive than the classical frequency-scanning operation of the ICR cell described above, in which only ions of a single *m/z* ratio can be determined at any given time.

The advantage of being able to detect all ions simultaneously across the mass spectrum by FTMS derives from the Fellgett advantage [169]. As with other Fourier transform methods, the FTMS spectrum can be acquired in a fraction of the time required by the corresponding frequency-scanning technique. In the frequency-scanning version of data acquisition from a classical ICR as described earlier, one samples the detector signal at each resolution element across the mass spectrum. If there are X resolution elements across the spectrum, the complete spectrum can be acquired by FTMS in 1/Xth of the time required by ICRMS for a complete scan of the frequency domain to obtain data at the same S/N ratio. Alternatively, the Fellgett advantage permits increased sensitivity in the same total measurement time if the overall signal is Fourier transformed rather than frequency scanned. The enhancement in overall sensitivity derives from an improvement in S/N which improves by the square root of **n**, where **n** is the number of ions detected. With the Fellgett advantage, **n** is increased because the complete mass spectrum is integrated during the total measurement time via Fourier transformation, rather than being timeshared at the detector as in the frequency-scanning technique [170].

FTMS is unique in that the increased measurement time increases both sensitivity and resolution [171]. This advantage derives from the fact that in FTMS the ions are not consumed during the detection process. The ICR mass spectrometer, when operated in the FTICR (or FTMS) mode, can achieve resolutions exceeding 100,000 without energy filtering [172]. Methodological refinements to permit simultaneous excitation and detection of ions have been proposed and implemented to minimize nonuniform power distribution over the excitation bandwidth for near-maximal resolving power in FTICR [173].

Figure 2-46 illustrates the resolving power possible with FTMS instrumentation. The mass difference between the positive and negative ^{35}Cl ions is 0.00109 mass unit, the mass of two electrons. The resolving power in this case is greater than 1,000,000 using the FWHM calculation. At present, achievement of this resolution requires several seconds, and therefore it is not possible to obtain a complete spectrum at this high resolving power on the chromatographic timescale. Mass resolution of 1,450,000 has been achieved by FTICRMS using a pulsed glow discharge source [174], and a resolution of 442,000 was illustrated in the analysis of complex mixtures of petroleum [175]; such resolving power allows the identification of oligomers in complex aerosols [176].

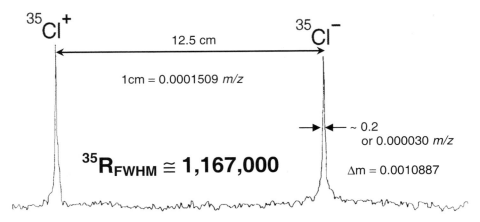

Figure 2-46. *Segment of mass spectrum in the region of the nominal m/z 35 showing a resolving power >1,000,000 (FWHM) using FTMS. The peaks represent the positive and negative ions of ^{35}Cl that have a difference in mass equal to the mass of two electrons (~0.15 millimass units). The instrument was switched from positive-ion detection mode to negative-ion detection mode during the scan between the two peaks.*
Courtesy of Spectrospin AG.

B. Operational Considerations

The operating pressure in the typical ICR cell must be in the 10^{-7} Pa range to permit optimal operation of the instrument. This operational pressure is some 2–3 orders of magnitude lower than that required in most other kinds of *m/z* analyzers. Special considerations must be given to providing for a vacuum system, especially one involving regions of differential pumping between the ion source and the ICR cell [18]. Differential pumping greatly reduces interference in the mass analyzer from the gas load associated with most external ionization sources [133, 177]. In spite of attention to using differentially pumped connections between the ion source and the mass analyzer in the ICR instrument, the high pressures associated with gas and liquid chromatographic inlets reduce the attractiveness of the FTMS approach for GC/MS and LC/MS applications. On the other hand, the approach of adding an axial AC potential to the Penning trap has been proposed and implemented in ICR technology to allow for stable ion trajectories in the face of ion–molecule collisions at relatively high pressures [178]. In addition, the integrating effect (Fellgett advantage) made possible through the Fourier transform approach is of particular advantage in dealing with the problem of mass spectral skewing due to changes in partial pressure of the analyte entering through a chromatographic inlet.

Because the frequency of ion cyclotron precession is directly proportional to the strength of the magnetic field, it is important to have a strong and uniform magnetic field for high-accuracy measurements. High-performance FTMS instruments use superconducting magnets that are large in size and expensive to maintain from the standpoint of cryogenic cooling. Because the operation of some ion sources is adversely affected by high magnetic fields, it is often necessary to separate the ion source and the ICR cell by as much as 1 meter [18, 179]. The transmission of ions over large distances between the ion source and the ICR cell requires complicated and extensive

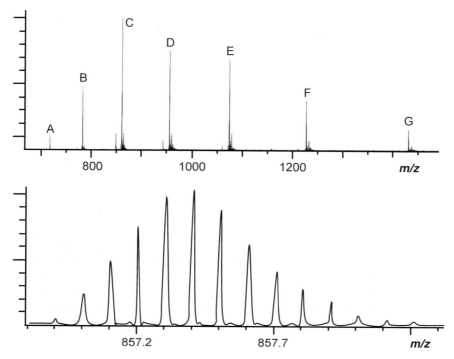

Figure 2-47. *(Top) ESI mass spectrum of ubiquitin obtained with an FTICR mass spectrometer at low resolving power (top) and at high resolving power (Bottom) in the vicinity of m/z 857 (peak C in top spectrum).* Data provided by I. Jon Amster, Department of Chemistry, University of Georgia, Athens, GA, USA.

ion optics to avoid severe transmission losses of the ions [13]. The dynamic range of stored selected ions in FTICR-MS can be increased by using stored waveforms to eject all but specified ions in order to utilize the limited storage capacity of the ICR cell efficiently [180].

C. Representative Applications

Although FTICR mass spectrometry has been used extensively with EI [177, 181], the great benefit of high resolving power and accuracy of mass determination is exemplified in its recent application to desorption/ionization sources [175, 182]; it is essential for applications using electron capture dissociation [183, 184]. Desorption/ionization permits nonvolatile macromolecules to be converted efficiently into ions that are amenable to analysis by mass spectrometry. The mass spectrum in Figure 2-47 shows the region of the mass spectrum with peaks representing the intact protonated molecule of ubiquitin. The notable feature in the bottom panel of Figure 2-47 is the fact that the FTICR mass spectrometer (R > 1,500,000) can resolve the individual isotope peaks representing the intact protonated molecule. Note that the individual isotope peaks are spaced at 0.1 *m/z*-unit intervals, indicating that the ion represented at C in the top spectrum has 10 charges; the ion represented by the peak D has nine charges, and so forth.

FTMS is useful for applications in forensic science [181], the petroleum industry [185, 186], and even the wine industry [187].

The FTICRMS instrument is ideally suited for ion/molecule reactions in the gas phase because of its operation as an ion trap. In the study of ion/molecule reactions, it is important to control the number and type of reactants. This can be done in FTMS by expelling all ions except those of interest for the ion/molecule reaction. For example, if it were of interest to study the interactions of ions of *m/z* 74 from methyl stearate with molecules of methanol, it would be important to remove all the other ions formed from methyl stearate by EI. That is to say, the complete EI mass spectrum of methyl stearate consists of molecular ions of *m/z* 298 together with many fragment ions of *m/z* 269, 267, 255, 199, 143, etc. For the above-stated ion/molecule reaction to be studied with little complication, the FTMS would be operated in such a way as to expel all ions from the ICR cell except those of *m/z* 74. Molecular vapors of methanol could then be admitted to the ICR cell in which ions of *m/z* 74 are trapped. As the ions of *m/z* 74 are allowed to interact with neutral molecules of methanol, the conditions in the ICR cell are adjusted so that ions of all masses once again will be trapped so that any product ions of the ion/molecule reaction will be retained and eventually *m/z*-analyzed for identification.

In summary, Fourier transformation (FT) of signals from the ion cyclotron resonance (ICR) mass spectrometer is the basis for FTMS, which has extraordinary attributes of high resolving power and mass accuracy. Recent developments in the theory and technology of FTICRMS have made it possible to correct for local aberrations in magnetic and electric fields within the ICR cell so that the potential benefits of high resolving power and high-mass accuracy of the instrument can be realized. For these reasons, FTICRMS instrumentation will continue to play a role in the analysis of ions formed by the wide variety of ionization techniques now available.

7. Ion Mobility Spectrometry (IMS)

An ion mobility spectrometer is *not* a mass spectrometer. Why then describe such technology in a chapter devoted to mass spectrometers? First, ion mobility spectrometry (IMS) has established itself as being complementary to mass spectrometry [188, 189]. Hybrid instrumentation has been designed and implemented by combining an ion mobility spectrometer with a mass spectrometer [190–193]; such instrumentation has been used for characterization of peptides [194–196], carbohydrates [197, 198], and drug products [199–201]. Therefore, familiarity with this promising complementary technology is very important. Second, at first glance, the ion mobility spectrometer might be confused with a TOF mass spectrometer. Both instruments have a flight tube in which ions drift in characteristic or diagnostically distinguishing ways. However, in the ion mobility spectrometer, the ions drift under the influence of an electric field; in the TOF mass spectrometer, the ions drift in a field-free region. In the ion mobility spectrometer, the ions drift through a collision gas that flows in an opposing direction at atmospheric pressure; in the TOF mass spectrometer, the ions drift in a vacuum where there is no collision with gaseous molecules. In the ion mobility spectrometer, the ions separate according to their size or cross-sectional area; in the TOF mass spectrometer, the ions separate according to their *m/z* values. These differences and other operating principles are described in more detail below.

A. Operating Principles of IMS

In the classical mode of ion mobility spectrometry, gas-phase ions migrate toward the detector, as accelerated by a constant electric field, through a drift gas that

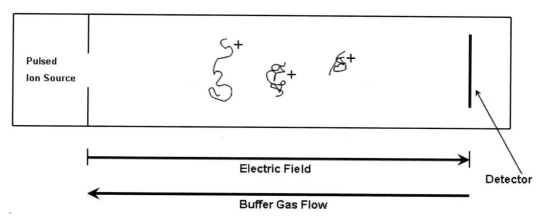

Figure 2-48.   Conceptual diagram of a conventional ion mobility spectrometer.

flows in the opposite direction, as shown in the conceptual diagram in Figure 2-48. Because the force on the ion is directly proportional to the magnitude of the electric field and the charge on the ion, and because the resistance to migration of the ion is proportional to its cross-sectional area, the process of IMS can be visualized as gas-phase electrophoresis to a first approximation. However, the resistive flow of the ion through the collision gas is a complicated function of the composition of the drift gas as well as its chemical and physical nature [189, 202]; in some cases, the order of drift times for certain analytes can be reversed by changing the drift gas [203]. The resolving power of IMS, in part, is dependent on the discrete dimensions of the ion packet admitted to the drift region through a gating potential downstream from the ion source; orthogonal injection of ions into the drift region significantly improves the resolving power of the technique [204].

B. FAIMS

In field-asymmetric ion mobility spectrometry (FAIMS), the gas-phase ions migrate toward the detector as transported by a drift gas through a periodic (nonconstant) orthogonal electric field as shown conceptually in Figure 2-49 [205, 206]. Thus, IMS differs from FAIMS in two important ways. In IMS, the ions drift collinearly with the electric field against the flow of drift gas. In FAIMS, the ions drift with the drift gas, which moves in a direction orthogonal to the electric field.

As illustrated in Figure 2-49, in the FAIMS instrument, the ions are carried toward the detector by the drift gas, but their path undergoes deviations from side to side as an alternating asymmetric electric field is applied in a direction orthogonal to the flow of the drift gas [205, 206]. The electric field is asymmetric in that it is applied in opposing directions for unequal periods (t_1 vs t_2 in Figure 2-49). The separation power of FAIMS is based on its capacity to discriminate small differences in the alpha factor of different analytes; note in Figure 2-49 that species with $\alpha = 0$ can be distinguished from those with positive and negative values under the experimental conditions. In this way, FAIMS provides improved resolution over that available in conventional IMS.

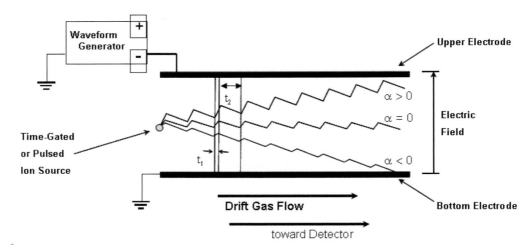

Figure 2-49. *Conceptual diagram of a field-asymmetric ion mobility spectrometer (FAIMS).* *Courtesy of G. Scherperel, Graduate Assistant, Department of Chemistry, Michigan State University, East Lansing, MI.*

As illustrated in Figure 2-49, in the FAIMS instrument, the ions are carried toward the detector by the drift gas, but their path undergoes deviations from side to side as an alternating asymmetric electric field is applied in a direction orthogonal to the flow of the drift gas [205, 206]. The electric field is asymmetric in that it is applied in opposing directions for unequal periods (t_1 vs t_2 in Figure 2-49). The separation power of FAIMS is based on its capacity to discriminate small differences in the alpha factor of different analytes; note in Figure 2-49 that species with $\alpha = 0$ can be distinguished from those with positive and negative values under the experimental conditions. In this way, FAIMS provides improved resolution over that available in conventional IMS as illustrated in the characterization of oligosaccharides [198].

The FAIMS instrument offers a greatly simplified drift tube design over those for conventional IMS because it does not require ion shutters, voltage dividers, or an aperture grid [189, 205-207].

C. Applications

Ion mobility spectrometry has been used to study and detect a variety of small molecules, such as dinucleotides [208] and tetrapeptides [209]. Greater selectivity is achieved when ion mobility is determined after mass/charge/selection, as illustrated in studies of small peptides [210] including bradykinin [211], tryptic peptides [212], amphetamines [213], amino acids [214, 215], opiates [216], and illicit drugs [217], also as detected by orthogonal TOFMS [188, 218]. Some structural information can be obtained by incorporating some means of CAD as has been demonstrated in the use of an orifice skimmer for CAD of drift-selected electrosprayed angiotension [219] or the use of an octapole collision cell for peptide [220] mixture analysis. Clemmer and coworkers [221] have designed an ingenious system for studying time-dependent conformational changes in electrosprayed ions as large as cytochrome-c using ion mobility spectrometry to determine the abundance of various conformers and TOF mass spectrometry to determine the charge state. Combining such a system with LC offers a powerful tool for mixture analysis as demonstrated with peptide mixture analysis [222–224]. IMS has been used to characterize large ions formed by MALDI, though not in combination with mass spectrometry [146]. IMS has also been used to determine the ionization pathways and drift behavior for sets of constitutional isomeric and stereoisomeric nonpolar terpene hydrocarbons [225].

A commercially available hybrid instrument called Synapt that integrates IMS with MS was introduced by Waters Corp. at Pittcon 2007. As described in more detail in Chapter 3, the Synapt incorporates an IMS module between a transmission quadrupole and an *oa*TOF analyzer, effectively replacing the standard collision cell in a conventional *tandem-in-space* instrument. The availability of the Synapt is expected to significantly increase the number of applications of combined IMS-MS/MS, especially in the area of proteomics and drug metabolism [196]. A key advantage of the IMS-MS/MS is that it allows better characterization of minor constituents of a complex mixture because the otherwise overwhelming isobaric major constituents have a different drift time, and therefore enter the CAD chamber or the *m/z* analyzer at a different time.

IV. Calibration of the *m/z* Scale

As was pointed out in the introduction to this chapter, manufacturers do not provide instruments with calibrated *m/z* axes. Calibration of the *m/z* axis is accomplished by introducing a sample of a compound whose mass spectrum is well known and adjusting the response of the mass spectrometer accordingly. Depending on the type of algorithm used, the investigator must enter the known *m/z* values for the peaks in the mass spectrum of the calibration compound; the computer will then make appropriate adjustments in the numerical presentation of the *m/z* scale. In other cases, the operator may adjust *m/z*-scale parameters until the computer output indicates the proper *m/z* values for known peaks in the mass spectrum of the calibration compound [226]. Specialized instrumentation as in single-stage reflection TOF-SIMS may require more attention to certain details [227]. Modern instruments that are used for GC/MS and LC/MS that are considered to be "unit-resolution" instruments have very stable *m/z* scales. The calibration may not need to be repeated for several days or even weeks; however, the calibration should be checked on at least a daily basis. Most of these types of instruments have a "mass-calibration stability" specification of 8 or 24 hours.

When performing accurate mass measurement to the nearest fraction of a millimass unit, it is not uncommon to use an internal standard, which is a substance introduced into the mass spectrometer's ion source with the analyte. The internal standard produces ions of known *m/z* values, usually close to the values of the ions whose accurate mass is to be determined. This allows for the accurate mass of the analyte ion to be calculated as a function of the exact mass of the internal standard's ion. This technique of using an internal standard is also used routinely in MALDI.

The internal standard technique can be used to determine the accurate *m/z* value of any mass spectral peak regardless of the instrument's resolving power. This internal standard technique has been used with QMF. It is important to remember that when this technique is applied to data obtained on a unit-resolution mass spectrometer that peak's accurate *m/z* value does not necessarily represent the mass of an ion of a specific elemental composition. Unless the peak represents a monoisotopic ion and/or an ion of single elemental composition, the returned accurate *m/z* value is a weighted average of the masses of the different species contributing to the unresolved peak. The reason instruments with high resolving power are usually used for accurate mass measurements is to assure that the ion being measured represents only one isotopic and/or elemental composition.

Because the substance being used to calibrate the *m/z* scale must be ionized and produce ions of multiple *m/z* values that will span the *m/z* scale of the instrument, no single substance is normally used as a calibrant. The choice of calibrant is dependent on the type of ionization and the *m/z* range of the mass spectrometer.

1. Electron Ionization

In practice for conventional EI mass spectrometry, perfluorinated compounds are preferred as calibration compounds for at least three reasons: (a) the mass spectra of these compounds have discrete peaks at regular intervals throughout the *m/z* range; e.g., from *m/z* 50 to 650 or higher (as in the case of perfluorokerosene); (b) these compounds are relatively volatile, and can be introduced and removed from the mass spectrometer and sample reservoir with ease; and (c) the exact masses of these ions (containing mostly carbon and fluorine) are almost equal to an integer value. Fluorine has a negative mass defect; therefore, ions containing fluorine can be resolved easily from the ions of compounds from biological or hydrocarbon sources by high-resolution mass spectrometry. The mass spectra of the two most common calibration compounds for EI, PFTBA (perfluorotributylamine, a.k.a. FC-43) and PFK (perfluorokerosene), are shown in Figures 2-50 and 2-51, respectively. The ions formed by PFTBA when it fragments under conditions of electron ionization are shown in the back matter of this book. Many perfluorinated compounds of this type are available commercially.

Calibration of the peak-intensity axis is less precise, but Eichelberger and coworkers [228] have recommended guidelines based on the mass spectrum of decafluorotriphenylphosphine (DFTPP) for helping to ensure uniform tuning of instruments in different laboratories. DFTPP is readily available in pure form, can be easily introduced via the GC inlet, and produces ions of high abundance at *m/z* 50, 198, and 442. The published spectrum of DFTPP can be used as a guide to adjust the focusing and tuning parameters of the ion optics to assure reasonable agreement with other laboratories for relative intensities throughout the mass scale. Various DFTPP tuning criteria have been established for many of the U.S. Environmental Protection Agency GC/MS methods. Another compound also used by the EPA in establishing the performance criteria of an instrument is bromofluorobenzene (BFB).

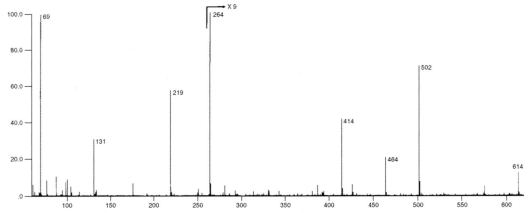

Figure 2-50.   EI mass spectrum of perfluorotributylamine used in the calibration of the m/z scale of mass spectrometers equipped with an EI source.

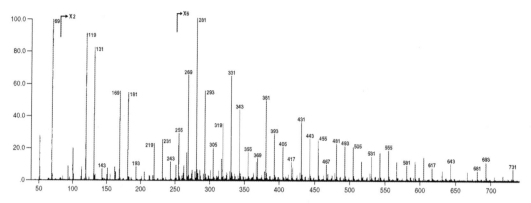

Figure 2-51. *EI mass spectrum of perfluorokerosene used for calibration of the nominal m/z scale of double-focusing instruments and as internal calibration for peak-matching accurate mass measurements in TOF and double-focusing instruments.*

The proliferation of different tune criteria mandated by different EPA methods for the same general set of analytes had to do with differences in relative intensities of peaks obtained on different types of instruments (mainly the differences obtained with the early commercial designs of the QMF vs the magnetic-sector instruments due to the mass discrimination observed in the QMF). As the QMF improved and manufacturers addressed the causes of mass discrimination, it became more difficult to meet the original tune criteria [229]. The more modern instruments had to be detuned in order to meet the EPA tune criteria. Different branches of the EPA began to promulgate a series of different criteria. The waste-water people had one set of criteria, while the drinking-water people had another, and the Super Fund group had yet a third for the same set of analytes. The DFTPP standards were for the class of compounds known as "semivolatiles".

Normally, during calibration of the *m/z* scale, the partial pressure of the calibration compound is kept constant, as is the temperature of the ion source and ion optics in order to avoid changes in the mass spectrum. In contrast, Eichelberger and coworkers proposed a dynamic calibration procedure in which the partial pressure of the analyte was constantly changing [228]. The rationale for the dynamic calibration scheme was to simulate the dynamic nature of the partial pressure of the analyte, which was continuously changing because it was introduced from the GC column. This meant that in order to meet the tune criteria, not only was it necessary to adjust various mass spectral instrument parameters, but also various GC parameters such as the carrier gas flow rate.

2. Chemical Ionization

Remember that there are two types of chemical ionization as differentiated by pressure: the type that is used with instruments designed for electron ionization (low pressure) and the type used with instruments primarily designed for atmospheric pressure ionization of LC eluates. The first type of CI uses PFTBA or PFK to calibrate the *m/z* scale in instruments that also have EI ion sources. These compounds do not provide a particularly strong CI signal when using most reagent gases recommended for positive ion formation, but the signal is strong enough to provide a good *m/z* scale calibration. PFTBA and PFK provide exceptionally good calibration of the *m/z* scale for

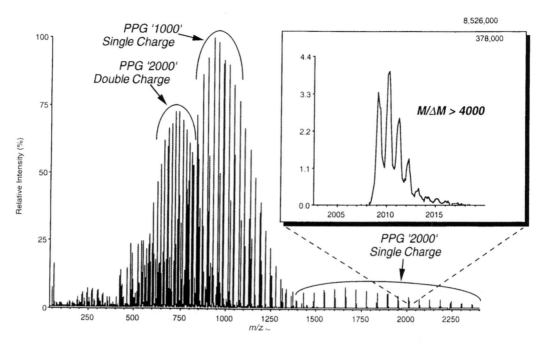

Figure 2-52.   Electrospray mass spectrum of PPG mixture.

resonance electron capture ionization.   Because of the way the internal QIT GC-MS performs CI, it is not necessary to recalibrate the instrument when switching between CI and EI.

3. Electrospray Ionization and Other APCI Techniques

APCI and ES instruments used with LC/MS do not easily produce a variety of ions over a broad range of m/z values from a single compound.   Calibration of these instruments is accomplished using a mixture of compounds.   A variety of different substances can be used to calibrate the LC-MS.   Until the introduction of the Agilent 1100 LCMSD, the substances most often used for calibrating the m/z scale of an LC/MS system were PPG (polypropylene glycol) and PEG (polyethylene glycol).   Today, a variety of substances are in use including mixtures of sugars and solvent clusters such as methanol clusters.

Mixtures of both polyethylene [PEG, $(HO-(CH_2CH_2O)_n-H)$] and polypropylene glycol [PEG, $(HO-(CH(CH_3)CH_2O)_n-H)$] have been used to calibrate the m/z scale.   Both of these materials are mixtures of different molar mass polymers.   Both have the same disadvantage that they are very sticky and remain in the system for a long period.   Often these materials are used to calibrate or check the calibration of the m/z scale at the end of the day so that the residual material can be purged during the night.   Use of post-acquisition data files is possible because many instruments store data as intensity:time pairs and then the m/z value is assigned using a calibration file as the data are viewed.

PPG is commercially available in a series of molar mass distributions:

PEG	Ave MW 425	Aldrich 20,233-4
PPG	Ave MW 1000	Aldrich 20,223-0
PPG	Ave MW 2000	Aldrich 20,233-9

An example of the mass spectrum obtained for PPG is shown in Figure 2-52. The following is a typical recipe used in a PPG calibration:

> Solvent: 2 mM NH_4OAc (from 1000 mM stock solution) in 0.1% HCO_2H of 1:1 $MeOH:H_2O$. Mix 500 mL of methanol (HPLC grade) with 500 mL of water (HPLC grade). Add 2 mL of 1000 mM stock NH_4OAc solution. Add 1 mL of 88% formic acid (Baker's analyzed reagent).

> To prepare PPG 10^{-4} M stock solution (2×10^{-4} M PPG 2000, 1×10^{-4} PPG 1000, and 3.3×10^{-5} M PPG 425). Weigh 0.4 g PPG 2000, 0.1 g PPG 1000, and 0.014 g PPG 425 into a 500-mL beaker. Dissolve the material with about 200 mL of the prepared solvent. Quantitatively transfer the contents of the beaker to a 1000-mL volumetric flask. Wash the beaker with several small aliquots of solvent into the volumetric flask. Dilute the contents of the volumetric flask to the volume mark.

Another typical solvent system used is 1×10^{-4} M in 80/20 CH_3CN/H_2O 10 mM NH_4OAc and 0.01% HOAc using PPG 1000 and 2000. This is introduced into the mass spectrometer at a rate of 5 $\mu L\ min^{-1}$. The ions formed in the positive mode are $[M + NH_4]^+$ and $[M + 2NH_4]^{2+}$. The incremental mass between polymer chains is 58 Da.

Another calibration solution used with electrospray for calibration of the *m/z* scale of the QMF mass spectrometer (over the range of 100–1500 *m/z*) is composed of a mixture of corn syrup, raffinose, maltose, and maltotetraose (500, 20, 100, and 20 ng μL^{-1}). The calibration is done using electrospray in the negative-ion detection mode; the system is then switched to the positive-ion mode for data collection. This technique of calibrating in negative-ion mode and switching back to positive-ion mode works well for transmission quadrupoles, but not for sector-based instruments. A mass spectrum generated by this mixture of sugars is shown in Figure 2-53.

This calibration solution is prepared in 50% aqueous acetonitrile by dissolving 1 mg of corn syrup in 0.5 mL, 1 mg of maltose in 5 mL, 1 mg of raffinose in 12 mL, and 1 mg maltotetraose in 12 mL; equal volumes of each solution are then mixed together.

A series of proprietary solutions have been employed by Agilent Technologies for the calibration of their QMF instrumentation out to *m/z* 3000. There are separate solutions for APCI and ESI. The materials used for the *m/z*-scale calibrations are perfluorinated phosphoazides, which are similar to a commercial calibration material called Ultra Mark that has been used in the calibration of magnetic-sector instruments with EI sources and high-*m/z* ranges (3–10K) for a number of years.

4. MALDI

Because MALDI produces principally protonated molecules of the analyte, a mixture of compounds (each producing essentially one peak in the mass spectrum) might be used to calibrate the mass spectrometer externally; i.e., the instrument is calibrated during an independent run before the sample of interest is analyzed. External calibration is a convenient procedure because the calibration process does not interfere with sample preparation procedures, but experience with a given instrument will determine whether external calibration is adequate for a given task.

The most reliable calibration procedure for MALDI is internal calibration, in which one or more compounds, similar in composition to the analyte, are mixed with the sample to provide a mass spectral peak on either side of the peak representing the protonated analyte for calibration of the *m/z* axis by interpolation.

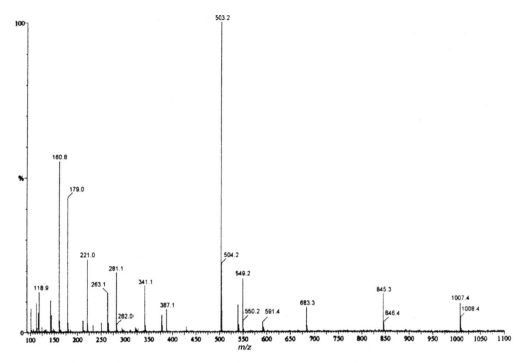

Figure 2-53.   Mass spectrum of sugar mixture obtained by electrospray ionization in the negative-ion detection mode.

V. Ion Detectors

1. General Considerations

Sensitivity, accuracy, and response time are important parameters that distinguish different ion detection systems [230]. High accuracy and fast-response times usually are mutually exclusive features [231]. The fast-response electron multiplier [232] required to follow rapid scans (e.g., m/z 50 to 500 in 0.1 sec) probably will not provide the accuracy desired (better than ±0.1%) in isotope ratio measurements. Such isotope ratio measurements are usually made with the Faraday cup detector [233]. However, with differential ion counting under computer control [234], the electron multiplier provides good precision (±0.5%) and 100 times greater sensitivity than the classical Faraday cup.

Ideally, a statement of sensitivity should explicitly define the ion current received by the detector for a specified rate of sample consumption. The definition should include (a) the name of the calibration compound, (b) the m/z value of the ion current that is measured, (c) the resolving power of the instrument, (d) the quantity of calibration compound consumed per second, (e) the intensity of the electron beam and pressure in the ion source (for CI), and (f) the ion current arriving at the detector (e.g., that impinging the conversion dynode of an electron multiplier). These practical units of sensitivity will permit a comparison of instruments, which is nearly impossible if manufacturers use different criteria, which is usually the case.

For scanning instruments, an increase in resolving power requires an increase in sample size or a decrease in scan rate to obtain a reasonable statistical representation of the peak shape at each *m/z* value in the spectrum; for example, for a resolving power of 10,000 and a scan time of 10 sec, each 0.0001 of the mass spectrum (along the *m/z* scale) is examined by the electron multiplier (detector) for only 10^{-3} sec. If the increments of the spectrum are further reduced (resolving power increased), the electron multiplier receives too few ions from this sample size to establish a representative mass spectral peak shape. Because the center of the true peak cannot be determined accurately under these conditions, the accuracy of the *m/z* measurement deteriorates. Analogously, sensitivity is inversely related to scan speed, because the higher the scan rate the less time does any given ion beam impinge the detector. A detectable signal requires a certain minimum number of particles to hit the detector; below this minimum (which depends on the accuracy desired), ion counting statistics are so poor that the signal cannot be distinguished from noise [235]. This means, for a given ion-beam current, the higher the scan rate, the fewer particles from the ion beam at each *m/z* value that impinge the detector, and the poorer the sensitivity.

2. Types of Detectors

Some of the common detectors are briefly described below; for more complete descriptions, refer to the authoritative works of Evans [236] and Geno [237]; modern detectors for mass spectrometry have been reviewed [230] by Koppenal *et al.* The detectors most often used with the modern mass spectrometers involve a secondary emission of electrons. This means that the detection device has a finite life, which is dependent on the number of ions that strike the device (amount of sample introduced and the number of samples analyzed). In many instruments that use a standard lead-doped glass electron multiplier (EM), this life is between 18 months and two years, depending on usage. When the EM dies, it must be replaced. This can be a very tedious and arduous task due to the requirement of the precise alignment of the device with the ion beam that exits the analyzer. In some cases, it may take several iterations of bringing the instrument to atmospheric pressure to install and position the EM, re-establishing the vacuum, and then testing the performance of the EM. In the case of instruments with oil diffusion pumps, each of these iterations can be quite long because of the need to cool the pumps before venting the instrument (increasing the pressure to atmospheric conditions) and then later having to reheat the pump oil before the vacuum can be reestablished.

In about 1989, K&M Electronics (now ITT Power Solutions) in West Springfield, Connecticut, developed a two-part EM shown in Figure 2-54. The first part is a permanently mounted holder which can be positioned once on the instrument platform so that the ion beam is in near-perfect alignment with the detector. The second part is the EM itself which is attached to the base using a compression lock. With this two-part device, the EM can be easily plugged and unplugged from the base without disturbing the orientation of the base or the alignment of the ion beam and the multiplier. Many mass spectrometer manufacturers have adopted this product

Figure 2-54. Plug-in continuous dynode EM assembly allowing for self-alignment during replacement.

and it finds widespread use. Several companies besides ITT Power Solutions now manufacture replacement EMs for these bases. The K&M product also led to a demand for such ease-of-use features and mass spectrometer companies that did not adopt this particular product designed their own devices to allow for self-positioning and self-alignment of the EM when it is inserted into the fixed base.

A. Faraday Cup

In this conventional electrical detector, positive ions impinging on the collector are neutralized by electrons drawn from ground after passage through a high-ohmic resistor (Figure 2-55). The voltage drop across the standard resistor is a measure of the ion current. The voltage signal from the resistor is then amplified by a DC amplifier or a vibrating-reed electrometer. The small metal electrode that the positive ions intercept (the collector) is mounted in a Faraday cage, as shown in Figure 2-55. The metal surface of the collector is inclined with respect to the ion trajectory so that reflected ions or ejected secondary electrons cannot escape from the cage.

The measured and amplified ion current from the Faraday cup is directly proportional to the number of ions and number of charges per ion; i.e., it is a charge detector. The response of the Faraday cup is independent of the energy, the mass, or the chemical nature of the ions. Faraday cup detectors are simple, inexpensive, rugged, and reliable. They have high accuracy, constant sensitivity, and low electrical noise. The principal disadvantage of the classical Faraday cup is its relatively long time constant inherent in the amplification system associated with the use of a high-ohmic resistor in the circuitry. Consequently, this classical version of the detector has been reserved for accurate measurement of slowly changing ion currents and not for recording rapidly scanned mass spectra such as those in GC/MS. However, replacement of the resistor-to-ground component in the electrical schematic shown in Figure 2-55 with a field-effect transistor has all but removed the amplification circuit as the limiting element in the dynamic performance of the Faraday cup [238].

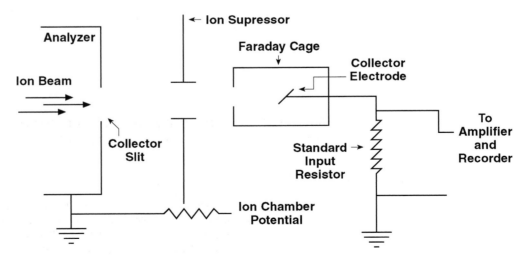

Figure 2-55. *Conceptual diagram of the Faraday cup detector.* From Watson JT "Mass Spectrometry Instrumentation" in Waller GR, Ed., Biomedical Applications of Mass Spectrometry, Wiley-Interscience, New York, 1972, with permission.

Building on earlier work by Cottrell and coworkers [239] and Hillenkamp and coworkers [240], Westphall and coworkers [241] have developed a charge detector based on the principle of the Faraday cup with an electronic signal amplification system fast enough to time- and mass-resolve the signals from a MALDI TOF mass spectrometer. Critical to this new design of the Faraday cup is placement of a transmission screen in close proximity to the collector plate to provide adequate time resolution [241, 242]. The transmission screen minimizes sluggishness of the detector response associated with image current charging at the collector surface, a problem second in importance to the circuitry issue described in the previous paragraph. The response of the Faraday cup detector is compromised by the dynamics of image charging that builds up as the approaching ion packet nears the collector surface (i.e., placing a charge near a conductor causes the charge carried in the conductor to move around until an equal but opposite charge distribution is achieved). The dynamics of image current formation results in a signal response that is proportional to the speed of the approaching ions. Insertion of a shielding element in close proximity to the collector surface blocks the electric field produced by the approaching ions until such time that the ions traverse the shielding element. The rise time of the signal is now proportional to the distance between the collector and the shield divided by the velocity of the ions; if this distance is made small enough, the importance of ion velocity becomes negligible [242].

A remaining problem in providing a Faraday cup detector with a fast response relates to the actual capacitance of the collector plate, which adds noise and reduces the slew rate of the amplification electronics. The capacitance of the collector scales with the physical dimensions of the device; therefore, the collector should be kept small. For applications requiring a large detection area as in TOFMS, an array of collectors will have to be used [242]. Segmented Faraday cups have recently been described [243, 244], which could be used for this purpose.

By combining the three corrective measures described above, a highly responsive Faraday cup detector can be designed that will be devoid of the mass discrimination frequently observed with electron multipliers and microchannel plates, caused by mass dependence of the bombarding particle in releasing secondary electrons [242, 245].

B. Electron Multiplier

1) Discrete-Dynode Version

The electron multiplier uses the principle of secondary-electron emission to effect amplification [232]. A discrete-dynode version of an electron multiplier is presented in Figure 2-56. The ion beam from the m/z analyzer is focused onto the conversion dynode, which emits electrons in direct proportion to the number of bombarding ions. The secondary electrons from the conversion dynode (ions to electrons) are accelerated and focused onto a second dynode, which itself emits secondary electrons (electrons to electrons). This means, amplification is accomplished through a "cascading effect" of secondary electrons from dynode to dynode, because the number of electrons ejected from each dynode surface is greater than the number of electrons impinging on it [246]. Each stage (copper–beryllium (CuBe) dynode) is connected to a successively higher potential by a voltage divider, and the final collector (or anode) is connected to a conventional amplifier. Versions of these electrostatic detectors with 10 to 20 stages are not uncommon. The efficiency of electron multipliers is dependent upon the velocity of the impinging particle [247], a feature responsible for the disappointing performance of channel electron multiplier arrays (CEMAs) in MALDI TOFMS for high-mass ions [245]. The CEMA is discussed later in this section.

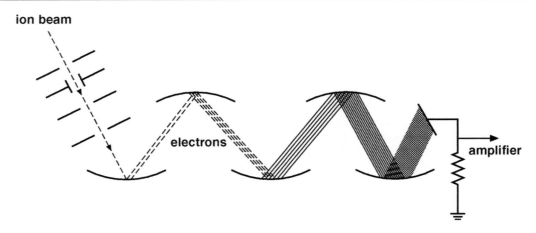

ion beam

electrons

amplifier

Figure 2-56.   Conceptual diagram of a discrete-dynode electron multiplier.
From McFadden, WH, Techniques of GC/MS, Wiley-Interscience, New York, 1973, with permission.

These ordinary copper–beryllium discrete-dynode multipliers were susceptible to poisoning by water vapor and oxygen. This meant that exposure to atmosphere would greatly shorten their useful life. Most early mass spectrometers using the EI techniques required that the vacuum be broken and the internal components be exposed to the atmosphere. This was usually done in conjunction with a need to replace the ionizing filament; however, with careful heating in an air-free oven at about 300 °C, the CuBe discrete-dynode electron multiplier could be rejuvenated. They can also be reconditioned by the original manufacturer, which is the recommended process. The CuBe discrete-dynode is no longer in widespread use. ETP, a wholly owned subsidiary of the Australian company SGE Analytical Sciences Pty Ltd, manufactures a discrete-dynode electron multiplier constructed from a lead-doped glass, which is the same material used in the continuous-dynode electron multipliers.

2) Continuous-Dynode Version

An alternative design of the electron multiplier (EM) uses a continuous dynode (CD) made of glass doped with lead. This lead-doped glass is electrically resistive, and thereby provides an electrical gradient when connected to a power supply as illustrated in Figure 2-57. The cornucopia shape illustrated in Figure 2-57 is a popular design for the modern EM. The large end of the cornucopia is adjusted at an angle that readily captures the continuous ion beam on the internal wall of the cornucopia and converts the ions to secondary electrons [246], making absolute alignment less of an issue. The secondary electrons are attracted along the positive electrical gradient farther on into the cornucopia. Each time these electrons collide with the wall, additional secondary electrons are expelled, thereby providing amplification. This form of the continuous dynode is curved so as to prevent positive ions from causing spurious signals or feedback signals due to secondary ionization of residual gas molecules.

This device was originally trademarked as the Channeltron® single-channel electron multiplier (CEM) by the Electro-Optics Division of Bendix Corporation (the first U.S. manufacturer of CEMs), which became Galileo Electro-Optics Corporation due to a management buyout of the Bendix Division. The Galileo Scientific Products Group was subsequently purchased by Burle Industries, Inc., who formed the wholly owned

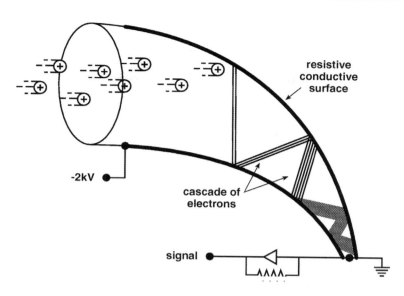

Figure 2-57. **Conceptual diagram of a continuous-dynode electron multiplier; the field gradient along the conductive internal surface of the cornucopia attracts the cascading electrons toward the preamplifier.**

subsidiary Burle Electro-Optics, Inc. Today, there are several other manufacturers of the lead-doped glass continuous dynode EM; however, all of these companies have their origins through people who were originally involved with Bendix or were part of one of its progeny. The first functioning CEMs were fabricated by William C. Wiley in 1958 at the Bendix Corporation at the same time he and McLaren were improving the TOF mass spectrometer [248]. This work was closely followed by developments at Mullard Research Laboratories in Leeds, UK, and the Laboratories d'Electronique et de Physique Appliquee (L.E.P.) in France.

The operational gain of the continuous-dynode EM is about 10^5. This device is much more tolerant of exposure to atmosphere than the CuBe devices. The CEM can produce a gain of $\sim 10^5$ with an initial operating voltage of ~ 1200 V; this is compared to an initial operating voltage of ~ 3000 V for CuBe multipliers. As the CEM ages, the voltage must be increased to about 1800–1900 V to achieve a 10^5 gain. At this point, the mass spectra exhibit a great deal of electrical noise and the multiplier should be replaced. The lead-doped glass multiplier does not have a particularly good shelf life. It is recommended only to purchase a replacement at the time it is needed.

The fast response (negligible time constant) and high sensitivity (gain of 10^5) make the CEM indispensable in GC/MS and LC/MS instruments. However, it does have some drawbacks: (a) it does not have the stability of an electrometer detector; the gain is variable, depending on usage and operating conditions; (b) it exhibits some mass discrimination [249] because secondary-electron emission at the conversion dynode is dependent on the velocity [250], mass, charge, and nature of the incident particles; that is to say, a given number of ions of m/z 500 will not generate an output signal as great as an equal number of ions of m/z 50; (c) overloading and saturation effects result when the output current exceeds 10^{-8} A; and (d) the multiplier is affected by variations in magnetic field and therefore must be located at least 50 cm away from sector magnets. It is also important to remember that the secondary emission of electrons can be initiated by neutrals. If a neutral particle strikes the detector at the same time that ions of a specific *m/z* value are supposed to be exiting the analyzer, the result could be a noise spike in the mass spectrum or an erroneous intensity. This can be determined by examining spectra on either side of the spectrum containing the suspected spurious peak.

C. Negative-Ion Detection

Modification of a magnetic mass spectrometer to detect negative ions instead of positive ions is relatively simple [251]. The polarity of the accelerating voltage and the magnet current must be reversed. The negative ions have sufficient injection kinetic energy (typically 3–4 kV) to impinge on the electron multiplier even though it has a negative bias on its first dynode so that secondary electrons will experience an overall positive gradient toward the last stage.

In mass spectrometers that use quadrupole fields no change is required in the analyzer part of the instrument because the field acts on positive and negative ions equally. In the transmission quadrupole, the repeller and acceleration voltages have to be changed from positive to negative. The kinetic energy of the ions exiting the quadrupole device is low (3–15 V) relative to that of magnetic instruments (typically, thousands of volts) as they approach the detector. If the quadrupole is equipped with a conventional electron multiplier, a negative bias on the first dynode will repel the negative ions, and they will not be detected.

Stafford *et al.* [252] designed a system with a special electrode that converts negative ions to positive ions; this conversion dynode is mounted near the entrance to a conventional electron multiplier, which then amplifies the secondary positive-ion current in the usual manner. Figure 2-58 illustrates the basic concept in these approaches to the detection of negative ions. Plate A is maintained at a positive potential of 1 to 3 kV to attract the negative ions and to provide them with sufficient kinetic energy for efficient conversion to positive (secondary) ions upon impact with plate B, which is well isolated

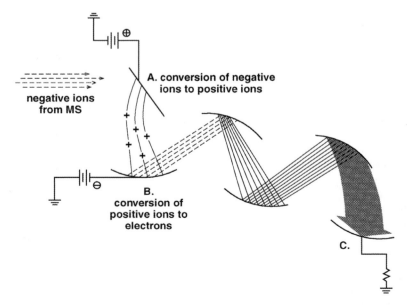

Figure 2-58. Conceptual diagram of EM for detection of negative ions.

from plate A. Plate B is maintained at a negative potential (1–3 kV) so that it will attract the secondary positive ions from plate A and supply them with sufficient kinetic energy (2–6 kV) for efficient conversion to secondary electrons upon impact with this conversion dynode (plate B). There is a positive gradient from plate B to plate C (from –3 kV or –1 kV up to ground potential) to cause a cascade of the secondary electrons through the electron multiplier to the collector (plate C). The conversion dynode also acts as a high-energy dynode to accelerate the nascent positive ions with sufficient energy to overcome the otherwise problematic effect of low velocity associated with ions having high *m/z* values.

Hunt *et al.* [253] modified a quadrupole mass spectrometer so that the polarity of the accelerating voltage could be changed at a rate of 10 kHz. Under these conditions, packets of positive and negative ions were ejected alternately from the ion source and into the *m/z* filter for mass analysis. Two electron multipliers were mounted on the instrument; one was floated above ground to detect the negative ions, the other maintained below ground potential to attract and detect the positive ions. With this system, they were able to obtain positive- and negative-ion mass spectra essentially simultaneously. This process, when used with CI GC/MS was trademarked by Finnigan Corp. as the PPNICI™ (pulsed-positive negative-ion chemical ionization); however, the technique has little practicality.

D. Post-Acceleration Detection and Detection of High-Mass Ions

Whether an ion produces a response, and if so the magnitude of that response when it hits a detector, is dependent upon its velocity [236, 237, 240, 250]. With the advent of the desorption/ionization technique MALDI, it has been possible to generate ions of extraordinarily high *m/z* values (single charge ions of very high mass); however, it is often difficult to detect such ions efficiently because their velocity is relatively low if it depends only on the accelerating voltage of the ion source [254–256].

One technique for improving the detection efficiency of high-*m/z* value ions is to provide a "post-acceleration" electric field to increase their velocity prior to impacting the conversion dynode of the detector [257]. An example of a post-acceleration detector (PAD) based on a CEMA chevron has been described by Cotter for use in the TOF mass spectrometer [258]. Although similar, the PAD should not be confused with the high-energy dynode used between the *m/z* analyzer and detector with instruments such as those based on the transmission quadrupole.

Detection of macro-ions (those heavier than hundreds of kDa having a single or double charge) has been a challenge in modern mass spectrometry. Conventional mass spectrometry has relied on ionization-based detectors, namely those producing and multiplying secondary electrons associated with particle bombardment of a metal surface; such detectors are restricted to ions having a mass less than 10^6 Da. An ion-to-photon conversion detector (IPD) shows promise for detecting macro-ions [259, 260]. The IPD functions by ions impinging a conversion surface coated with a fluorescent organic scintillator compound like Bu-PBD [2-(4-biphenyly)-5-(4-tert-butylphenyl)-1,3,4-oxadiazole], which emits photons in proportion to the ion current; the photons are collected by a head-on photomultiplier located behind the conversion surface [259–261]. Photosensitive detectors, relying on laser-induced fluorescence (LIF) or elastic light scattering (ELS), show promise in this area, as reviewed Peng *et al.* [262]. In principle, LIF has no size limitation, but its success relies on how well dye labeling can be accomplished and how well fluorescence can be collected from excited particles confined in the small laser probe volume. For this reason, ELS is currently the method of choice in detecting particles larger than 100 nm.

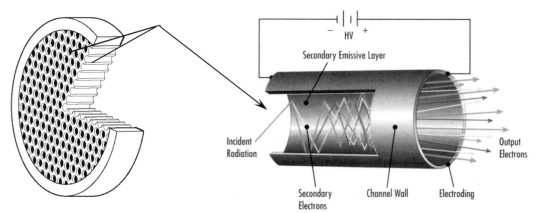

Figure 2-59.   *A channel electron multiplier array (CEMA). The device has a 40-mm diameter and a 0.6-mm thickness. Each channel has an inside diameter of 10 μm, and there is a gap of 12 μm between each channel.* Courtesy of JEOL USA AccuTOF.

E. Channel Electron Multiplier Array (CEMA)

The channel electron multiplier array (CEMA) consists of a honeycomb arrangement of many channels (up to several hundred per square inch), each having a diameter on the order of 10 μm made possible with fiber optic technology using metal-doped glass. A potential difference applied to opposite ends of the channels (top and bottom of the honeycomb) creates an electrical gradient along the resistive but conducting surface of each channel. The ions enter the individual channels slightly off-axis so that they impinge the wall of the channel as illustrated in Figure 2-59. This primary event (an ion colliding with the surface to expel electrons) has very poor efficiency; the efficiency is directly proportional to the velocity of the impacting ion, which explains the exceptionally poor response of the CEMA to massive ions. The secondary electrons resulting from ions impinging the wall continue to ricochet down the channel, producing more secondary electrons upon each impact, thereby amplifying the original ion beam.

The CEMA was designed to give a two-dimensional image of the impinging signal; an example of its use in electro-optical ion detection is described below. The CEMA is also employed in TOF mass spectrometers because it presents a simple flat surface that can be arranged perpendicular to the ion beam to preserve velocity resolution of the ion packets [263].

F. Electro-Optical Ion Detection

This novel approach [264] to ion detection employs a microchannel electron multiplier, a phosphor, an optical system, and an array of photodiodes or charge-coupled devices [265]. A block diagram of the individual components of this electro-optical ion detector (EOID) is shown in Figure 2-60. Individual (*m/z*-analyzed) ion beams impinge a chevron arrangement of two CEMAs; the offset angle in each CEMA (described earlier) plate oppose one another to minimize positive ion feedback. Each ion beam at unit resolution impinges two or more pixels of the CEMA; the electrons from the CEMA (gain

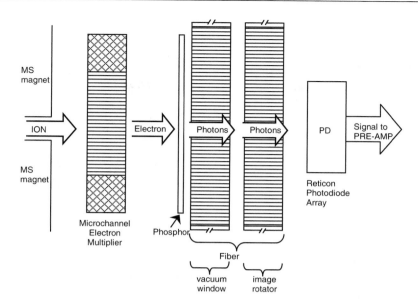

MS magnet

ION

MS magnet

Microchannel Electron Multiplier

Electron

Phosphor

Photons

Photons

vacuum window

Fiber

image rotator

PD

Signal to PRE-AMP

Reticon Photodiode Array

Figure 2-60. Conceptual diagram of EOID.

of 10^4) impinge a phosphor surface that in turn emits photons, which are transmitted via fiber optics to an optical window that serves as a vacuum barrier. The image from the vacuum window is transmitted to an array of photodiodes or charge-coupled devices (CCDs) that constitutes the final transducer in each of several hundred individual channels of the ion-beam sensor [266]. The EOID can be arranged along the focal plane of a magnetic-sector mass spectrometer so that the ion beams impinge the active surface.

The EOID was conceived as an electronic photoplate, which would have the advantage of integrating ion current during transitory excursions in partial pressure of the analyte in the ion source, as occurs with use of a chromatographic inlet. An early application of the EOID in GC/MS [267] did not significantly improve detection limits as they are usually a function of signal intensity relative to chemical background rather than sensitivity in two-sector instruments. However, considerable advantages of array detection were realized in tandem MS with magnetic-sector instruments [268–270]. Array detection is responsible for substantial advantages realized in discharge-source (ICP, etc.) mass spectrometry by allowing the simultaneous detection of multiple-ion currents of differing m/z values [271, 272]. When simultaneous detection was used in mass spectrometry, benefits such as improved detection limits and precision, reduced sample consumption and analysis time, and the elimination of correlated noise sources are realized [273].

G. The Daly Detector

The positive ion beam passing the detector slit is attracted toward an aluminized cathode (the Daly knob) held at a very large negative potential (e.g., −15,000 V) [274, 275]. Positive ions impacting the Daly Knob, which essentially serves as a conversion dynode, produces up to eight secondary electrons, which are attracted to scintillator unit held at ground potential as illustrated in Figure 2-61. When these electrons impact the scintillator, the secondary electrons create photons, which are then amplified in the usual manner in a conventional photomultiplier.

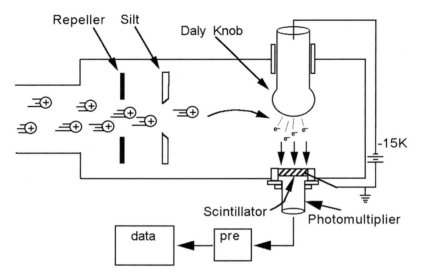

Figure 2-61. Daly knob used for ion detection in a mass spectrometer.

The Daly detector offers two significant advantages over other similar detectors. First, most of the mechanical components (all except the Daly knob) are located outside the vacuum chamber, and thus can be serviced without disrupting the pressure regime of the mass spectrometer. Second, the large potential difference used between the Daly knob and the slit is particularly advantageous for detecting ions of high mass; in this way, the Daly detector serves somewhat as a post-acceleration detector. In addition, a deceleration lens can be installed in front of the Daly knob to distinguish between stable ions and those that are ionic products of metastable decay; the latter have less kinetic energy than do the stable ions, which do not dissociate after acceleration out of the ion source.

A variation of the Daly detector, called the Dynolite detector, is illustrated schematically in Figure 2-62. This is the modern device used by Waters Corp. in their transmission quadrupole-based instruments and by JEOL in their double-focusing instruments. All other manufacturers use electron multipliers. These devices have one advantage over the use of the electron multipliers in that they do not deteriorate in performance over time. There have been detectors of this design in use for decades without a need to replace them.

H. Cryogenic Detectors

Cryogenic detectors are energy-sensitive calorimetric detectors that operate at low temperatures [276]. A cryogenically cooled $Nb-Al_2O_3-Nb$ superconductor–insulator superconductor tunnel junction detector operating at 1.3 kV has been used as an ion detector in a MALDI TOF mass spectrometer for detecting 66-kDa ions [277]. Compared to conventional, ionization-based detectors, which rely on secondary electron formation or the charge created in a semiconductor, cryogenic detectors measure low-energy solid state excitations created by particle impact. Cryogenic detectors reportedly have nearly 100% efficiency even for very large, slow-moving molecules, in contrast to microchannel plates whose efficiency drops considerably at large mass. Therefore, cryogenic detectors could contribute to extending the *m/z* range accessible by a TOF-MS by improving detection limits [276, 277].

I. Ion Detection in FTMS

All other detectors described in this chapter create an electrical signal based on the collision of ions with a specialized surface within the detector module. Whereas early versions of the ICR instrument, e.g., the omegatron, did use this type of detection system [278, 279], modern versions of the ICR, in the form of FTMS, use space-charge detection [162]. In this way, the ion current is not detected directly, but is sensed indirectly through an image current. Image current is that current that flows into a conductor to balance a charge induced by an approaching ion packet of opposite charge.

As presented earlier in Figure 2-45, the oscillating packet of ions in an ICR cell induces an oscillating image current in a pair of sensing plates, which serve as the detector. In this way, the ions in the ICR cell are detected indirectly. The frequency of oscillation relates to the *m/z* value of the ions and the magnitude of the image current relates to the abundance of ions in the packet.

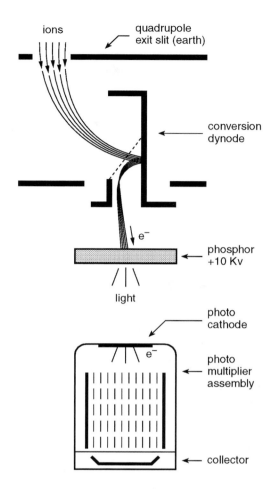

Figure 2-62.   The Dynolite® detector.

VI. Vacuum Systems

1. Introduction

The development of mass spectrometry would not have been possible without suitable vacuum systems. Low pressure is essential for production of free electrons and ions in the gas phase. Low pressure (i.e., infrequent ion/molecule or molecule/molecule collisions) is also essential for preservation of resolving power in *m/z* analyzers. In mass spectrometry, ions move from point A to point B. Once the direction of an ion's path has been established, any interaction with other matter can cause that direction to change. If an ion undergoes an inelastic collision with another particle, the ion can be neutralized or it can become collisionally activated and decompose into ions of a lesser mass. On the other hand, CI in the ionization chamber of some instruments is optimal at pressures 10^4 to 10^5 times higher than that recommended for EI; therefore, it is important to know how to achieve and measure low pressure. In the case of atmospheric pressure ionization, gas-phase ions are present with a large number of neutral molecules. The ions must be separated from the molecules without affecting the distribution of ions so that they can be *m/z*-analyzed. This separation is accomplished through the mass spectrometer's pumping system.

One of the primary reasons for the advancements in mass spectrometry has been the advancements in vacuum technology. It is unclear as to whether the demands for mass spectrometry forced the advancement in vacuum technology or the advancements in vacuum technology are responsible for the improved performance of the mass spectrometer in the last ten years. Just as advancements in vacuum technology during World War II led to the post-war proliferation of mass spectrometry, the advances in vacuum technology of the last ten years have closely followed the proliferation of better mass spectrometers.

Some of the basic principles of the more common components of vacuum systems are presented here. Much more detailed information is available in relevant books [280, 281].

2. Definitions

Vacuum technology, like all technologies, has a language of its own. Several terms and definitions will facilitate an understanding of vacuum systems, including the operating principles of pumps and measurement devices used to monitor the pressures of various components of the mass spectrometer. These are as follows:

pascal. The pascal (Pa) is equivalent to 7.5×10^{-3} mm Hg. The pascal has the dimensions of a newton per square meter, which is a force, whereas mm Hg is a linear measurement that is not a force. The pascal is the unit of pressure that is acceptable in the Systeme Internationale d'Unites (SI units) [282].

torr. A torr (Torr) is the pressure exerted by a 1-mm high column of mercury at 0 °C and at sea level at latitude 45°. A torr is equivalent to 133.32 Pa. Although not an acceptable SI unit because it is based on the weight, not the mass, of a column of mercury, it is still a widely used unit in vacuum technology. Two other nonstandard units of pressure are used: *atmosphere* and *bar* (*millibar*). The *atmosphere* and the *bar* are both equal to 760 Torr. The *millibar* is 1/1000 of bar or 1.32 Torr. The actual measurement of pressure may be in mm Hg, but this value is corrected to the definition of a torr when the pressure is reported.

molecular flow. The flow of gas through a channel under conditions such that the mean free path is greater by an order of magnitude than the largest dimension of a transverse section of the channel. At these pressures, the flow characteristics are determined by collisions of the gas molecules with the channel surfaces, and flow effects from molecular collisions are insignificant.

transition flow. The flow of a gas through a channel under conditions such that the mean free path is of the same order as the transverse dimension of the channel. In this pressure range, the flow characteristics are determined by collisions of the gas molecules with surfaces as well as with other gas molecules; also called **Knudsen flow**.

viscous flow. The flow of a gas through a channel under conditions such that the mean free path is very small in comparison with the smallest dimension of a transverse section of the channel. At these pressures, the flow characteristics are determined mainly by collisions between the gas molecules (i.e., the viscosity of the gas). The flow may be laminar or turbulent.

throughput. The volume of gas (**Q**) at a known pressure and temperature that passes a plane in a known time. In SI, throughput has the units of $Pa \times m^3 \, sec^{-1}$, which is equivalent to $J \, sec^{-1}$ or watts, because $Pa = N \, m^{-2}$ and $J = N \times m$.

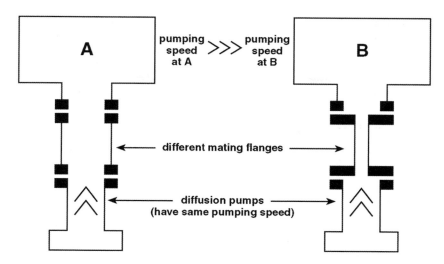

Figure 2-63. *Illustration of improper mating flanges used with high-vacuum oil diffusion pumps. If the potential speed of the pump is to be maintained, the pump must be as close to the volume being pumped as possible and the connection should not have a diameter less than the diameter of the pump's opening. These conductance issues are primary restricted to oil diffusion pumps. Turbomolecular pumps are closely coupled to the pumped volume.*

conductance. The throughput divided by the pressure drop along a channel or conduit is equal to the conductance: $C = Q / (P_2 - P_1)$. By analogy to an electrical circuit, conductance is analogous to electrical current divided by a potential difference.

resistance. The resistance to gas flow through a vacuum conduit causes a pressure drop in a fashion analogous to the voltage drop associated with the flow of electricity through a wire. The resistance (or impedance) to gas flow is related to the length, l, and diameter, d, of the conduit: $R = kl / d^4$.

pump speed. The volumetric rate of gas flow across a section at the pump inlet. Note that pumping speeds are quoted at the throat of a given pump; if comparable pumping rates are desired in other components of the system, they must be connected to the pump by short, large-diameter connections (see Figure 2-63 and the definition of resistance). Often cooling baffles are used in vacuum systems employing oil diffusion pumps to reduce the amount of background. These baffles can reduce the pumping speed by as much as 90%. The mesh of the screens used to protect the high-speed components of a turbomolecular pump will also diminish the pumping speed.

differential pumping. In an arrangement in which two volumes are connected by a small orifice and have separate vacuum pumps attached through conduits that are large compared with the common orifice (Figure 2-64), the two volumes are said to be differentially pumped [283].

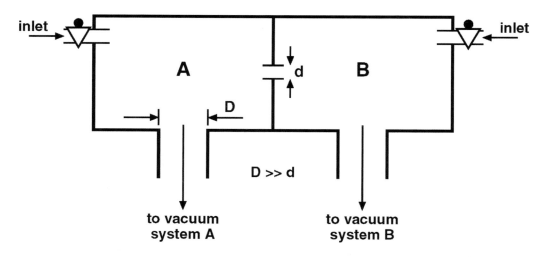

Figure 2-64.   *Illustration of differential pumping. A sudden increase in the gas load entering chamber A (e.g., an ion source) will result in little change in the pressure in chamber B (e.g., an m/z analyzer) because the gas flow through the communicating small orifice d is very low compared with that through D to the vacuum system A.*

3. Pressure Gauges

A. Thermal-Conductivity Gauges

This class of gauges measures pressure indirectly by responding to the capacity of gas molecules to remove energy from a heated wire. When a molecule collides with the wire, energy is transferred to the molecule; the result is loss of heat by the wire and a corresponding increase in kinetic energy of the gas and decrease in the temperature of the wire. The rate at which heat is removed from the wire is dependent on the frequency of collisions of gas molecules with the wire; in turn, the frequency of collisions is dependent on the concentration (pressure) of the gas molecules. Because the rate of energy removal is dependent on the type of molecules, the gauge must be calibrated for different gases.   These gauges are effective in the 66 to 0.06 Pa range [280], the pressure range referred to as "medium vacuum".

B. Pirani Gauge

When the heated wire serves as one arm of a Wheatstone resistance bridge, the gauge is called a Pirani gauge. The wire can be heated by electrical current. A change in gas pressure will cause a change in temperature of the wire and a corresponding change in its resistance, which is sensed through the Wheatstone bridge by a feedback circuit that controls the heating current such that a null condition is re-established. The null heating current can be calibrated in pressure units. A miniaturized version of the Pirani gauge has been developed [284].

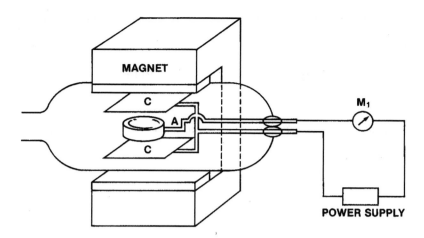

Figure 2-65.   *Schematic diagram of a Penning gauge. A is a center anode;*
C represents two cathodes (this can be replaced with a
cylindrical cathode). A strong permanent magnet surrounds
these elements. From Basic Vacuum Practice, 2nd ed., Varian, Inc., Palo Alto, CA,
USA, 1989.

C. Thermocouple Gauge

In this case, a thermocouple is attached directly to the heated wire, which is heated by a constant electrical current. At a given pressure, the wire will come to baseline temperature due to heat loss via the corresponding rate of collisions and via radiation. If the pressure increases, the rate of collisions increases, thereby withdrawing more heat from the wire so that the temperature of the wire drops and so does the electrical output from the thermocouple, and so forth. The output from the thermocouple can be calibrated in pressure units.

4. Ionization Gauges

At pressures below 10^{-3} Pa ($\sim 10^{-5}$ torr), ionization of gaseous molecules and measurement of the resulting ion current provide an effective means of measuring pressure.

A. Hot-Cathode Gauge

The operating principles involve ionization of gas molecules by EI and subsequent collection of the ions. The hot cathode generates the electrons for ionization. At constant temperature, the measured ion current is proportional to the number of gas molecules present, which is proportional to the pressure in the chamber.   The Bayard-Alpert ionization gauge was originally designed to measure pressures down to about 10^{-8} Pa [285], but there are reports of measurements down to 10^{-10} Pa [286]. The reasons for the upper and lower limits of measurement by these devices have been reviewed [287].

B. Cold-Cathode Gauge

The Penning gauge [288] consists of an anode ring held between two cathode electrodes, as illustrated in Figure 2-65. A potential of 2 kV to 10 kV is applied across the electrodes; this high field causes electrical breakdown, producing electrons and positive ions. An optional arrangement to increase the sensitivity involves use of a magnetic field (0.1–0.2 tesla) to cause the electrons to travel in spiral paths, thereby increasing the probability of collisions with gas molecules to generate additional ionization. The total current (ions plus electrons) is sufficiently large that no current amplifier is required; output currents of 10 to 50 mA Pa^{-1} are typical [280]. The pressure range of operation of the Penning gauge is 1 to 10^{-4} Pa. The Penning gauge has the advantage of spanning the ranges of operation of the thermal-conductivity gauges and the ionization gauges; however, it has the disadvantage of not operating below 10^{-4} Pa, primarily because of difficulty in maintaining the discharge [280]. The cold-cathode gauge is more rugged, but less accurate, than the hot-cathode ionization gauge [289].

5. Types of Pumps

A. Mechanical Pumps (Low Vacuum)

The first step toward creating a usable vacuum in a mass spectrometer is to reduce the pressure in a chamber from atmospheric to 10 to 10^{-1} Pa. This rugged task can be accomplished by a mechanical pump. The rotary mechanism of some types of mechanical pumps or fore pumps must be lubricated with oil; the vapor pressure of the hot oil is one of the factors that limits the ultimate vacuum that can be achieved by these pumps. The vapor from this hot oil can also be a source of chemical background in the mass spectrometer. The importance of a mechanical pump lies in its durability and throughput or high pumping speed at high pressure (above 100 Pa, a pressure region in which high-vacuum pumps are inoperable). By definition, a fore pump is a vacuum pump that maintains the *fore pressure* of another pump below its critical pressure. It is in this capacity that mechanical pumps serve as fore pumps to high-vacuum pumps such as the

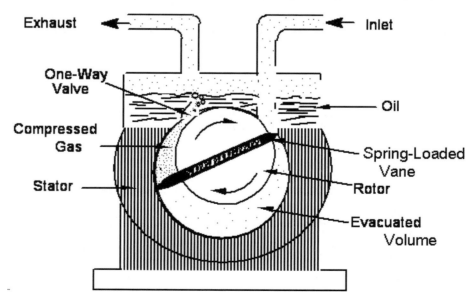

Figure 2-66.   *Cross-sectional sketch of a rotary vane pump showing how the spring-loaded vane sweeps the gas from one chamber and compresses it in another, from which it escapes through a one-way valve.*

diffusion pump or turbomolecular pump. Fore pumps used in mass spectrometry range in capacity from 350 to 50 liters min^{-1}, depending on whether they are part of the vacuum system for the ion source or that for the analyzer, where the throughput is not so great.

> **Warning!!!** The exhaust of the mechanical pump as well as the split vent of the gas chromatograph must be filtered (a carbon absorption filter) or vented to a chemical fume hood. Most of the sample introduced into a mass spectrometer ends up in the mechanical pump exhaust or (in the case of oil-based pumps) in the pump oil. The pump oil should be handled extremely carefully. This is a major hazardous waste.

1) Rotary Vane Pumps

One of the most common mechanical pumps is the direct-drive rotary vane pump illustrated in Figure 2-66. A revolving rotor with a spring-loaded vane sweeps some gas into a chamber where it is compressed and expelled into a separate exhaust chamber. This repetitive action depletes the number of gas molecules from the entrance throat to the pump, thereby reducing the pressure or creating a vacuum. Direct-drive mechanical pumps are popular because they eliminate problems due to faulty or broken belts.

When considering the venting of a rotary vane pump (especially with GC/MS systems) it is important to understand the difference between a filter that removes chemical components from the pump's exhaust and a mist eliminator supplied by many mass spectrometer manufacturers. The mist eliminator is designed to prevent atomized oil from being sprayed into the room when the pump starts or is turned off. The mist eliminator does not prevent toxic substances from being exhausted into the room. Figure 2-67 shows two different types of mist eliminators; a carbon filter is mounted above each mist eliminator. As can be seen in Figure 2-68, the components required to properly filter the pump exhaust are numerous and somewhat complex. For this reason, it is best to find a supplier that can provide a kit. It is also important to remember to change the carbon filter every 30 days.

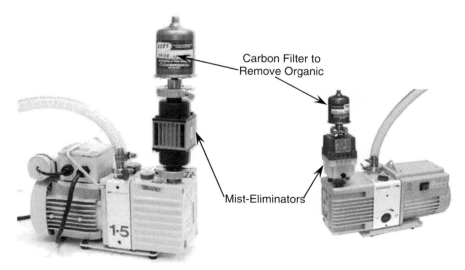

Figure 2-67.   Rotary vane pumps equipped with a mist eliminator and a carbon filter.

Figure 2-68.   Components in the pump filter kit sold by Scientific Instrument Services in Ringoes, NJ, USA.
(http://www.sisweb.com)

2)   Scroll Pumps

Another pump design that is becoming as popular as the fore pump in high-vacuum systems is the scroll pump. This is a dry pump that has no oil or other liquid lubricants, which are required in conventional rotary vane pumps and are responsible for some of the background in mass spectra. The pump is hermetically sealed to keep moisture from becoming a problem. This type of pump has its origin in the 1905 invention of a scroll-type fluid-displacement device patented by Leon Creux. The pump consists of two interleafed scrolls or spiral-shaped devices, one of which is fixed and the other is flexible. The fixed scroll consists of a spiral-shaped channel in a metal plate; the flexible scroll is essentially a belt made of PFTE that lies in the channel of the fixed scroll. The two scrolls have nonconcentric centers that are made to move in a nonrotational circular manner relative to one another (orbiting motion), which makes a pocket formed by the line contact between the flexible scroll and the wall of the fixed scroll shift along the spiral-curved surfaces as illustrated in Figure 2-69. The pumping action begins with the intake stroke, represented in (a) of Figure 2-69, in which a pocket of gas is captured between the flexible scroll and the wall of the fixed scroll. As the scrolls continue to orbit, the pocket becomes smaller and smaller as it moves toward the center of the interleafed scrolls (illustrated in (b) and (c) of Figure 2-69); this action compresses the original pocket of gas for expulsion through a one-way valve that is accessible at the end of the spiraling channel, as represented at the center of the diagram in (d) of Figure 2-69.

The scroll pumps are much more expensive than rotary vane pumps of the same performance, but they are much cleaner, quieter, and require less periodic maintenance (e.g., oil change, etc.) except for annual replacement of the PFTE membrane that is pressed against the outer edges of the channel in the fixed scroll and the edges of the

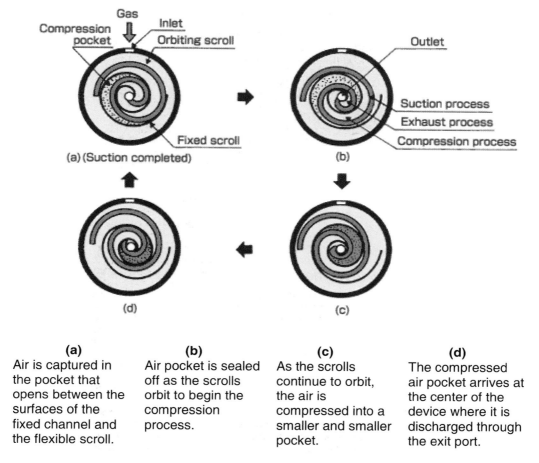

Figure 2-69.   Schematic flowchart of scroll pump operation.

(a)
Air is captured in the pocket that opens between the surfaces of the fixed channel and the flexible scroll.

(b)
Air pocket is sealed off as the scrolls orbit to begin the compression process.

(c)
As the scrolls continue to orbit, the air is compressed into a smaller and smaller pocket.

(d)
The compressed air pocket arrives at the center of the device where it is discharged through the exit port.

flexible scroll to form a nearly gas-tight seal between nontouching metal components that move orbitally in close proximity. The scroll pump can serve as a roughing pump to evacuate a chamber starting at atmospheric pressure to a final pressure of less than 10 Pa, or it can serve as a backing pump for a turbomolecular pump or other high-vacuum pumps.

3) Roots Pump

The Roots pump is a dry pump that has no touching metal or nonmetal surfaces, which might otherwise produce contamination by fine particles of material resulting from abrasive wear of mechanical surfaces as in the scroll pump. The operational components of the Roots pump are two lobed impellers (roughly shaped like a figure "8") that are mechanically linked via gears such that they rotate in opposite directions, as illustrated in Figure 2-70. The pumping action of the Roots pump derives from the compression of gas that is captured in cavities created between the convoluted surfaces of rotating impellers, the volume of which progressively decreases as the impellers rotate toward one another. The impellers never touch one another or the pump housing; thus there is no need for a lubricant. Because of the small gap allowing clearance of rotating

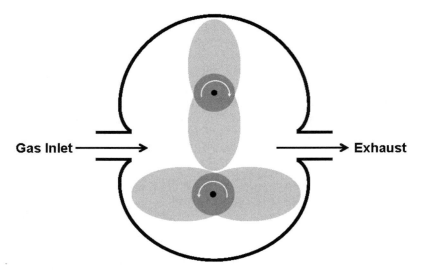

Figure 2-70.   Schematic representation of the Roots pump.

metal parts, the pressure differential achieved by a single stage is not great, but when several stages are arranged in series, the overall effect is diminution of the pressure from atmospheric to 10 Pa at the surface of the first of a contiguous string of four or five stages of impellers.   Isaiah Davies invented the design principle of the rotary lobe blower in 1848, about 20 years before Francis and Philader Roots applied it to commercial practice, which continues today (Wikipedia.com).

4) Diaphragm Pumps

A diaphragm pump is a positive-displacement pump that uses a combination of the reciprocating action of a rubber or PFTE diaphragm and suitable nonreturn check valves to pump a fluid.   Sometimes this type of pump is also called a membrane pump.

There are two main types of diaphragm pumps:

- In the first type, the diaphragm is sealed with one side in the fluid to be pumped and the other in air or hydraulic fluid.   The diaphragm is flexed, causing the volume of the pump chamber to increase, then decrease.   A pair of nonreturn check valves prevent reverse flow of the fluid.

- The second type of diaphragm pump has one or more unsealed diaphragms with the fluid to be pumped on both sides.   The diaphragm(s) again are flexed, causing the volume to change.

When the volume of a chamber of either type is increased (the diaphragm moving up), the pressure decreases, and fluid is drawn into the chamber.   As the diaphragm moves back down, the fluid is expelled from the chamber.   Another cycle begins by moving the diaphragm up, etc.   This pumping action is similar to that of the cylinder in an internal combustion engine.

B. High Vacuum

The high-vacuum system is what maintains the low pressures required by most mass spectrometers.   The pumps capable of maintaining these low pressures will not begin to function until the pressure is down to about 0.1 Pa.   It is in the area of high vacuum systems that some of the more dramatic advances have been made. One area of particular importance is the split-flow turbomolecular pump (as illustrated at the right). This device allows for differential pumping through the use of a single pump. In some cases, as many as three different chambers can be evacuated using a single pump. The oil diffusion pump, which dominated mass spectrometry for years, has largely been replaced by the turbomolecular pump, whose purchase price has dropped precipitously in the last few years due to increased manufacturing volumes and new designs. The oil diffusion pump is still a viable performer on some GC/MS systems, but is no longer found in the area of LC/MS.

It is important to realize that the pumping speed of a given high-vacuum pump as quoted by the mass spectrometer manufacturer is actually the specification quoted by the pump manufacturer, which in all cases except for Varian

Figure 2-71A. Split-flow turbomolecular pump with pumping from both the top and the side.

is different from the mass spectrometer manufacturer. This specification is usually based on the pump's ability to pump nitrogen or air (which is 80% nitrogen). In the case of LC/MS systems, this nitrogen-pumping performance specification is appropriate because the gases in LC/MS systems are primarily nitrogen. However, in the case of GC/MS, this nitrogen-pumping specification can be a problem because most GC/MS systems use helium as the GC mobile phase, which means that the gas load is due to helium. In some cases, hydrogen is used as a carrier gas. The pumping speed of oil diffusion and turbomolecular pumps drops by about 30% when pumping helium compared to pumping nitrogen. The pumping speed of turbomolecular pumps drops about another 15 to 20% when pumping hydrogen. Further details on this problem can be found in the GC/MS chapter.

1) Turbomolecular Pumps

The turbomolecular pump, based on original work by W. Becker, uses rapidly rotating blades to direct gas molecules through the system. The heart of the pump takes on the appearance of a turbine compressor in a jet engine (Figures 2-71A and 2-71B). Several sets of blades are mounted on a shaft that rotates at 20,000 to 60,000 rpm. When mounted in the pump, the rotating sets of blades are separated by sets of slotted stator (stationary) blades (Figure 2-72). Although the compression ratio across any one stage (a set of rotating and stationary blades) is not great, the compression across several sets (in series) is sufficient to cause efficient gas flow, as illustrated by the arrows in Figure 2-71B.

Because the blades impart momentum to the gas molecules most efficiently in the molecular-flow region, the turbomolecular pump, like the diffusion pump, must be backed up by a mechanical fore pump. If the fore-line pressure increases such that the rear blades are in transition or viscous flow, the rotor will sense additional torque due to viscous drag. It is also important to coordinate the operation of the fore pump (oil containing) and the turbomolecular pump to prevent back-streaming of oil vapors into the housing of the turbo. For example, the turbo and the backing pump should be started at the same time. The turbo should be vented to atmospheric pressure before it stops; never shut off the turbo and leave it at high vacuum because this will permit back-diffusion of oil vapors from the mechanical pump through the stator/rotor blades. Opening a nonoperating turbo to low pressure will cause the same problem; i.e., such a pressure gradient would force oil vapors from the mechanical pump "backwards" through the turbine, thereby contaminating it.   The turbo should be vented (exposed to atmospheric pressure with a dry gas) on the high-vacuum side or into the blade structure, but never on the fore-vacuum side. The fore line should be leak-tight. Accidental venting through the mechanical pump can be prevented by using only those equipped with safety valves that quickly close during power failure or shutoff. These operating procedures to maintain a hydrocarbon-free background in the turbomolecular pump are based on the simple principle of maintaining a decreasing pressure gradient toward the mechanical pump (J.A. Freeman, Balzer High Vacuum Corp., Personal Communication).

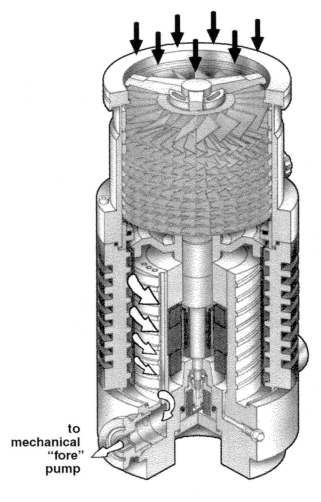

to
mechanical
"fore"
pump

Figure 2-71B.   *Cutaway view of a turbomolecular pump showing the rotating blades in the upper portion pushing gas-phase molecules to the bottom of the pump, where they are exhausted into a fore pump.* Courtesy of Balzer High Vacuum Corp., Hudson, NH, USA.

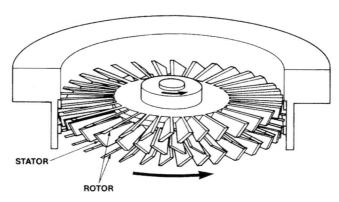

Figure 2-72. ***Close-up view of blades in a turbomolecular pump.*** *Courtesy of Balzer High Vacuum Corp., Hudson, NH, USA.*

In summary, the turbomolecular pump can rapidly attain high vacuum (pressures as low as 10^{-8} Pa in baked systems) with no hydrocarbon background [280]. However, mechanical failure of the rapidly rotating parts can be devastating (see Figure 2-73); maintenance can be expensive. A maintenance program that involves a trade-in for rebuilt systems will help reduce downtime in the case of a mechanical failure.

Figure 2-73. ***Catastrophic failure of 250 L sec^{-1} turbo pump.***

2) Oil Diffusion Pumps

The high-vacuum pumps cannot be used until the resident pressure has been reduced to 10 to 10^{-1} Pa. Oil diffusion pumps do not operate efficiently at higher pressures; most diffusion pumps become damaged (due to deposition of charred oil residue as described later) when exposed to pressure above 100 Pa for more than a few minutes. The outstanding feature of diffusion pumps is the high pumping speed that is attainable at low pressure. The pumping speed is proportional to the physical size of the diffusion pump. Mass spectrometry requires the use of diffusion pumps with pumping speeds in the range of 50 to 500 L sec^{-1}; the larger pumps were used on the ion source of instruments designed for chemical ionization or GC/MS operations where packed columns were employed, or both, that involve a relatively high throughput of gas. Today's modern GC/MS instrumentation (that usually provides only EI) uses oil diffusion pumps that are rarely larger than the 50 L sec^{-1}. This small size means that far less oil is used than in the larger versions, with a concomitant smaller chance of causing oil contamination to the analyzer in the event of an uncontrolled venting of the system to atmosphere.

The operating principles of an oil diffusion pump are illustrated schematically in Figure 2-74. There are at least two important points to be made concerning the operation of this type of pump. The first is that the pump cannot be used at pressures much above 1 to 10^{-2} Pa. Therefore, they should be turned on or allowed to operate only at pressures below 10^{-1} Pa; i.e., if a given device is to be evacuated to a very low pressure, the first step is to reduce the pressure in the device from atmospheric pressure down to >10^{-1} Pa with a mechanical pump such as the rotary vane pump. Once the pressure is reduced >10^{-1} Pa, the heater of the diffusion pump can be turned on. If the only mechanical pump available is one that is connected to the exit of the diffusion pump, as shown in Figure 2-74, the whole device can be evacuated by pumping with the mechanical pump through the cold diffusion pump; when the appropriate pressure is reached, the heater to the diffusion pump is turned on. High pressure does not affect the diffusion pump so long as it is cold; i.e., turned off.

The second point concerning operation of a diffusion pump is that there are no moving mechanical parts. The diffusion pump is relatively simple. Referring to Figure 2-74, note that there is a liquid in the bottom of the diffusion pump. Typically this liquid is a high-boiling oil, like a polyphenyl ether, which is heated, forcing its vapor up through the central chimney of the diffusion pump – thus the name oil diffusion pump. These vapors are expelled from the central chimney through radial openings at various stages or intervals along the chimney. In this way, molecular beams of oil are directed downward and toward the outer walls. The gas-phase oil molecules collide with gaseous molecules and knock them down toward the bottom of the pump. This pumping action compresses the gas causing the pressure at the bottom of the diffusion pump to increase (with a concomitant decrease in pressure at the top of the pump!), which is relieved by a mechanical or fore vacuum pump. When the oil molecules finally reach the cool outer walls of the pump, they condense, and after sufficient accumulation, form droplets that run down the cooled sides of the chamber to the reservoir at the bottom. Continuous heating of the reservoir recycles the oil through the pump. This means that there exists a pressure gradient along the vertical axis of the diffusion pump. At the top of the diffusion pump, the pressure might be 10^{-6} Pa or lower, depending on the vapor pressure of the pumping liquid; at the bottom of the diffusion pump, the pressure might be 10^{-1} Pa.

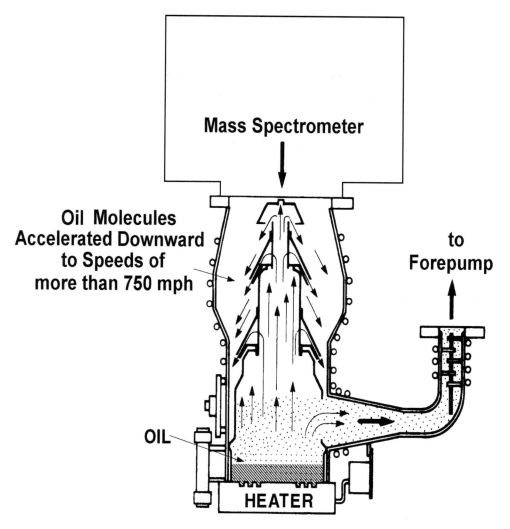

Figure 2-74.   *Schematic diagram of the oil diffusion pump.* *Courtesy of CVC Products, Rochester, NY, USA.*

The operating mechanism of the diffusion pump can be abused by a burst of gas or high pressure at the entrance to the diffusion pump, because the downward momentum of oil molecules can be overcome by the cumulative upward component of momentum of the high numbers of other gaseous molecules. As a consequence, the oil vapor may cascade above the top of the diffusion pump into the mass spectrometer, condensing on cool surfaces and thereby contaminating the instrument. The pump also may overheat because of a reduction in the flow rate of condensed oil vapor to the reservoir. Under these adverse conditions (in which some oil is shunted to volumes above the cool walls) the condensed oil vapor that eventually touches these overheated surfaces leaves a charred residue that is difficult to clean off. These pressure-related problems with the diffusion pump constitute the main reason that great care is taken in monitoring pressures during the pump's operation.

> Before turning off an oil diffusion pump, the temperature of the oil should be lowered to room temperature (20–25 °C) and maintained at this temperature for some time dependent on the volume of oil in the pump. During the cool-down time, the mechanical pump should continue to operate. The mechanical pump should be turned off as the mass spectrometer is being vented to atmospheric pressure with a dry gas such as nitrogen.

A baffle, cooled by an external source, can be installed at the throat of the diffusion pump to reduce the quantity of oil vapor that enters the mass spectrometer via diffusion or "back-streaming". Such baffles are only necessary on exceptionally large diffusion pumps; these baffles can cause as much as a 90% reduction in pumping speed. Most modern diffusion pump oils present no prohibitive background because their vapor pressure at ambient conditions is 10^{-5} Pa or lower. Diffusion pumps used on modern GC/MS instruments do not require baffles.

The pumping speed of a diffusion pump refers to the conductance or pumping rate of the system for nitrogen as measured at the throat of the diffusion pump. The effective pumping speed at other sites in the mass spectrometer is attenuated by the conductance of all intervening sections of conduit [280], as illustrated in Figure 2-63.

> One way to experience how a diffusion pump works is in a tub-based shower equipped with a flexible shower curtain. The water droplets streaming out of the shower-head will create a partial vacuum in the chamber formed by the walls of the enclosure and the nonrigid shower curtain. As the water droplets in the shower stream collide with some of the air molecules and force them down to the bottom of the shower enclosure, there are fewer air molecules remaining in the main volume of the shower stall so that the pressure is reduced. The shower curtain signals the partial vacuum being created within the shower enclosure as the atmospheric pressure in the bathroom forces it inward toward the parallel wall of the enclosure.

3) Sputter-Ion Pumps (Nonregeneratable Getter Pumps)

The operation of this pump is electrochemical in nature [280]. A fresh metal surface of titanium (the cathode of a pair of electrodes maintained at a potential difference of several thousand volts) is continually sputtered. Chemisorption of neutral gases at this activated surface removes them by conversion to low-vapor-pressure compounds.

The sputter-ion pump cannot be operated above 1 Pa, and it operates most efficiently at pressures below 10^{-3} Pa. This pump requires little maintenance and provides an oil-free vacuum (low background) and economical operation over a long lifetime (pump life is inversely proportional to gas throughput). These sputter-ion pumps have been used with several commercially available mass spectrometers such as the Siemens Quantra permanent magnet FTICR-MS and the Infinicon Hapsite transmission quadrupole GC-MS.

References

1. Sparkman OD, Mass Spectrometry PittCon 2007. *J Amer Soc Mass Spectrom* **18:** 1146-1159, 2007.
2. Sparkman OD, Mass Spectrometry PittCon 2006. *J Amer Soc Mass Spectrom* **17:** 873-884, 2006.
3. Cotter RJ, The New Time-of-Flight Mass Spectrometry. *Anal Chem* **71:** 445A-451A, 1999.
4. Price D and Milnes GJ, Renaissance of TOF-MS. *Int. J. Mass Spectrom. Ion Proc.* **99:** 1-39, 1990.
5. Loboda AV, Krutchinsky AN, Bromirski M, Ens W and Standing KG, A tandem Q/TOF MS with a MALDI source: design and performance. *Rapid Commun Mass Spectrom* **14:** 1047-1057, 2000.
6. Hao C and March RE, A survey of recent activity in quadrupole ion trap mass spectrometery. *International Journal of Mass Spec.* **212:** 337-358, 2001.
7. Ziegler Z, Ion traps come of age. *Anal Chem* **74**(17)**:** 489A-492A., 2002.
8. Mattauch J and Herzog R, Double-focusing MS. *Phys. Rev.* **50:** 617-623, 1936.
9. Johnson EG and Nier AO, Electromagnetic focusing. *Phys. Rev.* **9:** 10-17, 1953.
10. Duckworth HE and Ghoshal SN, High resolution MS. In: *Mass Spectrometry* (Ed. McDowell CA). McGraw-Hill, New York, 1963.
11. Finnigan RE, Quadrupole MS: A History. *Anal. Chem.* **66:** 969A-975A, 1994.
12. Chernushevich IV, Loboda AV and Thomson BA, An introduction to quadrupole-time-of-flight mass spectrometry. *J Mass Spectrom* **36**(8)**:** 849-65., 2001.
13. Guan S and Marshall AG, Stacked-Ring Electrostatic Ion Guide. *J. Am. Soc. Mass Spectrom.* **7:** 101-106, 1996.
14. Brubaker WM, Quadrupole MS. In: *Advances in Mass Spectrometry* (Ed. Kendrick E), pp. 293-300. Institute of Petroleum, London, 1968.
15. Barnett EF, Tandler WSW and Turner WR, Quadrupole mass filter with fringe field penetrating structure., 1971.
16. Berkout VD and Doroshenko VM, Improving the Quality of the Ion Beam Exiting a Quadrupole Ion Guide. *Journal of the American Society for Mass Spectrometry* **17**(3)**:** 335-340, 2006.
17. Chernushevich IV and Thomson BA, Collisional cooling of large ions in electrospray mass spectrometry. *Anal Chem* **76**(6)**:** 1754-1760, 2004.
18. Limbach P, Marshall A and Wang M, Electrostatic Ion Guide for FT-ICR-MS. *Int. J. Mass Spectrom. Ion Processes.* **125:** 135-143, 1993.
19. Kim T, Tang K, Udseth HR and Smith RD, A multicapillary inlet jet disruption electrodynamic ion funnel interface for improved sensitivity in API. *Anal Chem* **73:** 4162-70., 2001.
20. Ibrahim Y, Tang K, Tolmachev AV, Shvartsburg AA and Smith RD, Improving MS Sensitivity Using a High-Pressure Electrodynamic Ion Funnel. *J Amer Soc Mass Spectrom* **17:** 1299-1305, 2006.
21. Julian RR, Mabbett SR and Jarrold MF, Ion Funnels for the Masses: Experiments & Simulations with Simplified Ion Funnel. *J Amer Soc Mass Spectrom* **16:** 1708-1712, 2005.
22. Page JS, Tolmachev AV, Tang K and Smith RD, Variable low-mass filtering using an electrodynamic ion funnel. *Journal of Mass Spectrometry* **40**(9)**:** 1215-1222, 2005.
23. Page JS, Tolmachev AV, Tang K and Smith RD, Theoretical and Experimental Evaluation of Low m/z Transmission in Electrodynamic Ion Funnel. *J Amer Soc Mass Spectrom* **17:** 586-592, 2006.
24. Colburn AW, Giannakopulos AE and Derrick PJ, The ion conveyor. An ion focusing and conveying device. *European Journal of Mass Spectrometry* **10:** 149-154, 2004.
25. Heddle CWO, *Electrostatic Lens Systems, 2nd Ed.* Taylor & Francis, 2000.
26. Dawson PH, Acceptance of the quadrupole filter. *Int. J. Mass Spectrom. Ion Phys.* **17:** 423-445, 1975.
27. Jennings KR and Dolnikowski GG, Mass Analyzers. In: *Methods in Enzymology 193: Mass Spectrometry*, pp. 37-61. Academic Press, Nyc, 1990.
28. Farmer JB and McDowell CA, Types of MS. In: Mass Spectrometry. McGraw-Hill, New York., 1963.
29. Wiley WC and McLaren IH, TOF-MS. *Rev. Sci. Instrum.* **26:** 1150-1157, 1955.
30. Studier MH, Continuous Ion Source for TOF-MS. *Rev. Sci. Instrum.* **34:** 1367-1370, 1963.
31. Marable NL and Sanzone G, Impulse-focused TOF-MS. *Int. J. Mass Spectrom. Ion Phys.* **13:** 185-194, 1974.
32. Paul W and Raether M, Das Elektrische Massenfilter. *Z. Physik.* **140:** 262-271, 1955.
33. Paul W and Steinwedel H, *Z. Naturforsch, A* **8:** 448, 1953.
34. Dawson PH, Quadrupole Mass Spectrometry and Its Applications. Elsevier, New York, 1976.
35. Comisarow MB and Marshall AG, History of FT-ICR-MS. *J. Mass Spectrom.* **31:** 581-589, 1996.
36. Graff DK, Fourier and Hadamard Transforms. *J. Chem Educ.* **72:** 304-309, 1995.
37. Guilhaus M, Principles and Instrumentation for TOF-MS. *J. Mass Spectrom.* **30:** 1519-1532, 1995.
38. Wollnik H, TOF-MS. *Mass Spectrom. Rev.* **12:** 89-114, 1993.
39. Mamyrin BA, Time-of-flight mass spectrometry (concepts, achievements, and prospects). *International Journal of Mass Spectrometry* **206**(3)**:** 251-266, 2001.
40. Smythe WR and Mattauch JHE, A New Mass Spectrometer. *Phys. Rev.* **40:** 429-433, 1932.
41. Stephens WE, A Pulsed Mass Spectrometer with Time Dispersion. *Phys. Rev.* **69:** 691, 1946.
42. Wolff MM and Stephens WE, A Pulsed MS with Time Dispersion. *Rev. Sci. Instr.* **24:** 616-617, 1953.

43. Karas M, Historical Comment on reprint of 'Time-of-Flight Mass Spectrometer with Improved Resolution,' Wiley W. C;. McLaren, I. H. Rev. Sci. Instr. 1955, 26, 1150. *J. Mass Spectrom.* **32:** 1-3, 1997.

44. Hiroshi T, (tile not available). *Kage Kagaku Dzassi* **67:** 1769, 1964.

45. Kiser RW, *Introduction to Mass Spectrometry and Its Applications (Prentice-Hall International Series in Chemistry)*, 1965.

46. Karas M, Bachmann D, Bahr U and Hillenkamp F, Matrix-Assisted Laser Desorption Ionization Mass Spectrometry. *Int. J. Mass Spec. Ion Proc.* **78:** 53-68, 1987.

47. Stein R, A Time Focusing Equation for TOF-MS. *Int. J. Mass Spectrom. Ion Proc.* **132:** 29-34, 1994.

48. Bewig L, Buck U, Gandhi SR and Winter M, Pseudorandom tof analysis with a TOFMS. *Rev Sci. Instrum.* **67:** 417-422, 1996.

49. Ioanoviciu D, Ion optics in TOF-MS. *Int. J. Mass Spectrom. Ion Processes* **131:** 43-65, 1994.

50. Stein R, Time focusing and phase space dynamics in TOF-MS. *Int. J. Mass Spectrom. Ion Processes* **132:** 29-47, 1994.

51. Flory CA, Taber RC and Yefchak GE, Non-linear Ion Extraction Fields for Spatial Focusing in TOF-MS. *Int. J. Mass Spectrom. Ion Process.* **152:** 169-176, 1996.

52. Stein R, Space and velocity focusing in TOF-MS. *Int. J. Mass Spectrom. Ion Phys.* **14:** 205-218, 1974.

53. Browder JA, Miller RL, Thomas WA and Sanzone G, Impulse-field focusing for TOF-MS. *Int. J. Mass Spectrom. Ion Phys.* **37:** 99-108, 1981.

54. Juhasz P, Roskey MT, Smirnov IP, Haff LA, Vestal ML and Martin SA, Applications of Delayed Extraction MALDI TOF MS to Oligonucleotide Analysis. *Anal Chem* **68:** 941-6, 1996.

55. Juhasz P, Vestal ML and Martin SA, On the initial velocity of ions generated by MALDI and its effect on the calibration of delayed extraction TOF MS. *J Amer Soc Mass Spectrom* **8:** 209-217, 1997.

56. Takach EJ, Hines WM, Patterson DH, Juhasz P, Falick AM, Vestal ML and Martin SA, Accurate Mass measurements in MALDI-TOF with delayed extraction. *J Protein Chem* **16:** 363-369, 1997.

57. Kovtoun SV, Mass-correlated Delayed Extraction in Linear Time-of-Flight Mass Spectrometers. *Rap. Commun. Mass Spectrom.* **11:** 810-815, 1997.

58. Kovtoun SV and Cotter RJ, Mass-correlated pulsed extraction: theoretical analysis and implementation with a linear MALDI TOF MS. *J Am Soc Mass Spectrom* **11:** 841-53., 2000.

59. Pinkston JD, Rabb M, Watson JT and Allison J, New time-of-flight mass spectrometer for improved mass resolution, versatility, and MS/MS studies. *Rev. Sci. Instrum.* **57**(4): 583-9, 1986.

60. Bakker JMB, Beam-modulated TOF-MS. *J. Phys. E. Sci. Instrum.* **7:** 364-368, 1974.

61. Lubman DM and Jordan RM, TOF-MS of supersonic beams. *Rev. Sci. Instrum.* **56:** 373-376, 1985.

62. Hood R and Watson J, Realistic Expectations of Resolution and Mass Accuracy in MALDI-TOF for High-Mass Ions. In: *53rd Conf Amer Soc Mass Spectrom, San Antonio, 2005*, pp. MP 264.

63. Hood R and Watson J, Resolution and Mass Accuracy of High-Mass Ions in TOF MS: Applications to PAMAM Dendrimers. In: *54th Annual Conf Amer Soc Mass Spectrom, Seattle, 2006*, pp. MP 230.

64. Makarov AA, Raptakis and Derrick PJ, Pitfalls on the Road to the Ideal TOF Mirror. *Int. J. Mass Spectrom. Ion Proc.* **146/147:** 165-182, 1995.

65. Mamyrin BA, Reflectron TOF-MS. *Int. J. Mass Spectrom. Ion Processes* **131:** 1-19, 1994.

66. Short RT and Todd PJ, Improved Energy Compensation in TOF-MS. *J. Am. Soc. Mass Spectrom.* **5:** 779-787, 1994.

67. Scherer S, Altwegg K, Balsiger H, Fischer J, Jaeckel A, Korth A, Mildner M, Piazza D, Reme H and Wurz P, A novel principle for an ion mirror design in TOFMS. *Int J Mass Spectrom* **251:** 73-81, 2006.

68. Mamyrin BA, Karataev VI, Shmikk DV and Zagulin VA, Reflectron TOF-MS. *Sov. Phys. JETP* **37:** 45-48, 1973.

69. Mamyrin BA and Shmikk DV, Linear reflection. *Sov. Phys. JETP* **49:** 762-764, 1979.

70. Gohl W, Kutscher R, Laue HJ and Wollnik H, Time-of-flight mass spectrometry for ions of large energy spread. *International Journal of Mass Spectrometry and Ion Physics* **48:** 411-14, 1983.

71. Wollnik H, Gruener U and Li G, Time-of-flight mass spectrometers with grid-free ion mirrors and electron-impact ion sources. *Annalen der Physik (Berlin, Germany)* **48**(1-3): 215-28, 1991.

72. Grix R, Kutscher R, Li G, Gruener U and Wollnik H, A time-of-flight mass analyzer with high resolving power. *Rapid Communications in Mass Spectrometry* **2**(5): 83-5, 1988.

73. Cotter RJ, Gardner BD, Iltchenko S and English RD, Tandem Time-of-Flight Mass Spectrometry with a Curved Field Reflectron. *Analytical Chemistry* **76**(7): 1976-1981, 2004.

74. Cotter RJ, Iltchenko S and Wang D, The curved-field reflectron: PSD and CID without scanning, stepping or lifting. *International Journal of Mass Spectrometry* **240**(3): 169-182, 2005.

75. Warscheid B and Fenselau C, Characterization of Bacillus spore species and their mixtures using postsource decay with a curved-field reflectron. *Analytical Chemistry* **75**(20): 5618-5627, 2003.

76. Krutchinsky AN, Loboda AV, Spicer VL, Dworschak R, Ens W and Standing KG, Orthogonal injection of MALDI ions into a TOF MS. *Rapid Commun Mass Spectrom* **12:** 508-518, 1998.

77. Krutchinsky AN, Chernushevich IV, Loboda AV, Ens W and Standing KG, Measurements of protein structure and noncovalent interactions by TOF MS with orthogonal ion injection. *Mass Spectrometry in Biology & Medicine*: 17-30, 2000.

78. Nielsen ML, Bennett KL, Larsen B, Moniatte M and Mann M, Peptide end sequencing by orthogonal MALDI tandem mass spectrometry. *Journal of Proteome Research* 1(1): 63-71, 2002.
79. Laures AMF, Wolff J-C, Eckers C, Borman PJ and Chatfield MJ, Investigation into the factors affecting accuracy of mass measurements on aTOF MS using Design of Experiment. *Rapid Commun Mass Spectrom* 21: 529-535, 2007.
80. Hashimoto Y, Hasegawa H, Satake H, Baba T and Waki I, Duty Cycle Enhancement of an oaTOF Mass Spectrometer Using an Axially-Resonant Excitation Linear Ion Trap. *J Amer Soc Mass Spectrom* 17: 1669-1674, 2006.
81. Gonin M, Chen YH, Horvath T, Theiss M and Wollnik H, Ion-detectors for TOF MS. *Nuclear Instruments & Methods in Physics Research, Section B: Beam Interactions with Materials and Atoms* 136-138: 1244-1247, 1998.
82. Lincoln KA, Data acquisition in TOF-MS. *Dyn. Mass Spectrom.* 6: 111-119, 1981.
83. Holland JF, Enke CG, Allison J, Stults JT, Pinkston JD, Newcome B and Watson JT, MS in Chromatography. *Anal. Chem.* 55: 997A-1112A, 1983.
84. Gross JH, *Mass Spectrometry: A Textbook*. Springer-Verlag, Berlin/Heidelberg, 2004.
85. Brunelle A, Della-Negre S and LeBeyec Y, Progress in TOF-MS. *Analusis* 20: 417-420, 1992.
86. Welthagen W, Shellie RA, Spranger J, Ristow M, Zimmermann R and Fiehn O, Comprehensive 2D (GC * GC-TOF) for high resolution metabolomics. *Metabolomics* 1: 65-73, 2005.
87. Reichenbach SE, Kottapalli V, Ni M and Visvanathan A, Computer language for identifying chemicals with comprehensive 2D GC/MS. *Journal of Chromatography, A* 1071: 263-269, 2005.
88. Trapp O, Kimmel JR, Yoon OK, Zuleta IA, Fernandez FM and Zare RN, Continuous 2-channel TOFMS detection of electrosprayed ions. *Angewandte Chemie, International Edition* 43: 6541-6544, 2004.
89. Zare RN, Hadamard Transform TOFMS: More Signal, More of the Time. *Angewandte Chemie, International Edition* 42: 30-35, 2003.
90. Yoon OK, Zuleta IA, Kimmel JR, Robbins MD and Zare RN, Duty Cycle and Modulation Efficiency of Two-Channel Hadamard Transform TOFMS. *J Amer Soc Mass Spectrom* 16: 1888-1901, 2005.
91. Allison J and Stepnowski RM, Hows & whys of ion trapping. *Anal Chem* 59: 1072A-1088A, 1987.
92. Stafford GC, Kelley PE, Syka JEP, Reyonolds WE and J. TJF, Ion Trap Technology. *Int. J. Mass Spec. Ion Proc.* 60: 85-98, 1984.
93. March RE, Introduction to quadrupole ion trap MS. *J Mass Spectrom* 32: 351-369, 1997.
94. March RE, Quadrupole ion trap MS: theory, simulation, recent developments and applications. *Rapid Commun Mass Spectrom* 12: 1543-1554, 1998.
95. March RE, Quadrupole ion trap MS. A view at the turn of the century. *Intl J Mass Spectrom* 200: 285-312, 2000.
96. Hager JW, Trends in MS development. *Anal Bioanal Chem* 378: 845-850, 2004.
97. Syka JEP, Commercial Quadrupole Ion Trap. In: *Practicle Aspects of Ion Trap Mass Spectrometry: Fundamentals of Ion Trap MS*, Vol. 1 (Eds. March RE and J.. TJF). CRC, Boca Raton, FL, 1995.
98. Brodbelt J, Louris JN and Cooks RG, CI in an Ion Trap MS. *Anal. Chem* 59: 1278-1285, 1987.
99. Berberich DW and Yost RA, NCI in an Ion Trap MS. *J. Am. Soc. Mass Spectrom.* 5: 757-764, 1994.
100. Laughlin BC, Mulligan CC and Cooks RG, Atmospheric Pressure Ionization in a Miniature Mass Spectrometer. *Analytical Chemistry* 77(9): 2928-2939, 2005.
101. Gunawardena HP, Emory JF and McLuckey SA, Phosphopeptide Anion Characterization via Sequential Charge Inversion and ETD. *Anal Chem* 78: 3788-3793, 2006.
102. Syka JEP, Coon JJ, Schroeder MJ, Shabanowitz J and Hunt DF, Peptide and protein sequence analysis by ETD-MS. *Proc Natl Acad Sci USA* 101: 9528-9533, 2004.
103. Ottens AK, Arkin CR, Griffin TP, Palmer PT and Harrison WW, Ion-molecule reactions in quadrupole ion trap: implications for lightweight gas analysis. *Int J Mass Spectrom* 243: 31-39, 2005.
104. Douglas D, Frank A and Mao D, Linear ion traps in MS. *Mass Spectrom Rev* 24(1): 1-29, 2005.
105. Hager JW, A new linear ion trap MS. *Rapid Commun Mass Spectrom* 16: 512-526, 2002.
106. Michaud AL, Frank AJ, Ding C, Zhao X and Douglas DJ, Ion Excitation in a Linear Quadrupole Ion Trap with an Added Octopole Field. *J Amer Soc Mass Spectrom* 16: 835-849, 2005.
107. Londry FA and Hager JW, Mass selective axial ion ejection from a linear quadrupole ion trap. *Journal of the American Society for Mass Spectrometry* 14(10): 1130-1147, 2003.
108. Song Y, Wu G, Song Q, Cooks RG, Ouyang Z and Plass WR, Novel Linear Ion Trap Mass Analyzer Composed of Four Planar Electrodes. *J Amer Soc Mass Spectrom* 17(4): 631-639, 2006.
109. Song Q, Kothari S, Senko MA, Schwartz JC, Amy JW, Stafford GC, Cooks RG and Ouyang Z, Rectilinear Ion Trap MS with AP Interface & ESI Source. *Anal Chem* 78: 718-725, 2006.
110. Tabert AM, Goodwin MP and Cooks RG, Co-occurrence of Boundary and Resonance Ejection in a Multiplexed Rectilinear Ion Trap Mass Spectrometer. *J Amer Soc Mass Spectrom* 17: 56-59, 2006.
111. Hager JW and Le Blanc JCY, Product ion scanning using a Q-q-Qlinear ion trap (Q TRAP) mass spectrometer. *Rapid Communications in Mass Spectrometry* 17(10): 1056-1064, 2003.
112. Syka JEP, Marto JA, Bai DL, Horning S, Senko MW, Schwartz JC, Ueberheide B, Garcia B, Busby S, Muratore T, Shabanowitz J and Hunt DF, Novel linear quadrupole ion trap/FT MS: performance & use in analysis of histone H3 post-translational modifications. *J Proteome Res* 3: 621-626, 2004.

113. Gerlich D, Inhomogeneous RF fields: A versatile tool for the study of processes with slow ions. In: *Advances in chemical physics LXXXII.*, pp. 1-176. John Wiley and Sons., New York, 1992.
114. Ding C, Konenkov NV and Douglas DJ, Quadrupole mass filters with octopole fields. *Rapid Communications in Mass Spectrometry* **17**(22): 2495-2502, 2003.
115. Schwartz JC, Senko MW and Syka JEP, A two-dimensional quadrupole ion trap MS. *J Amer Soc Mass Spectrom* **13**: 659-669, 2002.
116. Zhan X and Desiderio DM, Linear ion-trap mass spectrometric characterization of human pituitary nitrotyrosine-containing proteins. *International Journal of Mass Spectrometry* **259**(1-3): 96-104, 2007.
117. Collings BA, Campbell JM, Mao D and Douglas DJ, A combined linear ion trap TOF system with improved performance and MSn capabilities. *Rapid Commun Mass Spectrom* **15**: 1777-1795, 2001.
118. Collings BA, Stott WR and Londry FA, Resonant excitation in a low-pressure linear ion trap. *Journal of the American Society for Mass Spectrometry* **14**(6): 622-634, 2003.
119. Collings BA, Increased Fragmentation Efficiency of Ions in a Low Pressure Linear Ion Trap with an Added dc Octopole Field. *Journal of the American Society for Mass Spectrometry* **16**(8): 1342-1352, 2005.
120. Reid GE, Data courtesy of Gavin Reid, Department of Chemistry, Michigan State University. 2006.
121. Ouyang Z, Wu G, Song Y, Li H, Plass WR and Cooks RG, Rectilinear Ion Trap: Concepts, Calculations, and Analytical Performance of a New Mass Analyzer. *Analytical Chemistry* **76**(16): 4595-4605, 2004.
122. Gao L, Song Q, Patterson GE, Cooks RG and Ouyang Z, Handheld Rectilinear Ion Trap Mass Spectrometer. *Anal Chem* **78**: 5994-6002, 2006.
123. Makarov A, Electrostatic axially harmonic orbital trapping: a high-performance technique of mass analysis. *Anal Chem* **72**(6): 1156-62, 2000.
124. Hardman M and Makarov AA, Interfacing the Orbitrap Mass Analyzer to ESI. *Anal Chem* **75**: 1699-1705, 2003.
125. Kingdon K, A method for the Neutralization of Electron Space Charge by Positive Ionization at Very Low Pressures. *Phys. Rev.* **21**: 408418, 1923.
126. Knight R, Storage of Ions from Laser-produced Plasmas. *Appl. Phys. Lett.* **38**: 221-223, 1981.
127. Hu Q, Noll RJ, Li H, Makarov A, Hardman M and Cooks RG, The Orbitrap: A new mass spectrometer. *Journal of Mass Spectrometry* **40**(4): 430-443, 2005.
128. Wu G, Noll RJ, Plass WR, Hu Q, Perry RH and Cooks RG, Ion trajectory simulations of axial a.c. dipolar excitation in the Orbitrap. *International Journal of Mass Spectrometry* **254**(1-2): 53-62, 2006.
129. Makarov A, Denisov E, Kholomeev A, Balschun W, Lange O, Strupat K and Horning S, Performance Evaluation of a Hybrid Linear Ion Trap/Orbitrap Mass Spectrometer. *Anal Chem* **78**: 2113-2120, 2006.
130. Olsen JV, de Godoy LM, Li G, Macek B, Mortensen P, Pesch R, Makarov A, Lange O, Horning S and Mann M, Parts per million mass accuracy on an Orbitrap mass spectrometer via lock mass injection into a C-trap. *Mol Cell Proteomics* **4**: 2010-21, 2005.
131. Marshall AG, Hendrickson CL and Emmett MR, Recent advances in Fourier transform ion cyclotron resonance mass spectrometry. *Advances in Mass Spectrometry* **14**: Chapter 10/221-Chapter 10/239, 1998.
132. Marshall AG, Hendrickson CL and Jackson GS, FT-ICR MS: a primer. *Mass Spectrom Rev* **17**: 1-35, 1998.
133. Senko MW, Hendrickson cL, Emmett MR, Shi SDH and Marshall AG, External accumulation of ions for enhanced ESI FT ICR MS. *J Amer Soc Mass Spectrom* **8**: 970-976, 1997.
134. Makarov A, Theory and Practice of the Orbitrap Mass Analyzer. In: *54th Annual Conf Amer Soc Mass Spectrom, Seattle, 2006*, pp. MOF 3:40.
135. Makarov A, Denisov E, Lange O and Horning S, Dynamic Range of Mass Accuracy in LTQ Orbitrap Hybrid MS. *J Amer Soc Mass Spectrom* **17**: 977-982, 2006.
136. Kholomeev A, Makarov A, Denisov E, Lange O, Balschun W and Horning S, Squeezing a Camel Through the Eye of a Needle: a Curved Linear Trap for Pulsed Injection into Orbitrap Analyzer. In: *54th Annual Conf Amer Soc Mass Spectrom, Seattle, 2006*, pp. TOB 10:35.
137. Hu Q, Makarov AA, Cooks RG and Noll RJ, Resonant ac Dipolar Excitation for Ion Motion Control in the Orbitrap Mass Analyzer. *Journal of Physical Chemistry A* **110**(8): 2682-2689, 2006.
138. Hu Q, Cooks RG and Noll RJ, Phase-Enhanced Selective Ion Ejection in an Orbitrap MS. *J Amer Soc Mass Spectrom* **18**: 980-983, 2007.
139. Thevis M, Makarov AA, Horning S and Schaenzer W, MS of stanozolol and its analogues by ESI-CAD with quadrupole-linear ion trap and linear ion trap-orbitrap hybrid mass analyzers. *Rapid Commun Mass Spectrom* **19**: 3369-3378, 2005.
140. Yates JR, Cociorva D, Liao L and Zabrouskov V, Performance of a Mass Analyzer with Orbital Trapping for Peptide Analysis. *Analytical Chemistry* **78**(2): 493-500, 2006.
141. Peterman SM, Duczak N, Kalgutkar AS, Lame ME and Soglia JR, Application of a Linear Ion Trap/Orbitrap in Metabolite Characterization Using Data-Dependent Accurate Mass Measurements. *J Amer Soc Mass Spectrom* **17**: 363-375, 2006.
142. Scigelova M and Makarov A, Orbitrap mass analyzer - overview and applications in proteomics. *Practical Proteomics* **1**(1-2): 16-21, 2006.

143. Thevis M, Krug O and Schaenzer W, Mass spectrometric characterization of efaproxiral (RSR13) and its implementation into doping controls using LC-API-MS/MS. *Journal of Mass Spectrometry* **41**: 332-338, 2006.

144. Venable JD, Wohlschlegel J, McClatchy DB, Park SK and Yates JR, III, Relative Quantification of Stable Isotope Labeled Peptides Using a Linear Ion Trap-Orbitrap Hybrid MS. *Anal Chem* **79**: 3056-3064, 2007.

145. Adachi J, Kumar C, Zhang Y, Olsen JV and Mann M, Human urinary proteome contains > 1500 proteins, with large proportion of membrane proteins. *GenomeBiology* **7**(9): No pp given, 2006.

146. Eiceman GA, Young D and Smith GB, Mobility spectrometry of amino acids and peptides with MALDI in air at ambient pressure. *Microchemical Journal* **81**: 108-116, 2005.

147. Miller PE and Denton MB, Operating Concepts of the Quadrupole MS. *J. Chem. Educ.* **63**: 617-622, 1986.

148. Dawson PH, Quadrupole Mass Analyzers. *Mass Spectrom. Rev.* **5**: 1-37, 1986.

149. Konenkov N, Londry F, Ding C and Douglas DJ, Linear Quadrupoles with Added Hexapole Fields. *J Amer Soc Mass Spectrom* **17**: 1063-1073, 2006.

150. Capron L, Rolando C and Sablier M, No Effect of Ion Structure on m/z Axis of Quadrupole MS. *Rapid Commun. Mass Spectrom.* **8**: 991-995, 1994.

151. Dawson PH and Yu B, 2nd Stability Region of the Quadrupole. *Int. J. Mass Spectrum. Ion Processes* **56**: 25-50, 1984.

152. Fite WL, Separation of fringe fields. *Rev. Sci. Instrum.* **47**: 326-330, 1976.

153. Wien WC, Die elektrostatische und magnetische Ablenkung der Kanalstrahlem. *Berlin Physikalische Gesellschaft Verhandlungen* **17**: 10-12, 1898.

154. Wien WC, Untersuchuugen über die elektrische Entladung in verdünnten Gasen. *Ann. Phys. Leipzig* **8**: 224-266, 1902.

155. Bartky W and Dempster AJ, Paths of Charged Particles in Electric and Magnetic Fields. *Phys. Rev.* **33**: 1019, 1929.

156. Trainor JR and Derrick PJ, Sectors and Tandem Sectors. In: *Mass Spectrometry in the Biological Sciences: A Tutorial* (Ed. Gross ML), pp. 3-27. kluwer Academic Publishers, 1992.

157. De Laeter J and Kurz MD, Alfred Nier and the sector field mass spectrometer. *J Mass Spectrom* **41**: 847-854, 2006.

158. Marshall AG, Lin Wang TC and Lebatuan Ricca T, ICR-MS. *Chem. Phys. Lett.* **105**: 233-236, 1984.

159. Marshall AG, Ion cyclotron resonance mass spectrometry: a brief history. *Actualite Chimique* (1): 18-22, 2001.

160. Marshall AG, Milestones in Fourier transform ion cyclotron resonance mass spectrometry technique development. *International Journal of Mass Spectrometry* **200**(1/3): 331-356, 2000.

161. Amster IJ, Fourier transform mass spectrometry. *Journal of Mass Spectrometry* **31**(12): 1325-1337, 1996.

162. Marshall AG and Hendrickson CL, FTICR Detection: Principles and Experimental Configurations. *International J Mass Spectrom* **215**: 59-75, 2002.

163. Jacoby CB, Holliman CL and Gross ML, FT-MS. In: *Mass Spectrometry in the Biological Sciences: A Tutorial* (Ed. Gross ML), pp. 93-116. Kluwer Academic Publishers, Netherlands, 1992.

164. Guan S, Kim HS, Marshall AG, Wahl MC, Wood TD and Xiang X, Shrink-wrapping an ion cloud for high-performance FT ICR MS. *Chemical Reviews (Washington, DC, United States)* **94**: 2161-82, 1994.

165. Easterling ML, Mize TH and Amster IJ, Routine part-per-million mass accuracy for high-mass ions: space-charge effects in MALDI FT-ICR. *Analytical Chemistry* **71**(3): 624-632, 1999.

166. Chan TW, Duan L and Sze TP, Accurate mass measurements for peptide and protein mixtures by using MALDI Fourier transform mass spectrometry. *Anal Chem* **74**: 5282-9., 2002.

167. Marshall AG, Accurate mass measurement: taking full advantage of nature's isotopic complexity. *Physica B: Condensed Matter (Amsterdam, Netherlands)* **346-347**: 503-508, 2004.

168. McLafferty FW, High Resolution FTMS at High Mass. *Acc. Chem. Res.* **27**: 379-386, 1994.

169. Fellgett P, A propos de la theorie du spectrometre interferentiel multiplex. *J. Phys. Radium* **19**: 187-191, 1958.

170. Bruce JE, Van Orden SL, Anderson GA, Hofstodler SA, Sherman MG, Rockwood AL and Smith RD, Selected Ion Accumulation in FTMS. *J. Mass Spectrom.* **30**: 124-133, 1995.

171. Vartanian VH and Laude DA, Optimization of Ion Cell for FT-ICR-MS. *Int. J. Mass Spectrom. Ion Proc.* **141**: 189-200, 1995.

172. Hendrickson CL, Drader JJ, Laude DA, Jr., Guan S and Marshall AG, FT ICR MS in a 20 T resistive magnet. *Rapid Communications in Mass Spectrometry* **10**: 1829-1832, 1996.

173. Beu SC, Blakney GT, Quinn JP, Hendrickson CL and Marshall AG, Broadband Phase Correction of FT-ICR Mass Spectra via Simultaneous Excitation and Detection. *Analytical Chemistry* **76**: 5756-5761, 2004.

174. Watson CH, Barshick CM, Wronka J, Laukien FH and Eyler JR, Pulsed-Gas Glow Discharge for FT-ICR-MS. *Anal. Chem.* **68**: 573-575, 1996.

175. Schaub TM, Hendrickson CL, Quinn JP, Rodgers RP and Marshall AG, Instrumentation and Method for Ultrahigh Resolution FD FT-ICR-MS of Nonpolar Species. *Analytical Chemistry* **77**: 1317-1324, 2005.

176. Reinhardt A, Emmenegger C, Gerrits B, Panse C, Dommen J, Baltensperger U, Zenobi R and Kalberer M, Ultrahigh Mass Resolution and Accurate Mass Measurements as a Tool To Characterize Oligomers in Secondary Organic Aerosols. *Anal Chem* **79**: 4074-4082, 2007.

177. Fu J, Purcell JM, Quinn JP, Schaub TM, Hendrickson CL, Rodgers RP and Marshall AG, External EI 7T FT ICR MS for resolution and ID of organics. *Rev Scientific Instruments* **77**: 025102/1-025102/9, 2006.

178. Huang Y, Li G-Z, Guan S and Marshall AG, A combined linear ion trap for mass spectrometry. *J Amer Soc Mass Spectrom* **8**: 962-969, 1997.

179. Anderson JS, Vartanian VH and Laude DA, Jr., Ion cells in FT-ICR-MS. *Trends Anal. Chem.* **13**: 234-239, 1994.

180. Bruce JE, Anderson GA and Smith RD, "Colored" Noise Waveforms and Quadrupole Excitation for the Dynamic Range Expansion of FT-ICR-MS. *Anal. Chem.* **68**: 534-541, 1996.

181. Rodgers RP, Blumer EN, Freitas MA and Marshall AG, Compositional analysis for identification of arson accelerants by EI FT-ICR high-resolution MS. *Journal of Forensic Sciences* **46**: 268-279, 2001.

182. Mize TH and Amster IJ, Broad-band ion accumulation with an internal source MALDI-FTICR-MS. *Analytical Chemistry* **72**(24): 5886-5891, 2000.

183. McFarland MA, Chalmers MJ, Quinn JP, Hendrickson CL and Marshall AG, Evaluation and Optimization of ECD Efficiency in FT ICR MS. *J Amer Soc Mass Spectrom* **16**: 1060-1066, 2005.

184. Cooper HJ, Hakansson K and Marshall AG, The role of electron capture dissociation in biomolecular analysis. *Mass Spectrometry Reviews* **24**(2): 201-222, 2005.

185. Marshall AG and Rodgers RP, Petroleomics: The Next Grand Challenge for Chemical Analysis. *Accounts of Chemical Research* **37**(1): 53-59, 2004.

186. Rodgers RP, Schaub TM and Marshall AG, Petroleomics: MS returns to its roots. *Analytical Chemistry* **77**(1): 20A-27A, 2005.

187. Cooper HJ and Marshall AG, Electrospray Ionization Fourier Transform Mass Spectrometric Analysis of Wine. *Journal of Agricultural and Food Chemistry* **49**(12): 5710-5718, 2001.

188. Steiner WE, Clowers BH, Fuhrer K, Gonin M, Matz LM, Siems WF, Schultz AJ and Hill HH, Jr., Electrospray ionization with ambient pressure ion mobility separation and mass analysis by orthogonal time-of-flight mass spectrometry. *Rapid Commun Mass Spectrom* **15**(23): 2221-6., 2001.

189. Eiceman GA and Karpas Z, *Ion Mobility Spectrometry, 2nd Edition*, 2004.

190. Creaser CS, Griffiths JR, Bramwell CJ, Noreen S, Hill CA and Thomas CLP, Ion mobility spectrometry: a review. Part 1. Structural analysis by mobility measurement. *Analyst (Cambridge, United Kingdom)* **129**(11): 984-994, 2004.

191. Khayamian T and Jafari MT, Design for ESI-ion mobility spectrometry. *Anal Chem* **79**: 3199-3205, 2007.

192. Koeniger SL, Merenbloom SI, Valentine SJ, Jarrold MF, Udseth HR, Smith RD and Clemmer DE, An IMS-IMS Analogue of MS-MS. *Analytical Chemistry* **78**(12): 4161-4174, 2006.

193. Tang X, Bruce JE and Hill HH, Jr., Design and performance of an atmospheric pressure ion mobility FT-ICR-MS. *Rapid Commun Mass Spectrom* **21**: 1115-1122, 2007.

194. Levin DS, Miller RA, Nazarov EG and Vouros P, Rapid Separation and Quantitative Analysis of Peptides Using a New NanoESI-Differential Mobility Spectrometer-MS System. *Anal Chem* **78**: 5443-5452, 2006.

195. Kaur-Atwal G, Weston DJ, Green PS, Crosland S, Bonner PLR and Creaser CS, Analysis of tryptic peptides using DESI combined with ion mobility spectrometry/MS. *Rapid Commun Mass Spectrom* **21**: 1131-1138, 2007.

196. Pringle SD, Giles K, Wildgoose JL, Williams JP, Slade SE, Thalassinos K, Bateman RH, Bowers MT and Scrivens JH, An investigation of the mobility separation of some peptide and protein ions using a new hybrid quadrupole/travelling wave IMS/oa-ToF instrument. *Int J Mass Spectrom* **261**: 1-12, 2007.

197. Clowers BH, Dwivedi P, Steiner WE, Hill HH and Bendiak B, Separation of Sodiated Isobaric Disaccharides & Trisaccharides by ESI-AP-IMS-TOFMS. *J Amer Soc Mass Spectrom* **16**: 660-669, 2005.

198. Levin DS, Vouros P, Miller RA and Nazarov EG, Using a Nanoelectrospray-Differential Mobility Spectrometer-MS for Analysis of Oligosaccharides with Solvent Selected Control Over ESI Aggregate Ion Formation. *J Amer Soc Mass Spectrom* **18**: 502-511, 2007.

199. Eckers C, Laures AMF, Giles K, Major H and Pringle S, Ion mobility separation in combination with HPLC/MS to facilitate detection of trace impurities in formulated drugs. *Rapid Commun Mass Spectrom* **21**: 1255-1263, 2007.

200. Budimir N, Weston DJ and Creaser CS, Analysis of pharmaceutical formulations using atmospheric pressure ion mobility spectrometry combined with LC and nano-ESI. *Analyst* **132**: 34-40, 2007.

201. Bollan HR, Stone JA, Brokenshire JL, Rodriguez JE and Eiceman GA, Mobility Resolution and Mass Analysis of Ions from Ammonia and Hydrazine Complexes with Ketones Formed in Air at Ambient Pressure. *Journal of the American Society for Mass Spectrometry* **18**(5): 940-951, 2007.

202. Shvartsburg AA, Tang K and Smith RD, Optimization of the design and operation of FAIMS analyzers. *J Am Soc Mass Spectrom* **16**(1): 2-12, 2005.

203. Asbury GR and Hill HH, Jr., Using different drift gases to change separation factors (alpha) in ion mobility spectrometry. *Anal Chem* **72**(3): 580-4., 2000.

204. Laiko VV, Orthogonal Extraction Ion Mobility Spectrometry. *J Amer Soc Mass Spectrom* **17**: 500-507, 2006.

205. Guevremont R, High-Field Asymmetric Waveform Ion Mobility Spectrometry (FAIMS). *Canadian Journal of Analytical Sciences and Spectroscopy* **49**(3): 105-113, 2004.

206. Guevremont R, High-field asymmetric waveform ion mobility spectrometry: a new tool for mass spectrometry. *Journal of Chromatography, A* **1058**(1-2): 3-19, 2004.
207. Krylov E, Nazarov EG, Miller RA, Tadjikov B and Eiceman GA, Field Dependence of Mobilities for Gas-Phase-Protonated Monomers and Proton-Bound Dimers of Ketones by Planar Field Asymmetric Waveform Ion Mobility Spectrometer (PFAIMS). *Journal of Physical Chemistry A* **106**(22): 5437-5444, 2002.
208. Gidden J, Bushnell JE and Bowers MT, Gas-phase conformations and folding energetics of oligonucleotides: dTG- and dGT. *J Am Chem Soc* **123**(23): 5610-1., 2001.
209. Kinnear BS, Hartings MR and Jarrold MF, Helix unfolding in unsolvated peptides. *J Am Chem Soc* **123**(24): 5660-7., 2001.
210. Wu C, Siems WF, Klasmeier J and Hill HH, Jr., Separation of isomeric peptides using electrospray ionization/high-resolution ion mobility spectrometry. *Anal Chem* **72**(2): 391-5., 2000.
211. Wyttenbach T, Kemper PR and Bowers MT, Design of new ESI Ion Mobility Mass Spectrometer. *Int. J. Mass Spectrom. Ion Proc.* **212**: 13-24, 2001.
212. Henderson SC, Valentine SJ, Counterman AE and Clemmer DE, ESI/ion trap/ion mobility/TOF MS for rapid and sensitive analysis of biomolecular mixtures. *Anal Chem* **71**: 291-301., 1999.
213. Matz LM and Hill HH, Jr., Evaluating the separation of amphetamines by electrospray ionization ion mobility spectrometry/MS and charge competition within the ESI process. *Anal Chem* **74**(2): 420-7., 2002.
214. Beegle LW, Kanik I, Matz L and Hill HH, Jr., Electrospray ionization high-resolution ion mobility spectrometry for the detection of organic compounds, 1. Amino acids. *Anal Chem* **73**(13): 3028-34., 2001.
215. Asbury GR and Hill HH, Jr., Separation of amino acids by ion mobility spectrometry. *J Chromatogr A* **902**(2): 433-7., 2000.
216. Matz LM and Hill HH, Jr., Evaluation of opiate separation by high-resolution electrospray ionization-ion mobility spectrometry/mass spectrometry. *Anal Chem* **73**(8): 1664-9., 2001.
217. Wu C, Siems WF and Hill HH, Jr., Secondary electrospray ionization ion mobility spectrometry/mass spectrometry of illicit drugs. *Anal Chem* **72**(2): 396-403., 2000.
218. Belov ME, Buschbach MA, Prior DC, Tang K and Smith RD, Multiplexed Ion Mobility Spectrometry-Orthogonal TOFMS. *Anal Chem* **79**: 2451-2462, 2007.
219. Lee YJ, Hoaglund-Hyzer CS, Taraszka JA, Zientara GA, Counterman AE and Clemmer DE, Collision-induced dissociation of mobility-separated ions using an orifice-skimmer cone at the back of a drift tube. *Anal Chem* **73**(15): 3549-55., 2001.
220. Hoaglund-Hyzer CS and Clemmer DE, Ion trap/ion mobility/quadrupole/time-of-flight mass spectrometry for peptide mixture analysis. *Anal Chem* **73**(2): 177-84., 2001.
221. Badman ER, Hoaglund-Hyzer CS and Clemmer DE, Monitoring structural changes of proteins in an ion trap over approx 10-200 ms: unfolding transitions in cytochrome c ions. *Anal Chem* **73**: 6000-7., 2001.
222. Srebalus CA and Clemmer DE, Assessment of purity and screening of peptide libraries by nested ion mobility-TOFMS: identification of RNase S-protein binders. *Anal Chem* **73**(3): 424-33., 2001.
223. Srebalus B, Hilderbrand AE, Valentine SJ and Clemmer DE, Resolving isomeric peptide mixtures: a combined HPLC/ion mobility-TOFMS analysis of a 4000-component combinatorial library. *Anal Chem* **74**(1): 26-36., 2002.
224. Matz LM, Dion HM and Hill HH, Jr., Evaluation of capillary liquid chromatography-electrospray ionization ion mobility spectrometry with mass spectrometry detection. *J Chromatogr A* **946**(1-2): 59-68., 2002.
225. Borsdorf H, Stone JA and Eiceman GA, Gas phase studies on terpenes by ion mobility spectrometry using different APCI techniques. *International J Mass Spectrom* **246**: 19-28, 2005.
226. Troost JR and Olavesen EY, GC-MS Calibration Bias. *Anal. Chem.* **68**: 708-711, 1996.
227. Green FM, Gilmore IS and Seah MP, TOF-SIMS: Accurate Mass Scale Calibration. *Journal of the American Society for Mass Spectrometry* **17**(4): 514-523, 2006.
228. Eichelberger JW, Harris LE and Budde WL, Reference Compound to Calibrate Ion Abundance in GC-MS. *Anal. Chem.* **47**: 995-1000, 1975.
229. Chu TZ and Finnigan RE, Thirty Years of a Mass Spectrometry Company. *Amer. Lab.* **October 1998:** 90s-109s, 1998.
230. Koppenaal DW, Barinaga CJ, Denton MB, Sperline RP, Hieftje GM, Schilling GD, Andrade FJ and Barnes JHIV, MS detectors. *Anal Chem* **77**: 418A-427A, 2005.
231. Nygren U, Ramebaeck H, Berglund M and Baxter DC, The importance of a correct dead time setting in isotope ratio MS. *International J Mass Spectrom* **257**: 12-15, 2006.
232. Harris FM, Trott GW, Morgan TG, Brenton AG, Kingston EE and Beynon JH, S/N in Electron multipliers at low ion currents. *Mass Spectrom. Rev.* **3**: 209-229, 1984.
233. McKinney CR, McCrea JM, Epstein S, H.A. A and Urey HC, Measurements of isotope abundance ratios. *Rev. Sci. Instrum.* **21**: 724-730, 1950.
234. Schoeller DA and Hayes JM, Ion Counting Isotope Ratio MS. *Anal. Chem.* **47**: 408-415, 1975.
235. Meili J, Walls FC, McPherron R and Burlingame AL, Design, implementation, and performance of a GC/high resolution MS data system for complex mixtures. *J Chrom Sci* **17**: 29-42, 1979.
236. Evans S, "Detectors" In: Methods in Enzymology: Mass Spectrometry. *J.A. McCloskey, ed.* **193**: 61-68, 1990.

237. Geno PW, *Ion Detection in MS. In: Mass Spectrometry in the Biological Sciences: A Tutorial.* Kluwer Academic Publ., Netherlands, 1992.
238. Fuerstenau SD and Benner WH, Molecular weight determination of megadalton DNA electrospray ions using charge detection TOF MS. *Rapid Commun Mass Spectrom* **9**: 1528-38, 1995.
239. Imrie DC, Pentney JM and Cottrell JS, A Faraday cup detector for high-mass ions in MALDI TOF MS. *Rapid Commun Mass Spectrom* **9**: 1293-6, 1995.
240. Bahr U, Röhling U, Lautz C, Strupat K, Schürenberg M and Hillenkamp F, A Charge Detector for TOF-MS of High-Mass Ions Produced by MALDI. *Int. J. Mass Spectrom. Ion Processes* **153**: 9-22, 1996.
241. Westphall MS, Scalf M and Smith LM, High-Speed Faraday Cup Detector for TOF MS. In: *54th Annual Conference of the American Society for Mass Spectrometry, Seattle,2006*, pp. MP 189.
242. Westphall MS and Smith LM, Personal Communication. Dept of Chemistry, University of Wisconsin, 2006.
243. Knight AK, Sperline RP, Hieftje GM, Young E, Barinaga CJ, Koppenaal DW and Denton MB, Development of a micro-Faraday array for ion detection. *Int J Mass Spectrom* **215**: 131-139, 2002.
244. Scheidemann AA, Darling RB, Schumacher FJ and Isakharov A, Faraday cup detector array with electronic multiplexing for multichannel mass spectrometry. *Journal of Vacuum Science & Technology, A: Vacuum, Surfaces, and Films* **20**(3): 597-604, 2002.
245. Chen X, Westphall MS and Smith LM, MS of DNA mixtures: Instrumental effects responsible for decreased sensitivity with increasing mass. *Anal Chem* **75**: 5944-5952, 2003.
246. Fies W, Jr., Measuring the Gain of an Electron Multiplier. *Int. J. Mass Spectrom. Ion Proc.* **82**: 111-129, 1988.
247. Westmacott G, Frank M, Labov SE and Benner WH, Using a superconducting tunnel junction detector to measure the secondary electron emission efficiency for a microchannel plate detector bombarded by large molecular ions. *Rapid Communications in Mass Spectrometry* **14**(19): 1854-1861, 2000.
248. Goodrich G and Wiley W, Continous Channel Electron Multiplier. *Rev. Sci. Instrum.* **33**: 761-762, 1962.
249. Holmes JL and Szulejko JE, Energy discrimination in ion detection systems: the performance of conversion dynodes. *Organic Mass Spectrometry* **18**(6): 273, 1983.
250. Beuhler RJ and Friedman L, Low noise, high voltage secondary emission ion detector for polyatomic ions. *International Journal of Mass Spectrometry and Ion Physics* **23**(2): 81-97, 1977.
251. Dougherty RC, Dalton J and Biros FJ, Negative chemical ionization mass spectra of polycyclic chlorinated insecticides. *Organic Mass Spectrometry* **6**(11): 1171-81, 1972.
252. Stafford GC, Reecher JR, Smith RD and Story MS, Neg Ion Detection. *Dyn. Mass Spectrom.* **5**: 55-57, 1978.
253. Hunt DF, Stafford GC, Jr., Crow FW and Russell JW, Pulsed Positive Negative Ion Chemical Detection. *Anal. Chem.* **48**: 2098-2105, 1976.
254. Beuhler RJ, Chu YY, Friedlander G, Friedman L, Alessi JG, LoDestro V and Thomas JP, Cluster-impact fusion: time-of-flight experiments. *Physical Review Letters* **67**(4): 473-6, 1991.
255. Matthew MW, Beuhler RJ, Ledbetter M and Friedman L, Large, energetic impacts on surfaces. *Nuclear Instruments & Methods in Physics Research, Section B: Beam interactions with Materials and Atoms* **B14**(4-6): 448-60, 1986.
256. Xu Y, Bae YK, Beuhler RJ and Friedman L, Secondary electron analysis of polymeric ions generated by an electrospray ion source. *Journal of Physical Chemistry* **97**(46): 11883-6, 1993.
257. Hedin A, Haakansson P and Sundqvist BUR, On the detection of large organic ions by secondary electron production. *International Journal of Mass Spectrometry and Ion Processes* **75**(3): 275-89, 1987.
258. Olthoff JK, Lys IA and Cotter RJ, A pulsed time-of-flight mass spectrometer for liquid secondary ion mass spectrometry. *Rapid Communications in Mass Spectrometry* **2**(9): 171-5, 1988.
259. Dubois F, Knochenmuss R and Zenobi R, An ion-to-photon conversion detector for mass spectrometry. *Int. J. Mass Spectrom. Ion Processes* **169/170**: 89-98, 1997.
260. Dubois F, Knochenmuss R, Zenobi R, Brunelle A, Deprun C and Le Beyec Y, A comparison between ion-to-photon and microchannel plate detectors. *Rapid Commun. Mass Spectrom.* **13**(9): 786-791, 1999.
261. Dubois F, Knochenmuss R and Zenobi R, Optimization of an ion-to-photon detector for large molecules in mass spectrometry. *Rapid Commun. Mass Spectrom.* **13**(19): 1958-1967, 1999.
262. Peng W, Cai Y and Chang H, Optical detection for MS of macroions. *Mass Spectrom Rev* **23**: 443-465, 2004.
263. Price D and Milnes GJ, Recent developments in techniques utilizing TOF MS. *International Journal of Mass Spectrometry and Ion Processes* **60**: 61-81, 1984.
264. Giffen CE, Boettger HG and Norris DD, EOID. *Int. J. Mass Spectrom. Ion Phys.* **15**: 437-449, 1974.
265. Sweedler JV and Ratzlaff KL, *Charge Transfer Devices in Spectroscopy.* VCH, NY, 1994.
266. Barnes JH, Schilling GD, Sperline R, Denton MB, Young ET, Barinaga CJ, Koppenaal DW and Hieftje GM, Focal Plane Camera for a Mattauch-Herzog Geometry Mass Spectrograph for ICP. *Anal Chem* **76**: 2531-2536, 2004.
267. Hedfjall B and Ryhage R, EOID for GC-MS. *Anal. Chem* **53**: 1641-1644, 1981.
268. Cottrell J and Evans S, Multichannel Electrooptical Detection in FAB-MS. *Anal. Chem.* **59**: 1990-1995, 1987.

269. Burlingame AL, *Array Detection in Sector MS. In: Biological Mass Spectrometry: Present and Future.* Wiley & Sons, NY, 1994.
270. Hill JA, Biller JE and Biemann K, A variable dispersion array detector for a tandem mass spectrometer. *International Journal of Mass Spectrometry and Ion Processes* **111:** 1-25, 1991.
271. Barnes JH and Hieftje GM, Detector-array technology for MS. *Int. J. Mass Spectrom.* **238:** 33-46, 2004.
272. Schilling GD, Andrade FJ, Barnes JH, Sperline RP, Denton MB, Barinaga CJ, Koppenaal DW and Hieftje GM, 2nd-Generation Focal-Plane Camera Coupled to an ICP Mattauch-Herzog Geometry MS. *Anal Chem* **78:** 4319-4325, 2006.
273. Sperline RP, Knight AK, Gresham CA, Koppenaal DW, Hieftje GM and Denton MB, Read-noise characterization of focal plane array detectors via mean-variance analysis. *Applied Spectroscopy* **59**(11)**:** 1315-1323, 2005.
274. Daly NR, McCormick A and Powell RE, Detector for the metastable ions observed in the mass spectra of organic compounds. *Review of Scientific Instruments* **39**(8)**:** 1163-7, 1968.
275. Daly NR, McCormick A and Powell RE, Metastable spectra of cis- and trans-butenes. *Organic Mass Spectrometry* **1**(1)**:** 167-8, 1968.
276. Frank M, Labov SE, Westmacott G and Benner WH, Energy-sensitive cryogenic detectors for high-mass biomolecule mass spectrometry. *Mass spectrometry reviews* **18**(3-4)**:** 155-86, 1999.
277. Frank M, Mears CA, Labov SE, Benner WH, Horn D, Jaklevic JM and Barfknecht AT, High-efficiency detection of 66,000-Da protein molecules using a cryogenic detector in a MALDI TOFMS. *Rapid Commun Mass Spectrom* **10:** 1946-1950, 1996.
278. Sommer H, Thomas HA and Hipple JA, *Phys. Rev.* **76:** 1877, 1949.
279. Ameri PV, Krouse HR and Fichtner H, *Rev. Sci. Instrum.* **72:** 2036–2042, 2001.
280. O'Hanlon JF, *A User's Guide to Vacuum Technology. 2nd Ed.* Wiley Interscience, New York, 1989.
281. Milonni PW and Shih M-L, The Quantum Vacuum. 1993.
282. Mosbacher CJ, Vacuum technology. *Res./Devel* **25:** 41-42, 1974.
283. Grimsrud E, Vacuum for High Pressure MS. *Anal. Chem.* **50:** 382-384, 1978.
284. Wilfert S and Edelmann C, Miniaturized vacuum gauges. *Journal of Vacuum Science & Technology, A: Vacuum, Surfaces, and Films* **22**(2)**:** 309-320, 2004.
285. Bayard RT and Alpert DA, Extension of the Low Pressure Range of the Ionization Gauge. *Rev. Sci. Instrum.* **21:** 571-572, 1950.
286. Kuo YH and Kanaji T, Measurement of uhv and xhv by hot cathode ionization gauge with higher sensitivity. *Vacuum* **44**(5-7)**:** 555-7, 1993.
287. Edelmann C and Wilfert S, Trends and problems in the development of hot filament ionisation gauges for total pressure measurements in the vacuum range. *Recent Research Developments in Vacuum Science & Technology* **3:** 271-290, 2001.
288. Penning FM, Ein neuer manometer für niedrige gasdruckes ins besondere zwischen 10-3 und 10-5 mm. *Physica* **4:** 71-75, 1937.
289. Jitschin W, Accuracy of vacuum gauges. *Journal of Vacuum Science & Technology, A: Vacuum, Surfaces, and Films* **8**(2)**:** 948-56, 1990.

"The mass spectrometer behaves at times in the most capricious and unaccountable manner...When by good fortune all is well, the arrangement is capable of good performance. Thus, after a favorable setting of the apparatus, six elements were successfully analyzed in as many working days. On the other hand, after dismantling became imperative, and it had to be cleaned and rebuilt, exactly as before as far as one could tell, no results of any value could be obtained during weeks of work."

~Francis William Aston

Chapter 3　Mass Spectrometry/Mass Spectrometry

Introduction to Mass Spectrometry, 4th Edition: Instrumentation, Applications, and Strategies for Data Interpretation; J.T. Watson and O.D. Sparkman, © 2007, John Wiley & Sons, Ltd

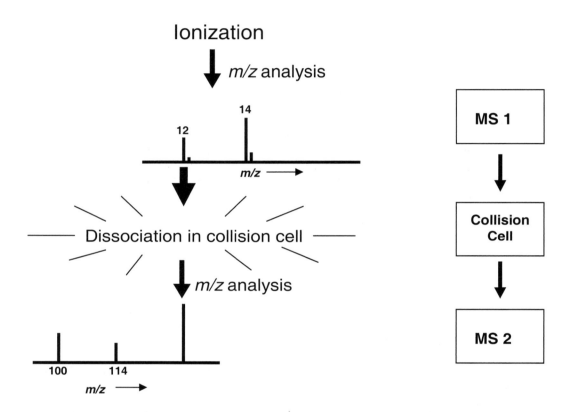

Figure 3-1. *Conceptual representation of the technique of MS/MS. In this case, ions of m/z 129 are selected to be precursor ions by the first mass spectrometer (MS1); these ions are directed into a collision chamber, and the decomposition products are analyzed by the second mass spectrometer (MS2) to produce the product-ion spectrum of m/z 129 (bottom).*

I. Introduction

1. History and the Evolution of the Technique

The first use of the acronym MS/MS was coined for "mass spectrometry/mass spectrometry" by William F. Haddon in a symposium organized by Michael L. Gross, presented at the University of Nebraska, Lincoln, Nebraska, November 3–5, 1976 (Haddon 1978). Haddon used the term to describe the technique of bringing about the decomposition of a stable ion by forcing it to collide with neutral gas atoms or molecules. The acronym was later defined by F. W. McLafferty and F. M. Bockhoff as "mass separation/mass spectral characterization" (MS/MS) by analogy to gas chromatography/mass spectrometry (McLafferty 1978); however, mass spectrometry/mass spectrometry is the definition of MS/MS that is used today. A synonym for MS/MS is *tandem mass spectrometry*. These two terms are used interchangeably throughout this book and throughout the scientific literature.

The use of metastable decay to determine structure goes back to 1945 when Verron H. Dibeler and Edward U. Condon published a paper entitled "Detection of Metastable Ions with the Mass Spectrometer" while both were at Westinghouse Research Laboratories, East Pittsburgh, Pennsylvania [1]. This was followed almost a year later by "Metastable Ions Formed by Electron Impact in Hydrocarbon Gases" [2]. Early in the 1960s the development of the acceleration-voltage scan by Michael Barber and R. M. Elliott [3] and the modification of time-of-flight mass spectrometers to use retarding grids and drift spaces to separate stable and metastable ions led to an increased interest in the subject [4, 5]. The next step in the evolution of MS/MS was the discovery that a collision gas present in various field-free regions of the mass spectrometer increased the intensity and number of metastable-ion peaks observed in the mass spectrometer [6, 7]. According to the book *Mass Spectrometry/Mass Spectrometry: Techniques and Applications of Tandem Mass Spectrometry,* it was not until publication of the book *Metastable Ions* [8] in 1973 that the field of MS/MS really took off. Probably the single event that caused MS/MS to become widely recognized in the late 1970s (ca. 1977–1978) with was the development of the triple-quadrupole mass spectrometer (described in detail later in this chapter) by Richard A. Yost and Christie G. Enke at Michigan State University (East Lansing, Michigan, USA) based on the earlier work of Donald McGilvery and James D. Morrison at La Trobe University (Melbourne, Australia). By the 1980s, commercial triple-quadrupole mass spectrometers were available from Finnigan Corp. (now ThermoFisher Scientific), Nermag, Extranuclear Laboratories, Inc., and Sciex. The three major magnetic-sector instrument manufacturers of the time (VG Micromass, Ltd, Kratos Scientific Instruments, Inc., and JEOL, Ltd) all produced different types of instruments for tandem mass spectrometry. VG had a reverse-geometry double-focusing instrument, the Micromass ZAB-2F; Kratos tried to get into the market with a one-of-a-kind instrument built for Michael Gross at the University of Nebraska, which was an MS50TA, a forward-geometry double-focusing instrument with a second electric sector following the magnetic sector; and JEOL built a four-sector instrument that was a reverse-geometry instrument in tandem with a forward-geometry instrument. However, the triple quadrupole quickly became the tool of choice until *tandem-in-time* was commercially introduced first by Varian in the Saturn quadrupole ion trap GC-MS (a 3D QIT), ca. 1992, followed soon after by the Finnigan external ionization 3D QIT GC-MS (the Polaris Q) and later by an LC-MS. The instruments that exploded into use in the late 1990s and continue today are the tandem transmission quadrupole-TOF instrument and the TOF-TOF instrument; more descriptions of all these technologies appear later in this chapter.

2. Concept and Definitions

Mass spectrometry/mass spectrometry [9–12] can be thought of as a means by which to obtain the mass spectrum of a mass spectrum. This is possible provided there is some means of causing ions in the primary mass spectrum to fragment further (or differently) so that the secondary mass spectrum will reveal new information. The most common device used to promote ion dissociation is a collision cell located between the first (MS1) and second (MS2) *m/z* analyzer. The conceptual representation of MS/MS (a.k.a. tandem mass spectrometry) in Figure 3-1 was developed in the context of an EI example in which a molecular ion (not represented by a peak in Figure 3-1) has spontaneously fragmented to produce ion current represented by the peaks at *m/z* 141 and *m/z* 129 in the primary mass spectrum at the top. In this contrived example, it is of interest to determine the extent to which ions of *m/z* 129 can be made to fragment further. This is done by using MS1 to select ion current at *m/z* 129 (precursor ions) for entry into the collision cell for collisionally activated dissociation (CAD).[1] The products of the dissociation process, which characterize the precursor ion, are then analyzed according to *m/z* values (a peak may still appear at *m/z* 129 because some of the ions can survive the CAD process). Because of the requirement for two *m/z*-selective processes as illustrated at the top and the bottom of Figure 3-1, MS/MS or *tandem mass spectrometry* is an obvious name for the technology.

There are other dissociative processes that can be used in conjunction with MS/MS, such as surface-induced dissociation [13–21] or photo-induced dissociation [22–28]. However, because CAD is by far the most commonly used, it will be covered exclusively in this chapter. In this book, the term MS/MS will be assumed to be the technique where dissociation comes about as a result of collisional activation. If ion dissociation results from some other type of dissociation, the process will be specified.

From a qualitative point of view, tandem mass spectrometry is used for two basic purposes: (1) identifying compounds relative to standards through pattern recognition [29–38] and (2) mapping fragmentation pathways [39–43] that lead to a given mass spectrum. Identification of targeted compounds by pattern recognition in product-ion mass spectra is common practice. Automated interpretation of product-ion mass spectra of peptides is the basis for amino acid sequencing and for identification of proteins in proteomic studies. Examples of such analyses following ionization by EI or ESI or other desorption/ionization techniques are described in later sections of this chapter as well as in Chapter 12 on biopolymers.

Many of the fragmentation pathways presented in Chapter 6 leading to EI mass spectra were validated by MS/MS. Such validation is often done for EI spectra in which fragment ions represented by a given peak are thought to undergo a second stage of decomposition leading to secondary ions represented by a peak at a lower *m/z* value in the mass spectrum. For example, consider the mass spectrum of *n*-propyl-1-propanamine (see Figure 6-65, EI mass spectrum of dipropylamine). The peak at *m/z* 30 represents fragment ions that arise by secondary decomposition of fragment ions with *m/z* 72. This pathway can be verified by introducing *n*-propyl-1-propanamine (nominal mass = 101 Da) into the ion source, and only allowing ions of *m/z* 72 to pass through MS1 for entry into the collision cell. The experiment is completed by adjusting MS2 so that it can transmit only ion current of *m/z* 30; a signal

---

[1] In MS/MS, the selected ion becomes energized (activated) by a collision. Some of these ions (precursor ions) then dissociate to form fragment ions (product ions) of the collisionally activated precursor ions. This process has also been called collision-induced dissociation (CID). Use of either term is acceptable. The authors use CAD because this term is more descriptive of the process.

detected under these conditions proves that ions of *m/z* 72 are the precursor of ions of *m/z* 30.

MS/MS is the only real way to obtain structural information from ions that represent the intact molecule produced by soft ionization techniques such as ESI and APCI. In this respect, MS/MS is an invaluable tool for qualitative analyses.

Probably the most common use of MS/MS is in quantitative applications using selected reaction monitoring (SRM). SRM involves monitoring the ion current associated with the transition of a given precursor ion to a particular product ion during collisionally activated dissociation of the precursor ion; the transition is called a selected reaction. The *m/z* transition established by MS1 and MS2 provides high specificity for the technique and the intensity of the ion current reaching the detector during SRM is a quantitative indicator of the targeted analyte for which the precursor and product ions are characteristic. The SRM technique produces a higher specificity for integer mass spectral data than selected ion monitoring and can achieve very low detection limits. SRM is described in more detail in the following section on nomenclature.

3. Nomenclature

There are numerous modes of ion dissociation including those by natural causes as in metastable dissociation as well as by stimulating the dissociation process as described in the next section [44–47]. Other relevant definitions in the technology of MS/MS follow immediately:

Metastable ions. There are three categories of ions in mass spectrometry: stable ions, unstable ions, and metastable ions. The distinguishing feature among the three groups is based on the half-life of the ions compared to some experimentally relevant time interval associated with the mass spectrometric experiment (residence time in an ion source, residence time in a flight tube, etc.). If the interval between ion formation and ion detection is one microsecond and the half-life of the ion is 3 msec, then the ion is considered to be stable because it survives as an intact ion as it passes through the *m/z* analyzer to reach the detector. If the half-life of the ion is 7 nsec in this hypothetical example, it is considered to be unstable because it decomposes within the ion source into fragment ions that will eventually leave the ion source and enter the *m/z* analyzer. If the half-life of the ion is 0.8 μsec, they are considered to be *metastable* because a significant number of the ions will decompose as they pass through the *m/z* analyzer; i.e., an ion that has a half-life that is of the same order of magnitude as the timescale of the experiment is said to be *metastable*. Metastable ions will exhibit peculiar behavior in the *m/z* analyzer because they change kinetic energy (because of decomposition into components of different *m/z* value) within the analyzer even though they travel through it with the same velocity as ions from which they are formed (i.e., the ion was accelerated out of the ion source as an intact ion, but decomposed into another ion or a neutral before it reached the detector) [8].

A **collision cell** is a small chamber mounted in the ion path of the mass spectrometer [48, 49]. The collision cell has two small openings, one to let the precursor ions in, the second to let the product ions and surviving precursor ions out. The chamber can be pressurized, usually 10^{-1} to 10^{-2} Pa (pressure of the collision gas will be a factor in how much energy is imparted to the ion), with a collision gas that serves as a target for the precursor ions, thereby activating them so that they undergo decomposition or fragmentation to product ions. The energy transferred during the collision can be controlled in part by the electric field seen by the ions [50]; another parameter that determines the extent of energy transfer during a collision is the mass of the ion relative to the mass of the target (collision) gas calculable through the center-of-mass formula. The collision cell is often differentially pumped so that the collision gas does not interfere with the proper operation of the two *m/z* analyzers in hardware tandem MS/MS (sometimes referred to as *tandem-in-space* mass spectrometry). The collision cell is mounted in an appropriate field-free region.

A **precursor ion** is any ion selected for analysis by CAD. The precursor ion can be a fragment ion from the first *m/z* analyzer. An ion representing the intact molecule such as a molecular ion ($M^{+\bullet}$ or $M^{-\bullet}$), a protonated molecule (MH^+ or $[M + H]^+$), a deprotonated molecule $[M - H]^-$, an ion formed by hydride abstraction ($[M - H]^+$, an adduct ion (e.g., $[M + Cl]^-$ or $[M + K]^-$), or a multiple-charge ion ($[M + nH]^{n+}$) can also be chosen as a precursor ion for a CAD experiment.

Product ions are those fragment ions produced when the precursor ion undergoes dissociation in the collision cell during CAD.

A **product-ion spectrum** is an array of product ions produced by decomposition of a given precursor. A product-ion spectrum is recorded by setting the first *m/z* analyzer to a fixed value and using a second *m/z* analyzer to acquire ion current over a range of *m/z* values. Product-ion mass spectra do not exhibit isotope peaks when a monoisotopic precursor ion is selected. This lack of isotope peaks in product-ion mass spectra makes it difficult to follow the fate of certain elements such as those that exhibit significant intensities at the X+2 position (Cl, S, K, etc.). The use of the term MS/MS or MS^n spectrum is inadequate and should not be used because it does not clarify which of the various modes of MS/MS described above was used to acquire the data.

A **precursor-ion spectrum** is an array of precursor ions each of which is capable of generating a given product ion of a particular *m/z* value. A precursor-ion spectrum is recorded by scanning the first *m/z* analyzer while setting the second *m/z* analyzer to a fixed value, namely the *m/z* value of the specified product ion. Implicit in this definition is the fact that ions of one *m/z* value at a time pass through MS1. This means that there must be a continuous beam of ions passing into MS1. A precursor-ion mass spectrum cannot be obtained when a TOF mass spectrometer is used as MS1 or when an ion trap (quadrupole or magnetic) is used as both MS1 and MS2 (but at different times).

A **neutral-loss spectrum** is an array of ions that undergoes a common loss; e.g., expulsion of a molecule such as H_2O [51] or a radical such as $^\bullet CH_3$. A neutral-loss spectrum is recorded by simultaneously scanning the first *m/z* analyzer and the second *m/z* analyzer at the same rate, but offset by a fixed number of *m/z* units, such as 18 in the case of the loss of H_2O. The resulting spectrum would show the *m/z* values for those ions that lost a molecule of H_2O in the collision cell. Again, as was the case with a precursor-ion spectrum, the neutral-loss spectrum requires a continuous ion source to provide a beam of ions consisting of multiple *m/z* values. Ions of a single *m/z* value will pass through each analyzer one *m/z* value at a time.

> The term MS/MS spectrum should not be used. As seen above, MS/MS can produce product-ion spectra, precursor-ion spectra, or neutral-loss spectra. Therefore, use of the term MS/MS spectrum would create confusion.

MS/MS or tandem mass spectrometry is most widely used for a technique called **selected reaction monitoring** (**SRM**). This is a technique that monitors the ion current associated with the *m/z* transition of a given precursor ion to a particular product ion during collisionally activated dissociation of the precursor; the transition is called a selected reaction. In most cases, much better selectivity can be obtained with SRM as compared to **selected ion monitoring** (**SIM**), and as a result lower detection limits can usually be obtained by using SRM; these points are illustrated in two examples in a later section devoted to SRM in this chapter.

SRM is not limited to a single transition pair; the technique is *selected reaction monitoring*, not *single reaction monitoring*. Because more than a single transition pair can be monitored during a specified time interval, the term *multiple reaction monitoring* (MRM) was a neologism created by an instrument manufacturer in an effort to promote their instruments for applications of the technique; unfortunately, this term has now become standard with several other companies. The rationale for choosing the term SRM for the technique of monitoring transition pairs as opposed to some of the other proposed synonyms is analogous to that presented in Chapter 10 for SIM.

Forward geometry describes the normal arrangement of the electric sector preceding the magnet in a double-focusing mass spectrometer. As illustrated in Figures 2-40 and 2-42, ions accelerated by the potential V pass through the electric sector for energy focusing at potential E before being momentum-analyzed (m/z-analyzed) by the magnetic field B. Note the position of the first field-free region (FFR) between the accelerator and the electric sector.

Reverse geometry is that configuration of a double-focusing mass spectrometer in which the ions are m/z-analyzed (momentum) before they are energy-focused in the electric sector; as shown in Figure 2-41, the magnet precedes the electric sector.

II. Ion Dissociation

1. Metastable Ions

The early study of metastable ions was carried out with a magnetic-sector mass spectrometer [52, 53]. Metastable ions [8, 54] are accelerated from the ion source as one species (precursor ion m_1), but decompose into another species (product ion m_2) before reaching the magnetic field. This metastable transition [54] produces a broad, diffuse "metastable peak" at an m/z value ($m*$) that is related to the initial mass (m_1) and final mass (m_2) by the relationship [8, 55]:

$$m* = (m_2)^2 / m_1$$ (Eqn. 3-1)

The metastable peak is detected at an m/z value lower than both m_1 and m_2, because the metastable ion, m_1, was accelerated as m_1, but was m/z-analyzed as m_2 while maintaining the velocity of m_1. Therefore, $m*$ does not represent the metastable ion *per se*; rather, $m*$ represents a metastable transition: $m_1 \rightarrow m_2$. The m/z value of $m*$ is determined experimentally in conventional mass spectrometry from the center of the broad, diffuse "metastable" peaks in an oscillographic recording of the mass spectrum; data systems ignore them because the peaks are not sharp. The metastable peak will usually cover 3 or 4 m/z units, and often its peak center will correspond to a nonintegral value. Because the m/z value of this metastable peak is related to ions of masses m_1 and m_2, the metastable peak can be cited as experimental evidence to link certain ions together in a fragmentation scheme describing unimolecular decomposition of a complex molecule. To facilitate determination of possible combinations of ions (m_1 and m_2) that could produce a given metastable peak, tables have been prepared that list plausible integral values of m_1 and m_2 that could be related to a given value of $m*$ [56].

Although it is not apparent from the conventionally recorded (oscillographic) mass spectrum, there are usually many more metastable transitions (or potentially detectable metastable peaks) than there are normal, discrete peaks [8, 57]. The peaks representing metastable ions in the conventional mass spectrum are weak and diffuse because the decomposition products from the metastable transitions possess a distribution of energies, and in a single-focusing mass spectrometer there is no facility for energy-focusing of the ions.

Metastable transitions can be detected in a TOF instrument if it is equipped with a reflectron that can supply a retarding potential at the end of the drift tube just before the detector. Under conventional TOF operation, the metastable ion of mass **m₁**, the product ion of mass m_2, and the neutral species ($m_1 - m_2$) will all reach the detector simultaneously if no translational energy is lost in the metastable transition occurring in the field-free region of the drift tube [9, 58]. Use of the retarding potential, however, can resolve these three species as follows: the flight time of the neutral species of mass $m_1 - m_2$ will not be altered, the metastable ion of mass m_1 will be retarded somewhat, and the product ion of lighter mass m_2 will be retarded even more and will reach the detector with the longest arrival time of the three species (further details on use of an electrostatic mirror with PSD for MALDI are provided in Chapter 7). Although observation of metastable transitions is feasible by TOF mass spectrometry [59–61], they have been studied most extensively with magnetic instruments [8]. The kinetics of metastable-ion decomposition greatly affect the detection limits that can be achieved for a given metastable transition [8].

2. Collisionally Activated Dissociation (CAD)

Structural information can be obtained from the product-ion spectrum of any ion by inducing its decomposition by collisional activation with a relatively high pressure of a collision gas [62, 63]. In collisionally activated dissociation (CAD), the ion in question is selected as the precursor ion and is directed into a collision cell, where it collides with atoms (molecules in the case of nitrogen) of a collision gas to convert some of its translation energy into internal energy [47, 64–69], which drives the decomposition of the precursor ion into product ions. The level of energy available in the CAD process significantly influences the type of fragmentation induced [70–72].

The maximal amount of translational energy that can be converted to internal energy, E_{com}, is given by Equation 3-2:

$$E_{com} = E_{lab} \ [m_c /(m_c + m_i)]$$

(Eqn. 3-2)

which is derived from the so-called center of mass (com) reference frame for inelastic collisions. In Equation 3-2, E_{lab} = kinetic energy of the ion provided by electric fields in the mass spectrometer, m_i = mass of the ion, and m_c = mass of the collision gas atom. As can be appreciated by inspection of Equation 3-2, a larger fraction of the ion's KE can be converted into internal energy when it collides with a Xe atom than when it collides with a He atom. Furthermore, the larger cross-sectional area of the larger atom increases the probability of a collision.

In addition, the efficiency of decomposition of a given precursor ion scales with its charge state because an ion gains energy from an electric field in proportion to the number of charges it carries; i.e., if two ions have the same mass, but one carries three charges whereas the other carries only one charge, the triple-charge ion will gain three times the kinetic energy in traversing a given electric field. For this reason, ESI-CAD MS/MS often provides analytical advantage because of the multiply protonated molecules of analyte produced by ESI [73]. The higher the charge state, the more likely an ion is to dissociate.

3. Electron Capture Dissociation

Electron capture dissociation (ECD) has been established as an effective means of stimulating extensive fragmentation of multiprotonated molecules of proteins [74, 75]. In general, the idea is to convert a multiprotonated molecule, an even-electron species, to a radical cation, an odd-electron species as represented by the following simple reaction:

$$MH_2^{2+} + e_{5\,eV}^- \rightarrow MH^{+\cdot} + H^\cdot$$

The radical cation formed at the time of electron capture readily fragments via a variety of pathways, but favors the formation of c- and z-type ions from proteins [76–78]. For this reason, the ECD has been of critical important in the development of top-down approaches to protein characterization and identification [76]. The source of low-energy electrons was originally just a simple heater filament [79], but additional control processes for injection of low-energy electrons has improved the technique [80, 81].

Whereas the movement of thermal electrons can be controlled within the ICR for ECD, the RF fields in a quadrupole ion trap cause most of the electron flux to be ejected from the electron/ion interaction chamber, thereby making ECD less effective in these instruments [82]. For this reason, most of the pioneering work with ECD has been accomplished with FTMS instrumentation [75, 83].

4. Electron-Transfer Dissociation

The advantages of ECD are not available to users of quadrupole ion trap instrumentation because, as stated above, the RF fields in such instruments expel most of the thermal electrons from the interaction region with protonated molecules of the analyte [82]. However, pioneering work by the McLuckey Group with ion/ion reactions [84, 85] and by the Hunt Group with electron-transfer dissociation (ETD) [86, 87] have resulted in effective dissociation techniques that can be used readily with quadrupole ion trap instrumentation [88].

Electron-transfer dissociation adds a low-energy electron to a multiprotonated molecule via an ion/ion reaction, thereby converting it from an even-electron species to a radical cation, which readily dissociates via a variety of pathways in a manner analogous to those observed in ECD [85–87]. The efficiency of dissociation is improved substantially by combining ETD with CAD in a procedure called ETcaD, which reportedly increases dissociation of the protein to permit nearly 90% sequence coverage [89].

The thermal electron of interest in ETD is transferred to the protonated analyte by interaction with an externally formed anion ($A^{-\bullet}$) as represented conceptually by the following reaction:

$$MH_2^{2+} + A^{-\bullet} \rightarrow MH^+ + AH$$

An analyte is typically ionized by ESI to form a multiprotonated molecule, which is guided into a reaction chamber for interaction with a beam of electron-rich anions formed in a separate ion source; ETD is readily accomplished due to electrostatic attraction between oppositely charged ions in the interaction chamber. Because the movement of both the reagent ions and the analyte ions can be controlled by electric fields, there are a variety of ways by which the ion/ion interaction can be conducted. Some of these are represented in Figure 3-2 as conceived by the McLuckey Group. One means of performing these methods of ion/ion interaction is by use of an axial-ejection linear quadrupole ion trap in the position of a *tandem-in-space* mass spectrometer. In such an implementation of ETD, a transmission quadrupole can be used as MS1 to select either analyte ions (multiple-protonated ions) or anions (which can be referred to as reagent ions). After the ETD and, in some cases, CAD is accomplished, the resulting

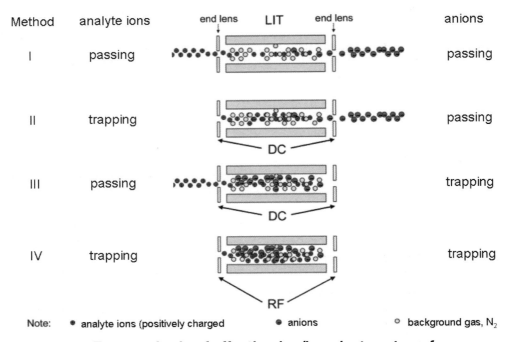

Figure 3-2. Four methods of effecting ion/ion electron-transfer dissociation experiments in an LIT: (I) passage of both polarity ions; (II) positive-ion storage/negative-ion transmission; (III) positive-ion transmission/negative-ion storage; (IV) mutual storage of both polarity ions. Reprinted from Liang X, Hager JW, and McLuckey SA "Transmission Mode Ion/Ion Electron-Transfer Dissociation in a Linear Ion Trap" Anal. Chem. 2007, 79(3), 1073–1081, with permission from the American Chemical Society.

product ions are analyzed according to their *m/z* values using a second transmission quadrupole (MS2).

One implementation in which quadrupole mass filters are used to control the input (MS1) and output (MS2) of ions for interaction in a linear ion trap is shown in Figure 3-3 [90]. Another implementation uses a pulsed triple ionization source, using a common atmospheric–vacuum interface and ion path, to generate different types of ions for sequential ion/ion reaction experiments in an LIT-based tandem mass spectrometer [91]. Unlike the low-mass electrons, the heavier anions (~100 Da) are not adversely affected by the RF fields in quadrupole ion traps, thereby allowing such relatively inexpensive instrumentation to be used for top-down analysis of proteins that was once exclusively reserved for those with FTMS instrumentation [92–95].

There have been two commercial implementations of ETD on QITs. One is the product introduced by Agilent Technologies/Bruker Daltonics at PittCon 2006 based on their high-capacity 3D QITs [96]. This implementation involves the isolation of multiple-protonated molecules of a single *m/z* value followed by the introduction of the negative-charge reagent ions. There is no isolation of reagent ions in this process, which means that if multiple reagent-ion species are formed, they compete with one another in their reactions with the analyte ions.

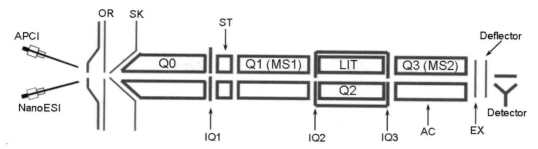

Figure 3-3. *Illustration of Applied BioSystems/MDS Sciex Qtrap tandem-in-space mass spectrometer arranged so that the LIT is in the position of the collision cell rather than being positioned as Q3 (MS2), as is the case with currently commercially available instruments. This illustration shows a homemade dual-ion source: nanospray (for formation of multiple-protonated analyte molecules) and APCI (for formation of reagent anions). Q1 can be set to alternately pass the two different polarity ions with storage of either in the LIT. Q3 is used for m/z analysis of the resulting product ions.* Reprinted from Liang X, Hager JW, and McLuckey, SA "Transmission Mode Ion/Ion Electron-Transfer Dissociation in a Linear Ion Trap" Anal. Chem. 2007, 79(3), 1073–1081, with permission from the American Chemical Society.

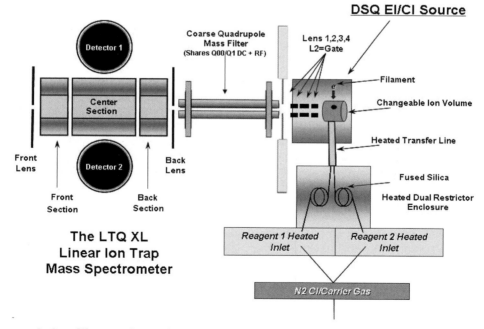

Figure 3-4. *Illustration of the Thermo Fisher Scientific segmented-linear quadrupole configured for electron-transfer dissociation (ETD).* Courtesy of Thermo Fisher Scientific.

The other implementation was introduced by Thermo Fisher Scientific based on their LTQ XL linear quadruple ion trap mass spectrometer at PittCon 2007 [97]. Because the LTQ XL is a segmented ion trap (see Figure 3-4), the analyte ions are isolated in the main body of the trap in the usual manner. The ions of a single *m/z* value can then be moved to the first segment (segment on the trap entry side). The reagent ions are then allowed to enter the trap from a conventional EI/CI ionization source. Reagent ions of a specific *m/z* value are isolated in the main body of the trap. The analyte ions are then allowed to return to the trap's main body where the ion/ion reaction and dissociation takes place with the storage of product ions. The flexibility offered by three segments of the LTQ XL translates into the capability for carrying out the combination of electron-transfer dissociation and collisionally activated dissociation (ETcaD), discussed above.

5. Illustrative Example of Qualitative Analysis by MS/MS

An example of the specificity achievable by MS/MS is illustrated in Figure 3-5 which was reported in the context of analyzing coal tar for dioxin [98]. Under the conditions for analysis, it had been established that a standard solution of 1,2,4-trichloro-dibenzo-*p*-dioxin would produce ion current at *m/z* 288 (the ^{37}Cl isotope peak of the M$^{+\bullet}$) in the conventional EI mass spectrum. The ion with *m/z* 288 contains two atoms of ^{35}Cl and one atom of ^{37}Cl. Under these same conditions, an aliquot of the coal tar produced the mass spectrum shown at the top of Figure 3-5. Whereas there does appear to be a peak at *m/z* 288 in this primary mass spectrum, it can hardly be claimed that such a spectrum is indicative of dioxin because there appears to be significant ion current at nearly every *m/z* value. In an effort to seek greater specificity in the analysis, the technique of MS/MS was employed to obtain the product-ion spectrum of *m/z* 288 from a standard of dioxin showing the loss of 63 Da (characteristic loss of CO^{35}Cl from chlorinated dibenzo-*p*-dioxins) to produce an ion represented by a peak at *m/z* 225, as seen in the middle spectrum of Figure 3-5. It should also be noted that there is a peak at *m/z* 223 which represents an ion formed by the loss of CO^{37}Cl. The intensity of this peak is about one-third the intensity of the peak at *m/z* 225, which would be expected.

Another aliquot of the coal tar sample spiked with 150 ppm dioxin was analyzed by MS/MS during which the ion current at *m/z* 288 (from a spectrum like that at the top of Figure 3-5) was selected as the precursor ion for CAD to produce the product-ion mass spectrum shown at the bottom of Figure 3-5. Based on the fact that most of the major peaks in the product-ion spectrum of standard dioxin (middle of Figure 3-5) are in agreement with those in the product-ion spectrum of *m/z* 288 from the coal tar (bottom), it appears that the coal tar does contain dioxin. This example illustrates the analytical specificity available in the technology of MS/MS.

III. Instrumentation for MS/MS

MS/MS can be accomplished in a spatial or temporal domain. The physical linking of *m/z* analyzers, one after the other, constitutes the spatial domain and is often called *tandem-in-space* mass spectrometry. The use of a single hardware device such as a 3D quadrupole ion trap (QIT) mass spectrometer to isolate an ion of a specific *m/z* value (the precursor ion) during one time interval, collisionally activate the precursor ion so that dissociation (CAD) takes place and collect (store) product ions during a second time interval, and obtain a product-ion mass spectrum by scanning the product ions out of the device during a third time interval is an example of MS/MS in a temporal domain, sometimes referred to as *tandem-in-time* mass spectrometry. The hardware requirements for *tandem-in-space* and *tandem-in-time* mass spectrometry uniquely characterize each category of MS/MS.

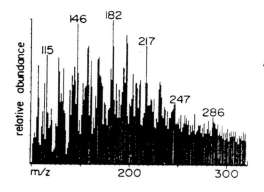

A. Direct Insertion Probe EI Mass Spectrum of Coal Tar Sample Spiked with 150ppm 1,2,4-Trichlorodibenzo-*p*-Dioxin Nominal Mass 286 Da

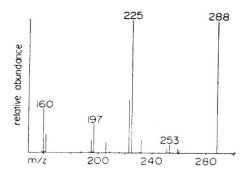

B. Product Ion Mass Spectrum of *m/z* 288 Precursor Ion (^{37}Cl isotope) of 1,2,4-Trichlorodibenzo-*p*-Dioxin – Pure Sample

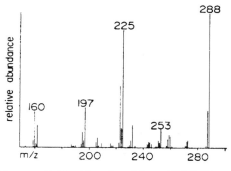

C. Product Ion Mass Spectrum of m/z 288 Precursor Ion Isolated from Ionization of Coal Tar Sample Spiked with 150 ppm 1,2,4-Trichlorodibenzo-*p*-Dioxin

Figure 3-5. ***Product-ion MS/MS analysis of coal tar for dioxin.*** *Reprinted from Singleton KE, Cooks RG and Wood KV, Anal. Chem. 1983, 55, 762–764, with permission from the American Chemical Society.*

A product-ion analysis (product-ion spectrum) can be performed using either *tandem-in-time* or *tandem-in-space* mass spectrometry. A precursor-ion analysis (precursor-ion spectrum) or a common neutral-loss analysis (neutral-loss spectrum) can only be performed using *tandem-in-space* mass spectrometry. As these two techniques are described in more detail later in the chapter, the limitations of *tandem-in-time* will become more obvious. *Tandem-in-time* mass spectrometry can be used for selected reaction monitoring; however, as the number of transitions increase, the efficiency of the operation severely suffers when compared with that based on *tandem-in-space* mass spectrometry.

1. *Tandem-in-Space* Mass Spectrometry (MS/MS)

Instruments designed for *tandem-in-space* use a mass-selective process to choose a given precursor ion at a given set of coordinates in space, transfer the ion to a different set of coordinates for ion dissociation, and then transfer the products of that dissociation to yet another set of coordinates in space for secondary mass analysis [58].

Tandem-in-space is the manifestation of MS/MS operation in which a given mass-selective device is used to select the precursor ion for the dissociation process and a second mass-selective device is used to analyze the products resulting from the dissociation [9]. The two mass-selective devices are placed on opposite sides of the ion-dissociation region, and thus are connected to one another across the ion-dissociation volume in a tandem arrangement. This arrangement of analyzers and the collision cell can be used for a *product-ion analysis*. The *tandem-in-space* mass spectrometer can also be used to carry out *product-ion analyses* and to conduct *neutral-loss analyses*.

Today, the most common *tandem-in-space* mass spectrometer is the triple quadrupole (two transmission quadrupole *m/z* analyzers separated by a collision cell that is a third quadrupole operated in the RF-only mode. Some of the first MS/MS experiments were carried out using double-focusing magnetic-sector mass spectrometers in which the collision cell was positioned between the ion source and the magnetic sector in a reverse-geometry double-focusing mass spectrometer. These instruments are still used for high-energy CAD studies.

The so-called *hybrid* tandem mass spectrometer has also become a popular *tandem-in-space* instrument. The most popular among this class of instruments is the transmission quadrupole-TOF mass spectrometer in which MS1 is a transmission quadrupole and MS2 is an orthogonal-injection reflectron time-of-flight (*oa*TOF) mass spectrometer. Several developmental iterations of this type of instrument are described in Chapter 2. Less popular today and considered to be in the category of legacy instruments are the multiple-magnetic and electric-sector instruments that were popular for early protein analyses by fast atom bombardment (FAB). There are also a few instruments that were designed to combine a double-focusing mass spectrometer as MS1 with either a transmission quadrupole (more than one was commercially manufactured) or a TOF (this was truly a one-of-a-kind instrument) as MS2. Recently, Shimadzu has introduced a 3D QIT that functions as MS1 and the collision cell with a TOF as MS2. This instrument really stretches the definition of *tandem-in-space* because it is not possible to perform either a precursor-ion or a common neutral-ion analysis in this instrument. The hybrid MS/MS instrument using a TOF as MS2 is not capable of common neutral-loss analyses.

The latest introduction to the *tandem-in-space* arena is the TOF-TOF, which is primarily being used for rapid identification of proteins in proteomics applications (see Chapter 12). This instrument, like those instruments that position the collision cell in front of an electric sector in a double-focusing instrument or after MS1 in a *tandem-in-space* instrument using a double-focusing instrument as MS1, performs high-energy CAD. The triple-quadrupole and the Q-TOF instruments perform only low-energy CAD.

An advantage of the *tandem-in-space* arrangement is the capability for controlling the pressure and the type of inert gas used in the collision cell. These instruments can use N_2 molecules or He, Ne, Ar, Kr, or Xe atoms as the collision gas, thereby providing some flexibility in energy deposition based on the mass of the collision gas; in addition, the pressure can easily be adjusted to control the number of collisions during the CAD process. As described in Chapter 2, the quadrupole ion trap (a *tandem-in-time*

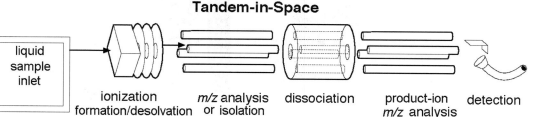

Figure 3-6. Concept of tandem-in-space MS/MS instrument.
Adapted with permission from Johnson JV and Yost RA "Tandem-in-Space and Tandem-in-Time Mass Spectrometry: Triple Quadrupoles and Quadrupole Ion Traps" Anal. Chem. 1990, 62, 2162–2172.

instrument) is usually restricted to the use of He[2] at a fixed pressure (required for dampening the amplitude of oscillations of ions so that they will remain in their individual orbitals).   If such an instrument is used, the operator cannot adjust the pressure or choose a different collision gas in an effort to adjust the degree of dissociation during CAD.   A hybrid of space/time instrumentation for MS/MS is manifest in the use of a collision cell in a TOF instrument [99].

A. Triple-Quadrupole Mass Spectrometer

The triple-quadrupole MS (TQMS) is a tandem arrangement, as illustrated in Figure 3-6, in which the first and third quadrupoles are mass-selective (**Q**) and the second quadrupole (**q_2**) is operated in the RF-only mode (transmits ions of all *m/z* values) to serve as a collision cell [100, 101].   The first quadrupole is used for selection of precursor ions following ionization by EI, CI, APCI, APPI, or ESI.   The precursor ion is transmitted to the collision cell (an RF-only quadrupole in this illustration, **q_2**, which collimates the residual precursor ions and all the product ions in an appropriate concentration of collision gas); typically, the pressure in the collision cell will be ~10^{-1} Pa due to the presence of a *collision gas*.   The third quadrupole (**Q_3**) provides a means of analyzing all the products of CAD that emerge from the collision cell (**q_2**).

The term triple quadrupole is somewhat misleading.   For example, in one commercial manifestation of the technology, the Quattro micro API produced by Waters Corp., the collision cell actually consists of a hexapole, rather than a quadrupole.   Other instruments use a lens stack or an octupole as a collision cell.   In all cases, the collision cell (**q_2**) is not *m/z*-selective; i.e., **q_2** is not a mass spectrometer.   Only the first and third quadrupoles (**Q_1** and **Q_3**) are *m/z* analyzers **MS1** and **MS2**, respectively.

The TQMS provides for selection of both precursor ions and product ions at unit resolution [102] throughout the *m/z* scale.   Operation of the TQMS for MS/MS is straightforward [103–105].   A precursor-ion spectrum is produced by scanning the first quadrupole while holding the third quadrupole at a constant value of *m/z*.   A product-ion spectrum is produced by holding the first quadrupole at constant *m/z* while scanning the

---

[2] The 3D QIT that is part of the Shimadzu QIT–TOF is designed to use Ar as the buffer gas; however, its pressure is fixed at a level needed to cool ions that are being stored. The Ar atoms make for a more energetic collision than do He atoms because of their increased mass.

Table 3-1. Advantages of the triple quadrupole for MS/MS.

- Simple to operate, rugged, and easy to maintain.
- More sensitive in selected ion monitoring (SIM) and selected reaction monitoring (SRM) than *tandem-in-time* mass spectrometry.
- Capable of product- and precursor-ion as well as common neutral-loss analyses.
- Allows for the use of different types of collision gases, which is not the case with commercially provided *tandem-in-time* instruments.
- Allows for the control (and variation) of the collision gas pressure.
- Can be used for selected ion monitoring.
- TQMS has a wide linear dynamic range for quantitation.
- No low-mass cutoff in MS/MS (explained later).
- Adequate CAD efficiencies (10–40%).
- High sample capacity and throughput.
- Very good for hyphenated-technique quantitation (SRM on a TQMS is at least an order of magnitude more sensitive than that for a quadrupole ion trap).
- Ideally suited for Open Access (Proven Track Record).
- Spectra obtained by MS/MS may not be as complex to interpret as those that are produced by a QIT.
- Easy-to-build spectral libraries.

third quadrupole. A neutral-loss spectrum is produced by simultaneously scanning the first and third quadrupoles at a constant offset of *m/z*. If both the second and third quadrupoles are operated in the RF-only mode, the TQMS behaves as a single analyzer in transmitting a normal mass spectrum (but at diminished intensity because of transmission losses (at least 10%) in each RF-only quadrupole) to the detector [106].

B. Q-TOF Hybrid Mass Spectrometer

The tandem transmission quadrupole-TOF mass spectrometer has the advantage of mass accuracy to the nearest 0.1 millimass unit. This high-mass accuracy results from the orthogonal arrangement between the longitudinal axis of the ion beam coming from the quadrupole and the direction in which the ion beam is extracted into the TOF instrument as described in detail in Chapter 2. This orthogonal extraction of ion packets into the TOF *m/z* analyzer reduces the effect of longitudinal (forward and backward) dispersion of ion motion in the ion beam just prior to entering the flight tube, thereby having an effect on the resolving power. The TOF is a reflectron, which results in another degree of resolving power enhancement to give these instruments an overall resolving power of 10,000–12,000. One manufacturer, Waters, has employed a double reflectron to produce a set of "W" optics that results in even higher resolving power.

Table 3-2. Disadvantages of the triple quadrupole for MS/MS.

- Limited to MS/MS, but some MS[3] results can be obtained in conjunction with in-source CAD.

- Low resolving power (1000–4000) – best *m/z* specifications for precursor and/or product ion is about 0.1 units – the mass measurement of a product ion can be stretched with internal calibration – most commercial instruments have a resolution of 0.3 *m/z*.

- Possibility of secondary product-ion fragmentation in the collision cell.

- Full-scale acquisition rate limited to about 1 spectrum/second; however, increased data acquisition rates are realistic because a narrow acquisition range is usually employed.

- Maximum *m/z* range to 4000 – most commercial LC/MS instruments have a maximum obtainable *m/z* value of 1500 to 3000.

- Initial cost is higher than entry-level quadrupole ion trap, although the price is dropping.

Figure 2-11 illustrates the Waters Q-TOF® [3] optics system. Most quadrupole-TOF instruments have a transmission quadrupole that offers unit resolution over an *m/z* range of 2000–4000 (there have been commercial instruments with maximum *m/z* transmissions of 32,000; however, this is at the sacrifice of resolution, which is ~3–4 *m/z* units). The tandem transmission quadrupole mass spectrometer–TOF mass spectrometer has been extensively used for proteomics; it has also found many uses in the area of small-molecule analytical chemistry.

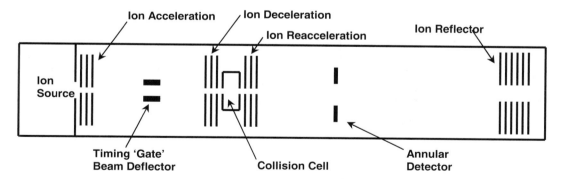

Figure 3-7. Schematic diagram of a TOF-TOF mass spectrometer.

---

[3] During recent times, the term Q-TOF has been applied to any manufacturer's mass spectrometer that is a *tandem-in-space* instrument employing a transmission quadrupole as MS1 and a reflectron TOF as MS2. This is inappropriate because the term Q-TOF is a registered trade mark of Waters Corporation. *Tandem-in-space* mass spectrometers using a transmission quadrupole as MS1 and a TOF as MS2 as manufactured by Agilent Technologies, Bruker Daltonics, and Applied BioSystems/MDS Sciex should not be referred to as a Q-TOF mass spectrometer.

C. TOF-TOF Mass Spectrometer

A schematic diagram of a TOF-TOF instrument equipped with a MALDI ion source is shown in Figure 3-7. As in normal MALDI-TOF operation, the extraction pulse from the ion source is triggered (with or without an adjustable delay) by the laser pulse. Ions then pass through a truncated field-free region to allow the ion packets of various *m/z* values to separate sufficiently for a discrete packet (typically, 1 to 3 *m/z* units in resolution) to be selected (gated) by the time gate (beam deflector unit) for passage on to the collision cell. As the gated packet of ions approaches the collision cell, it passes through a deceleration field to reduce its kinetic energy from several tens of thousands of eV to an operator-adjustable value (typically, a few hundred to a thousand eV) before entering the collision cell. As the ions continue through the collision cell (some have now been converted to product ions), they are reaccelerated to several tens of thousands of eV. The high-energy packet of ions extracted (actually, only reaccelerated) from the collision cell enters a longer field-free region to allow optimal separation of ion packets of various *m/z* values to produce a product-ion spectrum. The resolution of the product-ion spectrum is improved somewhat by the reflectron shown at the right-hand end of the schematic diagram. As explained in the description of reflectron TOF mass spectrometers in Chapter 9 on MALDI, the ion beam blooms radially within the reflectron so that the vast majority of the ions are reflected back to an annular detector back toward (but off-axis from) the ion source.

The TOF-TOF manifestation of MS/MS offers the analytical advantage of rapid data generation, which facilitates high-throughput analysis of large sample loads. The TOF-TOF instrument has been appreciated in proteomic analyses in which the analyst seeks some sequence information in addition to molecular mass of proteolytic fragments for protein identification [107–110] or in analyses for amino acids [111]; results of a recent study indicates that some fragmentation observed in TOF-TOF work is dependent on laser fluence [112]. The TOF-TOF technology has also proven useful in the analysis of carbohydrates [113–115]. An advantage of the TOF-TOF instrument is the possibility of controlling the energy of the CAD process in the collision cell; in particular, the possibility of high-energy CAD provides fragmentation processes that, in some cases, provide the basis for protein identification that is not attainable on quadrupole-based instrumentation. The mass accuracy of the product ions is not as good as that from a standalone TOF-MS equipped with a reflectron because of the space-focusing issues associated with the ions in the collision cell. As the *m/z*-selected ion packet traverses the collision cell, it suffers collisional broadening that degrades the resolving power of the second TOF instrument. The reflectron can only compensate for energy dispersion among the product ions; it cannot compensate for spatial dispersion of the ion packet. However, the overall resolving power of the TOF-TOF instrument is sufficient to provide mass accuracies better than 20 ppm, which translates into a few tenths of a dalton for most ions, and this is satisfactory for protein identification at a high level, mainly because of the sequence information provided in proteomic analyses. The TOF-TOF instrumentation is currently available from Applied Biosystems and Shimadzu.

D. BEqQ Hybrid Mass Spectrometer

A hybrid instrument [116] such as the BEqQ (magnetic sector (B), electric sector (E), RF-only quadrupole (q), quadrupole mass filter (Q) in series) offers the possibility of selecting precursor ions with high resolving power (e.g., R = 8000 or greater). The advantage with the BEqQ (see block diagram in Figure 3-8) is that the first quadrupole (**q**)

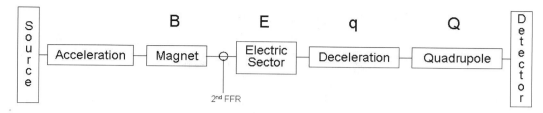

Figure 3-8. Conceptual illustration of the tandem BEqQ hybrid instrument.

is an efficient collision cell because the RF-only field ensures a broad acceptance angle for the diverging decelerated ion beam and also serves to contain the product ions as they tend to scatter during the CAD process. Although ions that are passed through MS2 (**Q**) are not to be resolved beyond unit values, due to the selectivity of a high resolving power, the product ions will most likely represent a precursor ion of a single *m/z* value. This increase in selectivity is not without a price. As will be recalled from the description of the way a double-focusing instrument achieves resolution in Chapter 2, the higher the resolving power, the fewer ions that will reach the collision cell.

A deceleration module is a necessary part of the BEqQ to reduce the translational energy of the ions from the BE sectors from several thousand electron volts to ~100 eV or less in the RF-only quadrupole (**q**). Like CAD in the triple quadrupole and the transmission quadrupole-TOF instrument, product ions are produced by low-energy collisional activation. The BEqQ hybrid instruments introduced in the 1980s suffered from compacted operation because of the lack of sophisticated electronics and microprocessors in the design of the double-focusing portion of the instrument and the high cost associated with the double-focusing portion. The deceased sensitivity and the higher degree of specificity bought very little in terms of reduced limits of quantitation; therefore, with the cost far exceeding that of the triple-quadrupole instruments, these systems quickly slipped into oblivion.

E. Double-Focusing Instruments

Double-focusing magnetic-sector instruments, which were designed to focus or select ions based both on energy (E) and on momentum (B), can be used in a variety of different linked-scan modes to allow products of dissociation in a field-free region to reach the detector [117]. Linked scanning is a coordinated use of the magnetic (B) and electric (E) sectors of a double-focusing instrument to select precursor ions and detect product ions in a single instrument; which means, this type of operation is not "true" tandem mass spectrometry. Depending on which field-free region is used for the collision cell, a variety of linked-scan modes are available including IKES (ion kinetic energy scan) [118, 119], MIKES (mass-analyzed ion kinetic energy scan) [120], and an arrangement that allows for direct analysis of product ions, constant B^2E [118, 121], which produces sharper peaks for product ions than does a B/E scan [55]. Triple- and four-sector instruments offer specialized opportunities for investigating ion dissociation [122]. One of the primary advantages of performing MS/MS experiments in a double-focusing instrument is the availability of high-energy CAD. In double-focusing instruments, the ions are accelerated out of the ion source with upwards of 1000 eV, whereas in a transmission quadrupole, the accelerating potential is on the order of 10–20 eV. The advantage of high-energy CAD is that the fragmentation processes involve fewer rearrangements and are, consequently, frequently more simple to interpret than those promoted by low-energy CAD.

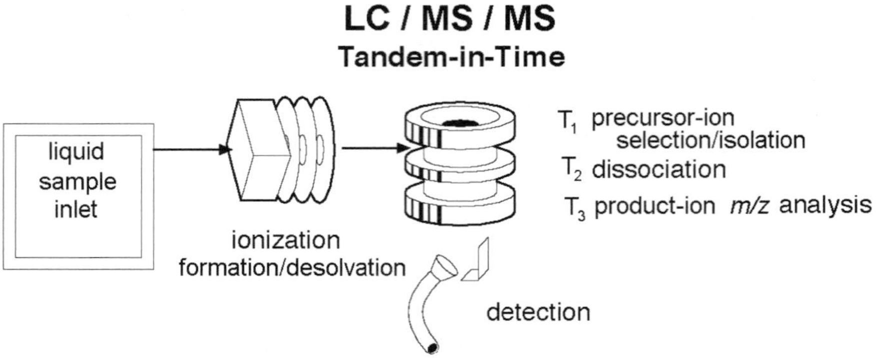

LC / MS / MS
Tandem-in-Time

T_1 precursor-ion
  selection/isolation

T_2 dissociation

T_3 product-ion *m/z* analysis

liquid
sample
inlet

ionization
formation/desolvation

detection

Figure 3-9. Conceptual illustration of tandem-in-time by 3D QIT *Adapted with permission from Johnson JV and Yost RA "Tandem-in-Space and Tandem-in-Time Mass Spectrometry: Triple Quadrupoles and Quadrupole Ion Traps" Anal. Chem. 1990, 62, 2162–2172.*

2. *Tandem-in-Time* Mass Spectrometry

MS/MS accomplished through *tandem-in-time* mass spectrometry means that the ions are injected into (or formed in) a given set of coordinates in space during a specified time interval; then, selected ions are expelled from that volume such that only ions of a specified *m/z* value remain in those same coordinates of space (during a second time interval). These stored precursor ions are then exposed to some dissociation process (during a third time interval), and finally an episode of *m/z* analysis of the products of dissociation is conducted (during a fourth time interval) in the very same coordinates of space (see conceptual diagram in Figure 3-9). That is to say, all the individual processes (ion accumulation, ion selection, dissociation, and *m/z* analysis) for MS/MS are carried out in the same coordinates of space, but in sequential periods of time. This concept was first reported by Robert B. Cody based on experiments that he did in Ben Freiser's laboratory at Purdue University as a doctoral student [123]. During this same period, Cody also introduced the concept of MS^n (multiple iterations of MS/MS), which can only be accomplished with *tandem-in-time* mass spectrometry [124]. The MS^n development also led to the concept of EIEIO (electron impact [sic] excitation of ions from organic compounds) while Cody was "working in a barn on Old McDonald's Farm" [125].

Instruments that allow this kind of analysis by MS/MS are the ion cyclotron resonance (FTMS) mass spectrometer, which has especially good resolving power [126, 127], the orbitrap (another trapping device capable of high resolving power), and the quadrupole ion trap (QIT)[4] [128, 129] mass spectrometer; a linear QIT-ICR hybrid instrument for this purpose improves dynamic range and detection limits for FTMS by use of the linear QIT for selective accumulation of the precursor ions of interest [130]. The concept of *tandem-in-time* is illustrated in Figure 3-9 in the context of using a 3D QIT ion trap for sequential (T_1, T_2, etc.) operational processes. The concept of "*tandem-in-time*" is that all three steps (precursor-ion isolation, precursor-ion dissociation, and product-ion *m/z* analysis) take place in the same position in space, but at different times. Unlike *tandem-in-space*, where ions are continuously passing from one operational step to the next, *tandem-in-time* applies only a single process (selection, dissociation, or analysis) to the ions in any given time interval.

---

[4] As explained in Chapter 2, there are two types of quadrupole ion trap mass spectrometers: the 3D QIT and the linear quadrupole ion trap (LIT) or 2D QIT. *Tandem-in-time* mass spectrometry is performed in the same way in both types of QITs.

Ions of all *m/z* values are stored in the trap during the time interval T_1. These ions can be injected into the trap, as is the case with LC/MS, or they can be formed in the trap, as can be the case with GC/MS. In the QIT (which is presently the most common of the *tandem-in-time* mass spectrometer), all of these ions are stored in concentric orbitals according to their *m/z* values, with the ions of higher *m/z* values being stored toward the center of the trap, as described in Chapter 2.

During the time interval T_2, ions of all *m/z* values other than that of the precursor ion are ejected from the trap.

During T_3, the precursor ions are excited by applying an increased amplitude of RF at their orbital frequency (resonant excitation) [131, 132]. The excitation process is carefully controlled so that the absorption of energy by the precursor ions is sufficient to cause dissociation upon collision with He atoms, which are present as a buffer gas to collisionally dampen the oscillatory amplitude of the ions to aid in keeping those of different *m/z* value in their individual orbitals. The amplitude of the applied RF potential is not great enough to increase the radii of ion orbitals to the point of premature ion ejection from the trap. Conditions in the QIT are adjusted so that the resulting product ions will be stored in their individual orbitals without being excited and without undergoing possible secondary dissociation.

During T_4, the ions (mostly product ions, but some residual precursor ions) now stored in the trap are scanned out one *m/z* value at a time to obtain the product-ion mass spectrum.

In summary, the QIT uses software for in-time replacement of multiple hardware in-space components. Several companies now offer QIT MS/MS instruments that can do "in-time" what the beam-type instruments do "in-space". An advantage of QIT mass spectrometers is the increased sensitivity because they do not suffer loss of ions during transfers between multiple segments of hardware in space.

An advantage of the 3D QIT and linear ion trap as an MS/MS device is its availability to perform subsequent iterations of MS on the product ions; i.e., steps T_2 through T_4 can be carried out on one of the product ions formed during the first stage of CAD; one of the secondary product ions can then be selected for yet a third stage of CAD to give MS^4 results, etc. This MS^n is accomplished with the addition of software routines. To perform MS^n with beam-type instruments requires the addition of expensive hardware for each degree of **n**. This capability of the QIT mass spectrometer to perform MS^n is its single strongest feature.

One of the main differences between *tandem-in-time* and *tandem-in-space* mass spectrometry is what is called the "low-mass cutoff" observed in the QIT. Product ions that have *m/z* values <1/3 of the *m/z* value of the precursor ion are not efficiently stored in the QIT during axial excitation that causes CAD. A dramatic example of this is seen in Figure 3-10. The top mass spectrum in Figure 3-10 was obtained using a 3D QIT and the lower spectrum was acquired using a triple-quadrupole mass spectrometer. At the low-*m/z* end of the spectrum are peaks at *m/z* 109 and 97. These two peaks represent ions that are characteristic of the CAD of protonated molecules of certain steroids. In the mass spectrum acquired with the triple-quadrupole MS/MS instrument, the peak at *m/z* 97 is the base peak and the peak at *m/z* 109 has a relative intensity of ~75%; however, both of these peaks have a relative intensity of less than 10% in the mass spectrum obtained with 3D QIT. The reason for the low relative intensity of these two peaks is the low-mass cutoff problem in the QIT when axial excitation RF is applied for CAD. During excitation of the precursor ion (having an *m/z* value of 331 in this case), ions of *m/z* values below

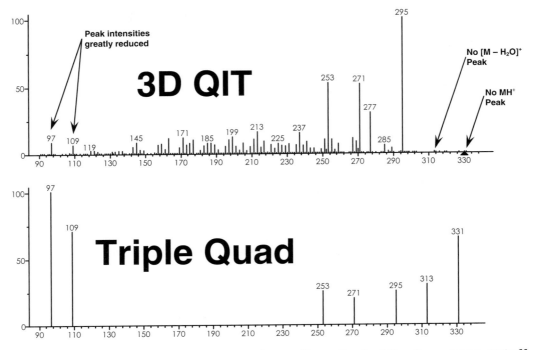

Figure 3-10. *(Top) Mass spectrum shows the effect of the low-mass cutoff problem with the QIT during a product-ion analysis. (Bottom) Spectrum results from excitation of precursor ion of same m/z value (nothing should be inferred from the difference in noise in the top spectrum and the bottom spectrum).*

110 are not efficiently stored, which explains why immonium ions are not observed in product-ion mass spectra of peptides as acquired with the QIT.

Figure 3-10 also reveals another important difference between *tandem-in-space* and *tandem-in-time* mass spectrometry. The product-ion mass spectrum acquired with the triple-quadrupole instrument has peaks of significant intensities at *m/z* 331 (representing the MH$^+$) and *m/z* 313 (representing the [M + H – H$_2$O]$^+$ ion). Neither of these peaks is significant in the mass spectrum obtained with the 3D QIT. Examination of the collision times and efficiency of the CAD process for the two instruments as shown in Table 3-5 reveals that both are far greater in the QIT mass spectrometer than in the triple-quadrupole instrument. This accounts for the lack of the MH$^+$ peak; however, it does not explain the lack of the [M + H – H$_2$O]$^+$ peak. The lack of the [M + H – H$_2$O]$^+$ peak could possibly be explained by the fact that the excitation energy covers a range of *m/z* values rather than a point value.

Tandem-in-time mass spectrometry in the FT-ICR is carried out in a similar manner to that accomplished in the QIT. Ion dissociation is accomplished by: (1) introduction of the collision gas into the ICR cell, (2) excitation of the precursor ion to promote collisionally activated dissociation, (3) evacuation of the collision gas, and (4) acquisition of the product-ion mass spectrum. Ion dissociation has also been accomplished with the use of lasers and through a process called electron capture dissociation (ECD) as well as electron-transfer dissociation ETD).

Table 3-3. Advantages of the quadrupole ion trap for MS/MS.

- Simple to operate, rugged, and easy to maintain
- Can perform MS^n, but entry-level instruments from most manufacturers limited to MS/MS
- Constant separation throughout the *m/z* scale
- Capable of selected reaction monitoring (SRM)
- Can be operated to higher *m/z* values than a triple quadrupole
- Initial cost for entry-level instrument is lower than for triple quadrupole
- High sample capacity and throughput
- Good sensitivity in the full-spectrum mode
- Higher CAD efficiency (60–80%) than the triple quadrupole
- Easy-to-build spectral libraries

Table 3-4. Disadvantages of the quadrupole ion trap for MS/MS.

- Limited to product-ion scanning in MS/MS—cannot do precursor-ion or common neutral-loss scanning
- No control of the pressure or type of collision gas
- Unwanted fragmentation possible, due to the high pressure of the required dampening gas
- Linear dynamic range limited in quantitation due to a maximum storage capacity of the trap
- Sensitivity not as great as that of a triple quadrupole in selected reaction monitoring
- The low-mass cutoff for product ions during MS/MS
- Low resolving power (1000–4000) – *m/z* accuracy about 0.1 units, but can stretch with internal calibration – most commercial instruments have a resolution of 0.3 *m/z*
- Spectra may be more complex than those obtained on a triple quadrupole due to the large number of ion/atom collisions resulting in a higher degree of fragmentation

In practice, MS/MS-in-time is much simpler to accomplish than MS/MS-in-space because the latter requires more points of ion focus that must be optimized simultaneously to achieve a successful experiment. The efficiency of CAD depends on several factors, but chief among them are the number of collisions that can be achieved within the collision cell and the translational energy available in the precursor ion. The frequency of collisions is dependent on the pressure of the collision gas (it is also dependent on the concentration of precursor ions, but that parameter is fixed at a given

Table 3-5. Comparison of CAD conditions for different instrument types.

Instrument Type	Number of Collisions*	eV, Ion Energy	Collision Gas	Collision Time, μsec	% Efficiency
Single Transmission Quadrupole; In-source CAD	M	1–400	N_2	5–50	80–95
Triple Quadrupole	S/M	10–100	N_2, Xe, N_2	5–50	10–40
3D and Linear Quadrupole Ion Trap	M	6–10	He	20,000–40,000	60–80
Double-Focusing Sector-Based	S	2000–8000	He	<5	1–3

* M = multiple collisions and S = only a single collision.
Adapted from Interpretation of CID Mass Spectra, Robert D. Voyksner, Research Triangle Institute, Research Triangle Park, NC, 2000, USA.

detection limit). The number of collisions achieved by a given precursor ion at a given pressure of collision gas is dependent on the residence time of the ion within the collision cell. The energy of the precursor ion depends on the accelerating potential of MS1 in combination with other potentials available in various components of the ion optics. Some of these parameters are summarized in Table 3-5 for some instrument configurations.

IV. Specialized Techniques and Applications

1. In-Source CAD

In-source CAD is a nondiscriminatory technique in which all ions near the entrance (inlet cone) are subjected to collisional activation. Unlike in MS/MS, where a particular precursor ion is selected for CAD, there can be a whole array of precursor ions participating in the in-source CAD process. Because there is no selection of a particular precursor ion with the technique of in-source CAD, the secondary mass spectrum of product ions may be difficult to interpret.

In-source CAD is accomplished by adjusting the potential between the ion-transfer region and the skimmer cone containing the aperture into the *m/z* analyzer as illustrated in Figure 3-11. In this way, the kinetic energy of the ions in this region can be adjusted to influence the efficiency of dissociation when the analyte ions collide with molecules of nitrogen in the region of the interface that is sometimes called the "first pumped region" where the pressure is about 10^{-1} Pa. The effect of collision energy during in-source CAD of sulfamethazine is illustrated in Figure 3-12. The energy available for in-source CAD is typically measured in somewhat instrument-dependent terms, but as can be seen in Figure 3-12 the degree of fragmentation increases in proportion to the applied potential difference between the skimmer cone and the opposite side of the interface; at a relative setting of 110 volts, the predominant peak represents the MH⁺, which diminishes in relative intensity as the potential difference increases. In the case of in-source CAD, the collision gas is essentially nitrogen from residual air, from use of a nitrogen nebulizer, or from use of a nitrogen drying or "curtain" gas in the interface.

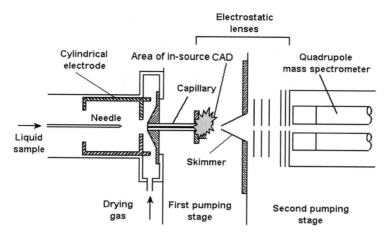

Figure 3-11. A schematic representation of the in-source CAD technique.

Because there is no selection of a specific precursor ion for in-source CAD, there may be several different precursor ions undergoing CAD, and there is no way to determine which product ions come from which precursor ions. Reasonable data interpretation from in-source CAD can only be done for pure samples. This can be accomplished by flow injection or infusion of standards during in-source CAD. In cases where the desired analyte is a component of a mixture, LC conditions must be established that allow the analyte fraction to pass through the in-source CAD region individually [133]. In spite of the shortcoming of no selection of a particular precursor ion, in-source CAD is frequently used to analytical advantage [38]. One advantage that in-source CAD has over selection of a monoisotopic precursor ion is that it is easy to follow the fate of X+2 elements like Cl, Br, and S.

2. CAD in Conjunction with Soft Ionization

Desorption/ionization (D/I) has advanced mass spectrometry significantly in several areas of application to thermally sensitive compounds. D/I techniques like electrospray ionization (ESI) and MALDI (matrix-assisted laser desorption/ionization) produce protonated molecules (MH^+) of the analyte; ESI can also produce $[M - H]^+$ and $[M - H]^-$ ions. None of these ions, which represent the intact molecule, tend to fragment; for this reason, such D/I techniques are said to be "soft" ionization techniques. Although these ions are useful in providing an indication of nominal or molar mass of the analyte, they are often disappointing in that they do not provide much, if any, structural information because they do not fragment readily. The same is true for the ions formed in atmospheric pressure chemical ionization (APCI) and electron capture negative ionization (ECNI) used with both GC/MS and LC/MS, CI used with GC/MS, and APPI used primarly with LC/MS. All of these ionization modes are also considered to be soft ionization techniques.

In some cases, these ionization techniques can also produce cationized molecules (MNa^+, MK^+, MAg^+, etc.) or anionized molecules (MCl^-, MBr^-, etc.). All ions formed by these soft ionization techniques are even-electron species, a feature that tends not to promote fragmentation. In most of these cases of soft ionization, there is little or no subsequent fragmentation of the even-electron ions that can aid in structural assignment of the analyte.

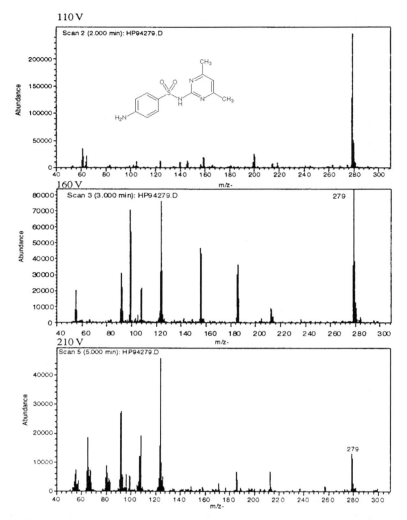

Figure 3-12. *Spectra acquired at different potentials applied to the ion volume between the API orifice (cone) and the entrance to the m/z analyzer (which increases the kinetic energy of the ions in this relatively high pressure region (10^{-1} Pa)). It should be noted that some of the product ions formed at 160 V shown in the middle spectrum appear to be converted to other product ions at higher potential (bottom spectrum). This formation of secondary product ions due to collisional activation of product ions also occurs in tandem-in-space MS/MS.*

Because CAD is a useful technique for promoting fragmentation of a selected ion, the combination of CAD with a soft ionization technique like ESI, APCI, CI, ECNI, or APPI has considerable analytical advantage [134–139]; a particularly illustrative application is in the use of CAD MS/MS to distinguish nine anabolic steroids that have structurally similar structures [140].

A. Data-Dependent Acquisition

The term "data-dependent analysis" was probably first coined by Finnigan-MAT (now Thermo Fisher) ca. 1997 [141]. The name Data-Dependent™ was trademarked by Thermo Fisher; therefore, Waters uses the term independent data-directed analysis (DDA™); Agilent uses the term data-directed acquisition; Applied BioSystems/MDS Sciex uses the term information-dependent acquisition (IDA); other companies may have different names for the same process. This software acquires a mass spectrum using one of the soft ionization techniques (usually ESI, APCI, or APPI); the resulting spectrum is analyzed to determine which mass spectral peaks represent various types of ions (charge states, MH^+ adducted molecules, etc.). Then, based on rules as to the ion type or ions of a specific *m/z* value, the next spectrum acquired will be in a product-ion or neutral-loss mode. This type of automated operation is very useful for high-throughput analyses such as those of proteolytic digests for purposes of protein identification as described in Chapter 12. However, caution should be exercised in using the data-dependent analytical strategy. When the precursor ion is selected solely as being represented by the most intense peak in the spectrum, the possibility of an abundant contaminant shifting results away from analytes of interest can become a significant problem.

> The term data-dependent analysis (acquisition, etc.) was originally used with respect to performing a single-stage mass spectral analysis followed by an MS/MS analysis. Recently, the term has been used with respect to performing an accurate mass measurement after the detection of a specified nominal *m/z* value by a 3D QIT mass spectrometer.

3. Selected Reaction Monitoring

In the technique of selected reaction monitoring (SRM), the product-ion current resulting from CAD of a selected precursor ion is recorded (monitored) during sample introduction into the MS/MS instrument. If the sample is introduced through a chromatographic inlet, the data record takes the form (profile) of a chromatogram.

A. An Illustrative Example Showing that SRM Has a Higher Specificity than SIM in Spite of a Lower Signal Strength

The selectivity available through SRM is illustrated here by comparison with selected ion monitoring (SIM) in the detection of targeted compounds. SIM is a technique in which the mass spectrometer is dedicated to monitoring ion current at a selected *m/z* value (as opposed to acquiring data over a specified range of *m/z* values) characteristic of the analyte as the sample is introduced into the instrument.

Figure 3-13 compares SIM chromatograms for *m/z* 255 obtained by injecting two different binary mixtures of ketoprofen and fenbufen into a Quattro-micro mass spectrometer equipped with an ESI source. Ketoprofen and fenbufen are constitutional isomers, each having a nominal mass of 254 Da; the ESI mass spectrum of each (data not shown) is dominated by a peak at *m/z* 255 representing the protonated molecule. The sample represented by the top SIM chromatogram contained equal amounts of the two compounds (60 ng mL^{-1}). The sample represented by the SIM chromatogram at the bottom of this figure contained 60 ng mL^{-1} ketoprofen and 6 ng mL^{-1} fenbufen. At 60 ng mL^{-1} each, the two compounds can easily be quantitated in the presence of one another. At a concentration of 60 ng mL^{-1} ketoprofen and 6 ng mL^{-1} fenbufen, it is clear based on

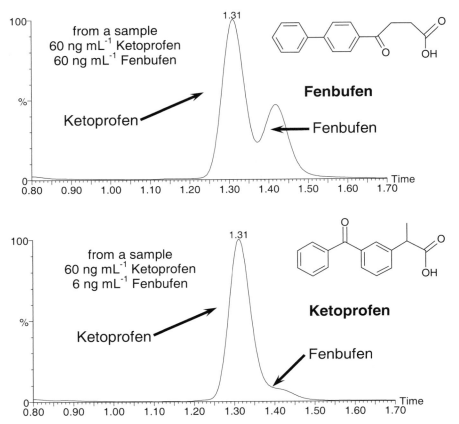

Figure 3-13. *Example of how a combination of response factor and close retention times of isobaric compounds can result in a limited range of quantitable concentration using SIM.* From Waters Corp. Applications Note, with permission.

an examination of the bottom chromatogram that the two cannot be quantitated properly in the presence of one another at these differing concentrations. The peak profile in the vicinity of 1.42 minutes does not make it possible to even confirm the presence of fenbufen. Because fenbufen has a lower ionization efficiency than ketoprofen and because it is present at only 10% of the concentration of ketoprofen, the signal-to-background ratio (S/B) at 1.42 min in the lower panel, the expected retention time for fenbufen, is not sufficient to show a discernible peak in the SIM profile.

However, when the instrument is set up for SRM under conditions for CAD MS/MS, readily discernible peaks are seen at 1.31 min for ketoprofen and at 1.42 min for fenbufen (in the top and bottom panels of Figure 3-14, respectively) following injection of a sample containing 60 ng mL^{-1} ketoprofen and 6 ng mL^{-1} fenbufen. The improvement in S/B for both compounds derives from monitoring a characteristic CAD reaction for each; the selected reaction for ketoprofen is m/z 255 → m/z 209, whereas that for fenbufen is m/z 255 → m/z 237. Note that even though the ion current signal (8.06e4 read as 8.06×10^4) in the bottom panel is much lower than that (1.43e6 read as 1.43×10^6) in the middle panel, consistent with the lower concentration and ionization efficiency of fenbufen, the S/B is excellent, allowing for a readily discernible peak at 1.42 min.

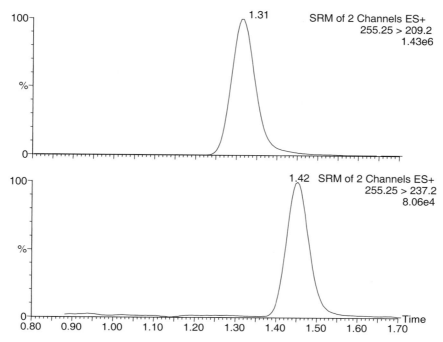

Figure 3-14. *Same analysis as shown in Figure 3-13 except that SRM is used for increased specificity for the two analytes.* From Waters Corp. Applications Note, with permission.

B. An Example Comparing the Specificity of SRM and SIM in the Context of Analyzing a Biological Sample for a Drug Metabolite

This example illustrates the typical analytical strategies used in setting up a selective assay for a targeted compound based on mass spectrometric techniques. The CI mass spectrum of the drug metabolite of interest is shown in Figure 3-15. The peak at m/z 308 in this mass spectrum would seem to be a good candidate to monitor during SIM because its high signal intensity should lead to low detection limits for the drug metabolite [142]. The top panel in Figure 3-16 shows the SIM chromatogram from monitoring the ion current at m/z 308 during the analysis of a urine extract by GC/MS using methane chemical ionization. As can be seen in the SIM profile in the top panel of Figure 3-16, the sample contains many compounds that are capable of generating ion current at m/z 308. The chromatographic peak representing the drug metabolite (indicated by an arrow at the expected retention time) lies on top of considerable background signal from other compounds associated with the sample matrix. Clearly, such S/B is not desirable because of the difficulty in recognizing such a chromatographic peak and in establishing a baseline for quantitation of the drug metabolite under these circumstances.

In an effort to seek better specificity in detecting the drug metabolite, the technique of SRM was considered. The CAD MS/MS product-ion mass spectrum of m/z 308 (the base peak in Figure 3-15) is shown in Figure 3-17. This product-ion spectrum shows that the precursor ion current at m/z 308 is readily converted to a product ion of

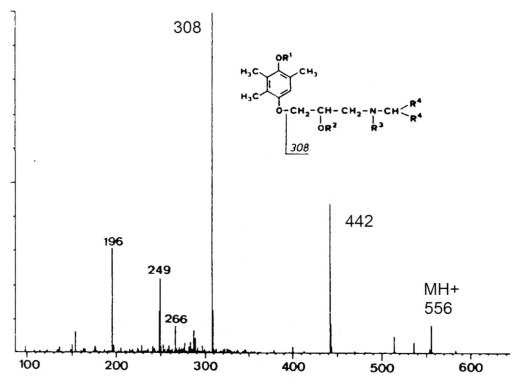

Figure 3-15.   Methane-CI mass spectrum of a drug metabolite.

m/z 266 under CAD MS/MS conditions. The transition of m/z 308 → m/z 266 during CAD is a good candidate CAD reaction to monitor during SRM, which is accomplished by adjusting MS1 to select ion current at m/z 308 to be transmitted to the CAD collision cell and adjusting MS2 to transmit ion current of m/z 266 from the collision cell to the detector.

Analysis of another aliquot of the urine extract again by GC/MS using methane chemical ionization, but using CAD MS/MS with SRM produces the chromatogram shown in the lower panel of Figure 3-16. The SRM chromatogram has fewer peaks than the SIM chromatogram (upper panel of Figure 3-16), indicating that most of the ion current of m/z 308 produced by compounds in the sample matrix cannot be converted to m/z 266 in the collision cell, which provides a much cleaner chromatogram. The much better S/B evident in the SRM profile (in the lower panel of Figure 3-16) makes it much easier to recognize the chromatographic peak at the expected retention time (indicated by the arrow) for the drug metabolite. Furthermore, quantitation of the drug metabolite will surely be more reliable from the SRM profile because of better peak area measurements against the flatter baseline in the vicinity of the peak representing the drug metabolite.

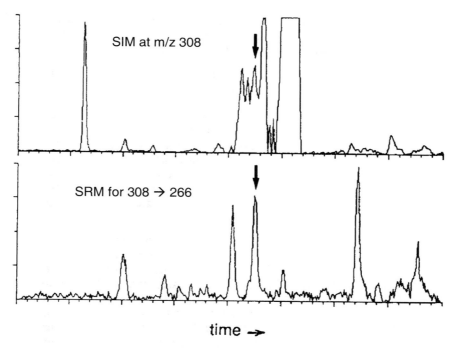

SIM at m/z 308

SRM for 308 → 266

time →

Figure 3-16. (Top) SIM chromatogram for m/z 308. The arrow points to RT of metabolite. (Bottom) SRM chromatogram of same sample using the transition of m/z 308 → m/z 266.

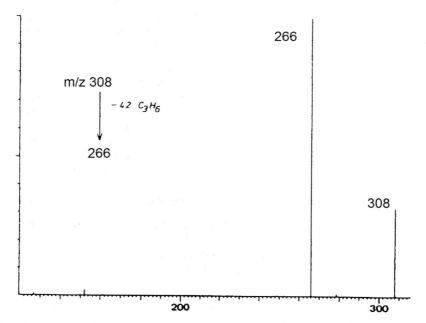

266

m/z 308

-42 C_3H_6

266

308

200   300

Figure 3-17. Product-ion mass spectrum resulting from CAD of the precursor ion with m/z 308.

> Often manufacturers will say that an SRM analysis is much more sensitive than an SIM analysis. The problem is in the choice of the word sensitivity. This term is incorrectly used (in lieu of detection limit) to mean the minimum number of ions required to result in a reproducible signal representing a specified amount of an analyte. Interestingly, the magnitude of the signal in SRM is less than that in SIM because the SRM transition does not necessarily result in a 100% conversion of the precursor ion to the product ion and ions are lost in the passage from MS1 to the collision cell and from the collision cell to MS2. However, as illustrated in the above two examples, SRM can result in a lower limit of detection than can be achieved in SIM because of the higher degree of specificity in SRM.

4. Precursor-Ion Analysis

The precursor-ion analysis (a.k.a. *common product-ion analysis*) reveals the *m/z* value of all ions that are capable of being transformed into a product ion of a given *m/z* value under CAD MS/MS conditions. Such a spectrum is acquired with CAD MS/MS instrumentation by keeping the *m/z* value in MS2 fixed while scanning MS1. A precursor-ion analysis can only be performed with *tandem-in-space* instruments such as the triple-quadrupole mass spectrometer, Q-TOF instruments,[5] and instruments with multiple magnetic and electric sectors. Ion trap instruments (both QIT and FT-ICR) are not capable of a precursor-ion analysis.

An example of a precursor-ion analysis is shown in Figure 3-18 in the context of simplifying data interpretation during analysis of a protein digest [143]. The complexity of the protein digest is represented by the conventional mass spectrum (from ESI with a triple-quadrupole instrument in which both q_2 and Q_3 were operated in the RF-only mode, a Q_1 scan) shown in panel A. The lettering of some peaks refers to specific tryptic fragments; note the appearance of peaks representing multiple-charge ions.

This experiment was designed to identify only those tryptic fragments that contain tyrosine. From previous work, it was known that tyrosine-containing peptides produce the immonium ion of tyrosine (136 Da). Thus, analyzing the digest using a precursor-ion-analysis mode in which Q_3 was adjusted to transmit only ions of *m/z* 136 results in the spectrum in B (middle panel) of Figure 3-18. This spectrum was obtained by scanning Q_1 (MS1) while holding Q_3 (MS2) at a fixed *m/z* value of 136. Inspection of the spectrum in panel B indicates that only six of the dozens of tryptic fragments represented in panel A contain tyrosine.

This procedure has been used to detect lysine- and glycosylated-containing peptides by performing precursor-ion scans of the immonium ion of lysine and of the oxonium ion of HexNAc, respectively [37, 144]; similar work has been done in the identification and quantitation of glycerophospholipids [145]. In an analogous manner, phosphopeptides can be recognized in the negative-ion mode from a precursor-ion scan for the phospho-group [144]. Once methionine-specific ions from CAD were well characterized by precursor-ion scans, they became useful as sequence-sensitive markers [146].

---

[5] When a precursor-ion analysis is performed using a Q-TOF mass spectrometer, a complete product-ion mass spectrum is obtained for each *m/z* value that passes through the scanned transmission quadrupole (MS1). The analysis is accomplished using a mass chromatogram for the *m/z* value corresponding to the desired product ion.

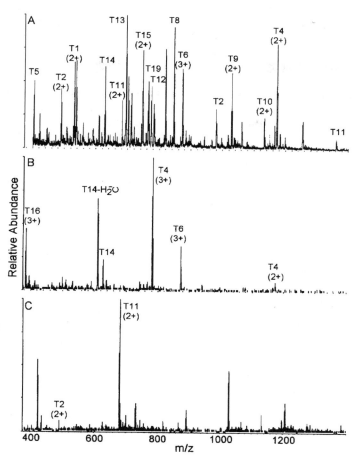

Figure 3-18. *(Top) ESI mass spectrum from direct infusion of a tryptic digest. (Middle) Mass spectrum obtained during a common product-ion analysis for m/z 136. (Bottom) Mass spectrum obtained during a common neutral-loss analysis looking for losses of 24 m/z units.*

5. Neutral-Loss (Common Neutral-Loss) Analysis

The neutral-loss analysis (a.k.a. *constant* or *common neutral-loss analysis*) provides a convenient means of recognizing ions that expel a given fragment (or at least lose a given mass) during the CAD process. The neutral-loss analysis can be used to recognize those components of a mixture that have common structural properties of molecules; it is this feature that can be used to detect some metabolites of a drug, which share some common structural features. To acquire the spectrum from a neutral-loss analysis, the two mass analyzers are scanned at the same rate; however, the second analyzer is offset by a fixed *m/z* value from the first. In this way, only ions that have a specific mass loss (dark matter) in the collision cell will be passed by the second analyzer and be detected (recorded).

An example of a neutral-loss analysis is shown in panel C of Figure 3-18. Recall from the description in the previous section that panel A in Figure 3-18 represents the complexity of a tryptic digest of a protein. The question that needed to be answered was which of the dozens of tryptic fragments represented in panel A contain methionine. From previous work, it was known that methionine loses CH_3SH from its side chain during CAD of double-protonated methionine-containing peptides. This represents a loss of 24 *m/z* units from a double-charge ion (mass of 48 divided by a charge of 2). Only those ions that exhibit this loss will be detected by scanning Q_1 (MS1) and Q_3 (MS2) simultaneously with a 24 *m/z* unit offset for Q_3. The neutral-loss spectrum in panel C of Figure 3-18 indicates that only a half dozen of the many peptides represented in panel A contain methionine by exhibiting this characteristic methionine side-chain loss.

Neutral-loss analyses are only possible with beam-type instruments such as transmission quadrupoles, Q-TOF instruments, and instruments with multiple magnetic and electric sectors.

6. Ion/Molecule Reactions

The MS/MS instrumentation has also been very useful in the study and application of ion/molecule reactions, where the first stage of mass selection in MS/MS can be used to control the recipe [147–156] of the reaction mixture. Ion/molecule reactions offer a means of providing structurally diagnostic information [157–161], including chirality [139, 162–167], or specificity as for nucleosides [168], distonic ions [169], organic nitrates [161], and in a variety of organometallic applications [170]. The feasibility of using beta5-cyclopentadienylcobalt ion (CpCo^{*+}) as a cationization reagent for saturated hydrocarbon analysis has been demonstrated in studies using FTMS [171]. The key to novel top-down approaches to analyses of macromolecules are charge permutation reactions, some of which are based on ion/molecule reactions [172].

Early applications of ion/molecule reactions for analytical advantage were based on the triple-quadrupole mass spectrometer in large measure because of the ease by which the GC inlet could be used for sample introduction [173, 174]. The pentaquadrupole as used by Rolando and coworkers [149] and by others allowed CAD to be used in the second RF-only collision cell to help identify and characterize products of ion/molecule reactions conducted in the first RF-only collision cell; an extension of this work used dimethyl ether in ion/molecule reactions to distinguish the 3-tolyl ion from the isomeric benzyl and tropylium cations [175]. The ICR or the FTMS [176] approach to ion/molecule reactions has been used extensively for multiple generations of ion/molecule reactions by conducting *tandem-in-time* procedures. The quadrupole ion trap mass spectrometer also allows for multiple stages of MS as a function of time [156], and it is readily compatible with the chromatographic inlet system for sample introduction, which continues to make this mode of instrumentation for MS/MS attractive for applications [177].

7. Hybrid Instrumentation for MS/MS and Ion Mobility Spectrometry (IMS)

As described in the previous chapter, ion mobility spectrometry (IMS) provides a useful means of characterizing ions by a process that might be conceptually thought of as gas-phase electrophoresis; i.e., separating ionic species based in part on their cross-sectional size (related to their structural conformation). Whereas there have been several proof of principle reports of combining IMS with mass spectrometry, these have usually been done in academic laboratories to capture the mass spectra of drift-time-resolved ions at the exit of an ion mobility spectrometer [178, 179] or the use of IMS to time-resolve conformationally distinct species of a given *m/z* value [180, 181].

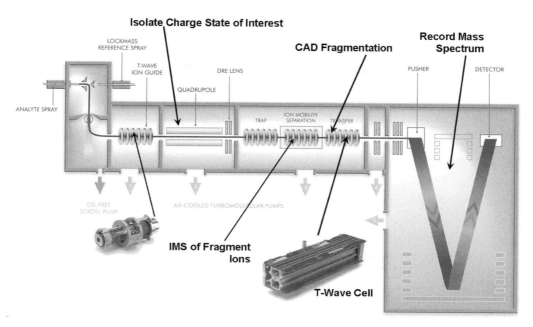

Figure 3-19. Illustration of Waters Synapt HDMS hybrid mass spectrometer.

A commercially available hybrid instrument, introduced as the Synapt HDMS by Waters Corp. at the American Society for Mass Spectrometry Annual Conference held in Seattle, WA in 2006, nicely integrates IMS into a *tandem-in-space* mass spectrometer, as illustrated schematically in Figure 3-19. The approach used by Waters Corp. was to replace the chamber usually reserved for a single collision cell in a conventional *tandem-in-space* instrument with a module consisting of three consecutive electrodynamic lensing elements shown in one of the central chambers in Figure 3-19. Each of these lensing elements is equipped with "traveling wave" technology (T-wave devices) which provides control of the electric field to either move ions through the lens or trap them within it for designated periods of time [182]. The first of these T-wave elements is designed to trap ions that are coming out of the transmission quadrupole in Figure 3-19. As indicated by the notation in Figure 3-19, CAD fragmentation can be performed in this trap; importantly, the T-wave technology allows the trapped ions to be expelled from this trap at a designated time, thereby serving as a gate to inject a bolus of ions into the second lensing element where IMS is performed. The third lensing element in this central chamber in Figure 3-19 is designated as a transfer element, which improves transmission of drift-time-resolved ion packets to the *oa*TOF-reflectron; but in other applications, another stage of CAD can be performed in this third lensing element [183]. The additional dimension of data that can be produced by this type of instrumentation is illustrated in Figure 3-20, which is a plot of *m/z* value vs ion mobility drift time, showing, for example, that ions of *m/z* 20,800 (see arrow at left) are mostly of an open conformation (longer drift time as encircled as group A) than those with a more compact conformation (shorter drift time as encircled as group B) [184]. A key advantage of this type of analyzer is that it allows better characterization of minor constituents of a complex mixture because the otherwise overwhelming isobaric major constituents have a different drift time, and thus enter the CAD chamber or the *m/z* analyzer at a different time. This instrument has applications in both in the field of biological polymers and small molecules.

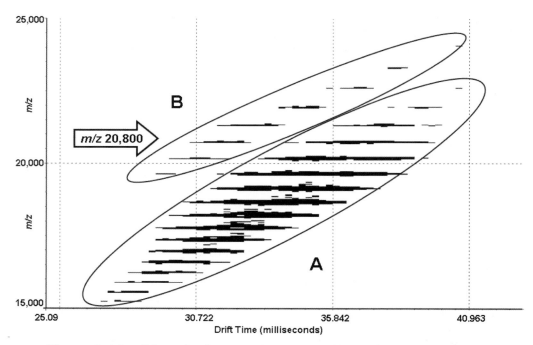

Figure 3-20.   Plot of m/z vs drift time (IMS) for CAD product ions.
As reprinted from research by Albert Heck, Utrecht University, and made available to ODS from Mark McDowall at Waters.

V. Analyte Identification from MS/MS Data

1. Introduction

Too often it is said, "Rules used for the interpretation of GC/MS data (fragmentation of molecular ions, odd-electron ions) cannot be used in the interpretation of data obtained by LC/MS/MS (fragmentation of even-electron ions)." This statement is completely untrue. There is much more to the interpretation of a mass spectrum than deducing a fragmentation mechanism. The odd or even m/z value of the nominal mass of the precursor ion allows a determination as to whether the analyte contains an odd-number of nitrogen atoms, the *Nitrogen Rule* (do not forget about mass defect and instrument resolving power – remember that the *Nitrogen Rule* is based on the *nominal mass*). Examination of the isotope cluster representing the precursor ion will reveal information as to whether there are X+2 elements like Cl, Br, Si, S, or K present (usually K is only present in ions formed by ESI). An accurate mass measurement of the precursor ion can lead to an exact elemental composition, depending on the accuracy of the measurement. Obvious neutral losses such as the loss of a water or ammonia molecule also provide information about the analyte. Tools to determine the presence of a substructure by comparison of a product-ion mass spectrum with a collection of EI mass spectra like the NIST05 Database will be described in the next section; this and other tools can facilitate proposals of fragmentation mechanisms, as is the case with the NIST MS Interpreter, Thermo Fisher's Xcalibur utility Mass Frontier, or the Advanced Chemical Development's (ACD/Labs) MS Fragmentor, in an effort to help deduce the correct structure.

All fragmentation of ions occurs as a result of the breaking of covalent bonds. Odd-electron ion fragmentation is often (but not always) initiated by a radical site, whereas even-electron ion fragmentation is initiated by a charge site. A radical will often be lost from an odd-electron ion to form an even-electron fragment ion. Because the energy of collisional activation is low in most MS/MS and in-source CAD experiments, there is a tendency for MH^+, $[M - H]^+$, $[M - H]^-$, and $[M + X]^-$ to undergo rearrangements that result in the expulsion of a molecule to form an even-electron fragment ion.

Another complicating factor in dealing with mass spectra obtained by CAD is the fact that the product ions produced can also undergo collisional activation and dissociate via in-source CAD and *tandem-in-space* MS/MS using a collision cell as in the case of the triple-quadrupole mass spectrometer. This secondary fragmentation is not so much of an issue when MS/MS is carried out in a QIT because only the precursor ion (and ions within a few *m/z* values of that of the precursor ion) becomes activated.

2. Identifying an Unknown Using a Product-Ion Mass Spectrum

Keeping the above principles in mind, look at the mass spectrum obtained using APCI and in-source CAD of an unknown analyte in Figure 3-21. The nominal *m/z* value of the MH^+ representing the analyte is 309, an odd number which means that the analyte does not contain an odd number of nitrogen atoms. The isotope cluster associated with the MH^+ peak does not indicate the presence of Cl, Br, S, or Si. Because the data were acquired using APCI, K would not be expected to be present and the appearance of the isotope cluster confirms this assumption to be true. There is a peak at *m/z* 213 representing an ion formed by a loss of 56 Da. Looking for elemental compositions that contain C, H, O, and/or N and that have a mass of 56 Da using the NIST Isotope Calculator Program (see Figure 3-22 for example) reveals:

C_2H_2NO for C, H, N, and O
C_3H_4O for C, H, and O
CH_2N_3, $C_2H_4N_2$, and C_3H_6N for C, H, and N
C_4H_8 for C and H

Only the C_3H_4O, $C_2H_4N_2$, and C_4H_8 formulas represent molecules. Because the peak at *m/z* 309 represents a protonated molecule, it is most likely that the ion represented by this peak will expel a molecule when it fragments, which means that one of these three elemental compositions could be the loss. The $C_2H_4N_2$ elemental composition has two rings and/or double bonds, and it has to represent one of these two structures, $H_2N-C{\equiv}C-NH_2$ or $HN{=}CH-CH{=}NH$, neither of which falls into the KIS (keep it simple) category. The C_3H_4O elemental composition also has two rings and/or double bonds and could be $H_2C{=}CHC(O)H$ (propenal), $HC{\equiv}C-CH_2-OH$ (another possible isomer $H_3C-C{\equiv}C-OH$), or a four-membered ring with a C=C and an O atom in the ring; again, these are structures that do not follow the KIS rule. Therefore, the best possibility is the loss of C_4H_8, which would strongly suggest the presence of a butyl moiety because a good structure for this elemental composition containing one ring or double bond is that of butene. A molecule of butene could have been expelled from the MH^+.

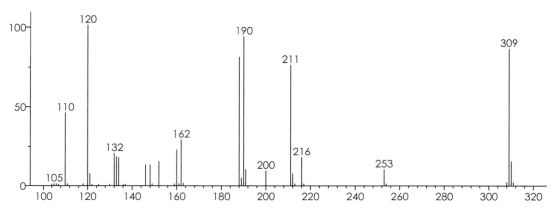

Figure 3-21. In-source CAD product-ion mass spectrum of a protonated molecule obtained by APCI.

Formula Generator (Element entry)

File View Options Help

Mass 93 Elements: C H N

Ions	O+E	RDB	Exact MW	C	H	N
$CH_{11}N_5$	Odd	0-4	93.10144	1	11	5
$C_2H_{13}N_4$	Even	0-2.5	93.11402	2	13	4
C_3HN_4	Even	5.5-9.5	93.02012	3	1	4
$C_3H_{15}N_3$	Odd	0-1	93.12659	3	15	3
$C_4H_3N_3$	Odd	5-8	93.03270	4	3	3
$C_4H_{17}N_2$	+	0	93.13918	4	17	2
$C_5H_5N_2$	Even	4.5-6.5	93.04527	5	5	2
C_6H_7N	Odd	4	93.05785	6	7	1

Number of different formulas is 8

Figure 3-22. Results of a formula search of mass of 93 Da containing C,H, and N.

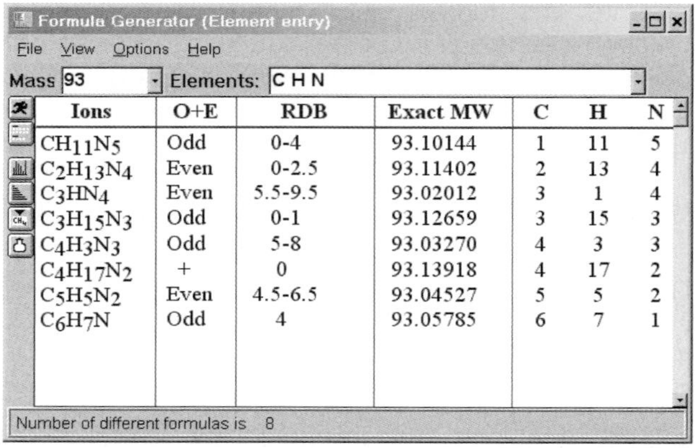

The peak at *m/z* 216 represents the loss of 93 Da. Keeping with the idea that the most likely loss from the MH^+ is a molecule, the only possible elemental composition must include an odd number of nitrogen atoms because the loss is represented by an odd number. Using the NIST Isotope Calculator Program for C, H, N, and O and C, H, and N, there are nine possible elemental compositions. Only four of these have only one nitrogen atom and only one of the compositions has one nitrogen atom and only C, H, and N, C_6H_7N. A formula search using the NIST MS Search Program (Figure 3-23) returns only seven compounds for this elemental composition: the three regioisomers of methyl pyridine, three compounds that contain triple bonds, and aniline. Aniline is the simplest of these possible structures and should be considered a good candidate for the neutral loss of 93 Da.

From this analysis, it can be concluded, based on the mass spectral data, that:

1. The analyte has a nominal mass of 308 Da.
2. There are no X+2 elements other than the possibility of oxygen.
3. There is probably a butyl moiety due to the loss of 56 Da (loss of a butene molecule).
4. There appears to be the loss of a molecule of aniline.
5. Based on the probable loss of aniline, the analyte contains an even number of nitrogen atoms (two is a good place to start), if N atoms are in fact present.

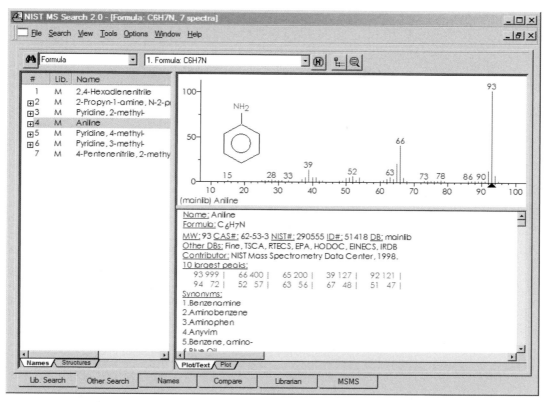

Figure 3-23. *Results of an NIST formula search for C_6H_7N with the best candidate highlighted.*

Losses of 98 (*m/z* 211), 109 (*m/z* 200), 119 (*m/z* 190), and 121 (*m/z* 118) do not conjure up any obvious structural features. The same is true for the peaks at *m/z* 162, 160, 134, 133, 132, 120, and 106. The peak at *m/z* 93 could represent an aniline OE ion and the peak at *m/z* 77 could represent a phenyl EE ion, but neither of these peaks provides additional information other than that already deduced.

The next step in trying to identify the compound that generated this product-ion mass spectrum is to see what information can be gained by an MS/MS search and the substructure identification routine using the NIST Mass Spectral Search Program (Figure 3-24). This search is carried out against the NIST05 Database of EI spectra of 163K+ compounds, and the resulting "hit list" is evaluated for the presence and absence of various substructures that are suggested to be present. Some interesting results are obtained. Information not previously known:

1. There is a heterocyclic ring that has nitrogen (.N. and het_ring).
2. There are atoms of oxygen present.
3. There are carbonyl groups present.
4. There is one or more aromatic rings bound to a nitrogen atom.
5. There is a (O)CN functionality (amide).
6. The analyte probably contains only C, H, O, and N.

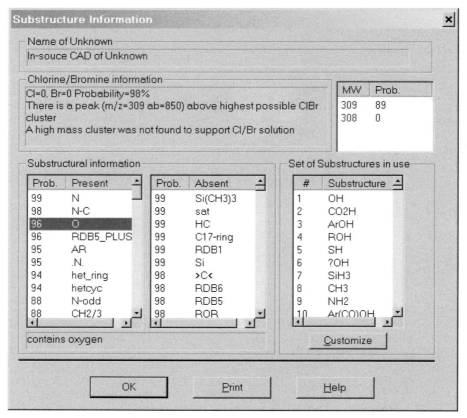

Figure 3-24. Results of substructure information after an MS/MS search.

There is some contradictory information, such as an 81% probability that the ion contains an odd number of nitrogen atoms. On the other hand, there is an 81% probability that there is a CH_3 group, which supports the presence of a C_4H_9 group.

Each spectrum in the NIST05 (NIST/EPA/NIH Mass Spectral) Database is indexed according to the substructures associated with the compound [185]. Substructures are ways of characterizing the structure of a compound. Substructures include heteroatoms, the number of rings and/or double bonds associated with the structure, structural moieties such as aromatic rings, alkyl radicals, carbonyl groups, ether moieties, etc. Currently, the NIST05 Database uses a set of 541 different substructures. The percent probability of the presence and probability of the absence of the various substructures is calculated by an evaluation of the "hit list" obtained during a search of an unknown spectrum against the Database. Best results come about when a neutral loss, hybrid similarity/neutral loss, or MS/MS search is performed on only the primary Database (excluding replicate spectra). Substructure indexing is only provided with the NIST05 Database.

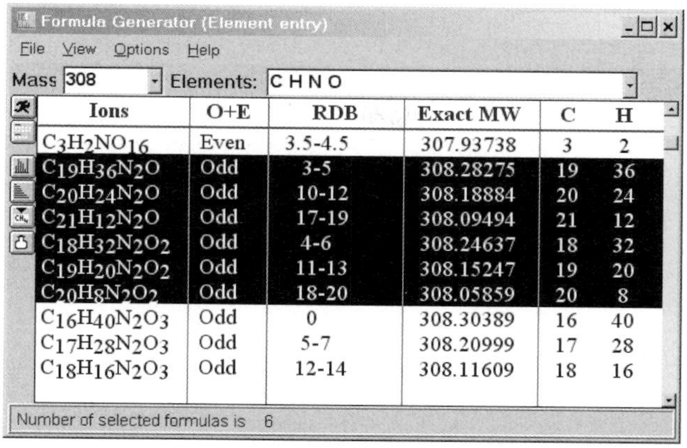

Formula Generator (Element entry)

File View Options Help

Mass 308 Elements: C H N O

Ions	O+E	RDB	Exact MW	C	H
$C_3H_2NO_{16}$	Even	3.5-4.5	307.93738	3	2
$C_{19}H_{36}N_2O$	Odd	3-5	308.28275	19	36
$C_{20}H_{24}N_2O$	Odd	10-12	308.18884	20	24
$C_{21}H_{12}N_2O$	Odd	17-19	308.09494	21	12
$C_{18}H_{32}N_2O_2$	Odd	4-6	308.24637	18	32
$C_{19}H_{20}N_2O_2$	Odd	11-13	308.15247	19	20
$C_{20}H_8N_2O_2$	Odd	18-20	308.05859	20	8
$C_{16}H_{40}N_2O_3$	Odd	0	308.30389	16	40
$C_{17}H_{28}N_2O_3$	Odd	5-7	308.20999	17	28
$C_{18}H_{16}N_2O_3$	Odd	12-14	308.11609	18	16

Number of selected formulas is 6

Figure 3-25. Results of formula search of mass of 308 Da containing C, H, N, and O.

Use the NIST Formula Calculator (Figure 3-25) to determine the possible elemental compositions for a compound that contains only C, H, O, and N and has a nominal mass of 308. Limit the possibilities to only those that have two atoms of nitrogen. There are three possibilities that have only one atom of oxygen: $C_{19}H_{36}N_2O$ (0, number of hits from formula search), $C_{20}H_{24}N_2O$ (16), and $C_{21}H_{12}N_2O$ (0). There are also three possibilities that have two atoms of oxygen: $C_{18}H_{32}N_2O_2$ (0), $C_{19}H_{20}N_2O_2$ (29), and $C_{20}H_8N_2O_2$ (0). Any one of these elemental compositions could generate multiple structures. Checking each of the six formulas against the NIST MS Formula Search of the NIST05 Database resulted in only $C_{20}H_{24}N_2O$ (16) and $C_{19}H_{20}N_2O_2$ (29) returning hits. Of these 45 hits, only two compounds were found to be listed in other databases. One of the hits from the Formula Search for the $C_{20}H_{24}N_2O$ elemental composition was listed in one other database. This hit was *N*-phenyl-*N*-[1-(phenylmethyl)-4-piperidinyl]-acetamide, which is also listed in the European Index of Commercial Chemical Substances; this compound has only three synonyms in the NIST05 Database, none of which are common names. Only one of the hits for the Formula Search (phenylbutazone) of the elemental composition $C_{19}H_{20}N_2O_2$ was in other databases. Phenylbutazone is in six of the possible nine other databases and has a total of 148 synonyms (many of which are common names) listed in the NIST05 Database compared with three synonyms for *N*-phenyl-*N*-[1-(phenylmethyl)-4-piperidinyl]-acetamide.

Based on the large number of common names and the entry in multiple other databases, phenylbutazone is the most probable candidate for the identity of the unknown. Examination of the structure of phenylbutazone (Figure 3-26) shows that all the assumptions from the examination of the spectrum and the results of the NIST MS/MS formula search followed by the substructure identification are supported. This analyte has a complex structure. As will be seen later in this section, when the EI and CAD mass spectra are re-evaluated, the results are far from intuitive; however, the peaks in the CAD mass spectrum can be rationalized by proposed fragmentation mechanisms of the phenylbutazone structure.

Comparing the EI mass spectrum of phenylbutazone in Figure 3-26 with the product-ion mass spectrum in Figure 3-21 reveals little if any similarities other than the loss of a molecule of butene from both the MH^+ and the $M^{+•}$. This lack of similarity points out the differences by which ions representing the same intact molecule can fragment. However, the example also points out that the identity can be obtained by methods other than rationalizations through fragmentation mechanisms.

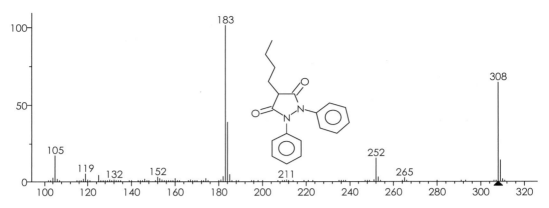

Figure 3-26. EI mass spectrum of phenyl butazone.

3. Similarities between EI and Product-Ion Mass Spectra

On the other hand, sometimes $M^{+\bullet}$ and MH^+ behave similarly when they fragment to produce similar mass spectra. Look at the product-ion and EI mass spectra in Figure 3-27. Both spectra exhibit intense peaks (the base peaks) representing the loss of a molecule of ketene (HC=C=O). Based on the *m/z* value of the peak representing the intact molecule in both spectra, it is obvious that the analyte contains an odd number of N atoms. If the $M^{+\bullet}$ has a tendency to undergo a rearrangement and expel a molecule, the MH^+ will undergo a similar rearrangement to expel the same molecule. Scheme 3-1 and Scheme 3-2 illustrate the fragmentation of the MH^+ and the $M^{+\bullet}$, respectively.

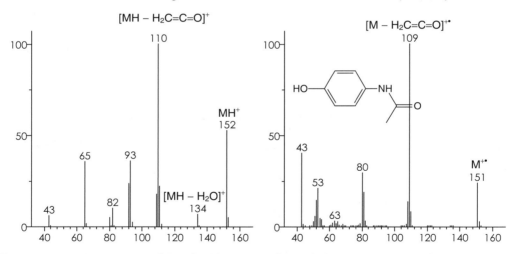

Figure 3-27. In-source CAD (left) and EI (right) fragmentation of acetaminophen.

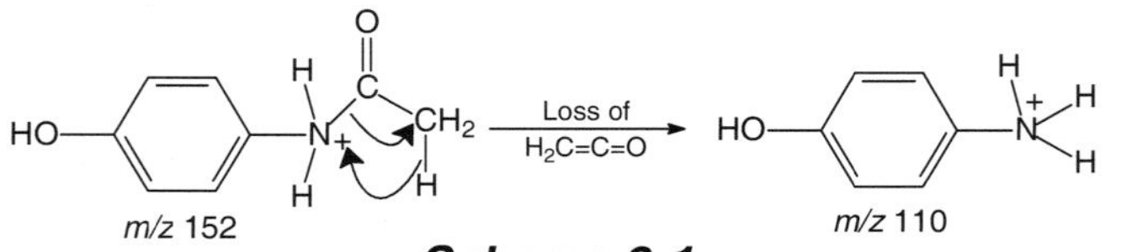

Scheme 3-1

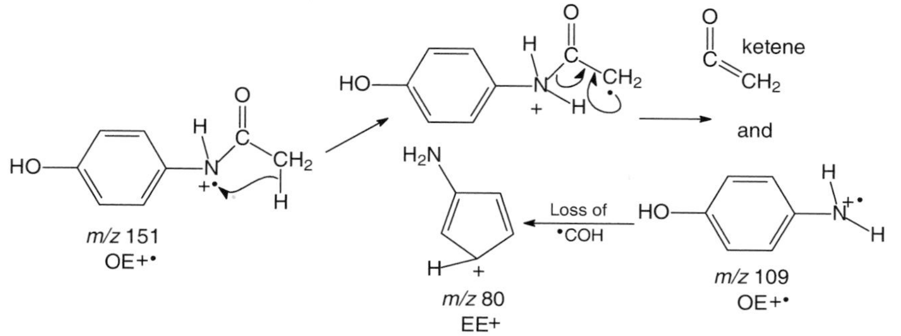

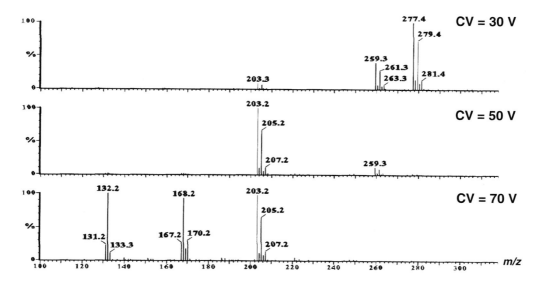

Scheme 3-2

4. Another Way of Using Substructure Identification

Another interesting situation involving CAD is the fragmentation of clenbuterol, a performance-enhancement drug. Figure 3-28 shows three spectra obtained at increasingly higher energy values. The top spectrum shows peaks representing a protonated molecule that clearly has two atoms of chlorine in its elemental composition and an ion formed by the loss of water (H_2O) from the protonated molecule. The middle mass spectrum, acquired at a higher cone voltage, has no peaks representing the protonated molecule, a weak set of peaks representing the $[MH - H_2O]^+$ ion, and a base peak 56 *m/z* units lower than the peak representing the $[MH - H_2O]^+$ ion. This pattern indicates that the ion (which obviously has retained the two atoms of chlorine) represented by the cluster of peaks 56 *m/z* units lower than the peaks representing the $[MH - H_2O]^+$ ion is formed by fragmentation of the $[MH - H_2O]^+$ ion. The third mass

Figure 3-28. ***Three in-source CAD mass spectra of clenbuterol at three successively higher cone voltages (CV).***

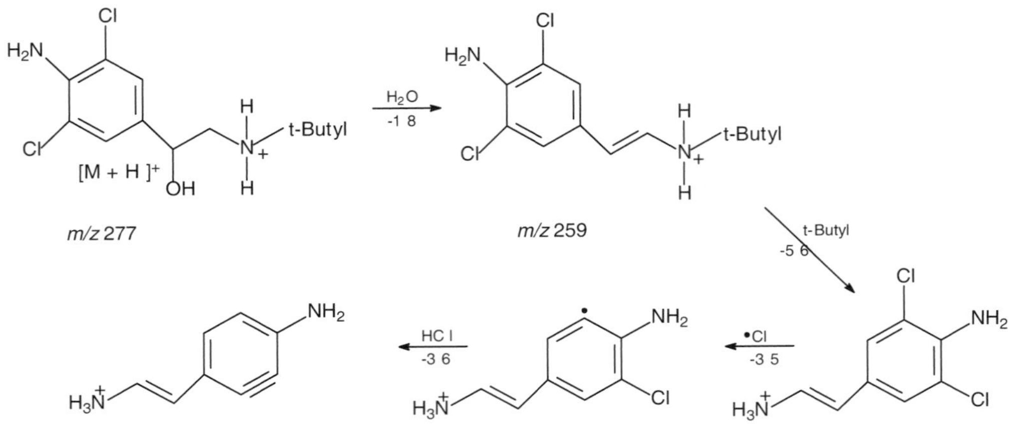

Scheme 3-3

spectrum, acquired at a yet higher cone voltage than the second spectrum, reveals that the $[M - 18 - 56]^+$ will become collisionally activated to lose a chlorine radical followed by the loss of a molecule of HCl, as supported by the isotope peak patterns at m/z 168 and m/z 132. Examination of Scheme 3-3 shows that it would be difficult to support these subsequent losses from the protonated molecule and its fragments with mechanisms.

There is nothing indicated by any of the mass spectral data that the analyte contains either nitrogen or oxygen. However, when an MS/MS search was carried out on the product-ion mass spectrum (Figure 3-29) obtained using a 3D QIT mass spectrometer with the protonated monoisotopic molecule designated as the precursor ion, the results showed that both nitrogen and oxygen were present based on the substructure identification report (Figure 3-30). It should not go unnoticed that the substructure identification is not infallible. The report also states that the nitrogen, and maybe the oxygen, is present in heterocyclic rings, which, of course, is not the case.

Using the NIST Isotope/Formula Calculator for a formula with a nominal mass of 276 Da (nominal mass of the MH$^+$ was 277 Da) that contained two atoms of chlorine, two atoms of nitrogen (because the MH$^+$ has an odd-m/z value, the number of N atoms has to be even according to the *Nitrogen Rule*), and one or two atoms of oxygen, in addition to carbon and hydrogen, resulted in five elemental compositions:

$$C_{12}H_{18}Cl_2N_2O$$
$$C_{13}H_6Cl_2N_2O$$
$$C_{10}H_{26}Cl_2N_2O_2$$
$$C_{11}H_{14}Cl_2N_2O_2$$
$$C_{12}H_2Cl_2N_2O_2$$

The first, second, and fourth formulas gave one and only one "hit" when a formula search was carried out. The hit for the second and fourth formulas were of compounds that were not in any other databases (explained in Chapter 6) besides the NIST and had no common names or synonyms. The hit for the first formula was the spectrum of clenbuterol, which is in three other databases (Registry of Toxic Effects of Chemical Substances, U.S. Pharmacopoeia/U.S.A.N. and European Index of Commercial Chemical Substances), and has four common (or trade names) and four chemical names. If this had been an unknown spectrum, clenbuterol would have been a very good candidate for the correct identity.

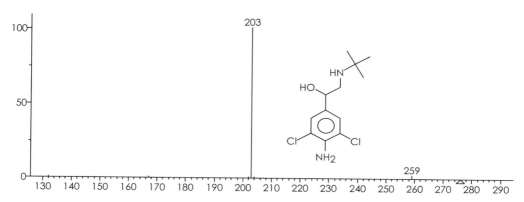

Figure 3-29. *Product-ion mass spectrum of MH⁺ of clenbuterol used to obtain substructure information.*

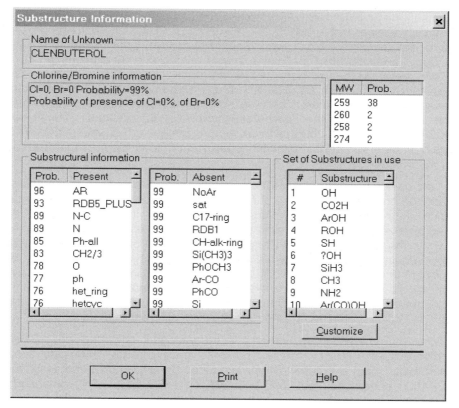

Figure 3-30. Result of substructure information after MS/MS search.

When trying to extract information from spectra produced by CAD the behavior exhibited in secondary fragmentations of aliphatic amines and alcohols in EI should always be kept in mind. Always look for the neutral losses of molecules, as opposed to radicals, in product-ion mass spectra. This is not to say that product-ion mass spectra cannot exhibit loss of radicals, but it is much rarer than in spectra produced by electron ionization.

5. Searching of Product-Ion Spectra against Standardized Databases

The question of why there are no commercially available databases of product-ion spectra is always raised in any discussion of CAD data. One of the reasons for the question is the misunderstanding of EI mass spectral databases and Library Search Programs that have for so long been with GC/MS. This misunderstanding often leads to the fatal statement, "The library search said this unknown was diethylbatguanna, so it has to be diethylbatguanna!" Library search programs (or for that matter, computers) do not identify compounds; analysts identify compounds from their mass spectra and, in most cases, other data and information.

The problem with developing a database of product-ion mass spectra is the variability of the ions formed and their abundances. Different instrument types (*tandem-in-time vs tandem-in-space*) have different efficiencies with respect to formation of product ions. The *tandem-in-space* instruments have the problem of fragmentation of product ions in the collision cell, which complicates abundance issues and the number of peaks that may appear in the spectrum. A search of *Chemical Abstracts* on December 30, 2006, of the research topic "Library Search of MS/MS Data" resulted in 17 citations where "Library" and "MS/MS Data" were closely related and 39 citations where "Library" and "MS/MS Data" were anywhere in the reference. A search of "Library Search of MS/MS Spectra" resulted in 66 citations where "Library" and "MS/MS Spectra" were closely associated and 87 references where the two concepts appeared anywhere in the reference. After removing the duplicates, there were 62 citations. These covered a period of 24 years (1982 to 2006) with six articles between 1982 and 1994. The first of this series of papers was concerned with establishing criteria for a standardized database [186].

As is obvious from the two product-ion spectra in Figure 3-10 (top acquired with a 3D QIT and the bottom acquired with a TQMS), there can be significant differences in the appearance of a given compound due to the type of instrument used. As NIST discovered in developing what is to date the largest database of product-ion mass spectra (included with the NIST MS Search Program, which is part of the NIST05 Database), not only is the instrument type a factor but the collision energy and number of collisions is very significant in spectral reproducibility. Product-ion mass spectra produced by CAD in a collision cell did not exhibit significant differences when data were acquired on the same type of instrument from different manufacturers. There is much more variation for in-source CAD spectra obtained on instruments from different manufacturers.

Other than the NIST product-ion spectra database, which includes ~5.2K spectra of ~2K different ions (1671 positive and 341 negative ions), there are some other product-ion mass spectral databases; one produced by John Halket and associates [187] contains ~1000 compounds obtained with ESI on a 3D QIT. This database can be downloaded from:

http://www.rijkswaterstaat.nl/rws/riza/home/projecten/spectrum_library/cicid_downloads.html

Table 3-6 is a list of known product-ion mass spectral databases provided in a private communication by John Halket, who is at the Bioanalytical Services, Centre for Chemical Sciences, Royal Holloway, University of London, Egham, Surrey, UK.

Table 3-6. Known product-ion mass spectral databases.

MS Type	Compound Type	Approx. Number Spectra	Approx. Number Compds.)	Polarity	Reference / Further Information
Sciex	*Drugs*	*1200*	*400*	*Pos./ Neg.*	*Gergov, 2000;* merja.gergov@helsinki.fi
MSD	Drugs	41	41	Pos./ Neg.	Lips, 2001; jos_lips@agilent.com
Quattro	Chemicals	88	88	Pos./ Neg.	Little, 2001; jllittle@eastman.com
Sciex	*Drugs/tox.*	*1200*	*1200*	*Pos./ Neg.*	*Marquet, 2001;* marquet@unilim.fr
Sciex	*Pesticides*	*90*	*90*	*Pos./ Neg.*	*Schreiber, 1999;* www.chemicalsoft.*de*
Sciex	*Explosives*	*50*	*50*	*Pos./ Neg.*	*Schreiber, 2000;* www.chemicalsoft.*de*
MSD	Pesticides	100	100	Pos./ Neg.	Hough, 2000
Sciex	*Drugs*	*1293*	*431*	*Pos / Neg.*	*Weinmann, 2000;* www.chemicalsoft.*de*

Some triple-quadrupole MS/MS libraries for LC/MS

Sciex	*Drugs*	*1200*	*400*	*Pos./Neg.*	*Gergov, 2000;* merja.gergov@helsinki.fi
TSQ	Drugs	70	N/a	Pos.	Josephs, 1995
TSQ	Pesticides	197	150	Pos./Neg.	Kienhuis,2000; p.kienhuis@riza.rws.minvenw.nl
Quattro	Pesticides	60	60	Pos.	Slobodnik, 1996
Sciex	*Drugs*	*2151*	*717*	*Pos./Neg.*	*Weinmann, 2003/Version 1..0;* www.chemicalsoft.de/ *(now available for sale!)*

Some ion trap MS/MS libraries

LCQ	Misc.	1200	1200	Pos.	Baumann, 2000; j.halket@rhul.ac.uk
LCQ	Drugs	80	80	Pos.	Franz.dussy@bs.ch
LCQ	Drugs	75	50	Pos.	Fitzgerald, 1999; rlfitzgerald@vapo.ucsd.edu
Esquire	Drugs	700	200	Pos.	Sander, 1998 (Bruker)
MSD/Trap	Drugs	(***)	(***)	Pos.	B. Wuest (Agilent)
LCQ	Drugs	43,000	29,000	Pos./Neg.	Josephs, 2003; jonathan.josephs@bms.com
LCQ	Nat.product	200	100	Pos./Neg	Kite (Kew, London)
LCQ	Nat.product	200	100	Pos./Neg.	Tolstikov (Max Planck, Potsdam)
LCQ	Nat.product	40	100	Pos./Neg.	Messner (Univ. Gent)
LCQ	Nat.product	2000	2000	Pos./Neg.	Stead, 2001 (GSK)
LCQ	Pesticides	100	60	Pos./Neg.	Stan, 2001 (TU Berlin)
MSD/Trap	Pesticides	3000	1000	Pos.	Robert_voyksner@lcmslimited.com

*** not available
LCQ is a 3D QIT and TSQ is a triple-quadrupole mass spectrometer manufactured by Finnigan Corp. (now Thermo Fisher); MSD/Trap is the 3D QIT sold by Agilent Technologies and is a jointly developed instrument between Bruker Daltonics and Agilent; Quattro is a triple quadrupole manufactured by Waters Corp. (formerly Micromass).

Most of the literature citations regarding product-ion mass spectral libraries are on data acquired on a single instrument (or at least a single type of instrument in the same laboratory) using a single set of conditions. Performance is evaluated on determinations made of unknowns on the same instrument used to collect the database. The only report of a standardized product-ion database is that of NIST, which has a database of spectra acquired on different types of instruments (including in-source CAD) and under multiple collision energies.

VI. Concluding Remarks about MS/MS

As was pointed out in another chapter, APCI was developed about ten years before MS/MS. APCI was never widely accepted for use with LC analytes because the only information provided for an unknown was its mass. When APCI was used with high-resolving power instruments, the mass measurement could result in an unambiguous elemental composition (if the resolving power was high enough and the mass was low enough). Use with transmission quadrupole mass spectrometers, which had exploded into use, APCI would only provide a nominal mass. The paradigm for mass spectrometry at that time was to determine structure from fragmentation data or at least compare fragmentation patterns of unknowns with fragmentation patterns of known compounds. APCI did not offer this possibility. When MS/MS with the transmission quadrupole (triple-quadrupole) mass spectrometer was developed, the lack of APCI of LC analytes could have been addressed, but it was not. Until the development of electrospray and ten years after the development MS/MS, there was only one company that offered an APCI MS/MS instrument and it was not designed for liquid sample introduction. That was the Targa from Sciex in Canada. Once ESI was introduced, and the similarity of being a system that produced gas-phase ions at atmospheric pressure was realized, the combination with MS/MS was a natural. MS/MS provided the ability to extract the structure based on fragmentation patterns.

Of course, MS/MS provides much more than just fragmentation. Through the power of selected reaction monitoring, MS/MS provides an unprecedented low level of detection. Just as chemical ionization is often viewed as a necessary companion to electron ionization in GC/MS (because EI spectra of all compounds capable of being ionized in the gas phase do not exhibit a molecular ion peak) to determine a mass for the analyte, MS/MS is a necessary companion for APCI or ESI. Much can be accomplished with a single-stage instrument and in-source CAD, but so much more can be accomplished with MS/MS.

An important aspect of designing an MS/MS experiment includes the potential role of isotope peaks. This is especially true for analytes containing the X+2 elements of Cl, Br, S, and Si when using a unit-resolution mass spectrometer. By choosing an *m/z* value range rather than the *m/z* value for a monoisotopic ion, the fate of the X+2 atoms can be more easily followed.

References

1. Dibeler V and EU C, Metastable Ions with the Mass Spectrometer. *Phys. Rev.* **68:** 54-55, 1945.
2. Dibeler V, Fox R and Condon E, Metastable Ions Formed by Electron Impact in Hydrocarbon Gases. *Phys. Rev.* **69:** 347-356., 1946.
3. Barber M and Elliott R, Comparsion of Metastable Spectra from Single and Double Focusing Mass Spectrometers. In: *ASTEM E 14 Conference on Mass Spectrometry and Allied Topics, Montreal, Canda, June 7 12, 1964 1964*, pp. 150.
4. Hunt W, Jr., Huffman R, Saari J, Wassel G, Betts J, Paufve E, Wyess W and Fluegge R, TOF-MS Adapted for Studying Charge Transfer, Ion Dissociation, and Photoionization. *Rev. Sci. Instrum.* **35:** 88-95, 1964.
5. Ferguson R, McCulloh E and Rosenstock H, Observations of the Products of Ionic Collision Processes and Ion Decomposition in a Linear, Pulsed TOF-MS. *J. Chem. Phys.* **42:** 100-106, 1965.
6. Haddon WF and McLafferty FW, Metastable Ion Characteristics. VII. Collision-induced Metastables. *J. Am. Chem. Soc.* **90:** 4745-6, 1968.
7. Jennings K, Collision-induced Decompositions of Aromatic Molecular Ions. *Inter. J. Mass Spectrom. Ion Phys.* **1:** 227-235, 1968.
8. Cooks RG, Beynon JH, Caprioli RM and Lester GR, *Metastable Ions.* Elsevier, New York., 1973.
9. De Hoffmann E, Tandem Mass Spectrometry: A Primer. *J. Mass Spectrom.* **31:** 129-137, 1996.
10. Busch KL, Glish GL and McLuckey SA, *Mass Spectometry/Mass Spectrometry.* VCH Publishers Inc., New York, 1988.
11. McLafferty FW, Tandem Mass Spectrometry. Wiley, New York., 1983.
12. McLafferty FW, 25 years of MS/MS. *Org. Mass Spectrom.* **28:** 1403-1406, 1993.
13. Cooks RG, Jo S-C and Green J, Collisions of organic ions at surfaces. *Applied Surface Science* **231-232:** 13-21, 2004.
14. Ouyang Z, Grill V, Alvarez J, Doerge CH, Gianelli L, Thomas P, Rohrs HW and Cooks RG, A multiquadrupole tandem mass spectrometer for the study of ion/surface collision processes. *Review of Scientific Instruments* **73**(6): 2375-2391, 2002.
15. Dongré AR, Somogyi Á and Wysocki VH, Surface-induced Dissociation: An Effective Tool to Probe Structure, Energetics and Fragmentation Mechanisms of Protonated Peptides. *J. Mass Spectrom.* **31:** 339-350, 1996.
16. Bier ME, Amy JW, Cooks RG, Syka JEP, Ceja P and Stafford G, A Tandem Quadrupole Mass Spectrometer for the Study of Surface-Induced Dissociation. *Int. J. Mass Spectrom. Ion Processes.* **77:** 31-47, 1987.
17. Bier ME, Schwartz JC, Schey KL and Cooks RG, Tandem Mass Spectrometry Using an In-Line Ion-Surface Collision Device. *Int. J. Mass Spectrom. Ion Processes* **103:** 1-19, 1990.
18. Ijames CF and Wilkins CL, Surface-Induced dissociation by Fourier Transform Mass Spectrometry. *Anal. Chem.* **62:** 1295-1299, 1990.
19. Schey K, Cooks RG, Kraft A, Grix R and Wollnik H, Ion/Surface Collision Phenomena in an Improved Tandem Time-of-Flight Instrument. *Int. J. Mass Spectrom. Ion Processes%V 94:* 1-14, 1989.
20. De Maaijer-Gielbert J, Gu C, Somogyi A, Wysocki VH, Kistemaker PG and Weeding TL, Surface-induced dissociation of singly and multiply protonated polypropylenamine dendrimers. *Journal of the American Society for Mass Spectrometry* **10**(5): 414-422, 1999.
21. Nikolaev EN, Somogyi A, Smith DL, Gu C, Wysocki VH, Martin CD and Samuelson GL, Implementation of low-energy surface-induced dissociation (eV SID) and high-energy collision-induced dissociation (keV CID) in a linear sector-TOF hybrid tandem mass spectrometer. *International Journal of Mass Spectrometry* **212**(1-3): 535-551, 2001.
22. Tecklenburg RE, Jr., Sellers-Hann L and Russell DH, The utility of a biased activation cell for laser-ion beam photodissociation studies. *International Journal of Mass Spectrometry and Ion Processes* **87**(1): 111-20, 1989.
23. Tecklenburg RE, Jr., Miller MN and Russell DH, Laser ion beam photodissociation studies of model amino acids and peptides. *Journal of the American Chemical Society* **111**(4): 1161-71, 1989.
24. Creaser CS, McCoustra MRS and and O'Neill KE, GC-MS/MS using Laser Photodissociation. *Org. Mass Spectrom.* **26:** 335-338, 1991.
25. Gimon-Kinsel ME, Kinsel GR, Edmondson RD and Russell DH, PID of Peptides and Proteins in 2-stage TOFMS. *J. Amer. Soc. Mass Spectrom.* **6:** 578-587, 1995.
26. Price WD, Schnier PD and Williams ER, MS/MS of Large Biomolecule Ions by Blackbody Infrared PID. *Anal. Chem.* **68:** 859-866, 1996.
27. Aben I, Pinkse FA, Nibbering NMM and Kistemaker PG, PID in a BEqQ Instrument. *J Rapid Commun. Mass Spectrom* **9:** 215-220, 1995.
28. Little DP and McLafferty FW, IR-Photodissociation of Non-Covalent Adducts of Electrosprayed Nucleotide Ions. *J. Am. Soc. Mass Spectrom.* **7:** 209-211, 1996.

29. Gross ML, Charge-remote fragmentation: an account of research on mechanisms and applications. *Int. J. Mass Spectrom.* **200**(1/3): 611-624, 2000.
30. Despeyroux D and Jennings KR, CID. In: Biological Mass Spectrometry: Present and Future. (Eds. Matsuo T and Caprioli RM), 1994.
31. Fenselau CC, Hunt DF, Shabanowitz J, Harvey TM and Coates M, Ion Chemistry of Biopolymers: Organics in the Environment by MS/MS. *Anal. Chem.* **57**: 129-146, 1994.
32. Gonzalez J, Besada V, Garay H, Reyes O, Padron G, Tambara Y and Foltz RL, MS/MS in Forensic Toxicology. *Int. J. Mass Spectrom. Ion Proc.%V 118/119*: 237-263, 1992.
33. Green MK and Lebrilla CB, Ion-molecule reactions as probes of gas-phase structures of peptides and proteins. *Mass Spectrom Rev* **16**(2): 53-71., 1997.
34. Kondrat RW and Cooks RG, Direct Analysis of Mixtures by MS. *Anal. Chem.%V 50*: 81A-92A, 1978.
35. McClellan JE, Quarmby ST and Yost RA, Parent and neutral loss monitoring on a quadrupole ion trap mass spectrometer: screening of acylcarnitines in complex mixtures. *Anal Chem* **74**(22): 5799-806., 2002.
36. Marcos J, Pascual JA, de la Torre X and Segura J, Fast screening of anabolic steroids and other banned doping substances in human urine by gas chromatography/tandem mass spectrometry. *J Mass Spectrom* **37**(10): 1059-73., 2002.
37. Rogalski JC and Kast J, Specific detection of O-linked N-acetylhexosamine modified peptides using multiple precursor ion scans. *Rapid Communications in Mass Spectrometry* **19**(1): 77-78, 2005.
38. Harrison AG, Characterization of a- and g-glutamyl dipeptides by negative ion collision-induced dissociation. *Journal of Mass Spectrometry* **39**(2): 136-144, 2004.
39. McLafferty FW, Bente PF, Kornfeld R, Tsai SC and Howe I, Collisional Activation Spectra of Organic Ions. *J. Mass Spectrom* **30**: 797-806, 1995.
40. Levsen L and Schwarz H, Gas-phase Chemistry by CAD. *Mass Spectrom. Rev.* **2**: 77-148, 1983.
41. Russell DH, Freiser BS, McBay EH and Canada DC, Benzyl versus Tropylium Ion Structures. *Org. Mass Spectrom.* **18**: 474-485, 1983.
42. Hunt DF, Giordani AB, Shabanowitz J and Rhodes G, Pathways in CAD of (M-1)–ions. *J. Org. Chem.* **47**: 738-741, 1982.
43. Harvey DJ, A new charge-associated mechanism to account for the production of fragment ions in the high-energy CID spectra of fatty acids. *Journal of the American Society for Mass Spectrometry* **16**(2): 280-290, 2005.
44. Sleno L and Volmer DA, Ion activation methods for tandem mass spectrometry. *Journal of Mass Spectrometry* **39**(10): 1091-1112, 2004.
45. Grill V, Shen J, Evans C and Cooks RG, Collisions of ions with surfaces at chemically relevant energies: Instrumentation and phenomena. *Rev. Sci. Instrum.* **72**(8): 3149-3179, 2001.
46. Ospina MP, Powell DH and Yost RA, Internal energy deposition in chemical ionization/tandem mass spectrometry. *Journal of the American Society for Mass Spectrometry* **14**(2): 102-109, 2003.
47. Vekey K, Internal Energy Effects in MS. *J. Mass Spectrom.* **31**: 445-463, 1996.
48. Glish GL and Todd PJ, Collision region for MS/MS. *Anal. Chem.* **54**: 842-843, 1982.
49. Cheng X, Wu Z, Fenselau C, Ishihara M and Musselman BD, Collision Cell Interface for a Four-Sector MS. *J. Am. Soc. Mass Spectrom.* **6**: 175-186, 1995.
50. Schneider BB and Chen DD, Collision-induced dissociation of ions within the orifice-skimmer region of an electrospray mass spectrometer. *Anal Chem* **72**: 791-9, 2000.
51. Eller K and Drewello T, Pitfalls in Reaction Intermediate SCANS. *Org. Mass Spectrom.* **28**: 1462-1466, 1993.
52. Beynon JH, Metastable Ions in MS. *Anal. Chem.* **42**: 97A-102A, 1970.
53. Budzikiewicz H and Grigsby RD, Half protons or doubly charged protons? The history of metastable ions. *Journal of the American Society for Mass Spectrometry* **15**(9): 1261-1265, 2004.
54. Beynon JH, Saunders RA and Williams AE, Dissociation of Metastable Ions with Release of Internal Energy. *J. Mass Spectrom* **30**: 793-796, 1995.
55. Jennings KR and Mason RS, MS/MS by Linked Scanning. In: *Tandem Mass Spectrometry* (Ed. McLafferty FW), pp. 197-222. Wiley, New York., 1983.
56. Beynon JH, Saunders RA and Williams AE, *Table of Metastable Transitions for Use in Mass Spectrometry*. Elsevier, London., 1965.
57. McLuckey SA, Glish GL and Cooks RG, KE Effects in MS/MS. *Int. J. Mass Spectrom. Ion Phys.* **39**: 219-230, 1981.
58. Jennings KR and Dolnikowski GG, Mass Analyzers. In: *Methods in Enzymology, Vol. 193: Mass Spectrometry* (Ed. McCloskey JA), pp. 37-61. Academic Press, Nyc, 1990.
59. Cotter RJ, Cornish TJ and Musselman B, Tandem Sector/TOF-MS with a Curved-field Reflectron. *Rapid Commun. Mass Spectrom.* **8**: 339-340, 1994.
60. Flory CA, Taber RC and Yefchak GE, One-Dimensional Mirror Potential for Energy Focusing in TOF-MS. *Int. J. Mass Spectrom. Ion Process.* **152**: 177-184, 1996.
61. Della-Negra S and Beyec Y, Metastable Ion Studies with TOF-MS. *Anal. Chem.* **57**: 2035-2040, 1985.

62. Hayes RN and Gross ML, Collision-Induced Dissociation. In: *Methods in Enzymology, Vol 193: Mass Spectrometry* (Ed. McCloskey JA), pp. 237-262. Academic Press, New York, NY., 1990.
63. Morgan DG and Bursey MM, *Energy Corrections FAB-MS/MS of Protonated N-Benzoylated Tripeptides: Tools for Probing Mchanisms of CAD Processes*, 1995.
64. Jennings KR, Collision-Induced Decompositions of Aromatic Molecular Ions. *Int. J. Mass Spectrom. Ion Phys.* **1:** 227-235, 1968.
65. Harrison AG and Lin MS, Energy dependence of the fragmentation of the n-butylbenzene molecular ion. *International Journal of Mass Spectrometry and Ion Physics* **51**(2-3)**:** 353-6, 1983.
66. Nacson S and Harrison AG, Energy transfer in collisional activation. Energy dependence of the fragmentation of n-alkylbenzene molecular ions. *International Journal of Mass Spectrometry and Ion Processes* **63**(2-3)**:** 325-37, 1985.
67. Liere P, Blasco T, March RE and Tabet JC, Effect of Delay Between Isolation and Excitation on Internal Energy of Ions in ITMS. *Rapid Commun. Mass Spectrom.* **8:** 953-958, 1994.
68. Doroshenko VM and Cotter RJ, Pulsed Gas Introduction for CID Efficiency for Peptides in a MALDI/Quadrupole Ion Trap MS. *Anal. Chem.* **68:** 463-472, 1996.
69. McLuckey SA, Glish GL and Cooks RG, KE effects in CAD by BEqQ. *Int. J. Mass Spectrom. Ion Phys.* **39:** 219-230, 1981.
70. Claeys M, Van den Heuvel H, Chen S, Derrick PJ, Mellon FA and Price KR, Comparison of High- and Low-Energy CID of Glycoalkaloids. *J. Am. Soc. Mass Spectrom.* **7:** 173-18, 1996.
71. Vachet RW, Winders AD and Glish GL, Correlation of Kinetic Energy Losses in High-Energy CID of Peptides. *Anal. Chem.* **68:** 522-526, 1996.
72. March RE, Li H, Belgacem O and Papanastasiou D, High-energy and low-energy CADof protonated flavonoids generated by MALDI and by ESI. *Int J Mass Spectrom* **262:** 51-66, 2007.
73. Loo JA, Edmonds CG and Smith RD, MS/MS of Multiply Charged ESI ions in TQMS. *Anal. Chem.* **63:** 2488-2499, 1991.
74. Cooper HJ, Håkansson K and Marshall AG, The role of electron capture dissociation in biomolecular analysis. *Mass Spec Rev* **24:** 201-222, 2005.
75. McFarland MA, Chalmers MJ, Quinn JP, Hendrickson CL and Marshall AG, Evaluation and Optimization of Electron Capture Dissociation Efficiency in Fourier Transform Ion Cyclotron Resonance Mass Spectrometry. *Journal of the American Society for Mass Spectrometry* **16**(7)**:** 1060-1066, 2005.
76. Ge Y, ElNaggar M, Sze SK, Oh HB, Begley TP, McLafferty FW, Boshoff H and Barry CE, Top down characterization of secreted proteins from Mycobacterium tuberculosis by ECD MS. *J Am Soc Mass Spectrom* **14:** 253-261, 2003.
77. Kjeldsen F, Haselmann KF, Budnik BA, Sorensen ES and Zubarev RA, Complete characterization of posttranslational modification sites in the bovine milk protein PP3 by tandem mass spectrometry with electron capture dissociation as the last stage. *Analytical Chemistry* **75**(10)**:** 2355-2361, 2003.
78. Syrstad EA and Turecek F, Toward a general mechanism of electron capture dissociation. *Journal of the American Society for Mass Spectrometry* **16**(2)**:** 208-224, 2005.
79. Tsybin YO, Hakansson P, Budnik BA, Haselmann KF, Kjeldsen F, Gorshkov M and Zubarev RA, Improved low-energy electron injection systems for high rate ECD in FT-ICR-MS. *Rapid Commun Mass Spectrom* **15:** 1849-54., 2001.
80. Tsybin YO, Ramstroem M, Witt M, Baykut G and Hakansson P, Peptide and protein characterization by high-rate ECD-FT-ICR-MS. *Journal of Mass Spectrometry* **39:** 719-729, 2004.
81. Kjeldsen F, Haselmann KF, Sorensen ES and Zubarev RA, Distinguishing of Ile/Leu amino acid residues in the PP3 protein by (hot) ECD in FT-ICR-MS. *Anal Chem* **75:** 1267-74., 2003.
82. Silivra OA, Kjeldsen F, Ivonin IA and Zubarev RA, Electron capture dissociation of polypeptides in a 3D quadrupole ion trap: Implementation and first results. *J Am Soc Mass Spectrom* **16:** 22-27, 2005.
83. Polfer NC, Haselmann KF, Zubarev RA and Langridge-Smith PR, Electron capture dissociation of polypeptides using a 3 Tesla FT-ICR-MS. *Rapid Commun Mass Spectrom* **16:** 936-43., 2002.
84. He M, Emory JF and McLuckey SA, Reagent Anions for Charge Inversion of Polypeptide/Protein Cations in the Gas Phase. *Anal Chem* **77:** 3173-3182, 2005.
85. Huang T-Y, Emory JF, O'Hair RAJ and McLuckey SA, Electron-Transfer Reagent Anion Formation via Electrospray Ionization and Collision-Induced Dissociation. *Analytical Chemistry* **78**(21)**:** 7387-7391, 2006.
86. Coon JJ, Ueberheide B, Syka JEP, Dryhurst DD, Ausio J, Shabanowitz J and Hunt DF, Protein identification using sequential ion/ion reactions and tandem mass spectrometry. *Proc Nat Acad Sci USA* **102:** 9463-9468, 2005.
87. Coon JJ, Shabanowitz J, Hunt DF and Syka JEP, Electron Transfer Dissociation of Peptide Anions. *Journal of the American Society for Mass Spectrometry* **16**(6)**:** 880-882, 2005.
88. Chi A, Bai DL, Geer LY, Shabanowitz J and Hunt DF, Analysis of intact proteins on a chromatographic time scale by ETD-MS/MS. *Int J Mass Spectrom* **259:** 197-203, 2007.

89. Swaney DL, McAlister GC, Wirtala M, Schwartz JC, Syka JEP and Coon JJ, Supplemental Activation Method for High-Efficiency Electron-Transfer Dissociation of Doubly Protonated Peptide Precursors. _Anal Chem_ **79:** 477-485, 2007.

90. Liang X, Hager JW and McLuckey SA, Transmission Mode Ion/Ion Electron-Transfer Dissociation in a Linear Ion Trap. _Anal Chem_ **79:** 1073-1081, 2007.

91. Liang X, Han H, Xia Y and McLuckey SA, A Pulsed Triple Ionization Source for Sequential Ion/Ion Reactions in an Electrodynamic Ion Trap. _J Amer Soc Mass Spectrom_ **18:** 369-376, 2007.

92. Chi A, Huttenhower C, Geer LY, Coon JJ, Syka JEP, Bai DL, Shabanowitz J, Burke DJ, Troyanskaya OG and Hunt DF, Analysis of phosphorylation sites on proteins from saccharomyces cerevisiae by ETD MS. _Proc Nat Acad Sci USA_ **104:** 2193-2198, 2007.

93. Mikesh LM, Ueberheide B, Chi A, Coon JJ, Syka JEP, Shabanowitz J and Hunt DF, The utility of ETD mass spectrometry in proteomic analysis. _Biochimica et Biophysica Acta, Proteins and Proteomics_ **1764**(12): 1811-1822, 2006.

94. Zarling AL, Polefrone JM, Evans AM, Mikesh LM, Shabanowitz J, Lewis S, Engelhard VH and Hunt DF, Identification of class I MHC-associated phosphopeptides as targets for cancer immunotherapy. _Proc Nat Acad Sci USA_ **103:** 14889-14894, 2006.

95. Catalina MI, Koeleman CAM, Deelder AM and Wuhrer M, Electron transfer dissociation of N-glycopeptides: loss of the entire N-glycosylated asparagine side chain. _Rapid Communications in Mass Spectrometry_ **21:** 1053-1061, 2007.

96. Sparkman OD, Mass Spectrometry PittCon 2006. _J Amer Soc Mass Spectrom_ **17:** 873-884, 2006.

97. Sparkman OD, Mass Spectrometry PittCon 2007. _J Amer Soc Mass Spectrom_ **18,** 2007.

98. Singleton KE, Cooks RG and Wood KV, Utilization of natural isotopic abundance ratios in tandem mass spectrometry. _Analytical Chemistry_ **55**(4): 762-4, 1983.

99. Cotter RJ, Gardner BD, Iltchenko S and English RD, Tandem Time-of-Flight Mass Spectrometry with a Curved Field Reflectron. _Analytical Chemistry_ **76**(7): 1976-1981, 2004.

100. Yost RA and Fetterolf DD, MS/MS instrumentation. _Mass Spectrom. Rev._ **2:** 1-46, 1983.

101. Yost RA and Enke CG, TQMS for Mixture Analysis. _Anal. Chem._ **51:** 1251A-1264A, 1979.

102. Crow FW, Tomer KB and Gross ML, Resolution in MS/MS. _Mass Spectrom. Rev._ **2:** 47-76, 1983.

103. Dawson PH, French JB, Buckley JA, Douglas DJ and Simmons D, TQMS Instrument Parameters. _Org. Mass Spectrom._ **17:** 205-211, 1982.

104. Dawson PH, French JB, Buckley JA, Douglas DJ and Simmons D, Use of TQMS. _Org. Mass Spectrom._ **17:** 212-219, 1982.

105. Wysocki VH, Triple Quadrupole MS. In: _Mass Spectrometry in the Biological Sciences: A Tutorial_ (Ed. Gross ML), pp. 59-77. Kluwer Academic Publishers, Netherlands, 1992.

106. Busch KL, Transmission in an RF-only Quadrupole for MS/MS. _Anal. Instrum. (N. Y.)_ **21:** 123-140, 1993.

107. Yergey AL, Coorssen JR, Backlund PS, Jr., Blank PS, Humphrey GA, Zimmerberg J, Campbell JM and Vesta IML, De novo sequencing of peptides using MALDI/TOF-TOF. _J Am Soc Mass Spectrom_ **13:** 784-91, 2002.

108. Liu Z and Schey Kevin L, Optimization of a MALDI TOF-TOF MS for intact protein analysis. _J Am Soc Mass Spectrom_ **16:** 482-90, 2005.

109. Mancone C, Amicone L, Fimia GM, Bravo E, Piacentini M, Tripodi M and Alonzi T, Proteomic analysis of human very low-density lipoprotein by two-dimensional gel electrophoresis and MALDI-TOF/TOF. _Proteomics_ **7**(1): 143-154, 2007.

110. Nachman RJ, Russell WK, Coast GM, Russell DH, Miller JA and Predel R, Identification of PVK/CAP2b neuropeptides from single neurohemal organs of the stable fly and horn fly via MALDI-TOF/TOF tandem mass spectrometry. _Peptides_ **27:** 521-526, 2006.

111. Gogichaeva NV, Williams T and Alterman MA, MALDI TOF/TOF Tandem Mass Spectrometry as a New Tool for Amino Acid Analysis. _J Amer Soc Mass Spectrom_ **18:** 279-284, 2007.

112. Campbell JM, Vestal ML, Blank PS, Stein SE, Epstein JA and Yergey AL, Fragmentation of Leucine Enkephalin as a Function of Laser Fluence in a MALDI TOF-TOF. _J Amer Soc Mass Spectrom_ **18:** 607-616, 2007.

113. Yu S-Y, Wu S-W and Khoo K-H, Distinctive characteristics of MALDI-Q/TOF and TOF/TOF tandem MS for sequencing of permethylated complex type N-glycans. _Glycoconjugate Journal_ **23:** 355-369, 2006.

114. Maslen S, Sadowski P, Adam A, Lilley K and Stephens E, Differentiation of isomeric N-glycan structures by LC-MALDI-TOF/TOF tandem MS. _Anal Chem_ **78:** 8491-8, 2006.

115. Maslen SL, Goubet F, Adam A, Dupree P and Stephens E, Structure elucidation of arabinoxylan isomers by HPLC-MALDI-TOF/TOF-MS/MS. _Carbohydrate Research_ **342:** 724-735, 2007.

116. Glish GL, McLuckey SA, McBay EH and Bertram LK, Design of a Hybrid QEB MS. _J Int. J. Mass Spectrom. Ion Proc_ **70:** 321-338, 1986.

117. Bruins AP, Jennings KR and Evans S, Metastables in Double Focusing MS. _Int. J. Mass Spectrom. Ion Phys._ **26:** 395-404, 1978.

118. Boyd RK, Porter CJ and Beynon JH, Linked Scan for Reversed Geometry MS. *Org. Mass Spectrom.* **16:** 490-494, 1981.
119. Beynon JH, Caprioli RM, Baitinger WE and Amy JW, IKES of aromatic hydrocarbons. *Org. Mass Spectrom.* **3:** 455-477, 1970.
120. Smith DH, Djerassi C and Rapp U, Metastable Molecular Ions. *J. Am. Chem. Soc.* **96:** 3482-3486, 1974.
121. Walther H, Schlunegger UP and Friedli F, B2E linked scan. *Org. Mass Spectrom.* **18:** 572-575, 1983.
122. Gross ML, Russell DH and McLafferty FW, Triple sectors in MS/MS. In: *Tandem Mass Spectrometry*. Wiley, New York, 1983.
123. Cody RB and Freiser BS, Collision-Induced Dissociation in a Fourier Transform Mass Spectrometer. *Int. J. Mass Spectrom. Ion Processes* **41:** 199-204, 1982.
124. Cody R, Burnier R, Cassady C and Freiser B, Consecutive Collision-Induced Dissociations in FTMS. *Anal. Chem* **54:** 2225-2228, 1982.
125. Cody RB, New Techniques in Ion Cyclotron Resonance Spectrometry. In: *Chemistry*. Purdue University, 1982.
126. Koster C, Kahr MS, Castoro JA and Wilkins CL, FT-MS. *Mass Spectrom. Rev.* **11:** 495-512, 1992.
127. Sleno L, Volmer DA and Marshall AG, Assigning product ions from complex MS/MS spectra: The importance of mass uncertainty and resolving power. *Journal of the American Society for Mass Spectrometry* **16**(2): 183-198, 2005.
128. Glish GL, Multiple Stage MS. *Analyst* **119:** 533-537, 1994.
129. March RE, Quadrupole ion trap mass spectrometry: theory, simulation, recent developments and applications. *Rapid Communications in Mass Spectrometry* **12**(20): 1543-1554, 1998.
130. Patrie SM, Charlebois JP, Whipple D, Kelleher NL, Hendrickson CL, Quinn JP, Marshall AG and Mukhopadhyay B, Construction of a hybrid quadrupole/Fourier transform ion cyclotron resonance mass spectrometer for versatile MS/MS above 10 kDa. *J Am Soc Mass Spectrom* **15**(7): 1099-108., 2004.
131. Murrell J, Despeyroux D, Lammert SA, Stephenson JL and Goeringer DE, \"Fast excitation\" cid in a quadrupole ion trap mass spectrometer. *Journal of the American Society for Mass Spectrometry* **14**(7): 785-789, 2003.
132. Gabelica V, Karas M and De Pauw E, Calibration of ion effective temperatures achieved by resonant activation in a quadrupole ion trap. *Analytical chemistry* **75**(19): 5152-9, 2003.
133. Gu J, Hiraga T and Wada Y, ESI MS of pyridylaminated oligosaccharide derivatives: sensitivity and in-source fragmentation. *Biological Mass Spectrometry* **23**(4): 212-7, 1994.
134. Ni J, Mathews MAA and McCloskey JA, Collision-induced dissociation of polyprotonated oligonucleotides produced by electrospray ionization. *Rapid Commun. Mass Spectrom.* **11**(6): 535-540, 1997.
135. Arnott D, Shabanowitz J, Hunt DF, Waugh RJ and Bowie JH, MS and MS/MS of Proteins and Peptides Deprotonated di- and tri-peptides by CID. *Clin. Chem.* **39:** 2005-2010, 1993.
136. Rozman E, Galceran MT and Albet C, Ebrotidine and its metabolites studied by mass spectrometry with electrospray ionization. Comparison of tandem and in-source fragmentation. *Rapid Communications in Mass Spectrometry* **9**(15): 1492-8, 1995.
137. Doerg DR, Bajic S, Blankenship LR, Preece SW and Churchwell MI, Determination of b-agonist residues in human plasma using liquid chromatography/atmospheric pressure chemical ionization mass spectrometry and tandem mass spectrometry. *Journal of Mass Spectrometry* **30**(6): 911-16, 1995.
138. Poon C, Kaplan H and Mayer PM, Methylating peptides to prevent adduct ion formation also directs cleavage in collision-induced dissociation mass spectrometry. *European Journal of Mass Spectrometry* **10**(1): 39-46, 2004.
139. Serafin SV, Maranan R, Zhang K and Morton TH, Mass Spectrometric Differentiation of Linear Peptides Composed of L-Amino Acids from Isomers Containing One D-Amino Acid Residue. *Analytical Chemistry* **77**(17): 5480-5487, 2005.
140. Guan F, Soma LR, Luo Y, Uboh CE and Peterman S, CID Pathways of Anabolic Steroids by ESI-MS/MS. *J Amer Soc Mass Spectrom* **17:** 477-489, 2006.
141. Josephs J, The Use of Triple Quadrupole and Ion Trap Mass Spectrometers for the Data Dependent LC/MS/MS Analysis of Bioactive Compounds. In: *213th ACS National Meeting, San Francisco,* April 13-17 1997.
142. Richter WJ, Blum W, Schlunegger UP and Senn M, Tandem mass spectrometry of pharmaceuticals. In: *Tandem Mass Spectrom.*, pp. 417-34, 1983.
143. Arnott D, Basics of Triple-Stage Quadrupole [and] Ion Trap Mass Spectrometry: Precursor, Product and Neutral-Loss Scanning. Electrospray Ionisation and Nanospray Ionisation. In: *Proteome Research: Mass Spectrometry* (Ed. James PE). Springer-Verlag, Berlin, 2001.
144. Wilm M, Neubauer G and Mann M, Parent Ion Scans of Unseparated Peptide Mixtures. *Anal. Chem.* **68:** 527-533, 1996.
145. Ejsing CS, Duchoslav E, Sampaio J, Simons K, Bonner R, Thiele C, Ekroos K and Shevchenko A, Automated Identification and Quantification of Glycerophospholipid Molecular Species by Multiple Precursor Ion Scanning. *Analytical Chemistry* **78**(17): 6202-6214, 2006.

146. Downward KM and Biemann K, Methionine-Specific Ions from CID Peptides. *J. Mass Spectrom.* **30:** 25-32, 1995.
147. Schmit JP, Dawson PH and Beaulieu N, Synthesis in Collision Cell of a MS/MS System. *Org. Mass Spectrom.* **20:** 269-275, 1985.
148. White EL, Tabet JC and Bursey MM, Reaction of Ammonia with Benzoyl Ions in TQMS. *Org. Mass Spectrom.* **22:** 132-139, 1987.
149. Dolnikowski GG, Heath TG, Watson JT, Scrivens JH and Rolando CH, A study of the gas-phase reaction between protonated acetaldehyde and methanol. *J. Am. Soc. Mass Spectrom.* **1**(6): 481-8, 1990.
150. Ni J and Harrison AG, Reactive Collisions of C5H8 Isomers with Metal Ions. *Rapid Commun. Mass Spectrom.* **10:** 220-224, 1996.
151. Kan SZ, Xu YC, Byun YG and Freiser BS, Ion-Molecule Reactions of C60Rh+ with CH3X, C6H5X(X=Cl, Br, I) and CH2I2. *J. Mass Spectrom.* **30:** 834-840, 1995.
152. Ranasinghe YA and Glish GL, Reactions of the Phenylium Cation with Small Oxygen- and Nitrogen-Containing Molecules. *J. Am. Soc. Mass Spectrom.* **7:** 473-481, 1996.
153. Bauerle GF, Hall BJ, Tran NV and Brodbelt JS, Ion-Molecule Reactions of Oxygenated CI Reagents with Vincamine. *J. Am. Soc. Mass Spectrom.* **7:** 250-260, 1996.
154. Ni J and Harrison AG, Reactive Collisions in Quadrulpole Cells 5. Reactions of Singly Charged Transition Metal Ions with Cyclopetene. *Int. J. Mass Spectrom. Ion Proc.* %V 146/147: 251-260, 1995.
155. Gruetzmacher H-F, Buechner M and Zipse H, Ion/molecule reactions of 2-chloro- and 2-bromopropene radical cations with methanol and ethanol-FT-ICR spectrometry and DFT calculations. *International Journal of Mass Spectrometry* **241**(1): 31-42, 2005.
156. O'Hair RAJ, The 3D quadrupole ion trap mass spectrometer as a complete chemical laboratory for fundamental gas-phase studies of metal mediated chemistry. *Chemical Communications (Cambridge, United Kingdom)* (14): 1469-1481, 2006.
157. Fetteroli DD, Yost RA and Eyler JR, Reactive Collisions in MS/MS. *Org. Mass Spectrom.* **19:** 104-107, 1984.
158. Dolnikowski GG, Allison J and Watson JT, Hydroxylation of selected hydrocarbon ions on reaction with methanol in the gas phase. *Org. Mass Spectrom.* **25:** 119-23, 1990.
159. Pachuta RR, Kenttämaa HI, Cooks RG, Zennie TM, Ping C, Ching CJ and Cassady JM, Reactive Collisions with Ethyl Vinyl Ether. *Org. Mass Spectrom.* %V **23:** 10-15, 1988.
160. McCarley TD and Brodbelt J, Structurally-diagnostic Ion-Molecule Rxns. *Anal. Chem.* **65:** 2380-2388, 1993.
161. Eberlin MN, Structurally diagnostic ion/molecule reactions: class and functional-group identification by mass spectrometry. *Journal of Mass Spectrometry* **41**(2): 141-156, 2006.
162. Tao WA, Clark RL and Cooks RG, Quotient ratio method for quantitative enantiomeric determination by mass spectrometry. *Anal Chem* **74**(15): 3783-9., 2002.
163. Wu L, Tao WA and Cooks RG, Ligand and metal-ion effects in metal-ion clusters used for chiral analysis of alpha-hydroxy acids by the kinetic method. *Anal Bioanal Chem* **373**(7): 618-27., 2002.
164. Tao WA and Cooks RG, Chiral analysis by MS. *Anal Chem* **75**(1): 25A-31A., 2003.
165. Seymour JL, Turecek F, Malkov AV and Kocovsky P, Chiral recognition in solution and the gas phase. Experimental and theoretical studies of aromatic D- and L-amino acid-Cu(II)-chiragen complexes. *Journal of Mass Spectrometry* **39**(9): 1044-1052, 2004.
166. Wu L, Meurer EC and Cooks RG, Chiral morphing and enantiomeric quantification in mixtures by mass spectrometry. *Analytical Chemistry* **76**(3): 663-671, 2004.
167. Moraes LAB, Sabino AA, Meurer EC and Eberlin MN, Absolute configuration assignment of ortho, meta, or para isomers by mass spectrometry. *Journal of the American Society for Mass Spectrometry* **16**(4): 431-436, 2005.
168. Alvarez EJ and Brodbelt JS, Nucleosides by Ion-Molecule Reactions in Ion Trap MS. *J. Mass Spectrom.* **30:** 625-631, 1995.
169. Beasley BJ, Smith RL and Kenttamaa HI, Dimethyl Diselenide: Ion-Molecule Reagent for Distonic Ions. *J. Mass Spectrom.* **30:** 384-385, 1995.
170. Operti L and Rabezzana R, Gas-phase ion chemistry in organometallic systems. *Mass spectrometry reviews* **25:** 483-513, 2006.
171. Byrd HC, Guttman CM and Ridge DP, Molecular mass determination of saturated hydrocarbons: reactivity of eta5-cyclopentadienylcobalt ion (CpCo*+) and linear alkanes up to C-30. *J Am Soc Mass Spectrom* **14**(1): 51-7., 2003.
172. He M and McLuckey SA, Charge permutation reactions in tandem mass spectrometry. *Journal of Mass Spectrometry* **39**(11): 1231-1259, 2004.
173. Schmit JP, Beaudet S and Brisson A, Ion/Molecule Reactions with TQMS. *Org. Mass Spectrom.* **21:** 493-498, 1986.
174. Dolnikowski GG, Kristo MJ, Enke CG and Watson JT, Ion Trapping in a TQMS. *Int. J. Mass Spectrom. Ion Process* **82:** 1-15, 1988.

175. Heath TG, Allison J and Watson JT, Selective detection of the tolyl cation among other [C7H7] isomers by ion/molecule reaction with dimethyl ether. *J Amer Soc Mass Spectrom* **2**: 270-7, 1991.
176. Gorman GS and Amster IJ, Kinetics and Thermodynamics of Bracketing Proton-transfer Reactions in FT-MS. *Org. Mass Spectrom.* **28**: 1602-1607, 1993.
177. Gal J-F, Maria P-C, Operti L, Rabezzana R and Vaglio GA, A quadrupole ion trap and FT-ICR concerted study of the kinetics of an ion/molecule association reaction: A chemometric approach. *European Journal of Mass Spectrometry* **9**: 435-444, 2003.
178. Steiner WE, Clowers BH, Fuhrer K, Gonin M, Matz LM, Siems WF, Schultz AJ and Hill HH, Jr., ESI with ambient pressure ion mobility separation and mass analysis by oaTOFMS. *Rapid Commun Mass Spectrom* **15**: 2221-6., 2001.
179. Laiko VV, Orthogonal Extraction Ion Mobility Spectrometry. *J Amer Soc Mass Spectrom* **17**: 500-507, 2006.
180. Clowers BH, Dwivedi P, Steiner WE, Hill HH and Bendiak B, Separation of Sodiated Isobaric Disaccharides and Trisaccharides Using ESI-AP Ion Mobility-TOFMS. *J Amer Soc Mass Spectrom* **16**: 660-669, 2005.
181. Eiceman GA, Young D and Smith GB, Mobility spectrometry of amino acids and peptides with matrix assisted laser desorption and ionization in air at ambient pressure. *Microchemical Journal* **81**(1): 108-116, 2005.
182. Giles K, Pringle SD, Worthington KR, Little D, Wildgoose JL and Bateman RH, Applications of a traveling wave-based rf-only stacked ring ion guide. *Rapid Commun Mass Spectrom* **18**: 2401-2414, 2004.
183. Pringle SD, Giles K, Wildgoose JL, Williams JP, Slade SE, Thalassinos K, Bateman RH, Bowers MT and Scrivens JH, The mobility separation of some peptide and protein ions using a new hybrid quadrupole/travelling wave IMS/oa-ToF instrument. *Int J Mass Spectrom* **261**: 1-12, 2007.
184. Heck A, paper in press according to ppt slides from Mark McDowell ala Pittcon 2007. 2007.
185. Stein SE, Chemical Substructures Identification by Mass Spectral Library Searching. *J. Am. Soc. Mass Spectrom.* **6**: 644-653, 1995.
186. Dawson P and Sun W, A Round Robin on the Reproducibility of Standard Operating Conditions for the Acquisition of Library MS/MS Spectra using Triple Quadrupoles. *Inter J Mass Spectrom Ion Proc* **55**: 155-70, 1984.
187. Cintora M, Lifante E, Eichler M, Rodriguez S, Cooke M, Przyborowska A, Down S, Patel R and Halket J, A 1000 compound library of API (ESI, APCI) daughter ion mass spectra and some LC/MS/MS applications . In: *Advances in Mass Spectrometry*, Vol. 15, pp. 609-610. Wiley, Chicester, UK, 2001.

"An early criticism of tandem mass spectrometry was that, because mass spectroscopists [sic] do not understand simple mass spectra, they are ill-advised to add to their difficulties."

R. Graham Cooks

Chapter 4   Inlet Systems

Introduction to Mass Spectrometry, 4th Edition: Instrumentation, Applications, and Strategies for Data Interpretation; J.T. Watson and O.D. Sparkman, © 2007, John Wiley & Sons, Ltd

"…but still try, for who knows what is possible…"

"Experiment is the servant of curiosity."

~Michael Farady

I. Introduction

Mass spectrometry, as a general analytical specialty, found its origins some 80 odd years ago, some time before the development of gas chromatography (GC) and long before the development of high-performance liquid chromatography (HPLC). The challenges in developing the early inlet systems were associated with the problems of transferring the sample from terrestrial conditions to the much lower operating pressure of the ion source (usually electron ionization) at about 10^{-1} Pa. As the development of mass spectrometry advanced, the primary inlets became gas and high-performance liquid chromatographs. The techniques of GC/MS and LC/MS are covered in separate chapters of this book.

Even with the broad utility of GC and HPLC interfaces, a major challenge in mass spectrometry stills involves the sample inlet. Some of the new ionization techniques require special means of ion transfer, and often such ion inlets have features that are not necessarily compatible with the temporal demands of GC and HPLC. In many mass spectrometers of today, the sample is ionized at atmospheric pressure and the resulting ions are introduced through a series of differentially pumped regions into the *m/z* analyzer, where the pressure is usually on the order of 10^{-4} Pa or less (the one exception being the quadrupole ion trap, where the pressure is about 10^{-1} Pa). In addition to the use of GC and HPLC as a mode of sample introduction, alternate sample introduction systems include a batch inlet (usually limited to pure samples that exhibit a discernible vapor pressure and are thermally stable); the technique known as membrane introduction mass spectrometry (MIMS) (for dissolved gases and other volatile components in aqueous or gaseous samples); the direct insertion probe (DIP) (usually reserved for pure samples, though these devices can be used for differential distillation of the components in an impure sample); a variant of the of DIP, the Chromatoprobe (a device for introducing vapors from volatile liquids and solids through a GC injection port into a CI or EI ion source); and the direct exposure probe (a.k.a. the desorption chemical ionization probe). Inlet systems for ions include the atmospheric pressure chemical ionization (APCI) interface, the electrospray ionization (ESI) interface, and the atmospheric pressure photoionization (APPI) interface. These interfaces are often used with liquid samples that flow directly into the interface, which is an integral part of the mass spectrometer. The APCI, ESI, and APPI interfaces are described in Chapter 11 on LC/MS and elsewhere in this book.

In addition to these atmospheric pressure interfaces, a series of new surface ionization techniques has been introduced in recent years (2005) by which ions are formed near or on the surface of material carrying the analyte, and the resulting ions then enter the *m/z* analyzer. These devices include desorption electrospray ionization (DESI), desorption atmospheric pressure chemical ionization (DAPCI), and a novel technique called direct analysis in real time (DART).

The inlet system has usually been designed to answer a sampling problem. In this category is the technique of selected ion flow tube mass spectrometry (SIFTMS), used specifically for the qualitative and quantitative analysis of gas-phase analytes in atmospheric air and breath in real time. Some inlet systems are designed to alter, rather than preserve, the sample in an effort to gain more information; a good example of such an interface is the chemical reaction interface mass spectrometry (CRIMS), which is used with both GC/MS and LC/MS to convert the analytes to specific gases to facilitate detection of isotopes that had been incorporated into the original intact analyte molecules, as in metabolic studies.

Even though not as popular today as they were during the ten years after their introduction, the techniques of fast atom bombardment (FAB) and liquid secondary ion mass spectrometry (LSIMS) are still in wide use, as can be seen from a review of any current issue of *CA Selects: Mass Spectrometry*. These two techniques offer a very good way to deal with analytes that are not volatile enough to be ionized using either electron or chemical ionization, or that may be too thermally unstable.

In most cases, these inlets are ancillary to the commercial instruments for GC/MS or LC/MS. In a few cases like SIFT, APCI, APPI, ESI, FAB, and LSIMS, the inlet is specifically designed for a given *m/z* analyzer. In some cases, such as some implementation of a direct insertion or direct inlet probe and in MIMS, an instrument designed for GC/MS can be used, but the gas chromatograph is replaced by the specialized inlet system.

II. Batch Inlets

1. Heated Reservoir Inlet

The heated reservoir was one of the first inlets used with electron ionization mass spectrometry. This device still has limited utility with instruments designed for EI and conventional chemical ionization; however, no commercial manufacturer of mass spectrometers any longer offers this as an option. The heated reservoir inlet played an important role in early analyses of petroleum and was used to obtain reference mass spectra of many compounds. It consists of a reservoir (ranging in size from 1 mL to approximately 1 liter) with two separate sets of valves. The reservoir can be evacuated by auxiliary vacuum and sampled through an orifice by the ion source, as illustrated schematically in Figure 4-1. The system can be heated to facilitate sample vaporization; care must be taken not to introduce samples that have relatively low volatility, as they may be difficult to evacuate from the reservoir and may produce a "memory" effect or a high background during subsequent analyses.

The reservoir is first evacuated, then the valve to the ion source is opened, and the mass spectrum is scanned to determine if any residue from a previous sample exists. When the reservoir is evacuated and isolated from all other components by closing all

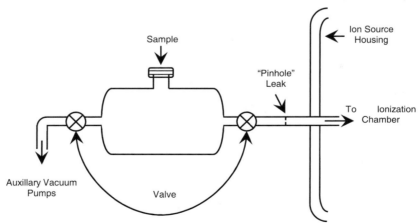

Figure 4-1. Schematic representation of a heated reservoir inlet for permanent gases or volatile liquids.

valves, the sample is introduced, usually through a septum with a microliter syringe, although other modes of introduction such as a gas sampling valve can be used [1]. The valve between the reservoir and the ion source is then opened, and the sample slowly diffuses through the viscous leak into the ion source. The viscous pinhole leak prevents the sample from being depleted during a period of as long as 15 to 30 min.

The heated reservoir is a convenient inlet system for pure, relatively volatile samples that primarily exist in the liquid phase. Some form of this inlet is used on most GC/MS instruments to introduce calibration compounds such as perfluorotributylamine (PFTBA, a.k.a. FC-43) that produce data for calibration of the *m/z* scale. The device has the advantage of maintaining a constant sample pressure in the ion source for prolonged periods, but the disadvantage of requiring relatively large sample sizes. When acquiring reference spectra, the sample should be pure to avoid the presence of extraneous and/or nonreproducible mass spectral peaks. Specialized versions of this type of inlet system have been used for high-temperature studies [2].

2. Direct Inlet Probe (DIP)

The mass spectra of pure compounds not suited for the heated reservoir inlet (such as solid samples that can be volatilized) can be obtained if the sample is placed close to the ion source and heated to cause vaporization. The direct inlet probe (DIP) was designed expressly for this purpose. Like the heated reservoir inlet, the DIP is used with an instrument designed for EI and/or conventional CI. The sample is placed in a small, clean glass ampoule (usually made from a standard melting-point capillary) that is mounted on the end of a rod or probe. As illustrated schematically in Figure 4-2, the sample probe can enter the vacuum of the ion source via a vacuum-lock assembly. The vacuum valve is normally closed. When the probe is introduced through the O-ring seals up to the initial position indicated by the broken line (in Figure 4-2), the vacuum lock is evacuated by an auxiliary vacuum system. When the pressures on either side of the vacuum valve are nearly equal, the vacuum valve is opened. The probe then can be pushed through the open valve, through a second set of O-ring seals, and into the ion source. The open end of the sample ampoule is positioned against the wall of the ionization chamber so that whatever vapor pressure is available can be used efficiently to move a stream of sample vapor directly into the ion source, as illustrated in Figure 4-3.

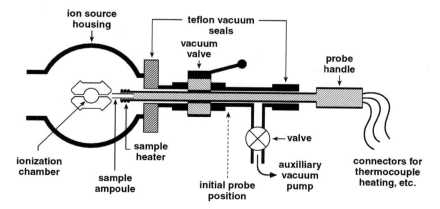

Figure 4-2. *Schematic diagram (showing involvement of vacuum lock) of the direct inlet probe for materials in the condensed phase (solids and liquids having low vapor pressure).*

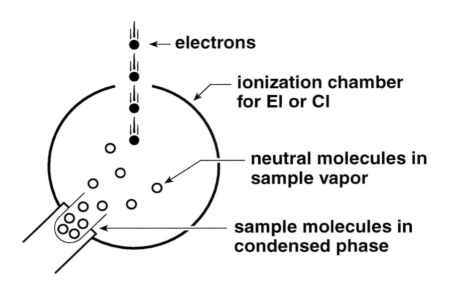

Figure 4-3. *Illustration of a conventional direct inlet probe in which the condensed-phase sample is placed in a tubular reservoir (e.g. melting point capillary) near the ionization chamber; the sample must be vaporized so that gas-phase molecules can be ionized.*

Most probes are designed with some means of heating the sample to increase its vapor pressure. In addition, heat from the ion-source block, which is usually maintained at a relatively high temperature to avoid condensation of samples vapors, will eventually warm the sample ampoule by conduction. Some probe models are designed with some means to cool the sample ampoule so that evaporation of relatively volatile compounds can be controlled; otherwise, such samples might be lost in the vacuum lock or as soon as the ampoule comes into contact with the hot ion-source block.

The principal advantage of the direct inlet probe is that it reduces the distance that sample vapors must travel to reach the ionization chamber to a few millimeters. It is this feature that permits the mass spectra of relatively involatile compounds to be obtained, provided sufficient vapor pressure can be generated at a temperature below that causing thermal degradation. Disadvantages of the direct probe are that (a) it increases the risk of venting (exposure to atmospheric pressure) the ion source, (b) it is likely to increase the rate at which the ion source becomes contaminated unless care is taken to avoid introduction of excessively large samples of relatively involatile compounds, and (c) the sample must be pure, or any contaminant in the sample must have a significantly different volatility from that of the compound of interest, so that they will evaporate from the probe at different times (see Chapter 5 for a discussion of the problem of mixed spectra and strategies on interpretation of spectra).

A DIP was a standard accessory supplied with early GC/MS systems. It was often referred to as a quick-and-dirty method of obtaining a mass spectrum, with the emphasis on the "dirty" part of this term. As manufacturers worked to reduce their after-sales support, the availability of the DIP went from a standard accessory to an optional accessory, and eventually to a no-longer-available feature. During typical improper use, most operators put about the same amount of sample into the sampling device as they might normally use for a melting-point determination. Such a huge quantity is thousands of times too much sample for the mass spectrometer. The result is usually contamination of the ion source. Another practice that became somewhat popular during use of the DIP was heating the unit to its maximum at the end of the analysis to clean the probe. This practice results in the volatilization of most of the sample, which then condenses on various components of the ion source and produces extensive contamination.

The DIP is best used in a manner similar to use of the direct exposure probe. Prepare a solution of the analyte. Place a drop of the solution in the sample device or on the outer surface; evaporate the solvent before introducing the DIP into the mass spectrometer. This will greatly reduce the amount of analyte introduced and result in far less contamination of the ion source. Never heat the DIP beyond the temperature required to obtain the desired mass spectrum. The sample holders or inserts are inexpensive and should be discarded after use. The sample should never be allowed to come into contact with any of the permanent parts of the DIP.

Refinements in DIP design, including use of a transducer to measure and control its temperature, have extended its versatility for qualitative and quantitative analyses of a variety of sample types, including simple mixtures. Quantitative mass spectral data can be obtained by imposing a linear increase in probe temperature. By plotting ion abundance versus temperature, elimination curves are produced; the areas of these elimination profiles are proportional to the amounts of the components present in a mixture [3].

A. The Chromatoprobe

An interesting variation on the DIP is the Chromatotoprobe shown in Figure 4-4. This is a device that is designed for insertion through a GC PTV (programmable temperature vaporization) injector. The device was developed by Aviv Amirav in the Chemistry Department of Tel Aviv University [4]. The Chromatoprobe uses the same type of sample container as the DIP. The sample in the container is lowered into the injector body where it is rapidly heated by the PTV. Sample vapors are swept into the mass spectrometer by a flow of GC carrier gas (usually helium through a short piece of 320 μm i.d. fused silica column (~1–3 m long). The Chromatoprobe was originally designed to be used with the QIT GC-MS. This arrangement for volatizing the analyte outside the ionization region greatly aids in preventing contamination of the ion source or the QIT. One of the most common sources of contamination when using a DIP is the sputtering of nonvolatile components of the sample. These substances remain in the fused silica column and do not create source contamination. Another advantage of the Chromatoprobe is that there is no necessity for a vacuum interface to atmospheric pressure.

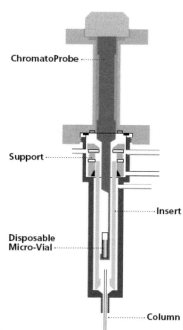

Figure 4-4. Cross-sectional view of the Chromatoprobe inside the body of a PTV injector.

Amirav also developed a variation of the Chromatoprobe called the SnifProbe [5]. This device uses a small piece of 530 μm i.d. capillary column packed with a porous polymer. Air containing analytes is passed through the device for several seconds, typically at a rate of ~30 mL min^{-1}. The analytes are then desorbed from the sampling device using the Chromatoprobe.

Details on the use of the Chromatoprobe and the SnifProbe can be found on Amirav's Web site at http://www.tau.ac.il/chemistry/amirav/dsi.shtml. Unfortunately, the Chromatoprobe is only available through Varian, Inc. for use with their QIT GC/MS systems. Made available on any GC/MS system, the Chromatoprobe would provide a simple system for the analysis of truly semivolatile analytes that are nearly pure.

3. Direct Exposure Probe (Desorption Chemical Ionization, DCI)

The direct exposure probe, in addition to being used to introduce samples deposited on a surface as illustrated in Figure 4-5, was refined to enhance the technique of desorption chemical ionization (DCI). The sample is dissolved in a volatile solvent, and an aliquot of the solution is placed on a convex surface or a DCI filament, which then is placed directly into the CI plasma in the vicinity of the optical focal point of the ion source. In this way, ions desorbed directly from the surface can be readily captured by the ion optics and sent to the *m/z* analyzer. This technique greatly reduces the requirement for sample volatility that in conventional EI or CI mass spectrometry provides "self-transport" of the analyte vapors into the ionization chamber of the instrument. The DCI technique is applicable to the analysis of nonvolatile compounds, as there is no

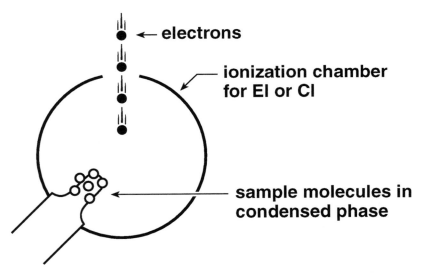

Figure 4-5. *Illustration of one type of direct exposure probe; the sample is in the condensed phase on the surface of the probe so that ionization (EI or CI) can take place on the surface of the probe. Desorption chemical ionization (DCI) is especially useful for nonvolatile analytes.*

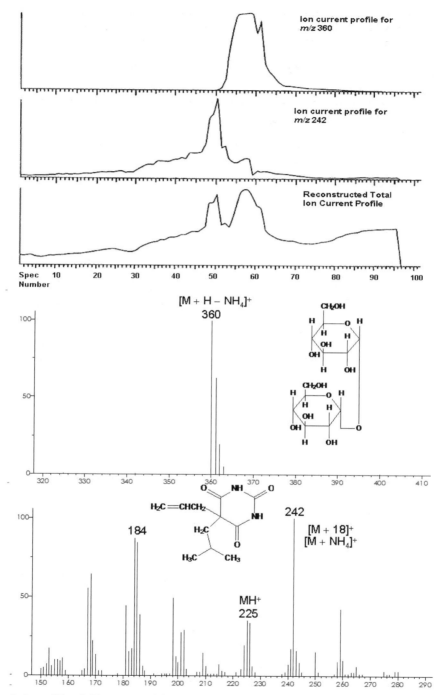

Figure 4-6. *(Top) Extracted ion profiles for m/z 360 and m/z 242 and TIC. (Middle) DCI mass spectrum of a carbohydrate diluent. (Bottom) DCI of street drug mixed with diluent.*

requirement for the sample to be in the vapor state [6, 7]. Results also can be obtained from thermally labile samples as illustrated in Figure 4-6. DCI from various surfaces ranging from gold [8] to Teflon [9] to polyimide-coated surfaces [10] has been described. The common design of modern DCI probes incorporates use of a sample filament constructed of a material such as thoriated tungsten (see Figure 7-8 in Chapter 7). A current is passed through the filament to activate the sample on the filament's surface. The activated analyte is ionized on the surface, and the resulting ions are desorbed from the condensed phase into the gas phase for m/z analysis. As seen in Figure 7-9, DCI is very useful for compounds that are obviously thermally labile and, therefore, not suited for normal EI or CI analyses. The DCI topic is more extensively covered in Chapter 7.

Both the DIP and DCI probes can be used as a device for gross separations (as by distillation), as seen in Figure 4-6 where the crude distillation of sample components of different condensed vapor-phase conversion points provides distinct analyte profiles.

4. Pyrolysis

Pyrolysis mass spectrometry (Py-MS) involves the decomposition or transformation of a nonvolatile material under controlled conditions to produce gas-phase analytes that can be used to characterize the original substance. Py-MS is also used for its selectivity and sensitivity in kinetic studies of high-temperature gas-phase reactions, and to detect and characterize highly reactive molecules as they form using a variety of methods developed for the study of gas-phase ion chemistry by mass spectrometry [11]. Unlike "decomposition by burning" as implied by the two Greek roots of the word pyrolysis, Py-MS involves sample decomposition in an inert atmosphere. The decomposition can be imposed by flash heating to temperatures as high as 1400 °C using a coiled platinum heater, a furnace to heat a quartz crucible, or a Curie-point device (inductive (RF) heating of a ferromagnetic material). Lasers have also been used to bring about such decompositions. The pyrolysis system can be linked with either an EI or CI mass spectrometer. The pyrolyzed material (the pyrolysate) is often separated into components using gas chromatography. If the pyrolytic conditions are carefully controlled, the results are reproducible from sample to sample. The data obtained using Py-MS or pyrolysis GC/MS (Py-GC/MS) must often be analyzed using sophisticated statistical computations and pattern recognition programs. In some cases, samples are extracted with an organic solvent to remove any monomers or low molecular mass materials prior to pyrolysis that otherwise might lead to complications during interpretation of the results; the residue from such preliminary extracts might be analyzed separately by GC/MS.

Many pyrolysis applications, especially in the analysis of polymers [12–15]and other organics [16–18], have been accomplished by modification of the DIP to accommodate a pyrolysis unit. In one application, a Curie-point pyrolysis/triple quadrupole and microtube pyrolysis furnace/quadrupole ion trap mass spectrometers were used to analyze bacterial spores [19].

Pyrolysis combined with GC/MS has been used to identify products of pyrolyzed soil samples [19], biological particulates from space missions [20], pyrolyzed polymers [21], coal hydrogenates [22], paints [23], bacteria [24], spruce needles [25], and lignin [26]. The GC/MS approach is especially useful in quantitative analyses. Pyrolysis GC/MS and Py-MS are indispensable tools in forensic science. In addition to the analysis of polymers and paints, Py-GC/MS and Py-MS are used extensively in the analysis of hair and fiber samples.

III. Continuous Inlets

1. Membrane Introduction MS (MIMS)

In MIMS, a thin membrane made of organic material such as a silicone elastomer (same type of material used for breast implants) provides a selective means of introducing low molecular weight volatile organic chemicals into an EI or CI ion source from samples such as flowing water [27–32]. Some selectivity is achieved by the differential permeability of the membrane for various compounds [33]. For example, the membrane is permeable to compounds that are soluble in the membrane material or that are adsorbed by the membrane, and which have a high diffusivity in the membrane material. The analytes must be sufficiently volatile to evaporate on the low-pressure side of the membrane to finally diffuse into the ion source of the mass spectrometer. Because the membrane is made of an organic material, the principle of "like dissolves like" holds to a first approximation such that the membrane excludes inorganic materials such as water, aqueous solutions of inorganic salts, fixed gases such as oxygen, nitrogen, and other inert gases. Hayward and Maden [32] evaluated ten polymer membrane materials for response time, detection limits, and water permeability for applications in environmental analyses and fermentation monitoring. A miniaturized MIMS instrument (12 kg) has been tested with several different membranes including a cellulose membrane to allow detection of organic contaminants and water at the 10–100 ppm level in organic solvents and with a standard silicone membrane to allow detection of volatile organic compounds in water at concentrations just below 1 ppm [34].

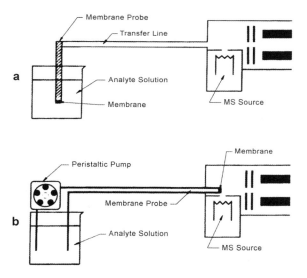

Figure 4-7.   ***Schematic diagrams of two possible implementations of a MIMS interface to a mass spectrometer: (a) involves submerging a membrane in an aqueous solution containing analytes or exposing it to air from which analytes are to be detected; analytes pass through the membrane which can act as a concentrating device; (b) a liquid or gas flows over the membrane surface that is mounted on the ion-source housing.*** *From Bauer JS and Cooks RG "MIMS for trace-level determination in on-line process monitoring and environmental analysis" American Laboratory, October 1993, p 38.*

Savickas *et al.* [35] developed a capillary tubular form of the thin membrane that showed remarkably good transmission of dissolved organics in water. In this way, aqueous solutions can be continuously transported through the membrane capillary, exposing the ionization chamber of the mass spectrometer to vapors of dissolved organics as shown in Figure 4-7. Bauer and Cooks [31] incorporated the membrane interface capillary flow-through interface into an ion trap mass spectrometer and demonstrated detection limits in the low PPT level for dozens of low molecular weight (i.e., below mass 300) compounds such as toluene, chloroform, xylenes, etc. LaPack *et al.* [36] developed a valved sample cell for MIMS; Mendes *et al.* [37] designed a cryotrap system to improve detection limits for volatile organics by MIMS.

The MIMS approach provides an efficient means by which to monitor aqueous streams continuously for dissolved organics [38–40]. The membrane assembly is particularly useful in monitoring realistic environmental samples [41, 42] such as river water, ocean water, or discharge fluids from industrial plants where suspended as well as high concentrations of inorganic materials that might interfere with more classical approaches offer no interference with the MIMS approach. Chemical modification of the membrane provides one means of adjusting the selectivity of the interface, called "affinity MIMS" [43]. An enzyme-modified membrane permits the analysis of carboxylic esters by the MIMS technique [44]. The MIMS method can also be set up with a pure water stream in which various samples can be analyzed for organic content by flow injection analysis. Other applications of MIMS have been described for monitoring VOCs in air [45], soil [46], and biological reactions [47]. The MIMS approach to sampling has revolutionized techniques for monitoring gases in blood [48, 49] and tissue [50].

2. Supercritical Fluid Chromatography (SFC)

Supercritical fluid chromatography (SFC) combined with mass spectrometry is a developing field [51, 52]; it is described here as one of the promising inlet systems for a mass spectrometer. Heating a gas or liquid to a temperature above its critical temperature while simultaneously compressing it to a pressure above its critical pressure produces a supercritical fluid. The solubilizing power of a supercritical fluid, which is directly related to its density, is much greater than that of a gas, a feature that makes SFC more suitable than GLC for the separation of nonvolatile and high molecular weight analytes [53]. Further, because the diffusion coefficients of many analytes are much greater in a supercritical fluid than in a liquid, and because the resistance-to-mass transfer term in SFC is much less than that in HPLC, the separation speed in SFC can be as much as an order of magnitude greater than in HPLC [54].

Although SFC has some unique attributes, open tubular-column SFC is not a particularly user friendly technique; the lack of general laboratory success with open tubular-column SFC is part of the reason that early attempts to perform SFC/MS were not widespread. Much of the problem was related to the difficulty in fabricating and using proper restrictors to maintain optimal separation conditions [55]. On the other hand, packed-column SFC is much more user friendly. Packed-column SFC has blossomed during the last decade in large measure because most of the hardware is similar, if not identical, to that used in conventional HPLC. Packed-column SFC has become very well accepted as a cost-effective technique for chiral analysis and preparative purification of pharmaceuticals; in fact, whole companies have been dedicated to developing and applying the technique, such as Accelapure. Attractive attributes of packed-column SFC include increased speed of analysis (usually three-fold faster than conventional HPLC), less use of organic solvents, and the fact that the isolated components of a mixture are readily available in concentrated form in a volatile solvent; e.g., methanol. A particularly impressive

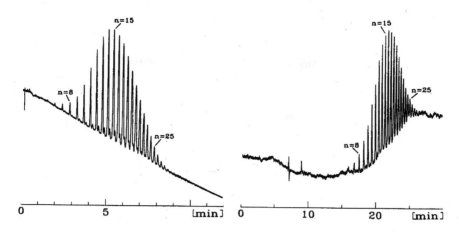

Figure 4-8.   Typical supercritical fluid chromatograms.

Adapted from Giorgetti et al. "Mixed Mobil Phases and Pressure Programming in Packed Column SFC"
J. Chromatogr. Sci. 1989, 27, 318, with permission from Preston.

application of packed-column SFC/MS for the rapid, high-throughput analysis of blood plasma for dextromethorphane reports the quantitative analysis of a 96-well array in 10 minutes [56].

In general, instrumental components (the injection and pump systems) for SFC are hybrids of those used in GLC and HPLC. A given sample dissolved in an organic solvent is injected into a sample loop and displaced into the column by a high-pressure liquid that subsequently becomes a supercritical fluid upon reaching the temperature of the column oven. Pressure control in SFC is more important than flow control because the density of a supercritical fluid varies with pressure, and because the solvent strength of a fluid is a function of its density, pressure control is important. However, from the standpoint of the vacuum system in the mass spectrometer, the gas load from the SFC column is an important operational consideration; typical flow rates (as liquids) in SFC correspond to 1–10 µL min^{-1} for open tubular columns.

A representative SFC separation on an open tubular column (10 m × 0.05 mm) containing immobilized SB-Biphenyl-30 is shown in Figure 4-8. The chromatogram results from a mixture of octylphenol poly(ethylene glycol) ethers (Triton X-165) eluted with 0.175 mL min^{-1} CO_2 (supercritical pressure = 7.15 × 10^6 Pa; supercritical temperature = 31.3 °C) containing 0.0265 mL min^{-1} of 2-propanol pressure programmed from 1.25 × 10^7 to 3.80 × 10^7 Pa at 170 °C.

Pinkston and Chester [52] reviewed the state-of-the-art in SFC/MS and suggested guidelines for its effective use in analytical chemistry. In particular, they emphasize the importance of the injection technique in obtaining the desired results from the SFC component of the combination and suggest optimum means for interfacing the SFC hardware for obtaining desirable mass spectrometric results. SFC/MS is particularly suitable for the analysis of complex mixtures of compounds in which the high molecular weight components are not sufficiently volatile for separation by GC and thus require analysis by HPLC, as, for example, silicone polymers. On the other hand, the lower-boiling fractions might otherwise be analyzed by GC; i.e., a complete analysis of a series of silicone polymers might require separate analysis by two different technologies, whereas the mixture could be completely analyzed in one run by SFC/MS.

There have been several reportsof combining SFC with MS via a simple interface [57, 58] or through a direct inlet [59]; in the latter application with the direct inlet, it was reported that the appropriate direct introduction of the SFC inlet does not adversely affect the operating pressure conditions in a differentially pumped mass spectrometer. Because of the high pressure used in SFC, appropriate use of a flow restrictor must be used, especially in the case of open tubular columns with the direct connection [55, 60, 61]. In some cases, the addition of an organic solvent and/or a volatile electrolyte as a modifier to CO_2 can be advantageous [62–64].

Most of the applications of SFC/MS have been made using a quadrupole mass spectrometer because of its tolerance to relatively high pressures in the analyzer; however, magnetic instruments [65, 66] have also been used, although more care is needed to avoid high pressures in the ion source where arcing may occur with the relatively high potentials used. The FTMS instrument [67] and the ion trap mass spectrometer [68, 69] can be used with external ionization to avoid problems with the high pressures associated with the SFC interface. SFC has also been combined with plasma sources of ionization [70] and used with NCI MS for the analysis of surfactants [71]. SFC/MS/MS has been used for high-throughput analyses (3 minutes per sample) to analyze mouse blood for enantiomers of propranolol and pindolol [72]. Chiral analysis of synthetic drug reactions can be monitored by SFC/MS [73]. SFC/MS using a packed column and CO_2 carrier can separate and quantitate 15 different estrogen metabolites in less than 10 min [74]. The separation of polypeptides up to 40-mers in CO_2/methanol has been demonstrated in applications of SFC/MS [75]; this demonstration prompted the use of liquid nitrogen to quench or eliminate back-exchange during separation of peptic peptides by SFC after experiments involving solution-phase HDX [76].

3. Electrophoretic Inlet

Electrophoresis is a separation technique based on the differential migration of charged analytes through a given medium under the influence of an electric field. One of the earliest forms of electrophoresis employs a gel impregnated with a buffer to provide a fixed medium for electrophoretic flow; polyacrylamide gel electrophoresis (PAGE) is still widely used today for separation of complex mixtures, such as the analysis of whole-cell lysates for proteins. Unlike gel electrophoresis, capillary electrophoresis is based on the differential migration of charged analytes through the free bulk solution under the influence of an applied electric field.

Capillary electrophoresis (CE) shares some mechanical features with HPLC in that the process involves fluid flow through a capillary with migration of well-defined concentration zones of analytes through a detector at characteristic times. The resolving power of CE can be much higher than that achieved by HPLC because of the discrete nature of "plug flow" of the analyte in CE as opposed to the band-broadening nature of laminar flow in HPLC as illustrated by the profiles in Figure 4-9. The resolving power of

A   **B**

Figure 4-9. Representation of concentration profiles of analyte in a flowing system under (A) laminar flow vs (B) electro-osmotic flow.

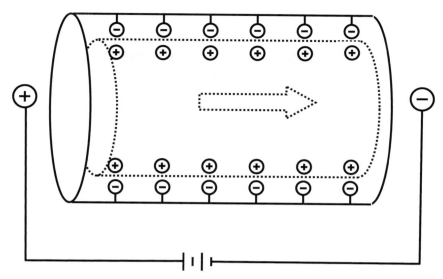

Figure 4-10. *Schematic diagram showing mechanism of EOF based on a double layer formed between fixed negative sites on the wall of capillary and positive counter ions in solution creating a cylindrical volume of solution that can then be moved by an applied electric field.*

CE is superior to conventional gel electrophoresis because the high lateral surface area of the capillary allows much greater heat dissipation associated with ohmic heating so that high electric fields (1000 V/m) can be used without causing band broadening due to thermally induced convection [77, 78].

Electro-osmotic flow (EOF) or plug flow in a fused silica capillary is promoted by the surface charge on the capillary wall. The silyl functional groups in the fused silica capillary are weak acids that tend to dissociate starting at pH 3, with complete dissociation at pH 10, to present a progressively more negative surface to the bulk solution. This negative charge on the fused silica capillary wall promotes counter-layer formation of a positive-charged cylindrical plug of solution (especially if it contains an electrolytic buffer), which moves as a discrete plug in response to the electric field applied across opposing ends of the capillary, as illustrated in Figure 4-10.

Injection of the sample into the capillary can be accomplished in several ways. Electrokinetic injection is achieved by simply immersing the capillary inlet into the sample solution, and then applying the electrophoretic potential for a few seconds. If the pH of the sample solution is below pH 3 such that no EOF occurs, analyte ions enter the capillary by electrophoretic mobility alone. If pH of the solution is such that EOF occurs, analyte ions enter the capillary under a combination of EOF and electrophoretic mobility. Electrokinetic injection offers the advantage that only species of like charge enter the capillary, discriminating against analytes of opposite charge, thereby simplifying the separation somewhat; however, under such conditions, ions enter the capillary based on mobility, thereby under sampling species of low mobility. Injection by displacement is preferred because analyte ions are present in the injected sample in direct proportion to their concentration in the bulk sample. Displacement injection can be accomplished by placing the capillary inlet into the sample solution and allowing a pressure gradient across the capillary, as imposed by a vacuum at the exit of the capillary, to force an aliquot of the bulk sample into the inlet of the capillary. Improved reproducibility and operation have been reported with internally tapered capillaries in the interface [79]; on the other hand, a tapered sheathless tip with a beveled edge has been reported to be especially rugged [80]. A multiplexed interface allows up to four capillary electrophoresis columns to be monitored sequentially during parallel operation [81].

On-line coupling to a mass spectrometer is most commonly configured via an electrospray interface [78, 82]; in such a configuration, the capillary outlet is grounded to the mass spectrometer. J. W. Jorgensen, who pioneered developments in CE, showed that separations can be accomplished on a very short timescale with peak widths of only a few milliseconds in the electrophoreogram [83]. Such rapid electrophoresis and short-lived samples in the ion source of the mass spectrometer will require heroic measures on the side of "scan" rates and approaches to data acquisition as were developed for capillary column GC/MS [84, 85]. The combination of CE with FTMS [86] may also offer an effective means of acquiring spectra from transient samples. A critical requirement in the CE/MS combination is proper selection of buffers that allow proper separation by CE, but which also are compatible with mass spectrometry. The reader should refer to the detailed description of volatile buffers in Chapter 11 on LC/MS. Whereas the volume flow rate of CE may be compatible with nanospray, ESI sources requiring higher flow rates can be fitted with a sheath interface, which essentially provides a make-up flow of solvent into the mass spectrometer interface. However, better detection limits can be obtained with a low-flow, sheathless interface that is relatively resistant to clogging and to adverse effects of salts, etc. [87, 88].

Combinations of CE with ESI [89–92] as well as with APCI [93, 94] have been reported. The CE column also offers the advantage of concentrating the analyte for subsequent analysis by MS [95, 96]. The combination of CE/MS has been used for the analysis of proteins [95–98] coenzymes [90], lipopolysaccharides [78], biogenic amines [99], and industrial alkylphosphonates [100].

IV. Ionization Inlet Systems

Prior to the development of atmospheric pressure chemical ionization (APCI) in the 1970s by the Horning research group at Baylor University, transfer of a gas-phase sample from atmospheric pressure outside the mass spectrometer into an ion source operated at 10^{-1} Pa had been a major problem in most applications of mass spectrometry [101]. As the utility of the atmospheric pressure ionization techniques of electrospray ionization and atmospheric pressure photoionization were realized and APCI gained popularity, the emphasis shifted from getting gas-phase molecules into the ion source of a mass spectrometer to getting gas-phase ions into the m/z analyzer from atmospheric pressure. Understandings in this area have led to developments that have produced new and easier ways of using mass spectrometry. During 2004 and 2005, sampling ionization techniques were introduced that will have as profound an impact on the use of mass spectrometry as did the gas chromatographic interface, the development of ESI and MALDI, and the refinement of the TOF m/z analyzer. Simplicity of analysis has always been a goal in the field of mass spectrometry. These new developments in ion production/transfer bring simplicity in application of mass spectrometry to a new level of realization. These new techniques are described in this section.

1. Direct Analysis in Real Time (DART)

DART is an analytical technique invented by Robert B. Cody and James Larameé [102] at JEOL while trying to develop an atmospheric pressure version of the tunable electron monochromator for electron capture LC/MS like that previously developed by Larameé and Deinzer [103] for use with GC/MS (see Chapter 10 on GC/MS). DART involves a comprehensive array of ionization mechanisms allowing atmospheric pressure ionization of fixed gases as well as molecules on the surface of a wide variety of materials (solids and liquids). Some of the mechanisms of ion formation in DART have not yet been fully explored. One of these mechanisms involves the use of excited-state species such as metastable helium atoms (2^3 S) that have an energy of 19.8 eV. These excited-state species can react with analytes on surfaces to produce molecular ions through a process of Penning ionization. The helium metastables can also react with atmospheric water to produce protonated water clusters $[(H_2O_{n-1})H]^+$ and hydroxyl ions. The presence of dopants will produce adduct ions and, in some cases, increase sensitivity. The ion/molecule reactions between protonated water clusters and analyte molecules can produce protonated molecules of analytes. The system can also produce thermalized electrons to form negative-charge molecular ions. Gases such as N_2 will produce positive-charge molecular ions through Penning ionization. Other ionization systems using a corona discharge in air, as is used by DART, produce nitrogen oxide anions that interfere with the analysis of nitrogen-based explosives. DART can detect ClO_3^- and ClO_4^- ions from sodium perchlorate deposited from solution on a glass rod as well as molecules of volatile solvents such as from an open bottle of acetonitrile 10–20 feet from the ionization source. Various dopants have been used to enhance analyses by the detection of adduct ions like $[M + NH_4]^+$ when vapors of NH_4OH are purged through the ion source [102]. Unlike in ESI, alkali adducts are never seen in conjunction with protonated molecules, nor are multiple-charge ions. Ionization takes place without the need for any solvents or sample preparation. This technique has been used with a high-resolution TOF mass spectrometer [102]. DART was introduced by JEOL at the ASMS Sanibel Conference in January 2005.

An interesting historical fact about DART is that at the time the technology was introduced, it was accompanied by a commercially available product. This may be the first occurrence of the simultaneous introduction of scientific technology and instrumentation. JEOL offers a commercial product with the DART interfaced to a *re*TOF mass spectrometer that produces accurate mass measurements (R > 7000 FWHM). These accurate mass measurements greatly add to the technique's utility. Another important feature of DART is its lack of reliance on sample position in the ionization interface, which is operated at atmospheric pressure. As can be seen in Figure 4-11, the DART interface is open to the atmosphere. What is not obvious from this figure is the fact that there is no electric potential between the DART exit and the ion inlet of the mass spectrometer. The gas used with DART can be temperature controlled from ambient to about 600 °C

Samples can be positioned in the interface while being held by hand. Samples such an intact orange have been used for the detection of a pesticide on its surface [102]. Detection of illicit drugs can be performed by placing a drop of urine on a melting point capillary and then holding the capillary in the DART ionization region. The versatility of DART is illustrated by the results shown in Figures 4-12 and 4-13. DART can also detect gas-phase molecules at very low concentrations in ambient air [102], as does APCI [101]. The spectrum in Figure 4-13 was acquired by Robert B. Cody (JEOL USA) on the Exposition floor at PittCon®2006 at the suggestion of pesticide researchers Imma Ferrer and Michael M. Thurman (Spain), using fruit purchased from a food vendor.

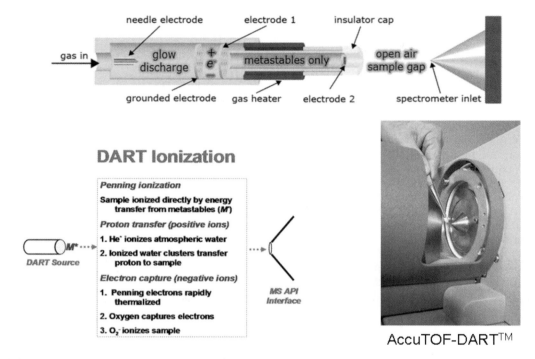

AccuTOF-DART™

Figure 4-11.   *Schematic representation of the DART interface showing ion inlet to mass spectrometer (top); DART ionization (bottom left); sample being held manually in the DART interface (bottom right).*

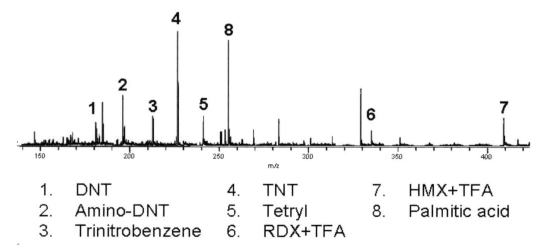

1.	DNT	4.	TNT	7.	HMX+TFA
2.	Amino-DNT	5.	Tetryl	8.	Palmitic acid
3.	Trinitrobenzene	6.	RDX+TFA		

Figure 4-12   *Mass spectrum (negative ion) resulting from explosives (3 ppm) in muddy pond water sampled from the surface of a glass rod using TFA dopant.* Adapted from Figure 4 of "Versatile New Ion Source for the Analysis of Materials in Open Air under Ambient Conditions" Cody RB, Larameé JA, and Durst HD Anal. Chem. 2005, 79, with permission from the American Chemical Society.

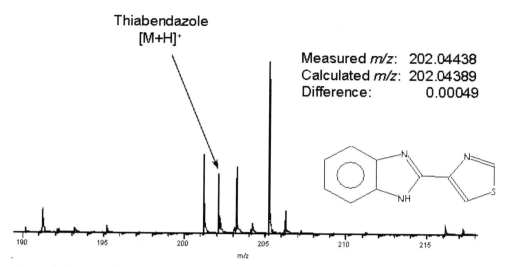

Thiabendazole
[M+H]$^+$

Measured *m/z*: 202.04438
Calculated *m/z*: 202.04389
Difference: 0.00049

Figure 4-13.   *Mass spectrum identifying thiabendazole (fungicide) from orange peel.*

DART has been used in studies of the deposition and release of fragrance on surfaces such as fabric and hair [104] and in methodology for detecting melamine in pet food [105].   The use of DART is not limited to the TOF mass spectrometer.   This technique could also be used with MS/MS instrumentation as has been illustrated with another related technique, DESI, described below. It could also be used with ion mobility spectrometry (IMS) in a manner analogous to the use of IMS with electrospray-equipped mass spectrometers [106, 107].

2. Desorption Electrospray Ionization (DESI)

Desorption ESI (DESI) is accomplished by directing electrospray-charged droplets and ions of solvent onto the surface to be analyzed, as illustrated in Figure 4-14 [108].   The impact of the charged microdroplets of electrosprayed solution onto the surface produces gaseous ions of the material originally adsorbed to the surface to give mass spectra similar in appearance to conventional ESI mass spectra in that they show mainly peaks corresponding to single or multiple protonated species or alkali-ion adducts of the analytes.   Studies of the DESI phenomenon indicate that accumulation of liquid on the surface being analyzed likely plays an important role in dissolution of some of the compounds deposited on the surface, thereby facilitating desorption/ionization during further bombardment of the nascent solution by additional droplets from the sprayer [109, 110]. A key feature of DESI is that it provides a means of keeping the sample outside the mass spectrometer; the objective is to allow only ions to enter the *m/z* analyzer via a capillary vacuum sweeper that carries ions from the sample bombardment region to the skimmer cone entrance [111]. As does ESI, DESI produces multiple-charge ions.

The DESI phenomenon was observed during the analysis of both conductive and insulating surfaces, and for compounds ranging from nonpolar small molecules such as lycopene, the alkaloid coniceine, small drugs, as well as polar compounds such as peptides and proteins; changes in the solution being electrosprayed can be used to selectively ionize particular compounds, including those in biological matrices [108]. In another application, DESI was used to characterize the active ingredients in

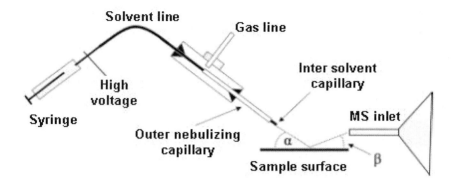

Figure 4-14. *Schematic illustration of a desorption electrospay ionization interface (DESI).* *From Venter A, Sojka PE, and Cooks RG "Droplet Dynamics and Ionization Mechanisms in Desorption Electrospray Ionization Mass Spectrometry" Anal. Chem. 2006, 78(24), 8549–8555, with permission.*

pharmaceutical samples formulated as tablets, ointments, and liquids [88]. DESI also has been used for trace detection of the explosives trinitrohexahydro-1,3,5-triazine (RDX), octahydro-1,3,5,7-tetranitro-1,3,5,7-tetrazocine (HMX), 2,4,6-trinitrotoluene (TNT), pentaerythritol tetranitrate (PETN), and their plastic components (Compn. C-4, Semtex-H, and Detasheet) directly from a wide variety of surfaces (metal, plastic, paper, polymer) without sample preparation or pretreatment [112–115]. Some industrial polymers also can be sampled successfully by DESI [116]. A variation of the technique, called "reactive" DESI, uses reactive chemicals in the bombarding droplets [117, 118]. In an application of reactive DESI to the analysis of specialized polymer containing Cu(II), a dilution solution of iodine as an oxidant enhances the detection limit by two orders of magnitude [119].

The DESI technique was used to demonstrate a novel manner of reading or visualizing the components of a mixture of aspirin, acetaminophen, and caffeine separated on a normal-phase silica gel plate by mass spectrometry [120]. DESI mass spectra can be obtained directly from the TLC plate by moving it incrementally under computer control while directing the charged-droplet plume from the stationary DESI emitter at the TLC plate surface [121] as demonstrated in the analysis of dietary supplements for alkaloids [122]. The DESI techniques appear to be much more effective and convenient to use than earlier techniques for analyzing components directly from a TLC plate, which were based on microelution and sampling systems that were manually directed to specific locations on the plate [120].

A somewhat related technique called matrix-assisted laser desorption electrospray ionization (MALDESI) generates multiprotonated proteins by an ESI process from proteins that are imbedded in an organic acid matrix [123]. At first glance, DESI appears to perform similarly to DART; however, there is one significant difference, namely the positioning of the sample. As described by Cooks, it appears that sample positioning is critical for the performance of DESI [124], however, recent results show that when the DESI process is conducted in an enclosed environment, the relative orientation of the sprayer and the sampling capillary becomes much less important [125].

3. Desorption Atmospheric Pressure Chemical Ionization (DAPCI)

DAPCI is a variation of DESI in which ionization can be carried out on surface-bound condensed-phase analytes or on gas-phase molecules. This technique has also been described by Cooks [126]. Dependence on source geometry as required for DESI apparently has not been studied.

There is also a technique called ASAP (atmospheric-pressure solids analysis probe) that involves the manual placement of the sample in an APCI interface [127]. The ASAP provides a more rapid method for analysis of volatile and semivolatile compounds than can be accomplished with the conventional solids probe inlet systems on commercially available API mass spectrometers. The ASAP interface can be installed with only a simple modification to either an ESI or APCI source [127, 128]. In general, ASAP provides fast solids/liquid sample introduction for ionization by corona discharge of analyte vapors generated by a stream of hot nitrogen gas flowing from an ESI or APCI probe [127]. Some caution is required with this technique of limited utility because of the high potentials associated with the APCI corona discharge.

V. Speciality Interfaces

1. Selected Ion Flow Tube Mass Spectrometry (SIFTMS)

A combination inlet and ionization system, selected ion flow tube (SIFT), of the same category of inlet system as the APCI interface, can be used in the analysis of gas samples for trace organic components such as pollutants in air and metabolic gases in breath using ionization that takes place in a fast flow tube. The SIFT mass spectrometry technique was conceived and developed 30 years ago [129, 130]. It quickly became a standard technique for the study of ion-neutral reactions at thermal energies [131].

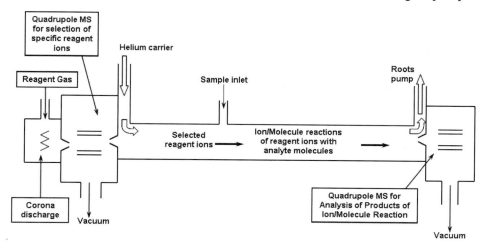

Figure 4-15.   *Schematic diagram of SIFT showing production of m/z-specified reagent ions on the left for injection into the central reaction chamber to react with analyte from the sample inlet to form characteristic ions to be analyzed by the MS system on the right.*

Reagent ions such as H_3O^+, NO^+, and $O_2^{+\bullet}$ are produced in a chemical ionization source using moist atmospheric air (p $\cong$ 50 Pa) as the reagent gas and a microwave gas discharge ion source (as opposed to the resistively heated filaments used to produce electrons in the conventional CI ion source) to provide an increased degree of ruggedness. The reagent (precursor) ions are selectively separated from other ions with a transmission quadrupole *m/z* analyzer. The reagent ions are injected into a tube (diameter = 1–2 mm, length = 30–100 cm) containing helium at a temperature of ~300 K and flowing at a linear velocity of 40–80 m sec^{-1} as maintained using a Roots pump (p $\cong$ 100 Pa). The term *fast flow tube* derives from the fact that a high velocity of gas flows through the tube. Samples are introduced into the flow through a heated sampling line (Figure 4-15). Because the sample gas also contains water molecules, the reagent ion complex becomes more complicated with the presence of protonated water clusters and water–NO^+ clusters, which along with the primary reagent ions can form adducts with analyte molecules. A second transmission quadrupole *m/z* analyzer is used to determine the ratio of product ions to precursor ions in the fast flow tube. The second quadrupole can be operated either in the full-scan mode or in the selected ion monitoring mode. Quantitative calculations are made using computer algorithms based on fluid dynamics and ion/molecule reactions that take place in the fast flow tube [132]; in this way quantitation can be accomplished without the time-consuming chore of calibration required for conventional GC methods [133]. The measurement error for metabolites in a single breath has been reported to be between ±5% and ±20% [130]. Other studies show that peroxides can be monitored in ambient air [134] and ion/molecule reactions of importance to outer space research can be studied [135]. A variation of SIFT that uses only the hydronium ion is called proton transfer reaction mass spectrometry (PTRMS) [136, 137].

2. Fast Atom Bombardment (FAB) and Liquid Secondary Ion Mass Spectrometry (LSIMS)

Prior to 1970, to contemplate the mass spectrum of an intact oligopeptide or even a single free amino acid was a fantasy because the negligible vapor pressure of such compounds precluded their analysis by EI or CI mass spectrometry. During the 1970s, desorption/ionization (D/I) techniques such as field desorption [138] and plasma (^{252}Cf) desorption [139] were introduced, and with special instrumentation, selected laboratories were able to conduct analyses of previously intractable compounds [140]. It was the introduction of fast atom bombardment [141, 142], however, that has had the greatest impact on analyses of nonvolatile and/or thermally labile compounds because the FAB source has been so easily adapted to all kinds of mass analyzers.

FAB was developed from attempts to analyze organic compounds mounted on metal surfaces by a well-developed technique in surface science called secondary ion mass spectrometry (SIMS). In SIMS, a primary beam of mass-selected ions is used to bombard the analyte and generate secondary ions that can be mass-analyzed to characterize the analyte [143, 144]. Early success in FAB (Figure 4-16) was achieved by bombarding the sample (dissolved in a liquid matrix) with a beam of atoms having a kinetic energy of several thousand eV, and thus the name "fast atoms". It was soon learned, however, that the charge on the bombarding particle was not important to the outcome of the desorption/ionization experiment [143, 145].

The analytical advantage provided by FAB is dependent upon the use of a liquid matrix to hold the analyte [146]. For this reason, analysis of a solution of analyte during bombardment by a beam of ions is often called liquid SIMS (or LSIMS) to connote the origin of the technology and to distinguish it from conventional SIMS, which does not employ a liquid matrix in analyzing the surface layer of solids.

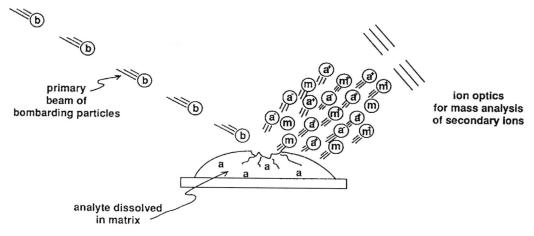

primary
beam of
bombarding particles

ion optics
for mass analysis
of secondary ions

analyte dissolved
in matrix

Figure 4-16. *Bombardment of a sample (dissolved in a liquid matrix (e.g. glycerin) by a primary beam of atoms or ions to produce secondary ions that are characteristic of the analyte (a, analyte; b, bombarding particle; m, matrix).*

The desorption/ionization process is not completely understood [143, 147], and the spectrum used by the analyst may very well be the composite result of several distinct physical and chemical processes [148, 149]. The primary event of momentum transfer by the high-energy bombarding particle causes nuclear excitation of some matrix molecules; as this impulse of energy decays with concomitant desorption of matrix and analyte molecules and ions due to high temperature (~1000 °C) in the local molecular environment, there is considerable "curve crossing" on the potential energy diagrams that explains the electron excitation leading to the formation of both odd- and even-electron ions. The impact of the bombarding particle is thereby responsible for direct desorption of some secondary ions, but the liquid matrix in the FAB or LSIMS experiment also provides important physical effects during the bombardment process. For example, the matrix is responsible for the long-lived (tens of minutes) FAB or LSIMS signal. The matrix facilitates refreshing the bombarded surface with analyte, while removing radiation-damaged species [150] that may otherwise polymerize and interfere with continued access of dissolved analyte to the bombarding beam. Solute transport in the matrix is a very important parameter in the FAB experiment; investigations of the transport process indicate that several mechanisms may play a role, including disruptive mixing due to abrupt changes in surface tension in the region of bombardment, as well as convective and diffusive processes [151, 152]. Several investigations have been made on the role of surface activity of solutes in the LSIMS experiment; for example, during the analysis of an equimolar mixture of alkyl quaternary amines, the longest chain, most surface-active amine dominates the spectrum [153].

Although no longer as popular as they once were, the soft ionization techniques of FAB and LSIMS are still in wide use, as can be seen by a review of the current literature. The FAB and LSIMS techniques performed best with the high acceleration voltages associated with double-focusing instruments. These techniques have largely been replaced by ESI and MALDI, both of which provide better sensitivity and are much easier to use.

3. Chemical Reaction Interface Mass Spectrometry (CRIMS)

Chemical reaction interface mass spectrometry (CRIMS) provides a means of selectively detecting the isotopic content of the analyte in a compound-independent manner [154–157]. The analyte is completely atomized in a helium plasma and the nascent atoms of the destroyed analyte are allowed to combine with a reactant gas such as SO_2 to form gases that can easily be transported to the mass spectrometer for analysis.

A schematic diagram of the CRIMS system is shown in Figure 4-17. The interface has been miniaturized to preserve temporal changes in analyte concentration as from a chromatographic column and to fit into existing instruments. Careful control of He flow is necessary to sustain the microwave discharge. The capacity of the CRIMS system is dependent upon the concentration of reactant gas, which must be kept at low levels to avoid quenching the discharge.

Qualitative and quantitative isotopic information is transformed from the original analyte into easily handled, simple gases by proper choice of the reactant gas [158]. For example, SO_2 readily captures the isotopic information for C, N, and Cl content by producing CO_2, NO, and HCl, respectively, from an analyte of composition $C_aH_bO_cN_dCl_f$ in the CRIMS system according to the following scheme [159]:

C $\rightarrow$ CO_2 + CO (CRIMS products of carbon)

N $\rightarrow$ NO + NO_2 (CRIMS products of nitrogen)

Cl $\rightarrow$ HCl   (CRIMS products of chlorine)

Other reactant gases must be used to obtain reliable information regarding other elements of interest. For example, H_2O is formed in the SO_2 reactant chamber with the hydrogen atoms from the destroyed analyte, but H_2O poses great problems in underreporting due to problems of adsorption and exchange.

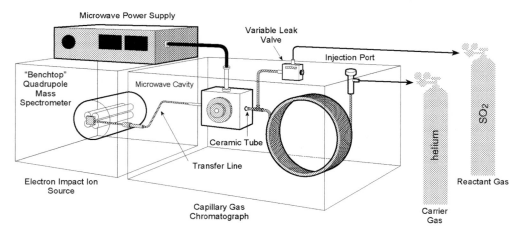

Figure 4-17. *Schematic illustration of the CRIMS interface as associated with the eluant of a GC column prior to introduction into the mass spectrometer.* Reprinted with permission of Donald Chase, Duke University Medical Center, Mass Spectrometry Facility, from 1996 ASMS Short Course entitled "Liquid Chromatography–Art and Practice".

The CRIMS approach is suitable as a detector for HPLC [160–163]; it rivals flow-through radioactivity detectors in sensitivity and avoids radiation hazards when the experiment is designed with stable isotopes. The CRIMS system has been used in many metabolism [164–167] and biological applications [168, 169].

4. Inductively Coupled Plasma Mass Spectrometry (ICPMS)

Physicists invented mass spectrometers to prove the existence of isotopes, so the field of elemental analysis is the oldest application of mass spectrometry [170, 171]. Thermal ionization [172–174] and spark sources [175] have been used for decades for elemental analysis of samples introduced in "batch" mode. There are several mass spectral techniques used today for elemental analysis, but the most widely used technique is inductively coupled plasma mass spectrometry (ICPMS) [64, 176–181]. It is particularly useful as a dynamic sampling system that can be coupled with separation processes, especially when using a low dead-volume nebulizer to preserve chromatographic resolution [182]; ICP also offers convenient sample introduction and lower detection limits for some elements than the batch inlet systems.

In this combination, the ICP serves both as a sample inlet system and as an ionization source for the mass spectrometer. The ICP-MS is ideally suited for elemental analysis of a wide variety of samples [170, 181, 183, 184], including those containing chlorinated hydrocarbons [185], long-lived radionuclides [186, 187], metals incorporated into bacteria [188], or containing platinum in the form of cisplatin and related drugs in human plasma [189]. The ICP technology has also been used in conjunction with laser ablation for imaging metals in human brain tumors [190].

The ICP provides an intense source of energy that achieves ionization of most elements in solutions of a wide variety of samples from biological to geological to environmental origins. The plasma formed in the ICP from an inert gas such as argon provides temperatures up to 10,000 °C. As in classical applications of ICP, the sample is aspirated into the torch assembly where the solutes are desolvated, atomized, and finally ionized in the nonoxidizing and energetic medium. The ICP is able to ionize approximately 80 of the elements in the periodic table, and when coupled to mass spectrometry, provides the possibility for multielement analysis at detection limits frequently two orders of magnitude lower than those achieved by emission spectroscopy.

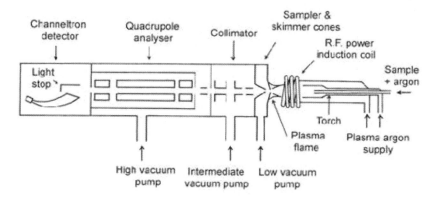

Figure 4-18. ***Schematic diagram of an ICP interface on a transmission quadrupole mass spectrometer.*** *Reprinted from Radiogenic Isotope Geology, Alan P. Dickin, with permission from Cambridge University Press (http://www.onafarawayday.com/Radiogenic/).*

A. Hardware Configuration

The ICP operates at atmospheric pressure, while the mass spectrometer operates at a considerably lower pressure of typically 10^{-3} to 10^{-5} Pa. A simplified illustration of the usual configuration is shown in Figure 4-18. Solutions of the analyte are introduced through aspiration into the ICP as is done in standalone operation of the classical ICP instrument. The plasma is sampled through an arrangement of skimmers that limit gas flow into the mass spectrometer, while still allowing a reasonable transfer of the ionized medium. The skimmer system consists of a small orifice, typically fractions of a millimeter in diameter, that leads into an intermediate chamber that is differentially pumped and connected with yet a second orifice leading to the ion optics of the m/z analyzer. The skimmer system allows most of the gas to be pumped away to avoid interference with operation of the mass spectrometer. The orifice leading into the mass spectrometer is off-axis from a line-of-sight to the detector so that background signals arising from photons in the ICP will be minimized.

As indicated in Figure 4-18, the quadrupole mass analyzer is frequently used in ICPMS because it is relatively insensitive to the somewhat elevated pressures associated with the skimmer interface; also, the quadrupole can accommodate a relatively large spread in kinetic energies of the ions as they are formed in the ICP. The quadrupole mass filter has good (i.e., 30%) ion transmission, and as with most mass spectrometers can easily achieve the mass resolution and mass range (e.g., up to m/z 300) needed for adequate mass analysis of elements. However, because the quadrupole m/z analyzer is a scanning instrument, only ions of a given m/z value may be detected at any given time. In this regard, an instrument capable of ionizing and storing or analyzing a wide variety of m/z values simultaneously provides better efficiency of detection. A small magnetic-sector mass spectrometer with an array detector offers the advantage of simultaneous detection of ion currents at several different m/z values across the spectrum [191, 192], but the magnetic mass spectrometer suffers from limitations in the pressure requirement (i.e., the higher voltages in the ion acceleration region of the magnetic mass spectrometer require a pressure lower than that for operation of a quadrupole mass spectrometer). The time-of-flight (TOF) mass spectrometer provides better detection limits and resolving power for applications with ICP than do the quadrupole instruments [193–195].

B. Operational Considerations

Fundamental studies of ICPMS are advancing the field [196–200], but it still suffers from many of the practical operating problems that are present in classical ICP operations [201]. Continuous operation of the ICPMS is frequently interrupted by clogging of the aspirator by insoluble components such as residual protein in a biological sample. Also, concentrations of analyte exceeding a few tenths of a percent often are high enough to cause problems by clogging the orifices with residual solute when the solvent is vaporized. On the other hand, the wide linear range (of up to six orders of magnitude) available with the ICPMS allows samples to be diluted, and offers a considerable analytical advantage in its own right.

Detection limits are 1–10 ppt (part per trillion by weight) with quadrupole instruments. These values improve to 10–50 ppq (part per quadrillion) with magnetic-sector devices because of the very low background. The reader should distinguish between the detection limit in the solution presented to the instrument and the detection limit in the original sample, usually a solid. The solute level in the final solution is typically 0.1%, so a detection limit of 1 ppt in the solution corresponds to 1 ppb in the original solid. The detection limits achievable by ICPMS are superior to those of any other multielement techniques and are rivaled only by highly selective single elemental methods like laser atomic fluorescence or resonance ionization [202]. At high signal levels, the precision is roughly 1–3% relative standard deviation (RSD) in the ion signal. At low signal levels, the precision is poorer and is limited by statistical variations in the number of ion counts observed [194]. Most quantification is done by ratio measurements, either with an internal standard of a different element or with a different stable isotope of the analyte element.

The simplicity of the mass spectra produced by the ICP greatly facilitates analysis of geochemical and nuclear samples. Traditionally, these materials have proven difficult for emission spectrometry because of the rich spectra emitted when elements like iron, rare earths, and actinides are excited. Considerable attention must be given to the background from the sample matrix and/or interferences from the ICPMS operation in formulating an assay and in interpreting results [201]. Many of the elements have isotopes that can be of advantage in confirming the detection of a particular element, but they also can be an interference in that they may provide a signal at an *m/z* value that is isobaric with another element. In other cases, polyatomic species may be formed in the plasma that interfere with detection of an element at a given *m/z* value. For example, Ar_2^+ at *m/z* 80 will interfere with detection of the major isotope of selenium. In samples that are aqueous solutions of a matrix prepared in hydrochloric acid, sea water, or unprocessed urine, which normally contains a fair amount of chloride ion, $^{40}Ar^{35}Cl^+$ will interfere with $^{75}As^+$. In these two examples of background interference, the principal constituent involves argon which, of course, is always present because it is the primary ingredient in the ICP torch. Interestingly, interference from argon can be greatly diminished by adding a small amount of xenon to the argon discharge gas. The use of $Ar-H_2$ mixed-gas plasma and cryogenic desolvation is also helpful in dealing with high levels of chloride in the sample. Another type of interference comes from the occasional occurrence of double-charge ions as, for example, $^{138}Ba^{+2}$; such a double-charge barium ion would provide interference for detection of $^{69}Ga^+$ at *m/z* 69. Resolving power of the mass spectrometer is the best way to deal with the problem of potentially overlapping signals from multiple elements in a given sample [203, 204].

C. Electrothermal Vaporization

When elemental composition is the primary goal, direct analysis of the sample using electrothermal vaporization provides a simple protocol [205, 206]. In electrothermal vaporization, a small sample (e.g., a few microliters of blood) is deposited on a sample probe that is rapidly heated electrothermally, and the vaporized sample is swept into the ICP by a flow of argon gas. This approach can be used for analyzing Hg in a wide variety of solid samples (e.g., hair [203], river sediment, meat) using ^{200}Hg as an internal standard [205].

D. Laser Ablation

In the laser ablation process, solid sample material with little or no sample preparation is irradiated with a laser pulse on the sample in the vicinity of the ICP aspirator [184, 207–210]. Because of the focusing power of the laser beam, spatially resolved analyses of the sample are possible. Quantitative calibration of the laser ablation experiment is difficult, however, because of the matrix-dependent nature of the ablation process.

E. Speciation

Whereas ICPMS is an excellent technique for elemental analysis of a given sample, it provides no means for speciation of the elements within the sample [211–214]. For example, in the case of an interest in arsenic content in a given sample, the arsenic may be present as arsenite, arsenate, methanearsonic acid (MMA), or dimethylarsenic acid (DMA). Upon entering the ICP, all the various species containing arsenic will be converted into atomic arsenic and, thus, all species will give a mass spectrometric signal at m/z 75. If speciation of the arsenic is of interest, the analyst may arrange to use HPLC in conjunction with ICPMS [215, 216]; alternatively, speciation has been accomplished by MALDI in monitoring $As_3O_5^-$ which is unique to As_2O_3 and $As_3O_8^-$ and is produced by As_2O_5. An example of assaying As in four different species as separated by ion chromatography and monitored by ICP [217] is provided in Figure 4-19; a more recent report uses a photo-oxidizer to allow detection of ten different species of As [218]. Speciation of Se [219–224], as well as Pb and Hg [225, 226], can be done in conjunction with GC; GC has also be used for speciation of organohalide and organometallic compounds [227]. Speciation of Fe in meat can be done by LC/ICPMS [228]. Quantitation of Cd and Zn in Alzheimer brain tissue can be done by ICPMS [229]. Quantitation of DNA adducts can be achieved based on detection of phosphorus by ICPMS [230].

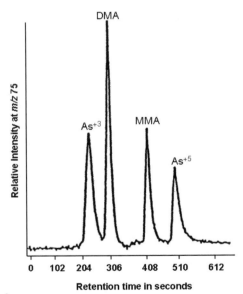

Figure 4-19. Mass chromatogram at m/z 75 showing the detection of arsenic in four compounds separated by HPLC. *Reprinted from Sheppard BS, Caruso JA, Heitkemper DT, and Wolnik KA, Arsenic speciation by ion chromatography with inductively coupled plasma mass spectrometric detection, Analyst (Cambridge, United Kingdom) 1992, 117(6), 971–975, with permission.*

F. Summary

ICPMS is ideally suited for elemental analysis of samples with high specificity and sensitivity. The sensitivity achieved by ICPMS is frequently two to three orders of magnitude better than that achieved by ICP in emission spectrometry. Analytical precision of results from ICPMS is on the order of 1% relative standard deviation. In general, the background in ICPMS is more manageable than that in standalone ICP emission spectrometry, but considerable attention must be given to avoiding isobaric interferences from both atomic and polyatomic species. Appropriate separation techniques such as HPLC must be combined with ICPMS if speciation of the various elements within the sample is of interest. Finally, all of the operational problems that plague conventional ICP operation, such as clogging of orifices with residues from solutions more concentrated than a tenth of a percent, or with other matrix-related solutes, such as residual proteins in a biological sample, are also a problem in ICPMS; the orifices in the skimmer interface between the ICP and the MS are also prone to clogging with residual sample residues. The ICPMS technology is well suited for stable isotope-tracer studies in various states of health.

VI. Final Statement

It should be remembered that mass spectrometry is a tool that can be used to separate mixtures of analytes as well as to identify individual compounds. To this end of using mass spectrometry as a separation tool, all of the inlets presented in this chapter can serve a purpose.

References

1. Biemann K, *Mass Spectrometry*. McGraw Hill, Boston, 1962.
2. Brutti S, Balducci G and Gigli G, A gas-inlet system coupled with a Knudsen cell mass spectrometer for high-temperature studies. *Rapid Commun Mass Spectrom* 21: 89-98, 2007.
3. Schronk LR, Grigsby RD and Scheppele SE, Microdistillation/MS for Petroleum. *Anal. Chem.* 54: 748-755, 1982.
4. Amirav A and Dagan S, A direct sample introduction device for mass spectrometry studies and gas chromatography mass spectrometry analyses. *European Mass Spectrometry* 3(2): 105-111, 1997.
5. Gordin A and Amirav A, SnifProbe: new method and device for vapor and gas sampling. *J Chromatogr A* 903(1-2): 155-72, 2000.
6. Dessort D, Bisseret P, Nakatani Y, Ourisson G and Kates M, Ammonia DCI-MS of Phospholipids. *Chem. Phys. Lipids* 33: 323-340, 1983.
7. Cotter RJ, Desorption MS with Extended Probes. *Anal. Chem.* 52: 1590A-1606A, 1980.
8. Bisseret P, Nakatani Y, Ourisson G, Hueber R and Teller G, Ammonia CI-MS of Lecithins on a Gold Support. *Chem. Phys. Lipids* 33: 383-392, 1983.
9. Hansen G and Munson B, Surface CI-MS. *Anal. Chem.* 50: 1130-1134, 1978.
10. Reinhold VN and Carr SA, DCI-MS with Polyimide-Coated Wires. *Anal. Chem.%V 54*: 499-503, 1982.
11. Turecek F, Pyrolysis Mass Spectrometry. In: *Encyclopedia of Mass Spectrometry*, Vol. 4, pp. 306-312. Wiley, 2004.
12. Schulten HR, Leinweber P and Theng BKG, Organics by Pyrolysis Methylation-MS. *Geoderma* 69: 105-118, 1996.
13. Qian K, Killinger WE, Casey M and Nicol GR, Polymers by Pyrolysis MS. *Anal. Chem.* 68: 1019-1027, 1996.
14. Schulten HR and Lattimer RP, MS of Polymers. *Mass Spectrom. Rev.* 3, 1984.
15. Smith CG, Pyrolysis of Polymers. *Anal. Pyrolysis* 1984: 428-452, 1984.
16. Galletti GC, Dinelli G and Chiavari G, Pyrolysis MS of Sulfonylureas. *J. Mass Spectrom.* 30: 333-338, 1995.
17. Meuzelaar HLC, Windig W, Harper AM, Huff SM, McClennen WH and Richards JM, Pyrolysis MS of Organic Materials. *Science* 226: 268-274, 1984.
18. Tsao T and Voorhees KJ, Smoke Aerosols by Pyrolysis MS. *Anal. Chem.* 56: 368-373, 1984.
19. Beverly MB, Basile F, Voorhees KJ and Hadfield TL, Dipicolinic Acid in Bacterial Spores by Pyrolysis-MS. *Rapid Commun. Mass Spectrom.* 10: 455-458, 1996.
20. Matney ML, Limero TF and James JT, Pyrolysis-GC-MS of Biological Particulates Collected during Space Shuttle Missions. *Anal. Chem.* 66: 2820-2828, 1994.
21. Roussis SG and Fedora JW, Thermal Extraction Unit for Polymers by Pyrolysis/GC-MS. *Rapid Comun. Mass Spectrom.* 10: 82-90, 1966.
22. Braekman-Danheux C, Pyrolysis GC-MS of Hydrocarbons. *J. Anal. Appl. Pyrolysis%V 6*: 195-200, 1984.
23. Wilcken H and Schulten HR, Resin-modified Paints by Pyrolysis-GC-MS. *Fresenius J. Anal. Chem.* 355: 157-163, 1996.
24. Engman H, Mayfield HT, Mar T and Bertsch W, Bacteria by Pyrolysis GC-MS. *J. Anal. Appl. Pyrolysis* 6: 137-156, 1984.
25. Schulter HR, Simmleit N and Müller R, Curie-Point Pyrolysis-GCMS of Spruce Needle. *Anal. Chem.* 61: 458-466, 1989.
26. Genvit W, Boon J and Faix O, Beech Milled Wood Lignin by Pyrolysis-GC/MS. *Anal. Chem.* 59: 508-513, 1987.
27. Bier M and Cooks R, Membrane Interface for Selective Introduction of Volatile Compounds Directly Into the ion source of MS. *Anal. Chem.* 59: 597-601, 1987.
28. Allen TM, Cisper ME, Hemberger PH and Wilkerson CWJ, Volatile, semivolatile organics, and organometallic compds in air and water by MIMS. *Int J Mass Spectrom* 212: 197-204, 2001.
29. Johnson, Cooks R and Allen, MIMS Trends & Applications. *Mass Spectrom. Rev.* 19: 1-37, 2000.
30. Ketola RA, Kotiaho T, Cisper ME and Allen TM, Environmental Appl of MIMS. *J. Mass Spectrom.* 37: 457-476, 2002.
31. Bauer SJ and Cooks RG, Membrane Interface for MS. *Amer. Lab* 25: 36-51, 1993.
32. Hayward MJ and Maden AJ, Sheet Materials as Membranes in MIMS. *Anal. Chem.* 68: 1805-1811, 1996.
33. Mansikka T, Kostiainen R, Kotiaho T, Ketola RA, Mattila I, Ojala M, Honkanen T, Wickstrom K, Waldvogel J and Pilvio O, Automatic MIMS system for online water monitoring. *Advances in Mass Spectrometry* 14: D021840/1-D021840/5, 1998.
34. Janfelt C, Frandsen H and Lauritsen FR, A mini MIMS for on-site detection of contaminants in both aqueous and liquid organic samples. *Rapid Commun Mass Spectrom* 20: 1441-1446, 2006.
35. Savickas PJ, LaPack MA, Tou JC, Hayward MJ, Kotiaho T, Lister AK, Cooks RG, Ausitn GD, Narayan R and Tsao GT, Hollow Fiber Membrane Probes for MIMS: Monitoring Bioreactions. *Anal. Chem.* 61: 2332-2336, 1989.

36. LaPack MA, Tou JC, Cole MJ and Enke CG, Valved Sampling Cell for MIMS. *Anal. Chem.* 68: 3072-3075, 1996.
37. Mendes MA, Pimpim RS, Kotiaho T and Eberlin MN, A Cryotrap MIMS System for Analysis of Volatile Organic Compounds in Water at the Low-PPT Level. *Anal. Chem.* 68: 3502-3506, 1996.
38. Soni M, Bauer S, Amy JW, Wong P and Cooks RG, PPQ Organics in H2O by MIMS. *Anal. Chem.* 67: 1409-1412, 1995.
39. Wong PSH, Srinivasan N, Kasthurikrishnan N, Cooks RG, Pincock JA and Grossert JS, On-Line Monitoring of the Photolysis of Benzyl Acetate and 3,5-Dimethoxybenyl Acetate by MIMS. *J. Org. Chem.* 61: 6627-6632, 1996.
40. Milagre CDF, Milagre HMS, Rodrigues JAR, Rocha LL, Santos LS and Eberlin MN, On-line monitoring of bioreductions via MIMS. *Biotechnology and Bioengineering* 90: 888-892, 2005.
41. Kotiaho T, On-site Environmental and In Situ Process Analysis by MS. *J. Mass Spectrom.* 31: 1-15, 1996.
42. Virkki VT, Ketola RA, Ojala M, Kotiaho T, Komppa V, Grove A and Facchetti S, On-Site Environmental Analysis via MIMS. *Anal. Chem.* 67: 1421-1425, 1995.
43. Xu C, Patrick JS and Cooks RG, Affinity MIMS. *Anal. Chem.* 67: 724-728, 1995.
44. Creba AS, Weissfloch ANE, Krogh ET and Gill CG, An Enzyme Derivatized Polydimethylsiloxane (PDMS) Membrane for MIMS. *J Amer Soc Mass Spectrom* 18: 973-979, 2007.
45. Cisper ME, Gill CG, Townsend LE and Hemberger PH, PPT VOCs in Air by MIMS. *Anal. Chem.* 67: 1413-1417, 1995.
46. Ojala M, Mattila I, Tarkiainen V, Saerme T, Ketola RA, Maeaettaenen A, Kostiainen R and Kotiaho T, Purge-and-membrane MS for analysis of VOCs in soil. *Anal Chem* 73: 3624-3631, 2001.
47. Lauritsen FR and Gylling S, Monitoring Biological Reactions via MIMS. *Anal. Chem.* 67: 1418-1420, 1995.
48. Lundsgaard JS and Gronlund J, Membrane Probe for Blood Gases by MS. *J. Appl. Physiol.* 48: 376-381, 1980.
49. Brodbelt J, Cooks R, Tou J, Kallos G and Dryzga M, In Vivo Mass Spectrometric Determination of Organic Compounds in Blood with a Membrane Probe. *Anal. Chem.* 59: 454-458, 1987.
50. Pinard E, Rigaud AS, Riche D, Naquet R and Seylaz J, Membrane for Tissue Gas by MS. *Neurosci* 23: 943-949, 1980.
51. Pinkston JD, Advantages and drawbacks of popular supercritical fluid chromatography/mass spectrometry interfacing approaches-a user's perspective. *European Journal of Mass Spectrometry* 11(2): 189-197, 2005.
52. Pinkston JD and Chester TL, Guidelines for SFC-MS. *Anal. Chem.* 67: 650A-656A, 1996.
53. Sheeley DM and Reinhold VN, SFC-MS of high-molecular-weight biopolymers. Instrumental considerations. *J Chromatog* 474: 83-96, 1989.
54. Poole CF and Poole SK, Supercritical Fluid Chromatography. In: *Chromatography today*, pp. 601-648. Elsevier, 1991.
55. Pinkston JD and Hentschel RT, Chromatography Columns with Cast Porous Plugs and Methods of Fabricating Them. *J. High Resolut. Chromatogr.* 16: 269-274, 1993.
56. Hoke SH, II, Tomlinson JA, Bolden RD, Morand KL, Pinkston JD and Wehmeyer KR, Increasing Bioanalytical Throughput Using pcSFC-MS/MS: 10 Minutes per 96-Well Plate. *Analytical Chemistry* 73(13): 3083-3088, 2001.
57. Smith R and Udseth H, MS Interface for Microbore SFC with Splitless Injection. *Anal. Chem.* 59: 13-22, 1987.
58. Olesik SV, SFC-MS. *High Resolut. Chromatogr.* 14: 5-9, 1991.
59. Huang EC, Jackson BJ, Markides KE and Lee ML, Heated Probe for Capillary SFC-MS. *Anal. Chem.* 60: 2715-2719, 1988.
60. Pinkston JD, Owens GD, Burkes LJ, Delaney TE, Millington DS and Maltby DA, Capillary SFC-MS Using a "High Mass" Quadrupole and Splitless Injection. *Anal. Chem.* 60: 962-966, 1988.
61. Pinkston JD and Bowling DJ, Cryopumping in SFC-MS. *Anal. Chem.* 65: 3534-3539, 1993.
62. Zheng J, Taylor LT, Pinkston JD and Mangels ML, Effect of ionic additives on the elution of sodium aryl sulfonates in SFC. *Journal of Chromatography, A* 1082(2): 220-229, 2005.
63. Zheng J, Glass T, Taylor LT and Pinkston JD, Study of the elution mechanism of sodium aryl sulfonates on bare silica and a cyano bonded phase with methanol-modified carbon dioxide containing an ionic additive. *Journal of Chromatography, A* 1090(1-2): 155-164, 2005.
64. Pinkston JD, Stanton DT and Wen D, Elution and structure-retention modeling of polar and ionic substances in SFC using volatile ammonium salts as mobile phase additives. *Journal of Separation Science* 27: 115-123, 2004.
65. Mertens MAA and et al., SFC Coupled to Magnetic MS. *J. High Resolut. Chromatogr.* 19: 17-22, 1996.
66. Reinhold VN, Sheeley DM, Kuei J and Her G, High Mass Samples on a Magnetic MS by SFC. *Anal Chem* 60: 2719-2722, 1988.
67. Baumeister ER, West CD, Ijames CF and Wilkins CL, Interface for SFC and FTMS. *Anal. Chem.* 63: 251-255, 1991.

68. Pinkston JD, Delaney TE, Morand KL and Cooks RG, SFC-MS Using a Quadrupole/Ion Trap Hybrid with External Source. *Anal. Chem.* 64: 1571-1577, 1992.
69. Garzotti M, Rovatti L and Hamdan M, Coupling of a SFC system to a hybrid (Q-TOF 2) mass spectrometer: on-line accurate mass measurements. *Rapid Commun Mass Spectrom* 15: 1187-90, 2001.
70. Carey J and Caruso JA, Plasma Spectrometric Detection for SFC. *Trends Anal. Chem.* 11: 287-293, 1992.
71. Kalinoski H and Hargiss L, SFC-MS of Non-ionic Surfactant Materials Using Chloride-Attachment NCI. *J. Chromatogr.* 505: 199-213, 1990.
72. Chen J, Hsieh Y, Cook J, Morrison R and Korfmacher WA, SFC-MS/MS for enantioselective determination of propranolol and pindolol in mouse blood by serial sampling. *Anal Chem* 78: 1212-1217, 2006.
73. Alexander AJ and Staab A, Achiral/Chiral SFC/MS for Profiling Isomeric Cinnamonitrile/Hydrocinnamonitrile Products in Chiral Drug Synthesis. *Anal Chem* 78: 3835-3838, 2006.
74. Xu X, Roman JM, Veenstra TD, Van Anda J, Ziegler RG and Issaq HJ, Analysis of 15 Estrogen Metabolites by Packed Column SFC MS. *Anal Chem* 78: 1553-1558, 2006.
75. Zheng J, Pinkston JD, Zoutendam PH and Taylor LT, Feasibility of SFC/MS of Polypeptides with Up to 40-Mers. *Anal Chem* 78: 1535-1545, 2006.
76. Emmett MR, Kazazic S, Marshall AG, Chen W, Shi SDH, Bolanos B and Greig MJ, Supercritical Fluid Chromatography Reduction of Hydrogen/Deuterium Back Exchange in Solution-Phase Hydrogen/Deuterium Exchange with Mass Spectrometric Analysis. *Analytical Chemistry* 78(19): 7058-7060, 2006.
77. Cunico RL, Gooding KM and Wehr T, *Basic HPLC and CE of Biomolecules*. Bay Bioanalytical Laboratory, Inc, 1998.
78. Li J and Richards JC, Application of capillary electrophoresis MS to the characterization of bacterial lipopolysaccharides. *Mass Spectrometry Reviews* 26: 35-50, 2007.
79. Zheng J, Norton D and Shamsi SA, Fabrication of Internally Tapered Capillaries for Capillary Electrochromatography ESI MS. *Anal Chem* 78: 1323-1330, 2006.
80. Tseng M-C, Chen Y-R and Her G-R, A Low-Makeup Beveled Tip Capillary Electrophoresis/Electrospray Ionization Mass Spectrometry Interface for Micellar Electrokinetic Chromatography and Nonvolatile Buffer Capillary Electrophoresis. *Analytical Chemistry* 76(21): 6306-6312, 2004.
81. Li F-A, Wu M-C and Her G-R, Development of a Multiplexed Interface for Capillary Electrophoresis-Electrospray Ion Trap Mass Spectrometry. *Analytical Chemistry* 78(15): 5316-5321, 2006.
82. Smith RD, Wahl JH, Goodlett DR and Hofstadler SA, CE-MS. *Anal. Chem.* 65: 574A-584A, 1993.
83. Monnig CA and Jorgenson JW, High Speed CE. *Anal. Chem.* 63: 802-807, 1991.
84. Holland JF, Enke CG, Allison J, Stults JT, Pinkston JD, Newcome B and Watson JT, Mass spectrometry on the chromatographic time scale: realistic expectations. *Anal. Chem.* 55: 997A-1012A, 1983.
85. Erickson ED, Enke CG, Holland JF and Watson JT, Time array detection in GC/TOFMS. *Anal. Chem.* 62: 1079-84, 1990.
86. Hofstadler SA, Swanek FD, Gale DC, Ewing AG and Smith RD, CE-ESI-FT-ICR for cellular proteins. *Anal. Chem.* 67: 1477-1480, 1995.
87. Moini M, Simplifying CE-MS Operation. 2. Interfacing Low-Flow Separation Techniques to MS Using a Porous Tip. *Anal Chem* 79: 4241-4246, 2007.
88. Chen H, Talaty NN, Takats Z and Cooks RG, DESI MS for High-Throughput Analysis of Pharmaceuticals in Ambient Environment. *Anal Chem* 77: 6915-6927, 2005.
89. Figeys D, van Oostveen I, Ducret A and Aebersold R, Proteins at Subfemtomole Level by CE-ESI-MS/MS. *Anal. Chem.* 68: 1822-1828, 1996.
90. Zhao Z, Wahl JH, Udseth HR, Hofstadler SA, Fuciarelli AF and Smith RD, On-line Capillary Electrophoresis-ESI-MS. *Electrophoresis* 16: 389-395, 1995.
91. Schmeer K, Behnke B and Bayer E, Capillary Electrochromatography-ESI. *Anal. Chem.* 67: 3656-3658, 1995.
92. Banks JF, Jr. and Dresch T, CE Peptide Separations Coupled to ESI-TOFMS. *Anal. Chem.* 68: 1480-1485, 1996.
93. Takada Y, Sakairi M and Koizumi H, APCI Interface for CE/MS. *Anal. Chem.* 67: 1474-1476, 1995.
94. Takada Y, Sakairi M and Koizumi H, CE-APCI-MS. *Anal. Chem.* 67: 1474-1476, 1995.
95. Tomlinson AJ, Benson LM, Oda RP, Braddock WD, Riggs BL and Katzmann JAANS, Preconcentration-CE-MS. *J. Cap. Electrophoresis* 2: 97-104, 1995.
96. Thompson T, Foret F, Vouros P and Karger B, Capillary Electrophoresis-ESI of Proteins. *Anal. Chem.* 65: 900-906, 1993.
97. Williams BJ, Russell WK and Russell DH, Utility of CE-MS Data in Protein Identification. *Anal Chem* 79: 3850-3855, 2007.
98. Hsieh YL, Cai J, Li YT, Henion JD and Ganem B, Protein/Drug Complexes by CE-ESI-MS. *J. Amer. Soc. Mass Spectrom.* 6: 85-90, 1995.

99. Vuorensola K, Kokkonen J, Siren H and Ketola RA, Optimization of CE-ESI-MS for catecholamines. *Electrophoresis* 22: 4347-4354, 2001.

100. Ortega-Gadea S, Bernabe-Zafon V, Simo-Alfonso EF, Ochs C and Ramis-Ramos G, Industrial alkylpolyphosphonates by infusion ESI-ion trap MS with ID of impurities by tandem CZE. *J Mass Spectrom* 41: 23-33, 2006.

101. Carroll DI, Dzidic I, Horning EC and Stillwell RN, API Review. *Appl. Spectrosc. Rev.* 17: 337-352, 1981.

102. Cody RB, Laramee JA and Durst HD, Versatile new ion source for the analysis of materials in open air under ambient conditions. *Analytical Chemistry* 77(8): 2297-2302, 2005.

103. Laramee JA and Deinzer ML, Capillary Gas Chromatographic Introduction of Environmental Compounds into a Trochoidal Electron Monochromator/Mass Spectrometer. *Analytical Chemistry* 66(5): 719-24, 1994.

104. Haefliger OP and Jeckelmann N, Direct mass spectrometric analysis of flavors and fragrances in real applications using DART. *Rapid Commun Mass Spectrom* 21: 1361-1366, 2007.

105. Vail TM, Jones PR and Sparkman OD, Rapid and Unambiguous Identification of Melamine in Contaminated Pet Food Based on Mass Spectrometry with Four Degrees of Confirmation. *J Anal Tox* (in press), 2007.

106. Pringle SD, Giles K, Wildgoose JL, Williams JP, Slade SE, Thalassinos K, Bateman RH, Bowers MT and Scrivens JH, An investigation of the mobility separation of some peptide and protein ions using a new hybrid quadrupole/travelling wave IMS/oa-ToF instrument. *Int J Mass Spectrom*, 2007.

107. Tang X, Bruce JE and Hill HH, Jr., Characterizing Electrospray Ionization Using Atmospheric Pressure Ion Mobility Spectrometry. *Analytical Chemistry* 78(22): 7751-7760, 2006.

108. Takats Z, Wiseman JM, Gologan B and Cooks RG, Mass Spectrometry Sampling Under Ambient Conditions with Desorption Electrospray Ionization. *Science* 306(5695): 471-473, 2004.

109. Costa AB and Cooks RG, Simulation of Atmospheric Transport in DESI. In: *55th Ann Conf Mass Spectrom, Indianapolis,2007.*

110. Ifa DR, Gumaelius LM, Eberlin LS, Manicke NE and Cooks RG, Forensic analysis of inks by imaging DESI mass spectrometry. *Analyst* 132(5): 461-467, 2007.

111. Cooks RG, Ouyang Z, Takats Z and Wiseman JM, Ambient Mass Spectrometry. *Science (Washington, DC, United States)* 311(5767): 1566-1570, 2006.

112. Cotte-Rodriguez I, Takats Z, Talaty N, Chen H and Cooks RG, DESI of Explosives on Surfaces: Sensitivity and Selectivity Enhancement by Reactive DESI. *Anal Chem* 77: 6755-6764, 2005.

113. Cotte-Rodriguez I, Chen H and Cooks RG, Rapid trace detection of triacetone triperoxide (TATP) by complexation reactions during DESI. *Chemical Communications (Cambridge, United Kingdom)* (9): 953-955, 2006.

114. Justes DR, Talaty N, Cotte-Rodriguez I and Cooks RG, Detection of explosives on skin using ambient ionization MS. *Chem Commun* (21): 2142-2144, 2007.

115. D'Agostino PA, Chenier CL, Hancock JR and Lepage CRJ, DESI of chemical warfare agents from solid-phase microextraction fibers. *Rapid Commun Mass Spectrom* 21: 543-549, 2007.

116. Nefliu M, Venter A and Cooks RG, DESI and electrosonic spray ionization for solid- and solution-phase analysis of industrial polymers. *Chemical Communications (Cambridge, United Kingdom)* (8): 888-890, 2006.

117. Sparrapan R, Eberlin LS, Haddad R, Cooks RG, Eberlin MN and Augusti R, Ambient Eberlin reactions via desorption electrospray ionization mass spectrometry. *Journal of Mass Spectrometry* 41(9): 1242-1246, 2006.

118. Nyadong L, Green MD, De Jesus VR, Newton PN and Fernandez FM, Reactive DESI linear ion trap MS of latest-generation counterfeit antimalarials via noncovalent complex formation. *Anal Chem* 79: 2150-2157, 2007.

119. Nefliu M, Cooks RG and Moore C, Enhanced Desorption Ionization Using Oxidizing Electrosprays. *Journal of the American Society for Mass Spectrometry* 17(8): 1091-1095, 2006.

120. Ford MJ, Deibel MA, Tomkins BA and Van Berkel GJ, Quantitative TLC MS of Caffeine by Surface Sampling Probe ESI-MS/MS System. *Anal Chem* 77: 4385-4389, 2005.

121. Van Berkel GJ and Kertesz V, Automated sampling and imaging of analytes separated on TLC plates using DESI-MS. *Anal Chem* 78: 4938-4944, 2006.

122. Van Berkel GJ, Tomkins BA and Kertesz V, TLC/DESI-MS: Investigation of Goldenseal Alkaloids. *Anal Chem* 79: 2778-2789, 2007.

123. Sampson JS, Hawkridge AM and Muddiman DC, Generation and Detection of Multiply-Charged Peptides and Proteins by Matrix-Assisted Laser Desorption ESI (MALDESI) FT_iCR MS. *J Amer Soc Mass Spectrom* 17: 1712-1716, 2006.

124. Takats Z, Wiseman JM and Cooks RG, Ambient MS by DESI: Instrumentation, mechanisms and applications in forensics, chemistry, and biology. *J Mass Spectrom* 40: 1261-1275, 2005.

125. Venter A and Cooks RG, DESI in a Small Pressure-Tight Enclosure. *Anal. Chem.* 79: in press, 2007.

126. Misharin AS, Laughlin BC, Vilkov A, Takats Z, Ouyang Z and Cooks RG, High-throughput using APCI MS with cylindrical ion trap array. *Analytical Chemistry* 77: 459-470, 2005.

127. McEwen CN, McKay RG and Larsen BS, Analysis of Solids, Liquids, and Biological Tissues Using Solids Probe Introduction at Atmospheric Pressure on Commercial LC/MS Instruments. *Anal Chem* 77: 7826-7831, 2005.

128. McEwen CN and Gutteridge S, Analysis of the Inhibition of the Ergosterol Pathway in Fungi Using the Atmospheric Solids Analysis Probe (ASAP) Method. *J Amer Soc Mass Spectrom* 18(7): in press, 2007.

129. Adams NG and Smith D, The selected ion flow tube (SIFT); a technique for studying ion-neutral reactions. *International Journal of Mass Spectrometry and Ion Physics* 21(3-4): 349-59, 1976.

130. Smith D and Spanel P, Selected Ion Flow Tube Mass Spectrometry (SIFT-MS) for On-Line Trace Gas Analysis. *Mass Spectrom. Rev.* 24: 661-700, 2005.

131. Smith D and Adams NG, SIFT: studies of ion-neutral reactions. *Advances in Atomic and Molecular Physics* 24: 1-49, 1988.

132. Spanel P and Smith D, Quantitative SIFT-MS: the influence of ionic diffusion and mass discrimination. *J Am Soc Mass Spectrom* 12: 863-872, 2001.

133. Kubista J, Spanel P, Dryahina K, Workman C and Smith D, Combined use of GC and SIFT MS for absolute trace gas quantification. *Rapid Commun Mass Spectrom* 20: 563-567, 2005.

134. Spanel P, Diskin AM, Wang T and Smith D, A SIFT study of the reactions of H3O+, NO+ and O2+ with hydrogen peroxide and peroxyacetic acid. *Intl J Mass Spectrom* 228: 269-283, 2003.

135. Jackson DM, Adams NG and Babcock LM, Ion-Molecule Reactions of Several Ions with Ethylene Oxide and Propenal in a Selected Ion Flow Tube. *J Amer Soc Mass Spectrom* 18: 445-452, 2007.

136. Lagg A, Taucher J, Hansel A and Lindinger W, Applications of proton transfer reactions to gas analysis. *Intl J Mass Spectrom Ion Proc* 134: 55-66, 1994.

137. Steeghs MML, Sikkens C, Crespo E, Cristescu SM and Harren FJM, Development of a proton-transfer reaction ion trap MS: Online detection and analysis of volatile organic compounds. *Int J Mass Spectrom* 262: 16-24, 2007.

138. Lattimer RP and Schulten HR, FI and FDMS: Past, Present and Future. *Anal. Chem.* 61: 1201A-1215A, 1989.

139. Macfarlane RD and Sundqvist BUR, Cf-252 PDMS. *Mass Spectrom Rev.* 4: 421-460, 1985.

140. Busch KL, Desorption ionization mass spectrometry. *Journal of Mass Spectrometry* 30: 233-40, 1995.

141. Barber M, Bordoli RS, Sedgwick RD and Tyler AN, F.A.B. *J. Chem. Soc. Chem. Commun.* 1981: 325-327, 1981.

142. Seifert Jr. WE and Caprioli RM, Fast Atom Bombardment MS. In: *Methods in Enzymology, High Resolution Separation of Macromolecules*, Vol. 270 (Eds. Hancock B and Karger BL), pp. 453-486, 1996.

143. Pachuta S and Cooks RG, Mechanisms in Molecular SIMS. *Chem. Rev.* 87: 647-669, 1987.

144. Benninghoven A, Organic SIMS and its relation to FAB. *Intl J Mass Spectrom Ion Phys* 46: 459-62, 1983.

145. Hamdan M, FIB, FAB, Liquid or Molecular SIMS? *Org. Mass Spectrom.* 27: 759-760, 1992.

146. De Pauw E, Matrix selection for LSIMS and FAB MS. *Methods in Enzymology* 193(Mass Spectrom.): 201-14, 1990.

147. Cook KD and Chan KWS, Energy deposition in desorpt/ionization. *Int. J. Mass Spectrom. Ion Processes* 54: 135-49, 1983.

148. Ligon WV, Jr., Extraction and analysis of organic cations from acid solution with strong electric fields and mass spectrometry. *Science (Washington, DC, United States)* 204(4389): 198-200, 1979.

149. Ligon WV, Jr., SIMS of glycerol-d5 shows radical cation intermediates. *Intl J Mass Spectrom Ion Phys* 52: 189-93, 1983.

150. Field FH, FAB study of glycerol: MS and radiation chemistry. *J Phys Chem* 86: 5115-23, 1982.

151. Kriger MS, Cook KD, Short RT and Todd PJ, Secondary ion emission from solutions: time dependence and surface phenomena. *Analytical chemistry* 64: 3052-8, 1992.

152. Ligon WV, Jr. and Dorn SB, Understanding the glycerol surface as it relates to the SIMS experiment. A review. *Intl J Mass Spectrom Ion Proc* 78: 99-113, 1987.

153. Ligon WV, Jr. and Dorn SB, Surfactants modify response factors in SIMS of liquid surfaces. *Intl J Mass Spectrom Ion Proc* 61: 113-22, 1984.

154. Abramson FP, Chemical Reaction Interface MS (CRIMS). *Mass Spectrom. Rev.* 13: 341-356, 1994.

155. Song H, Kusmierz J, Abramson F and McLean M, CRIMS with an MSD. *J. Amer. Soc. Mass Spectrom.* 5: 765-771, 1994.

156. Lecchi P and Abramson FP, An innovative method for measuring hydrogen and deuterium: CRIMS with nitrogen reactant gas. *J Am Soc Mass Spectrom* 11: 400-6., 2000.

157. Lecchi P and Abramson FP, Size-exclusion chromatography combined with chemical reaction interface mass spectrometry for the analysis of complex mixtures of proteins. *ACS Symposium Series* 893(Multiple Detection in Size-Exclusion Chromatography): 184-195, 2005.

158. McLean M, Vestal ML, Teffera Y and Abramson FP, Element and Isotope Specific Detection for HPLC Using CRIMS. *J. Chromatogr., A* 732: 189-199, 1996.

159. Chance DH and Abramson FP, CRIMS for Selective Detection of 13C- , 15N-, and 2H-labeled compounds. *Anal. Chem.* 61: 2724-2730, 1989.

160. Teffera Y, Kusmierz JJ and Abramson FP, Continuous-Flow Isotope Ratio MS Using CRIMS for either GC or LC. *Anal. Chem.* 68: 1888-1894, 1996.
161. Lecchi P and Abramson FP, Size exclusion chromatography-CRIMS: "a perfect match". *Anal Chem* 71: 2951-5., 1999.
162. Abramson FP, Black GE and Lecchi P, HPLC with isotope-ratio MS for measuring enrichment of underivatized materials. *J Chromatogr A* 913: 269-73., 2001.
163. Eckers C, Abramson FP and Lecchi P, Sulfur-containing impurities in pharmaceuticals by HPLC/CRIMS. *Rapid Commun Mass Spectrom* 15: 602-7., 2001.
164. Abramson FP, Teffera Y, Kusmierz J, Steenwyk RC and Pearson PG, Replacing 14C with Stable Isotopes in Drug Metabolism Studies. *Drug Metab. Dispos.* 24: 697-701, 1996.
165. Song H and Abramson FP, CRIMS for Detecting Isotope Enrichment in Metabolites. *Drug Metab. Disp.* 2: 868-873, 1993.
166. Abramson FP, The use of stable isotopes in drug metabolism studies. *Semin Perinatol* 25: 133-8., 2001.
167. Goldthwaite CAJ, Hsieh F-Y, Womble SW, Nobes BJ, Blair IA, Klunk LJ and Mayol RF, LC/CRIMS as an Alternative to Radioisotopes for Quantitative Drug Metabolism Studies. *Anal. Chem.* 68: 2996-3001, 1996.
168. Kusmierz JJ and Abramson FP, Tracing 15N-Amino Acids with CRIMS. *Biol. Mass Spectrom.* 23: 756-763, 1994.
169. Song H, and Abramson, F. P., CRIMS for Chlorinated Benzodiazenines Compounds. *Anal. Chem.* 65: 447-450, 1993.
170. Houk RS, Elemental and isotopic analysis by inductively coupled plasma mass spectrometry. *Accounts of Chemical Research* 27(11): 333-9, 1994.
171. Ramendik GI, Fatyushina EV, Stepanov AI and Sevast'yanov VS, New approach to the calculation of relative sensitivity factors in ICP MS. *J Anal Chem (Translation of Zhurnal Analiticheskoi Khimii)* 56: 500-506, 2001.
172. Cavazzini G, A method for determining isotopic composition of elements by thermal ionization source mass spectrometry. *International Journal of Mass Spectrometry* 240(1): 17-26, 2005.
173. Todd PJ, McKown HS and Smith DH, A new ion source for thermal emission mass spectrometry. *International Journal of Mass Spectrometry and Ion Physics* 42(3): 183-90, 1982.
174. Bencsath FA and Field FH, Electron, chemical, and thermal ionization mass spectra of alkali halides. *Analytical Chemistry* 58(4): 679-84, 1986.
175. Ramendik GI, Kryuchkova OI, Tyurin DA, McHedlidze TR and Kaviladze MS, Factors affecting the relative sensitivity coefficients in spark and laser plasma source MS. *Int J Mass Spectrom and Ion Proc* 63: 1-15, 1985.
176. Beauchemin D, Inductively coupled plasma mass spectrometry. *Analytical Chemistry* 78(12): 4111-4135, 2006.
177. Cottingham K, ICPMS: it's elemental. *Analytical Chemistry* 76(1): 35A-38A, 2004.
178. Beauchemin D, Inductively coupled plasma mass spectrometry. *Anal Chem* 76(12): 3395-416., 2004.
179. Lobinski R, Schaumloeffel D and Szpunar J, Mass Spectrometry in Bioinorganic Analytical Chemistry. *Mass Spectrometry Reviews* 25: 255-289, 2006.
180. Sturgeon R, *Inductively coupled plasma mass spectrometry handbook, Simon M. Nelms (ed)*, Vol. 20, 2005.
181. Hill SJ, *Inductively Coupled Plasma Spectrometry and its Applications.* Blackwell Publishing, UK, 2007.
182. Giusti P, Lobinski R, Szpunar J and Schaumloeffel D, Development of a Nebulizer for a Sheathless Interfacing of NanoHPLC and ICPMS. *Analytical Chemistry* 78(3): 965-971, 2006.
183. Axelsson MD, Rodushkin I, Ingri J and Ohlander B, Multielemental analysis of Mn-Fe nodules by ICP-MS: optimisation of analytical method. *Analyst* 127: 76-82., 2002.
184. Barnes JHIV, Schilling GD, Hieftje GM, Sperline RP, Denton MB, Barinaga CJ and Koppenaal DW, Novel array detector for analysis of solid samples by laser ablation ICP MS. *J Am Soc Mass Spectrom* 15: 769-776, 2004.
185. Van Acker MRMD, Shahar A, Young ED and Coleman ML, GC/Multiple Collector-ICPMS Method for Chlorine Stable Isotope Analysis of Chlorinated Aliphatic Hydrocarbons. *Analytical Chemistry* 78(13): 4663-4667, 2006.
186. Forte M, Rusconi R, Margini C, Abbate G, Maltese S, Badalamenti P and Bellinzona S, Uranium isotopes in food and environmental samples by different techniques: a comparison. *Radiat Prot Dosimetry* 97: 325-8., 2001.
187. Becker JS, ICP-MS and laser ablation for analysis of long-lived radionuclides. *Intl J Mass Spectrom* 242: 183-195, 2005.
188. Li F, Armstrong DW and Houk RS, Behavior of bacteria in the inductively coupled plasma: Atomization and production of atomic ions for mass spectrometry. *Analytical Chemistry* 77(5): 1407-1413, 2005.

189. Brouwers EEM, Tibben MM, Rosing H, Hillebrand MJX, Joerger M, Schellens JHM and Beijnen JH, Sensitive inductively coupled plasma mass spectrometry assay for the determination of platinum originating from cisplatin, carboplatin, and oxaliplatin in human plasma ultrafiltrate. *Journal of Mass Spectrometry* 41(9): 1186-1194, 2006.

190. Zoriy MV, Dehnhardt M, Reifenberger G, Zilles K and Becker JS, Imaging of Cu, Zn, Pb and U in human brain tumor resections by laser ablation inductively coupled plasma mass spectrometry. *International Journal of Mass Spectrometry* 257(1-3): 27-33, 2006.

191. Barnes JHIV, Hieftje GM, Denton MB, Sperline R, Koppenaal DW and Barinaga C, MS detector array that provides truly simultaneous detection. *American Laboratory (Shelton, CT, US)* 35: 15-16, 18, 20-22, 2003.

192. Barnes JHIV, Schilling GD, Stone SF, Sperline RP, Denton MB, Young ET, Barinaga CJ, Koppenaal DW and Hieftje GM, Simultaneous multichannel MS HPLC by array detector sector-field. *Anal and Bioanalytical Chem* 380: 227-234, 2004.

193. Hieftje GM, Barnes JH, Gron OA, Leach AM, McClenathan DM, Ray SJ, Solyom DA, Wetzel WC, Denton MB and Koppenaal DW, Evolution and revolution in instrumentation for plasma-source MS. *Pure and App Chem* 73: 1579-1588, 2001.

194. McClenathan DM and Hieftje GM, Absolute methods of quantitation in glow discharge TOFMS. *Journal of Analytical Atomic Spectrometry* 20: 1326-1331, 2005.

195. McClenathan DM, Ray SJ, Wetzel WC and Hieftje GM, Plasma source TOFMS. *Analytical Chemistry* 76: 158A-166A, 2004.

196. Niu H and Houk RS, Fundamental aspects of ion extraction in inductively coupled plasma mass spectrometry. *Spectrochimica Acta, Part B: Atomic Spectroscopy* 51B(8): 779-815, 1996.

197. McClenathan DM, Wetzel WC, Lorge SE and Hieftje GM, Effect of the plasma operating frequency on the figures of merit of an ICP TOFMS. *Journal of Analytical Atomic Spectrometry* 21: 160-167, 2006.

198. Gamez G, Bogaerts A and Hieftje GM, Temporal and spatially resolved laser-scattering plasma diagnostics for the characterization of a ms-pulsed glow discharge. *Journal of Analytical Atomic Spectrometry* 21(3): 350-359, 2006.

199. Ammann AA, ICP MS: a versatile tool. *J Mass Spectrom* 42: 419-427, 2007.

200. Butler OT, Cook JM, Harrington CF, Hill SJ, Rieuwerts J and Miles DL, Atomic spectrometry update. Environmental analysis. *J Anal Atomic Spectrom* 22: 187-221, 2007.

201. Houk RS and Praphairaksit N, Dissociation of polyatomic ions in the inductively coupled plasma. *Spectrochimica Acta, Part B: Atomic Spectroscopy* 56B(7): 1069-1096, 2001.

202. Myers DP, Ray SJ and Hieftje GM, Inorganic time-of-flight mass spectrometry. *Practical Spectroscopy* 23: 447-505, 2000.

203. Rodushkin I and Axelsson MD, Application of double focusing sector field ICP-MS for multielemental characterization of human hair and nails. Part I. Analytical methodology. *Sci Total Environ FIELD Full Journal Title:The Science of the total environment* 250(1-3): 83-100, 2000.

204. Milgram KE, White FM, Goodner KL, Watson CH, Koppenaal DW, Barinaga CJ, Smith BH, Winefordner JD, Marshall AG, Houk RS and Eyler JR, High-Resolution Inductively Coupled Plasma Fourier Transform Ion Cyclotron Resonance Mass Spectrometry. *Analytical Chemistry* 69(18): 3714-3721, 1997.

205. Resano M, Gelaude I, Dams R and Vanhaecke F, Solid sampling-electrothermal vaporization-ICP MS for Hg in materials using isotope dilution with a gaseous phase for calibration. *Spectrochimica Acta, Part B: Atomic Spectroscopy* 60B: 319-326, 2005.

206. Peschel BU, Andrade F, Wetzel WC, Schilling GD, Hieftje GM, Broekaert JAC, Sperline R, Denton MB, Barinaga CJ and Koppenaal DW, Electrothermal vaporization coupled with ICP array-detector MS for the multielement analysis of Al2O3 ceramic powders. *Spectrochimica Acta, Part B: Atomic Spectroscopy* 61B: 42-49, 2006.

207. Pickhardt C, Dietze H-J and Becker JS, Laser ablation ICP MS for direct isotope ratio measurements on solid samples. *Intl J Mass Spectrom* 242: 273-280, 2005.

208. Pickhardt C, Izmer AV, Zoriy MV, Schaumloeffel D and Sabine Becker J, On-line isotope dilution in laser ablation ICP MS by microflow nebulizer. *Int J Mass Spectrom* 248: 136-141, 2006.

209. Bajic SJ, Aeschliman DB, Saetveit NJ, Baldwin DP and Houk RS, Analysis of glass fragments by laser ablation-ICP-MS and principal component analysis. *J Forensic Sci* 50: 1123-1127, 2005.

210. Wang Z, Hattendorf B and Guenther D, Analyte Response in Laser Ablation ICP MS. *J Amer Soc Mass Spectrom* 17: 641-651, 2006.

211. Szpunar JaL, R., Hyphenated Techniques in Speciation Analysis. pp. 220. Royal Society of Chemistry, 2004.

212. Ray SJ, Andrade F, Gamez G, McClenathan D, Rogers D, Schilling G, Wetzel W and Hieftje GM, Plasma-source mass spectrometry for speciation analysis: state-of-the-art. *Journal of Chromatography, A* 1050(1): 3-34, 2004.

213. Rosen AL and Hieftje GM, ICP-MS & ESI-MS for speciation analysis: applications and instrumentation. *Spectrochimica Acta, Part B: Atomic Spectroscopy* 59B(2): 135-146, 2004.

214. Tu Q, Wang T, Welch CJ, Wang P, Jia X, Raab C, Bu X, Bykowski D, Hohenstaufen B and Doyle MP, ID and Speciation of Isomeric Intermediates in a Catalyst Formation Reaction by HPLC-ICPMS and HPLC-ESI-MS. *Anal Chem* 78: 1282-1289, 2006.

215. Francesconi KA, Tanggaar R, McKenzie CJ and Goessler W, Arsenic metabolites in human urine after ingestion of an arsenosugar. *Clin Chem* 48: 92-101., 2002.

216. Polatajko A and Szpunar J, Speciation of arsenic in chicken meat by anion-exchange LC with ICP-MS. *J Assoc Oil Anal Chem* 87(1): 233-7., 2004.

217. Sheppard BS, Caruso JA, Heitkemper DT and Wolnik KA, Arsenic speciation by ion chromatography with inductively coupled plasma mass spectrometric detection. *Analyst (Cambridge, United Kingdom)* 117(6): 971-5, 1992.

218. Nakazato T and Tao H, Photooxidation Reactor for Speciation of Organic Arsenicals by LC-Hydride Generation-ICPMS. *Anal Chem* 78: 1665-1672, 2006.

219. Devos C, Sandra K and Sandra P, Capillary GC-ICP-MS for D,L-selenomethionine in food supplements and urine. *J Pharm Biomed Anal* 27: 507-514., 2002.

220. McSheehy S, Yang L, Sturgeon R and Mester Z, Determination of Methionine and Selenomethionine in Selenium-Enriched Yeast by Species-Specific Isotope Dilution with LC/MS and ICP-MS. *Anal Chem* 77: 344-349, 2005.

221. Vonderheide AP, Mounicou S, Meija J, Henry HF, Caruso JA and Shann JR, Investigation of selenium-containing root exudates of Brassica juncea using HPLC-ICP-MS and ESI-qTOF-MS. *Analyst (Cambridge, United Kingdom)* 131(1): 33-40, 2006.

222. Munoz AHS, Kubachka K, Wrobel K, Gutierrez Corona JF, Yathavakilla SKV, Caruso JA and Wrobel K, Se-Enriched Mycelia of Pleurotus ostreatus: Distribution of Selenium in Cell Walls and Cell Membranes/Cytosol. *Journal of Agricultural and Food Chemistry* 54(9): 3440-3444, 2006.

223. Mounicou S, Shah M, Meija J, Caruso JA, Vonderheide AP and Shann J, Localization and speciation of selenium and mercury in Brassica juncea-implications for Se-Hg antagonism. *Journal of Analytical Atomic Spectrometry* 21(4): 404-412, 2006.

224. Gergely V, Kubachka KM, Mounicou S, Fodor P and Caruso JA, Selenium speciation in Agaricus bisporus and Lentinula edodes mushroom proteins using multi-dimensional chromatography coupled to inductively coupled plasma mass spectrometry. *Journal of Chromatography, A* 1101(1-2): 94-102, 2006.

225. Krupp EM and Donard OFX, Isotope ratios on transient signals with GC-ICP-MS. *Intl J Mass Spectrom* 242: 233-242, 2005.

226. Poperechna N and Heumann KG, Species-Specific GC/ICP-IDMS for Trimethyllead Determinations in Biological and Environmental Samples. *Analytical Chemistry* 77(2): 511-516, 2005.

227. Barnes JHIV, Schilling GD, Sperline RP, Denton MB, Young ET, Barinaga CJ, Koppenaal DW and Hieftje GM, Coupling of a GC to a simultaneous-detection ICP MS for speciation of organohalide and organometallic compounds. *J Anal Atomic Spectrom* 19: 751-756, 2004.

228. Harrington CF, Elahi S, Merson SA and Ponnampalavanar P, Iron speciation in meat by spectrophotometry & HPLC-ICP-MS. *Anal Chem* 73: 4422-7., 2001.

229. Panayi AE, Spyrou NM, Iversen BS, White MA and Part P, Cadmium and zinc in Alzheimer's brain tissue by ICP MS. *J Neurol Sci* 195: 1-10., 2002.

230. Edler M, Jakubowski N and Linscheid M, Quantitation of melphalan DNA adducts using HPLC - ICP MS. *J Mass Spectrom* 41: 507-516, 2006.

"Measure what is measurable and make measurable what cannot now be measured."

~Galileo

Chapter 5   Strategies for Data Interpretation (Other than Fragmentation)

Introduction to Mass Spectrometry, 4th Edition: Instrumentation, Applications, and Strategies for Data Interpretation; J.T. Watson and O.D. Sparkman, © 2007, John Wiley & Sons, Ltd

"The mass spectrum shows the mass of the molecule and the masses of pieces from it."

~Fred W. McLafferty

I. Introduction

A mass spectrum tells a story. The story is about the analyte that produced the mass spectrum. At first, the story may appear to be only about the mass of the analytes, but not always.[1] A determination of the mass of the analyte is especially reliable when the mass spectrum is generated by one of the soft ionization techniques. A closer examination of these spectra may allow the determination of whether the analyte has an odd or even number of nitrogen atoms. Further examination might reveal whether the analyte contains heteroatoms of Cl, Br, Si, or S or whether the ions that represent the intact molecule result from adduct formation with potassium and/or sodium ions. Usually, the conditions under which the mass spectrum is obtained are known, especially the ionization technique and whether the peaks in the spectrum represent ions with a positive or negative charge. Making a determination of an analyte's nominal mass and elemental composition from the spectrum may be all that is necessary to facilitate the identification of an unknown. Extraction of this information from the mass spectrum is the job of the mass spectrometrist.

In an effort to prepare for data interpretation, this chapter is devoted to key nomenclature relating to ions, peaks, isotope abundance, and various features of the mass spectrum. This information is defined in such a way as to be relevant for all types of ionization techniques in mass spectrometry. Examples of mass spectra are described to illustrate the influence of the natural abundance of stable isotopes on the multiplicity of peaks representing a given ion depending on its elemental composition. Attention is given to the electron configuration of ions formed by different types of ionization techniques and to the relative stability of different types of ions. The odd-electron molecular ions ($M^{+\bullet}$) produced during electron ionization (EI) tend to decompose or fragment extensively; several different mechanisms of fragmentation for odd-electron ions are described in the following chapter with emphasis on the relationship between the fragmentation pattern and the structure of the analyte. In some cases, this fragmentation, which can be the source of a wealth of structural information, can also be a curse because it is so extensive that there is no peak representing the intact molecule. In contrast, the even-electron ions resulting from the soft ionization techniques tend not to fragment, thereby providing only information relating to the mass of the analyte.

Table 5-1. Types of ions representing intact molecules formed by different ionization techniques.

Types of Ions	Produced by Ionization Technique
Positive molecular ions ($M^{+\bullet}$)	Electron ionization (EI), charge exchange CI, field ionization (FI)
Protonated molecules (MH^+) Adduct ions: (e.g., $[M + C_2H_5]^+$, MNa^+, and MCl^-) Hydride abstraction products ($[M - H]^+$) Deprotonated molecules ($[M - H]^-$)	Electrospray ionization (ESI), fast atom bombardment (FAB), field desorption (FD), field ionization (FI), matrix-assisted laser desorption/ionization (MALDI), and chemical ionization (CI), both in GC/MS and LC/MS, atmospheric pressure chemical ionization (APCI)
Negative molecular ions ($M^{-\bullet}$)	Resonance electron capture ionization (ECI)

---

[1] Only ~80% of the EI spectra in the NIST/EPA/NIH Mass Spectral Database exhibit a discernible molecular ion peak.

Later in this chapter, an example is described showing the way in which EI and CI provide complementary mass spectral information. The lack of fragmentation in the mass spectra resulting from soft ionization is due to the low amount of energy absorbed by the analyte molecule during ion formation. As seen in Chapter 3 on MS/MS, it is possible to put energy into these ions to induce fragmentation from which structural information may be inferred for identification. Strategies for interpretation of mass spectral data are presented in this chapter regardless of the ionization technique used; some aspects of data interpretation such as dealing with the influence of the natural abundance of stable isotopes on the multiplicity of peaks representing a given ion, applying the *Nitrogen Rule*, and determining the neutral loss (the dark matter) from an ion during the formation of another ion are the same for all types of analyte ionization. Other aspects such as the likelihood of multiple-charge ions or extensive fragmentation are peculiar to a given type of ion formation such as electrospray or electron ionization, respectively.

As mass spectrometry has developed and techniques have evolved for the formation of ions in both the condensed and vapor phases, there has been a great deal of discussion about the interpretation of "different types of mass spectra". A mass spectrum is a mass spectrum, regardless of how it is generated. The procedures for extracting information about the elemental composition, mass, and structure of an analyte are the same no matter how the mass spectrum is formed (the type of ionization technique used) or the *m/z* analyzer used to separate the ions.

There are several aspects of organic mass spectrometry that are important to keep in mind when trying to extract information about an analyte's structure from mass spectral data:

1. The technique of mass spectrometry only detects charged species, ions.

2. All fragmentations must involve the breakage of chemical bonds.

3. In most cases, ion fragmentation involves the formation of a new ion and a neutral. The masses of the new ion and the neutral will equal the mass of the original ion undergoing fragmentation.

4. Mass spectral fragmentation *almost never* results in an ion pair (a positive and negative ion).

5. Logical groupings of atoms (according to valence rules) are lost from ions as a radical or as a molecule.

6. The ions that are detected in a mass spectrometer behave as individuals, free from interaction with other matter (other ions or neutrals) because of the vacuum.

7. Analytes encountered will generally contain mostly atoms of carbon and hydrogen; therefore, certain assumptions can be made about the elemental composition of many ions.

II. Some Important Definitions

It is also a good idea to define the matter of mass spectrometry:

1. **Molecules** – neutral species composed of atoms connected according to valence rules, having an *even number* of electrons.

2. **Radicals** – neutral species composed of atoms connected according to valence rules, having an *odd number* of electrons.

3. **Ions** – species that have either a positive or negative charge, composed of atoms connected according to valence rules, having either an *odd number* or *even number* of electrons.

To the purist, the above three rather encompassing definitions may be outside the scope of more specific characterization of matter. However, these definitions will serve those trying to extract information from mass spectra. Another important term used throughout this book, and especially in this and the next chapter, is *dark matter*. The *dark matter* is the neutral loss that is observed as the difference in the *m/z* values of two peaks in a mass spectrum representing the original ion and the newly formed ion. The dark matter is often significant in deducing structural aspects from a mass spectrum. A peak at *m/z* 31 in the mass spectrum of 1-octanol obtained by EI represents an oxonium ion ($H_2C=O^+H$), indicating that the analyte is an aliphatic alcohol. As more experience is gained in interpreting EI mass spectra, a number of observed *m/z* values will elicit recognition of specific structural features. These specific *m/z* values will usually be low, e.g., <100. On the other hand, peaks at higher *m/z* values such as a peak at *m/z* 158 (the $M^{+\bullet}$ peak) or at *m/z* 143 may not elicit immediate structural insights; however, the difference between 158 and 143, which represents the mass of the neutral loss (the dark matter) from the molecular ion does reveal that the analyte has a methyl group located in a special position in its structure. This example and other observances of the dark matter are used as extensively as the recognition of peaks at specific *m/z* values in the spectrum to identify the structure of the analyte.

III. Possible Information That Can be Obtained from the Mass Spectrum

Mass spectrometry based on ionization promoted by interaction of vaporized analyte molecules with 70 eV electrons (electron ionization, EI) provides a wealth of information ranging from a possible determination of the mass of a molecule and an indication of its elemental composition based on isotope peak intensity ratios (the intensity of an isotope peak relative to the intensity of a monoistopic peak) to some structural information based on a fragmentation pattern. The fragmentation pattern principally results from the neutral losses of a variety of radical species and, occasionally, molecules of lesser mass from the intact analyte, which is present as both a radical and an ion (the molecular ion, $M^{+\bullet}$). These neutral losses are the dark matter of the mass spectrum (i.e., their mass is deduced, rather than being represented directly by a peak in the mass spectrum).

Soft ionization techniques such as chemical ionization (CI), electrospray ionization (ESI), matrix-assisted laser desorption/ionization (MALDI), photoionization (PI), field ionization (FI) and field desorption (FD), and fast atom bombardment (FAB) principally produce ions that represent the intact molecule, usually in the form of a protonated molecule (MH^+). Some energy is imparted in the formation of these ions that can result in fragmentation. As an example, when a protonated molecule is formed during an analysis by CI using GC/MS, the resulting ion has absorbed an amount of energy equivalent to the difference between the proton affinity of the analyte and that of the conjugate base of the reagent ion. This is why protonated molecules formed using ammonia as a reagent gas fragment less than those formed using methane as the reagent gas. The proton affinity of ammonia is close to that of most organic molecules, whereas the proton affinity of methane is much lower than that of ammonia (and that of most other organic molecules). Consequently, CI using methane results in much more energy being absorbed during formation of the protonated molecule, and sometimes this energy is sufficient to bring about fragmentation. For this reason, a certain degree of fragmentation of the protonated molecule may be observed in CI using GC/MS as well as during APCI using LC/MS.

Unlike the situation with other soft ionization techniques, ions of the analyte in electrospray ionization (ESI) have very little or no excess energy. This is because ions are formed in solution, where there is ample collisional cooling, and then they are desorbed into the gas phase from electrically charged droplets at atmospheric pressure where there is the opportunity for additional collisional cooling (shedding of excess energy). The limited fragmentation that is represented in spectra obtained during electrospray is a result of inelastic collisions of the ions and molecules of nitrogen that are in the interface between the atmospheric pressure region and the high-vacuum region of the mass spectrometer. As was illustrated in the chapter on MS/MS, the energy imparted to the analyte ions can be adjusted to control the fragmentation; although this fragmentation can be minimized, it cannot be eliminated.

Negative-charge molecular ions ($M^{-\bullet}$) (formed by ECI), deprotonated molecules ($[M - H]^-$) (formed by ESI and CI, including APCI), and adduct ions (e.g., $[M + Cl]^-$) (formed mainly in APCI and ESI) can also be produced. The production of these ions usually involves little or no transfer of energy.

CI (including APCI) and EI share a common requirement: the analyte molecules must be present in the gas phase at the point of ionization; i.e., the analytes must be thermally stable and present a discernible vapor pressure. The soft ionization techniques, those in which very little fragmentation of the analyte occurs, often produce a MH^+, which carries information on the mass of the analyte. Structural information can be obtained from imposed fragmentation of the MH^+, such as through the process of collisionally activated dissociation (CAD), also known as collision-induced dissociation (CID). When this CAD MS/MS technique is utilized to examine the MH^+, the fragmentation is of an even-electron ion (EE^+), unlike the fragmentation process that occurs in EI, which involves the fragmentation of an odd-electron ion ($OE^{+\bullet}$), the $M^{+\bullet}$. Regardless of the ionization technique or the mechanism of fragmentation of the ion representing the intact molecule, the information obtainable from the mass spectrum is the mass of the analyte and some of its structural features in the form of characteristic losses or "signature" fragmentation pattern.

Another important feature in the mass spectrum is the basis to determine whether the analyte has an odd number of nitrogen atoms based on whether the peak representing the ion containing the intact molecule has an odd or even *nominal m/z value*. The operative term here is *nominal m/z value*, which will become clear in the discussion of the *Nitrogen Rule* in this chapter.

IV. Elemental Composition of an Ion and the Ratios of Its Isotope Peaks

The *m/z* value of a mass spectral peak and the relative intensity of the peaks at adjacent higher *m/z* values (usually by 1, 2, 3, or a few more integer *m/z* units) can be used to extract a great deal of information about the elemental composition of the ion. The accurately determined *m/z* for these peaks will also reveal even more insight into the elemental composition of the ion. The first step is to establish some definitions or rules that will be used in the extraction of this information from a mass spectrum. It should be emphasized that the concern is with the ion's elemental composition, not how the ion was formed. These can be ions represented in mass spectra produced by the LC/MS techniques of electrospray or APCI, GC/MS techniques based on electron, chemical, or field ionization, or techniques such as fast atom bombardment and matrix-assisted laser desorption/ionization.

1. Definition of Terms Related to the Matter of Mass Spectrometry

Unified atomic mass unit (u). This term, established by IUPAC in 1962, is defined as 1/12 the mass of the most abundant naturally occurring stable isotope of carbon, ^{12}C. The term "dalton" (Da) is a synonym for the unified atomic mass unit (u). It is important to keep in mind that these terms are the units of mass on an atomic scale; they are not synonymous with the *m/z* of an ion.

Isotopes. It can be said that mass spectrometry deals with the masses of isotopes of the elements, rather than their atomic masses. Regardless of the ionization technique used, the elemental composition of an ion can sometimes be deduced from the ratio of isotope peak intensities that represent the ion in the mass spectrum [1]. Isotopes are species of the same element that have the same number of protons and electrons. Therefore, isotopes have the same chemical properties, but differ in mass because they contain different numbers of neutrons.

Atomic mass. The atomic mass of an element is the number that represents the element's mass based on the weighted average of the masses of its naturally occurring stable isotopes. For example, the integer atomic mass of bromine is 80 Da. This is because there are only two naturally occurring stable isotopes of bromine (^{79}Br and ^{81}Br), which exist in nature in about equal amounts. When the *relative mass* (M_r) of an ion, molecule, or radical is reported, it is based on the atomic masses of its elements.

Nominal mass. The nominal mass of an element is the *integer mass of its most abundant stable isotope*. Returning to the example of bromine in the previous paragraph, the nominal mass is 79 Da because ^{79}Br is more abundant (0.51) than ^{81}Br (0.49). The nominal mass of a molecule, radical, or ion is the sum of the nominal masses of all the atoms of its constituent elements. The nominal mass of $C_3H_6O^{+\bullet}$ is 58 Da. The nominal mass of $C_{67}H_{100}O_7^{+\bullet}$ is 1016 Da. With respect to the common elements encountered in organic mass spectrometry, the integer value of the lowest-mass isotope is the nominal mass. This is not true for all the elements. The transition element mercury (Hg) has nine stable isotopes over a range of 199–204 Da. The most abundant isotope is ^{202}Hg; therefore, the nominal mass of mercury is 202 Da, which is not the mass of the lowest-mass isotope.

Monoisotopic mass. The monoisotopic mass of an element is the ***exact*** mass of *the most abundant naturally occurring stable isotope* determined relative to the mass of ^{12}C, which is assigned an exact value of 12.0000 Da (all digits to the right of the decimal point are 0 regardless of how far the value is expanded). The monoisotopic mass of a molecule, ion, or radical is the sum of the monoisotopic masses of all the atoms of its constituent elements. The monoisotopic mass of $C_3H_6O^{+\bullet}$ is 58.04185 Da. The monoisotopic mass of $C_{67}H_{100}O_7^{+\bullet}$ is 1016.7468 Da (as compared to its nominal mass of 1016 Da). When using a mass spectrometer capable of reporting an accuracy of no greater than 1 *m/z* unit, the experimentally reported mass for $C_3H_6O^{+\bullet}$ would be 58 Da. The nominal mass and the monoisotopic mass are nearly the same for this low-mass ion within the operational limits of the instrument. With such an instrument, the reported value for the peak representing $C_{67}H_{100}O_7^{+\bullet}$ would be *m/z* 1017. In this case, the observed *m/z* value would be rounded to one greater than the nominal mass; this is because of the cumulative *mass defect* of a large number of hydrogen atoms.

Mass defect. The simplest way to consider mass defect (and the best way with respect to obtaining information from mass spectral data) is that it is the difference between the exact mass and integer mass of a nuclide.[2] The mass defect of an ion, molecule, or radical will be the sum of the mass defects of all atoms of its constituent nuclides. The most abundant stable isotope of carbon (^{12}C) has no mass defect because the atomic mass scale is defined relative to the mass of this nuclide. The mass of each of the other nuclides is measured relative to that of ^{12}C, and each of them does have a mass defect. The mass defect (δ) of the protonium isotope of hydrogen (^1H) is 0.0078 u or 0.78% of its mass. The mass defect of the only stable isotope of fluorine is –0.0016 u or 0.008% of its mass. The relative proportion represented by the mass defect of fluorine is three orders of magnitude less than that for ^1H. The cumulative effect of the ^1H mass defect becomes significant in hydrocarbons that have a nominal mass greater than 500 (more than 100 hydrogen atoms). No longer can the monoisotopic mass and the nominal mass of such large ions be considered the same when using instruments that report observed *m/z* values to the nearest integer. In the case of perfluorotributylamine, the compound often used to calibrate the *m/z* axis in EI transmission quadrupole and QIT mass spectrometers, if a $M^{+\bullet}$ peak is observed in the mass spectrum of this compound, it would have a nominal *m/z* value of 671. The calculated exact mass would be 670.9599 Da, representing a mass defect of 0.0401, or 0.006% of its mass, which is insignificant. This is why perfluorinated compounds make such good calibrants of the *m/z* axis. It should be noted that the mass defects of ^1H (protonium), ^2H (deuterium), ^{13}C, ^{14}N, and ^{15}N are positive numbers. The mass defect of the other common nuclides encountered in the mass spectrometry of organic compounds is negative. In dealing with biological compounds, the mass defects of ^{14}N and ^{16}O may cancel one another because nearly equal numbers of atoms of these two elements may be present in the ion.

Consider the mass spectral peak for $C_{50}H_{100}^{+\bullet}$. The nominal mass for this ion is 700 Da. The monoisotopic mass for the same ion to the nearest integer is 701 Da. This one-dalton difference is due to the mass defect of hydrogen. See the calculations below:

Nominal mass calculation	Monoisotopic mass calculation
50 C × 12 = 600	50 C × 12.0000 = 600.0000
100 H × 1 = 100	100 H × 1.0078 = 100.7800
700	700.7800 or 701

---

[2] A nuclide is characterized by its mass number (the number of protons and neutrons in the nucleus). Two different elements can be common nuclides; i.e., they can have the same mass number but different numbers of neutrons and protons.

2. The Nitrogen Rule

In mass spectrometry, the *Nitrogen Rule* serves two purposes: (1) it can aid in determining which peaks in the mass spectrum represent the ion that corresponds to the intact analyte molecule and, (2) it can be used to determine whether the intact molecule contains an odd number of nitrogen atoms. An examination of the nominal mass and valence of all the elements generally encountered in organic mass spectrometry (C, H, N, O, F, Na, Si, P, S, Cl, K, and Br) reveals that the numbers representing the nominal mass and the valence are either both odd or both even with one exception. Nitrogen, the exception, has an *odd valence* (+3) and an *even nominal mass* (14 Da). This means that *any molecule that contains an odd number of nitrogen atoms will have an odd nominal mass*. There are two very important restrictions that must be placed on the *Nitrogen Rule* [2] for it to be valid:

1. The above definition of a molecule must be accepted without condition.

2. The calculated *nominal mass* for the intact molecule must be considered, not necessarily the reported or experimentally measured *m/z* value for the peak representing the ion.

It is very important to be aware of the mass defect, especially when dealing with ions that have a mass greater than 500 Da, because the rounding function in the data system can easily cause a one-integer discrepancy between an assigned *m/z* value and a calculated nominal mass.

There are several ways to state the *Nitrogen Rule*. All will be useful when interpreting a mass spectrum to determine whether the corresponding analyte has an odd number of nitrogen atoms, or when considering a good candidate for a peak from which to determine the analyte's nominal mass. For now, these statements of the *Nitrogen Rule* will be restricted to single-charge ions:

1. All molecular ions ($M^{+\bullet}$ or $M^{-\bullet}$) of an analyte with an odd number of nitrogen atoms will have an odd nominal *m/z* value.

2. All odd-electron fragment ions ($OE^{+\bullet}$ or $OE^{-\bullet}$) that contain an odd number of nitrogen atoms will have an odd nominal *m/z* value.

3. All even-electron fragment ions (EE^{+} or EE^{-}) that contain an odd number of nitrogen atoms will have an even nominal *m/z* value.

4. All protonated molecules (MH^{+}) or other EE^{+} produced by the addition of an even-electron species (e.g., Na^{+}, $C_4H_7^{+}$, etc.) or $[M - H]^{+}$ produced by hydride abstraction that has an odd number of nitrogen atoms will have an even nominal *m/z* value.

5. All even-electron deprotonated molecules ($[M - H]^{-}$) or adduct ions produced by the addition of an even-electron species (e.g., Cl^{-}), and which contain an odd number of nitrogen atoms, will have an even nominal *m/z* value.

In all the above cases, if the species has an even number of nitrogen atoms (do not forget that zero is a number and that it is an even number) the odd/even values will be reversed. It is also important to remember that if an ion is formed by protonation, hydride abstraction, deprotonation, or adduct formation to yield a species that is a cluster of an analyte molecule containing an odd number of nitrogen atoms with another molecule that also has an odd number of nitrogen atoms (e.g., an acetonitrile molecule that is often found in the HPLC mobile phase), the resulting ion (due to protonation, Na^+ adduct formation, etc.) now has an even number of nitrogen atoms. Based on the *Nitrogen Rule*, once the peak that represents the intact molecule is identified, a determination can be made as to whether the analyte has an odd number of nitrogen atoms. It should also be noted that all the above definitions contain the term *nominal mass*. This is especially important for single-charge ions that have *m/z* values above 500.

The candidate peak for the molecular ion, protonated molecule, or deprotonated molecule must be in agreement with the *Nitrogen Rule* regarding what is known about the analyte. A compound containing an odd number of nitrogen atoms, in addition to any number of C, H, O, S, Si, P, or halogen atoms, will have an odd nominal mass. The nominal mass of the analyte will be an even number for any compound containing an even number of or no nitrogen atoms (again, do not forget zero).

3. Elemental Composition of an Ion Based on the Ratio of Isotope Peak Intensities

A. Isotope Peak Patterns Used to Determine the Elemental Composition of Ions

In all mass spectra, such as the EI mass spectrum of 1,3-bromocholorpropane (Figure 5-1), it can be seen that there are peaks that increase by one *m/z* unit beyond what will become known as the nominal *m/z* peak. These peaks usually represent ions of the same elemental composition as the nominal *m/z peak*, but of different isotopic compositions. In the mass spectrum in Figure 5-1, the peak at *m/z* 156 (the X peak) represents an ion with the elemental composition of C_3H_6BrCl. The isotopic composition of this ion is $^{12}C_3{}^1H_6{}^{79}Br^{35}Cl$. This is the nominal *m/z* peak and it is also the monoisotopic peak (where there is only one isotope of each element present, and it is the most abundant isotope, which is, in this case, the isotope of lowest mass). The peak at *m/z* 157 (the X+1 peak) represents an ion with the same elemental composition, but there is a possibility of two different isotopic compositions: $^{12}C_2{}^{13}C^1H_6{}^{79}Br^{35}Cl$ and $^{12}C_3{}^1H_5{}^2H^{79}Br^{35}Cl$. Both of these isotopic compositions have a mass that is 1 Da greater than that of the monoisotopic ion. The peak at *m/z* 158 (the X+2 peak), which has an intensity greater than the intensity of the previous two peaks, could represent ions of the following isotopic compositions:

$$^{12}C^{13}C_2{}^1H_6{}^{79}Br^{35}Cl$$

$$^{12}C_3{}^1H_4{}^2H_2{}^{79}Br^{35}Cl$$

$$^{12}C_2{}^{13}C^1H_5{}^2H^{79}Br^{35}Cl$$

$$^{12}C_3{}^1H_6{}^{79}Br^{37}Cl$$

$$^{12}C_3{}^1H_6{}^{81}Br^{35}Cl$$

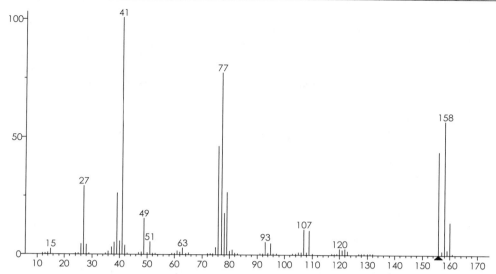

Figure 5-1. EI mass spectrum of 1,3-bromochloropropane.

Based on the abundance of the naturally occurring stable isotopes of each of the elements present (see the table on the inside front cover), the majority of the intensity at m/z 157 will be due to the isotopic composition $^{12}C_2{}^{13}C^1H_6{}^{79}Br^{35}Cl$ and the majority intensity of the peak at m/z 158 will result from the isotopic composition $^{12}C_3{}^1H_6{}^{81}Br^{35}Cl$; however, a significant amount of intensity of the peak at m/z 158 will come from the isotopic composition $^{12}C_3{}^1H_6{}^{79}Br^{37}Cl$. These peak intensities and their relation to the natural abundances of the various isotopes of the elements that compose the ions can be used to determine the elemental composition of the ion. As will be seen in the following example, once an elemental composition is established, it may even be possible to propose the ion's structure.

Some additional definitions need to be stated. The main elements of organic mass spectrometry can be divided into three categories: **X**, **X+1**, and **X+2**.

- **X elements** are those that exist only as a single nuclide (**F**, **Na**, **P**, **I**).

- **X+1 elements** are those that have only two naturally occurring stable isotopes (**H**, **C**, and **N**) and the mass of the more abundant is one dalton less than the other isotope. For all ions with a mass <1000 Da, H is treated like an **X** element because of the low abundance of deuterium compared to that of protonium 0.01 % vs 99.99%).

- **X+2 elements** are those that have a naturally occurring stable isotope 2 Da greater than the most abundant isotope (**O**, **S**, **Si**, **Cl**, **K**, and **Br**). All of the common X+2 elements encountered in organic mass spectrometry, except for Cl and Br, have three naturally occurring stable isotopes. The contributions of these elements (with the exception of oxygen) to the intensity of the X+1 peak must be considered before assigning the number of atoms of X+1 elements in the ion. This will become clear below.

The table on the inside front cover has X+1 and/or X+2 factors for all the elements that have heavier isotopes. These factors have to do with the probability that an ion with a heavier isotope will be present. The factors are based on the relative abundances of the heavier isotopes. The X+2 factor for chlorine is 32.5. This is derived from the fact that

75.77% of all chlorine atoms are ^{35}Cl and 24.23% are ^{37}Cl (24.23 / 75.77 × 100 = 32.5). The X+1 factor for carbon is 1.1, which is also the percent abundance of ^{13}C. This factor just turns out to be the same as the percent abundance for ^{13}C based on rounding (1.09 / 98.91 × 100 = 1.102, which rounds to 1.1). The abundances of ^{29}Si and ^{30}Si (abundances of the three isotopes being 92.2%, 4.7%, and 3.1%, respectively) give rise to an X+1 factor of 5.1 and X+2 factor of 3.4.

The factors at X+1 and at X+2 are additive for multiple elements according to the number of atoms of an element in an ion. If an ion's elemental composition has only a single atom of carbon, there is a 1.1% probability of finding one of these ions in which the carbon is ^{13}C. The probability of one ^{13}C being present in a given ion increases as the number of carbon atoms in the elemental composition increases. If an ion contains two atoms of an element, there is twice the probability of finding an ion with a heavy isotope of that element than there is in an ion containing only one atom of the element. These probabilities translate into the intensity of the X+1 peak relative to that of the X peak. In the case of an ion that contains a single atom of carbon, the relative intensity of the X+1 peak to that of the X peak is 1.1%. If there are two carbon atoms in the ion, the intensity of the X+1 peak relative to that of the X peak is 2.2% (the X+1 factor for carbon times the number of atoms of carbon). If the ion contains ten carbon atoms, the intensity of the X+1 peak relative to the X peak is 11%. If the ion contains ten carbon atoms and four sulfur atoms, the intensity of the X+1 peak relative to that of the X peak will be 14.2%; this number is derived from the number of C atom times the X+1 factor for carbon plus the number of S atoms times the X+1 factor for sulfur (10 C atoms × 1.1 + 4 S atoms × 0.8 (the X+1 factor for sulfur)).

A close examination of this same table shows that carbon also has an X+2 factor. Carbon does not have an X+2 isotope. This X+2 factor has to do with the probability that an ion contains two atoms of ^{13}C. Because there is likely to be an increasingly large number of carbon atoms in ions of higher and higher mass, there is an increasing probability of two atoms of ^{13}C being present in such an ion. These probabilities are actually determined from a binominal expansion of the equation $(a + b)^n$, where **a** is the abundance of the X isotope, **b** is the abundance of the heavier isotope, and **n** is the number of atoms of the element under consideration. In the case of carbon, this would be $(98.91 + 1.1)^n$. The coefficient for the third term in this expansion (which represents the intensity of the X+2 peak relative to the X peak) is approximately 0.006 times the number of carbon atoms squared ($0.006 × C_n^2$). The actual relationship of the abundances of the naturally occurring stable isotopes of the various elements follows a much more rigorous mathematical relationship. Details can be found in Jürgen Gross' book *Mass Spectrometry: A Textbook* (Springer-Verlag, 2004) and R. Martin Smith's *Understanding Mass Spectra: A Basic Approach* (John Wiley & Sons, Inc., 2004). Due to limitations in the capability of measuring the X+1 and X+2 peak intensities (except for isotope ratio mass spectrometers like those designed for geological applications), especially as the abundance of the ions represented by the X peak decreases, these more rigorous approaches are considered pedantic at best.

The X+1 and X+2 factors can be used to predict the relative intensity of the respective peaks when an ion's elemental composition is known. These factors can also be used in determining an unknown elemental composition based on relative peak intensities observed in a mass spectrum. When using these X+1 and X+2 factors in this latter way, the number of X+2 elements that also have a contribution at X+1 should be assigned before the number of C atoms (X+1 elements) is assigned. A practical example of this issue is described later in the context of the mass spectrum in Figure 5-5.

B. Isotope Peak Patterns for Ions Containing Various Combinations of Br/Cl

The influence of multiple chlorine and/or bromine atoms on the multiplicity of isotope peaks for any given ion is shown graphically in a figure found in the "back matter" of this book. These reference patterns are valuable for recognizing spectra of compounds containing Br and Cl that are commonly encountered during the analysis of environmental samples. To gain familiarity, examine the simple pattern at the upper left, which is that for any ion containing only one chlorine atom. It consists of two peaks separated by 2 *m/z* units; the second peak is only one-third as intense as the first because ^{37}Cl is only one-third as abundant as ^{35}Cl. When the ion contains more than one atom of Br and/or Cl, the multiplicity of peaks becomes more complex as shown by the second entry, which represents the isotope peak-intensity pattern for any ion containing two atoms of Cl. Further examination of the top row of spectra in this graphic shows that in addition to the X+2 peak becoming more intense, indicating the increased probability of the ion containing an atom of ^{37}Cl, the number of peaks separated by 2 *m/z* units increases. This is because of the greater number of possible combinations of ^{35}Cl and ^{37}Cl atoms. It is also important to note that for ions containing four or more atoms of chlorine, two or more atoms of bromine, or one or more atoms of both chlorine and bromine there is a higher probability of the ion containing an X+2 isotope than there is of an ion with no X+2 isotope. When looking at the accompanying tabular data associated with this graphic for ions with varying numbers of Cl and Br atoms, it will be seen that the data are reported relative to the most intense peak in the cluster (the base peak), <u>not</u> the intensity of the nominal *m/z* value peak. In order to obtain the relative intensity of X+1 to X, it will be necessary to normalize X+1 to X (intensity at X+1 / intensity at X × 100%). It is also worth noting that the peaks at X+3, X+5, and so forth are essentially the ^{13}C-isotope peaks relative to the X+2, X+4, etc., just as the X+1 is the ^{13}C-isotope peak relative to the X peak. This comes about because the intensity contribution resulting from two atoms of ^{13}C being present is small compared to the intensity of the X+2 isotope, and the intensity of three atoms of ^{13}C being present is miniscule compared to the intensity of one ^{13}C being present with the X+2 isotope, etc. For this reason, the analyst has several redundant sets of data from which to calculate the number of carbons present in the ion, namely, intensity at X+1 / intensity at X, intensity at X+3 / intensity at X+2, intensity at X+5 / intensity at X+4, and so forth.

In Figure 5-2, although the abscissa corresponds to the *m/z* scale, no numerical values are given so that these bar-graph representations of peak clusters are universally applicable to any group of atoms that comprise an ion together with the indicated number of chlorine and/or bromine atoms. This feature is illustrated in Figure 5-2; at the top, a portion of the reference bar-graph intensity pattern is reproduced from material in the back matter of this book. Compare the intensity pattern of the second entry of the reference bar graphs, namely that for Cl_2, with the three spectra at the bottom of Figure 5-2. The relative peak heights in the second group of peaks in the top row of the Cl/Br isotopic pattern shown in the back matter should represent the principal pattern of peak intensities in the mass spectrum of a molecular ion of chlorine (Cl_2) or a molecule of dichloromethane (CH_2Cl_2) or a molecule of dichlorobenzene ($C_6H_4Cl_2$). Notice the good agreement between the chlorine peak-intensity patterns regardless of which other elements are present. This is because the chlorine isotope peaks are separated at 2 *m/z* unit intervals, whereas the major isotopic contributions from carbon and hydrogen occur at 1 *m/z* unit intervals; thus, they do not overlap with the isotope peaks of Br and/or Cl.

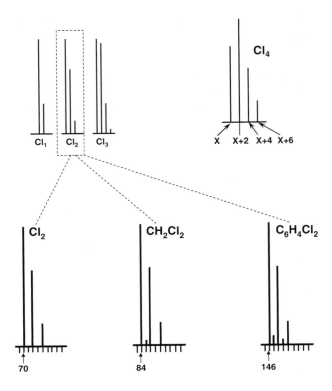

Figure 5-2. *Peak patterns for some ions that contain two atoms of chlorine.*

The *m/z* values of the peaks at the bottom of Figure 5-2 are different in each case because the mass of the two chlorine atoms is "riding" along with the cumulative mass of additional carbon and hydrogen atoms. In the mass spectrum of CH_2Cl_2, the discernible peak at *m/z* 85 is due to the presence of ^{13}C. There should also be a peak at *m/z* 87 representing an ion with the isotopic composition of $^{13}CH_2{}^{35}Cl^{37}Cl$, but it is not observed because of the resolution of the graphical display. This peak would probably be observable in the tabular display of the spectrum. There should also be a peak at *m/z* 89 which would represent the ion with the isotopic composition of $^{13}CH_2{}^{37}Cl_2$, but again this peak is not discernible in this compressed display. An examination of the tabular display of this spectrum may not exhibit the peak at *m/z* 89. This lack of a peak is probably due to very low ion abundances for this ion. In this case, the ion abundance is too low to be represented by a discernible peak in these realistic graphical displays of mass spectral data.

The peaks at *m/z* 86 and 88 in the mass spectrum of dichloroethane result from those molecular ions that contain one or two ^{37}Cl atoms, respectively. Similarly, in the bar-graph spectrum for dichlorobenzene, the peaks at *m/z* 147, 149, and 151 (not observed due to the graphic display resolution or due to low ion abundance) represent those molecular ions in which one of the six carbon atoms is present as ^{13}C. Again, the peaks at *m/z* 148 and 150 represent the abundance of those molecular ions that contain one or two ^{37}Cl isotopes, respectively, but no ^{13}C.

C. Constraint on the Number of Atoms Allowed for a Given Element

When interpreting an unknown mass spectrum, an assessment of the relative intensity of certain isotope peaks is made in an effort to establish limits on the number of atoms of a particular element that are contained in the ion. For example, by noting that the peak at *m/z* 147 (the X+1 peak) in the third abbreviated spectrum at the bottom of Figure 5-2 has an intensity of 6.6% (6 × 1.1%) relative to that at *m/z* 146 (the X peak), it is possible to deduce that no more than six atoms of carbon are present in the ion. Certain other combinations of elements can be eliminated from further consideration; for example, as many as 10 carbon atoms could not be present because (even though the mass of the ion would allow for this) in such a case, the intensity at *m/z* 147 would be 11% (10 × 1.1%) relative to that at *m/z* 146. By subtracting the nominal mass of the two obviously present chlorine atoms from the nominal mass of the ion, the value of the remaining mass and the X+1 to X intensity ratio, in this case, corroborate one another.

A potentially complicating problem in the use of the intensity ratio of an isotope peak to that of a monoisotopic peak in a determination of elemental compositions is that the X+1 contribution from three atoms of nitrogen (0.37 × 3 = 1.11) is approximately equivalent to the X+1 contribution of one atom of carbon. However, in such a case, use of the *Nitrogen Rule* would identify the ion containing an odd number of nitrogen atoms.

The isotope peak patterns for ions containing atoms of Cl and/or Br are very dramatic, as seen in the Cl/Br isotope patterns in the figure located in the back matter of this book. Although not as dramatic, the tell-tale isotope patterns for ions containing atoms of silicon, sulfur, and potassium can be just as revealing. Usually recognition of the presence of oxygen has to be obtained from examination of the tabular data because the abundances of the two heavier isotopes of oxygen (^{17}O and ^{18}O) are very low, 0.04% and 0.20%, respectively. In the case of higher-mass ions, such tabular data may be necessary to discern whether the ion contains silicon and sulfur. This concealed information about oxygen (and possibly sulfur and silicon) is one of the reasons that having a mass spectrum in both a tabular and a bar graph (or profile mode) display is requisite to a good interpretation.

D. Relationship of the Charge State of an Ion and the Spacing of the Corresponding Isotope Peaks

Although the description to this point was limited to single-charge ions, the role of isotope peaks in determining the charge state of an ion deserves some attention. When dealing with ions that have more than a single charge, isotopic species that differ in mass by one or more daltons will be represented by peaks separated by some fraction of an *m/z* unit. For example, the spectrum of a double-charge ion will exhibit X+1 (as for carbon) isotope peaks separated by 0.5 *m/z* units. For ions that have a charge state of 10, the X+1 isotope peaks will be separated by 0.1 *m/z* units. The mass of an ion can be determined from a peak at a given *m/z* value only if the charge state is known. In turn, the charge state of an ion can be computed only if the isotope peaks are resolved to the extent that the separation of the peaks can be ascertained to the nearest fraction of an *m/z* unit. This is why the technique of counting the number of peaks in a 1 *m/z* interval is limited to *m/z* analyzers that have very high resolving power.

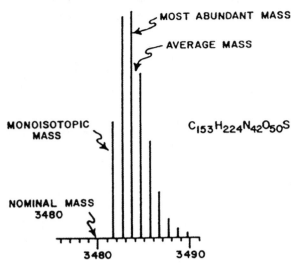

Figure 5-3. *Graphical illustration relating nominal, monoisotopic, and average mass for an ion containing a large number of atoms.* *Reprinted with permission from the American Chemical Society from Yergy J, Heller D, Hansen G, Cotter RJ and Fenselau C, Isotopic Distributions in MS of Large Molecules, Anal. Chem. 1983, 55, 353–356.*

1) Ions of High Mass-to-Charge Ratio

As mass spectrometry has advanced, it is possible to produce ions having high mass-to-charge ratios. This is especially true in techniques like MALDI, in which it is not uncommon to have peaks that represent single-charge ions of several thousand daltons and a single charge. It will be seen a little later in this chapter that the nominal *m/z* value peak is not always the base peak in the cluster of peaks representing the monoisotopic ion and its isotopic variants. This is especially true for ions that have multiple atoms of chlorine and bromine. High-mass ions are likely to contain a large number of carbon atoms. The more carbon atoms present, the higher the probability of finding an ion containing at least one atom of ^{13}C. Eventually, the probability of finding an ion with a single atom of ^{13}C is greater than the probability of finding an ion in which all the carbon atoms are ^{12}C. In the case of an ion containing 100 C atoms, the X+1 peak intensity is 110% of the intensity of X (1.1_C [X+1 factor for C] $\times$ 100 [number of C atoms] = 110%), or normalizing to the most intense peak in the cluster of peaks, the intensity of the X peak is 91% of that of the X+1 peak, which has become the base peak in the cluster of peaks. Figure 5-3 shows the appearance of the isotope peak pattern for an ion of moderately high mass [3]. The peak pattern is strikingly different from that shown in Figure 5-2 principally because the ion is composed of a greater number of atoms compared to the pattern for the ion of fewer atoms represented in Figure 5-2, although the ion represented in Figure 5-2 has special features associated with the Cl and/or Br peak pattern.

E. Steps to Assigning an Elemental Composition Based on Isotope Peak Intensities

Before applying these principles to an example of mass spectral data where isotope peak-intensity data will be used to obtain an elemental composition, some guidelines to such an undertaking must be established. A very important consideration with respect to the use of isotope peak intensities to the determination of an elemental composition is the precision with which peak intensity values can be measured with most mass spectrometers used in GC/MS and LC/MS. This precision is considered to be *about* ±10% relative standard deviation for most commercial mass spectrometers. Much better results can be obtained with instruments specifically designed for isotope ratio measurements, but these instruments are rarely used with hyphenated techniques because they require long data acquisition times (in some cases requiring several minutes per integer *m/z* unit, taking measurements every few hundredths of a unit). Because of the poor precision in acquiring peak-intensity data realized in most GC/MS and LC/MS analyses due to poor ion counting statistics during rapid data acquisition, it is not necessary to have an absolute accounting for the measured values during data interpretation. As a *general guideline*, the isotopic abundances will have been accounted for when *about* ±10% of the normalized experimentally determined peak intensity values have been accounted for. In the case of a very low signal intensity, a less rigorous application of this ±10% is acceptable.

The natural abundance of ^{18}O is low, approximately 0.2%. Because of the relatively poor precision of nonspecialized instruments in making intensity measurements, an error of ±1 or more in the determination of the number of oxygen atoms is a reasonable expectation.

Steps in Assigning an Elemental Composition

1. Assign the nominal *m/z* peak in the cluster. This will be the peak with the lowest *m/z* value in a cluster of peaks. Initially, it can be assumed that other peaks in the cluster for the next few *m/z* units represent the same ions, but of different isotopic compositions. The nominal *m/z* peak may or may not be the base peak in the isotope cluster (a group of peaks that includes the isotope peaks and the nominal *m/z* value peak). Often when multiple atoms of chlorine and/or bromine are present, the nominal *m/z* peak will not be the base peak. Examination of the table of tabular values for isotope peak intensities for ions containing these combinations of chlorine and bromine atoms in the back matter of this book reveals that the theoretical values are reported relative to the base peak in a given cluster.

2. Assign the number of atoms of X+2 elements with the exception of oxygen (Br, Cl, Si, and S). (Because these examples are going to relate to ions that are produced in electron ionization, no consideration will be given to the X element Na nor to the X+2 element K.)

3. Assign the number of atoms of X+1 elements (C and N); do not forget the *about* ±10% guideline in accounting for the observed peak intensities. It is important to keep in mind the fact that peak intensities in a cluster are reported relative to the base peak. As explained above, there are cases where the base peak and the nominal *m/z* value peak are not the same. In these cases, it is necessary to normalize X+1 intensity to that of X before assigning the number of X+1 elements.

4. Assign the number of atoms of oxygen present. (The reason for assigning the number of atoms of O [an X+2 element] after the number of C and N atoms is to account for the X+2 contribution from carbon first, and to use the residual value to

determine the number of O atoms. Depending on the number of carbon atoms present, the contribution at X+2 due to the presence of carbon (the probability of an ion containing two atoms of ^{13}C) can be significant.

5. Assign the number of X elements (F, I, and P). (Remember, when the mass of the ion is <1000 Da, consider H as an X element.)

At this point, a proposed elemental composition has been reached. There may be positive or negative remainders in the accounting of peak intensities at X+1 and X+2, but there cannot be any remainder in the accounting for mass (the observed *m/z* value). (Remember this is an elemental composition; it is *not* necessarily an empirical formula. The empirical formula relates to the proportion of the elements present, *not* to the number of atoms of the elements.) The next question to be answered is whether the elemental composition makes sense. One of the ways to do this is to calculate the number of rings and/or double bonds that could be associated with this elemental composition.

F. Validating the Putative Elemental Composition of an Ion

After an elemental composition has been proposed, its validity can be tested using the rings-plus-double-bonds (R + db) calculation. This calculation also has been called a determination of "double-bond equivalents" and "degrees of unsaturation" in organic chemistry. At least three assessments can be made from the numerical value of the R + db calculation. (1) If the value of the R + db computation is negative, the suggested elemental composition is invalid. (2) If the value of the R + db computation is larger than the number of carbons, the suggested elemental composition should be viewed with some suspicion. (3) Whether the isotope peak cluster represents an $OE^{+\bullet}$ or an EE^+ is related to the value of the R + db computation being integral or fractional, respectively:

$$R + db = \#C - \tfrac{1}{2}\#H + \tfrac{1}{2}\#N + 1 \ (\# = \text{number of atoms})$$

This formula treats elements of like valence the same; therefore, Si is treated as C (both have a valence of 4), the halogens are treated as H (all have a valence of 1), and P is treated as N (both have a valence of 3). The number of O and S atoms (both with a valence of 2) are not included in this calculation, but rings and/or double bonds associated with these elements can be calculated. If the R + db = 1 for an elemental composition believed to be represented by a $M^{+\bullet}$ peak and there is a single atom of oxygen in the elemental composition, then there is a good probability that the molecule contains a carbonyl group.

The R + db formula is only valid for elements in their lowest valence state. Any double bonds associated with a tetra- or hexa-valent sulfur or a penta-valent phosphorus will not be accounted for in the rings-plus-double-bonds calculation. If EI is used and the R + db formula computes to a fractional value (ends in 1/2), the ion is even-electron (EE^+), which would be a fragment ion produced by a single-bond cleavage in the $M^{+\bullet}$. If ESI or atmospheric pressure chemical ionization (APCI) is used as in an analysis by LC/MS, and the R + db formula computes to a fractional value (ends in 1/2), the peaks could represent either a protonated molecule or a fragment resulting from expulsion of a molecule from the protonated molecule. In the case of a fragment ion or an ion formed through proton or hydride abstraction, the actual number of rings and/or double bonds is obtained by rounding down. In the case of protonated molecules or adduct ions, the actual number of rings and/or double bonds is obtained by rounding up. The rounding direction can lead to an error in the value for the number of rings and/or double bonds based on the type of the ion; therefore, care must be taken in the rounding of these fractional values, especially with ESI and APCI data as described above.

G. An Illustrative Example of the Use of Isotope Peak Ratios to Determine an Elemental Composition

Once an elemental composition and the number of rings and/or double bonds are known, it may be possible to propose a structure. As with any conclusion drawn from mass spectral data, another important question that must be answered is whether the conclusions make sense. Is it possible and probable? Another useful guideline to follow is the *KIS Rule* (keep it simple – *law of parsimony*). There is a greater probability of a common structure being the proper match than a structure that is uncommon or bizarre. As will be seen in the illustrative example below, knowing the elemental composition can be suggestive of a possible structure.

Theoretically, the relative intensities of isotope peaks can be used to ascertain the elemental composition of an ion represented by a nominal *m/z* value peak in a mass spectrum. Consider Figure 5-4, which presents an abbreviated portion of a mass spectrum consisting of a group of peaks. It is important to realize that because this is not a complete mass spectrum, it cannot be known whether the *nominal m/z* peak represents a molecular ion, a protonated molecule, or a fragment ion produced by either the loss of a molecule or a radical from a molecular ion. Therefore, the potential answer should not be prejudiced by premature application of the *Nitrogen Rule*. The data will reveal some characteristics of the ion even though the *Nitrogen Rule* cannot be applied.

The data in Figure 5-4 are presented in graphic form as a cluster of peaks together with a numerical or a tabular listing of the peak intensities normalized to the most intense peak. The cluster of peaks starts at *m/z* 149, with isotope peaks at *m/z* 150 and 151. In the process of accounting for isotope peak intensity contributions from various elements, it is helpful to prepare a worksheet similar to that illustrated in Table 5-2. The worksheet for this example should contain three columns, "Mass", "X+1", and "X+2" for the

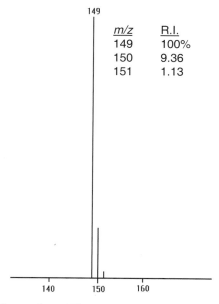

m/z	R.I.
149	100%
150	9.36
151	1.13

Figure 5-4. A portion of an EI mass spectrum of an unknown.

Table 5-2. Worksheet illustrating the process of determining an elemental composition from the intensity of the isotope peaks.

Accounting of observed peak intensities at:

The X+2 peak intensity would indicate that no X+2 elements other than O are present; i.e., no atoms of Br, Cl, Si, or S are present.

$\frac{X}{100}$	$\frac{X+1}{9.36}$	$\frac{X+2}{1.13}$	
	$\underline{-8.8}$	$\underline{-0.38}$	C_8 contributions
	0.56	0.75	
	$\underline{-0.12}$	$\underline{-0.6}$	O_3 contributions
	0.44	0.15	remainder

Accounting of mass for the nominal m/z peak:

$$C_8 = 8 \times 12 = 96 \text{ mass units}$$
$$O_3 = 3 \times 16 = \underline{48} \text{ mass units}$$
$$144 \text{ mass units}$$

However, the nominal mass peak was observed at *m/z* 149:

$$\underline{-144} \text{ mass units for } C_8O_3$$
$$5 \text{ mass units yet to be accounted for}$$

Consider five H atoms.

Is $C_8H_5O_3$ a reasonable elemental composition for *m/z* 149?

Consider rings + double bonds = R + db (see above for the definition of rule).
R + db = 8 − 5/2 + 1 = 13/2 = 6 1/2 = 6

Round off *low* for EE$^+$ fragment ions.

Proposed structure:

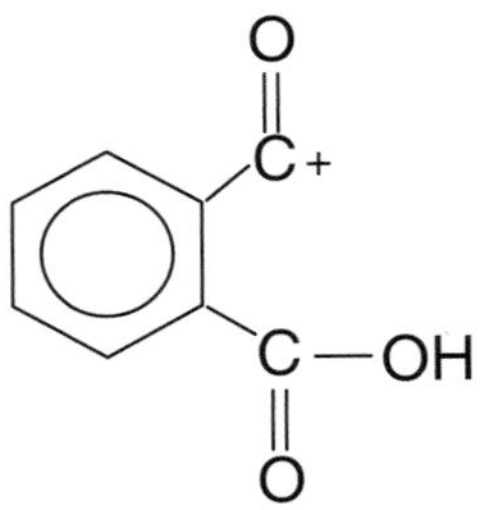

series of three peaks in this cluster, where "Mass" refers to the nominal *m/z* value peak at 149; "X+1" refers to the relative intensity of the X+1 peak (*m/z* 150 with a relative intensity of 9.36%); and "X+2" refers to the relative intensity of the X+2 peak (*m/z* 151 with a relative intensity of 1.13%). The first step in the process described above (*Step 1: Assign the nominal m/z value in the cluster*) has already been done in the process of generating the tabular form of the data. In this case, the nominal *m/z* value is assumed to be 149.

In the second step, (*Step 2: Assign the number of atoms of X+2 elements with the exception of oxygen (Br, Cl, Si, and S)*), there are no atoms of Br present. If only a single atom of Br were present, the X+2 peak would have a relative intensity almost as great as that of the X peak. If more than one atom of Br is present, the X+2 intensity would be greater than that of the X peak (see the reference patterns in the back matter). There are no atoms of Cl present. If one atom of Cl were present, the X+2 peak relative intensity would be 1/3 the intensity of the X peak and, as with the case for Br, it can be observed from the graphical presentation that neither is indicated. There are no atoms of S or Si indicated. Although not easily determined from the graphical representation of the mass spectrum, examination of the tabular presentation shows that the X+2 peak has a relative intensity of 1.13%. Examination of the X+2 factors for S and Si in the table on the inside front cover, it is seen that if a single atom of Si is present, X+2 would have a minimum relative intensity of 3.4%; if a single atom of S is present, the minimum X+2 relative intensity would be 4.4%.

There are no atoms of Br, Cl, S, or S present in the ion represented by the cluster of peaks in Figure 5-4. It is as important to know which elements are not present as it is to know which are present in the ion.

The third step in determining the elemental composition from the isotope peak intensity data is *Step 3: Assign the number of atoms of X+1 elements (C and N)*. There are a couple of important points to remember before assigning the number of atoms of C and/or N: (1) if the peak intensity at X+1 is *not* reported relative to that of X, it will be necessary to normalize the intensity of X+1 to that of X; (2) once the relative intensity at X+1 has been accounted for within *about* ±10% of the observed value, *stop!* To continue to rationalize the X+1 intensity could lead to overinterpretation. It is always possible to return for a further explanation of the intensity at X+1, if necessary.

The intensity at X+2 reported in the tabular data is 9.36% in Figure 5-4. Unfortunately, this experimental value falls between the theoretical values for eight and nine atoms of carbon as calculated using the carbon X+1 factors in the table on the inside front cover, or observed in the table giving the *X+1 and X+2 relative peak intensities for increasing numbers of C and H atoms* in the back matter. The X+1 value for eight C atoms is 8.8, and for nine C atoms, it is 9.9. In the case of nine C atoms, the theoretical and observed values differ by −0.54 percentage points. In the case of eight C atoms, the difference is 0.52 percentage points. Both values are well within the *about* ±10% limit. Which one should be chosen? There is a rational reason to choose eight C atoms over nine C atoms. Look at the observed value for X+2 in the experimental data. Examination of the X+2 column for eight and nine C atoms in the above referenced table in the back matter shows that the X+2 values for eight and nine C atoms are 0.38 and 0.49, respectively. In the case of eight C atoms, the X+2 remainder is 0.75, and for the case of nine C atoms, the remainder is 0.64. Both of these remainders are far greater than the *about* ±10%. Therefore, there must be a contribution from some source other than the possibility of an ion containing two atoms of ^{13}C. The only other possible contributor at X+2 could be oxygen (which was initially omitted from the above-described consideration for X+2). Oxygen also makes a contribution at X+1, which means that the X+1 value is not only going to arise from the ^{13}C isotope but also from the probable presence of an ^{17}O isotope.

Based on the above rationalization, eight atoms of C are selected. This results in a mass remainder of 53 Da, an X+1 relative intensity remainder of 0.56, and an X+2 remainder of 0.75. The "Mass" column would easily accommodate an atom of nitrogen. The "X+1" column would also accommodate a single nitrogen atom. The nominal *m/z* value is an odd number. There may be great temptation to consider a single atom of nitrogen. This temptation should be rejected. The X+1 value has been reconciled to within the *about* ±10% limit. It is unknown whether the nominal *m/z* value peak represents an $OE^{+\bullet}$ ion or an EE^+ ion. Therefore, it cannot be determined (from the *Nitrogen Rule*) whether the ion contains an odd number of N atoms.

The next step in this process of establishing an elemental composition based on intensity of an isotope peak is *Step 4: Assign the number of atoms of oxygen present.* The remaining intensity at X+2 to be reconciled is 0.75%. This value rounds nicely to 0.8%. This could account for as many as four atoms of O (at 0.2% each). However, 4 × 16 Da = 64 Da, which is greater than the unaccounted mass. The mass must always reconcile to zero. The maximum number of O atoms allowed based on the remaining mass is 3. Reconciling the three columns in Table 5-2 leaves 5 in the "Mass" column, 0.44 in the "X+1" column, and 0.15 in the "X+2" column.

> **Allow for some ambiguity.** The fact that the X+2 residual relative intensity value (after subtracting the contribution due to the probability of two atoms of ^{13}C being present) indicates that the presence of four atoms of oxygen in this exercise is a good example of how X+2 isotope values will have an accuracy of no better than ±1 atom of oxygen.

The remainder in the "X+2" column is slightly greater than *about* ±10% of its original value (0.15 relative to the initially observed value of 1.13); however, it is important to remember that as a relative intensity becomes smaller (and the actual ion abundance becomes less) deviations greater than the *about* ±10% are sometimes acceptable because the precision with which tiny ion currents can be measured is poor. Once again, the remaining X+1 contribution would allow for the single N atom, but the remaining mass would not.

The last step in assigning an elemental composition based on isotope peak intensities is: *Step 5: Assign the number of X elements (I, P, F, and H, for ions with a mass <1000 Da).* With only five remaining in the "Mass" column, the only possible X element to be considered is hydrogen. This will result in an elemental composition of $C_8H_5O_3$.

This elemental composition is then tested to make sure that it is reasonable (*Step 6* in the identification process). This is done using the *rings-plus-double-bonds* calculation. As seen in Table 5-2, the R + db calculation results in a value of 6½. The R + db calculation ends in a half, indicating an EE^+ ion. Now the *Nitrogen Rule* can be applied. An EE^+ ion with no (or an even number of) N atoms will have an odd *m/z* value. The nominal *m/z* value is odd (149). The elemental composition includes no atoms of nitrogen. The R + db calculation, nominal *m/z* value, and the elemental composition are in agreement with one another; therefore, there is a high probability that the ion represented by this cluster of peaks with nominal *m/z* 149 has the elemental composition of $C_8H_5O_3$ and a total of six rings and/or double bonds.

The next step is to attempt to assign a structure (*Step 7*). It is not always going to be possible to assign a structure based on an elemental composition and the results of the R + db calculation. In some cases, several structures may be possible, such as *meta* and *para* isomers versus the *ortho* isomer. In some cases, there will be one structure that has a higher possibility than all other structures based on the KIS rule described above. That is the situation with this particular illustration.

When the value of R + db is large, try a phenyl ring, which accounts for three double bonds and one ring. It also accounts for six atoms of carbon. In this case, choosing a phenyl ring will leave a remainder of two R + db, two atoms of C, three atoms of O, and five atoms of hydrogen. Surely, some of these atoms of hydrogen will be associated with the aromatic ring, but these single-valence elements should be used to fill out the final structure.

It is possible that two carbonyl groups could account for the remaining two R + db, the two remaining C atoms, and two of the three O atoms. As seen in the proposed structure, these two carbonyl groups can be positioned on adjacent atoms on the aromatic ring. Of course, it is possible that the two carbonyl groups could be positioned on any of the six carbon atoms (no more than one carbonyl per C atom), but it will become clear as to why adjacent atoms were chosen.

After assigning the carbonyl groups, the remaining O atom and one of the H atoms can be combined as a hydroxyl (OH) group. This can then be associated with one of the carbonyl groups to form a carboxylic acid. With two of the possible substitution points on the phenyl ring being occupied by the two carbonyl groups, this leaves a remainder of four positions, which is the exact number of remaining H atoms.

The last step in this process of trying to identify the analyte (*Step 8*) is to ask the question, "Given the information provided about the sample, does this answer make sense?"

In this case, the answer does make sense. The spectrum was obtained by electron ionization using GC/MS. This proposed structure is that of a phthalate ion, which is represented by the base peak in the EI mass spectrum of many common plasticizers that are esters of phthalic acid. These plasticizers are ubiquitous and are often encountered as common contaminants in both LC/MS and GC/MS. They are characterized by a peak at *m/z* 149 in the mass spectrum.

In most cases, the peak at *m/z* 149 will result from the presence of a plasticizer, a common contaminant. However, it is possible that *m/z* 149 does relate to the analyte (e.g., the EI mass spectra of some steroids do exhibit a significant peak at *m/z* 149). Before expending too much cerebral energy trying to rationalize a peak at *m/z* 149 as relating to the analyte, it is best to evaluate the mass spectral data to try to determine the source of the ion current at *m/z* 149. With data from a chromatography/mass spectrometry system, this determination can sometimes be made by a comparison of mass chromatographic peaks for *m/z* 149 and some other ion believed to represent the analyte. If the two mass chromatographic peaks do not rise and fall together, the two ions are not in common with the mass spectrum of the same compound. Often the generation of a mass chromatogram for *m/z* 149 results in a plot of constant signal above the zero line, and gives no discernible peak; this indicates that the ion is present in every spectrum and that it represents the constant chemical background.

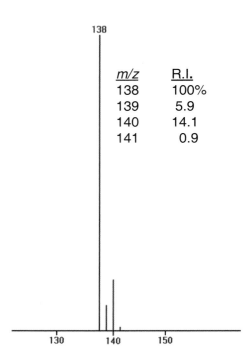

m/z	R.I.
138	100%
139	5.9
140	14.1
141	0.9

Figure 5-5.   EI mass spectrum with nominal m/z peak at 138, exhibiting the possible presence of an X+2 element other than Cl or Br.

To illustrate the importance of assigning contributions to X+1 from the X+2 elements Si, S, and K, consider the example shown in Figure 5-5.

From the intensity of the X+2 peak, it is obvious that there is an X+2 element other than Cl or Br present. However, there may be a tendency to assign the X+1 elements first. If the X+1 element of C is prematurely assigned based on the observed 5.9% relative intensity for the X+1 peak, five atoms of carbon would be the result. Due to the X+1:X+2 intensity ratio, it is very unlikely that Si is the X+2 element present because Si has a higher contribution at X+1 than at X+2 (X+1 = $5.2n_{Si}$ and X+2 = $3.4n_{Si}$). As indicated in the figure legend, the data were acquired from electron ionization; therefore, the presence of potassium can be eliminated. The number of oxygen atoms necessary for such an intense X+2 peak would not be very probable. This leaves S as the probable contributor to the X+2 peak. Dividing the observed X+2 relative intensity by the 4.4 factor for X+2 due to S results in 3.2 or a possibility of three atoms of S.

The contribution to the relative intensity of the X+1 peak from three atoms of S is 2.4. This would leave a residual of 3.5 as being due to an ion containing a single atom of ^{13}C, which indicates that the ion contains only three C atoms, *not* five!

It is clear from this example that the number of X+2 elements that also have a contribution at X+1 should be assigned before assigning the number of C atoms (X+1 elements).

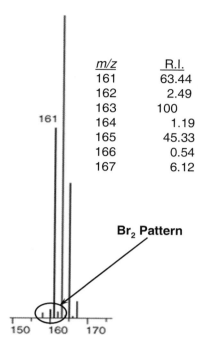

m/z	R.I.
161	63.44
162	2.49
163	100
164	1.19
165	45.33
166	0.54
167	6.12

Figure 5-6. EI mass spectrum with nominal m/z peak at 161 exhibiting possible overlay.

H. Potential Problems Arising from Adjacent Peaks

Another important consideration in the use of isotope peak intensities is the possibility that a peak may not only represent ions of different isotopic compositions but they may also represent ions of different elemental compositions. Consider the abbreviated mass spectrum shown as a bar graph and in tabular format in Figure 5-6. After assigning the number of atoms of X+2 elements, it is obvious, based on the remaining mass, that only a single atom of carbon can be present with no X elements or atoms of oxygen. The one carbon atom is confirmed by the intensity ratio of X+3 relative to that of X+2, the ratio of the intensity of X+5 relative to that of X+4, and the intensity of X+7 relative to that of X+6. All of these computations give a ratio of approximately 1.1%. However, when the peak intensity at X+1 is normalized relative to the peak intensity of X, it looks as though there could be as many as three or four atoms of carbon. Based on the remaining mass, this is not possible. Apparently, some other species is present that possibly has a different elemental composition. There is something more than the ion with the same elemental composition as represented by the peak at m/z 161 with an atom of ^{13}C and that contributes to the intensity of the X+1 peak. A closer examination of the mass spectrum shows that the peak at m/z 158 is followed by a peak at m/z 160 of almost twice the intensity. Now, the high intensity of the X+1 peak comes into focus. The peak at m/z 162 has an intensity that is almost equal to the intensity of the peak at m/z 158. The peak-intensity pattern of m/z 158, 160, and 162 is very similar to that of the peak pattern of an ion containing two atoms of Br. The mass of two atoms of Br (a Br molecule) is 158 Da. Therefore, the peak at m/z 162 could represent ions of two different elemental compositions (e.g., $^{13}C^{35}Cl_2{}^{79}Br$ and $^{81}Br_2$).

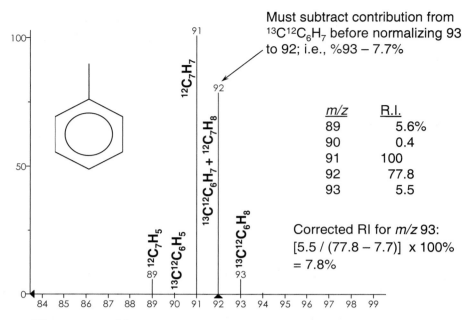

Must subtract contribution from
$^{13}C^{12}C_6H_7$ before normalizing 93
to 92; i.e., %93 − 7.7%

m/z	R.I.
89	5.6%
90	0.4
91	100
92	77.8
93	5.5

Corrected RI for m/z 93:
[5.5 / (77.8 − 7.7)] x 100%
= 7.8%

**Figure 5-7. Mass spectrum of toluene illustrating
interference with isotope peak intensities.**

R. Martin Smith in *Interpretation of Mass Spectra: A Basic Approach* (John Wiley & Sons, Inc., 2004) uses a more dramatic example of overlapping mass spectra, namely those involving the ^{13}C-isotope contribution of a tropylium ion to the intensity otherwise representing the M$^{+\bullet}$ peak of toluene. If the intensity of the M$^{+\bullet}$ peak is not adjusted for the isotope peak-intensity contribution of the tropylium ion, then the value for the relative intensity of the isotope peak for the molecular ion will appear to be in error (Figure 5-7).

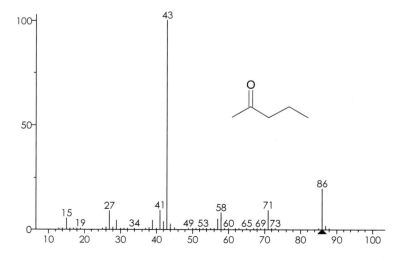

Figure 5-8. EI mass spectrum of 2-pentanone.

4. Elemental Composition as a Function of an Accurate Determination of the m/z Value of a Mass Spectral Peak

The *m/z* value of a mass spectral peak obtained using any type of *m/z* analyzer can be determined with a high degree of accuracy (within ±0.0001 mass units or better). If the peak is monoisotopic (representing an ion of a single elemental and isotopic composition), the accurately determined mass value can be used to ascertain the elemental composition of the ion [4]. Up to a mass of approximately 500 Da, an ion of a given elemental composition has a unique calculated exact mass. Comparison of an accurately determined mass of an unknown ion and the calculated exact mass of various elemental compositions will allow the elemental composition of the ion to be determined. The problem is in knowing whether the mass spectral peak represents an ion of single isotopic and elemental composition. It has already been illustrated in this chapter that in a cluster of peaks, only the peak of lowest *m/z* value represents an ion of a single isotopic composition. The X+1 peak can represent ions containing a single atom of ^{13}C, ^{15}N, ^{17}O, or some other element with an X+1 isotope. Ions of differing elemental composition that have the same nominal mass will have different calculated exact masses.

Experimental determination of the *m/z* value of a single-charge ion up to 500 Da by a mass spectrometer capable of high resolving power will reveal its elemental composition [2, 4, 5]. However, if the *m/z* analyzer is only capable of separating ions that differ by an integer *m/z* unit, then the accurately determined *m/z* value for the peak will be the weighted average of all the ions present (weighted based on the percent abundance of each isotopic composition represented in the integer *m/z* value peak).

Look at the mass spectrum of 2-pentanone (Figure 5-8). As will be seen in the next chapter, molecular ions of this compound can fragment to produce single-charge propyl ions (calculated exact mass of 43.0548 Da) and single-charge methyl-acylium ions ($H_3CC{\equiv}O^+$) with a calculated exact mass of 43.0184 Da. In order to separate these two ions by mass spectrometry, an instrument capable of *high resolving power* is necessary [2, 4, 5]. The resolving power necessary to separate these two ions can be calculated from the formula $R = M / \Delta m$, where $M = 43$ and $\Delta m = 43.0548 - 43.0184 = 0.0364$.

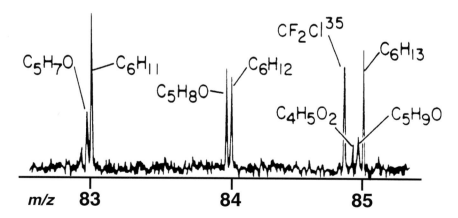

Figure 5-9. *Mass spectral data obtained with a resolving power (R) of 2500 (10% valley) for nominal m/z values of 83, 84, and 85.* Reprinted with permission from the American Chemical Society from Watson JT and Biemann K, Anal. Chem. 1964, 36, 1135–1137.

An example of the capability of such an instrument is seen in the mass spectrum in Figure 5-9. A close examination of these data begs the question, "Why are there no peaks representing $^{12}C_5{}^{13}CH_{11}$, $^{12}C_6{}^1H_{10}{}^2H$, and the three isotopic variants of C_5H_7O in the cluster of peaks with the integer m/z value of 84?" Based on the separation of the peaks representing the $^{12}C_5{}^1H_8{}^{16}O$ and $^{12}C_6{}^1H_{12}$ ions (the two peaks with nominal m/z 84, of about equal intensity, and separated so that the height from the baseline to the top of the valley between the two peaks is approximately 10% in Figure 5-9 [6]), the exhibited resolving power is 2309 ($M/\Delta m = 84.0575 / 0.0364 = 2309$). The resolving power required to separate $^{12}C_5{}^{13}CH_{11}$ or $^{12}C_6{}^1H_{10}{}^2H$ from $^{12}C_6H_{12}$ is 17,890 or 44,260, respectively, either of which is greater than the exhibited resolving power. The peak labeled as representing C_6H_{12} actually is a triplet, representing at least three ions: two of one elemental composition (but different isotopic compositions) and one of another elemental composition. A comparable resolving power would be required to separate C_5H_8O from the three isotopic variants of C_5H_7O ($^{13}CC_4H_7O$, $C_5{}^2HH_6O$, $C_5H_7{}^{17}O$).

As seen from Figure 5-9, at an m/z value of about 100, a resolving power of 2500 is more than sufficient to distinguish between two ions that are isobaric with respect to nominal mass and that have elemental compositions of C_xH_yO and C_zH_w. The accuracy of the measured m/z values for these ions is reported in parts per million. Measurements to within 5 ppm at the 100 m/z value are sufficient to achieve the experimental objective. This accuracy within so many ppm relates to the experimentally measured m/z value *vs* the calculated exact mass of a candidate elemental composition. As the m/z value increases, it is necessary to obtain a lower ppm accuracy to distinguish among several candidate elemental compositions with the assumption that the one closest in agreement is the correct one. Organic chemistry journals often state that an accuracy of ±5 ppm is required to confirm an elemental composition [7]. As was pointed out in an editorial by Michael Gross in *The Journal of the American Society for Mass Spectrometry* [8], an arbitrary ppm value cannot apply because the accuracy required is based on the elemental compositions to be distinguished and the mass of the ion under consideration. The following three examples demonstrate how quickly the difference in mass that distinguishes a given elemental composition from another diminishes with the nominal mass of (or more correctly, with the number of atoms in) the ion under consideration. (1) A molecule of nominal mass 118 Da, when considered with regard to the valence rules and under the constraints of candidate compositions of C_{0-100}, H_{3-74}, O_{0-4}, and N_{0-4}, will correspond to candidate molecular formulas that differ in mass by no more than 34 ppm (approximately 4 millimass units). (2) At a nominal mass of 500 Da under the same constraints, there are five elemental composition candidates that differ in mass by less than 5-ppm, which is 0.95 mmu. (3) A molecule with a mass of 750.4 Da (remember mass defect due to hydrogen) and under constraints of elemental compositions in the range of C_{0-100}, H_{25-110}, O_{0-15}, and N_{0-15} could correspond to any one of 626 elemental compositions within a 4-ppm window of mass. An error of less than 0.0018 ppm is required for an unambiguous elemental composition for this molecule.

Definition of ppm as used in mass spectrometry:

$ppm = [(\mathbf{E} - \mathbf{A}) / \mathbf{E}] \times 1 \times 10^6$

where $\mathbf{E}$ = calculated exact mass of an elemental composition and $\mathbf{A}$ = measured accurate mass of an ion of a given elemental composition

Before the ready availability of computers, books were published that listed the values of calculated exact mass for various elemental compositions of designated numbers of atoms of C, H, O, and N [5]. Today, the exact mass of these elemental compositions is calculated on-line by computer programs. Usually, a constrained number of atoms of suspected elements and a ppm-limit value are entered by the analyst.

A. Appearance of Mass Spectra of High-*m/z* Value Ions

Another important consideration of the resolving power of a mass spectrometer is the effect that the resolving power has on the appearance of a mass spectrum, especially when ions of high *m/z* values are involved. It should be recalled from the chapter on types of *m/z* analyzers (Chapter 2) that the quadrupole ion trap and the transmission quadrupole *m/z* analyzers operate at constant resolution, whereas the double-focusing and time-of-flight instruments operate at constant resolving power. This means that no matter what the *m/z* value, the isotope peaks can separated from one another with the transmission quadrupole or the quadrupole ion trap analyzer. However, both of these instrument types have upper *m/z* limits of well under 10,000. In contrast, there is no upper limit on the *m/z* range of the time-of-flight *m/z* analyzer, but these instruments usually have resolving powers of about 2000 to 3000 when operated in the linear mode, meaning that the isotope peaks can no longer be separated from one another once the *m/z* value passes the value for the resolving power. At *m/z* 100,000 on such an instrument, only a single peak will be observed, the centroid of which represents the weighted average of the *m/z* values of all the unresolved isotopic species. For single-charge ions, this numerical value will be equal to the *average mass* of the ion. This inability to separate the isotopic species is one of the reasons that data from such instruments are displayed in the profile mode. Figure 5-10 is a series of mass spectra that were obtained with a time-of-flight *m/z* analyzer for ions of specific nominal masses of increasing magnitude.

For an example of the use of accurate mass measurements in the interpretation of mass spectral data, consider the mass spectrum obtained by electron ionization (EI) in Figure 5-11, which has three peaks at nominal *m/z* values of 308, 279, and 238. The

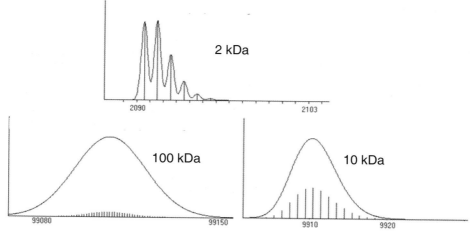

Figure 5-10. *TOF mass spectra (clockwise) acquired with a resolving power (R) of 2000 for ions with masses of ca. 2, 10, 100 kDa; the vertical lines correspond to the computed value for the center of isotope peaks even when not resolved experimentally.*

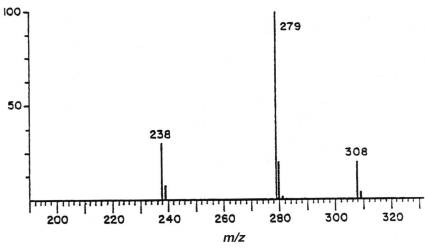

Figure 5-11. EI mass spectrum of an unknown.

Table 5-3. Elemental composition based on accurate mass measurement of ions represented by peaks in Figure 5-11.

Experimentally Determined Accurate Mass	Calculated Exact Mass	Deviation in mmu and (ppm)	Most Likely Corresponding Elemental Composition
308.1878	308.1888	−1 (3.2)	$C_{20}H_{24}N_2O$
279.1510	279.1497	1.3 (4.6)	$C_{18}H_{19}N_2O$
238.1249	238.1232	1.7 (7.1)	$C_{16}H_{16}NO$

peak at the highest m/z value is a candidate for a $M^{+\bullet}$ peak (the candidacy for a $M^{+\bullet}$ peak is discussed later in this chapter). The peak at m/z 308 is the highest m/z value peak in the mass spectrum that is not an isotope peak or a peak that is due to background. In this case, the m/z value of this peak is an even number (308), indicating that the ion does not contain an odd number of nitrogen atoms (the *Nitrogen Rule*). Unfortunately, this even m/z value does not rule out the presence of N atoms in the ion; there could be an even number of nitrogen atoms present.

The peak at m/z 279 supports the candidacy of the peak at m/z 308 as a $M^{+\bullet}$ peak because the loss of 29 corresponds to a logical loss of an ethyl radical or a CHO radical. If an even number of N atoms is present, one is not lost in the formation of the ion with m/z 279.

The peak at m/z 238 provides little information; it has an even m/z value, which could represent an odd-electron fragment ion formed by the loss of a molecule from the molecular ion or an even-electron ion with an odd number of N atoms formed by the loss of a radical containing an odd number of N atoms. The number reflecting the dark matter (70 Da) does not indicate an obvious grouping of atoms, although it could correspond to C_5H_{10}.

The isotope peak intensities associated with these three nominal m/z value peaks do not provide any obvious clues as to the elemental compositions of these ions, other than the fact that the X+2 elements are probably not present.

More information can be obtained with an accurate mass measurement [2, 4, 5]. The experimental accurate *m/z* values for the three nominal *m/z* peaks in the mass spectrum are seen in Table 5-3, which also contains the nearest calculated exact mass value found in a table of mass values for various combinations of C, H, O, and N atoms along with the matching elemental composition. It is now clear that the putative molecular ion contains two atoms of nitrogen; the peak at *m/z* 279 represents an even-electron ion formed by the loss of an ethyl radical; the peak at *m/z* 238 represents an even-electron ion formed by the loss of a C_4H_8N radical. In this illustration of data obtained by EI, the use of accurate mass measurements translates to elemental compositions of particular ions, thereby eliminating some of the ambiguity associated with the low-resolution data.

In this particular example, the accurate masses of the ions were determined using high resolving power mass spectrometry. It is worth noting that an accurate mass measurement can be made on any mass spectral peak regardless of the mass spectrometer's resolving power. However, in the absence of high resolving power, it will not be known whether the peak represents an ion of a single elemental composition. Unless a peak is known to represent a single elemental composition and a single isotopic composition for that elemental composition, the accurate mass measurement is a weighted average (an average weighted according to the abundance of each species) of the masses of each elemental and isotopic composition represented by the peak. In order to obtain an unambiguous elemental composition from an accurate mass measurement, the species being analyzed must consist of a single elemental and isotopic composition. The only ions of a given integer mass that will have a single elemental and isotopic composition are monoisotopic ions that have no interferences from ions of other elemental compositions.

5. Using EI Data to Identify Unknowns Detected During Analysis by LC/MS

When data are obtained using an LC/MS technique of electrospray or atmospheric pressure chemical ionization, only information about the intact analyte can be extracted from the spectrum unless some form of MS/MS is utilized. Sometimes, a determination of the suspected elements present can be made from the relative intensity of the isotope peaks. There are also computer programs to aid in this process. One advantage of modern time-of-flight (TOF) *m/z* analyzers is that an accurate mass measurement and isotope peak intensities are obtainable within a single spectrum. When a double-focusing *m/z* analyzer is used, it is necessary to obtain one set of spectra for the accurate *m/z* measurement and another set for the isotope peak-intensity information.

A suspicious peak at *m/z* 212 was observed in a mass spectrum (Figure 5-12) obtained during an analysis by LC/MS using electrospray and negative-ion detection. Generation of a mass chromatogram at *m/z* 212 resulted in no chromatographic peak. This result indicated that the mass spectral peak at *m/z* 212 represents something in the constant background of the system. Through a process of elimination it was determined that the contaminant was associated with the mass spectrometer and not the LC or the solvents being used. In an effort to identify the contaminant, an accurate mass measurement was made on the suspicious ions. Table 5-4 lists the elemental compositions that have a calculated mass within ±10 mmu of the experimentally measured mass and within the constraints of a designated number of atoms of C, H, N, O, and S for a [M – H]⁻ ion (*m/z* 212) believed to correspond to the suspected contaminant. Sulfur is a fairly unusual element to be found in an organic compound; however, the

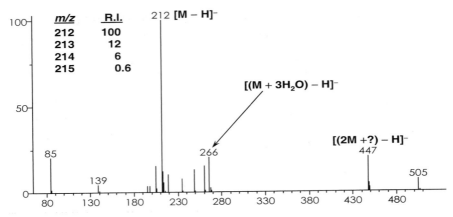

Figure 5-12. **Abbreviated mass spectrum obtained by negative-ion detection and electrospray exhibiting a [M – H]⁻ peak at nominal m/z 212.**

isotope peak-intensity pattern and tabular data are characteristic, and in this case, the data indicated the presence of a specific number of S atoms. Therefore, sulfur was included as a constraint in the automated computation of candidate elemental compositions based on an accurate mass measurement. The *m/z* value obtained for the deprotonated molecule using electrospray is an even number, which indicates that the analyte has an odd number of N atoms (the *Nitrogen Rule*). The presence of an odd number of N atoms and a single atom of S (based on the isotope peak-intensity data) narrowed the list to an elemental composition of $C_{10}H_{14}NO_2S$ for the deprotonated molecule or $C_{10}H_{15}NO_2S$ for the intact analyte.

Table 5-4. **Elemental compositions within ±10 mmu of the experimentally measured mass of the contaminant.**

Elemental Composition	R + db	Error (ppm)
$C_6H_{16}N_2O_2S_2$	0	8.1
$C_6H_{16}N_2O_4S$	0	−9.7
$C_9H_{12}N_2O_2S$	5	11
$C_9H_{12}N_2O_4$	5	−6.3
$C_{10}H_{14}NO_2S$	4.5	−1.1
$C_{11}H_{16}O_2S$	4	−14
$C_{11}H_{16}S_2$	4	4
$C_{12}H_8N_2O_2$	10	15
$C_{13}H_{10}NO_2$	9.5	2.2
$C_{14}H_{12}O_2$	9	−10
$C_{14}H_{12}S$	9	7.4

An elemental composition of $C_{10}H_{15}NO_2S$ could constitute any one of several different structures. One possible way of identifying the unknown is to use the *Formula Search* in the *NIST Mass Spectral Search Program* and the *NIST/EPA/NIH Database* (*NIST05*).

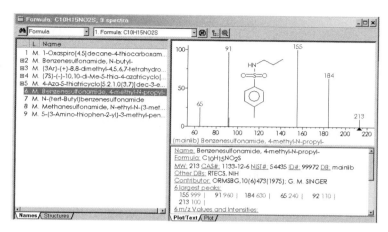

Figure 5-13. Results of a Formula Search using the NIST Mass Spectral Search Program and NIST/EPA/NIH NIST05 Database.

Although this *Database* contains only EI mass spectra, it also lists the nominal masses and elemental compositions of over 190,000 different compounds and contains hundreds of thousands of synonyms for "common compounds". The *NIST MS Search Program's Formula Search* for $C_{10}H_{15}NO_2S$, the elemental composition determined by the accurate mass calculation software, yielded a list of eight compounds (Figure 5-13). Only one of these compounds had any synonyms that included "common" or "trade" names, *N*-butylbenzenesulfonamide, a known plasticizer. This identification was confirmed by the use of CAD MS/MS, which indicated the loss of $HN=CH-C_3H_7$, a molecule that could only occur with one of the eight possible structures listed in Figure 5-13 [9].

6. Does the Result Make Sense?

One very important step in extracting information from mass spectral data is to make sure that the "Does the answer make sense?" question is asked. The data may lead to a conclusion that is logical from a computational point, but one that *does not make sense* after all factors are taken into consideration. A good example is seen in an attempt to identify a contaminant in a sample of a synthetic tripeptide using both isotope peak intensities and accurate mass measurements.

A sample of a putative tripeptide having the expected structure shown below was analyzed using electrospray LC/MS with positive-ion detection. The nominal mass of the peptide with an elemental composition of $C_9H_{17}N_3O_4$ is 231 Da (monoisotopic mass is 231.1219 Da); peaks were observed at *m/z* 232 representing the protonated molecule, at *m/z* 254 representing the Na⁺ adduct, and at *m/z* 270 representing the K⁺ adduct. In addition to these peaks and a number of other peaks in the mass spectrum, a peak was observed at a nominal *m/z* value of 308. It was believed that this latter peak represented a contaminant or a by-product of the synthesis.

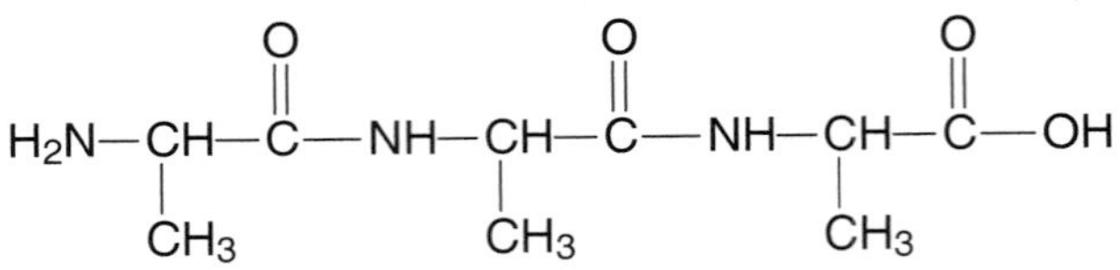

$$\underset{\underset{CH_3}{|}}{H_2N-CH}-\overset{\overset{O}{||}}{C}-\underset{\underset{CH_3}{|}}{NH-CH}-\overset{\overset{O}{||}}{C}-\underset{\underset{CH_3}{|}}{NH-CH}-\overset{\overset{O}{||}}{C}-OH$$

The isotope peak pattern associated with the peak at *m/z* 308 (assumed to be a protonated molecule) is represented by the following tabular listing:

308	100.00
309	11.17
310	15.62
311	1.73

The cluster of peaks starting at *m/z* 308 represents a single-charge ion (isotope peaks appear to be separated at 1 *m/z* unit intervals). An accurate mass measurement produced a value of 308.041496 Da for the monoisotopic mass peak at *m/z* 308.

Using the default elemental calculator in the mass spectrometer's data system, the closest match of the experimentally measured mass to a calculated exact mass corresponded to $C_6H_{18}N_3O_5S_3$, which has a calculated exact mass of 308.040860 Da (a difference of 0.64 millimass units or ~2 ppm). The theoretical peak pattern representing an ion of this elemental composition is:

308	100.00
309	10.40
310	15.07
311	1.42

All of the isotope peak relative intensity values, with the exception of the X+3 peak, are well within *the ±10% criterion* for agreement; the X+3 value is out-of-agreement beyond the 10% window, but remember that for peaks of very low intensity, it is permissible to expand the window of acceptable remainder up to perhaps 20% of the original value.

By analogy with the structure of the expected synthetic peptide, the following structure was proposed for the putative contaminant or by-product:

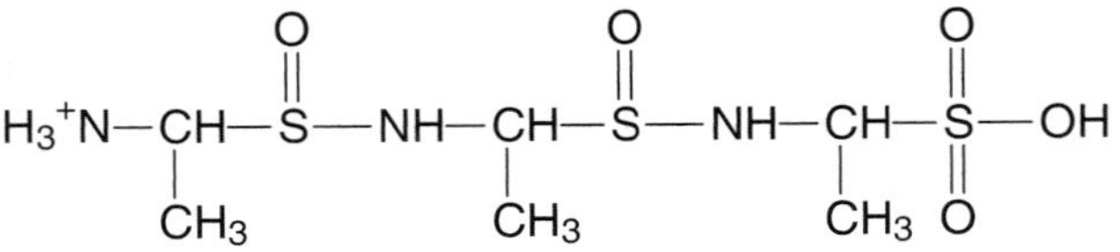

$$\underset{\underset{CH_3}{|}}{H_3{}^+N-CH}-\overset{\overset{O}{||}}{S}-\underset{\underset{CH_3}{|}}{NH-CH}-\overset{\overset{O}{||}}{S}-\underset{\underset{CH_3}{|}}{NH-CH}-\overset{\overset{O}{||}}{\underset{\underset{O}{||}}{S}}-OH$$

A detailed examination of the isotope peak intensity data reveals how easily the analyst might have been lulled into proposing a structure based on sulfur. Although a suspicious elemental composition may have been suggested in the first computational consideration shown below, it would be easy to arrive at what appears to be a valid computational solution when the second set of calculations is considered:

308	100.00
309	11.17
310	15.62
311	1.73

	Mass	X+1	X+2	
	308	11.17	15.62	
3S × 32 =	96	2.40	13.20	Based on the intensity of the X+2 peak
	212	8.77	2.42	
3N × 14 =	42	1.11	N/A	Based on the presence of 3 atoms of N
	170	7.67	2.42	in the analogous expected peptide
7C × 12 =	84	7.70	0.29	
	86	−0.03	2.13	
5O × 16 =	80	−0.12	0.80	Largest number of O atoms possible
	6	−0.15	1.33	based on the remaining mass

The remaining 6 mass units must be attributed to hydrogen.

Rings-plus-double-bonds calculation for $C_7H_6N_3S_3O_5$
$7C - 6/2H + 3/2N + 1 = 5\ 1/2 = 6$; round up for a protonated molecule

The number of R + db values is twice the number calculated for $C_9H_{17}N_3O_4$ as the synthetic peptide. Therefore, this elemental composition for the putative by-product is suspect.

A second computational consideration produces a more likely result.

	Mass	X+1	X+2	
	308	11.17	15.62	
3S × 32 =	96	2.40	13.20	No reason to change from original
	212	8.77	2.42	assumption
3N × 14 =	42	1.11	N/A	No reason to change from original
	170	7.66	2.42	assumption
6C × 12 =	72	6.60	0.22	6 C atoms will involve more H atoms
	98	1.17	2.20	causing the R + db count to be lower
5O × 16 =	80	0.12	0.80	
	18	1.05	1.40	

The remaining 18 mass units would have to be attributed to hydrogen.

Rings-plus-double-bonds calculation for $C_6H_{18}N_3S_3O_5$
$6C - 18/2H + 3/2N + 1 = -1/2 = 0$; round up for a protonated molecule

The results of the R + db calculation would support the proposed structure containing three atoms of sulfur because two of the sulfur atoms have a valence of 4 and the remaining sulfur atom has a valence of 6. Therefore, the double bonds associated with these sulfur atoms would not be accounted for in the R + db calculation.

This information was reported to the client. However, before reporting this structure, the analyst should have asked the very important question, "Does this result make sense?" The answer to this question would have to be "no!"

The large X+2 intensity value in the isotope peak pattern exhibited by the unknown ion led the analyst to accept the results of the data system's elemental calculator. Sulfur does make a significant contribution at the X+2 position. The presence of sulfur will also contribute significant intensity at X+1, and prompts the consideration of fewer carbon atoms than if sulfur had not been considered to be part of the elemental composition. The combined mass resulting from the three sulfur atoms and reduced number of carbon atoms (six as compared to the original nine) allows for, and supports, an increase in the number of oxygen atoms. The assumption that sulfur is present triggers a cascade of related assumptions that appear to produce a defensible computational solution. The solution is computationally logical, but it is *not likely in a chemical sense*.

The sample submission report stated that the sample contained a high concentration of a potassium salt. Potassium is an X+2 element with an X+2 factor of 7.22 per atom. Two potassium atoms could easily account for the observed X+2 intensity of 15.62 when contributions are also considered from carbon and oxygen. These facts along with the fact that the analysis was carried out by electrospray ionization (in which potassium adduct ions are commonplace) should have resulted in the inclusion of potassium in possible elemental compositions. There was no persuasive chemical rationale for including sulfur as a possible constituent in the elemental composition. The only reason sulfur should have been considered was because of the X+2 isotope peak intensity.

The software on the mass spectrometer used for this analysis was old (designed for EI data only) and did not include metal atoms (such as K and Na, the ions of which commonly form adducts with analyte molecules during electrospray) for elemental composition calculations. Therefore, the software did not calculate a possible formula containing potassium atoms. If the elemental composition calculation had been carried out using potassium as a possibility, it would have been clear that a more likely, and acceptable, result would be $C_9H_{16}N_3O_4K_2$, which could be represented by the following structure:

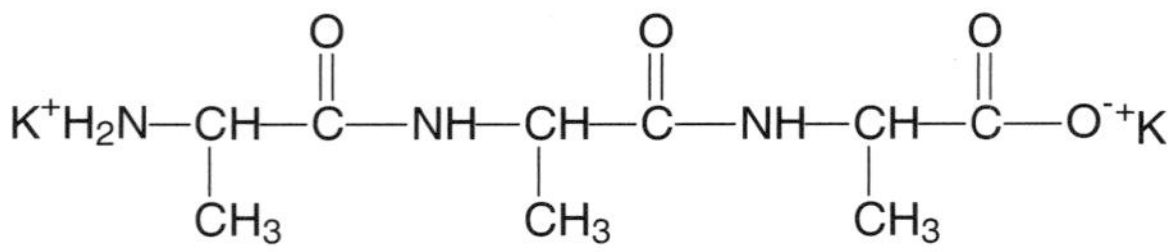

$$K^+H_2N-\underset{\underset{CH_3}{|}}{CH}-\overset{\overset{O}{\|}}{C}-NH-\underset{\underset{CH_3}{|}}{CH}-\overset{\overset{O}{\|}}{C}-NH-\underset{\underset{CH_3}{|}}{CH}-\overset{\overset{O}{\|}}{C}-O^{-+}K$$

This structure results from cationization of the amino terminus by a potassium cation and from ion-pair formation between another potassium cation and an anion of the analyte due to its loss of a proton from the carboxyl terminus. This solution is much more likely than the sulfur solution, especially when it is considered that two of the sulfur atoms replacing the three carbonyl carbon atoms have a valence of 4 and the third sulfur atom has a valence of 6.

This example illustrates the importance of the *"Does it make sense?"* part of the data interpretation procedure.

V. Identifying the Mass of an Analyte

Mass spectrometry is the only analytical instrumental technique that can be used to determine the mass of an analyte with any degree of accuracy, even to just an integer value. Instruments equipped for the soft ionization techniques that produce primarily ions representing the intact molecule have been referred to as *molecular mass machines*. At one time, even electron ionization (EI) was used primarily for analyte characterization based on nominal mass. The challenge in using mass spectrometry to determine mass is in recognizing the peak in the mass spectrum that represents the intact molecule. As demonstrated in an earlier section of this chapter, the elemental composition of an ion can be determined using accurate mass measurement or isotope peak intensities, but in order to determine the elemental composition of an analyte, it is necessary to identify the peak in the mass spectrum that represents the intact molecule. Therefore, recognition of the peak representing the intact analyte molecule is one of the most important aspects of interpreting mass spectra.

EI mass spectrometry can complicate the goal of determining the nominal mass of a molecule because the molecular ion of some compounds is so unstable that none survive to give rise to a discernible peak; i.e., for certain compounds, the molecular ions fragment completely. As pointed out earlier, only about 80% of the compounds in the NIST/EPA/NIH Mass Spectral Database produce EI mass spectra that exhibit a discernible peak for the molecular ion. Another 18% of the compounds represented in the library produce mass spectra in which the $M^{+\bullet}$ peak has a relative intensity of only 1 to 5%. In some cases, a low intensity $M^{+\bullet}$ peak may not be detected because of the operating mechanics of the instrument, such as the need for rapid data acquisition when using a narrow-bore capillary (i.d. of 32 μm or less) with GC/MS on scanning beam-type instruments such as the transmission quadrupole or magnetic double-focusing mass spectrometer.

In the case of liquid introduction of the sample under atmospheric pressure (ESI, APCI, APPI, etc.), the mass spectrum may be complicated by the presence of peaks that represent protonated or deprotonated species that are clustered with one or more molecules of one or more components of the solution containing the analyte. In the case of positive-ion ESI, there is the problem (or advantage) of adduct formation with alkali ions. There is also the possibility of forming multimers of the analyte. Furthermore, multimers of solvent molecules can give rise to significant peaks in the mass spectrum. The task of recognizing the peak that represents the intact molecule, and thus that allows a determination of the mass of the analyte, sometimes can be daunting at best.

Once the peak that represents the intact analyte molecule has been identified, the m/z value assigned to that peak must be interpreted. In the case of EI, the integer m/z value for the $M^{+\bullet}$ peak represents the analyte's nominal mass for all compounds having a nominal mass of <500 Da, which includes most of the analytes that can be ionized by EI. The NIST/EPA/NIH Mass Spectral Database, NIST05, has <6K compounds (out of 163K) that have a nominal mass of >500 Da.

It is easy to exceed the 500-Da level with ions formed by ESI and APCI. This means that mass defect becomes an issue and the word "nominal" in the definition of the *Nitrogen Rule* is very important. Mass defect can be an issue with GC/MS analytes because of compounds such as polybrominated diphenyl ethers (PBDEs). For example, the Br_7 congener of the PBDEs has a nominal mass of 716 Da and a monoisotopic mass of 715.4467, which means that the integer m/z value assigned for this analyte would be 715 due to a negative mass defect of approximately ~0.6 Da. In the case of a hydrocarbon with a nominal mass of >500 Da, the positive mass defect of hydrogen will cause the m/z value of the peak to be one unit higher than the nominal mass.

Until the development of desorption/ionization techniques, most mass spectrometry dealt with analytes that were <1000 Da. When this was the case, the m/z value of the detected ions was usually less than the resolving power of the mass spectrometer. As was pointed out above, with techniques like MALDI, FAB, and ESI, it is possible to detect ions whose mass far exceeds the resolving power of the instrument. The m/z value assigned to a mass spectral peak may not represent the mass of a single ion, but the mass of several ions, each having a different integer mass (e.g., a cluster of isotope peaks). The m/z value assigned to such a peak represents the average of this multitude of integer-value ions weighted by the abundance of each. If the m/z value is high enough and the resolution is low enough, this assigned m/z value will be the *average mass* of the ion (the mass of the ion based on the atomic masses of the elements that comprise it). Care must be taken in jumping to conclusions about these weighted-average m/z values. If the assigned m/z value is around 3K and the instrument has a resolving power of 3000, then the peak represents only three integers; i.e., if the assigned value is 3234, the m/z value of the next discernible peak would be 3237. Under these circumstances, the peak at m/z 3234 would not represent an ion whose *average mass* was this value (assuming a single-charge ion).

Another important consideration has to do with high-mass ions that carry multiple charges as in the case of ESI. In the case of an ion that has 1000 charges (e.g., addition of 1000 protons to a molecule), the mass of the ion will be 1000 times the m/z value of the ion. Assuming that this ion is detected with a unit resolution mass spectrometer, like the transmission quadrupole or the QIT, the m/z value assigned to the mass spectral peak multiplied by 1000 should be close to the *average mass* of the ion. To determine the mass of the analyte, the mass of the 1000 protons must be subtracted. However, what usually happens is that the mass of 1000 hydrogen atoms (a proton and an electron = 1.007825 Da) is subtracted, which can lead to an erroneous result. The mass of an electron at 0.55 mDa does not seem like much; however, 1000 electrons have a collective mass of 0.55 Da, which could be significant in the determination of the mass of the analyte. These issues must be considered when dealing with ions that are detected by mass spectrometers with high resolving power like the FT-ICR and the orbitrap.

1. Recognition of the Peak Representing the Molecular Ion in EI

The $M^{+\bullet}$ peak is expected at the high end of the spectrum. By convention, the nominal mass of the molecular ion is equal to the sum of the nominal masses of the constituent atoms. Depending on the type of compound, the molecular ion ($M^{+\bullet}$ or $M^{-\bullet}$) may represent only a small fraction of the total ion current. For example, the molecular ions of many aliphatic compounds are represented by peaks that correspond to only 1% or less of the total ion current, whereas the molecular ion of an aromatic compound might be represented by a peak corresponding to more than 25% of the total ion current. Compare the mass spectra of naphthalene and n-decane shown in Figure 5-14. Do not confuse the percentage of total ion current (relative ion current of ions of a specific m/z value divided by the sum of all exhibited ion currents) represented by a given mass spectral peak with the relative intensity of that peak (intensity of a peak at a specific m/z value reported as a percentage of the intensity of the base peak of the mass spectrum).

The $M^{+\bullet}$ peak will be at the highest m/z value in the mass spectrum that is neither an isotope peak nor a peak that represents a background ion. However, EI mass spectra can be frustrating in that the peak at the highest m/z value may not necessarily be the $M^{+\bullet}$ peak because some compounds do not produce a sufficiently stable $M^{+\bullet}$.

Once a candidate for the $M^{+\bullet}$ peak has been selected, its validity must be tested. The peak at the next lowest m/z value must represent a logical neutral loss from the molecular ion; the loss must consist of a grouping of atoms in the form of a radical or molecule that conforms to the nominal masses of and the valence rules of the constituent elements. This can be tested by use of the *Nitrogen Rule*. If the candidate $M^{+\bullet}$ peak represents an ion containing an odd number of nitrogen atoms, it will have an odd

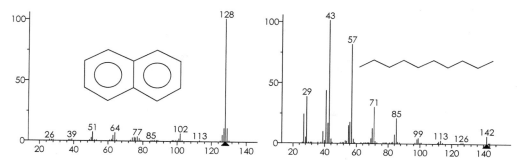

**Figure 5-14. EI mass spectra of naphthalene (left) and n-decane (right) with
M⁺• peaks at m/z 128 and 142, respectively.**

m/z value. If the peak at the next lower *m/z* value represents an ion that has retained the odd number of nitrogen atoms and it was produced by single-bond cleavage, then it must have an even value. If the molecular ion does not contain an odd number of nitrogen atoms, the *m/z* value of the M⁺• peak is even. If there are no atoms of nitrogen in the molecular ion, there can be no atoms of nitrogen in the fragment ions produced by that molecular ion. It is very important to keep in mind that the *Nitrogen Rule* relates to the nominal mass of the ion. When the mass of an ion is greater than about 500 Da, the nominal and the monoisotopic mass (the actual mass observed in the mass-to-charge ratio of the ion) can be significantly different (nominal mass will be less than the observed *m/z* value of the ion) due to the mass defect of hydrogen, as defined earlier.

> **Note.** No fragment ion will contain more atoms of an element than are present in the molecular ion.

A. Reasonable Losses from M⁺• in EI

Once a candidate for the molecular ion is selected, additional criteria involve simple arithmetic. Taking the nominal masses H = 1 Da, C = 12 Da, O = 16 Da, and so forth, the analyst contemplates combinations of elements that might be lost (the dark matter) from the molecular ion to generate the fragment ions represented in the spectrum. For example, the loss of 15 Da from the M⁺• is due to the loss of a methyl radical (•CH₃); M − 29 is due to loss of an ethyl radical (•CH₂CH₃) or a formal radical (HC•O); and M − 31 is due to loss of a methoxyl radical (•OCH₃). Losses such as M − 11 or M − 5 are not valid and would correspond to loss of an illogical group of atoms according to mass and valence rules. A list of reasonable losses of radical species of odd mass (if no nitrogen is included) and of valence-satisfied species (H₂O, methanol, etc.) is presented in a table in the back matter of this book.

2. Recognition of the Protonated Molecule (MH⁺) in Soft Ionization

The mass of a protonated molecule will be 1 dalton greater than the sum of the masses of the constituent atoms of the molecule from which it was produced. Soft ionization techniques (CI, APCI, FI, ESI, MALDI, etc.) can produce ions that have a mass greater than that of the protonated molecule and, in the case of multiple-charge ions, *m/z* values less than the mass of the ion. Unlike the M⁺• peak in an EI mass spectrum, the MH⁺ peak in a soft ionization mass spectrum may not be the peak at the highest *m/z* value;

this is because the analyte may also form adduct ions such as potassiated (MK^+) and sodiated (MNa^+) ions in the case of ESI, adducts with reagent gas ions like $C_2H_5^+$ and $C_3H_5^+$ in the case of methane CI, protonated adducts resulting from solvent/analyte clustering (like $[M + H_2O]H^+$) in ESI and APCI, and protonated multimers of analytes, like $(2M)H^+$, in ESI and APCI. Generally speaking, the soft ionization techniques of ESI and APCI in LC/MS, CI and field ionization in GC/MS, and MALDI and FAB used in the analysis of nonvolatile thermally labile compounds such as proteins do not produce significant ion current from fragmentation of the ion representing the intact molecule. However, due to adduct formation, multiple-charge states arising from different cationized species, and the formation of complexes with liquid chromatographic mobile-phase components, these spectra can be as complex as those produced by EI. In most cases, the peaks that represent these ions have high relative intensities, and it is fairly easy to recognize the peak for the MH^+ when it is accompanied by peaks 22 and 38 *m/z* units higher, representing MNa^+ and MK^+, respectively.

Note. Adduct ions and protonated species of analyte molecules clustered with mobile-phase components and/or other analyte molecules may contain more atoms of an element than does the protonated molecule.

The *Nitrogen Rule* also can be used to aid in establishing the peak that represents the protonated molecule. The *Nitrogen Rule* can be a powerful tool; its variations based on the ion type under consideration are very important ($M^{+\bullet}$, MH^+, or $[M - H]^{-\bullet}$). It is also important to obey the *nominal mass* part of the definition.

A. Probable Adducts Observed in the Mass Spectrum Produced by Soft Ionization

Some of the more common adducts that can be observed in spectra produced by the LC/MS techniques of ESI and APCI involve single and multiple molecules of water (H_2O; 18, 36, 54, 72 Da, etc.), acetonitrile (CH_3CN; 41 Da), ammonium (NH_4^+ or $[M + H + NH_3]^+$; 18 Da), acetic acid (H_3CCOOH; 60 Da), acetate ($H_3CCO_2^-$; 59 Da), formic acid ($HCOOH$; 46 Da), formate ($HCOO^-$; 45 Da), ethanol (H_3CCH_2OH; 46 Da), methanol (H_3COH; 32 Da), sodium (Na^+; 23 Da), potassium (K^+; 39 Da), and transition metal ions like Fe^+, Ag^+, Pt^+, and so forth. In order to discern whether isobaric species such as H_2O vs NH_4^+ or EtOH vs formic acid are a component of an adduct, the composition of the mobile phase will need to be known in detail. The problem is further exacerbated by the chance of adding multiple molecules or ions of various species to the analyte molecule.

3. Recognition of the Deprotonated Molecule ($[M - H]^-$) Peak in Soft Ionization

Soft ionization techniques can produce negative ions that have a mass less than that of the original analyte molecule. The mass spectrum produced by negative ions can be as complex as those involving protonated molecules. Adducts to anions of the analyte are represented by peaks at *m/z* values higher than those for the deprotonated molecule.

VI. Recognition of Spurious Peaks in the Mass Spectrum

Not all of the peaks in a mass spectrum will provide information about the analyte. Some peaks are from components of the sample other than the analytes. These can be due to ionization of contaminants in the chemical system introduced into the mass spectrometer, to the instrument itself, to functionality (or malfunction) of the instrument, or to coeluting components in the case of analyses involving chromatography. It is important to recognize or identify these peaks so that they may be excluded from the interpretation process. The best way to determine whether certain mass spectral peaks correspond to the same compound is through the use of mass chromatograms or mass chronograms.

When a plot of mass spectral peak intensity vs spectrum number for the *m/z* values of the mass spectral peaks in question (mass chromatograms) shows profiles (chromatographic peaks) that are superimposable, it provides evidence that the mass spectral peaks correspond to the same compound (or a single chromatographic component). However, if the plots of specified mass spectral peak intensities (mass chromatograms) show profiles that are not superimposable, the mass spectral ion currents (peaks) correspond to different compounds. Further, if the mass chromatograms show no peak profile (a flat profile), the mass spectral ion currents (peaks) correspond to constant background.

1. Noise Spikes

Noise spikes are characterized by the absence of isotope peaks. If a peak is suspected to be a noise spike, examination of the spectra acquired just before and just after the spectrum containing the suspicious peak usually will be devoid of this peak, confirming the identity of the suspicious peak as a noise spike. Noise spikes result from a neutral striking the ion detection device at the time an ion of a specific *m/z* value should be striking the device to produce a signal. At the next time interval (corresponding to the next *m/z* value) the neutrals probably will not be present; therefore, there will be no isotope peak associated with a noise spike. Noise spikes can also enhance the intensity of an isotope peak if the neutral strikes the detector at the same time as the isotope-containing ion strikes the detector. Noise spikes can have a wide variation in absolute intensities from small and almost unnoticeable to being several orders of magnitude greater than the peak representing the most abundant ion formed by the analyte. Noise spikes can appear in mass spectra obtained by all types of *m/z* analyzers and with all types of ionization techniques. Noise spikes are fairly uncommon when using modern instrumentation.

2. Peaks Corresponding to Contaminants in GC/MS and LC/MS

There are certain *m/z* values for peaks that should immediately raise suspicion when such data are acquired during GC/MS or LC/MS. These have been detailed in Chapters 10 and 11 on these two techniques under the heading of Common Contaminants and should be reviewed carefully before attempting to interpret a mass spectrum.

A. The Phthalate Ion Peak

As was pointed out during descriptions of using isotope peak-intensity data to determine the elemental composition of the corresponding ion, phthalate plasticizers (esters of phthalic acid) are common contaminants producing an ion of *m/z* 149 during an analysis by electron ionization using GC/MS and by ESI using LC/MS. Such a peak representing the same ion can be found in mass spectra obtained by all ionization techniques.

A peak at *m/z* 149 in mass spectra should always be considered with suspicion. The mass spectra of approximately 3000 of the 147K compounds in the NIST02 NIST/EPA/NIH Mass Spectral Database exhibit a peak at *m/z* 149 having a relative intensity of between 10 to 100% and have a $M^{+\bullet}$ peak with an intensity between 1 and 100%. There are 167 compounds in the NIST02 Database with the name phthalate. Of these, about half (81) have a mass spectrum that exhibits a peak at *m/z* 149 with a relative intensity between 10 and 100%; about two-thirds of these (60) are common plasticizers.

B. GC Column Bleed

There are two characteristic ions resulting from the bleed of the degraded GC column stationary phase represented by peaks at *m/z* 207 and 281 often observed in mass spectra obtained using GC/MS. These peaks are often identifiable by their intense X+1 and X+2 isotope peaks resulting from the presence of multiple atoms of silicon. Less-intense peaks at *m/z* 355, 429, etc. may also be detected, representing heavier fragments of truncated siloxane polymers that differ by 74 Da. More details on column bleed, other peaks resulting from it, and procedures to confirm the presence of column bleed are covered in Chapter 10.

C. Cluster Ions

One of the greatest potential inferences in LC/MS is formation of solvent clusters. These solvent clusters can be a problem when using either positive- or negative-ion detection. Solvent clusters are more of a problem in ESI than in APCI or APPI; however, they can be represented in spectra obtained by all the atmospheric pressure ionization techniques. Chapter 11 provides greater detail on these clusters and how to deal with them.

VII. Obtaining Structural Information from the Mass Spectrum

It may be possible to determine an elemental composition and/or the mass of an analyte from an EI mass spectrum. For the gas-phase analytes amenable to EI, it may be necessary to obtain another mass spectrum by some other technique such as chemical or field ionization to confirm (or establish) the mass and/or possibly the elemental composition. Both nominal and high resolving power data may be required by multiple techniques to establish an elemental composition and/or mass. Mass and elemental composition of the analyte do not ensure the establishment of the analyte's identity (structure).

Soft ionization techniques most often employed in LC/MS will not yield much structural information, and it may be necessary to "dig" into the mass spectrum to obtain the mass. More than the mass is necessary to ascertain the elemental composition of an ion. Although consideration of isotope peak-intensity ratios can provide a good approximation of the elemental composition of an ion, there are limitations to this procedure, as illustrated in this chapter. Often, the use of accurate mass measurement is useful in determining the elemental composition of an ion, but even this technique may not identify a specific elemental composition, especially when the ion has a mass greater than 500 Da [8].

Once an elemental composition and a mass have been determined, the task of *identifying* the analyte may still remain. The use of the *Formula Search* in the NIST Mass Spectral Search Program as an aid to an identification has already been illustrated. This will not always lead to *the answer*. The next step is to look at what the fragmentation reveals. In cases of little fragmentation, there may not be much information available.

This means that it will be necessary to promote fragmentation through a secondary process such as in-source CAD or CAD MS/MS. Care should be taken when embarking on such a venture. Often in CAD MS/MS, the selected precursor is the monoisotopic ion, meaning that it does not contain heavy isotopes. Because of no heavy isotopes, it will not be possible to monitor the fate of a heteroatom such as a chlorine, bromine, sulfur, or silicon because it will no longer be possible to use isotope peak intensities to recognize which fragment ions contain the tell-tale element.

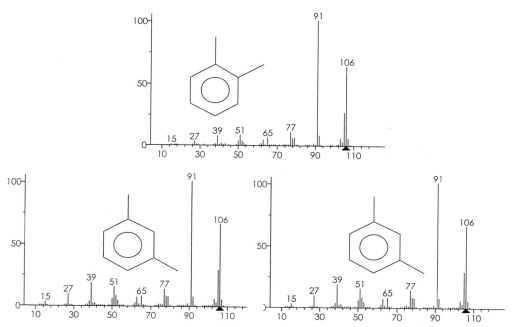

Figure 5-15. EI mass spectra of the three xylene isomers.

Even though a high degree of confidence might be developed as to the identity of an analyte, the only way to be sure is to obtain an authentic sample of the compound believed to be the analyte and determine whether it can produce the same mass spectral data as the unknown. Even this may not confirm the identity of the analyte. Not all compounds can be identified by mass spectrometry alone. A good example is seen with the EI mass spectra of the three regioisomers of xylene (*o*-, *m*-, and *p*-). These spectra, shown in Figure 5-15, are not distinguishable from one another. The only way these compounds can be identified is through a combination of mass spectrometry (to determine that they are xylenes) and gas chromatography, which will distinguish them according to their different retention times.

> **Note.** Mass spectrometry can usually be used to determine the mass of an organic analyte; however, structures and elemental composition can only be determined for some of the analytes.

Another fallacy relating to mass spectrometry is the belief that a molecule always can be broken apart. Looking at the mass spectra of six compounds in the NIST05 Mass Spectral Database that have the elemental composition of $C_{14}H_{10}$, five of the six have fairly similar structures (Figure 5-16). Five of the six mass spectra have about 80–90% of the ion current associated with the molecular ion and its isotopes, and the fragments are formed by the loss of either a hydrogen radical or a hydrogen molecule from the molecular ion. The spectra of all six compounds have a peak that represents the loss of a molecule of acetylene from the molecular ion (m/z 152). The intensity of the peak representing the ion resulting from this loss of acetylene is about 10% of the base peak, which is also the $M^{+\bullet}$ peak (m/z 178) in all the spectra, except for the spectrum of 6β,8α-dihydrocyclobut[α]acenaphthylene, which shows a peak at m/z 152 that is ~85% of the base peak. The limited fragmentation of these compounds is due to resonance stabilization of the molecules. These spectra were generated using 70 eV electrons. Any attempt to cause further fragmentation of these molecular ions using CAD will not be met with success because it is unlikely that more energy will be deposited in the molecules from CAD than from 70 eV electrons. It is also true that not all protonated molecules produce fragments when subjected to collisional activation.

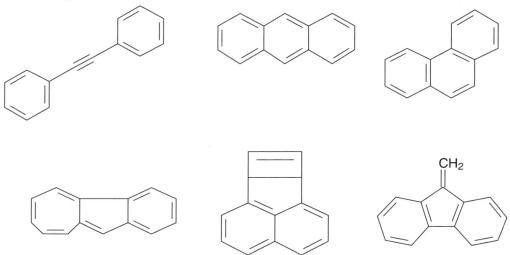

Figure 5-16.　　**Six structures in the NIST05 (2005 version of the NIST/EPA/NIH Mass Spectral Database) having the elemental composition $C_{14}H_{10}$.**

It can be said that mass spectrometry produces more information about the structure and composition of analyte from less sample than any other technique. However, it should be remembered that mass spectrometry has limitations; to try to push past these limitations can result in serious mistakes. Especially over the past 15 years, the field of mass spectrometry has grown exponentially; it has touched and been influenced by nearly every area of physical and biological science. With emerging techniques such as imaging in mass spectrometry, there is no reason to think that limits to the information obtainable from this technique have been reached.

> **Note.** Do not abuse mass spectral data. Use mass spectral data to glean as much information as possible, but do not overinterpret the data.

This chapter has concentrated on strategies for the general interpretation of a mass spectrum excluding features relating to fragmentation. However, the issue of the types of fragmentation processes is worth some commentary. Some analysts have the opinion that the strategies used in the evaluation of fragmentation data from EI (GC/MS) are not applicable to the fragmentation obtained by collisionally activated dissociation (CAD) of ions produced in LC/MS (usually by ESI or APCI). Although it is true that much of the fragmentation of the molecular ions (OE ions) in EI is initiated by a radical site, there are cases in which some of these OE ions do undergo heterolytic cleavage as driven by a charge site. Furthermore, the secondary fragmentation of EE fragment ions produced during primary fragmentation of OE ions in EI is sometimes analogous to the fragmentation or CAD of the EE ions produced with LC/MS ionization techniques. The fragmentation of EE ions in both cases is often dominated by processes involving the loss of a molecule, and these processes are charge-site-driven, often through rearrangements. In general, the "Even-Electron Rule" prevails: an EE ion tends to fragment to produce other EE ions, whereas OE ions tend to fragment to produce predominately EE ions and occasionally other OE ions.

Beyond the energetic differences in the fundamental processes involved in EI and non-EI techniques, the latter often used in conjunction with CAD, many of the differences in the appearance of the fragmentation patterns produced by these two quite different categories of analysis are likely to be due to the differences in the types of molecular structure being analyzed. Most of the analytes amenable to EI have fewer heteroatoms, are less polar, and might be referred to as being "less complex" than those analyzed by non-EI techniques in combination with CAD MS/MS. However, an understanding of the fragmentations exhibited by selected EE-fragment ions in EI (e.g., the hydride-shift rearrangement) can provide valuable clues to understanding the CAD and/or metastable fragmentation of protonated molecules, the principal EE ions of more than one type of non-EI technique.

References

1. Budzikiewicz H and Grigsby RD, Mass spectrometry and isotopes: A century of research and discussion. *Mass Spectrometry Reviews* **25**(1): 146-157, 2006.
2. Beynon J, *Mass Spectrometry and its Applications to Organic Chemistry*. Elsevier, Amsterdam, 1960.
3. Yergy J, Heller D, Hansen G, Cotter RJ and Fenselau C, Isotopic Distributions in MS of Large Molecules. *Anal Chem* **55**: 353-356, 1983.
4. McLafferty FW, Billionfold data increase from mass spectrometry instrumentation. *Journal of the American Society for Mass Spectrometry* **8**(1): 1-7, 1997.
5. Beynon J and Williams A, *Mass and Abundance Tables for Use in Mass Spectrometry*. Elsevier, NYC, NY, 1963.
6. Watson JT and Biemann K, High-resolution MS with GC. *Anal. Chem.* **36**: 1135-1137, 1964.
7. Editor, Guidelines for Authors. *J. Org. Chem.* **58**: 7A-12A, 1993.
8. Gross M, Accurate Masses for Structure Confirmation, Editorial. *J. Am. Soc. Mass Spectrom.* **5**: 57, 1994.
9. Vanferford B, Pearson R, Cody R, Rexing D and Synder S, "Determination of an Unknown System Contaminant Using LC/MS/MS". In: *Liquid Chromatography/Mass Spectrometry, MS/MS and Time-of-Flight MS: Analysis of Emerging Contaminants* (Eds. Frre I and Thurman E). American Chemical Society, Washington, DC, 2003.

Suggested Further Reading

1. Gross, JH *Mass Spectrometry: A Textbook*; Springer-Verlag, Heidelberg, Germany, 2004, ISBN:3540407391 reviewed *JASMS* 16:793)

2. Smith, RM *Understanding Mass Spectra: A Basic Approach*, 2nd ed.; Wiley: Hoboken, NJ, 2004, ISBN:047142949X reviewed *JASMS* 16:792). Smith, RM *Understanding Mass Spectra: A Basic Approach*; Busch, KL, Tech. Ed.; Wiley: New York, 1999, ISBN:0471297046 reviewed *JASMS* 11:664).

3. Snyder, AP *Interpreting Protein Mass Spectra: A Comprehensive Resource*; Oxford: New York, 2000, ISBN:0842135716 (reviewed *JASMS* 13:107).

4. Shrader, SR *Introductory Mass Spectrometry*, 2nd ed.; Shrader Laboratories: Detroit, MI, 1999 (originally published by Allyn and Bacon: Boston, MA, 1971).

5. Barker, J *Mass Spectrometry: Analytical Chemistry by Open Learning*, 2nd ed.; Ando, DJ, Ed.; Wiley: Chichester, UK, 1999, ISBN:0471967645. (Davis, R; Frearson, MJ, 1st ed., 1987, ISBN: 0471913898; reviewed *JASMS* 4:831).

6. Lee, TA *A Beginner's Guide to Mass Spectral Interpretation*; Wiley: Chichester, UK, 1998, ISBN:047197628 (hard) 0471976296 (paper) (reviewed *JASMS* 9:852).

7. Splitter, JS; Tureček, F, Eds. *Applications of Mass Spectrometry to Organic Stereochemistry*; VCH: New York, 1994, ISBN:089573303X (reviewed *JASMS* 6:152).

8. McLafferty, FW; Tureček, F *Interpretation of Mass Spectra*, 4th ed.; University Science: Mill Valley, CA, 1993, ISBN:0935702253 (reviewed *JASMS* 5:949).

9. ЭАИКИН, ВГ; МИКАЯ, АИ *ХИМИЧЕСКИЕ МЕТОДЫ В МАСС- СПЕКТРОМЕТРИИ ОРГАНИЧЕСКИХ СОЕДИНЕНИЙ*; Издательство "Наука": Moscow, 1987.

10. ТЕРЕНТЬЕВ, ПБ; СТАНКЯВИЧЮС, АП Масс-спектрометрический анализ биологически активных азотистых оснований; Иэдательство: Мокслас, 1987.

11. ВУЛьФСОН, НС; ЭАИКИН, ВГ; МИКАЯ, АИ *МАСС-СПЕКТРОМЕТРИЯ ОРГАНИЧЕСКИХ СОЕДИНЕНИЙ*; Издательство "Наука": Moscow, 1986.

12. Porter, QN *Mass Spectrometry of Heterocyclic Compounds*, 2nd ed.; Wiley–Interscience: New York, 1985.

13. McLafferty, FW; Venkataraghavan, R *Mass Spectral Correlations*, 2nd ed.; American Chemical Society: Washington, DC, 1982.

14. Sklarz, B, Ed. *Mass Spectrometry of Natural Products*, plenary lecturers presented at the International Mass Spectrometry Symposium on Natural Products, Rehovot, Israel, 28 August—2 September 1977; Pergamon: Oxford, UK, 1978.

15. Levsen, K *Fundamental Aspects of Organic Mass Spectrometry*; Verlag Chemie: Weinheim, Germany, 1978.

16. McLafferty, FW *Interpretation of Mass Spectra*, 2nd ed.; Benjamin: Reading, MA, 1973.

17. Hamming, MG; Foster, NG *Interpretation of Mass Spectra of Organic Compounds*; Academic: New York, 1972.

18. Hill, HC *Introduction to Mass Spectrometry*, 2nd ed.; Heyden: London, 1972 (1st ed., 1966; 2nd printing 1969; 2nd ed., revised by AG Loudon; 1st ed. translated into German, Italian, and Japanese; Hilson C. Hill died in 1967).

19. Seibl, J *Massenspektrometrie*; Akademische Verlagsgesellschaft: Frankfurt, Germany, 1970.

20. Beynon, JH; Saunders, RA; Williams, AE *The Mass Spectra of Organic Molecules*; Elsevier: Amsterdam, 1968.

21. Polyakova, AA; Khmel'nitskii, RA (Schmorak, J, Translator) *Introduction to Mass Spectrometry of Organic Compounds*; Israel Program For ScientificTranslations: Jerusalem, Israel, 1968 (original Russian language edition, *Vvedenie V Mass Spektrometriyu Organicheskikh Soedinenii*, Izdatel'stvo "Khimya," Moskva-Leningrad, 1966).

22. Budzikiewicz, H; Djerassi, C; Williams, DH *Mass Spectrometry of Organic Compounds*; Holden-Day: San Francisco, CA, 1967.Reed, RI *Applications of Mass Spectrometry to Organic Chemistry*; Academic: New York, 1966.

23. Spiteller, G *Massenspektrometrische Strukturanalyse Organischer Verbindungen*; Verlag Chemie: Weinheim, Germany, 1966.

24. Quayle, A; Reed, RI Interpretation of Mass Spectra. In *Interpretation of Organic Spectra*; Mathieson, DW, Ed.; Academic: New York, 1965.

25. Budzikiewicz, H; Djerassi, C; Williams, DH *Structure Elucidation of Natural Products by Mass Spectrometry*, Vol. I *Alkaloids*; Vol. II *Steroids, Terpenoids, Sugars, and Miscellaneous Natural Products*; Holden-Day: San Francisco, CA, 1964.

26. Budzikiewicz, H; Djerassi, C; Williams, DH *Interpretation of Mass Spectra of Organic Compounds*; Holden-Day: San Francisco, CA, 1964.

27. McLafferty, FW, Ed. *Mass Spectra of Organic Ions*; Academic: New York, 1963.

28. McLafferty, FW Mass Spectrometry. In *Determination of Organic Structures by Physical Methods*, Vol. II; Nachod, FC; Phillips, WD, Eds.; Academic: New York, 1962.

"Learning how to identify a simple molecule from its electron-ionization mass spectrum is much easier than from any other types of spectra."

~Fred W. McLafferty

Chapter 6   Electron Ionization

Introduction to Mass Spectrometry, 4th Edition: Instrumentation, Applications, and Strategies for Data Interpretation; J.T. Watson and O.D. Sparkman, © 2007, John Wiley & Sons, Ltd

Conventions Used in This Chapter and Throughout the Book

This is a typical graphical representation of a molecule, *n*-propanol:

Structure A

The lines between the symbols for each of the elements represent pairs of electrons shared by the two adjacent atoms. The two pairs of nonbonding electrons on the oxygen atom are represented by the two pairs of dots above and below its symbol; similar dots will be used to represent nonbonding electrons on other heteroatoms.

The *n*-propanol molecule can also be represented by this type of structure:

H₃C—CH₂—H₂C—O̤H

Structure B

The *n*-propanol molecule can also be represented by this type of structure:

Structure C

In a given formula, ions are designated as odd electron (OE) using the sign of the charge (+ or −) and a dot (•) for the unpaired electron; or as even electron using only the sign of the charge (+ or −). M⁺• is the symbol for a molecular ion.

A single-barbed arrow is used to indicate the movement of a single electron:

A double-barbed arrow is used to indicate the movement of a pair of electrons:

Sigma-bond ionization is indicated using the following symbolism: **C+•C** or **C•+C**.

The △ or ▲ symbol on the abscissa of a mass spectrum indicates the position of the molecular ion peak. The △ indicates that a molecular ion peak is not present. The ▲ indicates that an intensity value is in the mass spectrum for the molecular ion (although it may be so low that the molecular ion peak may not be observable in the graphic display due to the display resolution).

I. Introduction

Still the most widely used technique in mass spectrometry, electron ionization (EI) produces molecular ions from gas-phase analytes. These molecular ions then fragment in a reproducible way, which results in a "fingerprint" of the analyte. Because of the uniqueness of these "chemical fingerprints", commercially available libraries containing hundreds of thousands of standard EI mass spectra can be used to facilitate identification of unknown compounds.

At one time the abbreviation for EI meant "electron impact". This name gives an incorrect impression of the fundamental processes involved in this important ionization technique. As was pointed out by Ken Busch [1], if an ionizing electron (70 eV) actually collided with the nucleus of one of the atoms that compose the analyte molecule, the gain in internal energy by the molecule would be infinitesimal. The maximal fraction of translational energy (70 eV, in this case) that can be converted into internal energy of the molecule during an inelastic collision with an electron is related to a ratio of the masses of the colliding bodies via the "center-of-mass" formula; because of the disparity in mass between an electron (~1/2000 the mass of a proton) and any common atom in a molecule, such a fraction would be vanishing small. Furthermore, from the perspective of providing cross-sectional area for a target, the molecule is composed of mostly free space between the nuclei of its atoms, meaning that the probability of a "head-on" collision between an ionizing electron and one of the molecule's nuclei or one of its electrons is very small. Therefore, the analogy of a cue ball hitting a billiard ball (an elastic collision) is inappropriate for the excitation of a gas-phase molecule leading to ionization, and the term "electron impact" is no longer used.

Energy sufficient for ionizing and fragmenting gas-phase analyte molecules is acquired by interaction with 70 eV electrons produced from a hot filament and accelerated through a 70-V electric field. The ionization chamber is maintained at low pressure (~10^{-4} Pa). The pressure in the ion volume (1 cm^3, high gas conductance chamber) of the EI source is ~10^{-1} Pa, depending on the method of sample introduction (via GC, batch inlet, or direct insertion probe), to minimize ion/molecule collisions.

Some structural features of the analyte molecule can be deduced from the fragmentation pattern of the molecular ion ($M^{+\bullet}$). Knowledge gained from organic-solution chemistry concerning electron shifts together with a consideration of the relative stabilities of chemical bonds, odd- and even-electron ions, and radicals (neutral species that have an odd number of electrons) are used to suggest decomposition mechanisms of molecular ions *in vacuo* to rationalize observed fragmentation pathways.

II. Ionization Process

The source of electrons in nearly all EI mass spectrometers (regardless of the type of *m/z* analyzer) is a thin ribbon or filament of metal that is heated electrically to incandescence or to a temperature at which it emits free electrons (Figure 6-1). The emitted electrons are attracted to an anode (trap) situated on the opposite side of the ionization chamber from the filament or cathode. The electrons are forced by electric fields through the central space of the ionization chamber as a beam collimated by a slit near the filament. A superimposed magnetic field causes the electrons to move in a tight helical path, which increases the path length, and thus the probability of interaction with a molecule as they transverse the 15–20 mm gap between the slit and the anode.

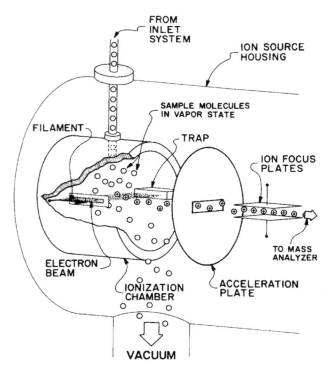

Figure 6-1. Schematic representation of an electron ionization source.

There are three indicators of direct current that can be monitored during operation of the ion source as illustrated conceptually in Figure 6-2. These indicators are:

1. The electrical heating current through the filament, which results in thermal emission of electrons; this will vary depending on the composition and condition of the filament, but will usually be on the order of 3 to 4 amperes.

2. The total-emission current, which provides an index of the total number of electrons emitted by the filament.

3. The trap or target current, which is the portion of the emitted electron current that flows directly from the filament across the central space of the ionization chamber to the anode. The trap current (that available for ionization) is smaller than the total emission of free electrons because some collide with the filament slit and housing. The magnitude of the trap current is usually on the order of 100 µA.

The energy of the electrons is controlled by the potential difference between the filament and the source block.

Other conditions being constant, the efficiency of ionization of molecules by the electron beam increases rapidly with electron energy from 10 eV to approximately 20 eV for most organic compounds (see Figure 6-3), which brackets the range of the ionization energy of most organic compounds. The energy of the electron beam is reported in electron volts; 1 eV is the energy gained (equivalent to 23 kcal mole^{-1}) by an electron in traversing an electric field maintained by a potential difference of 1 V. Most reference

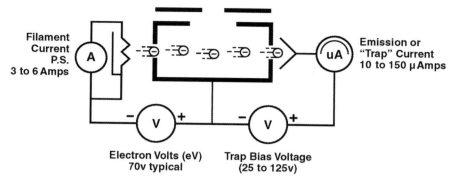

Figure 6-2. *Electrical circuit schematic of filament and trap in an EI source.*

spectra are obtained at 70 eV because at that level the perturbations in electron energy have negligible effects on ion production (Figure 6-3) and because reproducible fragmentation patterns are obtained that can be compared with thousands of reference spectra that also were acquired at 70 eV. The linear part of the curve in Figure 6-3 can be extrapolated to the abscissa to obtain the appearance potential for ions [2].

The efficiency of EI is also related to the degree of interaction between sample molecules and energetic electrons. This interaction can be effectively increased by (a) increasing the trap current, (b) increasing the cross section of the electron beam, or (c) increasing the sample pressure in the ionization chamber. However, each of these reasonable suggestions has practical limitations. Most often, the sensitivity or efficiency of ionization is increased by increasing the trap current. The limitation in this case relates

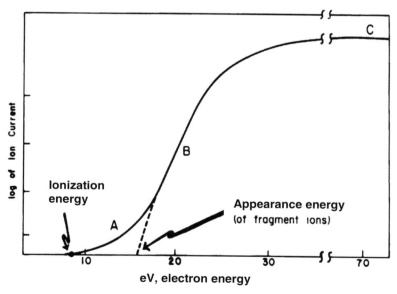

Figure 6-3. *Relationship between ion production and energy (electron volts) of ionizing electrons: A, threshold region, principally molecular ions produced; B, production of fragment ions becomes important; C, routine operation, mostly fragment ions.*

to the lifetime of the filament; usually, increasing the filament current above recommended levels to double the sensitivity will more than halve the filament lifetime. The other two suggestions for increasing ionization efficiency are limited by factors that may degrade the resolving power of the mass spectrometer. Some designs of ion optics are based on the formation of ions at a point source; in such designs, the dimensions of the ionizing medium (cross section of the electron beam) should be kept to a practical minimum. Finally, although an increase in sample pressure (short of that causing chemical ionization) in the ionization chamber would increase ionization efficiency, the accompanying increase in pressure in the analyzer would lead to a reduction in resolution due to band-broadening resulting from increased ion/molecule collisions and changes in fragmentation patterns due to collisional activation and subsequent dissociation of ions formed in the source. This approach is practical only with an instrument that has a powerful (250 L sec^{-1} or greater) pumping system or is differentially pumped using two high-vacuum pumps or a split-flow pump, as explained in Chapter 2 on Vacuum Systems.

The pressure in the ion-source housing is maintained at approximately 10^{-4} Pa so that the mean free path of the ions is approximately equal to the distance from the ion source to the detector. This relationship gives a satisfactory margin of protection against ion/molecule interactions in instruments that have lesser vacuum systems designed for low or unit resolution. If high-resolution data are required, the pressure in the analyzer must be lower than that reasonably achieved in the ion source; this can be accomplished most effectively with differential pumping of the ion source and analyzer (see Chapter 2). Alternatively, a differentially pumped instrument also provides the opportunity to increase sensitivity in special cases by increasing the sample pressure in the ion source by an order of magnitude over that permissible on a singly pumped instrument.

Regardless of the mode of sample introduction, the sample vapors should be channeled into the ion volume of the ionization chamber. This arrangement leads to maximum sensitivity because all of the sample must pass through the ionization chamber before it expands into the relatively large volume of the ion-source housing and escapes through the vacuum system.

EI might be considered something of an enigma. It is a process that, based upon close examination, appears not to work. Only about 0.01–0.001% of the analyte molecules that enter the ion volume will be ionized. The remaining unionized molecules are pumped away; i.e., 99.99% of the sample is wasted! However, even with this dismal ionization efficiency, a sample of 10 pg (some modern instruments claim even smaller amounts) of octafluoronaphthalene injected into a GC will yield a signal in a transmission quadrupole scanned from m/z 50 to m/z 300 at a rate of 3 scans sec^{-1} that can be quantitated using a mass chromatographic peak with a signal-to-background of >10:1 for the molecular ion. This quantity of material is sufficient to produce a full-scan mass spectrum that can be matched against the standard spectrum in the NIST/EPA/NIH Mass Spectral Database with a match factor of >90%. For this and other reasons, it has long been said, "More information about the structure of an analyte can be obtained from less sample by using mass spectrometry than by any other technique."

III. Strategy for Data Interpretation

1. Assumptions

When presented with a spectrum of peaks, the investigator assumes that all ions represented by the peaks are derived from a common source, namely the molecular ion of the compound of interest. This assumption may not be correct, as demonstrated by consideration of several criteria to be described shortly. Ions that may derive from contaminants should be identified and disregarded. Ions that result from traces of solvent are examples of those most easily recognized and disregarded. Elucidation of the structure of unknown compounds from their mass spectra should involve consideration of the structural significance of all ions of high mass or prominent abundance.

For those mass spectra that will be used for "fingerprinting" or comparison with reference spectra (mass spectral libraries), the interpreter will assume that the operator has obtained the mass spectra under conditions that reflect the true fragmentation pattern, not those that may distort peak intensities in the overall mass spectrum, such as scanning the *m/z* range of the mass spectrometer during GC elution when the sample concentration in the ion source may be changing rapidly during the scan (see Chapter 10).

During attempts to rationalize the losses observed from the molecular ion, it is generally assumed that the odd electron and positive charge are localized on a heteroatom or at the site of a double bond, if either of these is present. This assumption provides a rational starting point from which to consider various pathways of fragmentation of the molecular ion. The thermochemical basis for assuming that the odd electron in a molecular ion most likely resides on a heteroatom or at the site of a double bond is based on the fact that the ionization potential for an electron in a nonbonding orbital or a π bond is lower than for those in other orbitals (e.g., in a sigma bond), as explained in the following section.

2. The Ionization Process

A neutral molecule absorbs energy during the interaction of the ionizing electron with its electron cloud. As shown in Scheme 6-1, the analyte molecule interacts with an electron having 70 eV of kinetic energy, and through the interaction, absorbs approximately 14 eV as internal energy, which quickly causes the ejection of one of the electrons from the analyte (a thermal electron with <1 eV energy) and a residual energetic electron now having 56 eV of kinetic energy. Whereas some analyte molecules do absorb approximately 14 eV of energy from this interaction, there is actually a wide distribution of energies absorbed, such that a few molecular ions may absorb 30 eV, others only 9 eV of energy, etc. The greater the amount of energy absorbed from the ionizing electron, the greater the tendency for the nascent molecular ion to decompose into fragment ions. In principle, the 56-eV electron shown at the right in Scheme 6-1 can interact with yet another molecule of the analyte effecting ionization.

$$M + e^-_{\text{70 ev}} \longrightarrow M^+ + e^-_{\text{56 ev}} + e^-_{\text{thermal}}$$

Scheme 6-1

Because the analyte molecule has many electrons, it is reasonable to ask, "Which electron leaves during the ionization process?" Whereas sufficient energy may have been absorbed by the analyte molecule to eject any one of its electrons, the most probable ejection will be of an electron that is least tightly bound. Therefore, it is predictable that the most likely site of ionization will be where electrons are loosely bound; e.g., in the nonbonding orbital of a heteroatom.

Consideration of the ionization potential and electronic configuration of a few simple molecules provides some useful insight into the phenomenon of ionization. Imagine that you have a very basic EI mass spectrometer in which you can control the energy of the bombarding electrons so that you can measure the minimum energy necessary to cause ion current to be generated in the ion source, pass through the *m/z* analyzer, and strike the detector. This basic instrument would be equipped with a reservoir inlet system into which you could place the gaseous analyte, allowing a small quantity to continuously leak into the ion source while you increase the energy of the electrons in an effort to prepare a graph similar to that in Figure 6-3 to determine the minimum energy necessary to ionize the compound. Now place ethane in the reservoir and increase the energy of the ionizing electrons until the ionization potential is reached. For ethane, this will require approximately 11.5 eV to produce molecular ions, which would then be accelerated through the mass spectrometer to the detector.

In using the basic mass spectrometer to assess the ionization potential for each of the compounds listed in Table 6-1, it would be observed that only 10.5 eV is required to generate ion current from ethene rather than 11.5 eV for ethane [3]. Examination of the electronic configuration of these two compounds reveals that ethane consists of seven σ bonds, whereas ethene only has five σ bonds, but it also has two overlapping *p* orbitals to give a π-bond orbital. Because the principal difference between these two compounds is the presence of the π-bond orbital in ethene, it must be that an electron is easier to remove from a π-bond orbital than from a σ-bond orbital.

Table 6-1.   Ionization energy for selected compounds.

Compound	Ionization energy* in electron volts (eV)
$H_3C–NH–CH_3$	8.23
$H_3C–CH_2–NH_2$	8.86
$H_3C–O–CH_3$	10.03
$H_2C=CH_2$	10.51
$H_3C–CH_3$	11.52

*Data taken from Lias SG, Bartmess JE, Liebman JF, Holmes JL, Levin RD, and Mallard WG, Gas-phase ion and neutral thermochemistry, J. Phys. Chem. Ref. Data, 1988, 17(Suppl. 1).

Carrying out the experiment further, it will be found that only ~10 eV are required to cause ionization of dimethyl ether. An examination of the electronic configuration of dimethyl ether shows that it consists of eight σ bonds and two nonbonding orbitals. The implication from the experimental results so far is that it is easier to remove an electron from one of the nonbonding orbitals of oxygen than from the π-bond orbital in ethene or from a σ-bond orbital of ethane.

Continuing further with the analysis of ethylamine, it will be noted that it requires only approximately 8.9 eV to generate detectable ion current. The electronic configuration of ethylamine consists of nine σ-bond orbitals and one nonbonding orbital. Once again, the implication is that it requires less energy to remove an electron from a nonbonding orbital than from a π-bond orbital or from a σ-bond orbital, but, in addition, there is a disparity between the nonbonding orbital of nitrogen and that of oxygen. This can be explained by their relative positions in the periodic table, where oxygen is farther to the right, indicating a greater electronegativity, which is consistent with its greater reluctance to give up an electron from its nonbonding orbital relative to nitrogen relinquishing one of its nonbonding electrons.

Although it is certainly possible to generate a $M^{+\bullet}$ from ethene, dimethyl ether, or ethylamine by removing an electron from one of its many σ-bond orbitals, this is a more expensive process in terms of the amount of energy required than is removing an electron from either a π-bond orbital or a nonbonding orbital. Therefore, the probability of the electronic configuration of the $M^{+\bullet}$ of ethene, dimethyl ether, or ethylamine being in the form of a one-electron σ bond is much less than that for one in which the odd electron resides in a π-bond orbital or in a nonbonding orbital, respectively. It is important to remember that although there is a "most probable site" for the charge and for the odd electron in the $M^{+\bullet}$ of a specific analyte, in reality there can be a wide variety and distribution of molecular ions differing in charge-radical sites, especially in analytes containing a variety of heteroatoms, π-bond orbitals, and so forth.

When considering the fragmentation pattern of dimethyl ether or that of ethylamine, the fragmentation schemes in which the radical site or odd electron is on the heteroatom, namely oxygen or nitrogen, should dominate. In the case of the $M^{+\bullet}$ for ethene, it should be assumed that the odd electron resides in between the two carbon atoms. In the case of ethane, consideration of the electron being removed from any one of the seven σ-bond orbitals would be a good starting point for the fragmentation process.

The location of the charge and radical sites in a $M^{+\bullet}$ assumes that the molecule has a known structure. How can this knowledge of the charge-radical site be of benefit in determining the structure of an unknown? By understanding the way in which molecular ions of known compounds fragment, it is possible to deduce the structure of an unknown analyte from the presence of peaks in the mass spectrum and from the dark matter represented by the numerical difference between peaks representing fragment ions and molecular ions. Methods for interpreting a mass spectrum are provided in great detail in Chapter 5. This chapter describes the overall appearance of the EI mass spectra and mechanisms by which molecular ions fragment; these features assist in the deduction of what structural moieties are present in the corresponding analyte molecule.

As explained in Chapter 5, the first step in the interpretation of a mass spectrum is to determine whether a $M^{+\bullet}$ peak is present in the mass spectrum. In many cases, the elemental composition of the $M^{+\bullet}$ can be determined from the isotope peaks associated with it or from its accurately determined mass. The *Nitrogen Rule*, used in establishing the validity of a peak as representing the $M^{+\bullet}$, also provides some information about the elemental composition of the analyte. The elemental composition provides information as to whether heteroatoms are present and the possible number of π-bond orbitals that may be in the $M^{+\bullet}$ (as explained in Chapter 5). If no $M^{+\bullet}$ peak is present, it is still possible to identify the analyte. This can be done through an examination of other peaks in the mass spectrum or through the use of a different ionization technique such as

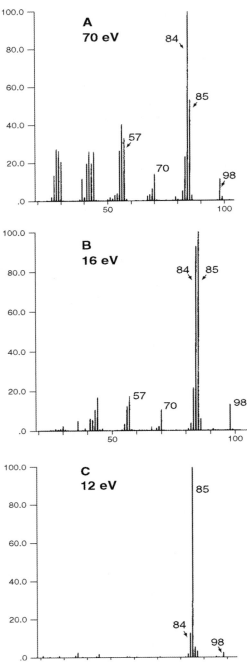

Figure 6-4. Three mass spectra illustrating the effect of ionization potential on the relative ion current of a M$^{+\bullet}$ and the fragment ions.

chemical ionization, which is designed to establish the analyte's nominal mass.

In some types of EI mass spectrometers, the M$^{+\bullet}$ peak candidate can be confirmed by low-electron-energy ionization. As the effective energy of the electron beam is reduced to a level near the ionization potential of the organic compound (typically 9–12 eV), fragmentation of the M$^{+\bullet}$ diminishes because less excess energy is absorbed, making the M$^{+\bullet}$ more stable and therefore more abundant relative to other ions in the spectrum. In this way, the validity of the choice for the M$^{+\bullet}$ peak can be tested by acquiring the mass spectrum at low electron energy, even though the detection limits are worse because the ionization efficiency also decreases with energy.

Consider the "unknown" mass spectrum in Figure 6-4A. Based on an initial examination, it appears that the peak at *m/z* 98 represents the nominal M$^{+\bullet}$. However, if that were the case, the ions with *m/z* 84 and 85 would have to be generated by losses of 14 Da and 13 Da, respectively, from the M$^{+\bullet}$. As such losses are not allowed, it is likely that the sample in the heated-inlet reservoir system, in this case, is impure. This means that Figure 6-4A is a mass spectrum of two or more analytes of different nominal masses. To prove this point, the electron energy is decreased to 16 eV for a second analysis (Figure 6-4B). Important changes in the relative intensities of several peaks are observed. Whereas in the original spectrum the base peak was *m/z* 84, the peak at *m/z* 85 is the base peak in this spectrum. The peak at *m/z* 98 now has a relative intensity equivalent to that of most of the peaks below *m/z* 84. Decreasing the electron energy still further to 12 eV for a third analysis gives the result in Figure 6-4C. Nearly all fragment ions represented by peaks below *m/z* 84 are insignificant; the peak at *m/z* 85 is not only the base peak, but it represents the majority of the ion abundance.

The peaks at *m/z* 85 and 98 in Figure 6-4C must represent the molecular ions of a two-component mixture, as these are the only two significant peaks remaining at low ionizing potential. Based on a logical-loss rational, it can be assumed that the peak at *m/z* 84 in all three mass spectra represents the loss of a hydrogen radical from the component that exhibits a $M^{+\bullet}$ peak at *m/z* 85. It is also clear from the decreased abundance that the component that has a nominal mass of 98 Da has a higher ionization potential than the component that has a nominal mass of 85 Da. This makes sense because based on the *Nitrogen Rule* (Chapter 5) the component with a nominal mass of 85 Da must contain an odd number of nitrogen atoms.

There are two important features to appreciate about obtaining spectra at low ionizing energy. The first is that as the ionizing energy is reduced to near the ionization potential of most organic molecules (10–11 eV), there will be very little excess energy transferred to the nascent molecular ions, which means that fragmentation of these molecular ions will be minimal. This feature is illustrated in the sequence in Figure 6-4 A–C, which shows a profound diminution in fragmentation in going to a lower ionizing potential. The second feature is not so obvious from viewing the sequence in Figure 6-4; the efficiency of the ionizing process also drops off dramatically with the reduction in ionizing energy (see Figure 6-3). The absolute magnitude of the $M^{+\bullet}$ current at *m/z* 85 and at *m/z* 98 in Figure 6-4C is approximately three orders of magnitude less than that in Figure 6.4A; however, the absolute intensity of the fragment ions diminishes by a factor of 4–5 orders of magnitude in going from 70 eV (Figure 6-4A) to 12 eV (Figure 6-4C). On a relative basis, the $M^{+\bullet}$ peak appears to grow with decreasing ionizing potential. Of course, it is the discrimination against fragmentation that is useful in this technique.

Finally, to complete the explanation of the spectra in Figure 6-4, the peak at *m/z* 85 represents the $M^{+\bullet}$ of piperidine, which contains one nitrogen atom; therefore, it has an odd nominal mass. Piperidine fragments readily by eliminating a hydrogen radical to form an ion of mass 84; because little energy is required to produce this fragment, a trace of it remains in the 12-eV spectrum (Figure 6-4C). The peak at *m/z* 98 represents the $M^{+\bullet}$ of cyclohexanone, a compound containing only carbon, hydrogen, and oxygen.

The general appearance of the data also indicates the type of compound that generated the mass spectrum. One important aspect of the mass spectrum is the percent total abundance of the $M^{+\bullet}$. This term "percent total abundance" expresses the abundance of a particular ion compared with the sum of the abundances of all ions in a specified *m/z* range. One convention that has been used to aid the determination of a total abundance scale is shown in Figure 6-5. In this convention, the scale for percentage total ionization is presented on the ordinate at the right or the high-mass end of the mass spectrum with a sigma (Σ) symbol, such as $\%\Sigma_{12}$ in Figure 6-5, where $\%\Sigma_{12}$ indicates that all of the ion current from *m/z* 12 up to and including the $M^{+\bullet}$ (and its isotope peaks, as described in Chapter 5) was used as the basis for determining the contribution of any given ion to the total ionization in the specified *m/z* range. For example, in Figure 6.5 the peak at *m/z* 43 (the base peak) in the specified *m/z* range is responsible for 47% of the total ion current (TIC) represented by all the peaks in the range *m/z* 12 to *m/z* 60; the peak at *m/z* 58 has a relative intensity of about 25%, which represents approximately 12% (25% of 47%) of the total ion current as estimated from the bar graph. Most data systems no longer provide this convenient graphic presentation, which is unfortunate. Percentage of total ion current ($\%\Sigma_n$) is the only expression of ion

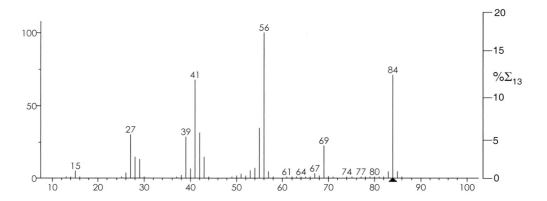

Figure 6-5. *Illustration of the concept of %Σ to show the percentage of the total ion current represented by any peak; in this presentation, the base peak is not assigned 100%, but it still represents the largest fraction of the total ion current.*

abundance that can be used in a valid comparison of the mass spectra of different types of compounds, such as those of isomers.

Depending on the type of compound, the $M^{+\bullet}$ may represent only a small fraction of the total ion current. For example, compare the abundance of the molecular ions of the several compounds in Table 6-2 [4]. Each of the compounds in Table 6-2 contains ten carbons, but in combination with heteroatoms or structures ranging from aromatic to aliphatic, normal chain to highly branched. Note that the $M^{+\bullet}$ of polyaromatic naphthalene resists fragmentation, and therefore the $M^{+\bullet}$ peak represents a large fraction of TIC (36.7%Σ). Although many aliphatic compounds have a $M^{+\bullet}$ peak of only 1%Σ, this peak is still readily discernible in the spectrum; however, the $M^{+\bullet}$ peak for long-chain alcohols or branched compounds may not be detectable (<0.01%Σ), especially at the data acquisition rates associated with techniques such as GC/MS using capillary GC columns.

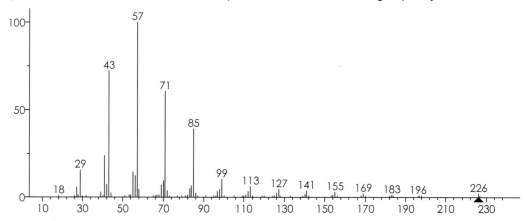

Figure 6-6. *Mass spectrum of hexadecane showing the ion series incremented by m/z 14 corresponding to –CH₂– homologs.*

Table 6-2. Abundance of molecular ions in mass spectra of selected compounds of various structures and elemental compositions.

Compound	Abundance (%Σ)
Naphthalene	44.3
Quinoline	39.6
n-Butylbenzene	8.26
trans-Decalin	8.22
tert-Butylbenzene	7.00
Alloocimene	6.40
Diamyl sulfide	3.70
n-Decane	1.41
n-Decylmercaptan	1.40
Diamylamine	1.14
Methyl nonanoate	1.10
Myrcene	1.00
Cyclododecane	0.88
3-Nonanone	0.50
n-Decylamine	0.50
Diamyl ether	0.33
cis-cis-2-Decalol	0.08
3-Nonanol	0.05
Linalool	0.04
3,3,5-Trimethylheptane	0.007
n-Decanol	0.002
Tetrahydrolinalool	0.000

From Biemann K, Mass Spectrometry: Organic Chemical Applications, McGraw-Hill, New York, 1962, with permission.

EI mass spectra that exhibit few peaks, several of which may have relatively high intensities, usually are of resonantly stabilized compounds such as aromatic compounds. Mass spectra that have a large number of peaks represent analytes that are more aliphatic in nature. One term that is used in the description of a mass spectrum is "ion series". An ion series is represented by peaks separated by a specific number of *m/z* units. The spectrum of hexadecane (Figure 6-6) shows periodic clusters of peaks; in most cases, there is a spacing of 14 *m/z* units between a peak at the highest *m/z* value (that does not represent an isotopic ion) in one cluster and that of an adjacent cluster.

These clusters of peaks represent ions composed of sequential increments of a $-CH_2-$ unit. That is not to say that a given ion is formed by the loss of a $-CH_2-$ unit from the ion represented by a peak 14 *m/z* units higher in the mass spectrum. These ions are formed by the loss of homologous radicals (of successively larger numbers of $-CH_2-$ units) from the molecular ion.

Another important feature of EI mass spectra is the presence of certain peaks that represent "diagnostic" ions in the mass spectrum. The *m/z* values for these peaks should indicate the possibility of a structural moiety in the analyte or a possible contaminant in the sample. The *m/z* values (or loss from the $M^{+\bullet}$) and structural formula for some common diagnostic ions are listed in Table 6-3.

Table 6-3. *List of significant peaks in EI mass spectra representing diagnostic ions of importance.*

Compound Type	Structure of Ions	*m/z* Value
Aliphatic amines	$R_2C=NH_2$ or $H_2N=CR_2$	30, 44, 58
Aliphatic alcohols	$R_2C=OH$	31, 45, 59
Aliphatic ketones	$(CH_3)_2C=OH$	58
Aliphatic acids	$H_3C(C=O)OH$	60
Methyl esters	$H_3C(C=O)OCH_3$ and $^\bullet OCH_3$ loss	74, M − 31
Aromatics	C_6H_5, C_7H_7, $H_3C-C_6H_4$, ring fragments	77, 91, 105, and 39, 51, 65
Phthalate	$(C_6H_4)(C=O)_2OH$	149
Column bleed	$[(CH_3)_2SiO]_n - CH_3$ where $n \geq 3$	207, 281
Trimethylsilyl derivative	$Si(CH_3)_3$	73

IV. Types of Fragmentation Pathways

Once an *m/z* value for the $M^{+\bullet}$ and its probable elemental composition has been established, it is possible to see which peaks may represent the various possible fragment ions resulting from cleavage of certain bonds in the $M^{+\bullet}$. The odd-electron (OE) site formed in the $M^{+\bullet}$ during EI can initiate cleavage of chemical bonds to form an even-electron (EE) species or another OE species. These species will be ions of the same charge sign or will be neutrals. When a $M^{+\bullet}$ fragments, it produces an ion with the same charge sign (positive in EI) and a lesser mass along with a neutral, which can be a molecule with a smaller mass (an even-electron species) or a radical with lesser mass than the molecular ion (an odd-electron species). EI fragmentation *does not* result in an ion pair (two ions of opposite charge sign).

After identifying the $M^{+\bullet}$ peak and ascertaining as much information as possible from this peak and using the general appearance of the mass spectrum to formulate some idea as to the type of compound that may have produced it, the next step is to identify any peaks that may represent odd-electron fragment ions (for more detail, see Chapter 5). Odd-electron fragment ions formed from molecular ions are very important because they require the presence of specific structural features. Odd-electron fragment ions result from rearrangements. Unlike the other three mechanisms for the $M^{+\bullet}$

fragmentation listed below, rearrangements involve breaking more than one chemical bond and the formation of new chemical bonds. The other three mechanisms involve only single-bond cleavage. Peaks representing OE ions can be recognized from the mass spectra according to whether they occur at even or at odd *m/z* values coupled with a knowledge of the elemental composition of the $M^{+\bullet}$ from which they derive. Remembering from the *Nitrogen Rule* that a molecule that contains an even number of nitrogen atoms (e.g., a molecule containing zero or two atoms of nitrogen) has an even nominal mass, a $M^{+\bullet}$ peak for such a species, an OE ion, would occur at an even *m/z* value in the spectrum. OOn the other hand, a fragment ion that contains an even number of nitrogen atoms, formed by loss of a radical from a $M^{+\bullet}$, would have an even number of electrons, but would be represented by a peak in the mass spectrum occurring at an odd *m/z* value. As a simple example, a molecule that does not contain any atoms of nitrogen will be represented by a $M^{+\bullet}$ peak at an even *m/z* value in the mass spectrum. When the odd-electron molecular ion (an odd-electron species) containing no nitrogen undergoes single-bond fragmentation through the loss of a radical, the resulting even-electron ion will have an odd *m/z* value (see Scheme 6-2). When a $M^{+\bullet}$ of the same compound fragments through a rearrangement, a molecule (an even-electron species) is the neutral loss (dark matter), and the resulting odd-electron ion will have an even mass.

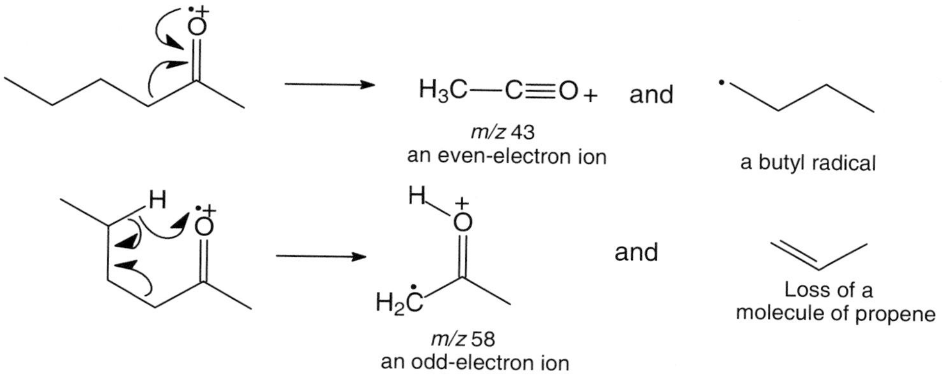

$H_3C-C\equiv O +$ and

m/z 43
an even-electron ion

a butyl radical

and

m/z 58
an odd-electron ion

Loss of a
molecule of propene

Scheme 6-2

Now look at a more complicated case of a $M^{+\bullet}$ that contains two atoms of nitrogen represented by a peak at an even *m/z* value. This $M^{+\bullet}$ (an odd-electron ion) fragments with the loss of a radical (an odd-electron species) that retains one of the two nitrogen atoms. The resulting fragment ion will have an odd number of nitrogen atoms; because it was formed by the loss of an odd-electron radical from an odd-electron ion, the fragment ion will have an even number of electrons, and because it contains an odd-number of nitrogen atoms, it is represented by a peak at an even *m/z* value (Scheme 6-3). There is an important structural significance of odd-electron fragment ions (in signaling the existence of a rearrangement i.e., the elimination of a molecule) that cannot be overemphasized.

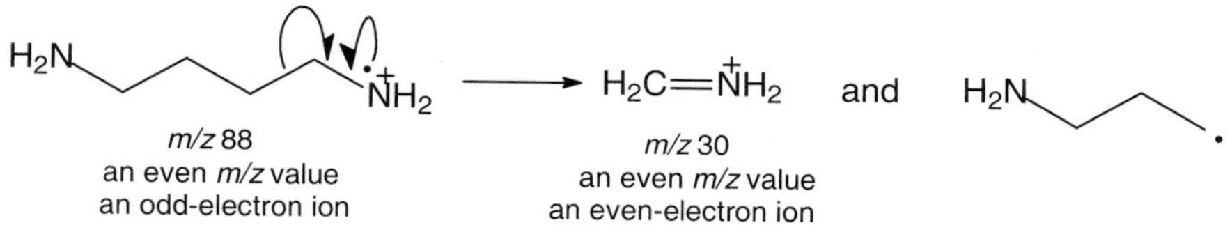

$H_2C=\overset{+}{N}H_2$ and

m/z 88
an even *m/z* value
an odd-electron ion

m/z 30
an even *m/z* value
an even-electron ion

Scheme 6-3

Fragmentation of the $M^{+\bullet}$ following formation by EI can occur by a variety of processes, most of which can be rationalized by one of the following four mechanisms:

    A. Sigma-Bond Cleavage

    B. Homolytic or Radical-Site-Driven Cleavage

    C. Heterolytic or Charge-Site-Driven Cleavage

    D. Rearrangements

1. Sigma-Bond Cleavage

The homolytic and sigma-bond cleavage pathways are initiated by the odd electron in OE ions, whereas the heterolytic cleavage pathways can be initiated by an electron pair in either OE ions or EE ions. Rearrangements are usually initiated by a radical site in an OE ion or by an electron pair in an EE ion. A given type of ion, whether it is an OE or an EE species, has the possibility of participating in more than one of the four different fragmentation processes.

In general, the driving force for fragmentation of an ion is somewhat dependent on the bond strength in the original ion, the stability of the resulting ion, *and* the stability of the neutral loss (radical or molecule) relative to the energetics of the original ionic species. That is to say, if the sum of the internal energies of ion B and radical C are less than the internal energy of $M^{+\bullet}$ in Scheme 6-4, then fragmentation will proceed spontaneously as it is

$$M^{+\bullet} \longrightarrow {}^+B \quad \text{and} \quad {}^\bullet C$$

Scheme 6-4

an overall exothermic process. Following this concept, "paper mechanisms" can be drawn to explain certain fragmentations if the site of electron deficiency and/or positive charge is presumed to be known. The paper mechanisms that are defined, described, and given as examples on the next several pages can be used to anticipate the kinds of data to expect during analysis of a given molecule by EI, or used to rationalize the data observed during analysis of a given compound by EI.

> A rigorous treatment of the thermodynamics and spectroscopic considerations associated with the fragmentation of ions during the electron ionization process can be found in *Mass Spectrometry: A Textbook* by Jürgen H. Gross (Springer-Verlag, Berlin, 2004).

It is important to realize that these paper mechanisms are generally attempts to extend the rationale of organic chemistry (in which solvents are usually involved) to an understanding of unimolecular decompositions *in vacuo*. These mechanisms should not be taken as an indication of detailed knowledge of actual electron movement within the ion. However, the general concepts described below have been used remarkably well in studies of the mechanisms of decomposition of OE ions [5–7]. The art and science of assigning structures to gas-phase ions have been reviewed [8].

Most of the bonds broken in a mass spectral fragmentation process are sigma bonds. The term "sigma bond cleavage" is reserved for fragmentation that occurs as a result of the loss of a σ-bond electron during the ionization process (sigma-bond ionization) that forms the M⁺•. As shown in Scheme 6-5, the site of sigma-bond ionization is represented by a single-electron bond; this is the weak link in the M⁺• in which all other bonds consist of two electrons. As shown in species A of Scheme 6-5, the M⁺• of *n*-butane consists of four carbon atoms held together by three σ bonds; two of the bonds consist of two electrons and one of the bonds consists of a single electron. The M⁺•, species A in Scheme 6-5, is represented by a peak at *m/z* 58 (an even value) because it does not contain an odd number of nitrogen atoms. Those molecular ions of *n*-butane illustrated in Scheme 6-5 will fragment at the bond between the methyl carbon and the other three carbons giving rise to a methyl radical (species C) that will not be detected (because it does *not* carry a charge) and a charged species (the propyl ion) that will be detected at *m/z* 43. In other words, the charge is retained by the propyl portion of the M⁺• and the radical is retained by the methyl portion (the neutral loss). If the other electron in this bond had been removed during ionization, sigma-bond cleavage of that one-electron bond would have left the odd electron on the propyl portion and the charge on the methyl group (a propyl radical and a methyl ion). A sigma-bond electron could have been lost from any of the carbon–carbon bonds or from a carbon–hydrogen bond in butane. In many cases (but not all), when sigma-bond cleavage occurs, ions of both possible *m/z* values will be represented in the mass spectrum, but rarely by peaks of equal intensity.

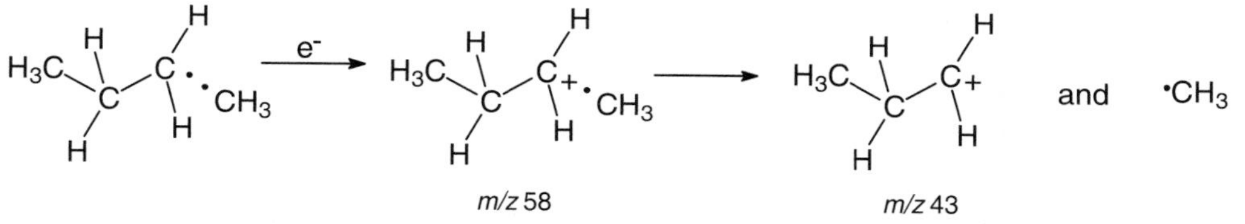

m/z 58   *m/z* 43

A   B   C

Scheme 6-5

Unlike charge sites resulting from the loss of an electron associated with heteroatoms and π bonds, charge sites resulting from the loss of sigma-bond electrons can migrate throughout the M⁺•, resulting in cleavage at any point along the carbon–carbon skeleton. There are a number of guidelines that can be applied to the species produced by the fragmentation of a M⁺•. As with many sets of guidelines, there are numerous exceptions:

- The charge site will be associated with the more stable of two possible ions.

- The radical site will be associated with the portion that has the highest ionization potential.

- There is a tendency to lose the largest possible neutral species.

- Fragmentation will occur at the weakest bond.

These guidelines not only apply to fragmentation resulting from sigma-bond cleavage but also to that resulting from the three other driving forces listed above.

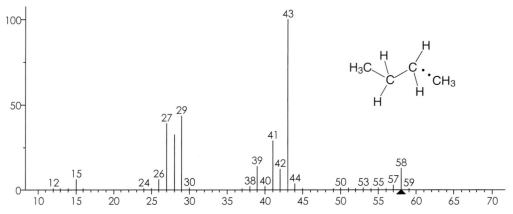

Figure 6-7.   EI mass spectrum of n-butane.

Examination of the mass spectrum of *n*-butane (Figure 6-7) shows that the most stable ion is the propyl ion, which is represented by the base peak; the intensity of this peak might also suggest that the methyl radical is the most stable radical. However, methyl radicals are considerably less stable than larger radicals. The disparity in the intensities of the peaks representing the methyl and propyl ions is also very noticeable, indicating that the propyl ion is formed in preference to the methyl ion even through both ions could be formed by cleavage of the carbon–carbon bond between the number 1 carbon and the number 2 carbon. Although ethyl ions could be formed by retention of the charge by either the number 2 or the number 3 carbon atom, the intensity of the peak at *m/z* 29 representing this ion is only about half that of the base peak. There is another important observation that can be made from this mass spectrum. Acylium ions of the form $C_nH_{(2n-1)}O^+$ have the same nominal mass as aliphatic ions of the form $^+C_{2n}H_{(4n+1)}$, i.e., both $H_3C-C^+H_2$ and $HC≡O^+$ have a nominal mass of 29 Da. In a mass spectrum where all *m/z* values are reported to the nearest integer, these two ions cannot be distinguished from one another; however, an examination of the mass spectrum of 3-pentanone (Figure 6-8) shows two fragment ion peaks, one at *m/z* 29 and the other at *m/z* 57, representing an ethyl ion and a methyl acylium ion, respectively. There are peaks of significant intensities at *m/z* 28, 27,

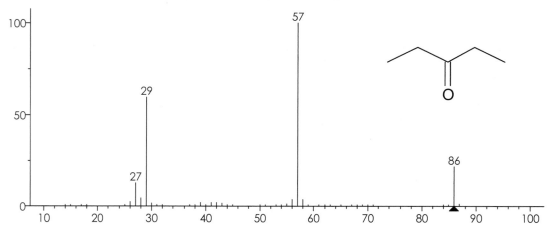

Figure 6-8.   EI mass spectrum of 3-pentanone exhibiting peaks
representing an acylium ion (m/z 57) and an ethyl ion (m/z 29).

and 26 preceding the peak at *m/z* 29, indicating secondary fragmentation of the aliphatic ion through the losses of hydrogen radicals and molecules, whereas such peaks representing secondary fragmentation ions are not associated with the peak at *m/z* 57, which represents the acylium ion. This tendency for aliphatic ions to undergo secondary fragmentation allows these types of ions to be differentiated from acylium ions of the same nominal mass in many cases. A number of factors determine the extent to which this secondary fragmentation of these aliphatic ions occurs. These factors include the pressure in the ion source, whether the aliphatic ion results from a fragmentation of the $M^{+\bullet}$ and whether the aliphatic ion was formed by the expulsion of a molecule from a larger aliphatic ion produced from a $M^{+\bullet}$.

2. Homolytic or Radical-Site-Driven Cleavage

Homolytic or radical-site-driven cleavage results from the tendency of the odd electron associated with a heteroatom or a π bond in the $M^{+\bullet}$ to pair with another electron for better stability; like cops and snakes, electrons travel in pairs. Typically, an initial consideration of the fragmentation mechanism assumes that the OE species has the "plus/dot" (the charge/the odd electron) at a single location, possibly in a nonbonding orbital of a heteroatom such as that of oxygen as illustrated in Scheme 6-6. The movement of single electrons in the paper mechanism (Scheme 6-6) is indicated by a single "fishhook" or single-barbed arrow [5]. The overall electron movement in homolytic cleavage will commonly be indicated with fishhooks **a** and **b** in Scheme 6-6; fishhook **c** is shown here for the sake of completeness. Fishhooks **a** and **b** indicate movement of two single electrons that form an additional bond between carbon and oxygen in this case.

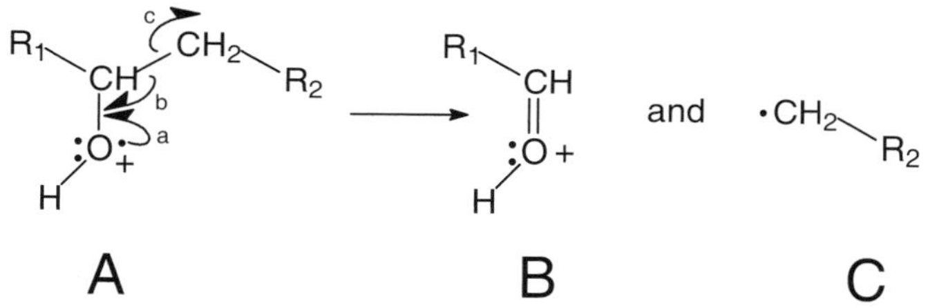

Scheme 6-6

The radical site on the OE molecular ion (A) stimulates transfer of a single electron from an adjacent orbital in an effort to cause a pairing of electrons on the oxygen to achieve a lower-energy state. In Scheme 6-6 it should be noted that the electron pairing with the radical-site electron is associated with a bond between the atom attached to the atom that bears the radical and charge sites and an adjacent atom. The "pairing" of the original odd electron on the heteroatom in Scheme 6-6 causes an OE site to occur elsewhere in the $M^{+\bullet}$. It may appear that no advantage is gained overall because an OE species still exists. However, depending on the element and its electronegativity, some radical species are more stable than other radical species, and if the fragmentation process proceeds, obviously a more favorable energetic state has been achieved. As illustrated in Scheme 6-6, the initial OE ion undergoes homolytic cleavage of the adjacent carbon–carbon bond by migration of a single electron (via fishhook **b**) to pair with the odd electron (via fishhook **a**) that was originally on the oxygen. This causes the formation of an EE oxonium ion (B) and the formation of an alkyl radical (via fishhook **c**). The sum of the internal energies of the oxonium ion and the alkyl radical are less than the energy of the original OE species and, therefore, the fragmentation process proceeds as shown.

During the homolytic cleavage process, the positive charge does not move; note that the positive charge is on the oxygen atom in both species A and species B in Scheme 6-6. There is no longer an odd electron on species B because the odd electron originally on oxygen has now been paired up with an electron that was taken from the bond between the two carbons. As the total number of electrons in the system is conserved, an odd electron now appears on another species, namely the methyl radical (species C) in this scheme. Species C is a neutral species, that is to say it has no net charge and is not detected by the mass spectrometer. In summary, species A in Scheme 6-6 is an OE ion and species B is an EE ion, whereas species C is an OE neutral (a radical). The oxonium ion can also be written in a resonance form with the charge on the carbon atom connected to the oxygen atom and the double bond moves to a position between the carbon atom and the R$_1$ group; this is another rationalization of stability borrowed from classical organic chemistry.

The example shown in Scheme 6-6 is called alpha cleavage because the bond that is cleaved is the bond between the carbon of a functional group and an adjacent carbon atom. A–H$_2$C–O– of an alcohol, ether, or ester and a –CO– of a ketone, aldehyde, acid, or ester are considered to be functional groups. The terminology describing this special case of homolytic cleavage was first suggested by Budzikiewicz, Djerassi, and Williams in 1964 [5]. McLafferty and Tureček [6] define "α-cleavage" as "cleavage of a bond adjacent to the atom alpha to that possessing the original radical site", which clearly indicates limitation of the use of the term "α-cleavage" to homolytic cleavages. Replacement of the oxygen with other heteroatoms (S, N, P, or a halogen) constitutes a similar functionality. Homolysis can also result in a beta cleavage, as will be seen later in this chapter in the discussion of the fragmentation of aliphatic amines. This alpha, beta, and gamma nomenclature convention in organic chemistry is illustrated in Scheme 6-7.

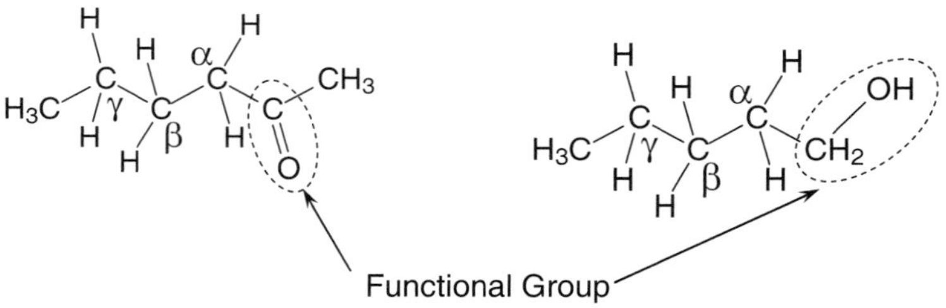

Functional Group

Scheme 6-7

Another example of homolytic cleavage is illustrated in Scheme 6-8, which shows the M$^{+\bullet}$ of diethyl ether with the "plus/dot" on the oxygen atom in a nonbonding orbital. The odd electron on the oxygen stimulates the migration of a single electron from the carbon–carbon bond alpha to the carbon–oxygen bond as illustrated in Scheme 6-8. Homolytic cleavage at the carbon–carbon bond forms an EE oxonium ion and expels a methyl radical as indicated in this scheme. Those molecular ions of diethyl ether that do not fragment would be detected at *m/z* 74, an even *m/z* value as the molecule contains an even number (namely 0) of nitrogen atoms, and the fragment resulting from homolytic fission of the carbon–carbon bond would be detected at *m/z* 59, an odd *m/z* value, which is consistent with an EE ion containing no atoms of nitrogen. The methyl radical would not be detected.

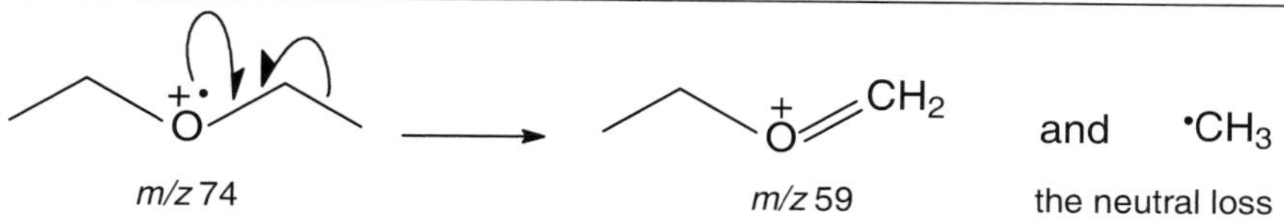

Scheme 6-8

3. Heterolytic or Charge-Site-Driven Cleavage

The positive charge on an ion, rather than the odd electron of an OE species, can stimulate fragmentation by inducing heterolytic capture of both electrons out of an adjacent bond as illustrated in Scheme 6-9. The driving force in heterolytic cleavage (so-called because the *electron pair* constituting the original chemical bond stays with one atom of the two originally connected) is the *induction* effect of the positive charge causing the electron pair to migrate toward the positive charge resulting in neutralization at the original site. The movement of both electrons from a given bond causes an electron deficiency on the other species, the first carbon of the propyl group in Scheme 6-9, which is reflected by the positive charge developing on that carbon atom as indicated on species B. Of course, the original bond from which both electrons migrated to the positive charge is broken as is shown with the carbon–chlorine bond of species A in Scheme 6-9; this movement of electrons is coupled with the formation of a radical species indicated by C.

The movement of two electrons is shown by a *double-barbed* arrow in the heterolytic fragmentation scheme. In inductive cleavage (synonymous with heterolytic cleavage) as illustrated in Scheme 6-9, the charge appears to migrate from its original locus in the M$^{+\bullet}$. In Scheme 6-9, the M$^{+\bullet}$, an OE species, that would be detected with an m/z 78 (an even value) has the positive charge on the chlorine. The fragment ion, species B, has the positive charge on the carbon of an EE ion that would be detected at m/z 43. The chlorine radical (an atom of chlorine) is quite stable, and because it is neutral will not be detected. In this scheme, the charge migrated from its original position on the chlorine atom to the carbon atom of the ethyl moiety.

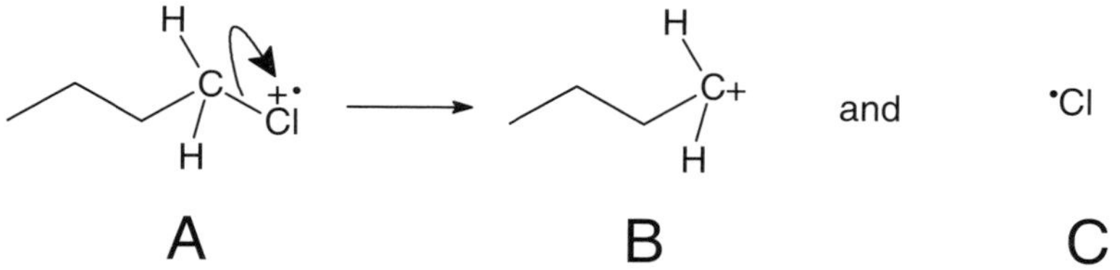

Scheme 6-9

Another example of charge-driven or heterolytic cleavage is illustrated in Scheme 6-10 for the M$^{+\bullet}$ of diethylether. Once again, note that the "+•" is located on the oxygen atom as this is the most likely locus from which to lose an electron during the ionization process. In this case, as opposed to that in Scheme 6-9, the charge site induces the movement of *both* electrons from an adjacent carbon–oxygen bond to neutralize the original charge on the oxygen atom to give rise to the radical species C shown in Scheme 6-10, a species with no charge, which therefore will not be detected. In addition to the radical species, the charged species B is formed with the positive charge now on the methylene carbon of the ethyl ion, which is detected at m/z 29, an odd m/z value. This is consistent with the ethyl ion being an EE species.

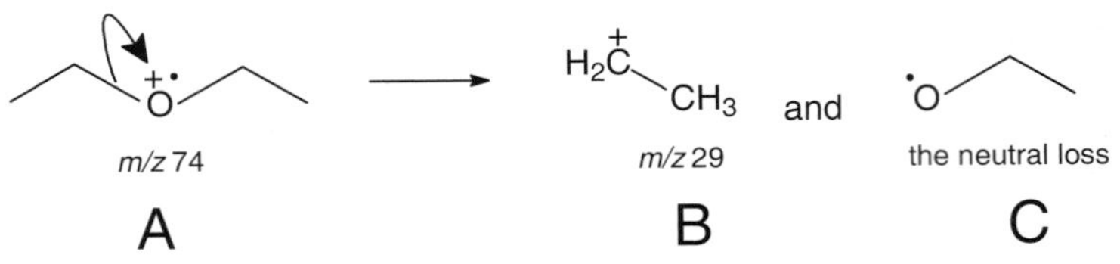

Scheme 6-10

It is interesting to compare the fragmentation processes of the diethylether $M^{+\bullet}$ illustrated in Schemes 6-8 and 6-10 that produce ions of different elemental composition and different mass. Even though the ions formed from diethylether via Schemes 6-8 and 6-10 have different m/z values, namely m/z 59 and m/z 29, both of these m/z values are odd because they both represent EE ions that do not contain an odd number of nitrogen atoms.

It is important to remember that heterolysis involves the movement of a pair of electrons from the bond that connects the atom with the charge and the adjacent atom. In some cases, this may not be clear. A good example involves heterolytic cleavage that takes place when the charge and the radical are associated with the carbonyl oxygen atom of a ketone (Scheme 6-11). The actual cleavage in this case occurs at the alpha carbon. However, this bond cleavage only occurs after a two-step process as illustrated in Scheme 6-11. Even though an alpha-carbon bond is broken in this process, it would be inappropriate to refer to this as alpha cleavage. The term alpha cleavage (as stated above) is reserved for the special case of homolytic cleavage that occurs at an alpha carbon.

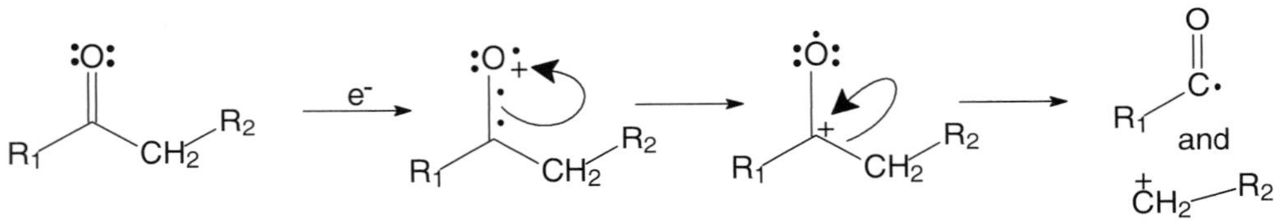

Scheme 6-11

Unlike homolytic cleavage, which always results in the formation of an ion and a radical, heterolytic cleavage can result in the formation of an ion and a molecule or an ion-radical pair. The ion and molecule formation associated with heterolytic cleavage usually occurs when the heterolytic cleavage proceeds as a secondary reaction in an ion formation. Secondary reactions involved with rearrangements of charges and/or radical sites followed by either homolytic or heterolytic cleavage are discussed in more detail below. A good example of heterolytic cleavage resulting in the formation of an ion–molecule pair is the common propensity for acylium ions (formed by homolytic cleavage of molecular ions) to lose a molecule of carbon monoxide as illustrated in Scheme 6-12. The propensity for this heterolytic cleavage to occur is based on the stability of the resulting ion versus that of the ion that undergoes the secondary fragmentation.

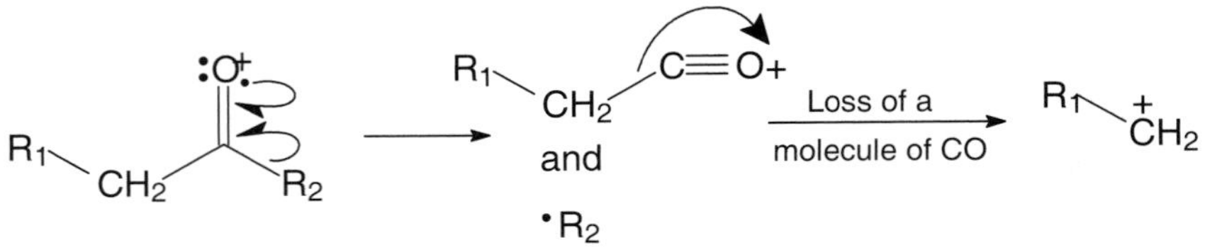

Scheme 6-12

The tendency for homolytic cleavage or heterolytic cleavage to occur is influenced by the chemical nature of the original electron removed during ionization. Homolytic cleavage will occur in preference to heterolytic cleavage when the charge and the radical site result from the loss of a nonbonding electron from a nitrogen or phosphorus atom. If the charge and radical site are associated with a nitrogen or phosphorus atom, it is very unlikely that ions resulting from heterolytic cleavages will be observed. If the charge and radical site are due to the loss of an electron from a nonbonding orbital of a halogen (F, Cl, Br, or I), the probability of heterolytic cleavage is more likely; however, ions resulting from both types of cleavage will be observed in the mass spectrum, though the abundances of ions produced by heterolytic cleavage will be significantly greater. When the charge and the radical sites are due to the loss of an electron from a π-bond orbital or a nonbonding orbital of an oxygen or sulfur atom, homolytic cleavage can be promoted by adjacent electron-withdrawing groups. Peaks due to both homolytic and heterolytic cleavages will be observed in the mass spectra of compounds that have π-bond orbitals or atoms of oxygen or sulfur, and these ions sometimes can have nearly equal abundances.

4. Rearrangements

Fragmentation as a result of a rearrangement involves multiple-bond cleavages and new bond formations. The neutral loss in a fragmentation involving a rearrangement is always a molecule, an EE species. Unlike the previously described mechanisms associated with single-bond cleavages, which result in EE ions being formed from OE ions and OE ions being formed from EE ions, rearrangement fragmentation processes produce the same type of ions as the precursor ion; i.e., EE ions produce EE fragment ions and OE ions produce OE fragment ions. In an EI mass spectrum, peaks representing OE ions are important because they signal the involvement of rearrangement fragmentations of OE species, which can provide insight about the location of heteroatoms and/or double/triple bonds in the analyte molecule, and/or to the overall structure of the analyte molecule.

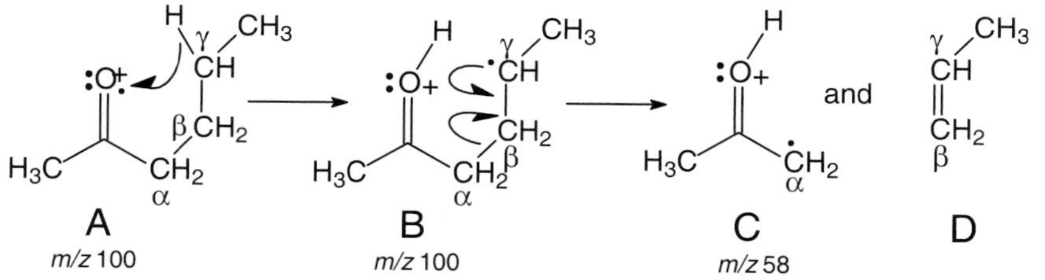

A	B	C	D
m/z 100	*m/z* 100	*m/z* 58	

Scheme 6-13

In EI, rearrangement fragmentation of EE ions is always associated with a secondary fragmentation of the $M^{+\bullet}$, a process that significantly affects the appearance of the spectrum as exemplified by the mass spectra of aliphatic alcohols and amines. Rearrangements involve a shift of a hydrogen atom or a hydride, and in some cases a small group of atoms in the form of a radical, which is analogous to a hydrogen atom. After a hydrogen or radical shift occurs, which involves the breaking of one bond and the forming of another, a distonic ion [9, 10] is formed with a new radical site, but with the charge remaining in its original position in the case of a rearrangement fragmentation of an OE ion; in the case of rearrangement fragmentations of an EE ion, the nascent ion has a different charge location after the shift occurs. Species C with *m/z* 58 in Scheme 6-13 is a distonic ion. The next step in the rearrangement fragmentation involves the movement of a single electron (a homolytic cleavage) in response to the new radical site or the movement of a pair of electrons (a heterolytic cleavage) in response to a new or the original charge site, both of which usually involve cleavage of a second bond and the formation of a second new bond, as illustrated in Schemes 6-13 and 6-14.

> The charge and radical sites in an odd-electron ion are usually associated with the same atom (e.g., as is the case when an electron is lost from a nonbonding orbital of an oxygen atom in the formation of a molecular ion of a ketone) or from a bond between two connected adjacent atoms (e.g., as is the case when a π-bond electron is lost in the formation of the molecular ion of an aromatic compound) in a nascent molecular ion. When the charge and radical sites are not associated with the same or two connected adjacent atoms, the ion is referred to as a *distonic ion*.

The formation of a primary oxonium ion, illustrated in Scheme 6-14, is a good example of a rearrangement mechanism involving the secondary fragmentation of an EE ion produced by the primary fragmentation of a $M^{+\bullet}$. The same process of multiple-bond cleavages and new bond formations described above for the rearrangement fragmentation of a $M^{+\bullet}$ (an OE ion) also occurs in these hydride-shift rearrangements of EE ions.

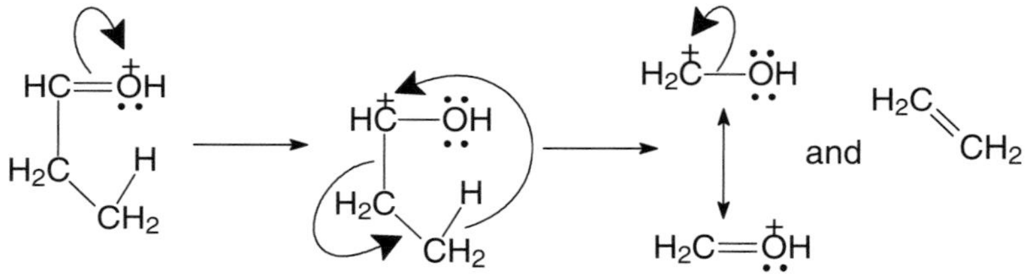

Scheme 6-14

A. Hydrogen-Shift Rearrangements

A commonly observed rearrangement occurs in the case of a $M^{+\bullet}$ containing a double bond and a γ hydrogen relative to the double bond as shown generally in Scheme 6-13. In species A, the γ hydrogen is positioned in such a way as to participate in a virtual six-membered ring involving the carbonyl oxygen, which provides the double bond and the radical site. This overall rearrangement process involves the shift of a hydrogen atom (a proton and an electron) from its original position on a carbon atom that is in the gamma position relative to the carbonyl group. This γ-hydrogen shift occurs because of the need for the radical electron on the oxygen atom to pair with another electron. The resulting odd-electron ion has the same mass (and elemental composition) and the same number of electrons as the original ion, but with a new geometry and a new radical site. The odd electron constituting the new radical site on the γ carbon initiates the subsequent cleavage of the β bond. This mechanism was originally described by Nicholson [11], who compared photolytic and EI decomposition products of aliphatic ketones; however, because of extensive studies by Fred W. McLafferty over the last five decades [12–15], this general rearrangement has affectionately taken on the name *McLafferty rearrangement*.

According to studies by McLafferty [13–15] and others [16, 17], the overall rearrangement proceeds according to the two discrete steps described above in Scheme 6-13. The first step in Scheme 6-13, involving the transfer of the γ hydrogen to the radical site on oxygen in species A, forms a new bond in a stable enol ion [18]. This exothermic step leads to generation of the distonic species B, which has the radical site on the γ carbon, and which is still detectable at *m/z* 100 because all of the atoms originally present are still connected to one another, albeit in a different order. If sufficient internal energy is available, the radical site on the γ carbon in some of species B stimulates homolytic fission of the bond between the α- and β-carbon atoms to form the OE species C appearing at *m/z* 58, together with elimination of the EE species D as a molecule of ethene.

As seen in Scheme 6-15, the γ-hydrogen shift can also be followed by a heterolytic cleavage. In this particular example, the driving force for the heterolytic cleavage is the formation of the resonantly stabilized substituted ketene OE ion.

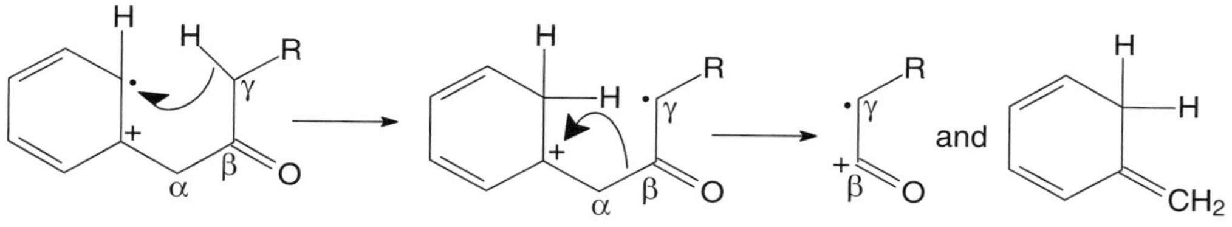

Scheme 6-15

In Scheme 6-15, the charge and radical sites resulting from the loss of one of the π electrons on the aromatic ring is shown. As described earlier in this chapter, the most probable electron to be lost during the formation of a M$^{+\bullet}$ for this compound is one of the nonbonding electrons associated with carbonyl oxygen. Although this is true, there may be some molecular ions formed with the charge and the radical associated with a π bond in the aromatic ring (as shown in Scheme 6-15) as opposed to those formed with the charge and the radical associated with the carbonyl oxygen. The ions with the charge and the radical associated with the ring will be far less abundant than those with the charge and radical sites associated with the carbonyl oxygen; none the less, there will be some molecular ions with the charge and radical sites associated with the ring.

> It is always important to keep in mind that even though there is a *most* probable position for the charge and radical sites in a molecular ion, some molecular ions with localized charge and radical sites in other positions might also exist; such a variety of possible positions for ionization provides a distribution of isomeric molecular ions.

An even more significant example of a hydrogen-shift rearrangement followed by a heterolytic cleavage is illustrated in the loss of water from an aliphatic alcohol as shown in Scheme 6-16. Again, a virtual six-membered ring plays a role in promoting the hydrogen shift from a γ carbon in response to the radical site created by the expulsion of an electron from a nonbonding orbital of the oxygen atom. The γ-hydrogen shift induces a heterolytic cleavage that comes about when the pair of electrons that constitute the bond between the carbon and oxygen atoms moves to neutralize the charge that is still on the oxygen atom (see Scheme 6-16). This heterolytic movement of electrons causes the charge to move to the carbon atom that was attached to the oxygen atom and releases a neutral molecule of water. As will be seen later on in this chapter during discussion of the fragmentation of aliphatic alcohols, the resulting distonic OE ion in Scheme 6-16 can undergo a series of secondary fragmentations, which account for the large number of peaks in the mass spectra of such compounds.

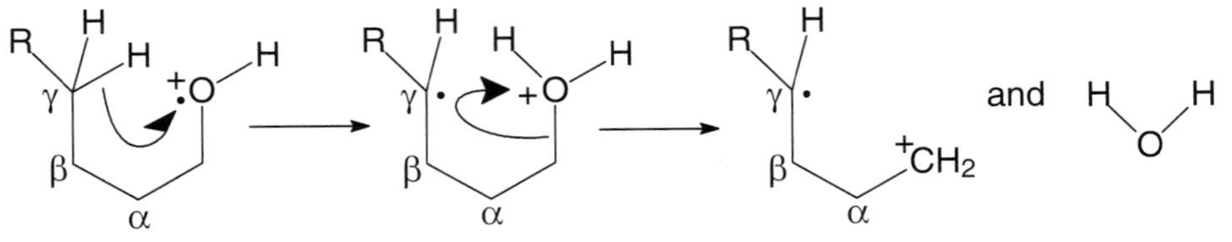

Scheme 6-16

The importance of the virtual six-membered ring in a rearrangement fragmentation cannot be overemphasized; however, eight- and ten-membered ring transitions are also observed, as will be seen later on in this chapter in the discussion of the fragmentation of methyl stearate. Tighter shifts (meaning shifts from carbons closer than five atoms away from the site of the charge or radical) can also take place, as seen with hydride and hydrogen shifts in aliphatic amines and alcohols. This process is illustrated in Scheme 6-17, which shows a hydrogen shift from the α carbon of ethyl amine in response to the radical site created by the expulsion of an electron from a nonbonding orbital of nitrogen during the formation of the M⁺•. Some of these distonic ions then undergo heterolytic cleavage with the loss of a molecule of ammonia and the formation of an OE ion of ethene.

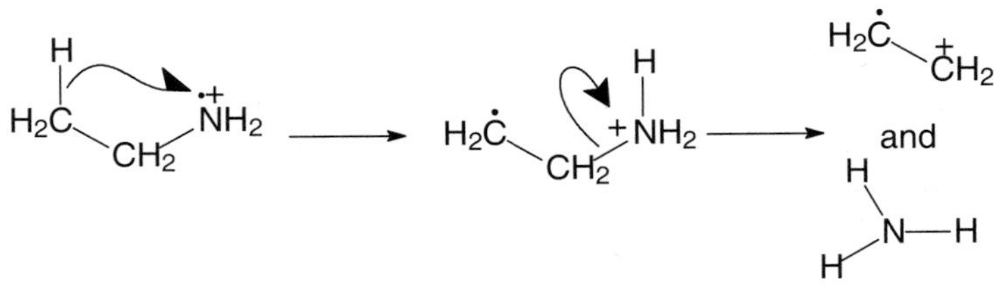

Scheme 6-17

Another important factor associated with the occurrence of a hydrogen shift and the associated virtual six-membered ring is the type of radical that is produced. If the γ carbon has three hydrogen atoms attached, the result of a hydrogen shift will be a primary radical (the radical-ion product in Scheme 6-17 is an example of a primary radical), which is not as stable as a secondary radical. The intensity of the peak representing the OE fragment ion formed by a γ-hydrogen shift involving intermediate formation of a primary radical in the mass spectrum of 2-pentanone (Figure 6-9) has a lower relative intensity than does the peak with the same *m/z* value (*m/z* 58) in the mass spectrum of 2-hexanone

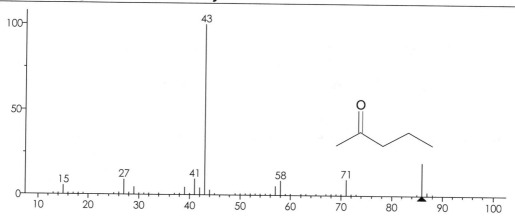

Figure 6-9.   EI mass spectrum of 2-pentanone.

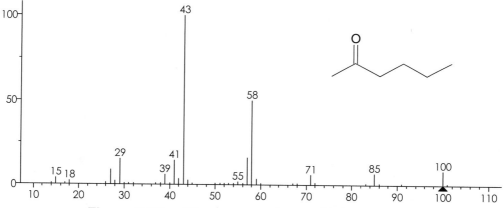

Figure 6-10.   EI mass spectrum of 2-hexanone.

(Figure 6-10) where the γ-hydrogen shift results in the intermediate formation of a secondary radical (like that shown with the second structure in Scheme 6-18). Another example of the influence of the intermediate formation of a primary vs secondary radical during a γ-hydrogen-shift rearrangement can be found later in Section V in the comparison of the mass spectra of *n*-propylbenzene (Figure 6-23) and *n*-pentylbenzene (Figure 6-25); for propylbenzene, formation of the OE ion of *m/z* 92 is virtually nonexistent in the mass spectrum (Figure 6-23), whereas, for *n*-pentylbenzene, formation of the same ion involves an intermediate secondary radical, and the peak at *m/z* 92 is 80% of the base peak (Figure 6-25).

The one exception to suppression of the γ-hydrogen-shift rearrangement by the intermediate formation of a primary radical is when the ion that is undergoing the rearrangement has just been produced by a preceding γ-hydrogen-shift rearrangement. The products of two successive γ-hydrogen shifts are represented by peaks at *m/z* 86 and *m/z* 58 in the mass spectrum of 4-decanone (Figure 6-11 and Scheme 6-18). The peak at *m/z* 86 represents the OE fragment ion formed by the loss of a molecule of 1-pentene, as shown early in Scheme 6-18. This OE ion will then undergo a γ-hydrogen shift that involves the intermediate formation of a primary radical, as shown in the second line of Scheme 6-18; the subsequent β cleavage of the distonic ion in the form of a primary radical

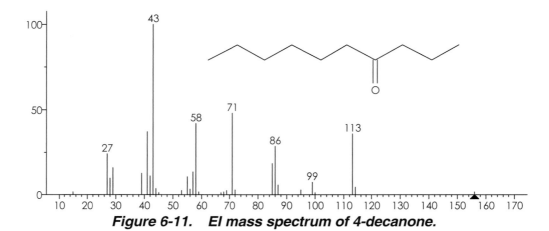

Figure 6-11.   EI mass spectrum of 4-decanone.

produces an OE ion represented by the peak at *m/z* 58. The intensity of the peak at *m/z* 58 indicates that there is a strong tendency for this secondary reaction (involving the intermediate formation of a primary radical) to take place.

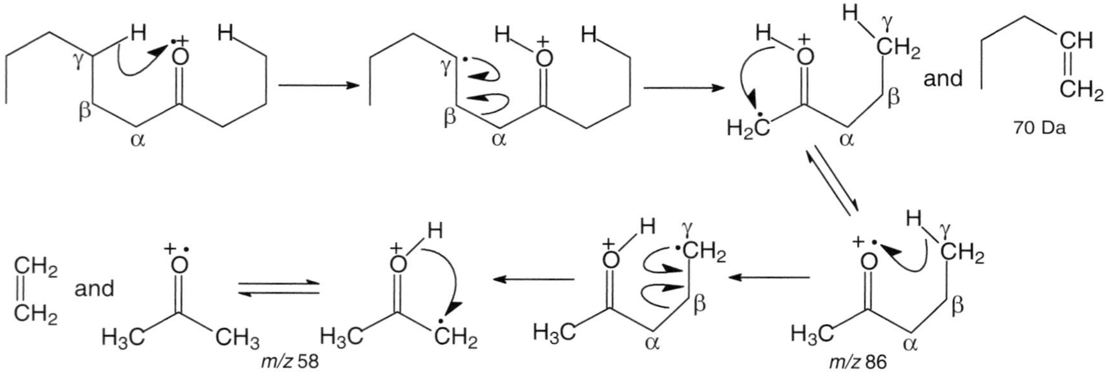

Scheme 6-18

Another factor that cannot be ignored in the formation of these OE ions resulting from γ-hydrogen shifts is the electronic configuration of the α, β, and γ carbons. If these carbons have an electronic configuration other than that of an sp^3 hybrid, the ion geometry and configuration may preclude formation of the virtual six-membered ring making the reaction less likely to occur, resulting in lower abundances of these corresponding OE fragment ions.

B. Hydride-Shift Rearrangements

Fragmentation occurring as a result of a hydride-shift rearrangement is mostly restricted to secondary fragmentation of EE ions formed by single-bond cleavage in a M$^{+\bullet}$. An example of a hydride-shift rearrangement fragmentation of a secondary oxonium ion produced by the alpha cleavage of a M$^{+\bullet}$ of a secondary alcohol was shown earlier in Scheme 6-14. Such rearrangements observed in an even-electron system have also been studied and explained by McLafferty [19].   As stated earlier, these hydride-shift rearrangements play a key role in the EI-induced fragmentation of aliphatic amines.

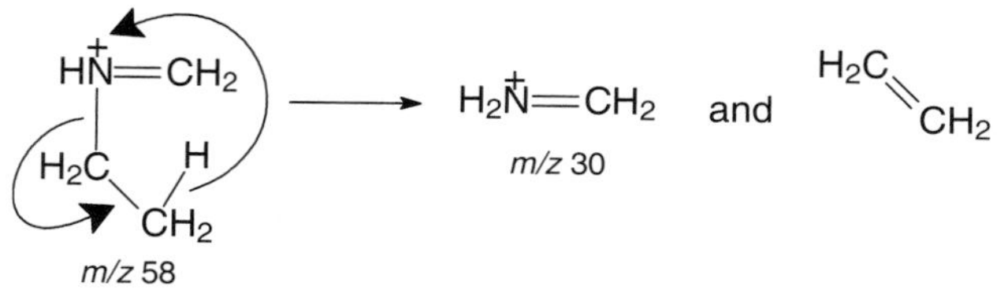

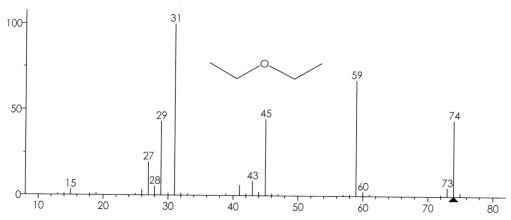

Scheme 6-19

An example of a hydride-shift rearrangement in the secondary fragmentation of an amine is shown in Scheme 6-19 as indicated by the double-barbed arrows in the species having an *m/z* value of 58. As shown in this scheme, the elimination of ethylene, an EE species of even mass, from an EE ion (containing one nitrogen) with *m/z* 58, produces another EE species (also containing the nitrogen) with *m/z* 30, another even number. The *m/z* values of these ions in Scheme 6-19 provide an example of a corollary to the *Nitrogen Rule*; namely that an EE fragment ion retaining an odd number of nitrogen atoms will have an even *m/z* value. In this case, the species at *m/z* 58 is a fragment ion (an EE species) from a larger OE species that contains at least one atom of nitrogen.

Another example of the significance of a hydride-shift rearrangement is seen in the mass spectrum of diethyl ether (Figure 6-12). This mass spectrum exhibits a M$^{+\bullet}$ peak at *m/z* 74 of significant intensity. There are major peaks at *m/z* 59, 45, 31, and 29. All the peaks in the mass spectrum appear to be consistent with the *Nitrogen Rule* and the formation of EE-fragment ions through single-bond cleavage.

Figure 6-12. EI mass spectrum of diethyl ether.

The M$^{+\bullet}$ peak in the mass spectrum of diethyl ether (Figure 6-12) is at an even *m/z* value which is consistent with the analyte not containing an odd number of nitrogen atoms. Because there are no atoms of nitrogen in the analyte, none of the fragment ions formed will contain an atom of nitrogen. If the fragment ions are EE ions, then according to the *Nitrogen Rule* they must have an odd *m/z* value. The peak at *m/z* 59 represents an oxonium ion produced by alpha cleavage in the M$^{+\bullet}$ of diethyl ether with the charge and radical sites on the oxygen atom as shown earlier in Scheme 6-8. Some of the molecular ions with the charge and radical sites on the oxygen atom will undergo heterolytic cleavage to produce aliphatic ions with *m/z* 29, as was illustrated in Scheme 6-10. The peak at *m/z* 29 obviously represents the aliphatic ethyl ion because of the presence of peaks at *m/z* 28, 27, and 26, which represent ions formed by the secondary fragmentation of the ethyl ion.   The peak at *m/z* 59 probably does not represent an alkyl ion because it is not accompanied by peaks at *m/z* 58, 57, and 56 that would correspond to secondary fragmentation of such an ion. Both peaks at *m/z* 29 and *m/z* 59 represent ions formed by single-bond cleavage. It is also possible to rationalize the peak at *m/z* 45 as representing an ion formed through a single-bond cleavage mechanism resulting in the loss of an ethyl radical from the $M^{+\bullet}$.

The origin of the ion represented by the peak at *m/z* 31 in Figure 6-12 cannot be rationalized by any single-bond cleavage mechanism relating to the fragmentation of diethyl ether. Ions with *m/z* 31 are listed in Table 6-3 as being a primary oxonium ion that is diagnostic for aliphatic alcohols. Therefore, in Figure 6-12, the only rationale for the formation of the ion represented by the peak at *m/z* 31 is a fragmentation of diethyl ether induced by a hydride-shift rearrangement as shown in Scheme 6-20. A hydride from the terminal carbon of the ethyl moiety of the oxonium ion with *m/z* 59 shifts in response to the charge on the oxygen atom in concert with the movement of the pair of electrons between the methylene carbon of the ethyl moiety and the oxygen in response to the new electron deficiency created by the hydride shift. As shown in Scheme 6-20, the charge remains on the oxygen atom, but now the oxonium ion with *m/z* 31 is composed of a hydrogen attached to the oxygen rather than to the ethyl group. The hydride-shift rearrangement, illustrated in Scheme 6-20, results in the formation of a molecule of ethene. Examination of a series of spectra of dialkyl ethers reveals that although an alkyl-substituted primary oxonium ion is formed regardless of the size of the chain on either side of the ether oxygen, the peak for the ion with *m/z* 31, while present, has a significant intensity only when one of the alkyl chains is an ethyl group.   This fact supports the supposition that some rearrangement fragmentations most likely involve the expulsion of "small" molecules.

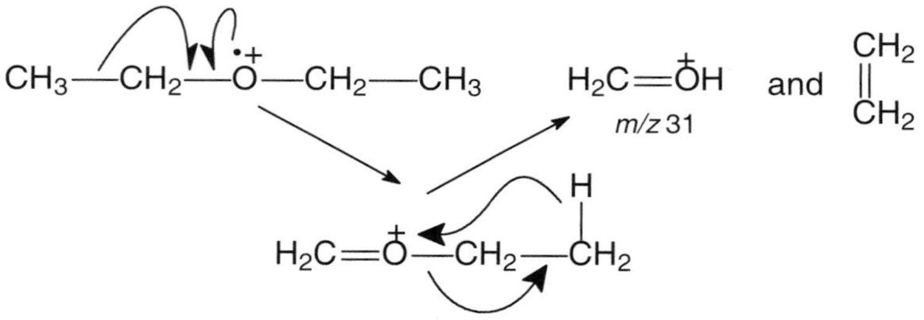

Scheme 6-20

Another important feature that distinguishes the mass spectra of ethers and aliphatic alcohols is that those of most aliphatic ethers exhibit a $M^{+\bullet}$ peak, whereas those of aliphatic alcohols containing more than four carbons do not. The intensity of the peak at *m/z* 31 in the mass spectra of all primary aliphatic alcohols always has a relative intensity of about 25%. The fragmentation of aliphatic alcohols and aliphatic ethers is discussed in more detail later in this chapter.

V. Representative Fragmentations (Spectra) of Classes of Compounds

This section compares the so-called "paper mechanisms" with the peaks observed in the mass spectra of a series of different types of compounds. By proposing paper mechanisms based on structures and observed peaks, it becomes easier to deduce a structure from an EI mass spectrum. Frequent practice with this strategy or procedure will make it easier to recognize the relationship between a structural moiety or feature in the molecule and observed peaks of specific *m/z* values or to specific neutral losses in the mass spectrum. Gaining familiarity with representative mass spectra of many different classes of compounds is another important aspect of learning to deduce a structure from an unknown mass spectrum. When the quality of a mass spectrum is being determined by the reviewers at NIST, they always rely on spectra of similar compounds where possible.

1. Hydrocarbons

There are four types of hydrocarbons: alkanes, alkenes, alkynes, and aromatics. Alkanes, alkenes, and alkynes can be broken down further into straight-chain, branched, and cyclic compounds. Aromatic hydrocarbons usually involve rings; however, a conjugated double-bond system such as found in butadiene also constitutes an aromatic compound. From a mass spectrometry standpoint, hydrocarbons can be defined a little differently. There are hydrocarbons that have π bonds (alkenes, alkynes, and aromatic) and there are hydrocarbons that do not have π bonds (saturated hydrocarbons: alkanes). There are linear hydrocarbons (those that have only two terminal carbon atoms and no carbons attached to more than two other carbon atoms) and there are cyclic hydrocarbons in which all the carbon atoms (no terminal carbons) are attached to two or more other carbon atoms. Branched hydrocarbons have more than two terminal carbon atoms, and contain some carbons that are attached to more than two other carbon atoms. An aromatic compound can have alkyl substituents. The fragmentation of saturated hydrocarbons is principally through σ-bond cleavage. Fragmentation of unsaturated hydrocarbons is initiated by the loss of an electron from a π-orbital electron in the formation of the $M^{+\bullet}$. Fragmentation of saturated cyclic compounds involves a two-step process. Because the driving force for the fragmentation of the various types of compound is different, the interpretation of their mass spectra will be discussed separately.

A. Saturated Hydrocarbons

1) Straight-Chain Hydrocarbons

The fragmentation of a saturated hydrocarbon results in the formation of carbenium ions. For straight-chain hydrocarbons, the mass spectrum is dominated by peaks that represent primary carbenium ions separated by 14 *m/z* units, as seen in Figure 6-13. Unlike molecular ions involved in charge or radical-site-driven cleavages, the location of the charge cannot be easily assigned in the $M^{+\bullet}$ of a saturated hydrocarbon. Even if an electron is ejected between two specific carbon atoms, the charge can migrate throughout the nascent $M^{+\bullet}$ because of the lack of highly localized electrons as exist in π orbitals of a double bond. It should also be mentioned that an electron in a carbon–carbon bond is more likely to be lost than an electron from a carbon–hydrogen bond. This is probably because the carbon–carbon bond consists of the overlay of two hybridized orbitals and a carbon–hydrogen bond is the overlay of a hybridized orbital and an *s* orbital; therefore, it is likely that the more highly hybridized bond can better accommodate the electron deficiency.

The abundance of the $M^{+\bullet}$ produced by straight-chain saturated hydrocarbons is about 2–5% of the total ion current. Sometimes, the rate of dissociation of molecular ions is high enough that the residence time of ions in the mass spectrometer can appreciably affect the intensity of the molecular ion peak. A long residence time, the interval between ion formation and detection, might be long enough for all sufficiently unstable molecular ions to decompose depending on the kinetics of the process. The residence time in a magnetic sector, TOF, or transmission quadrupole instrument is microseconds, whereas in a quadrupole ion trap it can be as much as 25 msec. The molecular ions of some

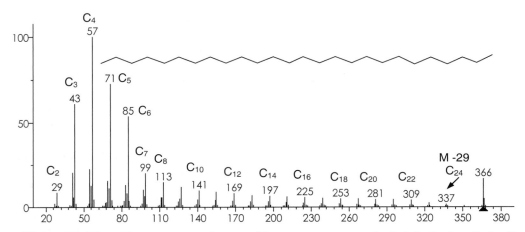

Figure 6-13.   EI mass spectrum of hexacosane, a straight-chain aliphatic hydrocarbon.

saturated hydrocarbons are stable enough to be detected even when the data are acquired using a quadrupole ion trap mass spectrometer.

There are several other important factors that should be noted about the mass spectrum of hexacosane (Figure 6-13). Looking at the spectrum as a whole, a characteristic pattern is observed. This pattern is sometimes referred to as the "ski-slope" pattern. This ski-slope pattern is the underlying pattern exhibited by the mass spectrum of compounds that have a straight-chain hydrocarbon backbone; it derives from a series of peaks that represent C_2, C_3, C_4, C_5, C_6 primary carbenium ions. These ions dominate because of their stability and the fact that they come from at least two different sources.

To some extent, the ski-slope pattern can be explained by the half-C_n rule (where C_n represents the number of C atoms in the molecular ion); this rule states that the majority (~80%) of the ions that contain half or more of the carbons comprising the complete chain result from direct fragmentation of the $M^{+\bullet}$. Only about 20% of the ions containing fewer than half of the carbon atoms result from direct fragmentation of the $M^{+\bullet}$ [20]; that is, most of the ions containing fewer than half the carbon atoms in the original molecule result from two or more stages of fragmentation. The source of most of the fragment ions in the mass spectrum of a straight-chain hydrocarbon (certainly down to the half-n position where n = total number of carbon atoms in the analyte molecule), are molecular ions; however, as described above, below the half-n position, the precursor is most likely a high-mass fragment ion. For example, the peak at *m/z* 281 in Figure 6-13 represents a fragment ion containing C_{20}^+ (as $C_{20}H_{41}^+$), which most likely was formed by the expulsion of $C_6^{\bullet}$ (as the radical $C_6H_{13}^{\bullet}$) from the C_{26} molecular ion. Furthermore, the peak at *m/z* 113 represents a fragment ion containing C_8^+ (as $C_8H_{17}^+$), which was likely formed by loss of C_{10} (as the olefin $C_{10}H_{20}$) from the C_{18} ion (as $C_{18}H_{37}^+$) or the loss of C_{11} (as an olefin) from the fragment ion comprising C_{19}^+ or the loss of C_9 from the C_{17}^+ ion, etc., as well as the loss of the radical $C_{18}^{\bullet}$ from the C_{26} molecular ion.

Another feature of note in the mass spectrum of hexacosane (Figure 6-13) is the pattern of peak clusters at intervals of 14 *m/z* units. Usually, the most intense peak in these clusters represents fragment ions with the elemental composition C_nH_{2n+1}. The serial clusters of peaks differing by 14 *m/z* units, up to the 1/2 C_n position in the spectrum, represent the integral effect of a large population of isomeric molecular ions losing a particular member of a homologous series of radical species, i.e., those containing different numbers of methylene (–CH_2–) groups; it does *not* represent the successive loss of methylene groups (which would suggest the loss of a diradical) from a particular fragment ion!

It should also be noted that in the mass spectra of straight-chain hydrocarbons there is the lack of a peak representing the [M – CH_3]$^+$ ion; this feature relates to the low stability of the methyl radical, especially as compared to radicals containing higher numbers of carbon atoms. The intensity of peaks representing ions formed from losses of successively larger radicals from the M$^{+\bullet}$ increases as the number of carbon atoms comprising the radical increases down to the half-n value of the molecular ion. After the half-n value is reached, the intensity of the peaks representing ions with successfully smaller numbers of carbon atoms increases at a much higher rate because these ions are also due to secondary fragmentation of larger fragment ions.

Although the loss of a methyl radical is used extensively in the explanation of some fragmentation mechanisms, it should always be remembered than the [M – 15]$^+$ peak indicates the presence of a *special* methyl (i.e., the methyl group is in a special location) in the M$^{+\bullet}$.

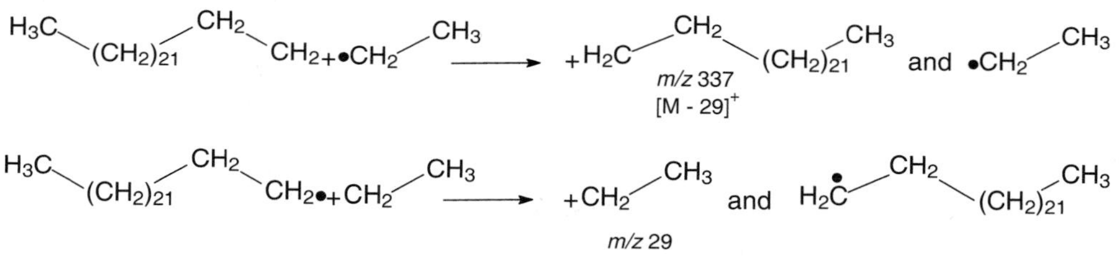

Scheme 6-21

As seen in Scheme 6-21, σ-bond cleavage can result in the formation of either an ethyl ion or the [M – 29]$^+$ ion. Formation of the ethyl ion is favored due to the increased stability accompanying the $C_{24}H_{49}$ radical compared to that of the ethyl radical that accompanies the [M – 29]$^+$ ion. This is the basis for the *Stevenson rule* [21], which states (paraphrased from McLafferty [6]):

> Cleavage of a carbon–carbon bond in an odd-electron ion due to the loss of a σ-bond electron can lead to two sets of ion-radical products; i.e., the M$^{+\bullet}$ ABCD$^{+\bullet}$ can produce A$^+$ + $^\bullet$BCD or A$^\bullet$ + $^+$BCD. The fragment with the higher ionization energy (IE) should be the fragment with the greater tendency to retain the unpaired electron. This means that there should be a higher probability for the formation of the ion corresponding to the species that has the lower ionization energy. The ions with the lower ionization energy are usually more stable and, therefore, will account for the more abundant of the two ions.

The Stevenson rule also corresponds to the *loss of the largest alkyl* radical statement often used in the description of EI mass spectra. Looking at mass spectra of compounds that form aliphatic ions through single-bond cleavage, it will be seen that these spectra are dominated by peaks that represent ions formed by the loss of the largest radical moiety. This *loss of the largest alkyl* will dominate over the *greatest stability of the ion formed* principle. It is a well-established fact that a tertiary carbenium ion is more stable than a secondary carbenium ion. Looking at the partial mass spectrum of 3-methyl-*n*-heptane (Figure 6-14), it is clear that the three secondary carbenium ions

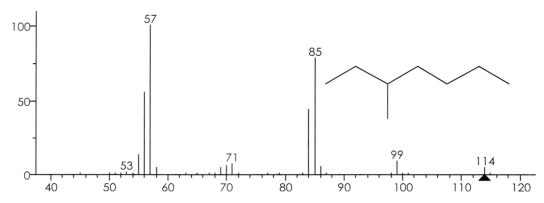

Figure 6-14.   Abbreviated EI mass spectrum of 3-methyl heptane.

formed by the respective loss of a butyl (*m/z* 57), ethyl (*m/z* 85), or methyl radical (*m/z* 99) from the M$^{+\bullet}$ are represented by peaks of decreasing intensity. This spectrum supports the *loss of the largest alkyl principle* and the Stevenson rule. The ion that should be the most stable (the tertiary carbenium ion formed by the loss of a hydrogen radical) is the least abundant; this observation is consistent the *loss of the largest group* and the instability of the hydrogen radical.   Although there are three independent factors in sigma-bond cleavage (stability of the ion formed, stability of the neutral loss, and tendency to lose the largest moiety), all three must be considered as a whole, rather than from just their individual contributions.

The series of peaks that starts with an *m/z* value of x and proceeds to about *m/z* x − 4 represents ions formed by successive losses of hydrogen radicals and hydrogen molecules from the ion with *m/z* x. These patterns are characteristic of aliphatic ions. There are oxonium and acylium ions that have *m/z* values corresponding to those of alkyl ions, but the mass spectrum representing these ions does not exhibit the secondary fragmentation pattern associated with the presence of alkyl ions.   The presence and absence (or minimization) of these secondary fragmentation patterns can be used to distinguish between alkyl ions and ions containing oxygen atoms that have the same nominal *m/z* values.

Not restricted to the mass spectra of hydrocarbons, alkyl ions formed in the mass spectra of many compounds can undergo secondary fragmentation through the loss of a molecule of methane, ethene, or larger olefins. These losses result from rearrangements that take place in the even-electron fragment ions to produce lighter even-electron product ions. A good example is seen in the spectrum of *n*-butylbromide (Figure 6-26 later in this chapter), which has peaks at *m/z* 41 and *m/z* 29 representing ions formed by the respective loss of methane (16 Da) and ethene (28 Da) from an even-electron butyl ion with *m/z* 57.

2)   Branched Hydrocarbons

The mass spectrum of a branched hydrocarbon is very similar to that of the straight chain saturated hydrocarbon:

1.   Low intensity, but discernible M$^{+\bullet}$ peak.

2.   Peaks every 14 *m/z* units beginning with [M − 29]$^{+}$.

3.   A recognizable "ski-slope" pattern with the spectrum dominated by peaks representing C_2, C_3, C_4, C_5, and C_6 ions.

4.   A moderate secondary fragmentation pattern associated with alkyl ions.

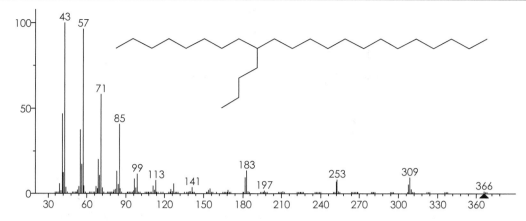

Figure 6-15.   EI mass spectrum of 9-butyldocosane, a branched aliphatic hydrocarbon.

The one significant difference in the mass spectra of a branched hydrocarbon, as seen in the mass spectrum of 9-butyldocosane in Figure 6-15, is the presence of what could be called extruding "bushes" along the "ski slope", peaks with far greater than expected intensities for a normal chain. These peaks represent secondary carbenium ions that are formed by fragmentation at the branch points in the backbone chain. The intensity of these peaks is greater than that of those representing the primary carbenium ions because secondary carbenium ions are more stable. Scheme 6-22 shows the formation of a secondary carbenium ion with the loss of one branch from the hydrocarbon backbone. Scheme 6-23 shows the formation of another secondary carbenium ion due to the loss of a different branch from the hydrocarbon backbone.   These two ions are represented by bushes sticking up above the ski-slope at *m/z* 182 and 253, respectively, in Figure 6-15.

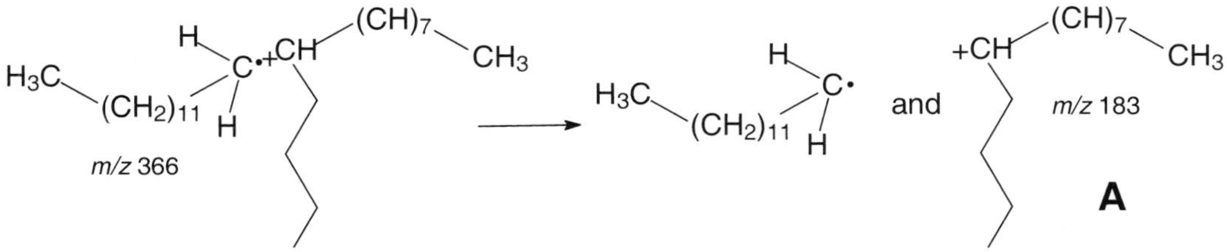

Scheme 6-22

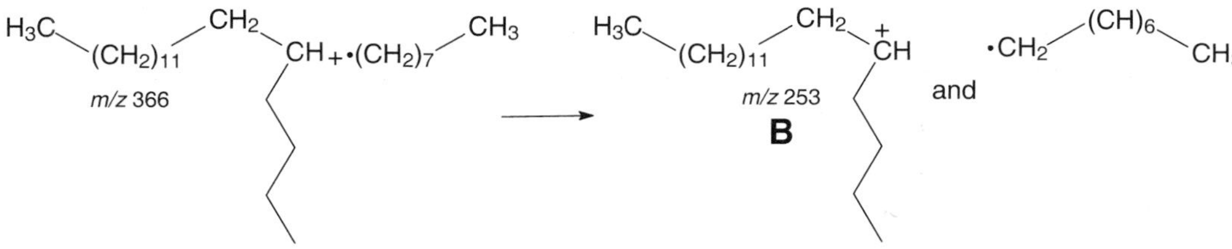

Scheme 6-23

Processing the mass spectra with Boolean logic allows determination of the location of the branch point on the hydrocarbon backbone. The branched hydrocarbon is schematically represented by Scheme 6-24. Let box A contain all the carbon atoms that make up the secondary carbenium ion in Scheme 6-22. Let box B contain all the carbon atoms that make up the secondary carbenium ion shown in Scheme 6-23. As illustrated in

Scheme 6-24, the total number of carbon atoms in the molecule subtracted from the sum of the carbon atoms in boxes A and B is equal to the number of carbon atoms in the branch plus one (the branch point carbon).

To determine the number of carbon atoms in a saturated hydrocarbon (branched or not), subtract two from the nominal mass ($366 - 2 = 364$ in this example) and divide by 14, which is the mass of a $-CH_2-$ unit. In this example (Figure 6-15), the total number of carbon atoms in the hydrocarbon that has a single branch point is 26 ($= 364 / 14$). To determine the number of carbon atoms in either box A or box B, subtract 1 from the mass of the carbenium ion represented by the particular box and then divide the resulting number by 14. For the current example, the result of this arithmetic process indicates that the number of carbon atoms in Box A is 13 ($= [183 - 1] / 14$) and the number of carbon atoms in Box B is 18. Substituting these numbers into the Boolean equation shown in Scheme 6-24 yields the number 5 ($= 13 + 18 - 26$), which is the sum of the carbon atoms in the branch plus the branch point carbon; this means that the branch consists of four carbon atoms, a butyl group. The next step is to determine the location of the branch point. By convention, the branch point is designated numerically from the nearest terminus of the main carbon chain; for this reason, box A was chosen for computation because it contained fewer carbons than box B. In this case, box A contains 13 carbon atoms; because 4 of these 13 carbon atoms are in the form of a butyl side chain, the other 9 carbon atoms must be in the main chain. Therefore, the butyl group is attached to carbon 9.

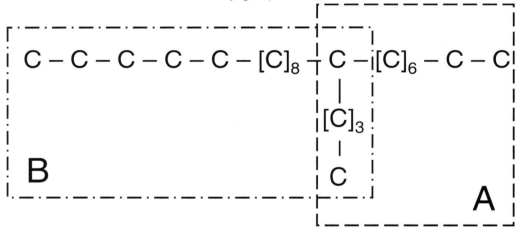

$$\text{Carbons}_A + \text{Carbons}_B - \text{Carbons}_{Total} = \text{Branch Point} + \text{Branch}$$

Scheme 6-24

The above computational example based on Boolean logic was contrived so that the number of carbon atoms in box A would be consistent with the correct name of the compound, namely 9-butyldocosane. Any two of the three high-intensity peaks representing secondary carbenium ions could have been assigned to box A and box B, and the resulting structure would be the same although the numbering would not be consistent with IUPAC nomenclature; e.g., 5-octyloctadecane. Whenever the *m/z* values for peaks representing secondary carbenium ions can be identified, the location of this secondary carbon atom can be established. Another example of the use of this Boolean logic is illustrated later (Figure 6-89) in this chapter in the context of using mass spectral data to establish the location of a double bond in the hydrocarbon backbone of an aliphatic ester.

Although the intensity of the M$^{+\bullet}$ peak is easily discernible in the mass spectrum of 9-butyldocosane, as the number of branches in a hydrocarbon increases, identification of a M$^{+\bullet}$ peak can be difficult. This is especially true in data acquired by rapid scanning (problem relates to poor ion-counting statistics) GC/MS instruments or in a quadrupole ion trap where the time between ion formation and ion detection can be quite long (the problem relates to a relatively short half-life of the M$^{+\bullet}$). Using chemical ionization in an internal ionization quadrupole ion trap mass spectrometer with acetonitrile as the reagent gas can be a good way to determine the nominal mass of these highly branched compounds [22].

3) Cyclic Hydrocarbons

The M$^{+\bullet}$ peak of a cyclic hydrocarbon is more intense than the M$^{+\bullet}$ peak for either straight-chain or branched saturated hydrocarbons (each of these latter compound types can be referred to as *linear hydrocarbons* to differentiate them from *cyclic hydrocarbons*). The reason for the greater intensity of the M$^{+\bullet}$ peak for the cyclic compound is due to the fact that cleavage of a single bond does not result in the production of a fragment ion. Breaking of a carbon–carbon bond in the cyclohexane M$^{+\bullet}$ results in an ion with the charge and radical sites on nonadjacent carbon atoms (see Scheme 6-25), but all of the atoms originally in the analyte are still connected to one another.

The distonic ion of cyclohexane produced by σ-bond cleavage (see Scheme 6-25) undergoes various reactions involving hydrogen-shift rearrangements, followed by homolytic or heterolytic cleavages resulting in single-bond fragmentations. Like the mass spectra of linear hydrocarbons, the mass spectra of cyclic hydrocarbons will exhibit peaks every 14 *m/z* units; however, the peak pattern and the appearance of the peak clusters will be more erratic than observed in the spectra of the linear hydrocarbons. A good example is the mass spectrum of cyclohexane shown in Figure 6-16.

As can be seen from the fragmentation pathways illustrated in Scheme 6-25, many of the ions in the cyclohexane spectrum can result from fragmentation of the M$^{+\bullet}$; however, these fragmentations may be induced by hydrogen shifts; e.g., the peaks at *m/z* 69 ([M – CH$_3$]$^+$), *m/z* 55 ([M – C$_2$H$_5$]$^+$), and *m/z* 41 ([M – C$_3$H$_7$]$^+$) represent ions resulting from homolytic cleavages induced by the respective shifts of a hydrogen atom from the number 4, 3, and 2 carbon atoms (the number 1 carbon atom is the site of the charge) to the radical site on the number 6 carbon. It is possible to have competing pathways for ion formation. Not only does the ion with *m/z* 41 result from a homolytic cleavage of M$^{+\bullet}$ induced by a hydrogen shift, this fragment ion can also be the result of a heterolytic cleavage of the ion

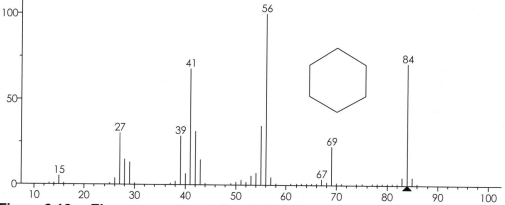

Figure 6-16. *El mass spectrum of cyclohexane, a cyclic saturated hydrocarbon.*

with *m/z* 69. Fragmentations of fragment ions contribute more to the appearance of the mass spectrum of cyclic saturated hydrocarbons than they do to the appearance of the spectra of aliphatic saturated hydrocarbons. The peaks at *m/z* 43 and *m/z* 29 represent ions formed by hydride-shift rearrangements, also shown in Scheme 6-25. It is clear from the spectrum in Figure 6-16 and the mechanism proposed in Scheme 6-25 that a number of different simultaneous reactions take place.

As the ring becomes increasingly substituted by alky groups, the relative intensity of the M$^{+\bullet}$ peak decreases; this is especially true when a given carbon has more than one substituent. Compare the spectra of 1-ethyl-1-methyl cyclohexane and 1-ethyl-2-methyl cyclohexane in Figure 6-17. The M$^{+\bullet}$ peak in the mass spectrum of 1-ethyl-2-methyl cyclohexane is obvious at *m/z* 126, whereas in the mass spectrum of 1-ethyl-1-methyl cyclohexane, for all practical purposes, there is no M$^{+\bullet}$ peak.

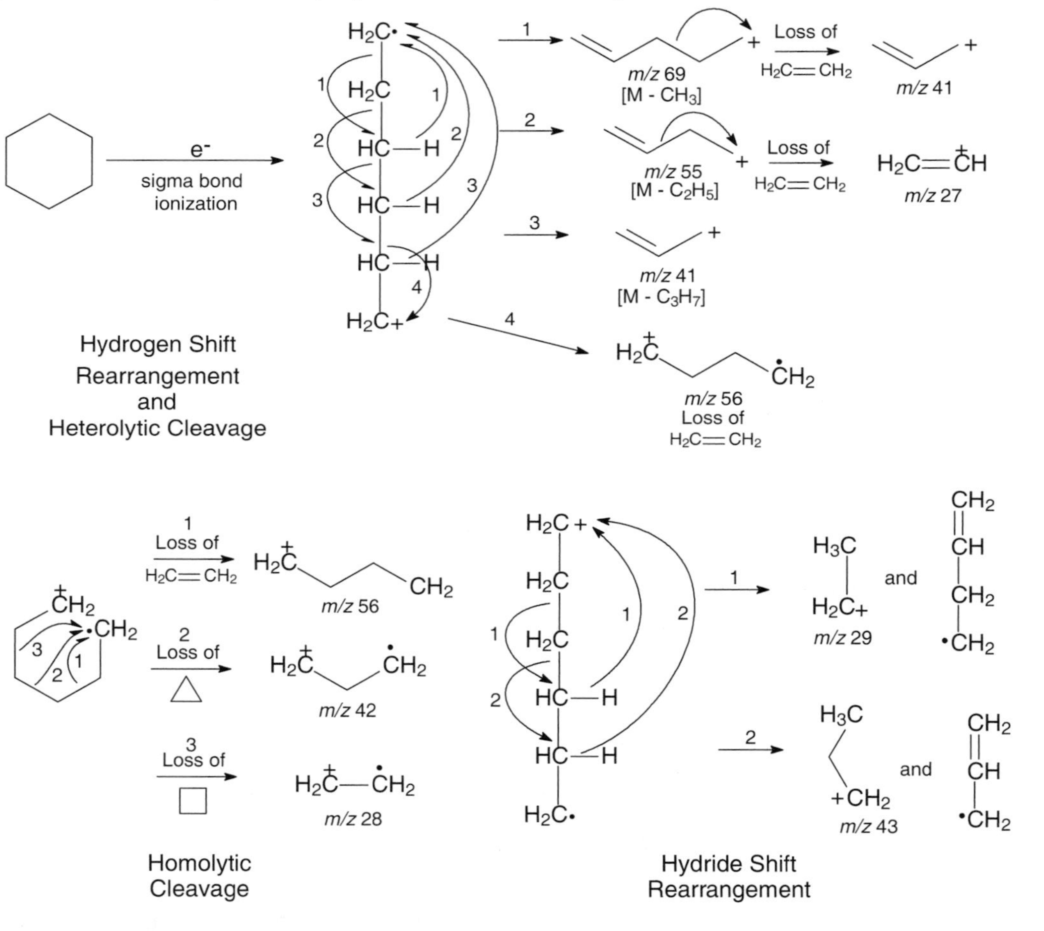

Scheme 6-25

Substitution of heteroatoms into the ring or attachment of functional groups to the ring will influence the fragmentation patterns. The principles associated with location of the charge and radical sites on heteroatoms, discussed for specific types of aliphatic compounds later in this chapter, provides a rationale for the formation of these patterns. It

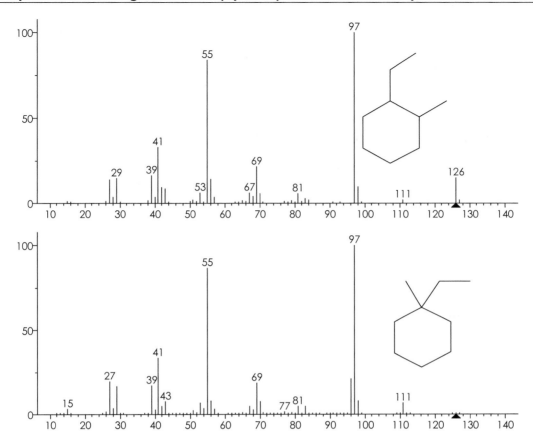

Figure 6-17. *EI mass spectrum of 1-ethyl-2-methyl cyclohexane (top) and 1-ethyl-1-methyl cyclohexane (bottom).*

is important to remember that an initial bond cleavage in a cyclic molecular ion usually does not result in the formation of a fragment ion.

> When the EI mass spectra of a given class of compounds becomes of interest, study spectra of as many compounds in that class as possible. Use the NIST Mass Spectral Database for this purpose.

B. Unsaturated Hydrocarbons

Unsaturated hydrocarbons can be straight-chain, branched, or cyclic hydrocarbons. These compounds are called alkenes because they have one or more double bonds. They do not have the resonance structure that simulates alternating double and single bonds as do aromatic compounds. They do have one or more pairs of π-bond electrons. The most probable site of the charge and radical sites in the M$^{+\bullet}$ of an alkene is between the two carbon atoms originally connected by the double bond. The M$^{+\bullet}$ is formed by the loss of a

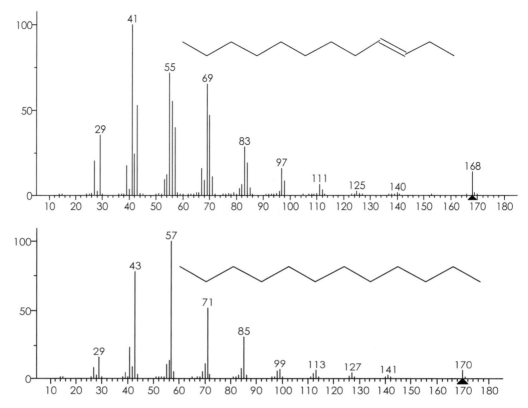

Figure 6-18. **EI mass spectra of 3-dodecene (bottom), a linear hydrocarbon with a single site of unsaturation, and n-dodecane (top).**

π-bond electron. A comparison of the mass spectra of *n*-dodecane and 3-dodecene in Figure 6-18 shows the similarities and differences in the mass spectra of the two types of compounds. Both compounds are straight chain with a total of 12 carbon atoms. Both mass spectra show a $M^{+\bullet}$ abundance of 2–5% of the total ion current. Both mass spectra exhibit a ski-slope pattern with peaks representing C_3, C_4, C_5, and C_6 ions dominating the spectrum and distinct clusters of peaks every 14 *m/z* units as they approach the $M^{+\bullet}$ peak. A significant difference in these two spectra is the $[M - 29]^+$ peak for *n*-dodecane vs the $[M - 28]^+$ peak for 3-dodecene. The $[M - 28]^+$ peak for 3-dodecene represents the loss of a molecule of ethene to form an odd-electron fragment ion. Another important difference in the two spectra is the increased abundance of the alkenyl and alkynyl ions for 1-dodecene that increment by two and four *m/z* units lower than the peak for the alkyl ions; e.g., *m/z* 41 and *m/z* 39 vs *m/z* 43. There are more peaks for OE ions in the mass spectrum of dodecene (e.g., *m/z* 42, 56, 70, etc.) than in that of dodecane; these represent odd-electron ions resulting from various olefin expulsions from the molecular ions of the alkene. The loss of ethene from an olefin is always observed when there is a terminal double bond; this loss is also sometimes observed when the double bond is deeper in the carbon chain.

C. Aromatic Hydrocarbons

Aromatic hydrocarbons are compounds composed of a ring structure that have a specific number of delocalized electrons so that the structure is consistent with alternating double and single bonds. The simplest such structure is a benzene ring. Aromatic compounds can be monocyclic or polycyclic. The most common monocyclic aromatic compounds are substituted benzenes. Some of the most studied polycyclic aromatic compounds are polycyclic aromatic hydrocarbons (PAHs), sometimes referred to as polynuclear aromatic hydrocarbons (PNAs) because of their health and environmental significance. The PAHs are composed of multiple fused aromatic rings. One of the most common PAHs is naphthalene, a compound used for control of moth infestation in clothing storage areas and as a disinfectant in urinals in public toilets.

The mass spectra of aromatic hydrocarbons (mainly those containing only carbon and hydrogen with all the carbon atoms arranged with conjugated double bonds and without substitutions) are very different from the mass spectra of aliphatic hydrocarbons. When molecular ions are formed from these aromatic molecules, it is due to the loss of a π-bond electron. This loss does not result in the cleavage of any bond. The resonance structure delocalizes the energy of the ionizing electron. This stabilization means that the mass spectra of these compounds are dominated by the $M^{+\bullet}$ peak, which is usually the base peak in their mass spectra. The mass spectrum of chrysene (a polycyclic aromatic hydrocarbon that has no commercial use and is only a laboratory curiosity) shown in Figure 6-19 is a good example of the spectra produced from aromatic hydrocarbons. The $M^{+\bullet}$ peak at m/z 228 is the base peak. There is a peak at m/z 226 that suggests that some of the $M^{+\bullet}$ ions expel a molecule of hydrogen. The peak at m/z 227 is too intense to be exclusively the ^{13}C-isotope peak related to the m/z 226 peak; it is likely that some of the ion current at m/z 227 represents the loss of a hydrogen radical from the $M^{+\bullet}$. There are peaks at m/z 114 and m/z 113 that represent ions having two charges [23], i.e., the nominal mass of chrysene is 228 Da; the ion with a mass of 228 Da and two charges will have an m/z value of 114, the mass divided by the number of charges. The one observation that may be somewhat confusing is the intensity ratio of the peaks at m/z 113 and 114, which is about the same as that of the peaks at m/z 226 and 228. The abundance ratio of $[M - H_2]^{+\bullet}$ and $M^{+\bullet}$ is about 1:4; the fact that the abundance ratio of the double-charge ions is nearly the same is not readily explained.

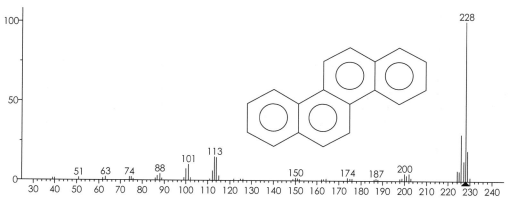

Figure 6-19. *El mass spectrum of chrysene, a polynuclear aromatic hydrocarbon.*

> A philosophical question. A molecular ion is formed by
> the addition or removal of an electron. Is a double-charge
> ion that retains all the atoms of the original molecule a
> molecular ion?

More likely to be encountered than the PAHs are compounds that contain a phenyl group. The mass spectra of compounds containing a phenyl group have the following characteristics:

1. $M^{+\bullet}$ peaks of significant intensity.

2. A dearth of peaks representing fragment ions.

3. Low-intensity peaks at *m/z* 39 and 65 representing secondary fragmentation of compounds with a benzyl moiety.

4. A low-intensity peak at *m/z* 51 for all phenyl-containing compounds (this includes compounds with a benzyl moiety).

5. Peaks at *m/z* 77 (of wildly varying intensity) for all phenyl-containing compounds.

6. Peaks of significant intensity at *m/z* 91 representing a tropylium ion for compounds containing a benzyl moiety.

7. An examination of the mass spectra of benzene and toluene shown in Figure 6-20 illustrates the difference a substituent on the phenyl ring makes in the appearance of the spectrum. The base peak in the mass spectrum of benzene is the $M^{+\bullet}$ peak, whereas in the toluene mass spectrum the $M^{+\bullet}$ peak is only about 20% of the intensity of the base peak at *m/z* 91 representing the tropylium ion.

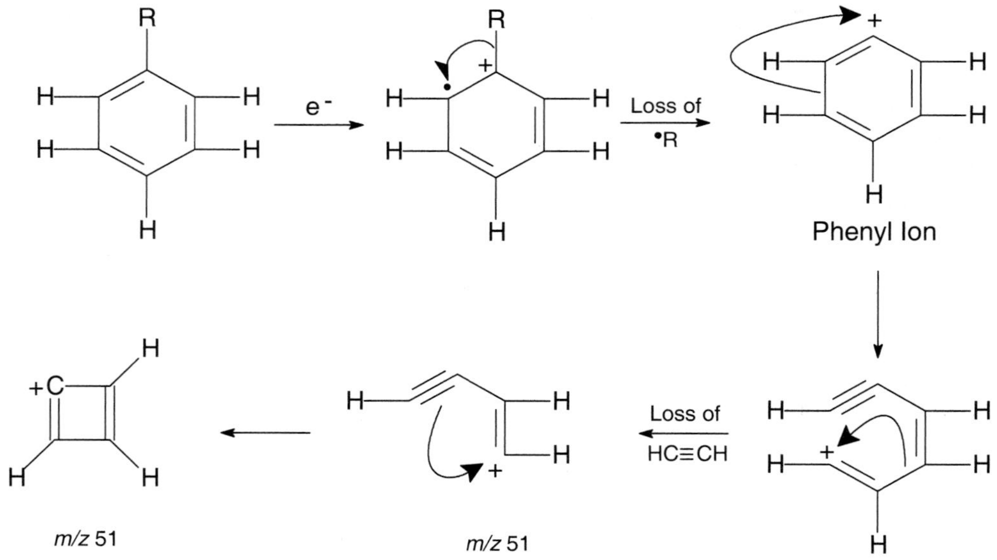

Scheme 6-26

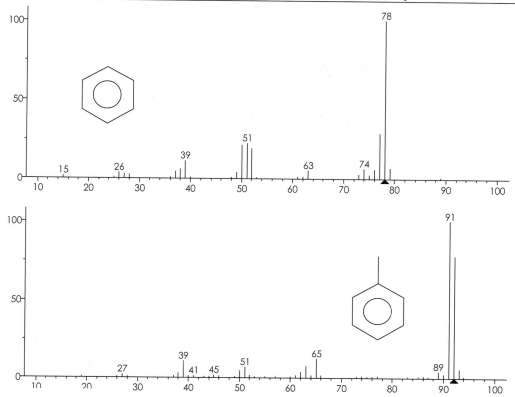

Figure 6-20. *El mass spectra of benzene (top) and toluene (bottom).*

Scheme 6-26 illustrates a fragmentation mechanism that results in the formation of the ion with *m/z* 51 from a phenyl ion. Scheme 6-27 illustrates the two-step fragmentation of the tropylium ion with successive losses of acetylene in the formation of ions with *m/z* values of 65 and 39. There is a peak at *m/z* 39 in the mass spectrum of benzene. However, benzene does not have a benzyl moiety, which, if present, could participate in the formation of a tropylium ion (see Scheme 6-28), which is the precursor of the ion with *m/z* 65. According to Scheme 6-27, the ion with *m/z* 65 is the precursor of the ion with *m/z* 39. However, there is no peak at *m/z* 65 in the mass spectrum of benzene. So where does the ion with *m/z* 39 come from in the mass spectrum of benzene? What is the number 39, besides the age of my coauthor's girlfriend? Thirty-nine is half of seventy-eight! The peak at *m/z* 39 in the mass spectrum of benzene represents a double-charge ion of the intact benzene molecule.

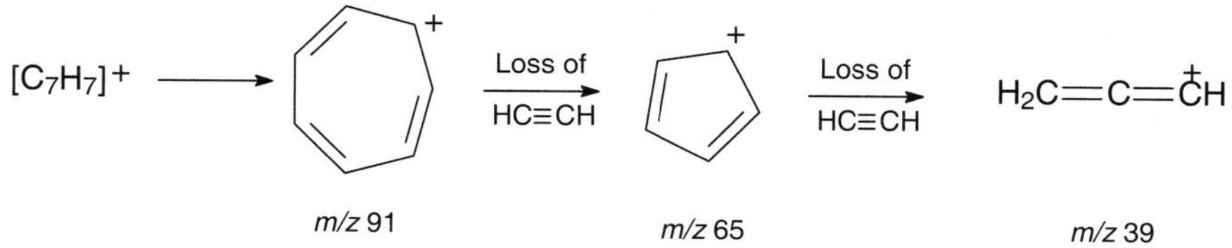

$[C_7H_7]^+$ ⟶ $\xrightarrow[\text{HC≡CH}]{\text{Loss of}}$ $\xrightarrow[\text{HC≡CH}]{\text{Loss of}}$ $H_2C{=}C{=}\overset{+}{C}H$

m/z 91   *m/z* 65   *m/z* 39

Scheme 6-27

Scheme 6-28 illustrates the formation of the tropylium ion [24] from the M⁺• of toluene. This cleavage of a hydrogen–carbon bond on the carbon attached to the ring with the charge and radical on the ring is called *benzylic cleavage*. Although the radical and charge sites are shown localized between carbons number 1 and number 2, it should be understood that in the actual M⁺•, the charge and radical are part of the ring structure and are not localized. The localization of the charge and radical sites is shown in Scheme 6-28 for didactic purposes only.

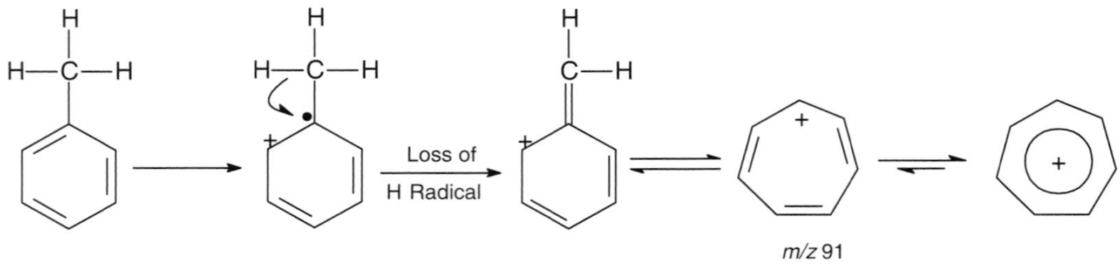

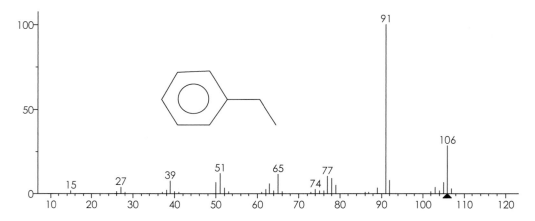

Scheme 6-28

Figure 6-21. EI mass spectrum of ethylbenzene.

Figure 6-21 is the mass spectrum of ethylbenzene and Figure 6-22 shows the spectra of the three regioisomers of xylene; the four compounds have the elemental composition C_8H_{10} configured on a single benzene ring. For all intents and purposes, these spectra are essentially identical, with the exception that the intensity of the peak at *m/z* 105 in the mass spectrum of ethylbenzene is far less than that in the mass spectra of the three regioisomers of xylene. As seen from the displayed structures on each of the spectra, molecular ions of all four compounds have structural features that can participate in benzylic cleavage. However, the three regioisomers of xylene can only lose a hydrogen radical in the formation of the tropylium ion, which in this case becomes a methyl-substituted tropylium ion as represented by the peak at *m/z* 105 of rather significant intensity. Ethylbenzene can lose either a hydrogen radical or a methyl radical. From a statistical standpoint, it would seem more likely that a hydrogen radical would be lost because there are two hydrogens and only one methyl group on the benzylic carbon. However, the low intensity of the peak at *m/z* 105 in Figure 6-21 indicates that this is not

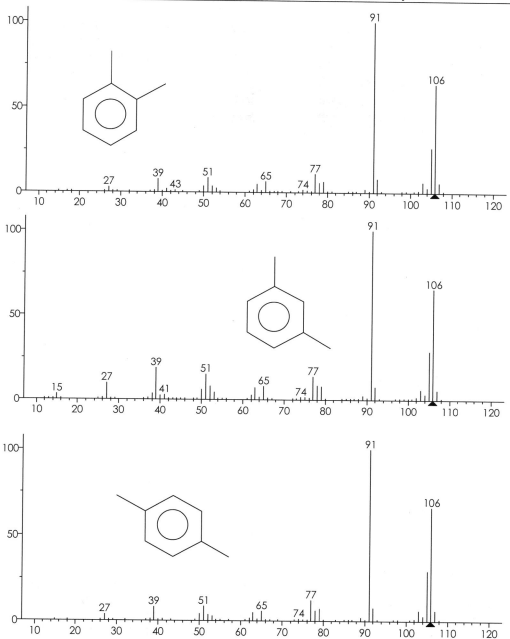

Figure 6-22. **El mass spectra of o-xylene (top), m-xylene (middle), p-xylene (bottom).**

the case. The favored loss of a methyl radical over a hydrogen radical during benzylic cleavage of the ethylbenzene $M^{+\bullet}$ relates to the greater stability of a methyl radical than a hydrogen radical. There is a peak at m/z 105 in the mass spectrum of ethylbenzene, albeit very small; therefore, it cannot be said that the $M^{+\bullet}$ will not lose a hydrogen radical. It is just that the loss of the hydrogen radical is *much less likely* than the loss of a methyl radical. It should also be noted that all four spectra are dominated by a peak at *m/z* 91. In the case of ethylbenzene, the tropylium ion is formed directly from M$^{+\bullet}$. In the case of the xylene isomers, the peak at *m/z* 91 represents a tolyl ion [24, 25] due to the loss of a methyl radical from the xylene M$^{+\bullet}$.

As would be expected, the base peak in the mass spectrum of *n*-propylbenzene, shown in Figure 6-23, is *m/z* 91, representing a tropylium ion. The mass spectrum in Figure 6-23 is typical of the mass spectra of many aromatic compounds:

1. Few peaks representing fragment ions.

2. An intense M$^{+\bullet}$ peak.

3. A dominant base peak representing a phenyl or tropylium ion, or a substituted phenyl or tropylium ion.

4. Minor peaks at *m/z* 39, 51, and 65.

A structural isomer with the same elemental composition as propyl benzene (C_9H_{12}), 1-ethyl-4-methylbenzene (4-ethyltoluene) produces a mass spectrum (Figure 6-24) very different from that of propylbenzene. Benzylic cleavage involving the ethyl moiety results in the loss of a methyl radical and the formation of a methyl-substituted tropylium ion with *m/z* 105 (the base peak). The methyl radical will be lost in preference to the loss of a hydrogen radical from either of the two alkyl groups attached to the ring because of the methyl radical's greater stability.

Careful examination of the spectra in Figures 6-21 and 6-22 can lead to a rather uncomfortable conclusion, which is that regioisomers of aromatic hydrocarbons cannot be distinguished by their mass spectra. It is clear that whereas the mass spectra of positional isomers such as propylbenzene and 4-ethyltoluene have very different appearances, the mass spectra of the three xylene regioisomers cannot be distinguished. However, because mass spectrometry for these types of compounds is often performed in conjunction with gas chromatography, these xylene isomers (with a unique mass spectrum as a group) can be distinguished from one another by their differences in retention times. As will be seen later in this chapter, when substitutions to the ring involve heteroatoms, the *ortho effect* can, in some cases, allow the *ortho* isomer to be distinguished from the *meta* and *para* isomers by the appearance of their mass spectra. The *meta* and *para* isomers can only be distinguished from one another by their retention time differences.

All of the mass spectra of aromatic hydrocarbons containing a phenyl ring exhibit an easily discernible M$^{+\bullet}$ peak at an even *m/z* value and peaks representing fragment ions at odd *m/z* values. These spectra are consistent with the *Nitrogen Rule*, which was discussed in Chapter 5.

The mass spectrum of *n*-pentylbenzene (Figure 6-25) is characterized by the unusual feature of a significant peak in the fragmentation region at an even *m/z* value, *m/z* 92. The peak intensity at *m/z* 92 is much greater than can be explained from the stable isotope contribution from the species represented by the peak at *m/z* 91. When such a situation as this is observed in the mass spectrum of an unknown there should be concern that this *m/z* 92 peak might indicate the presence of superimposed spectra of two compounds, where the peak at *m/z* 92 might represent the M$^{+\bullet}$ of a contaminant. As was the case for all the spectra viewed so far in this class of aromatic hydrocarbons (compounds that do not contain an odd number of nitrogen atoms), it would be expected that all fragment ions be represented by peaks at odd *m/z* values (consistent with the *Nitrogen Rule*).

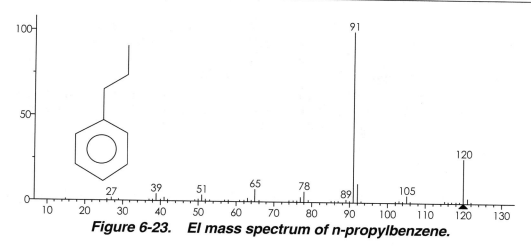

Figure 6-23.   EI mass spectrum of n-propylbenzene.

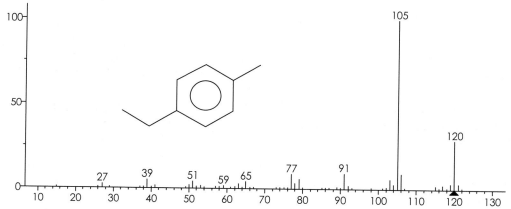

Figure 6-24.   EI mass spectrum of 1-ethyl-4-methylbenzene (4-ethyltoluene).

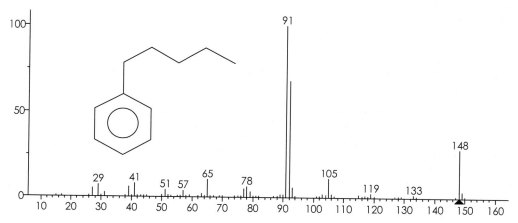

Figure 6-25.   EI mass spectrum of n-pentylbenzene.

One way to eliminate the possibility of superimposed spectra would be to examine each of several consecutively recorded mass spectra, assuming the sample was analyzed by GC/MS, to determine whether the peak intensity at *m/z* 92 is consistently 40% of that at *m/z* 91 in all spectra. Another way would be to prepare mass chromatograms (Chapter 1) for *m/z* 91 and *m/z* 92 to determine whether the chromatographic peaks rise and fall together (which would suggest a pure sample). Having ruled out the possibility of superimposed spectra associated with a contaminant or coeluting analyte, observing a peak at an even *m/z* value in the fragmentation region of the mass spectrum is an indicator that a rearrangement has occurred during fragmentation of the M⁺•.

Recalling Scheme 6-13 earlier in this chapter, a rearrangement can occur when there are hydrogen atoms on a moiety that is gamma in position to the sites of the charge and the radical is an odd-electron ion. These hydrogen atoms are called γ hydrogens. The peak at *m/z* 92 in the mass spectrum of *n*-pentylbenzene can be explained as a rearrangement of a γ hydrogen, which subsequently promotes a homolytic cleavage. The nascent radical site on the γ carbon formed by the γ-hydrogen shift in response to the radical site in the M⁺• stimulates cleavage at the β carbon (the bond between the β- and α-carbon atoms) of the pentyl side chain as illustrated in Scheme 6-29. Observing the peak representing a rearrangement in the mass spectrum of *n*-pentylbenzene is good evidence that most of the carbons in the side chain are in a normal chain. It is clear, for example, that there is no branch at the α carbon because if there were, the branch would be carried along with the α carbon in the rearrangement fragmentation product, and the peak at *m/z* 92 would be shifted to *m/z* 106 if a methyl group were on the α carbon or to *m/z* 120 if an ethyl group were on the α carbon. No evidence would be given in the mass spectrum for branching at the β carbon in the side chain as no fragment ions containing the β carbon and its substituents carry a positive charge. It is unlikely that any branching occurs at the γ carbon, although the rearrangement would still be possible if at least one γ hydrogen remained.

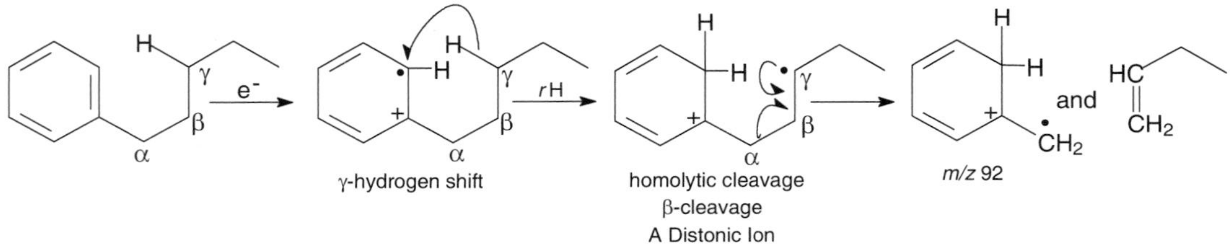

Scheme 6-29

Indirect evidence for all five carbons being in a straight chain is given by the fact that *m/z* 92 is quite intense in the mass spectrum. Compare the intensity of the peak at *m/z* 92 in the spectrum of *n*-propylbenzene in Figure 6-24 (top) with the intensity at *m/z* 92 in the mass spectrum of *n*-pentylbenzene (Figure 6-25). Propylbenzene has three γ hydrogens; therefore, the rearrangement is possible. However, notice that the intensity at *m/z* 92 in the mass spectrum of *n*-propylbenzene is only slightly higher than would be expected from the isotopic composition of the ion with *m/z* 91. This low intensity means that the rearrangement process is not highly favored in the case of *n*-propylbenzene because the intermediate radical produced by the γ-hydrogen shift is a primary radical as opposed to a secondary radical in the rearrangement of *n*-pentylbenzene, as shown in the first step in Scheme 6-28. It can be concluded that when the γ hydrogens are on a terminal carbon (i.e., an ω carbon), the rearrangement process is frequently suppressed. Alternatively, as will be seen in the discussion of rearrangement fragmentations associated with aliphatic ketones later in this chapter where there are two sets of γ hydrogens and one set is on an ω carbon, this suppression effect is not a factor.

2. Alkyl Halides

Alkyl halides (a.k.a. haloalkanes) are composed of a saturated hydrocarbon (straight-chain or branched) and one or more halogen (fluorine, chlorine, bromine, or iodine) atoms. One-, two-, three-, and four-carbon alkyl halides are important environmental compounds that are found as chlorine disinfection by-products in drinking water; they are also used as solvents (CCl_4), aerosol propellants (CCl_3F – Freon 11), fire extinguishers (CF_3Br – Halogen 1301), anesthetics ($CF_3CHClBr$ – Halothane or Fluothane), and in the production of plastics such as polyvinyl chloride (PVC) and polytetrafluorethene (PTFE – Teflon®).[1] Larger carbon chain alkyl halides are used in the chemical industry; e.g., lauryl chloride (1-chlorododecane) is used in the manufacture of photographic chemicals, pharmaceuticals, and surfactants. These higher nominal mass compounds, as a class, pose less of a health hazard than short-chain alkyl chlorides.

The most probable location of the charge and the radical sites in the M⁺• of an alkyl halide is associated with the halogen moiety because of the three pairs of nonbonding electrons of the halogen. As was pointed out earlier in this chapter, there is a greater tendency for heterolytic cleavage to occur than homolytic cleavage when the charge and the radical sites are on a halogen atom.

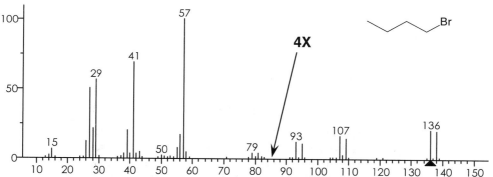

Figure 6-26. *EI mass spectrum of n-butylbromide with peak intensities above m/z 85 magnified by a factor of four.*

As seen in Figure 6-26, the base peak in the mass spectrum of *n*-butylbromide is at *m/z* 57, which represents the loss of a bromine radical from the M⁺•. The peak at *m/z* 41 represents a secondary fragment ion resulting in the loss of a molecule of methane (16 Da) from the ion with *m/z* 57. The peak at *m/z* 29 represents another secondary fragmentation product of the butyl ion resulting from the loss of a molecule of ethene (28 Da). Therefore,

---

[1] Vinyl chloride and polytetrafluorethene are not really alky halides because they contain a double bond; however, most references include these compounds in discussions about alkyhalides. Halogenated aromatic compounds will be discussed later in this chapter in the section on Multiple Heteroatoms or Heteroatoms and a Double Bond.

it can be concluded that the major fragmentation of the *n*-butylbromide $M^{+\bullet}$ occurs via heterolytic cleavage following the initial loss of a bromine radical.

The fragmentation pathways for the *n*-butylbromide $M^{+\bullet}$ are shown in Scheme 6-30. When viewing the mass spectrum of *n*-butylbromide in Figure 6-26, it should be noted that the intensities of all peaks above m/z 86 have been magnified by a factor of four.

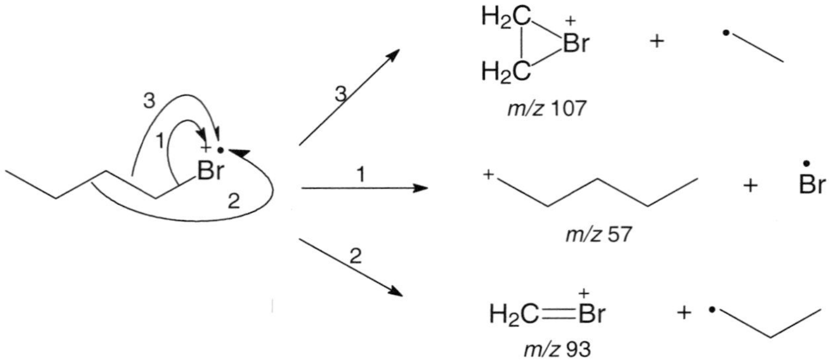

Scheme 6-30

When the mass spectra of the four possible *n*-halobutanes (Figures 6-26 and 6-27) are compared, one feature stands out. The mass spectra of the fluoro- and chloro-analogs show a base peak at m/z 56 in contrast to one at m/z 57 in the spectra of the bromo- and iodo-analogs. There are peaks of significant intensity at m/z 43 in both the spectra of the fluoro- and chloro-analogs. There is essentially no peak at m/z 43 in the mass spectra of the bromo- and iodo-analogs. The peaks at m/z 56 represent odd-electron ions produced by the loss of H–X from the $M^{+\bullet}$. The peaks at m/z 43 represent the alternative loss of a H_2C–X radical from the $M^{+\bullet}$ as a result of a heterolytic cleavage brought about by the movement of the pair of electrons between the number 1 and the number 2 carbon atoms in response to the charge site on the halogen atom.

The hydrohalogen loss to form the odd-electron ion with m/z 56 (which must involve a hydrogen shift in response to the radical site on the halogen) decreases as the size of the halogen increases. It should be noted that there is a peak of measurable intensity at m/z 56 in the mass spectrum of *n*-butylbromide, but there is no m/z 56 peak in the mass spectrum of *n*-butyliodide. These differences in the mass spectra of halocarbons where the only variable is the halogen are probably due to both the electronegativity and the steric factor associated with the size of the halogen.

In many cases, alkyl halides do not exhibit a $M^{+\bullet}$ peak. The lack of a $M^{+\bullet}$ peak is probably due to the high propensity for the loss of H–X, which will produce a distonic odd-electron fragment ion. If the hydrogen shift resulting in this distonic ion were from the number 4 carbon, a subsequent heterolytic cleavage involving the loss of a molecule of ethene would result in an odd-electron terminal olefin $M^{+\bullet}$, as was the case involving the loss of water from aliphatic alcohols. This means that the first possible peak in the mass spectrum of an alkyl chloride could correspond to [M – 64] (the loss of HCl followed by the loss of $H_2C=CH_2$). This process is especially prominent as the number of carbon atoms increases. Both chloromethane and chloroethane produce mass spectra with an obvious and intense $M^{+\bullet}$ peak, which is the base peak. The next member in this homologous series,

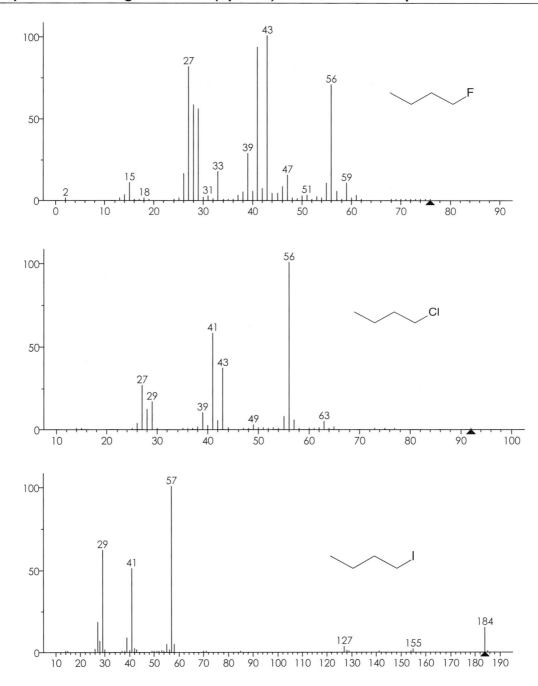

Figure 6-27. **Mass spectra of fluoro- (top), chloro- (middle), and iodo-n-butane (bottom).**

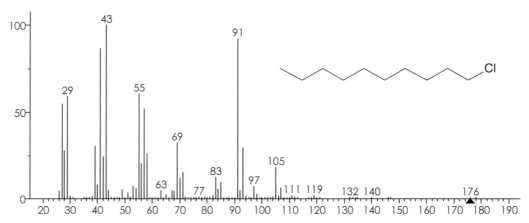

**Figure 6-28.   *El mass spectrum of 1-chlorodecane. The M⁺• peak is
vanishingly small and there are peaks of significant
intensity at m/z 91 and m/z 105.***

chloropropane, produces a mass spectrum with the base peak representing [M – HCl]⁺.
The M⁺• peak in chloropropane is vanishingly small. This trend of a very intense [M – HCl]⁺
peak continues among alkylchlorides up to C_6, at which point a peak at *m/z* 91 representing
an ion that contains an atom of chlorine (obvious because of the isotope peak at *m/z* 93,
which has an intensity about one-third that of the *m/z* 91 peak) becomes a prominent peak
in the mass spectrum. A good example of a prominent *m/z* 91 peak is seen in the mass
spectrum of 1-chlorodecane in Figure 6-28.   There is also a pair of peaks with a 3:1
intensity ratio at *m/z* 105 and *m/z* 107, which represents the six-membered ring analog of
the ion with a nominal *m/z* value of 91 (Scheme 6-31). Examination of the mass spectrum
of 1-bromodecane shows pairs of peaks with almost equal intensity at *m/z* 135 and *m/z* 137
and at *m/z* 149 and *m/z* 151, representing the bromine analogs of the $C_4H_8Cl^+$ and $C_5H_{10}Cl^+$
ions formed by alkyl chlorides. In the mass spectrum of 1-iodododecane there are peaks at
m/z 155 and 169, which could possibly represent three- and four-membered ring analogs
with the iodine being one of the ring members.

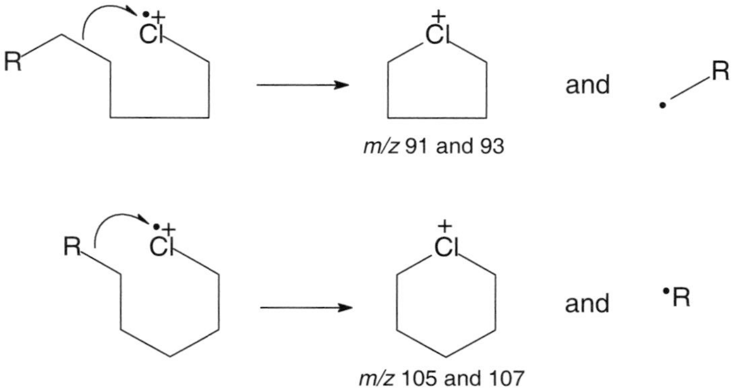

m/z 91 and 93

and

m/z 105 and 107

and   •R

Scheme 6-31

These two competing reactions, the formation of the halogen-containing rings and the loss of H–X followed by the loss of $H_2C=CH_2$ involving the formation of a ring containing the halogen, both of which have very high tendencies to occur, diminishes the possibility of a $M^{+\bullet}$ peak in the mass spectrum of an alkyl halide.

At first inspection of the mass spectrum of 1-chlorodecane there is no obvious pattern; however, if the spectrum is closely examined, excluding the peaks at *m/z* 91 and 93 and at *m/z* 105 and 107, representing the chlorine-containing ions, the ski-slope pattern of straight-chain hydrocarbons is apparent (see Figure 6-29).

In cases where the halogen (or multiple halogens) is (are) chlorine or bromine, the characteristic X+2 isotope peak patterns are obvious indicators of the number of atoms of such elements. The spectra will often have clusters of peaks separated by 14 *m/z* units in the C_2–C_6 region, as seen in the mass spectrum of 1-chlorododecane (Figures 6-27 and 6-28). Because of the lack of a $M^{+\bullet}$ peak and the similarity between the mass spectra of many species of a homologous series of 1-chloroalkanes, an identification, even using a library search, can be difficult. The same is true for other compounds composed of different halogens in this class, and retention time becomes very important for an unambiguous identification.

A library search was conducted using the NIST Main Library mass spectrum of 1-chlorododecane against the NIST05 Database. Nine of the first 20 hits were for seven different members of a $C_nH_{2n+1}Cl$ homologous series. There were four hits in this group of twenty for α-ω-dichloro-*n*-alkanes; all of these hits had *Match Factors* that were in the range of 884–720.

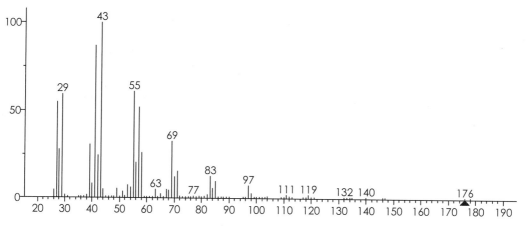

Figure 6-29. ***EI mass spectrum of 1-chlorodecane with the peaks representing the chlorine-containing ions at m/z 91 and 93 and at m/z 105 and 107 removed. Notice that the spectrum's pattern is characteristic of a straight-chain hydrocarbon.***

3. Oxygen-Containing Compounds

A. Aliphatic Alcohols

The EI mass spectra of aliphatic alcohols are some of the most widely studied of all classes of compounds in mass spectrometry.   This is because these compounds are volatile over a large molecular-mass range and they are used in a wide range of applications from disinfectants to ingredients of perfumes. The most probable location for the charge and the radical sites in the $M^{+\bullet}$ will be on the oxygen atom due to a loss of an electron from one of the two pairs of nonbonding electrons. The fragmentation of aliphatic alcohols is somewhat analogous to the fragmentation of alkyl halides in that the $M^{+\bullet}$ of alkyl halides of a certain minimum size have a tendency to lose H–X; the $M^{+\bullet}$ of alkyl alcohols of a certain minimum size (larger than the corresponding alkyl halide) have a tendency to lose water. The distonic ions formed by the loss of a molecule of water will then undergo one or more heterolytic cleavages to expel one or more molecules of $H_2C=CH_2$ to produce a terminal olefin $M^{+\bullet}$, which accounts for the unsaturated-hydrocarbon appearance of the mass spectrum. Because of the lack of a $M^{+\bullet}$ peak, it is difficult to use the mass spectra of aliphatic alcohols for identification without the concomitant use of GC retention times.

There is a peak at m/z 31 in the mass spectra of all 1°, 2°, and 3° alcohols that can be considered diagnostic for aliphatic alcohols. This peak at m/z 31 represents a primary oxonium ion ($H_2C=O^+H$) formed by homolytic cleavage of the $M^{+\bullet}$ at the α carbon (the number 2 carbon atom) as illustrated in Scheme 6-32.

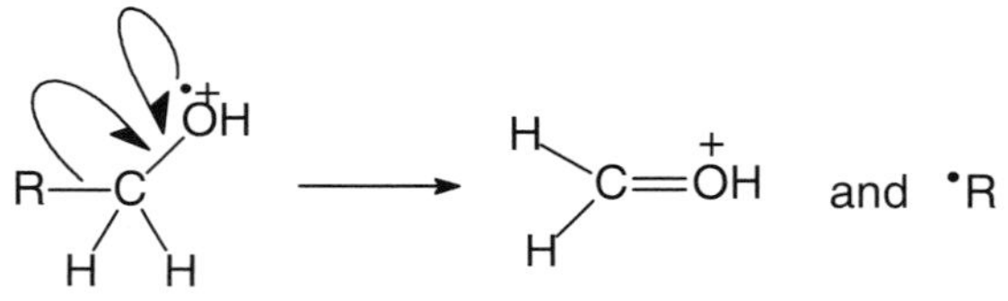

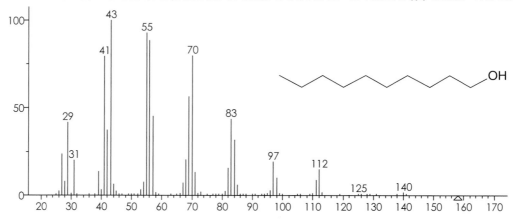

Scheme 6-32

The mass spectrum of 1-decanol, shown in Figure 6-30, is a good representation of the spectra produced by primary alcohols. There is no $M^{+\bullet}$ peak. The peak representing the ion formed by the loss of a molecule of water from the $M^{+\bullet}$ is vanishingly small. The first

Figure 6-30.   *EI mass spectrum of 1-decanol. The peak at m/z 31 represents the diagnostic primary oxonium ion.*

easily distinguishable peak, at m/z 112, represents the ion formed by the loss of $H_2C=CH_2$ from the $[M - H_2O]^{+\bullet}$ ion. The peak at m/z 31, representing the primary oxonium ion, is readily discernible in this mass spectrum. The remainder of the mass spectrum looks very much like the spectrum of 1-decene with the characteristic pattern seen in the spectra of terminal olefins.

Formation of the primary oxonium ion with an m/z value of 31 from 2° and 3° alcohols is explained in Scheme 6-33. The intensity of the peak at m/z 31 decreases as the chain length increases. The intensity of the m/z 31 peak in the mass spectrum of n-butanol is ~98% of the base peak, which represents the $[M - H_2O]^{+\bullet}$ ion, whereas the m/z 31 peak in the mass spectrum of n-decanol is ~20% of the base peak at m/z 43, which represents a propyl ion. There is also a decrease in the intensity of the m/z 31 peak in a series of mass spectra ranging from primary to tertiary alcohols. The m/z 31 peak in the mass spectrum of *tert*-butanol is only ~30% of the intensity of base peak at m/z 59, which represents the dimethyl-substituted oxonium ion. The ratio of the relative intensities of the m/z 31 peaks remains constant for primary and tertiary alcohols regardless of the main chain length.

The relative intensity of peaks in the mass spectrum is important; however, it is also important to be aware that as the abundance of an ion of a specific m/z value increases while that of another remains constant, the *relative intensity* of the second ion may appear to decrease if the first is represented by the base peak.

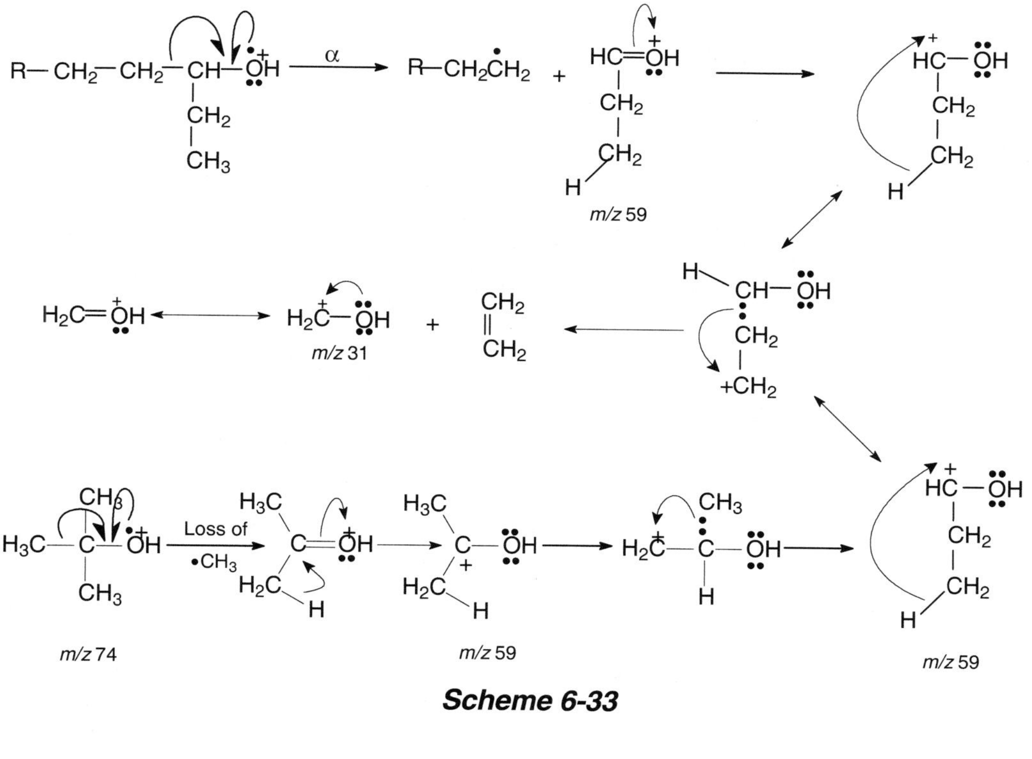

Scheme 6-33

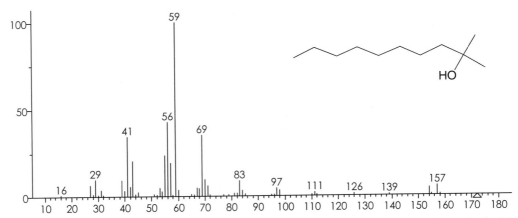

Figure 6-31. *El mass spectrum of 2-methyl-2-decanol. The peak at m/z 31 is barely distinguishable.*

Fred McLafferty and František Tureček, on page 241 of the 4th Edition of *Interpretation of Mass Spectra* [26], describes the formation of primary oxonium ions from mono- and disubstituted oxonium ions using a series of mechanisms involving the pairing of single electrons with no apparent driving force. The mechanisms in Scheme 6-33 all have a driving force related to a charge site attracting a pair of electrons.   Such cause–effect mechanisms appear more logical than the *hand-waving* associated with odd-electron pairing for no apparent reason. However, it should be remembered that all of these schemes illustrate so-called *paper mechanisms*, and a degree of faith is required to accept them as "truth".

Peaks representing mono- and disubstituted oxonium ions will be present in the mass spectra of 2° and 3° alcohols. The principle of the *loss of the largest moiety* will always be the driving force for the initial homolytic cleavage that results in formation of the oxonium ion. This is illustrated in Scheme 6-33 and corroborated by the fact that the base peak in the mass spectrum of 2-methyl-2-decanol (Figure 6-31) and 3-decanol (Figure 6-32) is at *m/z* 59.

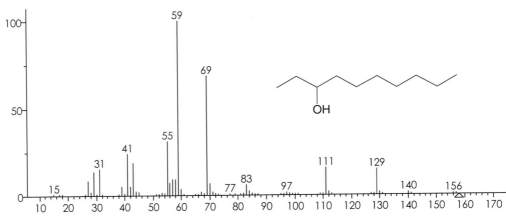

Figure 6-32.   *El mass spectrum of 3-decanol.*

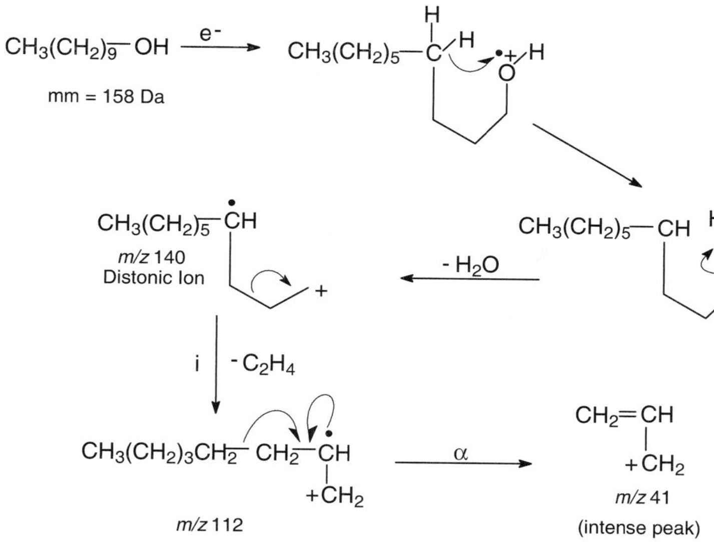

Scheme 6-34

Scheme 6-34 illustrates the loss of water from the $M^{+\bullet}$ of decanol to produce a distonic fragment ion, some of which then undergo heterolytic cleavage to expel a molecule of ethene. The first step involves a hydrogen shift from the γ carbon (in aliphatic alcohols, the number 4 carbon is the γ carbon because the number 1 carbon is a part of the *carbonyl* functional group) as facilitated by the putative existence of a virtual six-membered ring involving the oxygen atom that carries both the radical and charge sites. The second step in Scheme 6-34 involves the actual loss of water via heterolytic cleavage of the distonic molecular ion.

As illustrated in Scheme 6-35, the ion with m/z 112 produced by the 1,2-elimination of water, and then ethene, from the $M^{+\bullet}$ of 1-decanol can undergo fragmentation via multiple pathways. This is the same type of fragmentation expected for a terminal olefin such as 1-octene.

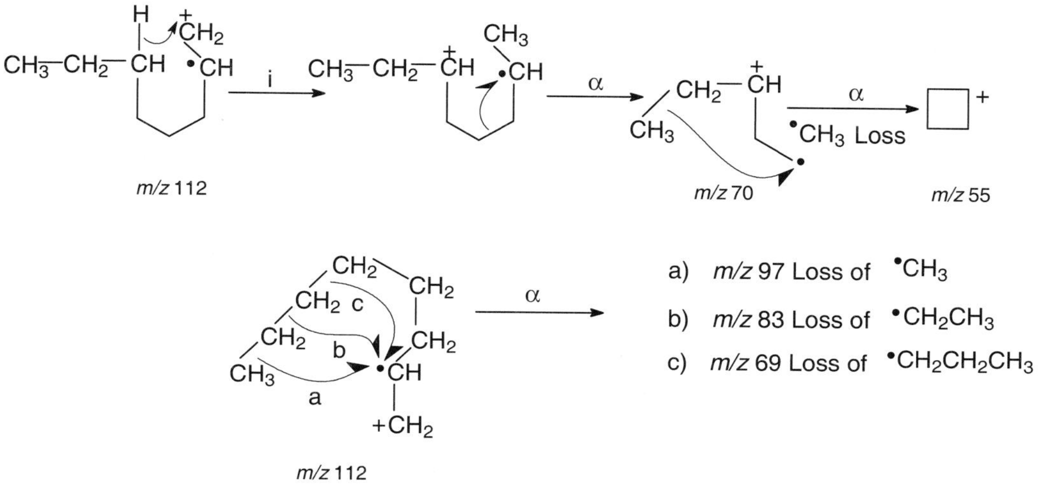

Scheme 6-35

An important consideration with respect to peaks representing substituted oxonium ions is that a peak at *m/z* 59 could represent an ethyl-monosubstituted oxonium ion or a dimethyl-disubstituted oxonium ion. It should also be kept in mind that these mono- and disubstituted oxonium ions are the source of the primary oxonium ions with *m/z* 31.

As pointed out earlier, the absence of a $M^{+\bullet}$ peak makes it difficult to identify the exact structure of an aliphatic alcohol from its mass spectrum. It is fairly easy to recognize a mass spectrum as being that of an aliphatic alcohol because of the characteristic peak at *m/z* 31. Once it is suspected that the analyte is an aliphatic alcohol, an educated guess can be made as to its nominal mass. The nominal mass will be either 28 or 46 Da greater than the highest *m/z* value peak representing an odd-electron ion of a significant abundance. To confirm the validity of this guess, it is possible to prepare the trimethylsilyl derivative (discussed later in this chapter) of the analyte for reanalysis by EI or to reanalyze the original sample by chemical ionization. It is also possible to confirm the nominal mass of an aliphatic alcohol by increasing the ion source pressure to promote the formation of an $[M + H]^{+}$ ion. When analyses are carried out in an internal ionization quadrupole ion trap *m/z* analyzer, the $[M + H]^{+}$ peak is often observed because of the high propensity for analyte molecules to react with proton-donating ions formed from ion/molecule reactions involving the analyte in this type of instrument. In this case, the proton donor would be the primary or substituted oxonium ion.

One last comment about the mass spectra aliphatic alcohols: it should not have gone unnoticed that the numerical difference between the nominal mass of an aliphatic alcohol and the value of the highest *m/z* peak of a reasonable intensity in the mass spectrum is often 46. The nominal mass of a natural alcohol, ethanol, is 46 Da. For some time it was thought that the $M^{+\bullet}$ of higher nominal mass alcohols expelled a molecule of ethanol; however, this is not the case. The apparent $[M - 46]^{+}$ peak is actually an $[(M - H_2O) - H_2C=CH_2]^{+}$ peak.

B. Aliphatic Ethers

As was the case with aliphatic alcohols, the most probable site for the charge and the radical will be on the oxygen atom due to the loss of an electron from one of its two pairs of nonbonding electrons. As long as the two alkyl branches consist of three carbons or less, there will be an easily observable $M^{+\bullet}$ peak in the mass spectrum. When one of these branches consists of four or more carbon atoms, the $M^{+\bullet}$ peak may be very weak, although more intense than the (nearly nonexistent) $M^{+\bullet}$ peak in the mass spectra of aliphatic alcohols. Another feature of the mass spectra of aliphatic ethers (where one of the branches has two or more carbon atoms) that is similar to the mass spectra of aliphatic alcohols is a peak at *m/z* 31 representing a primary oxonium ion. Aliphatic ethers and aliphatic alcohols are the only compounds that produce ions with *m/z* 31. In the fragmentation of aliphatic ethers, this ion is produced by a hydride-shift rearrangement that occurs in a primary oxonium ion with an alkyl substituent on the oxygen. This is illustrated in Scheme 6-36.

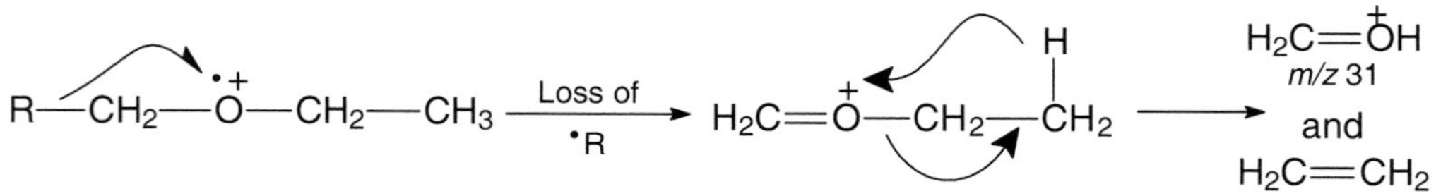

Scheme 6-36

The disubstituted oxonium ion from an asymmetric ether will be formed by loss of a radical (via homolytic cleavage) from the longer chain (based on the *loss of the largest moiety principle*). Even if the short chain contains as few as three carbon atoms, the intensity of the *m/z* 31 peak is less than 5% of the base peak. Regardless of the number of carbon atoms in the long chain, if the short chain has only two carbon atoms, the intensity of the *m/z* 31 peak will be well above 50% of the base peak. This aspect probably is related to the stability of the ethene molecule compared with that of larger olefins that would be formed when this shorter chain contains more than two carbon atoms. Even though the mass spectra of alkyl ethers and alkyl alcohols both exhibit peaks at *m/z* 31, its intensity is far greater in the mass spectra of ethers than in the mass spectra of alcohols. Another very important fact is that the mass spectra of ethers do not exhibit the obvious [M − H$_2$O]$^{+\bullet}$ or [(M − H$_2$O) − H$_2$C=CH$_2$]$^{+\bullet}$ peak seen in the mass spectra of the aliphatic alcohols. Care must be taken with this assertion, because the mass spectra of some alkyl ethers (ones that have a chain length that will supply hydrogens on a γ carbon that is not an ω carbon) do exhibit peaks at even *m/z* values that represent odd-electron ions. However, these spectra of aliphatic ethers lack the clusters of peaks every 14 *m/z* units, distinguishing them from the spectra produced by the alcohols.

In addition to the intense peak representing the primary oxonium ion with *m/z* 31, spectra of ethers containing an ethyl moiety will have a discernible peak representing the disubstituted oxonium ion (H$_2$C=O$^+$–CH$_2$–CH$_3$) at *m/z* 59. Peaks can also be observed in the mass spectra of aliphatic ethers that represent ions produced by heterolytic cleavage due to the shift of the pair of electrons between the number 1 carbon and the oxygen atom in response to the charge on the oxygen atom. This heterolytic cleavage reaction is quite prevalent in symmetrical aliphatic ethers. In some cases, the spectra of ethers will exhibit peaks representing ions formed by cleavage of the carbon–oxygen bond of one branch where the charge is retained by the oxygen atom still attached to the other branch; i.e., the loss of a radical representing the entire chain (see Scheme 6-39 later).

As stated above, in cases where one of the chains has hydrogens on a γ carbon that is not an ω carbon, peaks representing odd-electron ions will be present in the mass spectra. Although not thoroughly documented in the literature, the formation of these odd-electron ions separated by 28 *m/z* units can be explained by a mechanism analogous to that for the loss of water from aliphatic alcohols followed by the subsequent loss of a molecule of ethene as shown in Scheme 6-36, which shows a possible mechanism for the formation of the ions represented by the peaks at *m/z* 84 and *m/z* 56 in the mass spectrum of ethyl hexyl ether (Figure 6-33).

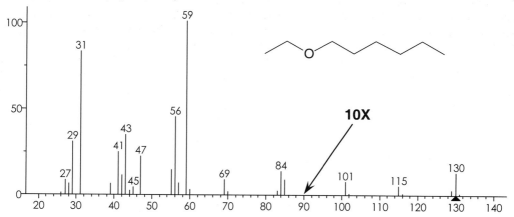

Figure 6-33. *EI mass spectrum of ethyl hexyl ether; nominal mass of 130 Da.*

An explanation of the peaks representing the ions formed by the fragmentation of ethyl hexyl ether (Figure 6-33) is a good example of the various fragmentation pathways that can be involved in the mass spectrometry of aliphatic ethers. The origin of the ions represented by the peaks at *m/z* 84 and *m/z* 56 is explained in Scheme 6-37. The peaks at *m/z* 59 and *m/z* 115 represent oxonium ions with an alkyl substituent on the oxygen formed by homolytic (α cleavage) taking place on either side of the oxygen in the M$^{+\bullet}$(see Scheme 6-38). The peak at *m/z* 31 represents a primary oxonium ion formed by the loss of an olefin through a process involving a hydride-shift rearrangement that occurs in the alkyl portion of the ion with *m/z* 59 or *m/z* 115 (Scheme 6-38).

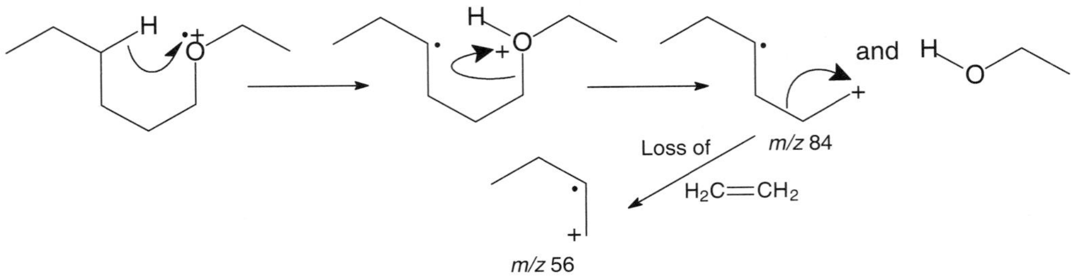

Loss of / *m/z* 84

$H_2C{=}CH_2$

m/z 56

Scheme 6-37

m/z 59

m/z 115

$H_2C{=}\overset{+}{O}H$
m/z 31

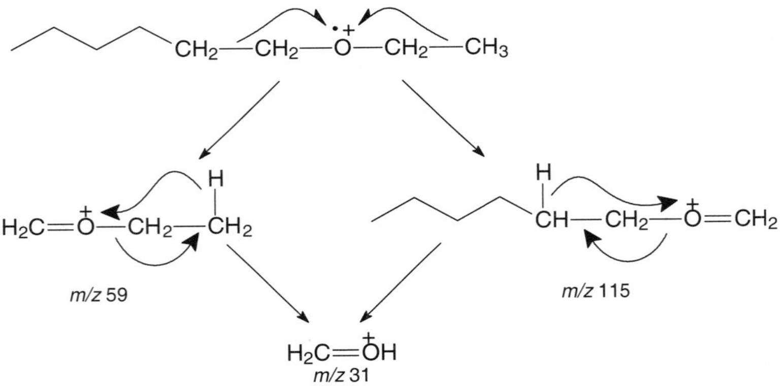

Scheme 6-38

The peaks at *m/z* 29 and *m/z* 85 in Figure 6-33 could represent alkyl ions formed by heterolytic cleavage of the M$^{+\bullet}$ at one of the two carbon–oxygen bonds with the charge and the radical sites on the oxygen atom (Scheme 6-39). The peak at *m/z* 69 could represent an ion formed by the loss of a molecule of methane (16 Da) from the hexyl ion (85 Da). The ions with *m/z* 45 and *m/z* 101 could be formed by the cleavage of the appropriate one-electron C–O bond created on either side of the oxygen by one-electron shifts within the M$^{+\bullet}$ (mm = 130 Da), as illustrated in Scheme 6-40.

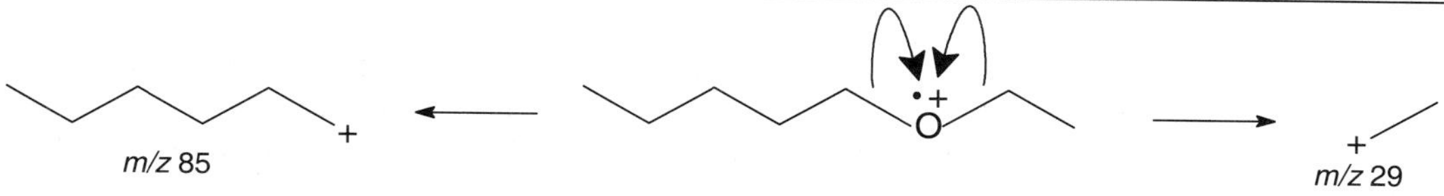

Scheme 6-39

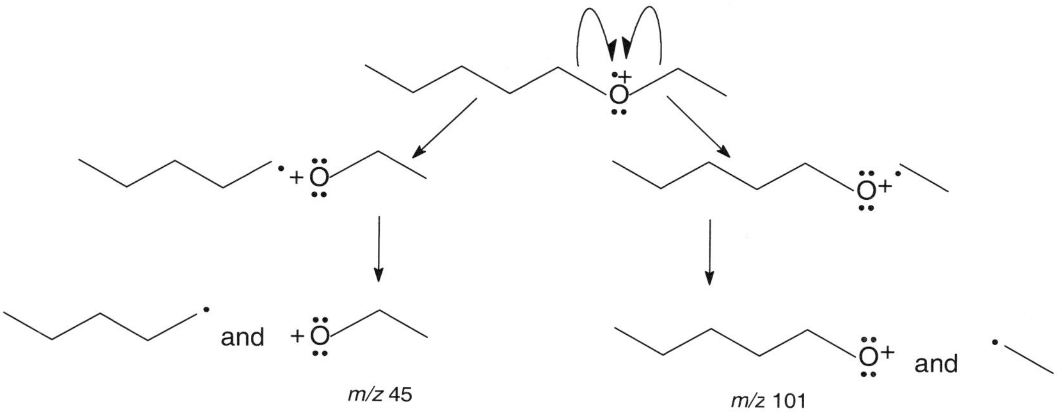

Scheme 6-40

The peak at *m/z* 47 could represent a $CH_3-CH_2-O^+H_2$ ion formed by a double hydrogen rearrangement [27] described in Scheme 6-41. As pointed out in the source of this mechanism, "[the mechanism was] subject to verification by deuterium labeling experiments" in 1967 [28]. It is not known whether that verification was ever accomplished.

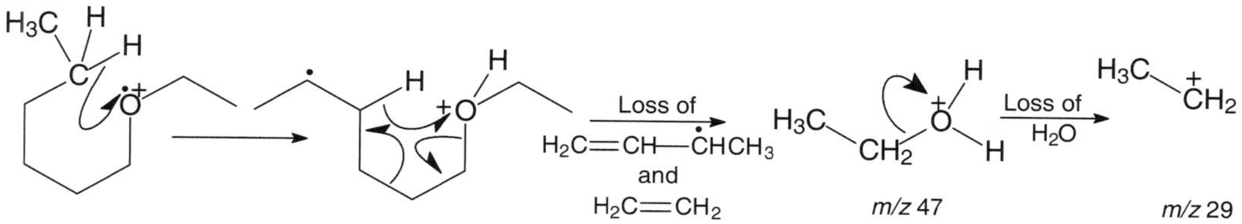

Scheme 6-41

The pattern formed by the peaks at *m/z* 43, 42, 41, and 39 in Figure 6-33 indicate that the *m/z* 43 peak represents a propyl ion. It is clear from the described possible origins of the ion with *m/z* 29 that this ion is an ethyl ion; the pattern of the peaks at *m/z* 29, 28, and 27 supports this assertion.

A possible explanation for the formation of the propyl ion is shown in Scheme 6-42. This ion is produced by the secondary fragmentation of the hexyl ion with *m/z* 85 formed by the heterolytic cleavage of the M⁺•.

An interesting point about the mechanism proposed in Scheme 6-42 is that it involves another of these virtual six-membered ring mechanisms. In this case, the mechanism was inspired by the mechanism shown for the formation of the ion with *m/z* 43 in the mass spectrum of cyclohexane (Figure 6-16 and Scheme 6-25). As was also seen in the formation of the ion with *m/z* 29 in the fragmentation of cyclohexane, these hydride-shift rearrangement fragmentations of an EE ion can also occur through five-membered ring transitions.

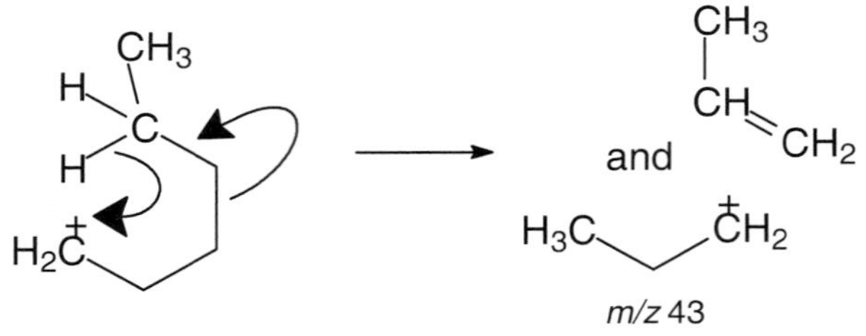

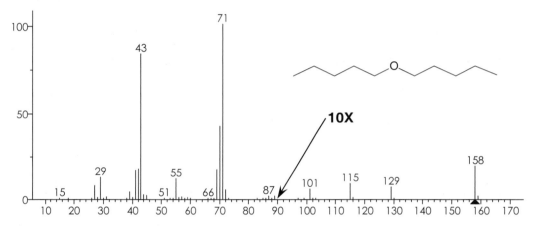

Scheme 6-42

A good example of the mass spectrum of a symmetrical ether is shown in Figure 6-34. This spectrum also illustrates the effect of chain length on the formation of the primary oxonium ion represented by the peak at m/z 31, which is vanishingly small, but still perceptible. Also notice the peaks at m/z 70 and m/z 42 (neither of which is labeled) for odd-electron ions resulting from the molecular ion of dipentyl ether participating in a rearrangement mechanism analogous to that illustrated in Scheme 6-37.

Figure 6-34. EI mass spectrum of dipentyl ether; nominal mass of 158 Da.

C. Aromatic Alcohols

Aromatic alcohols are phenols. These are compounds that have a hydroxyl group attached to a phenyl ring. Like other aromatic compounds, the mass spectra of phenols exhibit intense molecular ion peaks. These spectra will often exhibit a peak representing the loss of a molecule of carbon monoxide; a mechanism for this loss of CO is illustrated in Scheme 6-43.

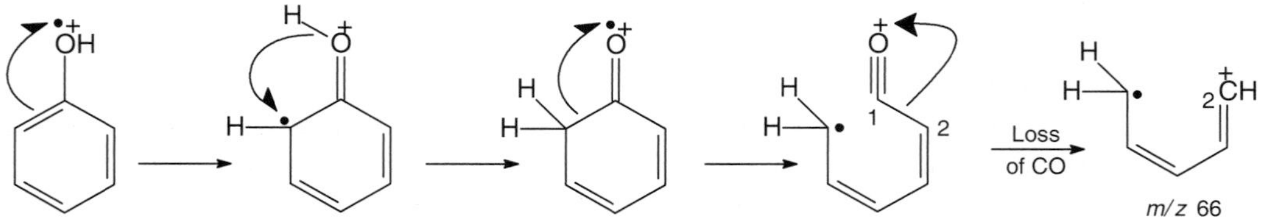

Scheme 6-43

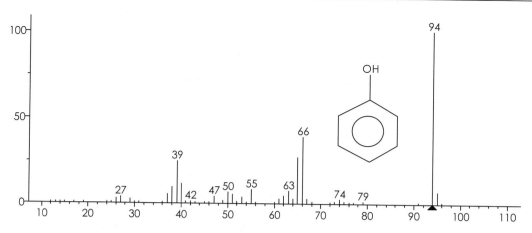

Figure 6-35.   El mass spectrum of phenol; nominal mass of 94 Da.

Also as shown in Scheme 6-43, some of the distonic ions with *m/z* 66 resulting from the expulsion of CO from ionized phenol continue to fragment by losing a molecule of ethene via another heterolytic cleavage driven by the nascent charge site on carbon atom 2 that was adjacent to the carbon atom 1 (lost as CO). As can be seen in Figure 6-35, in addition to the intense M$^{+\bullet}$ peak, the mass spectrum of phenol also is characterized by peaks at *m/z* 65, 51, and 39 representing secondary fragmentations of the aromatic ring discussed earlier in this chapter.

An important differentiator between the mass spectra of aromatic alcohols and aliphatic alcohols is the fact that phenols do not lose water under normal circumstances.

In cases where the phenolic hydroxyl is on a carbon atom adjacent to a carbon atom with a non-alkyl substitution, there will be an *ortho effect,* which promotes the loss of a molecule from the M$^{+\bullet}$. The *ortho effect* is a rearrangement that will cause a molecule to be expelled from the M$^{+\bullet}$ of an aromatic compound that has two functionalities on adjacent carbon atoms.   One of the functionalities can be a heteroatom.   When one of the substituents is an alkyl group (CH$_3$, C$_2$H$_5$, etc.), the *ortho effect* is minimal.   In the mass spectrum of *o*-chlorotoluene, there is an obvious [M – HCl]$^{+\bullet}$ peak (although very small) whereas no peak of significant intensity at the same *m/z* value appears in the mass spectrum of either the *m*- or *p*-isomer of this compound.   The mass spectra of the *o*-isomers of chloroethylbenzene and chloroisopropylbenzene do not exhibit a peak representing the [M – HCl]$^{+\bullet}$ ion (none of these spectra is shown). The *ortho effect* was first reported in 1959 by McLafferty and Golke in a publication dealing with hydroxylated aromatic acids [29].   There are three other references in the literature that give a great deal more information on the *ortho effect* [30–32].

The *ortho effect* is pronounced with phenols in which the substituent is anything other than an alkyl group. All three regioisomers of methylphenol exhibit an [M – H$_2$O]$^{+\bullet}$ peak at *m/z* 90. None of the mass spectra of ethylphenol exhibits such a peak. The mass spectra of the regioisomers of dichlorophenol show the influence of the *ortho effect;* there are six possible isomers. For each isomer in which the chlorine moiety is on a carbon atom adjacent to the carbon atom with the hydroxyl group, the spectrum exhibits an [M – HCl]$^{+\bullet}$ and an [M – HCl – CO]$^{+\bullet}$ peak. In the mass spectra of the isomers that do not have at least one of the two chlorine atoms adjacent to the OH, these OE ion peaks do not exist (Figures 6-36 and 6-37). There is more on the *ortho effect* later in this chapter.

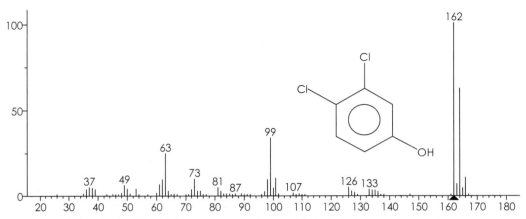

Figure 6-36.   EI mass spectrum of 3,4-dichlorophenol.

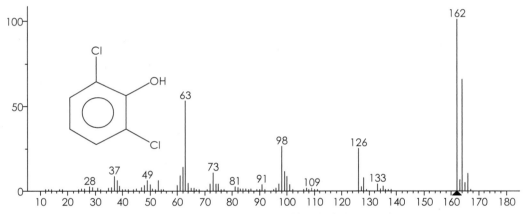

Figure 6-37.   EI mass spectrum of 2,6-dichlorophenol.

Returning to the point that phenols do not lose water as do aliphatic alcohols and cycloalkanols, a good example is available in a comparison of the mass spectra of two positional isomers, one constituting a cycloalkanol (1,2,3,4-tetrahydro-1-naphthalenol) (Figure 6-38), and the other a phenol (5,6,7,8-tetrahydro-1-naphthalenol) (Figure 6-39). Scheme 6-44 is a possible mechanism showing the loss of water from the M$^{+\bullet}$ to justify the peak at *m/z* 130 in the mass spectrum (Figure 6-38) of the cyclic aliphatic alcohol (1,2,3,4-tetrahydro-1-naphthalenol).

m/z 148   *m/z* 148   *m/z* 130

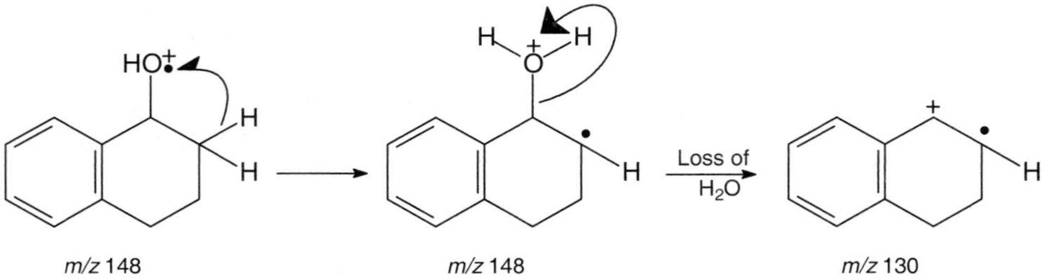

Scheme 6-44

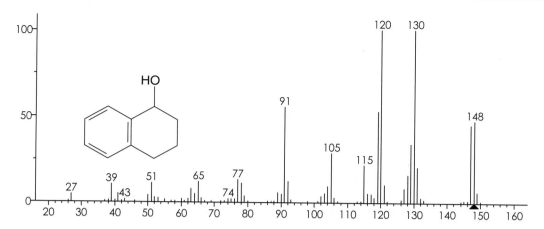

Figure 6-38.   *EI mass spectrum of 1,2,3,4-tetrahydro-1-naphthalenol.*

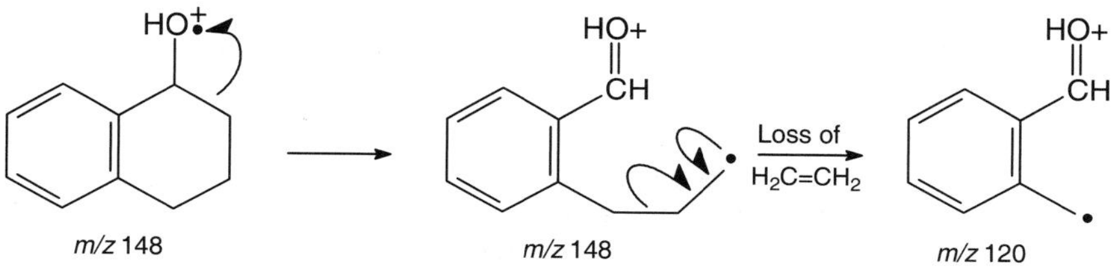

Scheme 6-45

An intense peak at *m/z* 120 is observed in the spectra of both of these compounds; however, whereas this ion has the same elemental composition as the two possible ions that have *m/z* 120 formed by the phenolic compound, the structure of the *m/z* 120 ion formed by the cyclic alkanol is very different than the two proposed structures for the ion with the same *m/z* value formed by the phenolic compound. This is seen in Scheme 6-45.

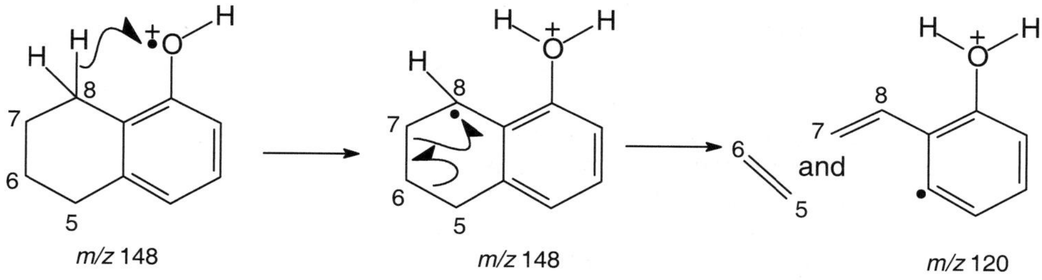

Scheme 6-46

The phenolic isomer (5,6,7,8-tetrahydro-1-naphthalenol) exhibits a mass spectrum with a significantly different appearance (Figure 6-39). This is because the phenolic compound has two possible mechanistic pathways by which to lose ethene (Schemes 6-46 and 6-47).

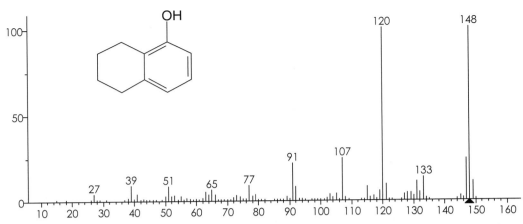

Figure 6-39.   El mass spectrum of 5,6,7,8-tetrahydro-1-naphthalenol.

The two ions with *m/z* 120 in Schemes 6-46 and 6-47 have the same elemental composition and thus the same exact mass, but each of the structures could be validated by the use of stable isotope labeling by incorporating ^{13}C for carbons 6 and 7.   The formation of the ion via the loss of ethene involving carbons 5 and 6 (as shown in Scheme 6-46) would result in an $[M - 29]^+$ peak at *m/z* 121 (loss of $H_2^{13}C{=}CH_2$ from a $M^{+\bullet}$ with *m/z* 150).   The other ion resulting from a loss of a molecule of ethene involving carbons 6 and 7 would have an *m/z* value of 120 (loss of $H_2^{13}C{=}^{13}CH_2$).

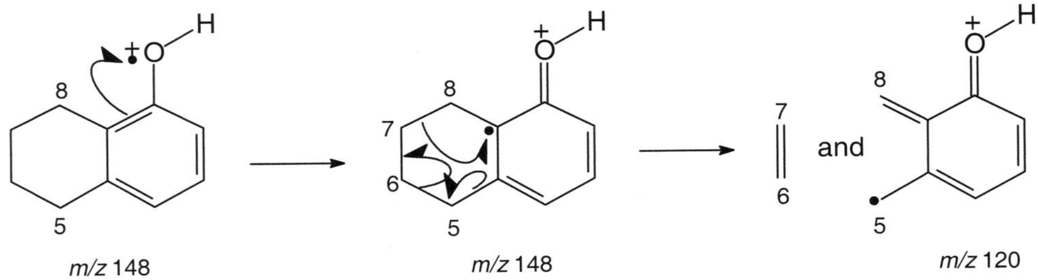

m/z 148   *m/z* 148   and   *m/z* 120

Scheme 6-47

In all of the examples of ionization and fragmentation of hydroxyl groups attached to phenyl rings described above, the mechanisms are based on the charge and radical sites being on the oxygen atom due to the loss of a nonbonding electron.   It is also possible to have molecular ions in which the charge and radical sites are associated with the aromatic ring due to the loss of a π-bond electron.   Because there are competing fragmentation pathways of the different types of molecular ions resulting from the loss of different types of electrons, both sets of mechanisms must be considered.

Two features differentiate the mass spectra of aromatic alcohols and aliphatic alcohols.   First, the intense $M^{+\bullet}$ peak in the mass spectra of aromatic alcohols is in stark contrast to the vanishingly small $M^{+\bullet}$ peak in the mass spectra of aliphatic alcohols.   Second, aromatic alcohols do not expel water, whereas aliphatic alcohols do.   Even though the mass spectrum of 1,2,3,4-tetrahydro-1-naphthalenol exhibits a molecular ion peak, its intensity is only about half that of the molecular ion peak exhibited in the mass spectrum of 5,6,7,8-tetrahydro-1-naphthalenol.

D. Cyclic Ethers

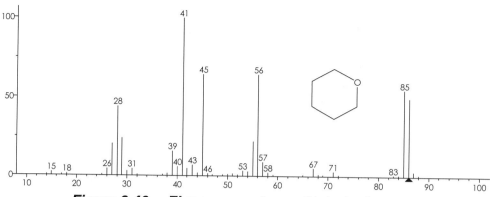

Figure 6-40.   EI mass spectrum of tetrahydropyran.

The mass spectrum of tetrahydropyran is a good example of the complex nature of the mass spectra of cyclic ethers. The source of all the ions represented by the major peaks in this mass spectrum (Figure 6-40) is shown in Schemes 6-48 and 6-49. Because of the simplicity of the mechanisms in Scheme 6-49 and the stability of the resulting ions, it might be expected that the peak at m/z 31 and the one corresponding to $[M - 28]^{+\bullet}$ (m/z 58) would be more intense than observed in the mass spectrum (Figure 6-40). The peak at m/z 58 could be mistaken for an isotope peak associated with m/z 56; however, its intensity is too great to be explained away as representing a species containing two atoms of ^{13}C or an atom of ^{18}O. Also, the ion with m/z 58 appears to be a precursor of the ion represented by the peak at m/z 30 in the mass spectrum. Although the peaks at m/z 57 and 71 represent O-substituted oxonium ions of the form $H_2C=O-C_nH_{2n-1}$ (where n = 2 and 3, respectively), the peak at m/z 43 probably represents a propyl ion rather than a substituted oxonium ion (the fragmentation pathway is not shown).

The mass spectrum of tetrahydropyran, along with the associated fragmentation schemes, is a good illustration of the complexity that is introduced by multiple EE-ion rearrangements that can occur in such compounds. The fragmentation of tetrahydropyran is consistent with the fragmentation of tetrahydrofuran reported by Smakman and De Boer [33].

E. Ketones and Aldehydes

Ketones and aldehydes are characterized by a carbonyl (C=O) moiety positioned between two R groups. In aldehydes, one of the R groups is a hydrogen atom; this hydrogen is located in a "special place" which promotes its expulsion to form a stable acylium ion as represented by the characteristic $[M - H]^+$ peak of significant intensity in the mass spectra of many aldehydes, regardless of whether the other R group is aliphatic or aromatic. Ketones have two non-hydrogen moieties attached to the carbonyl.

> An intense $[M - 1]^+$ peak (>5% of the base peak) in the mass spectrum of an aliphatic compound indicates the possibility that the compound is an aldehyde. $[M - 1]^+$ peaks are also observed when an aldehydic group is attached to an aromatic ring.

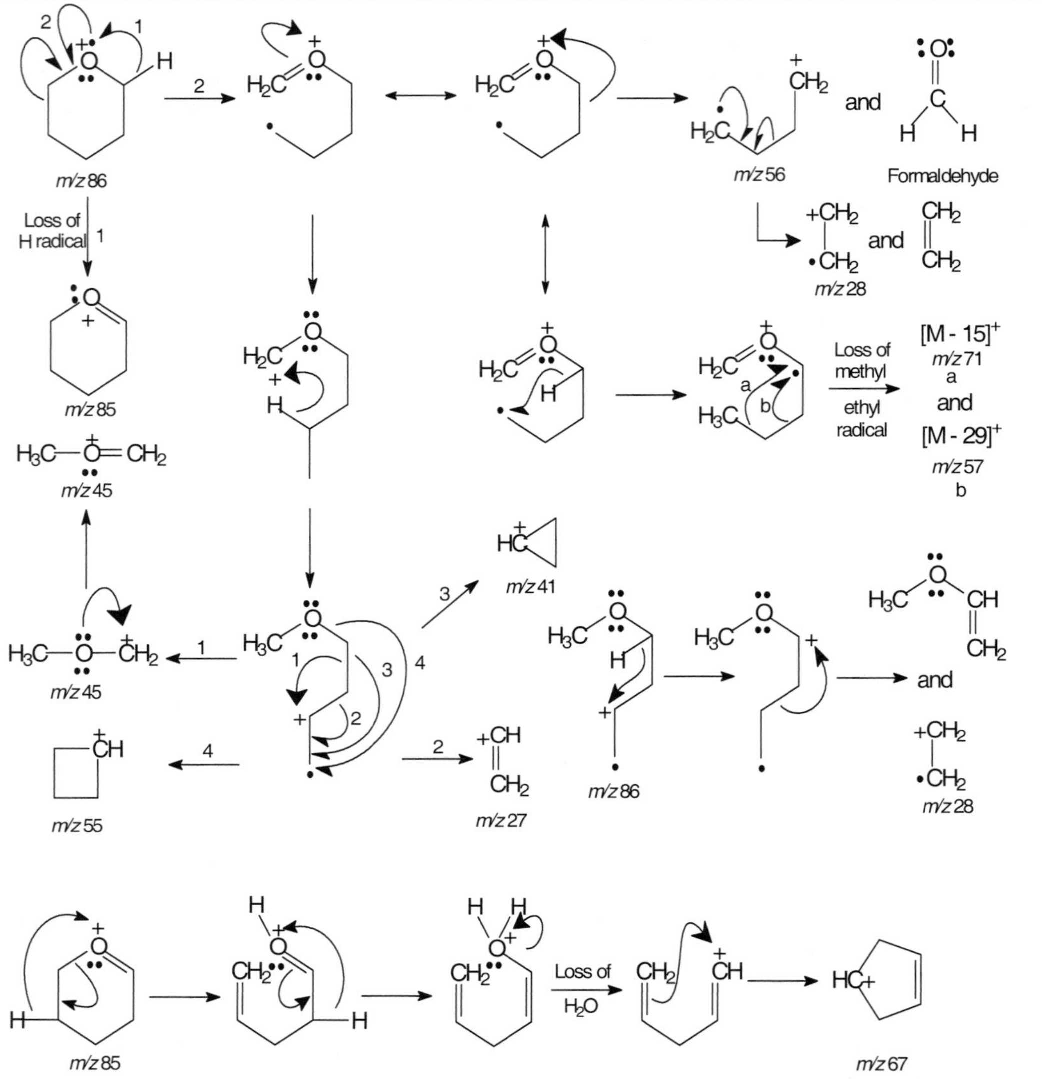

Scheme 6-48

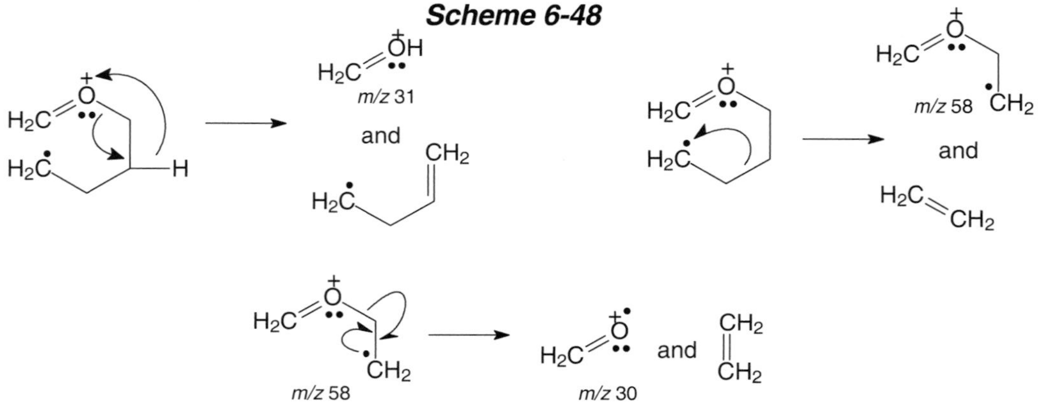

Scheme 6-49

Straight-chain aliphatic aldehydes are similar in their mass spectral behavior to aliphatic alcohols. The mass spectra of straight-chain aliphatic aldehydes, through C_4, exhibit a relatively intense $M^{+\bullet}$ peak (>50% of the base peak) and the signature $[M - 1]^+$ peak. Figure 6-41 is the mass spectrum of butyl aldehyde showing both an intense $M^{+\bullet}$ peak and a readily discernible $[M - H]^+$ peak. The mass spectra of straight-chain aldehydes larger than C_4 show no observable $M^{+\bullet}$ peak. The mass spectra of straight-chain aliphatic aldehydes larger than C_3 exhibit a peak at *m/z* 44, which represents an OE ion formed by a γ-hydrogen-shift rearrangement fragmentation (a McLafferty rearrangement). This *m/z* 44 peak is the base peak in the mass spectra of C_4 through C_6 straight-chain aliphatic aldehydes. The intensity of this peak declines to about 25% of the base peak in the mass spectrum of octadecyl aldehyde.

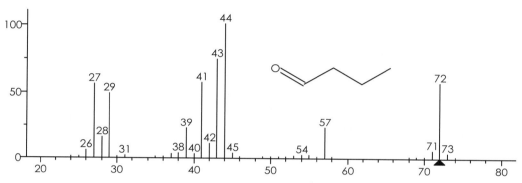

Figure 6-41.   *EI mass spectrum of butyl aldehyde (Wiley Registry 8th Ed).*

Figure 6-42 is the mass spectrum of decyl aldehyde, which exhibits no discernible $M^{+\bullet}$ peak. However, there are several OE ion peaks in the $M^{+\bullet}$ peak region that appear in the mass spectra of all straight-chain aliphatic aldehydes that contain more than four carbon atoms. These are peaks representing $[M - 18]^{+\bullet}$ (loss of H_2O), $[M - 28]^{+\bullet}$ (loss of $H_2C=CH_2$), and $[M - 44]^{+\bullet}$ ions. The loss of ethene from the $M^{+\bullet}$ is explained in Scheme 6-50. The loss of water from the $M^{+\bullet}$ can occur via two different pathways, as illustrated in Scheme 6-51. The loss of 44 Da from the $M^{+\bullet}$ is more complex. This ion could be formed by the loss of a molecule of HC≡CH from the ion formed by the loss of a water molecule via an enol form of the aldehyde as shown in Scheme 6-51; alternatively, this $[M - 44]^{+\bullet}$ ion could result from loss of the aldehydic H$^\bullet$ radical followed by the loss of CO followed by the loss of a $^\bullet CH_3$ radical from the $M^{+\bullet}$.

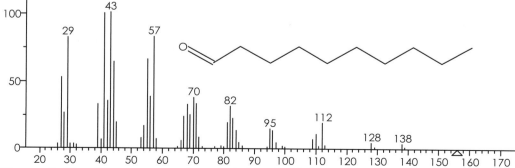

Figure 6-42.   *EI mass spectrum of decyl aldehyde (Wiley Registry 8th Ed).*

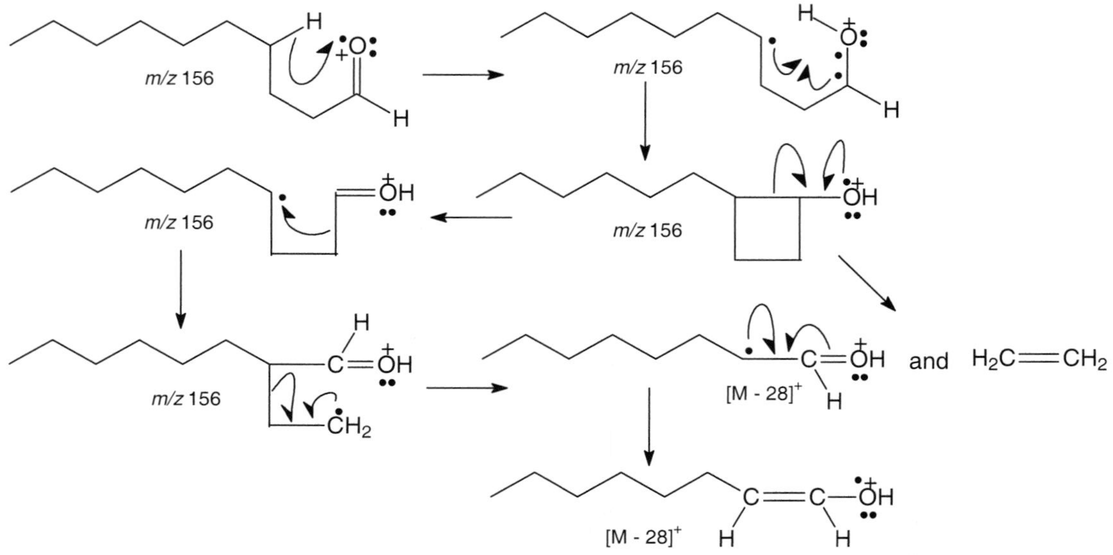

Scheme 6-50

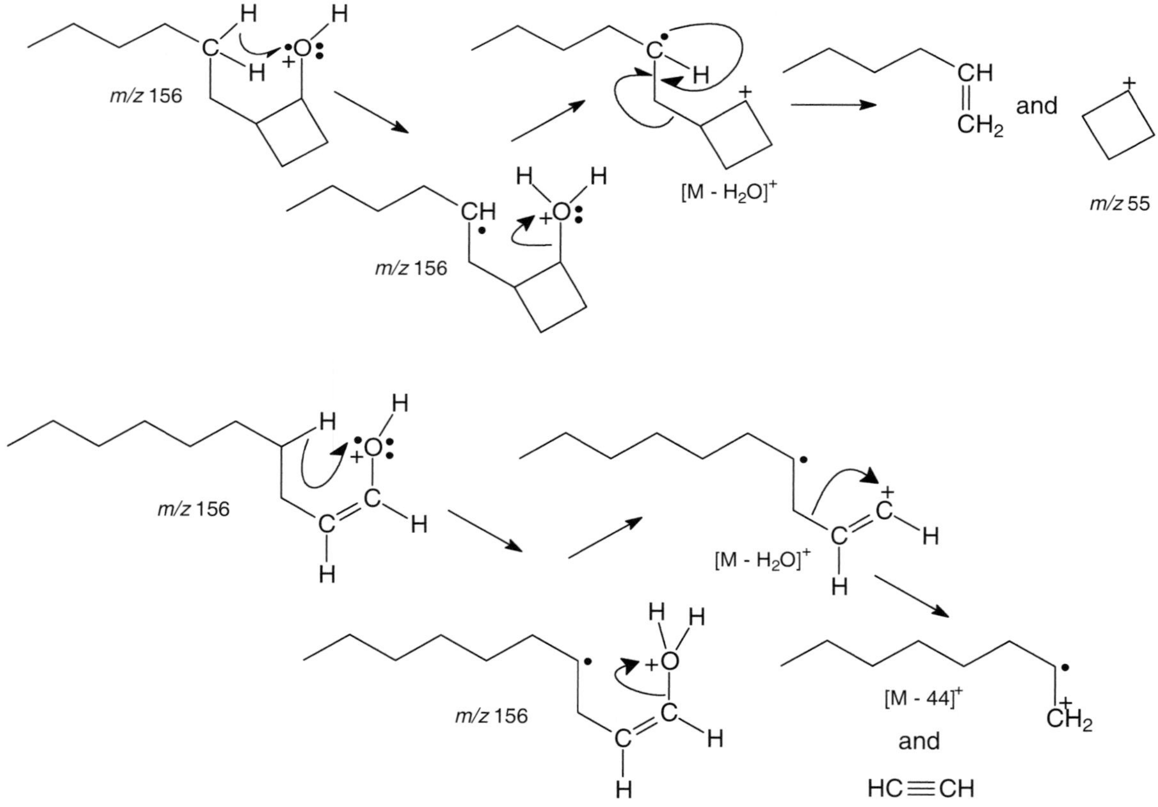

Scheme 6-51

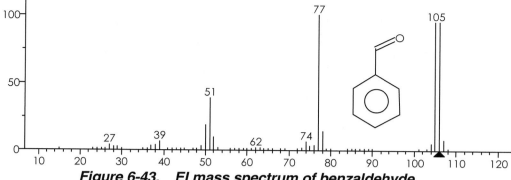

Figure 6-43.　El mass spectrum of benzaldehyde.

The mass spectrum of benzaldehyde (Figure 6-43) exhibits a very intense $M^{+\bullet}$ peak and an $[M - 18]^{+\bullet}$ peak of almost equal intensity. The high intensity of the $M^{+\bullet}$ and $[M - 1]^+$ peaks reflect the resonance stability of the corresponding ions. The peak at m/z 77 represents a phenyl ion formed by heterolytic cleavage of the molecular ion that was formed by the loss of one of the nonbonding electrons on oxygen.

Compared to the mass spectrum of benzaldehyde, the spectrum of cyclohexanol, the saturated analog (Figure 6-44), presents a very different picture. In the mass spectrum of cyclohexanal, there is an easily discernible $M^{+\bullet}$ peak, although the intensity is not a large value (<25% of the base peak). The intensity of the $[M - 1]^+$ peak is very low, barely detectable in the graphics display. The intensity of the $M^{+\bullet}$ peak is as great as it is largely due to delocalization of the charge and radical sites among several structures, all of which have the same m/z value as rationalized in Scheme 6-52 through β cleavage resulting first in a cyclic oxonium ion with subsequent cleavage through a hydride-shift rearrangement, all of which may be reversible.

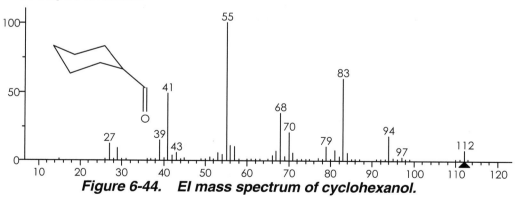

Figure 6-44.　El mass spectrum of cyclohexanol.

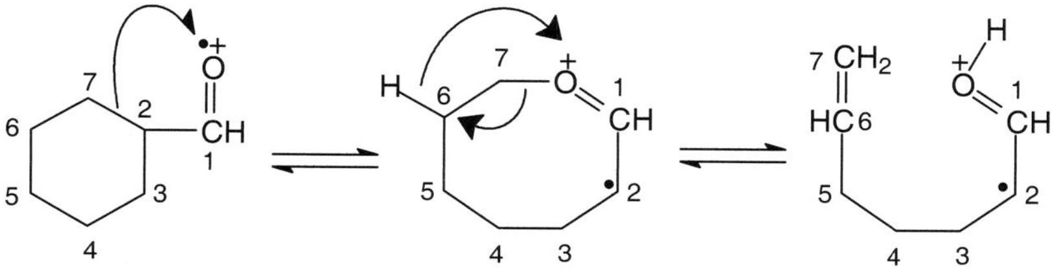

Scheme 6-52

A good example of the fragmentation of aliphatic ketone molecular ions is the mass spectrum of 4-decanone (Figure 6-11). In addition to the formation of OE fragment ions with m/z 86 and 58, through the rearrangement fragmentation of the $M^{+\bullet}$ followed by the rearrangement fragmentation of that product (as illustrated in Scheme 6-18), a number of other fragmentations are initiated by the charge and radical sites on the carbonyl oxygen atom due to the loss of one of the nonbonding electrons. All $M^{+\bullet}$ fragmentations of 4-decanone can be rationalized through mechanisms starting with $M^{+\bullet}$ having the "plus/dot" on the oxygen, as will become clear in the next few pages.

One very important aspect of the rearrangement fragmentation of ketones that should be pointed out is the formation of the OE ion with m/z 58. This ion is present in the mass spectrum of all aliphatic ketones except those that have a substituent on the α carbon. As was stated in the discussion on primary and secondary radicals in relation to formation of OE ions through γ-hydrogen-shift β-cleavage rearrangements, the intensity of the peak at m/z 58 will be significant as long as the process involves intermediate formation of a secondary radical. When the carbonyl is the number 3 carbon atom, the diagnostic peak at m/z 58 will be shifted by 14 m/z units to m/z 72. This arrangement can be considered as a special case of α-substitution.

> A nonisotopic, nonbackground peak at m/z 58 in the mass spectrum of a compound believed to contain no atoms of nitrogen strongly suggests that the compound is an aliphatic ketone. There are spectra of 575 compounds in the NIST05 Mass Spectral Database with a peak at m/z 58 having a relative intensity of 10–100% and containing no atoms of nitrogen, but one or more atoms of oxygen, and the string "ONE" in one of its names.

Schemes 6-53 and 6-54 illustrate the pathways of α cleavage (a special case of homolytic cleavage) for the $M^{+\bullet}$ of 4-decanone to account for the ion current with m/z 113 and m/z 71 in Figure 6-11. Some of the resulting acylium ions subsequently expel a molecule of CO via a heterolytic-cleavage mechanism (not shown) to account for some of the ion current at m/z 85 and m/z 43.

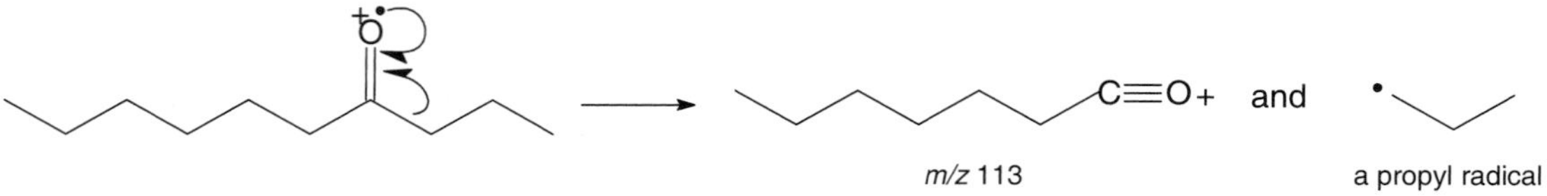

m/z 113   a propyl radical

Scheme 6-53

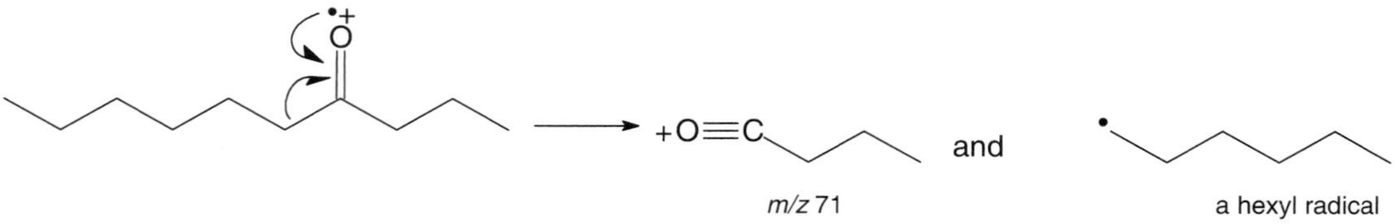

m/z 71   a hexyl radical

Scheme 6-54

> The terms α cleavage, β cleavage, or γ cleavage refer to special cases of homolytic cleavage where the carbon bond next to the α-C, β-C, or γ-C atom, on the side closest to the moiety carrying the charge, is broken.

Some of the other 4-decanone molecular ions will undergo fragmentation via heterolytic cleavage to produce ions with m/z 43 and m/z 85, as illustrated in Schemes 6-55 and 6-56. The base peak at m/z 43 in the mass spectrum (Figure 6-11) of this compound represents an ion formed by the $M^{+\bullet}$ losing a seven-carbon carbonyl radical, which is the largest possible moiety that can be lost. Although there are competing pathways for the formation of the ions with m/z 43 and m/z 85, the ion with m/z 43 dominates because of the loss of the largest moiety during heterolytic cleavage.

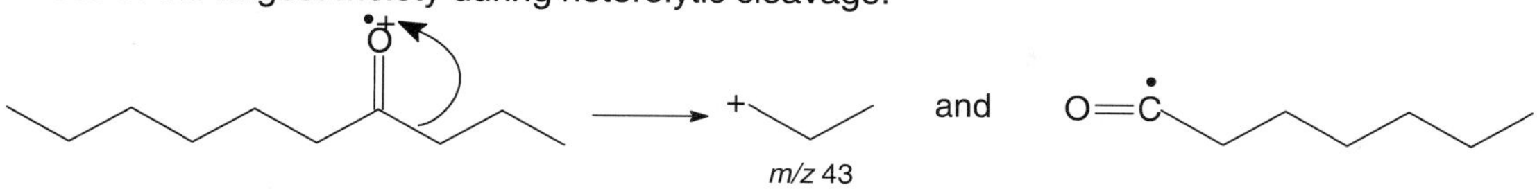

Scheme 6-55

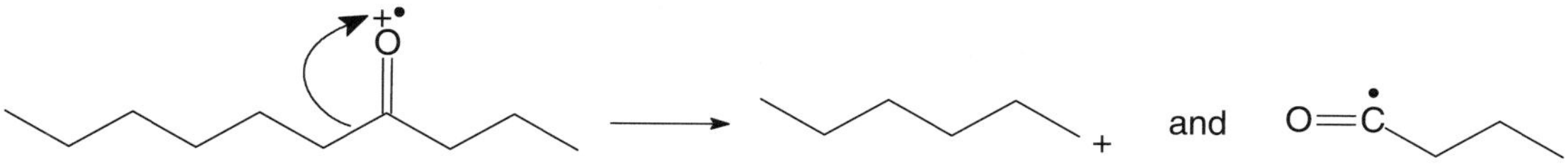

Scheme 6-56

The peaks at the lower m/z values (42, 41, and 39) preceding the peak at m/z 43 represent the secondary fragmentation ions so often associated with aliphatic ions. There are peaks with nominal m/z values of 99, 57, and 29 that cannot be accounted for through the fragmentation mechanisms of homolytic cleavage, heterolytic cleavage, or rearrangements. From the peak pattern between m/z 53 and m/z 57, it appears that the peak at m/z 57 represents a butyl ion. The difference between 156 (the nominal mass of 4-decanone) and 57 is 99, which is the m/z value of one of the other yet unexplained peaks in the mass spectrum. It is possible that these two peaks represent ions produced by σ-bond fragmentation at a site of σ-bond ionization. This process of σ-bond fragmentation could also be an explanation for the peak at m/z 29, although the peak at m/z 27 appears to be too intense to be rationalized as representing an ion formed by the loss of a hydrogen molecule from an ethyl ion, which could be represented by the peak at m/z 29.

Continuing with the interpretation of the mass spectrum (Figure 6-11) of 4-decanone, there is a problem with the relative intensities of the yet-to-be-explained peaks at m/z 99, 29, and 27. The intensity of these peaks would suggest that a significant number of the molecular ions formed were due to σ-bond ionization, which is not reasonable based on the much lower ionization potential associated with the loss of one of the nonbonding electrons on the oxygen atom.

There are some other possibilities for fragmentation of molecular ions that have the charge and radical sites on the oxygen atom that could relate to the origin of the so-far-unexplained peaks. The peak at m/z 57 could represent a butyl ion formed by the loss of a molecule of ethene via a secondary heterolytic cleavage of hexyl ion as shown in Scheme 6-57. The *n*-butyl ion would then expel another molecule of ethene to produce the ethyl ion with m/z 29, also shown in Scheme 6-57.

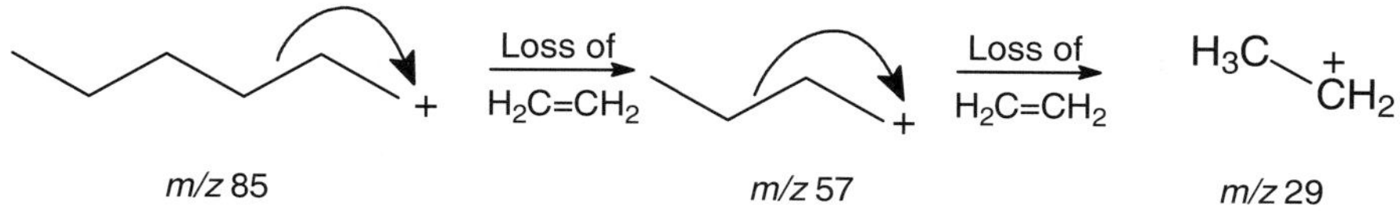

Scheme 6-57

The above description for the formation of the ion with m/z 29 does not explain why the intensity of the peak at m/z 27 is as great as it appears in Figure 6-11. The origins of the vinyl ion represented by the peak at m/z 27 could be from both the expulsion of a hydrogen molecule from the ethyl ion and the expulsion of a molecule of methane from the propyl ion. This leaves only the peak at m/z 99 in Figure 6-11 to be explained.

A possible source of the ion with m/z 99 is through a γ cleavage as illustrated in Scheme 6-56. According to Reference 26, page 138, this reaction may be the result of the rupture of an allylic bond in the enol form of the M$^{+\bullet}$ or, as illustrated in Scheme 6-58, may involve the formation of a cyclic oxonium ion. The reference points out that ions formed by this γ cleavage have a far greater abundance than those formed either by β cleavage or by δ cleavage; however, the presence of these products would be masked by peaks at m/z 113 and at m/z 85 for the isobaric ions formed by homolytic and heterolytic cleavage, respectively. The dominance of γ cleavage (m/z 113) over β cleavage (m/z 99) and δ cleavage (m/z 127) is reflected in the mass spectrum of di-n-butyl ketone (Figure 6-45).

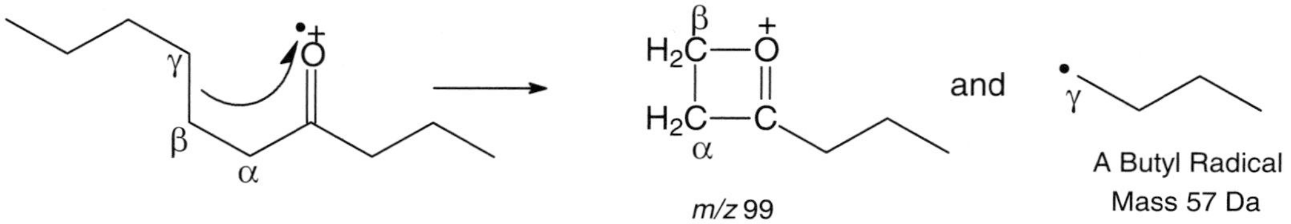

Scheme 6-58

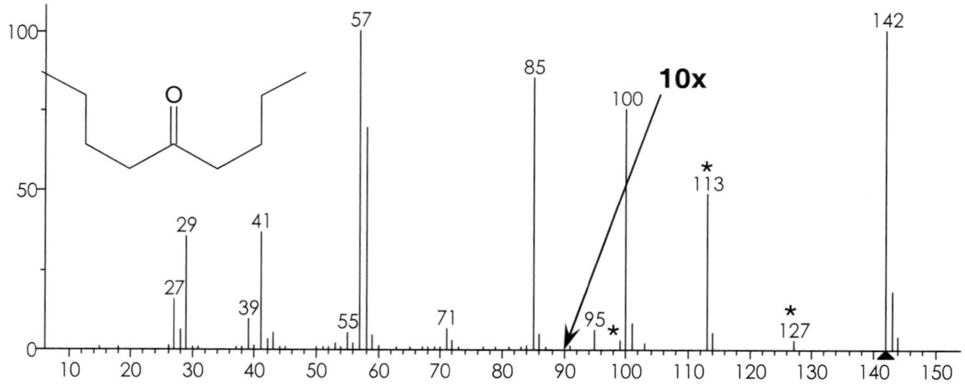

Figure 6-45.   EI mass spectrum of di-n-butyl ketone with peaks representing β, γ, and δ cleavages (left to right) marked with an *.

The fragmentation of aromatic ketones can usually be explained by a mechanism starting with the most probable site for the charge and the odd electron in the M$^{+\bullet}$. The peaks at m/z 105 and m/z 77 in the mass spectrum of diphenyl ketone (Figure 6-46) represent ions formed by homolytic and heterolytic cleavages, respectively. It should be noted that the high abundance of the ion with m/z 105 is a result of resonance stabilization of the acylium moiety resulting from its being attached directly to the phenyl ring.

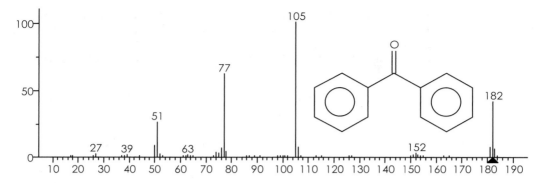

Figure 6-46. *EI mass spectrum of diphenyl ketone (a.k.a. benzophenone).*

Peaks resulting from the charge and radical sites being only associated with the carbonyl oxygen are also seen in the mass spectrum of ethyl phenyl ketone (Figure 6-47). A significant change in the mass spectrum of a compound consisting of $C_{10}H_{10}O$ results from a slight change in the arrangement of the atoms to position the carbonyl between a benzyl carbon and a methyl carbon (in going from ethyl phenyl ketone to phenyl acetone); compare Figures 6-47 and Figure 6-48.

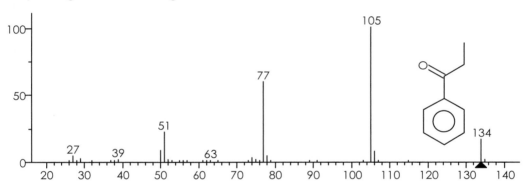

Figure 6-47. *EI mass spectrum of ethyl phenyl ketone.*

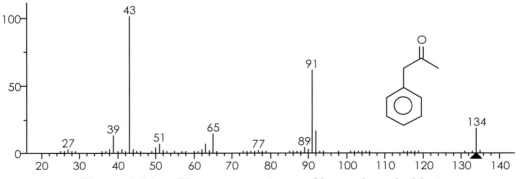

Figure 6-48. *EI mass spectrum of benzyl methyl ketone.*

The change in fragmentation pattern illustrated in Figures 6-47 vs Figure 6-48 results from changing structural features in going from a resonantly stabilized M$^{+\bullet}$ to an aliphatic ketone M$^{+\bullet}$; in both cases, the charge and radical sites are on the carbonyl oxygen. From the structure of benzyl methyl ketone (a.k.a. phenyl acetone), a peak in the mass spectrum representing a benzyl acylium ion might be predicted. However, no such peak is observed. If the benzyl-acylium ion is formed, it will quickly lose CO to produce a tolyl ion,

An ion with the elemental composition of C_7H_7 originating from a $CH_2C_6H_5$ moiety can have three structural forms [24]: the tolyl ion in which the charge is on the carbon of the CH_2 moiety; the tropylium ion, which is a C_7H_7 resonantly stabilized seven-membered ring; and the benzyl ion, in which the charge is on the ring and in which the carbon atom of the CH_2 moiety can be considered to be attached to a ring carbon through a double bond. The benzyl and tropylium ions are considered to be in equilibrium, with the tropylium ion dominating.

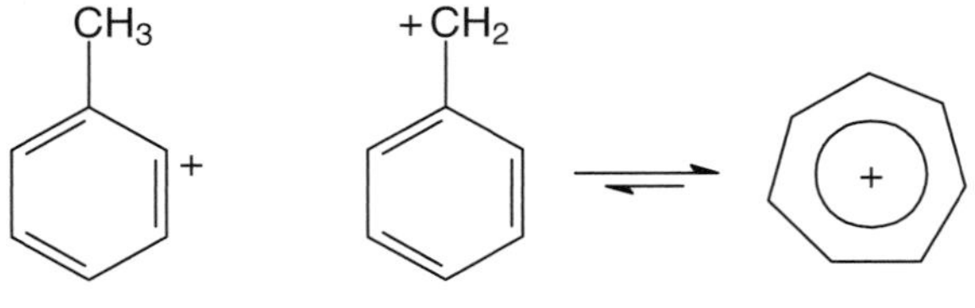

Tolyl Ion   Benzyl Ion   Tropylium Ion

which has an *m/z* value of 91. The mass spectrum of benzyl methyl ketone (Figure 6-48) also exhibits a peak at *m/z* 43, which represents the very stable methyl acylium ion. There is also a peak at *m/z* 92, which is far more intense than would be expected for a ^{13}C-isotope peak of the ion with *m/z* 91. This peak represents an ion formed by a rearrangement that can only be explained by the charge and radical sites resulting from the loss of an electron on the ring (Scheme 6-59).

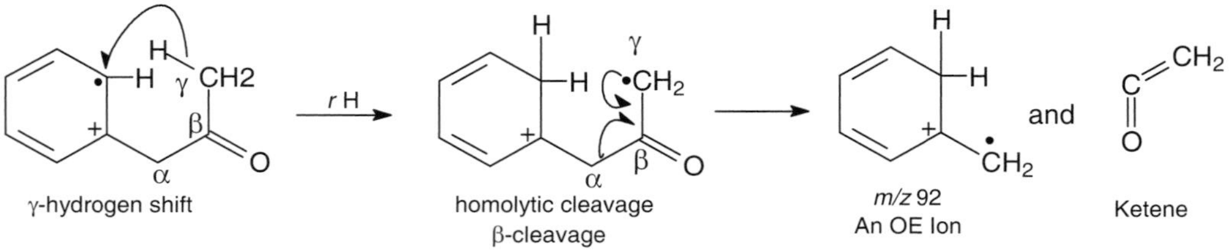

γ-hydrogen shift   homolytic cleavage   *m/z* 92   Ketene
                           β-cleavage   An OE Ion

Scheme 6-59

Although ethyl phenyl ketone also has hydrogens on a γ-carbon atom, the lack of the carbonyl next to this ω carbon makes the formation of the primary radical much less likely than is the case with benzyl methyl ketone. The mass spectrum of benzyl methyl ketone is a good example of the effect of having competing sites for the radical and the charge in molecular ions. The most probable sites of the charge and radical will be the result of the loss of a nonbonding electron from the oxygen; however, there will also be molecular ions formed by the loss of a π electron from the phenyl ring.

It is also worth noting that the peak at *m/z* 91 in the mass spectrum of benzyl methyl ketone (Figure 6-48) probably represents tropylium ions formed through multiple fragmentation pathways [24]. A tropylium ion is formed via benzylic cleavage when the charge and radical sites are initially on the ring. Benzyl ions will be formed by the loss of CO from the benzyl-acylium ions (with *m/z* 119 in Scheme 6-60) formed via homolytic cleavage of those molecular ions produced with the charge and radical sites on the oxygen atom (analogous to that shown in Scheme 6-60). Benzyl ions can also be produced directly by heterolytic cleavage of the bond between the carbonyl carbon and the benzyl carbon in molecular ions with the charge and radical sites on the oxygen atom (not shown).

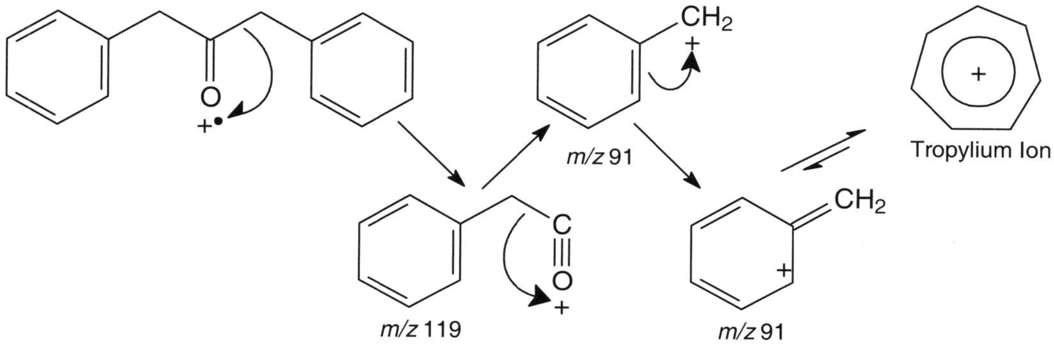

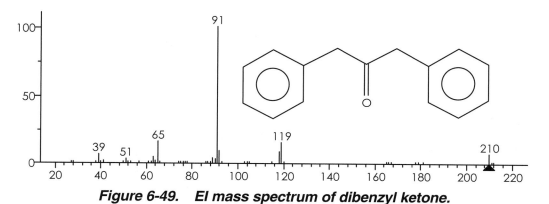

Scheme 6-60

Another interesting property of an alkyl-aryl ketone is reflected in the mass spectrum of dibenzyl ketone (Figure 6-49), which exhibits a peak at *m/z* 119 that is of much lower intensity than might be predicted. Also, there is an OE ion peak at *m/z* 118.

Figure 6-49.   El mass spectrum of dibenzyl ketone.

Scheme 6-60 illustrates the formation of an acylium ion of *m/z* 119 from homolytic fission of a molecular ion of dibenzyl ketone having the charge and radical sites on the carbonyl oxygen. Because this acylium ion is not resonantly stabilized through conjugation with the phenyl ring, it readily loses carbon monoxide to form the tolyl ion, which likely isomerizes to the tropylium ion of *m/z* 91 [24].

In addition to being formed via the loss of CO from the benzyl-acylium ion, the ion with *m/z* 91 may also be formed directly by heterolytic cleavage of the benzylic carbon–carbonyl carbon bond in the M$^{+\bullet}$ (Scheme 6-60). A third possibility for the formation of the ion with *m/z* 91 is through benzylic cleavage of molecular ions produced by loss of a π electron from the phenyl ring. This type of M$^{+\bullet}$ would also offer a possible explanation for the formation of the ion with *m/z* 118; this OE ion could be the result of a heterolytic cleavage following an initial γ-hydrogen shift, as illustrated in Scheme 6-61.

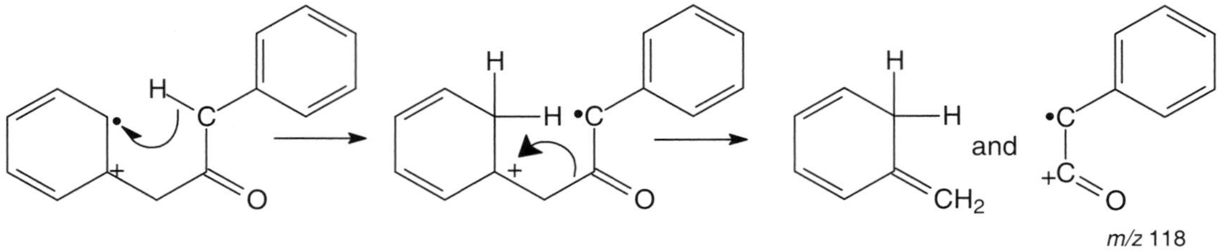

Scheme 6-61

The fragmentation of dibenzyl ketone (Schemes 6-60 and 6-61) supports the premise that isomeric molecular ions can be produced during the ionization process. It is obvious from the above explanation and the mass spectrum that molecular ions of dibenzyl ketone with the charge and radical sites on the carbonyl oxygen are formed as well as those with the charge and radical sites on the ring.

F. Aliphatic Acids and Esters

Like ketones and aldehydes, acids and esters are characterized by the presence of a carbonyl group. Just as an aldehyde can be viewed as an oxidation product of an alcohol, an organic acid (carboxylic acid) can be viewed as an oxidation product of an aldehyde. Organic acids are seldom encountered in EI mass spectrometry because of their polarity. These highly polar molecules are also difficult to separate by gas chromatography. When these acids also contain hydroxyl and amino groups, the resulting increase in polarity and thermal lability make vaporization of the compound difficult; for example, amino acids decompose upon heating. The analysis of an organic acid by EI usually involves preliminary conversion to a less polar, more volatile derivative such as methyl ester or a trisubstituted silyl ester (discussed later in this chapter).

The mass spectra of aliphatic carboxylic acids and esters of carboxylic acids are very similar. Like the EI mass spectra of aliphatic aldehydes and ethers, those of carboxylic acids and esters exhibit discernible M$^{+\bullet}$ peaks unless the molecule contains a γ carbon (i.e., a γ carbon on the carboxylic acid moiety). The presence of this γ carbon allows for β cleavage induced by a γ-hydrogen shift (a McLafferty rearrangement) in molecular ions of both acids and esters where the γ carbon is on the acid side. The resulting OE ion is characteristic of both of these compound types. The McLafferty rearrangement ion from a methyl ester has an *m/z* value of 74 and that for carboxylic acids has an *m/z* value of 60. If the alcohol moiety of an ester has a carbon that is gamma to the alkoxy oxygen, a rearrangement can occur when the charge and radical sites are associated with this alkoxy oxygen.

Even though *n*-butyl acetate and methyl butanoate both have a carbon atom that is a γ carbon, the geometry of the two molecular ions is different in that the virtual six-member ring is formed via the ether oxygen (sp^3 hybrid) in the *n*-butyl acetate and the carbonyl oxygen (sp^2 hybrid) in the methyl butanoate. The γ-hydrogen shift in the *n*-butyl acetate M$^{+\bullet}$ in response to the charge and radical sites on the alkoxy oxygen of the ester linkage results in the inductive cleavage of the sp^3 hybrid oxygen–aliphatic carbon bond to form the OE ion with *m/z* 56 (Scheme 6-62), whereas the γ-hydrogen shift in the methyl butanoate M$^{+\bullet}$ is in response to the charge and radical sites being on the carbonyl oxygen resulting in β cleavage to form the ion of *m/z* 74 that characterizes the McLafferty rearrangement (Scheme 6-63).

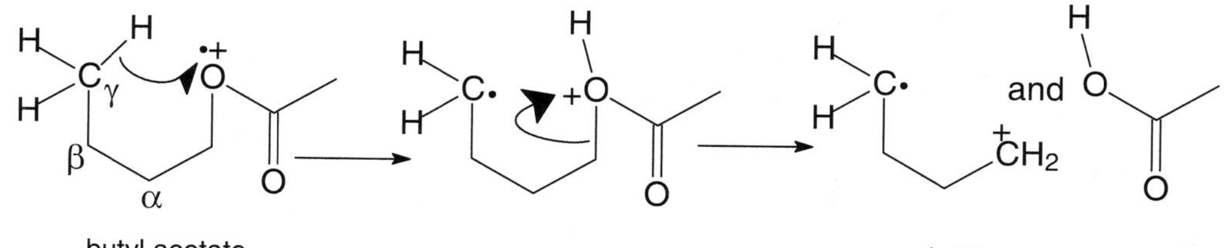

butyl acetate   *m/z* 56

Scheme 6-62

methyl butanoate   *m/z* 74

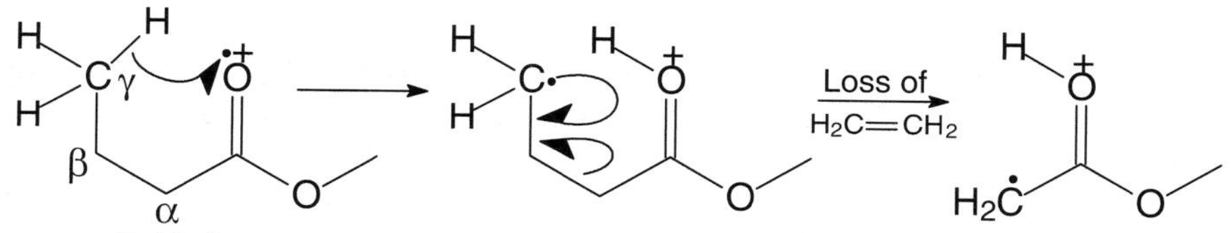

Scheme 6-63

The intensity of the peaks representing these OE ions formed from these rearrangement fragmentations is quite high, even though the γ hydrogens are associated with terminal methyl groups and result in intermediate formation of primary radicals during the γ-hydrogen shift process. Examination of the mass spectra in Figures 6-50 and 6-51 shows that the peaks representing OE fragment ions account for more than 50% of the intensity of the base peak.

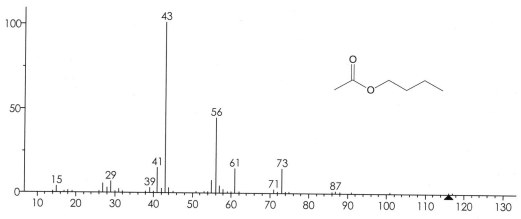

Figure 6-50.   *EI mass spectrum of n-butyl acetate; the peak at m/z 56 represents an odd-electron ion formed by a γ-hydrogen shift through a virtual six-membered ring transition.*

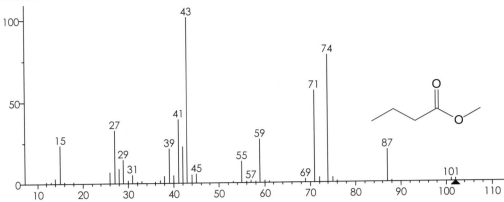

Figure 6-51.   EI mass spectrum of methyl butanoate; the peak at m/z 74 represents an OE ion formed by a γ-hydrogen shift.

The mass spectra of acetates formed by esterification of acetic acid with a homologous series of straight-chain alcohols illustrate how fragmentation mechanisms can be supported. Scheme 6-62 proposes that the peak at *m/z* 56 in the mass spectrum of *n*-butyl acetate represents a distonic aliphatic ion with the charge on one terminal carbon atom and the radical site on the other terminal carbon atom. Examine the four spectra in Figures 6-50, 6-52, 6-53, and 6-54.

Unlike the mass spectrum of *n*-butyl acetate (Figure 6-50), the mass spectrum of *n*-pentyl acetate (Figure 6-52) does not exhibit a significant peak at *m/z* 56; however, there is an OE fragment ion peak at *m/z* 70 that is 14 *m/z* units greater than *m/z* 56. This means that a distonic ion containing one more –CH₂– is formed via a six-membered ring transition analogous to that shown in Scheme 6-62 with the radical site still located on the C-4.

In the mass spectrum of *n*-hexyl acetate (Figure 6-53), the OE ion peak at *m/z* 70 is no longer present; however, there are OE fragment ion peaks at *m/z* 84 and *m/z* 56. The ion with *m/z* 84 is a homolog of the ion with *m/z* 56 (contains two more –CH₂– units than does the ion with *m/z* 56) that was seen in the mass spectrum of *n*-butyl acetate. The ion with *m/z* 86, like the ion at *m/z* 56 formed from *n*-butyl acetate and the ion with *m/z* 70 formed from *n*-pentyl acetate, has the radical site located on C-4. The peak at *m/z* 56 observed in the mass spectrum of *n*-hexyl acetate represents an ion formed by the loss of a molecule of H₂C=CH₂ from the OE fragment ion with *m/z* 84 through heterolytic cleavage.

In the mass spectrum of *n*-heptyl acetate (Figure 6-54), there are OE ion peaks at *m/z* 98, *m/z* 70, and *m/z* 56. In the mass spectrum of *n*-octyl acetate (data not shown), there are OE ion peaks at *m/z* 112, *m/z* 84, *m/z* 70, and *m/z* 56. From the examination and description of these mass spectra of homologous aliphatic acetates, it is apparent that the distonic ion (homologs of the ion with *m/z* 56 in Scheme 6-62) represented by the high-mass OE ion peak undergoes secondary fragmentation to eliminate an olefin through inductive cleavage. For example, the ion with *m/z* 70 in the mass spectrum of *n*-heptyl acetate (Figure 6-54) results from inductive cleavage of the initial distonic ion with *m/z* 98 in response to the charge site to lose a molecule of H₂C=CH₂ as illustrated in Scheme 6-64. The OE ion peak at *m/z* 56 in Figure 6-54 represents an ion formed by hydrogen-shift-induced homolytic cleavage of the initial distonic ion with *m/z* 98 to expel propene, through the mechanism illustrated in Scheme 6-64. The OE fragment ion represented by the peak at *m/z* 56 in the mass spectrum of *n*-hexyl acetate (Figure 6-53) could also be formed as a result of the loss of a molecule of ethene from the ion with *m/z* 84 as a result of inductive cleavage (analogous to Scheme 6-64).

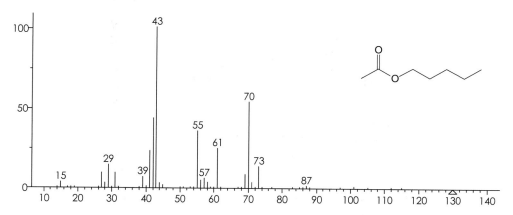

Figure 6-52. *EI mass spectrum of n-pentyl acetate.*

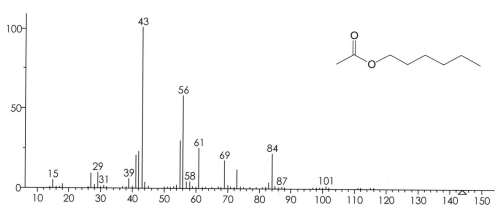

Figure 6-53. *EI mass spectrum of n-hexyl acetate.*

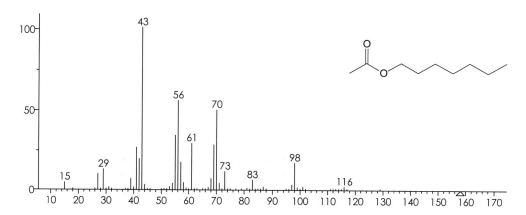

Figure 6-54. *EI mass spectrum of n-heptyl acetate.*

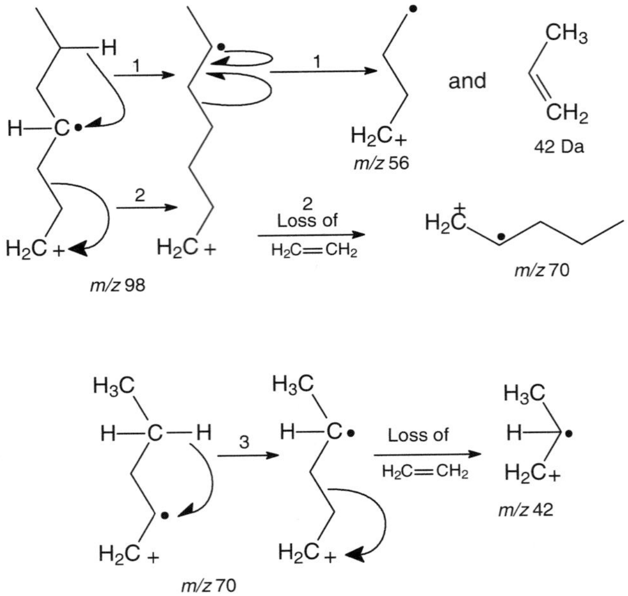

Scheme 6-64

The EI mass spectra of the propionates of the same homologous series of alcohols show the same OE ion peaks. Even as the size of the carbon chain on the acid side of the ester grows, the rearrangement driven by the charge and radical sites on the alkoxy oxygen dominates. These same series of OE ion peaks dominate the spectra of the corresponding propionate compounds. The utility of these mass spectra in identifying the corresponding compounds is limited by the lack of a $M^{+\bullet}$ peak. Therefore, GC retention time becomes an important factor in distinguishing these compounds. Determination of nominal mass through derivatization is not an option in helping to distinguish this series of compounds because esters do not contain active hydrogens to react with derivatizing reagents.

Aliphatic carboxylic acids with 12–20 carbon atoms are known as fatty acids. These compounds have a great deal of biological significance and have been extensively studied by mass spectrometry [34–36]. As was stated at the beginning of this section, because of the polarity of organic acids, they are usually analyzed by gas chromatography and/or mass spectrometry in the form of their methyl esters.

A comparison of the mass spectra of stearic acid (Figure 6-55) and methyl stearate (Figure 6-56) shows a great number of similarities. Both spectra exhibit an intense peak at *m/z* 60 and *m/z* 74, respectively, representing the ion formed through a γ-hydrogen shift-induced β cleavage (McLafferty rearrangement). In each spectrum, there is a peak 13 *m/z* units higher (at *m/z* 73 and *m/z* 87, respectively) representing a structurally diagnostic EE ion. Among the subsequent series of peaks appearing at 14 *m/z* unit intervals; i.e., 87, 101, 115, etc., there are those of more significant intensity appearing at intervals 56 *m/z* units; i.e., 73, 129, 185, etc. in Figure 6-55 and 87, 143, 199, etc. in Figure 6-56.

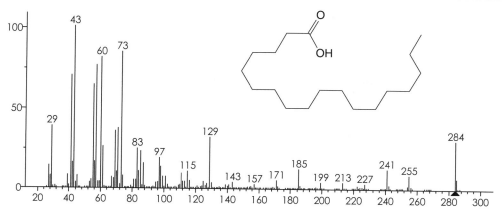

Figure 6-55. *EI mass spectrum of octadecanonic acid; peaks at m/z 73, m/z 129, m/z 185, and m/z 241 separated by 56 m/z units.*

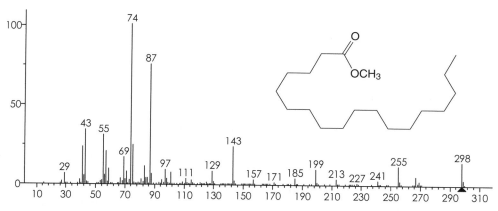

Figure 6-56. *EI mass spectrum of methyl octadecanonate (methyl stearate); peaks at m/z 87, m/z 143, m/z 199, and m/z 255 separated by 56 m/z units.*

Spectra of both the acid and the ester also exhibit $[M - 29]^+$ and $[M - 43]^+$ peaks that have greater-than-expected relative intensities. The origin of the ions represented by all of these notable peaks is the same for both molecular ions. The formation of these ions has been extensively studied for methyl stearate. Based on stable isotope-labeling studies using both deuterium and ^{13}C, Ryhage and Stenhagen proposed an explanation for the formation of the ions represented by the peaks at *m/z* 87, *m/z* 143, and *m/z* 199 in the mass spectrum of methyl stearate as shown in Scheme 6-65 [34, 37, 38].

Even though the logic of the mechanisms involving several distonic ions proposed in Scheme 6-65 is irrefutable, it has been argued that these proposed mechanisms follow from an initial six-membered ring transition [39] as opposed to the eight-membered ring transition shown here. Sparkman and Ren used the B3LYP/6-31+G(d) method for geometry and energy calculation with the Windows® version of *Gaussian 98* to model methyl heptanoate. Based on the results of these calculations, the ion with *m/z* 87 can be formed via an initial eight-membered ring hydrogen atom transfer with less energy requirement than that for the initial six-membered ring transition [40].

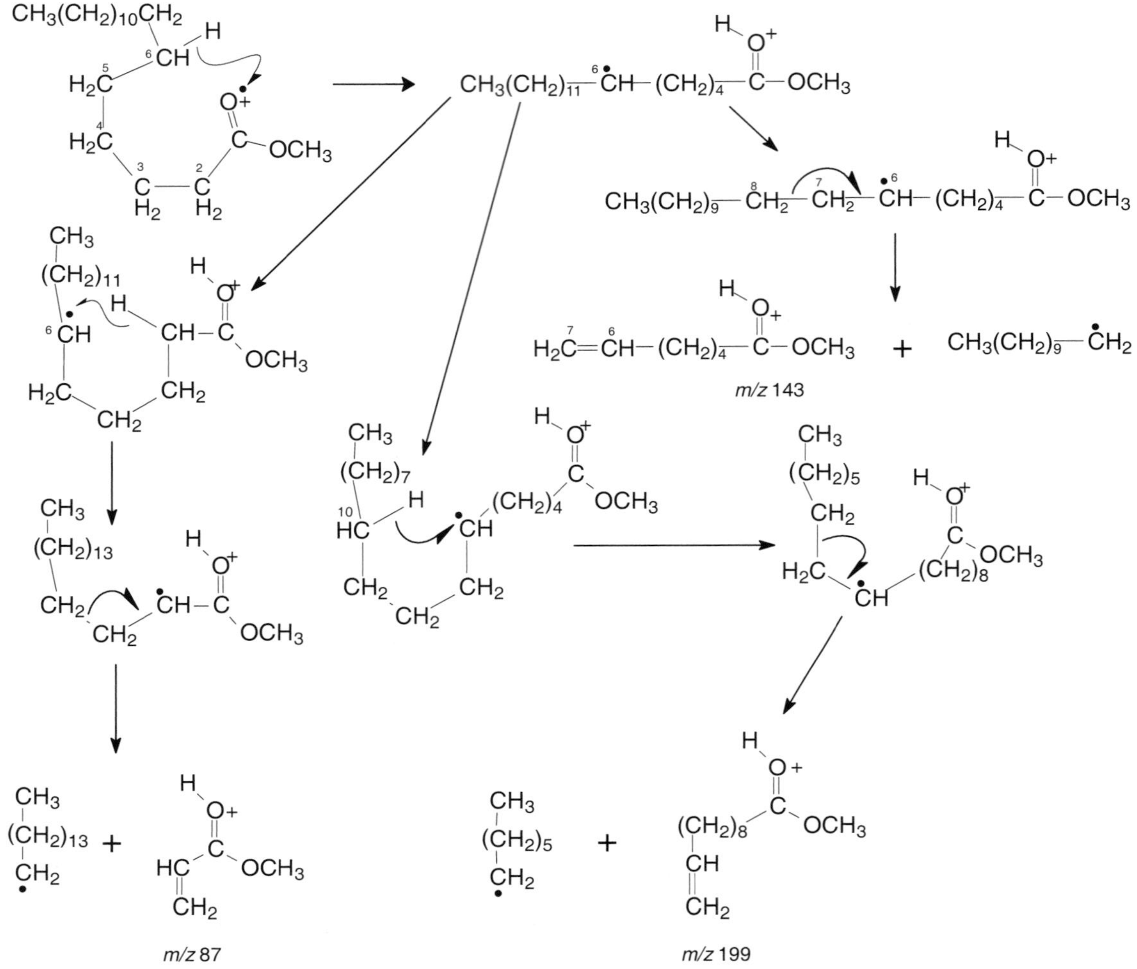

Scheme 6-65

However, if the six-membered ring secondary transition philosophy shown in Scheme 6-65 for the formation of the ions with m/z 87, m/z 143, and m/z 199 is extended one more step, an alternative justification for the formation of the ion with m/z 255 can be made by proposing the mechanism in Scheme 6-66.

The species at the left in Scheme 6-66 is formed starting with the distonic ion (shown as the second structure in Scheme 6-65) initially formed through a hydrogen shift in response to the radical site associated with the carbonyl oxygen occurring through an eight-membered ring transition; the radical will be on C-6. Just as was the case with the formation of the ion with m/z 199 in Scheme 6-65, a six-membered transition involving a hydrogen atom on C-10 occurs, causing the radical site to now be located on C-10.

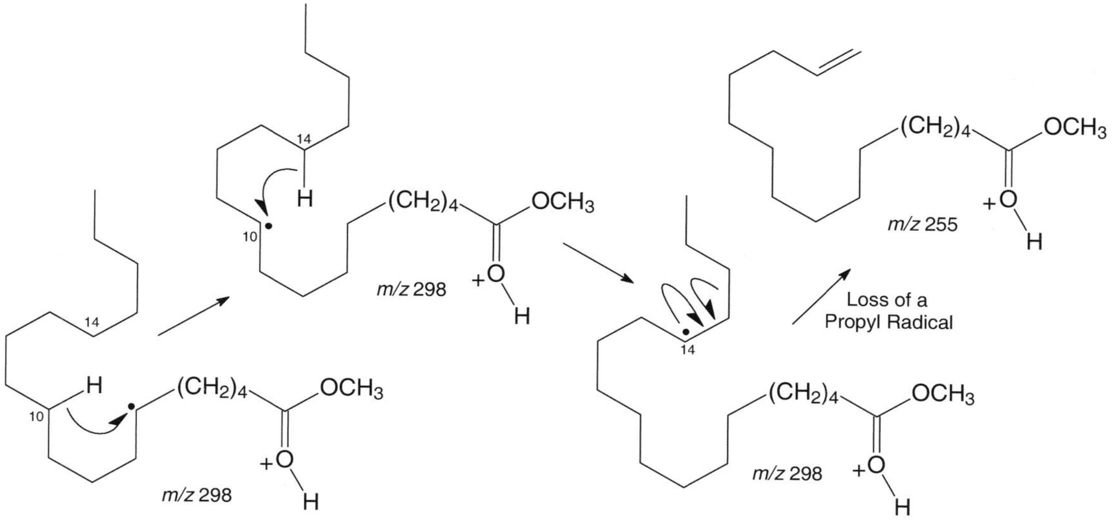

Scheme 6-66

However, rather than a homolytic cleavage being initiated by the movement of one of the two electrons between C-11 and C-12 in response to the radical site on C-10, another six-membered transition involving a hydrogen shift from C-14 could occur. The new radical site at C-14 initiates homolysis by having one of the two electrons forming the bond between the C-15 and C-16 pair with the radical site at C-14. This would result in the loss of a propyl radical.

Hubert Budzikiewicz and colleagues [28] in Carl Djerassi's laboratory at Stanford University suggested the mechanism shown in Scheme 6-67 based on the stable isotope-labeling studies carried out by Stenhagen. The Stenhagen studies showed that the $[M - 43]^+$ peak in the mass spectrum of methyl stearate was due to the loss of carbon atoms 2, 3, and 4. This same work also showed that the $[M - 29]^+$ ion was formed through the loss of carbon atoms 2 and 3. The Budzikiewicz *et al.* mechanism proposes the loss of the methylene units comprised of C-2, C-3, and C-4 plus an additional hydrogen atom that could come from any position along the chain beyond C-5. The $[M - 29]^+$ ion is formed through a five-membered ring transition rather than the six-membered transition shown in Scheme 6-67.

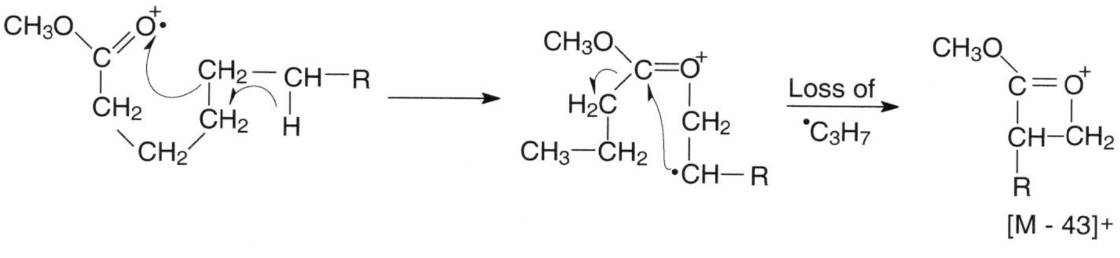

Scheme 6-67

One of the reasons for showing these somewhat bizarre mechanisms is to emphasize the point that with such ionic reactions and/or isomerizations apparently taking place inside the EI ion source, it is amazing that it is possible to extract so much information about the structure of an analyte from its EI mass spectrum. As was stated early on, all of these schemes are paper mechanisms. They are based on what is known about the behavior of organic molecules. However, even with the results of stable isotope-labeling studies, it is not always possible to describe fragmentation behavior accurately.

A significant difference in the mass spectra of methyl esters and those of the corresponding free fatty acids is the presence of a $[M - 31]^+$ ($[M - OCH_3]^+$) peak for the methyl esters, but not a $[M - 17]^+$ ($[M - OH]^+$) peak for the free acids. Alkyl esters lose an alkoxy radical via homolytic cleavage when the charge and radical sites are on the carbonyl oxygen atom (Scheme 6-68). The analogous $M^{+\bullet}$ of the free acid does not lose a hydroxy radical. A peak at m/z 74, together with a $[M - 31]^+$ peak, is a strong indication that the analyte is a methyl ester in which the acid moiety consists of at least four carbon atoms.

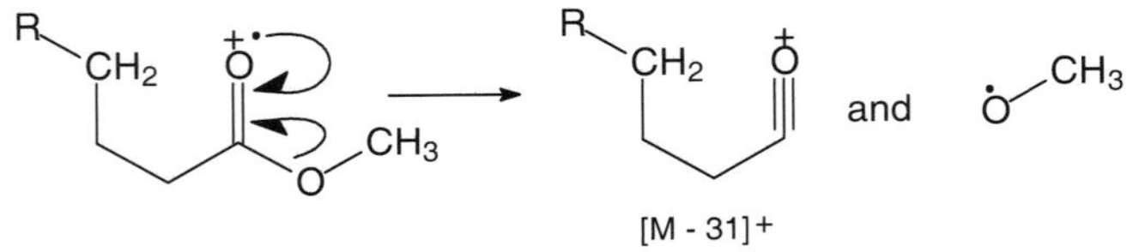

$$R\diagdown CH_2 \quad \overset{+\bullet}{O} \longrightarrow R\diagdown CH_2 \quad \overset{+}{O} \quad \text{and} \quad \overset{\bullet}{O}\diagdown CH_3$$

$$[M - 31]^+$$

Scheme 6-68

A periodic series of peaks every 56 m/z units are observed in the mass spectra of all long-chain fatty acids and esters of higher mass alcohols. There are also peaks that stand out from the basic spectral pattern of peaks separated by 14 m/z units including those that designate the position of branch points and other functional groups. It is important to be able to recognize the periodic peaks that correspond to the above-described rearrangements and to differentiate these peaks from those that are structurally diagnostic.

The mass spectra of esterified fatty acids provide a good example of using general observations and knowledge of specific fragmentation mechanisms to elucidate the structure of an unknown from its EI mass spectrum.

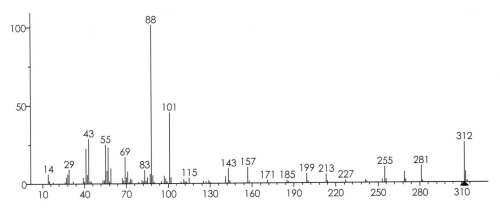

Figure 6-57.   EI mass spectrum of an unknown.

Consider the mass spectrum in Figure 6-57 as that of an unknown. Reading the spectrum from the right to the left, the first observation is that the highest m/z value peak that is not due to background or an isotope peak is at m/z 312. This is an even number indicating, according to the *Nitrogen Rule*, that the analyte does not contain an odd number of N atoms. The base peak in the mass spectrum is at m/z 88, which could well represent an OE fragment ion. There is an obvious series of peaks separated by 56 m/z units (see m/z 101, 157, 213, etc.). It is known that the mass spectra of methyl esters of fatty acids exhibit a $M^{+\bullet}$ peak, a peak that represents an OE fragment ion at m/z 74, and periodic peaks separated by 56 m/z units that represent other rearrangement fragmentations of the $M^{+\bullet}$. Having noted such clues that are ester-like, the peak at m/z 88 might be expected to correspond to an ethyl ester considering the homologous shift of 14 from m/z 74. The formation of an ion of m/z 88 could be produced as illustrated in Scheme 6-69 from the $M^{+\bullet}$ of an ethyl ester.

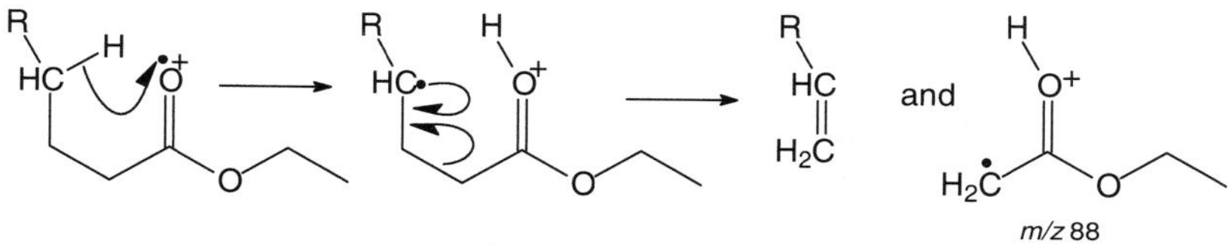

Scheme 6-69

Further evidence that the unknown is likely to be an ester comes from noting that the difference between the OE ion peak (m/z 88) and the first peak (m/z 101) representing the 56-m/z unit periodic rearrangement ion is 13 m/z units just as in the mass spectrum of methyl stearate the peak (m/z 74) for the OE ion and the first (m/z 87) of the periodic series of peaks separated by 56 m/z units is 13 m/z units. Because the difference between the nominal mass of methyl stearate and the apparent nominal mass of the unknown is 14 Da, the evidence described to this point suggests that the unknown could be the ethyl ester of stearic acid.

However, examination of the high-mass end of the mass spectrum in Figure 6-57 shows no evidence of a peak at M – 45 (m/z 267) corresponding to loss of an ethoxy radical, as would be expected for an ethyl ester. Recall the important characteristic fragmentation of esterified fatty acids where some of the molecular ions tend to lose the alkoxy radical as shown in Scheme 6-68. Instead, there is a peak at m/z 281 in Figure 6-57 that could correspond to the loss of a methyl radical! This evidence suggests that the unknown is a methyl ester. A structural possibility that would be consistent with all the mass spectral evidence is a methyl ester of a C_{18} acid, which is substituted on the C-2 position (the α carbon) with a methyl group (α-methyl methyl stearate). Some of the molecular ions of this compound could produce an OE fragment ion with m/z 88 as illustrated in Scheme 6-70. The mass spectrum of ethyl stearate is shown in Figure 6-58.

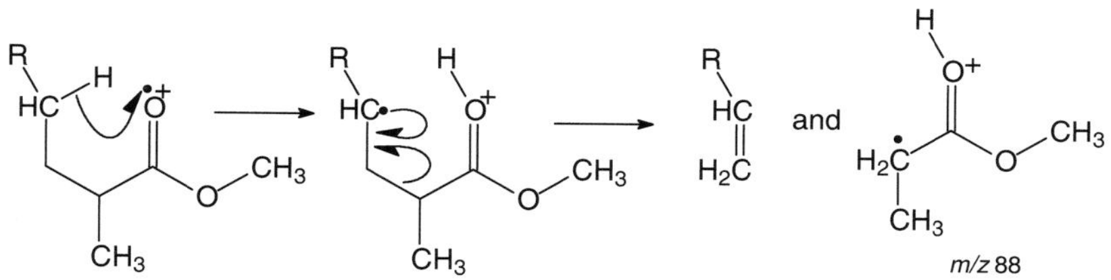

Scheme 6-70

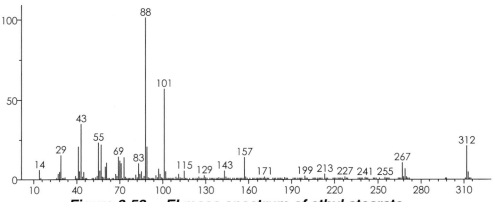

Figure 6-58.   EI mass spectrum of ethyl stearate.

The above example illustrates the importance of extracting several pieces of information from a mass spectrum to support the identification of an unknown. As was seen in this particular case, even though there was a low-*m/z* value OE ion peak that represented a specific behavior, its meaning was less than conclusive. It was necessary to look at the high-*m/z* end of the mass spectrum to obtain the unambiguous information.

G. Aromatic Acids and Esters

Aromatic acids and esters provide some of the best examples of the *ortho effect*. Two commercially significant compounds that fall into this category are acetylsalicylic acid (a.k.a. Aspirin) and methylsalicylate (2-hydroxymethylbenzoate, a.k.a oil of wintergreen). The mass spectra of these compounds have the same general characteristics as the mass spectra of all aromatic compounds: (1) M$^{+\bullet}$ peaks of significant intensity, (2) ion current distributed throughout a few *m/z* values, and (3) peaks representing secondary fragmentation ions with *m/z* 39, 51, and 65.

The best illustration of the behavior of these types of compounds is the fragmentation of the regioisomers of hydroxybenzoic acid. Examination of Scheme 6-71 shows how the *ortho effect* results in the expulsion of a molecule of water from the M$^{+\bullet}$ of salicylic acid. Some of these [M – H$_2$O]$^{+\bullet}$ ions with *m/z* 120 undergo two successive losses of carbon monoxide to form the OE ions with *m/z* 92 and 64 (see the corresponding peaks in Figure 6-59). Without the *ortho effect*, most of the molecular ions of the *meta* and *para* isomers lose a hydroxyl radical to form an EE ion (*m/z* 121), some of which lose CO to form another EE ion (*m/z* 93); some of these ions then lose another molecule of CO to form yet a third EE ion (*m/z* 65) (see the corresponding peaks in Figure 6-60).

If the acid moiety is esterified, the *ortho effect* will cause the M$^{+\bullet}$ to expel an alcohol molecule. If the hydroxyl moiety of 2-hydroxybenzoic acid is replaced with an –NH$_2$ group, the M$^{+\bullet}$ will still expel a molecule of water; however, if the –COOH group is replaced with a –CONH$_2$, the M$^{+\bullet}$ will expel a molecule of NH$_3$. There is no *ortho effect* observed in the mass spectrum of 2-chlorobezoic acid (data not shown).

2-acetyloxybenzoic acid (acetylsalicylic acid, Aspirin) is another interesting compound in that it has both alkyl and aryl moieties. The mass spectra of both the 2- and 4- regioisomers of acetyloxybenzoic acid (Figures 6-61 and 6-62) exhibit discernible M$^{+\bullet}$ peaks. These two mass spectra also exhibit a peak at *m/z* 138 (the same elemental composition and possibly the same structure as 2-OH benzoic acid), corresponding to the loss of a molecule of ketene from the M$^{+\bullet}$ (Scheme 6-72). Some of the resulting OE ions then fragment in the same way as the two isomers of hydroxybenzoic acid whose mass spectra are shown in Figures 6-59 and 6-60. In this case, the *ortho effect* functions during a secondary fragmentation process.

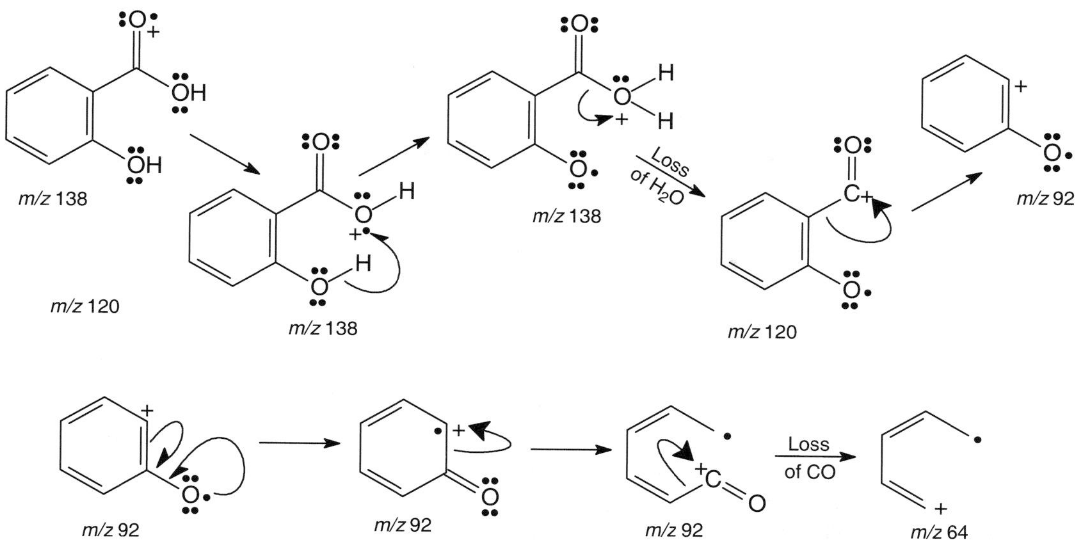

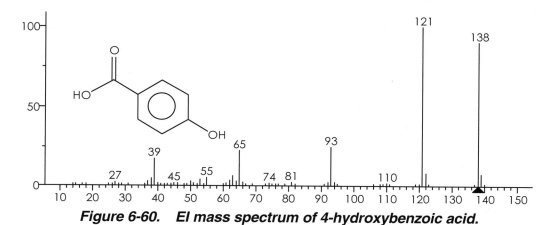

Scheme 6-71

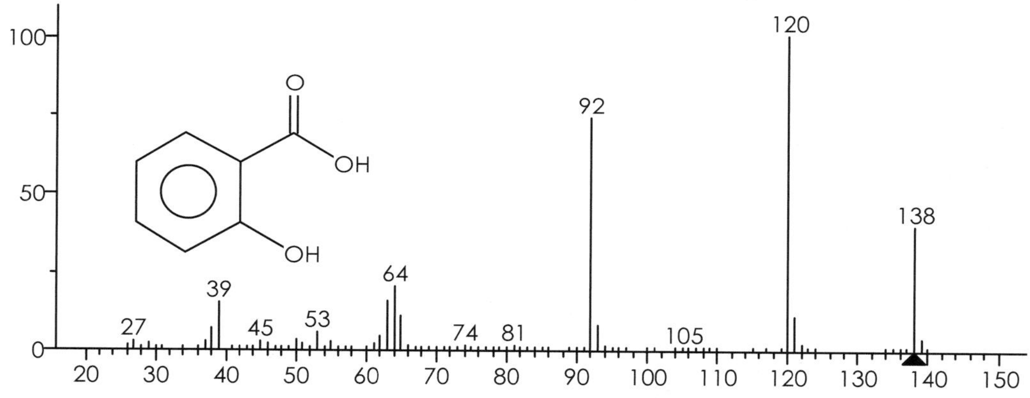

Figure 6-59. EI mass spectrum of salicylic acid (2-hydroxybenzoic acid).

Figure 6-60. EI mass spectrum of 4-hydroxybenzoic acid.

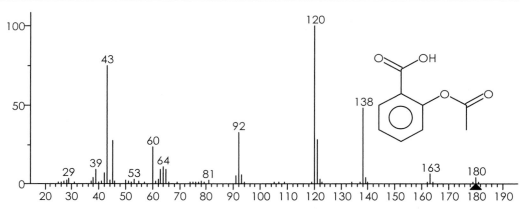

Figure 6-61.   EI mass spectrum of 2-acetyloxybenzoic acid.

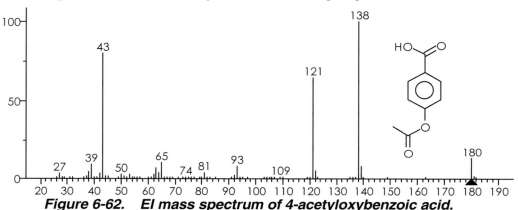

Figure 6-62.   EI mass spectrum of 4-acetyloxybenzoic acid.

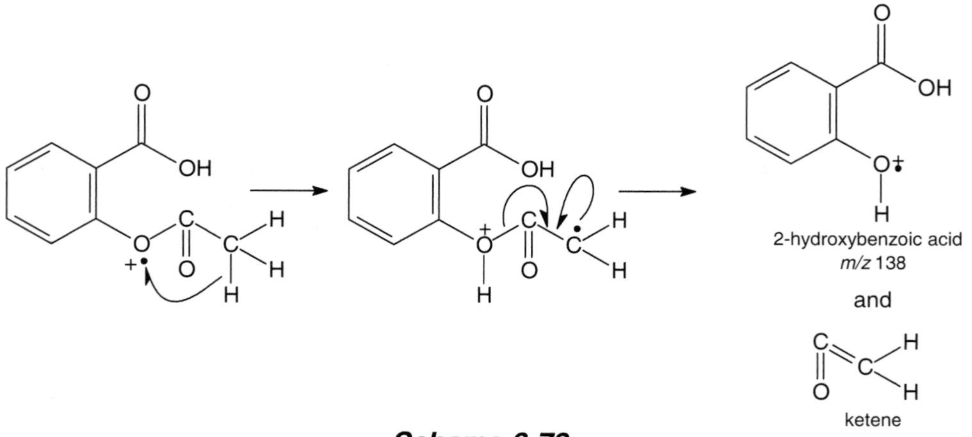

Scheme 6-72

Because of the resonance structure of acetylsalicylic (acetyloxybenzoic) acid, the charge and radical sites may be transitory.   The mass spectra of these compounds (Figures 6-61 and 6-62) both show a peak at *m/z* 138 corresponding to loss of ketene from the M⁺•. To support a mechanism for the loss of ketene, the charge and radical sites in the M⁺• need to be on the position shown in Scheme 6-72.

4. Nitrogen-Containing Compounds

A. Aliphatic Amines

Aliphatic amines behave a lot like aliphatic alcohols in that the initial fragmentation occurs by cleavage of the bond between C-1 and C-2 due to homolysis driven by the charge and radical sites on the heteroatom (α cleavage). Structurally, aliphatic amines differ from aliphatic alcohols in that 1°, 2°, and 3° amines are constituted by the number of non-hydrogen substituents on the N atom, whereas the number of non-hydrogen substituents on the hydroxylated carbon determines whether the alcohol is 1°, 2°, or 3°.

Depending on the number of carbon atoms in substituents on the nitrogen atom, a number of hydride-shift rearrangements resulting in secondary fragmentations can occur in the nascent EE fragment ions formed by the initial homolytic cleavage of the $M^{+\bullet}$. Once a radical site is no longer a factor due to its departure with the dark matter (the neutral loss), all of these secondary fragmentations are charge-site-driven with the charge still on the N atom.

The electrophilic character of oxygen is higher than that of nitrogen; therefore, in the mass spectra of ketones and ethers, peaks are observed representing ions that are due to heterolytic cleavage as well as homolytic cleavage. In the mass spectra of aliphatic amines, there are no peaks that represent ions formed by heterolytic cleavage of the $M^{+\bullet}$, only peaks representing ions formed by homolytic cleavage of the $M^{+\bullet}$.

The EI mass spectra of five isomers of the elemental composition $C_6H_{15}N$ will be examined on the following pages. The first three isomers represent 1°, 2°, or 3° amines (1-hexylamine, Figure 6-64; dipropylamine, Figure 6-65; and triethylamine, Figure 6-66). Members of the second group are all primary amines, differing only in the position of the $-NH_2$ group on the hydrocarbon chain (1-hexylamine; 2-hexylamine, Figure 6-67; and 3-hexylamine, Figure 6-68); 1-hexylamine is common to both groups.

The mass spectra of 1-hexylamine, dipropylamine, and triethylamine shown in Figures 6-64, 6-65, and 6-66, respectively, illustrate an interesting property of the mass spectral display. Reading each spectrum from right to left, the relative intensity of the $M^{+\bullet}$ peak obviously increases from the primary (Figure 6-64) to the tertiary amine (Figure 6-66). The relative intensity of the peak at m/z 30 is the same for the primary and secondary amines (the base peak). In reality, what is happening is that the abundance of the ion with m/z 30 is decreasing. Therefore, the intensity of the $M^{+\bullet}$ peak relative to the base peak (at m/z 30) in Figure 6-65 is greater than the intensity of the $M^{+\bullet}$ peak relative to the base peak (also at m/z 30) in Figure 6-64. A determination of the percent abundance of the $M^{+\bullet}$ in all three spectra reveals that it is about the same (this numerical determination is not possible from the data provided in these three figures). It is important to remember that the physical dimensions of the mass spectral display are fixed. The *y*-axis shows the intensity of all the peaks *relative* to that for the most abundant ion (the base peak) in the mass spectrum. The separation of peaks along the *x*-axis is determined by the number of m/z units required to show all the peaks within a selected display range.

The mass spectra of all five of these amines exhibit a peak at m/z 30 representing a characteristic ion formed by aliphatic amines, just as the ion with m/z 31 is a characteristic ion for aliphatic alcohols. Both of these ions are EE ions and both conform to the *Nitrogen Rule* in that the EE ion with m/z 30 (an even number) has an odd number of N atoms ($H_2C=N^+H_2$) and the EE ion with m/z 31 (an odd number) does not contain an odd number of N atoms ($H_2C=O^+H$). These important peaks will not be observed in a mass spectrum unless the data acquisition starts at a sufficiently low *m/z* value. It is common practice to start data acquisition at *m/z* 35 to avoid ions due to air (*m/z* 28 and *m/z* 32) and water (*m/z* 18); therefore, special attention is required in performing mass spectral analyses of amines and alcohols. Compare the mass spectrum of 1-hexylamine where the acquisition began at *m/z* 35 (Figure 6-63) and that of the same compound where the data acquisition began at *m/z* 15 (Figure 6-64).

Returning to the comparison of the mass spectra of aliphatic alcohols and amines, another important difference between the oxonium ions formed by aliphatic alcohols and the immonium ions formed by aliphatic amines is their abundance. The primary immonium ion formed by an initial fragmentation of the M$^{+\bullet}$ of amines is far more abundant than the primary oxonium ion formed by the initial fragmentation of an aliphatic alcohol M$^{+\bullet}$.

The mass spectrum of 1-hexylamine (Figure 6-64) is dominated by a single peak at *m/z* 30. There is a peak for the M$^{+\bullet}$ at *m/z* 101 (consistent with the *Nitrogen Rule*); however, it is of very low intensity. There is also a low intensity peak at *m/z* 100, which represents the loss of a hydrogen radical from the M$^{+\bullet}$. The most probable positions for the charge and radical sites in an amine will be on the nitrogen atom as shown in Scheme 6-73, due the loss of an electron from the nonbonding orbital of nitrogen. The ion of *m/z* 30 is formed by homolytic cleavage of the bond between C-1 and C-2 (the α carbon) as illustrated in Scheme 6-73. The mass spectra of all aliphatic amines in which the –NH$_2$ group is located on a terminal carbon have the same general appearance: a base peak at *m/z* 30 (representing a primary immonium ion formed by α cleavage) and a barely discernible M$^{+\bullet}$ peak. When data are acquired using rapid scanning in GC/MS to accommodate narrow chromatographic peaks, the M$^{+\bullet}$ peak may not be observed due to poor ion counting statistics in the detection system.

The ion with *m/z* 30 is observed in the mass spectra of all aliphatic amines (1°, 2°, and 3°) and regardless of the position of the –NH$_2$ group on the chain. Ions with *m/z* 30 are not only formed by initial fragmentation of the M$^{+\bullet}$, but also by secondary fragmentation of EE ions of higher *m/z* values produced by initial fragmentation of the M$^{+\bullet}$, as described in detail in some of the following examples.

Aliphatic amines also undergo β cleavage (cleavage of the bond between C-2 and C-3) to produce ions with *m/z* 44 as illustrated in Scheme 6-74. This will be illustrated in more detail in the discussion of the fragmentation of dipropyamine and triethylamine. Like the ion with *m/z* 30, the ion with *m/z* 44 also can result from the secondary fragmentations of EE ions formed by an initial β cleavage.

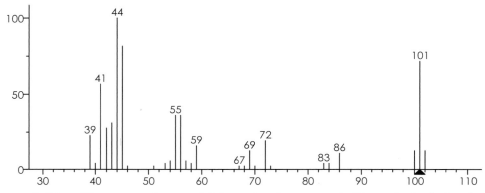

Figure 6-63.   *Mass spectrum of 1-hexylamine; acquisition started at m/z 35.*

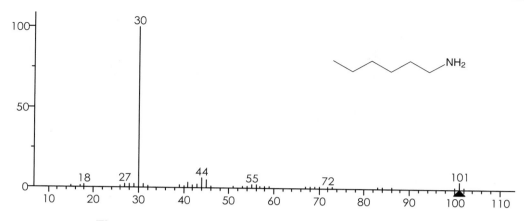

Figure 6-64.   EI mass spectrum of 1-hexylamine.

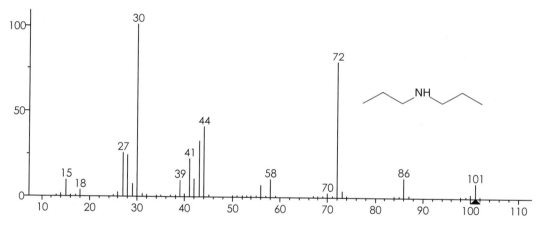

Figure 6-65.   EI mass spectrum of dipropylamine.

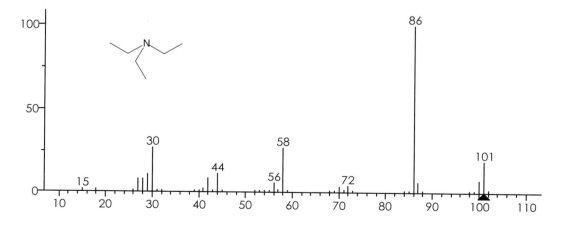

Figure 6-66.   EI mass spectrum of triethylamine.

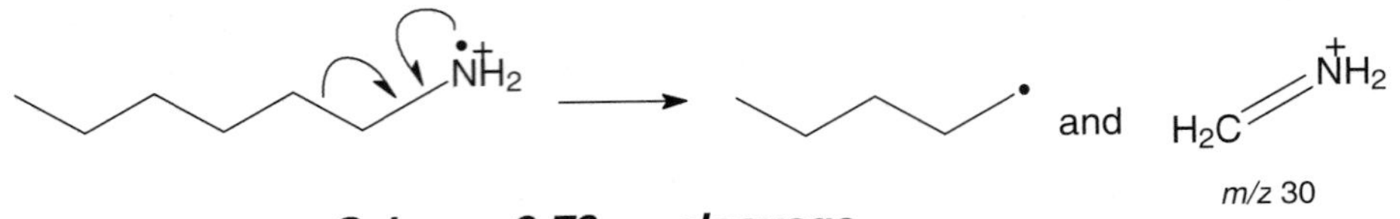

Scheme 6-73. α *cleavage.*

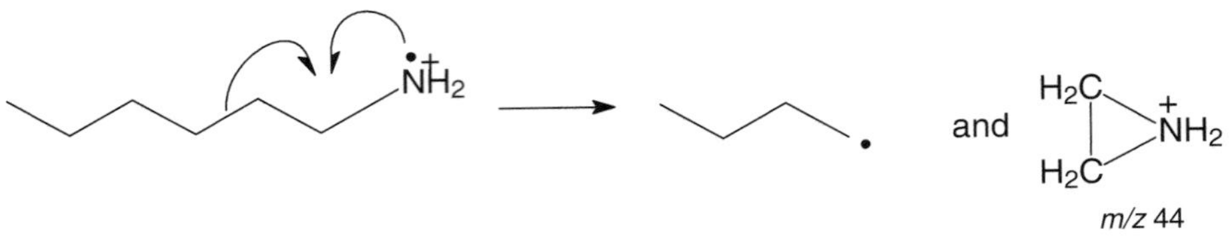

Scheme 6-74. β *cleavage of a primary amine.*

Although β cleavage is an important factor in the explanation of the fragmentation of aliphatic amines, it is not the only source of ions with *m/z* 44. Peter Derrick and associates [41] used deuterium labeling to show that aliphatic amines with γ hydrogens on nonterminal carbon atoms form ions with *m/z* 44 through what Fred McLafferty refers to as a "pseudo α cleavage" involving a multistep hydrogen and skeletal rearrangement to produce an α-methyl-alkylamine, which then undergoes α cleavage to produce the methyl-substituted immonium ion (H₃C–CH=N⁺H₂) as illustrated in Scheme 6-75.

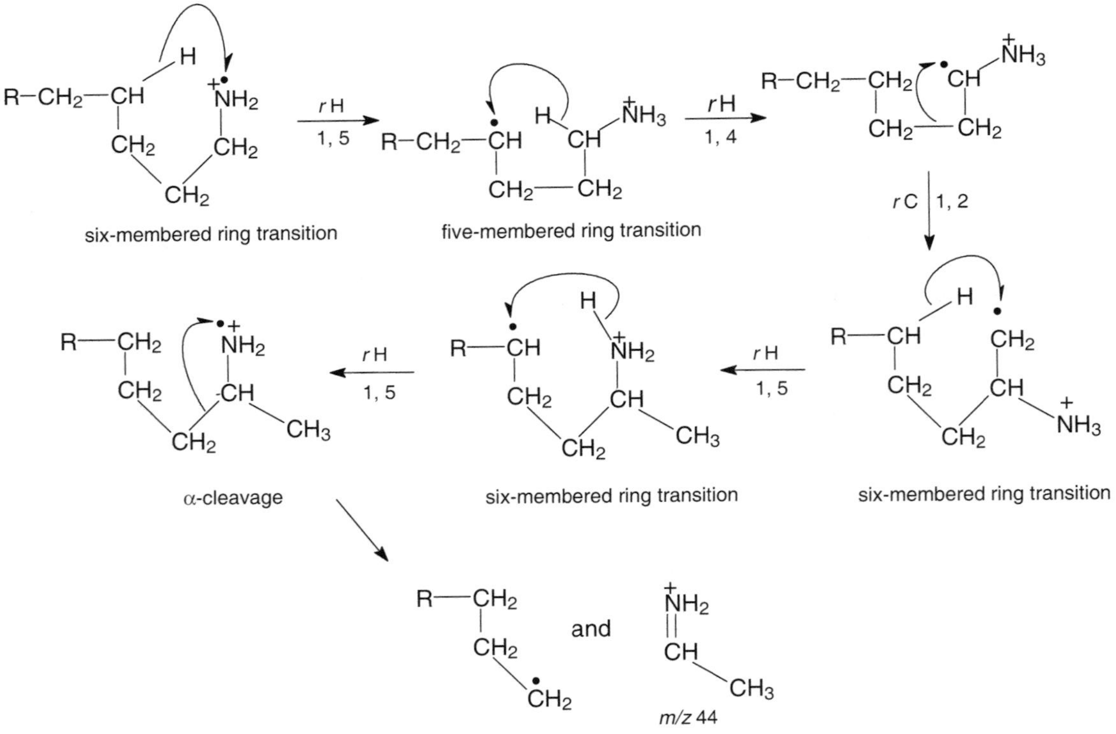

Scheme 6-75

The ion with *m/z* 44 probably results both from the rearrangement (Scheme 6-75) and from β cleavage (Scheme 6-74). Examination of the mass spectra of a homologous series of aliphatic amines with the –NH$_2$ group located on a terminal carbon reveals that the relative intensity of the peak at *m/z* 44 increases with increasing chain length.

The mass spectrum (Figure 6-65) of the secondary aliphatic amine, dipropylamine, appears to be much more complex than that of 1-hexylamine. Initial fragmentation of the M$^{+\bullet}$ of dipropylamine via α cleavage results in formation of the ion with *m/z* 72, a propyl-substituted immonium ion, and the loss of an ethyl radical (Scheme 6-76). Some of these EE ions of *m/z* 72, with the charge still on the N atom, will undergo a hydride-shift rearrangement fragmentation to produce the primary immonium ion with *m/z* 30 (Scheme 6-77).

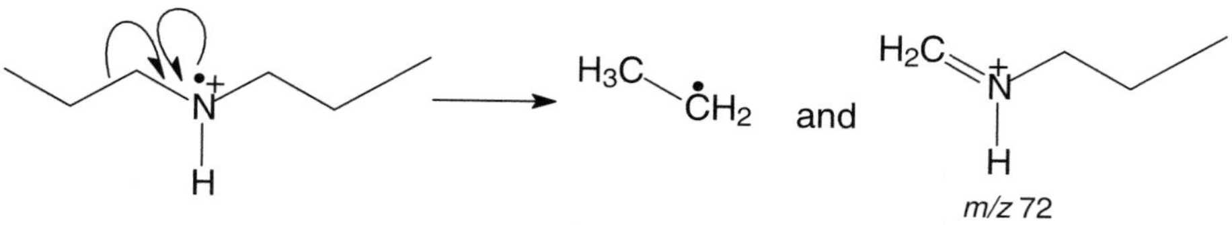

Scheme 6-76

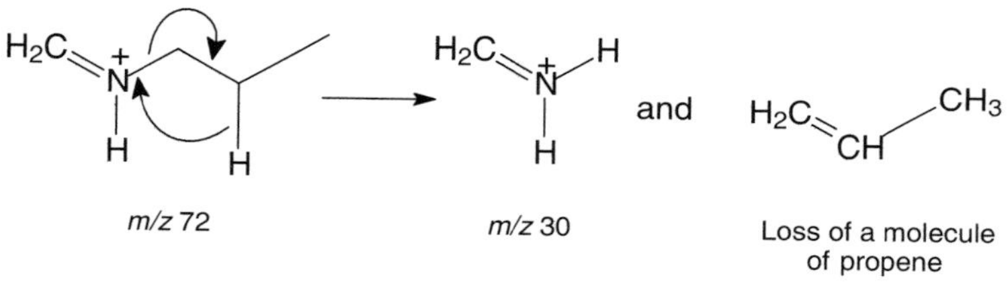

Scheme 6-77

Some dipropylamine molecular ions will undergo β cleavage to produce ions with *m/z* 86 through the loss of a methyl radical (a methyl group in a special location) as shown in Scheme 6-78. Some of the ions with *m/z* 86 will then undergo a hydride-shift rearrangement to produce the ion with *m/z* 44 (Scheme 6-79).

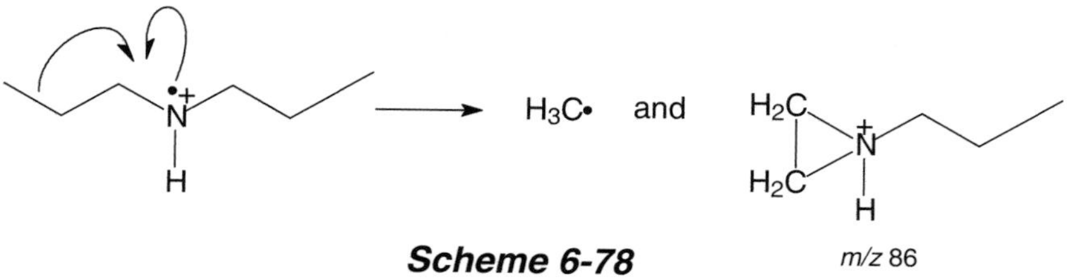

Scheme 6-78

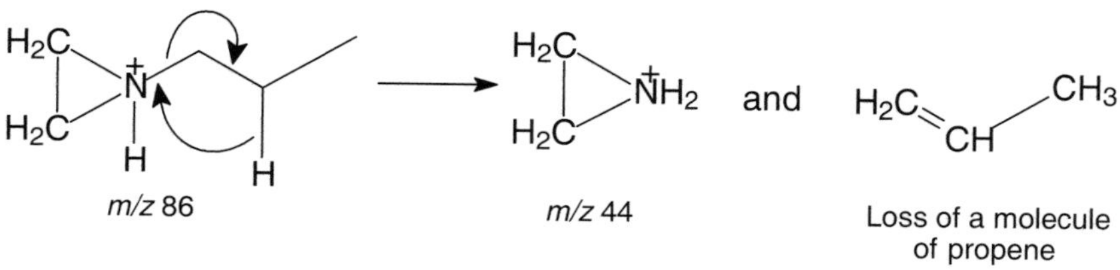

Scheme 6-79

Examination of the starting ions in Schemes 6-75 (*m/z* 72) and 6-77 (*m/z* 86) reveals a strong structural similarity. Both ions have a propyl moiety attached to the nitrogen atom that carries the charge. Just as this propyl moiety can initiate a hydride-shift rearrangement fragmentation of the primary α-cleavage product (Scheme 6-77), it can do the same in the primary β-cleavage product as shown in Scheme 6-79. In both cases, a molecule of propene is expelled to produce a primary immonium ion.

Another interesting behavior of the M⁺• ion of dipropylamine is the fragmentation of the ions represented by the peaks at *m/z* 58 and *m/z* 56 in Figure 6-65. This involves the loss of a hydrogen radical through a five-membered ring transition to form a propyl-substituted four-membered ring immonium ion (*m/z* 100) as shown in Scheme 6-80. Many of these ions lose a molecule of propene through a hydride-shift rearrangement forming the ion with *m/z* 58; some of these ions then lose a molecule of hydrogen (H_2) to form the ion with *m/z* 56.

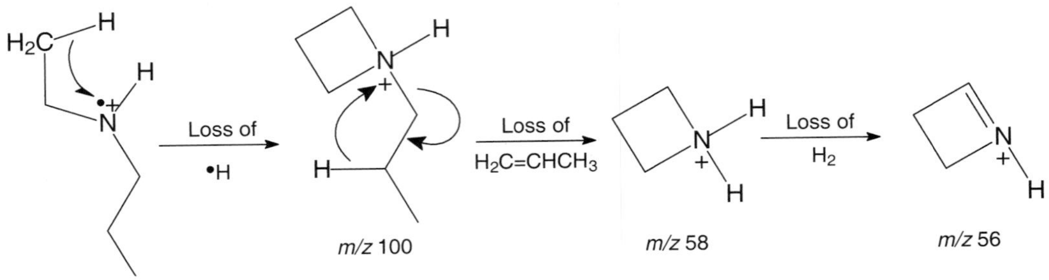

m/z 100   *m/z* 58   *m/z* 56

Scheme 6-80

Just as the secondary amine required a two-step process to form the primary immonium ion with *m/z* 30, as illustrated through Schemes 6-76 and 6-77 in the explanation of the fragmentation of dipropylamine, the formation of this primary immonium ion with *m/z* 30 in the mass spectrum of a tertiary amine requires a three-step process as illustrated in Scheme 6-81. Not all of the initially formed ions are converted to the primary immonium ion; in fact, the more steps in the conversion process, the lower the abundance of the primary immonium ion formed as seen by the intensity of the peak at *m/z* 30 in the mass spectrum of the primary amine relative to the mass spectrum of the tertiary amine.

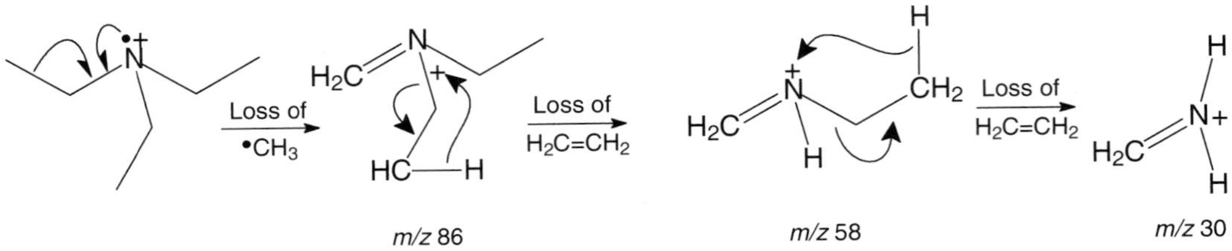

m/z 86   *m/z* 58   *m/z* 30

Scheme 6-81

The M⁺• of triethylamine first undergoes α cleavage resulting in the loss of a methyl radical to produce the EE ion with *m/z* 86 as shown in Scheme 6-81. The base peak at *m/z* 86 in the mass spectrum of triethylamine (Figure 6-66) represents the only α-cleavage EE product ion regardless of which ethyl moiety loses a methyl radical. Some of the resulting EE ions with *m/z* 86 then undergo a hydride-shift rearrangement fragmentation to lose a molecule of ethene to produce the EE ion with *m/z* 58; some of these ions will also undergo a hydride-shift rearrangement to produce the primary immonium ion with *m/z* 30.

Some molecular ions of triethylamine will undergo β cleavage to produce an EE disubstituted immonium ion with *m/z* 100 through the loss of a hydrogen radical as illustrated in Scheme 6-82. The fact that there are eight other β hydrogens as well as six α hydrogens that could be expelled as a radical explains why the peak for [M − 1]⁺ has such a high intensity relative to that of the M⁺• peak in Figure 6-66 (compare with that in Figure 6-65). Most of the disubstituted immonium ions will undergo two iterations of hydride-shift rearrangements, each resulting in the loss of a molecule of ethene to finally form the ion with *m/z* 44 as shown in Scheme 6-82.

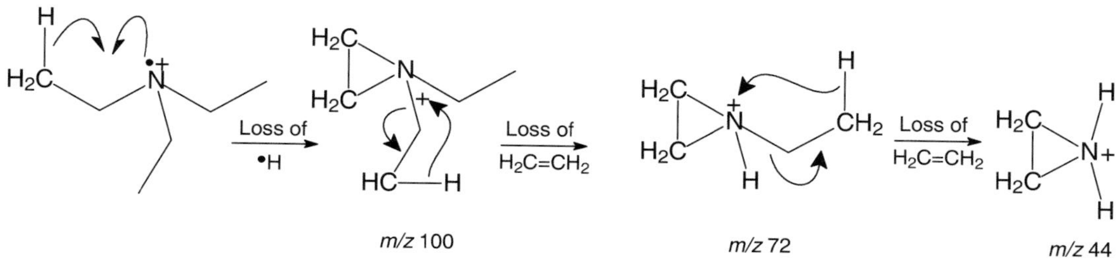

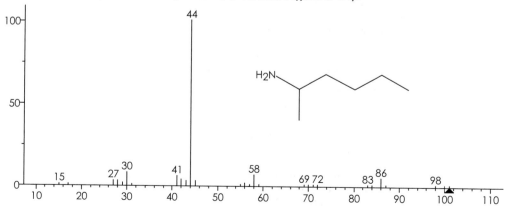

m/z 100   *m/z* 72   *m/z* 44

Scheme 6-82

Another interesting observation in the mass spectra of aliphatic amines is the consistent presence of an [M − 1]⁺ peak. The ion represented by this peak was explained for the spectrum of triethylamine, but what about the [M − 1]⁺ peak in the mass spectra of dipropylamine (Figure 6-65), 1-hexylamine (Figure 6-64), 2-hexylamine (Figure 6-67), and 3-hexylamine (Figure 6-68)? The source of the [M − 1]⁺ ion in all of these mass spectra is the loss of a hydrogen radical through α cleavage from the carbon atom bonded to the nitrogen atom. The hydrogens attached to the carbon bonded to the nitrogen are in the α-position. The tendency of the radical site on a nitrogen atom to pair with another electron is so great that even though the hydrogen radical is not very stable, there will be some hydrogen loss through α cleavage.

Comparison of the mass spectra of the three primary amines, 1-, 2-, and 3-hexylamine, reveals how the position of the –NH₂ group on the carbon chain can cause significant differences in the appearance of the mass spectrum. The mass spectrum of 1-hexylamine was discussed earlier in this chapter. Its mass spectrum is dominated by the α-cleavage product with *m/z* 30 (Scheme 6-73 and Figure 6-64).

Figure 6-67.   *El mass spectrum of 2-hexylamine.*

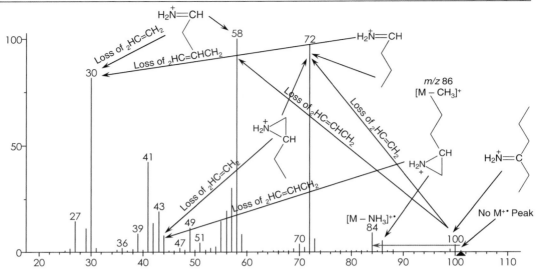

Figure 6-68. *EI mass spectrum of 3-hexylamine showing the structure of various ions and the origin of the ions represented by selected peaks.*

The mass spectrum of 2-hexylamine (Figure 6-67) is dominated by a single peak at m/z 44. This mass spectrum exhibits the characteristic peaks of a diminishingly small $M^{+\bullet}$ peak followed by an $[M - 1]^+$ peak and a peak at m/z 30. The $M^{+\bullet}$ of 2-hexylamine will undergo fragmentation along three different pathways of α cleavage as shown in Scheme 6-83. Most of the ions with m/z 100 will undergo a hydride-shift rearrangement to produce ions of m/z 44 analogous to the mechanism shown in Scheme 6-84 for the formation of the ion with m/z 44 from the ion with m/z 86. The same mechanism is employed by some of the EE ions with m/z 86 in the formation of ions with m/z 30 (Scheme 6-84). The resonance structures of the EE ions with m/z 100 (Scheme 6-83) and m/z 86 (Scheme 6-84) are keys to understanding these hydride-shift rearrangements that result in the primary immonium ions.

β cleavage (not shown) of some of the 2-hexylamine molecular ions will result in $[M - 1]^+$ ions due to the loss of a hydrogen radical from C-1 and other ions with m/z 58 having a methyl-substituted cyclic immonium structure. This latter ion, like the methyl-substituted primary immonium ion, cannot undergo further fragmentation. The $[M - 1]^+$ ion can undergo a hydride-shift rearrangement to produce an ion with m/z 58 in a manner analogous to that of the α-cleavage product that produced the ion with m/z 44 from the $[M - 1]^+$ ion as shown in Scheme 6-82. Mechanisms for these fragmentations are purposefully omitted to challenge the reader, who may wish to retreat to review appropriate schemes described on the last few pages.

Molecular ions of 3-hexylamine can lose hydrogen, ethyl, or propyl radicals by α cleavage and methyl or ethyl radicals by β cleavage. Some of the resulting EE ions will then undergo hydride-shift rearrangements as summarized in Figure 6-68. The mechanisms associated with the primary fragmentation of molecular ions of 3-hexylamine and subsequent secondary fragmentations of the resulting EE ions have been purposefully omitted to challenge the reader's grasp of the subject, as was the case with the secondary fragmentations associated with β-cleavage products during interpretation of the mass spectrum of 2-hexylamine.

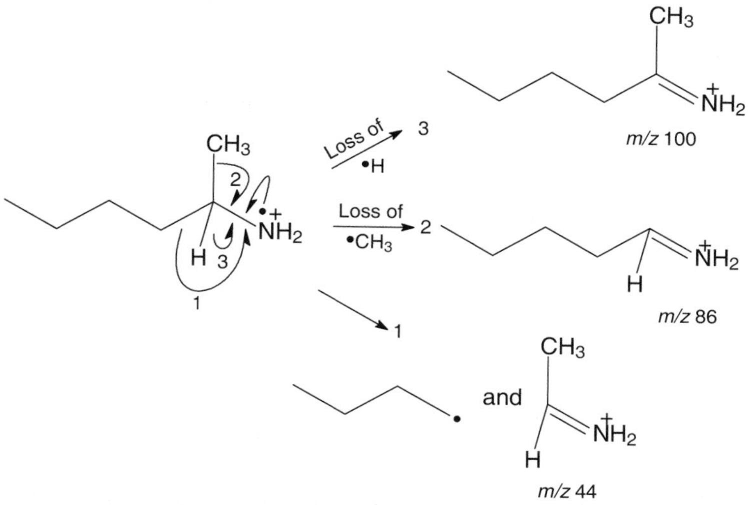

Scheme 6-83

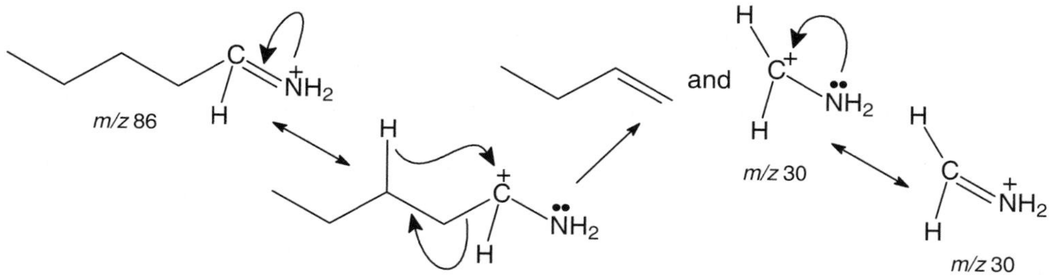

Scheme 6-84

Not seen in many of the mass spectra of aliphatic amines shown in Figures 6-63 through 6-70, there is sometimes a peak that represents a distonic OE ion formed by the loss of a molecule of ammonia (NH_3) from the $M^{+\bullet}$. The mechanism for this rearrangement is shown in Scheme 6-85. The peak for $[M - NH_3]^{+\bullet}$ is clearly visible in the mass spectrum of 3-hexylamine (Figure 6-68).

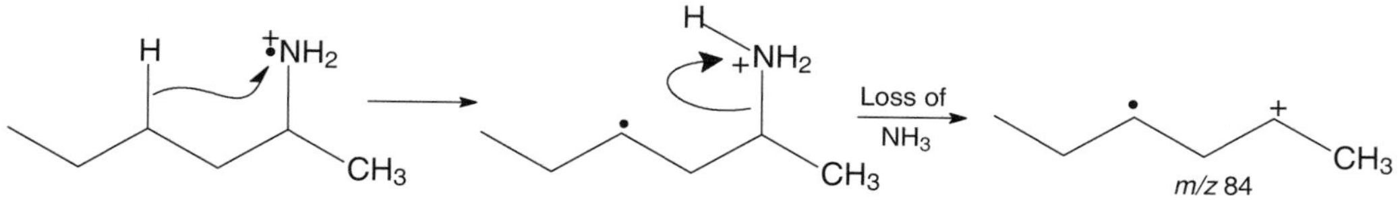

Scheme 6-85

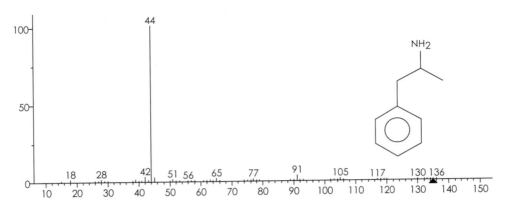

Figure 6-69. **EI mass spectrum of amphetamine.**

There is one exception to the statement, "The mass spectra of all aliphatic amines exhibit a peak at *m/z* 30." This exception is when the –NH₂ group is on C-2. The methyl radical in the form of C-1 is lost, and there are no β hydrogens present to participate in a hydride-shift rearrangement. The smallest ion that can be formed from such a structure has an *m/z* value of 44. The ion with *m/z* 44 results from an α-cleavage process that forms the methyl-substituted immonium ion in which the methyl substitution is C-1. A good example of this phenomenon is seen in the mass spectrum of amphetamine (Figure 6-69).

B. Aromatic Compounds Containing Atoms of Nitrogen

There are two types of aromatic amines. There are aromatic amines where an amino group is attached to an aromatic ring (aniline) and there are aromatic compounds with one or more nitrogen atoms in the ring (pyridine). Like the mass spectra of all aromatic compounds, these are characterized by intense peaks for the $M^{+\bullet}$, and not much fragmentation.

Molecular ions of compounds with an amine group attached to an aromatic ring, like aniline (Figure 6-70), behave somewhat like those of the corresponding aromatic alcohols. The mass spectra of both types of these compounds have intense $M^{+\bullet}$ peaks

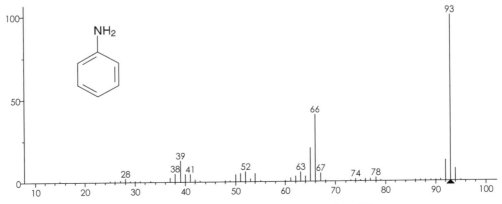

Figure 6-70. **EI mass spectrum of aniline.**

reflecting the high resonance stability of the molecular ion. The molecular ions of aromatic alcohols do not lose water and those of aromatic amines do not lose ammonia. Just as the $M^{+\bullet}$ of phenol loses CO, the $M^{+\bullet}$ of aniline loses HCN. This loss of HCN from the $M^{+\bullet}$ of aniline results in an OE ion of cyclopentadiene, which has an *m/z* value of 66. Some of these ions will then lose a hydrogen radical to form the ion with *m/z* 65. This loss of HCN is not observed in the mass spectra of ring-substituted anilines. The peak representing the $[M - 1]^+$ ion in the mass spectrum of aniline is formed by the loss of a hydrogen radical from the amino group [42]. When there are alkyl substituents on the ring of aniline, the $M^{+\bullet}$ behaves as if the charge and radical sites were on the ring. Benzylic cleavage occurs, which results in the formation of an amino-tropylium ion.

When a substituent is on the nitrogen atom of an aniline analog, the $M^{+\bullet}$ behaves as if the charge and radical sites were on the nitrogen atom. However, the mass spectra of *o*-, *m*-, and *p*-methylaniline and *N*-methylaniline do not allow for differentiation between N-substituted and ring-substituted compounds. This is not true for dimethyl-substituted anilines in which the substituents are both on the ring, one on the ring and one on the nitrogen, or both on the nitrogen. Examine the three mass spectra in Figures 6-71, 6-72, and 6-73. It is also interesting to note that the *ortho effect* is not observed for haloanilines, 2,6-dihaloanilines, hydroxyaniline, or amino aniline; however, a strong *ortho effect* is observed in the mass spectrum of 2-aminobenzoic acid. This lack of the *ortho effect* in the mass spectra of certain anilines, which have analogous structures to phenols whose mass spectra do exhibit the *ortho effect*, is a good example of the necessity to look at appropriate spectra before jumping to a conclusion of similar behavior.

Once there is an alkyl group on the nitrogen that can undergo a hydride-shift rearrangement fragmentation of a substituted immonium ion with the loss of an alkene, differentiation between ring-substitution and N-substitution is possible. An example is the mass spectrum of *N,N*-diethylaniline (Figure 6-74). In addition to the $M^{+\bullet}$ peak at *m/z* 149, there is an $[M - CH_3]^+$ peak at *m/z* 134 due to the loss of a methyl radical from a $M^{+\bullet}$ with the charge and radical sites on the nitrogen atom. Some of these EE ions with *m/z* 134 will then undergo a hydride-shift rearrangement fragmentation to lose a molecule of ethane, as shown in Scheme 6-86, to produce the EE ion with *m/z* 106, which still contains the single atom of nitrogen.

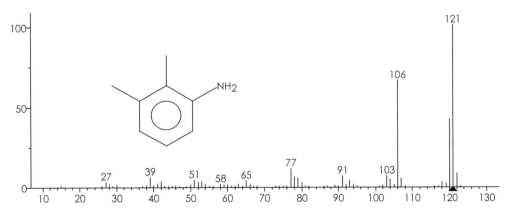

Figure 6-71.   El mass spectrum of 2,3-dimethylaniline.

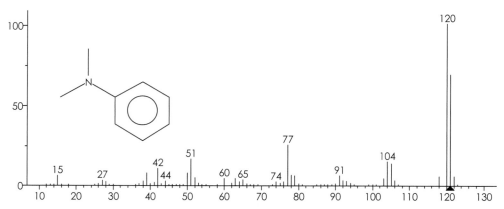

Figure 6-72.   *EI mass spectrum of N,N-dimethylaniline.*

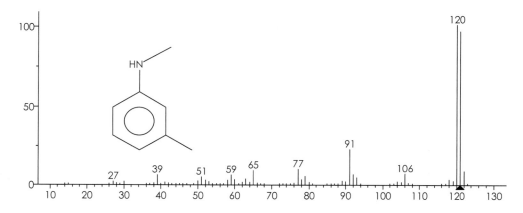

Figure 6-73.   *EI mass spectrum of 3-methyl-N-methylaniline.*

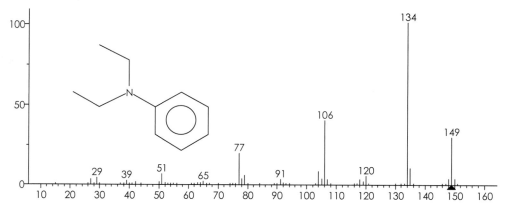

Figure 6-74.   *EI mass spectrum of N,N-diethylaniline.*

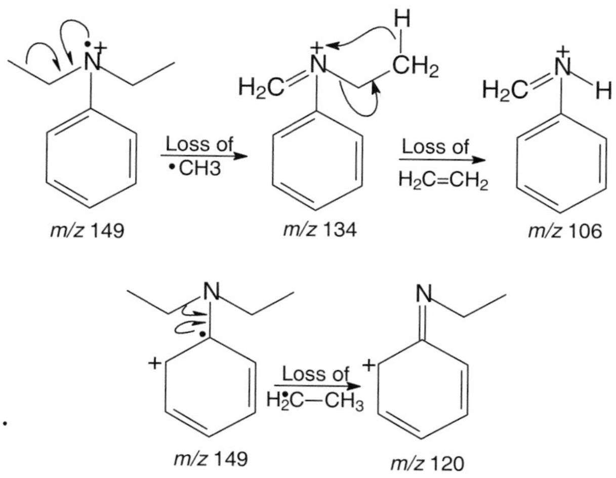

Scheme 6-86

The peak at m/z 120 in the mass spectrum of N,N-diethylaniline (Figure 6-74) illustrates that fragmentation can also be initiated with the charge and radical sites associated with the ring (also shown in Scheme 6-86).

Like the mass spectrum of aniline, the mass spectrum of pyridine has the $M^{+\bullet}$ peak as the base peak and exhibits a strong $[M - HCN]^{+\bullet}$ peak. There can be no substituents on the nitrogen in pyridine compounds, only on the carbon atoms in the ring; there are three regioisomers of a substituted pyridine. The mass spectra of three methyl isomers are almost identical and cannot be differentiated; however, the spectra of the three ethyl isomers are distinguishable. The spectrum of 2-ethylpyridine shows no peak at m/z 92 ($[M - CH_3]^+$ or $[M - 15]^+$). The base peak in the mass spectrum of 3-ethylpyridine is at m/z 92. The peak at m/z 92 in the mass spectrum of 4-ethylpyridine is about 55% of the intensity of the base peak, which is the $M^{+\bullet}$ peak at m/z 107. All three spectra exhibit a weak $[M - HCN]^{+\bullet}$ peak and a strong $[M - 1]^+$ peak, which is the base peak in the mass spectrum of 2-ethylpyridine. (No spectra of the methyl- or the ethyl-substituted pyridines are shown.)

The mass spectra of the three regioisomers of butylpyridine exhibit significant differences (Figures 6-75, 6-76, and 6-77). All three spectra exhibit an OE ion peak at m/z 93, which represents the product of a γ-hydrogen-shift rearrangement (McLafferty rearrangement) resulting in a β cleavage with the loss of a molecule of propene (analogous to Scheme 6-29). This reaction is common for the $M^{+\bullet}$ of aromatic compounds where there are γ hydrogens associated with a nonterminal carbon atom. The intensity of the $M^{+\bullet}$ peak is almost nonexistent in the mass spectrum of the 2-butyl isomer, but it is >70% of the intensity of the base peak in the mass spectrum of the 3-butyl isomer; the $M^{+\bullet}$ peak has an intensity of ~30% of the base peak in the mass spectrum of the 4-butyl isomer. Of the mass spectra compared here, that of the 3-butyl isomer (Figure 6-76) shows the most intense peak (m/z 92) representing loss of a propyl radical via benzylic cleavage. All three spectra have peaks representing $[M - 15]^+$ and $[M - 29]^+$ ions. Another challenge for the reader is to propose mechanisms to explain the formation of these ions.

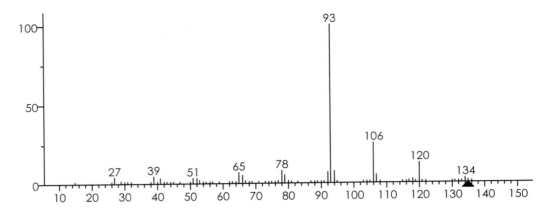

Figure 6-75.   EI mass spectrum of 2-butylpyridine.

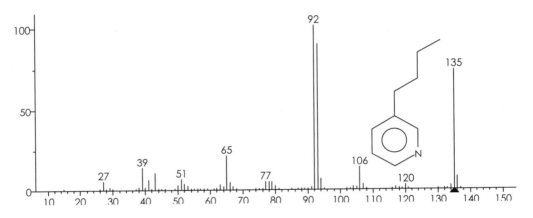

Figure 6-76.   EI mass spectrum of 3-butylpyridine.

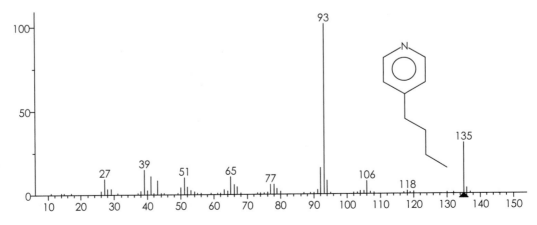

Figure 6-77.   EI mass spectrum of 4-butylpyridine.

The mass spectra of substituted pyridines and other aromatic compounds that contain two and three atoms of nitrogen in the ring often reveal much more information about the position of substitution than is forthcoming from the mass spectra of alkylbenzenes. Any time a class of compounds is to be analyzed, it is a good idea to look for examples of the mass spectra of these compounds in a mass spectral database like the NIST/EPA/NIH (NIST05) Database. Compare various spectra in a class such as those provided in Figures 6-75, 6-76, and 6-77.

C. Heterocyclic Nitrogen-Containing Compounds

There are saturated heterocyclic compounds that contain nitrogen atoms. The molecular ions of these compounds behave much like those of cyclic alkyl ethers. Compare the spectra in Figure 6-40 (tetrahydropyran) and Figure 6-78 (piperidine). The mass spectra of both compounds exhibit an intense $M^{+\bullet}$ peak because α cleavage results only in the opening of the ring. There is a strong $[M - 1]^+$ peak because α cleavage can also result in the loss of a hydrogen radical. To understand the origins of the other peaks in the mass spectrum of piperidine, look at Scheme 6-48 and draw the necessary parallels.

Many compounds that contain atoms of nitrogen are too nonvolatile or thermally labile for analysis by EI. These compounds, such as dimethyltryptamine, require derivatization. Because of rapid data acquisition rates used in GC/MS, a detectable molecular ion peak may not be observed in the mass spectra of nitrogen-containing compounds. Such compounds may require analysis by CI or by EI following derivatization or both to determine their nominal mass.

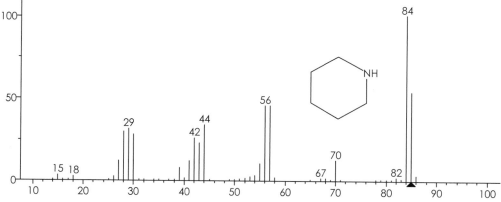

Figure 6-78.   *EI mass spectrum of piperidine.*

D. Nitro Compounds

Nitro compounds are unusual in that the nitrogen atom has a valence state greater than three. As would be expected, some the molecular ions of these compounds will fragment with the loss of the $-NO_2$ group. However, there are also peaks that represent the loss of oxygen and the loss of NO from other molecular ions; see peaks at *m/z* 107 and *m/z* 93, respectively, in the mass spectrum of nitrobenzene (Figure 6-79). The mechanisms for these losses are shown in Scheme 6-87. It is because of the high valence state of nitrogen that the loss of the $-NO_2$ group results in an EE-fragment ion. The NO_2 loss (even though NO_2 has an even mass and contains an odd number of nitrogen atoms) is not the loss of a molecule, but the loss of a radical. NO_2 is a radical, two of which exist as the molecule N_2O_4. In the mass spectra of both aromatic- and aliphatic-nitro compounds, the base peak is the $[M - NO_2]^+$ peak.

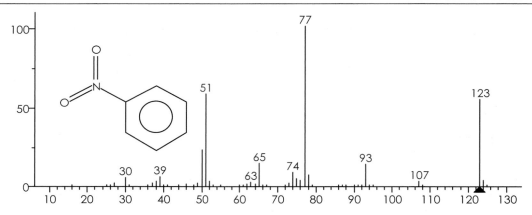

Figure 6-79. EI mass spectrum of nitrobenzene.

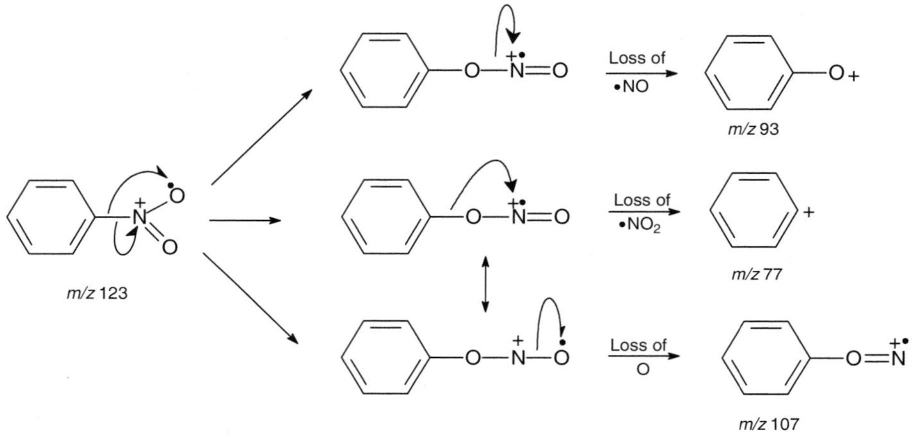

Scheme 6-87

E. Concluding Remarks on the Mass Spectra of Nitrogen-Containing Compounds

When a mass spectrum is examined, it is always prudent to be mindful of the *Nitrogen Rule*. Many of the mass spectra of compounds containing a single atom of nitrogen as presented here exhibit the classic *Nitrogen Rule* spectrum with a M$^{+\bullet}$ peak at an odd *m/z* value and several peaks at even *m/z* values representing EE ions containing the single nitrogen atom. However, it must be remembered that if an EE-fragment ion is formed that does not contain the nitrogen atom, its *m/z* value will be odd. Also be heedful of the fact that molecular ions containing an even number of nitrogen atoms (2, 4, 6, etc.) can fragment to form EE ions containing an odd number of nitrogen atoms, which will have an even *m/z* value.

> Earlier in this chapter, it was pointed out a peak at *m/z* 58 indicated the presence of an OE ion that could well be the result of a rearrangement fragmentation of an aliphatic ketone. However, a peak at *m/z* 58 also could possibly represent an EE fragment ion containing an odd number of nitrogen atoms.

5. Multiple Heteroatoms or Heteroatoms and a Double Bond

Throughout this chapter, the emphasis has been placed on whether the fragmentation is charge-site or radical-site-driven and on designating the probable sites of the charge and odd electron in the $M^{+\bullet}$. In compounds that contain a single functional group like ketones, aromatic hydrocarbons, chloroalkanes, etc., designating these sites is easy. When dealing with analytes that have more than one functional group that is readily ionizable, the situation becomes more complex. During the discussion of ketones, it was clear from an examination of the mass spectrum of benzyl methyl ketone (Figure 6-48) that there were peaks that represented ions formed by a fragmentation initiated with the charge and radical sites in the $M^{+\bullet}$ on the carbonyl oxygen (peak at m/z 43) and those with the charge and radical sites in the $M^{+\bullet}$ on the aromatic ring (m/z 92) as shown in Scheme 6-59. Ions formed by the charge and radical sites being associated with either the aromatic ring or the carbonyl oxygen in the $M^{+\bullet}$ were also pointed out as being represented in the mass spectrum of dibenzyl ketone (Figure 6-49 and Schemes 6-60 and 6-61).

There are some examiners of mass spectra who say that there can be only one site from which an electron can be lost in the formation of the $M^{+\bullet}$ and that site is the site having the lowest ionization potential. Even when there are heteroatoms, double bonds, and areas of resonance due to double-bond conjugation, any type of electron can be lost to form a molecular ion. This includes the possibility of initially losing a sigma-bond electron (unlikely, but possible) in the formation of a $M^{+\bullet}$. Thus, the molecular ion peak is likely to represent a distribution of different molecular ions, which are distinguished by the location of the charge and radical sites.

Ethanol amine is a simple compound that is both a primary alcohol and a primary amine. Examination of the mass spectrum (Figure 6-80) of this compound clearly illustrates that some molecular ions are formed with the charge and radical sites on the oxygen of the alcohol and other molecular ions are formed with the charge and radical sites on the nitrogen. The mass spectrum of ethanol amine has the base peak at m/z 30, which represents the primary immonium ion $H_2C=N^+H_2$. The peak at m/z 31 is far too intense relative to the peak at m/z 30 to represent just the ^{13}C-isotopic contribution of the immonium ion. This peak at m/z 31 also (and primarily) represents the primary oxonium ion $(H_2C=O^+H)$ characteristic of aliphatic alcohols.

There are many other examples of the role of multiple possible sites for the charge and odd electron. An area in which this phenomenon is especially important is the mass spectral analysis of illicit drugs. R. Martin Smith, in his book *Understanding Mass Spectra:*

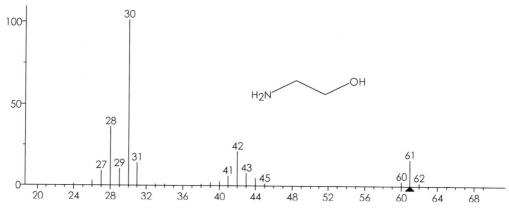

Figure 6-80. *EI mass spectrum of ethanol amine (Wiley Registry 8th Ed).*

A Basic Approach, 2nd ed. (John Wiley & Sons, Inc., New York, 2004), uses the mass spectra of illicit drugs to show that multiple fragmentation pathways exist and that not all are initiated by the charge and radical sites being on the same atom.

Another interesting aspect of the Smith book is that a significant number of the mass spectral displays have amplified peak intensities for clarity. Many of the mass spectra presented are of aliphatic amines with oxygen- and aromatic-containing moieties. Such mass spectra often show no peak for the $M^{+\bullet}$, but are dominated by a single peak representing a product of an initial homolytic cleavage promoted by the charge and radical sites being on the nitrogen atom. This use of expanded-intensity spectra is an important reminder that a significant amount of important information may be present in the low-intensity peaks that are often dismissed as being part of the background (the grass).

6. Trimethylsilyl Derivative

Derivatization can be important in EI mass spectrometry. As already mentioned, fatty acids are usually analyzed by GC/MS as the methyl ester derivative due to the high polarity of the free acids. Derivatives are prepared to make the analyte more volatile (by chemical modification to replace active hydrogens that would otherwise participate in hydrogen bonding) or, as in the case of aliphatic amines and alcohols, to aid in a determination of the analyte's nominal mass. Besides methylating agents, other useful derivatizing reagents are those that form trimethylsilyl (TMS) derivatives [43, 44]. The TMS reagents shown in Figure 6-81 react with compounds that have reactive hydrogen atoms, like all alcohols (including phenols), organic acids ($R–CO_2H$), all 1° and 2° amines, and some amides. Other alkyl-silicon compounds also can be used as derivatization reagents.

As seen in Figure 6-82, these silylating reagents do not react with ketones or esters. One way of quickly determining the classification of an unknown analyte is to analyze some of it by GC/MS before and after reacting it with a silylating reagent. If there is no change in the mass spectral data or retention time of the chromatographic peaks, the analyte is probably a ketone or an ester. New retention times and differences in the mass spectra mean that the analyte could be a compound with reactive hydrogens.

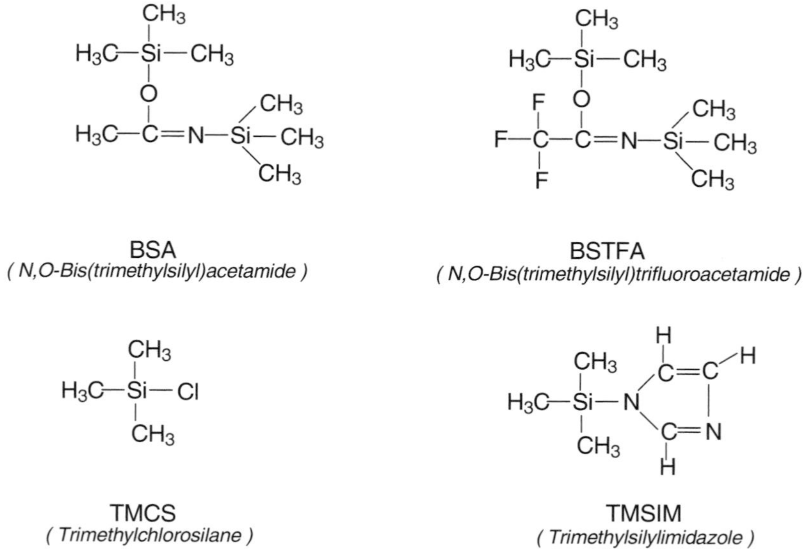

BSA
(*N,O-Bis(trimethylsilyl)acetamide*)

BSTFA
(*N,O-Bis(trimethylsilyl)trifluoroacetamide*)

TMCS
(*Trimethylchlorosilane*)

TMSIM
(*Trimethylsilylimidazole*)

Figure 6-81. Four common silylating reagents.

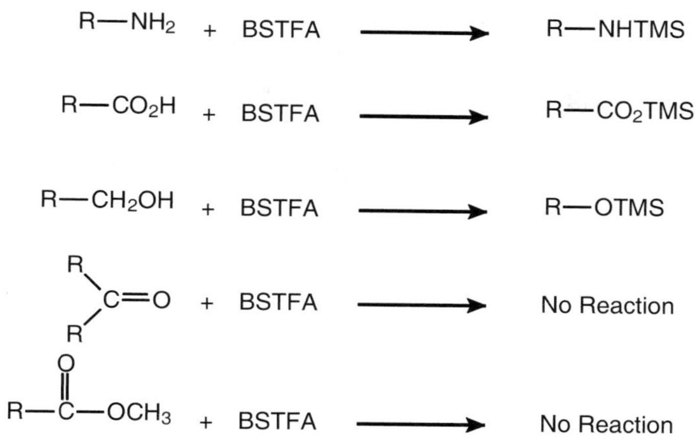

R—NH₂ + BSTFA ⟶ R—NHTMS

R—CO₂H + BSTFA ⟶ R—CO₂TMS

R—CH₂OH + BSTFA ⟶ R—OTMS

$\underset{R}{\overset{R}{>}}$C=O + BSTFA ⟶ No Reaction

R—C(=O)—OCH₃ + BSTFA ⟶ No Reaction

Figure 6-82. *Reactions of a silylating reagent.*

The mass spectra of TMS derivatives usually do not exhibit a $M^{+\bullet}$ peak. The highest m/z value peak observed in the mass spectrum may be the $[M - CH_3]^+$ ($[M - 15]^+$) peak. There will always be a peak at m/z 73, which represents the $(CH_3)_3Si^+$ EE ion. If the derivatized moiety is on a nonterminal carbon atom in an aliphatic structure, there will be ions representing secondary carbenium ions with the $-Si(CH_3)_3$ group attached. Formation of the TMS derivative of a primary aliphatic alcohol and subsequent fragmentation of the $M^{+\bullet}$ to form the $[M - 15]^+$ ion is illustrated in Scheme 6-88. The silyloxonium, such as the $[M - 15]^+$ ion, are very stable and unless there is some overriding factor such as the possibility of secondary carbenium ion formation, there always will be a discernible $[M - 15]^+$ peak in the mass spectrum. The analogous silylimmonium ions formed by fragmentation of the $M^{+\bullet}$ of the TMS derivative of amines are also quite stable and is likely to be represented by discernible peaks in the mass spectrum.

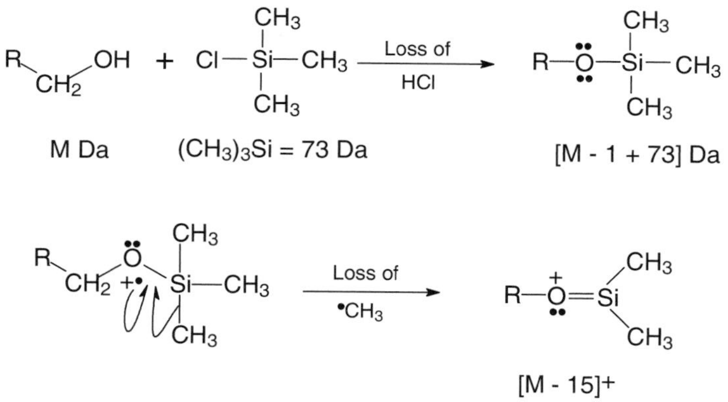

Scheme 6-88

As can be seen in the mass spectrum of the TMS ether formed from *n*-decanol, there are other peaks worth noting (Figure 6-83). The peak at *m/z* 75 is found in the mass spectra of TMS derivatives of all aliphatic alcohols. Peaks at *m/z* 89 and 103 are found in the mass spectra of TMS derivatives of all primary alcohols. The structures of these three ions are:

$$H^+O=Si(CH_3) \qquad\qquad H_2C=SiO^+(CH_3)_2 \qquad\qquad H_2C=O^+Si(CH_3)_3$$
$$\textit{m/z}\ 75 \qquad\qquad\qquad\quad \textit{m/z}\ 89 \qquad\qquad\qquad\quad \textit{m/z}\ 103$$

The peaks at *m/z* 89 and *m/z* 103 will shift to higher *m/z* values in the mass spectra of TMS derivatives of 2° and 3° alcohols consistent with the mass of the substituents on the alcohol carbon. For example, the peak at *m/z* 89 in Figure 6-83 will shift to *m/z* 145 in the mass spectrum of the TMS derivative of 5-tetradecanol (a 2° alcohol with an *n*-butyl group on the alcohol carbon); similarly, the peak at *m/z* 103 in Figure 6-83 will shift to *m/z* 159. The peaks at *m/z* 73 and *m/z* 75 in Figure 6-83 would not shift to higher *m/z* values in the mass spectrum of the TMS derivative of 5-tetradecanol.

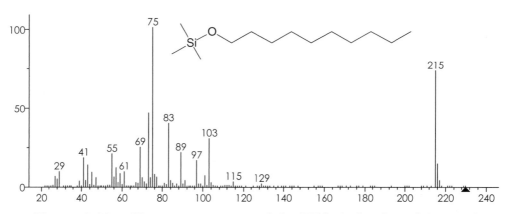

Figure 6-83.   *EI mass spectrum of the TMS derivative of decanol.*

Examination of the mass spectrum of the TMS derivative of 1,10-decandiol (Figure 6-84) reveals an interesting, and what turns out to be a characteristic, pair of peaks at *m/z* 147 and *m/z* 149 [45]. Peaks at these *m/z* values are common in the mass spectra of TMS derivatives of *vic*-diols; i.e., compounds in which there are hydroxyl groups on adjacent carbon atoms. Even though the two –OH groups are well separated in this particular example, the molecular ions are in the gas phase and the structure is free to fold so that the two trimethylsilylether groups can approach one another with sufficient proximity to form these diagnostic ions indicating the presence of at least two silyl groups in the analyte molecule [45]. The formation of these two diagnostic ions from their precursors is shown in Scheme 6-89. The presence of these diagnostic peaks in a mass spectrum only indicates that more than one group was derivatized; it does not indicate the total number of groups that were derivatized.

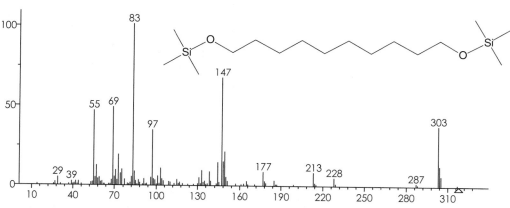

100
50
0
83
147
55
69
97
303
29
39
177
213
228
287
10
40
70
100
130
160
190
220
250
280
310

Figure 6-84. EI mass spectrum of the TMS derivative of 1,10-decandiol.

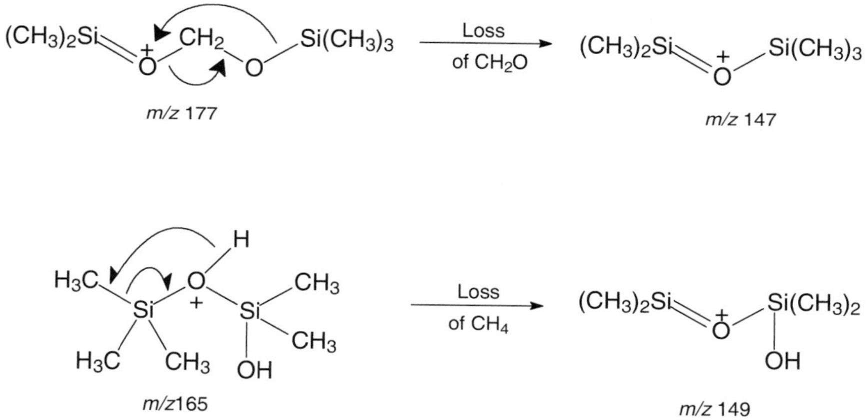

(CH3)2Si
CH2
Si(CH3)3
Loss of CH2O
(CH3)2Si
Si(CH3)3
m/z 177
m/z 147
H
H3C
CH3
Si
Si
CH3
H3C
CH3
OH
Loss of CH4
(CH3)2Si
Si(CH3)2
OH
m/z165
m/z 149

Scheme 6-89

The formation of the TMS derivative results in a mass change of 72 Da. The trimethylsilyl group has a mass of 73 Da, but it replaces an active hydrogen atom in the analyte molecule. Therefore, the derivatization process results in a net change of 72 Da between the mass of the underivatized and mass of the derivatized analyte.

The number of sites derivatized in a given analyte molecule can be determined by analyzing derivatives prepared in parallel using unlabeled and stable isotope-labeled analogs of the derivatization reagent [45]. A commercially available stable isotope-labeled analog of the *N,O*-bis(trimethylsilyl)acetamide reagent has eighteen deuterium atoms, three on each of the three methyl groups in each of the two TMS groups, and is designated as d_{18}-BSA. The mass spectrum in Figure 6-85 is that of an unknown derivatized with d_0-BSA; i.e., the unlabeled reagent containing no deuterium atoms. The mass spectrum in Figure 6-86 is that of another aliquot of the same unknown after derivatization using d_{18}-BSA.

> The upper left-hand structure in Figure 6-81 is that of BSA. When BSA reacts with an analyte with one active hydrogen, only one TMS ($(CH_3)_3Si$) group is added.

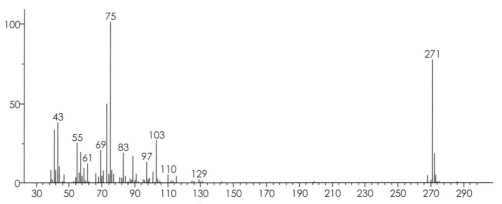

Figure 6-85. *EI mass spectrum of an unknown in the form of a TMS derivative prepared with d_0-BSA.*

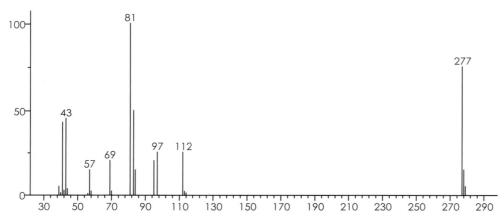

Figure 6-86. *EI mass spectrum of an unknown in the form of a TMS derivative prepared with d_{18}-BSA.*

The net change in mass between the derivatized and the underivatized analyte is again 72 Da when the –Si(CH$_3$)$_3$ group replaces the active hydrogen by the use of d_0-BSA. However, when d_{18}-BSA is used, each hydrogen atom (mass 1 Da) in the –Si(CH$_3$)$_3$ group is replaced with a deuterium atom (D or ^{2}H, mass 2 Da), resulting in a net increase between the mass of the underivatized analyte and that of the derivatized form of 81 Da: Si (28 Da) plus that of three –CD$_3$ groups (18 Da each or 54 Da) minus that of the replaced active hydrogen atom (–1 Da) in the analyte molecule ($28 + 54 - 1 = 81$). Note that the net shift in mass of the derivatized analyte molecule prepared using d_{18}-BSA is 9 Da more for each derivatizable functional group than that effected using d_0-BSA ($81 - 72 = 9$).

Examination of the mass spectrum (Figure 6-85) of the d_0-BSA-derivatized unknown shows that there are peaks at m/z 271 (which can be presumed to be the peak representing [M – CH$_3$]$^+$) and at m/z 73, 75, 89, and 103, indicating that the unknown is a primary alcohol. What is not known is how many –Si(CH)$_3$ groups were added.

In the spectrum (Figure 6-86) of the d_{18}-BSA-derivatized unknown, it appears that the (CH$_3$)$_3$Si$^+$ EE ion with m/z 73 in the spectrum of the unlabeled derivative prepared with d_0-BSA has shifted to m/z 82. The other diagnostic ions represented in the mass spectrum of a primary-alcohol TMS derivative have also shifted to new m/z values as follows:

$H^+O=Si(CH_3)_2$
m/z 75

$H_2C=SiO^+(CH_3)_2$
m/z 89

$H_2C=O^+Si(CH_3)_3$
m/z 103

$H^+O=Si(CD_3)_2$
m/z 81

$H_2C=SiO^+(CD_3)_2$
m/z 95

$H_2C=O^+Si(CH_3)_3$
m/z 112

The peak at the highest *m/z* value in the mass spectrum (Figure 6-86) of the derivative prepared with the d_{18}-BSA is at *m/z* 277. It is presumed that this peak at *m/z* 277 represents $[M - CD_3]^+$ (therefore $M^{+\bullet} = 296 = 277 + 18$) just as it is presumed that the peak at *m/z* 271 in the mass spectrum (Figure 6-85) of the derivative prepared with d_0-BSA represents $[M - CH_3]^+$ (therefore $M^{+\bullet} = 286 = 271 + 15$). This means that the nominal mass of the derivative formed with the labeled reagent has a nominal mass 9 Da greater than the derivative formed with the unlabeled reagent. This difference in the nominal mass of the labeled and unlabeled derivative is the same as the difference in mass of a single $-Si(CD_3)_3$ unit and a single $-Si(CH_3)_3$ unit. This difference of 9 Da indicates that only a single active hydrogen existed and that only a single TMS group exists in the derivatized compound. If two TMS groups had been incorporated into the derivative, the difference in the nominal mass of the labeled and unlabeled derivative would be 18 Da; if three TMS groups had been incorporated, the mass difference would be 27 Da, and so forth.

What feature in the mass spectrum reveals the number of sites that have been derivatized? Because there is a shift of 6 *m/z* units ($277 - 271 = 6$) for a high-mass peak in the mass spectra of the unknown after parallel treatments with d_0- and d_{18}-BSA, it can be inferred that the unknown contains only one derivatizable functional group. This inference is based on the fact that the peak at *m/z* 271 represents a dimethylsilyloxy ion (similar in structure to the ions with *m/z* 89 as shown above and having the structure shown in the second line of Scheme 6-88); the ion with *m/z* 271 is shifted by only 6 Da in the mass spectrum of the labeled derivative because the fragment ion contains only two $-CD_3$ (d_3-CH_3) groups. If the unknown had contained more than one derivatizable functional group, this high-mass dimethylsilyloxy ion, corresponding to $[M - CD_3]^+$, would contain another $-Si(CD_3)_3$ group for each additional functional group. In an example where the unknown had two derivatizable functional groups, the high-mass dimethylsilyloxy ion, corresponding to $[M - CD_3]^+$, would contain one intact $-Si(CD_3)_3$ group (net shift of 9 Da) in addition to the two CD_3 groups (net shift of 6 Da) comprising the dimethylsilyloxy moiety (analogous to the structure with *m/z* 95 directly above); in such a case, the peaks representing $[M - CH_3]^+$ and $[M - CD_3]^+$ in the corresponding mass spectra of the unknown after parallel treatments with d_0- and d_{18}-BSA would be separated by 15 *m/z* units ($6 + 9 = 15$).

Based on the information obtained from the peaks at the low end of the *m/z* scale of the mass spectrum (Figure 6-85) of the d_0-BSA-derivatized unknown, the analyte is a primary alcohol. Only one $-SI(CH_3)_3$ group was added. This means that there is only one OH group and it is on a $-CH_2$ moiety. The nominal mass of the TMS derivative is 286 Da; therefore, the nominal mass of the unknown is 214 Da ($286 - 72 = 214$). To determine the mass of hydrocarbon backbone, subtract 17 Da for the $-OH$ group and 1 Da for a hydrogen on the terminal carbon atom. The remainder is 196 Da. This number is evenly divisible by 14, which means that the unknown consists of 14 $-CH_2-$ units. Neither the mass spectrum of the deuterated nor of the unlabeled TMS derivative of the unknown showed any peaks that would indicate branch points in the chain or that the $-CH_2OH$ group was at a location other than at the end of the chain. Therefore, there is a good probability that the unknown is tetradecanol:

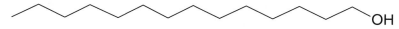

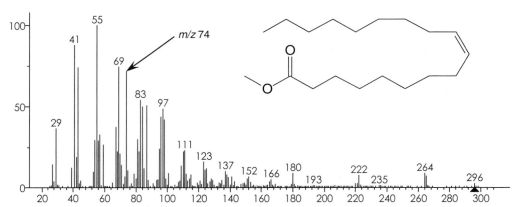

Figure 6-87. El mass spectrum of methyl oleate (methyl 9-octadecenoate).

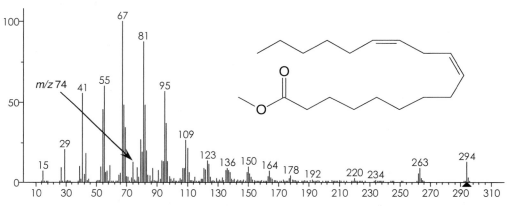

Figure 6-88.   El mass spectrum of methyl linoleate (methyl 9,12-octadecdienoate).

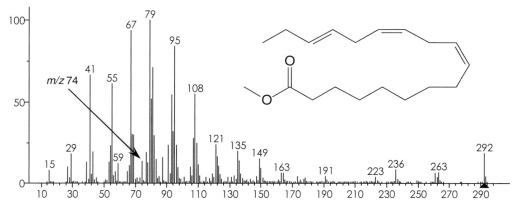

Figure 6-89.   El mass spectrum of methyl linolenate (methyl 9,12,15-octadectrienoate).

7. Determining the Location of Double Bonds

The use of derivatization can facilitate a determination of the nominal mass of an analyte when there is no $M^{+\bullet}$ peak in the EI mass spectrum of the underivatized compound. Derivatization also can solve the issue of high polarity, nonvolatility, and thermal lability. Derivatization can be used to determine the position of substitution or sites of unsaturation in aliphatic compounds. However, care must be taken with regard to relying on the results of a derivatization reaction. Derivatizing an analyte is a chemical synthesis and it is necessary to practice this process in order to develop the confidence required for an unambiguous result.

Another use of derivatization is in the determination of the location of double bonds in a molecule. In some cases, interpretation of an EI mass spectrum can reveal the position of functional groups in a molecule; however, recognition of the position of a double bond in a molecule through the use of EI mass spectrometry requires some chemical means of marking the original site of unsaturation. Whereas the mass spectra of unsaturated fatty acid esters differ significantly from those of their saturated analogs, the mass spectra of isomeric unsaturated fatty esters exhibit no distinguishing peaks [34, 35, 46]. Derivatives must be prepared that mark the location of the original double bond and that generate fragmentation to facilitate recognition of its position in the carbon chain [47–52]. McCloskey [35] has reviewed several derivatization procedures that have been used to characterize unsaturated esters by GC/MS.

The mass spectra of esters of unsaturated fatty acids usually exhibit discernible $M^{+\bullet}$ peaks. Some of these molecular ions usually produce an abundant ion resulting from the loss of CH_3OH ($[M - 32]^{+\bullet}$) [46] and, in many cases, the important $[M - 31]^+$ ion corresponding to the loss of the methoxy radical. Unfortunately, as can be seen from the three mass spectra in Figures 6-87, 6-88, and 6-89, peaks in the remainder of the mass spectrum provide very little information. None of the peaks in these three mass spectra corresponds to fragmentation at the locus of a double bond in the long carbon chain of an unsaturated methyl ester. The peak at m/z 222 in the spectrum of methyl oleate (Figure 6-87) represents an OE ion formed by the loss of $H_3CC(O)OCH_3$ (methyl acetate) which is expelled from the ester moiety [46]. The other obvious OE ion peaks in this mass spectrum are at m/z 180, 166, and 152, none of which would correspond to the charge and radical site existing between C-9 and C-10, which is the site of the double bond. Scheme 6-90 illustrates fragmentation pathways that produce the ions that would be expected regardless of which of the two carbon atoms that originally constituted the double bond carry the charge and radical site. Close examination of the mass spectrum reveals that there are peaks representing the ions with m/z 143 and possibly at m/z 157; however, there is no appreciable ion current evidenced at m/z 99, other than that due to a ^{13}C-isotope ion of the ion with m/z 98, or at m/z 197. The ion current at m/z 143 could well be due to a rearrangement fragmentation brought about by a hydrogen-shift rearrangement in the $M^{+\bullet}$.

It is also interesting to note that unlike the mass spectra of their saturated analogs, these unsaturated straight-chain methyl esters do not have m/z 74 as the base peak. In some cases, finding the peak at m/z 74 can be difficult because of the significant ion current at other m/z values in that region. It should also be noted that the intensity of the m/z 74 peak decreases as the degree of unsaturation increases.

An explanation for the lack of discrete fragmentation relating to the position of the double bond in the molecule has been postulated by Biemann [4]. Formation of the $M^{+\bullet}$ with the site of electron deficiency at the original location of the double bond (as illustrated in Scheme 6-91) produces species A, which is interconvertible with species B via hydrogen

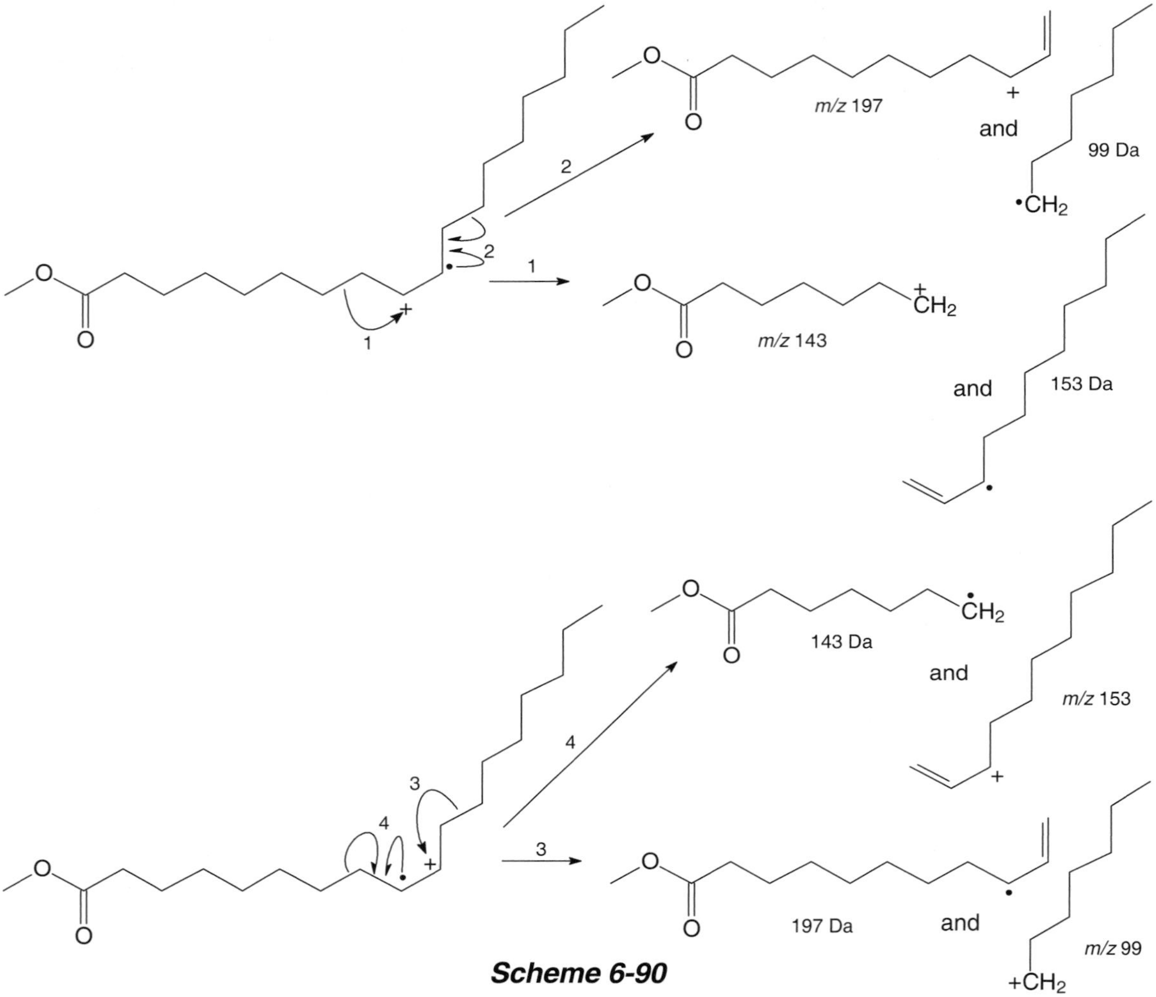

Scheme 6-90

and hydride shifts. This interconversion of molecular ion species corresponds to shifts or migration of the original double bond; thus, no discrete fragmentation relating to the original position of the double bond is observed.

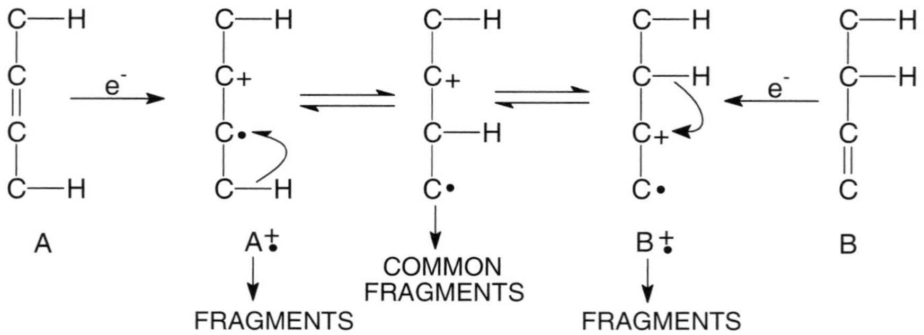

Scheme 6-91

The position of a double bond can be "marked" by oxidizing it with OsO_4 to a vicinal diol followed by silylation (with bis-trimethylsilylacetamide, BSA) of the hydroxyl groups to produce a derivative (as shown in Scheme 6-92) suitable for analysis by GC/MS [53]. The mass spectrum (Figure 6-90) of the bis-trimethylsilyloxy derivative of the vicinal diol derived from methyl oleate shows two intense peaks that can be related to the position of the double bond in the original molecule.

Based on the nominal mass of the derivative and the *m/z* values of the two ions represented by the intense peaks in Figure 6-90, it can be deduced that the two peaks represent moieties that can be joined to form the intact molecule. Although fragmentation studies with labeled compounds have shown that production of the ion of *m/z* 259 involves preliminary migration of a trimethylsilyl moiety to the ester moiety [53], the simple scheme of bond fission in Figure 6-90 is essentially correct. It would be expected that fission of the C-9–C-10 bond would produce two secondary carbenium ions that would be more stable than isobaric primary carbenium ions produced through cleavage resulting from sigma-bond ionization.

If the location of the double bond were not known in the example relating to Figure 6-90, the following logic would result in an answer. One of the two secondary carbenium ions represented by the two peaks at *m/z* 215 and *m/z* 259 will have the structure $(CH_3)_3Si^+O=CH(CH_2)_xCO_2CH_3$ and the other structure will be $H_3C(CH_2)_yCH=O^+Si(CH_3)_3$. Both ions have an $-OSi(CH_3)_3$ moiety [nominal mass of 89 Da (16 +28 + 3(15) = 89)]. Subtract the nominal mass of the $-OSi(CH_3)_3$ moiety from the *m/z* value of each of the two secondary carbenium ions and divide by 14 (the mass of a $-CH_2-$ unit) to determine the number of methylene groups. In this case, 215 – 89 divided by 14 = 9 (an integer) and 259 – 89 divided by 14 = 12.14 (a fractional number). The integer (9) equals the number of methylene units in the hydrocarbon-containing TMS moiety, $H_3C(CH_2)_yCH=O^+Si(CH_3)_3$. The noninteger value (12.14) corresponds to the ester-containing TMS moiety, $(CH_3)_3Si^+O=CH(CH_2)_xCO_2CH_3$; if the mass of the $-CO_2CH_3$ moiety (59 Da) is subtracted along with the mass of the $-OSi(CH_3)_3$ moiety from 259, the resulting value will be one less than the sum of the masses of the internal $-CH_2-$ units in the ester-containing TMS moiety; i.e., 259 – (89 + 59) + 1 = 112, which divided by 14 = 8. This means that the double bond is between carbon atoms 9 and 10. This solution was arrived at using the same logic that was used to determine a branch point in a branched alkane as shown earlier in Figure 6-15.

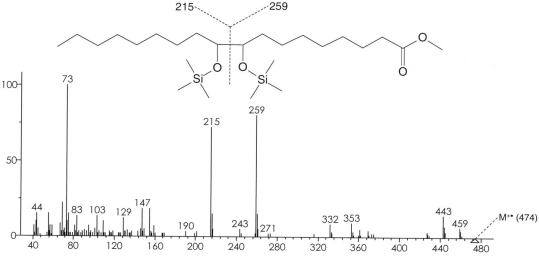

Figure 6-90. *EI mass spectrum of methyl 9,10-bis[(trimethylsilyl)oxy]octadecanoate.*

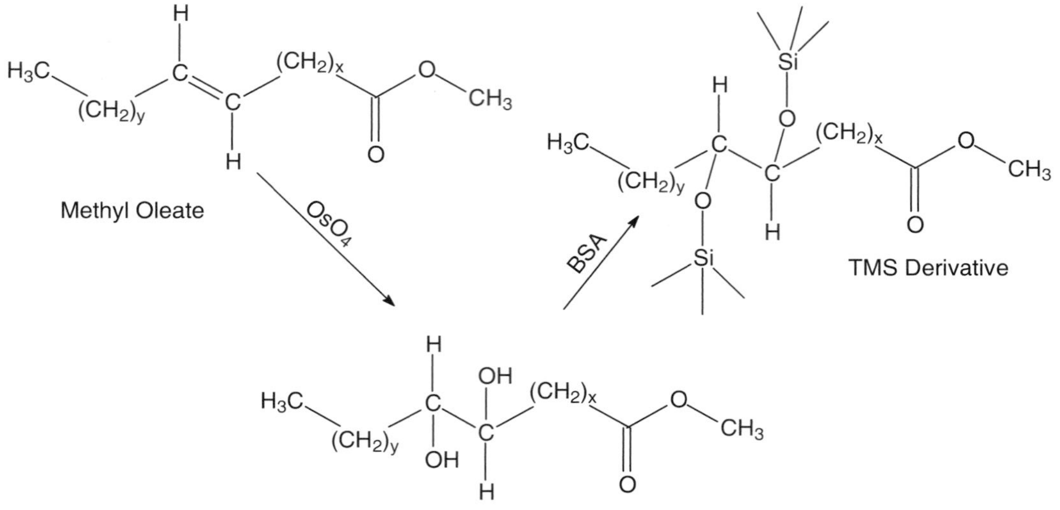

Scheme 6-92

The derivatization approach to marking the site of unsaturation can be extended to polyunsaturated fatty acids. The spectra of derivatives of fatty acids having more than two double bonds are much more complex than that shown in Figure 6-90; however, with careful interpretation, the fragmentation pattern can be related to the locations of the double bonds in the original molecule [54, 55].

Similar approaches can be used to determine the positions of cyclopropane [35] or cyclopropene [35, 56] moieties in the main chain. The presence and positions of oxygen functions (hydroxy and keto groups) also can be determined by shifts in mass spectral data obtained from normal-chain fatty acid derivatives [35, 57]. The mass spectra of pyrrolidide [58, 59] and picolinyldimethylsilyl ether [60, 61] derivatives of unsaturated fatty acids also provide a generally reliable means of recognizing the site of the double bond. The use of substituted cyclopropylimines [62, 63], halocarbenes [64], or Diels–Alder adduct formation [65] also have been employed in determining the position and geometry of olefins.

Dimethyldisulfide (DMDS) is an example of a reagent that reacts directly with a double bond as shown in Scheme 6-93. The covalently modified hydrocarbon backbone of analyte (after reaction with dimethyldisulfide) produces a mass spectrum characterized by a discernible $M^{+\bullet}$ peak and fragment ions that can be correlated with the position of the double bond in the original olefin [48, 66–68]. Formation of both prominent fragment ions is shown in Scheme 6-93; homolytic cleavage of the sulfur–carbon bond with charge retention on either sulfur atom is equally probable. From the mass spectrum of the dimethyldisulfide reaction product of 3-dodecene in Figure 6-91, it appears that fragmentation only occurs to produce ions with a single sulfur atom. Although theoretically possible, it appears that there is no cleavage that would result in a secondary carbenium ion that contains both sulfur atoms.

Reaction of olefins with 5,5-dimethoxy-1,2,3,4-tetrachlorocyclopentadiene modifies the double bond in a manner similar to that shown in Scheme 6-93. However, in this case, the mass spectrum provides information not only on the position of the double bond in the olefin, but also on the geometry [66].

Double-bond location in fatty acids can also be accomplished by collisionally activated dissociation (CAD) of ions formed during chemical ionization (CI). When using acetonitrile as the CI reagent gas in an internal ionization 3D QIT mass spectrometer, an ion is formed by the reaction of the protonated adduct of acetonitrile with the methyl ester of

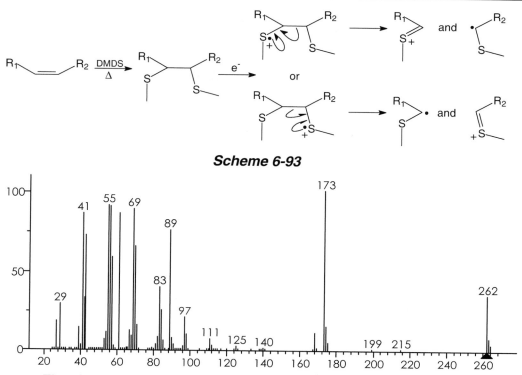

Scheme 6-93

Figure 6-91. EI mass spectrum of DMDS-modified 3-dodecene.

the fatty acid. When this adduct ion is subjected to CAD, fragments related to the location of double bonds are produced [69]. The locations of as many as eight double bonds in a single analyte have been reported [69]. It is also possible to determine whether the analytes are *cis* or *trans* isomers using this technique, which is more of a direct-analysis method than those involving formation of derivatives.

VI. Library Searches and EI Mass Spectral Databases

1. Databases

One of the most powerful aids in the identification of an unknown compound is the comparison of the compound's mass spectrum with those in a database of mass spectra acquired under the same conditions. There are four large commercially available databases of EI mass spectra: NIST05 (NIST/EPA/NIH Database of 190,825 spectra), Wiley Registry 8th Ed. (399,383 spectra), the Wiley/NIST (combination of the Wiley 8th Ed. and NIST05) (532,573 spectra), and the Palisade Complete Mass Spectral Library (composed of all the spectra in the Wiley Registry 7th Ed., NIST02, and another ~150K spectra for an advertised total of over 600,000 spectra). The NIST and Wiley databases are provided with chemical structures in a format that can be used in various structure search routines. The Palisade Complete Mass Spectral Library has a limited number of structures in a bit-map format that makes them unusable for structure searching.

There are several much smaller commercially available databases of EI mass spectra that are application specific, such as the Wiley Database of Designer Drugs and the Allured 4th Ed. of the Robert P. Adams database of Essential Oil Compounds. These "boutique" databases, as well as the major ones listed above, are available in multiple formats to accommodate various proprietary Database Search Engines. Table 6-4 contains a list of the known commercially available mass spectral databases.

There are also some EI databases that can be downloaded without charge. These are usually in the Agilent GC/MS ChemStation format; however, the NIST Mass Spectral (MS) Search Program has a utility that will convert mass spectral databases in the ChemStation format to the NIST MS Search Program's format. Two of the better known free-downloadable databases are the ones from the American Academy of Forensics Sciences, Toxicology Section (AAFS) "Comprehensive Drug Library", and The International Association of Forensic Toxicologists (TIAFT) "Derivatives of Drugs".

The American Academy of Forensics Sciences (AAFS), Toxicology Section and "Comprehensive Drug Library" (http://www.ualberta.ca/~gjones/mslib.htm) of 2200 spectra, including replicates, was last updated in 2004. This database does not include structures, but it does include CAS registry numbers, molecular weights (nominal mass), and elemental compositions. If an NIST structure is present for a given CASrn, that structure will be displayed with that spectrum.

The current TIFF collection contains 205 EI mass spectra, including 122 spectra of TMS derivatives of drugs, provided by Aldo Polettini and 82 examples of various derivatives of relevant doping substances provided by Klaus Müller and Detlef Thieme. The majority of these spectra are of TMS derivatives. This database can be downloaded from (http://www.tiaft.org//main/mslib.html).

Table 6-4. *Commercially available databases sold for use with GC/MS data systems.*

Publisher	Database Name	Number of Spectra
National Institute of Standards and Tech.	NIST05	190,825 spectra of 163,198 compounds
John Wiley & Sons	Wiley Registry 8th Ed.	399,383 spectra
John Wiley & Sons	Wiley 8N combination of Wiley 8th Ed. and the NIST05	532,573 spectra
Palisade MS	Palisade Complete	>600,000 spectra
Allured Publishing	Mass Spectra of Essential Oil Compounds, 4th Ed.	2205 spectra
John Wiley & Sons	Designer Drugs 2005	3437 spectra of 2959 compounds
John Wiley & Sons	Geochemical, Petro Chemical, and Biomarkers	1110 spectra
John Wiley & Sons	Volatiles in Food	1620 spectra
John Wiley & Sons	Androgens, Estrogens and Other Steroids	2979 spectra
John Wiley & Sons	Alexander Yarkov Organic Cpd	37,055 spectra
John Wiley & Sons	Chemical Concepts Mass Spectra Data, 4th Ed.	>40,000 spectra
Agilent Technologies	Stan Pesticide Database	340 spectra
Agilent Technologies	RTL Pesticide Database	927 spectra
Agilent Technologies	RTL Hazardous Chemical DB	731 spectra
Not applicable	Pfleger/Maurer/Weber (PMW)	>6300 spectra

2. Library Search Programs

The comparison made between the sample spectrum and the spectra in the database is carried out by a Database Search Engine. The various search engines are the underlying structure of the Library Search Program. Even though the look and feel of various Library Search Programs may be different, the underlying search engine can be the same. The most widely used search engine today is that of the NIST MS Spectral Search Program, developed under the direction of Steve Stein at the Mass Spectrometry Data Center of the National Institute of Standards [70], part of the United States Department Commerce. This search engine is incorporated into the mass spectral application software of Varian, Thermo Electron, Perkin Elmer, Shrader/JEOL, and LECO. Agilent Technologies and Waters Corporation have direct links to the NIST MS Search Program through their software as well as their proprietary search systems. With some manipulation on the part of the user, spectra can be imported into the NIST MS Search Program from the Applied Biosystems/MSD Sciex, Shimadzu, and Bruker Daltonics mass spectral data applications. The Waters proprietary search engine uses the same base INCOS algorithm [71] used as the foundation of the NIST MS Search Program. In addition to their own library search routines Varian, Thermo Electron, Perkin Elmer, Shrader/JEOL, LECO, Agilent Technologies, and Waters Corporation provide the NIST MS Search Program as a standard addition to their software or when the NIST/EPA/NIH Mass Spectral Database is purchased for use with their software.

A copy of the NIST Mass Spectral Search Program along with a database of about 2,400 spectra can be downloaded from http://www.nist.gov/srd/nist1a.htm by clicking on the link "**The NIST 05 Demo version may be downloaded here**", which is just under the graphic display of the NIST MS Search Program on this cited Web page. This download includes MS Interpreter and the Lib2NIST program used to create NIST MS Search user databases from GC/MS ChemStation libraries and JACMP and text files.

The search engine used by Shimadzu and Agilent in their proprietary mass spectral software is the Probability Based Matching (PBM) algorithm developed by Fred McLafferty and Bockhoff [72]. The problem with this search engine is that it is dependent on the base peak for an identified spectrum in the database, being at the same *m/z* value as the base peak of the unknown's spectrum.

The remainder of the discussion regarding the Library Search of mass spectra will use examples and terminology relating to the NIST Mass Spectral (MS) Search Program.

In order to obtain the best results from a library search, a good-quality spectrum should be submitted for identification. Issues of spectral skewing and background subtraction discussed in Chapter 10 (GC/MS) are very important in preparing the spectrum that will be used in the search. As pointed in Chapter 10 under the discussion of AMDIS (Automated Mass spectral Deconvolution and Identification System), it is best to use spectra that have been deconvoluted by AMDIS for a *Library Search* because AMDIS addresses both background subtraction and spectral skewing.

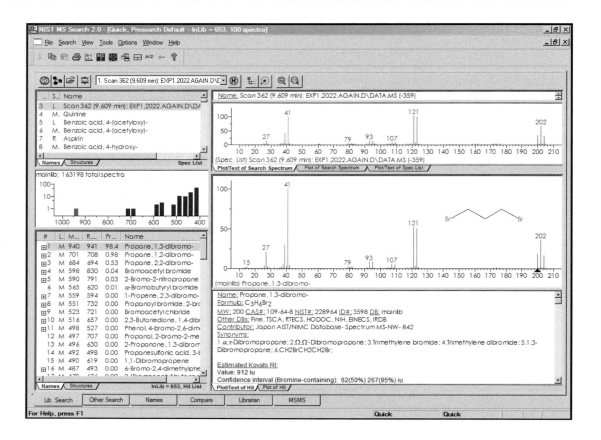

Figure 6-92. **The NIST MS Search Program Lib Search tab. There are a total of eight visible panes; the Compare pane is not displayed. The vertical and horizontal slider bars can be adjusted according to user preference. The orientation of the bar between the display of the bar-graph spectrum and the text data can be switched from horizontal (the displayed orientation) to vertical.**

Figure 6-92 is an example of a typical result obtained when an *Identity Search* is performed using the NIST MS Search Program. An *Identity Search* involves a presearch based on a limited number of the most intense peaks in the sample spectrum along with some other features explained in detail in the User's Manual for the NIST MS Search Program. The presearch is followed by a peak-by-peak comparison between the sample spectrum and a subset of the database determined by the presearch. The peaks are rated according to their *m/z* values and relative intensities with the peaks at the high end of the *m/z* scale given the most weight. The best 100 matches (*Hits*) are listed according to the degree of match in the *Hit List*.

The *Hit List* (a list of the database spectra that most closely match the sample spectrum) is found in the lower left section of Figure 6-92. The first column on the left side of this panel is the ranking value, with "1" being the best match. The second column is an up-to-two letter code (first two letters) of the database's name that contains the *Hit*. The single letter M and R are reserved for the NIST/EPA/NIH "main library" (*mainlib*) and "replicate library" (*replib*), respectively. Designation of the library corresponding to the *Hit* is necessary because multiple databases can be searched simultaneously. The next three columns relate to how closely the sample spectrum matches that of the *Hit*. From the left, the first of these columns has the Match Factor (Match). This is a number between 0 and 999 that indicates how closely the sample spectrum matches the library spectrum of the *Hit* based on a peak-by-peak comparison of the two spectra. Values greater than 700 are considered to be good matches. However, even very high values (>900) do not guarantee that the sample spectrum and the *Hit* spectrum were produced by the same compound.

The next value is the Reverse Match Factor (R.Match). This is a match factor that is calculated disregarding any peaks in the sample spectrum that are not in the spectrum of the *Hit*. The Reverse Match Factor has the same numeric range as the Match Factor. These nonconforming peaks in the sample spectrum are considered to be from a source other than a single analyte. These spurious peaks may be due to background or a substance that coeluted during the chromatographic process.

All library search programs have numeric indicators with respect to how good a match has been achieved; however, reliance on these numeric values should be approached with caution. More reliance should be placed on a visual comparison of the graphical presentation of the two spectra. This visual comparison of the graphical displays may be enhanced by a comparison of the two spectra in a tabular format.

The third value with respect to the quality of the *Hit* is the Probability (Prob), which is unique to the NIST MS Search Program. This number is expressed as a percentage. This number represents the probability that the sample spectrum and the spectrum of the *Hit* come from the same compound if a spectrum of the analyte is in the database. This value is dependent on the uniqueness of the spectrum of a particular compound. For example, if the sample spectrum is that of *m*-xylene and the Match Factor for *m*-xylene is over 900, the Probability of a match may be low because of the similarity of the mass spectra of the xylene isomers. For example, in the case where the sample spectrum is *m*-xylene, which is compared (searched) against the NIST05 database, the top five *Hits* will be the spectra of *o*-xylene, *p*-xylene, and ethylbenzene, all of which are nearly the same as the spectrum of *m*-xylene. On the other hand, if the sample is phenylbutazone, the Probability value will be high because of the uniqueness of this compound's mass spectrum compared to all the mass spectra in the NIST05 database. The degree of uniqueness of a spectrum can be determined by comparing the appearance of a spectrum in the *Hit List* with other spectra close to it.

Match Factors and Reverse Match Factors are calculated based on individual spectra regardless of how many spectra of a particular compound are found in a search of multiple databases. The Probability is based on the compound, and will be the same for each spectrum of the same compound in a multiple-database search or in a search of a database with multiple spectra for the same compound.

It is very important to remember that chemists (people) identify unknown analytes, not Library Search Programs. This is why much more than just reliance on the numerical values for Match Factors must be used. First, it should be clearly understood that the various commercially available databases do not contain all the mass spectra of all analytes that can pass through a gas chromatograph and/or that can undergo electron ionization in the gas phase. This is why companies like Eastman Chemical maintain their own databases of mass spectra, which may contain mass spectra of tens of thousands of compounds that are not in any commercially available database. It may even be that there are not mass spectra of compounds similar to the specific analyte under investigation in any of the available databases.

If the analyte has what can be referred to as a "public origin" (e.g., a contaminant in a packaged substance such as a packaged food, a compound in a substance found to be associated with a crime against a person or property, a material associated with an equipment failure, etc.), that material is probably going to be known and there could well be a mass spectrum of it in the NIST/EPA/NIH Mass Spectral Database, if it is suitable for electron ionization mass spectrometry. Such substances, compounds, and materials will also probably be listed in some other database, like the one maintained by the U.S. EPA or the European Index of Commercial Chemical Substances. Compounds in the NIST/EPA/NIH Database are indexed according to their inclusion in the nine databases listed in Table 6-5. Given a choice for the identity of a substance of a public origin between a compound that is not listed in any of these nine databases and one that is in several of them, the latter would be the best choice. Another indicator as to which of several possibilities should be selected from a list is whether the Synonyms List contains several common or trade names for the compound. Compare the entries of the two suggested compounds (*Hits*) with very similar mass spectra that appear in the top 15 *Hits* of a mass spectral database search (comparison) of an unknown against the NIST05 Database: 3-aminoisonicotinic acid (CASrn 7579-20-6), which is not listed in any of the nine other databases and has no synonyms other than one other chemical name and aspirin (CASrn 50-78-2), which is also found in seven of the nine other databases and has 141 synonyms, many of which are common names. It is obvious which of these would be the most likely candidate to be a compound of public origin.

Table 6-5. ***Other databases that may contain NIST/EPA/NIH database compounds.***

NIST MS Search Program Designation	Database Name
Fine	Commercially Available Fine Chemical Index
TSCA	Toxic Substances Control Act Inventory
RTECS	Registry of Toxic Effects of Chemical Substances
EPA	EPA Environmental Monitoring Methods Index
USP	U. S. Pharmacopoeia/U.S.A.N.
HODOC	CRC Handbook of Data of Organic Compounds
NIH	NIH-NCI Inventory File
EINECS	European Index of Commercial Chemical Substances
IR	NIST/EPA Gas Phase IR Database

Only 28,297 of the 163,198 compounds found in the NIST05 Mass Spectral Database are in one or more of the nine other databases. The remaining compounds are not so common and may have only been seen inside a mass spectrometer once under some special circumstances. Therefore, if substances associated with natural products, new drug metabolites, etc., are being analyzed, the "public origin" hypothesis may not be applicable.

3. When the Spectrum of the Unknown Is Not in the Database(s)

When there is a high confidence that a mass spectrum of the analyte is not present in one or more of the available mass spectral databases, the NIST MS Search Program has two other search modes. These are the Similarity Search (which compares the sample spectrum against the spectra of the database(s) to see which spectra are similar, even if they do not match) and the Neutral Loss Search (which looks for spectra in the database(s) that exhibit the same neutral losses from the $M^{+\bullet}$). A hydride Neutral Loss and Similarity Search can also be carried out. The *Hit List* from any one of these three searches can be evaluated using the Substructure Identification Option accessed from the Tools menu on the Main Menu Bar of the Lib Search tab of the NIST MS Search Program. The functionality of this Substructure Identification utility [73] is explained in more detail in the Library Search section of Chapter 3 on MS/MS.

It should be noted that when this Substructure Identification utility is used with El mass spectra of compounds whose spectra are not in the database(s), the only information that can be developed is the probable presence and probable absence of various substructures. The analyst will have to construct a probable structure. Once a structure has been proposed and drawn using some structural drawing program, the structure can be associated with the sample spectrum. The spectrum, along with the proposed structure, can be sent to MS Interpreter Program.

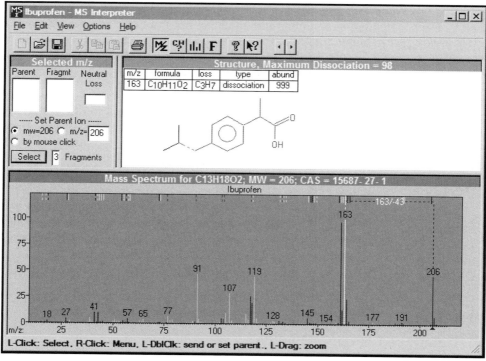

Figure 6-93. *There is a peak at m/z 163 in this mass spectrum. Through the use of alternating colors in the program's actual display (shown in grayscale), the loss of an isopropyl radical is shown from the structure in the top portion of the display. The resulting ion and the neutral loss are explained in the text boxes of the display.*

The MS Interpreter Program predicts at what *m/z* values there will be a mass spectral peak based on the structure. The position of these peaks can be compared with those in the spectrum that was imported with the proposed structure. If there is too much variation, then the structure can be adjusted, reassociated with the spectrum, and the two reimported into the program. An example of the MS Interpreter Program applied to the structure and mass spectrum of ibuprofen is shown in Figure 6-93.

In addition to the MS Interpreter Program that is provided with the NIST MS Search Program and is also part of the free download of the NIST MS Search Program, there are some other commercially available programs that perform similarly. There are the Mass Spec Calculator Pro from ChemSW (http://www.chemsw.com), the Mass Frontier (a Thermo Fisher product that requires Xcalibur Mass Spectral Software and components of Microsoft Office, http://www.highchem.com), and the ACD/MS Fragmentor Program from Advanced Chemistry Development (ACD Labs), all of which can be integrated with ACD/MS Manager: (http://www.acdlabs.com/products/spec_lab/exp_spectra/ms_fragmentor/).

4. Searching Multiple Databases

Care must be taken in decisions to search multiple mass spectral databases at the same time. When multiple databases are searched using the NIST MS Search Program, the 100 best matches are listed in the *Hit List* without indicating the database source(s). Each database is searched sequentially, but the order of presentation in the *Hit List* is based on the Match Factor or the Reverse Match Factor, depending on how the search was structured. This is not the case with PBM Search in the Agilent Technologies GC/MS ChemStation, which is one of the most widely used GC/MS library searches. Multiple libraries (up to three) are listed along with a "Match Quality" (a number between 0 and 100) for the first two. If the search finds a *Hit* that has a Match Quality greater than the minimum specified for the first library, the other two libraries are not searched. If the search program goes to the next library and the search finds a *Hit* that is greater than the minimum specified Match Quality for that library, that *Hit* will be reported and the third library will not be searched; however, a *Hit* with a higher Match Quality than that of the reported spectrum from the first database, but below that of the first database's minimum Match Quality, will not be reported.

This unusual behavior of the GC/MS ChemStation internal library search means that if the minimum Match Quality for the first database is set at 90 and the minimum Match quality for the second database is set at 70 and no *Hits* are found in the first library with >90 Match Quality and no *Hits* are found in the second library with a Match Quality >70, the third library will be searched where the *Hit* with the best Match Quality may be only 50. This *Hit* will be listed as the best match for the search even though there was a possible *Hit* in the first database that would have had a Match Quality of 89, which will not be reported at all.

5. Database Size and Quality

It certainly has not gone unnoticed that the Wiley Registry 8th Ed. has about twice as many spectra as does the NIST05, and the Palisade Complete has over three times as many spectra as are in the NIST05 Database. Is bigger better? If the bigger comes from quality spectra of more compounds, then there is no question that bigger is better when it comes to a library of mass spectra. Although the Wiley Registry and Palisade Complete do contain spectra of compounds that are not in the NIST Database, both also contain many more replicate[2] spectra and they also have a number of duplicate[3] spectra. As an example, the Wiley Registry has 13 spectra with the name "phenylbutazone" and at least 13 spectra for "hexachlorobenzene", whereas the NIST05 has only four spectra for these two common and public origin compounds. The NIST05 can be searched without inclusion of the replicate spectra, whereas the Wiley Registry and the Palisade Complete cannot. The importance of being able to search a database separately without replicates (or duplicates) is in trying to identify compounds whose spectra are not in the database. Too many spectra for a certain compound can limit the number of different *Hits* that may provide the necessary clues for elucidation of a structure.

Another important factor about a mass spectral database is spectral quality. All the spectra in the NIST05 have been individually examined to assure their correctness with the name, provided structure, and the spectrum itself [74]. Spectra in both the Wiley Registry and the Palisade have been added as provided to the publisher without any evaluation. As an example, there two spectra in the Wiley Registry that are labeled Maneba, which is a herbicide that is a polymer of manganese containing a polymer of an ethylene carbanoditioate. Examination of these two spectra shows that there are no similarities between them. Even though the listed chemical names for the two spectra are exactly the same and the elemental compositions are the same (no structure provided), they have different CAS registry numbers. Examination of the spectrum of CASrn 12427-38-2 reveals that it is clearly a mass spectrum of elemental sulfur (S_8). When the mass spectrum of elemental sulfur is searched against the Wiley Registry, Maneba is one of the *Hits*.

The search of the NIST05 mass spectrum of elemental sulfur against the Wiley Registry using the NIST MS Search Program points out another problem. Many of the top 15 *Hits* are spectra that are reported to be of S_8; however, five of these top 15 *Hits* are for compounds whose spectra only have peaks that represent the molecular ion and its isotopes. The Wiley Registry (and the Palisade Complete because it incorporates the 7th Ed. of the Wiley Registry) contain a significant number of these single-peak spectra. These spectra can cause problems with a search because these single peaks are also the base peak for these spectra and there is significant weighting given to them based on their high *m/z* values and the fact that this peak has 100% relative intensity.

Because there are compounds in the Wiley Registry that are not in the NIST05 Database, it is probably best to have both. The advantage of having both, as opposed to the Wiley/NIST combination, is that NIST05 can be searched without the possible interferences from duplicate and replicate spectra. This is not the case with the Wiley/NIST combination.

6. Concluding Remarks on the NIST Mass Spectral Search Program

The reason for the popularity of the NIST MS Search Program with so many mass spectrometer manufacturers and third-party database publishers like John Wiley & Sons, Inc. is its versatility. Not only can sample spectra be searched against the NIST05

---

[2] A replicate spectrum is another spectrum of the same compound; this spectrum is usually acquired by a different laboratory.

[3] A duplicate spectrum is a second copy of the same spectrum of the same compound. Duplicate spectra originate during the combining of various databases, more than one of which contains the same spectrum.

and other mass spectral databases (up to 12 databases in addition to the NIST mainlib and replib) using several different search algorithms but constraints regarding the number of specific elementals, name fragments, molecular (nominal) mass ranges, presence of peaks based on their type (normal or loss), and *m/z* value can be placed on these searches. The databases (NIST05, other commercial databases, and user-generated databases) can be searched according to the ID number (position in the database) of a spectrum, the CAS registry number of the spectrum, and the spectrum's NIST number (a unique number given to each spectrum in the NIST archive), as well as by any of the previously mentioned constraints. It is also possible to submit a structure for purposes of retrieving its spectrum or spectra of similar structures from the database. Spectra can easily be exported to other applications like Microsoft Word. The NIST MS Search Program can also be linked to structure-drawing programs to allow the easy import and export of structures in MOL format.

Used in combination with AMDIS (described in Chapter 10), it is easy to import spectra from nearly all commercial mass spectrometer data file formats into the NIST MS Search Program for searching or building user databases.

VII. Summary of Interpretation of EI Mass Spectra

There are many excellent reference books on the interpretation of EI mass spectra and rationalizations of fragmentation schemes for molecular ions of organic compounds. Budzikiewicz *et al.* [28] and Porter [75] offer systematic approaches to the interpretation and rationalization of the mass spectra of many classes of organic compounds; several graphic mass spectra and mechanisms of fragmentation are presented for representative compounds in each class.

Introductory approaches to the interpretation of mass spectra of organic molecules have been written by Biemann [4], McLafferty and Tureček [26], Hill [76], and Shrader [77]. Many other systematic studies of the mass spectra and fragmentation mechanisms of different classes of compounds are available in the literature, such as those of sulfur-containing compounds [78]. For example, much of the information characterizing the mass spectra of steroids gathered from two decades of studies of fragmentation pathways has been reduced to computer algorithms for investigations of marine sterols [79, 80]. In other exhaustive studies of steroids [81], deuterium labeling has been used to delineate fragmentation in Δ^4- and $\Delta^{1,4}$-3-ketosteroids [82, 83] and stereochemical correlations in other steroids [84, 85]. Gaskell [86] has reviewed the mass spectrometry of steroids. General methods based on mass spectrometry have been reported for bile acids [87] and glycerolipids [88]. Splitter and Tureček [7] have addressed the general application of mass spectrometry to stereochemical features of molecules; a specific example of stereochemical differentiation has been reported for 1,3-amine acids [89]. Smith [90] has provided a great deal of information on general fragmentation and specific data regarding the fragmentation of molecular ions of illicit drugs.

Mass spectrometry of fatty acids [35, 36] has been well documented, including that of the eicosanoids [91, 92]. Other fields of application in natural products include studies of terpene derivatives [93] and alkaloids [94]. Mass spectrometry has also been useful in environmental [95–98] chemistry and toxicology studies of dioxins [99]. Fundamental studies of fragmentation mechanisms of other compounds, such as those of alkylbenzenes [100, 101], were important contributions to the current understanding of mass spectra. Detailed investigations of specific types of ions such as aliphatic oxonium ions [102], ion–dipole complexes [103], and 1,2-hydrogen shifts in allylic ions [104] help to advance the field and lay the groundwork for applications such as recognizing the location of double bonds [105] or branch points [106] in hydrocarbons and fatty acids.

The best approach to understanding fragmentation mechanisms in EI or those associated with any other type of ionization is to continue to practice the interpretation process. Draw the structure of the molecular ion of the compound (or putative analyte) with the sites of the charge and the odd electron at the most probable location (heteroatom, in a π bond, or lastly in a sigma bond). Try to identify driving forces that would cause electrons to move in a way that would result in the apparent bond cleavage, and always keep in mind the valence rules that allow for logical losses.

References

1. Busch KL, Electron ionization, up close and personal. *Spectroscopy (Eugene, OR)* **10:** 39-42, 1995.
2. Ipolyi I, Cicman P, Denifl S, Matejcik V, Mach P, Urban J, Scheier P, Maerk TD and Matejcik S, Electron impact ionization of alanine: Appearance energies of the ions. *International Journal of Mass Spectrometry* **252**(3): 228-233, 2006.
3. Lias SG, Bartmess JE, Liebman JF, Holmes JL, Levin RD and Mallard WG, Gas-Phase Ion and Neutral Thermochemistry. *J. Phys. Chem. Ref. Data.* **17:** Supplement No. 1, 1988.
4. Biemann K, *Mass Spectrometry: Organic Chemical Applications.* McGraw Hill, New York, 1962.
5. Budzikiewicz H, Djerassi C and Williams DH, *Interpretation of Mass Spectra of Organic Compounds.* Holden - Day, San Francisco, 1964.
6. McLafferty FW and Turecek F, *Interpretation of Mass Spectra.* University Science Books, Mill Valley, Calif, 1993.
7. Splitter JS, *Applications of Mass Spectrometry to Stereochemistry.* VCH Publishers Inc., NY, 1994.
8. Holmes JL, Assigning Structures to Ions in the Gas Phase. *Org. Mass Spectrom.* **20:** 169-183, 1985.
9. Hammerum S, Reactions of Distonic Ions in the Gas Phase. *Mass Spectrom Rev* **7:** 123-202, 1988.
10. Yate GF, Bouma WJ and Radom L, Ylidions as Distonic Ions in the Gas Phase. *J. Amer. Chem. Soc.* **106:** 5805-5808, 1984.
11. Nicholson AJC, Photochemical Decomposition of Aliphatic methyl Ketones. *Trans Faraday Soc.* **50:** 1067-1073, 1954.
12. McLafferty FW, Analysis by MS. *Anal. Chem.* **28:** 306-316, 1956.
13. McLafferty FW, Molecular Rearrangements in MS. *Anal. Chem.* **31:** 82-87, 1959.
14. McLafferty FW, Decompositions and Rearrangements of Organic Ions. In: *Mass Spectrometry of Organic Ions* (Ed. McLafferty FW), pp. 309-342. Academic Press, New York, 1963.
15. McLafferty FW, *Mechanisms of Ion Decomposition Reactions. In: Topics in Organic Mass Spectrometry.* Wiley Interscience, New York, 1970.
16. Turecek R, Drinkwater DE and McLafferty FW, Stepwise Nature of the gamma-H Rearrangement. *J. Amer. Chem. Soc.* **112:** 993-997, 1990.
17. Van Baar BLM, Terlouw JK, Akkok S, Zummack W and Schwarz H, Questions on the Mechanism of the McLafferty Rearrangement. *Int. J. Mass Spectrom. Ion Processes* **81:** 217-225, 1987.
18. Turecek F and (Ed) ZR, The Chemistry of Ionized Enols in the Gas Phase. In: The Chemistry of Enols. *J. Wiley and Sons Ltd:* 95-146, 1990.
19. McLafferty A, Rearrangement in an Even-Electron System: C3H6 Elimination from the a-Cleavage Product of Tri-n-Butylamine. *Org. Mass Spectrom.* **26:** 709-712, 1991.
20. Lavanchy A, Houriet R and Gaümann, The mass spectrometric fragmentation of n-alkanes. *Org. Mass Spectrom.* **14:** 79-85, 1979.
21. Stevenson D, Ionization and Dissociation by EI: Ionization Potentials and Energy of Formation of sec-Propyl and tert-Butyl Radicals; Limitations on the Method. *Discuss. Faraday Soc.* **10:** 35-45, 1951.
22. Moneti G, Pieraccinni G, Dani F, Catanella S and Traldi P, Acetonitrile as an Effective Reactant Species for Positive-ion Chemical Ionization of Hydrocarbons by Ion Trap Mass Spectrometry. *Rapid Comm. Mass Spectrom.* **10:** 167-170, 1996.
23. Harris FM, Review on Energetics of Doubly Charged Ions. *Int. J. Mass Spectrom. Ion Proc.* **120**(1/2): 1-44, 1992.
24. Schulze S, Paul A and Weitzel K-M, Formation of C7H7+ ions from ethylbenzene and o-xylene ions: Fragmentation versus isomerization. *International Journal of Mass Spectrometry* **252**(3): 189-196, 2006.
25. Olesik S, Baer T, Morrow JC, Ridal JJ, Buschek J and Holmes JL, Dissociation dynamics of halotoluene ions. Production of tolyl, benzyl and tropylium ([C7H7]+)ions. *Org Mass Spectrom* **24:** 1008-16, 1989.
26. McLafferty FW and Turecek F, *Interpretation of Mass Spectra, 4th Ed.* University Science Books, Mill Valley, CA, 1993.
27. Spiteller-Friedmann M and Spiteller G, Mass Spectra of Slightly Excited Molecules. III.Aliphatic Ethers. *Chem Ber* **100:** 79-92, 1967.
28. Budzikiewicz J, Djerassi C and Williams DH, *Mass Spectrometry of Organic Compounds.* Holden-Day, San Francisco, 1967.
29. McLafferty F and Golke R, Mass Spectrometric Analysis: Aromatic Acids and Esters. *Anal. Chem.* **31:** 2076-2082, 1959.
30. Schwarz H, Some new aspects of mass spectrometric orth effects. *Top. Curr. Chem.* **73:** 231, 1978.
31. Barkow A, Pilotek S and Grützmacher H, Ortho effects: A Mechanistic Study. *Eur. Mass Spectrom.* **1:** 525, 1995.
32. Benezra S and Bursey M, ortho-Effects on Ordering Factors in Mass Spectral Rearrangements. Loss of Keten [sic] from Halogenated Phenyl Acetates and Acetanilides. *J. Chem. Soc.(B).* 1515, 1971.
33. Smakman R and De Boer TJ, Mass spectra of some cyclic ethers. *Org Mass Spectrom* **1:** 403-16, 1968.
34. Ryhage R and Stenhagen E, Mass spectra of Methyl Esters of Saturated, Normal Chain Carboxylic Acids. *Journal* **13:** 523-534, 1959.

35. McCloskey JA, *Mass spectrometry of Fatty Acid Derivatives. In: Topics in Lipid Chemistry.* Logos Press, London, 1970.
36. Sweeley CC, Analysis of lipids by combined gas chromatography-mass spectrometry. *Fundam. Lipid Chem.*: 119-69, 1974.
37. Ryhage R and Stenhagen E, Low-energy spectra of methylesters of dibasic acids. *E Arkiv Kemi* **23**: 167, 1964.
38. Ryhage R and Stenhagen E, Mass spectra of long-chain fatty esters. In: *Mass Spectrometry of Organic Ions* (Ed. McLafferty F), 1963.
39. Vidavsky I, Chorush R, Longevialle P and McLafferty F, Functional Group Migration in Ionized Long-Chain Compounds. *J. Am. Chem. Soc.* **116**: 5865-5872, 1994.
40. Sparkman O and Ren J, Understanding Electron Ionization Fragmentation Mechanisms of Long-Chain Aliphatic Compounds with Carbonyl Functional Groups. In: *16th International Mass Spectrometry Conference, Edinburgh, UK,2003.*
41. Hammerum S, Christensen J, Egsgaard H, Larsen E, Derrick P and Donchi K, Slow Alkyl, Alkene, and Alkenyl Loss from Primary Alkylamines: Ionization of the Low-energy Molecular Ions Prior to Fragmentation in the - sec Timeframe. *Int J. Mass Spectrom Ion Phys* **47**: 351-354, 1983.
42. Rylander P, Meyerson S, Eliel E and McCollum J, Organic Ions in the Gas Phase XII: Aniline. *J. Am. Chem. Soc.* **85**: 2723-2725, 1963.
43. Halket JM and Zaikin VG, Derivatization in MS-1. Silylation. *Eur J Mass Spectrom* **9**: 1-21, 2003.
44. Pierce A, *Silylation of Organic Compounds.* Pierce Chemical Co, Rockford, IL, 1968.
45. McCloskey JA, Stillwell RN and Lawson AM, Use of deuterium-labelled trimethysilyl derivatives in mass spectrometry. *Anal. Chem.* **40**: 233-236, 1968.
46. Hallgren B, Ryhage R and Stenhagen E, Mass Spectra of Methyl Oleate, Methyl Linoleate, and Methyl Linolenate. *Journal* **13**: 845-847., 1959.
47. Attygalle AB, Jham GN and Meinwald J, Double-Bond Position in Unsaturated Terpenes by Alkylthiolation. *Journal* **65**(18)**:** 2528-2533, 1993.
48. Carlson D, Roan C, Yost R and Hector J, Dimethyl Disulfide Derivatives of Long Chain Alkenes, Alkadienes, and Alkatrienes for GC-MS. *Journal* **61**: 1564-1571., 1989.
49. Janssen G, Verhulst A and Parmentier G, Double Bonds in Polyenic Long-Chain Carboxylic Acids Containing a Conjugated Diene Unit. *Biomed. Environ. Mass Spectrom.* **15**: 1-6, 1988.
50. Yu Q, Zhang J and Huang Z, Double bond location in long chain olefinic acids:the mass spectra of 2-alkylbenzoazoles. *Biomed. Environ. Mass Spectrom* **13**: 211-216, 1986.
51. Voinov VG and Elkin YN, ECNI for Double Bond and Hydroxy Group Location in Fatty Acids on Methyl Esters or Pyrrolidides. %j *Org. Mass Spectrom.* **29**: 641-646, 1994.
52. Horiike M, Oomae S and Hirano C, Determination of the Double Bond Positions in Three Dedecenol Isomers. *Biomed. Environ. Mass Spectrom* **13**: 117-120, 1986.
53. Capella P and Zorzut CM, Double Bond Position in Monounsaturated Fatty Acid Esters by MS. *Anal. Chem.* **40**, 1968.
54. Eglington G, Hunneman DH, McCormick A, McCloskey JA and Law JH, GC-MS of Long Chain Hydroxy Acids: Ring Location in Cyclopropane Fatty Acid Esters by MS. *Org. Mass Spectrom.* **1**: 593-611, 1968.
55. Niehaus WG, Jr. and Ryhage R, Double Bond Positions in Polyunsaturated Fatty Acids using GC-MS. *Tetrahedron Lett.*: 5021-5026, 1967.
56. Hooper NK and Law JH, MS of Derivatives of Cyclopropene Fatty Acids. *J. Lipid Res.*, 1968.
57. Johnson JA, Bull AW, Welsch CW and Watson JT, Separation and ID of Linoleic Acid Oxidation Products in Mammary Gland Tissue from Mice Fed Low- and High-fat Diets. *Lipids* **32**: 369-375, 1997.
58. Anderson BA and Holman RT, Pyrrolidides for MS of the Double Bond in Monounsaturated Fatty Acids. *Lipids* **9**: 185-190, 1974.
59. Eagles J, Fenwick GR and Self R, MS of Pyrrolidide Derivatives of Oxygen-Containing Fatty Acids. *Biomed. Mass Spectrom.* **6**, 1979.
60. Harvey DJ, MS of Picolinyldimethylsilyl (PICSI) Ethers of Unsaturated Fatty Alcohols by G.C./M.S. *Biomed. Environ. Mass Spectrom.* **14**: 103-109, 1987.
61. Harvey DJ, Picolinyl esters for the structural determination of fatty acids by GC/MS. *Mol Biotechnol* **10**(3)**:** 251-60., 1998.
62. Lopez JF and Grimalt JO, Alkenone Distributions in Natural Environment by Better Method for Double Bond Location by GC-MS Cyclopropylimines. *J Amer Soc Mass Spectrom* **17**: 710-720, 2006.
63. Lopez JF and Grimalt JO, Phenyl- and cyclopentylimino derivatization for double bond location in unsaturated C37-C40 alkenones by GC-MS. *Journal of the American Society for Mass Spectrometry* **15**(8)**:** 1161-1172, 2004.
64. Schwerch S, Howald M, Gfeller H and Schlunegger UP, Halocarbenes as Diagnostic Tools for Mass Spectral Localization of Double Bonds. *Rapid Commun. Mass Spectrom.* **8**: 248-251, 1994.
65. Kidwell DA and Biemann K, Determination of double bond position and geometry of olefins by mass spectrometry of their Diels-Alder adducts. *Analytical Chemistry* **54**(14)**:** 2462-5, 1982.
66. Batt BD, Ali S and Prasad JV, GC/GC-MS studies on determination of position & geometry of double bonds in straight-chain olefins by derivatization. *J Chromatog Sci* **31**: 113-19, 1993.

67. Yuan G and Yan J, ID of double-bond position in isomeric linear tetradecenols and related compounds by MS dimethyl disulfide derivatives. *Rapid Commun Mass Spectrom* **16:** 11-14, 2002.
68. Mejanelle L, Laureillard J and Saliot A, Novel marine flagellate fatty acid: structural elucidation by GC-MS analysis of DMOX derivatives and DMDS adducts. *J Microbiol Methods* **48**(2-3): 221-237., 2002.
69. Michaud A, Yarawecz M, Delmonte P, Corl B, DE B and Brenna J, ID of Conjugated Fatty Acid Methyl Esters of Mixed Double Bond Geometry by Acetonitrile CI-MS/MS. *Anal. Chem.* **75:** 4925-4930, 2003.
70. Stein SE, Estimating Probabilities of Correct Identification from Results of Mass Spectral Library Searches. *J. Am. Soc. Mass Spectrom.* **5:** 316-323, 1994.
71. Sokolow SK, J.; Gustafson, P., The Finnigan Library Search Program. *Finnigan Application Report No. 2, March 1978,* 1978.
72. McLafferty FW and Bockhoff FM, Separation/Identification Systems for Complex Mixtures Using Mass Separation and Mass Spectral Characterization. *Anal. Chem.* **50:** 69-78, 1978.
73. Stein SE, Chemical Substructures Identification by Mass Spectral Library Searching. *J. Am. Soc. Mass Spectrom.* **6:** 644-653, 1995.
74. Ausloos P, Clifton CL, Lias SG, Mikaya AI, Stein S, D T, D. SO, Zaikin V and Zhu D, The Critical Evaluation of a Comprehensive Mass Spectral Library. *J. Am. Soc. Mass Spectrom.* **10:** 287-299, 1999.
75. Porter QN, *Mass Spectrometry of Heterocyclic Compounds.* Wiley Interscience, New York, 1985.
76. Hill HC, *Introduction to Mass Spectrometry.* Heyden and Son, London, 1972.
77. Shrader SR, *Introductory Mass Spectrometry.* Allyn & Bacon, Boston, 1971.
78. Bortolini O and Fogagnolo M, Mass Spectrometry of Sulfur-Containing Compounds in Organic and Bioorganic Fields. *Mass Spectrom. Rev.* **14:** 117-162, 1995.
79. Gray NAB, Buchs A, Smith DH and Djerassi C, Computer assisted structural interpretation of mass spectra data. *Helv. Chim. Acta* **64:** 458-470, 1981.
80. Lavanchy A, Varkony T, Smith DH, Gray NAB, White WC, Carhart RE, Buchanan BG and Djerassi C, Rule-based mass spectrum prediction and ranking: Applications to structure elucidation of novel marine sterols. *Org. Mass Spectrom.* **15:** 355-366, 1980.
81. Budzikiewicz H, MS of Amino Steroids and Steroidal Alkaloids-an Update. *Mass Spec. Rev.* **10:** 79-88, 1991.
82. Brown P and Djerassi C, Fragmentation of D4 and D1,4-3-Ketosteroids. *J. Am. Chem. Soc.* **102:** 807-817, 1980.
83. Brown P and Djerassi C, Effect of Additional Conjugation on Fragmentation of D4-3-Ketosteroids. *J. Org. Chem.* **46:** 954-963, 1981.
84. Patterson DG and Djerassi C, Stereochemical Correlations in Steroid Mass Spectra. *Org. Mass Spectrom.* **15:** 41-50, 1980.
85. Patterson DG, Haley MJ, Midgley I and Djerassi C, The Effect of Stereochemistry on D-Ring Fragmentation of Steroid Hydrocarbons. *Org. Mass Spectrom.* **19:** 531-538, 1984.
86. Gaskell SJ, Mass Spectrometry of Steroids. *Methods Biochem. Anal.* **29:** 385-434, 1983.
87. Setchell KDR and Matsui A, Bile Acids by GC-MS. *Clin. Chim. Acta* **127:** 1-17, 1983.
88. Kuksis A, Marai L and Myher JJ, Rev of Methodology on Glycerolipids. *J Chromatogr* **273:** 43-66, 1983.
89. Partanen T, Vainiotalo P, Stajer G, Bernard G and Pihlaja K, EI and CI MS in Stereochemical Differentiation of Some 1,3-Amine Acids. *Org. Mass Spectrom.* **29:** 126-132, 1994.
90. Smith RM, *Understanding Mass Spectra: A Basic Approach, 2nd Ed.* Wiley, NYC, 2004.
91. Murphy RC, Leukotrienes by MS. *J. Mass Spectrom* **30:** 5-16, 1995.
92. Murphy RC and Zirrolli JA, *Lipids Mediators, Eicosanoids, and Fatty Acids. In: Biological Mass Spectrometry Present and Future.* J. Wiley & Sons, New York, 1994.
93. Budzikiewicz H, Djerassi C and Williams DH, *Structure Elucidation of Natural Products by Mass Spectrometry. II. Steroids, Terpenoids, Sugars.* Holden-Day, San Francisco, 1964.
94. Budzikiewicz H, Djerassi C and Williams DH, *Structure Elucidation of Natural Products by Mass Spectrometry. Alkaloids.* Holden-Day, San Francisco, 1964.
95. Richardson SD, Environmental mass spectrometry. *Analytical chemistry* **72**(18): 4477-96, 2000.
96. Richardson SD, The role of GC-MS and LC-MS in the discovery of drinking water disinfection by-products. *J Environ Monit* **4**(1): 1-9., 2002.
97. Garcia de Oteyza T and Grimalt JO, GC and GC-MS characterization of crude oil transformation in sediments and microbial samples after 1991 spill in Saudi Arabian Gulf. *Environ Pollut (Amsterdam)* **139:** 523-531, 2006.
98. Richardson SD, Environmental mass spectrometry: Emerging contaminants and current issues. *Analytical Chemistry* **78**(12): 4021-4045, 2006.
99. Ryan JJ, Lau BP-Y and Boyle MJ, *Dioxin-Like Compounds in Human Blood. In: Biological Mass Spectrometry Present and Future.* J. Wiley & Sons, New York, 1994.
100. Kuck D, Mass Spectrometry of Alkylbenzenes and Related Compounds. Part II Gas Phase Ion Chemistry of Protonated Alkylbenzenes (Alkylbenzenium Ions). *Mass Spec. Rev.* **9:** 583-630, 1990.
101. Childs RF, The Homotropylium Ion and Homoaromaticity. *Acc. Chem. Res* **17:** 347-352, 1984.
102. Zahorszky VI, Unimolecular Reactions of Bifunctional Aliphatic Oxonium ions. *Mass Spec. Rev.* **11:** 343-388, 1992.

103. Postma R, vanHelden SP, vanLenthe JH, Ruttink PJA, Terlouw JK and Holmes JL, The [CH2 = CHOH/H2O]+×System = A Theoretical Study of Distonic, Hydrogen-bridged Ions and Ion-Dipole Complexes". *Org. Mass Spectrom.* **23:** 503-510, 1988.

104. Bowen RD, Gallagher R and T.Meyerson S, Structure and Mechanism of Formation of C5H9O+ from Ionized Phytyl Methyl Ether. *J. Am. Soc. Mass Spectrom.,* **7:** 205-208, 1996.

105. Yruela I, Barbe A and Grimalt JO, Detmn double bond position and geometry in linear & branched hydrocarbons and fatty acids by GCMS of epoxides and diols by stereospecific resin hydration. *J Chromatog Sci* **28:** 421-7, 1990.

106. Simon E, Kem W and Spiteller G, Localization of the Branch in Monomethyl Branched Fatty Acids. *Biomed. Environ. Mass Spectrom.* **19:** 129-136, 1990.

"Most people would rather die than think; in fact, they do so."

~Bertrand Russell

Chapter 7   Chemical Ionization

Introduction to Mass Spectrometry, 4th Edition: Instrumentation, Applications, and Strategies for Data Interpretation; J.T. Watson and O.D. Sparkman, © 2007, John Wiley & Sons, Ltd

"The spectra which are produced by this chemical ionization technique are very different from those which are produced by conventional electron impact mass spectrometry and are sometimes much more useful for both qualitative and quantitative analysis, particularly of high molecular weight and polyfunctional compounds."

~Burnaby Munson
~Frank Field

I. Introduction

Chemical ionization[1] (CI) is one of the "soft" ionization techniques; i.e., an ionization process that produces an ion representing the intact analyte that does not fragment significantly. Thus, the CI mass spectrum consists chiefly of peaks representing ions composed of the intact analyte molecule [1]. CI achieves ionization of the analyte without transferring excessive energy to the nascent analyte ions; this means that there is little subsequent fragmentation of the intact analyte ion. By definition, CI involves an *ion/molecule* reaction (see Scheme 7-1) in which the analyte (M), in the form of a gas-phase *molecule*, collides with a gas-phase *ion* (RH^+), a reagent ion (often a proton-rich ion). The result is the formation of abundant adduct ions (usually, the most abundant is a protonated analyte molecule, MH^+) that contain the intact molecular species of the analyte. Chemical ionization is the subject of a book [2] and several reviews [3–6].

$$M + RH^+ \longrightarrow MH^+ + R$$

Scheme 7-1

CI was originally developed by Burnaby Munson and Frank Field at Esso Research and Engineering Company, Bay Town Research and Development in 1996 as an ancillary capability for the EI mass spectrometer [7]. The original CI source was similar in design to the EI source except that there was less gas conductance out of the ion volume, a feature that allowed the pressure in the ionization chamber to rise to an abnormally high value from the standpoint of normal EI conditions. The CI source required a more powerful vacuum system than was typically available in conventional instruments of the time because of the necessity for maintaining a partial pressure of 10–100 Pa of the reagent gas (source of reagent ions) in the source while still being able to achieve the low pressures of ~10^{-4} Pa required for *m/z* analyzers in use at the time. The high pressure of reagent gas in a CI source is necessary to achieve a high concentration of reagent ions and to minimize the direct ionization of the analyte by EI, as explained in the following sections. The CI source used in concert with EI mass spectrometers today is very much like that originally designed by Field and Munson.

Approximately ten years later in the mid 1970s, Evan Horning's Group at the Baylor College of Medicine (Houston, TX) developed a variation of chemical ionization, which they originally called atmospheric pressure ionization (API) and which today is known as atmospheric pressure chemical ionization (APCI) [8–13]. Both CI as developed by Field and Munson and APCI as developed by the Horning Group required gas-phase analyte–reagent ion reactions. The main difference was the operating pressure and the sensitivity. As the name implies, APCI involves reactions that take place at atmospheric pressure rather than at the reduced pressures used in the Field–Munson technique. This atmospheric pressure operation means that the mass spectrometer must be designed to take in ions at atmospheric pressure. Mass spectrometers designed to operate with electrospray ionization (ESI), another atmospheric pressure technique, are well suited for

---

[1] As a point of clarification, *chemi-ionization* and *chemical ionization* describe two different processes. *Chemi-ionization* refers to the phenomenon in which two neutral particles collide to effect ionization, such as metastable helium He* and a sample molecule M:

$$He^* + M \rightarrow He + M^{+\bullet} + e^-$$

Chemical ionization is based on an ion/molecule reaction involving collision of a charged species, the reagent ion, with a neutral analyte molecule, resulting in ionization of the latter by transfer of a charged particle (proton or larger species) to it. *Chemi-ionization* is not considered further in this book.

APCI. Like ESI, APCI is a good technique for use with HPLC provided that the analyte has sufficient volatility to exist in the vapor phase after it exits the LC.

The sensitivity of the Field–Munson CI technique (which is the type of chemical ionization referred to when the general term "CI" or "chemical ionization" is used) was less than that obtained with EI. The low limits of detection sometimes achieved by CI are usually due to the high abundance of ions representing the intact analyte, which are often at *m/z* values above those of the ions that may produce interference from the sample matrix. Furthermore, limits of detection are often lower using the Field–Munson CI technique due to judicious choice of reagent ions that would discriminate against potential interferences in the sample matrix. On the other hand, APCI proved to be the most sensitive mass spectrometry technique ever developed. Horning's Group demonstrated the applicability of APCI to LC eluates, but the technique of APCI LC/MS did not gain popularity until the development of electrospray ionization. As described in the chapter on LC/MS (Chapter 11), the first commercialization of APCI was in the area of analyzing atmospheric gases for trace pollutants. APCI has proved to be a valuable tool in both GC/MS and LC/MS as well as in the somewhat esoteric area of supercritical fluid chromatography/mass spectrometry (SFC/MS).

The ionization techniques of CI and APCI can be used with any type of *m/z* analyzer. One of the problems with CI is the difficulty in reproducibly establishing the optimal pressure of reagent gas in the ion source.[2] This is because of the difficulty of measuring the pressure in the ionization volume and the requirement for a pressure that is two to three orders of magnitude greater than that required for the EI source. This inability to obtain reproducible pressure settings in the CI source is one of the reasons that CI is not an ionization technique of choice for quantitative analyses. The quadrupole ion trap (QIT) GC/MS that is designed for internal rather than external ionization is an exception to this shortcoming. During operation of the QIT with internal ionization as described in Chapter 2, the partial pressure of the reagent gas is maintained (at about 10^{-3} Pa) with feedback control from monitoring current associated with the reagent ions, thereby providing reproducible conditions for quantitation. The requirement for a relatively low partial pressure of reagent gas in the QIT increases the variety of reagent gases that can be used; e.g., rather than using ammonia gas from a high-pressure gas cylinder, the vapor from the headspace of a small vial of ammonium hydroxide is adequate for NH_3–CI in the internal ionization QIT. The QIT designed for internal ionization has other advantages. For example, it is possible to acquire alternate CI and EI spectra in the same analysis with such an instrument. During acquisition of the EI spectrum, CI reagent ions are not stored; therefore, no products of the ion/molecule reaction can be formed that would otherwise interfere with the EI spectrum.

The process of reagent ion formation is very complex in both CI and APCI. This process for the formation of methane reagent ions in CI is described below in Section III entitled "Production of Reagent Ions from Various Reagent Gases". The process of forming reagent ions that react with the gas-phase analyte molecules is even more complex in APCI and is described in Chapter 11.

Most discussions of CI used in modern instrumentation involve introduction of

---

[2] Most commercial mass spectrometers that provide a capability for monitoring the "ion source" pressure actually measure the pressure in the large ion source housing, not the ionization volume, which is tiny compared to the overall vacuum housing of the ion source. For this reason, the numerical value of the "ion source pressure" reading in an EI source is often reported to be 10^{-4} Pa, whereas the pressure quoted in this book is 10^{-1} Pa.

gas-phase analytes through a GC column. As found in Chapter 4 on inlets, condensed-phase samples can be introduced into the ion source of an EI mass spectrometer; as the sample is heated to increase its vapor pressure, some gas-phase molecules of analyte penetrate the low-pressure region of the ionization chamber. The same procedure is used in CI mass spectrometry. A variation of this technique is desorption chemical ionization (DCI). Unlike the solid probe as described in the previous sentences, which places the heated condensed-phase analyte close to the ionization chamber so that some equilibrated gas-phase molecules can enter the chamber and be ionized, the DCI probe promotes desorption of ions that appear to be formed in the condensed phase. This technique has been called desorption chemical ionization, direct chemical ionization, in-beam ionization, and surface ionization [14]. Unlike in CI and APCI, the analyte apparently does not have to be in the gas phase for ionization to occur. This technique was first reported by Fred McLafferty involving an underivatized peptide (VAAF) to produce a mass spectrum characterized by an intense peak representing the protonated molecule $[M + H]^+$ or MH^+ [15].

A great deal of the discussion on CI in this chapter is focused on the formation and detection of positive ions; however, unlike in EI, negative ions can also be formed easily during CI. Unfortunately, what is often referred to as negative-ion CI is actually the technique of resonance electron capture ionization, which does not involve an ion/molecule reaction, but rather the interaction of a thermal electron to produce a negative-charge molecular ion; this technique, properly called ECNI, is the basis of many important applications in environmental chemistry and pharmacology. The apparent reason for the confusion between negative-ion CI (NCI) and electron capture negative ionization (ECNI) is explained in Section V of this chapter, which describes the operating principles of both processes.

CI and EI are complementary in that CI produces mostly information relating to the molecular mass of the analyte by producing adduct ions that contain the intact analyte molecule, whereas EI promotes extensive fragmentation of a molecular ion to yield a

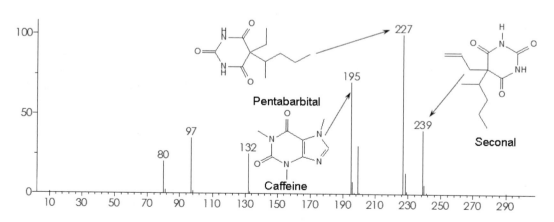

Figure 7-1. *Mass spectrum of an extract from a gastric lavage obtained by chemical ionization (CI) using isobutane as the reagent gas to generate protonated molecules of the named compounds.*

fragmentation pattern from which structural information can be deduced. The simplicity of a CI spectrum is illustrated in Figure 7-1, which is the CI mass spectrum of a sample of gastric lavage associated with a drug overdose case [16].   The three intense peaks in Figure 7-1 represent the protonated molecules of caffeine (nominal mass = 194 Da; peak for $[M + H]^+$ at m/z 195), pentobarbital (nominal mass = 226 Da; peak for $[M + H]^+$ at m/z 227), and secobarbital (nominal mass = 238 Da; peak for $[M + H]^+$ at m/z 239).   The CI spectrum of each of these individual compounds (data not shown) consists essentially of only a peak for the protonated molecule; i.e., no peaks relating to fragmentation of the $[M + H]^+$.   For this reason, CI is especially good for mixture analysis because there is essentially no overlap of mass spectral data relating to the individual components. Another advantage of CI is its tendency to discriminate against the sample matrix; this feature is also illustrated in Figure 7-1, where there are very few extraneous peaks (of remarkably low intensity) for such a complex mixture as a gastric lavage.   Later in this chapter, the complementary nature of CI and EI is emphasized in a descriptive application to the pesticide Dursban.   In addition, the complementary nature of EI and CI is described in Example 1-4 in Chapter 1.

II. Description of the Chemical Ionization Source

In CI, a reagent gas is converted into reagent ions, which react with the analyte to achieve its ionization, usually by the formation of a protonated molecule (which is an EE ion represented by MH^+ or $[M + H]^+$).   The CI source is similar to that described in the previous chapter for EI, except that the ionization chamber is "tighter" (allowing for less conductance) so that the local pressure can be elevated to approximately 10–100 Pa. This relatively high local pressure is due to the reagent gas, some of which is ionized by EI to promote production of constant flux reagent ions (except for the ion trap [17] or ICR) as explained below.

As in the conventional EI source, the source of electrons in a CI source is a heated filament mounted just outside the ionization chamber.   The emitted electrons are accelerated into the ionization chamber with energies up to several hundred electron volts [2, 5]; the electron energy is greater than that (70 eV) used in conventional EI so that there will be effective penetration of the electrons into the region of relatively high pressure reagent gas.   Also, in contrast to EI, the filament current in CI is regulated from total emission of electrons rather than from the electron current reaching the trap or anode; in CI, most of the electrons do not reach the anode because of the high probability of collision with the reagent gas at high pressure.

It is important to realize that the high pressure under discussion is required only in the ion volume of the ion source, a volume on the order of 1 cm^3 in the middle of the ion-source housing.   As the gas from the ion volume spills out into the much larger volume of the ion-source housing, the pressure there approaches 10^{-2} to 10^{-1} Pa, compared with 10^{-5} to 10^{-4} Pa during EI.   A pressure of this magnitude in the ion flight path in the m/z analyzer and detector degrades the sensitivity (ion transmission) and resolving power of the instrument.   These undesirable effects of high pressure can be eliminated or minimized with differential pumping, as explained in the section on vacuum systems in Chapter 2.

Other important features of CI source design are the measures taken to prevent electrical discharge in the relatively high pressures near the high-voltage acceleration region in the magnetic-sector instrument.   This is not a problem in the quadrupole mass spectrometer because the accelerating potential is only on the order of 15 V, but it can be a problem with the TOF mass spectrometer, which also uses high accelerating potentials.

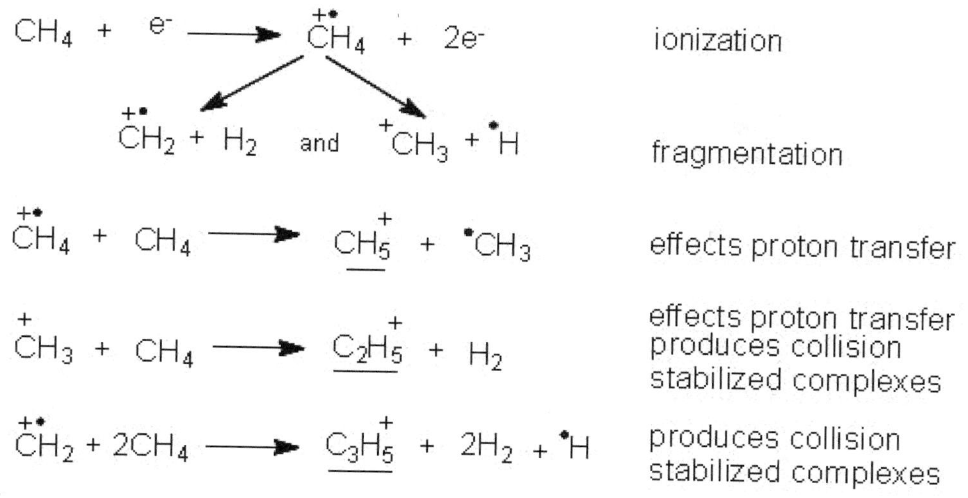

$$CH_4 + e^- \longrightarrow \overset{+\bullet}{C}H_4 + 2e^- \qquad \text{ionization}$$

$$\overset{+\bullet}{C}H_2 + H_2 \quad \text{and} \quad \overset{+}{C}H_3 + \overset{\bullet}{H} \qquad \text{fragmentation}$$

$$\overset{+\bullet}{C}H_4 + CH_4 \longrightarrow \underline{CH_5^+} + \overset{\bullet}{C}H_3 \qquad \text{effects proton transfer}$$

$$\overset{+}{C}H_3 + CH_4 \longrightarrow \underset{}{C_2}\overset{+}{H_5} + H_2 \qquad \begin{array}{l}\text{effects proton transfer} \\ \text{produces collision} \\ \text{stabilized complexes}\end{array}$$

$$\overset{+\bullet}{C}H_2 + 2CH_4 \longrightarrow \underline{C_3H_5^+} + 2H_2 + \overset{\bullet}{H} \qquad \begin{array}{l}\text{produces collision} \\ \text{stabilized complexes}\end{array}$$

Scheme 7-2

III. Production of Reagent Ions from Various Reagent Gases

Production of a steady state concentration of reagent ions is initiated by EI of the reagent gas. Even though both the analyte and the reagent gas are present in the ionization chamber, the vast majority of ion current derives from the reagent gas, which is present in great excess; typically, the partial pressure of analyte will be 10^{-3} Pa, whereas that of the reagent gas will be 10–100 Pa. As illustrated in Scheme 7-2 for methane, once some of the methane molecules are converted to molecular ions of methane, some of them decompose into fragment ions. The residual molecular ions and the nascent fragment ions of methane react via ion/molecule reactions with neutral molecules of methane to form proton-rich ions such as CH_5^+, $C_2H_5^+$, and $C_3H_5^+$. These proton-rich ions are represented by RH^+ in Scheme 7-1. The concentration of these proton-rich reagent ions is high relative to that of the analyte molecules, which means, an analyte molecule is more likely to interact with one of the proton-rich reagent ions than a 200-eV electron; for this reason, essentially no molecular ions of the analyte are formed. During collision with a reagent ion, if the analyte molecule has a higher proton affinity than methane, a proton is transferred from the proton-rich reagent ion to the analyte, which is thereby ionized by protonation as was shown in Scheme 7-1.

Although the gas-phase reactions in Scheme 7-2 are presented in a stepwise fashion, all these processes (EI of reagent gas, reagent gas molecule–ionized reagent gas interaction, and reagent ion–analyte molecule reactions) proceed simultaneously in the ion source [1].

Principal reagent ions of some other reagent gases are found in Table 7-1. Ion/molecule reaction schemes for the production of these reagent ions can be found in the indicated references. In addition, general surveys of reagent gases for high-pressure ionization mass spectrometry have been published [18]. Clean sources of the reagents H_3O^+, NH_4^+, NO^+ and O_2^+ can be obtained without any mass selection using technology borrowed from proton-transfer MS [19]. More details on the use of the protonating reagent gases can be found in a later section on data interpretation. Nitrogen, helium, nitric oxide, argon, and other inert, aprotic gases are discussed in a later section entitled "Ionization by Charge Exchange".

Electron ionization of isobutane provides a source of *t*-butyl carbenium ions ($C_4H_9^+$) that readily transfer protons to heteroatoms in a molecule [20]. The technical details of using high-pressure isobutane and the reactivity of the *t*-butyl carbonium ion have been described by Field [21].

Table 7-1. Characteristics of reagent gases for CI.

Reagent Gas	Predominant Reactant Ions	Proton Affinity* (kJ mol^{-1})	Hydride Affinity (kJ mol^{-1})
He/H$_2$	HeH$^+$	176	—
H$_2$	H^{3+}	424	1256
CH$_4$	CH$_5^+$	551	1126
	C$_2$H$_5^+$	666	1135
H2O	H$_3$O$^+$	697	—
CH$_3$CH$_2$CH$_3$	C$_3$H$_7^+$	762	1130
CH$_3$OH	CH$_3$OH$_2^+$	762†	—
(CH$_3$)$_3$CH (isobutane)	C$_4$H$_9^+$	821‡	1114
NH$_3$	NH$_4^+$, (NH$_3$)$_2$H$^+$, (NH$_3$)$_{33}$H$^+$	854	—
(CH$_3$)$_2$NH	(CH$_3$)$_2$NH$_2^+$, (CH$_3$)$_2$H$^+$, C$_3$H$_8$N$^+$	921	—
(CH$_3$)$_3$N	(CH$_3$)$_3$NH$^+$	9431	—

* Lias SG, Bartmess JE, Liebman JF, Holmes JL, Levin RD, and Mallard WG *J. Phys. Chem. Ref. Data* 17, Suppl. 1, 1988.
† from reference [1].
‡ Proton affinity of isobutylene, which is the conjugate base of isobutane. All values converted from kcal to kJ.

Tetramethylsilane [22, 23] as a reagent gas produces a single reagent ion, (CH$_3$)$_3$Si$^+$, which forms adducts with a variety of compounds providing analytically useful CI mass spectra, especially in molecular weight determinations, because of the prominent (M + 73)$^+$ ion. Tetramethylsilane reacts as a Lewis acid with analyte molecules and therefore effects ionization by association or adduct formation [24, 25]; this is a low-energy-release reaction, and fragmentation of the adduct ion is not a major process. A variety of organic gases have been studied for analytical advantage in producing reagent ions for CI, including dimethyl ether [26], benzene [27, 28], nitromethane [29], methanol [30], diisopropyl ether [31], acrylonitrile [32, 33], ethylene oxide, a.k.a. oxiran [34], and other unusual reagents [32, 35]. Inorganic gases such as nitric oxide [36] and the direct use of metal ions such as K^+ [37–39], Fe^+ [40], and other elements [41, 42] have been used to ionize organic molecules in the gas phase. Chemical ionization with NH_3 [43–45] offers the advantage of selective ionization via protonation because of its high proton affinity. Interesting experiments can be pursued with an isotopically labeled reagent gas [46], including the use of ND_3 for H–D exchange in peptides [47]. Some of the more esoteric substances that have been used as reagent gases in beam-type instruments are used as mixtures in methane. A few percent of ammonia in methane gives a performance in CI equal to that obtained with pure ammonia.

Internal ionization 3D QIT mass spectrometers require only a low partial pressure of reagent gas, which allows the use of some reagent gases that would be difficult to use in the high-pressure CI source of a beam-type mass spectrometer. As an example, the head space from a vial containing acetonitrile provides sufficient partial pressure of this reagent to allow for CI in this type of QIT.

IV. Positive-Ion Formation under CI

As a result of ionization (protonation) by interaction with proton-rich ions in a CI source, the analyte is distinguished in the mass spectrum by an intense peak representing the intact molecule at an *m/z* value one higher than the value of its molecular mass. The reason for the high abundance of the protonated molecule (MH^+) is due in large measure to the fact that most of the excess energy transferred to the protonated molecule during the ionization process [48] is lost due to collisional cooling in the relatively high-pressure environment.

It is proper to refer to the protonated form of the analyte as a *protonated molecule* (MH^+) and *not as a protonated molecular ion*. The latter name suggests some sort of combination of a proton (positively charged) and a molecular ion (also positively charged as described in the previous chapter on EI or negatively charged as in the case of electron capture ionization), which does not happen because of the repulsive forces between $M^{+\bullet}$ and H^+. If a positive-charge molecular ion could be protonated, such a species would have two positive charges associated with it! If a negative-charge molecular ion (as in the case of electron capture ionization) were to be protontated, neutralization of the $M^{-\bullet}$ by the H^+ would result.

1. Fundamentals

To appreciate the selectivity and energetics of CI, some fundamental concepts of physical chemistry must be recalled. Proton affinity is an important parameter to consider [49–52]. The proton affinity [53–55] of a molecule M is defined as the negative of the enthalpy for the following general protonation reaction:

$$H^+ + M \rightarrow MH^+$$

(Eqn. 7-1)

The higher the value of proton affinity, the more tightly bound is the proton in the substrate species (see the trend in Table 7-1); this parameter is inversely related to the gas-phase acidity of the analyte [56–59]. The relative acidities of substituted silanes, germanes, and stannanes have been measured by the equilibrium method using FTICR techniques [60]; similar studies have been conducted with allylphosphine [61].

The relative gas-phase basicity (ΔGB, defined as $-\Delta G^0$ of proton transfer) can be computed as shown in Equation 7-2 from the equilibrium constant of the equilibrium reaction shown in Equation 7-3 [58]. R is the universal gas constant and T is the temperature of the medium in which the equilibrium is established. If the equilibrium is found to favor the forward reaction in Equation 7-3 with a constituent composition shifted to the right, it can be concluded that the gas phase basicity of Y is greater than the gas phase basicity of X; further, if the entropy of proton transfer is negligible, it can also be concluded that the proton affinity of Y is greater than the proton affinity of X [62]. From the measured value of $-\Delta G$ of protonation, the $-\Delta H$ of protonation, the proton affinity [59, 63], can be derived.

$$\Delta GB = -RT\,ln(K_{eq}) = -RT\,ln([X]\,[YH^+])\,/\,([Y]\,[XH^+]) \qquad \text{(Eqn. 7-2)}$$

$$XH^+ + Y \rightleftharpoons X + YH^+ \qquad \text{(Eqn. 7-3)}$$

One of the most successful experimental procedures for measuring the relative gas-phase basicity of compounds of interest is to set up an experiment in which the compound of interest is given the opportunity to exchange a proton with a compound of known gas-phase basicity [64]. By setting up several such experiments, a compound of known gas-phase basicity can eventually be found that will protonate the unknown material, as well as another compound that will not protonate the compound. By this bracketing procedure, the experimental limits can be determined for the gas-phase basicity of the compound of interest relative to known values for carefully chosen standard compounds.

As an example of the bracketing approach for determining gas-phase basicities, consider the dipeptide alanylvaline as the compound of interest to be compared in protonation reactions with the compounds ethyl amine and 3-fluoropyridine as compounds of known gas-phase basicity [63]. The molecular mass of alanylvaline is 188 Da, whereas the molecular mass of 3-fluoropyridine is 97 Da and has a gas phase basicity of 885 kJ mol^{-1}, and ethyl amine has a molecular mass of 45 Da and a gas phase basicity of 213.3 kcals mol^{-1} [63]. The dipeptide will be represented by the symbol Y in Equation 7-3 to be tested in a bracketing manner between the first reaction with protonated 3-fluoropyridine as represented by X in Equation 7-3 and subsequently by a similar reaction with protonated ethyl amine. Protonated species of the volatile base of known gas-phase basicity were prepared by standard CI techniques. Neutral species of the dipeptide (Y in Equation 7-3) were introduced into the ion source of a mass spectrometer through the specialized technique of substrate-assisted laser desorption [65]. The results of the first step in the bracketing method to test the forward reaction in Equation 7-3 are shown in Figure 7-2, where the peak at *m/z* 98 represents the protonated molecule of 3-fluoropyridine and the peak at *m/z* 189 represents the protonated molecule of alanylvaline in panel A. Because alanylvaline was admitted to the ion source of the mass spectrometer as a neutral, it became protonated only by capturing a proton from some of the protonated 3-fluoropyridine, thereby indicating that the gas-phase basicity of alanylvaline is greater than that of 3-fluoropyridine. The results of the second experiment in the bracketing test for the forward reaction of Equation 7-3 are shown in panel B of Figure 7-2 in which the peak at *m/z* 46 represents the protonated molecule of ethyl amine. No peak is observed at *m/z* 189 representing protonated alanylvaline, indicating that this dipeptide has a gas-phase basicity lower than that of ethyl amine.

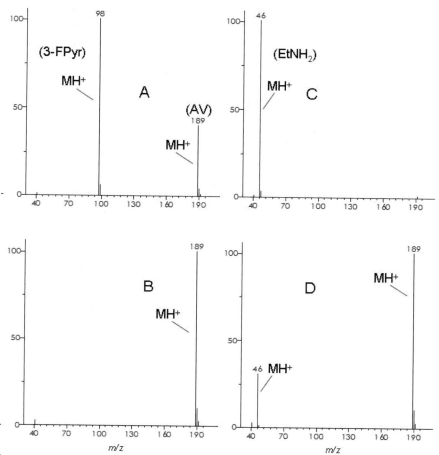

Figure 7-2. *Mass spectral data supporting the bracketing experiment to determine relative proton affinities.*

The reverse reaction in Equation 7-3 can be assessed by using MALDI to produce protonated molecules of the dipeptide for experiments in which the volatile vapors of the neutral base of known gas-phase basicity are admitted to the mass spectrometer. Panel C of Figure 7-2 shows the results of allowing the protonated molecules of the dipeptide to interact with the neutral vapors of 3-fluoropyridine. Because there is no peak at m/z 98, it is clear that the gas-phase basicity of 3-fluoropyridine is not sufficient to pull one of the protons away from the protonated dipeptide. Therefore, the gas-phase basicity of alanylvaline is greater than 885 kJ mol^{-1}. Panel D shows the results of interaction between neutral vapors of ethyl amine and the protonated molecule of the dipeptide. The peak at m/z 46 shows that ethyl amine has a sufficiently high gas-phase basicity to capture or remove a proton from some of the protonated molecules of the dipeptide to enable it to become protonated and give the peak at m/z 46. These results indicate that the gas-phase basicity of ethyl amine is greater than that of the dipeptide.

A more controlled method for obtaining thermochemical data has become known as the kinetic method [64, 66, 67], in which the competitive fragmentation of cluster ions is used to infer thermochemical properties. The relative rates of fragmentation of a selected precursor, such as a protonated dimer, are represented by the relative abundances of the corresponding fragment ions. An advantage of the kinetic method is that it is sensitive to small differences in thermochemical values. For example, pyridine and 3-aminopentane have similar proton affinities (differ by less than 0.5%) as determined by the equilibrium method [68], but the relative abundances of the two protonated monomers generated by metastable decay of the proton-bound dimer are significantly different [69]. To a considerable extent, the bracketing method and the kinetic method are related in that the intermediate involved is likely to be a proton-bound dimer. The kinetic method has been extended to estimate the ionization energies of polycyclic aromatic hydrocarbons [70] and of substituted anilines [71] as well as the heterolytic bond dissociation energy of inorganic salts [72], and to determine the electron affinities of polycyclic aromatic hydrocarbons [73]. The kinetic method has also been used for chiral recognition by dissociation of suitable complexes [74–76]. The gas-phase acidities of 52 alkanols and cycloalkanols have been determined from studies of the unimolecular dissociation and by CAD of proton-bound cluster ions [77].

2. Practical Considerations of Proton Affinity in CI

Two general, practical trends of predictability in CI reactions are derived from the relative proton affinities of the analyte and the reagent gases [1, 78]. First, the greater the difference in proton affinities of the analyte and the reagent gas, the more energy will be transferred to the protonated molecule, and the greater will be the degree of subsequent fragmentation of the analyte. Second, the predictability of the principal type of ion/molecule reaction is less reliable, but this is also based on a consideration of the proton affinities. If the reagent ion is a good Brønsted acid or if the analyte is a Brønsted base, then proton transfer to the analyte is a likely event resulting in formation of the MH^+ ion, a protonated molecule of the analyte. A Brønsted acid is any molecular or ionic species that can act as a proton donor; a Brønsted base is one that can act as a proton acceptor. In ion/molecule reactions that involve hydride transfer from the analyte to the reagent ion, the reagent ion acts as a Lewis acid. A Lewis acid is any species that can accept or attach to a pair of unshared electrons, such as that on a hydride ion (H^-). The relative strength of a Lewis acid is measured as its hydride affinity (see Table 7-1).

The ion/molecule reaction must be exothermic or no protonated molecule will be observable in conventional CI; i.e., the proton affinity of the analyte must be greater than that of the reagent gas (the conjugate base of the reagent ion) for a proton to be transferred from the reagent ion to the analyte. The proton affinity of most organic molecules containing only C, H, and O is around 200 kcal mol^{-1} [78]. This means that many compounds can be ionized under CI conditions using isobutane or methane, but not with ammonia, according to the data in Table 7-1.

When the proton affinities of the analyte and the reagent gas are nearly equal, adduct ions, formed by association of the reagent ion with the compound, may be observed instead of the protonated molecule. The gas-phase basicities and proton affinities of amino acids and peptides as measured by CI and related techniques have been reviewed [79]. Affinities between other particles (e.g., $SiCl^+$ and pyridine) can be assessed by the kinetic method [80]. Application of the kinetic method to dynamic systems (those involving a changing concentration of analyte) allow the experiment to be carried out more quickly using flow-injection analysis or in the context of LC as demonstrated with chiral detection [81].

3. Selective Ionization

The concept of selective ionization of the compound of interest has obvious analytical advantages [82, 83]. It can be achieved to some extent in CI by choosing a reagent gas that has a proton affinity just slightly lower than that of the analyte. Although all compounds having proton affinities greater than that of the analyte will also be ionized, at least those with proton affinities less than that of the reagent gas will not be ionized and, therefore, will not interfere with the analysis.

The utility of selective ionization in CI has been studied for several classes of compounds using various reagent gases [27, 82]. Ammonia is a reagent gas of relatively high proton affinity (Table 7-1), but nitrogen-containing compounds are sufficiently good Brønsted bases (weak Brønsted acids) to accept proton transfer from NH_4^+ [3, 84]. This means that by using a reagent gas such as ammonia [47, 85, 86] with a relatively high proton affinity, CI conditions can be established in which nitrogenous compounds or other analytes with unusually high proton affinities will be selectively ionized. Ordinary molecules containing only C, H, and O will usually not be ionized when ammonia is the CI gas, but nitrogenous compounds will be [3, 84]. Positive-ion chemical ionization methodology was used in the detection and quantification of amphetamine, opiates, and cocaine and metabolites in human postmortem brain relative to stable isotope-labeled internal standards [87].

4. Fragmentation

A characteristic feature of most CI mass spectra is that they contain fewer fragment ion peaks than the corresponding EI mass spectra. The extent of this difference is proportional to the excess energy transferred to the analyte molecule during the process of protonation. The magnitude of this excess energy is equal to the difference in the proton affinity of the analyte and that of the reagent gas. The greater this difference in proton affinities, the greater the excess energy available to the nascent protonated molecule of the analyte, and therefore the more extensive the subsequent fragmentation of the protonated analyte. The "rules" for fragmentation in CI [88, 89] are different from those in EI, as the principal ions are even-electron species.

The relationship between fragmentation of the RH$^+$ and the type of reagent gas is illustrated in Figure 7-3, which shows the CI mass spectra of 5α-dihydrotestosterone obtained using three different reagent gases. The spectrum in the top panel was obtained using isobutane. Assuming that the proton affinity of 5α-dihydrotestosterone is only 837 kJ mol^{-1}, then excess energy of 21 kJ mol^{-1} will be available to the protonated steroid from reaction with the *t*-butyl reagent ion. This small amount of energy causes only a minor dehydration of the protonated molecule, as indicated by the peak at *m/z* 273. The spectrum in the middle panel was obtained using methane as the reagent gas. Referring to Table 7-1, methane has an average proton affinity of 600 kJ mol^{-1}; this results in an excess energy on the order of 230 kJ mol^{-1}, which causes a greater degree of dehydration of MH$^+$ as indicated by the increase in intensity of the peak at *m/z* 273 relative to that of the protonated-molecule peak at *m/z* 291. Finally, the bottom panel in Figure 7-3 shows the CI spectrum of 5α-dihydrotestosterone obtained with hydrogen as the reagent gas. In this case, the excess energy available to the protonated molecule of the steroid is 440 kJ mol^{-1}. The effect of this excess energy is reflected by the appearance of several fragment peaks in the spectrum in addition to the further-enhanced peak intensity at *m/z* 273. A CI spectrum obtained with a mixture of helium and hydrogen will sometimes show as much fragmentation as a spectrum obtained under EI conditions, with the addition of a MH$^+$ peak. Of the reagent gases commonly used, the *t*-butyl ion from isobutane produces the least fragmentation, and protonated hydrogen

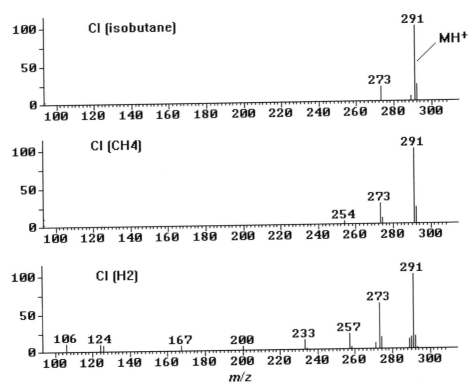

Figure 7-3. *Influence of the type of reagent gas on energy transfer to the nascent protonated molecule of 5α–hydroxytestosterone as illustrated by the extent of fragmentation: (top) isobutane, (middle) methane, and (bottom) hydrogen.*

produces the most extensive fragmentation, with the reagent ions from methane being of intermediate potency in this regard.

It has been a general experience in CI to observe fragmentation involving elimination of a functional group **X** together with a hydrogen atom from the protonated molecule [90]. It is generally assumed that this mode of fragmentation is initiated by transfer of a proton from the reagent ion to a specific site, usually a heteroatom, on the sample molecule. This process is summarized in Scheme 7-3 for the analyte molecule (AX) where X is a heteroatom or functional group and RH^+ is the reagent ion:

$$AX + RH^+ \longrightarrow AXH^+ + R \longrightarrow A^+ + HX$$

Scheme 7-3

The extent to which the protonated molecule can eliminate a functional group together with a hydrogen atom is illustrated in the CI mass spectra of free androsterone and the trimethylsilyl ether of androsterone in Figure 7-4. CI of silyl derivatives has been studied in detail [91]. Rearrangement reactions observed in CI beyond functional group eliminations have been reviewed [92].

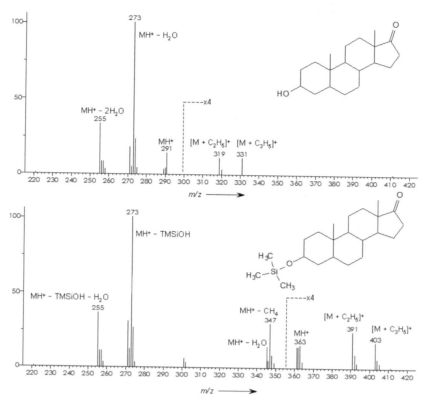

Figure 7-4. **Methane CI mass spectrum of (top) free androsterone and (bottom) TMS ether.** *Courtesy of Her GR, Graduate Assistant, Department of Chemistry, Michigan State University, East Lansing, MI.*

Hydride abstraction from the sample molecule, generating an $[M - 1]^+$ ion, is seen most often in CI of hydrocarbons [93]. Although there is some suggestion that hydride abstraction may be the net effect of the loss of H_2 from MH^+, there is also some evidence to suggest direct abstraction of H^- from the molecule [2]. Hydride abstraction is also observed in CI of alcohols; hydride abstraction is site-specific for the hydroxyl-bearing carbon for short-chain primary and secondary alcohols, but is less specific for long-chain primary alcohols [2, 20].

There are many factors that influence the stability of the MH^+ ion beyond the enthalpy of protonation. An increase in temperature of the ion source will increase the extent of fragmentation observed; this effect is more pronounced with isobutane than with methane [2]. Functional-group interaction sometimes permits intramolecular proton sharing [94], and the steric configuration of the protonated molecule sometimes plays an important, but rarely predictable, role in the route and extent of fragmentation [2, 95].

V. Negative-Ion Formation under CI

1. True Negative Chemical Ionization

First there are a few words on nomenclature. Just as the words positive chemical ionization connote the ion/molecule reaction illustrated in Scheme 7-1, the words negative chemical ionization (NCI) connote ion/molecule reactions similar to those illustrated in Scheme 7-4. Namely, NCI is meant to refer to an ion/molecule reaction in which the analyte molecule (M) interacts with an anion. The product of such an ion/molecule reaction carries a negative charge because of forming an adduct with the anion or by losing a proton as illustrated in Scheme 7-4.

$$M \ + \ Cl^- \longrightarrow [M + Cl]^-$$

$$M \ + \ OH^- \longrightarrow [M - H]^- \ + \ H_2O$$

Scheme 7-4

.The abbreviation NCI is frequently misused in the literature by using it to refer to electron capture negative ionization (ECNI), a technique described below that often employs a gas at high pressure under conditions similar to those used for CI to produce thermal electrons as a by-product during ionization of some of the gas molecules. The words "employs a gas at high pressure under conditions similar to those used for CI" in the preceding statement are probably the basis for the confusion that results in the incorrect use of the term NCI to describe ECNI. Whereas both NCI and ECNI share operational conditions, the reactions involved in the actual ionization process are quite different; therefore care should always be taken to distinguish the difference. NCI (negative-ion CI) involves an ion/molecule reaction either to add an anion to a gas-phase analyte molecule or to abstract a proton from the analyte molecule as shown in Scheme 7-4. ECNI does not involve an ion/molecule reaction; see below for more details on ECNI.

Operational conditions required for CI also provided an excellent environment for producing negative ions that could be detected by mass spectrometry. Previously, negative-ion mass spectrometry [96] had not enjoyed the success of positive-ion mass spectrometry, mainly because the energies of the electrons available in conventional EI sources were so high that the analyte molecule could not capture the electron, meaning that analytically useful negative ions were not likely to be formed [97]. As illustrated in Scheme 7-5, the process of negative-ion formation involving direct interaction of an analyte molecule with an electron can follow any of several pathways, depending on the energy of the available electrons. One of the most desirable pathways from the analytical viewpoint involves resonance electron capture (see the first reaction in Scheme 7-5) because this produces a negative molecular ion ($M^{-\bullet}$). This process of forming a $M^{-\bullet}$ requires a thermal electron; i.e., one having an energy of 0.1 eV or less. In a conventional low-pressure EI source normally operated at 70 eV, there are very few thermal electrons. Thus, in a world committed to the use of conventional EI mass spectrometers prior to the discovery of CI, applications based on negative-ion detection under EI conditions were scarce because the sensitivity was so poor.

Resonance Electron Capture

$$AB \; + \; e^- \; (\sim 0.1 \text{ eV}) \; \longrightarrow \; AB^{-\bullet}$$

Dissociative Electron Capture

$$AB \; + \; e^- \; (0\text{--}15 \text{ eV}) \; \longrightarrow \; A^\bullet \; + \; B^-$$

Ion Pair Formation

$$AB \; + \; e^- \; (>10 \text{ eV}) \; \longrightarrow \; A^- \; + \; B^+ \; + \; e^-$$

Scheme 7-5

Applications of true NCI that involve the generation of OH^- have been reviewed by Harrison [98]. NCI and fragmentation pathways for fluoroaromatic compounds have been reported [99]. Negative-ion formation due to proton abstraction $[M - H]^-$ have been used in the NCI of hydrocarbons [100]. Using ammonia to produce NH_2^- [101], which serves as a Brønsted base, is effective in analyzing samples by NCI for triacylglycerols [102]. Bowie [97] has reviewed the formation and fragmentation of negative ions derived from organic compounds. Bursey [103] has reviewed charge inversion of negative ions during MS/MS. NCI using O^- [104–106] or OH^- [107–109] and other anions have been reviewed [35]. Studies of molecular anions with O_2 have also been described [106, 110].

2. Resonant Electron Capture Negative Ionization

The prospect of high sensitivity from electron capture negative ionization (ECNI) is a major reason for the growth in applications of this technique; the key process for ECNI is illustrated by the first reaction listed in Scheme 7-5. The sensitivity in ECNI for compounds that have high electron affinity is greater than that in EI or in positive CI for reasons of kinetics. The rate constants for electron capture may be as high as 10^{-7} cm^3 molecule^{-1} sec^{-1} compared to maximal rate constants for particle-transfer (e.g., a proton) ion/molecule reactions of $3 \times 10^{-9} \text{ cm}^3$ molecule^{-1} sec^{-1} [2, 111]. The low mass and high mobility of the electron are responsible for an enhancement factor of nearly 100 in the sensitivity of ENCI compared to that of positive CI for a suitably electrophilic compound.

The most common source of thermal electrons needed for ECNI is that consisting of the secondary electrons produced in a high-pressure EI process otherwise used as the source of reagent ions for classical chemical ionization. The three mechanisms in Scheme 7-5 do not represent CI by definition; however, they do represent some of the pathways of forming negative ions in a high-pressure ion source. The high number of collisions in the high-pressure CI source helps produce a large flux of low-energy electrons (by collisional cooling) including thermal electrons [2]. This means that the pathways of dissociative electron capture and resonance electron capture are important in a CI source operated at approximately 100 Pa of reagent gas.

Alternatively, production of thermal electrons can be achieved with an electron monochromator [112, 113]. The electron monochromator offers the advantage of exact control of the energy of the electrons, but it requires an additional piece of equipment for the standard mass spectrometer. Another advantage of the electron monochromator is its capacity to provide a beam of electrons of continuously variable kinetic energy to bombard molecules of the analyte [114]; in this way, an energy spectrum of ECNI can be

Table 7-2.   Examples of successful ECNI methods.

Compounds	Derivative Compound	Amount Injected (fg)	S/B
Dopamine	(CH₃)₃SiO—[benzene ring]—CH₂CH₂—N=C(H)—[benzene ring]—F₅ ; (CH₃)₃SiO—	25	4
TH-cannabinol	[cannabinol ring system]—O—C(=O)—[benzene ring]—F₅ ; C₅H₁₁	10	1
Amphetamine	[benzene ring]—C(=O)—N(H)—C(H)(CH₃)—C(H)(H)—[benzene ring]; F₅	10	4
Amphetamine	[phthalimide ring system, F₄]—N—C(H)(CH₃)—C(H)(H)—[benzene ring]	10	12

From Hunt DF and Crow FW "ECNI Mass Spectrometry" Anal. Chem. 1978, 50, 1781–1784, with permission.

obtained to characterize the analyte, as was demonstrated in the analysis of isomeric tetrachlorodibenzo-*p*-dioxins, which show distinct differences in the electron energies needed to produce the corresponding molecular and fragment anions [115]. Related work in dissociative electron attachment has been reported for families of monosubstituted bromo- and chloroalkanes [116]. An electron monochromator system was also used to detect and identify three dinitroaniline pesticides, flumetralin, pendimethalin, and trifluralin, in both mainstream and sidestream tobacco smoke [117]. ECNI can be used to detect triclosan as its pentafluorobenzoyl ester in human plasma and milk, with an improvement of detection limit to 0.018 ng g^{-1} [118].

During early development work in this field, Dougherty *et al.* [119] and later Hunt and colleagues [120, 121] discovered that the sensitivity of negative-ion detection under CI conditions can be extraordinary. Some representative detection limits are reported in Table 7-2; a comparison of the S/B ratio obtained for selected drugs by ECNI, positive-ion CI (PCI), and EI is presented in Table 7-3.

Ideally, for applications to quantitative analysis, the negative molecular ion will be stable or will undergo only minor decomposition to a high-mass fragment ion; this will generate a mass spectrum with peaks at high mass, a feature that helps to uniquely

Table 7-3. *Comparison of S/B achieved with different ionization methods.*

Trade Name of Benzodiazepine	ECNI	PCI	EI
Nordiazepam-TMS	115	25	22
Termazepam-TMS	2686	25	53
α-HO-alphazolam-TMS	2048	25	4
α-HO-triazolzm-TMS	2314	1	2

From Fitzgerald RL et al. Detection limits for selected drugs by ECNIMS J. Anal. Toxicol. 1993, 17, 341–347, with permission.

characterize a given compound. Unfortunately, some negative molecular ions decompose to low-mass fragment ions that are representative of only the electrophilic moiety on the original molecule [97]. For example, in work represented in Figure 7-5, ECNI was used to assay biological samples for testosterone as the pentafluorobenzyloxime-TMS derivative [122]. The ion current at *m/z* 181 can be monitored by selected ion monitoring to achieve high sensitivity; however, because *m/z* 181 could represent the pentafluorobenzyl moiety from any compound, the selectivity of the assay is put in jeopardy.

It is preferable to use a derivative that under ECNI conditions would produce a stable molecular ion or fragment ion consisting of a major portion of the molecule of interest. Although this ideal is rarely achieved for compounds of interest, some notable successes have been reported in assays for pesticides [123], dioxins [124, 125], chlordanes [126], steroids [127–129], long-chain alcohols [130], LSD [131] and other drugs [132], leukotrienes [133], and other bioactive materials [134–136] as well as for

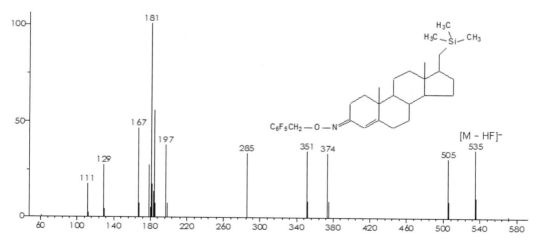

Figure 7-5. *ECNI spectrum of the pentafluorobenzyloxime-trimethylsilyl ether derivative of testosterone.* From Gaskell SJ "Analysis of Steroids" in Methods of Biochemical Analysis, Vol. 29; Glick D, Ed., Wiley-Interscience, New York, 1983, pp 385–432, with permission.

trichothecenes [137], PAHs [138], glucuronides [139], triazines [140], and chlorine-containing compounds [141]. The ECNI GC/MS approach has been used to determine levels of toxaphene in environmental samples [142]. Beyond the quantitative application cited above, the oxygen-induced dechlorination of PCBs and DDT [143] and oxidation of dibenzothiophene [106, 144] by oxygen have been reported. Also, ECNI has been used to analyze fatty acids in an effort to recognize the location of double bonds and hydroxy groups [145, 146].

Most applications of ECNI involve the preparation of an electrophilic derivative that also gives all other derivatizable materials in the sample the opportunity for enhanced detectability; this means that some of the chemical selectivity for such an assay is lost. A novel approach to sample processing capitalizes on the latent electrophilic nature of the compound of interest, when possible [147]. An example of this strategy is represented in Scheme 7-6. It has been observed that unsaturated-ketogenic steroidal nuclei produce a high response to ECNI [148]; the high response is apparently due to the nonclassical conjugation possible in these steroidal systems [149]. In this way, if suitably unsaturated, corticosteroids can be oxidized to a ketogenic steroidal nucleus (see Scheme

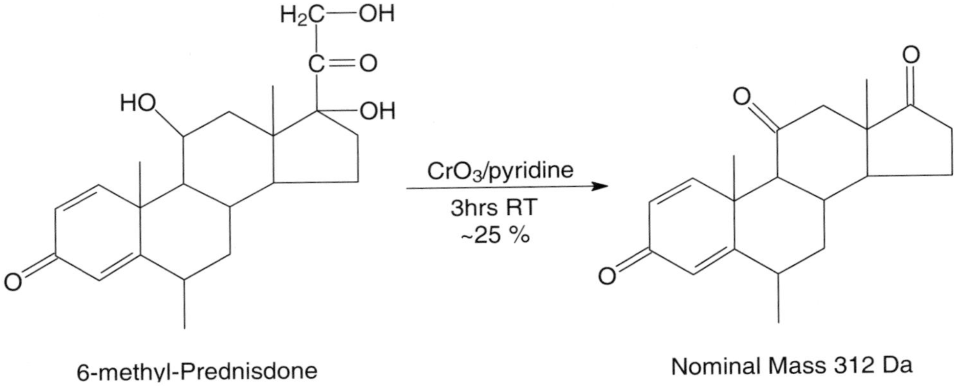

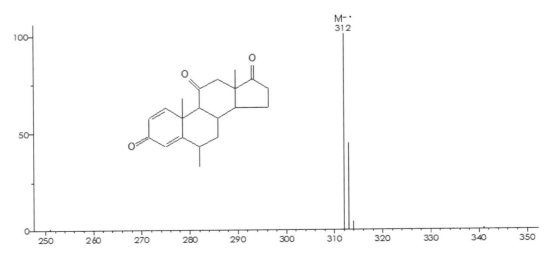

6-methyl-Prednisdone   Nominal Mass 312 Da

Scheme 7-6

Figure 7-6. **ECNI mass spectrum of 6-methylprednisolone.** *Courtesy of Her GR, Graduate Assistant, Department of Chemistry, Michigan State University, East Lansing, MI.*

7-6). They produce simple ECNI mass spectra containing only a few intense peaks as shown in Figure 7-6. Considerable analytical advantage can be gained in selectively detecting the oxidized product of synthetic steroids, because most other material (including most endogenous steroids) in the biological sample does not undergo such a transformation, at least not to a product that can be so easily detected [150].

The operational aspects of ECNI mass spectrometry have been reviewed [151] and an interlaboratory comparison of detection limits has been published [152].

VI. Data Interpretation and Systematic Studies of CI

The first question to be answered might be whether the peak in the nominal-mass region of the mass spectrum represents MH^+ or $[M - H]^+$. If the sample is known to be a hydrocarbon and the reagent gas is methane, then an ion in the molecular-mass region is probably $[M - H]^+$; the $[M - H]^+$ ion predominates in the methane CI spectra of *n*-alkanes [153]. Non-hydrocarbons, such as esters [2], may also produce an $[M - H]^+$ ion in methane, but it will usually be accompanied by an MH^+ ion, as in Figure 7-7. Methane as a reagent gas often produces adduct ions of $[M + C_2H_5]^+$ and $[M + C_3H_5]^+$ that can be used as a parity check on a peak for the MH^+ ion. For example, in Figure 7-7 the peaks at *m/z* 306 and *m/z* 318 fall in the series $[M + C_2H_5]^+$ and $[M + C_3H_5]^+$ as "satellite" peaks to the peak for the MH^+ at *m/z* 278. If the spectrum in Figure 7-7 had been of an unknown compound, the series of ions would have been realized only after recognizing that the separation between peaks at *m/z* 278 and *m/z* 306 was 28 *m/z* units and that from *m/z* 278 to *m/z* 318 was 40 *m/z* units, which would suggest the sequence of peaks for $[M + 1]^+$, $M + 29]^+$, $[M + 41]^+$. The peak at *m/z* 276 in Figure 7-7 represents the $[M - H]^+$ ion.

If assignment of the molecular-mass-related ions in the methane CI spectrum is ambiguous, then a reagent gas of lower strength as a Brønsted acid (higher proton affinity) should be used; e.g., isobutane or possibly ammonia (see Example 1-4 in Chapter 1). The peak at the high-mass end of the spectrum is much more likely to represent MH^+ when a reagent gas of relatively high proton affinity is used; a comparison of the effects of methane and isobutane as reagent gases can be seen in Figure 7-3.

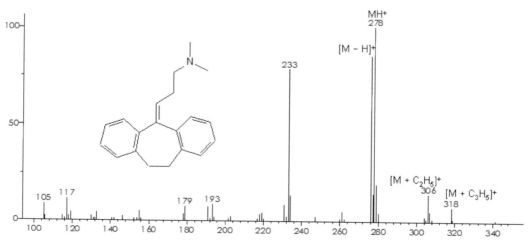

Figure 7-7. *Methane CI mass spectrum of amitriptyline.* From Finks BS, Foltz RL, Taylor DM et al. "A Comprehensive GC/MS reference Data System for Toxicological and Biomedical Purposes" J. Chromatogr. Sci. 1974, 12, 304-328, with permission.

Systematic studies of CI MS for characterization PAHs [154], sugars [45, 155, 156], peptides [157, 158] including stereochemical differentiation [159], steroids [160], carbamate pesticides [161], PCBs [162], additives to polypropylene [163], and many other classes of compounds have been reviewed by Harrison [2].

VII. Ionization by Charge Exchange

Charge-exchange mass spectrometry [2, 164–166], which sometimes produces mass spectra similar to those from EI, can be conducted in any instrument that is equipped for high-pressure ionization such as CI. The principal difference between CI and charge exchange is that aprotic gases are used in the latter technique.

1. Mechanism of Ionization

In the charge-exchange reaction, molecular ions of the sample are generated by transfer or exchange of an electron from the analyte molecule M to the reagent-gas ions $(G^{+\bullet})$ as shown in Scheme 7-7. As in CI, the charge-exchange reagent ions are produced by EI of the reagent gas at pressures of 0.5–1 torr with an electron energy up to 400–500 eV [164]. For some aprotic reagent gases (e.g., Ar, N_2, Kr, CO_2, and Xe), two or more major ions are generated that are capable of charge-exchange reactions.

$$M \; + \; G^{+\bullet} \longrightarrow M^{+\bullet} \; + \; G$$

Scheme 7-7

Charge-exchange ionization can be achieved between the analyte molecule and reagent-gas ion if the recombination energy of the reagent-gas ion (with an electron) is greater than the ionization potential of the analyte molecule. Because charge exchange is a resonance process, the recombination energy of the reagent-gas ion strictly defines the amount of energy transferred to the product ion during charge exchange. Therefore, following charge exchange with high-recombination-energy species such as He^+ or Ne^+, extensive fragmentation of complex product ions is likely unless frequent collisions in the high-pressure source can quench long-lived excited states of these ions.

Most reagent gases produce ions with more than one recombination energy. Furthermore, molecular ions such as $N_2^{+\bullet}$ or $CO^{+\bullet}$ have several electronic and vibrational excited states that are available to the charge-transfer process. Consequently, the amount of energy available for transfer during charge exchange is more diffuse with molecular ions (e.g., $N_2^{+\bullet}$) than with atomic ions (e.g., $Ne^{+\bullet}$). This fact may explain the observation that $N_2^{+\bullet}$ charge-exchange mass spectra resemble those from EI.

The resonance rule stipulates that charge exchange will occur only if the product ion can occupy an energy state that will accommodate a quantum of energy equal to the recombination energy of the reagent ion. For complex organic molecules, the number of energy levels available is sufficiently large that resonance is usually possible and ionization by charge exchange is observed. For small molecules, especially diatomic systems, and definitely for atomic systems, ionization by charge exchange is less likely because there are fewer energy levels available to accommodate the quantized recombination energy according to the resonance rule [165, 166].

2. Fragmentation and Appearance of Mass Spectra

Einolf and Munson [164] have studied the basic charge-exchange reactions of several aprotic gases: He, Ne, Ar, Kr, Xe, N_2, CO, CO_2, CF_4, and NO. The mass spectra of most compounds produced by reaction with ions from He, Ne, or Ar result principally from dissociative charge exchange. The molecular ions that appear in these spectra appear to be produced by direct EI, not by charge-exchange reactions. The mass spectra obtained by O_2 charge exchange contain both molecular ions and fragment ions, but the high pressure of oxygen in the ion source greatly reduces the lifetime of the filament. Nitric oxide (NO) or CH_4 will give simple mass spectra that are usually dominated by $M^{+\bullet}$ and $[M + NO]^+$ or $[M + CH_3]^+$ ions [164]. Charge exchange offers some advantage in the detection of polyaromatic hydrocarbons [167] and in the study of substituent effects [168].

VIII. Atmospheric Pressure Chemical Ionization

Ionization in this case is accomplished in an ionization chamber at atmospheric pressure, as compared with a pressure of 100 Pa for CI or 10^{-2} Pa for EI. In the atmospheric-pressure chemical ionization (APCI) source, ionization is initiated by low-energy electrons from a radioactive beta source or a corona discharge [169]. (Because the ion source operates at atmospheric pressure, it is not possible to use a heated filament to produce electrons as the primary source of ionization.) The low-energy electrons ionize a reagent gas (e.g., N_2, O_2, H_2O, etc.) that, through a complex series of ion/molecule reactions, efficiently produces positive and negative ions of the analyte. The basic processes of ionization by APCI have been reviewed [2, 170–174].

During operation, the APCI source is purged with N_2 or another suitable carrier gas. The high ionization efficiency (and thus high sensitivity of the technique) is due to the short mean free path at 760 torr, which translates into an increased number of collisions between the sample molecules and reagent ions. As H_2O has the highest proton affinity (PA) of those gases normally present at ambient conditions, it captures any free protons, and upon collisions with other molecules of H_2O forms a series of small clusters $H_3O^+(H_2O)_n$, where n = 0–4 [175]. Once a molecule more basic (higher PA) than H_2O enters the APCI source, the proton is transferred to it, and subsequent collisions form clusters of the newly protonated species.

The high pressure or high collision rate that gives rise to efficient ionization interferes with efficient transfer of the ionized species to the *m/z* analyzer. The challenge of sampling the APCI source is usually met by using a skimmer interface similar to that described for electrospray (Chapter 8) to maintain the vacuum in the mass spectrometer. Heat from or collisions with the drift tube surface provides a means of breaking up the protonated clusters described above.

McEwen and coworkers have modified a commercially available LC/MS instrument to accommodate both APCI LC/MS and GC/MS processes [176]. They cite advantages of this additional capability vs those of LC/MS alone to include higher chromatographic resolution in the GC vs LC mode, greater peak capacity for complex mixture analysis, better detection limits for a variety of volatile compounds, and the capability for observing compounds of low polarity that are not readily observed in LC/MS. Advantages of the combined system over those of conventional GC/MS include the capacity for higher carrier gas flow and shorter columns for passing less-volatile materials through the gas chromatograph, selective ionization, and rapid switching between positive- and negative-ion modes in combination with cone voltage-induced fragmentation together with MS^n. Disadvantages of the combined system include the fact that APCI

GC/MS provides no or limited ionization of saturated hydrocarbons and other highly nonpolar compounds; therefore, there are fewer EI-like fragmentation patterns for computer-aided library searching. On the other hand, the lack of a response from nonpolar compounds can be advantageous for certain analyses, as is the simplified mass spectrum and easy determination of molecular weight that results from less fragmentation in the AP ionization mode. Simple modification of a commercially available ESI or APCI instrument to provide for a hot stream of nitrogen gas to vaporize the analytes sufficiently for transport into a corona discharge for ionization provides a system comparable to the DESI and/or DART systems for rapid analysis of solid or liquid samples [177].

APCI has been used for low-level detection of sugars [178] by chloride attachment in the negative-ion mode, and for dimethylenedihydrothiophene [179], alkanes [180], metalloporphyrins [181], and pollutants [182, 183], and for structure elucidation of triacylglycerols [184] and detection of chemical warfare agents [183] in the positive-ion mode. Reviews by Harrison [2] and Bruins [170] provide guidance to applications of APCI. One of the attractive features of the APCI source is its simplicity of design. A microchip system incorporating a capillary channel for a nebulizer gas inlet fabricated on the silicon wafer has been described as the APCI interface between a capillary column gas chromatograph and a mass spectrometer [185]. The high sensitivity of the APCI source places high demands on solvent and sample purity. The APCI source, like the ESI source, can accommodate capillary electrophoresis [186]; more on capillary electrophoresis/mass spectrometry can be found in Chapter 4.

IX. Desorption Chemical Ionization

The foregoing description of CI has focused on those compounds for which the analyte molecules were clearly in the gas phase prior to interaction with reagent ions. To overcome restrictions of volatility, nonvolatile samples have been successfully analyzed from various surfaces that are exposed directly to the reagent-ion plasma. This technique has been variously called direct chemical ionization, in-beam ionization, desorption chemical ionization (DCI), and surface ionization [187]. Whether the analyte is ionized in the condensed phase directly on the surface by the CI plasma or whether the overall procedure of DCI merely utilizes more efficiently what little analyte vapor is available has never been convincingly established.

Today, DCI is carried out using direct-exposure probes that are constructed with a filament that can be rapidly heated. The physical shape, material of construction, and treatment of these filaments all contribute to the results obtained. The procedure involves dissolving the sample in an appropriate volatile solvent, distributing a drop of the solution evenly over the filament, and evaporating the solvent (Figure 7-8). The DCI sample probe is passed through a vacuum lock into the ion source of the mass spectrometer so that the heatable filament enters the ionization chamber. Once a CI plasma has been established in the ionization chamber, the filament is rapidly heated by electrical current to promote the DCI process. An analogous technique using the direct-exposure probe is called direct electron ionization (DEI), which results when only a beam of electrons is used for ionization.

Figure 7-8. Desorption chemical ionization (DCI) filament used in the Nermag R10-10.

The exact mechanism of DCI has never been clearly explained; there is evidence that indicates that the intact analyte is desorbed from the surface of the filament and that ionization takes place in the gas phase [188]. This evidence supports an approach that uses rapid heating. However, the difference in ion current obtained with different reagent gases indicates that the ionization takes place on the surface of the filament, and the ions are then desorbed into the gas phase (Devant G, Private Communication from Nermag S.A., Rueil Malmaison, France, 1979). DCI is especially applicable to use with transmission quadrupole mass spectrometers. It is unclear whether it was a combination of filament physical design and manufacturing material or factors pertaining to ion-source potentials, but results obtained using transmission quadrupole instruments appear to be significantly better than those obtained with sector-based instruments.

DCI can produce mass spectra from underivatized sugars and other organic compounds that are thermally labile. DCI has produced spectra for many compounds that had been unsuccessfully analyzed by field desorption. One of the more significant examples was the mass spectrum of Brevetoxin B (Figure 7-9), the toxic substance associated with Red Tide, as obtained by Koji Nakanishi at Columbia University in the early-1980s [189]. The mass spectrum in Figure 7-9 shows a peak at m/z 895 for the protonated molecule and a peak at m/z 912 for the ammonium adduct of Brevetoxin B. When NH_3 is used as the reagent gas for the analysis of hydroxalated compounds, the adduct ion is readily formed. In some cases, these NH_4^+ adduct ions will lose water (H_2O, which also has a nominal mass of 18 Da like the ammonium ion), leaving a peak that could be characterized as a *pseudomolecular ion peak*. That this mass spectrum of Brevitoxin B was acquired during the installation of a Nermag R10-10 transmission quadrupole mass spectrometer is a testament to the simplicity of the DCI technique.

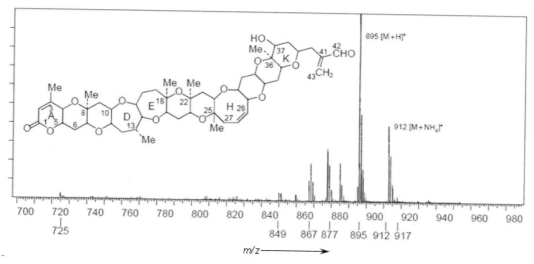

Figure 7-9. *DCI mass spectrum of Brevitoxin B obtained by Koji Nakanishi at Columbia University in the mid-1980s on a Nermag R10-10 mass spectrometer.* Courtesy of Koji Nakanishi.

DCI originated and matured during the same timeframe as FAB. Both techniques were suitable for the analysis for compounds that were thermally labile and/or nonvolatile. DCI has been a very good technique for use with transmission quadrupole mass spectrometers, which are more limited in their maximum *m/z* detectability than double-focusing instruments. FAB was the preferred technique for proteins and peptides until the development of MALDI and electrospray, which have now displaced FAB, whereas DCI continues to work nicely with 1 to 5 saccharide sugars, coordination complexes, and other such nonvolatile thermally labile analytes.

Useful mass spectra have been obtained by DCI from peptides [190] and carbohydrates [191–193] deposited on surfaces. The type and composition of the surface [192] is particularly important in avoiding sample decomposition when energy is transferred relatively slowly as in resistive heating of the sample. Abrupt transfer of energy to the sample as with a pulsed laser [157] seems to be most effective. Atmospheric sampling of vapors from explosives into a glow discharge ion source has been reported [194].

X. General Applications

A good example of the use of chemical ionization in GC/MS is the analysis of the pesticide Dursban on the skins of various commodities (fruits and vegetables), where the objective is to obtain as low a detection limit as possible. As can be seen from the EI mass spectrum of Dursban shown in Figure 7-10(a), the ion current is distributed over a

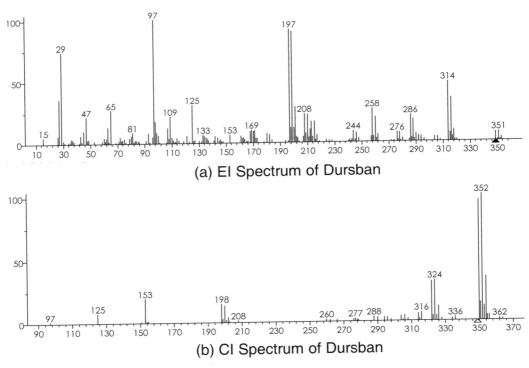

(a) EI Spectrum of Dursban

(b) CI Spectrum of Dursban

Figure 7-10. *EI and methane CI mass spectra of Dursban showing concentration of ion current in ions with high m/z values for CI as compared to EI.* Courtesy of Varian, Inc., Applications Note.

large number of *m/z* values from the peak at *m/z* 349 for the M$^{+\bullet}$ down to peaks at *m/z* 29 (representing an ethyl ion) and the OE ion peak at *m/z* 28 (representing a H$_2$C$^{\bullet}$—$^{+}$CH$_2$ ion). The base peak is at *m/z* 97. The next most intense peak, which represents an ion with three atoms of chlorine, is at *m/z* 197. Most of the ion current in the EI mass spectrum is below *m/z* 300. Many endogenous substances associated with commodities produce extensive ion current in the range below *m/z* 300. Sometimes it is impossible to separate these endogenous substances from the analyte using gas chromatography. Even when using selected ion monitoring for better specificity and sensitivity, the interferences from the natural compounds found in the skins of fruits and vegetables can make a satisfactory detection limit impossible. Using the peaks that represent the M$^{+\bullet}$ could improve the specificity, but doing so would mean that only about 10% of the total ion current would be used for quantitation

---

The labeled peak in the clusters of isotope peaks for M$^{+\bullet}$ and MH^{+} of the Dursban in the respective EI and CI spectra is not the nominal *m/z* peak in either case. This is an important illustration. It should always be kept in mind that the computer system does not necessarily label the important or significant peaks. It usually labels only the most intense peak within a cluster in order to produce an attractive display. Another important point is illustrated by these two spectra which show that the X+2 peak is the most intense peak within the isotope cluster for either the M$^{+\bullet}$ or MH^{+}; this is contrary to the expectation that the X peak should dominate for an ion containing three chlorine atoms. This "deviation" can be explained by the fact that the ion also contains sulfur, which makes a substantial contribution at X+2. These are two important points that show why it is important to perform a careful examination of the mass spectrum before reaching any conclusions.

Examination of the methane CI mass spectrum in Figure 7-10(b) shows that most of the ion current is associated with the cluster starting at *m/z* 350 for the nominal MH^{+}. Using the three chlorine isotope peaks at *m/z* 350, 252, and 354 in an SIM analysis carried out with CI will utilize most of the analyte ion current to optimize the detection limit. The fact that the ion current being monitored is above *m/z* 300 should minimize the likelihood of interference from matrix endogenous substances.

A better method with a decreased limit of detection and greater specificity in the analysis of Dursban would be to use ECNI because of the presence of the three atoms of chlorine in Dursban making it electronegative and ideally suitable for ECNI. The sample matrix components likely to be encountered in commodities do not have a highly electrophilic nature and are not likely to be ionized by ECNI.

When describing a mass spectrum obtained using chemical ionization, it is always a good idea to include the name of the reagent gas in the description; i.e., a CI mass spectrum acquired using isobutane as a reagent gas would be called an "isobutane CI" mass spectrum.

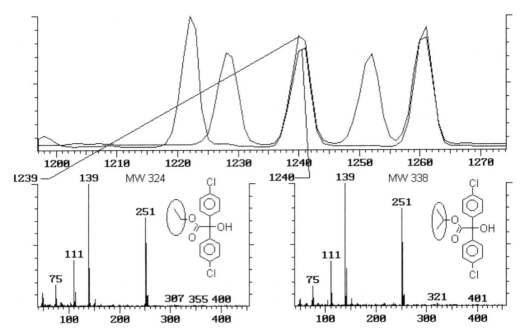

Figure 7-11A. *EI mass spectra of chlorobenzylate (lower left) and chloropropylate (lower right) along with superimposed RTIC chromatograms for samples containing each analyte (top).*

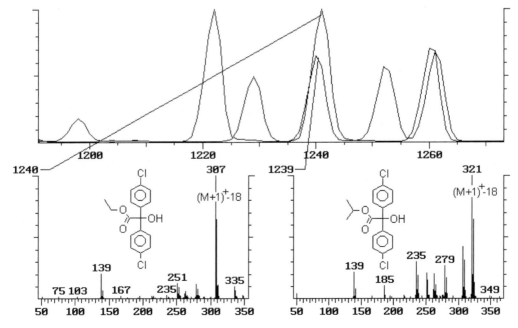

Figure 7-11B. *Same as in Figure 7-11A except that the spectra were acquired with methane CI.* Data courtesy of Varian, Inc., Applications Note.

Another good example of the utility of chemical ionization is illustrated in the analysis of two pesticides (chloropropylate, nominal mass 338 Da and chlorobenzylate, nominal mass 324 Da) that have identical GC retention times and EI mass spectra as shown in Figure 7-11A. These two compounds have the same structure with the exception of a difference in the ester portion of the molecule. The chloropropylate is an isopropyl ester, whereas the chlorobenzylate is an ethyl ester. These two compounds produce very different mass spectra under methane CI. The methane CI mass spectra for the two compounds (Figure 7-11 B) shows a characteristic ion for the chloropropylate (ion represented by a cluster of peaks with nominal m/z 321) that is not present in the methane CI spectrum of chlorobenzylate. There is a peak at m/z 307 in both mass spectra; however, the two could be quantitated in the presence of one another by first determining the amount of chloropropylate present and then determining the amount of chlorobenzylate attributing a portion of the mass chromatographic peak for m/z 307 to the presence of chloropropylate. The two characterizing ions in this example are not the MH^+. In fact, there is no peak representing the MH^+. The two characterizing ions in the methane CI mass spectra are formed by loss of a water molecule from the MH^+ formed by addition of a proton to the hydroxyl moiety of both compounds.

This detection limit for the assay in this example could also have been improved by using ECNI because the $M^{-\bullet}$ (negative charge molecular ions of the two compounds have different m/z values (i.e., m/z 338 for chloropropylate and m/z 324 for chlorobenzylate).

Other applications of CI include studies of chemical reactions in the upper troposphere and lower stratosphere that require monitoring concentrations of HNO_3, HCl, and chlorine nitrate ($ClONO_2$) [195]. A new method for detecting gas-phase hydroperoxides and other weak and strong acids in the atmosphere is based on the clustering action of CF_3O^- [196].

XI. Concluding Remarks

As was pointed out early in this chapter, CI in all instruments except internal ionization QIT m/z analyzers requires a powerful and complicated vacuum system to deal with the amount of gas necessary to maintain the proper plasma of reagent ions. This high ion concentration in the "tight" ion source also results in contamination that can quickly degrade performance. Some CI reagent gases are dirtier than others; e.g., isobutane contaminates the ion source much quicker than either methane or ammonia.

Although the need for complicated vacuum systems greatly adds to the cost of instruments, many analysts insisted on having both EI and CI capabilities when purchasing instruments. Because CI usually requires changes in the ion-source geometry and the use of CI meant that the instrument would have to be cleaned, CI was not readily embraced. The CI accessory often went unused, even when it would have been greatly beneficial in solving a problem involving the identity of an analyte that did not exhibit a molecular ion peak in its mass spectrum.

New vacuum systems such as the split-flow turbomolecular pumps and secure-design changes that allowed for changing the geometry of the ion source without having to break vacuum have led to wider use of the technique. Several manufacturers now offer ion sources that can be used for both EI and CI without any changes to their geometry. Agilent Technologies has published an applications note illustrating how to tune the CI source of the 597X mass spectrometers to allow for acquisition of EI spectra. These combination sources or use of a CI source for EI often result in poorer EI sensitivity than can be obtained with a dedicated EI source. However, the use of a combination source provides for convenience and, in some cases, even more detectable molecular ions due to reactions of analytes with helium ions.

The easiest (and cleanest) CI is performed using the internal ionization QIT *m/z* analyzer. Even one of the negative aspects of these instruments, namely the tendency for ion/molecule reactions to take place between highly acidic fragment ions of the analyte and unionized analyte molecules (so-called "self–CI"), has a useful outcome in that molecular mass information may be provided.

Although there have been no commercial manufacturers of APCI GC/MS instruments since Sciex discontinued the Targa series of instruments, APCI is still probably the most sensitive of all GC/MS ionization methods. There is a company (M&M Mass Spec Consulting, LLC [http://www.asap-ms.com]) that sells hardware to interface a GC to a conventional APCI interface designed for use with LC/MS instruments. This instrumental interface can provide a good way to obtain APCI data using GC/MS.

References

1. Harrison AG, Chemical ionization. *Journal de Chimie Physique et de Physico-Chimie Biologique* **90**(6): 1411-32, 1993.
2. Harrison AG, *Chemical Ionization Mass Spectrometry*. CRC Press, Boca Raton, Fla., 1992.
3. Westmore JB and Alauddin M, M., Ammonia CI-MS. *Mass Spec. Rev* **5**(4): 381-466, 1986.
4. Munson B, CI-MS: 10 years later. *Anal. Chem.* **49**: 772A-778A, 1977.
5. Harrison AG, Chemical ionization mass spectrometry. *Advances in Mass Spectrometry* **11A**: 582-95, 1989.
6. Munson B, Development of chemical ionization mass spectrometry. *International Journal of Mass Spectrometry* **200**(1/3): 243-251, 2000.
7. Munson B and Field FH, Chemical Ionization Mass Spectrometry. *J Amer Chem Soc* **88**: 2621-2630, 1966.
8. Horning EC, Horning MG, Carroll DI, Dzidic I and Stillwell RN, New Picogram Detection System Based on MS with an External Ionization Source at Atmospheric Pressure. *Anal. Chem.* **45**: 936-943, 1973.
9. Horning EC, Carroll DI, Dzidic I, Haegele KD, Horning MG and Stillwell RN, Atmospheric Pressure Ionization (API) Mass Spectrometry; Solvent-Mediated Ionization of Samples Introduced in Solution and a Liquid Chromatograph Effluent Stream. *J. Chromatogr. Sci.* **12**: 725-729, 1974.
10. Horning EC, Carroll DI, Dzidic I, Haegele KD, Horning MG and Stillwell RN, Liquid Chromatography-Mass Spectrometry Computer Analytical Systems; Continuous-Flow Systems Based on API MS. *J. Chromatogr.* **99**: 13-21, 1974.
11. Horning EC, Carroll DI, Dzidic I, Haegele KD, Stillwell RN, Lin SN and Oertli CU, Development and Use of Bioanalytical Systems Based on Mass Spectrometry. *Clin. Chem.* **23**: 13-21, 1977.
12. Carroll DI, Dzidic I, Horning MG, Stillwell RN and Horning EC, Subpicogram Detection System for Gas-Phase Analysis Based upon Atmospheric Pressure Ionization (API) Mass Spectrometry. *Anal. Chem.* **46**: 706-710, 1974.
13. Carroll DI, Dzidic I, Haegele KD, Stillwell RN and Horning EC, API MS: Corona Discharge Ion Source for Use In LC/MS Computer Analytical System. *Anal. Chem.* **47**: 2369-2373, 1975.
14. Cotter RJ, Desorption from Extended Probes. *Anal. Chem.* **52**: 1589A-1599A, 1980.
15. Baldwin MA and McLafferty FW, Direct chemical ionization of relatively involatile samples. Application to underivatized oligopeptides. *Org. Mass Spectrom.* **7**: 1353, 1973.
16. Milne GWA, Fales HM and Axenrod T, Identification of Dangerous Drugs by Isobutane CI MS. *Anal. Chem.* **43**(1815-1820), 1971.
17. Brodbelt J, Louris JN and Cooks RG, CI in an Ion Trap MS. *Anal. Chem* **59**: 1278-1285, 1987.
18. Hunt F, Reagent gases for CI. *Adv. Mass Spectrom.* **6**: 517-522, 1974.
19. Blake RS, Wyche KP, Ellis AM and Monks PS, CI reaction TOFMS: Multi-reagent analysis for determination of trace gas. *Int J Mass Spectrom* **254**: 85-93, 2006.
20. Munson B, Feng TM, Ward HD and Murray RK, Jr., Isobutane chemical ionization mass spectra of unsaturated alcohols. *Organic Mass Spectrometry* **22**(9): 606-9, 1987.
21. Field FH, CI using isobutane. *J. Am. Chem. Soc.* **91**: 6334-6341, 1969.
22. Odiorne TJ, Harvey DJ and Vouros P, CI using Tetramethylsilane. *J. Phys. Chem* **76**: 3217-3220, 1972.
23. Orlando R, Ridge DP and Munson B, Tetramethylsilane CI of Ethers. *Org. Mass Spectrom.* **23**: 527-534, 1988.
24. Lin Y, Ridge DP and Munson B, Association reactions of trimethylsilyl ions. *Organic Mass Spectrometry* **26**(6): 550-8, 1991.
25. Orlando R, Ridge DP and Munson B, Radiative stabilization of trimethylsilyl adduct ions. *Journal of the American Society for Mass Spectrometry* **1**(2): 144-8, 1990.
26. Burrows EP, Dimethyl Ether CI MS. *Mass Spectrom. Rev.* **14**: 107-116, 1995.
27. Allgood C, Lin Y, Ma Y-C and Munson B, Benzene for CI gas. *Org. Mass Spectrom.* **25**: 497-502, 1990.
28. Allgood C, Ma YC and Munson B, Quantitation using benzene in gas chromatography/chemical ionization mass spectrometry. *Analytical Chemistry* **63**(7): 721-5, 1991.
29. Vairamani M, Nitromethane in CI. *Org. Mass Spectrom.* **25**: 271-273, 1990.
30. Buchanan MV, Methanol for CI. *Anal. Chem.* **56**: 546-549, 1984.
31. Barry R and Munson B, CI with Diisopropyl ether. *Anal. Chem.* **59**: 466-471, 1987.
32. Vairamani M, Mirza UA and Srinivas R, Unusual reagents in CI. *Mass Spectrom. Rev.* **9**: 235-258, 1990.
33. Lawrence P and Brenna JT, Acetonitrile Covalent Adduct CI MS for Double Bond Localization in Non-Methylene-Interrupted Polyene Fatty Acid Methyl Esters. *Anal Chem* **78**: 1312-1317, 2006.
34. Lange C, Oxiran (Ethylene Oxide) as a CI Gas. *Org. Mass Spectrom.* **28**: 1285-1296, 1993.
35. Budzikiewicz H, Negative Chemical Ionization (NCI) of Organic Compounds. *Mass Spec. Rev.* **5**: 345-380, 1986.
36. Malosse C and Einhorn J, Nitric Oxide CI. *Anal. Chem.* **62**: 287-293, 1990.
37. Bombick D, Pinkston JD and Allison J, Potassium Ion CI. *Anal. Chem.* **56**: 396-402, 1984.

38. Bombick DD and Allison J, Desorption/Ionization MS of Thermally Labile Compounds Based on Thermoinic Emission. *Anal. Chem.* **59:** 458-466, 1987.
39. Kassel DB and Allison DB, Potassium Ion Ionization of Desorbed Species (K+IDS) Organic Acidemias. *Biomed. Environ. Mass Spectrom.* **17:** 221-228, 1988.
40. Peake DA and Gross ML, Iron(I) for CI. *Anal. Chem.* **57:** 115-120, 1985.
41. Eller K and Schwarz H, Organometallic Chemistry in the Gas Phase. *Chem. Rev.* **91:** 1121-1177, 1991.
42. Chowdhury AK, Cooper JR and Wilkins CL, Indium-Alkene Ions for CI. *Anal. Chem.* **61:** 86-88, 1989.
43. Cody RB, Ammonia CI. *Anal. Chem.* **61:** 2511-2515, 1989.
44. Rudewicz P and Munson B, Effect of Ammonia Partial Pressure in CI-MS. *Anal. Chem.* **58:** 2903-2907, 1986.
45. Rudewicz P and Munson B, CI vs EI for the Determination of Position and Extent of Labeling for 18O- and 13C-Containing Permethylated Alditols by GC-MS. *Anal. Chem.* **58:** 358-361, 1986.
46. Ligon WV, Jr and Girade H, CI with an Isotopically Labeled Reagent Gas. *J. Amer. Soc. Mass Spectrom.* **5:** 596-598, 1994.
47. Dookeran NN and Harrison AG, H-D Exchange in Peptides using ND3. *J. Mass Spectrom.* **30:** 666-674, 1995.
48. Nakata H, Arakawa N and Mizuno R, Signifant Differences in Site of Protonation and Extent of Fragmentations in CI and FAB MS of Simple Bifunctional Compounds. A Mechanistic Implication for Formation of Protonated Molecules. *Org. Mass Spectrom.* **29:** 192-196, 1994.
49. Gal J-F, Maria P-C and Raczynska ED, Thermochemical aspects of proton transfer in the gas phase. *Journal of Mass Spectrometry* **36**(7): 699-716, 2001.
50. Fossey J, Mourgues P, Thissen R and Audier HE, Proton affinity of some radicals of alcohols, ethers and amines. *International Journal of Mass Spectrometry* **227**(3): 373-380, 2003.
51. Meot-Ner M, The proton affinity scale, and effects of ion structure and solvation. *International Journal of Mass Spectrometry* **227**(3): 525-554, 2003.
52. Deakyne CA, Proton affinities and gas-phase basicities: theoretical methods and structural effects. *International Journal of Mass Spectrometry* **227**(3): 601-616, 2003.
53. Bojesen G, The Order of Proton Affinities of the 20 Common L-a-amino acids. *J. Amer. Chem. Soc.* **109:** 5557-5558, 1987.
54. Bliznyuk AA, Schaefer HF and Amster IJ, Proton Affinity of Lys and His by the Bracketing Method. *J. Amer. Chem. Soc.* **115:** 5149-5154, 1993.
55. Traeger JC, The Absolute Proton Affinity of Isobutene. *Rapid Commun. Mass Spectrom* **10:** 119-122, 1996.
56. Bartness JE, Gas Phase Acidity. *Mass Spectrom. Rev* **8:** 297-344, 1989.
57. Bursey MM, Comments to readers = On Acids and Bases. *Mass Spec. Rev* **9:** 503-504, 1990.
58. Gorman GS and Amster IJ, Gas Phase Basicity of Dipeptides. *J. Amer. Chem. Soc.* **115:** 5729-5735, 1993.
59. Bleiholder C, Suhai S and Paizs B, Revising the Proton Affinity Scale of the Naturally Occurring a-Amino Acids. *Journal of the American Society for Mass Spectrometry* **17**(9): 1275-1281, 2006.
60. Gal J-F, Decouzon M, Maria P-C, Gonzalez AI, Mo O, Yanez M, Chaouch SE and Guillemin J-C, Acidity Trends in a,b-Unsaturated Alkanes, Silanes, Germanes, and Stannanes. *Journal of the American Chemical Society* **123**(26): 6353-6359, 2001.
61. Sicilia MdC, Mo O, Yanez M, Guillemin J-C, Gal J-F and Maria P-C, Is allylphosphine a carbon or a phosphorus base in the gas phase? *European Journal of Mass Spectrometry* **9**(4): 245-255, 2003.
62. Lias SG, Liebman JF and Levin RD, Thermochemical Reference Data for Gaseous Ions and Radicals. *J. Chem. Phys. Ref. Data.* **13:** 695-808, 1984.
63. Gorman GS, Speir JP, Turner CA and Amster IJ, Proton Affinities of the 20 Common a-Amino Acids. *J. Amer. Chem. Soc.* **114:** 3986-3988, 1992.
64. Cooks RG and Wong PSH, Kinetic Method of Making Thermochemical Determinations: Advances and Applications. *Accounts of Chemical Research* **31**(7): 379-386, 1998.
65. Speir JP and Amster IJ, Substrate Assisted Laser Desorption. *Anal. Chem.* **64:** 1041-1045, 1992.
66. Cooks RG, Patrick JS, Kotiaho T and McLuckey SA, Thermochemical determinations by the kinetic method. *Mass Spectrometry Reviews* **13**(4): 287-339, 1994.
67. Decouzon M, Gal JF, Herreros M, Maria PC, Murrell J and Todd JFJ, On the use of the kinetic method for the determination of proton affinities by FT-ICR MS. *Rapid Commun Mass Spectrom* **10:** 242-5, 1996.
68. Kebarle P, Ion Thermochemistry and Solvation from Gas Phase Ion Equilibria. *Ann Rev Physical Chem* **28:** 445-476, 1977.
69. Cooks RG and Kruger TL, Intrinsic Basicity Determined Using Metastable Ions. *J. Amer. Chem. Soc.* **99:** 1279-81, 1977.
70. Chen G and Cooks RG, Estimation of ionization energies of polycyclic aromatic hydrocarbons using the kinetic method. *Journal of Mass Spectrometry* **32**(3): 333-335, 1997.

71. Denault JW, Chen G and Cooks RG, Estimation of ionization energies of substituted anilines by the kinetic method. *International Journal of Mass Spectrometry and Ion Processes* **175**(1,2): 205-213, 1998.

72. Chen G and Cooks RG, Estimation of heterolytic bond dissociation energies by the kinetic method. *Journal of Mass Spectrometry* **32**(11): 1258-1261, 1997.

73. Chen G and Cooks RG, Electron affinities of polycyclic aromatic hydrocarbons determined by the kinetic method. *Journal of Mass Spectrometry* **30**(8): 1167-73, 1995.

74. Tao WA, Zhang D, Wang F, Thomas PD and Cooks RG, Kinetic Resolution of D,L-Amino Acids Based on Gas-Phase Dissociation of Copper(II) Complexes. *Analytical Chemistry* **71**(19): 4427-4429, 1999.

75. Tao WA and Cooks RG, Chiral analysis by MS. *Analytical Chemistry* **75**(1): 25A-31A, 2003.

76. Gal JF, Stone M and Lebrilla CB, Chiral recognition of non-natural a-amino acids. *International Journal of Mass Spectrometry* **222**(1-3): 259-267, 2003.

77. Haas MJ and Harrison AG, The fragmentation of proton-bound cluster ions and the gas-phase acidities of alcohols. *International Journal of Mass Spectrometry and Ion Processes* **124**(2): 115-24, 1993.

78. Kabli S, van Beelen ESE, Ingemann S, Henriksen L and Hammerum S, The proton affinities of saturated and unsaturated heterocyclic molecules. *International Journal of Mass Spectrometry* **249/250**: 370-378, 2006.

79. Harrison AG, The gas-phase basicities and proton affinities of amino acids and peptides. *Mass Spectrometry Reviews* **16**(4): 201-217, 1997.

80. Yang SS, Wong P, Ma S and Cooks RG, $SiCl_3+$ and $SiCl+$ Affinities for Pyridines Determined by Using the Kinetic Method with MSn. *J. Am. Soc. Mass Spectrom.* **7**: 198-204, 1996.

81. Lemr K, Ranc V, Frycak P, Bednar P and Sevcik J, Chiral analysis by mass spectrometry using the kinetic method in flow systems. *Journal of Mass Spectrometry* **41**(4): 499-506, 2006.

82. Barry R and Munson B, Selective Reagents in CI-MS. *Anal. Chem.* **59**: 466-471, 1987.

83. Watkins MA, Price JM, Winger BE and Kenttaemaa HI, Ion-Molecule Reactions for Mass Spectrometric Identification of Functional Groups in Protonated Oxygen-Containing Monofunctional Compounds. *Analytical Chemistry* **76**(4): 964-976, 2004.

84. Buchanan MV, Nitrogen-containing Compounds by NH_3-CI. *Anal. Chem.* **54**: 570-574, 1982.

85. Leinonen A and Vainiotalo P, CIMS of Simple 1,3-Dioxolanes and their Sulphur Analogues Recorded with Methane, Isobutane, Ammonia and Acetone as Reagent Gas. *Org. Mass Spectrom.* **29**: 295-302, 1994.

86. Lehtelä P-L and Vainiotalo P, Unexpected Isotope Exchange Reaction of Oxazolidines Under Ammonia Chemical Ionization. *Org. Mass Spectrom.* **29**: 205-206, 1994.

87. Lowe RH, Barnes AJ, Lehrmann E, Freed WJ, Kleinman JE, Hyde TM, Herman MM and Huestis MA, A validated CI GC/MS method for ID and quantification of amphetamine, opiates, cocaine, and metabolites in human postmortem brain. *J Mass Spectrom* **41**: 175-184, 2006.

88. Uggerud E, Reactions of Protonated Molecules in the Gas Phase. *Mass Spectrom. Rev.* **11**: 389-430, 1992.

89. Cairns T, Siegmund EG and Stamp JJ, A Basic Mechanistic Fragmentation Concept in CI. *Org. Mass Spectrom.* **21**: 161-162, 1988.

90. Nakata H, Protonation Susceptibility of Functional Groups in CI. *Org. Mass Spectrom.* **27**(6): 686-688, 1992.

91. Odiorne TJ, Harvey DJ and Vouros P, Reactions of Alkyl Siliconium ions in CI. *J. Org. Chem.* **38**: 4274-4278, 1973.

92. Kingston EE, Shannon JS and Lacey MJ, Rearrangements in CI-MS. *Org. Mass Spectrom.* **18**: 183-192, 1983.

93. Houriet R, Parisod G and Gäumann T, Hydride Abstraction in CI of n-Paraffins. *J. Am. Chem. Soc.* **99**: 3599-3602, 1977.

94. Wilson MS and McCloskey JA, CI of Nucleosides. *J. Am. Chem. Soc* **97**: 3436-3444, 1975.

95. Weisz A, Cojocaru M and Mandelbaum A, Stereochemical Effects in CI. *J. Chem. Soc. Chem. Commun*: 331-332, 1989.

96. Bowie JH, 25 Years of Negative-Ion Studies at Adelaide. *Org. Mass Spectrom* **28**: 1407-1413, 1993.

97. Bowie JH, Fragmentation of Even-Electron Organic Negative Ions. *Mass Spec. Rev.* **9**: 349-379, 1990.

98. Marshall A, Tkaczyk M and Harrison AG, NCI of Carbonyl Compounds. *J. Am. Soc. Mass Spectrom.* **2**: 292-298, 1991.

99. Merrett K, Young AB and Harrison AG, O-.bul. chemical ionization mass spectra of fluoroaromatic compounds. *Organic Mass Spectrometry* **28**(10): 1124-8, 1993.

100. Bosma NL, Young AB and Harrison AG, Negative ion mass spectrometry of some C6H10 isomers. *Canadian Journal of Applied Spectroscopy* **39**(4): 91-6, 1994.

101. Kallio H and Currie G, NH- (from NH3) in NCI-MS. *Lipids* **28**: 207-215, 1993.

102. Laakso P and Kallio H, Optimization of the mass spectrometric analysis of triacylglycerols using negative-ion chemical ionization with ammonia. *Lipids* **31**: 33-42, 1996.

103. Bursey MM, Charge Inversion of Negative Ions in MS/MS. *Mass Spec. Rev.* **9**: 555-574, 1990.

104. Annan IY and Vouros P, Reactions of the Oxide Radical Anion (O-.) with Simple Aromatic Compounds. *J. Amer. Soc. Mass Spectrom.* **5**: 357-376, 1994.
105. Bruns G and Birkholz D, *Toxaphene by GC-MS with Oxygen ECNI MS. Chemosphere* **27**: 1873-1878, 1993.
106. Drabner G and Budzikiewicz H, Analysis of Benzannelated Thiophene Derivatives by Negative-ion Chemical Ionization Using Oxygen as a Reagent Gas. *J. Mass Spectrom.* **30**: 893-899, 1995.
107. Cheung M, Young AB and Harrison AG, O- and OH- Chemical Ionization of Some Fatty Acid Methyl Esters and Triacylglycerols. *J. Amer. Soc. Mass Spectrom.* **5**: 553-557, 1994.
108. Roy TA, Field FH, Lin YY and Smith LL, Hydroxyl ion negative chemical ionization mass spectra of steroids. *Analytical Chemistry* **51**(2): 272-8, 1979.
109. Smit ALC and Field FH, [OH]- negative chemical ionization mass spectrometry. Reactions of [OH]- with methadone, l-a-acetylmethadol and their metabolites. *Biomedical Mass Spectrometry* **5**(10): 572-5, 1978.
110. Knighton WB, Bognar JA and Grimsrud EP, Reactions of Selected Molecular Anions with Oxygen. *J. Mass Spectrom.* **30**: 557-562, 1995.
111. Leyh B and Hautot D, Mechanisms of Single-Electron Capture by the Dichlorocarbene Dication. *J. Am. Soc. Mass Spectrom.* **7**: 266-275, 1996.
112. Laramee JA, Mazurkiewicz p, Berkout V and Dienzer ML, Electron monochromator-mass spectrometer instrument for negative ion analysis of electronegative compounds. *Mass Spectrometry Reviews* **15**(1): 15-42, 1996.
113. Laramée JA and Deinzer ML, A Trochoidal Electron Monochromator for GC-MS. *Anal. Chem.* **66**: 719-724, 1994.
114. Havey CD, Eberhart M, Jones T, Voorhees KJ, Laramee JA, Cody RB and Clougherty DP, Theory and application of dissociative electron capture in molecular identification. *Journal of Physical Chemistry A* **110**(13): 4413-4418, 2006.
115. Laramee JA, Kocher CA and Deinzer ML, Application of a trochoidal electron monochromator/mass spectrometer system to the study of environmental chemicals. *Analytical Chemistry* **64**(20): 2316-22, 1992.
116. Pshenichnyuk SA, Pshenichnyuk IA, Nafikova EP and Asfandiarov NL, Dissociative electron attachment in selected haloalkanes. *Rapid Communications in Mass Spectrometry* **20**(7): 1097-1103, 2006.
117. Dane AJ, Havey CD and Voorhees KJ, The Detection of Nitro Pesticides in Mainstream and Sidestream Cigarette Smoke Using Electron Monochromator-Mass Spectrometry. *Analytical Chemistry* **78**(10): 3227-3233, 2006.
118. Allmyr M, McLachlan MS, Sandborgh-Englund G and Adolfsson-Erici M, Determination of Triclosan as Its Pentafluorobenzoyl Ester in Human Plasma and Milk Using ECNI MS. *Anal Chem* **78**: 6542-6546, 2006.
119. Dougherty RC, Dalton J and Biros FJ, Negative Ionization of Chlorinated Insecticides. *Org. Mass Spectrom.* **6**: 1171-1181, 1972.
120. Hunt DF and Crow FW, ECNI. *Anal. Chem.* **50**: 1781-784, 1978.
121. Hunt DF, Stafford GCJ, Crow FW and Russell JW, Pulsed Positive Negative CI. *Anal. Chem.* **48**: 2098-2105, 1976.
122. Gaskell SJ, *Analysis of Steroids. In: Methods of Biochemical Analysis*, Vol. 29. Wiley Interscience, New York, 1983.
123. Laramée JA, Eichinger PCH, Mazurkiewicz P and Deinzer ML, ECNI Organophosphate Pesticides Using Electron Monochromator. *Anal. Chem.* **67**: 3476-3481, 1995.
124. Laramée JA, Arbogast BC and Deinzer ML, ECNI of 1,2,3,4-Tetrachlorodibenzo-p-Dioxin. *Anal. Chem.* **58**(14): 2907-2912, 1990.
125. Focant J-F, Pirard C, Eppe G and De Pauw E, Recent advances in mass spectrometric measurement of dioxins. *Journal of chromatography. A* **1067**(1-2): 265-75, 2005.
126. Offenberg JH, Naumova YY, Turpin BJ, Eisenreich SJ, Morandi MT, Stock T, Colome SD, Winer AM, Spektor DM, Zhang J and Weisel CP, Chlordanes in the indoor and outdoor air of three U.S. Cities. *Environmental Science and Technology* **38**(10): 2760-2768, 2004.
127. Hofmann U, Holzer S and Meese CO, Synthetic Corticosteriods Using Isotope Dilution GC-MS with ECNI. *J. Chromatogr.* **508**: 349-356, 1990.
128. Murray S and Watson D, Bis-Trifluoromethylbenzoyl Derivatives for Steroids by ECNI with GC-MS. *J. Steroid Biochem.* **25**: 255-264, 1989.
129. Her G-R and Watson JT, ECNI of Chemically Oxidized Corticosteroids. *Biomed. Mass Spectrom.* **13**: 57-63, 1986.
130. Wolf BA, Condrad-Kessel W and Turk J, Long-chain Fatty Alcohol Quantitation by GC-MS with ENCI. *J. Chromatogr.* **25**: 255-265, 1990.
131. Lim HK, Andrenyak D, Francom P, Foltz RL and Jones RT, LSD and N-Demethyl-LSD in Urine by GC-MS with ECNI MS. *Anal. Chem.* **60**: 1420-1425., 1988.
132. Fitzgerald RL, et al., Detection Limits for Selected Drugs by ECNIMS. *J. Anal. Toxicol.* **17**: 342-347, 1993.

133. Balazy M and Murphy RC, Sulfidopeptide Leukotrienes in Biological Fluids by GC-MS. *Anal. Chem.* **58:** 1098-1101, 1986.
134. Lewy AJ and Markey SP, Melatonin by ECNI. *Science* **201:** 741-743, 1978.
135. Morrow JD, Minton TA, Badr KF and Roberts LJ, F2-isoprostane by ECNI-MS. *Biochim. Biophys. Acta,* **1210:** 244-248, 1994.
136. Strife R and Murphy R, Pentafluorobenzyl Esters by ECNI. *J. Chromatogr.* **305:** 3-12, 1984.
137. Roach JAF, Sphon JA, Easterling JA and Colvey EM, ECNI of Trichothecenes via SFC. *Biomed. Environ. Mass Spectrom.* **18:** 64-70, 1989.
138. Sim PG and Elson CM, PAHs by ECNI. *Rapid Commun. Mass Spectrom.* **2:** 137-138, 1988.
139. Brown SY, Garland WA and Fukuda EK, NCI of Glucuronides. *Biomed. Environ. Mass Spectrom.* **19:** 32-36, 1990.
140. Matting MJJ and Huang LQ, Adduct Ion Formation in ECNI MS of 2-(Alkylthio)- and 2-Alkoxy-5-Triazines. *Anal. Chem.* **62:** 602-609, 1990.
141. Kassel DB, Kayganich KA, Watson JT and Allison J, Ion Source Pretreatment with Chlorine-containing Compounds for Enhanced Performance in GC-ECNI-MS. *Anal. Chem.* **60:** 911-917, 1988.
142. Raff JD and Hites RA, Transport of Suspended-Sediment-Bound Toxaphene in the Mississippi River. *Environmental Science and Technology* **38**(10): 2785-2791, 2004.
143. Lepine FL, Milot S and Mamer OA, Regioselectivity of the Oxygen Addition-Induced Dechlorination of PCBs and DDT Metabolites in ECNI. *J. Am. Soc. Mass Spectrom.* **7:** 66-72, 1996.
144. Drabner G and Budzikiewicz H, Oxidation Reactions of Dibenzothiophene Subjected to Negative CI with Oxygen. *J. Am. Soc. Mass Spectrom.* **4:** 949-954, 1993.
145. Voinov VG, Boguslavskiy VM and Elkin YN, ECNI for Determining Double Bond and Hydroxy Group Location in Fatty Acids. *Org. Mass Spectrom.* **29:** 641-646, 1994.
146. Voinov VG, Van den Heuvel H and Claeys M, Resonant electron capture mass spectrometry of free fatty acids: examination of ion structures using deuterium-labeled fatty acids and collisional activation. *J Mass Spectrom* **37**(3): 313-21., 2002.
147. Watson JT and Kayganich K, Novel Sample Preparation for Analysis by ECNI. *Biochem. Soc. Trans.* **17:** 254-257, 1989.
148. Kayganich K, Watson JT, Kilts C and Ritchie J, Plasma Dexamethasone by Chemical Oxidation and ECNI. *Biomed. and Environ. Mass Spectrom.* **19:** 341-347, 1990.
149. Mayer HK, Reusch W and Watson JT, Effect of Structure on the ECNI Response of Steroids. *Org. Mass Spectrom.* **27:** 560-566, 1992.
150. Ritchie JC, Owens MJ, Mayer H, Watson JT, Kilts C and Carroll BJ, Preliminary Studies of 6b-Hydroxydexamethasone and its Importance in the DST. *Biol. Phychiatry* **32:** 825-833, 1992.
151. Ong VS and Hites RA, Why is ECNI-MS not Reproducible. *J. Am. Soc. Mass Spectrom.* **4:** 270-277, 1993.
152. Arbogast B, Budde WL, Deinzer M, Dougherty RC, Eichelberger J, Foltz RL and Grimm CC, Limits of Detection in ECNI. *Org. Mass Spectrom.* **25:** 191-196, 1990.
153. Field FH, CI-MS. *Accounts Chem. Res.* **1:** 42-49, 1968.
154. Van Orden SL, Malcolmson ME and Buckner SW, CI-MS of PAHs. *Anal. Chim. Acta.* **246:** 199-210, 1991.
155. Hedin PA and Phillips VA, CI of Sugars. *J. Agric. Food Chem.* **39:** 1106-1111, 1991.
156. Wang G and Sha Y, Acetone CI-MS of Monosaccharides. *Anal. Chem.* **57:** 2283-2286, 1985.
157. Speir JP, Borman GS, Cornett DS and Amster IJ, Peptides by laser DCI. *Anal. Chem.* **63:** 65-69, 1991.
158. Speir JP, Gorman GS, Cornett DS and Amster IJ, Dissociation of Peptide Ions using Laser Desorption/CI FT-MS. *Anal. Chem.* **63:** 55-60, 1991.
159. Partanen T, Vainiotalo P, Stajer G, Bernard G and Pihlaja K, EI and CI MS in Stereochemical Differentiation of Some 1,3-Amino Acids. *Org. Mass Spectrom.* **29:** 126-132, 1994.
160. Prome D, Prome J-C and Stahl D, 17-Epimeric Hydroxy Steroids by CI. *Org. Mass Spectrom.* **20:** 525-529, 1986.
161. Stamp JJ, Siegmund EG, Cairns T and Chan KK, CI-MS of Carbamate Pesticides. *Anal. Chem.* **58:** 873-881, 1986.
162. Voyksner RD, Bursey JT, Pack TW and Proch RL, Positive CI of Polychlorinated Biphenyls. *Anal. Chem.* **58:** 621-626, 1986.
163. Rudewicz P and Munson P, Additives in Polypropylene by CI-MS. *Anal. Chem.* **58:** 358-361, 1986.
164. Einolf N and Munson B, Charge Exchange-MS. *Int. J. Spectrom. Ion Phys.* **9:** 141-160, 1972.
165. Harrison AG and Cotter RJ, Methods of ionization. *Methods in Enzymology* **193**(Mass Spectrom.): 3-37, 1990.
166. Harrison AG, Charge exchange mass spectrometry. *NATO ASI Series, Series C: Mathematical and Physical Sciences* **118**(Ionic Processes Gas Phase): 23-40, 1984.
167. Simonsick WJ, Jr and Hites RA, PAHs by Charge Exchange CI MS. *Anal. Chem.* **58:** 2114-2121, 1986.

168. Simonsick WJ, Jr and Hites RA, Substituent Effects in Charge Exchange CI-MS. *Anal. Chem.* **58**: 2121-2126, 1986.
169. Bruins AP, MS with Ion Sources Operating at Atmospheric Pressure. *Mass Spec. Rev.* **10**: 53-78, 1991.
170. Bruins AP, API-MS. *Trends Anal. Chem.* **13**: 37-43, 1994.
171. Proctor CJ and Todd JFJ, Atmospheric Pressure Ionization MS. *Org. Mass Spectrom.* **18**: 509-516, 1983.
172. Carroll DI, Dzidic I, Horning EC and Stillwell RN, API Review. *Appl. Spectrosc. Rev.* **17**: 337-352, 1981.
173. Kambara H, API-MS. *Anal. Chem.* **54**: 143-146, 1982.
174. Sunner J, Nicol G and Kebarle P, Factors for Sensitivity in API. *Anal. Chem.* **60**: 1300-1307, 1988.
175. Zook DR and Grimsrud EP, Ion-Molecule Reactions in API. *J. Phys. Chem.* **92**: 6374-6378, 1988.
176. McEwen CN and McKay RG, A Combination AP LC/MS:GC/MS Ion Source. *J Amer Soc Mass Spectrom* **16**: 1730-1738, 2005.
177. McEwen CN, McKay RG and Larsen BS, Analysis of Solids, Liquids, and Biological Tissues Using Solids Probe at Atmospheric Pressure on Commercial LC/MS Instruments. *Anal Chem* **77**: 7826-7831, 2005.
178. Kato Y and Numajiri Y, CI- ion API of sugars. *J. Chromatogr., (in press)* **562**: 81-86, 1991.
179. Shiea J, Wang W-S, Wang C-H, Chen P-S and Chou C-H, Dimethylenedihydrothiophene in CH2Cl2 by Low-Temperature API-MS. *Anal. Chem.* **68**: 1062-1066, 1996.
180. Bell SE, Ewing RG, Eiceman GA and Karpas Z, Atmospheric Pressure CI of Alkanes, Alkenes, and Cyelvalkanes. *J. Amer. Soc. Mass Spectrom.* **5**: 177-185, 1994.
181. Rosell-Mele A and Maxwell JR, Metallo Porphyrins by Gel Permeation Chromatography/API CIMS. *Rapid Commun. Mass Spectrom.* **10**: 209-213., 1996.
182. Ketkar SN, Dulak JG, Fite WL, Buckner JD and Dheandhanoo S, API-MS/MS of low-level Pollutants in Air. *Anal. Chem.* **61**: 260-264, 1989.
183. Cotte-Rodriguez I, Justes DR, Nanita SC, Noll RJ, Mulligan CC, Sanders NL and Cooks RG, Analysis of gaseous toxic industrial compounds and chemical warfare agent simulants by APCI MS. *Analyst (Cambridge, United Kingdom)* **131**(4): 579-589, 2006.
184. Xu Y and Brenna JT, Atmospheric Pressure Covalent Adduct CI Tandem MS for Double Bond Localization in Monoene- and Diene-Containing Triacylglycerols. *Anal Chem* **79**: 2525-2536, 2007.
185. Oestman P, Luosujaervi L, Haapala M, Grigoras K, Ketola RA, Kotiaho T, Franssila S and Kostiainen R, GC-Microchip APCI-MS. *Anal Chem* **78**: 3027-3031, 2006.
186. Takada Y, Sakairi M and Koizumi H, CE-APCI-MS. *Anal. Chem.* **67**: 1474-1476, 1995.
187. Cotter RJ, Desorption from Extended Probes. *Anal. Chem.* **52**: 1589A-1599A, 1980.
188. Beuhler RJ, Flanagan E, Greene LJ and Friedman L, Proton Transfer MS of Peptides. *J. Am. Chem. Soc.* **96**: 3990, 1974.
189. Lin YY, Rick M, Ray S, DV E, Clardy J, Golik G, James J and Nakanishi K, Isolation and Structure of Brevetoxin B from the "Reded Tide" Dinoflagellate Ptychodiscus brevis (Gymnodinium breve). *J. Am. Chem. Soc.* **103**: 6773-6775, 1981.
190. Beuhler RJ, Flanigan E, Greene LJ and Friedman L, Proton transfer MS of peptides. *J. Am. Chem. Soc.* **96**: 3990, 1974.
191. Reinhold VN, *DCI-MS of carbohydrates, in Complex Carbohydrates, Part E. Methods in Enzymology.*, Vol. 138. Academic Press, Orlando, 1987.
192. Reinhold VN and Carr SA, DCI with poly-imide coated wires. *Anal. Chem.* **54**: 499-503, 1982.
193. Demirev PA, Handjieva N, Saadi H, Popov SS, Reshetova OS and Rozynov BV, Ammonia DCI of glycosides. *Org. Mass Spectrom.* **26**: 151-153, 1991.
194. McLuckey SA, Goeringer DE, Asano KG, Vaidyanathan G and Stephenson JLJ, High Explosives Vapor Detection by Glow Discharge-Ion Trap MS. *Rapid Commun. Mass Spectrom.* **10**: 287-298, 1996.
195. Marcy TP, Gao RS, Northway MJ, Popp PJ, Stark H and Fahey DW, Using CI MS for detection of HNO3, HCl, and ClONO2 in the atmosphere. *Int J Mass Spectrom* **243**: 63-70, 2005.
196. Crounse JD, McKinney KA, Kwan AJ and Wennberg PO, Measurement of Gas-Phase Hydroperoxides by Chemical Ionization Mass Spectrometry. *Analytical Chemistry* **78**(19): 6726-6732, 2006.

Chapter 8   Electrospray Ionization

Introduction to Mass Spectrometry, 4th Edition: Instrumentation, Applications, and Strategies for Data Interpretation; J.T. Watson and O.D. Sparkman, © 2007, John Wiley & Sons, Ltd

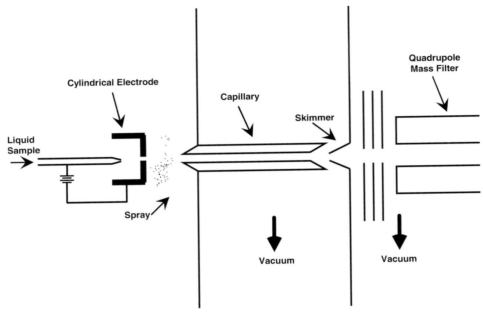

Figure 8-1. *Schematic diagram of a typical electrospray ionization interface mounted on a transmission quadrupole m/z analyzer.*

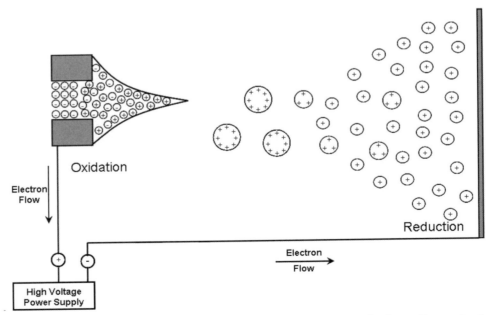

Figure 8-2. *Illustrative representation of the electrolytic cell created between the electrospray needle (left side of the circuit) and the counter-current electrode (right side of the circuit).*

I. Introduction

Electrospray ionization (ESI) brings many new features to classical mass spectrometry, which has traditionally relied on gas-phase analytes for ionization [1–4]:

(a) ESI is a means of producing ions from nonvolatile, thermally labile compounds.

(b) Because ESI can produce multiple-charge ions, the mass-to-charge (m/z) value of ions of macromolecules may fall within the range of most commonly used mass spectrometers.

(c) Due to the redundant assessment of the mass of these macromolecular ions (usually as variably protonated molecules) through detection of electrosprayed ions differing in charge state, it is possible to determine the mass of the analyte to 1 part in 10,000 or better, depending on the type of m/z analyzer.

(d) ESI serves as one of the most effective and successful interfaces for LC/MS (Chapter 11) that has been developed.

(e) ESI is a soft ionization technique that permits investigations of noncovalent associations of macromolecules such as proteins. Also, soft ionization allows the accurate mass of small molecules to be determined, using an instrument such as the TOF m/z analyzer.

(f) ESI allows direct analysis of inorganic cations and anions, providing information on the valence state and molecular formulation.

In addition to allowing for mass spectral measurements of ions in the mass range common to classical mass spectrometry produced from nonvolatile compounds, ESI has opened the field of mass spectrometry to analytes whose mass far exceeds that which can be measured by classical mass spectrometry. However, classical instruments now can be employed to this end.

II. Operating Principles

ESI is generally accomplished by forcing a solution of the analyte through a small capillary such that the fluid sprays into an electric field, thereby generating very fine droplets as illustrated conceptually in Figure 8-1. The presence of an electric field is essential for effective spraying action and, in many cases, for the ionization itself through electrochemistry [3–8]. The electric field is imposed between the tip of the spraying capillary and a counter electrode as indicated in Figure 8-2. The imposed electric field is important for several reasons, one being that it keeps the droplets from freezing (during endothermic loss of solvent by evaporation) by causing the charged droplet to endure many collisions through which some translational energy is converted to internal energy, thereby warming the droplet [9, 10].

Electrosprayed droplets will possess an excess of positive or negative charges depending on the capillary bias polarity. Redox chemistry [5, 11] at the interface between the solution and the electrode plays an important role in the charging of the droplets [12]. The initial "fine" droplets formed at atmospheric pressure in the spray orifice are enormous on the molecular scale, even though they may not be visible to the naked eye. These droplets diminish in size due to simple evaporation of the solvent in a drying gas at atmospheric pressure [13] or in a heated chamber at somewhat reduced pressure [14]. As the droplets shrink, ions (which are involatile) are retained and the concentration of the analyte increases, which can affect some equilibrium processes [15]. The decreasing droplet size also increases the repulsive forces between the excess charges in the droplet

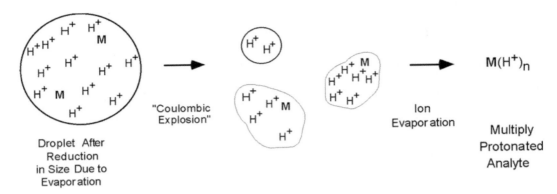

Figure 8-3. *Illustration showing a microscopic droplet containing the analyte and an excess of positive charges (protons). After size reduction due to evaporation, the droplet disintegrates into smaller droplets due to coulombic repulsion of the charges; this process repeats until finally the analyte molecule (M) is ejected into the gas phase along with some of the excess charge (proton[s]).*

eventually promoting electrohydrodynamic disintegration into many smaller droplets as illustrated in Figure 8-3 [16]. Asymmetry in the distribution of charge at the droplet surface may stimulate "coulombic explosion" of the droplet before reaching the Rayleigh limit (the point at which repulsive forces between like charges in an electrolytic solution overcome the cohesive forces of the solvent) for stability of a spherical charged droplet [16–19]. Finally, when the macroscopic solvent has evaporated, the analyte molecule finds itself with residual charges attached; for positive ESI, protons presumably attach at sites of high Lewis basicity. Another school of thought on the ESI mechanism champions a field desorption model [20], sometimes called an atmospheric pressure ion-evaporation process [21]. Whether the source of protons involved in ionization comes from the original sample solution or whether they enter the solution during the ESI process continues to be debated [22, 23]. However, it is clear that the maximum charge states and charge-state distributions of analyte ions formed by ESI can be influenced by the solvent composition [24].

Whereas some operational features of ESI are reminiscent of those of electrohydrodynamic MS [9, 25, 26], the mechanism of ionization in ESI is different [3, 4, 27–31]. A systematic study of instrument settings has shown that the ESI signal can be compound dependent, a feature that has both advantages and disadvantages [32]. Microscopic examination [33] of the spraying process indicates that a drastic change in the shape of the meniscus of the sprayed liquid and droplets occurs as the electric field reaches its optimum value. At the onset of electrospray, the electric field establishes an electrostatic force sufficient to pull liquid and droplets out of the nozzle toward the ground plate. The occurrence of this phenomenon indicates the electrostatic force becoming equal to the surface tension of the liquid. At that moment, the emanating liquid changes in shape from an elliptically shaped fluid cone into the so-called "Taylor cone", which is drawn out to a relatively sharp point along the axis of fluid flow through the spraying orifice [33]. After a flight path of approximately a millimeter, the jet spreads into a plume of fine droplets, which soon becomes invisible to the naked eye due to solvent evaporation and further subdivision of the droplets. Phase–Doppler particle analysis (PDPA), based on the interference pattern in crossed laser beams, measures droplet sizes from ESI sources [34]. PDPA measurements show that the charged droplets rupture when the charging levels reach between 70% and 80% of the Rayleigh limit [35]. Results of another PDPA study showed that during a given coulombic explosion as much as 15–20% of the charge in a droplet of methanol or acetonitrile and 20–40% of the charge on a droplet of water is lost with only relatively little accompanying loss in mass; the principal mechanism for diminishing droplet size is evaporation [36]. High-speed imaging of droplet formation during the burst mode, the pulsating Taylor-cone mode, and the cone-jet mode of ESI indicate that the primary drops are produced by varicose waves and lateral kink instabilities on the liquid jet emerging from the Taylor cone, whereas secondary drops are formed by fission [37].

In the more macroscopic conventional ESI sources [3, 4, 29] the droplets formed initially are on the order of one to ten micrometers, and diminish in size on the order of 100 nm due to the processes depicted in Figure 8-3. Results of a study of the evaporation and discharge dynamics of highly charged droplets of some common hydrocarbons may provide the basis for optimizing ESI in the presence of such solvents [38]. Results of a mechanistic study show that use of a high-velocity gas flow increases the rate of desolvation, decreases pH in the droplet, and increases ion current signals [39]; comparisons among the ESI spectra of several model proteins over a range of pH values, with and without the high-velocity gas, indicate that use of the high-velocity gas promotes a shift in charge-state distribution that is consistent with the native-like conformation of the protein as driven by changes in pH within the droplets. A variation of this technique using a supersonic nebulizing gas, called electrosonic spray ionization [40], greatly promotes desolvation as manifested by peak widths for various multiple-charge protein ions that are an order of magnitude narrower than those achieved with nanospray; under optimum conditions, the vast majority of the protein ions can be accumulated into the lowest charge state achievable by ESI or nanospray, indicating that folded protein ions are generated [41]. Recent studies indicate that energetic ions are formed in the ESI process, but that collisions with the curtain gas serves to cool the nascent ions and prevents cluster formation by shielding them from solvent vapors [42–44]. Other studies involving the use of evaporation inhibitors favor an ion evaporation mechanism for ultimate ion emission for relatively small molecules, by which ions are "field emitted" from a highly charged droplet surface [45]. A clever way to monitor the pH within the droplets using a fluorescent dye has been reported [46].

Electrochemical processes accompanying the ESI process involve the continuous withdrawal of ions from the ion source, and thus require that a current of electrons of equal magnitude be maintained (by oxidation in positive ESI as illustrated in Figure 8-2 or by reduction in negative-ion ESI) [47–49]. The ESI sprayer tip is one electrode of a conventional controlled-current electrolytic cell [50, 51]; the actual interfacial potential will depend on which chemical species are present [52]. In the conventional use of ESI, in which the sprayer capillary is effectively a cylindrical electrode, very little of the analyte actually touches the electrode (i.e., most of the redox chemistry is suffered by the solvent) because such interaction is controlled by the diffusivity of the analyte for a given ESI flow rate [47]. The brunt of the electrolytic process in ESI is suffered by water molecules; in the positive mode, a high local concentration of protons (low pH) is generated by electrolysis of water residing in the sprayer tip, and conversely, in the negative mode, a high local concentration of hydroxide ions (high pH) occurs in the sprayed solution. Solvent composition within the ESI capillary affects the extent of oxygen adsorption by the spraying liquid within the Taylor cone, thereby affecting the electrochemical potential near the capillary exit [53]. Whether the analyte is electrochemically affected in the ESI process depends on the electrode material and the ESI flow rate [49]; some artifact formation deriving from electrochemical processes has been reported, though it can be avoided with proper adjustment with a redox buffer [54]. A porous flow-through electrode emitter has been designed to improve mass transport of the analyte to the electrode for complete oxidation of the analyte at ESI flow rates approaching 1 mL min^{-1} [11]. Electrochemical control of the analyte solution in ESI provides a way to observe oxidized and reduced forms of metalloproteins [55] or to monitor other complex redox reaction pathways [56]. Electrochemical preconcentration has been used to advantage in an assay of tamoxifen and its metabolites by ESI [57]. A rotating ball inlet to the mass spectrometer facilitates studies of electrochemical effects [58]. Oxidation of steroid sulfates possessing a reactive double bond between C-5 and C-6 due to electrochemical processes in a distal metal HPLC capillary may occur [59]; peptides containing sites of unsaturation may also be susceptible to similar modification promoted by free radical chemistry following on-column electrolysis of the HPLC mobile phase.

Charbonnier *et al.* [60] measured the electrolytic current of the ESI process directly on a fast oscilloscope and found it to consist of discontinuous bursts of approximately 10^{-10} C corresponding to individual droplets. They observed that whereas the overall current increased with applied potential (2800 V to 4000 V), the instantaneous charge bursts remained at approximately 10^{-10} C, and their frequency increased. Overall, their results showed a typical droplet emission frequency of approximately 300 kHz.

Particularly in negative-ion ESI, ion emission is often accompanied by electron emission directly from the emitter or counter electrode. This discharge current will typically be larger than the spray current. Thus, through an arrangement similar to that used in electrochemistry, the electrospray process can be optimized through the observation of the compensatory current drawn from ground to the ESI source. In some cases, gases such as SF_6 can be used to suppress the electron current to achieve an optimum spectrum; for example, it was possible to achieve negative ion ESI of nucleotides using SF_6 for discharge suppression [61]. The use of a copper capillary suppresses (by maintaining the electrode at 0.34 V vs a standard hydrogen electrode) the oxidation of *N*-phenyl-1,4-phenylenediamine as occurs when using a stainless steel capillary (electrode potential = 0.46 V vs a standard hydrogen electrode) in the electrospray interface [62].

Some failures of a sheathless interface may be due to loss of the sputter-coated gold conductive coating from mechanical stress caused by gas evolution accompanying redox reactions of the solvent [63]; in an application to CE, hydroquinone (more easily oxidized than water) was used to suppress evolution of O_2 [64]. In some cases, derivatization of the analyte improves the assay, as was demonstrated by the use of ferrocene-based derivatives of simple alcohols, sterols, and phenols [65]. Formation of a charged hydrophobic derivative improves the ionization efficiency of some proteins and especially small peptides [66].

Although some of the general concepts of ESI are represented in Figure 8-3, other aspects remain to be clarified. In any case, the analyte molecule **M** together with a charged adduct becomes "airborne" and, without the bulk solvent, acquires a mass-to-charge ratio, *m/z*, which can be related to its molecular weight. The ion is accelerated through a small aperture (skimmer) in a differentially pumped region into the mass analyzer.

III. Appearance of ESI Mass Spectra and Data Interpretation

Most reported applications of ESI MS involve biomolecules, in which case a charge is most often generated by protonation of basic sites (e.g., amino groups) [67] or deprotonation of acidic sites (e.g., phosphates). In these cases, a series of peaks in the electrospray mass spectrum represents multiple-charge variants [68, 69] of the analyte; these multiple-charge ions are almost always protonated [70], but sometimes they include

$$[M + H]^+$$

$$[M + nH]^{+n}$$

$$[M - H]^-$$

$$[M - nH]^{-n}$$

$$[M + Na]^+$$

$$[M + nNa]^{+n}$$

$$[M + K]^+$$

$$[M + nK]^{+n}$$

$$[M + nH + xNa + yK]^{+(n+x+y)}$$

$$[nM + H]^+$$

Figure 8-4. *Some of the possible combinations of clusters of the analyte molecule (M) with various alkali ions and/or protons.*

species in which Na or K have replaced hydrogen, which may occur during ion/molecule reactions in the ESI source [71]. Some possible clusters formed during ESI are represented in Figure 8-4.

Because these multiple-charge species differ only by a consecutively decreasing number of charges as *m/z* increases, it is possible to deduce mathematically the molecular weight of the intact molecule. Several algorithms [72–76] have been developed to compute the mass of the analyte; a related algorithm has been developed to process LC/MS data [77]. In Figure 8-5, a general scheme is presented where the *m/z* values of the peaks in the spectrum are represented by m_1, m_2, etc. Note that the *m/z* value for m_2 is greater than that for m_1. The molecular weight of the analyte is designated M and the mass of the charge carrier (e.g., H$^+$) is designated **X**. The number of charges on a given ion is designated n. It is also important to note that the charge state (n_1) of the ion represented by m_1 is greater than the charge state (n_2) of the species represented by m_2. By using these defined relationships, the equations shown at the bottom of Figure 8-5 can be established. Solving these equations algebraically for the charge state and using the experimentally determined values for m_1 and m_2, it is possible to compute a value for M, the molecular weight of the analyte. This process can be repeated for different pairs of analyte ions in the spectrum, giving averagable multiple estimates of M.

The reduced mass-to-charge value of a massive ion containing several charges makes *m/z* analysis possible by inexpensive instruments such as the quadrupole MS (maximum *m/z* range of approximately 4000). Given that each peak in Figure 8-5 represents ions containing the same M, but with different values of z, the reinforcement of parallel calculations results in mass accuracies as good as 0.01% or better with most *m/z* analyzers. This means that the molecular weight of a macromolecule at, for example, 38,842 daltons can be measured routinely with an accuracy of ± 4 daltons.

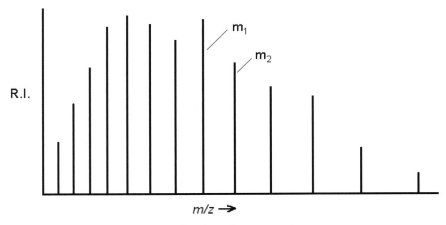

$m = m/z$ of a given peak
$M =$ mol. wt of analyte
$X =$ mass of adduct (e.g., H^+ or Na^+)
$n =$ number of charges

$$m_2 > m_1 \quad \text{and} \quad n_1 > n_2$$

Adjacent peaks represent ions differing by 1 charge state

$$m_2 = (M + n_2 X)/n_2 \tag{a}$$
$$m_2 = (M + (n_2 + 1)X)/(n_2 + 1) \tag{b}$$

Solve Equation (a) for M:

$$m_2 = (M + n_2 X)/n_2 \tag{a}$$

Multiply both sides by n_2

$$m_2 n_2 = M + n_2 X$$

Subtract $n_2 X$ from both sides

$$M = m_2 n_2 - n_2 X = n_2(m_2 - X) \tag{a}$$

Substitute the value for M into Equation (b)

$$m_1 = (M + (n_2 + 1)X)/(n_2 + 1) \tag{b}$$
$$m_1 = (n_2(m_2 - X) + (n_2 + 1)X)/(n_2 + 1) \tag{b}$$
$$m_1(n_2 + 1)\ (n_2(m_2 - X)\ ((n_2\ (\ 1)X$$
$$n_2 = (m_1 - X)/(m_2 - m_1) \tag{c}$$

       Guess at X
       Measure m_1 and m_2
       Calculate n_2 from Equation (c)
       Compute M from Equation (a) or Equation (b)

Figure 8-5. *Scheme for mathematical relationship between charge state and mass of analyte and charge carrier (adduct) in rationalizing the m/z value of peaks observed in the ESI mass spectrum.*

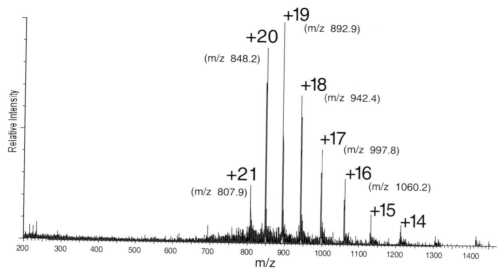

Figure 8-6. **ESI mass spectrum of 10 pmol horse heart myoglobin (reported average mol. wt = 16,951.5 Da; experimentally determined as 16,951.0 Da) obtained with a low-resolution quadrupole mass spectrometer.** Data provided courtesy of James Bradford, Graduate Assistant, in the Macromolecular Structure Facility at Michigan State University, East Lansing, MI.

Figure 8-6 shows an ESI mass spectrum obtained with a quadrupole MS that is capable of unit resolution on the *m/z* axis. Because the peaks at *m/z* 848.2, 892.9, and 942.4 indicate 20, 19, and 18 charges, respectively, they represent much larger masses. Analysis by ESI can be accomplished with remarkable sensitivity; in some cases, the molecular weight of a protein can be determined using only a few femtomoles (10^{-15}) of analyte.

IV. ESI with an *m/z* Analyzer of High Resolving Power

When an ESI source is used in combination with ion cyclotron resonance (ICR), superior resolution relative to the ESI quadrupole combination can be achieved as illustrated by the spectrum in Figure 2-47 in the context of FTICR in Chapter 2. When the isotope peaks in a given mass spectrum can be resolved, the calculated monoisotopic mass of candidate ions should be used during data interpretation [78–80]. As defined in Chapter 1, the monoisotopic peak corresponds to the ion containing the most abundant isotope of each element present. If the isotope peaks cannot be resolved, the centroid of the mass spectral peak should correspond well with the calculated average mass of the ion. The *m/z* values of any two multiple-charge ions corresponding to a single compound are unique numbers that enable the charge states for each ion to be unequivocally identified [81]. In contrast to conventional deconvolution processes for ESI data, the charge ratio analysis method (CRAM) identifies the charge states of multiple-charge ions without any prior knowledge of the nature of the charge-carrying species. In the case of high-resolution ESI mass spectral data, in which multiple-charge ions are resolved to their isotopic components, the CRAM is capable of correlating the isotope peaks of different multiple-charge ions that share the same isotopic composition with ESI mass spectral data of lysozyme and oxidized ubiquitin [81]. Peculiar to FTMS, coalescence can adversely affect the resolution, but special operational measures [82] as well as isotopically pure analytes [83] can greatly improve the situation.

The charge state of an ion can be determined by visual inspection of isotope peaks in the spectrum (if resolvable). For example, does a peak at *m/z* 486 in an unknown spectrum represent a single-charge ion of mass 486 Da or a double-charge ion of mass 962 Da? If a single-charge species is represented, the first isotope peak will be at *m/z* 487; if a double-charge species is present, the isotope peak will appear at *m/z* 486.5 (963/2). The inverse relationship between isotope peak spacing and charge state may lead to problems when more than 10 charges are present, and it is best to use some sort of automated data system [84] to avoid subjective errors when the spectrum cannot be resolved as well as that in Figure 2-47. Alternatively, derivatization of the analyte to a Schiff base has been reported as a means of determining the charge state of ions [85].

V. Conventional ESI Source Interface

Many of the conventional ESI sources (or those of early design) use a somewhat "macro scale" interface in which the liquid containing the analyte is sprayed through an aperture having an inner diameter (ID) on the order of tens of micrometers; recent designs of spray tips use lithographic approaches with high geometrical fidelity [86]. In some cases, the spraying process is ultrasonically assisted or pneumatically assisted using a high-velocity annular flow of gas at the capillary terminus. In some cases where the solvent does not spray effectively, a second "sheath" (often composed of water, methanol, or acetonitrile, sometimes modified by acids, bases, or other reagents) flows through the annular space between a 200-µm o.d. capillary and a larger fused-silica or stainless steel capillary (generally greater than 250 µm i.d.), mixing with the analyte solution (flowing through the inner capillary) at the terminus where spraying occurs [30]. In some cases, the analyte can form large clusters with solvent molecules, a situation that makes data interpretation difficult or at least difficult, especially when clusters include combinations of cations and solvent molecules as indicated in Figure 8-4; the clusters can be removed by collisional dissociation, which is dependent on both the energy and the pressure in or near the ion source [87]. In some cases, intermolecular cluster reactions in aging solutions can cause artifact formation as demonstrated in an ESI MS study of solutions of diacylglycerophosphocholine phospholipids [88].

The RG Cooks group showed that nebulizing the protein solution with a high-velocity gas during electrosonic ESI or electrosonic spray ionization resulted in principally a single peak in the mass spectrum representing nine protons attached to the protein molecule as opposed to obtaining an array of five or six differently protonated species in the conventional ESI mass spectrum [39–41]. Because all of the ion current from ionizing the protein contributes to a single peak in the mass spectrum in such a case, the signal–to-background ratio improves. This observation has other implications, such as an indication that conventional ESI may substantially denature most of the protein molecules as represented by an array of different species that capture a number of protons proportional to their degree of denaturation (i.e., the more denatured a protein, the greater the number of basic sites available to capture a proton) [89]. These results indicate that ESI may be a harsher ionization technique than originally thought [39, 40, 90].

VI. Nanoelectrospray and Microelectrospray Ionization

In general, the ionization efficiency of the electrospray ion source is comparable to that of the APCI source and can be several orders of magnitude greater than that achieved by more conventional ion sources, like the EI source. Although the response of ESI is concentration-dependent for a given analyte, the exact relationship between ESI signal strength and analyte concentration is not straightforward [69, 91]. Hill and coworkers found that although miniaturization of the ESI ion source leads to greater ionization efficiency results of a study of the effect of flow rates on a nanoelectrospray emitter by ion mobility spectrometry (IMS) seemed contradictory in that a greater ion abundance was observed with increasing flow rate [92]. Whereas it is generally well accepted that decreasing the flow rate increases both ionization and transmission efficiency [3], the results of the ESI IMS revealed that a decreased flow rate decreases the ion signal because the ion transfer is constant [92].

Wilm and Mann [93] developed a mathematical and theoretical model of the ESI process, which allowed them to optimize the dimensions of their ion source, leading to the production of microdroplets on the order of 100 nm in diameter. In efforts to optimize the ionization efficiency of the electrospray process, some research groups have miniaturized the device with a spraying aperture on the order of 1 μm that can accommodate flows of nL min^{-1} when a potential of approximately 800 V is impressed between the counter electrode and the spraying orifice; these miniaturized devices have an efficiency approaching 10% [33, 94]. The analytical advantages of nanoelectrospray are not due to the miniaturization of the device, per se, but rather by the capability of miniaturized devices to produce much smaller droplets. Smaller initial droplets in the ESI process means fewer coulombic explosions are required to expose the analyte to sampling by the mass spectrometer, which translates into lower detection limits [95]. In addition, because the droplets of solution are smaller in nanoelectrospray than in conventional ESI, there is less evaporative loss of bulk solvent, thereby reducing the tendency to concentrate contaminants in the sample.

Nanoelectrospray offers practical advantages of low sample consumption and sufficient time to conduct detailed analyses (CAD, etc.) on a given component even during LC/MS. The consumption of analyte solution is only nanoliters per minute, so a few microliters of sample solution can provide useful mass spectral signals for roughly half an hour, providing ample time for a variety of real-time experiments including parent-ion scans [96]. Flow rates approaching 25 nL min^{-1}, usually achieved with an aperture approaching 1 μm in diameter, are needed to minimize signal suppression and adduct formation [97, 98]. A dramatic difference in performance can be achieved in reducing the aperture from 10 to 1 μm [99].

From a practical point of view, nanoelectrospray operation is often prone to difficulties associated with clogging of the fine conduits and apertures required for minuscule flow rates. Le Gac and Rolando have developed and implemented a nib-like nanoelectrospray emitter in the form of a capillary slot fabricated into a silicon wafer [100]. Prototypes of the device have shown good reproducibility in applications [101, 102], and the most recent version offers the advantage of electrical connection to the silicon wafer without use of an external wire [100]. Experiments using standards of gramicidin S and Glu-fibrinopeptide B indicate detection limits to 10^{-8} M [100]. Such microfabricated devices are likely to facilitate the development of high-throughput applications of nanospray MS.

Microelectrospray differs from nanoelectrospray in that the flow in a microelectrospray system is dependent on the use of a nebulizing gas to force the liquid through the electrospray needle; also, the potential applied to the needle is higher. The nanospray technique developed by Wilm and Mann differs from the microelectrospray technique developed by Emmett who was the first to talk about these low-flow techniques in 1993; also Caprioli pointed out that the flow microelectrospray is dependent on a mechanical pump. Another differentiating factor is the diameter of the needle; in microelectrospray, the needle i.d. is ~50 μm, whereas for nanoelectrospray, it is ~1–3 μm. The flow rates are also different for the two techniques; microelectrospray typically uses a flow rate of ~25 nL min^{-1}, whereas microelectrospray uses 300–800 μL min^{-1}. Emmett first described his technique in May 1993 at the San Francisco ASMS meeting; he used the term "nanoliter flow LC/ES/MS". The next year, at the May ASMS meeting in Chicago, IL, Emmett used the term "micro-MS" and Wilm and Mann used the term "microMAS"; however, in a 1994 publication by Wilm and Mann, they used the term "nano-ES".

Although the terms nanoelectrospray and microelectrospray are often used interchangeably, they should not be, as can be seen from the above descriptions. It is also worth noting that use of the terms nanospray and microspray could be in violation of trademark regulations. The terms MicroIonSpray™ and NanoFlow™ ES are trade names that refer to the techniques marketed by Applied Biosystems/Sciex and Micromass, respectively. NanoSpray™ is a trademark belonging to Bruker Daltonics. These three trade names should only be used to describe the trademarked products.

VII. Desorption Electrospray Ionization (DESI)

Desorption ESI (DESI) is accomplished by directing electrospray-charged droplets (due to protonated solvent in most cases) on to the surface to be analyzed as illustrated in Figure 8-7 [103, 104]. The interaction of the charged particles with materials on the surface produces gaseous ions of some of the compounds originally adsorbed to the surface to give mass spectra similar in appearance to conventional ESI mass spectra in that they show mainly peaks corresponding to single- or multiproton or alkali-ion adducts of the analytes. The proximity of the spraying device to the sample is important, and can be optimized with simple adjustments while observing the signal of the mass spectrometer. There is some dependence on the angle of incidence that is related to the size of the droplets; large droplets have more momentum and are able to leave the surface at an angle close to the angle of incidence, whereas smaller droplets are swept along the plane of the surface by the bulk gas flow [104, 105]. At the microscopic level, the action of the DESI beam of sprayed droplets appears to simulate that of a "power scrubber" in that accumulation of liquid on the surface being analyzed seems to dissolve some of the compounds deposited on the surface, thereby facilitating desorption/ionization during further bombardment of the nascent solution by additional droplets from the sprayer [106, 107]. The ions (possibly trapped in small droplets) emanating from the sample surface are effectively "vacuumed up" by a relatively large capillary through which a high volume rate of air carries the secondary ions (sometimes as far as 3 m) to a skimmer cone for entry into the *m/z* analyzer [108]. For this reason, the proximity of the entrance to the "sweeper" capillary to the point of impact of the bombarding spray is important.

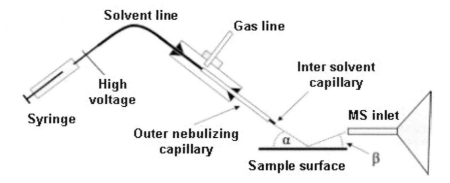

Figure 8-7. *Schematic illustration of a desorption electrospray ionization interface (DESI).* *From Venter A, Sojka PE, and Cooks RG "Droplet Dynamics and Ionization Mechanisms in Desorption Electrospray Ionization Mass Spectrometry" Anal. Chem., 2006, 78(24), 8549–8555, with permission.*

The results of a study of droplet dynamics associated with the DESI process provide some clues to the mechanism of secondary ionization [104]. The electrohydrodynamically produced droplets have an average diameter of 2–4 µm and a velocity of 120 m sec^{-1} at a distance of 2 mm downstream from the spraying capillary made of 100 µm i.d. fused silica [104]. The velocity of the droplets rapidly drops to 40 m sec^{-1} at a distance of 10 mm from the sprayer tip to approximately 10 m sec^{-1} at a distance of 40 mm; most DESI experiments are conducted at distances of 2–10 mm from the sprayer tip. Evaporation times of droplets of methanol were estimated to be 2.6 millisec (droplets of water:methanol of 1:1 take longer); this means, some ionization of fairly volatile compounds may occur by direct interaction with free ions or with protonated clusters of water molecules [105]. The principal means of ionizing nonvolatile compounds appears first to involve a droplet pick-up scavenging process that takes place during the brief contact time when droplets collide with sample surfaces with the analytes being extracted into the departing droplet. After the analyte is consumed by the droplet, ionization apparently occurs by the standard putative ESI mechanism according to the field-desorption model [20] or the charge-residue model [16]. The resulting ions corresponding to attomole quantities of analyte [109], whether free in the gas phase or still contained in residual microdroplets, are swept up by the ion-capturing capillary for transport to the mass spectrometer.

Given the size and velocity of the bombarding droplets, the kinetic energy per impacting water molecule is less than 0.6 meV; meaning, sputtering through momentum transfer during collisions or ionization through other electronic processes is unlikely [104], although some sort of coherent phonon excitation at the molecular level has been proposed [110]. Because droplets arrive at the surface of samples with velocities well below the speed of sound, the possibility of ionization by shockwave can be excluded [104].

The experimental setup for DESI as represented in Figure 8-7 calls for a liquid flow rate of 2 µL min^{-1} and high-velocity nebulizing gas maintained at a headpressure of 1130 kPa (~150 psi) at an undisclosed flow rate that is restricted by the annular cross-sectional area between the 190-µm o.d. of the inner capillary and the 250 µm i.d. of the outer fused silica capillary comprising the electrohydrodynamic sprayer that is maintained at a potential of 5 kV [104]. The extremely small droplets are produced by the shearing force of the high-velocity nebulizing gas, which prevents formation of a stable Taylor cone that is usually produced when a high potential is applied to conducting liquid as it flows out of a capillary [111]. There is some evidence that decreasing the size of the bombarding droplets increases the signal strength in the DESI spectrum [104]. The strength of the DESI spectrum will surely also depend on the efficiency with which the secondary ions can be collected; therefore, a rather large collection capillary should be employed to serve as a "vacuum cleaner" to capture a relatively large volume of the reflected flux of secondary ions.

Unlike SIMS, the DESI phenomenon causes no damage to polymeric or metallic surfaces, although physical ablation of soft sample material such as biological material does occur [104]. DESI has been used to analyze organic materials from both conductive and insulator surfaces, and for compounds ranging from nonpolar small molecules such as lycopene, the alkaloid coniceine, small drugs, as well as polar compounds such as peptides and proteins; changes in the solution being electrosprayed can be used to ionize particular compounds selectively, including those in biological matrices [103] or in inks for forensic purposes [106]. In another application, DESI was used to characterize the active ingredients in pharmaceutical samples formulated as tablets, ointments, and liquids [112]; it has also been used to sample industrial polymers [113]. DESI has also been used for trace detection of the explosives trinitrohexahydro-1,3,5-triazine (RDX), octahydro-1,3,5,7-tetranitro-1,3,5,7-tetrazocine (HMX), 2,4,6-trinitrotoluene (TNT), pentaerythritol tetranitrate (PETN), and their plastic components (Compn. C-4, Semtex-H, Detasheet) directly from a wide variety of surfaces (metal, plastic, paper, polymer), including skin [114] and SFE microfibers [115], without sample preparation or pretreatment [115, 116]; in combination with complexation reactions with alkali metal ions in the DESI spray, triacetone triperoxide (TATP) can be detected on surfaces [117]. Urinary metabolites [118], peptides [119], proteins [120], as well as intact bacteria [121] have been characterized by DESI. The DESI technique can be used to read or visualize the components of a mixture separated on a thin-layer chromatography (TLC) plate by mass spectrometry [122]. DESI mass spectra were obtained directly from the TLC plate by moving it incrementally under computer control while directing the stationary DESI emitter charged droplet plume at the TLC plate surface. Fundamentals and practical applications of the technique were demonstrated in both positive- and negative-ion modes from dyes separated on reversed-phase plates [123], and from a mixture of aspirin, acetaminophen, and caffeine separated on a normal-phase silica gel plate [122].

The DESI technique for analyzing TLC plates appears to be much more effective and convenient to use than earlier techniques for analyzing components directly from a TLC plate, which were based on microelution and sampling systems that were manually directed to specific locations on the plate [124–126]. DESI has also been proposed as part of a 2D chemical imaging system with a lateral resolution of a few hundred meters, as demonstrated in imaging a model system composed of an ink pattern on photographic paper and thin coronal sections of rat brain fixed to a glass microscope slide [127]; ion current characterizing the various ink dyes were used to image the ink pattern, and ions corresponding to deprotonated lipid molecules were monitored in the negative-ion mode to prepare patterns of various lipid sections in the coronal slices.

A variation of the technique, called reactive DESI, uses reactive chemicals in the bombarding droplets [128]. By adding an organic salt (e.g., a phosphonium chloride) to the spraying solution, the bombarding droplets can be used to direct selected chemical modification of the analyte to improve specificity of the analysis. By adding boronic acid to the spraying solution in DESI, it is possible to promote covalent reactions with analytes containing the *cis*-diol functionality as in some carbohydrates, catecholamines, etc. [129]. The addition of alkylamines to the DESI spray is the basis for a direct assay for artesunate, the active ingredient in some antimalarial drugs [130]. Again, such use of the DESI experiment is exceptional in that it keeps unnecessary chemistry and sample preparation out of the mass spectrometer. In an application of reactive DESI to the analysis of a specialized polymer containing copper(II), a dilute solution of iodine as an oxidant enhances the detection limit by two orders of magnitude [131].

A hallmark of the DESI experiment, as with the DART experiment [132], is that the sample is not introduced into the mass spectrometer, only the ions are introduced through the skimmer cone [108]. In this way, there is minimal insult to the *m/z* analyzer from the materials in the sample matrix that would otherwise contaminate the instrument. The newly described technique of matrix-assisted laser desorption electrospray ionization (MALDESI) is somewhat related to DESI in that the ions of proteins that are imbedded in an organic acid matrix have charge states that can be highly correlated to those obtained via nano-ESI, providing evidence that the ESI process dictates the observed charge-state distribution [133].

VIII. Effect of Composition and Flow Rate of an Analyte Solution

Sample preparation prior to analysis by ESI, as in MALDI, is an important determinant of the quality of the resulting spectrum. Great care must be given to choosing additives that can help solubilize the analyte without interfering with the ionization process. For example, in dealing with the classical problem of solubilizing hydrophobic proteins, detergents [134] may interfere with the electrospray process; however, acid-labile surfactants have been reported that have minimal interference with the ionization process [135]. One successful approach is to dissolve the protein in 90% formic acid [136]; prompt injection of the dissolved sample on to a disposable column of reversed-phase particles allows separation of salts and detergents [137] and avoids problems associated with formylation of the protein [138]. In some cases, using dimethylsulfoxide and dimethylformamide as solvents in the analysis of hydrophobic compounds has been suggested [139]. The usual problem of limited dynamic range apparently is related to some complex interplay of ion formation/instrumental/space-charge effects [140]. The impact of other characteristics and properties of the solvent, such as surface tension, on the electrospray process are addressed in Chapter 11 on LC/MS.

Salt or other compounds of fixed charge often dominate the ESI spectrum when present in the sample; for this reason, it is important to take some measure to remove or diminish the salt content of the sample [141]. Fibers made of capillary-channeled polymers of polypropylene can be used as sorbents in micro-SPE tips to remove both inorganic and organic buffers from proteins solutions [142]. A novel and very rapid (fraction of a second) procedure for on-line sample desalting is based on the differential diffusion of macromolecular analytes and low molecular weight contaminants [143]. Rather than using semi-permeable membranes, this on-line system uses a two-layered laminar flow geometry for a brief exposure of the sample to the rinsing medium so that 70% of the analyte is retained in the stream entering the ESI source together with only 10% of the original salt or low-mass contaminant; this on-line system is effective only if the diffusion coefficient of the contaminant is at least an order of magnitude greater than that of the analyte [143]. Another interesting ion source, in which aqueous droplets from an aerosol of the analyte solution is fused with charged droplets of electrosprayed methanol, has been reported to be 10 times more tolerant than a conventional ESI source in producing useful spectra of proteins; this source uses a single capillary to electrospray fused droplets containing multiprotonated proteins (e.g., cytochrome c) into the *m/z* analyzer [144].

Optimum conditions for obtaining ESI mass spectra vary significantly depending on the solvent system used to deliver the analyte to the interface [4, 69, 145–147]; in some cases, interference from the sample matrix can be removed by HPLC [148]. In a study of mobile-phase additives for LC/MS of polysulfonated compounds six di- and tri-alkylammonium acetates were compared with tetraalkylammonium salts and ammonium acetate; the concentration of the additives in the range 0–20 mM has a significant impact on the degree of signal suppression [149]. In another approach to dealing with the problem of signal suppression, the investigators used a nanosplitting device to drastically reduce the flow of sample into the ESI source [150]. The composition of the buffer solution also can greatly influence the appearance of the mass spectrum [151]. For example, Neubauer and Anderegg [152] used sodium acetate in the mobile phase during LC/MS to promote the formation of sodiated ions [153] in addition to the normally observed protonated species. The sodium-induced change in spacing of peaks allowed unambiguous determination of the charge state of the ions and hence their actual mass. Such a procedure is particularly useful when the analyte is unknown or represents a modified peptide under analysis. Solutions of mM Na^+ do not cause deterioration of chromatographic performance or fouling of the ion source, although such additives may attenuate the ESI signal. For example, if both proton and sodium adducts are present, the available analyte signal in either form is "diluted" by the other. In such cases, the strength of the ESI spectrum may depend on the degree to which adducted species have been removed (e.g., by ion exchange) or dissociated. Solvent adducts also can diminish the analyte signal by dispersing it among several different ionic species. Increasing the electric field between electrodes in the interface often breaks up ion/solvent adducts. It is generally necessary to seek a balance between dissociation of solvent adducts and conditions under which covalent bonds in the analyte can undergo dissociation. The higher the charge state of the adduct ion, the greater the efficiency of dissociation of ions due to collision; this feature can be used to seek structural information by MS/MS [154, 155].

The pH of the sample solution has important effects on the analysis as it can be used to control the degree of protonation of the analyte. Charge normalization of a given analyte can be made on the basis of the weighted average charge of the analyte as reflected in ESI mass spectrum. When proteins can be regarded as a collection of equivalent charge sites, the signal response from one protein can be used to predict the responses for other proteins [89, 156]. In using the concept of charge normalization, it is possible to predict the dependence of the signal response for a particular protein in a mixture on the concentration of other proteins in the mixture; this feature is especially important in top-down analysis of proteins [156].

There are reports of analyses of oligonucleotides [157, 158] and polyols [159] as metal ion adducts, however, in general, electrolyte solutions more concentrated than 100mM adversely affect an analysis by ESI. Some workers have found that adding pyridine to the analyte solution suppresses Na^+ adduct formation without removing the electrolytes [160]. However, cleaner spectra can be obtained by removing the electrolytes as with an on-line micro-dialysis unit [161]. An algorithm designed to reduce the level chemical noise and to increase the intensity of analyte ion peaks has been described [162].

IX. Special Applications

1. Direct Analysis of Ions in Solution by ESI

The molecular form and valence state of inorganic anions and cations have been analyzed directly by ESI MS [163–171]. However, because of easily driven fragmentation of some species in the ion source, the tendency to form clusters in the ion source, and the effects of pH on free and bound ions, the ESI spectrum should be interpreted with informed caution [172]. The "softness" of the ESI process preserves the valence and molecular speciation of the elements during the analysis of anions and cations in solution [173–175], unlike ICP which destroys such information in the harsh thermal environment of the plasma. Ru and Zr can be determined directly from urine after complexation with diethyldithiocarbamate, thereby avoiding interferences associated with such analyses using ICPMS [176]. Some copper(II) and copper(I) salts have been studied by ESI with in-source CID [177]. The stoichiometry of two polyfunctional mercurial Lewis acids has been studied by ESI MS/MS [178]; pentacoordinated bisaminoacylspirophosphoranes have also been characterized by ESI [179].

A method based on pressure-assisted capillary electrophoresis coupled to ESI MS that allows the simultaneous and quantitative analysis of multivalent anions, such as citrate isomers, nucleotides, nicotinamide-adenine dinucleotides, flavin adenine dinucleotide, and coenzyme A compounds, is key to the analysis and uses a noncharged polymer, poly(dimethylsiloxane), coated on the inner surface of the capillary to prevent anionic species from adsorbing on to the capillary wall [180].

2. Cold-Spray Ionization

A variation of the theme of ESI, cold-spray ionization (CSI), operates at a substantially lower temperature, namely in the range of 10–80 °C, and which allows for the analysis of more thermally labile compounds [181]. For example, one of the key structures of a Grinard reagent has been analyzed by CSI [181]; the more gentle nature of the CSI process suggests that it may be important in the study of noncovalent interactions and complexes.

3. Negative-Ion Detection

In some circumstances, it may be advantageous to operate the instrument in the negative-ion mode to achieve selectivity in detecting analytes that can be converted from neutral to negative simply by loss of a proton. During ESI in the negative mode, the pH of the sample solution at the spray tip is high due to the local production of hydroxide ions during the electrolytic process associated with ESI. Thus, in this application in which the HPLC conditions called for low pH (as achieved by formic acid) for good retention of certain analytes, deprotonation of the acidic analytes could still be achieved as they passed through the sprayer capillary due to the local production of hydroxide ions. The reconstructed total ion current chromatogram in Figure 8-8 illustrates the complexity of an extract of the field cotton plant as analyzed by ESI with LC/MS; some of the major peaks correspond to 15 of the known constituents of the plant. It was of interest to know whether the cotton plant extract also contained any chlorogenic acid (mol. wt = 354 Da). Although the presence of chlorogenic acid was not represented by any of the peaks in the total ion current chromatogram, its presence is clearly indicated by a major peak in the extracted mass chromatogram at *m/z* 353 (corresponding to the deprotonated carboxylate) as shown in the top panel of Figure 8-8 (Doctoral Dissertation, Bart A. O'Brien, Penn State University, University Park, PA, 2004; communicated by A. Daniel Jones, Department of Biochemistry, Michigan State University, East Lansing, MI). Of the three discernible peaks in the mass chromatogram, the one at a retention time of 14.44 min corresponded to a mass spectrum and a retention time identical to those for authentic chlorogenic acid, giving proof that chlorogenic acid was a constituent of the field cotton plant.

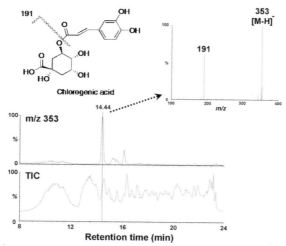

Figure 8-8. ***ESI negative-ion mass spectrum of chlorogenic acid.***
Data courtesy of Professor A. D. Jones, Dept of Biochemistry, Michigan State University, East Lansing, MI.

In other cases, it may be advantageous to form negative ions of the analyte intentionally because subsequent CAD of such ions provides structural information. For example, when neutral oligosaccharides were exposed to acetate, chloride, or fluorides, CAD of the negative adducts provided structurally informative fragment ions [182]. Direct infusion ESI MS can distinguish must derived from six different varieties of grapes during the fermentation process; the addition of unfermented must or sugar to wine can also be clearly detected [183]. In another application of negative ESI, the unsaturated fatty acid substituents of some phosphatidic acid, phosphatidylserine, phosphatidylinositol, and phosphatidylglycerol species were converted to their 1,2-di-hydroxy derivatives by OsO4; subsequent CAD MS/MS of the deprotonated species allowed positional determination of the double bonds by the production of specific product ions [184]. A study has been made of modifiers for detection of negative ions [185].

4. Secondary Electrospray Ionization (SESI)

Secondary electrospray ionization (SESI) was introduced by Chen *et al.* [186] who demonstrated that ESI could produce reactant ions, from electrospraying a solvent, which in turn could ionize neutral gaseous analytes. Later, Wu *et al.* [187] evaluated the analytical figures of merit of SESI, in comparison with those of ESI, in the analysis of illicit drugs, and concluded that SESI was a more sensitive ionization technique than ESI.

The ionization mechanism for SESI is similar to that of APCI, in which gas-phase reactant ions are produced in a conventional ESI source and subsequently reacted with the sample analyte in the gas phase forming a charged analyte, which can then be separated and detected by ion mobility spectrometry (IMS). Wu *et al.* [187] reported that SESI had higher ionization efficiency than ESI in the analysis of illicit drug samples by IMS. In other work [188], the nonradioactive feature of SESI has been stressed, and it has been used as a convenient ion source for IMS in the analysis of both vapor-phase and aqueous-phase explosives. Results from the analysis of RDX in the presence of the nonvolatile nitrate dopant, rather than the volatile chloride dopant, produced a lower detection limit, greater sensitivity, and better linearity [188].

It must be emphasized that to date SESI has been used only in conjunction with ion mobility spectrometry (IMS), rather than with mass spectrometry.

5. Kinetic Measurements of Chemical Reactions

A novel interface connecting a stopped-flow mixing instrument to an ESI mass spectrometer allows the real-time kinetics of chemical reactions to be monitored [189]. Immediately after mixing two reactant solutions, the mixture can be transferred to a reaction tube from which aliquots can be analyzed by ESI mass spectrometry. This experimental setup was used to study the kinetics of acetylcholine hydrolysis under alkaline conditions as a function of pH; the intensities of reactant (acetylcholine) and product (choline) ions were monitored simultaneously as a function of time to allow determination of rate constants that compare favorably with literature data [189]. A method for enzyme kinetic measurements and inhibitor screening of yeast hexokinase has also been described [190].

6. ESI Generation of Ions for Ancillary Experiments

Besides serving as an ion source for identification of compounds, ESI provides a convenient means for generating ions of various types to be used for other kinds of experiments. For example, ESI of aqueous iron(III) nitrate provides an easy route for the generation of bare iron oxide and hydroxide cations such as FeO^+, $FeOH^+$, and

Fe(OH)$^{2+}$ in the gas phase [191]. For experiments involving electron-transfer dissociation (ETD), it is sometimes inconvenient to vaporize compounds that are known to provide the appropriate reagent ions for the ion–ion reactions involved in such technology. McLuckey and coworkers use ESI to prepare ETD reagent anions by CAD of ESI-produced ions of arenecarboxylic acids [192]; by producing the anion reagents at the stage of CAD, undesirable premature reactions with highly protonated species associated with the ESI can be avoided.

X. General Applications of ESI

Because an ESI spectrum of a single compound usually consists of an extensive envelope of peaks, analysis of a mixture of compounds can be confusing, if not impossible to interpret; some means of charge reduction, as with electrons from a corona discharge [193], can simplify the problem. ESI MS continues to be useful in providing molecular weight information and with CAD MS/MS to help characterize a variety of proteins [194–198] including those containing metal ions [199–203]. CAD of multiply protonated molecules is especially efficient, but in some circumstances, e.g., studies of protein–protein reactions, it is important to reduce the charge state of the precursor ion; special instrumentation to manipulate the charge state of ESI ions facilitates such studies [204, 205]. ESI CAD is also useful in peptide mapping [206–208] including recognition of phosphorylation sites [209–212]. The LOD for certain proteins by ESI can be reduced by the use of antibodies immobilized on CNBr-activated sepharose to isolate the corresponding proteolytic peptides for the sample [213]. Dopamine and 6-hydroxydopamine [214] have been studied by ESI-CAD MS/MS, which promotes characteristic fragmentation of free peptides [70, 215] and those that have been chemically modified with a fixed charged [216] or as the trimethylsilylethyl ester [217]; the fragmentation of T-rich oligodeoxynucleotides has also been studied [218]. Derivatization of amino acids with an *N*-hydroxysuccinimide ester of *N*-alkylnicotinic acid enhances their detection by ESI by more than an order of magnitude in some cases [219]. Good synopses of the applications of ESI MS to biological compounds are available in several reviews [95, 220, 221].

ESI-CAD MS/MS has been used in the detection, characterization, and analysis of serine clusters [222], oligonucleotides [223–227], ergot alkaloids [228, 229], the bioactive phenolic compounds in rhubarb [230], glutathione conjugates [231], a series of chalcones [232], carbohydrates [233–236], glycoproteins [237], anabolic steroids [238], the dimethylglycine derivatives of oxysterols [239], phospholipids [240, 241], leukotriene A4 and prostaglandin H4 in the form of the [M – H]$^-$ ion [242], polyunsaturated fatty acids [243] including leukotrienes [244, 245], and novel lipids [246]. Underivatized isomeric monosaccharides can be distinguished by positive-ion ESI tandem MS when analyzed from an aqueous solution containing 1 mM of ammonium acetate [247]. Fatty acid positions in native lipid A can be determined by positive and negative ESI [248]. Precursor scan modes using ESI MS/MS have been used to detect the acyl species of the major classes of lipids, such as monogalactosyldiacylglycerols and phosphatidylglycerols, found in plant chloroplasts [249]. Acylphosphatidylglycerol, a polar lipid class containing three fatty acyl groups, was isolated from *Salmonella* bacteria and characterized by ESI-CAD MS/MS [250]. The detection of oxidized squalene may enhance characterization of forensic fingerprinting [251]. A high-throughput method for single nucleotide polymorphisms (SNPs) requiring no off-line purification of the polymerase chain reaction product was achieved by simple addition of reagent solution into a single sample well, thereby eliminating tedious and time-consuming steps of sample preparation [252]. ESI MS/MS was used to help identify some of the biomarkers of beeswax in an Etruscan cup at the parts-per-million level to provide the first evidence for the use of this material by the

Etruscans as fuel or as a waterproof coating for ceramics [253]. The fragmentation pathways of the $[M + H]^+$ and $[M + Li]^+$ ions of phosphatidylcholine have been examined by ESI-CID MS/MS [254]. ESI MS has been used to characterize PVDF telomers [255], cyclodextrin [256] and other complexes [257], components in waste water [258], complex humic and fulvic acid mixtures [259, 260], industrial polymers [261–263], C_{60} fullerenes [264, 265], oxidative degradation products of vitamin C [266], and insecticides [267]. During analyses of zinc complexes of diaryl bis(*p*-nitrophenyl)porphyrins by ESI-CID MS/MS, loss of two nitro groups was observed [268]. Polyesters prepared from different phthalic acid isomers could be distinguished by their fragmentation behavior during ESI-CID MS/MS [269]. Other studies of polyesters by ESI MS/MS have shown that formation and analysis of anions are often preferable to cations for determination of block length, end-group structure, and copolymer sequence [270]. Industrial alkylpolyphosphonates have been characterized by negative-ion ESI under both acidic and alkaline conditions in conjunction with capillary electrophoresis [271]. Calibration for determining monomer ratios in copolymers of poly(styrene sulfonate-co-maleic acid) can be achieved as a function of the cone voltage [272]. Direct analysis of plasma for drugs using nanoelectrospray infusion from a silicon chip under conditions of ESI MS/MS has been demonstrated to be a viable high-throughput technology [273].

The gentle nature [274] of the ESI process permits some noncovalent [69, 275–282] intermolecular interactions to be studied, including those involving deuterium exchange [70, 283, 284]. A huge noncovalent assembly of 18 monomers of hemocyanin was detected as the native protein with a measured molecular mass of 1,354,940±480 Da [285]; other noncovalent measurements of peptides and proteins can be found in Chapter 12. The interactions of cobalt(III) hexammine, $Co(NH3)_6^{3+}$, with five RNA hairpins have been studied by ESI MS [286]. The interactions between a novel enediyne (a potential anti-tumor agent) and various cytosine-containing oligonucleotides were studied using ESI MS [287]. Iron(II) has been used as a di-cation to enhance the detection of cyclodextrin plus aromatic-compound complexes in ESI MS [288]. The interaction between guanidinium and sulfonate- and carboxylate-functionalized binding partners can be studied by ESI [289]. The formation of complexes between tetanus toxin and carbohydrates containing NeuAc groups as studied by ESI MS may play an important role in ganglioside binding and molecular recognition [290].

The higher charge density of multiple-charge ions increases the effectiveness of CID [19, 291, 292] so that MS/MS spectra of appropriate ions may reveal sequence information for oligonucleotides [293], peptides [294, 295], proteins [296–298], and structural information on eicosanoids [299, 300], hydroxylated lipoxygenase products [301], and triacylglycerols [302]. The detection of relatively small molecules can be enhanced by preparation of the dansyl derivative, which provides a basic site for protonation; this derivatization approach was used to advantage in quantitative methodology for ethinylestradiol [303].

The chirality of a given compound can be determined by the extent of its participation in appropriate ion/molecule reactions. In the quotient ratio method [304], the chiral analyte and a chiral reference compound are simultaneously coordinated to a transition metal ion (e.g., Cu^{+2}) to form a single-charge deprotonated cluster ion; the trimeric complex ion (three chiral ligands – 2 mol of the analyte and 1 mol of the reference compound) is collisionally activated and undergoes dissociation by competitive loss of either a neutral reference or a neutral analyte. The ratio of the product ion branching ratios measured when pure chiral reference compounds are employed in separate experiments is related via the kinetic method to the enantiomeric composition of the chiral mixture; this quotient ratio is logarithmically related to enantiomeric purity. Determinations of enantiomeric excess for amino acids and the chiral drug, DOPA, have been reported using the quotient ratio method [304]. A new version of the kinetic method for chiral analysis, which employs a fixed (nondissociating) ligand as well as the usual analyte and chiral reference ligands, has been reported as applied to the analysis of penicillamine [305]. An alternative approach used enantioselectivite gas-phase guest exchange reactions to assess the chirality of nonnatural α-amino acids by complexation with permethylated β-cyclodextrin [306].

The ESI source can accommodate the eluant from capillary electrophoresis. One version of a rugged CE-ESI interface integrates the separation column, an electrical porous junction, and the spray tip into a single piece of a fused-silica capillary [307]. Another robust version of the capillary interface consists only of a small section of porous tubing, and therefore no liquid junction is needed to secure the porous section; in this case, there is no dead volume and the electrical connection is achieved simply by inserting the capillary outlet containing the porous junction into the existing ESI needle [308]; another report indicates that an overt electrical connection in the sheathless interface is not necessary for successful combination of CE and ESI [309]. The CE-ESI MS technique shows promise for detecting peptides with post-translational modifications [310]. A method based on CE-ESI MS has been developed for the separation and characterization of the D/L-enantiomers of underivatized amino acids [311]. The CE-ESI MS/MS combination has been used in the analysis of liposaccharides and glycans [312] as well as protein glycoforms [313].

Pinkston and coworkers have used SFC-ESI to characterize a variety of complex, low molecular weight polymers, in some cases with quantitation of distinct component classes of the polymer mixtures [314]. Bacteria can be characterized and distinguished by their enzymatic release of a bacterium-specific tag from an array of substrates; the tags released into the supernatant after incubation of substrates with the immobilized bacteria can be directly analyzed by ESI MS, without chromatographic separation [315]. Identification of constituent dyes and the relative amounts of those dyes present on a single fiber of nylon can be established by ESI MS/MS [316].

A method has been developed for the analysis of humic and fulvic acids by size-exclusion chromatography-ESI MS/MS using a completely volatile eluent [317]. A novel interface allows normal-phase liquid chromatography to be coupled with ESI; a make-up solution of 60 mM of ammonium acetate in methanol is infused at 5 ml min^{-1} through the tip of the electrospray probe, providing a sheath liquid that though poorly miscible with the chromatographic effluent, promotes efficient ionization of targeted analytes [318].

References

1. Fenn JB, Mann M, Meng CK, Wong SK and Whitehouse CM, ESI of Large Biomolecules. *Science* **246:** 64-71, 1989.
2. Hofstadler SA, Bakhtiar R and Smith RD, Instruments &Spectral Interpretation for ESI. *J. Chem. Ed.* **73:** A82-A88, 1996.
3. Kebarle P, Overview of mechanisms involved in ESI. *J Mass Spectrom* **35:** 804-817, 2000.
4. Kebarle P, Gas phase ion thermochemistry based on ion-equilibria. *Int J Mass Spectrom* **200:** 313-330, 2000.
5. Van Berkel GJ, Asano KG and Schnier PD, Electrochemical processes in a wire-in-a-capillary bulk-loaded, nona-electrospray emitter. *J Am Soc Mass Spectrom* **12**(7): 853-62., 2001.
6. Van Berkel GJ, McLuckey A and Glish GL, Electrochemical Origin of Radical Cations Observed in ESI-MS. *Anal. Chem.* **64:** 1586-1593, 1992.
7. Van Berkel GJ, Insights into analyte electrolysis in an electrospray emitter from chronopotentiometry experiments and mass transport calculations. *J Am Soc Mass Spectrom* **11:** 951-960, 2000.
8. Mora JF, Van Berkel GJ, Enke CG, Cole RB, Martinez-Sanchez M and Fenn JB, Electrochemical processes in electrospray ionization mass spectrometry. *J Mass Spectrom* **35**(8): 939-52., 2000.
9. Dulcks T and Roellgen FW, Electrohydrodynamic MS. *J. Mass Spectrom* **30:** 324-332, 1995.
10. Hoxha A, Collette C, De Pauw E and Leyh B, Mechanism of Collisional Heating in Electrospray Mass Spectrometry: Ion Trajectory Calculations. *Journal of Physical Chemistry A* **105**(31): 7326-7333, 2001.
11. Van Berkel G, Kertesz V, Ford M and Granger M, Efficient analyte oxidation in an electrospray ion source using a porous flow-through electrode emitter. *J Am Soc Mass Spectrom* **15**(12): 1755-66, 2004.
12. Van Berkel GJ and Zhou F, Gas-Phase Molecular Dications from Electrolytic Process Inherent to ESI. *J. Am. Soc. Mass Spectrom.* **7:** 157-162, 1996.
13. Covey T, Bonner R, Sushan B and Henion J, Ion-Spray MS. *Rapid Comm. Mass Spectrom.* **2:** 249-256, 1988.
14. Chait BT and Kent SBH, Weighing Naked Proteins. *Science* **257:** 145-164, 1992.
15. Wortmann A, Kistler-Momotova A, Zenobi R, Heine MC, Wilhelm O and Pratsinis SE, Shrinking Droplets in ESI and Their Influence on Chemical Equilibria. *J Amer Soc Mass Spectrom* **18:** 385-393, 2007.
16. Dole M, Mack LL, Hines RL, Mobley RC, Ferguson LD and Alice MB, Molecular beams of macroions. *Journal of Chemical Physics* **49**(5): 2240-9, 1968.
17. Duelcks T and Roellgen FW, Ionization conditions and ion formation in electrohydrodynamic mass spectrometry. *International Journal of Mass Spectrometry and Ion Processes* **148**(1/2): 123-44, 1995.
18. Duelcks T and Roellgen FW, Ion source for electrohydrodynamic MS. *J Mass Spectrom* **30:** 324-32, 1995.
19. Cerda BA, Breuker K, Horn DM and McLafferty FW, Charge/radical site initiation versus coulombic repulsion for cleavage of multiply charged ions. Charge solvation in poly(alkene glycol) ions. *J Am Soc Mass Spectrom* **12**(5): 565-70., 2001.
20. Iribarne JV and Thomson BA, On the evaporation of small ions from charged droplets. *J Chem Phys* **64:** 2287-94, 1976.
21. Iribarne JV, Dziedzic PJ and Thomson BA, Atmospheric pressure ion evaporation-MS. *Int J Mass Spectrom Ion Phys* **50:** 331-47, 1983.
22. Zhou S and Cook KD, Protonation in ESI: wrong-way- or right-way-round? *J. Am. Soc. Mass Spectrom.* **11:** 961-966, 2000.
23. McMahon T, Thermochemistry and solvation of gas phase ions. *Int J Mass Spectrom* **227:** vii-viii, 2003.
24. Lavarone AT, Jurchen JC and Williams ER, Effects of solvent on the maximum charge state and charge state distribution of protein ions produced by ESI. *J Amer Soc Mass Spectrom* **11:** 976-985, 2000.
25. Cook KD, Callahan JH and Man VF, Factors Affecting the Sampling of Poly (ethylenimines) by Electrohydrodynamic MS. *Anal. Chem.* **60:** 706-713, 1988.
26. Hamdan M and Curcuruto O, Development of the electrospray ionization technique. *International Journal of Mass Spectrometry and Ion Processes* **108**(2-3): 93-113, 1991.
27. Blades AT, Ikonomou MG and Kebarle P, Mechanism of ESI. *Anal. Chem.* **63:** 2109-2114, 1991.
28. Chillier XFD, Monnier A, Bill H, Guelacar FO, Buchs A, McLuckey SA and Van Berkel GJ, *Rapid Commun. Mass Spectrom.* **10:** 467-479, 1996.
29. Kebarle D and Tang L, Ions in Solution to Ions in Gas Phase: ESI Mechanism. *Anal. Chem.* **65:** 972A-986A, 1993.
30. Smith RD, Light-Wahl KJ, Winger BE and Goodlett DR, Electrospray Ionization MS. In: *Biological MS: Present and Future.* (Ed. Matsuo T, Caprioli, R.M., Gross, M.L. and Seyana, Y.), pp. 41-74. John Wiley & Sons Ltd, 1994.
31. Cech NB and Enke CG, Practical implications of ESI fundamentals. *Mass Spectrom Rev* **20:** 362-87., 2001.

32. Vaidyanathan S, Kell DB and Goodacre R, Selective detection of proteins in mixtures using ESI MS: Influence of instrumental settings and implications for proteomics. *Analytical Chemistry* **76**(17): 5024-5032, 2004.
33. Wilm M and Mann M, Analytical properties of the nanoelectrospray ion source. *Anal Chem* **68**(1): 1-8., 1996.
34. Olumee Z, Callahan JH and Vertes A, Velocity Compression in Cylindrical Capacitor Electrospray of Methanol-Water Mixtures. *Analytical Chemistry* **71**(18): 4111-4113, 1999.
35. Gomez A and Tang K, Charge and fission of droplets in electrostatic sprays. *Phys. Fluids* **6**: 404-414, 1994.
36. Smith JN, Flagan RC and Beauchamp JL, Droplet Evaporation and Discharge Dynamics in Electrospray Ionization. *Journal of Physical Chemistry A* **106**(42): 9957-9967, 2002.
37. Nemes P, Marginean I and Vertes A, Spraying Mode Effect on Droplet Formation and Ion Chemistry in Electrosprays. *Anal Chem* **79**: 3105-3116, 2007.
38. Grimm RL and Beauchamp JL, Evaporation and discharge dynamics of highly charged droplets of heptane, octane, and p-xylene generated by electrospray ionization. *Anal Chem* **74**(24): 6291-7., 2002.
39. Yang P, Cooks RG, Ouyang Z, Hawkridge AM and Muddiman DC, Gentle Protein Ionization Assisted by High-Velocity Gas Flow. *Analytical Chemistry* **77**(19): 6174-6183, 2005.
40. Takats Z, Wiseman JM, Gologan B and Cooks RG, Electrosonic Spray Ionization: Gentle Technique for Folded Proteins & Complexes in the Gas Phase. *Anal Chem* **76**: 4050-4058, 2004.
41. Wiseman JM, Takats Z, Gologan B, Davisson VJ and Cooks RG, Direct characterization of enzyme-substrate complexes by using electrosonic spray ionization MS. *Angewandte Chemie, International Edition* **44**: 913-916, 2005.
42. Takats Z, Drahos L, Schlosser G and Vekey K, Formation of hot ESI ions. *Anal Chem* **74**: 6427-9., 2002.
43. Naban-Maillet J, Lesage D, Bossee A, Gimbert Y, Sztaray J, Vekey K and Tabet J-C, Internal energy distribution in electrospray ionization. *Journal of Mass Spectrometry* **40**(1): 1-8, 2005.
44. Drahos L, Heeren RMA, Collette C, De Pauw E and Vekey K, Thermal energy distribution observed in electrospray ionization. *Journal of Mass Spectrometry* **34**(12): 1373-1379, 1999.
45. Siu KWM, Guevremont R, Le Blanc JCY, O'Brien RT, Berman SS, LeBlanc J and Sui K, Droplet Evaporation in ESI MS: Ethylene Glycol as a Solvent and Its Effects on Ion Desorption. *Org. Mass Spectrom.* **28**: 579-583, 1993.
46. Zhou S, Prebyl BS and Cook KD, Profiling pH changes in the electrospray plume. *Anal Chem* **74**(19): 4885-8., 2002.
47. Zhou F and Van Berkel GJ, Electrochemistry On-Line with ESI-MS. *Anal. Chem.* **67**: 3643-3649, 1995.
48. Van Berkel GJ, Electrolytic deposition of metals on to the high-voltage contact in an electrospray emitter: implications for gas-phase ion formation. *J Mass Spectrom* **35**(7): 773-83., 2000.
49. Van Berkel GJ, Asano KG and Kertesz V, Enhanced study and control of analyte oxidation in electrospray using a thin-channel, planar electrode emitter. *Anal Chem* **74**(19): 5047-56., 2002.
50. Jackson GS and Enke CG, Electrical Equivalence of Electrospray Ionization with Conducting and Nonconducting Needles. *Analytical Chemistry* **71**(17): 3777-3784, 1999.
51. De la Mora JF, Van Berkel GJ, Enke CG, Cole RB, Martinez-Sanchez M and Fenn JB, Electrochemical processes in electrospray ionization mass spectrometry. *Journal of Mass Spectrometry* **35**(8): 939-952, 2000.
52. Van Berkel GJ and Zhou F, Electrospray as a Controlled-Current Electrolytic Cell: Electrochemical Ionization of Neutral Analytes for Detection by Electrospray Mass Spectrometry. *Analytical Chemistry* **67**(21): 3958-64, 1995.
53. Pozniak BP and Cole RB, Ambient Gas Influence on ESI Potential As Revealed by Potential Mapping within the ESI Capillary. *Anal Chem* **79**: 3383-3391, 2007.
54. Chen M and Cook KD, Oxidation Artifacts in the Electrospray MS of Ab Peptide. *Anal Chem* **79**: 2031-2036, 2007.
55. Johnson KA, Shira BA, Anderson JL and Amster IJ, Chemical and on-line electrochemical reduction of metalloproteins with high-resolution electrospray ionization mass spectrometry detection. *Anal Chem* **73**(4): 803-8., 2001.
56. Kertesz V and Van Berkel GJ, Ionic adducts to elucidate reaction mechanisms: reduction of tetracyanoquinodimethane and oxidation of triphenylamine in electrochemistry/ESI MS. *J Solid St Electrochem* **9**: 390-397, 2005.
57. Pretty JR, Deng H, Goeringer DE and Van Berkel GJ, Electrochemically modulated preconcentration and matrix elimination for organic analytes coupled on-line with ESI MS. *Anal Chem* **72**: 2066-74., 2000.
58. Degn H, Christensen T and Sorensen R, Rotating ball inlet mass spectrometry enhanced by electrochemical effects. *Rapid Commun Mass Spectrom* **16**(6): 527-30., 2002.
59. Liu S, Griffiths WJ and Sjovall J, On-column electrochemistry in ESI. *Anal Chem* **75**: 1022-30., 2003.
60. Charbonnier F, Rolando C, Saru F, Hapiot P and Pinson J, Short Time Scale Observation of ESI Current. *Rapid Commun. Mass Spectrom.* **7**: 707-710, 1993.

61. Wampler RM, Blades AT and Kebarle P, Negative ESI of Nucleotides. *J. Amer. Soc. Mass Spectrom.*, 1992.
62. Van Berkel GJ and Kertesz V, Redox buffering in ESI with Cu emitter. *J Mass Spectrom* **36**: 1125-32., 2001.
63. Nilsson S, Svedberg TM, Pettersson J, Bjorefors TF, Markides K and Nyholm L, Evaluations of the stability of sheathless ESI MS emitters using electrochemical techniques. *Anal Chem* **73**: 4607-16., 2001.
64. Moini M, Cao P and Bard AJ, Hydroquinone as a buffer additive for suppression of bubbles formed by electrochemical oxidation of the CE buffer at the outlet electrode in capillary electrophoresis/ESI MS. *Anal Chem* **71**: 1658-61., 1999.
65. Van Berkel GJ, Quirke JM, Tigani RA, Dilley AS and Covey TR, Derivatization for electrospray ionization mass spectrometry. 3. Electrochemically ionizable derivatives. *Anal Chem* **70**(8)**: 1544-54., 1998.
66. Mirzaei H and Regnier F, Enhancing ESI of Peptides by Derivatization. *Anal Chem* **78**: 4175-4183, 2006.
67. Amad MH, Cech NB, Jackson GS and Enke CG, Importance of gas-phase proton affinities in determining the electrospray ionization response for analytes and solvents. *J Mass Spectrom* **35**(7)**: 784-9., 2000.
68. Meng CK, Fenn JB, Ancleto JF, Pleasance S and Boyd RK, Formation of Charged Clusters in ESI for Calibration. *Org. Mass Spectrom.* **26**: 542-549, 1991.
69. Peschke M, Verkerk UH and Kebarle P, Features of ESI mechanism multi-charged noncovalent protein complexes. *J Amer Soc Mass Spectrom* **15**: 1424-1434, 2004.
70. Buijs J, Hagman C, Hakansson K, Richter JH, Hakansson P and Oscarsson S, Inter- and intra-molecular migration of peptide amide hydrogens during electrospray ionization. *J Am Soc Mass Spectrom* **12**(4)**: 410-9., 2001.
71. Rodriqueza CF, Guoa X, Shoeiba T, Hopkinsona AC and Siua KWM, Formation of [M -nH + mNa](m-n)+ and [M -nH + mK](m-n)+ ions in ESI MS of peptides and proteins. *J Am Soc Mass Spectrom* **11**: 967-975, 2000.
72. Mann M, Meng C and Fenn J, Interpreting MS of Multiply Charged Ions. *Anal. Chem.* **61**: 1702-1708, 1989.
73. Labowsky M, Whitehouse C and Fenn JB, Mass Computations in ESI. *Rapid Commun. Mass Spectrom.*: 71-84, 1993.
74. Hagen JJ and Monnig CA, Algorithm for Mass in ESI. *Anal. Chem.* **66**: 1877-1883, 1994.
75. Reinhold BB and Reinhold VN, Entropy-based Algorithm for ESI. *J. Amer. Soc. Mass Spectrom.* **3**: 207-215, 1992.
76. Williams JD, Weiner BE, Ormand JR, Brunner J, Thornquest AD, Jr. and Burinsky DJ, Automated molecular weight assignment of electrospray ionization mass spectra. *Rapid Commun Mass Spectrom* **15**(24)**: 2446-55, 2001.
77. Pearcy JO and Lee TD, MoWeD, a computer program to rapidly deconvolute low resolution ESI LC/MS runs to determine component molecular weights. *J Am Soc Mass Spectrom* **12**: 599-606., 2001.
78. Zubarev RA, Demirev PA, Håkansson P and Sundqvist BUR, Limits to Accurate Mass Characterization of Large Biomolecules. *Anal. Chem.* **67**: 3793-3798, 1995.
79. Beavis RC, Chemical Mass of Carbon in Proteins. *Anal. Chem.* **65**: 496-497, 1993.
80. Senko MW, Beu SC and McLafferty FW, Monoisotopic Mass in FTICR. *J. Am. Soc. Mass Spectrom.* **6**: 229-233, 1995.
81. Maleknia SD and Downard KM, Charge Ratio Approach to Deconvolution of ESI Spectra. *Anal Chem* **77**: 111-119, 2005.
82. Stults JT, Minimizing Peak Coalescence: High Resolution Separation of Isotope Peaks in MALDI FT-ICR MS. *Anal. Chem.* **69**: 1815-19, 1997.
83. Tang K, Shahgholi M, Garcia BA, Heaney PJ and Cantor CR, Improvement in the Apparent Mass Resolution of Oligonucleotides by Using 12C/14N-Enriched Samples. *Anal. Chem.* **74**: 226 -231, 2002.
84. Senko MW, Beu SC and McLafferty FW, Automated Assignment of Charge States in ESI. **6**: 52-56, 1995.
85. Guan Z, Campbell VL and Laude DA, Charge State Determination by Schiff-Bass Formation in ESI. *J. Mass Spectrom.* **30**: 119-123, 1995.
86. Kameoka J, Orth R, Ilic B, Czaplewski D, Wachs T and Craighead HG, An electrospray ionization source for integration with microfluidics. *Anal Chem* **74**(22)**: 5897-901., 2002.
87. Schmidt A, Bahr U and Karas M, Influence of pressure in the first pumping stage on analyte desolvation and fragmentation in nano-ESI MS. *Analytical Chemistry* **73**(24)**: 6040-6046, 2001.
88. James PF, Perugini MA and O'Hair RAJ, Artefacts in ESI of Saturated Diacylglycerophosphocholines: Condensed Phase Hydrolysis to Gas Phase Intercluster Reactions. *J Amer Soc Mass Spectrom* **17**: 384-394, 2006.
89. Peschke M, Verkerk UH and Kebarle P, Prediction of the charge states of folded proteins in electrospray ionization. *European Journal of Mass Spectrometry* **10**(6)**: 993-1002, 2004.
90. Myung S, Wiseman JM, Valentine SJ, Takats Z, Cooks RG and Clemmer DE, Coupling DESI with Ion Mobility/MS for Protein Structure: Evidence for Desorption of Folded and Denatured States. *J Phys Chem B* **110**: 5045-5051, 2006.

91. Tang L and Kebarle P, Dependence of ESI Intensity on Analytes Concentration. *Anal. Chem.* **65**: 3654-3668, 1993.
92. Tang X, Bruce JE and Hill HH, Jr., Characterizing Electrospray Ionization Using Atmospheric Pressure Ion Mobility Spectrometry. *Analytical Chemistry* **78**(22): 7751-7760, 2006.
93. Wilm MS and Mann M, ESI and Taylor Cone Theory. *Int. J. Mass Spectrom. Ion Proc.* **136**: 167-180, 1994.
94. El-Faramawy A, Siu KWM and Thomson BA, Efficiency of Nano-ESI. *J Amer Soc Mass Spectrom* **16**: 1702-7, 2005.
95. Wickremsinhe ER, Singh G, Ackermann BL, Gillespie TA and Chaudhary AK, A review of nanoelectrospray ionization applications for drug metabolism and pharmacokinetics. *Current Drug Metabolism* **7**(8): 913-928, 2006.
96. Wilm M, Neubauer G and Mann M, Parent Ion Scans of Unseparated Peptide Mixtures. *Anal. Chem.* **68**: 527-533, 1996.
97. Tang K and Smith R, Physical/chemical separations in the break-up of highly charged droplets. *J Am Soc Mass Spectrom* **12**(3): 343-7, 2001.
98. Cech N and Enke C, Effect of affinity for droplet surfaces on the fraction of analyte. *Anal Chem* **73**(19): 4632-9, 2001.
99. Schmidt A, Karas M and Dulcks T, Effect of different solution flow rates on analyte ion signals in nano-ESI MS, or: when does ESI turn into nano-ESI? *Journal of the American Society for Mass Spectrometry* **14**(5): 492-500, 2003.
100. Le Gac S, Rolando C and Arscott S, Open Design Microfabricated Nib-Like NanoESI Emitter Tip on Conducting Silicon for the Application of Ionization Voltage. *J Amer Soc Mass Spectrom* **17**: 75-80, 2006.
101. Le Gac S, Cren-Olive C, Rolando C and Arscott S, A novel nib-like design for microfabricated nanospray tips. *Journal of the American Society for Mass Spectrometry* **15**(3): 409-412, 2004.
102. Le Gac S, Arscott S and Rolando C, Novel microfabricated nano-ESI sources for proteomics applications. *Spectra Analyse* **34**(245): 20-25, 2005.
103. Takats Z, Wiseman J, Gologan B and Cooks R, Mass spectrometry sampling under ambient conditions with desorption electrospray ionization. *Science* **306**(5695): 471-3, 2004.
104. Venter A, Sojka PE and Cooks RG, Droplet Dynamics and Ionization Mechanisms in Desorption Electrospray Ionization Mass Spectrometry. *Analytical Chemistry* **78**(24): 8549-8555, 2006.
105. Takats Z, Wiseman JM and Cooks RG, Ambient MS by DESI: Instrumentation, mechanisms and applications in forensics, chemistry, and biology. *J Mass Spectrom* **40**: 1261-1275, 2005.
106. Ifa DR, Gumaelius LM, Eberlin LS, Manicke NE and Cooks RG, Forensic analysis of inks by imaging DESI mass spectrometry. *Analyst* **132**(5): 461-467, 2007.
107. Costa AB and Cooks RG, Simulation of Atmospheric Transport in DESI. In: *55th Ann Conf Mass Spectrom, Indianapolis, 2007.*
108. Cooks RG, Ouyang Z, Takats Z and Wiseman JM, Ambient Mass Spectrometry. *Science (Washington, DC, United States)* **311**(5767): 1566-1570, 2006.
109. Bereman MS and Muddiman DC, Detection of Attomole Amounts of Analyte by DESI-MS Determined Using Fluorescence Spectroscopy. *J Amer Soc Mass Spectrom* **18**: 1093-1096, 2007.
110. Hiraoka K, Mori K and Asakawa D, Fundamental aspects of ESI droplet impact/SIMS. *J Mass Spectrom* **41**: 894-902, 2006.
111. Cole RB, Some tenets pertaining to ESI MS. *J Mass Spectrom* **35**: 763-72, 2000.
112. Chen H, Talaty NN, Takats Z and Cooks RG, Desorption Electrospray Ionization Mass Spectrometry for High-Throughput Analysis of Pharmaceutical Samples in the Ambient Environment. *Analytical Chemistry* **77**(21): 6915-6927, 2005.
113. Nefliu M, Venter A and Cooks RG, DESI and electrosonic spray ionization for solid- and solution-phase analysis of industrial polymers. *Chemical Communications (Cambridge, United Kingdom)* (8): 888-890, 2006.
114. Justes DR, Talaty N, Cotte-Rodriguez I and Cooks RG, Detection of explosives on skin using ambient ionization MS. *Chem Commun* (21): 2142-2144, 2007.
115. D'Agostino PA, Chenier CL, Hancock JR and Lepage CRJ, DESI MS of chemical warfare agents from solid-phase microextraction fibers. *Rapid Commun Mass Spectrom* **21**: 543-549, 2007.
116. Cotte-Rodriguez I, Takats Z, Talaty N, Chen H and Cooks RG, DESI of Explosives on Surfaces: Sensitivity and Selectivity Enhancement by Reactive DESI. *Anal Chem* **77**: 6755-6764, 2005.
117. Cotte-Rodriguez I, Chen H and Cooks RG, Rapid trace detection of TATP by complexation reactions during DESI. *Chemical Communications (Cambridge, United Kingdom)* (9): 953-955, 2006.
118. Pan Z, Gu H, Talaty N, Chen H, Shanaiah N, Hainline BE, Cooks RG and Raftery D, Principal component analysis of urine metabolites detected by NMR and DESI-MS in patients with inborn errors of metabolism. *Analytical and Bioanalytical Chemistry* **387**(2): 539-549, 2007.

119. Kaur-Atwal G, Weston DJ, Green PS, Crosland S, Bonner PLR and Creaser CS, Analysis of tryptic peptides using DESI combined with ion mobility spectrometry/MS. *Rapid Commun Mass Spectrom* **21:** 1131-1138, 2007.

120. Shin Y-S, Drolet B, Mayer R, Dolence K and Basile F, DESI-MS of Proteins. *Anal Chem* **79:** 3514-3518, 2007.

121. Song Y, Talaty N, Tao WA, Pan Z and Cooks RG, Rapid ambient mass spectrometric profiling of intact, untreated bacteria using DESI. *Chem Commun* (1): 61-63, 2007.

122. Van Berkel G, Ford M and Deibel M, TLC & MS coupled by desorption ESI. *Anal Chem* **77:** 1207-15, 2005.

123. Ford M, Kertesz V and Van Berkel G, TLC/ESI/MS system: analysis of rhodamine dyes separated on reversed-phase C(8) plates. *J Mass Spectrom* **22:** 22, 2005.

124. Van Berkel GJ, Sanchez AD and Quirke JM, Thin-layer chromatography and electrospray mass spectrometry coupled using a surface sampling probe. *Anal Chem* **74**(24): 6216-23., 2002.

125. Hsu F-L, Chen C-H, Yuan C-H and Shiea J, TLC Interface to ESI MS. *Anal Chem* **75:** 2493-2498, 2003.

126. Van Berkel GJ, Ford MJ, Doktycz MJ and Kennel SJ, Evaluation of surface-sampling probe ESI MS for the analysis of surface-deposited & affinity-captured proteins. *Rapid Commun Mass Spectrom* **20:** 1144-1152, 2006.

127. Ifa DR, Wiseman JM, Song Q and Cooks RG, Development of capabilities for imaging MS under ambient conditions with desorption electrospray ionization (DESI). *Int J Mass Spectrom* **259:** 8-15, 2007.

128. Sparrapan R, Eberlin LS, Haddad R, Cooks RG, Eberlin MN and Augusti R, Ambient Eberlin reactions via desorption electrospray ionization mass spectrometry. *Journal of Mass Spectrometry* **41**(9): 1242-1246, 2006.

129. Chen H, Cotte-Rodriguez I and Cooks RG, cis-Diol functional group recognition by reactive desorption electrospray ionization (DESI). *Chemical Communications (Cambridge, United Kingdom)* (6): 597-599, 2006.

130. Nyadong L, Green MD, De Jesus VR, Newton PN and Fernandez FM, Reactive DESI linear ion trap MS of latest-generation counterfeit antimalarials via noncovalent complex formation. *Anal Chem* **79:** 2150-2157, 2007.

131. Nefliu M, Cooks RG and Moore C, Enhanced Desorption Ionization Using Oxidizing Electrosprays. *Journal of the American Society for Mass Spectrometry* **17**(8): 1091-1095, 2006.

132. Cody RB, Laramee JA and Durst HD, Versatile new ion source for the analysis of materials in open air under ambient conditions. *Analytical Chemistry* **77**(8): 2297-2302, 2005.

133. Sampson JS, Hawkridge AM and Muddiman DC, Generation and Detection of Multiply-Charged Peptides and Proteins by MALDESI FT-ICR MS. *J Amer Soc Mass Spectrom* **17:** 1712-1716, 2006.

134. Breaux GA, Green-Church KB, France A and Limbach PA, Surfactant-aided, matrix-assisted laser desorption/ionization mass spectrometry of hydrophobic and hydrophilic peptides. *Anal Chem* **72**(6): 1169-74., 2000.

135. Yu Y-Q, Gilar M, Lee PJ, Bouvier ESP and Gebler JC, Enzyme-friendly, mass spectrometry-compatible surfactant for in-solution enzymatic digestion of proteins. *Analytical Chemistry* **75**(21): 6023-6028, 2003.

136. le Coutre J, Whitelegge JP, Gross A, Turk E, Wright EM, Kaback HR and Faull KF, Proteomics on full-length membrane proteins using mass spectrometry. *Biochemistry* **39**(15): 4237-42., 2000.

137. Tsai C-Y, Pai P-J, Ho Y-H, Lu J-F, Wang J-S, Lin W-Y and Her G-R, Rapid protein identification using a disposable on-line clean-up/concentrating device and ESI MS. *Rapid Commun Mass Spectrom* **21:** 459-465, 2007.

138. Whitelegge JP, Gundersen CB and Faull KF, Electrospray-ionization mass spectrometry of intact intrinsic membrane proteins. *Protein Sci* **7**(6): 1423-30., 1998.

139. Szabo PT and Kele Z, Electrospray mass spectrometry of hydrophobic compounds using dimethyl sulfoxide and dimethylformamide as solvents. *Rapid Commun Mass Spectrom* **15**(24): 2415-9, 2001.

140. Zook DR and Bruins AP, On cluster ions, ion transmission, and linear dynamic range limitations in electrospray (ionspray) mass spectrometry. *International Journal of Mass Spectrometry and Ion Processes* **162**(1-3): 129-147, 1997.

141. Rist W, Mayer MP, Andersen JS, Roepstorff P and Jorgensen TJD, Rapid desalting of protein samples for on-line microflow ESI MS. *Anal Biochem* **342:** 160-162, 2005.

142. Fornea DS, Wu Y and Marcus RK, Capillary-Channeled Polymer Fibers as a Stationary Phase for Desalting of Protein Solutions for Electrospray Ionization Mass Spectrometry Analysis. *Analytical Chemistry* **78**(15): 5617-5621, 2006.

143. Wilson DJ and Konermann L, Ultrarapid Desalting of Protein Solutions for Electrospray Mass Spectrometry in a Microchannel Laminar Flow Device. *Analytical Chemistry* **77**(21): 6887-6894, 2005.

144. Chang DY, Lee CC and Shiea J, Detecting large biomolecules from high-salt solutions by fused-droplet electrospray ionization mass spectrometry. *Anal Chem* **74**(11): 2465-9., 2002.

145. Cole RB and Harrata AK, Solvent Effect on Negative ion ESI. *J. Am. Soc. Mass Spectrom.* **4:** 546-556, 1993.

146. Cech NB and Enke CG, ESI response vs nonpolar character of small peptides. *Anal Chem* **72:** 2717-23., 2000.

147. Charles L, Pepin D, Gonnet F and Tabet JC, Effects of liquid phase composition on salt cluster formation in positive ion mode ESI MS: implications for clustering mechanism in ESI. *J Am Soc Mass Spectrom* **12:** 1077-84., 2001.

148. Pascoe R, Foley JP and Gusev AI, Reduction in matrix-related signal suppression effects in electrospray ionization mass spectrometry using on-line two-dimensional liquid chromatography. *Anal Chem* **73**(24)**:** 6014-23., 2001.

149. Holcapek M, Volna K, Jandera P, Kolarova L, Lemr K, Exner M and Cirkva A, Effects of ion-pairing reagents on ESI signal suppression of sulphonated dyes and intermediates. *Journal of Mass Spectrometry* **39**(1)**:** 43-50, 2004.

150. Gangl ET, Annan M, Spooner N and Vouros P, Reduction of Signal Suppression Effects in ESI-MS Using a Nanosplitting Device. *Analytical Chemistry* **73**(23)**:** 5635-5644, 2001.

151. Kralj B, Kocjan D and Kobe J, ESI MS of ammonium ion complexes with anomeric 2,3-O-isopropylidene-1alpha- and -1beta-D-ribofuranosyl azides: anomeric and kinetic isotope effects. *Rapid Commun Mass Spectrom* **16:** 1-10., 2002.

152. Neubauer G and Anderegg RJ, Identifying Charge States in ESI-MS. *Anal. Chem.* **66:** 1056-1061, 1994.

153. Kish MM, Ohanessian G and Wesdemiotis C, The Na+ affinities of a-amino acids: side-chain substituent effects. *International Journal of Mass Spectrometry* **227**(3)**:** 509-524, 2003.

154. Schnier PD and Gross DS, Maximum Charge State of Peptide and Protein Ions in ESI. *J. Am. Soc. Mass Spectrom.* **6:** 1086-1097, 1995.

155. Fabris D, Kelly M, Murphy C, Wu Z and Fenselau C, High-Energy CID of Multiply Charged Peptides from ESI. *J. Am. Soc. Mass Spectrom.* **4:** 652-661, 1993.

156. Pan P and McLuckey SA, Electrospray ionization of protein mixtures at low pH. *Anal Chem* **75**(6)**:** 1491-9., 2003.

157. Wang Y, Taylor JS and Gross ML, Fragmentation of electrospray-produced oligodeoxynucleotide ions adducted to metal ions. *J Am Soc Mass Spectrom* **12**(5)**:** 550-6., 2001.

158. Wang Y, Taylor JS and Gross ML, Fragmentation of photomodified oligodeoxynucleotides adducted with metal ions in an electrospray-ionization ion-trap mass spectrometer. *J Am Soc Mass Spectrom* **12**(11)**:** 1174-9., 2001.

159. Chen R and Li L, Lithium and transition metal ions enable low energy collision-induced dissociation of polyglycols in electrospray ionization mass spectrometry. *J Am Soc Mass Spectrom* **12**(7)**:** 832-9., 2001.

160. Greig M and Griffey RH, Organic Bases for ESI of Oligonucleotides. *Rapid Commun. Mass Spectrom.* **9:** 97-102, 1995.

161. Liu C, Wu Q, Harms AC and Smith RD, On-Line Micro-dialysis for Sample Cleanup in ESI-MS. *Anal. Chem.* **68:** 3295-3299, 1996.

162. Kast J, Gentzel M, Wilm M and Richardson K, Noise filtering techniques for electrospray quadrupole time of flight mass spectra. *Journal of the American Society for Mass Spectrometry* **14**(7)**:** 766-776, 2003.

163. Cheng ZL, Siu KWM, Guevremont R and Berman SS, ESI of Aqueous Salt Solutions. *J. Am. Soc. Mass Spectrom.* **3:** 281-288, 1992.

164. Agnes GR and Horlick G, Determination of Solution Ions by ESI. *Appl. Spectrosc.* **48:** 655-661, 1994.

165. Lamb JH and Sweetman GMA, Organogermaniums by ESI-MS/MS. *Rapid Commun. Mass Spectrom.* **10:** 594-596, 1996.

166. Hop CE and Bakhtiar R, ESI in Inorganic Chem and Synthetic Polymer Chem. *J. Chem. Educ.* **73:** A162, A164-A169, 1996.

167. Colton R, D'Agostino A and Traeger JC, ESI Applied to Inorganic and Organometallic Chemistry. *Mass Spectrom. Rev.* **14:** 79-106, 1995.

168. Cardwell TJ, Colton R, Lambropoulos N, Traeger JC and Marriott PJ, ESI-MS of Zinc Dithiophosphates. *Anal. Chim. Acta* **280:** 239-244, 1993.

169. Ketterer ME and Guzowski JP, Isotope Ratio Measurement in Elemental-Mode ESI-MS. *Anal Chem* **68:** 883-887, 1996.

170. Di Marco VB and Bombi GG, ESI-MS in the study of metal-ligand solution equilibria. *Mass Spec Rev* **25:** 347-379, 2006.

171. Sigman ME and Armstrong P, Analysis of oxidizer salt mixtures by electrospray ionization mass spectrometry. *Rapid Commun Mass Spectrom* **20:** 427-432, 2005.

172. Ross AR, Ikonomou MG and Orians KJ, Electrospray ionization of alkali and alkaline earth metal species. Electrochemical oxidation and pH effects. *J Mass Spectrom* **35**(8)**:** 981-9., 2000.

173. Arakawa R, Sasao A and Sonoda N, ESI MS of chemical reactions of Se in strongly basic amines. *J Mass Spectrom* **40:** 66-69, 2005.

174. Borrett VT, Colton R and Traeger JC, ESI-MS of Phosphoric/phosphonic Acid-ester with Alkali Ions. *Eur. Mass Spectrom.* **1:** 131-140, 1995.

175. Lover T, Bowmaker GA, Henderson W and Cooney RP, ESI of Cadmium Thiophenolate Complexes. *Chem. Commun. (Cambridge)* **5:** 683-685, 1996.

176. Minakata K, Suzuki M and Suzuki O, Determination of molybdenum and/or ruthenium in urine using electrospray ionization mass spectrometry. *Analytical Biochemistry* **348**(1)**:** 148-150, 2006.

177. Schroder D, Weiske T and Schwarz H, Dissociation behavior of Cu(urea)+ complexes generated by electrospray ionization. *International Journal of Mass Spectrometry* **219**(3)**:** 729-738, 2002.

178. Koomen JM, Lucas JE, Haneline MR, Beckwith King JD, Gabbai F and Russell DH, NanoESI MS and MS-MS of polydendate Lewis acids, (C6F4Hg)3 and o-C6F4(HgCl)2: binding selectivity. *Int J Mass Spectrom* **225:** 225-231, 2003.

179. Liu Z, Yu L, Chen Y, Zhou N, Chen J, Zhu C, Xin B and Zhao Y, Differences between the protonated and sodium adducts of pentacoordinated bisaminoacylspirophosphoranes in ESI. *J Mass Spectrom* **38:** 231-233., 2003.

180. Soga T, Ueno Y, Naraoka H, Matsuda K, Tomita M and Nishiokat T, Pressure-assisted capillary electrophoresis electrospray ionization mass spectrometry for analysis of multivalent anions. *Anal Chem* **74**(24)**:** 6224-9., 2002.

181. Yamaguchi K, Cold-spray ionization MS: Principle and applications. *J Mass Spectrom* **38:** 473-490, 2003.

182. Jiang Y and Cole RB, Oligosaccharide analysis using anion attachment in negative mode electrospray mass spectrometry. *Journal of the American Society for Mass Spectrometry* **16**(1)**:** 60-70, 2005.

183. Catharino RR, Cunha IBS, Fogaca AO, Facco EMP, Godoy HT, Daudt CE, Eberlin MN and Sawaya ACHF, Characterization of must and wine of 6 varieties of grapes by infusion ESI MS. *J Mass Spectrom* **41:** 185-190, 2006.

184. Moe MK, Anderssen T, Strom MB and Jensen E, Structure characterization of unsaturated acidic phospholipids by vicinal di-hydroxylation of fatty acid double bonds and negative ESI MS. *J Amer Soc Mass Spectrom* **16:** 46-59, 2005.

185. Wu Z, Gao W, Phelps MA, Wu D, Miller DD and Dalton JT, Favorable Effects of Weak Acids on Negative-Ion Electrospray Ionization Mass Spectrometry. *Analytical Chemistry* **76**(3)**:** 839-847, 2004.

186. Chen YH, Hill HH, Jr. and Wittmer DP, Analytical merit of electrospray ion mobility spectrometry as a chromatographic detector. *J. Microcolumn Sep.* **6**(5)**:** 515-524, 1994.

187. Wu C, Siems W and Hill H, Jr., Secondary electrospray ionization ion mobility spectrometry/mass spectrometry of illicit drugs. *Anal Chem* **72**(2)**:** 396-403, 2000.

188. Tam M and Hill HH, Jr., Secondary Electrospray Ionization-Ion Mobility Spectrometry for Explosive Vapor Detection. *Analytical Chemistry* **76**(10)**:** 2741-2747, 2004.

189. Kolakowski BM, Simmons DA and Konermann L, Stopped-flow electrospray ionization mass spectrometry: a new method for studying chemical reaction kinetics in solution. *Rapid Commun Mass Spectrom* **14**(9)**:** 772-6., 2000.

190. Gao H and Leary JA, Multiplex inhibitor screening and kinetic constant determinations for yeast hexokinase using mass spectrometry based assays. *Journal of the American Society for Mass Spectrometry* **14**(3)**:** 173-181, 2003.

191. Schroeder D, Roithova J and Schwarz H, Electrospray ionization as a convenient new method for the generation of catalytically active iron-oxide ions in the gas phase. *International Journal of Mass Spectrometry* **254**(3)**:** 197-201, 2006.

192. Huang T-Y, Emory JF, O'Hair RAJ and McLuckey SA, Electron-Transfer Reagent Anion Formation via Electrospray Ionization and Collision-Induced Dissociation. *Analytical Chemistry* **78**(21)**:** 7387-7391, 2006.

193. Ebeling DD, Westphall MS, Scalf M and Smith LM, Corona Discharge in Charge Reduction Electrospray Mass Spectrometry. *Anal. Chem.* **72:** 5158 -5161, 2000.

194. Shevchenko A, Chernushevich I, Wilm M and Mann M, De Novo peptide sequencing by nanoelectrospray tandem mass spectrometry using triple quadrupole and quadrupole/time-of-flight instruments. *Methods Mol Biol* **146:** 1-16., 2000.

195. Feistner GJ, Faull KF, Barofsky DF and Roepstorff P, Peptide & Protein Charting. *J. Mass Spectrom.* **30:** 519-530, 1995.

196. Bakhtiar R, Hofstadler SA and Smith RD, Peptides and Proteins by ESI. *J. Chem. Ed.* **73:** A118-A123, 1996.

197. Arnott D, Henzel WJ and Stults JT, Rapid Identification of Co-Migrating Gel-Isolated Proteins Using Ion Trap Mass Spectrometry. **19:** 968-980, 1998.

198. Arnott D, O'Connell KL, King KL and Stults JT, An Integrated Approach to Proteome Analysis: Identification of Proteins Associated with Cardiac Hypertrophy. *Anal. Biochem.* **258:** 1-18, 1998.

199. Allen MH and Hutchens TW, *Rapid Commun. Mass Spectrom.* **6:** 308-312, 1992.

200. Wang J, Guevremont R, Siu KWM, Ball LJ and et al., ESI-MS/MS of Alkali Ion Complexes of Tripeptides: Zinc Co-ordination in Yeast Transcriptional Activator PPR1. *Eur. Mass Spectrom.* **358:** 171-181, 1995.

201. Yu X, Wojciechowski M, Fenselau C, Moreau S, Awade AC, Molle D, LeGraet Y and Brule G, Metal-containing Proteins by ESI: Hen Egg White Lysozyme-Metal Ion Interactions. *Anal. Chem.* **65:** 1355-1359, 1993.

202. Wells JM, Reid GE, Engel BJ, Pan P and McLuckey SA, Dissociation reactions of gaseous ferro-, ferri-, and apo-cytochrome c ions. *J Am Soc Mass Spectrom* **12**(7): 873-6., 2001.

203. Johnson KA and Amster IJ, Mass spectrometry of a 3+ oxidation state for a [4Fe-4S] metalloprotein: ESI-FTICR MS of the high potential iron-sulfur protein from Chromatium vinosum. *J Am Soc Mass Spectrom* **12**: 819-25., 2001.

204. Badman ER, Chrisman PA and McLuckey SA, A quadrupole ion trap mass spectrometer with three independent ion sources for the study of gas-phase ion/ion reactions. *Anal Chem* **74**(24): 6237-43., 2002.

205. Xia Y, Chrisman PA, Erickson DE, Liu J, Liang X, Londry FA, Yang MJ and McLuckey SA, Implementation of Ion/Ion Reactions in a QTOF MS. *Anal Chem* **78**: 4146-4154, 2006.

206. Burdick DJ and Stults JT, Analysis of Peptide Synthesis Products by Electrospray Ionization Mass Spectrometry. *Meth. Enzymol.* **289**: 499-519, 1997.

207. Champion KM, Arnott D, Henzel WJ, Hermes S, Weikert S, Stults JT, Vanderlaan M and Krummen L, A Two-Dimensional Protein Map of Chinese Hamster Ovary Cells. *Electrophoresis* **20**: 994-1000, 1999.

208. Chalkley RJ and Burlingame AL, Identification of GlcNAcylation sites of peptides and alpha-crystallin using Q-TOF mass spectrometry. *J Am Soc Mass Spectrom* **12**(10): 1106-13., 2001.

209. Nuwaysir LM and Stults JT, ESI of Phosphopeptides. *J. Am. Soc. Mass Spectrom.* **4**: 662-669, 1993.

210. Kassel DB, Consler TG, Shalaby M, Sekhri P, Gordon N, Nadler T, Wickham G, Iannitti P, Boschenok J and Sheil MM, Mapping of Phosphorylation Sites: ESI of an Antibiotic-Oligonucleotide Adduct. *FEBS Lett.* **360**: 39-46, 1995.

211. Cao P and Stults JT, Phosphopeptide Analysis by On-Line Immobilized Metal-Ion Affinity Chromatography-Capillary Electrophoresis-Electrospray Ionization Mass Spectrometry. *J. Chromatogr. A* **853**: 225-235, 1999.

212. Cao P and Stults JT, Mapping the Phosphorylation Sites of Proteins Using On-line Immobilized Metal Ion Affinity Chromatography/CE/ESI-MS/MS. *Rapid Commun. Mass Spectrom.* **14**: 1600-1606, 2000.

213. Fenaille F, Tabet JC and Guy PA, Immunoaffinity purification and characterization of 4-hydroxy-2-nonenal- and malondialdehyde-modified peptides by ESI-MS/MS. *Anal Chem* **74**: 6298-304., 2002.

214. Hao C, March RE, Croley TR, Chen S, Legault MG and Yang P, Study of dopamine and the neurotoxin 6-hydroxydopamine by ESI-MS/MS. *Rapid Commun Mass Spectrom* **16**: 591-9., 2002.

215. Brinkworth CS, Dua S, McAnoy AM and Bowie JH, Negative ion fragmentations of deprotonated peptides: backbone cleavages directed through both Asp and Glu. *Rapid Comm. Mass Spectrom.* **15**: 1965-1973, 2001.

216. Sadagopan N and Watson JT, Mass spectrometric evidence for mechanisms of fragmentation of charge-derivatized peptides. *J Am Soc Mass Spectrom* **12**(4): 399-409., 2001.

217. Lejeune V, Cavelier F, Enjalbal C, Martinez J and Aubagnac JL, ESI of Amino Acid Trimethylsilylethyl Esters. *J Mass Spectrom* **37**(11): 1168-1170., 2002.

218. Wan KX and Gross ML, Fragmentation mechanisms of oligodeoxynucleotides: effects of replacing phosphates with methylphosphonates and thymines with other bases in T-rich sequences. *J Am Soc Mass Spectrom* **12**(5): 580-9., 2001.

219. Yang W-C, Mirzaei H, Liu X and Regnier FE, Enhancement of Amino Acid Detection and Quantification by Electrospray Ionization Mass Spectrometry. *Analytical Chemistry* **78**(13): 4702-4708, 2006.

220. Burlingame AL and Carr SA, Mass Spectrometry in the Biological Sciences. Humana Press, Totowa, NJ., 1996.

221. Costello CE, Time, life ... and mass spectrometry. New techniques to address biological questions. *BIOPHYSICAL CHEMISTRY* **68**(1-3): 173-188, 1997.

222. Nanita SC, Sokol E and Cooks RG, Alkali Metal-Cationized Serine Clusters Studied by Sonic Spray Ionization MS/MS. *J Amer Soc Mass Spectrom* **18**: 856-868, 2007.

223. Chenna A and Iden CR, DNA Adducts by ESI-MS. *Chem. Res. Toxicol.* **6**: 261-268, 1993.

224. Ding J and Anderegg RJ, ESI-MS of Oligonucleotides. *J. Am. Soc. Mass Spectrom.* **6**: 159-164, 1995.

225. Habibi-Goudarzi S and McCluckey SA, CAD of Nucleosides after ESI in Ion Trap. *J. Am. Soc. Mass Spectrom.* **6**: 102-113, 1995.

226. Kowalak JA, Dalluge JJ, McCloskey JA, Stetter KO, Limback PA and Crain PF, Transfer RNA from Hyperthemophiles: Improved Mass Accuracy (Salt Removal) in ESI of Nucleotides. *Biochemistry* **33**: 7869-7876, 1994.

227. Flora JW and Muddiman DC, Complete sequencing of mono-deprotonated peptide nucleic acids by sustained off-resonance irradiation collision-induced dissociation. *J Am Soc Mass Spectrom* **12**(7): 805-9., 2001.

228. Lehner AF, Craig M, Fannin N, Bush L and Tobin T, Fragmentation patterns of selected ergot alkaloids by electrospray ionization tandem quadrupole mass spectrometry. *Journal of Mass Spectrometry* **39**(11): 1275-1286, 2004.

229. Wang D, Liu Z, Guo M and Liu S, Structural elucidation and identification of alkaloids in rhizoma coptidis by electrospray ionization tandem mass spectrometry. *Journal of Mass Spectrometry* **39**(11): 1356-1365, 2004.

230. Ye M, Han J, Chen H, Zheng J and Guo D, Analysis of Phenolic Compounds in Rhubarbs Using LC Coupled with ESI MS. *J Amer Soc Mass Spectrom* 18: 82-91, 2007.
231. Davis MR, Kassahun K, Jochheim CM, Brandt KM, Baillie TA, Hua Y, Lu W, Henry MS, Pierce RH and Cole RB, Glutathione Conjugates by ESI-MS: Brevetoxins in "Red Tide" Algae. *Chem. Res. Toxicol.* 6: 376-383, 1993.
232. Zhang J and Brodbelt JS, Structural characterization and isomer differentiation of chalcones by electrospray ionization tandem mass spectrometry. *Journal of Mass Spectrometry* 38(5): 555-572, 2003.
233. Reinhold VN, Reinhold BB and Costello CE, Carbohydrate Molecular Weight Profiling Sequence, Linkage, and Branching Data from ESI-MS/MS. *Anal. Chem.* 67: 1772-1784, 1995.
234. Rozaklis T, Ramsay SL, Whitfield PD, Ranieri E, Hopwood JJ and Meikle PJ, Determination of Oligosaccharides in Pompe Disease by Electrospray Ionization Tandem Mass Spectrometry. *Clin Chem* 48(1): 131-139., 2002.
235. Metelmann W, Peter-Katalinic J and Muthing J, Gangliosides from human granulocytes: a nano-ESI QTOF MS fucosylation study of low abundance species in complex mixtures. *J Am Soc Mass Spectrom* 12: 964-73., 2001.
236. Harvey DJ, Ionization and collision-induced fragmentation of N-linked and related carbohydrates using divalent cations. *J Am Soc Mass Spectrom* 12(8): 926-37., 2001.
237. Roberts GD, Johnson WP, Burman S, Anumula KR and Carr SA, Strategy for Analysis Protein and Carbohydrate Components of Monoclonal Antibodies. *Anal. Chem.* (67): 3613-3625, 1995.
238. Guan F, Soma LR, Luo Y, Uboh CE and Peterman S, CID of Anabolic Steroids by ESI-MS/MS. *J Amer Soc Mass Spectrom* 17: 477-489, 2006.
239. Jiang X, Ory DS and Han X, Characterization of oxysterols by ESI-MS/MS after one-step derivatization with dimethylglycine. *Rapid Commun Mass Spectrom* 21: 141-152, 2007.
240. Larsen s, Uran S, Jacobsen PB and Skotland T, Collision-induced dissociation of glycero phospholipids using electrospray ion-trap mass spectrometry. *Rapid Commun Mass Spectrom* 15(24): 2393-8, 2001.
241. Zemski Berry KA and Murphy RC, Cell membrane aminophospholipids as isotope-tagged derivatives. *J Lipid Research* 46: 1038-1046, 2005.
242. Dickinson JS and Murphy RC, Mass spectrometric analysis of leukotriene A4 and other chemically reactive metabolites of arachidonic acid. *J Am Soc Mass Spectrom* 13(10): 1227-34., 2002.
243. Wheelan P, Zirrolli JA and Murphy RC, ESI-CID of Polyhydroxy Unsaturated Fatty Acids. *J. Am. Soc. Mass Spectrom.* 7: 140-149, 1996.
244. Wheelan P, Zirrolli JA and Murphy RC, Negative Ion ESI-MS/MS LTB4-Derived Metabolites. *J. Am. Soc. Mass Spectrom.* 7: 129-139, 1996.
245. Hevko JM and Murphy RC, Electrospray ionization and tandem mass spectrometry of cysteinyl eicosanoids: leukotriene C4 and FOG7. *J Am Soc Mass Spectrom* 12(7): 763-71., 2001.
246. Murphy RC, Krank J and Barkley RM, LC/MS methodology in lipid analysis and structural characterization of novel lipid species. *Functional Lipidomics*: 17-55, 2006.
247. Zhu X and Sato T, The distinction of underivatized monosaccharides using ESI MS. *Rapid Commun Mass Spectrom* 21: 191-198, 2007.
248. Sforza S, Silipo A, Molinaro A, Marchelli R, Parrilli M and Lanzetta R, Determination of fatty acid positions in native lipid A by positive and negative electrospray ionization mass spectrometry. *Journal of Mass Spectrometry* 39(4): 378-383, 2004.
249. Welti R, Wang X and Williams TD, Electrospray ionization tandem mass spectrometry scan modes for plant chloroplast lipids. *Analytical Biochemistry* 314(1): 149-152, 2003.
250. Hsu F-F, Turk J, Shi Y and Groisman EA, Characterization of acylphosphatidylglycerols from salmonella typhimurium by tandem mass spectrometry with ESI. *Journal of the American Society for Mass Spectrometry* 15: 1-11, 2004.
251. Mountfort KA, Bronstein H, Archer N and Jickells SM, Identification of Oxidation Products of Squalene in Solution and in Latent Fingerprints by ESI-MS and LC/APCI-MS. *Anal Chem* 79: 2650-2657, 2007.
252. Zhang S, Van Pelt CK, Huang X and Schultz GA, Detection of single nucleotide polymorphisms using ESI MS: validation of a one-well assay and quantitative pooling studies. *J Mass Spectrom* 37: 1039-50., 2002.
253. Garnier N, Cren-Olive C, Rolando C and Regert M, Characterization of archaeological beeswax by electron ionization and electrospray ionization mass spectrometry. *Anal Chem* 74(19): 4868-77., 2002.
254. Hsu F-Fu and Turk J, Electrospray ionization/tandem quadrupole mass spectrometric studies on phosphatidylcholines: the fragmentation processes. *Journal of the American Society for Mass Spectrometry* 14: 352-363, 2003.
255. Marie A, Fournier F and Tabet JC, Collision-Induced Dissociation Studies of Poly(vinylidene) Fluoride Telomers in an Electrospray-Ion Trap Mass Spectrometer. *Anal. Chem.* 74: 3213 -3220, 2002.
256. Cunniff JB and Vouros P, False Positives and the Detection of Cyclodextrin Inclusion Complexes by ESI MS. *J. Amer. Soc. Mass Spectrom.* 6: 437-447, 1995.
257. Young D-S, Hung H-Y and Liu LK, An Easy and Rapid Method for Determination of Stability Constants by Electrospray Ionization Mass Spectrometry. *Rapid Comm. Mass Spectrom.* 11: 769-773, 1997.

258. Hughes BM, McKenzie DE and Duffin KL, Monitoring Waste Streams by ESI-MS/MS. *Anal. Chem.* **67:** 1824-1830, 1993.

259. Stenson AC, Landing WM, Marshall AG and Cooper WT, Ionization and fragmentation of humic substances in ESI FT-ICR MS. *Anal Chem* **74:** 4397-409., 2002.

260. Stenson AC, Marshall AG and Cooper WT, Exact masses and chemical formulas of individual Suwannee River fulvic acids from ultrahigh resolution ESI FT-ICR MS. *Anal Chem* **75:** 1275-84., 2003.

261. Jasieczek CB, Buzy A, Haddleton DM and Jennings KR, ESI-MS of Poly(styrene). *Rapid Commun. Mass Spectrom.* **10:** 509-514, 1996.

262. Cai Y, Peng W-P, Kuo S-J, Lee YT and Chang H-C, Single-Particle Mass Spectrometry of Polystyrene Microspheres and Diamond Nanocrystals. *Anal. Chem.* **74:** 232 -238, 2002.

263. Keki S, Nagy L, Deak G and Zsuga M, Multiple charging of poly(propylene glycol) by binary mixtures of cations in electrospray. *Journal of the American Society for Mass Spectrometry* **16**(2)**:** 152-157, 2005.

264. Liu TY, Shiu LL, Luh TY and Her GR, Direct Analysis of C60 by ESI-MS. *Rapid Commun. Mass Spectrom.* **9:** 93-96, 1995.

265. Wilson SR and Wu Y, View of Fullerene Chemistry via ESI-MS. *Org. Mass Spectrom.* **29:** 186-191, 1994.

266. Schulz A, Trage C, Schwarz H and Kroh LW, ESI MS of a-dicarbonyl compounds-Probing intermediates during nonenzymatic browning reaction of -ascorbic acid. *Int J Mass Spectrom* **262:** 169-173, 2007.

267. Fleet IA and Monaghan JJ, Comparison of Electrospray MS of Chrysanthemic Acid Ester Pyrethroid Insecticides with EI and Positive-ion Ammonia CI Methods. *Rapid Comm. Mass Spectrom.* **11:** 796-802, 1997.

268. Silva EMP, Domingues MRM, Barros C, Faustino MAF, Tome JPC, Neves MGPMS, Tome AC, Santana-Marques MG, Cavaleiro JAS and Ferrer-Correia AJ, Dinitroporphyrin zinc complexes by ESI MS/MS. fragmentations of b-(1,3-dinitroalkyl) porphyrins. *J Mass Spectrom* **40:** 117-122, 2005.

269. Laine O, Laitinen T and Vainiotalo P, Characterization of polyesters prepared from three different phthalic acid isomers by CID-ESI-FT-ICR and PSD-MALDI-TOF mass spectrometry. *Anal Chem* **74**(16)**:** 4250-8., 2002.

270. Arnould MA, Vargas R, Buehner RW and Wesdemiotis C, Tandem mass spectrometry characteristics of polyester anions and cations formed by electrospray ionization. *European Journal of Mass Spectrometry* **11:** 243-256, 2005.

271. Ortega-Gadea S, Bernabe-Zafon V, Simo-Alfonso EF, Ochs C and Ramis-Ramos G, Characterization of industrial alkylpolyphosphonates by infusion ESI-MS with ID of impurities by tandem CE. *J Mass Spectrom* **41:** 23-33, 2006.

272. Prebyl BS, Johnson JD and Cook KD, Calibration for determining monomer ratios in copolymers by electrospray ionization mass spectrometry. *International Journal of Mass Spectrometry* **238**(3)**:** 207-214, 2004.

273. Dethy JM, Ackermann BL, Delatour C, Henion JD and Schultz GA, Demonstration of direct bioanalysis of drugs in plasma using nanoelectrospray infusion from a silicon chip coupled with MS/MS. *Anal Chem* **75:** 805-11., 2003.

274. Gabelica V and De Pauw E, Internal energy and fragmentation of ions produced in ESI. *Mass Spec Rev* **24:** 566-587, 2005.

275. Anderegg R, Wagner D, Stevenson C and Borchardt R, The MS of Helical Unfolding in Peptides. *J. Amer. Soc. Mass Spectrom.* **5:** 425-433, 1994.

276. Stephansson M, Sjoberg PJR and Markides KE, Regulation of Multimer Formation in ESI-MS. *Anal. Chem.* **68:** 1792-1797, 1996.

277. Leize E, Jaffrezic A and Van Dorsselaer A, Correlation of Solvation Energy and ESI Response for Supramolecular Complexes. *J. Mass Spectrom.* **31:** 537-544, 1996.

278. Knight WB, Swiderek KM and Sakuma Tea, Enzyme-inhibitor Complex by ESI-MS. *Biochemistry* **32:** 2031-2035, 1993.

279. Kapur A, Beck JL, Brown SE, Dixon NE and Sheil MM, Use of electrospray ionization mass spectrometry to study binding interactions between a replication terminator protein and DNA. *Protein Sci* **11**(1)**:** 147-57., 2002.

280. Schmidt A and Karas M, The influence of electrostatic interactions on the detection of heme-globin complexes in ESI-MS. *J Am Soc Mass Spectrom* **12**(10)**:** 1092-8., 2001.

281. Lorenz SA, Maziarz EP, 3rd and Wood TD, Using solution phase H/D exchange to determine origin of non-covalent complexes in ESI MS: in solution or in vacuo? *J Am Soc Mass Spectrom* **12:** 795-804., 2001.

282. Gabelica V, Vreuls C, Filee P, Duval V, Joris B and De Pauw E, Advantages and drawbacks of nanospray for noncovalent protein-DNA complexes by MS. *Rapid Commun Mass Spectrom* **16:** 1723-1728, 2002.

283. Johnson RS, Changes in Protein Hydrogen Exchange Rates that Result from Point Mutations. *J. Am. Soc. Mass Spectrom.* **7:** 515-521, 1996.

284. Wagner DS and Anderegg RJ, Conformation of Cytochrome-c by D/H Exchange. *Anal. Chem.* **66:** 706-711, 1994.
285. Zal F, Chausson F, Leize E, Van Dorsselaer A, Lallier FH and Green BN, Quadrupole Time-of-Flight Mass Spectrometry of the Native Hemocyanin of the Deep-Sea Crab Bythograea thermydron. *Biomacromolecules* **3**(2)**:** 229-231., 2002.
286. Kieltyka JW and Chow CS, Probing RNA Hairpins with Cobalt(III)hexammine and Electrospray Ionization Mass Spectrometry. *Journal of the American Society for Mass Spectrometry* **17**(10)**:** 1376-1382, 2006.
287. Sherman CL, Pierce SE, Brodbelt JS, Tuesuwan B and Kerwin SM, Identification of the Adduct Between a 4-Aza-3-ene-1,6-diyne and DNA Using Electrospray Ionization Mass Spectrometry. *Journal of the American Society for Mass Spectrometry* **17**(10)**:** 1342-1352, 2006.
288. Cai Y, Tarr MA, Xu G, Yalcin T and Cole RB, Dication induced stabilization of gas-phase ternary beta-cyclodextrin inclusion complexes by ESI. *J Amer Soc Mass Spectrom* **14:** 449-459, 2003.
289. Schug K and Lindner W, Using electrospray ionization-mass spectrometry/tandem mass spectrometry and small molecules to study guanidinium-anion interactions. *International Journal of Mass Spectrometry* **241**(1)**:** 11-23, 2005.
290. Conway MCP, Whittal RM, Baldwin MA, Burlingame AL and Balhorn R, Electrospray Mass Spectrometry of NeuAc Oligomers Associated with the C Fragment of the Tetanus Toxin. *Journal of the American Society for Mass Spectrometry* **17**(7)**:** 967-976, 2006.
291. Chernushevich IV, Verentchikov AN, Ens W and Standing KG, Effect of Ion-Molecule Collisions in the Vacuum Chamber of an ESI-TOFMS on Mass Spectra of Proteins. *J. Am. Soc. Mass Spectrom.* **7:** 342-349, 1996.
292. Adams J, Strobel FH, Reiter A and Sullards MC, Charge-Separation Reactions in ESI-MS/MS of Doubly Protonated Angiotensin II. *J. Amer. Soc. Mass Spectrom.* **7:** 30-41, 1996.
293. Boschenok J and Sheil MM, ESI-MS/MS of Nucleotides. *Rapid Commun. Mass Spectrom.* **10:** 144-149, 1996.
294. Gross DS and Williams ER, Coulomb Energy and Polarizability of a Multiply Protonated Peptide using ESI-FTMS. *J. Am. Chem. Soc.* **117:** 883-890, 1995.
295. Kolli VSK and Orlando R, Sequence of Large Peptides ESI-MS/MS of Multiply Protonated Ions. *J. Am. Soc. Mass Spectrom.* **6:** 234-241, 1995.
296. Cassady CJ and Carr SR, Elucidation of Isomeric Structures for Ubiquitin M+12H 12+ Ions Produced by ESI. *J. Mass Spectrom* **31:** 247-254, 1996.
297. Shevchenko A, Wilm M and Vorm O, Mass Spectrometric Sequencing of Proteins from Silver-Stained Polyacrylamide. *Anal. Chem.* **68:** 850-858, 1996.
298. Körner R, Wilm M, Morand K, Schubert M and Mann M, Peptides and Protein Digests by ESI-Ion Trap MS. *J. Am. Soc. Mass Spectrom.* **7:** 150-156, 1996.
299. Murphy RC, Barkley RM, Zemski Berry K, Hankin J, Harrison K, Johnson C, Krank J, McAnoy A, Uhlson C and Zarini S, Electrospray ionization and tandem mass spectrometry of eicosanoids. *Analytical Biochemistry* **346**(1)**:** 1-42, 2005.
300. Zemski Berry KA and Murphy RC, Analysis of polyunsaturated aminophospholipid molecular species using isotope-tagged derivatives and MS/MS/MS. *Anal Biochem* **349:** 118-128, 2006.
301. Griffiths WJ, Yang Y, Sjovall J, Lindgren JA, Ashton DS, Beddell CR, Green BN and Oliver RWA, ESI-MS/MS of Hydroxylated Lipoxygenase Products by ESI-MS. *Rapid Commun. Mass Spectrom.* **10:** 183-196, 1996.
302. McAnoy AM, Wu CC and Murphy RC, Direct Qualitative Analysis of Triacylglycerols by ESI MS Using a Linear Ion Trap. *Journal of the American Society for Mass Spectrometry* **16**(9)**:** 1498-1509, 2005.
303. Anari MR, Bakhtiar R, Zhu B, Huskey S, Franklin RB and Evans DC, Derivatization of ethinylestradiol with dansyl chloride to enhance ESI in analysis of ethinylestradiol in rhesus monkey plasma. *Anal Chem* **74:** 4136-44., 2002.
304. Tao WA, Clark RL and Cooks RG, Quotient ratio method for quantitative enantiomeric determination by mass spectrometry. *Anal Chem* **74**(15)**:** 3783-9., 2002.
305. Wu L and Cooks RG, Chiral analysis using the kinetic method with optimized fixed ligands: applications to some antibiotics. *Anal Chem* **75**(3)**:** 678-84., 2003.
306. Gal JF, Stone M and Lebrilla CB, Chiral recognition of non-natural alpha-amino acids. *International Journal of Mass Spectrometry* **222**(1-3)**:** 259-267, 2003.
307. Janini GM, Conrads TP, Wilkens KL, Issaq HJ and Veenstra TD, A sheathless nanoflow electrospray interface for on-line capillary electrophoresis mass spectrometry. *Anal Chem* **75**(7)**:** 1615-9., 2003.
308. Whitt JT and Moini M, CE-MS interface using a porous junction. *Anal Chem* **75:** 2188-91., 2003.
309. Wu Y-T and Chen Y-C, Sheathless Capillary Electrophoresis/Electrospray Ionization Mass Spectrometry Using a Pulled Bare Fused-Silica Capillary as the Electrospray Emitter. *Analytical Chemistry* **77**(7)**:** 2071-2077, 2005.
310. Kim J, Zand R and Lubman DM, Electrophoretic mobility for peptides with post-translational modifications in capillary electrophoresis. *Electrophoresis* **24**(5)**:** 782-93., 2003.

311. Schultz CL and Moini M, Analysis of underivatized amino acids and their D/L-enantiomers by sheathless capillary electrophoresis/electrospray ionization mass spectrometry. *Anal Chem* **75**(6): 1508-13., 2003.

312. Kelly J, Masoud H, Perry MB, Richards JC and Thibault P, O-Deactylated Lipooligosaccharides and Glycans by CE-ESI-MS/MS. *Anal. Biochem.* **233**: 15-30, 1996.

313. Kelly JF, Locke SJ, Ramaley L and Thibault P, Protein Glycoforms by CE-ESI-MS. *J. Chrom A.* **720**: 409-427, 1996.

314. Pinkston JD, Marapane SB, Jordan GT and Clair BD, Characterization of low molecular weight alkoxylated polymers using long column SFC/MS and image analysis based quantitation. *J Am Soc Mass Spectrom* **13**: 1195-208., 2002.

315. Basile F, Ferrer I, Furlong ET and Voorhees KJ, Simultaneous multiple substrate tag detection with ESI-ion trap MS for in vivo bacterial enzyme activity profiling. *Anal Chem* **74**: 4290-3., 2002.

316. Tuinman AA, Lewis LA and Lewis SA, Trace-Fiber Color Discrimination by Electrospray Ionization Mass Spectrometry: A Tool for the Analysis of Dyes Extracted from Submillimeter Nylon Fibers. *Analytical Chemistry* **75**(11): 2753-2760, 2003.

317. Reemtsma T and These A, On-line coupling of size exclusion chromatography with electrospray ionization-tandem mass spectrometry for the analysis of aquatic fulvic and humic acids. *Anal Chem* **75**(6): 1500-7., 2003.

318. Charles L, Laure F, Raharivelomanana P and Bianchini J-P, Sheath interface for LC ESI MS for analysis of neoflavonoids. *J Mass Spectrom* **40**: 75-82, 2005.

"Well, in homely terms, we learned how to make elephants fly, as it were. These huge big molecules cannot be put into the gas phase as vapors or gases without catastrophic decomposition. If you've ever tried to distill an egg out of a frying pan, you know you can't do it."

~John J. Fenn

Chapter 9   MALDI

Introduction to Mass Spectrometry, 4th Edition: Instrumentation, Applications, and Strategies for Data Interpretation; J.T. Watson and O.D. Sparkman, © 2007, John Wiley & Sons, Ltd

"The great tragedy of science – the slaying of a beautiful hypothesis by an ugly fact."

~Thomas Huxley

Historical Perspective and Introduction

Prior to 1970, to contemplate the mass spectrum of an intact oligopeptide was a fantasy. Because the negligible vapor pressure of high molecular weight, thermally labile compounds precluded their analysis by EI or CI mass spectrometry, no mass spectrometer had been designed to obtain mass spectra beyond m/z 1000. During the 1970s, desorption/ionization (D/I) techniques such as field desorption [1] and plasma (^{252}Cf) desorption [2] were introduced, and with special instrumentation, selected laboratories were able to conduct analyses of previously intractable compounds. However, it was the introduction of fast atom bombardment (FAB) [3] that had the greatest impact on analyses of nonvolatile and/or thermally labile compounds because the FAB source was so easily adapted to all kinds of m/z analyzers [4]. As a point on recent development, a continuous-flow version of field desorption has been described that provides improved spectral quality, higher sample throughput, and a simpler interface to sample handlers and chromatographic equipment [5].

FAB had been developed during attempts to analyze organic compounds mounted on metal surfaces by the then well-established technique of secondary ion mass spectrometry (SIMS). In SIMS, a primary beam of mass-selected ions is used to bombard the analyte and generate secondary ions that can be mass-analyzed to characterize the analyte. A FAB mass spectrum was acquired by bombarding the sample (dissolved in a liquid matrix) with a beam of atoms having a kinetic energy of several thousand eV – thus the name "fast atoms". It was soon learned, however, that the charge on the bombarding particle was not important to the outcome of the desorption/ionization experiment [6]. The analytical advantage provided by FAB was dependent upon the use of a liquid matrix [4], most often glycerol, to hold the analyte; for this reason, analysis of a solution of analyte during bombardment by a beam of ions was often called liquid SIMS (or LSIMS) to connote the origin of the technology and to distinguish it from conventional SIMS, which does not employ a liquid matrix in analyzing the surface layer of solids.

In early applications of lasers in mass spectrometry, the neat analyte (no matrix) was irradiated directly with intense pulses of laser light for short durations [7]. Energy transfer to the analyte was achieved with UV lasers to cause electronic excitation or IR lasers to cause vibrational excitation and ionization of the analyte. This technique of direct laser desorption (LD) had limited application and was effectively eclipsed by the development of MALDI in the late 1980s; some applications of laser desorption provide analytical advantage, however, such as in the detection of aromatic contaminants [8]. The thesis of MALDI was published in 1988 [9], for which K. Tanaka was awarded a portion of the Nobel Prize in chemistry in 2002 [10]. The routine techniques for MALDI as used today were developed by Karas and Hillenkamp and coworkers [11–14] as described in the following sections.

I. Operating Principles

1. The Matrix

In MALDI, the sample is mixed with an organic matrix compound; e.g., dihydroxybenzoic acid (see Figure 9-1), in a convenient solvent to achieve a molar ratio of analyte to matrix of approximately 1:5000 [11, 12, 15–18]. One role of the matrix is to separate the analyte molecules (by dilution) to prevent analyte–analyte molecular (or ionic) interactions during the ionization process. The key role of the matrix is in absorbing the radiation, thereby protecting the analyte from radiation damage. However, the matrix is selected not only for its capacity to absorb laser energy absorption, but for solubility characteristics similar to those of the analyte. The high energy density at the spot illuminated by the laser causes a phase transition from solid to gas; the explosive expansion of the localized segment of matrix (containing a trace of the analyte) mostly disperses neutrals, but also some ions, with initial velocities of 400 to 800 msec^{-1}. Interestingly, very little increase in the internal energy of the analyte occurs as there is little or no fragmentation of the analyte, making MALDI one of the soft ionization techniques.

Matrix	Application	Structure	Matrix Solution
α-cyano-4-hydroxy cinnamic acid (α-CHCA)	First choice for small proteins and peptides (<10 kDa)	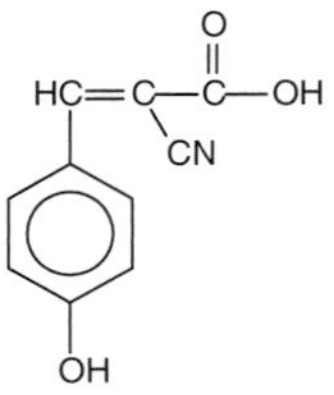	10 mg mL^{-1} in 50:50:0.3 water/ acetonitrile/TFA
3,5-Dimethoxy-4-hydroxy cinnamic acid (Sinapinic acid)	First choice for heavy proteins (>10 kDa)	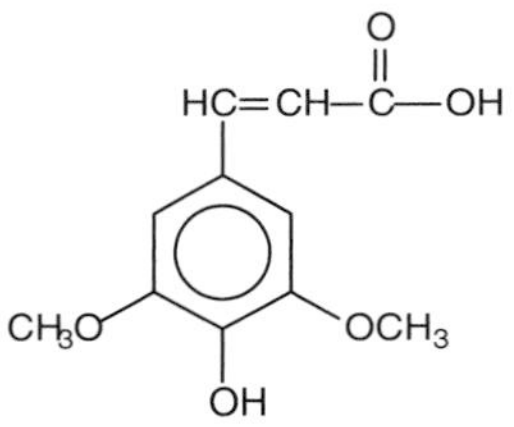	10 mg mL^{-1} in 70:30:0.3 water/ acetonitrile/TFA 10 mg mL^{-1} in 50:50:0.3 water/ acetonitrile/TFA if sample contaminated
2-Mercapto-benzothiazole (MBT)	Unusually fine crystals/even distribution on plate Good for peptidoglycans Also good for heavy proteins	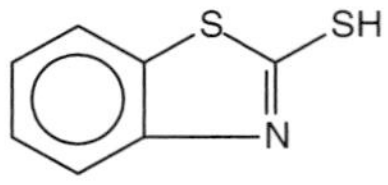	10 mg mL^{-1} in 1:1:1 ethanol (EtOH)/ tetrahydrofuran (THF)/water

Figure 9-1.   *Structures and properties of representative compounds used as MALDI matrices.*

When a solid matrix is to be employed, the sample and matrix are mixed together in an organic or aqueous solvent of mutual solubility. Once the solvent is removed, the matrix and analyte molecules codeposit [19] onto a sample planchet, as illustrated in Figure 9-2 in which the magnified inset shows an artist's conception of the deposition of matrix molecules and a protein analyte. Incorporation of the analyte into the crystalline matrix is critical to the

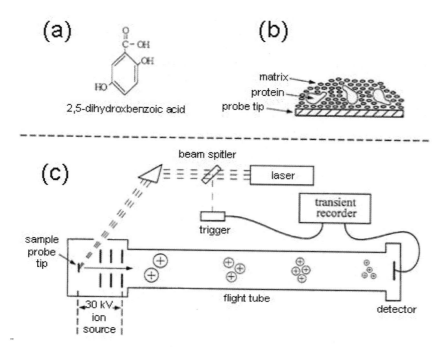

Figure 9-2. *Schematic diagram of a MALDI TOF instrument (c). A solid deposit of analyte/matrix mixture is conceptually represented on a sample plate (b). Laser radiation is focused onto the sample spot on the plate to effect ionization.*

success of the MALDI experiment; studies of protein ground up in dry crystalline dihydroxybenzoic acid failed to produce MALDI spectra [20], though such results are in conflict with those of a study of the solvent-free method of sample preparation [21]. The structures of some representative matrices are shown in Figure 9-1. Many other compounds have been used as the MALDI matrix [22, 23], including aquatic fulvic acids, which show some advantage for the analysis of carbohydrates and peptides [24]. A statistical approach to choosing a particular matrix compound has been proposed in the context of analyzing synthetic polymers [25].

Sample preparation is perhaps the most time-consuming aspect of analysis by MALDI MS; a few of the literature methods are briefly described below in Section III on Sample Handling.

Desirable attributes of the matrix are at least two-fold: (1) the matrix molecules must have high absorptivity for the laser radiation and (2) the matrix must be capable of forming a fine crystalline solid during co-deposition with the analyte.

The suitability of a matrix compound for MALDI is not completely dependent upon its capacity to absorb radiation from the laser. For example, Liao and Allison [26] found that the UV (337 nm) absorptivity of sinapinic acid (13,000 L cm^{-1} mol^{-1}) is nearly an order of magnitude greater than that of α-cyano-4-hydroxy cinnamic acid (1600 L cm^{-1} mol^{-1}), but both serve efficiently as a matrix during the analysis of insulin by MALDI. The molar absorptivity of 1,4-diphenyl-1,3-butadiene was 27,000 L cm^{-1} mol^{-1}, but it was not an effective matrix. Further, they noted that each of these compounds decomposes (as do most of the successful matrix compounds) at its melting point, a feature that may facilitate the physical desorption process. Many of the commonly used matrix compounds are organic acids; in some cases, it may be important to use a matrix compound that has basic properties [27]. The proton affinity of various matrix compounds has been determined by the kinetic method [28].

The more homogeneous and fine-grained the morphology [29, 30] of the crystal formation within the matrix, the more intense is the MALDI mass spectrum of the analyte. The morphology of irradiation damage to crystals of DHB by laser ablation during MALDI has been described [31], but with correlation between ionization and degree of ablation. 2,5-Dihydroxyacetophenone (DHAP) shows great promise as a matrix for proteins and glycoproteins in the molecular mass range of 8–100 kDa [32]; the use of 2,4,6-trihydroxyacetophenone also facilitates the analysis of glycoproteins and glycans [33], lipids [34], and polysaccharides [35]. The use of a self-assembled monolayer as matrix may lead to new approaches to MALDI [36] as certainly it has a substrate for preparing polymer-modified surfaces [37, 38], as described in some detail later in this chapter.

A new ionic liquid MALDI mass spectrum matrix 2,5-dihydroxybenzoic acid butylamine was found to provide substantially more consistent analytical responses than the conventional matrices of alpha cyano, etc. especially during the analysis of oligosaccharides and synthetic polymers [39]. Other ionic matrices are reported to facilitate quantitation of tryptic digests [40]. Oxidized carbon nanotubes have been found to offer advantages as a matrix, in part because its constituent carboxyl groups can serve as a source of protons during the analysis of oligosaccharides [41]. Immobilization of carbon nanotubes in polyurethane improves their performance as demonstrated in the analysis of small neutral carbohydrates [42]. In other applications of these carbon nanotubes, they served as the adsorbent for solid-phase extraction of small molecules such as drugs, as well as the matrix material during subsequent analysis by MALDI [43]. Solid-phase extraction of peptides is facilitated by the large surface area on C18 functionalized silica nanoparticles, which are compatible with subsequent analysis by AP MALDI MS [44]. Carboxylated/oxidized diamond nanoparticles (nominal size 100 nm) exhibit exceptionally high affinity for proteins through both hydrophilic and hydrophobic forces and their small size allows them to be incorporated into the MALDI matrix with no adverse effects on resolution or accuracy due to their presence during experimental measurements [45, 46]; the use of these diamond nanoparticles to accumulate proteolytic peptides from protein digests for analysis by AP MALDI has resulted in detection limits that rival those achievable by ESI [46].

Depending on the application, choice of the matrix compound can be important. For example, an excellent matrix for photodissociation consists of 1:1 α-cyano-4-hydroxycinnamic acid/fructose because $[M + H]^+$ ions are formed with low internal energies [47]. Often, a mixture of matrices (e.g., α-cyano-4-hydroxycinnamic acid combined with 2,5-dihydroxybenzoic acid) gives improved sequence coverage in protein mass mapping, especially after digesting with a mixture of enzymes and cleanup of the proteolytic peptides via ZipTip prior to analysis [48]. The reductive properties of a matrix like 1,5-diaminonaphthalene may be useful in analyses of disulfide linkages to narrow down the number of isomeric possibilities as demonstrated with two simple model cystinyl proteins [49]. In the case of halogenated fullerenes, sulfur serves well as the matrix [50]; metal powders have been recommended as a matrix for the analysis of small molecules [51]. Anionic dopants allow neutral oligosaccharides to be examined in the same mixture as acidic oligosaccharides [52]; special attention may be required in choosing a matrix for nucleic acid polymers because of their high negative charge [53, 54].

Adduct formation by the matrix to the analyte sometimes occurs, especially with α-cyano-4-hydroxycinnamic acid (α-CHCA), although this phenomenon is somewhat analyte-dependent. In some cases, adduct formation by the matrix can be suppressed by adding surfactants to the sample/matrix mixture; results of one study indicate that cationic surfactants, particularly cetyltrimethylammonium bromide, are useful in the analysis of acidic analytes, while the anionic surfactant, sodium dodecyl sulfate, showed promise for peptide analyses [55]. The possibility of forming Na^+ or K^+ ion adducts of the analyte also exists. In other cases, matrix complexes with sodium and potassium ions can be annoying as they may interfere with detection of low-mass ions of the analyte. On the other hand, because these matrix–alkali ion complexes have been so well characterized, they can be used as internal calibrants of the *m/z* axis [56]. If desirable, these matrix–alkali ion complexes can be suppressed by adding ammonium salt, such as citrates or phosphates, directly to the MALDI matrix [57]. Use of iodine as a matrix additive has been reported to reduce adduct formation with the matrix, as well as diminish adduct formation with Na^+ and K^+ [58]. Addition of 7 mM nitrilotriacetic acid to the matrix solution significantly reduces background due to matrix and alkali–metal clusters and enhances signals due to peptides [59].

In general, MALDI is recognized as a soft ionization technique (i.e., there is little, if any, fragmentation of the protonated molecule formed during the MALDI process). On the other hand, some matrix compounds do promote some fragmentation of the protonated molecule, as nicely summarized by Brown and coworkers in their efforts to promote fragmentation of peptides [60], albeit a few percent of the total ionization. A more recent study on the influence of the matrix on fragmentation of the analyte during analysis by MALDI reports that bumetamide used alone or in combination with α-cyano-4-hydroxycinnamic acid can significantly promote fragmentation of peptides [61]. The fragmentation-promoting effect of common matrix compounds can be suppressed when a saccharide is used as an additive, which participates as a nonenergy absorber in crystal formation; the net result is production of $[M + H]^+$ with low internal energies [47, 62]. In an effort to help classify "hot" and "cold" matrix compounds, the use of so-called "thermometer molecules" has been proposed [63]. Alpha-cyano-4-hydroxy-cinnamic acid (α-CHCA) and 2,5-dihydroxybenzoic acid (DHB) each control in different ways the nature and extent of subsequent CAD (TOF/TOF) fragmentation during the characterization of synthetic oligosaccharides [64]. Common pencil lead is a useful matrix for several groups of analytes, including peptides, polymers, and actinide metals [65-68]; an advantage of pencil lead is the absence of abundance background ions observed with the more common matrix compounds. The problem of interference from abundant background ions from the matrix can sometimes be controlled by reducing the molar ratio of matrix to analyte [69]. The temperature dependence of ion yields from the matrix may provide some insight into the mechanism of laser desorption/ionization [70]. The addition of phosphoric acid to 2,5-DHB matrix can facilitate the detection of phosphopeptides [71].

2. The Laser, *m/z* Analyzer, and Representative Mass Spectra

The operating principles and a typical configuration of a MALDI instrument with a TOF-MS are represented in Figure 9-2. Bursts of radiation at 10 Hz from the laser are focused on the sample probe tip through appropriate optics as illustrated in Figure 9-2. A UV laser, based on nitrogen, at 337 nm is most commonly used for MALDI; however, IR lasers have been used in some applications [72, 73]. Because the MALDI process is inherently a pulsed procedure, it is a logical application to use a TOF-MS to analyze the ions formed. A high repetition rate (1 kHz) laser reportedly increases the ion yield by a factor of 80 over that achieved with the conventional 30 Hz laser [74]. As indicated in Figure 9-1, a fraction of the laser pulse initiates the timing circuitry for measuring the *tof* for the ions produced by each laser pulse. As illustrated by the smallest circles in the flight tube of the TOF MS, the lightest ions formed during each ionization pulse reach the detector first, as all of the ions are given the same energy in the 30-kV ion acceleration region. Because $KE = 1/2 \ mv^2$, the ions of lightest mass have the highest velocity and thus reach the detector first. This transient (duration of approx. 300 microseconds) mass spectrum is recorded by a transient recorder at the detector. It is common practice to sum many (typically 10–100) of these transient mass spectra to produce a usable mass spectrum. Although TOF instruments are most often used for MALDI, ICR and ion trap instruments are being used with greater frequency. A clever technique for accumulating ions from multiple laser shots, including those from a calibration compound located separate from the analyte, has been reported based on an ICR instrument for improved signal reproducibility and mass accuracy [75]. In addition, certain hybrid instruments, such as the Qq-TOF with orthogonal acceleration, offer special applications or effects, such as the possibility of CID [76].

A representative MALDI mass spectrum obtained with a TOF *m/z* analyzer is shown in Figure 9-3. This MALDI TOF mass spectrum results from the analysis of approximately five picomoles of a disulfide-linked pair (heterodimer) of peptides (calculated mass or molecular weight = 3351.74 Da) utilizing α-cyano-4-hydroxycinnamic acid as the matrix. As can be seen in Figure 9-3, the most significant peak in the MALDI mass spectrum usually represents the protonated molecules of the analyte. In this case, the protonated molecules of the disulfide-linked peptides (a) and (b), the structure of which is illustrated at the upper left in Figure 9-3, are represented by a peak at *m/z* 3352.88.

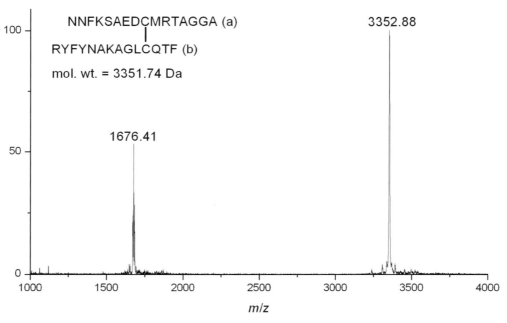

Figure 9-3. MALDI TOF mass spectrum of disulfide-linked peptides; α-cyano-4-hydroxcinnamic acid used as the matrix. Peak at m/z 3352.88 represents the protonated molecule. Does the peak at m/z 1676.41 represent a double-protonated molecule? See Figure 9-4 for the answer. *Data obtained by Xue Li, Graduate Assistant in the Department of Chemistry, Michigan State University, East Lansing, MI.*

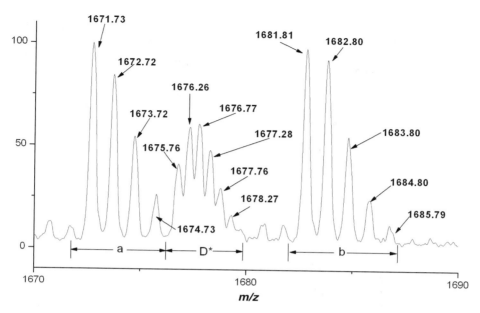

Figure 9-4. Expanded segment of the MALDI TOF mass spectrum in Figure 9-3 in the vicinity of m/z 1680 as obtained with a reflectron; see text for explanation of peaks within the brackets as a, b, and D* along the m/z axis. Data obtained by Xue Li, Graduate Assistant in the Department of Chemistry, Michigan State University, East Lansing, MI.

Does the peak at *m/z* 1676.41 in Figure 9-3 represent the double-protonated molecule of the analyte? The calculated *m/z* value for the average mass of the double-protonated disulfide-linked peptides is (3351.74 + 2.02) / 2 = 1676.88. Because the mass accuracy of the mass measurement was not particularly good, this could be considered to be a reasonable agreement between experimental and calculated values of mass. However, careful examination of an expanded version of the analog data along the *m/z* axis suggests otherwise. Consider the expanded view of the *m/z* axis around *m/z* 1680 as shown in Figure 9-4. Because the isotope peaks have been resolved in Figure 9-4, it is now appropriate to consider the monoisotopic mass (*vide infra*) of each of the ions, rather that the average mass of all the isotopic variants.

Note that there are three distinct clusters of peaks in Figure 9-4. The first cluster, designated **a** along the *m/z* axis starts at *m/z* 1671.73 and ends at *m/z* 1674.73. The second cluster **D*** begins at *m/z* 1675.76 and ends at *m/z* 1678.27, while the third cluster **b** begins at *m/z* 1681.81. Furthermore, note that the peaks in clusters **a** and **b** are separated by 1 *m/z* unit, whereas those in cluster **D*** are separated by 0.5 *m/z* units. The peak separations mean that clusters **a** and **b** represent single-protonated species, while cluster **D*** represents a double-protonated species.

The mass spectrum in Figure 9-4 illustrates the often-observed phenomenon of reduction of a disulfide bond by laser irradiation [77] during analysis by MALDI. The peak at *m/z* 1671.73 represents the monoisotopic protonated molecule of peptide (a) (the calculated monoisotopic mass is 1671.71 Da) and the peak at *m/z* 1681.81 represents the monoisotopic protonated molecule of peptide (b) (the calculated monoisotopic mass is 1681.81 Da). The peak at *m/z* 1675.76 represents the monoisotopic double-protonated molecule of the disulfide-bonded peptides (calculated monoisotopic mass = (3349.51 + 2.02) / 2 = 1675.77 Da). Results of other studies have shown that laser fluence can affect the minimal degree of fragmentation observed in MALDI MS [78].

In summary, the mass spectrum in Figure 9-4 indicates that laser irradiation during MALDI cleaves a significant fraction of the disulfide-bonded peptide molecules and yields protonated molecules of each cysteinyl peptide. On the other hand, a majority of the disulfide-bonded heterodimer resists reduction, and some of this intact species captures two protons during analysis by MALDI; approximately 20% of the peak intensity at *m/z* 1676.41 in Figure 9-3 represents the double-protonated molecule based on interpretation of the spectrum in Figure 9-4. In other applications, the reducing properties of 1,5-diaminonaphthalene (1,5-DAN) as a MALDI matrix cleave disulfide-bonded peptides, which can then be more effectively sequenced as individual species during CAD [49].

A representative MALDI TOF mass spectrum of bovine serum albumin (calculated mass is 66,431 Da) is shown in Figure 9-5. The peak at *m/z* 60,432 represents the protonated molecules, while the peaks at *m/z* 30,216 and *m/z* 22,144 represent double- and triple-protonated molecules, respectively. The mass spectrum was collected in the reflectron mode (*vide infra*), yet the resolution apparent in the peak at *m/z* 60,431 is quite poor due in large measure to the broad distribution of high initial velocities of all ions resulting from the explosive nature of the MALDI process. These deleterious effects are even more prominent at higher mass as illustrated in the analysis of the tenth generation of some dendrimers [79]. The deleterious effect of high initial velocity [80] of the ions is exaggerated at high mass because such ions express the lowest velocities resulting from the acceleration potential in the ion optics.

The commonly observed dropoff in MALDI signal with mass should not be blamed on the MALDI process *per se*, but rather on the common electron multiplier detector system used with the mass spectrometer [81]. As addressed in greater detail in the chapter on detectors, the response of an electron multiplier is dependent on the velocity of the incoming ion; the velocity of ions of a given energy drops off inversely with the square root of their mass.

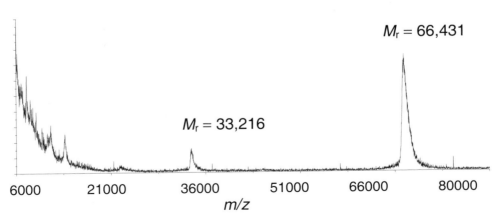

Figure 9-5. *MALDI TOF mass spectrum of bovine serum albumin from a matrix of sinapinic acid.* Data obtained by Dr. Robin Hood, Visiting Scientist in the Department of Biochemistry, Michigan State University, East Lansing, MI.

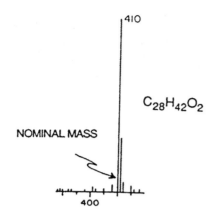

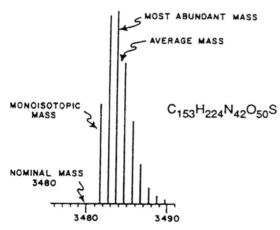

Figure 9-6. Comparison of isotope peak patterns for ions of low mass vs high mass.

The resolving power of a typical linear TOF MS is around 5000 [82]. With a reflectron, the resolving power approaches 10,000 or sometimes even greater. The peaks above m/z 30,000 in Figure 9-5 are more than 100 μm wide. Because of the problem of resolution, it is important to try to avoid contaminants that have a molecular weight close to that of the analyte of interest. The mass accuracy is typically about 0.01% [79, 83, 84], as illustrated in the results in Figure 9-3 through Figure 9-5.

As in the mass spectrometry of any compound containing very high numbers of atoms of carbon, hydrogen, nitrogen, and oxygen, the concept of nominal mass does not apply. The peak representing those ions containing only the lightest isotopes (^{12}C, ^{14}N, ^{1}H, etc.) is called the monoisotopic peak; it will lie at the low m/z edge of a cluster of peaks representing a particular ion. The monoisotopic peak will not be the most intense peak in such a cluster; the isotopic peaks are more intense for those species representing inclusions of ^{13}C and/or ^{15}N and/or ^{2}H because the probability of the ion including a few of these heavy isotopes is very high given the large number of atoms of these elements present. As can be seen in Figure 9-6, the monoisotopic mass can differ from the average mass (the weighted average of all the isotopic variants) by several daltons. If the mass spectrometer is capable of separating the isotope peaks for a given ion, then the monoisotopic mass of the ion should be used in calculations. If the resolving power of the instrument is not adequate, the peaks observed in the MALDI mass spectrum will represent the average distribution of the isotopic composition; in this case, the average mass, rather than the monoisotopic mass, of the ions should be used in calculations involved in data interpretation.

3. The Ionization Process

The mechanism for ionization of the analyte during MALDI is still under investigation [85–96]. The ionization and dissociation dynamics of molecules in superexcited states as produced through interaction with UV photons has been surveyed [97]. Liao and Allison [26] reported some curious observations several years ago as the need for better understanding of fundamentals of MALDI was being realized. For example, some analytes produce principally alkaline metal ion adducts of the analyte in one particular matrix, but in a different matrix, only protonated molecules of the analytes are formed. Liao and Allison observed that the MALDI mass spectrum of hexatyrosine showed a peak for the protonated molecule of the analyte even when the matrix of sinapinic acid was doped with a large concentration of NaCl (ratio of peptide:matrix:NaCl = 1:50:125); on the other hand, analysis of hexatyrosine from a matrix such as 2,5-dihydroxybenzoic acid yielded principally $[M + Na]^+$ even when no NaCl was added to the matrix. This phenomenon is probably due to the greater water solubility of DHB compared to sinapinic acid; because of the high solubility of DHB, the alkali ions tend to co-crystalize, possibly as salts of DHB, as the residual water evaporates during the sample crystallization process (R. Brown, Personal Communication, 2003). Similar results have been observed when trifluoroacetic acid (TFA) is added to a weakly acidic matrix like hydroxyacetophenone; the addition of TFA not only suppresses cationization of the analyte (due to competition between H^+ and Na^+), but it increases the abundance of multiprotonated species [98].

Contrary to early assessments, a carboxylic proton from the matrix is not essential as even mercaptobenzothiazole hydrogens (see structure at bottom of Figure 9-1) appear to be sufficiently acidic to protonate analytes [99]. The gas-phase basicities of selected matrix compounds [100] as well as of some analytes [87] have been measured, a property that has been correlated with internal energy of the nascent ions formed during the MALDI process [101]. *Ab initio* calculations for Na^+ interacting with several types of matrix molecules have been made [102]. Some results [26, 91, 103] indicate that a gas-phase mechanism may be involved in the formation of $[M + Na]^+$ ions; a recent report demonstrated that attachment of free gas-phase cations, rather than cation transfer from the cationized matrix, is the predominant process in cationization. Other preliminary results indicate that gas-phase protonation of peptides is unlikely, while formation of Na^+ or K^+ ion adducts of peptides in the gas phase proceeds readily (R. Brown, Utah State University, Personal Communication); on the other hand, many neutrals are apparently available in the laser plume, as Coon and coworkers have recently reported an increase of three orders of magnitude in producing protonated molecules when a corona discharge in used in conjunction with IR laser desorption [104]. The possible involvement of acoustic mechanisms of ion formation and desorption have been discussed in the context on nonresonant MALDI of oligonucleotides [105].

From a practical viewpoint, the analyst notes that matrix–adduct formation correlates with the energy of the photon, whereas the extent of sodium- or potassium-adduct formation is inversely related to the energy of the laser photon; i.e., cationization of the analyte is more prominent with an IR laser than with the standard UV laser. These observations are consistent with the premise that organic matrix adducts to the analyte are photochemically generated reaction products, a process in which IR is not sufficiently energetic; cationization (sodium or potassium), a thermal process, is promoted by IR, which can generate a high flux of inorganic ions for attachment to the analyte, possibly via charge exchange in the plume [106]. A combination of two laser pulses, one in IR [92, 107], the second in UV, has been reported to improve ion yield at 337 nm due to transient heating by the IR pulse [108]. It has been reported that shortening the pulse duration of an IR laser from 80 to 6 nsec reduces the threshold fluence by a factor of 1.2–1.9 [73], although another report suggests that MALDI mass spectra are unaffected by laser wavelength and pulse duration [109]. Yet another report states that the ion yield in MALDI is a high-power exponential function of laser fluence [110]. A recent report indicates that particle density in the plume increases dramatically with laser fluence above threshold, a feature that could shed light on the mechanism of the MALDI process [111]. Iodine as a matrix additive has been reported to lower the threshold for ionization of the analyte [58]. The general consistency of the MALDI mass spectrum; i.e., $\pm100\%$ ionization efficiency, regardless of variables associated with the laser pulse and/or fluence, when above the ionization threshold, implies a dominant role by secondary chemical processes in the energized matrix [88]. In general, it seems that much of the emphasis on laser pulse duration and energy has been on physical phenomena related to ablation for imaging [112, 113], rather than on mechanistic aspects of MALDI [114].

Knochenmuss and coworkers have long proposed that in-plume ion/molecule reactions play a major role in the MALDI process [86, 88, 114, 115]. Results of fluorescence experiments using an aromatic dye to trap excitions thereby diminishing the efficiency of ionizing the matrix DHB lend credence to the proposed role of energy pooling in their quantitative model for MALDI [116]. The equilibrium nature of proton transfer [117] as well as electron-transfer reactions [118] in the plume have been studied. Consistent with molecular orbital calculations based on this model showing that a sequential 2-photon mechanism is possible near a stainless steel interface, the Knochenmuss group shows enhanced MALDI signals from thin sample deposition on SS sample plates [119].

Photofragmentation of desorbed molecules shows a dependence on duration of the laser pulse; a ps-laser pulse can complete the ionization process so quickly that intermediate relaxation processes are not important [120, 121]. Results of other studies indicate that some fragmentation processes are dependent on laser fluence [78]. When dications $(MH_2)^{+2}$, formed during secondary ionization by 10-eV electrons of MH^+ produced in MALDI, are allowed to capture low-energy electrons, subsequent fragmentation results similar to that observed in photofragmentation [122]. The fragmentation patterns obtained with fs-laser pulses relate well to structure of the vaporized molecule, a feature used for analytical advantage in characterizing explosive compounds [123]; the photofragmentation of PAHs [124] as well as nitro-PAHs [125] have also been studied.

The nature of the matrix-analyte interaction is becoming better defined. Sample preparation must provide some means for the analyte to become incorporated into the matrix crystal [126]; experiments in which solid protein was milled in powdered dihydroxybenzoic acid failed to produce MALDI spectra [20]. Karas and coworkers [85, 127, 128] have proposed cluster formation between the analyte and matrix molecules as a critical prerequisite in the MALDI process. A model for preionization interactions between analyte (protein, in this case) and matrix molecules is shown in Scheme 9-1 based on ideas described by Karas *et. al* [127]. Hydrogen bonding of the analyte and matrix may be important (R. Brown, Utah State University, personal communication), and so may residual stoichiometric quantities of solvent [129]. In the model portrayed in Scheme 9-1, proton transfer is assumed to take place by disproportionation of matrix and analyte in the solid deposit or early in the lifetime of ejected clusters; the resulting cationic (protonated) and anionic (deprotonated) species now constitute an ionic, rather than a molecular, structure. Explosive separation of some of these ionic species illustrated in Scheme 9-1 could be responsible for protonated and deprotonated species being available in the gas phase for *m/z* analysis following pulsed illumination of the sample by the laser.

The mechanism of proton transfer to the analyte during MALDI has been of continuing interest since pioneering reports of the technology [14, 130–132]. In a recently described theoretical model for analyte ionization, matrix-to-analyte charge-transfer reaction kinetics are described by a hard-sphere Arrhenius expression [86]. In an experimental study, it has been found that clusters of the analyte with dihydroxbenzoic acid

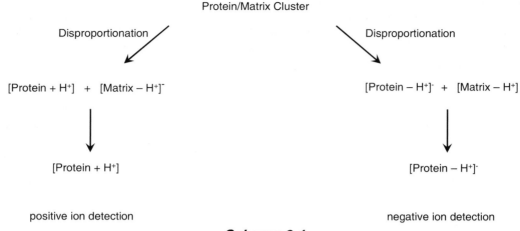

Scheme 9-1

are easier to ionize (reduction in ionization potential by 1 eV) than either compound taken separately [133, 134]. MALDI proton transfer product-ion distributions for a small peptide have been predicted by the energetics of possible secondary ion/molecule reactions at all laser fluences sufficient to generate a dense plume when using sinapinic acid [135].

Kruger and Karas [85] continue to document evidence for cluster formation and its critical role in charge separation as part of the MALDI process; they have found that residual stoichiometric quantities of solvent in the sample may also be important, especially as it controls the local pH at the site of photon absorption [129]. Several other studies report results that support clustering of the analyte with DHB [133, 134, 136, 137], and similar results have been observed with ferulic acid, vanillic acid, or sinapinic acid [128]. Other researchers propose a "poisoning" model to explain matrix–adduct formation [138]. Electrons can be formed by the photoelectric effect on the metal/organic matrix interface [139]; other studies of matrices doped with copper(II) salts indicate the availability of free electrons in the MALDI plume [94, 140]; electron capture by the matrix has also been reported [141].

Whereas explosive separation of charge pairs in the analyte/matrix clusters (caused by abrupt phase transition of the matrix upon absorption of UV radiation) as illustrated in Scheme 9-1 leads to protonated (or deprotonated) molecules of the analyte (as well as the matrix), multiple steps of disproportionation in a more extensive cluster (not just the heterodimer in Scheme 9-1) and subsequent charge separation would be likely to lead to multiple-protonated molecules of the analyte, etc. However, most of the multiple-protonated species are likely to be neutralized by reacting with electrons or negative ions; those species not completely neutralized (a neutral species would not be detected) are the "lucky survivors" and are detected as single-protonated molecules [127]. The continuous chemical background in the typical MALDI spectrum derives from clusters of matrix molecules and fragments therefrom [127, 142]; this chemical background is often more pronounced in atmospheric pressure MALDI (AP MALDI), possibly as a result of more efficient ablation of sample than in conventional MALDI. The related greater tendency of matrix–adduct formation with the analyte in AP MALDI can be controlled to some extent by increasing the cone voltage skimmer (causing dissociative collisions of the matrix-adducted analyte molecules) or by increasing the temperature of surfaces available to these species [143].

Some insight into the ionization process in MALDI can be gleaned from a study of the effect of esterification of peptides on their response to MALDI [144]. Esterification of the peptides greatly diminished their detection in the negative-ion mode, and substantially increased their response in the positive-ion mode. These results can be rationalized by the diminished acid character of the peptide upon esterification; because there would be a lesser proportion of the peptide molecules being deprotonated, there would be a larger proportion available for protonation and thus, detection in the positive-ion mode.

4. High-Pressure (HP) MALDI and Atmospheric Pressure (AP) MALDI

While the original development of MALDI was conducted at the typical low pressures (approximately 10^{-6} Torr) found in most mass spectrometer ion sources, recent reports indicate successful MALDI operation at higher pressures, including atmospheric pressure [145–151]. According to the preliminary report [146], atmospheric pressure (AP) MALDI at a laser frequency of 20 Hz gives an almost continuous ion current similar to that in ESI. Due to the thermalization of the ions, there is little or no metastable decay, which preserves larger ion currents for MS/MS experiments, an especially important issue in FTMS [152]; this feature has been documented during the analyses of proteins [152] and of RNA [153]. Because of the highly effective relaxation of metastable ion formation due to collisional cooling at atmospheric pressure, AP MALDI is considered to be a softer ionization technique than conventional vacuum MALDI [146, 154]. The nearly continuous beam of ion current from the AP MALDI source affords some flexibility in the nature of the *m/z* analysis technique and makes AP MALDI readily compatible with an orthogonal acceleration TOF mass spectrometer, for example.

The AP MALDI unit is compact (laser radiation via fiber optic as shown in Figure 9-7) and, in some cases, designed to be interchangeable with an ESI source for some instruments. Poor transfer efficiency of the ions to the analyzer in this prototype, a problem destined for an early solution, leads to a high LOD of 100 fmol of a 2.4-kDa peptide; special arrangements of ion optics, such as the ion funnel [155–157] or the AP ion lens [158], are likely candidates for improved design of the system. The use of counter-current heated gas streams [159] and the development of pulsed dynamic focusing [160], allowing for some entrainment of ions into the *m/z* analyzer, improves the ion transmission efficiency by an order of magnitude over that achieved with early versions of the AP technique [147, 150].

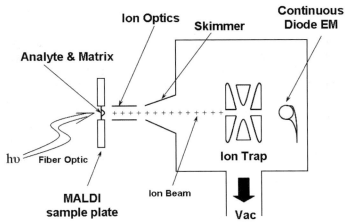

Figure 9-7. Conceptual diagram of an AP MALDI source in transmission geometry for a QIT instrument. *Diagram by G. Boyd, Professorial Assistant, Michigan State University, East Lansing, MI; after Galicia et al. Anal. Chem. 2002, 74, 1891.*

Other preliminary experiments with the prototype AP MALDI unit [146] involved the use of a corona discharge placed near the AP MALDI target to produce reagent ions to protonate compounds containing basic sites. Whereas the intensities of the matrix ion signals were increased 20–40 times by use of the corona discharge, the peptide ion peaks were largely unaffected; this suggests that the majority of the analyte molecules were ionized during the AP MALDI process (possibly within the condensed phase or at least within a few molecular dimensions of the surface of the matrix, or early in the plume) without the benefit of corona discharge. Application of the IR AP MALDI technique has been extended to analysis of frozen aqueous samples for peptides, carbohydrates, and glycolipids [161] and to noncovalent complexes of peptides with oligosaccharides [162]. This technique has yet to be reconciled with the report by Coon and coworkers in using a corona discharge and an IR laser for signal enhancement in MALDI [104]. A high repetition rate (1 kHz) laser has been reported to increase the ion yield from angiotensin in AP MALDI by a factor of 80 over that achieved with a 30-Hz laser [74]. The AP MALDI source with transmission geometry, illustrated in Figure 9-7, offers advantages described by Galicia *et al.* [163].

Extensive sample consumption during AP MALDI lends itself to automated spectral measurement, in which the laser beam could be rastered across the entire sample. A substantial number of microcrystals of the inhomogeneous target material would contribute to a final spectrum that was relatively independent of sample preparation, unlike the case in conventional MALDI, which relies more on identifying "sweet spots" on the target surface, a challenging process to automate [146]. High-throughput applications benefit from ready access to the sample chamber at ambient pressure, free from the encumbrance of vacuum locks required in conventional MALDI operation. A MALDI source has been designed for high-speed analyses on a triple quadrupole, and its performance compared with that of an ESI source [164]. High-throughput analyses also benefit from a high (2 kHz) pulse repetition rate, which translates to a data acquisition rate 10 times faster than currently available commercial instrumentation, as demonstrated in the analysis of tryptic digests of an *Escherichia coli* lysate [165].

In other cases, particularly those using ICR as the *m/z* analyzer, high pressure is limited to pulsed excursions to an estimated 1–10 mbar in the region of the MALDI sample during desorption, with the result of significantly decreased fragmentation compared to that in similar systems operating with pressures of <0.1 mbar [166]; a newer design allows better positioning of the sample plate with provision for collisional cooling [167]. More recent results with high-pressure (HP) MALDI for FTMS indicate enhanced production of multiple-charged proteolytic peptides, which may fragment more effectively and thus provide more definitive identification of the corresponding protein [168]. Galicia designed an AP MALDI source with transmission geometry using a transparent sample plate to allow laser illumination from the backside of the plate [163]. Feasibility studies with an AP laser desorption prototype showed many exciting possibilities including analysis of proteolytic peptides directly from aqueous solution (using water as the MALDI matrix) or by DIOS (no matrix) to achieve twice the sequence coverage obtained while using a matrix [169]. Use of a 3-µm IR laser showed good performance for AP MALDI, allowing a much broader choice of matrices including glycerol and liquid water for the analysis of peptides up to 2000 Da [170].

Advantages of the AP MALDI approaches have been reported in the context of analyzing carbohydrates [171] and sulfonated derivatives of peptides [172]. A recent report indicates that AP MALDI is ideal for producing a high flux of precursor ions from O-linked and N-linked complex glycans for subsequent structural studies by CAD MS/MS with FTICR technology [173]. AP MALDI has been used in the detection of cyclic lipopeptides during the characterization of whole-cell bacteria [174]. AP MALDI with an IR laser was used in an investigation of sugar–sugar noncovalent complex fragmentation in an ion trap mass spectrometer [154].

II. Sample Handling

1. Sample Preparation of the Conventional Plate

Proper sample preparation is critical for successful analyses by MALDI MS [175]. Matrix solutions are prepared in water, water–acetonitrile, or water–alcohol mixtures at a concentration of 5–10 mg mL^{-1} depending on the solubility properties of the matrix. The analyte is prepared at a concentration of about 0.1 g L^{-1}, and in a solvent that is miscible with the matrix solution (for peptides or proteins, aqueous 0.1% trifluoroacetic acid (TFA) is frequently used). The matrix and analyte solutions are mixed to give a final matrix-to-analyte molar ratio of approximately 5000:1, in a convenient final volume.

Several factors affect the sensitivity of a MALDI MS experiment. In some cases, analysis of smaller amounts of sample yields a better MALDI signal; this may simply be the result of approaching a more optimum molar ratio between the analyte and matrix, especially with macromolecules whose mass exceeds 10 kDa. Additionally, dilution of a contaminant, which can interfere with analyte signal acquisition, may also lead to improved spectra; in some cases, signal enhancement can be achieved by doping with copper(II) chloride [176]. A typical exploratory procedure involves making serial dilutions (10- to 1000-fold) of the sample to be analyzed prior to mixing with the matrix solution in an effort to reach an efficacious analyte-to-matrix ratio. As part of this exploratory procedure, simply changing the matrix may yield a favorable response if none was detected from an initial sample–matrix combination.

In conventional operation, aliquots of matrix and analyte solutions are applied to a stainless steel MALDI sample plate, and allowed to dry by either ambient evaporation, by heating with a stream of warm air, or under vacuum. During the drying process, the matrix co-deposits from solution with the analyte on the MALDI sample probe (as shown in Figure 9-2). Several procedures for sample preparation have been reported [177–179]; one comparative study of matrix/sample preparation focused on proteins of high mass [180]. Another study found that use of the two commonly used matrixes, DHB and α-CHCA, gave results leading to increased sequence coverage and spot-to-spot reproducibility for peptide mass mapping compared to the use of the single matrix components [181]. Some representative methods of sample preparation are briefly described below.

The "dried droplet" method [11] consists of mixing microliter quantities of an analyte solution and a matrix solution on a sample plate such that the two miscible solutions deposit approximately 10–100 nmol of matrix compound together with 1–10 pmol of analyte as the solvent(s) evaporate. A common problem with this method of sample preparation is the accumulation of most of the analyte/matrix crystals as a circular ridge around the edge of the deposit; this bulky ridge is usually heterogenous and irregularly distributed. Another problem often observed during formation of the solid deposit is segregation, which refers to the exclusion of salts and some of the analyte from the matrix as it crystallizes during evaporation of the solvent(s). Component segregation generates a heterogeneous mixture of analyte throughout the solidified sample, which results in highly variable production of analyte ions as the laser moves across the sample surface. Despite the many disadvantages of this earliest described method of sample preparation, it remains one of the most popular, probably because of its straightforward simplicity.

The "fast evaporation" method is a variant of the dried droplet method; it was developed in 1994 with the main goal of improving the resolution and mass accuracy of MALDI measurements [177]. First, a thin layer of the matrix compound is deposited onto the sample plate. A droplet of the analyte solution is then applied to the solidified matrix bed. Before the solvent evaporates from the newly applied droplet, it dissolves some of the matrix that then recrystallizes onto the residual underlying matrix bed on the surface of the sample plate. Because this method does not provide reproducible sample-to-sample data for peptide and protein mixtures, it is not recommended for multicomponent protein/peptide samples. There is some tendency for methionine and tryptophan to oxidize during the fast evaporation method [182], but not so with the dried droplet method [183], as reported by others [184].

The "ultrathin layer" method [98] consists of first applying a saturated matrix solution over the whole surface of a sample plate; before the solvent evaporates, most of the applied matrix solution is gently wiped away with a tissue, leaving only a faintly visible layer of matrix crystals. The sample is dissolved directly in a saturated matrix solution, an aliquot (usually 0.5 µL) of which is spotted onto the precoated sample plate. A protein–matrix crystalline film (cloudy in appearance) soon forms, and before the solvent evaporates, the residual solution containing most of the salts, detergents, and other components of the sample are removed by vacuum aspiration. The sample spot can be further cleaned by washing with dilute acidic solution to remove residual salts and detergents. This ultrathin layer method presents a homogeneous matrix–protein co-crystallized solid surface, allowing for reproducible data acquisition over the entire surface area. Compared with the conventional "dried droplet" method, this method shows better detection limits, mass accuracy, and resolution. It is a robust, detergent-friendly method for a variety of proteins and mixtures including membrane proteins. A three-layer variant of this procedure reportedly offers advantages for samples containing salts [185].

Even today, the choice of matrix and sample preparation method remains an empirical process due to a lack of understanding of the desorption/ionization mechanism. In some cases, an acid-sensitive analyte may require the use of a neutral or aprotic matrix, like the benzthiazole shown at the bottom of Figure 9-1; another aprotic matrix, 2-[(2E)-3-(4-*tert*-butylphenyl)-2-methylprop-2-enylidene]malononitrile (DCTB) shows promise for analyses of coordination compounds, organometallics, conjugated organic compounds (including porphyrins and phthalocyanines), carbohydrates, calixarenes, and macrocycles [186]. The quality of MALDI MS results are controlled by many factors, such as the choice of matrix and solvent, the additive used, and the crystallization conditions employed. Parallel analyses under different conditions may be required for optimization of the MALDI MS experimental conditions.

Samples to be analyzed by MALDI should *not* be stored in glass, as trace levels of the analyte can easily pick up stoichiometric amounts of alkaline earth cations. Any exposure of an analyte to glassware can promote Na^+ or K^+ adduct formation during subsequent analysis by MALDI or FAB, etc. Such adduct formation may predominate giving intense peaks representing $M + Na^+$ or $M + K^+$, rather than $M + H^+$. The trace levels of Na^+ or other metal ions that may interfere with the analysis can be removed by chelation or by ion exchange [187]. It is often best to store these types of samples in polycarbonate vessels, or even in some types of plastic! The analytes to be analyzed by MALDI will most often be soluble in aqueous media, and not in organic solvents that would dissolve the plasticizers from plastic vessels.

The demand for high-throughput analyses has led to some modifications to the conventional MALDI sample plate; an acoustical droplet ejector has been introduced to improve the reproducibility and speed of some aspects of sample preparation [188]. A modified MALDI probe consists of twin chambers, one for the sample, the other for internal standards; it has been reported that use of this twin-anchor device reported improves mass accuracy and reduces signal suppression for the analyte [189]. High-throughput procedures have gained momentum from specialized microfluidic use of otherwise traditional compact disks (CDs) [190, 191]. The success rate of the CD technology in protein identification is reported to be about twice that achieved with C18 ZipTips and standard MALDI steel targets. The CDs can be operated using robotics to transfer samples and reagents from microcontainers to the processing inlets on the disposable CD with spinning for microfluidic control of liquid movement through the microstructures. Interference from matrix ions has reportedly been reduced when using a pre-prepared MALDI target plate coated with a thin layer of α-cyano-4-hydroxycinnamic acid (CHCA) and nitrocellulose [192, 193]; this offers advantage for analyses of compounds having a molecular weight less than 1000 Da. Using a mixture of CHCA and nitrocellulose on an AnchorChip™ gives more comprehensive detection of peptides and with less laser power than using either CHCA or DHB alone as the matrix [194].

2. The Problem of Analyte Solubility

The poor solubility of some analytes in solvents that are suitable for the matrix can cause problems during sample preparation. In particular, hydrophobic membrane proteins have required special techniques of sample preparation. Cadene and Chait [98] used a high concentration of formic acid after solubilizing the protein in a nonionic detergent at a minimal concentration above the critical micellar concentration; others have reported that *n*-octylglucopyranoside minimizes hydrophobic adsorptive sample loss due [175, 184]. Significant improvement in detecting hydrophobic proteins can be accomplished by using matrix formulations consisting of perfluorooctanoic acid and sorbitol [195]. Breaux *et al.* [196] evaluated a wide variety of anionic, cationic, zwitterionic, and nonionic surfactants to solubilize a hydrophobic protein without interferring with subsequent analysis by MALDI; other than SDS, *n*-octylglucoside shows great promise for analyses of large (ca. 50 kDa) hydrophobic proteins [197]. Results of other studies show that a so-called "temperature-leap tactic" [198] can be helpful in maintaining protein solubility, while promoting proper crystal formation with the matrix for an optimal MS signal for large hydrophobic proteins [199].

Industrial polymers are another class of problem samples [183, 200, 201]; in some cases, it has been found that grinding the polymer and matrix crystals as dry solids produces a fine powder, which adheres to the MALDI plate, and with proper stoichiometry generates a good signal for the polymer.

3. The Problem of Sample Purity

The signal generated by a given compound during MALDI is affected by other components in the sample. In particular, any salts, surfactants, or even other similar compounds can cause severe attenuation or suppression of the signal expected from a given compound [202, 203], although concomitant use of a mid-IR free-electron laser [204] has been reported to eliminate or minimize otherwise adverse affects of 8 M guanidine. Specialized devices have been designed to combine the operation of coarse chromatographic separation of analyte and contaminants with the process of sample deposition on the MALDI target [205]. The detrimental effect of certain contaminants may be due to interference associated with either matrix crystal formation and/or the necessary interactions of the analyte with the growing matrix crystals. The matrix suppression effect can be controlled to some extent to facilitate detection of the analyte [206]. Frequently, detergents are required to solubilize proteins during the sample preparation procedure. Detergents such as SDS (sodium dodecylsulfate) are known to severely attenuate the signal expected from peptides and proteins, although exceptions have been reported. For this reason, considerable attention has been given to removal of such detergents, or the use of nonionic [196, 207, 208] or acid-labile [209, 210] or otherwise cleavable [211] detergents has been emphasized. Some matrix compounds are more forgiving of such contamination than others. For example, 2,4,6-trihydroxyacetophenone containing diammonium citrate reduces the need for extensive washing of ZipTip-bound peptides or other on-plate cleaning procedures [212]; a novel MALDI matrix, 3,4-diaminobenzophenone (DABP), permits a discernible ion signal for insulin in the presence of 2 M guanidine hydrochloride and 1.5 M urea [213].

A clever and useful approach to the problem of the adverse effect of detergents on MALDI response has been the use of detergents that can be degraded to innocuous compounds once they have served their purpose in solublizing the analyte. The use of acid-labile detergents, which can be degraded to a form that does not interfere with subsequent analysis by MALDI, has been reported [214]. Caprioli and coworkers have developed hybrid detergent molecules in which the polar head group is the α-cyano moiety or a sinapinic acid moiety that is linked to the nonpolar chain via an acid-labile functionality; at pH 2, the hybrid molecule decomposes into α-cyano cinnamic acid or sinapinic acid and a long-chain alcohol that is transparent in the MALDI process [215].

4. On-Probe Sample Purification and/or Modification

If the use of detergents and/or salts cannot be avoided in sample preparation, some means of removing these offending compounds before analysis by MALDI must be employed. In some cases, rinsing the solid deposits of the matrix/analyte on the sample plate with distilled water, or better in other cases with monoammonium phosphate, greatly suppressed matrix and alkali ion clustering, thereby improving detection limits [57]. A wide variety of surfaces has been used as a means of retaining particular analytes, while washing away MALDI-offending compounds.

A. SAMs and Polymer-Modified Surfaces

Some means to desalt the sample may lead to successful analysis by MALDI [216]. Watson and coworkers [37, 38] have developed a polymer-modified MALDI sample plate surface that facilitates both concentrating and cleaning the analyte during the sample preparation process. The general strategy for on-plate sample cleanup is illustrated in Scheme 9-2, where the polymer-modified metal surface selectively binds the analyte and retains it as the salty contaminant is washed away with water; subsequent addition of matrix releases the analyte from the polymer and allows co-crystallization prior to analysis by MALDI. Further, Watson and coworkers found that use of 200 μm hydrophilic gold spots improved reproducibility of the MALDI signal from sample to sample because the relatively small spot allowed a large fraction of the total sample area to be illuminated by the laser beam [38]; similar results with the use of small gold spots exposed through a hydrophobic coating have been reported in the context of analyzing brain tissue for neuropeptides [217]. MALDI sample plates have also been modified with polymethyl methacrylate in an effort to minimize sample handling during the analysis of nucleic acids [218]. A cationic polymer-modified surface on the MALDI sample probe has been shown to improve analyses of single strands of DNA by facilitating removal of salt from the sample,

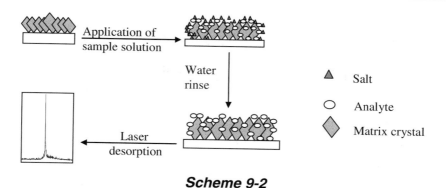

Scheme 9-2

thereby eliminating problems associated with multication adducts [37]. In other developing work for selective detection of phosphorylated peptides, metal-affinity complexes were chemically grafted onto the surface of a gold MALDI plate by coupling a derivative of nitrilotriacetate (NTA) to immobilized poly(acrylic acid) and subsequently forming the iron(III)–NTA complex, which accumulated phosphopeptides from protein digests [38]. Zirconium-modified stainless steel targets are effective in accumulating phosphopeptides from protein digests for washing prior to direct analysis by MALDI [219].

Ready-made sample plates [220] have been described for simplifying the sample preparation procedure, including those using a microfilm of α-CHCA on a prestructured sample support as an affinity surface for peptides [184]. Other prestructured sample supports have been designed to serve as an interface to CE/MS for analysis of proteolytic peptides [221] or to allow on-probe digestion of the protein prior to analysis by MALDI [222]. MALDI of liquid solutions from a graphite surface has been reported [223].

In a different approach, a thin layer of Teflon was deposited on the MALDI plate and then this surface was sputter-coated with an array of 200 μm diameter gold spots [224, 225]. Because of the difference in hydrophobicity between gold and Teflon, sample droplets attach to the surface only at the hydrophilic gold spots. In this way, large droplets often evaporate to the size of the gold spot, allowing concentration of the sample. Because a large fraction of the compact solid deposit can be illuminated by the laser, reproducibility of the signal improves. Paraffin-coated sample plates have been used for convenient and dependable preparation for high-throughput analyses of protein digests [226]. Disposable hydrophobic surfaces also have been proposed to facilitate localization of the sample to a small area to eliminate or minimize "sweet spots" on the sample plate [227]. In a somewhat analogous approach, an elastomeric device is reversibly sealed onto the MALDI target to form a multiwell plate to accommodate relatively large sample volumes for desalting, concentrating, and digesting the sample directly on the sample plate [228, 229].

Cationic and anionic nanoparticles (2 nm core diameter) can selectively accumulate peptides with low and high isoelectric points, respectively. Use of these nanoparticle scaffolds to extract and concentrate peptides from dilute solutions greatly improves the detection limits for assays based on MALDI MS [230]; targeted peptides can be detected from 250 mL aliquots of solution concentrations as low as 500 pM.

There is considerable interest in the direct analysis of proteins adsorbed onto a transfer membrane by MALDI because this membrane procedure is used routinely with gel electrophoresis, which is a dominant technique in separating complex mixtures of proteins [231]. Applications of MALDI MS to samples adsorbed onto transfer membranes include

(a) direct analysis of protein and peptide samples transferred from PAGE gels [72, 232–236] and (b) manipulations of membrane-adsorbed proteins such as washings to remove MALDI contaminants or chemical derivatization [237]. Cellulose precoated with matrix has been used for continuous deposition of the eluant from capillary electrophoresis for eventual analysis by MALDI [238]. Nylon-66 and positive charge-modified nylon (ZETABIND) membranes are suitable for analyzing picomole levels of a protein that has been immobilized by adsorption onto the nylon membrane, washed to remove MALDI contaminants, and digested enzymatically and/or chemically prior to adding matrix [237]. In other cases, on-target endoglycosidase digestion of glycoproteins has been described [239]. Both chemical and enzymatic digestions have been reported for proteins on PVDF and glass fiber [237]; nitrocellulose [240] also shows comparable advantages. For digestion of proteins electroblotted to nitrocellulose (NC), a new protocol dissolves the NC membrane in a solution of 70% acetonitrile–30% methanol (rather than acetone by other protocols) [241]; this new protocol requires only half the time of in-gel digestion and results in fewer missed cleavages and better protein coverage. Deposition of peptides and proteins onto nonporous ether-type polyurethane membranes allowed the sample to be washed prior to analysis by MALDI [242]. The use of sample probes modified with polyethylene membranes reportedly minimizes signal suppression during the analysis of large (100 kDa) proteins [243]. Coating the MALDI probe surface with a film consisting of a perfluorosulfonated ionomer (Nafion) prior to the addition of the sample/matrix mixture reportedly improves the quality of the spectra of highly acidic oligosaccharides [244]. In related work, a glass-chip-based sample preparation method using a sol-gel-derived material covalently bound to the surface of a glass chip was hybridized with 2,5-dihydroxybenzoic acid as the MALDI matrix to provide a medium for on-probe digestion of proteins prior to analysis by MALDI MS [245].

A novel liquid–liquid extraction procedure using ethyl acetate as the water-immiscible solvent was used to segregate hydrophobic and hydrophilic polypeptides on the MALDI sample probe in an effort to reduce the complexity of the sample prior to analysis; addition of the MALDI matrix to the organic solvent enhanced the efficiency of the technique for analysis of hydrophobic peptides and proteins [246].

B. Affinity Surfaces

In the affinity surface approach, an analyte-binding molecule, such as an antibody, lectin, or receptor, is covalently attached to the surface of a MALDI probe. In this way, the analyte of interest can be selectively captured and concentrated on the probe surface prior to analysis by MALDI MS [247, 248]. A general method for preparing bioaffinity probes has been reported [249]; these types of plates have been useful for screening urines for selected proteins [250], and in some cases adapted to chips [251]. However, the binding of the antibody to the surface of the conventional MALDI sample plate is not optimal, mainly because the antibody can bind to the template in various orientations, many of which block antigen recognition; strategies for orienting the antibody during immobilization to improve the situation have been described [252]. On the other hand, the thrombin-binding DNA aptamer has been covalently attached to a MALDI probe surface for successful affinity capture of thrombin during analyses of human plasma [253]. Antibodies have been used to characterize the nature of target-unrelated peptides recovered in the screening of phage-displayed random peptide libraries [254]. In other work, a monolayer of fourth-generation poly(amidoamine) dendrimers was adapted to construct the immunoaffinity surface of an antibody layer [255]; the general applicability of mass spectrometry to dendrimers has been addressed [256].

The surface of MALDI targets can be modified for the capture of biotin-tagged proteins by first binding a poly-L-lysine poly(ethylene glycol)-biotin polymer to the target; tetrameric neutravidin is then applied, which serves as a bridging molecule to capture biotinylated proteins [257]. Use of this immobilized biotin–neutavidin–biotin "sandwich interface" can be used to capture biotinylated proteins from complex mixtures, for subsequent on-probe cleanup, and finally analysis by MALDI. Another promising approach is based on a polymer thin film, produced by pulsed RF-plasma polymerization of allylamine on a metal surface that can be subsequently biotinylated to develop a bioaffinity-capture MALDI probe [258].

Surface-enhanced laser desorption/ionization (SELDI) probes [259–261] are useful for capturing and detecting specific biochemicals or biomarkers [262, 263], as in screening for transcription factors [264]; methodology to generate ordered nanocavity arrays on a Si wafer may offer additional selectivity to this approach [265]. A variation of this technique involves use of antibody-coated magnetic beads that bind specific peptides/proteins for subsequent analysis by MALDI, a technique that should be helpful in bridging the technology gap between biomarker discovery and clinical analysis [266]. A somewhat related approach called material-enhanced laser desorption/ionization (MELDI) uses derivatized carrier materials (cellulose, silica, poly(glycidyl methacrylate/divinylbenzene) particles, and diamond powder) to adsorb specific analytes for fast and direct analysis by MALDI TOFMS as tested in protein profiling of human serum samples [267]. Diamond nanoparticles seem promising as a general method for scavenging proteins from dilute solutions for subsequent analysis by MALDI MS [268]. A so-called mass spectrometric immunoassay in which MALDI is used to detect beta-2-microglobulin off-line after affinity capture by specialized pipettor tips proved useful in screening urine samples without interference from glycosylated forms of the protein [269]. A lab-on-a-chip approach, using a pressure-driven pumping mechanism based on the vacuum of the instrument to drive reacting solutions through microchannels, can monitor organic syntheses or biochemical reactions by MALDI TOFMS in real time [270]. A related procedure for on-probe cleanup of glycoprotein-released carbohydrate samples involves mixing small amounts of chromatographic media with matrix for codeposition onto a MALDI probe [271]. This simple procedure was used successfully to minimize interferences from cations, anions, and/or detergents and increase the mass spectral signal-to-background, improve mass accuracies, and better equalize the MALDI signal response for diverse carbohydrate structures released with multiple glycosidases. Ion exchange ProteinChip arrays have been used to remove SDS and concentrate proteins eluted from SDS-PAGE gels [272]. A wax-coated MALDI target plate has been used to allow the recovery of proteolytic peptides from the MALDI sample plate for subsequent analysis by ESI LC/MS [273].

5. Direct Analysis from Gels

Encouraging results have been obtained in the direct analysis of proteins from thin PAGE preparations [274, 275]; however, the conventional 2D gel warps and cracks when dehydrated, leaving a nonplanar surface which precludes accurate mass measurements during direct analysis. Much thinner 1D isoelectric-focusing gels provide a more planar surface when placed in the mass spectrometer, allowing useful mass spectra to be acquired by scanning the surface of the gel; Andrews and coworkers then assimilated the data into a "virtual" 2D gel, which is analogous to a "classical" 2D gel, except that the molecular weight information was acquired by MS [274, 275]. Preliminary results from IR laser desorption and ionization mass spectrometry of peptides and proteins directly from a polyacrylamide gel (i.e., without the addition of a matrix) have been reported [276].

A promising method for the analysis of in-gel digests is based on laser desorption only, as opposed to MALDI; the technique utilizes an IR laser pulse to desorb intact neutral molecules, followed by ionization via reagent ions produced by APCI using a corona discharge [104, 277]. Another non-MALDI approach uses an IR laser for desorption and ionization of peptides from a gel [278]. A recent report describes performing SDS PAGE on a chip fabricated from poly(methylmethacrylate) with a poly(di-Me siloxane) cover; after electrophoresis, the cover was removed and either the chip or the cover was subjected to IR MALDI, allowing a decrease in the volume of material required for analysis, thereby enabling improvement in the detection limit to the pmol level [279].

6. Hydrogen/Deuterium Exchange

Special sample handling is required to achieve measurement of hydrogen/deuterium exchange (HDX) by MALDI [280–282]. The principal concern in sample handling is to minimize back-exchange of any deuterons that have been incorporated into the analyte during the HDX experiment. In most cases, back-exchange is minimized by processing the analyte in acidic solution (pH = 3) at low temperature (at least 4 °C) [280, 282]. The use of fast microchromatography for sample handling has also been reported to help minimize back exchange [281]. MALDI-based HDX measurements have been used to determine folding free energies for proteins with the accuracy and precision of conventional spectroscopy-based methods [283].

III. Special Instrumental Techniques

There is very little fragmentation associated with the conventional MALDI MS [284] and thus this type of information is often not readily available without specialized instrumental techniques as described below.

1. Post-Source Decay (PSD)

In general, there are three types of ions in mass spectrometry: (1) stable ions, (2) unstable ions, and (3) metastable ions. The distinction between these three types of ions is made based on a comparison of the half-life for decay or fragmentation of the ions relative to the timeframe of a particular experiment. Consider the situation in TOF mass spectrometry; in this experiment, the relevant timeframe is the flight time (or residence time in the flight tube). If the flight time is on the order of 100 μsec, ions with a $t_{1/2}$ of 10 msec or more would be considered "stable" ions. Ions with a $t_{1/2}$ of 10 picosec or less would be considered "unstable" ions; those with a $t_{1/2}$ of 100 μsec would be "metastable" ions. In TOF mass spectrometry, a metastable ion is extracted from the ion source before it fragments, but it does fragment in the flight tube (post source) before it enters the detector region [285, 286]. Those ions that fragment before they leave the ion source (a phenomenon frequently observed in EI) are said to originate by in-source decay (ISD) in the context of MALDI [287].

Post-source decay [285] is a name to remind you that the ions of interest are those that fragment or dissociate in the flight tube of the TOFMS after they leave the ion source. If the ion had dissociated within the source, the nascent fragment and residual parents would obtain the same energy during acceleration, and because they have different masses, would travel with different velocities and have different flight times. However, for those ions that undergo PSD (metastable decay) in the flight tube, the resulting product ions continue toward the detector with the same velocity as the residual precursors, and thus are detected as a single peak in a conventional linear TOFMS. Differential detection of precursor and product ions requires an ion mirror as illustrated in Figure 9-8. A common technique available on some commercial MALDI TOF instruments for detecting PSD fragments involves selecting a given precursor ion with a "timed gate"; because a reflectron with a linear electric field provides optimal focus only over a limited *m/z* range, the results of several PSD experiments must be "stitched" together via the data system. Use of a curved-field reflectron [288, 289] for PSD (as commercialized by Shimadzu for the *Kratos AXIMA* series of instruments) optimally accommodates the complete *m/z* range, thereby offering the advantage of speed and some gain in performance. In a novel approach to PSD [290], ions are matched to their corresponding precursor ions by comparing spectra acquired at slightly different reflectron electric fields, and by measuring the difference in time-of-flight between the two spectra for each fragment, the mass of the fragment ion can be calculated; this approach saves analysis time and requires less sample than previous approaches.

Consider the ionized molecule AB$^+$ that dissociates into fragment ion A$^+$ and neutral B during its flight from the source to the detector. As illustrated in Figure 9-8, the ions AB$^+$ and A$^+$ can be distinguished by use of an ion mirror. Assume that AB$^+$ was accelerated with 30 kV; when A$^+$ ions were formed in flight, they maintained the same velocity as other AB$^+$

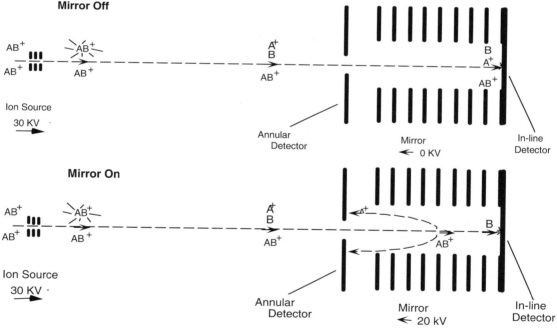

Figure 9-8. *Diagram of a linear electrostatic mirror on a TOF mass spectrometer. With no voltage on the mirror (top panel), ion AB$^+$ dissociates to ion A$^+$ and neutral B in the field-free drift tube ("post-source decay"); B and A$^+$ maintain the initial velocity of precursor AB$^+$ and all three species reach the in-line detector simultaneously. With opposing voltage on the mirror (lower panel), B is unaffected and hits the in-line detector as before, but A$^+$ is reflected to the annular detector.*

ions that remained intact, but the kinetic energy of A^+ diminished in direct proportion to the ratio of the masses of A^+ to AB^+. Thus, in the absence of any retarding potential on the mirror, both ions AB^+ and A^+ as well as neutral B arrive at the in-line detector simultaneously as shown in upper panel of Figure 9-8. If a retarding potential, say +20 kV, is impressed across the mirror, the ions AB^+ which entered the mirror with 30 kV will traverse the mirror with a net energy of 10 kV. Assume that the product ions, A^+, enter the reflector with a KE of only 18 kV; they will be repelled by the mirror with a net KE of +2 kV, which will cause them to reverse direction and hit the annular detector. Under these conditions, the neutrals B would be unaffected and hit the in-line detector as before, and while the ions AB^+ would eventually hit the detector, they would trail behind the neutrals as the ions AB^+ will lose 20 kV in traversing the mirror and hit the detector with only 10 kV of kinetic energy.

The repelled ions "fan out" by virtue of their inherent radial velocity, which is constant and unaffected by the mirror. Because the ions slow down in the mirror, the radial velocity operates during their relatively long residence time in the mirror to cause the ions to disperse or fan out, so that if they are repelled by the mirror, they will hit the annular-shaped detector on their reverse flight path, as shown in lower panel of Figure 9-8.

A comparative study was made of the negative- and positive-ion PSD spectra of nine peptides, which produced significant levels of deprotonated ions, $[M - H]^-$, although more abundant protonated molecules, $[M + H]^+$ [291]. In one application, PSD was used to determine the N-termini of proteins blotted on a PVDF membrane [292]. When a peptide contained a mixture of acidic and basic residues, the negative PSD spectra were more complex and the locations of acidic residues dictated some fragmentations. Although some structurally useful fragmentation is available through PSD [293], it is somewhat difficult to obtain, and the analyst has relatively little control over the extent of fragmentation produced. A more efficient analysis can be obtained by MALDI with a Q-TOF hybrid instrument [294], a TOF/TOF hybrid instrument [295], or a modified quadrupole ion trap mass spectrometer [296].

2. Ion Excitation

As MALDI becomes increasingly integrated into non-TOF instrumentation, such as quadrupole ion traps and specially designed hybrid-TOFs, PSD is being eclipsed by techniques for ion excitation. PSD relies on metastable decay of MALDI ions, leaving the analyst with little or no control over the outcome of the experiment. The incorporation of MALDI into FTMS, quadrupole ion trap, and certain hybrid-TOF instruments allows the operator to control the extent of ion dissociation as well as take advantage of classical MS/MS procedures involving mass selection of a precursor ion. The results of a preliminary study using an in-house assembled MALDI quadrupole ion trap mass spectrometer shows that *de novo* sequencing of peptides is possible from fragmentation by CAD MS/MS (single-charge MALDI ions within the ion trap appear to fragment at nearly every peptide bond along the peptide backbone) [296].

The nonresonant excitation technique of boundary-activated dissociation (BAD) has been used to obtain MS/MS spectra for peptide ions generated by MALDI in a quadrupole ion trap [297]. BAD MS/MS spectra for model peptides such as proctolin, des-Arg9-bradykinin, and substance P are qualitatively similar to those for which resonant excitation has been used and can be obtained with the same activation time. The conditions for optimal product ion formation are easily established when BAD is used because of its dependence upon a single activation parameter; consequently, MS/MS spectra of MALDI-generated ions can be obtained with simpler electronic equipment when single-frequency resonant excitation is used in comparison with broadband excitation [297].

3. Delayed Extraction (DE)

The technique of delayed extraction [298–300] dramatically improves the resolution achievable in MALDI TOFMS. Delayed extraction in MALDI TOFMS is designed to mimic [301] the processes used in time-lag focusing as described in Chapter 2 for TOFMS for gas-phase analysis [302]. The "delay" in delayed extraction is an operator-selectable period (typically 400 nanoseconds) between the laser pulse and the application of the ion extraction field to the nascent ions. The fundamental idea is to allow those ions having the greatest initial kinetic energy to travel farther into the ion source than ions having a lesser kinetic energy, so that when the extraction field is turned on, the ion having the lesser *initial* kinetic energy will traverse a greater distance through the electric field, thereby picking up more kinetic energy in a compensatory effort to give all ions of the same mass the same kinetic energy.

In principle, delayed extraction, a.k.a. time-lag focusing (TLF), causes ions of the same mass to arrive at the detector at nearly the same time, giving rise to a sharp peak or greater resolution. The benefit of delayed extraction in MALDI TOFMS is illustrated in Figure 9-9 which compares the MALDI mass spectrum of a fragment of ACTH obtained by conventional use of a linear TOF-MS (top panel) with those obtained by delayed extraction into a linear (middle panel) and a reflectron (bottom panel) TOF-MS.

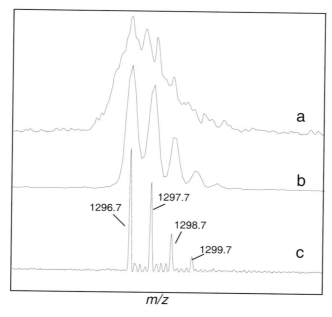

Figure 9-9.   *Comparison of MALDI mass spectra of angiotensin obtained during three different modes of operation: (a) linear mode with continuous extraction, (b) linear mode with delayed extraction, and (c) reflectron mode with delayed extraction.*
Data collected by Yingda Xu, Graduate Assistant, Department of Chemistry, Michigan State University, East Lansing, MI.

Some improvement in resolution with delayed extraction may be due to dissipation of the very high local pressure (tens of atmospheres) in the laser plume following (for picoseconds) the laser pulse. However, use of delayed ion extraction in MALDI TOF mass spectrometry distorts the linear relationship between *m/z* and the square of the ion flight time with the consequence that, if a mass accuracy of 10 ppm or better is to be obtained, the calibrant signals must lie close to the analyte signals. If this is not possible, systematic errors arise; to eliminate these, a higher-order calibration function and additional calibration peaks are necessary [303].

4. Desorption Ionization On Silicon (DIOS)

Porous silicon (pSi) constitutes a unique surface from which to conduct desorption/ionization as mediated by photons; it is included in this chapter as a MALDI technique without a matrix [304]. It is believed that the high surface area, low thermal conductivity, and high UV absorptivity of pSi has enabled its successful application to desorption/ionization on silicon (DIOS) mass spectrometry [305, 306]. Because DIOS involves no exogenous matrix compound to accomplish desorption/ionization of the analyte, the technique offers the distinct advantage of no (or very little) background ions in the mass spectrum [307–310], although the problem of interference from abundant background ions from the usual matrix can sometimes be controlled by reducing the molar ratio of matrix to analyte [69]. DIOS from pSi has been used for quantitation of small synthetic polymers [311]. Although some applications have been reported for analytes heavier than a few thousand daltons; e.g., in the characterization of selected enzymes by detecting the products of substrate modification [312], most applications of pSi have been to analytes below 1000 Da. Useful fragmentation has been reported when using TOF/TOF in combination with DIOS from pSi [313]. Parameters and procedures for the use of DIOS with pSi have been published [306, 314], but the mechanism of ionization remains in question [315]. Quantification of pure peptide standards relative to stable isotope-labeled standards by DIOS has been described [316]. Pure peptides (below 500 Da) have been detected at the pmol level, and good-quality mass spectra of polyethylene glycol (below 2000 Da) have been obtained by DIOS [306]. Remarkably, DIOS MS from pSi has been used to detect peptides (up to *m/z* 4000) directly from invertebrate exocrine tissue blots (after cellular lysis by a methanol-citrate solution directly on the silicon substrate) and also from neurons cultured directly on pSi substrates [317]. IR soft desorption/ionization without matrix has been attempted on stainless steel, aluminum, copper, silicon, pSi, and polyethylene; silicon surfaces gave the best performance in terms of signal level and low-mass interference [318].

Silylation of pSi provides a more hydrophobic surface allowing for preferential adsorption of analytes from salty samples as demonstrated in the analysis of protein digests [319]; a detection limit of approximately 500 molecules of *des*-Arg[9]-bradykinin was achieved with the use of such silanized pSi. Perfluorinated surfactants dramatically enhance detection of amino acids, carbohydrates, and other small organic compounds during analysis by fluorinated DIOS mass spectrometry [320]; whereas perfluorooctanesulfonic acid improved the detection limits and sequence coverage of tryptic digests, the use of conventional surfactants like nonfluorinated sodium dodecyl sulfate had no effect. In analysis of proteolytic digests using DIOS, the addition of diammonium hydrogencitrate facilitated the detection of cysteine sulfonic acid-containing peptides [321].

Single-crystal silicon nanowires (SiNWs) share many of the basic properties of pSi and appear to be a good platform for surface-based mass spectrometry [322]. In contrast to pSi, SiNWs are grown on the surface of a substrate and their physical dimensions, composition, density, and position can be precisely controlled at the nanoscale level, thereby offering potential for designing analyte-sampling surfaces for mass spectrometry. Silylation of SiNWs is necessary to achieve low detection limits, as demonstrated in the analysis of 5 pmol of a tryptic digest of bovine serum albumin containing peptides approaching 2000 daltons [322]. The wicking action of the silanized SiNW surface provides a chromatographic surface, which was demonstrated in analysis of simple binary mixtures (1mg mL^{-1}) of drugs around 400 daltons with detection by DIOS MS [322]. Although this preliminary report did not declare the quantities applied to the SiNW chromatographic surface, it is possible that this technology may ultimately provide analytical advantage over the more conventional combinations of TLC and mass spectrometry [323–328], even as combined with desorption ESI [329]. Laser desorption/ionization from silicon nanopowder (5–50 nm) shows some of the advantages of DIOS, such as the absence of background from the conventional MALDI matrix and low detection limits for small molecules [330]. DIOS continues to show advantages in the analysis of nonvolatile small molecules such as ribose and other aldopentose isomers [331]. Atmospheric pressure DIOS has been reported recently in an application to drug metabolism [332]. In a comparative study with MALDI, DIOS with and without CAD MS/MS is more sensitive in the detection of amphetamines and fentenyls during analysis of putative drug seizures [333].

A related technique, desorption/ionization on self-assembled monolayer (SAM) surfaces, generates gas-phase ions directly from samples deposited onto a SAM surface in a matrix-free manner [334, 335]; the hydrophobic nature of the SAM surface is suitable for analyses of lipids as well as peptides. Desorption/ionization from hydrophobic polymeric monoliths like poly(benzyl methacrylate-co-ethylene dimethacrylate) [336] or from alumina [337] have also been used to detect small molecules (explosives and drugs) and peptides, respectively.

5. Tissue Profiling or Imaging

The possibility of directly analyzing frozen tissue samples for proteins, in which ice serves as the matrix, had been explored nearly a decade ago [338]. The use of MALDI TOFMS spatially to map peptides, proteins, lipids, or xenobiotics (drugs) [339–349] or ICP to map selected elements [350] directly from a tissue is becoming a well-established application of imaging mass spectrometry. Specific protocols have been proposed for the correlation of 3D ion images with histological features observed by optical microscopy [351, 352]; in one case, such coordination of images has been reported to facilitate a more comprehensive understanding of healthy and pathological brain functions in rodents [353]. Related methodology allows direct analysis of intact tissue to detect and image pharmaceutical compounds [349, 354, 355].

Recent advances in instrumentation for MS have resulted in instruments capable of achieving spatial resolution to several micrometers, while acquiring high-resolution mass spectra. The quality of these mass spectrometric images is highly dependent on the techniques and protocols for sample preparation as reviewed recently [203, 356–358]. Instrumentation capable of MS/MS provides another dimension of specificity to the images [359]. The magnifying ion optics of an ion microscope allows ion images to be obtained with a lateral resolution of 4 mm; this approach obviates the need for tight laser focus and the accompanying sensitivity losses, features that offer an improvement of several orders of magnitude in the speed of acquisition compared to the conventional (microprobe) approach to MALDI MS imaging [360]. Use of mixtures of matrices improves many operational aspects of the imaging process, including better crystal formation on the tissue, all leading to improved resolution in the resulting image [361]. Improved resolution and sensitivity in chemical imaging has also been reported by involving matrix ions in the process [362]. Matrix-enhanced SIMS-TOF instrumentation has been used to image nervous tissue in the snail at micrometer spatial resolution [363, 364]; a $^{60}Ce^+$ ion beam causes minimal chemical damage to the tissue, as demonstrated in the imaging via lipid-derived ions from a freeze-dried single cell using a SIMS-TOF instrument [365]. The feasibility of MALDI MS imaging of features smaller than the laser beam size has been demonstrated [366]. Ion current that constitutes the image emanates from molecules as far as 40 mm below the surface of the tissue sample [367]. In related work, the use of gold cluster focused ion beams produced by a liquid metal ion gun for SIMS/TOFMS enhanced secondary ion emission of phospholipids and peptides during analyses of mouse brain samples [368, 369]. The combination of ion mobility spectrometry with an IR laser allows multiple-charged monomers and multimers to be resolved, thereby yielding pure spectra of the single-charge protein ion that are virtually devoid of chemical noise in the direct analysis of rat brain tissue for phospholipids [370]. Use of an IR laser in imaging by MALDI allows the water in the sample tissue to serve as the matrix, thereby permitting analysis of the constituent molecules of the tissue to be analyzed in their natural environment, i.e., without addition of the usual MALDI matrix solution [371]. The debut of DESI into the arena of molecular imaging is based on the use of negative-ion mass spectra of specific lipids (deprotonated molecules) from coronal sections of rat brain tissue [372].

A variety of sample preparation and matrix deposition protocols have been evaluated in the context of spatially profiling *Aplysia californica* exocrine gland and neuronal tissues; electrospray matrix deposition and a variety of freezing methods were found to be optimum for these invertebrate tissues [373]. A new approach to sample preparation involves placing a thin tissue section on an array of tiny glass beads that are mechanically attached to an elastic, hydrophobic membrane, which when stretched, causes rapid division of the tissue into thousands of pieces, each about the size of a glass bead; the array of glass beads maintains the relative morphological organization of the intact tissue and facilitates extraction of analytes from the tissue for efficient interaction with the MALDI matrix, resulting in a good-quality image [374]. A "spray droplet" method for applying the matrix solution to tissue samples improves the image resolution, as demonstrated in the analysis of rat brain sections [375]. While most imaging applications focus on proteins, lipids are also an important component of most tissues from an anatomical point of view, and because they tend to control the distribution of pharmaceutical agents and other small exogenous compounds [376–378]. Other applications involve direct analysis of rat brain tissue for phospholipids [193] and cell membranes for cholesterol [379]; such analyses using a MALDI TOF/TOF mass spectrometer allow assignment of specific phosphatidylcholine species [380]. An alternative technology for tissue imaging has been demonstrated, based on the excellent spatial resolution of secondary ion mass spectrometry (SIMS) combined with the sample preparation protocols of MALDI [381, 382]. Attempts to gain 3D information is often thwarted by damage to organic molecules during particle bombardment [383], though this seems to less problematic with bombardment by fullerene ions [384]. SIMS-TOF images of macrophages and glial cells are possible with use of a trehalose-glycerol matrix [385].

With related technology, SIMS/TOFMS can be used for profiling polymers, especially as facilitated with sample treatment involving deposition of Au and Ag metal nanoparticles [364, 386].

IV. Representative Applications

1. Proteins and Peptides

MALDI has been used extensively in the structural characterization of proteins [387–390], including isoenzymes [391]. MALDI PSD is quite effective in providing sequence information on proteins and peptides [392, 393], including those with phosphorylated serines and threonines [394]; this technique has been used to identify tryptic fragments, especially after charge derivatization [395, 396]. Prompt fragmentation in combination with CID in a Q-TOF has been used to extend the sequencing analytical strategy to larger proteins and peptides, including phosphorylated species [339]. Prompt fragmentation of DABMI derivatives greatly facilitates identification of cysteine-containing peptides during the analysis of UV-absorbing chromatographic components of a proteolytic digest [397]. Protein–tannin complexes can be determined by MALDI MS [398]. The relative binding strength of a series of terpyridine metal complexes of the type $[M(II)L_2]^+$ was investigated by using variable laser intensities for a series of transition metal ions including cadmium, cobalt, copper, iron, manganese, nickel, and ruthenium [399]. MALDI has also been used to characterize carbohydrates and glycoconjugates [400] as well as protein–protein interactions [401, 402]. An operational approach called "intensity fading" facilitates the use of MALDI to study some noncovalent interactions [403]; intensity fading shows a reduction in the relative signal intensity of low-molecular-mass binding partners (e.g., protease inhibitors) when their target protein (i.e., protease) is added to the sample, as illustrated in experiments with carboxypeptidase bound to potato carboxypeptidase inhibitor [404].

2. Microbes

MALDI has been used for the characterization and recognition of various bacteria by direct detection of proteins [405–415] or lipopeptides as characterized by subsequent CAD MS/MS [174, 416]. Viruses have also been characterized by MALDI TOF by detection of glycoproteins [197] and proteolytic products [417–419] and fungal spores [420, 421]. There is some indication that "shotgun" proteomic procedures offer analytical advantage [422]. Results of a recent study of analyzing whole cells of 37 strains of *Mycobacteria* indicated that all but one of the isolates could be differentiated, giving hope that the characteristic mass spectral fingerprint based on desorbed ions from the cell surface can provide the basis for rapid and reproducible identification and characterization of *Mycobacterium* species [415]. The usefulness of non-culture-based diagnosis of infectious diseases using mass spectrometry has been reviewed recently [423]. (Results of a study using ESI to identify PCR products distinguished 47 of 48 *Acinetobacter baumannii* genotypes in good agreement with results obtained by parallel analysis by pulse-field gel electrophoresis indicating that the MS-based methodology has great promise for rapid and reliable analysis to support efforts to track transmission during infectious outbreaks in hospitals, etc. [424].)

Sample preparation of the bacteria greatly affects the quality and reproducibility of the mass spectra obtained [180, 425]. Treatment of both whole-cell gram-positive and gram-negative bacteria with *N*-octyl-*B-D*-glactopyranoside enhances the detection of characteristic high molecular weight (up to 80 kDa) proteins using MALDI TOFMS with ferulic acid as matrix [426]. Flow field–flow fractionation of whole bacterial cells gave various fractions according to size and possibly different growth stages of bacteria for analysis by MALDI MS [427]. Analysis of whole cells of *E. coli* for various proteins, using FTMS for high-accuracy mass measurements and high-resolution isotope profile data in the range of *m/z* 5000 to 10,000, has also been reported as a means by which to characterize bacteria [428]. Intact *Bacillus* spore species can be identified from MALDI TOF/TOF fragment ion spectra obtained from whole (undigested) protein biomarkers [429]. Lectins immobilized to glass slides with activated surfaces have been used to concentrate and purify agglutinated *Bacillus* spores for subsequent direct analysis by MALDI [430]. On-probe digestion of bacterial proteins facilitates rapid identification [431]; other investigations of protein mass mapping approaches also seem promising [432]. A method to distinguish fungal infections and nonfungal affections is based on analysis of tryptic digests of tissue by MALDI [433]. MALDI PSD has been used for the rapid identification of *Bacillus* spores based on detection of a selectively released family of proteins [289]. In a related non-MALDI method, characteristic mass spectra from individual bacterial endospores of *Bacillus subtilis* var. *niger* were obtained in a bipolar aerosol TOF-MS using a pulsed 266-nm laser for molecular desorption and ionization [434, 435]. A useful demonstration that microbial mixtures can be analyzed by MALDI MS was based on a double-blind analysis of 50 different simulated mixed bacterial cultures [436]. In another test of similar methodology with automated spectral collection from whole-cell bacteria and subsequent data processing by specialized algorithms, fingerprints from three different laboratories were constructed and compared [437].

A new approach to identifying bacteria is based on using ribosomal protein sequences as biomarkers, rather than mass spectral fingerprints; a systematic statistical assessment is described using two different matrices and two different ion modes for this proteome-based approach (providing at least seven biomarkers) to characterizing bacteria [438]. Identification of bacteria based on detection of DNA and PCR products also has been reported [439].

3. Biomarkers

The detection of biomarkers for oral cancer from analysis of saliva by MALDI TOF mass spectrometry has been reported [440]. Intact protein biomarkers of *Bacillus cereus* T-spores, detected initially by MALDI, have been more completely analyzed by ESI-CAD FTMS/MS [441]. Because of the more comprehensive information available from CAD MS/MS, the AP MALDI approach to analyzing microbial samples would appear to be preferred [174]. Screening for diabetes based on high-throughput analyses by MALDI MS with quantitation of glutathionylated hemoglobin in blood samples has been reported [442]. The use of antibody-coated magnetic beads to bind specific peptides for subsequent analysis by MALDI should be helpful in bridging the technology gap between biomarker discovery and clinical analysis [266]. Results from model studies using MALDI PSD or CAD to analyze trypsinized human T47D breast cancer cells and cryo-preserved sections of murine brain tissue, followed by bioinformatics, suggest that these techniques can be used for the direct identification of proteins from cells and tissues in clinical applications of quantitative proteomics [443].

4. Synthetic Polymers

The status of mass spectrometry for the analysis of synthetic polymers has been summarized recently [444–450]. An update on studies of the mechanism of cationization of polymers indicates that separation of species in the initial acceleration region of a TOF instrument is not involved in the process [451]. Quantitation based on internal standards has been reported [452]. Electrospray deposition reduces the analyte segregation that can occur during traditional dried droplet deposition for MALDI [453]; in other developments, the solvent-free sample preparation method for synthetic polymers has been improved to minimize cross-contamination of samples [454–456]. The solvent-free sample preparation methodology has also been used successfully in the analysis of proteins [21, 457]. MALDI has been used to determine chain end groups introduced during functionalization [458]. Polystyrene nanoparticles approaching 1 MDa have been reported [459]; chromatographic processing of highly branched polystyrenes improves their analysis involving off-line MALDI MS [460]. The properties and relative ionization efficiencies of a series of polystyrenes containing hydroxyl, hydrogen, tertiary amine, and quaternary amine end functionalities have been studied [461]. The effect of matrix and laser energy on the molecular mass distribution of polystyrene, poly(ethylene glycol), poly(methyl methacrylate) and poly(tetrahydrofuran) has been reported [462]; these parameters can be optimized for quantitative purposes as demonstrated in the analysis of polystyrenes [463]. Buckminsterfullerene (C60) can be used as a MALDI matrix for thermal studies of polymers [464]. Other industrial polymers [201, 465–469] and blends thereof [183, 470], including copolymers of ethylene oxide and propylene oxide [471], are amenable to analysis by MALDI; the esters of such polymers have also been characterized by MALDI [472]. MALDI PSD has been used to characterize low molecular weight ethoxylated polymers [473]. An interlaboratory comparison of MALDI results on molecular weight distribution and mass fraction has been reported on mixtures of synthetic polymers having the same repeat unit [474]. Several kinds of polymers have been analyzed successfully by MALDI using pencil lead as the matrix [65–67].

Copolymers, such as methyl methacrylate and butyl methacrylate, can be determined by accurate assignment of a unique composition, especially when cesium adducts can be formed allowing higher resolution [475]; polyether copolymers have also been characterized [476]. Poorly soluble polyamides(<4600 Da), Nomex and Kevlar, were successfully analyzed after sample preparation involving wet grinding with 1,8-dihydroxyanthrone (dithranol) and 3-aminoquinoline as matrices and potassium trifluoroacetate as the cationizing agent [477, 478]. In some cases, it may be helpful to form metal ion adducts of the polymer [479, 480]. In an effort to understand polymer sample segregation in solid-phase MALDI samples better, matrix-enhanced secondary ion mass spectrometry has been used to investigate the solid-phase solubility of a variety of low molecular weight polymer materials and their interaction with various MALDI matrices [481]. In one report, the results of on-line GPC-ESI MS have been compared with those from automated GPC-MALDI TOFMS for poly(dimethylsiloxane) [482]. The analysis of perfluoropolyethers is facilitated by the use of fluorinated derivatives of benzoic acid and cinnamic acid as matrix compounds [483]. A deposition device for the off-line combination of supercritical fluid chromatography and MALDI has been used to characterize synthetic silicone oils [484]. In some cases, substrate-assisted laser desorption, rather than MALDI, has been used to analyze polyethylene using cobalt, copper, nickel, or iron metal powders as a sample substrate and silver nitrate as the cationization reagent [485].

5. Small Molecules

The usefulness of analyzing low molecular weight compounds, including pesticides [182], by MALDI should not be overlooked [486–488]. The matrix suppression effect is also an important consideration in the analysis of small molecules [206]. In an effort to extend more generally the utility of MALDI to small molecules, an exploratory study has been made of the effectiveness of converting classes of organic compounds to various derivatives (e.g., carboxylic acids to amides, aldehydes to hydrazones, etc.) to reduce their volatility and increase their mass; in general, it was found that the LOD of representative derivatized compounds by MALDI was about the same as that for the unmodified compound [489]. Charge derivatization of small amine molecules facilitates qualitative and quantitative analysis by MALDI MS [490]. MALDI has also been used to analyze tar pitch [491], creosote [492], glycosphingolipids [493], polysulfonated azo dyestuffs [494], and samples of forensic interest [495]. The structural analysis of phosphatidylcholine has been accomplished by MALDI PSD; sodiated precursors produced many fragment ions including those derived from the loss of fatty acids, whereas protonated molecules showed little fragmentation [496]. Some structural information can be obtained from negative ions of mildly acidic saccharides during MALDI PSD [497]. Sialylated oligosaccharides can be stablilized by amidation for structural analysis by MALDI PSD or CAD MS/MS [498]. 3-Hydroxy and oxosteroids derivatized with Girard P (GP) hydrazine give GP hydrazone cations that upon fragmentation produce structurally informative high-energy CAD MS/MS (TOF/TOF) spectra following MALDI, as demonstrated by the analysis of rat brain extracts for oxysterols [499]. Derivatized fullerenes can be analyzed by MALDI following solvent-free sample preparation with a variety of unconventional matrix compounds [500].

6. Quantitation

The issue of quantitation by MALDI has been addressed [501]. A systematic study of parameters affecting quantitation in the parent field of laser desorption/laser ionization found that factors such as laser power, alignment, pulse delay, etc. could affect ratios of signal response by an order of magnitude during the analysis of PAHs, underscoring the need for carefully chosen internal standards [502]. The practical utility of automated MALDI TOFMS as a tool for quantifying a diverse array of biomolecules, including growth hormone, insulin, defensins, homovanillic acid, and epinephrine has been assessed in the context of analyzing biological samples [503, 504].

A quantitative method for fatty acids from five triacylglycerol products by MALDI has been verified by independent analysis by GC/MS [505]. A high-throughput method for serum sulfatides in the form of lysosulfatides appears to be quantitatively suitable for clinical analytes [506]. Quantitative detection of free fatty acids can be achieved by MALDI when a *meso*-tetrakis porphyrin matrix is used in combination with cesium acetate as a cationizing agent to give a dominating reproducible signal by the cesiated cesium carboxylates $[RCOOCs + Cs]^+$ corresponding to a mass shift of 264.8 Da for each fatty acid [507]; in demonstrating this method, a linear relationship between fatty acid concentration and the ratio of the corresponding fatty acid to internal standard peak intensity was achieved for fatty acids ranging from 14 to 20 carbons across a concentration range from 4.40 to 150 mM in rat plasma. MALDI has been used to quantitate low molecular weight (<400 Da) substrates and products of enzyme-catalyzed reactions relative to isotope-labeled and fluorinated internal standards to assess enzyme activity [508], as well as other small molecules [489]. Quantitation of peptides and oligonucleotides is facilitated by the use of ionic liquid matrices [40, 509]. The use of antibody-coated magnetic beads to capture specific peptides (biomarkers) for analysis by MALDI can provide quantitative high-throughput clinical results when used with a spiked internal standard [266]. In fast screening of new enzymes (or the search for related substrates or inhibitors), quantitation of tryptic peptides can be accomplished without the need for an internal standard provided when using an ionic liquid matrix [40]. Various forms of hemoglobin, including importantly glycated Hb which is used to gauge the long-term glycemic state in diabetic patients, can be quantitated successfully by MALDI as validated in parallel with an HPLC method [442]. Synthetic polymers [452], including polystyrene [463], can be quantitated by MALDI when using an internal standard. The results of a study of optimizing instrument parameters for the quantitation of synthetic polymers show that detector potential and delay time were most important for analyzing polystyrene in a matrix of all-*trans* retinoic acid, whereas laser energy was more important when dithranol was the matrix [463].

7. Combined with Liquid Chromatography

Clever arrangements of sampling have been proposed to monitor chromatographic [510] processes. A new LC/MS interface, the laser spray, utilizes a 10.6-mm IR laser to irradiate the vapor and mist formation as an aqueous solution effuses from the tip of a stainless-steel capillary; when a high voltage (3–4 kV) was applied to the capillary, strong ion signals (orders of magnitude greater than obtained by ESI) appeared as the plume was sampled through the ion sampling orifice [511]. Another interface for off-line, but automated, LC/MS analysis is based on partial evaporation of the mobile phase from a hanging droplet at the outlet of the HPLC unit [512, 513]; the concentrated droplet is periodically transferred to a MALDI sample plate to accommodate flow rates up to 50 mL min^{-1}. A one-piece elastomeric device affixed to a MALDI target creates a prestructured 96-well sample array that processes high-flow HPLC fractions of a protein digest by collecting them directly into the elastomeric device, which concentrates the peptides, allows desalting, and causes matrix/analyte co-crystalization for analysis by MALDI MS [514]. Electrostatic-guided transfer of tiny volumes of eluant from monolithic LC columns to a 576-well SS MALDI plate, precoated by air-brushing with a standard matrix solution and incremented on an *x-y* platform, provides a well-defined deposit of analytes for subsequent analysis by MALDI; this off-line LC-MALDI approach has been shown to double the sequence coverage achieved by conventional "batch" MALDI analysis in simulated shotgun proteomic analyses using BSA, myglobin, and cytochrome c [515]. In an LC-MALDI TOF/TOF application, the chromatographic separation on a MALDI target yields adequate time to obtain MS/MS spectra for analyzing protein complexes [516]. A rotating ball is key to a clever design of a CE-MALDI interface [517, 518].

References

1. Lattimer RP and Schulten HR, FI and FDMS: Past, Present & Future. *Anal. Chem.* **61:** 1201A-15A, 1989.
2. Macfarlane RD and Sundqvist BUR, Cf-252 PDMS. *Mass Spectrom Rev.* **4:** 421-460, 1985.
3. Barber M, Bordoli RS, Sedgwick RD and Tyler AN, F.A.B. *J. Chem. Soc. Chem. Commun.* **1981:** 325-327, 1981.
4. Seifert Jr. WE and Caprioli RM, Fast Atom Bombardment MS. In: *Methods in Enzymology, High Resolution Separation of Macromolecules*, Vol. 270 (Eds. Hancock B and Karger BL), pp. 453-486, 1996.
5. Schaub TM, Linden HB, Hendrickson CL and Marshall AG, Continuous-flow sample introduction for field desorption/ionization mass spectrometry. *Rapid Commun Mass Spectrom* **18**(14)**:** 1641-4., 2004.
6. Pachuta S and Cooks RG, Mechanisms in Molecular SIMS. *Chem. Rev.* **87:** 647-669, 1987.
7. Cotter R, An Overview of Laser MS. *Journal* **195:** 45-59, 1987.
8. Weickhardt C, Tonnies K and Globig D, Detection and quantification of aromatic contaminants in water and soil samples by means of laser desorption laser MS. *Anal Chem* **74:** 4861-7., 2002.
9. Tanaka K, Waki H, Ido H, Akita S and Yoshida T, Protein and polymer analysis up to 100,000 by laser ionization time-of-flight mass spectrometry. *Rapid Commun Mass Spectrom* **2:** 151-153, 1988.
10. Tanaka K, The Origin of Macromolecule Ionization by Laser Irradiation (Nobel Lecture). *Angew. Chem. Int. Ed.* **42:** 3861-3870, 2003.
11. Karas M, Bachmann D, Bahr U and Hillenkamp F, Matrix-Assisted Laser Desorption Ionization Mass Spectrometry. *Int. J. Mass Spec. Ion Proc.* **78:** 53-68, 1987.
12. Karas M and Hillenkamp F, Laser Desorption of Proteins with Molecular Masses Exceeding 10,000 Daltons. *Anal. Chem.* **60**(20)**:** 2299-2301, 1988.
13. Karas M, Bahr U and Giessman U, MALDI-MS. *Mass Spectrom. Rev.* **10:** 335-357, 1991.
14. Hillenkamp F, Karas M, Beavis RC and Chait BT, Matrix-assisted laser desorption/ionization mass spectrometry of biopolymers. *Anal Chem* **63**(24)**:** 1193A-1203A., 1991.
15. Karas M and Hillenkamp F, Matrix-assisted laser desorption ionization mass spectrometry - fundamentals and applications. *AIP Conf. Proc.* **288**(Laser Ablation: Mechanisms and Applications--II)**:** 447-58, 1993.
16. Hillenkamp F, Karas M, Beavis RC and Chait BT, MALDI-MS. *Anal. Chem.* **63:** 1193A-1203A., 1991.
17. Hillenkamp F, T. Matsuo RM, Caprioli, Gross ML and Seyama Y, MALDI. In: *Biological Mass Spectrometry: Present and Future.* J. Wiley & Sons Ltd., Ny, 1994.
18. Hillenkamp F and Karas M, MALDI, an experience. *Int. J. Mass Spectrom.* **200:** 71-77, 2000.
19. Bogan MJ, Bakhoum SFW and Agnes GR, Promotion of matrix/peptide cocrystallization within levitated droplets with net charge. *J Amer Soc Mass Spectrom* **16:** 254-262, 2005.
20. Horneffer V, Glueckmann M, Krueger R, Karas M, Strupat K and Hillenkamp F, Matrix-analyte-interaction in MALDI-MS: Pellet and nano-electrospray preparations. *Int J Mass Spectrom* **249/250:** 426-432, 2006.
21. Trimpin S, Raeder HJ and Muellen K, Investigations of theoretical principles for MALDI-MS derived from solvent-free sample preparation. *Int J Mass Spectrom* **253:** 13-21, 2006.
22. Beavis RC and Chait BT, Matrix-assisted laser-desorption mass spectrometry using 355 nm radiation. *Rapid Communications in Mass Spectrometry* **3**(12)**:** 436-9, 1989.
23. Beavis RC and Chait BT, Cinnamic acid derivatives as matrices for ultraviolet laser desorption mass spectrometry of proteins. *Rapid Communications in Mass Spectrometry* **3**(12)**:** 432-5, 1989.
24. Mugo SM and Bottaro CS, Aquatic fulvic acid as a matrix for MALDI-TOFMS. *Rapid Commun Mass Spectrom* **21:** 219-228, 2007.
25. Meier MAR, Adams N and Schubert US, Statistical Approach To Understand MALDI-TOFMS Matrices: Discovery and Evaluation of New MALDI Matrices. *Anal Chem* **79:** 863-869, 2007.
26. Liao PC and Allison J, Ionization Processes in MALDI. *J. Mass Spectrom.* **30:** 408-423, 1995.
27. Parr GR and Smith LM, Basic Matrices for MALDI of Proteins and Oligonucleotides. *Anal. Chem.* **65:** 3204-3211, 1993.
28. Mirza SP, Raju NP and Vairamani M, Estimation of the proton affinity values of 15 MALDI matrices by ESI using the kinetic method. *J Amer Soc Mass Spectrom* **15:** 431-435, 2004.
29. Westman A, Huth-Fehre T, Demirev P and Sundqvist BUR, Sample Morphology in MALDI. *J. Mass Spectrom.* **30:** 206-211, 1995.
30. Williams TI, Saggese DA, Wilcox RJ, Martin JD and Muddiman DC, Effect of matrix crystal structure on ion abundance of carbohydrates by MALDI FTICR MS. *Rapid Commun Mass Spectrom* **21:** 807-811, 2007.
31. Fournier I, Marinach C, Tabet JC and Bolbach G, Irradiation effects in MALDI, ablation, ion production, & surfaces II 2,5-dihydroxybenzoic acid monocrystals. *J Amer Soc Mass Spectrom* **14:** 893-9, 2003.
32. Wenzel T, Sparbier K, Mieruch T and Kostrzewa M, 2,5-dihydroxyacetophenone: a matrix for highly sensitive MALDI of proteins. *Rapid Commun Mass Spectrom* **20:** 785-789, 2005.
33. Fenaille F, Le Mignon M, Groseil C, Siret L and Bihoreau N, Combined use of 2,4,6-trihydroxyacetophenone as matrix and enzymatic deglycosylation for characterization of complex glycoproteins and N-glycans by MALDI-TOFMS. *Rapid Commun Mass Spectrom* **21:** 812-816, 2007.
34. Stuebiger G and Belgacem O, Analysis of Lipids Using 2,4,6-Trihydroxyacetophenone as a Matrix for MALDI MS. *Anal Chem* **79:** 3206-3213, 2007.

35. Hsu N-Y, Yang W-B, Wong C-H, Lee Y-C, Lee RT, Wang Y-S and Chen C-H, MALDI of polysaccharides with 2',4',6'-trihydroxyacetophenone as matrix. *Rapid Commun Mass Spectrom* **21**: 2137-2146, 2007.
36. Mouradian S, Nelson CM and Smith LM, A Self-Assembled Matrix Monolayer for UV-MALDI. *J. Am. Chem. Soc.* **118**, 1996.
37. Xu Y, Bruening ML and Watson JT, Use of polymer-modified MALDI-MS probes to improve analyses of protein digests and DNA. *Analytical Chemistry* **76**(11): 3106-3111, 2004.
38. Dunn JD, Watson JT and Bruening ML, Detection of Phosphopeptides Using Fe(III)-Nitrilotriacetate Complexes Immobilized on a MALDI Plate. *Anal Chem* **78**(5): 1574-1580, 2006.
39. Mank M, Stahl B and Boehm G, 2,5-Dihydroxybenzoic acid butylamine and other ionic liquid matrixes for enhanced MALDI-MS analysis of biomolecules. *Anal Chem* **76**(10): 2938-50., 2004.
40. Tholey A, Zabet-Moghaddam M and Heinzle E, Quantification of Peptides for the Monitoring of Protease-Catalyzed Reactions by MALDI Using Ionic Liquid Matrixes. *Anal Chem* **78**: 291 - 297, 2006.
41. Ren S-f and Guo Y-l, Oxidized carbon nanotubes as matrix for MALDI TOF MS of biomolecules. *Rapid Communications in Mass Spectrometry* **19**: 255-260, 2005.
42. Ren S-f, Zhang L, Cheng Z-h and Guo Y-l, Immobilized carbon nanotubes as matrix for MALDI-TOF-MS of neutral small carbohydrates. *J Amer Soc Mass Spectrom* **16**: 333-339, 2005.
43. Pan C, Xu S, Zou H, Guo Z, Zhang Y and Guo B, Carbon nanotubes as adsorbent for SPE and matrix for laser desorption/ionization MS. *J Amer Soc Mass Spectrom* **16**: 263-270, 2005.
44. Turney K, Drake TJ, Smith JE, Tan W and Harrison WW, Functionalized nanoparticles for liquid AP-MALDI MS peptide analysis. *Rapid Commun Mass Spectrom* **18**: 2367-2374, 2004.
45. Kong XL, Huang LCL, Hsu CM, Chen WH, Han CC and Chang HC, High-Affinity Capture of Proteins by Diamond Nanoparticles for Mass Spectrometric Analysis. *Analytical Chemistry* **77**: 259-265, 2005.
46. Chen W-H, Lee S-C, Sabu S, Fang H-C, Chung S-C, Han C-C and Chang H-C, Solid-Phase Extraction and Elution on Diamond (SPEED): A Fast and General Platform for Proteome Analysis with MS. *Anal Chem* **78**: 4228-4234, 2006.
47. Hettick JM, McCurdy DL, Barbacci DC and Russell DH, Optimization of sample preparation for peptide sequencing by MALDI-TOF photofragment mass spectrometry. *Anal Chem* **73**(22): 5378-86., 2001.
48. Wa C, Cerny R and Hage DS, High sequence coverage in MALDI TOFMS for studies of protein modification: human serum albumin as a model. *Analytical Biochemistry* **349**: 229-241, 2006.
49. Fukuyama Y, Iwamoto S and Tanaka K, Sequencing and disulfide mapping of peptides containing disulfide bonds in 1,5-diaminonaphthalene as a reductive matrix. *J Mass Spectrom* **41**: 191-201, 2006.
50. Streletskiy AV, Kouvitchko IV, Esipov SE and Boltalina OV, Sulfur as a matrix for laser desorption/ionization of halogenated fullerenes. *Rapid Commun Mass Spectrom* **16**: 99-102., 2002.
51. Kinumi T, Saisu T, Takayama M and Niwa H, MALDI TOFMS using an inorganic particle matrix for small molecule analysis. *J Mass Spectrom* **35**: 417-22., 2000.
52. Wong AW, Wang H and Lebrilla CB, Selection of anionic dopant for quantifying desialylation reactions with MALDI-FTMS. *Anal Chem* **72**(7): 1419-25., 2000.
53. Distler AM and Allison J, 5-Methoxysalicylic acid and spermine: a new matrix for MALDI MS of oligonucleotides. *J Am Soc Mass Spectrom* **12**: 456-62., 2001.
54. Distler AM and Allison J, Additives for the stabilization of double-stranded DNA in UV-MALDI MS. *J Am Soc Mass Spectrom* **13**(9): 1129-37., 2002.
55. Grant DC and Helleur RJ, Surfactant-mediated MALDI-TOFMS of small molecules. *Rapid Commun Mass Spectrom* **21**: 837-845, 2007.
56. Neubert H, Halket JM, Fernandez Ocana M and Patel RKP, MALDI PSD and LIFT-TOF/TOF study of a-cyano-4-hydroxycinnamic acid cluster interferences. *J Amer Soc Mass Spectrom* **15**: 336-343, 2004.
57. Smirnov IP, Zhu X, Taylor T, Huang Y, Ross P, Papayanopoulos IA, Martin SA and Pappin DJ, Suppression of Matrix Clusters and Chemical Noise in MALDI-TOF MS. *Anal Chem* **76**: 2958-2965, 2004.
58. Bashir S, Burkitt WI, Derrick PJ and Giannakopulos AE, Iodine-assisted matrix-assisted laser desorption/ionization. *International Journal of Mass Spectrometry* **219**(3): 697-701, 2002.
59. Kim J-S, Kim J-Y and Kim H-J, Suppression of Matrix Clusters and Enhancement of Peptide Signals in MALDI-TOF MS Using Nitrilotriacetic Acid. *Anal Chem* **77**: 7483-7488, 2005.
60. Reiber DC, Brown RS, Weinberger S, Kenny J and Bailey J, Unknown peptide sequencing using matrix-assisted laser desorption/ionization and in-source decay. *Anal Chem* **70**(6): 1214-22., 1998.
61. Lavine G and Allison J, Bumetanide as a matrix for prompt fragmentation MALDI and demonstration of prompt fragmentation/PSD MALDI MS. *J Mass Spectrom* **34**: 741-8., 1999.
62. Distler AM and Allison J, Improved MALDI-MS analysis of oligonucleotides through the use of fucose as a matrix additive. *Anal Chem* **73**(20): 5000-3., 2001.
63. Luo G, Marginean I and Vertes A, Internal energy of ions generated by matrix-assisted laser desorption/ionization. *Anal Chem* **74**(24): 6185-90., 2002.
64. Stephens E, Maslen SL, Green LG and Williams DH, Fragmentation Characteristics of Neutral N-Linked Glycans Using a MALDI-TOF/TOF Mass Spectrometer. *Analytical Chemistry* **76**: 2343-2354, 2004.
65. Black C, Poile C, Langley J and Herniman J, The use of pencil lead as a matrix and calibrant for MALDI. *Rapid Commun Mass Spectrom* **20**: 1053-1060, 2006.

66. Berger-Nicolet E, Wurm F, Kilbinger A and Frey H, Pencil Lead as a Matrix for MALDI-TOF MS of Sensitive Functional Polymers. *Macromolecules* **40**: 746-751, 2007.

67. Price PC, Use of Graphite as a Simple MALDI Matrix for Synthetic Polymer Applications. In: *55th Annual Meeting of the Amer. Soc. for Mass Spectrom., Indianapolis,2007*, pp. TP 116.

68. Langley GJ, Herniman JM and Townell MS, 2B or not 2B, that is the question: further investigations into the use of pencil as a matrix for MALDI. *Rapid Commun Mass Spectrom* **21**: 180-190, 2007.

69. Vaidyanathan S, Gaskell S and Goodacre R, Matrix-suppressed MALDI MS and its suitability for metabolome analyses. *Rapid Commun Mass Spectrom* **20**: 1192-1198, 2006.

70. Wallace WE, Arnould MA and Knochenmuss R, 2,5-Dihydroxybenzoic acid: laser desorption/ionisation as a function of elevated temperature. *International Journal of Mass Spectrometry* **242**(1)**: 13-22, 2005.

71. Kjellstroem S and Jensen ON, Phosphoric Acid as a Matrix Additive for MALDI MS Analysis of Phosphopeptides and Phosphoproteins. *Analytical Chemistry* **76**(17)**: 5109-5117, 2004.

72. Strupat K, Eckerskorn C, Karas M and Hillenkamp F, IR-MALDI-MS of proteins electroblotted onto polymer membranes after SDS-PAGE separation. *Mass Spectrom. Biol. Sci.*: 203-16, 1996.

73. Menzel C, Dreiseverd K, Berkenkamp S and Hillenkamp F, The role of the laser pulse duration in infrared MALDI MS. *J Am Soc Mass Spectrom* **13**: 975-84., 2002.

74. McLean JA, Russell WK and Russell DH, A high repetition rate (1 kHz) microcrystal laser for high throughput atmospheric pressure MALDI-oaTOF MS. *Anal Chem* **75**: 648-54., 2003.

75. Mize TH and Amster IJ, Broad-Band Ion Accumulation with an Internal Source MALDI-FTICR-MS. *Anal. Chem.* **72**: 5886 -5891, 2000.

76. Baldwin MA, Medzihradszky KF, Lock CM, Fisher B, Settineri TA and Burlingame AL, MALDIcoupled with oaTOF MS for protein discovery, ID, and structural analysis. *Anal Chem* **73**: 1707-20., 2001.

77. Jones MD, Patterson SD and Lu HS, Determination of disulfide bonds in highly bridged disulfide-linked peptides by MALDI MS with PSD. *Anal Chem* **70**: 136-43., 1998.

78. Campbell JM, Vestal ML, Blank PS, Stein SE, Epstein JA and Yergey AL, Fragmentation of Leucine Enkephalin as a Function of Laser Fluence in a MALDI TOF-TOF. *J Amer Soc Mass Spectrom* **18**: 607-616, 2007.

79. Mueller R and Allmaier G, Molecular weight determination of ultra-high mass compounds on a standard MALDI-TOFMS: PAMAM dendrimer generation 10 and immunoglobulin M. *Rapid Commun Mass Spectrom* **20**: 3803-3806, 2006.

80. Berkenkamp S, Menzel C, Hillenkamp F and Dreiseverd K, Measurements of mean initial velocities of analyte and matrix ions in IR MALDI MS. *J Amer Soc Mass Spectrom* **13**: 209-20., 2002.

81. Chen X, Westphall MS and Smith LM, MS of DNA mixtures: Instrumental effects responsible for decreased sensitivity with increasing mass. *Anal Chem* **75**: 5944-5952, 2003.

82. Ingendoh A, Karas M, Hillenkamp F and Giessmann U, Factors affecting the resolution in matrix-assisted laser desorption-ionization mass spectrometry. *Int. J. Mass Spectrom. Ion Processes* **131**: 345-54, 1994.

83. Edmondson RD and Russell DH, Evaluation of MALDI-TOF Mass Measurement Accuracy by Using Delayed Extraction. *J. Am. Soc. Mass Spectrom.* **7**: 995-1001, 1996.

84. Clauser KR, Baker P and Burlingame AL, Accurate mass measurement (+/- 10 ppm) in protein ID strategies employing MS or MS/MS and database searching. *Anal Chem* **71**: 2871-82., 1999.

85. Kruger R and Karas M, Formation and fate of ion pairs during MALDI analysis: anion adduct generation as an indicative tool to determine ionization processes. *J Am Soc Mass Spectrom* **13**(10)**: 1218-26., 2002.

86. Knochenmuss R, A Quantitative Model of UV MALDI Including Analyte Ion Generation. *Anal Chem* **75**: 2199-2207, 2003.

87. Zenobi R and Knochenmuss R, Ion formation in MALDI MS. *Mass Spectrom. Rev.* **17**: 337-366, 1999.

88. Knochenmuss R, Stortelder A, Breuker K and Zenobi R, Secondary ion-molecule reactions in MALDI. *J. Mass Spectrom.* **35**(11)**: 1237-1245, 2000.

89. Lehmann E, Diederich F, Zenobi R, Salih B and Gomez-Lopez M, Do MALDI MS reflect solution-phase formation of cyclodextrin inclusion complexes? *Analyst (Cambridge, U. K.)* **125**: 849-854, 2000.

90. Lehmann E, Knochenmuss R and Zenobi R, Ionization mechanisms in MALDI MS: contribution of pre-formed ions. *Rapid Commun. Mass Spectrom.* **11**: 1483-1492, 1997.

91. Lehmann E and Zenobi R, Investigations of the cationization mechanism in matrix-assisted laser desorption/ionization. *Adv. Mass Spectrom.* **14**: B063490/1-B063490/11, 1998.

92. Dreiseverd K, Berkenkamp S, Leisner A, Rohlfing A and Menzel C, Fundamentals of MALDI mass spectrometry with pulsed infrared lasers. *Int J Mass Spectrom* **226**: 189-209, 2003.

93. Karas M, Bahr U, Fournier I, Gluckmann M and Pfenninger A, Initial-ion velocity as a marker for different desorption-ionization mechanisms in MALDI. *Int J Mass Spectrom* **226**: 239-248, 2003.

94. Knochenmuss R and Zenobi R, MALDI: role of in-plume processes. *Chem Rev* **103**(2)**: 441-52., 2003.

95. Derrick PJ, Symposium in print on mechanisms of MALDI. *European Journal of Mass Spectrometry* **12**(6)**: v, 2006.

96. Dreiseverd K, The desorption process in MALDI. *Chemical Reviews (Washington, DC, United States)* **103**(2)**: 395-425, 2003.

97. Hatano Y, Interaction of photons with molecules--cross-sections for photoabsorption, photoionization, and photodissociation. *Radiat Environ Biophys* **38**(4)**: 239-47., 1999.

98. Cadene M and Chait BT, A robust, detergent-friendly method for mass spectrometric analysis of integral membrane proteins. *Anal Chem* **72**(22)**:** 5655-8., 2000.

99. Xu N, Huang Z-H, Watson JT and Gage DA, Mercaptobenzothiazoles: a new class of matrixes for laser desorption ionization mass spectrometry. *J. Am. Soc. Mass Spectrom.* **8**(2)**:** 116-124, 1997.

100. Steenvoorden RJJM, Breuker K and Zenobi R, The gas-phase basicities of matrix-assisted laser desorption/ionization matrixes. *Eur. Mass Spectrom.* **3**(5)**:** 339-346, 1997.

101. Stevenson E, Breuker K and Zenobi R, Internal energies of analyte ions generated from different matrix-assisted laser desorption/ionization matrices. *J. Mass Spectrom.* **35**(8)**:** 1035-1041, 2000.

102. Ohanessian G, Interaction of MALDI matrix molecules with Na+ in the gas phase. *International Journal of Mass Spectrometry* **219**(3)**:** 577-592, 2002.

103. Kish MM, Ohanessian G and Wesdemiotis C, The Na+ affinities of a-amino acids: side-chain substituent effects. *International Journal of Mass Spectrometry* **227**(3)**:** 509-524, 2003.

104. Coon JJ, Steele HA, Laipis PJ and Harrison WW, Laser desorption-APCI: a novel ion source for direct coupling of PAGE to MS. *J Mass Spectrom* **37:** 1163-7., 2002.

105. Golovlev VV, Lee SH, Allman SL, Taranenko NI, Isola NR and Chen CH, Nonresonant MALDI of oligonucleotides: mechanism of ion desorption. *Anal Chem* **73**(4)**:** 809-12., 2001.

106. Leisner A, Rohlfing A, Roehling U, Dreisewerd K and Hillenkamp F, Time-Resolved Imaging of the Plume Dynamics in IR MALDI with a Glycerol Matrix. *Journal of Physical Chemistry B* **109:** 11661-11666, 2005.

107. Menzel C, Dreisewerd K, Berkenkamp S and Hillenkamp F, Mechanisms of energy deposition in IR MALDI MS. *International Journal of Mass Spectrometry* **207:** 73-96, 2001.

108. Little MW, Kim J-K and Murray KK, Two-laser infrared and ultraviolet matrix-assisted laser desorption/ionization. *Journal of Mass Spectrometry* **38**(7)**:** 772-777, 2003.

109. Papantonakis MR, Kim J, Hess WP and Haglund RF, Jr., What do MALDI mass spectra reveal about ionization mechanisms? *J Mass Spectrom* **37:** 639-47., 2002.

110. Westmacott G, Ens W, Hillenkamp F, Dreisewerd K and Schurenberg M, The influence of laser fluence on ion yield in MALDI mass spectrometry. *Int J Mass Spectrom* **221:** 67-81, 2002.

111. Jackson SN, Kim J-K, Laboy JL and Murray KK, Particle formation by IR laser ablation of glycerol: implications for ion formation. *Rapid Commun Mass Spectrom* **20:** 1299-1304, 2006.

112. Hanley L, Kornienko O, Ada ET, Fuoco E and Trevor JL, Surface mass spectrometry of molecular species. *J Mass Spectrom* **34**(7)**:** 705-23., 1999.

113. Georgiou S and Hillenkamp F, Introduction: laser ablation of molecular substrates. *Chem Rev* **103**(2)**:** 317-20., 2003.

114. Knochenmuss R, A quantitative model of ultraviolet matrix-assisted laser desorption/ionization. *J Mass Spectrom* **37**(8)**:** 867-77., 2002.

115. Knochenmuss R and Zhigilei LV, Molecular Dynamics Model of UV MALDI Including Ionization Processes. *Journal of Physical Chemistry B* **109:** 22947-22957, 2005.

116. Setz PD and Knochenmuss R, Exciton Mobility and Trapping in a MALDI Matrix. *Journal of Physical Chemistry A* **109:** 4030-4037, 2005.

117. Kinsel GR, Yao D, Yassin FH and Marynick DS, Equilibrium conditions in laser-desorbed plumes: thermodynamic properties of a-cyano-4-hydroxycinnamic acid and protonation of amino acids. *European Journal of Mass Spectrometry* **12**(6)**:** 359-367, 2006.

118. Hoteling AJ, Nichols WF, Giesen DJ, Lenhard JR and Knochenmuss R, Electron transfer reactions in MALDI: factors influencing matrix and analyte ion intensities. *European Journal of Mass Spectrometry* **12**(6)**:** 345-358, 2006.

119. McCombie G and Knochenmuss R, Enhanced MALDI Efficiency at the Metal-Matrix Interface. *J Amer Soc Mass Spectrom* **17:** 737-745, 2006.

120. Weinkauf R, Aicher P, Wesley G, Grotemeyer J and Schlag EW, Femtosecond versus Nanosecond Multiphoton Ionization and Dissociation of Large Molecules. *J Phys Chem* **98:** 8381-91, 1994.

121. Lockyer NP and Vickerman JC, Multiphoton ionization mass spectrometry of small biomolecules with nanosecond and femtosecond laser pulses. *Int J Mass Spectrom* **176:** 77-86, 1998.

122. Nielsen ML, Budnik BA, Haselmann KF and Zubarev RA, Tandem MALDI/EI tandem FTICR MS of polypeptides. *Int J Mass Spectrom* **226:** 181-187, 2003.

123. Hankin SM, Tasker AD, Robson L, Ledingham KW, Fang X, McKenna P, McCanny T, Singhal RP, Kosmidis C, Tzallas P, Jaroszynski DA, Jones DR, Issac RC and Jamison S, Fs laser TOFMS of labile molecular analytes: laser-desorbed nitro-aromatics. *Rapid Commun Mass Spectrom* **16:** 111-6., 2002.

124. Robson L, Tasker AD, Ledingham KWD, McKenna P, McCanny T, Kosmidis C, Tzallas P, Jaroszynski DA and Jones DR, Ionization and fragmentation dynamics of laser desorbed PAHs using fs and ns post-ionization. *Int J Mass Spectrom* **220:** 69-85, 2002.

125. Tasker AD, Robson L, Ledingham KWD, McCanny T, McKenna P, Kosmidis C and Jaroszynski DA, Fs ionization & dissociation of laser desorbed nitro-PAHs. *Int J Mass Spectrom* **225:** 53-70, 2003.

126. Horneffer V, Dreisewerd K, Ludemann HC, Hillenkamp F, Lage M and Strupat K, Is the incorporation of analytes into matrix crystals a prerequisite for MALDI MS? A study of five positional isomers of dihydroxybenzoic acid. *Int J Mass Spectrom* **185/186/187:** 859-870, 1999.

127. Karas M, Gluckmann M and Schafer J, Ionization in matrix-assisted laser desorption/ionization: singly charged molecular ions are the lucky survivors. *J Mass Spectrom* **35**(1): 1-12., 2000.

128. Meffert A and Grotemeyer J, Dissociative proton transfer in cluster ions: clusters of aromatic carboxylic acids with amino acids. *International Journal of Mass Spectrometry* **210/211**(1-3): 521-530, 2001.

129. Krueger R, Pfenninger A, Fournier I, Glueckmann M and Karas M, Analyte incorporation and ionization in MALDI visualized by pH indicator molecular probes. *Anal Chem* **73**: 5812-5821, 2001.

130. Hillenkamp F and Karas M, Mass spectrometry of peptides and proteins by matrix-assisted ultraviolet laser desorption/ionization. *Methods Enzymol* **193**: 280-95., 1990.

131. Beavis RC and Chait BT, Rapid, sensitive analysis of protein mixtures by mass spectrometry. *Proc Natl Acad Sci U S A* **87**(17): 6873-7., 1990.

132. Beavis RC and Chait BT, MALDI MS of proteins. *Methods Enzymol* **270**: 519-51., 1996.

133. Land CM and Kinsel GR, The mechanism of matrix to analyte proton transfer in clusters of 2,5-dihydroxybenzoic acid and the tripeptide VPL. *J Am Soc Mass Spectrom* **12**(6): 726-31., 2001.

134. Kinsel GR, Knochenmuss R, Setz P, Land CM, Goh SK, Archibong EF, Hardesty JH and Marynick DS, Ionization energy reductions in small DHB/acid-proline clusters. *J Mass Spectrom* **37**: 1131-40., 2002.

135. Breuker K, Knochenmuss R, Zhang J, Stortelder A and Zenobi R, Thermodynamic control of final ion distributions in MALDI: in-plume proton transfer reactions. *Int J Mass Spectrom* **226**: 211-222, 2003.

136. Lin Q and Knochenmuss R, Two-photon ionization thresholds of matrix-assisted laser desorption/ionization matrix clusters. *Rapid Commun Mass Spectrom* **15**(16): 1422-6., 2001.

137. Ermer DR, Baltz-Knorr M and Haglund RF, Jr., Intensity dependence of cation kinetic energies from 2,5-dihydroxybenzoic acid near the IR MALDI threshold. *J Mass Spectrom* **36**: 538-45., 2001.

138. Loboda AV and Chernushevich IV, Investigation of the mechanism of matrix adduct formation in MALDI at elevated pressure. *International Journal of Mass Spectrometry* **240**(2): 101-105, 2005.

139. Frankevich V, Knochenmuss R and Zenobi R, The origin of electrons in MALDI and their use for sympathetic cooling of negative ions in FTICR. *Int J Mass Spectrom* **220**: 11-19, 2002.

140. Zhang J, Frankevich V, Knochenmuss R, Friess SD and Zenobi R, Reduction of Cu(II) in matrix-assisted laser desorption/ionization mass spectrometry. *J Am Soc Mass Spectrom* **14**(1): 42-50., 2003.

141. Pshenichnyuk SA and Asfandiarov NL, The role of free electrons in maldi: Electron capture by molecules of a-cyano-4-hydroxycinnamic acid. *Eur J Mass Spectrom* **10**: 477-486, 2004.

142. Krutchinsky AN and Chait BT, On the nature of the chemical noise in MALDI mass spectra. *J Am Soc Mass Spectrom* **13**(2): 129-34., 2002.

143. Harvey DJ, Bateman RH, Bordoli RS and Tyldesley R, Ionisation and fragmentation of complex glycans in MALDI oaTOF MS. *Rapid Commun Mass Spectrom* **14**: 2135-42., 2000.

144. Lecchi P, Olson M and Brancia FL, The Role of Esterification on Detection of Protonated and Deprotonated Peptide Ions in MALDI MS. *J Amer Soc Mass Spectrom* **16**: 1269-1274, 2005.

145. Laiko VV, Moyer SC and Cotter RJ, AP MALDI/Ion Trap MS. *Anal. .Chem.* **72**: 5239 -5243, 2000.

146. Laiko VV, Baldwin MA and Burlingame AL, Atmospheric Pressure Matrix-Assisted Laser Desorption/Ionization Mass Spectrometry. *Anal. Chem.* **72**: 652 -657, 2000.

147. Moyer SC and Cotter RJ, Atmospheric pressure MALDI. *Anal Chem* **74**: 468A-476A., 2002.

148. Doroshenko VM, Laiko VV, Taranenko NI, Berkout VD and Lee HS, Recent developments in atmospheric pressure MALDI mass spectrometry. *International Journal of Mass Spectrometry* **221**(1): 39-58, 2002.

149. Creaser CS and Ratcliffe L, Atmospheric pressure MALDI MS: A review. *Current Analytical Chemistry* **2**(1): 9-15, 2006.

150. Berkout VD, Kryuchkov SI and Doroshenko VM, Modeling of ion processes in atmospheric pressure MALDI. *Rapid Commun Mass Spectrom* **21**: 2046-2050, 2007.

151. Moyer SC, Marzilli LA, Woods AS, Laiko VV, Doroshenko VM and Cotter RJ, AP MALDI on a quadrupole ion trap mass spectrometer. *Int J Mass Spectrom* **226**: 133-150, 2003.

152. Kellersberger KA, Tan PV, Laiko VV, Doroshenko VM and Fabris D, Atmospheric Pressure MALDI-Fourier Transform Mass Spectrometry. *Analytical Chemistry* **76**(14): 3930-3934, 2004.

153. Kellersberger KA, Yu ET, Merenbloom SI and Fabris D, Atmospheric pressure MALDI-FTMS of normal and chemically modified RNA. *J Amer Soc Mass Spectrom* **16**: 199-207, 2005.

154. Von Seggern CE and Cotter RJ, Fragmentation studies of noncovalent sugar-sugar complexes by IR AP-MALDI. *J Amer Soc Mass Spectrom* **14**: 1158-1165, 2003.

155. Kim T, Tang K, Udseth HR and Smith RD, Multicapillary inlet jet disruption electrodynamic ion funnel for improved sensitivity using atmospheric pressure ion sources. *Anal Chem* **73**: 4162-70., 2001.

156. Tang K, Tolmachev AV, Nikolaev E, Zhang R, Belov ME, Udseth HR and Smith RD, Control of ion transmission in jet disrupter dual-channel ion funnel ESI MS interface. *Anal Chem* **74**: 5431-7., 2002.

157. Page JS, Bogdanov B, Vilkov AN, Prior DC, Buschbach MA, Tang K and Smith RD, Automatic gain control in MS by jet disrupter in ion funnel. *J Amer Soc Mass Spectrom* **16**: 244-253, 2005.

158. Schneider BB, Douglas DJ and Chen DDY, An atmospheric pressure ion lens that improves nebulizer assisted electrospray ion sources. *J Amer Soc Mass Spectrom* **13**: 906-913, 2002.

159. Miller CA, Yi D and Perkins PD, An AP-MALDI ion trap with enhanced sensitivity. *Rapid Communications in Mass Spectrometry* **17**: 860-868, 2003.

160. Tan PV, Laiko VV and Doroshenko VM, Atmospheric pressure MALDI with pulsed dynamic focusing for high-efficiency transmission of ions into a mass spectrometer. *Anal Chem* **76**: 2462-2469, 2004.
161. Von Seggern CE, Gardner BD and Cotter RJ, IR AP MALDI Ion Trap MS of Frozen Samples Using a Peltier-Cooled Sample Stage. *Anal Chem* **76**: 5887-5893, 2004.
162. von Seggern CE and Cotter RJ, Study of peptide-sugar non-covalent complexes by infrared atmospheric pressure matrix-assisted laser desorption/ionization. *Journal of Mass Spectrometry* **39**(7): 736-742, 2004.
163. Galicia MC, Vertes A and Callahan JH, Atmospheric pressure matrix-assisted laser desorption/ionization in transmission geometry. *Anal Chem* **74**(8): 1891-5., 2002.
164. Corr JJ, Kovarik P, Schneider BB, Hendrikse J, Loboda A and Covey TR, Design Considerations for High Speed Quantitative MS with MALDI Ionization. *J Amer Soc Mass Spectrom* **17**: 1129-1141, 2006.
165. Moskovets E, Preisler J, Chen HS, Rejtar T, Andreev V and Karger BL, High-Throughput Axial MALDI-TOF MS Using a 2-kHz Repetition Rate Laser. *Analytical Chemistry* **78**(3): 912-919, 2006.
166. O'Connor PB and Costello CE, A high pressure MALDI FTMS ion source for thermal stabilization of labile biomolecules. *Rap. Commun. Mass Spectrom.* **15**: 1862-1868, 2001.
167. O'Connor PB, Budnik BA, Ivleva VB, Kaur P, Moyer SC, Pittman JL and Costello CE, High pressure MALDI for FTMS for large targets with diverse surfaces. *J Amer Soc Mass Spectrom* **15**: 128-132, 2004.
168. Budnik BA, Moyer SC, Pittman JL, Ivleva VB, Sommer U, Costello CE and O'Connor PB, High pressure MALDI-FTMS: implications for proteomics. *Int J Mass Spectrom* **234**: 203-212, 2004.
169. Laiko VV, Taranenko NI, Berkout VD, Musselman BD and Doroshenko VM, Atmospheric pressure laser desorption/ionization on porous silicon. *Rapid Commun Mass Spectrom* **16**(18): 1737-42., 2002.
170. Laiko VV, Taranenko NI, Berkout VD, Yakshin MA, Prasad CR, Lee HS and Doroshenko VM, D/I of biomolecules from solution at AP by IR laser at 3 microm. *J Am Soc Mass Spectrom* **13**: 354-61., 2002.
171. Creaser CS, Reynolds JC and Harvey DJ, Structural analysis of oligosaccharides by AP MALDI MS. *Rapid Commun Mass Spectrom* **16**: 176-84., 2002.
172. Keough T, Lacey MP and Strife RJ, AP MALDI MS of sulfonic acid derivatized tryptic peptides. *Rapid Commun Mass Spectrom* **15**: 2227-2239, 2001.
173. Zhang J, LaMotte L, Dodds ED and Lebrilla CB, Atmospheric Pressure MALDI FTMS of Labile Oligosaccharides. *Anal Chem* **77**: 4429-4438, 2005.
174. Madonna AJ, Voorhees KJ, Taranenko NI, Laiko VV and Doroshenko VM, Detection of cyclic lipopeptide biomarkers from Bacillus species using AP MALDI MS. *Anal Chem* **75**: 1628-37., 2003.
175. Cohen L and Chait BT, Influence of Matrix on the MALDI-MS Analysis of Peptides and Proteins. *Anal. Chem.* **68**: 31-37, 1996.
176. Dashtiev M, Frankevich V and Zenobi R, Signal enhancement in MALDI by doping with Cu(II) chloride. *Rapid Communications in Mass Spectrometry* **19**: 289-291, 2005.
177. Vorm O, Roepstorff P and Mann M, Improved Analysis by MALDI of Matrix Surfaces from Fast Evaporation. *Anal. .Chem.* **66**: 3281-3287, 1994.
178. Karbach V, Knochenmuss R and Zenobi R, Matrix-assisted filament desorption/ionization mass spectrometry. *J. Am. Soc. Mass Spectrom.* **9**(11): 1226-1228, 1998.
179. Kussmann M and Roepstorff P, Sample preparation techniques for peptides and proteins analyzed by MALDI-MS. *Methods Mol Biol* **146**: 405-24., 2000.
180. Vaidyanathan S, Winder CL, Wade SC, Kell DB and Goodacre R, Sample prep in MALDI MS of whole bacterial cells fOR high mass proteins. *Rapid Commun Mass Spectrom* **16**: 1276-86., 2002.
181. Laugesen S and Roepstorff P, Combination of two matrices results in improved performance of maldi ms for peptide mass mapping and protein analysis. *J Amer Soc Mass Spectrom* **14**: 992-1002, 2003.
182. Cheng Y and Hercules DM, Studies of pesticides by CAD, PSD, MALDI TOF MS. *J Am Soc Mass Spectrom* **12**: 590-8., 2001.
183. Alicata R, Montaudo G, Puglisi C and Samperi F, Influence of chain end groups on the MALDI spectra of polymer blends. *Rapid Commun Mass Spectrom* **16**: 248-60., 2002.
184. Gobom J, Schuerenberg M, Mueller M, Theiss D, Lehrach H and Nordhoff E, a-cyanohydroxycinnamic acid affinity sample prep for MALDI-MS of peptides. *Anal Chem* **73**: 434-8., 2001.
185. Keller BO and Li L, Three-Layer Matrix/Sample Preparation Method for MALDI MS of Proteins. *J Amer Soc Mass Spectrom* **17**: 780-785, 2006.
186. Wyatt MF, Stein BK and Brenton AG, MALDI MS Using 2-[(2E)-3-(4-tert-Butylphenyl)-2-methylprop-2-enylidene]malononitrile Matrix. *Mass Spec Rev* **78**: 199 - 206, 2006.
187. Liao PC, Leykam J, Gage DA and Allison J, Recognition of Phosphorylation Sites by MALDI Mass-Mapping. *Anal. Biochem.* **219**: 9-20, 1994.
188. Aerni H-R, Cornett DS and Caprioli RM, Automated Acoustic Matrix Deposition for MALDI Sample Preparation. *Analytical Chemistry* **78**(3): 827-834, 2006.
189. Sjoedahl J, Kempka M, Hermansson K, Thorsen A and Roeraade J, Chip with Twin Anchors for Reduced Ion Suppression and Improved Mass Accuracy in MALDI-TOF MS. *Anal Chem* **77**: 827-832, 2005.
190. Gustafsson M, Hirschberg D, Palmberg C, Joernvall H and Bergman T, Integrated sample preparation and MALDI MS on a microfluidic compact disk. *Analytical Chemistry* **76**: 345-350, 2004.
191. Hirschberg D, Tryggvason S, Gustafsson M, Bergman T, Swedenborg J, Hedin U and Jornvall H, ID of endothelial proteins in MALDI-MS by compact disc microfluidic system. *Protein J* **23**: 263-71., 2004.

192. Donegan M, Tomlinson AJ, Nair H and Juhasz P, Controlling matrix suppression for MALDI analysis of small molecules. *Rapid Commun. Mass Spectrom.* **18:** 1885-1888, 2004.
193. Jackson SN, Wang H-YJ, Woods AS, Ugarov M, Egan T and Schultz JA, Direct analysis of phospholipids in rat brain by MALDI-TOF and -ion mobility-TOF. *J Amer Soc Mass Spectrom* **16:** 133-8, 2005.
194. Wu Y-C, Hsieh C-H and Tam MF, MALDI of peptides on AnchorChipTM targets with CHCA and nitrocellulose as matrix. *Rap. Commun. Mass Spectrom.* **20:** 309–312, 2006.
195. Loo RRO and Loo JA, MALDI MS of Hydrophobic Proteins in Mixtures Using Formic Acid, Perfluorooctanoic Acid, and Sorbitol. *Anal Chem* **79:** 1115-1125, 2007.
196. Breaux GA, Green-Church KB, France A and Limbach PA, Surfactant-aided, MALDI MS of hydrophobic and hydrophilic peptides. *Anal Chem* **72:** 1169-74., 2000.
197. Kim YJ, Freas A and Fenselau C, Analysis of viral glycoproteins by MALDI-TOF mass spectrometry. *Anal Chem* **73**(7): 1544-8., 2001.
198. Xie Y and Wetlaufer DB, Control of aggregation in protein refolding: the temperature-leap tactic. *Protein Sci* **5**(3): 517-23., 1996.
199. Bird GH, Lajmi AR and Shin JA, Manipulation of Temperature to Improve Solubility of Hydrophobic Proteins for Cocrystallization in MALDI-TOF MS. *Anal. Chem.* **74:** 219-225, 2002.
200. Hoteling AJ, Mourey TH and Owens KG, Importance of Solubility in the Sample Preparation of Poly(ethylene terephthalate) for MALDI TOFMS. *Analytical Chemistry* **77**(3): 750-756, 2005.
201. Skelton R, Dubois F and Zenobi R, A MALDI Sample Preparation Method Suitable for Insoluble Polymers. *Anal. Chem.* **72**(7): 1707-1710, 2000.
202. Dubois F, Knochenmuss R, Steenvoorden RJJM, Breuker K and Zenobi R, Mechanism and control of salt-induced resolution loss in MALDI. *Eur. Mass Spectrom.* **2:** 167-172, 1996.
203. Monroe EB, Koszczuk BA, Losh JL, Jurchen JC and Sweedler JV, Measuring salty samples without adducts with MALDI MS. *Int J Mass Spectrom* **260**(2-3): 237-242, 2007.
204. Naito Y, Yoshihashi-Suzuki S, Ishii K, Kanai T and Awazu K, MALDI of protein samples containing concentrated denaturant by mid-IR free-electron laser. *Int J Mass Spectrom* **241:** 49-56, 2005.
205. Chen VC, Cheng K, Ens W, Standing KG, Nagy JI and Perreault H, rp-Separation and On-Target Deposition of Peptides by Hydrophobic Barrier for MALDI MS. *Anal Chem* **76:** 1189-1196, 2004.
206. McCombie G and Knochenmuss R, Small-Molecule MALDI Using the Matrix Suppression Effect To Reduce or Eliminate Matrix Background Interferences. *Analytical Chemistry* **76**(17): 4990-4997, 2004.
207. Barthelemy P, Ameduri B, Chabaud E, Popot JL and Pucci B, Use of Et-terminated perfluoroalkyl nonionic surfactants derived from tris(hydroxymethyl)acrylamidomethane. *Org Lett* **1:** 1689-92., 1999.
208. Prata C, Giusti F, Gohon Y, Pucci B, Popot JL and Tribet C, Nonionic amphiphilic polymers keep membrane proteins soluble and native in absence of detergent. *Biopolymers* **56:** 77-84., 2000.
209. Meng F, Cargile BJ, Patrie SM, Johnson JR, McLoughlin SM and Kelleher NL, Processing Complex Mixtures of Intact Proteins for Direct Analysis by Mass Spectrometry. *Anal. Chem.* **74:** 2923 -2929, 2002.
210. Yu Y-Q, Gilar M, Lee PJ, Bouvier ESP and Gebler JC, Enzyme-friendly, mass spectrometry-compatible surfactant for in-solution enzymatic digestion of proteins. *Analytical Chemistry* **75**(21): 6023-6028, 2003.
211. Norris JL, Porter NA and Caprioli RM, Mass Spectrometry of Intracellular and Membrane Proteins Using Cleavable Detergents. *Analytical Chemistry* **75**(23): 6642-6647, 2003.
212. Oehlers LP, Perez AN and Walter RB, MALDI TOFMS of 4-sulfophenyl isothiocyanate-derivatized peptides on AnchorChip sample supports using the sodium-tolerant matrix 2,4,6-trihydroxyacetophenone and diammonium citrate. *Rapid Commun Mass Spectrom* **19:** 752-758, 2005.
213. Xu S, Ye M, Xu D, Li X, Pan C and Zou H, Matrix with High Salt Tolerance for Analysis of Peptide and Protein Samples by Desorption/Ionization TOF MS. *Analytical Chemistry* **78:** 2593-2599, 2006.
214. Nomura E, Katsuta K, Ueda T, Toriyama M, Mori T and Inagaki N, Acid-labile surfactant improves in-SDS PAGE protein digestion for MALDI MS peptide mapping. *J Mass Spectrom* **39:** 202-207, 2004.
215. Norris JL, Porter NA and Caprioli RM, Combination Detergent/MALDI Matrix: Functional Cleavable Detergents for Mass Spectrometry. *Analytical Chemistry* **77**(15): 5036-5040, 2005.
216. Zhang H, Andren PE and Caprioli RM, Desalting Procedure for High Sensitivity MALDI. *J. Mass Spectrom.* **30:** 1768-1771, 1995.
217. Wei H, Dean SL, Parkin MC, Nolkrantz K, O'Callaghan JP and Kennedy RT, Microscale sample deposition onto hydrophobic target plates for trace level detection of neuropeptides in brain tissue by MALDI-MS. *Journal of Mass Spectrometry* **40**(10): 1338-1346, 2005.
218. Berhane BT and Limbach PA, Functional Microfabricated Sample Targets for MALDI MS of Ribonucleic Acids. *Anal Chem* **75:** 1997-2003., 2003.
219. Blacken GR, Volny M, Vaisar T, Sadilek M and Turecek F, In Situ Enrichment of Phosphopeptides on MALDI Plates Functionalized by Reactive Landing of Zirconium(IV)-n-Propoxide Ions. *Anal Chem* **79**(in press), 2007.
220. Miliotis T, Kjellstrom S, Nilsson J, Laurell T, Edholm LE and Marko-Varga G, Ready-made MALDI targets coated with matrix for automated sample deposition. *Rapid Commun Mass Spectrom* **16:** 117-26., 2002.
221. Johnson T, Bergquist J, Ekman R, Nordhoff E, Schurenberg M, Kloppel KD, Muller M, Lehrach H and Gobom J, Prestructured sample support in CE-MALDI. *Anal Chem* **73:** 1670-5., 2001.
222. Kleno TG, Andreasen CM, Kjeldal HO, Leonardsen LR, Krogh TN, Nielsen PF, Sorensen MV and Jensen ON, MALDI peptide mapping by on-probe in-gel digestion. *Anal Chem* **76:** 3576-3583, 2004.

223. Sunner J, Dratz E and Chen YC, Graphite Surface Assisted Laser Desorption/Ionization TOFMS of Peptides and Proteins from Liquid Solutions. *Anal. Chem.* **67**: 4335-4342, 1995.

224. Schuerenberg M, Luebbert C, Eickhoff H, Kalkum M, Lehrach H and Nordhoff E, Prestructured MALDI-MS sample supports. *Anal Chem* **72**(15): 3436-42., 2000.

225. Nordhoff E, Schurenberg M, Thiele G, Lubbert C, Kloeppel K-D, Theiss D, Lehrach H and Gobom J, Prestructured surface MALDI of peptides & oligonucleotides. *Int J Mass Spectrom* **226**: 163-180, 2003.

226. Tannu NS, Wu J, Rao VK, Gadgil HS, Pabst MJ, Gerling IC and Raghow R, Wax-coated plates for MALDI in high-throughput ID of proteins. *Anal Biochem* **327**: 222-232, 2004.

227. Gundry RL, Edward R, Kole TP, Sutton C and Cotter RJ, Disposable Hydrophobic Surface on MALDI Targets for Enhancing MS and MS/MS Data of Peptides. *Analytical Chemistry* **77**(20): 6609-6617, 2005.

228. Chen X, Murawski A, Kuang G, Sexton DJ and Galbraith W, Sample Preparation for MALDI MS Using an Elastomeric Device Reversibly Sealed on the MALDI Target. *Anal Chem* **78**: 6160-6168, 2006.

229. McComb ME, Perlman DH, Huang H and Costello CE, Evaluation of an on-target sample preparation system for MALDI-TOFMS in conjunction with normal-flow peptide HPLC for peptide mass fingerprint analyses. *Rapid Commun Mass Spectrom* **21**: 44-58, 2007.

230. Vanderpuije BNY, Han G, Rotello VM and Vachet RW, Mixed Monolayer-Protected Gold Nanoclusters as Selective Peptide Extraction Agents for MALDI-MS Analysis. *Analytical Chemistry* **78**(15): 5491-5496, 2006.

231. Vestling MM and Fenselau C, Surfaces for Interfacing Protein Gel Electrophoresis with MS. *Mass Spectrom. Rev.* **14**: 169-178, 1995.

232. Blais JC, Nagnan-Le-Meillour P, Bolbach G and Tabet JC, MALDI-MS of Proteins Electroblotted to PVDF Membranes. *Rapid Commun. Mass Spectrom.* **10**: 1-4, 1996.

233. Schleuder D, Hillenkamp F and Strupat K, IR-MALDI-MS of electroblotted proteins directly from different membranes, on-membrane digestion, & protein ID. *Anal Chem* **71**: 3238-47., 1999.

234. Eckerskorn C, Strupat K, Schleuder D, Hochstrasser D, Sanchez JC, Lottspeich F and Hillenkamp F, Proteins by direct IR-MALDI MS after 2D-PAGE & electroblotting. *Anal Chem* **69**: 2888-92., 1997.

235. Schmidt F, Lueking A, Nordhoff E, Gobom J, Klose J, Seitz H, Egelhofer V, Eickhoff H, Lehrach H and Cahill DJ, Minimal protein identifiers in 2D gels & recombinant proteins. *Electrophor* **23**: 621-5., 2002.

236. Nordhoff E, Egelhofer V, Giavalisco P, Eickhoff H, Horn M, Przewieslik T, Theiss D, Schneider U, Lehrach H and Gobom J, 2D electrophoresis-MALDI MS for protein mixtures. *Electrophoresis* **22**: 2844-55., 2001.

237. Zaluzec EJ, Gage DA, Allison J and Watson JT, MALDI of proteins immobilized on nylon-based membranes. *J. Am. Soc. Mass Spectrom.* **5**: 230-237, 1994.

238. Zhang H and Caprioli RM, Precoated Membranes for MALDI. *J. Mass Spectrom.* **31**: 690-692, 1996.

239. Colangelo J and Orlando R, On-target endoglycosidase digestion MALDI MS of glycopeptides. *Rapid Commun Mass Spectrom* **15**: 2284-9., 2001.

240. Mock KK, Sutton CW and Cottrell JS, Nitrocellulose as a Surface to Immobilize Proteins for MALDI. *Rapid Commun. Mass Spectrom.* **6**: 233-238, 1992.

241. Luque-Garcia JL, Zhou G, Sun T-T and Neubert TA, Use of Nitrocellulose Membranes for Protein Characterization by Matrix-Assisted Laser Desorption/Ionization Mass Spectrometry. *Analytical Chemistry* **78**(14): 5102-5108, 2006.

242. McComb ME, Oleschuk RD, Manley DM, Donald L, Chow A, O'Neil JD, Ens W, Standing KG and Perreault H, Polyurethane membrane as a sample support in MALDI MS of peptides and proteins. *Rapid Commun Mass Spectrom* **11**: 1716-22., 1997.

243. Blackledge JA and Alexander AJ, Polyethylene membrane as a sample support for direct MALDI MS of high mass proteins. *Anal Chem* **67**: 843-8., 1995.

244. Jacobs A and Dahlman O, Enhancement of the quality of MALDI mass spectra of highly acidic oligosaccharides by using a nafion-coated probe. *Anal Chem* **73**(3): 405-10., 2001.

245. Lin YS, Yang CH and Chen YC, Glass-chip-based sample preparation and on-chip trypic digestion for MALDI MS with sol-gel/DHB hybrid matrix. *Rapid Commun Mass Spectrom* **18**: 313-8., 2004.

246. Kjellstroem S and Jensen ON, In Situ Liquid-Liquid Extraction as a Sample Preparation Method for MALDI MS of Polypeptide Mixtures. *Analytical Chemistry* **75**: 2362-2369, 2003.

247. Brockman AH and Orlando R, New immobilization chemistry for probe affinity mass spectrometry. *Rapid Commun Mass Spectrom* **10**(13): 1688-92., 1996.

248. Griesser HJ, Kingshott P, McArthur SL, McLean KM, Kinsel GR and Timmons RB, Surface-MALDI mass spectrometry in biomaterials research. *Biomaterials* **25**(20): 4861-75., 2004.

249. Wang H, Tseng K and Lebrilla CB, A general method for producing bioaffinity MALDI probes. *Anal Chem* **71**(10): 2014-20., 1999.

250. Nelson RW, Nedelkov D and Tubbs KA, Biomolecular Interaction Analysis MS to detect proteins in biological fluids at femtomole level. *Anal Chem* **72**: 404A-411A., 2000.

251. Nedelkov D and Nelson RW, Analysis of human urine protein biomarkers via biomolecular interaction analysis mass spectrometry. *Am J Kidney Dis* **38**(3): 481-7., 2001.

252. Neubert H, Jacoby ES, Bansal SS, Iies RK, Cowan DA and Kicman AT, Affinity capture MALDI-TOF MS: immunoglobulin G using recombinant protein G. *Anal Chem* **74**: 3677-83., 2002.

253. Dick LW, Jr. and McGown LB, Aptamer-enhanced laser desorption/ionization for affinity mass spectrometry. *Analytical Chemistry* **76**(11): 3037-3041, 2004.

254. Menendez A and Scott JK, The nature of target-unrelated peptides recovered in the screening of phage-displayed random peptide libraries with antibodies. *Analytical Biochemistry* **336**(2): 145-157, 2005.

255. Seok H-J, Hong M-Y, Kim Y-J, Han M-K, Lee D, Lee J-H, Yoo J-S and Kim H-S, MS of affinity-captured proteins on a dendrimer for on-chip proteolytic digestion. *Anal Biochem* **337**: 294-307, 2005.

256. Baytekin B, Werner N, Luppertz F, Engeser M, Brueggemann J, Bitter S, Henkel R, Felder T and Schalley CA, How useful is MS of dendrimers? Fake defects. *Int J Mass Spectrom* **249/250**: 138-148, 2006.

257. Koopmann JO and Blackburn J, High affinity capture surface for MALDI compatible protein microarrays. *Rapid Commun Mass Spectrom* **17**: 455-62., 2003.

258. Li M, Timmons RB and Kinsel GR, Radio Frequency Plasma Polymer Coatings for Affinity Capture MALDI Mass Spectrometry. *Analytical Chemistry* **77**(1): 350-353, 2005.

259. Merchant M and Weinberger SR, Recent advancements in surface-enhanced laser desorption/ionization-time of flight-mass spectrometry. *Electrophoresis* **21**(6): 1164-77., 2000.

260. Issaq HJ, Conrads TP, Prieto DA, Tirumalai R and Veenstra TD, SELDI-TOF MS for diagnostic proteomics. *Anal Chem* **75**(7): 148A-155A., 2003.

261. Tang N, Tornatore P and Weinberger SR, Current developments in SELDI affinity technology. *Mass Spectrom Rev* **23**(1): 34-44., 2004.

262. Huang Y-F and Chang H-T, Analysis of ATP and Glutathione through Gold Nanoparticles Assisted Laser Desorption/Ionization MS. *Anal Chem* **79**: 4852-4859, 2007.

263. Paweletz CP, Wiener MC, Sachs JR, Meurer R, Wu MS, Wong KK, Yates NA and Hendrickson RC, Surface enhanced laser desorption ionization spectrometry reveals biomarkers for drug treatment but not dose. *Proteomics* **6**(7): 2101-2107, 2006.

264. Forde CE, Gonzales AD, Smessaert JM, Murphy GA, Shields SJ, Fitch JP and McCutchen-Maloney SL, Capture of transcription factors by SELDI MS. *Biochem Biophys Res Commun* **290**: 1328-35., 2002.

265. Finkel NH, Prevo BG, Velev OD and He L, Ordered Silicon Nanocavity Arrays in Surface-Assisted Desorption/Ionization Mass Spectrometry. *Analytical Chemistry* **77**(4): 1088-1095, 2005.

266. Whiteaker JR, Zhao L, Zhang HY, Feng L-C, Piening BD, Anderson L and Paulovich AG, Antibody-based enrichment of peptides on magnetic beads for MS-based quantification of serum biomarkers. *Anal Biochem* **362**: 44-54, 2007.

267. Feuerstein I, Najam-ul-Haq M, Rainer M, Trojer L, Bakry R, Aprilita NH, Stecher G, Huck CW, Bonn GK, Klocker H, Bartsch G and Guttman A, Material-Enhanced Laser Desorption/Ionization (MELDI)-Profiling Tool Utilizing Specific Carrier Materials for TOFMS. *J Amer Soc Mass Spectrom* **17**: 1203-1208, 2006.

268. Kong XL, Huang LCL, Hsu CM, Chen WH, Han CC and Chang HC, High-Affinity Capture of Proteins by Diamond Nanoparticles for Mass Spectrometric Analysis. *Analytical Chemistry* **77**(1): 259-265, 2005.

269. Tubbs KA, Nedelkov D and Nelson RW, Detection and quantification of beta-2-microglobulin using mass spectrometric immunoassay. *Anal Biochem* **289**(1): 26-35., 2001.

270. Brivio M, Fokkens RH, Verboom W, Reinhoudt DN, Tas NR, Goedbloed M and van den Berg A, Microfluidic system for biochemical reactions in on-line MALDI-TOF MS. *Anal Chem* **74**: 3972-6., 2002.

271. Rouse JC and Vath JE, On-probe sample cleanup strategies for glycoprotein-released carbohydrates prior to MALDI TOFMS. *Anal Biochem* **238**: 82-92., 1996.

272. Jorgensen CS, Jagd M, Sorensen BK, McGuire J, Barkholt V, Hojrup P and Houen G, Efficacy of elution of proteins from SDS PAGE and polyvinyldifluoride membranes for MS. *Anal Biochem* **330**: 87-97, 2004.

273. Terry DE, Umstot E and Desiderio DM, Optimized sample-processing time and peptide recovery for MS of protein digests. *J Amer Soc Mass Spectrom* **15**: 784-794, 2004.

274. Loo RR, Cavalcoli JD, VanBogelen RA, Mitchell C, Loo JA, Moldover B and Andrews PC, Virtual 2D gel electrophoresis: visualization & analysis of E. coli proteome by MS. *Anal Chem* **73**: 4063-70., 2001.

275. Walker AK, Rymar G and Andrews PC, Mass spectrometric imaging of immobilized pH gradient gels and creation of "virtual" two-dimensional gels. *Electrophoresis* **22**(5): 933-45., 2001.

276. Xu Y, Little MW, Rousell DJ, Laboy JL and Murray KK, Direct Analysis from Polyacrylamide Gel Infrared Laser Desorption/Ionization. *Analytical Chemistry* **76**(4): 1078-1082, 2004.

277. Coon JJ and Harrison WW, Laser desorption-APCI MS for the analysis of peptides from aqueous solutions. *Anal Chem* **74**: 5600-5., 2002.

278. Baltz-Knorr M, Ermer DR, Schriver KE and Haglund RF, Jr., Infrared laser desorption and ionization of polypeptides from a polyacrylamide gel. *J Mass Spectrom* **37**(3): 254-8., 2002.

279. Xu Y, Little MW and Murray KK, Interfacing Capillary Gel Microfluidic Chips with IR Laser Desorption MS. *J Amer Soc Mass Spectrom* **17**: 469-474, 2006.

280. Mandell JG, Falick AM and Komives EA, Measurement of amide hydrogen exchange by MALDI-TOF mass spectrometry. *Anal Chem* **70**(19): 3987-95., 1998.

281. Nazabal A, Laguerre M, Schmitter J-M, Vaillier J, Chaignepain St and Velours J, HDX on yeast ATPase supramolecular protein complex by MALDI MS. *J Amer Soc Mass Spectrom* **14**: 471-481, 2003.

282. Li X, Chou YT, Husain R and Watson JT, Integration of hydrogen/deuterium exchange and cyanylation-based methodology for conformational studies of cystinyl proteins. *Anal Biochem* **331**(1): 130-7., 2004.

283. Powell KD, Wales TE and Fitzgerald MC, Thermodynamic stability measurements on multimeric proteins using a new H/D exchange- & MALDI MS-based method. *Protein Sci* **11**: 841-51., 2002.

284. Brown RS, Carr BL and Lennon JJ, Factors that influence the observed fast fragmentation of peptides in matrix-assisted laser desorption. *J. Am. Soc. Mass Spectrom.* **7**(3): 225-32, 1996.

285. Spengler B, Kirsch D, Kaufmann R and Jaeger E, Sequencing of Peptides by MALDI and PSD. *Rapid Comm. in Mass Spectrom.* **6**: 105-108, 1992.

286. Stahl-Zeng J, Hillenkamp F and Karas M, Metastable fragment-ion analysis in a reflectron instrument with a gridless ion mirror. *Eur. Mass Spectrom.* **2**(1): 23-32, 1996.

287. Takayama M, In-source decay characteristics of peptides in matrix-assisted laser desorption/ionization time-of-flight mass spectrometry. *J Am Soc Mass Spectrom* **12**(4): 420-7., 2001.

288. Cotter RJ, Iltchenko S and Wang D, The curved-field reflectron: PSD and CID without scanning, stepping or lifting. *International Journal of Mass Spectrometry* **240**(3): 169-182, 2005.

289. Warscheid B and Fenselau C, Characterization of Bacillus spore species and their mixtures using postsource decay with a curved-field reflectron. *Analytical Chemistry* **75**(20): 5618-5627, 2003.

290. Kenny DJ, Brown JM, Palmer ME, Snel MF and Bateman RH, A Parallel Approach to Post Source Decay MALDI-TOF Analysis. *Journal of the American Society for Mass Spectrometry* **17**(1): 60-66, 2006.

291. Clipston NL, Jai-nhuknan J and Cassady CJ, Comparison of negative and positive ion TOF PSD MS for peptides containing basic residues. *Int J Mass Spectrom* **222**: 363-381, 2003.

292. Yamaguchi M, Nakazawa T, Kuyama H, Obama T, Ando E, Okamura T, Ueyama N and Norioka S, High-Throughput N-Terminal Sequencing of Proteins by MALDI MS. *Anal Chem* **77**: 645-651, 2005.

293. Kaufmann R, Spengler B and Lutzenkirchen F, MS peptide sequencing by CID in rTOFMS using MALDI. *Rapid Commun Mass Spectrom* **7**: 902-10., 1993.

294. Shevchenko A, Sunyaev S, Loboda A, Bork P, Ens W and Standing KG, Charting proteomes by unsequenced genomes by MALDI-oaTOF & BLAST homology search. *Anal Chem* **73**: 1917-26., 2001.

295. Medzihradszky KF, Campbell JM, Baldwin MA, Falick AM, Juhasz P, Vestal ML and Burlingame AL, Peptide CID in MALDI-TOF/TOF MS. *Anal Chem* **72**: 552-8., 2000.

296. Zhang W, Krutchinsky AN and Chait BT, \"De novo\" peptide sequencing by MALDI-quadrupole-ion trap MS: a preliminary study. *J Amer Soc Mass Spectrom* **14**: 1012-1021, 2003.

297. Ray KL and Glish GL, Matrix-assisted laser desorption/ionization-boundary-activated dissociation of peptide ions in a quadrupole ion trap. *International Journal of Mass Spectrometry* **222**(1-3): 75-83, 2003.

298. Brown RS and Lennon JJ, Mass Resolution Improvement by Incorporation of Pulsed Ion Extraction in a MALDI TOF MS. *Anal. Chem.* **67**: 1998-2003, 1995.

299. Vestal M, Juhasz P and Martin S, Delayed Extraction MALDI-MS. *Rapid Commun. Mass Spectrom.* **67**: 3990-3999, 1995.

300. Whittal R and Li L, High-Resolution MALDI-MS. *Anal. Chem.* **67**: 1950-1954, 1995.

301. Gleitsmann E and Karas M, Computer simulation and practical performance of DE MALDI-TOF mass spectrometer. *Adv. Mass Spectrom.* **14**: B073480/1-B073480/10, 1998.

302. Karas M, Commentary on "Time-of-flight mass spectrometer with improved resolution" W. C. Wiley and I. H. McLaren, Rev. Sci. Instrum., 26, 1150 (1955). *J. Mass Spectrom.* **32**(1): 1-3, 1997.

303. Gobom J, Mueller M, Egelhofer V, Theiss D, Lehrach H and Nordhoff E, Calibration to simplify & improve accurate determination of peptide masses by MALDI-TOF MS. *Anal Chem* **74**: 3915-23., 2002.

304. Peterson DS, Matrix-free methods for laser desorption/ionization mass spectrometry. *Mass Spectrometry Reviews* **26**: 19-34, 2007.

305. Kruse RA, Li X, Bohn PW and Sweedler JV, Experimental factors controlling analyte ion generation in laser desorption/ionization MS on porous silicon. *Anal Chem* **73**(15): 3639-45, 2001.

306. Shen Z, Thomas JJ, Averbuj C, Broo KM, Engelhard M, Crowell JE, Finn MG and Siuzdak G, Porous silicon as a versatile platform for laser desorption/ionization MS. *Anal Chem* **73**(3): 612-9., 2001.

307. Lewis WG, Shen Z, Finn MG and Siuzdak G, Desorption/ionization on silicon (DIOS) mass spectrometry: background and applications. *International Journal of Mass Spectrometry* **226**(1): 107-116, 2003.

308. Budimir N, Blais J-C, Fournier F and Tabet J-C, The use of DIOS MS of negative ions for fatty acids. *Rapid Commun Mass Spectrom* **20**: 680-684, 2005.

309. Vaidyanathan S, Jones D, Ellis J, Jenkins T, Chong C, Anderson M and Goodacre R, DIOS for metabolome analyses: influence of surface oxidation. *Rapid Commun Mass Spectrom* **21**: 2157-2166, 2007.

310. Budimir N, Blais J-C, Fournier F and Tabet J-C, Desorption/ionization on porous silicon mass spectrometry (DIOS) of model cationized fatty acids. *J Mass Spectrom* **42**: 42-48, 2007.

311. Okuno S, Wada Y and Arakawa R, Quantitative analysis of polypropyleneglycol mixtures by DIOS MS. *International Journal of Mass Spectrometry* **241**(1): 43-48, 2005.

312. Thomas JJ, Shen Z, Crowell JE, Finn MG and Siuzdak G, DIOS: a diverse MS platform for protein characterization. *Proc Natl Acad Sci U S A* **98**: 4932-7., 2001.

313. Go EP, Prenni JE, Wei J, Jones A, Hall SC, Witkowska HE, Shen Z and Siuzdak G, Desorption/ionization on silicon time-of-flight/time-of-flight mass spectrometry. *Analytical Chemistry* **75**(10): 2504-2506, 2003.

314. Kruse RA, Li X, Bohn PW and Sweedler JV, Experimental factors controlling analyte ion generation in laser desorption/ionization MS on porous silicon. *Anal Chem* **73**: 3639-45., 2001.

315. Thomas JK and Ellison EH, Various aspects of the constraints imposed on the photochemistry of systems in porous silica. *Adv Colloid Interface Sci* **89-90**: 195-238., 2001.

316. Go EP, Shen Z, Harris K and Siuzdak G, Quantitative analysis with desorption/ionization on silicon mass spectrometry using electrospray deposition. *Analytical Chemistry* **75**(20): 5475-5479, 2003.

317. Kruse RA, Rubakhin SS, Romanova EV, Bohn PW and Sweedler JV, Direct assay of Aplysia tissues by DIOS. *J Mass Spectrom* **36**: 1317-22., 2001.

318. Rousell DJ, Dutta SM, Little MW and Murray KK, Matrix-free infrared soft laser desorption/ionization. *Journal of Mass Spectrometry* **39**(10): 1182-1189, 2004.

319. Trauger SA, Go EP, Shen Z, Apon JV, Compton BJ, Bouvier ESP, Finn MG and Siuzdak G, Analyte Capture with Silylated DIOS MS. *Anal Chem* **76**: 4484-4489, 2004.

320. Nordström A, Apon JV, Uritboonthai W, Go EP and Siuzdak G, Surfactant-Enhanced DIOS Mass Spectrometry. *Anal Chem* **78**: 272 - 278, 2006.

321. Kinumi T, Shimomae Y, Arakawa R, Tatsu Y, Shigeri Y, Yumoto N and Niki E, Peptides containing cysteine sulfonic acid by MALDI & DIOS MS. *J Mass Spectrom* **41**: 103-112, 2006.

322. Go EP, Apon JV, Luo G, Saghatelian A, Daniels RH, Sahi V, Dubrow R, Cravatt BF, Vertes A and Siuzdak G, Desorption/Ionization on Silicon Nanowires. *Analytical Chemistry* **77**(6): 1641-1646, 2005.

323. Mehl J and Hercules D, Direct TLC-MALDI coupling using a hybrid plate. *Anal Chem* **72**(1): 68-73, 2000.

324. Hayen H and Volmer DA, ID of siderophores by TLC/MALDI TOFMS. *Rapid Commun Mass Spectrom* **19**: 711-720, 2005.

325. Dreisewerd K, Koelbl S, Peter-Katalinic J, Berkenkamp S and Pohlentz G, Milk Oligosaccharides from TLC by MALDI oaTOFMS. *J Amer Soc Mass Spectrom* **17**: 139-150, 2006.

326. Salo PK, Salomies H, Harju K, Ketola RA, Kotiaho T, Yli-Kauhaluoma J and Kostiainen R, Small molecules by ultra TLC-APMALDI. *J Amer Soc Mass Spectrom* **16**: 906-915, 2005.

327. Dreisewerd K, Muething J, Rohlfing A, Meisen I, Vukelic Z, Peter-Katalinic J, Hillenkamp F and Berkenkamp S, Gangliosides Directly from TLC Plates by IR MALDI oaTOF MS in Glycerol. *Anal Chem* **77**: 4098-4107, 2005.

328. Salo PK, Vilmunen S, Salomies H, Ketola RA and Kostiainen R, 2D Ultra-TLC and AP-MALDI MS in Bioanalysis. *Anal Chem* **79**: 2101-2108, 2007.

329. Van Berkel GJ, Ford MJ and Deibel MA, Thin-Layer Chromatography and Mass Spectrometry Coupled Using Desorption Electrospray Ionization. *Analytical Chemistry* **77**(5): 1207-1215, 2005.

330. Wen X, Dagan S and Wysocki VH, Small-Molecule Analysis with Silicon-Nanoparticle-Assisted Laser Desorption/Ionization MS. *Anal Chem* **79**: 434-444, 2007.

331. Li Q, Ricardo A, Benner SA, Winefordner JD and Powell DH, DIOS MS Studies on Pentose-Borate Complexes. *Analytical Chemistry* **77**: 4503-4508, 2005.

332. Steenwyk RC, Hutzler JM, Sams J, Shen Z and Siuzdak G, Atmospheric pressure DIOS for quantitation of midazolam in rat plasma for kinetic studies. *Rapid Commun Mass Spectrom* **20**: 3717-3722, 2006.

333. Pihlainen K, Grigoras K, Franssila S, Ketola R, Kotiaho T and Kostiainen R, Amphetamines and fentanyls by AP DIOS and MALDI MS for forensics of drug seizures. *J Mass Spectrom* **40**: 539-545, 2005.

334. Seino T, Sato H, Yamamoto A, Nemoto A, Torimura M and Tao H, Matrix-Free Laser Desorption/Ionization-MS Using Self-Assembled Germanium Nanodots. *Anal Chem* **79**: 4827-4832, 2007.

335. Sanguinet L, Aleveque O, Blanchard P, Dias M, Levillain E and Rondeau D, Desorption/ionization on self-assembled monolayer surfaces. *Journal of Mass Spectrometry* **41**(6): 830-833, 2006.

336. Peterson DS, Luo Q, Hilder EF, Svec F and Frechet JMJ, Porous polymer monolith for surface-enhanced laser desorption/ionization TOF MS of small molecules. *Rapid Commun Mass Spectrom* **18**: 1504-1512, 2004.

337. Nayak R and Knapp DR, Effects of Thin-Film Structural Parameters on Laser Desorption/Ionization from Porous Alumina. *Anal Chem* **79**: 4950-4956, 2007.

338. Berkenkamp S, Karas M and Hillenkamp F, Ice as a matrix for IR-MALDI MS from a protein single crystal. *Proc Natl Acad Sci U S A* **93**(14): 7003-7., 1996.

339. Raska CS, Parker CE, Huang C, Han J, Glish GL, Pope M and Borchers CH, Pseudo-MS3 in a MALDI orthogonal quadrupole-time of flight mass spectrometer. *J Am Soc Mass Spectrom* **13**(9): 1034-41., 2002.

340. Todd PJ and Schaaff TG, A secondary ion microprobe ion trap mass spectrometer. *J Am Soc Mass Spectrom* **13**(9): 1099-107., 2002.

341. Todd PJ, McMahon JM and McCandlish, Carl A., Secondary ion images of the developing rat brain. *Journal of the American Society for Mass Spectrometry* **15**(7): 1116-1122, 2004.

342. Spengler B and Hubert M, Scanning microprobe MALDI (SMALDI) for sub-micrometer resolved LDI and MALDI surface analysis. *J Am Soc Mass Spectrom* **13**: 735-48., 2002.

343. Pacholski ML and Winograd N, Imaging with Mass Spectrometry. *Chemical Reviews (Washington, D. C.)* **99**(10): 2977-3005, 1999.

344. Sanni OD, Wagner MS, Briggs D, Castner DG and Vickerman JC, Classify protein static TOF-SIMS spectra by principal component & neural networks. *Surface and Interface Analysis* **33**: 715-728, 2002.

345. Cliff B, Lockyer NP, Corlett C and Vickerman JC, Development of instrumentation for routine ToF-SIMS imaging analysis of biological material. *Applied Surface Science* **203-204**(Complete): 730-733, 2003.

346. Rubakhin SS, Jurchen JC, Monroe EB and Sweedler JV, Imaging mass spectrometry: fundamentals and applications to drug discovery. *Drug Discovery Today* **10**(12): 823-837, 2005.

347. Monroe EB, Jurchen JC, Lee J, Rubakhin SS and Sweedler JV, Vitamin E Imaging and Localization in the Neuronal Membrane. *Journal of the American Chemical Society* **127**(35): 12152-12153, 2005.

348. McDonnell LA and Heeren RMA, Imaging Mass Spectrometry. *Mass Spectrom Rev* **26**: 606-643, 2007.

349. Khatib-Shahidi S, Andersson M, Herman JL, Gillespie TA and Caprioli RM, Direct Molecular Analysis of Whole-Body Animal Tissue Sections by Imaging MALDI MS. *Anal Chem* **78**: 6448-6456, 2006.

350. Becker JS, Becker JS, Zoriy MV, Dobrowolska J and Matusch A, Imaging MS in biological tissues by laser ablation ICP MS. *Eur J Mass Spectrom* **13**: 1-6, 2007.

351. Chaurand P, Schwartz SA, Billheimer D, Xu BJ, Crecelius A and Caprioli RM, Integrating histology and imaging mass spectrometry. *Analytical Chemistry* **76**(4): 1145-1155, 2004.

352. Chaurand P, Schwartz SA and Caprioli RM, Assessing protein patterns in disease using imaging mass spectrometry. *Journal of Proteome Research* **3**(2): 245-252, 2004.

353. Crecelius AC, Cornett DS, Caprioli RM, Williams B, Dawant BM and Bodenheimer B, 3D Visualization of Protein Expression in Mouse Brain by Imaging MS. *J Amer Soc Mass Spectrom* **16**: 1093-1099, 2005.

354. Reyzer ML, Hsieh Y, Ng K, Korfmacher WA and Caprioli RM, Direct analysis of drug candidates in tissue by MALDI MS. *Journal of Mass Spectrometry* **38**: 1081-1092, 2003.

355. Hsieh Y, Casale R, Fukuda E, Chen J, Knemeyer I, Wingate J, Morrison R and Korfmacher W, MALDI imaging MS of clozapine in rat brain tissue. *Rapid Commun Mass Spectrom* **20**: 965-972, 2006.

356. Schwartz SA, Reyzer ML and Caprioli RM, Direct tissue analysis by MALDI MS: Practical sample preparation. *J Mass Spectrom* **38**: 699-708, 2003.

357. Lemaire R, Wisztorski M, Desmons A, Tabet JC, Day R, Salzet M and Fournier I, MALDI-MS Direct Tissue Analysis of Proteins: Improving Signal Sensitivity Using Organic Treatments. *Analytical Chemistry* **78**(20): 7145-7153, 2006.

358. Sherrod SD, Castellana ET, McLean JA and Russell DH, Spatially dynamic laser patterning using advanced optics for imaging MALDI MS. *Int J Mass Spectrom* **262**: 256-262, 2007.

359. Garrett TJ, Prieto-Conaway MC, Kovtoun V, Bui H, Izgarian N, Stafford G and Yost RA, Imaging of small molecules in tissue sections by an intermediate-pressure MALDI linear ion trap MS. *Int J Mass Spectrom* **260**: 166-176, 2007.

360. Luxembourg SL, Mize TH, McDonnell LA and Heeren RMA, High-Spatial Resolution Mass Spectrometric Imaging of Peptide and Protein Distributions on a Surface. *Analytical Chemistry* **76**(18): 5339-5344, 2004.

361. Lemaire R, Tabet JC, Ducoroy P, Hendra JB, Salzet M and Fournier I, Solid Ionic Matrixes for Direct Tissue Analysis and MALDI Imaging. *Analytical Chemistry* **78**(3): 809-819, 2006.

362. McDonnell LA, Mize TH, Luxembourg SL, Koster S, Eijkel GB, Verpoorte E, de Rooij NF and Heeren RMA, Using Matrix Peaks To Map Topography. *Anal Chem* **75**: 4373-4381, 2003.

363. Altelaar AFM, Van Minnen J, Jimenez CR, Heeren RMA and Piersma SR, Imaging of Lymnaea stagnalis Nervous Tissue at Subcellular Spatial Resolution by MS. *Anal Chem* **77**: 735-741, 2005.

364. Altelaar AFM, Klinkert I, Jalink K, de Lange RPJ, Adan RAH, Heeren RMA and Piersma SR, Gold-Enhanced Imaging of Cells and Tissue by SIMS and MALDI MS. *Anal Chem* **78**: 734-742, 2006.

365. Fletcher JS, Lockyer NP, Vaidyanathan S and Vickerman JC, TOF-SIMS 3D Biomolecular Imaging of Xenopus laevis Oocytes Using Buckminsterfullerene (C60) Primary Ions. *Anal Chem* **79**: 2199-2206, 2007.

366. Jurchen JC, Rubakhin SS and Sweedler JV, MALDI-MS Imaging of Features Smaller than the Size of the Laser Beam. *Journal of the American Society for Mass Spectrometry* **16**(10): 1654-1659, 2005.

367. Crossman L, McHugh NA, Hsieh Y, Korfmacher WA and Chen J, Investigation of the profiling depth in MALDI imaging MS. *Rapid Communications in Mass Spectrometry* **20**: 284-290, 2005.

368. Touboul D, Halgand F, Brunelle A, Kersting R, Tallarek E, Hagenhoff B and Laprevote O, Tissue molecular ion imaging by gold cluster ion bombardment. *Analytical Chemistry* **76**(6): 1550-1559, 2004.

369. Sjoevall P, Lausmaa J and Johansson B, Mass Spectrometric Imaging of Lipids in Brain Tissue. *Analytical Chemistry* **76**(15): 4271-4278, 2004.

370. Woods AS, Ugarov M, Jackson SN, Egan T, Wang H-YJ, Murray KK and Schultz JA, IR-MALDI-LDI Combined with Ion Mobility oaTOFMS. *J Proteome Res* **5**: 1484-1487, 2006.

371. Li Y, Shrestha B and Vertes A, Atmospheric Pressure Molecular Imaging by IR MALDI MS. *Anal Chem* **79**: 523-532, 2007.

372. Ifa DR, Wiseman JM, Song Q and Cooks RG, Development of capabilities for imaging MS under ambient conditions with DESI. *Int J Mass Spectrom* **259**: 8-15, 2007.

373. Kruse R and Sweedler JV, Spatial profiling invertebrate ganglia using MALDI MS. *Journal of the American Society for Mass Spectrometry* **14**(7): 752-759, 2003.

374. Monroe EB, Jurchen JC, Koszczuk BA, Losh JL, Rubakhin SS and Sweedler JV, Massively Parallel Sample Preparation for the MALDI MS Analyses of Tissues. *Analytical Chemistry* **78**(19): 6826-6832, 2006.

375. Sugiura Y, Shimma S and Setou M, Two-Step Matrix Application Technique To Improve Ionization Efficiency for MALDI in Imaging MS. *Anal Chem* **78**: 8227-8235, 2006.

376. Jackson SN, Wang H-YJ and Woods AS, Direct Profiling of Lipid Distribution in Brain Tissue Using MALDI-TOFMS. *Analytical Chemistry* **77**(14): 4523-4527, 2005.

377. Amaya KR, Monroe EB, Sweedler JV and Clayton DF, Lipid imaging in the zebra finch brain with secondary ion mass spectrometry. *International Journal of Mass Spectrometry* **260**(2-3): 121-127, 2007.

378. Boerner K, Malmberg P, Mansson J-E and Nygren H, Molecular imaging of lipids in cells and tissues. *International Journal of Mass Spectrometry* **260**(2-3): 128-136, 2007.

379. Ostrowski SG, Kurczy ME, Roddy TP, Winograd N and Ewing AG, Secondary Ion MS Imaging To Relatively Quantify Cholesterol in the Membranes of Individual Cells from Differentially Treated Populations. *Anal Chem* **79:** 3554-3560, 2007.

380. Jackson SN, Wang H-YJ and Woods AS, In Situ Structural Characterization of Phosphatidylcholines in Brain Tissue by MALDI-MS/MS. *J Amer Soc Mass Spectrom* **16:** 2052-2056, 2005.

381. McDonnell LA, Piersma SR, Altelaar AFM, Mize TH, Luxembourg SL, Verhaert PDEM, Van Minnen J and Heeren RMA, Subcellular imaging MS of brain tissue. *J Mass Spectrom* **40:** 160-168, 2005.

382. Brunelle A, Touboul D and Laprevote O, Biological tissue imaging with time-of-flight secondary ion mass spectrometry and cluster ion sources. *Journal of Mass Spectrometry* **40**(8): 985-999, 2005.

383. Debois D, Brunelle A and Laprevote O, Attempts for molecular depth profiling directly on a rat brain tissue section using fullerene and bismuth cluster ion beams. *Int J Mass Spectrom* **260:** 115-120, 2007.

384. Jones EA, Lockyer NP and Vickerman JC, MS analysis and imaging of tissue by ToF-SIMS-The role of buckminsterfullerene, C60 +, primary ions. *Int J Mass Spectrom* **260:** 146-157, 2007.

385. Parry S and Winograd N, High-Resolution TOF-SIMS Imaging of Eukaryotic Cells Preserved in a Trehalose Matrix. *Analytical Chemistry* **77**(24): 7950-7957, 2005.

386. Marcus A and Winograd N, Metal Nanoparticle Deposition for TOF-SIMS Signal Enhancement of Polymers. *Analytical Chemistry* **78**(1): 141-148, 2006.

387. Molloy MP, Phadke ND, Maddock JR and Andrews PC, Two-dimensional electrophoresis and peptide mass fingerprinting of bacterial outer membrane proteins. *Electrophoresis* **22**(9): 1686-96., 2001.

388. Baskakov IV, Legname G, Baldwin MA, Prusiner SB and Cohen FE, Pathway complexity of prion protein assembly into amyloid. *J Biol Chem* **277**(24): 21140-8., 2002.

389. Jiang Y, Lee A, Chen J, Cadene M, Chait BT and MacKinnon R, The open pore conformation of potassium channels. *Nature* **417**(6888): 523-6., 2002.

390. Rappsilber J, Moniatte M, Nielsen ML, Podtelejnikov AV and Mann M, Experiences and perspectives of MALDI MS and MS/MS in proteomic research. *Int J Mass Spectrom* **226:** 223-237, 2003.

391. Hardouin J, Hubert-Roux M, Delmas AF and Lange C, Identification of isoenzymes using MALDI MS. *Rapid Commun Mass Spectrom* **20:** 725-732, 2005.

392. Chaurand P, Luetzenkirchen F and Spengler B, Peptide and protein identification by MALDI TOF PSD MS. *J Am Soc Mass Spectrom* **10:** 91-103., 1999.

393. Fournier I, Chaurand P, Bolbach G, Lutzenkirchen F, Spengler B and Tabet JC, Sequencing of a branched peptide by MALDI TOF MS. *J Mass Spectrom* **35:** 1425-33., 2000.

394. Hoffmann R, Metzger S, Spengler B and Otvos L, Jr., Sequencing of peptides phosphorylated on serines and threonines by PSD MALDI TOF MS. *J Mass Spectrom* **34:** 1195-204., 1999.

395. Huang ZH, Shen T, Wu J, Gage DA and Watson JT, Protein sequencing by MALDI PSD MS of N-Tris(2,4,6-trimethoxyphenyl)phosphine-acetylated tryptic digests. *Anal Biochem* **268:** 305-17., 1999.

396. Keough T, Youngquist RS and Lacey MP, Sulfonic acid derivatives for peptide sequencing by MALDI MS. *Anal Chem* **75**(7): 156A-165A., 2003.

397. Borges CR and Watson JT, Recognition of cysteinyl peptides by prompt fragmentation of 4-dimethylaminophenylazophenyl-4'-maleimide derivative in MALDI-MS. *Protein Sci* **12:** 1567-72., 2003.

398. Mane C, Sommerer N, Yalcin T, Cheynier V, Cole RB and Fulcrand H, Assessment of the Molecular Weight Distribution of Tannin Fractions through MALDI-TOF MS. *Anal Chem* **79:** 2239-2248, 2007.

399. Meier MAR, Lohmeijer BGG and Schubert US, Relative binding strength of terpyridine model complexes under MALDI MS conditions. *Journal of Mass Spectrometry* **38:** 510-516, 2003.

400. Harvey DJ, MALDI MS of carbohydrates and glycoconjugates. *Int J Mass Spectrom* **226:** 1-35, 2003.

401. Farmer TB and Caprioli RM, Determination of protein-protein interactions by matrix-assisted laser desorption/ionization mass spectrometry. *J Mass Spectrom* **33**(8): 697-704., 1998.

402. Loo JA, Studying noncovalent protein complexes by electrospray ionization MS: The tools of proteomics. *Mass Spectrom Rev* **16**(1): 1-23., 1997.

403. Song F, A Study of Noncovalent Protein Complexes by MALDI. *J Amer Soc Mass Spectrom* **18:** 1286-1290, 2007.

404. Yanes O, Aviles FX, Roepstorff P and Jorgensen TJD, Exploring the \"Intensity Fading\" Phenomenon in the Study of Noncovalent Interactions by MALDI-TOF MS. *J Amer Soc Mass Spectrom* **18:** 359-367, 2007.

405. Bright JJ, Claydon MA, Soufian M and Gordon DB, Rapid typing of bacteria by MALDI TOFMS and pattern recognition software. *J Microbiol Methods* **48**(2-3): 127-138., 2002.

406. Smole SC, King LA, Leopold PE and Arbeit RD, Sample preparation of Gram-positive bacteria for ID by MALDI TOFMS. *J Microbiol Methods* **48:** 107-115., 2002.

407. Fenselau C and Demirev PA, Intact Microorganisms by MALDI MS. *Mass Spectrom. Rev.* **20:** 157-171, 2001.

408. Lay JO, Jr, MALDI-TOF mass spectrometry of bacteria. *Mass Spectrom. Rev.* **20:** 172-194, 2001.

409. Bundy JL and Fenselau C, Lectin and carbohydrate affinity capture surfaces for mass spectrometric analysis of microorganisms. *Anal Chem* **73**(4): 751-7., 2001.

410. Ryzhov V, Bundy JL, Fenselau C, Taranenko N, Doroshenko V and Prasad CR, MALDI TOFMS of Bacillus spores using a 2.94 μm infrared laser. *Rap. Commun. Mass Spectrom.* **14:** 1701-1706, 2000.

411. Ryzhov V and Fenselau C, Characterization of the protein subset desorbed by MALDI from whole bacterial cells. *Anal Chem* **73**(4): 746-50., 2001.

412. Wang Z, Dunlop K, Long SR and Li L, Mass Spectrometric Methods for Generation of Protein Mass Database Used for Bacterial Identification. *Anal. Chem.* **74**: 3174 -3182, 2002.

413. Demirev P, Lin J, Pineda F and Fenselau C, Bioinformatics & MS to ID microorganisms: proteome-wide post-translational modifications and database search of intact H. pylori. *Anal Chem* **73**: 4566-4573, 2001.

414. Du Z, Yang R, Guo Z, Song Y and Wang J, Identification of Staphylococcus aureus and determination of its methicillin resistance by MALDI TOFMS. *Anal Chem* **74**: 5487-91., 2002.

415. Pignone M, Greth KM, Cooper J, Emerson D and Tang J, Identification of Mycobacteria by MALDI TOFMS. *Journal of Clinical Microbio* **44**: 1963-1970, 2006.

416. Dworzanski JP, Snyder AP, Chen R, Zhang H, Wishart D and Li L, ID of bacteria using MS/MS combined with a proteome database and statistical scoring. *Anal Chem* **76**: 2355-2366, 2004.

417. Yao ZP, Demirev PA and Fenselau C, Mass spectrometry-based proteolytic mapping for rapid virus identification. *Anal Chem* **74**(11)**: 2529-34., 2002.

418. Yao ZP, Afonso C and Fenselau C, Rapid microorganism ID with on-slide proteolytic digestion followed by MALDI MS &database searching. *Rapid Commun Mass Spectrom* **16**: 1953-6., 2002.

419. Pribil P and Fenselau C, Characterization of Enterobacteria Using MALDI-TOF Mass Spectrometry. *Analytical Chemistry* **77**(18)**: 6092-6095, 2005.

420. Valentine NB, Wahl JH, Kingsley MT and Wahl KL, Direct surface analysis of fungal species by matrix-assisted laser desorption/ionization MS. *Rapid Commun Mass Spectrom* **16**: 1352-7., 2002.

421. Pribil PA, Patton E, Black G, Doroshenko V and Fenselau C, Rapid characterization of Bacillus spores targeting species-unique peptides AP MALDI MS. *J Mass Spectrom* **40**: 464-474, 2005.

422. Verberkmoes N, Hervey W, Shah M, Land M, Hauser L, Larimer F, Van Berkel G and Goeringer D, Evaluation of shotgun proteomics for ID of biological threat agents. *Anal Chem* **77**: 923-32, 2005.

423. Fox A, Mass spectromery for species of strain identification after culture or without culture: past, present, and future. *Journal of Clinical Microbiology* **44**(8)**: 2677-2680, 2006.

424. Ecker JA, Eshoo MW and etal, Identification of Acinetobacter species and genotyping of Acinetobacter baumannii by multilocus PCR and ESI MS. *J Clinical Microbiology* **44**: 2921-2932, 2006.

425. Williams TL, Andrzejewski D, Lay JO and Musser SM, Experimental factors affecting the quality and reproducibility of MALDI TOF MS of whole bacteria cells. *J Amer Soc Mass Spectrom* **14**: 342-351, 2003.

426. Meetani MA and Voorhees KJ, MALDI MS of High Molecular Weight Proteins from Whole Bacterial Cells: Pretreatment of Samples with Surfactants. *J Amer Soc Mass Spectrom* **16**: 1422-1426, 2005.

427. Lee H, Williams SKR, Wahl KL and Valentine NB, Analysis of whole bacterial cells by flow field-flow fractionation and MALDI TOFMS. *Analytical Chemistry* **75**(11)**: 2746-2752, 2003.

428. Jones JJ, Stump MJ, Fleming RC, Lay JO, Jr. and Wilkins CL, Investigation of MALDI-TOF and FT-MS techniques for analysis of Escherichia coli whole cells. *Anal Chem* **75**(6)**: 1340-7., 2003.

429. Demirev PA, Feldman AB, Kowalski P and Lin JS, Top-Down Proteomics for Rapid Identification of Intact Microorganisms. *Analytical Chemistry* **77**(22)**: 7455-7461, 2005.

430. Afonso C and Fenselau C, Use of bioactive glass slides for matrix-assisted laser desorption/ionization analysis: application to microorganisms. *Anal Chem* **75**(3)**: 694-7., 2003.

431. Harris WA and Reilly JP, On-probe digestion of bacterial proteins for MALDI-MS. *Anal Chem* **74**(17)**: 4410-6., 2002.

432. Tao L, Yu X, Snyder AP and Li L, Bacterial Identification by Protein Mass Mapping Combined with an Experimentally Derived Protein Mass Database. *Analytical Chemistry* **76**(22)**: 6609-6617, 2004.

433. Hollemeyer K, Jager S, Altmeyer W and Heinzle E, Proteolytic peptide patterns as indicators for fungal infections and nonfungal affections of human nails by MALDI TOFMS. *Anal Biochem* **338**: 326-331, 2005.

434. Steele PT, Tobias HJ, Fergenson DP, Pitesky ME, Horn JM, Czerwieniec GA, Russell SC, Lebrilla CB, Gard EE and Frank M, Laser power in bacterial spore bioaerosol MS. *Anal Chem* **75**: 5480-5487, 2003.

435. Steele PT, Srivastava A, Pitesky ME, Fergenson DP, Tobias HJ, Gard EE and Frank M, Desorption/Ionization Fluence Thresholds and Improved MS Consistency Using a Flattop Laser Profile in Bioaerosol MS of Single Bacillus Endospores. *Analytical Chemistry* **77**: 7448-7454, 2005.

436. Wahl KL, Wunschel SC, Jarman KH, Valentine NB, Petersen CE, Kingsley MT, Zartolas KA and Saenz AJ, Analysis of microbial mixtures by MALDI TOFMS. *Anal Chem* **74**: 6191-9., 2002.

437. Wunschel SC, Jarman KH, Petersen CE, Valentine NB, Wahl KL, Schauki D, Jackman J, Nelson CP and White E, Bacteria by MALDI: inter-lab comparison. *J Amer Soc Mass Spectrom* **16**: 456-462, 2005.

438. Pineda FJ, Antoine MD, Demirev PA, Feldman AB, Jackman J, Longenecker M and Lin JS, Microbe ID by MALDI MS & Model-Derived Ribosomal Biomarkers. *Anal Chem* **75**: 3817-3822, 2003.

439. Taranenko NI, Hurt R, Zhou JZ, Isola NR, Huang H, Lee SH and Chen CH, Laser desorption mass spectrometry for microbial DNA analysis. *J Microbiol Methods* **48**(2-3)**: 101-6., 2002.

440. Chen YC, Li TY and Tsai MF, Analysis of oral cancer patient saliva by MALDI TOF MS. *Rapid Commun Mass Spectrom* **16**: 364-9., 2002.

441. Demirev PA, Ramirez J and Fenselau C, Tandem mass spectrometry of intact proteins for characterization of biomarkers from Bacillus cereus T spores. *Anal Chem* **73**(23)**: 5725-31., 2001.

442. Biroccio A, Urbani A, Massoud R, di Ilio C, Sacchetta P, Bernardini S, Cortese C and Federici G, Quantitate glycated & glutathionylated Hb by MALDI TOF MS. *Anal Biochem* **336**: 279-288, 2005.

443. Pevsner PH, Naftolin F, Hillman DE, Miller DC, Fadiel A, Kogus A, Stern A and Samuels HH, Direct ID of proteins from T47D cells and murine brain tissue by MALDI-PSD or CAD. *Rapid Commun Mass Spectrom* **21**: 429-436, 2007.
444. Peacock PM and McEwen CN, MS of Synthetic Polymers. *Anal Chem* **78**: 3957-3964, 2006.
445. McEwen CN, Recent developments in polymer characterization using mass spectrometry. *Advances in Mass Spectrometry* **16**: 215-227, 2004.
446. Peacock PM and McEwen CN, MS of synthetic polymers. *Anal Chem* **76**(12): 3417-28., 2004.
447. Lattimer GMaR, Mass Spectrometry of Polymers. CRC Press, Inc., Boca Raton, FL, 2002.
448. Murgasova R and Hercules DM, MALDI of synthetic polymers-an update. *International Journal of Mass Spectrometry* **226**(1): 151-162, 2003.
449. Montaudo G, Samperi F, Montaudo MS, Carroccio S and Puglisi C, Current trends in matrix-assisted laser desorption/ionization of polymeric materials. *European Journal of Mass Spectrometry* **11**(1): 1-14, 2005.
450. Wesdemiotis C, MALDI-TOF Mass Spectrometry of Synthetic Polymers. *Analytical and Bioanalytical Chemistry* **381**(7): 1317-1318, 2005.
451. Hanton SD, Owens KG, Chavez-Eng C, Hoberg A-M and Derrick PJ, Updating evidence for cationization of polymers in the gas phase during MALDI. *Eur J Mass Spectrom* **11**: 23-29, 2005.
452. Chen H and He M, Quantitation of synthetic polymers using an internal standard by MALDI TOF MS. *J Amer Soc Mass Spectrom* **16**: 100-106, 2005.
453. Hanton SD, Hyder IZ, Stets JR, Owens KG, Blair WR, Guttman CM and Giuseppetti AA, Investigations of electrospray sample deposition for polymer MALDI MS. *J Amer Soc Mass Spectrom* **15**: 168-179, 2004.
454. Hanton SD and Parees DM, Extending the solvent-free MALDI sample preparation method. *Journal of the American Society for Mass Spectrometry* **16**(1): 90-93, 2005.
455. Trimpin S, Keune S, Raeder HJ and Muellen K, Solvent-Free MALDI-MS: Analysis of Synthetic Polymers and Giant Organic Molecules. *J Amer Soc Mass Spectrom* **17**: 661-671, 2006.
456. Trimpin S and McEwen CN, Multisample Preparation Methods for the Solvent-free MALDI-MS Analysis of Synthetic Polymers. *J Amer Soc Mass Spectrom* **18**: 377-381, 2007.
457. Trimpin S and Deinzer ML, Solvent-Free MALDI-MS for the Analysis of a Membrane Protein via the Mini Ball Mill Approach: Case Study of Bacteriorhodopsin. *Anal Chem* **79**: 71-78, 2007.
458. Arnould MA, Polce MJ, Quirk RP and Wesdemiotis C, Probing chain-end functionalization reactions in living anionic polymerization via MALDI TOF MS. *Int J Mass Spectrom* **238**: 245-255, 2004.
459. Cai Y, Peng WP, Kuo SJ, Sabu S, Han CC and Chang HC, Optical detection and charge-state analysis of MALDI-generated particles with molecular masses larger than 5 MDa. *Anal Chem* **74**(17): 4434-40., 2002.
460. Im K, Park S, Cho D, Chang T, Lee K and Choi N, HPLC and MALDI-TOF MS Analysis of Highly Branched Polystyrene: Resolution Enhancement by Branching. *Anal Chemi* **76**: 2638-2642, 2004.
461. Cox FJ, Johnston MV and Dasgupta A, Characterization and relative ionization efficiencies of end-functionalized polystyrenes by MALDI MS. *J Amer Soc Mass Spectrom* **14**: 648-657, 2003.
462. Wetzel SJ, Guttman CM and Girard JE, The influence of matrix and laser energy on the molecular mass distribution of synthetic polymers obtained by MALDI-TOF-MS. *Int J Mass Spectrom* **238**: 215-225, 2004.
463. Wetzel SJ, Guttman CM, Flynn KM and Filliben JJ, Significant Parameters in the Optimization of MALDI-TOF-MS for Synthetic Polymers. *J Amer Soc Mass Spectrom* **17**: 246-252, 2006.
464. Wallace WE and Blair WR, MALDI MS of covalently cationized polyethylene as a function of sample temperature. *Int J Mass Spectrom* **263**: 82-87, 2007.
465. Meyer T, Kunkel M, Frahm AW and Waidelich D, Residue mass plot and abundance plot: isobaric interferences in DE-MALDI-TOF MS of polymer mixtures. *J Am Soc Mass Spectrom* **12**: 911-25., 2001.
466. McEwen CN and Peacock PM, MS of chemical polymers. *Anal Chem* **74**(12): 2743-8., 2002.
467. Liu HMD and Schlunegger UP, MALDI of Synthetic Polymers with Azo Compound Matrices. *Rapid Commun. Mass Spectrom.* **10**: 483-489, 1996.
468. Williams JB, Chapman TM and Hercules DM, MALDI MS of Discrete Mass Poly(butylene glutarate) Oligomers. *Analytical Chemistry* **75**: 3092-3100, 2003.
469. Chen H, He M, Pei J and He H, Quantitation of Synthetic Polymers by MALDI TOF MS. *Anal Chem* **75**: 6531-6535, 2003.
470. Murgasova R and Hercules DM, Quantitative Characterization of a Polystyrene/Poly(a-methylstyrene) Blend by MALDI MS and Size-Exclusion Chromatography. *Anal Chem* **75**: 3744-3750, 2003.
471. Terrier P, Buchmann W, Cheguillaume G, Desmazieres B and Tortajada J, Analysis of Poly(oxyethylene) and Poly(oxypropylene) Triblock Copolymers by MALDI-TOF MS. *Anal Chem* **77**: 3292-3300, 2005.
472. Keki S, Nagy M, Deak G, Miklos Z and Herczegh P, MALDI MS study of bis(imidazole-1-carboxylate) endfunctionalized polymers. *J Amer Soc Mass Spectrom* **14**: 117-123, 2003.
473. Hanton SD, Parees DM and Owens KG, MALDI PSD of low molecular weight ethoxylated polymers. *International Journal of Mass Spectrometry* **238**(3): 257-264, 2004.
474. Guttman CM, Wetzel SJ, Flynn KM, Fanconi BM, VanderHart DL and Wallace WE, MALDI TOF MS Interlab Comparison of Polystyrenes with Different End Groups. *Anal Chem* **77**: 4539-4548, 2005.
475. Kaufman JM, Jaber AJ, Stump MJ, Simonsick WJ and Wilkins CL, Interference from multiple cations in MALDI-MS spectra of copolymers. *International Journal of Mass Spectrometry* **234**(1-3): 153-160, 2004.

476. Jackson AT, Scrivens JH, Williams JP, Baker ES, Gidden J and Bowers MT, Microstructural and conformational studies of polyether copolymers. *Int J Mass Spectrom* **238**: 287-297, 2004.

477. Gies AP and Nonidez WK, A Technique for Obtaining MALDI TOF Mass Spectra of Poorly Soluble and Insoluble Aromatic Polyamides. *Anal Chem* **76**: 1991-1997, 2004.

478. Gies AP, Nonidez WK, Ellison ST, Ji H and Mays JW, A MALDI-TOF MS Study of Oligomeric Poly(m-phenyleneisophthalamide). *Analytical Chemistry* **77**(3): 780-784, 2005.

479. Mowat IA and Donovan RJ, Metal-ion Attachment to Non-polar Polymers during MALDI. *Rapid Commun. Mass Spectrom.* **9**: 82-90, 1995.

480. Knochenmuss R, Lehmann E and Zenobi R, Polymer cationization in MALDI. *Eur. Mass Spectrom.* **4**: 421-427, 1998.

481. Hanton SD and Owens KG, Using MESIMS to Analyze Polymer MALDI Matrix Solubility. *Journal of the American Society for Mass Spectrometry* **16**(7): 1172-1180, 2005.

482. Liu XM, Maziarz EP, Heiler DJ and Grobe GL, Comparison of poly(dimethyl siloxanes) using automated GPC-MALDI-TOF MS and on-line GPC-ESI-TOF MS. *J Amer Soc Mass Spectrom* **14**: 195-202, 2003.

483. Marie A, Alves S, Fournier F and Tabet JC, Fluorinated matrix approach for the characterization of hydrophobic perfluoropolyethers by MALDI TOFMS. *Anal Chem* **75**: 1294-9., 2003.

484. Planeta J, Rehulka P and Chmelik J, Sample deposition device for off-line combination of SFC and MALDI TOFMS. *Anal Chem* **74**: 3911-4., 2002.

485. Yalcin T, Wallace WE, Guttman CM and Li L, Metal powder substrate-assisted laser desorption/ionization mass spectrometry for polyethylene analysis. *Anal Chem* **74**(18): 4750-6., 2002.

486. Lidgard R and Duncan MW, MALDI of Low Molecular Weight Compounds. *Rapid Commun. Mass Spectrom.* **9**: 128-132, 1995.

487. Goheen SC, Wahl KL, Campbell JA and Hess WP, Mass Spectrometry of Low Molecular Mass Solids by Matrix-assisted Laser Desorption/Ionization. *J Mass Spectrom* **32**(8): 820-828, 1997.

488. Debre O, Budde WL and Song X, Negative ion electrospray of bromo- and chloroacetic acids and an evaluation of exact mass measurements TOF MS. *J Am Soc Mass Spectrom* **11**: 809-21., 2000.

489. Tholey A, Wittmann C, Kang MJ, Bungert D, Hollemeyer K and Heinzle E, Derivatization of small biomolecules for optimized MALDI MS. *J Mass Spectrom* **37**: 963-73., 2002.

490. Lee PJ, Chen W and Gebler JC, Qualitative and quantitative analysis of small amine molecules by MALDI-TOF mass spectrometry through charge derivatization. *Anal Chem* **76**(16): 4888-4893, 2004.

491. Herod AA and et al., MALDI-MS of Pitch Fractions. *Rapid Commun. Mass Spectrom.* **10**: 171-177, 1996.

492. Fye JL, Nelson HH, Mowery RL, Baronavski AP and Callahan JH, Scanning UV Two-Step Laser MS of PAH Distributions on Creosote-Contaminated Soil Particles. *Anal Chem* **74**: 3019 -3029, 2002.

493. Hunnam V, Harvey DJ, Priestman DA, Bateman RH, Bordoli RS and Tyldesley R, Neutral & acidic glycosphingolipids by Q-TOF MS & MALDI. *J Am Soc Mass Spectrom* **12**: 1220-5., 2001.

494. Sullivan AG and Gaskell SJ, The Analysis of Polysulfonated Azo Dyestuffs by MALDI and Electrospray Mass Spectrometry. *Rap. Commun. Mass Spectrom.* **11**: 803-809, 1997.

495. Grim DM, Siegel J and Allison J, Evaluation of desorption/ionization mass spectrometric methods in the forensic applications of the analysis of inks on paper. *J Forensic Sci* **46**(6): 1411-20., 2001.

496. Al-Saad KA, Siems WF, Hill HH, Zabrouskov V and Knowles NR, Structural analysis of phosphatidylcholines by MALDI-TOFMS PSD. *J Amer Soc Mass Spectrom* **14**: 373-382, 2003.

497. Cai Y, Jiang Y and Cole RB, Anionic adducts of oligosaccharides by matrix-assisted laser desorption/ionization time-of-flight mass spectrometry. *Anal Chem* **75**(7): 1638-44., 2003.

498. Sekiya S, Wada Y and Tanaka K, Derivatization for Stabilizing Sialic Acids in MALDI-MS. *Analytical Chemistry* **77**(15): 4962-4968, 2005.

499. Wang Y, Hornshaw M, Alvelius G, Bodin K, Liu S, Sjövall J and Griffiths WJ, MALDI High-Energy CAD of Steroids: Oxysterols in Rat Brain. *Anal Chem* **78**: 164 - 173, 2006.

500. Kotsiris SG, Vasil'ev YV, Streletskii AV, Han M, Mark LP, Boltalina OV, Chronakis N, Orfanopoulos M, Hungerbuhler H and Drewello T, Application and evaluation of solvent-free MALDI MS of derivatized fullerenes. *European Journal of Mass Spectrometry* **12**(6): 397-408, 2006.

501. Tang K, Alliman SL, Jones RB and Chen CH, Quantitative MALDI-MS with Internal Standard. *Anal. Chem.* **65**: 2164-2166, 1993.

502. Elsila JE, de Leon NP and Zare RN, Factors affecting quantitative analysis in laser desorption/laser ionization mass spectrometry. *Analytical Chemistry* **76**(9): 2430-2437, 2004.

503. Bucknall M, Fung KY and Duncan MW, Practical quantitative biomedical applications of MALDI-TOF mass spectrometry. *J Am Soc Mass Spectrom* **13**(9): 1015-27., 2002.

504. Jorgensen M, Bergkvist KSG and Welinder KG, Quantification of defensins by MALDI-TOFMS. *Anal Biochem* **358**: 295-297, 2006.

505. Hlongwane C, Delves IG, Wan LW and Ayorinde FO, Comparative fatty acid analysis of triacylglycerols using MALDI TOF MS and GC. *Rap. Commun. Mass Spectrom.* **15**: 2027-2034, 2001.

506. Li G, Hu R, Kamijo Y, Nakajima T, Aoyama T, Inoue T, Node K, Kannagi R, Kyogashima M and Hara A, Establishment of a quantitative, qualitative, and high-throughput analysis of sulfatides from small amounts of sera by MALDI TOFMS. *Anal Biochem* **362**: 1-7, 2007.

507. Yu H, Lopez E, Young SW, Luo J, Tian H and Cao P, Quantitative analysis of free fatty acids in rat plasma using matrix-assisted laser desorption/ionization time-of-flight mass spectrometry with meso-tetrakis porphyrin as matrix. *Analytical Biochemistry* **354**(2): 182-191, 2006.

508. Bungert D, Heinzle E and Tholey A, Quantitative matrix-assisted laser desorption/ionization mass spectrometry for the determination of enzyme activities. *Analytical Biochemistry* **326**(2): 167-175, 2004.

509. Li YL and Gross ML, Ionic-liquid matrices for quantitative analysis by MALDI-TOF mass spectrometry. *Journal of the American Society for Mass Spectrometry* **15**(12): 1833-1837, 2004.

510. Fei X and Murray KK, On-Line Coupling of Gel Permeation Chromatography with MALDI Mass Spectrometry. *Anal. Chem.* **68**: 3555-3560, 1996.

511. Hiraoka K, Laser spray: electric field-assisted MALDI. *J Mass Spectro* **39**: 341-350, 2004.

512. Zhang B, McDonald C and Li L, Combining Liquid Chromatography with MALDI Mass Spectrometry Using a Heated Droplet Interface. *Analytical Chemistry* **76**(4): 992-1001, 2004.

513. Zhang N, Li N and Li L, Liquid chromatography MALDI MS/MS for membrane proteome analysis. *Journal of Proteome Research* **3**(4): 719-727, 2004.

514. Perlman DH, Huang H, Dauly C, Costello CE and McComb ME, Coupling of Protein HPLC to MALDI-TOF MS Using an On-Target Device for Fraction Collection, Concentration, Digestion, Desalting, and Matrix/Analyte Cocrystallization. *Anal Chem* **79**: 2058-2066, 2007.

515. Ro KW, Liu J and Knapp DR, Plastic microchip LC-MALDI MS using monolithic columns. *J Chromatography, A* **1111**(1): 40-47, 2006.

516. Zhen Y, Xu N, Richardson B, Becklin R, Savage JR, Blake K and Peltier JM, Development of an LC-MALDI method for the analysis of protein complexes. *J Am Soc Mass Spectrom* **15**(6): 803-22., 2004.

517. Musyimi HK, Narcisse DA, Zhang X, Stryjewski W, Soper SA and Murray KK, Online CE-MALDI-TOF MS Using a Rotating Ball Interface. *Analytical Chemistry* **76**(19): 5968-5973, 2004.

518. Musyimi HK, Guy J, Narcisse DA, Soper SA and Murray KK, Direct coupling of polymer-based microchip electrophoresis to online MALDI-MS using a rotating ball inlet. *Electrophoresis* **26**(24): 4703-4710, 2005.

Chapter 10   Gas Chromatography/Mass Spectrometry

Introduction to Mass Spectrometry, 4th Edition: Instrumentation, Applications, and Strategies for Data Interpretation; J.T. Watson and O.D. Sparkman, © 2007, John Wiley & Sons, Ltd

I. Introduction

Gas chromatography/mass spectrometry (GC/MS) is a combination of two microanalytical techniques: gas chromatography (GC), a separation technique, and mass spectrometry (MS), an identification technique. This combination has several advantages [1, 2]. First, it separates components of a complex mixture so that mass spectra of individual compounds can be obtained for qualitative purposes; second, it can provide quantitative information on these same compounds. Because sample volatility is a requirement for GC, ionization techniques for MS are restricted to those that require gas-phase analytes. These techniques are electron ionization (EI), chemical ionization (CI), reduced-pressure and atmospheric pressure CI (APCI), electron capture negative ionization (ECNI), and field ionization (FI). GC/MS can provide a complete mass spectrum from a few femtomoles of an analyte; ideally, this spectrum gives direct evidence for the nominal mass and, in the case of EI, provides a characteristic fragmentation pattern or "chemical" fingerprint that can be used as the basis for identification.

The GC/MS combination overcomes certain deficiencies or limitations caused by using each technique individually. For example, in using MS alone to analyze an impure sample, superimposed mass spectra of two or more compounds are obtained; the result can be confusing, if not uninterruptible. On the other hand, analysis of an impure sample by GC/MS will separate the impurities from the analyte and permit the mass spectrum of each of the components to be obtained individually. The use of GC alone may lead to questionable results because of ambiguities in relying only on a retention index for identification of various sample components.

Even though the technique of GC/MS is considered to be a combination of gas chromatography and mass spectrometry, it should always be kept in mind that GC/MS is as different from either gas chromatography or mass spectrometry as gas chromatography and mass spectrometry are from one another. The gas chromatograph-mass spectrometer (GC-MS) should not be considered as just a GC with an MS as the detector or as an MS with the GC as an inlet system. As will be discussed in this chapter in more detail, the differences in pressure requirements of the two instruments is a paramount consideration in the operation of the GC-MS.

As a reminder from Chapter 1, when discussing the data obtained with GC/MS, the word "peak" should never be used without the modifying adjective of "chromatographic" or "mass spectral". Following this rule will avoid confusion during discussions or in descriptions of GC/MS data.

GC/MS had its beginning shortly after Anthony Trafford James and Archer John Porter Martin, at the National Institute for Medical Research in London, UK, introduced gas chromatography in 1951 [3, 4] based on work done by Martin and Richard Laurence Millington Synge in 1941 [5, 6] for which they were awarded the Nobel Prize in chemistry in 1952 for the invention of partition chromatography. There has always been a controversy as to who first coupled a gas chromatograph to a mass spectrometer. Joseph C. Holmes and Francis A. Morrell at Philip Morris, Inc. (a cigarette manufacturer) in Richmond, Virginia, published a paper in 1957 on the interfacing of a GC to a Consolidated Engineering Corporation (CEC) Model 2-103B magnetic-sector mass spectrometer [7]. This work had been presented at the Fourth Annual Meeting of ASTM Committee E-14 on *Mass Spectrometry* in Cincinnati, Ohio, in May of 1956. In a 1993 remembrance of the first use of GC/MS [8], Fred McLafferty states that he and Roland S. Gohlke (both at Dow Chemical Company in Midland, Michigan, at that time) first presented their work on interfacing a GC to a time-of-flight mass spectrometer at the

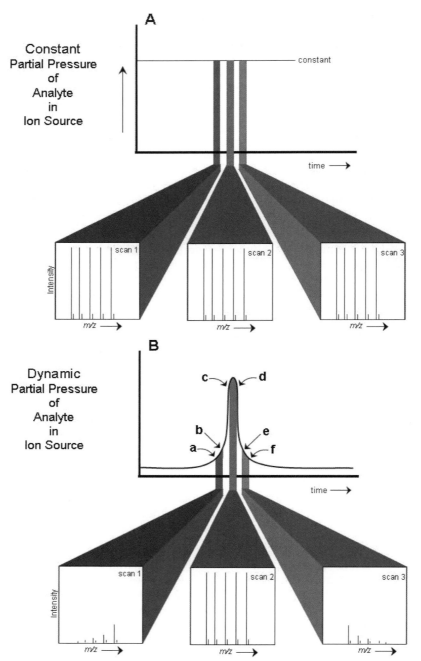

Figure 10-1.   Conceptual illustration of the effect of the chromatographic process on the mass spectra obtained by GC/MS. The top panel illustrates the spectra being acquired while the partial pressure of the analyte is held constant in the ion source. The bottom panel shows how the spectra change in appearance as the partial pressure of the analyte changes during elution of the sample into the source.

129th National American Chemical Society Meeting in April of 1956 in a Symposium on *Vapor Phase Chromatography*; however, the first real report in the literature appeared in the April 1959 issue of *Analytical Chemistry* almost a year after its receipt on May 31, 1958 [9], by the journal and three years after McLafferty claims it was first presented. This issue of Gohlke/McLafferty vs Holmes/Morrell is a little reminiscent of the Tanaka vs Hillenkamp/Karas situation with MALDI and the Nobel Prize in chemistry. The work reported by Gohlke was much more extensive than that reported by Holmes and Morrell. Gohlke and McLafferty went on to publish extensively about the application of GC/MS, whereas Holmes and Morrell never published additional information after their original report.

One of the primary differences between the mass spectral aspect of GC/MS and that of conventional mass spectrometry is the continuous variation in partial pressure of the analyte during the acquisition of all the spectra representing the analyte and the individual spectra. These differences are illustrated in Figure 10-1. These differences can be a significant factor in using EI mass spectra for the identifications of analytes and should always be considered.

II. Introduction to GC

The operating principles of GC [10, 11] are summarized in Figure 10-2. The sample represented at the left of the diagram is a mixture of A and B dissolved in solvent X. The sample solution is vaporized when it is injected into the GC. The mobile phase carries the vapor onto the column where it is partitioned between an organic stationary phase [12] and the mobile phase. The mobile phase (an inert carrier gas) moves the sample components through the column. The solvent X interacts very little (ideally not at all) with the stationary phase; thus it is the first component to emerge from the column. As A and B travel through the column, B has a greater solubility in the stationary phase than does A; this means that B is retained longer by the column than A. As a consequence, A emerges from the column before B (i.e., A has a shorter retention time than B). The detector at the end of the column produces an electrical response when an organic substance, in addition to the carrier gas, passes through it. A recording of the detector response vs time produces a chromatogram (an analog recording), as shown at the right in Figure 10-2. In the case of GC/MS, a pure component exits from the column into the ion source of the mass spectrometer.

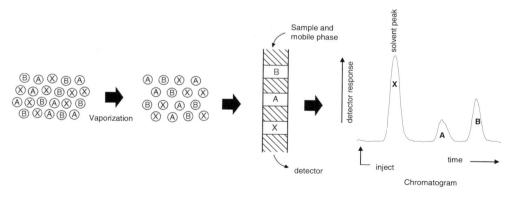

Figure 10-2. Conceptual operational features of a gas chromatograph.

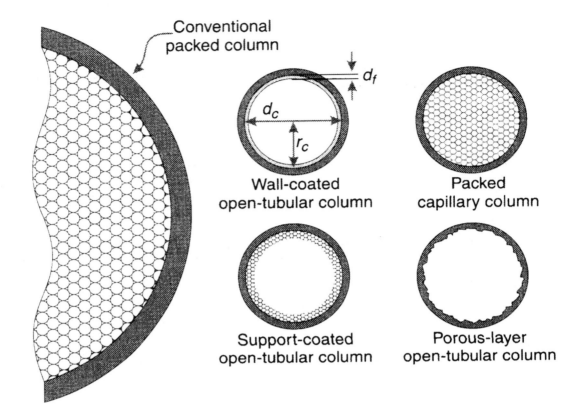

Figure 10-3. *A comparative representation of a 2-mm i.d. packed column to 0.53 mm i.d. open-tubular columns.* Reproduced with permission from *Introduction to Open-Tubular Column Gas Chromatography*, Hinshaw JV and Ettre LS Advanstar Communications, Cleveland, OH, 1994.

The stationary phase in GC is either a solid such as a molecular sieve (naturally occurring zeolites or synthetic substances such as alkali metal aluminosilicates), silica gel, alumina, or porous polymers like the Porapaks® or viscous high-boiling liquids such as organopolysiloxanes SE-30 or substances such as Carbowax or Apiezon. The solid stationary phases are used in what is sometimes referred to as gas–solid chromatography (GSC). GSC is the technique used for separation of permanent gases in packed columns or the porous-layer open-tubular (PLOT) columns. The mechanism for GSC usually involves *adsorption* (sorption of the analyte on the surface of the stationary phase). When high-boiling liquids are used as the stationary phase, the term gas–liquid chromatography (GLC) is sometimes used. The mechanism for GLC usually involves *absorption* (sorption of the analyte into the stationary phase). The liquid stationary phase can be applied to the walls of the column as in the case of the wall-coated open-tubular (WCOT) or capillary column or it can be applied to a solid coating the walls of a capillary column as in the case of the support-coated open-tubular (SCOT) column. The liquid phase can also be coated onto an inert support such as diatomaceous earth, and this material is then put into a packed column. A graphical illustration of the different types of columns is provided in Figure 10-3.

When the term GC is used, it is probably referring to the absorption process of GLC. That is how the term GC is used in this book.

Some of the features of a typical gas chromatogram are illustrated in Figure 10-4. The time required for an unretained compound (the solvent in the above description) to pass through the column is represented by t_0 where t_0 is related to the volume of the column and the flow rate of the carrier gas. The retention time, t_{RA}, is the time required for substance A to pass through the column. The adjusted retention time, t'_{RA}, is the difference between t_{RA} and t_0, and it is an indication of the degree of solubility of substance A in the stationary phase. There are a number of terms used with respect to relative retention times such as *Kovats Retention Index* and *Rohrschneider–McReynolds Constants*. A detailed discussion of these terms is beyond the scope of this book. There are several very good introductory references for modern gas chromatography [10, 13, 14].

The vertical axis of a gas chromatogram represents the magnitude of the detector response, which is proportional to the amount of substance present. The horizontal axis shows the retention time, which can be used as a qualitative characteristic for a given compound.

Modern GC/MS instrumentation mostly uses capillary columns sometimes called open-tubular columns or WCOT columns as pointed out above. These columns are narrow-diameter (100–750 µm) tubes constructed of glass, fused silica, or an inert metal (e.g., nickel). The most common material is the fused silica developed by Hewlett-Packard ca. 1977. These tubes are of varying length (2–60 m or more). The minimum length used in GC/MS is about 30 m. Shorter lengths cause problems due to

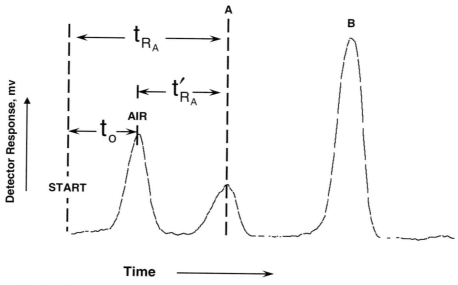

Figure 10-4. *Format of the gas chromatogram, the data record in GC.*

low pressure along a major fraction of the partitioning path because the exit of the column is under vacuum. In the WCOT column, the walls are coated with a high-boiling liquid or a chemically immobilized polymer, often an organosiloxane polymer that serves as the stationary phase. The organosiloxane polymer can have a large number of methyl groups mixed with other functional groups to give separation specificity for various types of compounds.

Columns with internal diameters of 320 μm or less typically have optimized flow rates of less than 2 mL min^{-1}. These columns can be directly inserted into the mass spectrometer [15]. Larger diameter columns (530 μm and 750 μm) require optimized flow rates much higher than 2 mL min^{-1} and must have some type of interface device. Packed columns are tubes that have internal diameters between 2 and 6 mm, sometimes as large as 12 mm. The length is between 1 and 3 meters, but may be as long as 15 m. The same stationary phase (high-boiling liquid) as used on the walls of capillary columns is used to coat a solid support at a level of 1% (wt/wt) or greater; this coated solid is then packed into the column to serve as the stationary phase. The diameter of the solid support particles is usually a function of the internal diameter of the tubing being used. The conventional packed GC columns require flow rates in excess of 20 mL min^{-1}. These types of flow rates are too high for modern pumping systems, and the packed columns are no longer used in modern GC/MS. Today, GC/MS applications are based on the use of WCOT, SCOT, or packed capillary columns that use flow rates that are compatible with modern vacuum systems. Packed capillary columns usually have an internal diameter greater than 500 μm.

There are many factors that go into the selection of the GC column. Among these are the stationary phase, the diameter and length of the column, and the thickness of the stationary phase on the column (stationary-phase loading). All of these parameters should be reported with GC/MS results.

Optimal column performance is dependent on the flow rate of carrier gas, the diameter of the column, and the film thickness of the stationary phase. The flow rate is a function of the column's inner diameter and the linear velocity of the mobile phase. The efficiency of a chromatographic column is expressed in terms of the height equivalent to a theoretical plate (HETP); i.e., that length of column required to achieve equilibrium in the partitioning of the analyte between the stationary and mobile phases. The shorter the segment (HETP) of column required to achieve this equilibrium, the more efficient the column. A typical plot of HETP as a function of linear velocity of the carrier gas is shown in Figure 10-5. Fortunately, not only do helium or hydrogen give the most efficient operation of a GC column (lower values for HETP) but they exhibit a more gradual change in the HETP with changing linear velocity than nitrogen; therefore, the two primary mobile phases in GC/MS are helium and hydrogen. It is important to keep in mind that most GC separations in GC/MS are accomplished with the use of temperature programming of the GC column. As the temperature of the mobile phase increases, the linear velocity of the gas through the column decreases because its viscosity increases. This means that the HETP will change during a temperature-programmed analysis. Examination of the van Deemter–Golay curves in Figure 10-5 shows that the efficiency of the column is less sensitive to changes in linear velocity when using hydrogen rather than helium; therefore, it would appear that hydrogen would be preferred over helium. However, because helium can be pumped better than hydrogen by turbomolecular pumps, helium is the preferred mobile phase in GC/MS because of consideration for the vacuum system in maintaining the appropriate pressure in the mass spectrometer. The pumping speeds for these three gases as listed in Table 10-1 indicate the basis for potential problems in maintaining a proper vacuum.

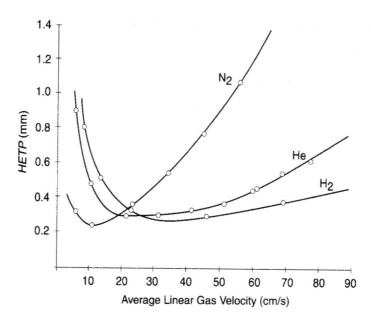

Figure 10-5. *Van Deemter–Golay curves for three common GC mobile phases.* Reproduced with permission from Hinshaw JV; Ettre LS Introduction to Open-Tubular Column Gas Chromatography, Advanstar Communications, Cleveland, OH, 1994, page 113.

Table 10-1. *Pumping speeds for various gases by the Alcatel MDP5030 CP 27 L sec^{-1} pump.*

Gas	Pumping Speed	Compression Ratio
N_2	27 L sec^{-1}	1×10^7
He	18 L sec^{-1}	1×10^3
H_2	15 L sec^{-1}	2×10^2

Although a ready reference for the pumping speed for He compared to those for N_2 and H_2 in an oil diffusion pump is not available, it is reasonable to assume that the speed is about the same for H_2 as for He based on the molar mass of the two gases and the operating principles of the oil diffusion pump. The pumping speed for H_2 appears to be about 35% greater than that for N_2 in oil diffusion pumps (200 vs 135, 500 vs 280, 1300 vs 700, and 3000 vs 2000 L sec^{-1} H_2 vs N_2). This is the reverse of the situation with the turbomolecular pump. Because the oil diffusion pump exhibits similar performance with both He and H_2, there should not be a problem using H_2 as a carrier gas with GC/MS instruments equipped with an oil diffusion pump.

Edwards manufactures a turbomolecular pump (the EXT series) that shows less than a 5% difference in pumping speed for N_2 and He (He > N_2) in the 40, 60, and 250 L sec^{-1} (N_2) models. These same pumps show a greater than 15% reduction in speed for H_2 compared to He or N_2. For 240 L sec^{-1} (N_2) pumps in this same series, the pumping speed for N_2 is ~4% less than for He, but over 20% greater for N_2 than H_2. A 295 L sec^{-1} (N_2) pump has 10% better pumping speed for He than N_2 and almost 30% poorer for H_2 vs N_2.

The design of the pump appears to have a lot to do with variations in pumping speeds based on the chemical composition of the gas. A pump that is reported to be equivalent to the 295 L sec^{-1} (with the exception of a shorter rotor) has a 300 L sec^{-1} speed for N_2, 220 L sec^{-1} for He (~27% less), and 170 L sec^{-1} for H_2 (~43% less). The reductions in speed are even greater for He and H_2 compared to N_2 for a 450 L sec^{-1} pump with this same change in rotor design (>50% for H_2).

Looking at pumping speed vs pressure curves for N_2, He, and H_2 in a turbomolecular pump, it appears that the maximum pumping speeds for N_2 and He are achieved well before a pressure of 10^{-1} Pa is established, whereas the maximum pumping speed for H_2 is not achieved until the pressure drops to at least 7.5×10^{-2} Pa.

Fluctuations in operating pressure, especially an increase, can affect the performance of the mass spectrometer. For example, a pressure increase in a quadrupole ion trap mass spectrometer due to decreased pumping speed could be somewhat problematic. In order to obtain unit resolution of ions, it is necessary to buffer the ions with a helium pressure of about 10^{-1} Pa. The decrease in pumping speed based on the chemical composition of the carrier gas could cause high pressures, which would in turn further reduce the pumping speed. It should be remembered that most commercial GC/MS systems are designed for optimum performance with helium as the carrier gas.

Another important consideration of the mobile phase is its purity. Even the highest purity commercially available carrier gas can be a problem. Always use some type of filter to remove water and hydrocarbons. The example showing the effect of a good filter (Figure 10-6A) is rather dramatic. This particular filter (Figure 10-6B) is now being provided with new instruments during installation by most of the major manufacturers. The filter is manufactured by Valco Instrument Company (VICI®) and is sold under the brand name *VICI Mat/Sen* (http://www.VICI.com).

A potential problem that has been reported with ultra-high-purity helium (5 and 6 nines; i.e., 99.999 or 99.9999%) is the presence of argon as a contaminant. If argon were present, the chromatogram shown in Figure 10-6B for the filter gas would show a much higher background. Each time a new cylinder of He is installed on a GC/MS system, use the mass spectrometer to test for the presence of argon.

Another solution to gas purity is the Valco heat-activated nonevaporative gettering alloy purifier. The getter alloy is enclosed in a replaceable sealed tube which is mounted in a heater. There are specific models for purification of helium (GC/MS) and nitrogen (LC/MS). The getter is activated by heating, which eliminates an oxide film on the surface. The gas to be purified diffuses through the getter particles. The heater is self-regulating in order to maintain an optimum temperature for the getter material and to prevent thermal runaway. As the gas filters through, the getter material adsorptively captures any contaminants. This heated getter system is considered to be the best of all gas purification systems; however, the results from the VICI Mat/Sen filters are pretty spectacular (Figure 10-6B).

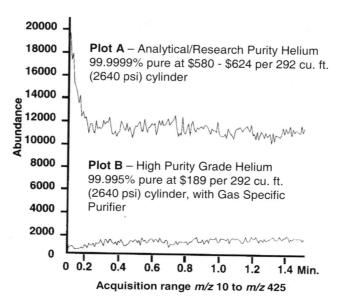

Plot A – Analytical/Research Purity Helium
99.9999% pure at $580 - $624 per 292 cu. ft.
(2640 psi) cylinder

Plot B – High Purity Grade Helium
99.995% pure at $189 per 292 cu. ft.
(2640 psi) cylinder, with Gas Specific
Purifier

Acquisition range *m/z* 10 to *m/z* 425

Figure 10-6A. *Comparative illustration of the effect of filtering GC carrier gas to reduce a background signal in the GC detector.*

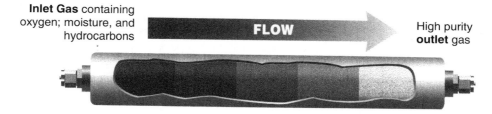

Inlet Gas containing oxygen; moisture, and hydrocarbons

FLOW

High purity **outlet** gas

Figure 10-6B. *Schematic representation of components in the gas filter.*

In GC/MS, the column is connected to a vacuum source, which can affect column performance compared to its use at atmospheric pressure with a GC detector such as flame ionization (FID). The effect of a vacuum at the exit of the column can alter retention times and increase chromatographic peak widths (i.e., decrease chromatographic resolution).

Frequently, the compound(s) of interest represent 10% or less of a mixture of substances that survive isolation or purification procedures (e.g., solvent extraction followed by a variety of liquid–liquid or liquid–solid partitioning systems, or both) and derivatization. After adequate preliminary purification and chemical modification (derivatization), if necessary, the GC can serve as the final purification step by resolving the various components in time and presenting them individually to the mass spectrometer for analysis. Although the chromatograph and mass spectrometer function at vastly different pressures, a compatible feature of these vapor-phase analytical techniques is that if the sample is sufficiently volatile for analysis by GC, it is also suitable for analysis by one of the gas-phase ionization techniques used in MS. Furthermore, the sample quantities required by the two instruments are approximately the same. The GC inlet is probably the most convenient and efficient means of fully utilizing the available sensitivity of a mass spectrometer.

1. Basic Types of Injectors

It is important to remember that the true inlet system in GC/MS is the injection port on the gas chromatograph. This is a frequently overlooked aspect of the integrated GC/MS system and deserves some attention here from the standpoint of definitions and comments on its usage. Injection techniques [16, 17] in capillary GC can dramatically affect qualitative and quantitative aspects of the results, especially if misused.

Split injection is the classical approach to introducing samples onto capillary GC columns [16]. Liquid samples are vaporized in a heated glass tube. A small portion of the combined solvent and solute vapor passes onto the capillary column. The majority of the vaporized injectate is discharged from the injector through the split vent. The split injection technique is an excellent means by which to insure initial placement of a discrete band of solute onto the stationary phase, and it is an effective way of introducing small samples that cannot reliably be manipulated with syringes designed for 0.1–1-µL volumes. Split ratios range from 20:1 to as high as 500:1 (1 being the part placed on the column). Analyte concentration should be high (around 1 µg µL^{-1}) when using split injections; this is not a high-sensitivity technique.

Splitless injection is a means of placing all or, at least most, of the sample into the column [16]. Usually, the GC injector is designed for either split or splitless operation. During the splitless mode of operation, the split exit port is closed so that the vaporized sample (i.e., solvent and solute) can be forced onto the column by carrier gas sweeping through the injection volume, a process that may take tens of seconds. Near the end of this process, the split outlet is reopened to purge the vaporization chamber after about 90% or more of the sample material has entered the column. This splitless injection technique is required for very low concentrations of analytes.

On-column injection deposits the entire solution from the syringe directly onto the entrance to the column or into an uncoated precolumn. During the injection process, the carrier gas redistributes the liquid from the syringe onto a film on the tubing wall. This mode of injection provides the most reliable or reproducible means of sample introduction to the GC column [16, 17]. Reproducibility is established by the fact that evaporation of the solvent occurs from the layer of sample liquid on the surface of the column. Both split and splitless on-column injections can be made. Often, on-column injections are made into cold injectors. The injector temperature is then programmed to first evaporate the solvent and then vaporize the analytes. In the cold-split injection mode, sample volumes as large as 100 µL can be accommodated; this allows for extremely low detection limits in terms of ppm [18].

Figure 10-7.   Schematic representation of a split/splitless injector.
This injector can be operated in the split or splitless mode.
There is no such thing as a "split/splitless injection".

2. Injection Considerations and Syringe Handling

There are a number of considerations that must be made with respect to the injection. Just because there appears to be liquid between the end of the syringe barrel and the tip of the plunger does not mean that the syringe contains the volume of liquid indicated by the marks on its calibrated barrel; remember that the needle has a significant dead volume. The most reliable way to assure reproducible injection volumes is through the use of the "sandwich" technique as described below.

After thoroughly cleaning the syringe with an appropriate solvent, pull the plunger back until a small volume of air is observed in the barrel. Then place the empty needle into the sample solution. Pull the plunger back further until its terminus passes through the desired sample volume (i.e., if the terminus is on the 0.5-µL mark when the needle is immersed into the sample, and the desired sample volume is 1.0 µL, pull the plunger terminus to the 1.5-µL mark). Next, remove the needle from the sample solution and pull the plunger tip back further until a second air gap (empty needle) is observed below the plug of sample already in the syringe barrel. This completes the "sandwich" of pure solvent, air, solution of the analyte, and air as illustrated in Figure 10-8. When the injection is made, the pocket of air in the lower part of the barrel and the needle ensure no fractionation of the sample (due to premature vaporization) during the time the needle penetrates the injector and before the plunger pushes the sample into the injector. The air bubble between the solvent plug and the plug of sample solution minimizes mixing and helps to ensure that no sample is left in the needle. The final solvent plug "washes" residual sample solution out of the needle. A variation of this technique is to place an air plug between the tip of the plunger and the plug of pure solvent. The "sandwich" injection technique is available on most modern GC autosamplers.

Important in all gas chromatography, but especially important in GC/MS, is the injector port cap. This device holds the septum, a disk of silicone rubber through which the syringe needle, as a means of maintaining an airtight seal, passes. It is important to keep air out of the GC column. The stationary phase can be damaged (oxidized) in the presence of air. Because the GC column is under a partial vacuum in a GC/MS system, if the septum becomes damaged, air can enter the column. After repeated injections, the septum will lose its structural integrity and allow the passage of air into the column. With repeated injections, the septum can be cored, a process by which a small piece of the septum material can fall into the injection port liner or even into the head of the column. When this happens, an increase in background signal is observed, the column can be further damaged, and the analytical integrity of the system becomes compromised. For this reason, it is important that the septum be changed after 50–100 injections (closer to 50 rather than 100). Septa also

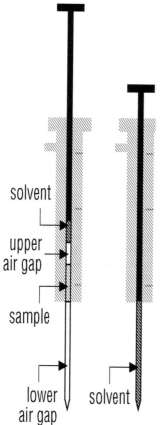

solvent

upper air gap

sample

lower air gap

solvent

Figure 10-8. *Conceptual illustration of "sandwich" injection technique.*

contribute to other problems. Because of inconsistencies in their manufacture, septa can "bleed", thereby contributing to background. Materials that do not readily volatilize can condense on the septum and contribute to the background through continuous bleed. Septa used with autosamplers are especially vulnerable to coring. For all of these reasons, it is best not to use a silicone septum in a GC/MS system. There are two types of septumless injector caps: the Merlin Microseal septumless injector cap and the JADE septumless injector cap (Figure 10-9).

Another important consideration is the type of sample. This will determine which type of injection liner should be used or whether a precolumn should be used. When the "sandwich" technique is not used, slow displacement of the plunger of the syringe may leave some liquid hanging at the needle-tip in the hot injection volume where the more volatile components preferentially evaporate. However, the higher-boiling solutes will be maintained in the hanging drop of solution and some may be removed from the sampling volume as the syringe is withdrawn from the injection port. This phenomenon results in a negative bias to the high-boiling components [16, 17].

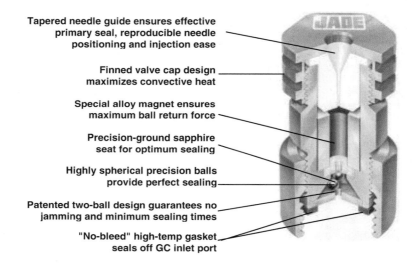

Tapered needle guide ensures effective primary seal, reproducible needle positioning and injection ease

Finned valve cap design maximizes convective heat

Special alloy magnet ensures maximum ball return force

Precision-ground sapphire seat for optimum sealing

Highly spherical precision balls provide perfect sealing

Patented two-ball design guarantees no jamming and minimum sealing times

"No-bleed" high-temp gasket seals off GC inlet port

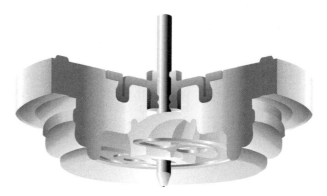

Figure 10-9.   Illustration of the Jade and Merlin Microseal septumless injector caps.

Injection speed is a very important factor in the type of results obtained in an analysis by GC/MS. The time the syringe needle resides in the injector before the plunger displaces the sample and the time the needle remains within the injector are other factors that can affect the quality of results. These and other injection techniques are described elsewhere in more detail [16, 19]. A split injection should be made as quickly as possible; this will avoid sample fractionation that can cause a disproportionate loss of high-boiling components. Fast-injection techniques are required for packed columns. Injections into hot columns (on-column injections) and splitless injections should be made slowly to avoid flashback through the syringe, which results in loss of low-boiling components. Most modern autosamplers offer some type of injection-speed control. Modern fast chromatography systems frequently rely on syringeless injection techniques [20].

Split injections give the best qualitative results in capillary column GC. Because the sample is deposited in a discrete volume at the head of the column, the analytes are partitioned into discrete bands resulting in well-resolved chromatographic peaks. Splitless and on-column injection techniques cause the solute to be distributed over a relatively large segment of the early column. This lack of discreteness results in a less-efficient separation and broader peaks in the chromatogram.

3. Syringeless Modes of Sample Injection for Fast GC

In the context of fast GC (complete chromatographic analyses in tens of seconds), the mechanics of conventional syringe operation become limiting to the overall performance. In such fast GC systems, modes of sample injection based on sample cryofocusing adsorption/thermal desorption show favorable temporal figures of merit [21].

Novel approaches to sample injection via microneedles [22] should lead to advances in sample handling and automated analyses. In this regard, the use of multiwalled carbon nanotubes show great promise for microscale solid-phase sampling and sample introduction for GC/MS [23]. Modern GC/MS instrumentation more often than not includes an autosampler, which allows for adjustment of all the parameters of sample size, rise steps, use of sandwich injection, and so forth.

III. Sample Handling

1. Proper Sample Container

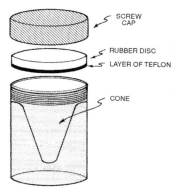

SCREW CAP

RUBBER DISC
LAYER OF TEFLON

CONE

Impurities in a sample may interfere with proper interpretation of the mass spectrum of the analyte. Some sources of contamination can be avoided by proper choice of storage vessels. One of the best means for storing samples for vapor-phase analysis is in a vessel like that illustrated schematically in Figure 10-10. There are two important features in the design of this container. First, a thin layer of Teflon is bonded to the side of the rubber septum that faces the interior of the container. The Teflon prevents most organic solvents from leaching any plasticizers or other contaminants

Figure 10-10. *Schematic illustration of a proper sampling vial.*

from the rubber disk. The rubber disk is soft and flexible to ensure a uniform seal around the top of the vessel when the screw-cap lid is attached. The seal is sufficient to allow heating or refluxing without loss of the vaporized sample or solvent. Second, the internal volume of the vial is in the shape of a cone to facilitate access to microliter quantities of sample.

Samples that are oily or that are dissolved in an organic solvent should not be stored in plastic containers or in conventional glass screw-cap bottles that have a plastic-coated liner glued to the inside of the lid. The problem is that lipid-soluble plasticizers may be extracted from the plastic container or the liner or the adhesive in the cap of the glass bottle. Under such conditions, these plasticizers can become major constituents in a sample; plasticizers such as dioctylphthalate have good vapor-phase properties and will produce attractive peaks in the gas chromatogram, and also produce a spectrum like that in Figure 10-11.

Sample losses caused by adsorption on glass surfaces can be minimized or eliminated by silanization.[1] Silanization [24] involves a chemical reaction between a silanizing reagent and polar groups (e.g., a hydroxy group, as indicated in Scheme 10-1). The silanizing reagent dimethyldichlorosilane (DMCS) can be used as a 5–10% solution in toluene to modify the glass surface chemically as it is poured into or through the vessel. After 5 to 10 min, the excess reagent is removed from the glass surface by rinsing it with toluene. Methanol is then added to convert the monochlorodimethylsilyl derivative to a methoxydimethylsilyl derivative; the methanol also assists in removing HCl liberated during the silylation reaction. Finally, the glass surface should be rinsed with toluene and dried. The silyl derivative is much less polar than the original hydroxyl group; thus, sample losses due to adsorption on silanized surfaces are greatly reduced.

Many problems in sample handling, such as contamination or losses of low-level samples, or both, can be avoided or reduced if there is good liaison between the user and the operator of the mass spectrometer. This liaison can also be important in avoiding misinterpretation of mass spectral data if basic assumptions concerning sample handling are understood and agreed upon.

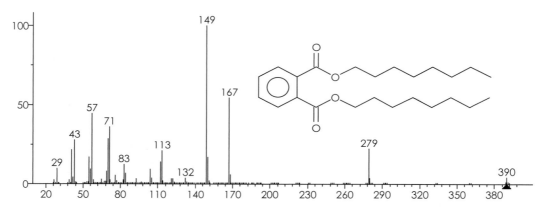

Figure 10-11. Mass spectrum of dioctyl phthalate obtained by electron ionization.

---

[1] *Silanization* involves surface inactivation by means of a chemical reaction. *Siliconization* involves surface inactivation by a process of physically coating the surface with a thin film of silicone oil.

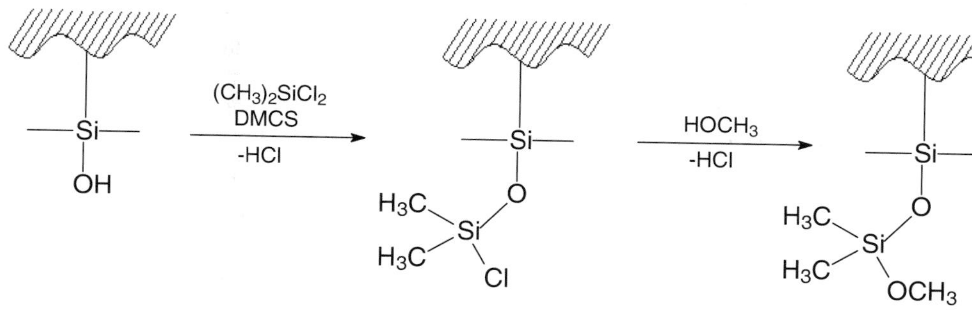

Scheme 10-1

2. Analyte Isolation and Purification

The analytical reliability of a method usually parallels the degree of isolation of the analyte from a complex mixture. At a minimum, a simple solvent extraction will often reduce the amount of nonvolatile material coinjected with the analyte for analysis by GC/MS; preliminary processing of the sample with solid-phase extraction (SPE) can also be helpful [23, 25]. Accumulation of nonvolatile material near the injection port of the GC inlet can present a reactive and/or adsorbent surface that may adversely affect transfer of the desired compounds into the mass spectrometer. The extent of sample purification required is often inversely related to the relative abundance of the analyte in the original sample. For example, identification of the drugs involved in an overdose case can usually be achieved following solvent extraction because of the high levels of the toxicant in biological samples [26, 27], whereas samples for the analysis of prostaglandins, which are only very minor constituents in biological samples, require extensive purification [28]. Fundamental aspects of chemical separations, phase equilibria and extractions, and chromatography are well presented in most standard chemistry textbooks.

3. Derivative Formation

Preparation of a derivative of a given compound is usually carried out to increase the thermal stability and volatility of the precursor compound. Derivatives can also be chosen to shift the mass of an ion that results from characteristic fragmentation of the precursor molecule to a different region of the mass spectrum to facilitate its detection. In other cases, a derivative with good thermal stability, but low volatility, may be required for manageable sample introduction via the direct inlet probe.

The following are a few examples of various types of derivatives that are commonly used to increase the volatility and thermal stability of polar compounds. Books by Blau and Halket [29] and Knapp [30] and major reviews by Vouros [31] and Halpern [32] provide many more details on a wide variety of derivatization reactions. Specialized treatment of carbohydrates for analysis by mass spectrometry have been summarized by Hellerqvist and Sweetman [33], Dell and coworkers [34, 35], and Hanisch [36]. Recent reviews are available on derivatization by alkylation [37], by formation of cyclic structures [38], on the formation of mixed derivatives of polyfunctional compounds [39], as well as specific derivatization of monofunctional compounds [40]. There are reviews on on-line (e.g., on-column or during transfer from a column) derivatization [41] and on chemical derivatization for metabolic profiling [42].

A. Silyl Derivatives

The most common silyl derivative consists of the trimethylsilyl (TMS) moiety. Silylation is a chemical reaction in which an active hydrogen (from an amino group, carboxylic acid, hydroxyl group, etc.) is replaced by a silyl group [43, 44]. The TMS derivative of a compound can usually be prepared by simple addition of any one of several "ready-made" commercial reagents, (see Figure 6-80 in Chapter 6), such as bis-trimethylsilylacetamide (BSA) bis-trimethylsilyltrifluoroacetamide (BSTFA), trimethylchlorosilane (TMCS), or trimethylsilylimidazole (TMSI). In most cases, BSTFA and BSA have comparable reactivities, but BSTFA is more volatile because of the fluorine, and thus it is easier to remove by evaporation from the sample matrix. TMCS is often added to BSA or BSTFA to serve as a catalyst. TMSI is a potent silyl donor, but it has relatively low volatility compared with BSA/BSTFA.

The reagents described above can be used in various combinations to effect different overall reactivities or potencies as silyl donors. For example, with regard to steroids containing hydroxyl groups in both the 11- and tertiary 17-positions, a mixture of BSA and 1% TMCS will react with the 11-hydroxyl, but not the tertiary 17-hydroxyl group. When a reagent mixture consisting of BSA/TMSI/TMCS (in the ratio 3:3:2) is heated with pyridine and the steroid at 60 °C for 12 hours, both the 11- and the tertiary 17-hydroxyl groups are converted to TMS ethers. Reaction conditions with BSTFA and TMSI that effect aromatization and catecholization of certain steroids have also been described [45]. In some cases, it may be necessary to heat the sample and reagent in a sealed vial for only 15 min at 60 °C to drive the reaction to completion. Similar reagents are available for preparation of bulky silyl derivatives [46]. A method for analyzing lauric, myristic, and palmitic acids and their ω and ω–1 hydroxylated metabolites from *in vitro* incubations of cytochrome P 450 CYP4A1, involving solid-phase extraction and trimethysilyl derivatization, has been developed [47]. Compounds containing multiple double bonds can be analyzed by oxidizing (with OsO_4) the double bond to a diol, which is then converted to the TMS ether [48]; fragmentation associated with cleavage at the site of the TMS ethers marks the site of the original double bond as described in detail for an example illustrated in Figure 6-85. In some cases, the reagents are sufficiently volatile that the reaction mixture can be injected directly onto a GC column. Deuterium-labeled TMS reagents are useful in special cases. The mass spectra of these derivatives are discussed extensively in Chapter 6.

B. Esters of Carboxylic Acids

The use of diazomethane to prepare methyl esters is simple, rapid, quantitative, and clean. However, the reagent is poisonous and explosive, and care must be exercised in its preparation from *N*-nitrosomethylurea; preparation of diazomethane from *p*-toluenesulfonylmethylnitrosamide is considered a safer procedure. Diazomethane is a gas under ambient conditions; however, it can be kept as a solution in cold (−5°C) diethyl ether in a screw-cap container with a Teflon-lined rubber septum. The yellow color of the diazomethane stock solution provides a convenient marker for the end point of the methylation reaction. Provided that the carboxylic acid sample is originally colorless, it will remain colorless so long as the added diazomethane is consumed in the esterification reaction. The appearance of the yellow color indicates that an excess of reagent has been added. The reaction is rapid and can be regarded as quantitative after 2 or 3 min.

An alternative method for preparing methyl esters involves mixing a benzene solution of the carboxylic acid with a solution of boron trifluoride (BF_3) in methanol (14% w/v). After heating on a steam bath, the reaction is stopped by adding water, which causes a phase separation; the methyl esters are in the upper phase of benzene. The reagent is stable and is commercially available in ready-made ampoules. Further details on the reaction have been reported by Metcalfe and Schmitz [49]. Longer-chain (ethyl, propyl, butyl) esters of carboxylic acids can be prepared conveniently by reaction with the corresponding dimethylformamide dialkyl acetals. Chloroformate reagents provide a useful one-step procedure for derivatizing fatty acids [50, 51] and amino acids [52, 53] for vapor-phase analysis.

C. Oxime Derivatives

In some cases, preparation of the oxime derivative of a carbonyl function improves the stability of an entire functional group. For example, the E-type prostaglandins contain a β-ketol ring structure (I) that is susceptible to thermal or chemical dehydration [54]. However, preparation of the methoxamine derivative (II) stabilizes the ring system so that the hydroxyl group can be converted to the TMS or acetate derivative without dehydration when the corresponding reagents are added as shown in Scheme 10-2.

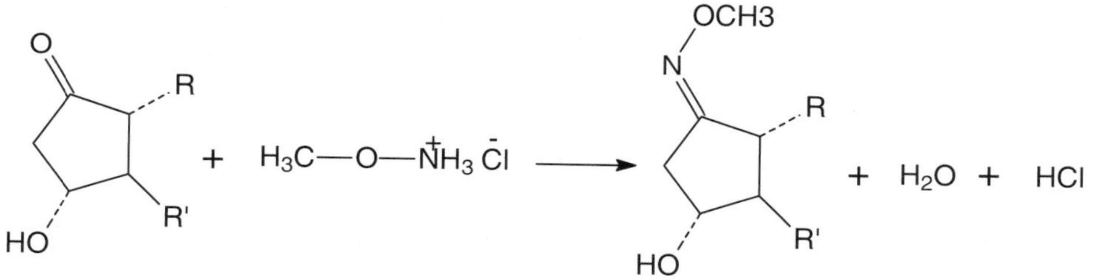

Scheme 10-2

Similarly, the dihydroxyacetone functional group (III) in C_{21} corticosteroids can be stabilized by preparation of the methoxime derivative (IV) of the C_{20} carbonyl function as shown in Scheme 10-3. This derivative (IV) can then be further derivatized as with TMSI to produce structure V [55]. Without stabilization by an oxime derivative, the dihydroxyacetone side chain of the corticosteroid decomposes under most GC conditions. Ketogenic steroids can be converted to hydrazones to facilitate their detection as in the analysis of urine [56].

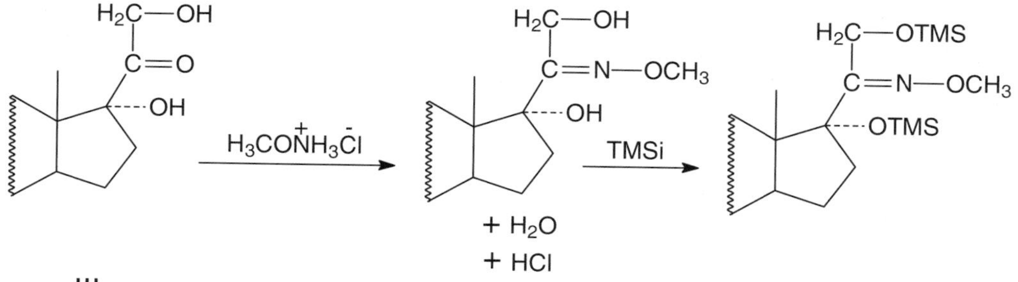

Scheme 10-3

D. Acyl Derivatives

The acyl derivatives of hydroxyl and amino groups have good vapor-phase characteristics [57]. The pentafluoropropionyl (PFP) derivatives of biogenic amines have excellent vapor-phase properties [58]; the PFP derivative of normetanephrine is described in a following section on selected ion monitoring in the context of an introductory qualitative example of SIM. PFP derivatives are especially useful for increased specificity and lower detection limits when using electron capture negative ionization (ECNI). The presence of fluorine in these derivatives makes them very good for ECNI techniques.

In most cases, preparation of the acyl derivatives is very simple. For example, for preparation of the acetate derivative, in a vial similar to that in Figure 10-10, 10 µL pyridine and 10 µL acetic anhydride are added to the sample; after 1 or 2 hours at room temperature, an aliquot of the reaction mixture containing the acetate derivative and excess reagent can be injected directly into the gas chromatograph. The PFP derivatives of biogenic amines can be prepared directly from a stable salt of the amine. For example, to a few crystals of normetanephrine as the hydrochloride salt, 20 µL ethyl acetate and 20 µL pentafluoropropionic acid anhydride are added. After 15 to 30 min at room temperature, the crystals will have dissolved and the reaction will be complete; the ethyl acetate facilitates the reaction because the PFP derivatives are more soluble in it than in neat excess reagent. An aliquot of the reaction mixture containing the PFP derivatives and excess reagent can be injected directly into the GC inlet.

E. Derivatives for Characterizing Double Bonds

Dimethyldisulfide is an example of a reagent that reacts directly with a double bond as shown in Scheme 6-93 in Chapter 6 on EI. The covalent adduct with dimethyldisulfide (DMDS) produces a mass spectrum characterized by a discernible molecular ion and fragment ions that can be related to the position of the double bond in the original olefin [59]. Formation of the two prominent fragment ions is also shown in Scheme 6-93. The DMDS derivative can also be used in the analysis of linear tetradecenols [60].

Reaction of olefins with 5,5-dimethoxy-1,2,3,4-tetrachlorocyclopentadiene also produces a covalent adduct; however, in this case, the mass spectrum provides information not only on the position of the double bond in the olefin but on the geometry as well [61]. Epoxide formation by oxidation with *m*-chloroperbenzoic acid of linear as well as branched hydrocarbons and fatty acids prior to analysis by GC/MS facilitates a determination of the geometry of double-bond position [62]. *n*-Propyl-, isopropyl-, and cyclopropylamines can be used to derivatize double bonds in aliphatic compounds to facilitate analysis by MS [63]. As described in Chapter 6, the location of double bonds can be determined by chemical ionization using products of the electron and chemical ionization of acetonitrile as the reagent ions in an internal ionization 3D QIT mass spectrometer.

IV. Instrument Requirements for GC/MS

1. Operating Pressures

The maximum permissible operating pressure in most mass spectrometers (10^{-4} to 10^{-3} Pa) is generally incompatible with the normal operating pressure at the exit of a GC column (1 atm). Pressures higher than 10^{-3} Pa in the region of the ion optics, ion accelerator, and *m/z* analyzer lead to deterioration of resolution or broadening of ion beams due to excessive ion–molecule collisions. The effects of high pressure in

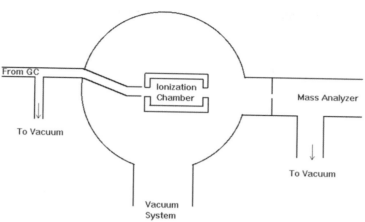

Figure 10-12. *Illustration showing relative arrangement and approximate relative sizes of conduits and orifices for gas flow through the ion source. Note that the GC inlet and the ion source, as well as the ion source and the analyzer, are differentially pumped.*

the ion source on resolution and ion abundance are greatly reduced in differentially pumped instruments in which the ion source and *m/z* analyzer are evacuated by separate pumping systems, as illustrated in Figure 10-12; however, most EI instruments are not differentially pumped. In some cases, instruments designed for CI in addition to EI will be differentially pumped. Differential pumping can also be accomplished using a split-flow turbomolecular pump. These pumps, introduced in the past five years, have greatly improved the performance of GC/MS instrumentation.

Modern instruments designed for GC/MS work have vacuum systems that can accommodate flow rates of common carrier gases (helium or hydrogen) up to 2 mL atm min^{-1} and maintain a general ion-source and analyzer pressure of 10^{-3} Pa or lower to avoid deleterious pressure effects. This is accomplished through the conductance of a single pump, permitting the direct connection of capillary columns to the mass spectrometer. Instruments that are designed only for EI may employ an oil diffusion pump with a pumping speed as low as 50 L sec^{-1} rather than a turbomolecular pump. Most instruments designed for both EI and CI are equipped with a high-capacity turbomolecular pump with a speed of 200 L sec^{-1} or greater. If columns larger than 320 μm i.d. are used, the instrument will employ an interface such as a molecular separator to remove most of the carrier gas, but transmit most of the sample to the mass spectrometer, or an open-split interface to divert a portion of the flow from the exit of the column. Both of these devices are described in a following section.

GC/MS requires a compromise in the optimum operating conditions of either GC or MS. In MS, it is important to have as little variation in the partial pressure of the analyte as possible during the acquisition of a spectrum, especially in an instrument that generates the spectrum by scanning across the *m/z* range; see later in Section V on mass spectral skewing. Spectral skewing is not a consideration in array detection TOF or in quadrupole ion traps because they are pulsed instruments, and they effectively integrate the spectrum over time.

The elution of the sample from the chromatograph is such that the partial pressure of the analyte is never constant. If the rate of change is reduced too much (through manipulation of mobile-phase flow rates or rates of heating the column) to allow the MS to spend more time measuring the ion current of each *m/z* value in the scan range to produce a better quality mass spectrum, chromatographic peak broadening can occur to the point that no distinguishable chromatographic peak will be observed and the mass spectrum may not be interpretable due to poor ion counting statistics (i.e., poor signal strength). A balance must be achieved between the scan speed of the mass spectrometer and the sharpness of the chromatographic peak.

The best chromatographic results are indicated by the narrowest GC peaks; this relates to the definition of chromatographic peak resolution. In GC/MS, the vacuum at the exit of the column may adversely affect chromatographic resolution. Some molecular separators and direct interfacing of the column to the ion source produce this adverse vacuum effect. However, results of a study of the low-pressure behavior of open-tubular columns revealed that this deviation from normal GC-column operation results in no significant loss in chromatographic efficiency [64, 65]. Use of a membrane separator or an open-split interface leaves the exit of the column at atmospheric pressure.

In GC/MS, the chromatographic peak is not produced by continuously monitoring an analog signal as is the case when using an FID with a conventional GC. In GC/MS, the chromatographic peak is a reconstructed profile of total ion current (RTIC or TIC) or, in some cases, a mass chromatographic peak, which is also called an extracted ion current (EIC) chromatogram. As defined in Chapter 1, the RTIC is generated from the mass spectral data (the only data acquired) by summing the ion current for each *m/z* value and plotting this value against spectrum number. In the case of the EIC, the ion current for a given *m/z* value, a range of *m/z* values, or the sum of several *m/z* values is plotted against spectrum number to obtain the "reconstructed" chromatographic peak. Because the mass spectra are acquired at a constant rate, the spectrum number can be used as a measure of the analyte's retention index (a function of retention time).

The number of points necessary to define a chromatographic peak to be used in quantitation and for assessment of mass spectral quality is an important factor in data evaluation. The best results for quantitation are achieved when 20 data points are used to define the chromatographic peak. This will produce a precision of ±2%. As few as five points across the chromatographic

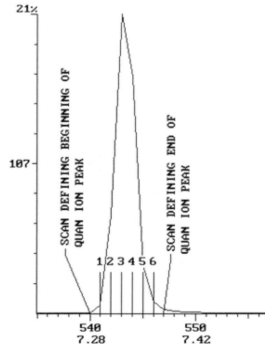

Figure 10-13.   Illustration of the minimum number of spectra required to define a chromatographic peak.

peak (Figure 10-13) can still result in a precision of ±10%. Trying to acquire 20 spectra during elution of a chromatographic peak while scanning through an *m/z* range of 20–400 will result in poor-quality mass spectral data. Manufacturers of GC/MS instrumentation have made great improvements in the electronics of their instruments in recent years that reduce the electrical noise during rapid acquisitions; however, this does not improve the signal strength generated by a specific number of ions. The signal-to-noise ratio is improved, but the signal itself remains the same. Only small incremental improvements in the quality of the mass spectrum have been achieved by these increased scan rate capabilities in modern instrumentation.

Another feature in some modern instruments is the capability of acquiring data in the scan and selected ion monitoring mode alternately during a given period of data acquisition. The selected ion monitoring (SIM) mode is explained in more detail later in this chapter; but, for now, it can be defined as a mode in which only ions of a few *m/z* values are monitored rather than monitoring all the *m/z* values in a specified interval. These monitored ions do not have to be at contiguous *m/z* values. The SIM mode allows for more time to be spent measuring the ion current of ions of interest rather than expending measurement time at *m/z* values where there may be no ion current. The SIM mode is usually used when performing target-compound analyses. The one drawback to using the SIM mode is that unambiguous confirmation of the analyte from a full scan spectrum will not be possible and contaminants that do not produce ions of the same *m/z* values as the analyte will be missed. Therefore, the ability to acquire data in both modes (alternating data points in SIM and scan modes) can be beneficial. Instruments such as the Agilent 5975B store data from these two modes used in the same data acquisition in separate files. The SIM data file can be used for quantitation offering good chromatographic peak definition and the scan file can be used for unambiguous confirmation of the analyte's presence and identifying any nontargeted analytes that may be present.

It should be pointed out that the SIM mode provides the basis for obtaining a lower limit of quantitation as well as better quantitative precision. If the goal of an experiment is to achieve the lowest possible detection limit by using SIM, then the alternating SIM/scan process cannot be used for purposes of providing unambiguous analyte confirmation. It has been pointed out that in most cases when using EI GC/MS, unambiguous identification can be achieved by use of the peak-area ratios for three monitored ion currents [66]. Regarding the debate over minimum data requirements for demonstrating that signals from an unknown sample are identical to those from a known compound, the results of a large-scale intercomparison between thousands of spectra and a library of one hundred thousand EI mass spectra indicate that the confidence in providing the correct identification increases by about an order of magnitude for each additional ion current monitored by SIM [67].

2. Typical Parameters for a Conventional GC/MS Interface

An example of a GC/MS interface between a capillary column and the ion source of a mass spectrometer is shown in Figure 10-14. The capillary column runs from the GC on the left directly into the ionization chamber within the ion source block of the MS [15].

Typically, a 25-m by 250–320-μm i.d. fused silica column containing a chemically bonded partitioning phase (the stationary phase) allows a flow of about 1–2 mL min^{-1} (carrier gas linear velocity of 30 cm sec^{-1}) into the ionization chamber [68]. Care should be taken during column installation to avoid problems of fine particles occluding the end of the column when it is passed through the mounting nut and ferrule; it is good

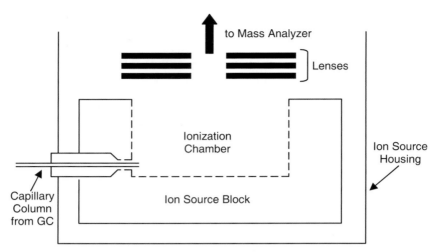

Figure 10-14. ***Illustration of the inlet system configuration showing an example of inserting a capillary GC column directly into the ionization chamber of the mass spectrometer.***

practice to cut off a few mm of the column after pushing it through the ferrule. An alterative is to install permanently a short piece (500 mm or less) of fused silica tubing of the same i.d. as the columns being used into the ion source and connect the GC column to this inside the GC oven. It is important to make sure all connections are airtight so that these connections are not another source of air leaks.

It is also good practice to inquire about manufacturing processes used for the fused-silica column. For example, the organic platting on the outside of the GC column may outgas when heated to 250 °C under vacuum.

A simple, efficient device may be used to divert the column effluent to a vacuum pump during the emergence of the solvent [69, 70]. This effluent diverter (also called a "vacuum diverter") is heatable and has a low dead-volume so that the chromatographic resolution is not affected. In fact, the appearance of the total ion chromatogram is improved, because tailing of the solvent or excess reagent is greatly diminished. Use of the diverter offers a convenient and efficient means of limiting the amount of material (excess reagents, uninteresting GC components, long-term column bleed, etc.) entering the ion source, thereby reducing the rate at which the ion source becomes contaminated. In recent years, similar devices have been made available for direct interfacing of capillary columns to ion sources that also allow for changing columns without having to break the vacuum of the mass spectrometer. The use of these devices is highly recommended and their availability or nonavailability should be considered as an important factor in making a decision on a new instrument.

3. Supersonic Molecular Beam Interface for GC/MS

The conventional interface for a gas chromatograph-mass spectrometer (GC-MS) is designed to maintain the vapor-phase status of the sample achieved in the GC that is usually operated at an elevated temperature. A new approach to interfacing the GC with a mass spectrometer employs a supersonic molecular beam, which provides several analytical advantages [71, 72]. The high-volume flow rate that can be accommodated by this interface allows use of a high linear velocity through the GC column, which translates to lower retention temperatures, which, in turn, allows many thermally labile compounds to be analyzed that are otherwise refractory to conventional GC/MS [73]; this interface is also compatible with flow-modulation GC applications [74]. Second, diminishing the internal energy (lost during gas expansion through a jet orifice) of the analyte molecules increases the stability of the molecular ions formed subsequently by EI as reflected by significantly enhanced molecular ion peaks in the mass spectra [75]. Third, combining a high-flux electron source with the molecular beam interface allows achievement of limits of detection for thermally labile compounds that are more than an order of magnitude lower than those achieved by conventional GC/MS [71, 72].

Two of the key features of the molecular beam GC-MS are shown in Figure 10-15. The eluate from the GC is forced through an expansion jet (left side of Figure 10-15), which cools the molecules to a level that promotes abundant molecular ion formation. Some of the expanded gas passes through the skimmer (central section of Figure 10-15) into the cylindrical "fly-through" EI source constructed with a very large thoriated-tungsten ribbon filament (30 mm long, 0.75 mm wide, and 0.025 mm thick) that produces a 10-mA emission current; this flux of electrons is approximately two orders of magnitude greater than that produced in a conventional EI source. Interaction of the analyte molecules with the very high flux of electrons promotes efficient electron ionization, which translates into extraordinarily low detection limits for many kinds of thermally labile small molecules [71, 72]. Because of the cold molecules in the supersonic molecular beam, the mass spectra dependably include a significant peak for the molecular ion, a feature that favors identification of the analyte [75].

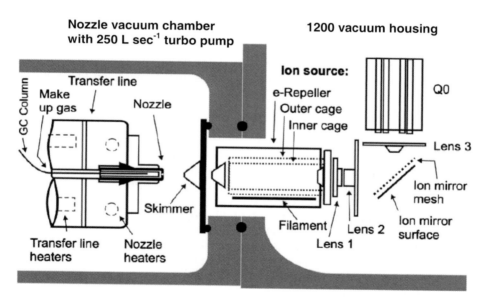

Figure 10-15. **Schematic representation of the Amirav supersonic molecular beam interface employed in a GC-MS.** *Fialkov AB, Steiner U, Jones L and Amirav A "A new type of GC-MS with advanced capabilities" Int. J. of Mass Spectrom. 2006, 251(1), 47–58, with permission.*

Impressive limits of detection (LODs) can be achieved with the supersonic molecular beam approach to GC/MS, especially when using a GC-MS/MS equipped with such an interface [72]. LODs typically listed by major vendors of GC/MS instruments are based on the analysis of octafluoronaphthalene (OFN) under standardized conditions. Whereas most current GC/MS instruments equipped with a conventional GC-MS interface can achieve LODs in the low femtogram range for OFN, the LODs for analytes in realistic samples are often a few orders of magnitude higher. Results from the laboratory of Amirav and coworkers show that LODs for a variety of small molecules approaches that for OFN when analyses are performed on a GC-MS/MS instrument equipped with the supersonic molecular beam interface [72]. Whereas the MS/MS operation of the instrument is principally responsible for reducing background signals, there is little question that attributes of the supersonic molecular beam interface contribute substantially to the greatly improved LODs. Major advantages of the supersonic molecular beam are manifested in the enhanced abundance of the molecular ion of the analyte (which becomes the precursor ion in the MS/MS experiment) by accepting large-volume injections of sample, by reducing the need for elevated temperature in the GC column and thereby reducing column bleed, etc. In this way, the Amirav group achieved a LOD of 2 fg for diazinon, the lowest value reported by GC/MS to date for this common pesticide [72]. In addition, the Amirav group has reported improvement factors in the LOD of 24 for dimethoate, 30 for methylstearate, 50 for cholesterol, 50 for permethrin, >400 for methomyl, and >2000 for $C_{32}H_{66}$ [72].

Importantly, the lower operating temperature of the supersonic molecular beam interface extends the range of thermally labile and low-volatility compounds that are amenable to analysis by GC/MS when using a supersonic molecular beam interface. For example, the Amirav group achieved an LOD of 10 fg for underivatized testosterone, which is not amenable to analysis by conventional GC/MS [72]. In general, it seems that the more difficult a given analysis is by conventional GC/MS, the greater the improvement in LOD that can be realized in performing the same analysis by using the supersonic molecular beam interface.

4. Open-Split Interface

The open-split interface [70, 76] illustrated conceptually in Figure 10-16 offers a convenient connection between the GC column and the mass spectrometer. The dimensions of the capillary leading to the mass spectrometer fix the flow rate into the ion source. Depending on the flow rates of the GC column and the purge gas, all or only a fraction of the GC column eluant enters the ion source. The open-split interface permits most of the solvent and other eluants to be shunted away from the ion source merely by increasing the flow rate of purge gas. It also permits the GC column to be changed without venting the MS or using an isolation valve; the open-split interface ensures normal operation of the capillary column with the exit at atmospheric pressure.

In the open-split interface, the column end is connected to an open sleeve, which has a flow of helium through it. The column end is butted against a restrictor line going into the ion source. This restrictor allows a fixed amount of the column eluate to enter the ion source. The restrictor typically allows 1 mL min^{-1} to enter the ion source. If a 530-μm i.d. Megabore™ column is connected to the interface at a flow rate of 5 mL min^{-1}, only 20% of the eluate and, therefore, only 20% of the analyte would be introduced to the ion source. This limiting of the amount of analyte reaching the ion source should be considered as an issue when sensitivity is important.

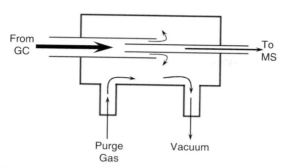

*Figure 10-16. **Representation of the open-split interface.***

When using the open-split interface, the flow rate of the column can be changed without affecting the conditions in the ion source. Only a fixed amount of sample enters the mass spectrometer as established by the conductance of the smaller capillary (restrictor) in Figure 10-16. This means that the same ionization conditions will exist even if GC column flow rates change during temperature programming.

The open-split interface gives the best chromatographic integrity of all the various types of interfaces. Its yield is dependent on the flow rate and the size of the restrictor. There is no enrichment of the analyte. The open-split interface has been found to discriminate against higher-mass and higher-boiling compounds, a feature that will limit its usability.

5. Molecular Separators

Separators were designed to meet the demands of GC/MS instrumentation in the 1960s when packed columns having flow rates on the order of 20–30 mL min^{-1} were being used.[2] The separators substantially reduced the pressure of the eluant stream from the GC that eventually entered the mass spectrometer. This was accomplished by enriching the analyte in the residual carrier gas entering the ion source either by acting directly on the carrier gas, as with the effusion [77, 78] and diffusion-type [79] separator, or by acting directly on the analyte as with the membrane separator [80]. The effusion separator had the virtue of inertness because it was fabricated from all-glass components, but suffered from fragility and complications in construction. The jet-orifice separator originally built of metal was eventually transformed into an all-glass design and was used extensively in several commercial GC/MS interfaces, as was the membrane separator. Because of the diminishing need for separators in applications of GC/MS, only the jet-orifice separator and the membrane separator will be described below from the standpoint of operating principles.

A. Jet-Orifice Separator

The most popular of the various enrichment devices used with packed columns was the jet-orifice separator invented, and later patented, by Einar Stenhagen (Swedish medical scientist) and perfected by Ragnar Ryhage, also of Sweden. The jet-orifice separator continues to be used by modern instrumentation, which is dominated by use of the capillary column. The operation of this separator is based on a diffusion principle [79, 81]. As shown in Figure 10-17, the effluent from the GC comes in from the left, expands

---

[2] The GC/MS industry was dominated by U.S. manufacturers during the 1960s and early 1970s. Although the capillary column was in widespread use outside the United States, its use in the U.S. was almost nonexistent because Perkin-Elmer Corp. held U.S. patents on the GC capillary column technology and required a licensing fee that made manufacture by other companies economically unviable. The use of the capillary GC column did not become a factor in GC/MS until the Perkin-Elmer patents expired at the end of the 1970s.

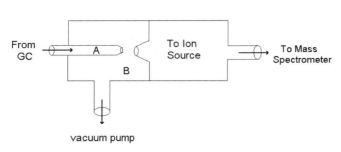

Figure 10-17. Representation of the jet separator.

through a nozzle at position A, and shoots toward an orifice in the wall of an adjoining chamber on its way to the ion source. Across the gap between points A and B, there is a tremendous expansion of the gases. Compounds that have high diffusivity will diffuse at right angles much more than those having a lower diffusivity. Thus, if a mixture of materials, such as 400 Da organic molecules in helium, which has a mass of 4 Da, is forced through the nozzle at position A, the helium will have a greater tendency to diffuse at right angles into chamber B than will the molecules with a mass of 400 Da, and a greater number of organic molecules of mass 400 Da will remain in the beam to enter the ion source than atoms of helium. This action imposes a discrimination against the carrier gas, which means that most of it diffuses into chamber B and is removed by a vacuum pump. Most of the organic material is preferentially shot into the ion source, thereby achieving an enrichment of the analyte molecules among residual molecules of the uninteresting carrier gas. The jet-orifice separator works in the so-called viscous-flow pressure region, as opposed to the molecular-flow pressure region as required for the effusion separator. A single-stage separator can accommodate flow rates from 5 to 30 mL min^{-1}; a double-stage separator is usually required for flow rates above 30 mL min^{-1}. The single-stage jet separator has an enrichment factor of approximately 50% and an efficiency of 30–40%. The jet-orifice separator tends to discriminate against low-mass components in the GC eluate. In viscous flow, the flow rate through a pore or orifice is proportional to the square of the total pressure. The jet-orifice separator can accommodate flow rates up to 40 mL min^{-1} at an efficiency of 40% [82]. The jet-orifice separators are prone to clogging, most often with particles of GC packing, but occasionally with deposits of organic material.

One important disadvantage to the jet-orifice separator is that it must have a flow rate of ~20 mL min^{-1} to perform properly. The largest of capillary columns (720 µm i.d.) have optimum flow rates of less than 15 mL min^{-1} which means that, for the jet-orifice separator to function properly, a makeup gas must be added at the exit of the GC column. This will further dilute the analyte in the gas stream going to the ion source and have a deleterious effect on sensitivity. With this added complication for the use of the jet separator, the results obtained in terms of sensitivity are not that much better than those that can be achieved with the open-split interface.

B. Membrane Separator

Duane Littlejohn (who went on to develop the Siemens/Applied Automation permanent magnet FTICR mass spectrometer in the late 1990s) and Peter Llewellyn, at the Varian Research Center in Palo Alto, CA, developed the membrane separator in the mid-1960s. No longer in commercial use, this device was the only separator offered by Hewlett-Packard on their GC/MS instruments during a good part of the 1970s. The membrane separator operates preferentially on the organic analyte [80, 83]. As illustrated in Figure 10-18, a thin membrane (a few square millimeters), typically made of silicone

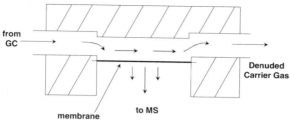

Figure 10-18. *Schematic representation of a membrane separator.*

elastomer, is the interface between the gas chromatograph and the mass spectrometer. The eluant from the GC is directed onto this membrane. The membrane, being organic, adsorbs and dissolves the organic constituents from the gas stream [83] in an analogous manner to those used in MIMS today (see Chapter 4). The carrier gas, such as helium or hydrogen, is not soluble in the organic membrane and is discharged through the exit of the separator module. Those organic compounds that dissolve in the elastomer rapidly diffuse through the very thin membrane into the vacuum of the ion source. In this way, the separator preferentially allows passage of the organic materials across the membrane into the MS with an enrichment factor of approximately 100.

Even though the membrane separator exhibited excellent performance with respect to analyte enrichment and had a high efficiency, it was never very popular because of the perceptions that it would cause chromatographic peak broadening and memory effects. Neither of these fears was ever realized and it was probably the best performing of all the molecular separators.

6. Inertness of Materials in the Interface

Prolonged exposure to hot metal surfaces [84, 85] can result in degradation of sensitive compounds, such as steroids, prostaglandins, modified sugars, pesticides, etc., during vapor-phase analysis. Modern GC/MS systems have very few, if any, metal surfaces. However, as was pointed out earlier in this chapter, glass surfaces may adsorb analytes. It has been demonstrated that silanized glass components in GC/MS systems can provide an inert surface [85]. The general rule seems to be, "When in doubt, silanize! And silanize all mechanical components that the sample will encounter *en route* between the injector and ion source." One caveat to silylation is that this process may cause problems with the analysis of basic compounds such as amines due to the presence of residual acidic sites.

Assessment of the degree of inertness of the GC interface has been described in experiments using free cholesterol [85]. The evaluation is based on the extent to which cholesterol is dehydrated, a process that is sensitive to active surfaces. Under inert conditions, the $[M - 18]^+$ peak in the mass spectrum of free cholesterol (Figure 10-19) will be approximately equal to or less than the $M^{+\bullet}$ peak; when the $[M - 18]^+$ peak is larger than the $M^{+\bullet}$ peak, a reactive surface is indicated. The $M^{+\bullet}$ peak should be the base peak for cholesterol when data are acquired over a range of m/z 30 to m/z 400. When the peak representing the ion with m/z 43 is the base peak (Figure 10-19), either there is a problem with mass discrimination or there is a reactive surface between the injection port cap and the ion source. Fales *et al.* [86] described the use of a pyrrolidine to test interface surfaces of sample-inlet systems. In this case, the pyrrolidine undergoes facile dehydrogenation in the presence of active surfaces in the interface. Another test for inertness used by the GC industry is the shape of the chromatographic peak produced by an injection of a solution of the pesticide endrin (1,2,3,4,10,10-hexachloro-6,7-epoxy-1,4,4a,5,6,7,8,8a-octahydro-exo-1,4-exo-5,8-dimethanonaphthalene). If this chromatographic peak is not symmetrical (exhibits tailing or fronting), then there is reactivity in the system. This test has been used to assess the effect of the electrode

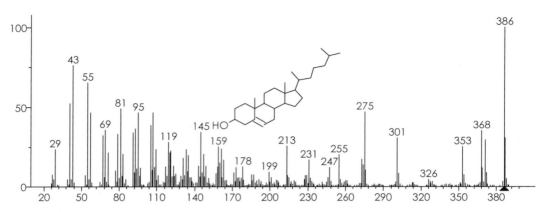

Figure 10-19. EI mass spectrum of cholesterol with the M⁺• peak as the base peak indicating the inertness of the GC/MS system.

surfaces in the internal ionization 3D QIT GC/MS system on chromatographic peak shapes. Like cholesterol, endrin has a significant molar mass for a gas chromatographic analyte with a number of structural features that make it particularly vulnerable (as can be seen from an examination of its structure) to reactive sites in an analytical instrument.

The composition of the materials used to construct the interface is important if the sample vapor is likely to come in contact with internal surfaces *en route* to the ionization chamber in cases where the tip of the capillary column is not inserted directly into the ionization chamber. Because the interface is maintained at a temperature of 200–280 °C, the possibility of reactive metal surfaces should be avoided. The active sites on such surfaces may catalyze chemical modification of the analyte. In some interfaces, glass-lined stainless steel tubing has been used, even though there may be some difficulties with the fragile nature of this material. In any case, it may be good practice to silanize these surfaces. Most manufacturers now recommend that a length of fused silica tubing be inserted into the ion source and the column connected to the other end of this tubing in the GC oven. This tubing should be as short as possible (no more than 500 mm in length) to avoid chromatographic peak broadening due to excessive dead volume.

> Glass-lined stainless steel can be deceptive. Stainless steel tubing can be easily bent by applying force with a tubing bender. This should not be done with glass-lined stainless steel unless it is heated to a temperature sufficient to soften the glass lining while force is being applied to bend the steel. If the heating is not done properly, the glass will break and the sample vapor will be as vulnerable as if the glass were not present; unfortunately, the problem is not obvious from external examination of the tubing.

V. Operational Considerations

1. Spectral Skewing

Rapid acquisition of the mass spectrum is required in GC/MS to avoid distortion of relative mass spectral peak intensities (spectral skewing) in instruments that acquire data by scanning the *m/z* scale. This spectral skewing is due to transient changes in the partial pressure of the sample in the ion source. The very nature of the GC output (namely, a changing sample concentration in the GC eluate with time) violates one of the cardinal rules in mass spectrometry, namely, that the sample pressure should remain constant during the time the mass spectrum is recorded. The relative peak heights in a mass spectrum will be distorted significantly if the required time for spectrum acquisition is comparable to the duration of the emerging gas chromatographic fraction. An extreme example of this mass spectral peak-intensity distortion is illustrated in Figure 10-20. In this case, imagine a hypothetical compound whose idealistic mass spectrum consists of only three peaks of equal intensities, as indicated in the inset spectrum at the upper left quadrant of Figure 10-20. The mass spectrum inset at the upper left represents the case in which there was no spectral skewing because it was acquired while the sample pressure in the ion source was constant.

Several other spectra were acquired as this hypothetical compound eluted from the GC column, which means that its partial pressure in the ion source was changing as a function of time as illustrated by the temporal profile (a GC peak) in the central portion of Figure 10-20. Three mass spectra acquired at designated intervals along the GC elution profile are also displayed in Figure 10-20. During scan #1, the mass spectrum was

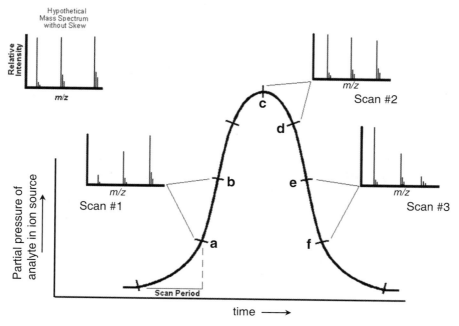

Figure 10-20. **Illustration of spectral skewing due to dynamics in partial pressure of the analyte in an instrument that is scanned from a low m/z value to a high m/z value during GC elution.**

acquired at the leading edge of the GC peak (between points **a** and **b**) when the partial pressure of the analyte in the ion source was changing (increasing) rapidly, and it shows significant peak-intensity distortion; namely, the peaks at low-m/z values are less intense (because the sample pressure was lower) than those at high-m/z values (when the partial pressure of the analyte was higher) compared with those in the spectrum acquired at constant sample pressure (inset at upper left).

The mass spectrum acquired during scan #2 (between points **c** and **d** on the chromatographic profile in Figure 10-20) shows less skewing of the peaks because the relative change in partial pressure of the analyte during this period was less.

The mass spectrum resulting from scan #3 (between points **e** and **f** on the chromatographic profile in Figure 10-20) shows different skewing of the peak intensities with discrimination at high-m/z values because now the partial pressure of the analyte is dropping as the scan function reaches the higher values of the m/z scale. In this case, the peaks at low-m/z values are more intense (because the sample pressure was relatively higher) than those at higher-m/z values (when the sample partial pressure was lower) compared with those in the spectrum acquired at constant sample pressure (inset at upper left).

Skewed spectra cause obvious problems in the use of pattern recognition, especially as manifested by failures in library searches. If the RTIC chromatographic peak is symmetrical and represents only a single component, spectral averaging can be used to obtain a spectrum that has close to "correct" relative intensities of the mass spectral peaks. From the chromatographic point of view, a repetitive slow mass spectral scan relative to fast elution provides too few data points from which to reconstruct the chromatographic profile, leading to poor quantitation and errors in retention time determination. Also, with fewer spectra throughout a GC peak, there is less of a basis for confirming the purity of a GC component or for deconvoluting unresolved components.

Attempts have been made to correct slowly recorded (and therefore distorted) mass spectra from a GC/MS system [87]. By simultaneously monitoring the total ion current (which is dependent on sample concentration in the ion source), the recorded peak intensities should be amenable to normalization or correction [88]. However, this procedure has not been well received, and rapid scanning of the mass spectrum appears to be the most desirable approach. Spectral skewing is not an issue with the TOF or QIT analyzers because these instruments acquire spectra in a pulsed "batch" mode as opposed to a scan mode.

2. Background/Bleed

Vapors from the GC liquid phase [12] or the septum, or both, enter the ion source, at which point they undergo continuous ionization and fragmentation; thus, their characteristic fragmentation pattern will be represented in each acquired mass spectrum. The mass spectrum of column bleed from a 2-m by 2-mm i.d. column of 3% OV-1 (methylsilicone) at 260 °C is shown in Figure 10-21. The "column-bleed" peaks in the mass spectrum can be used to advantage, in some respects, in verifying the correctness of the m/z scale. For example, in the mass spectrum of silicone GC stationary phases, discernible peaks can usually be found at m/z 207, 281, and 355 (Figure 10-21); more complete background spectra for common GC phases have been published [89]. The background peaks at m/z 207, etc., represent $[M - 15]^+$ ions from thermal degradation products of the siloxane polymer [90] as illustrated in Scheme 10-4.

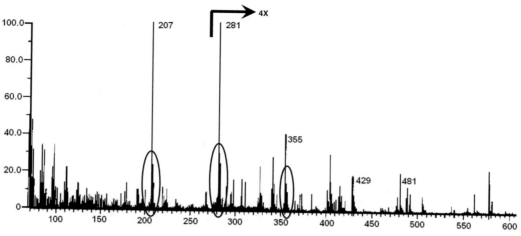

Figure 10-21. *EI mass spectrum of column bleed from the OV-1 stationary phase; note that the intensities from m/z 270 are magnified by a factor of four.*

The column bleed can have a detrimental effect on the mass spectrometer. The continuous bathing of the ion source with vapors from the relatively nonvolatile GC phase gradually contaminates nearly all the critical components of the ion optics, causing nonoptimal ion focusing and, in some cases, electrical breakdown or discharges. These problems may occur more frequently in ion sources that lack auxiliary heaters or in those

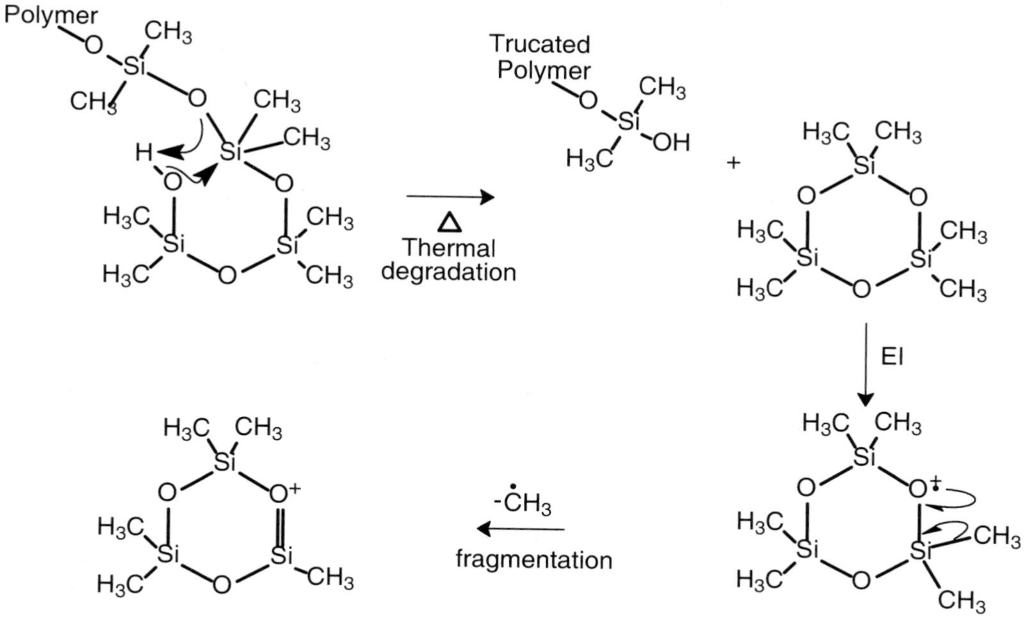

Scheme 10-4

that are operated "cold" (i.e., at 150 °C or lower). Because the vapor pressure of the stationary phase is a function of temperature, the rate of contamination of an ion source connected to a GC column will increase in proportion to the duration of high-temperature GC applications.

Contamination of the ion source may be a special problem if the column must be "baked out" or conditioned (at 20–25 °C above the expected operating temperature) while connected to the MS. The advent of chemically bonded stationary phases has helped to minimize many of these problems caused by column bleed, though some of the peaks in Figure 10-21 may still be detected.

> "Baking out" a GC column connected to the ion source of a mass spectrometer can be problematic. The material being purged from the column possibly will contaminate the ion source. For this reason, it is best to use a divert system that is now available for capillary GC columns from several different manufacturers. This means that the GC column is connected to the ion source through a length of fused silica tubing rather than extending the active column directly into the ion source.

The presence of peaks in a mass spectrum due to column bleed can often be identified by the intensity of the isotope peaks. The ions represented by these peaks contain several atoms of silicon. As will be recalled from Chapter 5, silicon is an X+2 element with an X+1 isotope contribution of >5% and an X+2 contribution of >3% per atom of Si. This means that X+1 and X+2 peaks associated with the nominal mass peaks for these ions will be much greater than those associated with ions that do not contain several atoms of Si. In Figure 10-21, short segments of the column-bleed spectrum that contain these intense isotope peaks in the vicinity of *m/z* 207, 281, and 355 have been circled.

3. The Need for Rapid Acquisition of Mass Spectra

Whereas rapid data acquisition (in some cases, scanning of the mass spectrometer) was a necessary innovation in applications to GC, use of a mass spectrograph for GC/MS demonstrated the advantage of its photographic recording system, which integrates any variation (resulting from changing sample concentration) in ion abundance throughout the exposure time [77]. Unfortunately, these photographs are discontinuous in time and produce data records of limited dynamic range.

As fast chromatography makes further demands for increasingly rapid acquisition of mass spectra, continued innovation in overall instrument performance must continue [91–93]; considerable improvement in this area has been made since the earliest experiments in GC/MS [8].

Rapid acquisition of mass spectra should be employed in GC/MS to avoid or minimize the possibility of distorting relative intensities (spectral skewing) of mass spectral peaks as described earlier in this chapter. As a rule, the acquisition time for a complete mass spectrum should require no more than one-fifth of the duration of the GC peak and preferably less. Applications to capillary columns offers the greatest challenge [65, 92]. From a capillary column, the duration of a chromatographic peak may be on the order of one second, thereby ideally requiring an acquisition rate of ten or more mass spectra per second. Most commercial GC/MS instruments can acquire a mass spectrum (*m/z* 50–500 at unit resolution, 10% valley) in less than 1 sec [91] while maintaining good mass spectral quality. Although some degree of peak-intensity distortion in mass spectra can be tolerated in most applications, data obtained by GC/MS should be used with caution in attempts to distinguish isomers (e.g., *cis* and *trans* isomers) or other compounds that may produce very similar mass spectra.

A. Performance Trade-Offs of Conventional Instruments for GC/MS

Three parameters—spectral acquisition rate, sensitivity, and useful dynamic range—must be considered simultaneously to describe the compatibility of a mass spectrometer with a high-resolution chromatograph [91]. The principal trade-off is that for a given sample size, a high acquisition rate (in a scanning instrument) acquires fewer ions (weaker signals), which translates to poorer ion counting statistics and lower S/N in the data.

The operating mechanics of a given instrumental design can establish an upper limit on the effective scan rate of the instrument. In the magnetic-sector (or double-focusing) instrument, the spectral acquisition rate, in large measure, is limited by the reluctance of the magnet; with state-of-the-art laminated magnets, the repetition rate for the magnetic sector is currently 1–3 Hz (or less) per decade.[3] At repetition rates higher than 10 Hz, problems may arise with the ion optics, which are effectively defocused by changes in the magnitude of the magnetic field during the transit time of an ion through the field; the result is a degradation of spectral quality from magnetic-sector instruments scanned at rates exceeding 10 Hz per decade. The mass filtering processes in the transmission quadrupole instrument also depend on the residence time of the ion in the quadrupole field. When the RF and DC fields change from one *m/z* stability value to another while an ion of a given *m/z* value is in transit from the ion source to the detector, that ion will not reach the detector. Thus, at unit resolution for a quadrupole filter consisting of poles that have a length of 20 cm, maximum useful spectral acquisition repetition limits of 5–10 Hz might be achievable. As was pointed out above, the QIT mass spectrometer, related to the quadrupole mass filter (see Chapter 2), is a pulsed instrument, but its spectral acquisition is based on operating principles that effectively limit it to producing complete mass spectra at a rate of 5–10 Hz [94].

B. Time-Array Detection

Innovations in data acquisition technology for a time-of-flight (TOF) mass spectrometer permit the technique of time-array detection (TAD) to solve many of the trade-offs in performance-specification problems forced onto conventional scanning instruments in an effort to meet the demands of chromatographic resolution. In the concept of TAD, all of the ion current information from an ion source extraction can be utilized [95] to substantially improve ion counting statistics. TAD provides improved sensitivity in TOF MS, while producing complete mass spectra at a rate greater than that required by modern capillary GC techniques. In essence, TOF MS with the use of TAD provides complete mass spectra at rates that can reach 100 Hz, at a sensitivity that otherwise could be reached only by the technique of selected ion monitoring in instruments that scan the *m/z* scale.

The TOF mass spectrometer is a pulsed instrument, but its spectral generation rate is effectively limited only by the transit time of the highest *m/z* value (slowest) ion in the mass spectrum. There are several variables that affect the transit time of a given ion, but under

---

[3] A decade is a segment of a mass spectrum in which the *m/z* value at the end point is an order of magnitude greater than the *m/z* value at the starting point. For example, the segment from *m/z* 30 to *m/z* 300 constitutes one decade; the segment from *m/z* 300 to *m/z* 3000 constitutes another decade. A mass spectrum over a range of *m/z* 20 to *m/z* 700 covers more than a decade, but less than two decades.

typical operation for ions over an *m/z* range of 1–1000 a spectral generation rate of 10 kHz can be easily accomplished. Although the information in any one of these 10,000 transient spectra is not particularly useful, the summation processes available with an integrating transient recorder (ITR) have the capacity to permit tens to hundreds of transients to be summed to produce individual mass spectra at an effective rate of 20–1000 Hz, each having suitable S/N. An example of high-speed GC-TOFMS involves the analysis of oils rendered from the peel of five diverse species of orange; a 14-m by 180-μm i.d. column (containing 5% phenyl dimethyl polysiloxane stationary phase) was temperature-programmed at 50 °C min^{-1} to achieve analysis times of under 140 sec with a spectral acquisition rate of 25 spectra sec^{-1} [96]. Other examples of rapid acquisition of mass spectra in TOF GC/MS involve the assay of pesticides in drinking water [97, 98].

4. Selected Ion Monitoring (SIM)

A. Definition and Nomenclature

Selected ion monitoring (SIM) refers to the dedicated on-line use of a mass spectrometer to acquire and record ion current only at certain selected *m/z* values. The resulting data records, selected ion current profiles (chromatograms), represent transitory profiles of particular compounds through the ion source. These data records have the appearance of a gas chromatogram if a GC inlet to the mass spectrometer is used and may be called selected ion current chromatograms; however, just like mass and TIC chromatograms, these chromatograms are reconstructed from mass spectral data. If the direct inlet probe is used, the data record can more accurately be called a selected ion current chronogram.   A more comprehensive or generic term would be selected ion current profile at a given *m/z* value; this name may be cumbersome, but it has the advantage of being explicit and descriptive.

Although almost two dozen names have been used in the scientific literature for this technique [99], SIM is used in this book for the following reasons: (1) the term *selected* is appropriate because it implies both choice and specificity; furthermore, it imposes no restriction as to the number of ions involved; (2) the word *ion* accurately describes the species being monitored; and (3) the term *monitoring* is preferred because it connotes the element of time in this specialized technique, which records profiles of ion currents as a function of time [99]. The data records are referred to as ion current profiles at a given *m/z* value.

B. Development of the Technique

The development of ion monitoring techniques has been reviewed by Falkner [100]. The modern version of ion monitoring was introduced by Sweeley *et al.* [101], who made it a practical technique for "simultaneously" monitoring ion currents at two or three *m/z* values with a magnetic mass spectrometer equipped with an accelerating-voltage alternator.   Hammar, Holmstedt, and Ryhage [102] refined the technique, which they preferred to call "mass fragmentography", and played a major role in demonstrating its analytical potential to the biomedical community.

A review article [103] covers applications of SIM into the 1990s; earlier reviews may also be useful [32, 104]. Even with all of these various names for the technique, the term SIM is still the most appropriate. Care should be taken to avoid calling this technique by the word "sim" or using the term "sim mode" when speaking about the technique; each letter should be clearly pronounced (i.e., S – I – M).

C. Qualitative Example of SIM

To illustrate the effectiveness of SIM as a selective means of detecting an analyte, some of the difficulties encountered in analyzing urine for normetanephrine (NMN) by gas chromatography using an electron capture detector (ECD) are first presented.

NMN, a metabolite of norepinephrine, must be chemically modified by acylation to the pentafluorylpropionyl (PFP) derivative as shown in Scheme 10-5 in order to improve its thermal stability for vapor-phase analysis. The PFP derivative of NMN is readily detectable by ECD because the derivative contains 15 fluorine atoms per molecule. Standard solutions of NMN-PFP were used to demonstrate that 50 pg of this derivative injected into the column could be readily detected by ECD, as illustrated in the chromatogram on the left side of Figure 10-22. However, when an aliquot of the derivatized urine extract was injected into the column under the same GC conditions, the electron capture detector was nearly saturated, and much lower settings of sensitivity had to be used in order to keep the detector response (to the interfering materials) to scale as shown in the chromatogram on the right side of Figure 10-22. There is no discernible discrete response for NMN-PFP at the expected retention time (arrow) in the chromatogram on the right side of Figure 10-22.

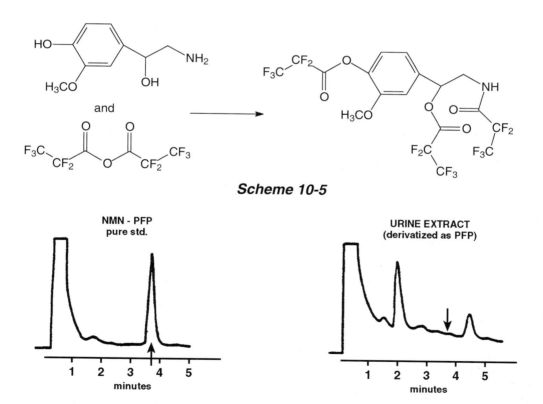

Scheme 10-5

Figure 10-22. **Gas chromatogram based on the response of pure standard of NMN-PFP (left) to an ECD and extract of a spiked urine.**

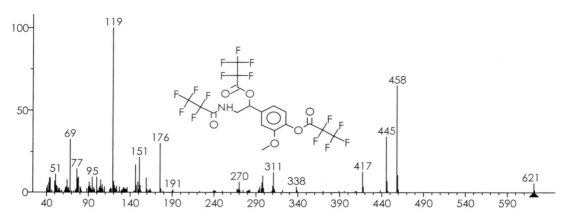

Figure 10-23.   EI (70 eV) mass spectrum of NMN as the pentafluoropropionyl derivative (NMN-PFP).

As an alterative, an SIM acquisition was set up on a magnetic-sector GC-MS to analyze the same samples. A cursory examination of the mass spectrum of authentic NMN-PFP (Figure 10-23) shows two major peaks at *m/z* 445 and *m/z* 458; the mass spectrometer was adjusted to monitor only the ion currents at these two *m/z* values throughout the gas chromatographic analysis of the sample.

The specificity of the SIM technique is illustrated in Figure 10-24, which shows two different detection profiles from the GC-MS during analysis of an aliquot of the derivatized biological extract as described below. Under the GC conditions, it had been established that NMN-PFP would emerge from the column at a retention time of 3.7 min. The upper chromatogram in Figure 10-24 is a profile of the total ion current (TIC) during the analysis; the TIC is non-selective in that it shows a response for any material that is ionized upon entering the ion source. As illustrated in the upper chromatogram of Figure 10-24, the TIC provides no discernible chromatographic peak at 3.7 min, the expected retention time for NMN-PFP.   In contrast, the bottom chromatogram in Figure 10-24 clearly shows the selected ion current chromatograms for *m/z* 445 and *m/z* 458. Although there is a substantial peak at a retention time of approximately 2 min, this compound produces only ion current at *m/z* 458 (nothing at *m/z* 445). However, at 3.7 min, the ion currents at *m/z* 445 and *m/z* 458 rise and fall

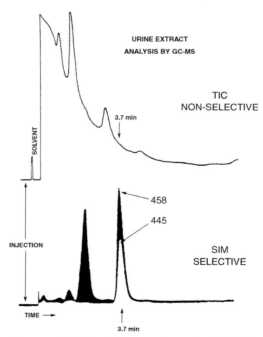

Figure 10-24.   Reconstructed total ion current chromatogram (top) and selected ion current chromatograms for m/z 445 and m/z 458 (bottom).

simultaneously, indicating that they have the same origin. Furthermore, the ratio of the areas of the two peaks in these selected ion current chromatograms ($I_{458}:I_{445}$) is approximately 2:1, as expected from the ratio of these major peaks in the mass spectrum of authentic NMN-PFP (Figure 10-23).

Therefore, observation of simultaneous peaks in the two selected ion current chromatograms in the expected abundance ratio and at the expected retention time provides good evidence that NMN-PFP is present in the derivatized urine extract. The aliquot of urine extract injected into the GC-MS to produce the chromatographic peaks in Figure 10-24 contained approximately 2 ng of NMN-PFP. This relatively large aliquot of the extract was injected so that a discernible signal could be observed in the relatively insensitive TIC profile. For routine analyses, sample aliquots containing a few pg of NMN-PFP gave readily discernible peaks by SIM [105].

D. Quantitative Example of SIM

Ideally, quantitative determination of an analyte by SIM is based on comparison between the ion current obtained from the analyte and the ion current from another compound chosen as an internal standard in the sample matrix; stable isotope-labeled analogs serve as ideal internal standards [106]. In this example based on normetanephrine (NMN), the internal standard is deuterium-labeled normetanephrine-PFP (d_3-NMN-PFP); the structure and mass spectrum of d_3-NMN-PFP are shown in Figure 10-25. The spectrum of d_0-NMN-PFP is shown in Figure 10-23. Note that the α carbon of the internal standard contains two deuterium atoms, and the β carbon contains a single atom of deuterium. These three deuterium atoms shift the mass of the molecule to 624 Da. As can be seen in the bar-graph mass spectrum in Figure 10-25, the $M^{+\bullet}$ of the deuterated analog of NMN-PFP appears at *m/z* 624.

The fragment ion with *m/z* 445 results from homolytic fission of the carbon-carbon bond between the α and β carbons as shown in Scheme 10-6. d_3-NMN-PFP fragments the same way to produce a fragment ion with *m/z* 446 because of the deuterium atom on the β carbon. d_3-NMN-PFP has virtually the same chemical and physical characteristics as d_0-NMN-PFP; only the mass spectrometer senses the difference between the d_0- and d_3-molecular ions and fragments formed by each. The ion with *m/z* 458 in the mass

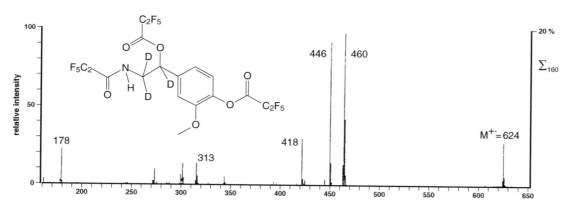

Figure 10-25. EI (70 eV) mass spectrum of *d₃*-NMN-PFP.

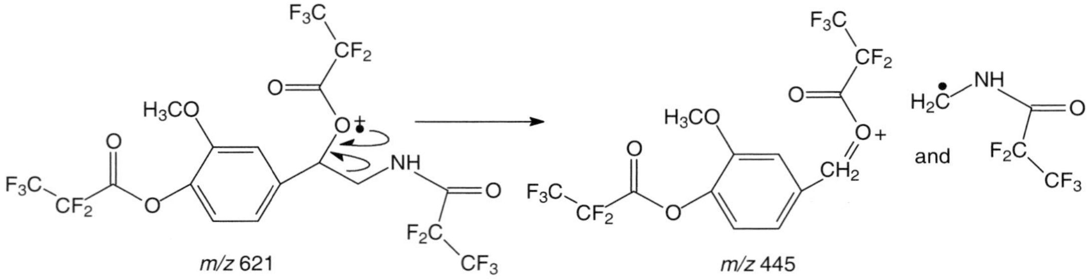

Scheme 10-6

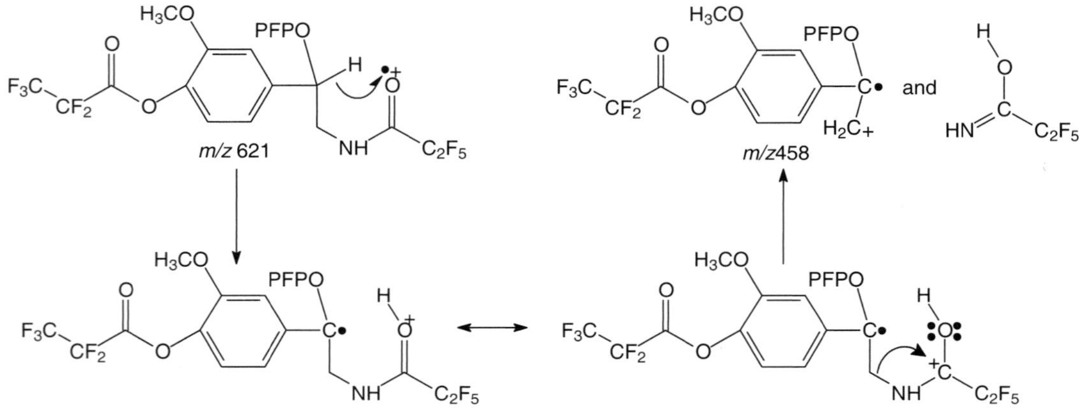

Scheme 10-7

spectrum of d_0-NMN-PFP is generated through a rearrangement as shown in Scheme 10-7. By transposing the molecule containing the three deuterium atoms into Scheme 10-7, it should become clear that the peak at m/z 458 shifts to m/z 460 rather than to m/z 461 because the deuterium atom on the β carbon is retained in the neutral loss radical.

Continuing with the quantitative viewpoint, it can be seen from the mass spectra in Figures 10-23 and 10-25 that the d_0-compound generates some ions that are distinctly different in mass from those generated by the d_3-compound. There is also some difference in relative intensities of the major peaks, such as those at m/z 458 and 460; this is due to an isotope effect of the deuterium atom on the α carbon during a γ-H rearrangement. Calibration plots made from data acquired during analyses of standard quantities of the d_3- and d_0-material will take this difference in fragmentation intensities into account. As can be seen in Figure 10-24, an important point is that the ion current at m/z 445 and m/z 458 can be monitored as two indices of the quantity of the d_0-compound present in any case; similarly, the ion current at m/z 446 and m/z 460 can be monitored as two indices of the quantity of the internal standard, namely the d_3-NMN-PFP, present in any case. Quantitative relationships can be based on the ratio of ion current at m/z 445 to that at m/z 446, or that at m/z 458 to that at m/z 460.

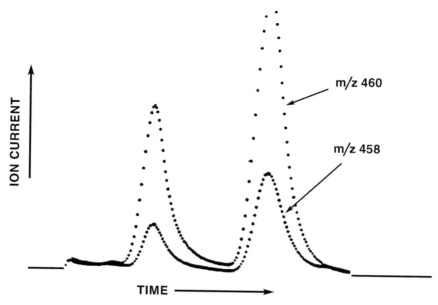

Figure 10-26. Selected ion current chromatograms for m/z 460 and m/z 458.

Consider the simple example of monitoring only one ion current from the deuterium-labeled material, the internal standard, and only one ion current from the analyte, namely the d_0-compound. Specifically, consider the ion current at m/z 458 as a quantitative index for the d_0-molecule, the analyte, and consider the ion current at m/z 460 as a quantitative index of the amount of internal standard present. If these two ion currents are monitored as a function of time as these two compounds are introduced through a gas chromatograph, it is expected that the selected ion current profiles at m/z 458 and m/z 460 will have the appearance of conventional-looking gas chromatograms. Figure 10-26 illustrates such a set of selected ion current chromatograms. The data are discontinuous or somewhat discrete in their presentation, as indicated by the dots; the discontinuities result from time-sharing the detector between the two ion currents. If these dots were connected for each ion current, smooth curves would result, showing the envelope of elution for each of the compounds. The reason that there are two sets of peaks of different retention times is because these results are from an analysis of a derivatized human urine extract, which also contains metanephrine (MN). The first pair of peaks in Figure 10-26 corresponds to MN-PFP and the second set of peaks correspond to NMN-PFP, the material under discussion [105].

For quantitative purposes, it is necessary to calibrate the mass spectrometer for its response to standard quantities of d_0- and d_3-NMN-PFP. This can be done quite simply by injecting replicate quantities of known amounts of the d_0- and d_3-NMN-PFP while monitoring ion currents at m/z 445 and 446, and at m/z 458 and 460, respectively. Then the ratio of those ion currents for the corresponding ratio of quantities of d_0- and d_3-labeled materials that were injected into the GC-MS is plotted against one another. Figure 10-27 is a calibration plot for the SIM results of standard quantities of NMN-PFP. The important feature in this plot is that the ratios of ion currents are linear functions of the ratios of quantities of material used to generate the ion currents. The fact that the two plots have different slopes is related to the isotope effect during the fragmentation

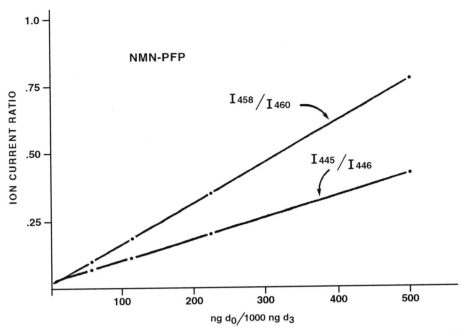

Figure 10-27. *Calibration plot for indicated ratios of ion currents monitored and ratios of the amounts of d_0-NMN-PFP (I_{458} or I_{445}) and d_3-NMN-PFP (I_{460} or I_{446}) injected into the GC-MS.*

mechanism involving shift of a γ H described earlier through Scheme 10-7. These different slopes have no bearing on the final quantitative answer so long as the relationship of the ratio of ion currents to the ratio of quantities is linear. The calibration plot in Figure 10-27 can be used to determine the ratio of an unknown quantity of analyte to the standard quantity of internal standard added at the beginning of the assay (i.e., the ratio of unlabeled to labeled material can be determined in the sample from the ratio of their corresponding ion currents defined earlier). With this ratio and knowledge of the quantity of labeled material (namely, the internal standard) that was added to a known quantity of biological sample in the first place, the quantity of unlabeled material or analyte that must have been in the sample at the outset can be calculated.

This example introduces quantitative relationships based on SIM. An important strategy when using an internal standard is to add a standard aliquot of this material to the sample at an early stage of processing; in this way, the internal standard accounts for losses of the analyte suffered during the sample processing. The ion current profiles in Figure 10-26 result from monitoring only one ion current from each of the two compounds, namely that at m/z 458 and at m/z 460. This procedure offers no alternative for quantitative computations if coeluting materials interfere with one ion current or the other. To prepare for such contingencies, it is a good idea to monitor a second set of ion currents so that if interference occurs in one set, the second set can be relied on for a good result. Even if there is no interference, it is always good to have a second set as a confirming ratio of ion currents. In the case of NMN, the ion currents at m/z 445 and at m/z 446 could also be monitored to represent the d_0- and d_3-compounds, respectively.

E. Mechanics of Ion Monitoring

Most mass spectrometers have only one detector. This means that to monitor more than one ion current, some provision must be made to focus the selected ions of interest alternately onto the detector for short periods of time (e.g., 50–250 msec). This is achieved by abruptly changing an instrument parameter (accelerating voltage or amplitudes of RF and DC voltages, which are being maintained at a static ratio to one another, depending on the type of mass spectrometer; see Chapter 2) to effect the necessary shift in the *m/z* scale (*vide infra*).

The benefits of increased sensitivity from SIM are only realized with instruments that operate by scanning a continuous ion beam through the *m/z* analyzer. Because the TOF and 3D QIT mass spectrometers operate in a pulsed fashion, there is no benefit of operation in the SIM mode. TOF instruments routinely acquire the complete spectrum 10,000 times per second, each spectrum requiring 100 μsec. An integrating transient recorder (ITR) has been developed [91, 107] to record all the mass spectral information from each pulse of the ion source of a TOF instrument. The ITR allows for time-array detection, which provides full-spectrum sensitivity at a level otherwise possible only by using SIM with a scanning instrument. The 3D QIT has all the ions of all *m/z* values available for detection in each pulse. These instruments cannot acquire data as rapidly as the TOF, but their full-spectrum sensitivity rivals that of the TOF.

1) Adjustment of the Mass Scale

Before ion monitoring can be used effectively, it is good practice to ascertain that the preset instrument parameters are indeed monitoring the ions of interest. Fine adjustment of the mass scale is best achieved during analysis of a standard aliquot of the compound of interest. During the preliminary runs, several different channels of slightly different fractional *m/z* value (e.g., 325.2, 325.3, 325.4, etc.) can be assigned to the same integer *m/z* (e.g., *m/z* 325 in this case) value to determine which will allow the detector to sample the ion beam from the "top" of the mass spectral peak.

2) Mass Range

Ideally, access to any combination of selected ions, regardless of *m/z* value throughout the mass spectrum, would be preferred. This is possible with a transmission quadrupole mass spectrometer, but with some magnetic instruments, the range of *m/z* values monitored may be limited to 10% of the upper value. Operational features as they relate to SIM are described here; more details on the operating principles of these mass spectrometers can be found in Chapter 2.

3) Magnetic Mass Spectrometer

As described in Chapter 2, either the accelerating voltage or the magnetic field can be used to control the mass scale. Because adjustments of the accelerating voltage can be achieved more rapidly and precisely than adjustments of the magnetic field, the former parameter is most often used for ion switching. The procedure commonly used for SIM involves focusing the ion of lowest *m/z* value onto the detector by adjustment of the magnetic field, which is then kept constant. The other ions of interest that have higher *m/z* values are then focused alternately onto the detector by attenuating the accelerating voltage to appropriate levels with a mechanical voltage divider [108] or via a programmable power supply [109]. Because the efficiency of ion transmission is a function of the magnitude of the accelerating potential, there is often a limit to the extent of attenuation, and hence the range of *m/z* values that can be monitored with the magnetic instrument.

4) Transmission Quadrupole Mass Spectrometer

The transmission quadrupole is very well suited for SIM [110] because selected ions from any region of the mass spectrum can be monitored without altering optimum conditions in the ion source or mass analyzer. Further, the parameters (superimposed RF and DC fields) that control the m/z axis can be changed rapidly, with good response and with no drift throughout the mass range for several hours.

5) Number of Ion Currents (Masses)

Although high specificity is a hallmark of SIM, high sensitivity is another. The highest sensitivity achievable by the beam-type mass spectrometer occurs when the recording system is dedicated to acquiring the ion current at only a single m/z value. In this situation, ion counting statistics of detection are optimized at the center of a discrete ion beam. In repetitive scanning, the data-acquisition system spends the majority of the analysis time sampling the barren region of the mass spectrum in between the maxima of individual ion beams. For this reason, the SIM mode of operation is approximately 1000 times more sensitive than the scanning mode.

To a limited extent, the more ion currents (m/z values) in a given mass spectrum that are monitored, the greater the confidence that the corresponding compound is present in the sample. Alternatively, monitoring a large number of ion currents leads to a poorer statistical representation of an ion current at any individual m/z value. Because SIM is ordinarily employed in cases in which high sensitivity is important, only two or three ion currents for each compound of interest are usually monitored. It is preferable to monitor at least two ion currents for each compound so that any ambiguity on one ion current profile can be qualified by comparison with another ion current profile, which should indicate simultaneous peaks at the appropriate retention time. In quantitative SIM, it is preferable to use an internal standard for each compound of interest; therefore, at a minimum, it is desirable to monitor two ion currents from each compound of interest and its internal standard. It was also pointed out earlier in this chapter that by comparing the intensity ratios of (or areas under the chromatographic peaks for) ions of three different m/z values, confirmation of the analyte was assured in most cases when using EI GC/MS.

F. Programmable SIM

In programmable SIM (a term unique to this book), the specificity of detection is changed as a function of time (e.g., as during the development of a gas chromatogram). This technique permits optimum detection of several sample components that can be separated by GC. Because a given sample may contain several different compounds of interest, it may be desirable to monitor as many as 8 to 12 or even more different ion currents so that all components can be quantitated. However, it would not be desirable to monitor all ion currents contemporaneously, because the greater the number monitored during a given timeframe, the poorer the S/N available for recording any given ion current. Therefore, an optimum compromise might involve monitoring only two to four ion currents during a specified time interval when the corresponding compounds are expected to emerge from the gas chromatograph.

The type of data record obtained with this technique is illustrated in Figure 10-28. In the first time period of this selected ion current chromatogram, the ion currents at m/z 236, 232, and 260 were acquired. At 1.9 min after sample injection (start of the second time period), the computer monitored ion currents at m/z 270, 261, and 290 for 1.2 min. Finally, during the third time period, the computer-controlled quadrupole monitored the ion currents at m/z 269, 237, and 266.

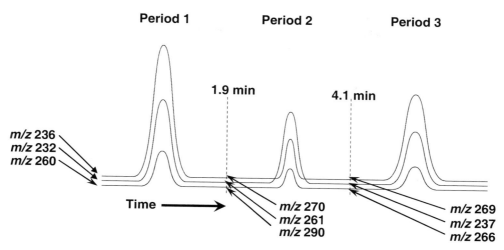

Figure 10-28. Data record from the use of programmable SIM to monitor three different sets of ion currents during three specified time periods.

In summary, programmable SIM offers an optimum compromise between operating the GC-MS for maximum sensitivity and obtaining high confidence in analytical results.   By monitoring more than one ion current for each compound of interest, the probability of reporting false positives is reduced. The programmable feature provides the desired specificity of detection at several points in a gas chromatogram without prohibitive loss in sensitivity resulting from monitoring a large number of ion currents.

G. SIM at High Resolving Power

The specificity of SIM can be increased by monitoring ion current only at the *m/z* value equal to the *accurate m/z* value of the selected ion.   For example, in analyzing plasma extracts for androst-4-en-3,17-dione, Millington and coworkers [111, 112] monitored the ion current at *m/z* 286.193. Figure 10-29 illustrates the narrow *m/z* window of acceptance used in the vicinity of nominal *m/z* 286 to achieve high selectivity against chemical noise.   This approach has been used to analyze a variety of biomedical [111–113] and environmental [109] samples.

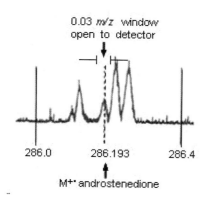

Figure 10-29. Oscillographic recording of isobars at a nominal m/z value of 286 when scanned with a R of 10,000. This figure illustrates the effective window or slit open to the detector at this R that permits detection of the ion beam centered at m/z 286.193 for androst-4-en-3,17-dione without interference from the other nominal isobaric ion currents; the effective m/z window to the detector is only 0.03 m/z units wide. From Millington DS "New Techniques in Quantitative Mass Spectrometry" J. Reprod. Fert. 1977, 51, 303–308, with permission.

VI. Sources of Error

1. Errors Relating to Equipment or Procedure

In qualitative work, a common source of error may be improper identification of the *m/z* scale. The validity of the *m/z* scale should be verified daily by obtaining the mass spectrum of a calibration compound as described in Chapter 2. This procedure may require a few minutes, but it can save hours of wasted time in attempting to interpret bad data.

It is important to be familiar with the software algorithm used in identifying a mass spectral peak and assigning an *m/z* value. If the software has the feature of rounding off the *m/z* value to the nearest integral unit, it is possible that errors can accumulate due to disparity in the magnitude of the mass defect (see Chapter 5 for more information about mass defect) between elements in the calibration compound and those in the analyte, or from failure to account for these differences properly.

Errors in library searches can result from mass spectral skewing (i.e., a distortion in the relative intensities of peaks). Skewing can be imposed by improper matching of timeframes involved in mass spectral data acquisition (slow scan speeds) and in analyte delivery (narrow peaks in GC) as described earlier in this chapter. Mass spectral skewing can also result from improper ion-abundance tuning of the instrument; this issue has been addressed officially by the EPA [114].

Quantitative applications are prone to a number of sources of error [115]. A good starting place for error prevention is in establishing a detection limit for the analyte. Some good questions then might be, "Why is the detection limit only one nanogram?", "Are low levels of the analyte being irreversibly adsorbed before they reach the ion source?", "Has the column been silanized?", "Is the peak shape symmetrical, or is there tailing, which might suggest sample adsorption to surfaces?". Other derivatives might be evaluated for vapor-phase characteristics, fragmentation pattern, and *m/z* of abundant ions.

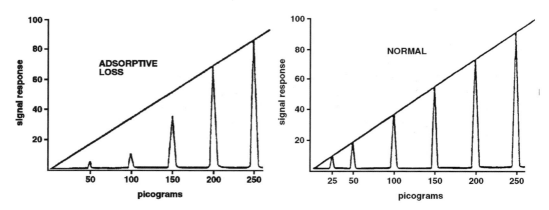

Figure 10-30. *Plot of instrument response to a series of individual injections of increasing amounts of analyte for the expected or normal case (right) and for the case in which an aliquot of each injection is lost to surface adsorption (left).* From Millard BJ, *Quantitative Mass Spectrometry, Wiley-Heyden, London, 1978, with permission.*

The problem of adsorptive loss of low levels of analyte can be recognized from the shape of the quantitative calibration plot. The adsorptive surface responsible for the problem will usually adsorb a given quantity of material each time the sample passes the offending site; when the concentration of analyte is high in a given sample, the diminution in signal response may not be noticeable because the relative proportion of the sample is small, but when the concentration of analyte is low in a given sample, the relative proportion of analyte adsorbed is likely to be significant. This scenario is illustrated in Figure 10-30. As shown in the right panel, the signal response decreases linearly with quantity injected; however, in the left panel, the signal response for low levels of analyte injected is lower than expected (due to adsorptive loss of the analyte). Such adsorptive loss of analyte can occur on the walls of the sample container or on any surfaces that the sample "sees" *en route* to the ion source. Because the surface of glassware presents a potential adsorptive surface, it is good practice to silanize it prior to using it for processing or storing samples containing low levels of polar analytes, which are likely to be prone to adsorptive loss. Introducing an excess of a stable isotope-labeled analog of the analyte (to act as a carrier) can also be used to alleviate this problem of low-level adsorption.

Other errors in quantitation can be similar to those incurred in peak area determination in GC [116]. Area measurements are subject to error in selected ion monitoring as well as in the reconstruction of mass chromatograms.

Accumulation of nonvolatile materials at the entrance to the GC column (injector liner) often has a detrimental effect because it provides an activated surface for sample adsorption or thermal decomposition. If the GC column is made of glass, this possible source of error can be evaluated by visual inspection for dark oily deposits or discolored column packing. This problem can often be remedied by removing the discolored packing, removing oily deposits from the injector liner with a solvent-moistened cotton swab, or simply cutting off the first few centimeters of a fused silica capillary column, and/or treating the column with a silanizing reagent such as bis-trimethylsilyltrifluoroacetamide (BSTFA), if compatible with the liquid phase.

2. Errors Relating to Contamination

Plasticizers are common contaminants of samples that have been stored in organic solvents in bottles or ampoules with plastic or rubber seals (proper containers for storage of organic liquids are described earlier in this chapter). Dioctyl phthalate has good vapor-phase properties; it elutes from a nonpolar GC column at approximately 200 °C as a sharp peak and produces a mass spectrum similar to that in Figure 10-11. The peak at *m/z* 149 in Figure 10-11 is the base peak and the most commonly encountered containment peak with phthalates; peaks at *m/z* 207, 281, 355, and 429 (along with their dominant isotope peaks due to the large number of Si atoms present) are due to column bleed from OV-1 as shown in greater detail in Figure 10-21.

A mass spectral peak at *m/z* 149 should always be considered with suspicion because it can represent an ion produced by any dialkyl phthalate other than the dimethyl ester [117, 118]. Some of these phthalates are components of pump oil, the vapors of which can be a source of contamination. It is easy enough to test the ions of *m/z* 149 to see if they are a part of the mass spectrum to be interpreted or are due to the background or to an almost coeluting substance. Plot a mass chromatogram for *m/z* 149 and a mass chromatogram for another *m/z* value that is most certainly a part of the spectrum to be interpreted. If the two mass chromatographic peaks rise and fall together, the peak at *m/z* 149 is part of the spectrum under consideration and must be rationalized. If the mass chromatogram for *m/z* 149 is a more or less flat line, then the source of this ion is some substance(s) that is (are) in the background of the GC/MS system (e.g., a contaminant in the pump oil). If the mass chromatographic peak of the ion with m/z 149 is a true peak (raises and falls), but does not follow the profile of the mass chromatographic peak of the other ion, then the ion with m/z 149 is produced by a near-coeluting substance.

3. Sources of Interference

The stationary phase in GC columns has an appreciable vapor pressure that at most operating temperatures is sufficient to produce a recognizable mass spectrum when the column is connected to the mass spectrometer's inlet. The mass spectrum of column bleed from a 2-m by 2-mm column of 3% OV-1 (methylsilicone) at 260 °C was described earlier as shown in Figure 10-21. It is useful to obtain a background mass spectrum of each GC stationary phase to be used so that ions associated with the GC column can be easily recognized and disregarded during the interpretation of mass spectra. Of course, the predominance of the column-bleed mass spectrum is proportional to the column temperature. In most cases the interfering ions will not be a problem in mass spectra obtained from 10 pmol or more of sample.

4. Dealing with Background in a Mass Spectrum

Background has been the bane of GC/MS since the technique's origin. Background subtraction is a process in which a spectrum is subtracted from another spectrum in an effort to remove mass spectral peaks that have their origin from a chemical source other than the analyte. The result of the background subtraction is presumably a pure spectrum, the interpretation of which will lead to an unambiguous structure for the analyte. Should the spectrum to be used as the background be selected from a part of the data file just before or just after the chromatographic peak or an average of spectra from both areas? Whether the background spectrum should be a single spectrum from the data set or an average of a number of spectra from a contiguous range or from different areas of the data file is another question that is always important in discussions of background subtraction. There is not a single answer to these questions.

> Column bleed is not only a result of stationary-phase decomposition, but is also due to high-boiling substances from the injected samples that elute very slowly and never can be represented by a distinguishable chromatographic peak. If sources other than stationary phase are believed to be bleeding off the GC column, a data file should be acquired using the temperature program used for analyzing the samples, but without injection or with an injection of pure solvent. These data files should be compared with data files acquired during an analysis from a clean column with no injection or with only solvent injected.

Because the origin of the background is often column bleed, the abundance of ions due to the background will increase with increasing column oven temperature. In such cases, good candidate background spectra are those acquired during an interval just after the chromatograph peak. However, if another chromatographic peak precedes the one of interest such that the two peaks nearly overlap, the spectrum that was acquired during the valley between the two chromatographic peaks should be considered as a candidate for the background spectrum. It may be desirable to use this "valley" spectrum averaged with a few spectra that follow the chromatographic peak of interest.

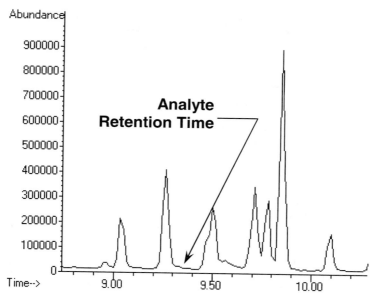

**Figure 10-31.   Partial RTIC chromatogram of data acquired from
m/z 40 to m/z 450.**

Rather than using RTIC chromatographic data in selecting a background spectrum, it may be best to use mass chromatograms.   Consider the partial RTIC chromatogram shown in Figure 10-31.   It turns out that there is no discernible RTIC chromatographic peak at the expected retention time for the analyte of interest (see arrow in Figure 10-31).   As can be seen from examining the spectrum in Figure 10-32, there appears to be a number of mass spectral peaks due to background (e.g., *m/z* 207, *m/z* 281, *m/z* 355, and *m/z* 429).   Also, about half of all the peaks between *m/z* 40 and *m/z* 340 are about the same size, which is consistent with these peaks representing electronic noise that is quantized at the same level through the analog-to-digital converter between

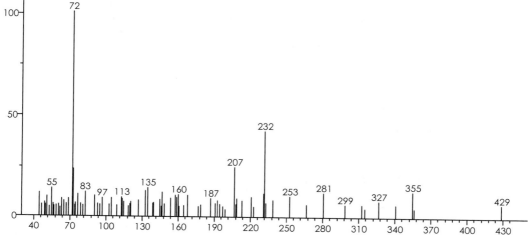

**Figure 10-32.   Mass spectrum from the position indicated by the arrow
in the RTIC chromatogram in Figure 10-31.**

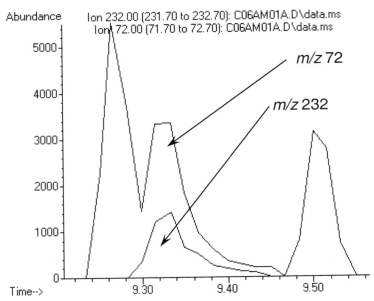

Figure 10-33. *Overlay of mass chromatograms of m/z 72 and m/z 232.*

the detector and the data system. However, there are two relatively intense mass spectral peaks in Figure 10-32 that may possibly be used to characterize the analyte: the base peak at *m/z* 72 and the peak at *m/z* 232, which could possibly represent a $M^{+\bullet}$.

An overlay of mass chromatograms for *m/z* 72 and *m/z* 232 is shown in Figure 10-33. From a comparison of the two mass chromatograms, it is obvious that the two ions represent the same compound because the two mass chromatograms have the same temporal profile. The compound of interest has an ion in common (*m/z* 72) with the preceding chromatographic component. Therefore, an apparently good choice for the background spectrum is the spectrum acquired at approximately 9.30 minutes during the chromatographic valley between the two components that have ions of *m/z* 72. This is a case in which a single spectrum is justified as the background spectrum. Figure 10-34 is the background-subtracted spectrum resulting from subtracting the spectrum acquired at the valley (approximately 9.30 min in Figure 10-33) from the spectrum at the apex of the mass chromatographic peak at *m/z* 232 in Figure 10-33 (two spectra later than the background spectrum).

The nonbackground-subtracted spectrum (Figure 10-32) and the spectrum that was background-subtracted (Figure 10-34) produced different and rather interesting results when searched against the NIST05 Mass Spectral Database of 163K compounds using the NIST MS Search Program. The NIST MS Search Program has two modes for an Identity Search (a search whose purpose is to identify the unknown spectrum if it is in the Database): Quick Search and Normal Search. The difference between these two searches is the pre-search algorithm employed. The Quick Search of the nonbackground-subtracted (raw data) spectrum (Figure 10-32) did not result in a *Hit* for the correct compound. The Normal Search did produce a *Hit* for the compound as the first *Hit*. Obviously the compound's spectrum was not one of the selected spectra in the pre-search when the Quick Search was carried out. Both the Quick Search and the Normal Search of the background-subtracted spectrum yielded the compound as the first *Hit* with identical Match and Reverse Match Factors (466 and 698, respectively). These

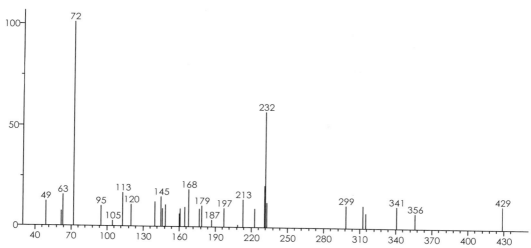

Figure 10-34.   Same spectrum as Figure 10-32, but background-subtracted using spectrum between two mass chromatographic peaks.

numbers were similar to the values obtained for the nonbackground-subtracted spectrum (437 and 784, respectively).   The higher Reverse Match Factor for the nonbackground-subtracted spectrum is understandable because background subtraction may have removed completely, or at least reduced in intensity, certain peaks representative of the compound.

Although it appeared obvious where to select the background spectrum, it may not have been the best spectrum to use for background subtraction. A Library search of the last spectrum (shown in Figure 10-35) acquired on the tailing edge (right side) of the mass chromatographic peak for *m/z* 232 resulted in a Match Factor of 629 and a Reverse Match Factor of 790. This Search yielded a Probability Value of 90.9% that the analyte had been identified, whereas the Probability for the Normal Identity Search of the

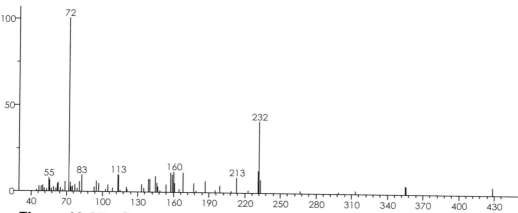

Figure 10-35.   Same spectrum as Figure 10-32, but with spectrum on tailing edge of the chromatographic peak used as the background spectrum.

spectrum shown in Figure 10-34 was below 40%. This is a good illustration of the fact that when selection of the background spectrum is based on human judgment, guesswork may play as big a role in the outcome as logic and rules. Computer algorithms for background subtraction employed by most commercial mass spectrometer data systems mainly use a background based on the average of several spectra. These spectra, which are averaged, are often acquired during intervals far away from the elution point of the analyte and therefore do not represent the mass spectra of coeluting substances, as would have been the case with the just-presented example. A special program has been designed to address the issue of background subtraction; this is AMDIS (Automated Mass spectral Deconvolution and Identification System) as described below.

A. AMDIS (Automated Mass spectral Deconvolution and Identification System)

AMDIS was developed by the National Institute of Standards and Technology Mass Spectrometry Data Center for the United States Army [119]. This program is designed to support compliance with the *Convention on the Prohibition of the Development, Production, Stockpiling and Use of Chemical Weapons and on their Destruction*, to which the United States is now a party to since April 29, 1997. The program is used by the Organization for the Prohibition of Chemical Weapons (OPCW) to determine if any of the Convention's listed compounds are present in samples analyzed by GC/MS. AMDIS is designed to separate near-coeluting analytes observed in GC/MS data files that are in widely varying concentrations compared to one another and other materials in the GC eluate. AMDIS deconvolutes components in RTIC chromatographic peaks. The program automatically compensates for spectral variation based on such factors as spectral skewing that is observed with beam-type instruments, based on the origin of the data file (instrument manufacturer's data file format). The spectral skewing in an Agilent GC/MS ChemStation data file will be different than that in a Thermo Fisher Xcalibur data file because the Agilent transmission quadrupole mass spectrometer scans from high to low *m/z* values and the Thermo Fisher instrument (and all other manufacturers of transmission quadrupoles) scans from low to high *m/z* values. TOF and QIT GC/MS systems do not produce skewed data; therefore spectral skewing correction should not be applied to these data.

> When describing an RTIC or mass chromatographic peak by stating the number of spectra that define the peak, it is important to consider only spectra of the compound. A chromatographic peak can have a starting spectrum that is primarily of the background. This is followed by a spectrum that represents the analyte. The third spectrum, like the first spectrum, could represent primarily the background. If so, this chromatographic peak would have to be called a single-spectrum reconstructed chromatographic peak. As a general rule, each spectrum chosen to represent the chromatographic peak must contain signals, corresponding to the most intense mass spectral peaks for the putative analyte, which are >10% of the signal due to the background.

AMDIS recognizes the instrument that produced a data file when the file is loaded based on the data file's extension and information contained in the data file's header. AMDIS is specific to every type of commercially manufactured GC/MS with the exception of the LECO Pegasus GC/TOFMS instrument, which has its own proprietary software for GC/MS RTIC chromatographic peak deconvolution. AMDIS can also be used with generic data files in the NetCDF format, a universal GC/MS data file format. LECO files can be processed with AMDIS. AMDIS functions best when an RTIC chromatographic peak is defined by at least four spectra. The peak area function will not work if there are more than 20 spectra on either side of the spectrum representing the apex of the RTIC chromatographic peak.

A simplistic explanation of how AMDIS performs deconvolution is that it plots mass chromatograms for each integer m/z value in the range of m/z values defined in the data acquisition parameters of the data file. These mass chromatographic peaks are then analyzed to determine which ones rise and fall together. This is the same principle applied by the original Biller–Biemann algorithm [120]. However, unlike the Biller–Biemann algorithm, which assigns the ion current of a given m/z value to a single component, AMDIS will proportionally distribute the ion current to analytes that have isobaric ions. These features can be appreciated in the context of a problem like the following involving the analysis of a mixture of benzene (mass spectrum shown in Figure 10-36) and 1,2-dichloro-2-propene (mass spectrum shown in Figure 10-37), in which the two components nearly coelute from a particular GC column as represented by a single peak in the RTIC chromatogram (not shown). Both compounds have a mass spectral peak at m/z 77 (for the $[M - H]^+$ ion of benzene and the ^{37}Cl-isotope peak for the $[M - Cl]^+$ ion of 1,2-dichloro-2-propene). Because of differences in the distribution of the total ion current in the mass spectra of the two compounds, the ion current at m/z 77 from benzene will be greater than that from the 1,2-dichloro-2-propene when equal amounts are present or even when benzene is the minor component, down to about 25% of the amount of 1,2-dichloro-2-propene. Use of the Biller–Biemann algorithm will result in all of the ion current of m/z 77 being applied to the benzene-deconvoluted mass spectrum (not shown). However, because AMDIS examines the shapes of the mass chromatographic profiles, it can assign proportions of the ion current at m/z 77 to each of the two

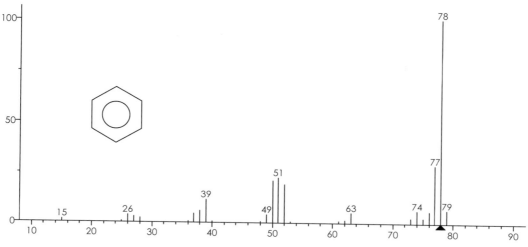

Figure 10-36. EI mass spectrum of benzene.

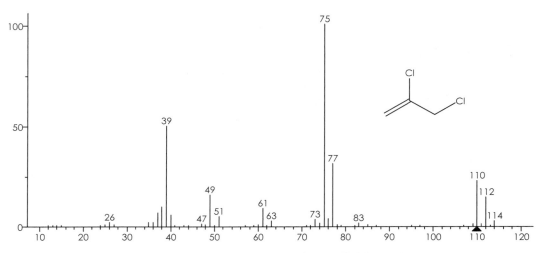

Figure 10-37. EI mass spectrum of 1,2-dichloro-2-propene.

deconvoluted spectra.   In this way, AMDIS can be applied to mass spectra of all components in a data file, to spectra of selected target compounds (with or without specified retention times), or to data files related to both target and nontarget analytes.

Reconsider the example described above involving Figure 10-31 in which the background spectrum was chosen manually. In the earlier-described example relating to Figures 10-33 to 10-35, it was assumed that mass spectral evidence was present at the expected retention time even though, as can be seen in Figure 10-31, there was no discernible peak in the RTIC chromatogram to indicate the appropriate series of mass spectra representing the analyte. However, if the same data file as represented in Figure 10-31 is processed automatically by AMDIS, the presence of the analyte (characterized by mass spectral peaks at m/z 72 and m/z 232) is clearly indicated by the output in Figure 10-38. The top panel in Figure 10-38 is the RTIC chromatogram over a relatively broad retention-time interval from approximately 8.8 to 10.0 minutes. The presence of the component of interest is indicated in the RTIC chromatogram window (top panel in Figure 10-38) by the ▼ symbol with the bar on top at a position above the RTIC chromatogram that corresponds to a retention time of approximately 9.334 minutes. The graphic portion of the middle panel of Figure 10-38 shows an overlay of the RTIC (here labeled as TIC) with two mass chromatograms (one at m/z 72, the other at m/z 232) from an expanded interval of retention time between approximately 9.29 and 9.38 minutes. The bottom panel of Figure 10-38 shows two superimposed, but slightly off-set, mass spectra; the dark lines represent the raw data, the light lines represent the background-subtracted (extracted) mass spectrum. Although not visible in the spectrum display, there are peaks that are represented by dashed light lines; these peaks are peaks that AMDIS has identified as questionably belonging to the deconvoluted spectrum (these are referred to as questionable (mass spectral) peaks or *uncertain peaks*).

An examination of the AMDIS background-subtracted spectrum in Figure 10-39 indicates that many mass spectral peaks from the "raw data" spectrum (Figure 10-32) have been removed in this deconvoluted spectrum. A search of the AMDIS-deconvoluted spectrum resulted in a Match Factor of 710, a Reverse Match Factor 737, and a Probability of 92. Other than the Match Factor, these values are not very different from the values obtained when the spectrum constituting the tailing edge of the

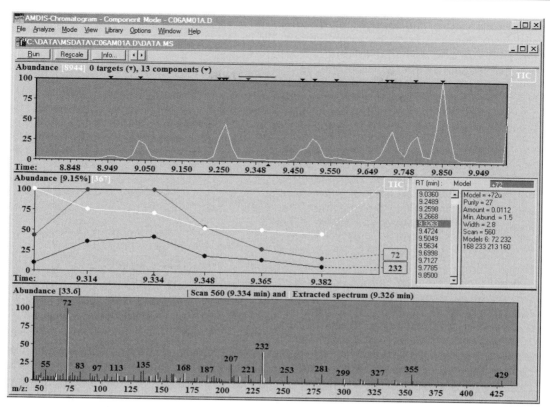

Figure 10-38. Result of an AMDIS deconvolution analysis.

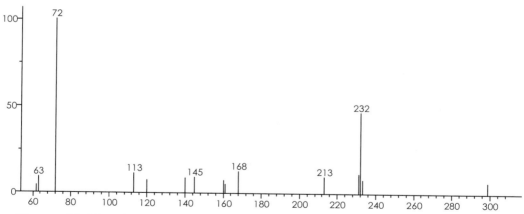

Figure 10-39. Mass spectrum resulting from an AMDIS deconvolution background subtraction.

component in the mass chromatogram was used as the background spectrum (~6% or less). It should be noted that selection of this spectrum for the manual background subtraction was not intuitive and the non-AMDIS process did not involve selecting a spectrum for the analyte. The results (Match Factor, Reverse Match Factor, and Probability) of the library search of the spectrum resulting from the logical choice for a background spectrum (Figure 10-34) were exceeded by the results of the library search of the AMDIS deconvoluted spectrum, in all three categories by 10–50%; the most notable of which was the Probability value of 92.0% vs 38.5%.

The above-described example involved AMDIS processing of mass spectra of what could be considered a nontarget analyte. AMDIS is even more impressive in processing the mass spectral data from analyzing target analytes. The RTIC chromatographic peak in Figure 10-40 looks like it could easily represent a single component; it is symmetrical and defined by seven data points from seven consecutively recorded mass spectra. The spectrum (Figure 10-40) taken at the apex of the RTIC chromatographic peak (Figure 10-41) appears to be very clean with little in

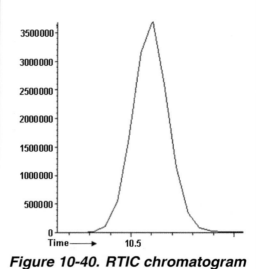

Figure 10-40. RTIC chromatogram of an apparent single component.

terms of background. Based on information from Chapters 5 and 6, it appears that the peak at *m/z* 128 could be a M$^{+\bullet}$ peak. If the peak at *m/z* 128 is a M$^{+\bullet}$ peak, the analyte is probably an aromatic compound based on the intensity of the M$^{+\bullet}$ peak and the fact that there is little apparent fragmentation of the M$^{+\bullet}$. The intensity of the peak at X+2 (*m/z* 130) indicates that the analyte may contain more than a single atom of sulfur, especially when compared to the intensity of the peak at *m/z* 129 (the X+1 peak). A closer examination of the seven consecutively recorded mass spectra used to reconstruct this RTIC chromatographic peak reveals that they represent at least two different compounds. The spectrum (Figure 10-42) at the beginning (leading edge) of the chromatographic peak (Figure 10-40) exhibits an X+2 peak with a much greater intensity

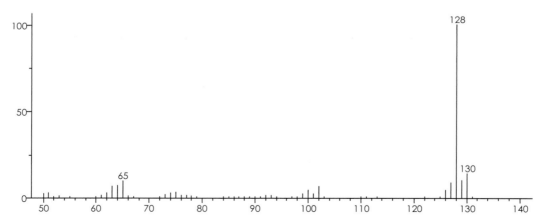

Figure 10-41.   Mass spectrum at the apex of the RTIC chromatographic peak shown in Figure 10-40.

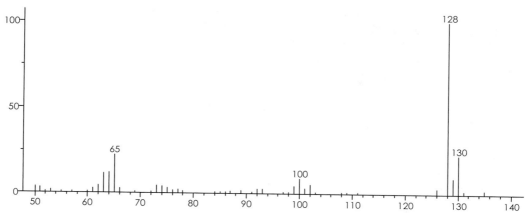

Figure 10-42. Spectrum at the leading edge of the RTIC chromatographic peak shown in Figure 10-41. Note the increased intensity of the X+2 peak at m/z 130.

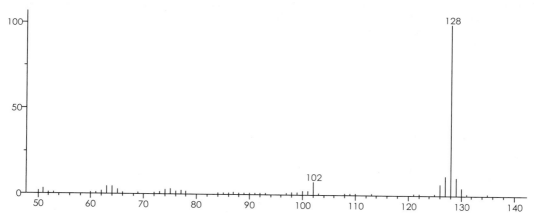

Figure 10-43. Spectrum at the tailing edge of the RTIC chromatographic peak shown in Figure 10-42. Note the intensity of the X+2 peak at m/z 130 is less than that observed in Figure 10-42 or Figure 10-41.

than that in the spectrum (Figure 10-41) at the apex of the chromatographic peak, whereas the spectrum (Figure 10-43) at the end (tailing edge) of the chromatographic peak shows an X+2 peak with an intensity expected for an ion that contains only carbon and hydrogen, and maybe oxygen. When this data file is submitted to AMDIS using an AMDIS target-compound library that contains spectra of the two analytes that compose this RTIC chromatographic peak (Figure 10-40), AMDIS identifies both compounds; the first is 3-chlorophenol and the second is naphthalene (Figure 10-44).

Figure 10-44 is the display obtained when a Target Compound Analysis is carried out using AMDIS. The top panel is the reconstructed total ion current chromatogram. Below and to the left are the mass chromatograms that characterize the naphthalene

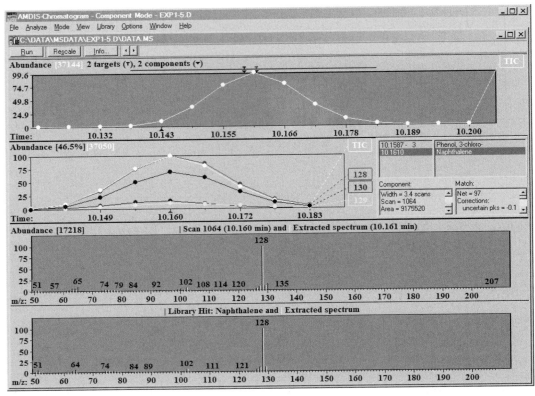

Figure 10-44. AMDIS Target Analysis display identifying 3-chlorophenol and naphthalene as two components that comprise the RTIC chromatographic peak shown in the top panel of the display.

component. To the right of that is the report window identifying the two components. The information in the lower part of the report window concerns the quality of the match. The window just below the mass chromatogram and report windows contains the sample spectrum and the deconvoluted spectrum superimposed on one another. The bottom window contains the deconvoluted spectrum and the AMDIS library spectrum for naphthalene superimposed on one another.

Naphthalene and 3-chlorophenol have essentially the same mass spectrum with the exception of the ^{37}Cl-isotope peak in the spectrum of 3-chlorophenol. Both compounds maximize in the same spectrum number (scan 1064) as indicated in Figure 10-44, but they have retention times that are 0.002 minutes (0.12 seconds) different (10.159 min vs 10.161 min) as determined by deconvolution using AMDIS. Because of the similarities in chromatographic and mass spectral behavior, it would not have been possible to perform a manual background subtraction to identify these two components. An automatic background subtraction routine does nothing with respect to component deconvolution. Therefore, it was necessary to use AMDIS to deconvolute the data to allow identification of these two compounds. AMDIS also produces a deconvoluted-component chromatographic peak area that can be used for quantitation of these two compounds. Conventional quantitation software associated with the usual commercial data system cannot quantitate these two components under these elution conditions.

These examples illustrate just some of the tasks that can be performed with AMDIS. More details can be obtained by working with the program and by using the tutorial that can be found in the manual. AMDIS is a free program provided with the NIST MS Search Program and the NIST05 Mass Spectral Database. It can also be downloaded from http://chemdata.nist.gov/mass-spc/amdis/.

It should be noted that AMDIS does produce two text files that contain sufficient information to perform quantitation; however, AMDIS does not produce quantitation reports and does not provide the type of quantitation reporting capabilities that are available in the form of proprietary software from some instrument companies. Custom programming is necessary to view the data created by AMDIS or to output it as text files. Another important limitation of AMDIS, especially when it comes to quantitation, is that chromatographic peaks reconstructed from more than twenty spectra on either side of the spectrum representing the apex are not properly processed by the program. The spectra beyond the twentieth spectrum are not considered in determinations of chromatographic peak area or for purposes of deconvolution.

B. Other Software Techniques

Although AMDIS is discussed in detail, it should not go unmentioned that the LECO Pegasus GC/TOF-MS is provided with a deconvolution routine that is optimized for the fast-GC (use of short, narrow bore (<2 μm i.d.), low stationary phase-loading columns with rapid temperature ramps) capabilities of this instrument. The LECO deconvolution system is also designed to work with the GCxGC (Pegasus 4D) capabilities that are provided as an option for their TOF GC/MS system. Details of the LECO ChromaTOF® software can be found on their Web site at http://www.leco.com under the Products topic. Like AMDIS, the LECO deconvolution software is designed to be used with target analytes or to identify components in a mixture even though they are not target analytes. The LECO deconvolution software is ideally suited for the data produced by the Pegasus GC/MS system and has been reported to provide results that rival those of AMDIS.

A third system that is designed to work with target compounds and unknowns like AMDIS and the LECO deconvolution system software is AnalyzerPro from SpectralWorks Ltd in the United Kingdom (http://www.spectralworks.com). Like AMDIS, the program will read a number of different instrument manufacturers' proprietary data file formats. Unlike AMDIS, the program treats all data files the same regardless of what instrument produced them. The data files are converted from the instrument company's file format to an AnalyzerPro format. These files in the converted format are permanently stored until they are deleted by the user. Also, unlike AMDIS, AnalyzerPro does have a very detailed reporting feature for quantitation. AnalyzerPro has many of the features of AMDIS, but with a simpler interface and a capability for generating quantitation reports. AnalyzerPro requires more interaction to obtain the best set of Deconvolution parameters than does AMDIS, but it also makes forming and editing target compound libraries much easier. Both AMDIS and AnalyzerPro allow spectra to be sent to the NIST MS Search Program or to be searched internally using the NIST dll. AnalyzerPro has made the inclusion of the NIST MS Search Program's dll very straightforward.

There is another third-party software package that provides a degree of deconvolution for target compounds. Unlike AMDIS, AnalyzerPro, and the LECO deconvolution software, if the analyte has not been registered according to its retention time (RT) and the relative intensity of certain mass spectral peaks, it will not be detected by the *Quantitative Deconvolution Software for Mass Spectrometry* produced by Ion Signature Technology (IST–http://www.ionsingnature.com). The IST software uses an algorithm based on mass chromatograms and the ratio of the areas of these mass chromatographic peak areas for ions that are characteristic of the analyte. These data are used in conjunction with retention time data. Quantitation is performed based on a single mass chromatogram. Confirmation of the analyte is a function of RT and area ratios of up to nine different mass chromatograms used to characterize the analyte. Each target analyte must be characterized by a minimum of two ions. There is nothing in the description of the algorithm that states how the program handles ions from contaminants that are isobaric with ions used to characterize a target analyte. Quantitation is always carried out against an internal standard.

Another very convenient program that allows for accurate mass measurements for data acquired using a transmission quadrupole GC/MS system is Cerno Bioscience's MassWorks (http://www.cernobioscience.com). Provided that the mass spectral data are stored in a profile mode (multiple data points for each mass spectral peak), this program will return accurate mass data for ions. MassWorks is based on a calibration that is carried out using a compound that produces mass spectral peaks of known exact masses. The calibration compound can be in the data file containing the analytes or in an external data file. The calibration is performed much like the normal *m/z* scale calibration procedure for an instrument, but differs based on the exact masses of ions rather than their nominal mass values. The software claims a mass accuracy of 5 ppm for data acquired using a unit-resolution instrument.

There is a third-party program that uses a unique automated library search routine on databases of <10K spectra and retention times for the analysis of GC/MS data. This is MassFinder produced by Dr. Hochmuth Scientific Consulting in Germany (http://massfinder.com). What makes this program unique is that it displays the preceding and subsequent spectrum along with the spectrum of interest. As a spectrum of interest is displayed, it is automatically searched against a specified library. Another unique feature of this particular program is the capability for including structures with chirality features. This can be done in user libraries with the NIST MS Search Program and the ACD Labs software, but is lacking in most other mass spectral database-building software.

Most commercial mass spectral data system software is covered by a restrictive license that limits its use to a single computer. Purchasing additional licenses for use on other computers can be price prohibitive. There is a program called Wsearch32 (http://www.wsearch.com.au/) that is a free download and will read the native format for most commercially available chromatography/mass spectrometry data systems. This software allows for the display of RTIC and mass chromatograms, raw spectra and background-subtracted spectra, averaged spectra, library search results, etc. Wsearch32 will work with Agilent .L (Agilent ChemStation libraries are actually folders with a dot-L extension) mass spectral libraries (Wsearch32 has its own internal PBM mass spectral library search algorithm) and will export spectra to the NIST MS Search Program. It is important that commercially provided mass spectral databases like the NIST05 and the Wiley Registry 8th Ed. are provided with licenses that restrict their use to a single computer. Therefore, when using a database with a program like Wsearch32, the libraries that are to be searched must be user libraries or databases that are in the public domine, not a copy of a commercially distributed library.

Third-party software such as Wsearch32, AMDIS, AnalyzerPro, MassWorks, MassFinder, the Ion Signature software, software from ACD Labs, etc. can be very beneficial in the extraction of information from chromatographic/mass spectral data.

VII. Representative Applications of GC/MS

Applications of GC/MS to environmental chemistry are common, such as in the detection of diphenyl ethers [121], dioxins [122, 123], PAHs [124], and other volatile organics that affect air quality [125, 126]. A method for chlorinated fatty acids from fish, after conversion to pentafluorobenzyl esters, has been developed based on negative chemical ionization [127]; data interpretation was facilitated by mass chromatogram generation at *m/z* 35 and 37. The precision of quantitation of tributyltin (TBT) in sediments has been improved by using solid-phase microextraction (SPME) with isotope dilution procedures [128]. Other applications range from forensic uses for drugs [129–132] to screening urines for steroids [25, 133] to characterizing oil spills [134, 135] to kinetic studies of biosynthesis using stable isotope-labeled molecules [136, 137]. GC/MS is the mass spectral method of choice for analytes that have a mass of less than 1000 Da, are volatile, and nonthermally labile. Compounds that are thermally labile or may be nonvolatile for other reasons can be made amenable to GC/MS through derivatization. As was emphasized at the beginning of this chapter, GC/MS is not only for EI analyses but has applications in CI, ECI, and FI. FI and CI can be used in conjunction with EI for identification of unknowns. ECI, CI, and FI can be used for specificity and, especially in the case of ECI, for increased sensitivity for targeted analyte analyses.

VIII. Special Techniques

1. Purge and Trap

The purge and trap technique [138–140] is used to capture volatile organic compounds (VOCs) from water or other liquid samples for subsequent analysis by GC or by GC/MS [141, 142]. Basically, the technique involves disruptively displacing the VOCs from solution by bubbling an inert gas through the sample; in this way, the VOCs are driven into the gas phase where they are transported by the inert gas into a canister of adsorptive material that accumulates them for later analysis [143]. The adsorptive material is usually Tenax, a synthetic polymer that is not reactive with the VOCs, but efficiently binds them under ambient conditions and releases them at elevated temperature without chemical modification; other adsorptive materials include XAD-2, Porapak, and Chromasorb [144]. In particular, Tenax is more suitable than classical adsorptive material like activated carbon, which requires use of an organic solvent to efficiently remove most VOCs, thereby diluting the accumulated analytes with undesirable organic solvent. Volatiles collected in the Tenax trap are then heated and transferred to a glass-bead cryogenic focusing trap to facilitate more discrete injection onto the GC or GC-MS.

The purge and trap technology has become an essential component for the standard methodology for monitoring estuarine waters for public health and safety [142, 145, 146]. In this way, the purge and trap technique is the initial step in analyses of water for targeted compounds including haloforms [147], trihalomethanes [148], phenols [149], and dimethyl sulfide [150]. The purge and trap technique has also been used in the analysis of urine for benzene, toluene, and xylene [151], of human milk for benzene and toluene [152], of the pulp of fruit at various stages of maturation for volatile organic compounds [153], of wine for volatile components [154], of honey for 1,2-dibromoethane, 1,4-dichlorobenzene, and naphthalene residues [155], and of other foodstuffs [156].

2. Thermal Desorption

This technique is used to accumulate gas-phase organic compounds from ambient air or any other gas that might occupy a given volume (head space) [157]. This technique differs from "purge and trap" in that no inert gas is used to carry the analytes to the canister of adsorptive material; in "thermal desorption" the gas sample itself transports the organic analytes to the adsorptive collector as it is forced through the canister [158]; this technique has become part of the standard methodology for monitoring atmospheric contamination by polyaromatic hydrocarbons (PAHs) [159]. As with the purge and trap technique, the analytes in the thermal desorption technique are eventually desorbed thermally from the canister of adsorptive material for analysis by GC [159] or by GC/MS. In other cases, volatile compounds are driven out of solid-phase samples by carefully controlled heating as, for example, in the analysis of pharmaceuticals for free acids [160], of virgin olive oil for volatiles [161], of cheddar cheese for volatiles [162], of tomato leaves for methyl salicylate [163], or of ballpoint pen inks for volatile organic components that can be used for classifying and dating [164].

In other cases, thermal desorption is used in conjunction with ancillary sampling procedures such as the "rolling stir bar" sampling procedure [165, 166] for VOCs. Briefly, the stir bar technique simply uses the capacity of the hydrophobic surface of the typical stir bar to adsorb organic compounds, which can then be thermally desorbed when the stir bar in warmed in a controlled manner in a glass desorption liner [166]. This stir bar sampling procedure has been used in the quantitative profiling of volatiles on the surface of human skin [167], in the analysis of fruits and vegetables for pesticides [165], and in the analysis of water samples for 17β-estradiol [168] or benzophenone [169] or other semivolatiles [170]. In a somewhat new development, semivolatiles are captured from environmental water samples while pumping them through a length of a porous-layer open-tubular capillary column, which is then thermally desorbed onto a GC or GC-MS [171]. In another application, glass-fiber-reinforced poly(acrylate)-based sorptive material is used to enrich micropollutants from aqueous samples, which are then thermally desorbed for analysis by GC/MS [172].

References

1. Kitson FG, Larsen BS, McEwen CN and Editors, *Gas Chromatography and Mass Spectrometry*, 1996.
2. Abian J, The coupling of gas and liquid chromatography with MS. *J Mass Spectrom* **34:** 157-168, 1999.
3. James AT and Martin AJP, Liquid-Gas Partition Chromatgraphy. *Biochem. J. Proc.* **48:** VII, 1951.
4. James AT and Martin AJP, Gas Liquid Partition Chromatography: A Technique for the Analysis of Volatile Material. *Analyst* **77:** 915-932, 1952.
5. Martin A and Synge R, Separation of the Higher Monoamino Acids by Counter-Current Liquid-Liquid Extraction: The Amino Acid Composition of Wool. *Biochem. J.* **35:** 91-121, 1941.
6. Martin A and Synge R, A New form of Chromatography Employing Two Liquid Phases. I. A Theory of Chromatography. II. Applications to the Microdetermination the Higher Monoamino Acids in Proteins. *Biochem. J.* **35:** 1358-1368, 1941.
7. Holmes J and Morrell F, Oscillographic Mass Spectrometric Monitoring of Gas Chromatography. *Applied Spectrscopy* **11:** 86-87, 1957.
8. Gohlke RS and McLafferty FW, Early GC-MS. *J. Amer. Soc. Mass Spectrom.* **4:** 367-371, 1993.
9. Gohlke R, Time-of Flight Mass Spectrometry and Gas Liquid Partition Chromatography. *Anal. Chem.* **31:** 535-541, 1959.
10. Grob RL, Modern Practice of Gas Chromatography, Third Edition. pp. 888 pp, 1995.
11. Fowlis IA, *Gas Chromatography. Second Edition*, 1995.
12. Yancey JA, Review of liquid phases in GC, Part II: applications. *J Chrom Sci* **32:** 403-13, 1994.
13. McNair H and Miller J, *Basic Gas Chromatography: Tecniques in Analytical Chemistry*. Wiley, New York, 1997.
14. Ettre L and JV H, *Basic Relationships of Gas Chromatography*. Advanstar, Cleveland OH, 1993.
15. Hites RA, Gas chromatography mass spectrometry. *Handb. Instrum. Tech. Anal. Chem.*: 609-626, 1997.
16. Grob K, *Split and Splitless Injection for Quantitative Gas Chromatography: Concepts, Processes, Practical Guidelines, Sources or Error, 4th Edition*, 2000.
17. Grob K, Injection Technique in GC. *Anal. Chem.* **66:** 1009A-1019A, 1994.
18. Whiting JJ and Sacks RD, Evaluation of split/splitless operation and rapid heating of a multi-bed sorption trap used for GC of large-volume air samples. *Journal of Separation Science* **29**(2): 218-227, 2006.
19. Hinshaw JV and Ettre LS, *Introduction to Open-Tubular Column GC*. Advanstar Communications, Cleveland, OH, 1994.
20. Wittrig RE, Dorman FL, English CM and Sacks RD, High-speed analysis of residual solvents by flow-modulation gas chromatography. *Journal of Chromatography, A* **1027**(1-2): 75-82, 2004.
21. Sacks RD, Nowak ML and Smith HL, New gas chromatography technologies for high-speed, online monitoring. *Advances in Instrumentation and Control* **51**(Pt. 1): 97-107, 1996.
22. Tirillini B and Stoppini AM, Injection of a Sample by Means of Microneedles for Capillary GC. *J. Chromatogr. Sci.* **33:** 139-142, 1995.
23. Basheer C, Alnedhary AA, Rao BSM, Valliyaveettil S and Lee HK, Development and Application of Porous Membrane-Protected Carbon Nanotube Micro-Solid-Phase Extraction Combined with GC/MS. *Analytical Chemistry* **78**(8): 2853-2858, 2006.
24. Supina WR, Henly RS and Kruppa RF, Silane treatment of solid supports for gas chromatography. *Journal of the American Oil Chemists' Society* **43:** 202A-204A, 1965.
25. Basheer C, Jayaraman A, Kee MK, Valiyaveettil S and Lee HK, Polymer-coated hollow-fiber microextraction of estrogens in water samples with analysis by GC/MS. *Journal of Chromatography, A* **1100**(2): 137-143, 2005.
26. Costello CE, Hertz HS, Sakai T and Biemann K, GC/MS to identify drugs and their metabolites in body fluids of overdose victims. *Clinical chemistry* **20:** 255-65, 1974.
27. Köppel C and Tenczer J, Scope and Limitations of GC-MS in Acute Poisoning. *J. Am. Soc. Mass Spectrom.* **6:** 995-1003, 1995.
28. Daniel VC, Minton TA, Brown NJ, Nadeau JH and Morrow JD, Simplified assay for 2,3-dinor-6-keto-PGF1 alpha by GCMS. *Journal of chromatography. B, Biomedical applications* **653:** 117-22, 1994.
29. Blau K, Halket JM and Editors, *Handbook of Derivatives for Chromatography. 2nd Ed*, 1996.
30. Knapp DR, *Handbook of Analytical Derivatization Reactions*. John Wiley & Sons, NYC, 1979.
31. Vouros P, Chemical derivatization in GC/MS. In: *Practical Spectroscopy*, Vol. 3, pp. 129-251, 1980.
32. Halpern B, Biomedical Applications of GC-MS. *CRC Crit. Rev. Anal. Chem.* **11:** 49-78, 1981.
33. Hellerqvist CG and Sweetman BJ, Mass spectrometry of carbohydrates. In: *Methods of Biochemical Analysis*, Vol. 34 (Ed. Suelter CaW, J.T.), pp. 91-143, 1990.
34. Dell A, Preparation and desorption mass spectrometry of permethyl and peracetyl derivatives of oligosaccharides. In: *Methods in enzymology*, Vol. 193 (Ed. McCloskey JA), pp. 647-60, 1990.
35. Dell A, Reason AJ, Khoo KH, Panico M, McDowell RA and Morris HR, Mass spectrometry of carbohydrate-containing biopolymers. In: *Methods in enzymology*, Vol. 230, pp. 108-32, 1994.

36. Hanisch FG, Methylation analysis of complex carbohydrates: overview and critical comments. *Biological Mass Spectrometry* **23**: 309-12, 1994.
37. Halket JM and Zaikin VG, Derivatization in mass spectrometry- 3. Alkylation (arylation). *European Journal of Mass Spectrometry* **10**(1): 1-19, 2004.
38. Zaikin VG and Halket JM, Derivatization in mass spectrometry-4. Formation of cyclic derivatives. *European Journal of Mass Spectrometry* **10**(4): 555-568, 2004.
39. Zaikin VG and Halket JM, Derivatization in mass spectrometry. 6.: Formation of mixed derivatives of polyfunctional compounds. *European Journal of Mass Spectrometry* **11**(6): 611-636, 2005.
40. Halket JM and Zaikin VG, Derivatization in MS-5. Specific derivatization of monofunctional compounds. *Eur J Mass Spectrom* **11**: 127-160, 2005.
41. Halket JM and Zaikin VG, Derivatization in MS-7. On-line derivatization/degradation. *Eur J Mass Spectrom* **12**: 1-13, 2006.
42. Halket JM, Waterman D, Przyborowska AM, Patel RKP, Fraser PD and Bramley PM, Chemical derivatization and MS libraries in metabolic profiling by GC/MS and LC/MS/MS. *Journal of Experimental Botany* **56**(410): 219-243, 2005.
43. Pierce AE, *Silylation of Organic Compounds*. Pierce Chemical Co., Rockford, IL, 1968.
44. Halket JM and Zaikin VG, Derivatization in MS-1. Silylation. *Eur J Mass Spectrom* **9**: 1-21, 2003.
45. Schwarz E, Abdel Baky S, Lequesne PW and Vouros P, Aromatization and Catecholization of the A ring of nor 19 Methyl Steroidal 3 keto Epoxides by Trimethylsilylation. *Int. J. Mass Spectrom. Ion Phys.* **47**: 511 514, 1983.
46. Harvey DJ, Dimethoxymethylsilyl Ethers for GC/MS. *J. Chromatogr.* **196**: 156-159, 1980.
47. Holmes VE, Bruce M, Shaw PN, Bell DR, Qi FM and Barrett DA, GC/MS for the measurement of fatty acid w and w-1 hydroxylation kinetics by CYP4A1 using an artificial membrane system. *Anal Biochem* **325**: 354-363, 2004.
48. Rontani JF, Prahl FG and Volkman JK, Characterization of unusual alkenones and alkyl alkenoates by EI GC/MS. *Rapid Communications in Mass Spectrometry* **20**(4): 583-588, 2005.
49. Metcalfe LD and Schmitz AA, Rapid Preparation of Esters of Fatty Acids for GC. *Anal. Chem.* **33**: 363-364, 1961.
50. Husek P, Simek P and Matucha P, Smooth esterification of di- and tricarboxylic acids with methyl and ethyl chloroformates in GC profiling of urinary acidic metabolites. *Chromatographia* **58**(9/10): 623-630, 2003.
51. Husek P, Improved Derivatization of Hydroxycarboxylic Acids with Chlorformates for GC. *J. Chromatogr.* **630**: 429-437, 1993.
52. Huang ZH, Wang J, Gage DA, Watson JT, Sweeley CC and Husek P, Characterization of N-ethoxycarbonyl ethyl esters of amino acids by MS. *J Chromatogr* **635**: 271-81, 1993.
53. Wang J, Huang ZH, Gage DA and Watson JT, Analysis of amino acids by GC-FID and GC/MS: simultaneous derivatization of functional groups by an aqueous-phase chloroformate-mediated reaction. *J Chromatogr A* **663**: 71-8, 1994.
54. Green K, GC/MS of O-Methyl Oxime Derviatives of Prostaglandins. *Chem. Phys. Lipids* **3**: 254-272, 1969.
55. Thenot JP and Horning EC, MO-TMS Derivatives of Steroids for GC/MS. *Anal. Lett.* **5**: 21-33, 1972.
56. Kirk JM, Tarbin J and Keely BJ, Analysis of androgenic steroid girard P hydrazones using multistage tandem mass spectrometry. *Rapid Communications in Mass Spectrometry* **20**(8): 1247-1252, 2006.
57. Zaikin VG and Halket JM, Derivatization in MS-2. Acylation. *Eur J Mass Spectrom* **9**: 421-434, 2003.
58. Anggard E and Sedvall G, GC of Catecholamine Metabolites using ECD and MS. *Journal* **41**: 1250-1256, 1969.
59. Bhatt BD, Ali S and Prasad JV, Position and Geometry of Double Bond in Olefins by Derivatization and Analysis by GC-MS. *J. Chrom. Sci.* **31**: 113-119, 1993.
60. Yuan G and Yan J, ID of double-bond position in isomeric linear tetradecenols and related compounds by MS dimethyl disulfide derivatives. *Rapid Commun Mass Spectrom* **16**: 11-14, 2002.
61. Kidwell D, Biemann K, Brauner A, Budzikiewicz H and Boland W, Determination of Double Bond Position and Geometry of Olefins by MS of Their Diels-Alder Adducts: CI of Homoconjugated Triene and Tetraene Units in Aliphatic Compounds. *Anal. Chem.* **54**: 2462-2465, 1982.
62. Yruela I, Barbe A and Grimalt JO, Determination of double bond position and geometry in linear and highly branched hydrocarbons and fatty acids from GC/MS of epoxides and diols generated by stereospecific resin hydration. *Journal of Chromatographic Science* **28**(8): 421-7, 1990.
63. Lopez JF and Grimalt JO, Alkenone Distributions in Natural Environments by Improved Method for Double Bond Location Based on GC-MS Cyclopropylimines. *J Amer Soc Mass Spectrom* **17**: 710-720, 2006.
64. Teranishi R, Buttery RG, McFadden WH, Mon TR and Wasserman J, GLC efficiencies in GC-MS. *Anal. Chem.* **36**: 1509-1512, 1964.
65. Trehy M, Yost R and Dorsey J, Short Open Tubular Columns in GC-MS. *Anal. Chem.* **58**: 14-19, 1986.
66. Sphon J, Use of Mass Spectrometry for Confirmations of Animal Drug Residues. *J. Assoc. Off. Anal. Chem.* **61**: 1247-1252, 1978.

67. Stein SE and Heller DN, On the Risk of False Positive Identification Using Multiple Ion Monitoring in Qualitative Mass Spectrometry: Large-Scale Intercomparisons with a Comprehensive Mass Spectral Library. *Journal of the American Society for Mass Spectrometry* **17**(6): 823-835, 2006.
68. Jensen TE, Kaminsky R, McVeety BD, Wozniak TJ and Hites RA, Direct Connection of Capillary Column to MS. *Anal. Chem.* **54**: 2388-2390, 1982.
69. Wolen RL and Pierson HE, Diverter valve for GC-MS. *Anal. Chem.* **47**: 2068-2069, 1975.
70. Ligon WV, Jr. and Grade H, Adjustable open-split interface for GC/MS providing solvent diversion and invariant ion source pressure. *Analytical Chemistry* **63**: 2386-90, 1991.
71. Fialkov AB, Steiner U, Jones L and Amirav A, A new type of GC-MS with advanced capabilities. *International Journal of Mass Spectrometry* **251**(1): 47-58, 2006.
72. Fialkov AB, Steiner U, Lehotay SJ and Amirav A, Sensitivity and noise in GC-MS: Achieving low limits of detection for difficult analytes. *International Journal of Mass Spectrometry* **260**(1): 31-48, 2007.
73. Fialkov AB, Gordin A and Amirav A, Extending the range of compounds amenable for gas chromatography-mass spectrometric analysis. *Journal of Chromatography, A* **991**(2): 217-240, 2003.
74. Kochman M, Gordin A, Alon T and Amirav A, Flow modulation comprehensive 2D GC/MS with a supersonic molecular beam. *Journal of Chromatography, A* **1129**: 95-104, 2006.
75. Alon T and Amirav A, Isotope abundance analysis methods and software for improved sample identification with supersonic gas chromatography/mass spectrometry. *Rapid Communications in Mass Spectrometry* **20**(17): 2579-2588, 2006.
76. Bourne S and Croasmun W, Cross-Bore Open-Split Interface for GC-MS. *Anal. Chem.* **60**: 2172-2174, 1988.
77. Watson JT and Biemann K, High resolution MS of GC effluents. *Anal. Chem.* **37**: 844-851, 1965.
78. Watson JT and Biemann K, High-resolution MS with GC. *Anal. Chem.* **36**: 1135-1137, 1964.
79. Ryhage R, MS as a Detector for GC. *Anal. Chem.* **36**: 759-764, 1964.
80. Llewellyn PM and Littlejohn DP, Membrane Separator for GCMS. *Proc. Pittsburgh Conference on Analytical Chemistry and Applied Spectroscopy*, 1966.
81. Reis VH and Fenn JB, Gas Separation in Supersonic Jets. *J. Chem. Phys.* **39**: 3240-3250, 1963.
82. Ryhage R, Efficiency of Separators in GC-MS. *Arkiv Kemi* **26**: 305-316, 1967.
83. Black DR, Flath RA and Teranishi R, Membrane Separators in GC-MS interfaces. *J. Chromatogr. Sci.* **7**: 284-289, 1969.
84. Foltz RL, Neher MB and Hennenkamp ER, Labile TMS Derivatives in GC-MS. *Anal. Chem.* **39**: 1338-1339, 1967.
85. Arnold JE and Fales FM, Interface Materials and Labile Compounds. *J. Gas Chromatogr.* **3**: 131-133, 1965.
86. Fales HM, Comstock W and Jones TH, Dehydrogenation Test for GC-MS systems. *Anal. Chem.* **52**: 980-982, 1980.
87. Pool WG, de Leeuw JW and van de Graaf B, Correction for Skewing in GC-MS. *J. Mass Spectrom.* **31**: 213-215, 1996.
88. Kennett BH, Ratio recording in GC-MS. *Anal. Chem.* **39**: 1506-1508, 1967.
89. Ende M and Spiteller G, Contaminants in MS. *Mass Spectrom. Rev.* **1**: 29-62, 1982.
90. Grassie N and MacFarlane IG, Thermal Degradation of Apolar Polysiloxanes. *Europ. Polym. J.* **14**: 875, 1978.
91. Holland JF, Enke CG, Allison J, Stults JT, Pinkston JD, Newcome B and Watson JT, Mass spectrometry on the chromatographic time scale: realistic expectations. *Anal. Chem.* **55**: 997A-1012A, 1983.
92. Leclercq PA and Cramers CA, GC/MS & Chromatographic Challenge. *J. High. Resolut. Chromatogr.* **11**: 845-848, 1988.
93. Veriotti T and Sacks R, Characterization and quantitative analysis with GC/TOFMS comparing enhanced separation with tandem-column stop-flow GC and spectral deconvolution of overlapping peaks. *Anal Chem* **75**(16): 4211-6., 2003.
94. Tiebach R and Blass W, Direct Coupling of GC to an Ion Trap Detector. *J. Chromatogr.* **454**: 372-381, 1988.
95. Holland JF, Newcome B, Tecklenburg RE, Jr., Davenport M, Allison J, Watson JT and Enke CG, Design, Construction and Evaluation of an Integrating Transient Recorder for GC-TOFMS. *Rev. Sci. Instr.* **62**: 69-76, 1991.
96. Veriotti T and Sacks R, High-speed characterization and analysis of orange oils with tandem-column stop-flow GC and time-of-flight MS. *Anal Chem* **74**(21): 5635-40., 2002.
97. Hada M, Takino M, Yamagami T, Daishima S and Yamaguchi K, Trace analysis of pesticide residues in water by narrow-bore GC/MS with programmable temperature vaporizer. *Journal of Chromatography, A* **874**: 81-90, 2000.
98. Hirsch R, Ternes TA, Bobeldijk I and Weck RA, Determination of environmentally relevant compounds using fast GC/TOF-MS. *Chimia* **55**(1-2): 19-22, 2001.

99. Watson JT, Falkner FC and Sweetman BJ, Ion monitoring nomenclature. *Biomed. Mass Spectrom.* **1**(2): 156-7, 1974.

100. Falkner FC, Historical Account of Ion Monitoring. *Biomed. Mass Spectrom.* **4**: 66-67, 1977.

101. Sweeley CC, Elliott WH, Fries I and Ryhage R, MS Determination of Unresolved Components in GC. *Anal. Chem.* **38**: 1549-1553, 1966.

102. Hammar CG, Holmstedt B and Ryhage R, Mass Fragmentography of Chlorpromazine. *Anal. Biochem.* **25**: 532-548, 1968.

103. De Leenheer AP and Thienpont LM, Applications of Isotope Dilution-MS in Clinical Chemistry, Pharmacokinetics and Toxicology. *Mass Spectrom. Rev.* **11**: 249-307, 1992.

104. Garland WA and Powell ML, Quantitative SIM of Drugs in Biological Matrices. *J. Chrom. Sci.* **19**: 392-434, 1981.

105. Robertson D, Heath E, Falkner F, Hill R, Brilis G and Watson JT, Assay for Urinary Normetanephrine using GC-MS with SIM. *Biomed. Mass Spectrom.* **5**: 704-708, 1978.

106. Liu RH, Lin DL, Chang WT, Liu C, Tsay WI, Li JH and Kuo TL, Isotopically labeled analogues for drug quantitation. *Anal Chem* **74**(23): 618A-626A., 2002.

107. Erickson ED, Enke CG, Holland JF and Watson JT, Application of time array detection to GC/MS. *Anal. Chem.* **62**(10): 1079-84, 1990.

108. Pagura C, Daolio S, Traldi P and Doretti L, Peak Matching Unit for Double Ion Monitoring. *Org. Mass Spectrom.* **19**: 204-205, 1984.

109. Tondeur Y, Hass JR, Harvan DJ and Albro PW, Computer-Assisted Determination of Masses in High Resolution SIM. *Anal. Chem.* **56**: 373-376, 1984.

110. Caprioli RM, Fies WF and Story MS, Direct Analysis of Stable Isotopes with a Quadrupole MS. *Anal. Chem.* **46**: 453A-462A, 1974.

111. Millington D, Jenner DA, Jones T and Griffiths K, Steroid in Human Breast Tumours by High Resolution SIM. *Biochem. J.* **139**: 473-475, 1974.

112. Millington DS, Methods for Improving Selectivity in Quantitative GC-MS. In: *Applications of Mass Spectrometry to Trace Analysis* (Ed. Facchetti S), pp. 189-202. Elsevier, Amsterdam, 1981.

113. Thorne GC, Gaskell SJ and Payne PA, Improved Quantitative Precision in High Resolution SIM. *Biomed. Mass Spectrom.* **11**: 415-420, 1984.

114. Eichelberger JW, Harris LE and Budde WL, Reference Compound to Calibrate GC-MS. *Anal. Chem.* **47**: 995-1000, 1975.

115. De Brabander HF, Batjoens P, Vanden Braembussche C, Dirin P, Smets F and Pottie G, Pitfalls in SIM: A Theoretical Example. *Anal. Chim. Acta.* **275**: 9-15, 1993.

116. Meyer VR, Errors in the Area Determination of Incompletely Resolved Chromatography Peaks. *J. Chromatogr. Sci.* **33**: 26-33, 1995.

117. Friocount MP, Picart D and Flock HH, Fragmentation of Phthalic Acid Esters. *Biomed. Mass Spectrom.* **7**: 193-200, 1980.

118. McLafferty FW, and Gohlke, R. S., MS of Aromatic Acids and Esters. *Anal. Chem.* **31**: 2076-2082, 1959.

119. Stein SE, An Integrated Method for Spectrum Extraction and Compound Identification from Gas Chromatography/Mass Spectrometry. *J. Am. Soc. Mass Spectrom.* **10**: 770-781, 1999.

120. Biller J and K B, Reconstructed Mass Spectra, ANovel Approch for the Utilitzation of Gas Chromatograph-Mass Spectrometer Data. *Anal Lett.* **7**: 515-528, 1974.

121. Eljarrat E, Lacorte S and Barcelo D, Optimization of congener-specific analysis of 40 polybrominated diphenyl ethers by gas chromatography/mass spectrometry. *J Mass Spectrom* **37**(1): 76-84., 2002.

122. Eljarrat E and Barcelo D, Congener-specific determination of dioxins and related compounds by gas chromatography coupled to LRMS, HRMS, MS/MS and TOFMS. *J Mass Spectrom* **37**(11): 1105-17., 2002.

123. Focant J-F, Pirard C, Eppe G and De Pauw E, Recent advances in mass spectrometric measurement of dioxins. *Journal of chromatography. A* **1067**(1-2): 265-75, 2005.

124. Hafner WD and Hites RA, Effects of wind and air trajectory directions on atmospheric concentrations of persistent organic pollutants near the Great Lakes. *Environmental Science and Technology* **39**(20): 7817-7825, 2005.

125. Tumbiolo S, Gal J-F, Maria P-C and Zerbinati O, Determination of benzene, toluene, ethylbenzene and xylenes in air by solid phase micro-extraction/gas chromatography/mass spectrometry. *Analytical and Bioanalytical Chemistry* **380**(5-6): 824-830, 2004.

126. Tumbiolo S, Gal J-F, Maria P-C and Zerbinati O, SPME sampling of BTEX before GC/MS analysis: outdoor and indoor air quality measurements in public and private sites. *Annali di Chimica (Rome, Italy)* **95**(11-12): 757-766, 2005.

127. Zhuang W, McKague AB, Reeve DW and Carey JH, ID of chlorinated fatty acids in fish by GC/MS with negative ion CIof pentafluorobenzyl esters. *J Mass Spectrom* **39**: 51-60, 2004.

128. Bancon-Montigny C, Maxwell P, Yang L, Mester Z and Sturgeon RE, Improvement of measurement precision of SPME-GC/MS determination of tributyltin using isotope dilution calibration. *Anal Chem* **74**(21): 5606-13., 2002.
129. Polettini A, Montagna M, Segura J and De La Torre X, b-Agonists in Hair by GC-MS. *J. Mass Spectrom* **31**: 47-54, 1996.
130. Lurie IS, Moore JM, Kram TC and Cooper DA, Isolation, Identification and Separation of Isomeric Truxillines in Illicit Cocaine. *J. Chromatogr.* **504**: 391-401, 1990.
131. Kidwell DA, Kidwell JD, Shinohara F, Harper C, Roarty K, Bernadt K, McCaulley RA and Smith FP, Comparison of daily urine, sweat, and skin swabs among cocaine users. *Forensic Science International* **133**(1-2): 63-78, 2003.
132. Campora P, Bermejo AM, Tabernero MJ and Fernandez P, Use of GC/MS with positive chemical ionization for the determination of opiates in human oral fluid. *Rapid Commun Mass Spectrom* **20**: 1288-1292, 2006.
133. Marcos J, Pascual JA, de la Torre X and Segura J, Fast screening of anabolic steroids and other banned doping substances in human urine by GC/MS/MS. *J Mass Spectrom* **37**(10): 1059-73., 2002.
134. Wang Z and Fingas M, Oil Spill Tracking and Weathering via GC-MS. *LC-GC* **13**: 950-958, 1995.
135. Garcia de Oteyza T and Grimalt JO, GC and GC-MS of oil in sediments and microbials after 1991 Saudi Arabian oil spill. *Environ Pollution (Amsterdam)* **139**: 523-531, 2006.
136. Kapetanovic I, Yonekawa W and Kupferberg H, Use of Stable Isotopes and GC-MS in the Study of Different Pools of Neurotransmitter Amino Acids in Brain Slices. *J. Chromatogr.* **500**: 387-394, 1990.
137. Nieto R, Calder AG, Anderson SE and Lobley GE, 15NH3 Enrichment in Biological Samples by GC-MS with EI. *J. Mass Spectrom.* **31**: 289-294, 1996.
138. Bianchi A, Varney MS and Phillips J, Modified analytical technique for trace organics in water using dynamic headspace and GC/MS. *J Chromatog A* **467**: 111-28, 1989.
139. Ettre LS, Headspace-gas chromatography: An ideal technique for sampling volatiles present in non-volatile matrices. *Advances in Experimental Medicine and Biology* **488**(Headspace Analysis of Foods and Flavors): 9-32, 2001.
140. Martinez E, Lacorte S, Llobet I, Viana P and Barcelo D, Multicomponent analysis of volatile organic compounds in water by automated purge and trap coupled to GC/MS. *J Chromatography, A* **959**: 181-190, 2002.
141. Huybrechts T, Dewulf J and Van Langenhove H, State-of-the-art of GC-based methods for analysis of anthropogenic volatile organic compounds in estuarine waters, illustrated with the river Scheldt as an example. *J Chromatog, A* **1000**: 283-297, 2003.
142. Santos FJ and Galceran MT, Modern developments in GC/MS-based environmental analysis. *Journal of Chromatography, A* **1000**: 125-151, 2003.
143. Wang DKW and Austin CC, Determination of complex mixtures of volatile organic compounds in ambient air: canister methodology. *Analytical and Bioanalytical Chemistry* **386**(4): 1099-1120, 2006.
144. Pollmann J, Helmig D, Hueber J, Tanner D and Tans PP, Evaluation of solid adsorbent materials for cryogen-free trapping-gas chromatographic analysis of atmospheric C2-C6 non-methane hydrocarbons. *Journal of Chromatography, A* **1134**(1-2): 1-15, 2006.
145. Huybrechts T, Dewulf J and Van Langenhove H, State-of-the-art of gas chromatography-based methods for analysis of anthropogenic volatile organic compounds in estuarine waters, illustrated with the river Scheldt as an example. *Journal of Chromatography, A* **1000**(1-2): 283-297, 2003.
146. Wang DKW and Austin CC, Determination of complex mixtures of volatile organic compounds in ambient air: an overview. *Analytical and Bioanalytical Chemistry* **386**(4): 1089-1098, 2006.
147. Loos R, Analytical methods for determination of haloforms in drinking water. *Handbook of Environmental Chemistry* **5**(Pt. G): 175-192, 2003.
148. Culea M, Cozar O and Ristoiu D, Methods validation for the determination of trihalomethanes in drinking water. *J Mass Spectrom* **41**: 1594-1597, 2006.
149. Zhao R-S, Cheng C-G, Yuan J-P, Jiang T, Wang X and Lin J-M, Sensitive measurement of ultratrace phenols in natural water by purge-and-trap with in situ acetylation coupled with GC/MS. *Anal and Bioanal Chem* **387**: 687-694, 2007.
150. Sakamoto A, Niki T and Watanabe YW, Establishment of Long-Term Preservation for Dimethyl Sulfide by the Solid-Phase Microextraction Method. *Anal Chem* **78**: 4593-4597, 2006.
151. Brcic I and Skender L, Determination of benzene, toluene, ethylbenzene, and xylenes in urine by purge and trap GC. *Journal of Separation Science* **26**: 1225-1229, 2003.
152. Fabietti F, Ambruzzi A, Delise M and Sprechini MR, Monitoring of the benzene and toluene contents in human milk. *Environment International* **30**: 397-401, 2004.
153. Narain N, Galvao MdS and Madruga MS, Volatile compounds captured through purge and trap technique in caja-umbu (Spondias sp.) fruits during maturation. *Food Chemistry* **102**: 726-731, 2007.
154. Simmons C and Bezoari MD, Analysis of volatile components of wines by GC/MS and NMR. *Journal of Undergraduate Chemistry Research* **5**: 41-47, 2006.

155. Tananaki C, Zotou A and Thrasyvoulou A, Determination of 1,2-dibromoethane, 1,4-dichlorobenzene and naphthalene residues in honey by GC/MS using purge and trap thermal desorption extraction. *Journal of Chromatography, A* **1083:** 146-152, 2005.

156. Fleming-Jones ME and Smith RE, Volatile Organic Compounds in Foods: A Five Year Study. *Journal of Agricultural and Food Chemistry* **51:** 8120-8127, 2003.

157. Hays MD and Lavrich RJ, Developments in direct thermal extraction gas chromatography-mass spectrometry of fine aerosols. *TrAC, Trends in Analytical Chemistry* **26**(2): 88-102, 2007.

158. Wei M-C, Chang W-T and Jen J-F, Monitoring of PAHs in air by collection on XAD-2 adsorbent then microwave-assisted thermal desorption coupled with headspace solid-phase microextraction and GC/MS. *Analytical and Bioanalytical Chemistry* **387**(3): 999-1005, 2007.

159. Poster DL, Schantz MM, Sander LC and Wise SA, Analysis of polycyclic aromatic hydrocarbons in environmental samples: a critical review of GC methods. *Anal and Bioanal Chem* **386:** 859-881, 2006.

160. Kemp EA, Nelson ED and Seburg RA, Use of Thermal Desorption MS for the Detection of Free Acid Impurities in Pharmaceutical Products. *Anal Chem* **78:** 6595-6600, 2006.

161. Kanavouras A and Hernandez RJ, The analysis of volatiles from thermally oxidized virgin olive oil using dynamic sorption-thermal desorption and solid phase micro-extraction techniques. *International Journal of Food Science and Technology* **41:** 743-750, 2006.

162. Gogus F, Ozel MZ and Lewis AC, Analysis of the volatile components of Cheddar cheese by direct thermal desorption-GC * GC-TOF/MS. *Journal of Separation Science* **29**(9): 1217-1222, 2006.

163. Deng C, Qian J, Zhu W, Yang X and Zhang X, Rapid determination of methyl salicylate, a plant-signaling compound, in tomato leaves by direct sample introduction and thermal desorption followed by GC-MS. *Journal of Separation Science* **28**(11): 1137-1142, 2005.

164. Buegler JH, Buchner H and Dallmayer A, Characterization of ballpoint pen inks by thermal desorption and GC/MS. *Journal of Forensic Sciences* **50**(5): 1209-1214, 2005.

165. Kende A, Csizmazia Z, Rikker T, Angyal V and Torkos K, Combination of stir bar sorptive extraction-retention time locked GC/MS and automated MS for pesticide identification in fruits and vegetables. *Microchemical Journal* **84**(1-2): 63-69, 2006.

166. Ochiai N, Sasamoto K, Kanda H and Nakamura S, Fast screening of pesticide multiresidues in aqueous samples by dual stir bar sorptive extraction-thermal desorption-low thermal mass GC/MS. *Journal of Chromatography, A* **1130**(1): 83-90, 2006.

167. Soini HA, Bruce KE, Klouckova I, Brereton RG, Penn DJ and Novotny MV, In Situ Surface Sampling of Biological Objects and Preconcentration of Their Volatiles for Chromatographic Analysis. *Anal Chem* **78:** 7161-7168, 2006.

168. Kawaguchi M, Ito R, Sakui N, Okanouchi N, Saito K and Nakazawa H, Dual derivatization-stir bar sorptive extraction-thermal desorption-GC/MS for determination of 17b-estradiol in water sample. *Journal of Chromatography, A* **1105:** 140-147, 2006.

169. Kawaguchi M, Ito R, Endo N, Sakui N, Okanouchi N, Saito K, Sato N, Shiozaki T and Nakazawa H, Stir bar sorptive extraction and thermal desorption-GC/MS for trace analysis of benzophenone and its derivatives in water sample. *Analytica Chimica Acta* **557:** 272-277, 2006.

170. Leon VM, Llorca-Porcel J, Alvarez B, Cobollo MA, Munoz S and Valor I, Analysis of 35 priority semivolatile compounds in water by stir bar sorptive extraction-thermal desorption-GC/MS. *Analytica Chimica Acta* **558**(1-2): 261-266, 2006.

171. Pyle SM, Sovocool GW and Riddick LA, Analysis of volatiles and semivolatiles in drinking water by microextraction and thermal desorption. *Talanta* **69**(2): 494-499, 2006.

172. Rodil R, von Sonntag J, Montero L, Popp P and Buchmeiser MR, Glass-fiber reinforced poly(acrylate)-based sorptive materials for the enrichment of organic micropollutants from aqueous samples. *Journal of Chromatography, A* **1138**(1-2): 1-9, 2007.

Chapter 11   Liquid Chromatography/Mass Spectrometry

Introduction to Mass Spectrometry, 4th Edition: Instrumentation, Applications, and Strategies for Data Interpretation; J.T. Watson and O.D. Sparkman, © 2007, John Wiley & Sons, Ltd

"Combined liquid chromatography mass spectrometry: a difficult courtship."

~Patrick Arpino

I. Introduction

The mass spectrometer has the potential to serve as the universal detector for the high-performance liquid chromatograph (HPLC). As outlined in reviews [1, 2] and books [3] on liquid chromatography/mass spectrometry (LC/MS), there are two major obstacles in achieving such a universal LC/MS combination. First, there is the problem of dealing with the liquid solvent; even though the flow rates can be small (10 µL min^{-1} for microcolumns), they still represent a major problem in maintaining an operational vacuum in the mass spectrometer. The more common flow rates used with the analytical column (4.6 mm i.d.) HPLC are 1–2 mL min^{-1}, which are even a greater problem. An example of how enormous the solvent removal problem can be is seen in the fact that 1 mL min^{-1} of liquid acetonitrile (a common LC mobile-phase component) produces >0.5 L min^{-1} of gas at atmospheric pressure and room temperature.

The second consideration in the use of LC/MS is that LC is usually reserved for those compounds that are so nonvolatile and/or polar that they are not amenable to analysis by GC and, in many cases, these compounds cannot be converted from the condensed to the gas phase, meaning that they are not suitable for analysis by mass spectrometry with electron ionization (EI) or chemical ionization (CI). Most of the original techniques for LC/MS addressed only the issue of disposing of the liquid solvent. Only the technique of electrospray ionization (ESI) has addressed the issue of ionization as well as the problem of dealing with the relatively high flow rate of liquid entering the sampling interface; as described in detail in Chapter 8, the vast majority of the sprayed liquid in the first differentially pumped chamber of the ESI interface is shunted away from the "line-of-sight" passage leading to the *m/z* analyzer.

The modern suite of techniques available through atmospheric pressure ionization (API), which includes electrospray ionization (ESI), atmospheric pressure chemical ionization (APCI), and atmospheric pressure photoionization (APPI), has allowed LC/MS to become an effective methodology in the analysis of organic compounds that are important in environmental, pharmaceutical, forensic chemistry, and many other application areas. Of the three techniques, ESI is especially important for samples that cannot be analyzed by GC or GC/MS. APCI and APPI are limited to the analysis of volatile compounds; however, the thermal stability required for GC is not important for these two techniques. All three of these techniques are described either in this chapter or in other designated parts of this book. Although the introduction of analyte solutions into a mass spectrometer is often associated with HPLC, many analyses using these liquid–sample interfaces are carried out without the use of an LC column. The modern LC/MS instrument is often supplied without the liquid chromatograph. The matching of the HPLC and the *m/z* analyzer is less exacting than the matching of the gas chromatograph and the *m/z* analyzer in GC/MS instrumentation. Even taking into consideration the statements below about the mass spectrometrist requiring fragmentation information, the most common objective in analyzing a sample solution by mass spectrometry is simply to determine the molecular mass of the analyte. To that end, there are now API interfaces that simultaneously perform APCI and ESI (introduced by Agilent Technologies at PittCon 2005) as well as interfaces that can perform APCI and ESI in alternating modes. The LC-MS interface is as much a tool for the analysis of pure analytes in solution as a tool for the analysis of analytes separated by HPLC. The API interface, for the most part, generates only ions representing the intact molecule, but these ions can be plentiful and result in mass spectra that have a number of peaks representing products of interactions between the analyte and the solvent molecules as well as between analyte molecules and reactive materials in the sample matrix and in the analytical system itself. The spectrum produced by LC/MS, although lacking the fragmentation information available in a mass spectrum obtained using EI, can be just as challenging during interpretation.

II. Historical Milestones in the Development of the Interface

1. Introduction

Soon after the first reports of what has become known as high-performance liquid chromatography (HPLC) in the 1970s attempts to interface the technique to a mass spectrometer began. (The first use of high-performance liquid chromatography to describe chromatography involving the passage of a liquid mobile phase through a column containing a solid stationary phase using a pump to pressurize the liquid is credited to Csaba Horváth of Yale University at the 1970 Pittsburgh Conference on *Applied Spectroscopy and Analytical Chemistry* [4].) One of the early methods was based on the technique now known as atmospheric pressure chemical ionization (APCI) developed in Evan Horning's laboratory at Baylor College of Medicine in Houston, TX [5, 6]; Horning originally called this technique atmospheric pressure ionization (API). Today, the term API is an abbreviation for the comprehensive technology conducted at atmospheric pressure including APCI as well as other modes such as electrospray ionization and atmospheric pressure photoionization. Horning's technique was largely rejected by the mass spectrometry community at the outset because only molecular mass information could be obtained. It was not until ESI exploded on to the scene that APCI as an LC/MS technique finally gained popularity. Until that time, mass spectrometrists had to have mass spectra with peaks representing fragment ions of the intact analyte for it to be "mass spectrometry". To many, "no structural information meant no information". Originally, the only commercially manufactured APCI instruments were those used for the analysis of air for trace organic components; APCI is still the most sensitive technique for these types of analyses. This fact is supported by the recent description of the modification of an APCI interface to a commercial LC/MS to accommodate the eluent from a GC column [7]. Another factor that resulted in eventual acceptance of APCI, especially as an LC/MS technique, was the development of MS/MS as a practical technique: ion dissociation coupled with the second stage of MS provided the requisite fragmentation data.

During the early era of LC/MS there were a number of techniques that allowed for the introduction of sample solutions into the *m/z* analyzer, including the Direct Inlet, the Moving-Belt Interface, the Thermospray Interface, and Continuous-Flow FAB, all of which are briefly described below. Like APCI, the Particle Beam Interface was also developed in this pre-ESI era. However, the particle beam interface is considered by many as a currently viable interface. Lessons learned about mobile-phase cocktails and their impact on mass spectrometry in these early attempts at interfacing proved to be beneficial for the currently viable techniques.

2. The Direct Inlet

In the early efforts to provide continuous monitoring in HPLC, approximately 10 µL of the eluent were channeled directly into the ion source of the mass spectrometer through a heated capillary. This direct inlet approach required that the eluent from the conventional LC be split [8], as illustrated in Figure 11-1. In these early applications, the pressure in the mass spectrometer's ion source was approximately 100 Pa because of the residual vaporized mobile phase, which was used as the reagent gas for chemical ionization. The interface developed by Arpino in the mid-1970s [9] was commercialized by several manufacturers.

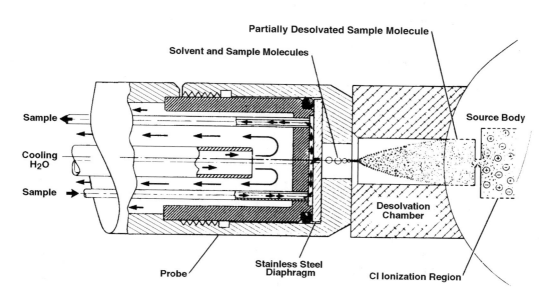

Figure 11-1. *Schematic illustration of a direct inlet interface as distributed by Hewlett-Packard (now Agilent Technologies). A portion of the flow from the LC enters the interface and it flows past an orifice at the end of the probe tip. The other side of the orifice is at reduced pressure causing a portion of the flow to be forced into the vacuum region of the mass spectrometer. The LC flow entering the mass spectrometer evaporates due to the reduced pressure; adiabatic cooling is countered by the heating block. The solvent is ionized by high-energy electrons. These ionized solvent molecules then function as reagent ions to bring about chemical ionization of the gas-phase analyte molecules.* Courtesy of Hewlett-Packard Corp.

Modern vacuum systems can accommodate direct introduction of LC solvent at up to 50 μL min^{-1} while maintaining proper operating pressure in the mass spectrometer; the microbore HPLC column is particularly well suited for this direct introduction of liquid without splitting the eluent [10, 11]. Miniaturized versions of the direct liquid inlet are now making a comeback; the nanoscale version of the direct liquid inlet produces an ultra-fine spray in the LC-MS interface, allowing the use of EI in some cases [12]. Commercial offerings of nanoscale EI direct inlets have not been forthcoming as of 2005. It should be remembered that the EI technique requires that the analyte have a finite vapor pressure and be thermally stable. The use of EI is a significant deviation from operation of the original direct liquid inlet technique in that the early efforts were only concerned with ion formation through chemical ionization (CI).

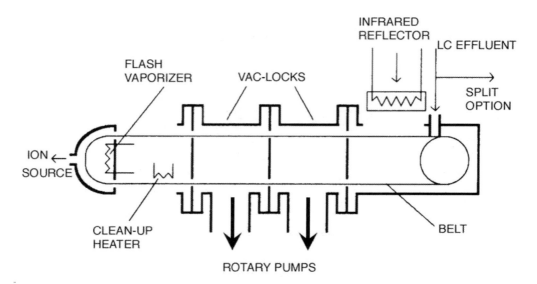

Figure 11-2. *Schematic diagram of the moving-belt interface for LC/MS. Note the two stages of differential pumping to remove the solvent and to cope with air leaks at the belt–vacuum housing interface.* Courtesy of WH McFaddon, Finnigan Corp. (now Thermo Fisher Scientific).

3. The Moving-Belt Interface

The moving-belt interface developed by McFadden *et al.* [13] provided a convenient means of introducing about 40% of the eluate from the HPLC column into the mass spectrometer under conditions that allowed EI [14]. This interface consisted of a series of auxiliary vacuum chambers (Figure 11-2) through which a continuous train (belt) carried the column eluate (or an aliquot), evaporated the solvent, and subsequently vaporized the solute in the ion source of the mass spectrometer for analysis by EI or CI using a conventional reagent gas. This arrangement was able to accommodate volatile nonpolar solvents such as hexane at rates up to 2 mL min^{-1}, but flow rates of aqueous solutions had to be kept below 0.3 mL min^{-1}. The moving belt did not address the issue of nonvolatile and/or thermally labile analytes. The analytes still had the requirement of being in the gas phase before they could be analyzed. As awkward as this apparatus appears, it did obtain a degree of commercial success in that more than a single company marketed versions of the hardware and there were extensive publications on the technique [15].

4. The Thermospray Interface

The thermospray [16] approach to LC/MS addressed two fundamental problems: (a) reduction of the pressure from residual vaporized HPLC solvent and (b) generation of ions of the analyte from the condensed phase, which were frequently of nonvolatile compounds [17]. The liquid stream from the LC was completely or partially vaporized by applying heat – hence the name thermospray. The eluate from the vaporizer was transported as a superheated mist in a supersonic vapor jet; the majority of the vaporized solvent and mist went to an auxiliary vacuum pump, though some of it traversed a skimmer aperture with the ions into the *m/z* analyzer. The primary advantage of the

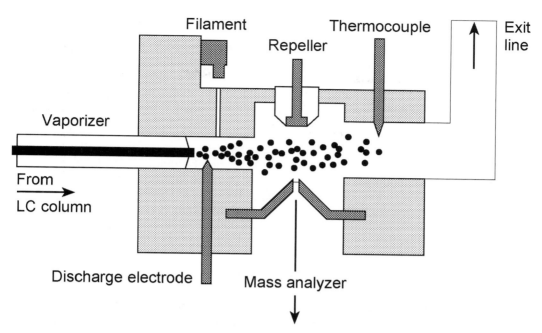

Figure 11-3. *Schematic diagram of the thermospray interface for LC/MS. Ionization is accomplished during evaporation of the solvent in the reduced-pressure region either by a process of chemical ionization (filament-on operation) or transfer of charge from charged droplets similar to electrospray (filament-off operation).*

thermospray interface was its capacity to generate ions of nonvolatile analytes in aqueous effluents at rates up to 2 mL min^{-1} [18]. In some ways, thermospray can be considered as a precursor to electrospray ionization [19]. The original thermospray device, developed by Marvin Vestal [20] and shown as a schematic representation in Figure 11-3, was designed to vaporize eluate from the LC column in a heated vacuum. A corona discharge or an EI filament was used to ionize the volatilized mobile phase. These ions became the reagent ions for chemical ionization. Serendipitously, during a demonstration of the apparatus, a disappointingly low ion current was observed for adenosine 5'-monophosphate (AMP), which was the analyte being used. As is usually the case when an important demonstration is being carried out, something went wrong. The filament failed (burned-out). However, this "something-went-wrong" turned out to be a "something-went-very-right"! When the filament went out, the ion current for the analyte increased significantly. This caused the developers to coin the term "filament-off" thermospray as opposed to the original mode of operation which then became known as "filament-on" thermospray. Not only were peaks observed that represented the protonated molecule but there were also peaks that represented potassium adducts [20]. It was not long after the development of thermospray that John Fenn began to show the advantages of ion desorption at atmospheric pressure (electrospray ionization) as opposed to ion desorption in a vacuum (filament-off thermospray).

5. Continuous-Flow FAB

Fast atom bombardment (FAB) is a desorption/ionization technique in which the analyte, dissolved in glycerol, is exposed to a beam of high-energy (fast) atoms of argon [21]. When impacted by the fast atoms, some of the glycerol molecules are degraded in such a way as to produce protons, electrons, and other products; during entrainment of the analyte into the plasma that exists within several molecular dimensions of the site of bombardment, ion/molecule reactions result in protonation of the analyte [22]. In early manifestations of FAB, great efforts were made to remove ions emanating from the "fast atom gun"; in later developments, it was learned that protonation of the analyte was just as effective under ion bombardment as under atom bombardment [23]. In many cases, ion sources were equipped with "ion guns" rather than the more cumbersome "fast atom guns"; in these cases, the technique was called "liquid SIMS", not FIB (fast ion bombardment) in deference to the technique of secondary ion mass spectrometry (SIMS), but with use of glycerol as a liquid matrix [24, 25]. In general, FAB was developed as a batch analytical technique, rather than a technique allowing on-line chromatographic separation of the sample [21].

In the most popular version of continuous-flow fast atom bombardment (CF-FAB), a microbore capillary is incorporated into a typical direct inlet probe assembly to continuously supply a dilute aqueous solution of glycerol to the bombardment platform [26–30]. A schematic diagram of the tip of a continuous-flow probe (highly magnified) is shown in Figure 11-4. The bombardment platform is machined of copper metal, and in the center of the metal probe a fused silica capillary is connected in such a manner as to

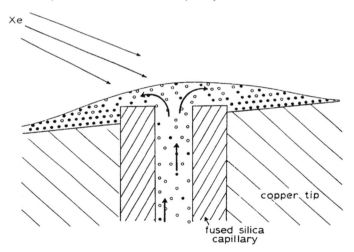

Figure 11-4. *Artist's conception of the appearance of the bombardment platform during "stabilized" operation in the continuous-flow FAB mode. Arrows in the center of the sketch indicate the direction of liquid flow through the fused silica capillary; the open circles represent zones containing high water content (low viscosity) and the closed circles, high glycerol content (high viscosity).* From Seifert WE Jr. and Caprioli RM in Methods in Enzymology, Vol. 270: High Resolution Separation Methods of Biological Macromolecules (Hancock B and Karger BL, Eds.) Academic Press, San Diego, CA, 1996, pp 453–486.

supply a continuous flow of the analyte in dilute aqueous glycerol. The fast atom or ion beam bombards this continuously refreshed solution surface during the analysis. The situation might be likened to a swamp, in which water is slowly, but continuously, supplied to a wetland area from an underground spring.

As illustrated in Figure 11-4, the fluid coming up through the copper tip at the terminus of the fused silica capillary is mostly water (5% glycerol). As the aqueous solution of the glycerol containing the analyte comes to the bombardment platform, it is exposed to the high vacuum of the mass spectrometer; in this low-pressure region, the vast majority of the water evaporates, leaving behind a glycerol-rich residue of the solution. This interface was useful in allowing FAB, which was designed as a batch inlet, to monitor the dynamic system of reversed-phase (rp) HPLC.

Continuous-flow FAB found its primary success with the double-focusing mass spectrometer, which uses tandem electric and magnetic fields to separate ions according to very narrow differences in *m/z* values. Although the CF-FAB technique, especially for peptides, has largely been replaced by ESI and ESI tandem mass spectrometry, many papers using CF-FAB still appear in current literature, ca. 2006.

III. Currently Viable Versions of the Interface

1. Atmospheric Pressure Ionization

A. Electrospray Ionization Interface

Electrospray ionization (ESI) is the most widely used interface [31] for LC/MS. The ionization mechanism involved with ESI is widely debated [32–34]; however, one concept is based on desorption of ions from the liquid surface, so-called "ion evaporation". The eluate from the LC column passes through a tube (the ESI needle), which serves as one of the electrodes in the constant-current electrolytic cell that establishes the ESI condition. A potential of 3–5 kV is impressed between the sprayer needle (primary electrode) and a counter electrode. The spray is primarily due to the difference in electrical potential between the needle and the counter electrode, in much the same way as an electrical paint sprayer works. In some cases, a nebulizing gas is used to facilitate the spraying of the LC eluate (see Chapter 8 for a more detailed explanation). Through the use of a drying gas (usually nitrogen, which is sometimes heated), the charged droplets of the spray will lose solvent to the gas phase. These diminishing droplets contain analyte ions that have existed in solution from the beginning of the LC process. These smaller droplets will experience an increase in the number-of-charges/surface area ratio, which results in a "coulombic explosion" that produces more charged droplets of a smaller size. This process continues until the analyte ions are now in the gas phase free from the solvent molecules. A typical ESI interface is shown in Figure 11-5.

The capacity to form ions in solution rather than in the gas phase removes the requirements for the analyte to have thermal stability and a discernible vapor pressure. For example, a mixture of carbohydrates is used to calibrate some ESI instruments in the negative-ion mode as used with the Micromass instrumentation. The most significant advantage of ESI for LC/MS is the ability to form ions with multiple charges, thereby allowing for the analysis of analytes that have masses that are higher numerically than the upper *m/z* limit of the mass spectrometer.

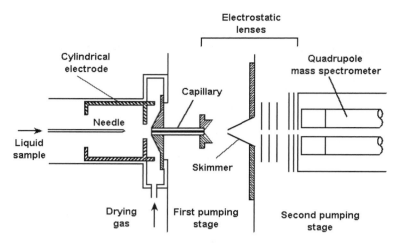

Figure 11-5. Schematic diagram of a typical electrospray interface to a transmission quadrupole mass spectrometer.

One of the features that distinguishes ESI from other ionization techniques is that the signal strength is a function of the *concentration* of the analyte, not the *amount* of analyte, that is introduced into the instrument. If a solution having a concentration of 1 ng mL^{-1} flows into an ESI interface at a rate of 1000 µL min^{-1}, the signal will be no greater than that obtained when the same sample is introduced at a flow rate of 100 µL min^{-1}. This concentration-dependent property of ESI is one of the reasons for its success. Because of this feature, it was possible to accommodate the low flow requirements of the original ESI interface designs by splitting the flow from the LC column and not sacrificing signal strength.

To gain increases in sensitivity, HPLC methods that result in higher analyte concentrations have been improved and enhanced. One way to increase the concentration of the analyte is to reduce the diameter of the LC column, thereby reducing dilution of the analyte. Prior to the development of ESI, most HPLC was carried out using 4.6 mm i.d. columns with flow rates of about 1 mL min^{-1}. Today, there is increasing use of techniques that use 2 and 1 mm i.d. columns. Specialized micro-HPLC systems are needed for capillary columns and nanocolumns (e.g., for i.d. ≤250 µm). More care must be taken in assembling the components to avoid leaks and, in general, a different attitude is required in setting up the equipment involving closer attention to detail.

1) Optimization for Analyses by HPLC

In general, the mass spectra obtained by ESI during LC/MS operations are identical to those obtained by ESI with sample introduction by infusion as described in Chapter 8. With proper planning and use of HPLC parameters [35], the adverse effects of certain solvents, salts, and easily ionizable compounds in a given sample can be avoided. Of course, care must be taken in performing HPLC in a manner that is compatible with ESI operation, as described in detail in a following section.

Invariably, difficulties will be encountered during the analysis of realistic samples that degrade the detection limits achieved with pure standards [36]. King and coworkers [37] have designed a clever and practical means of recognizing the chromatographic behavior of components of complex biological samples that interfere with detection of the analyte. They simply combine the HPLC eluate from a blank biological sample with a

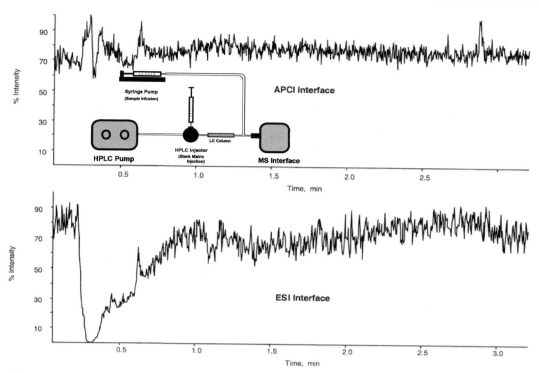

Figure 11-6. *Comparison of signal profiles from APCI and ESI sources set up to test for signal suppression as described in the text. The chromatograms show the effect of components eluting from the analytical column on the response to a post-column infusion of 10 µM urapidil. Diminution of the signal in the lower profile starting at 0.25 min indicates that the ESI source is more sensitive to signal suppression from the LC effluent from an injected protein sample than the API source.*
From King R, Bonfiglio R, Fernandez-Metzler C, Miller-Stein C and Olah T "Mechanistic investigation of ionization suppression in electrospray ionization" J. Am. Soc. Mass Spectrom. 2000, 11, 942–950, with permission.

constant infusion of their analyte via a T-connection as shown in the sketch accompanying the mass chromatogram of the analyte in Figure 11-6. A diminution in the mass chromatogram (it should constitute a level plot under ideal conditions) indicates the intervals during which sample components emerge from the column that interfere with detection of the analyte. Whereas this approach by King *et al.* does not solve the problem, it clearly indicates the chromatographic and temporal features of the offending components, and it provides a means by which to monitor progress in efforts to eliminate these components during method development [38]. Others have used a nanosplitting device to investigate how ionization and ion-transfer efficiencies are affected by drastically reducing the sample flow into the MS [39]. The proper collection and storage of biological samples can have a significant effect on the analytical result by ESI LC/MS [40].

2) Capillary Electrophoresis Interface

A variation of the ESI inlet accommodates a capillary electrophoresis (CE) column [41–44]. The CE interface is especially useful in the analysis of peptides [45] and oligonucleotides [46, 47], including those with adducts of PAHs [48, 49]. The mass spectra obtained with the LC-MS interfaces described above are similar in appearance to those shown in Chapter 8; a summary of applications follows in the last section. Problems associated with bubble formation (as associated with the redox chemistry at electrode surfaces) might be eliminated with the use of hydroquinone, which is more easily oxidized than water, to suppress evolution of O_2 [50].

B. APCI Interface

The atmospheric pressure chemical ionization (APCI) system can be used for continuous monitoring of an HPLC column as per Figure 11-6; the APCI interface is similar to that shown in Figure 11-7 for LC-ESI MS except that in APCI there is no strong electric field in the spray region, but there is a needle point with an electrical potential to promote and maintain the corona discharge. Primary electrons from the corona discharge interact with both the nitrogen nebulizing gas and the vaporized HPLC solvent to produce a corona of ions formed through direct ionization as shown in Scheme 11-1. Though nitrogen is the principal gas in the APCI chamber, nitrogen ions feed into a cascade of ion/molecule reactions in which water molecules (usually part of the mobile

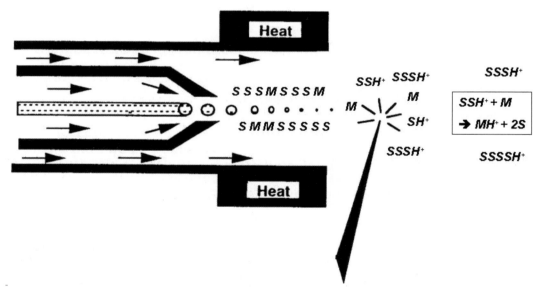

Figure 11-7. *Schematic illustration of an APCI interface. Analyte molecules (M) and solvent molecules (S) are nebulized into a heated zone at atmospheric pressure where both pass into the gas phase. The solvent molecules are converted to reagent ions through a series of reactions initiated by the ionization of gas-phase water molecules in the corona discharge; the reagent ions then ionize the analyte molecules through ion/molecule reactions.*

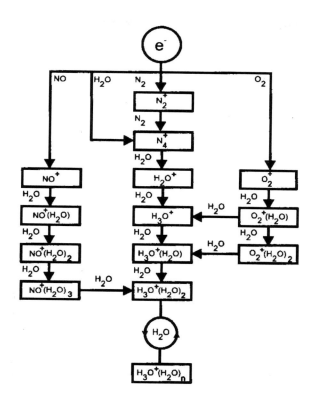

Scheme 11-1. *Mechanism by which ambient gases and solvent vapor are ionized to form reagent ions (protonated water molecules) in APCI.* From Willoughby R, Sheehan E, Mitrovich S *A Global View of LC/MS, Global View, Pittsburgh, PA, 1998, with permission.*

phase) capture a proton and react with additional water molecules to form a high flux of protonated clusters of water molecules. Because of the high proton affinity of water molecules, even a trace of water vapor in the APCI source causes the principal reagent ion eventually produced from a complicated series of ion/molecule reactions to be a protonated cluster of water molecules. Scheme 11-1 shows various pathways in which N_2, O_2, NO, and water vapor undergo ionization in positive-ion operation. In negative-ion operation, air is recommended as the nebulizing gas because oxygen is necessary to start the ionization process (e.g., O_2^-), and a different cascade of ion/molecule reactions results in the accumulation of hydroxylated water clusters as the primary reagent ions; pure nitrogen is not appropriate as a nebulizing gas for negative-ion work because it does not form negative ions itself to start the ionization cascade. Because these reactions take place at atmospheric pressure, a high number of collisions takes place between all species in the gas phase. Therefore, a very dense cloud of reagent ions is formed, which, in turn, promotes nearly complete ionization of the analyte.

Analyte molecules are ionized (usually by protonation to form single-charge ions) as detected in positive-ion APCI through ion/molecule reactions with the protonated water clusters. Similarly, negative ions of the analyte are formed (as detected in the negative-ion mode) through ion/molecule reactions with the accumulated hydroxylated

Proton Transfer: $H_3O^+ +$ M $\longrightarrow$ $[M + H]^+ +$ H_2O

$\quad\quad\quad\quad\quad\quad\quad\quad\quad$ $CH_3OH_2^+$

$\quad\quad\quad\quad\quad\quad\quad\quad\quad$ CH_3CNH^+

$\quad\quad\quad\quad\quad\quad\quad\quad\quad$ NH_4^+

Proton Abstraction: $HO^- +$ M $\longrightarrow$ $[M - H]^- +$ H_2O

$\quad\quad\quad\quad\quad\quad\quad\quad\quad$ CH_3O^-

$\quad\quad\quad\quad\quad\quad\quad\quad\quad$ CH_2CN^-

Adduct Attachment: $NH_4^+ +$ M $\longrightarrow$ $[M + NH_4]^+$

$\quad\quad\quad\quad\quad\quad\quad\quad\quad$ CH_3CNH^+

$\quad\quad\quad\quad\quad\quad\quad\quad\quad$ $CH_3OH_2^+$

$\quad\quad\quad\quad\quad\quad\quad\quad\quad$ H_3O^+

$\quad\quad\quad\quad\quad\quad\quad\quad$ $Cl^- +$ M $\longrightarrow$ $[M + Cl]^-$

$\quad\quad\quad\quad\quad$ $CH_3COO^- +$ M $\longrightarrow$ $[M + CH_3COO]^-$

Charge Exchange: $C_6H_6^{+\bullet} +$ M $\longrightarrow$ $M^{+\bullet} +$ C_6H_6

$\quad\quad\quad\quad\quad\quad\quad\quad$ $O_2^{-\bullet} +$ M $\longrightarrow$ $M^{-\bullet} +$ O_2

Electron Capture: $e^- +$ M $\longrightarrow$ $M^{-\bullet}$

Scheme 11-2. *Representative gas-phase ion/molecule reactions for the production of negative and positive ions during APCI.*

water clusters. It is important to keep in mind that because APCI involves gas-phase ion/molecule reactions, the analyte must have sufficient vapor pressure to exist in detectable quantities in the gas phase. Compounds such as sugars, which have an insufficient vapor pressure and are thermally labile, cannot be detected using APCI. APCI is limited to applications with relatively volatile analytes, typically those having a molecular weight less than 2000 Da. It should also be pointed out that identification of very low mass analytes or fragment ions (≤100 Da) is likely to be difficult because of interference from the residual protonated/deprotonated solvent clusters.

In positive-ion APCI operation, the hydronium ion (protonated water molecule) formed by the mechanism shown in Scheme 11-1 is the source of protons for formation of CI reagent ions from solvent molecules [3]. Some of these reagent ions (e.g., $CH_3OH_2^+$) are shown in Scheme 11-2. Note that the hydronium ion itself is a reagent ion in high-aqueous mobile phases; alternatively, methanol, acetonitrile, and other mobile-phase components may form protonated molecules and act as reagent ions. A similar mechanism occurs in negative-ion operation with the hydroxyl ion acting as the primary proton acceptor to form negative reagent ions.

Scheme 11-2 also shows ion/molecule reactions between reagent ions and analytes, leading to the formation of analyte ions in APCI. The primary reactions producing analyte ions in APCI involve proton transfer, forming $[M + H]^+$, for the positive-ion mode and proton abstraction, forming $([M - H]^-)$, for the negative-ion mode. The reactions are driven by the gas-phase acidity and basicity of the analytes and of the precursors of the reagent ions, as well as those of the mobile-phase components. For usable results in the positive-ion mode, the analytes must have a higher proton affinity than any of the mobile-phase components. In the negative-ion mode, the analyte should have a higher gas-phase acidity (i.e., proton donor capability) than any of the mobile-phase components.

A specialized and selective application of negative-ion detection involves resonance electron capture. This application is selective for highly electrophilic compounds; e.g., dioxin. Resonance electron capture requires the use of thermal electrons; i.e., those with very low energy, e.g., <0.1 eV, for efficient production of molecular anions; the use of electrons having energies >1 eV leads to dissociative electron capture, in which case only negative fragment ions (no molecular anions) are detected. For the resonance electron capture experiment, a high flux of thermal-energy electrons can be created in the conventional CI ion source by subjecting a reagent gas such as methane to a beam of energetic electrons as in EI. As illustrated in Scheme 11-3, thermal electrons are a by-product of the process of ionizing the methane molecules by EI; these thermal electrons can then react with an electrophilic molecule to form a molecular anion, $M^{-\bullet}$, as illustrated in the last reaction shown in Scheme 11-2.

$$CH_4 + e^-_{70eV} \rightarrow CH_4^{+\bullet} + e^-_{thermal} + e^-_{58eV}$$

Scheme 11-3

Whereas methane is commonly used in the conventional CI source for purposes of resonance electron capture negative ionization, any gas could be used for this purpose. In the case of APCI, atmospheric gases are the source of thermal electrons as they are ionized by the corona discharge. While APCI is now known as an LC/MS technique, it was originally used as a gas analysis technique after it was commercialized by Sciex in Toronto in the mid-1970s (called the Targa™); the Targa was used for the detection of air pollutants. APCI instruments have also been used as the ionization technique for eluates in supercritical fluid chromatography [51, 52].

Recent reviews [53] summarize the status, design, and operational characteristics of the APCI source for HPLC applications. APCI has been used in the analysis of lutein monoesters from the marigold plant [54]. In the analysis of canola oil, results from ESI MS and APCI MS were found to provide complementary information, which together contributed to a better understanding of the identities of the products formed by oxidation of triacylglycerols [55].

Some compounds can be reduced during analysis by APCI, as evidenced by an increase in mass of 2 Da, presumably by hydrogen radical production [56]. This observation raises some concern about artifact formation during such analyses; the prevalence of such reactions may be diminished by the limited use of protic[1] solvents. In some cases, the redox conditions can be controlled by the choice of metal used in fabricating the interface connections; e.g., the use of a copper capillary suppresses (by maintaining the electrode at 0.34 V vs a standard hydrogen electrode) the oxidation of *N*-phenyl-1,4-phenylenediamine as occurs when using a stainless steel capillary (electrode potential = 0.46 V versus a standard hydrogen electrode) in an electrospray interface [57].

C. APPI Interface

Atmospheric pressure photoionization (APPI) is the newest of the API techniques used in LC/MS [58, 59]. Photoionization (PI), which has long been used in GC [60, 61] and ion mobility spectrometry [62], is now being used for LC/MS [63–66]. For this purpose, a UV source has been incorporated into a module that nebulizes or sprays the LC effluent into the interface. In some cases, APPI augments APCI in that it can ionize some analytes that are refractory to ESI, though signal response from the three somewhat related techniques is compound-dependent [67]. There are two general approaches to APPI, one using UV radiation directly to ionize the analyte [64, 68] and the other involving a dopant [58, 63], which after being ionized by UV, converts the analyte to a protonated molecule as described below.

1) Operating Principles of APPI

The primary ionization process in APPI [68–71] involves absorption of a photon (as from a hydrogen discharge lamp; hv = 10.2 eV) by molecules having the appropriate chromophore; e.g., a π-electron system in an analyte or in a dopant like acetone or toluene. An attractive feature of using a hydrogen discharge is that the emitted photon is less energetic than the ionization potential (IP) of the principal components of the mobile phase in rp HPLC: water (IP = 12.6 eV), acetonitrile (IP = 12.2 eV), and methanol (IP = 10.8 eV). This means ionization of the major components of the mobile phase does not occur, thereby allowing detection of the analyte without interference.

When a photon from the hydrogen lamp interacts with an analyte molecule (M), it can be converted to a molecular ion ($M^{+\bullet}$) according to the mechanism illustrated in Scheme 11-4. Subsequently, the nascent molecular ion ($M^{+\bullet}$) can be converted to a protonated molecule (MH^+) via collision with a protic solvent molecule (S) from which it can withdraw a hydrogen atom according to the mechanism illustrated in Scheme 11-5.

$$M + hv \rightarrow M^{+\bullet} + e^-$$

Scheme 11-4

---

[1] In chemistry, solvents are classified as protic (a solvent that has a hydrogen attached to an electronegative group oxygen and is represented by the general formula ROH; e.g., H_2O, CH_3CO_2H, and CH_3OH), aprotic (a.k.a. dipolar aprotic solvents that do not contain an O–H bond, but do contain a bond that has a large bond dipole like a carbon–oxygen or carbon–nitrogen bond; e.g., acetone and acetonitrile), and nonpolar (solvents that have a low dielectric constant and are not miscible with water; e.g., benzene, carbon tetrachloride, and diethyl ether).

$$M^{+\bullet} + S \rightarrow MH^+ + [S-H]^{-\bullet}$$
Scheme 11-5

Depending on the nature of a protic solvent, variable quantities of both $M^{+\bullet}$ and MH^+ may be observed, as illustrated in the array of mass spectra in Figure 11-8. Also, as indicated in Figure 11-8, use of aprotic solvents as the stationary phase for LC ensures preservation of the $M^{+\bullet}$ in mass spectra produced by APPI during analyses by LC/MS [65].

In the mode of APPI based on dopants (D), a compound like acetone (IP = 9.70 eV) or toluene (IP = 8.83 eV) is ionized by the 10.2-eV photon from the hydrogen lamp [58]. The nascent molecular ions of the dopant ($D^{+\bullet}$) are soon converted to protonated molecules of the dopant as illustrated in Scheme 11-6. In turn, the protonated molecule of the dopant (acetone or toluene as D) can ionize the analyte molecule by proton transfer as illustrated in line (b) of Scheme 11-6 (provided that the proton affinity of D is less than that of M). Whether a dopant should be used depends on the ionization potentials of the mobile-phase solvent(s) and solvent complexes, and on the photon energies of VUV lamps [72].

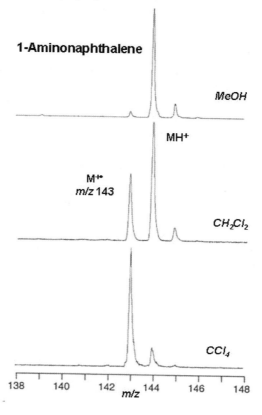

Figure 11-8. *Three APPI mass spectra of 1-aminonapthalene showing the influence of the protic nature of the solvent on the degree of formation of a protonated molecule vs a molecular ion.*
From Syage JA "Mechanism of [M + H]+ formation in photoionization mass spectrometry" J. Am. Soc. Mass Spectrom. 2004, 15, 1521–1533, with permission.

(a) $D^{+\bullet}$ + D → $(D - H)^{\bullet}$ + DH^+

(b)   M + DH^+ → D + MH^+

Scheme 11-6

The ionization of the dopant and subsequent reactions are indicated separately in Scheme 11-6, but in reality these simultaneous processes produce a dynamic flux of ions, resulting in protonation of the analyte provided that the dopant (D) has a proton affinity lower than that of the analyte (M). The reason that this suite of reactions leads to analytical advantage is that a relatively large flux of reagent ions, DH^+ in this case, can be produced by the relatively inefficient photoionization of a relatively high concentration of the dopant, D. The analyst can and should play an active role in the analytical scheme by choosing an appropriate compound as the dopant, the photoionization of which will lead to optimal ionization of the analyte (by protonation as shown in line (b) of Scheme 11-6 or by charge exchange, etc.). Furthermore, the analyst can optimize the signal-to-background for the analysis by choosing a solvent (for dissolution or chromatography of the analyte) that has an IP lower than the energy of the photons used in the primary photoionization event. The validity of the reaction shown in line (a) of Scheme 11-6 has been demonstrated in the results of related studies by Syage with a variety of protic and aprotic solvents [68].

Under these conditions, naphthalene and diphenyl sulfide produce only $M^{+\bullet}$ radical cations, while carbamazepine and acridine (each of which contains a nitrogen atom) produce protonated molecules MH^+, which are 100 times more abundant than the $M^{+\bullet}$ radical cations produced by naphthalene and diphenyl sulfide [63]. Anisole used as a dopant can increase the APPI signal response for analytes with low proton affinities in acetonitrile by two orders of magnitude [73].

2) Operating Mechanics for APPI

From the standpoint of the vacuum system, the APPI source is comparable to that used for ESI and APCI [72]. In APPI as in APCI, there is not a high electric field as there is in ESI. However, some electrostatic means, such as an "off-set potential", is necessary to move the ions toward the sampling orifice of the *m/z* analyzer. A conceptual diagram for APPI based on direct photoionization of the analyte is shown in Figure 11-9 [64].

Whereas, in principle, APPI discriminates against the principal components of the LC mobile phase, the ionization efficiency for the analyte can be disappointingly low. The high concentration of solvent vapor and entrained droplets in the APPI source absorbs most of the radiation before it reaches the center of the sample chamber; therefore, little radiation is available for photoionization in the center of the sprayer plume, which is highly sampled by the orifice leading to the mass spectrometer. Because of the problem of poor penetration of radiation from the hydrogen lamp into the APPI chamber, the use of dopants is often necessary to make APPI a viable technique (Bruce Bell, Dow Chemical, Midland, MI, Personal Communication, June 2005). The high concentration of dopant (relative to the concentration of analyte) allows good ionization of the dopant at the edge of the plume near the hydrogen lamp; protonated dopant diffuses into the core of the plume to promote ionization of the analyte near the sampling orifice leading into the mass

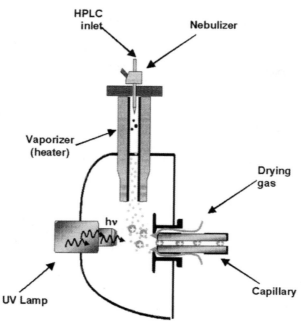

Figure 11-9. *Schematic illustration of an APPI source as implemented by Agilent Technologies on their various LC/MS instruments.*
From Figure 2 of Hanold KA, Fischer SM, Cormia PH, Miller CE and Syage JA "Atmospheric Pressure Photoionization. 1. General Properties for LC/MS" Anal. Chem. 2004, 76(10), 2842–2851, with permission.

spectrometer, thereby improving the detection limit of the APPI technique. Other work shows that the efficiency of APPI drops off with increasing solvent flow rate, probably through diminution of the photon flux due to absorption in the solvent vapor [74].

3) Signal Suppression

Signal suppression is the name given to the phenomenon of one compound affecting the signal response of another; the phenomenon seems to be more complicated than simple differences in ionization efficiency, although it is likely to be a significant factor. In the case presented in Figure 11-10, APPI shows good resistance to signal suppression that is otherwise suffered by ESI [72]; however, APPI is not always immune to signal suppression [75].

It has been recognized that analysis of an equimolar mixture of compounds is unlikely to produce equal responses for each component, it is especially frustrating to deal with signal suppression from some unknown matrix component of a complex "real world" sample (see additional descriptive detail on signal suppression and a schematic diagram for such study later in Section IV on Special Operation of LC under MS Conditions). The effect of a complex biological sample matrix on the ionization efficiency of APPI, APCI, and ESI as used in LC/MS is shown in Figure 11-10. The performance of each ionization technique is "visualized" by infusing a constant input of a test compound, fluphenazine in this case, into the LC effluent before it enters the ion source; ideal performance by a given ion source is represented by a constant signal or "flat line" in the output recording. For the test represented in Figure 11-10, the extract from a rat plasma sample was injected into the LC; the RTIC corresponding to the LC/MS run is shown in the top

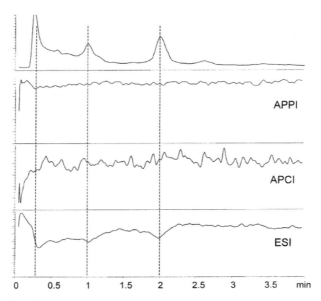

Figure 11-10. Comparison of signal suppression in APPI, APCI, and ESI.
From Figure 7 of Syage JA, Hanold KA, Lynn TC, Horner JA and Thakur RA
"Atmospheric pressure photoionization. II. Dual source ionization" J. Chromatogr. A
2004, 1050(2), 137–149, with permission.

panel of Figure 11-10, in which discernible peaks can be seen at 0.25, 1.0, and 2.0 minutes. The second panel indicates that the APPI signal for constant infusion of the test compound was unaffected by the components of the rat plasma entering the APPI source at 0.25, 1.0, and 2.0 minutes. In marked contrast, the fourth panel shows a diminution in the ESI signal at 0.25, 1.0, and 2.0 minutes, indicating that signal suppression of the test compound occurred as certain components of the rat plasma extract entered the ESI source. In other work, the impact of matrix effects from rat plasma on the performance of the APPI interface has been evaluated in pharmacokinetic analyses involving 42 different compounds [59, 75]. The technologies of ESI, APCI, and APPI have been assessed in the context of analyzing rat plasma for cyclosporine A [76].

4) Applications of APPI

Bruins and coworkers [63] have reported some early figures of merit for the use of dopants in APPI for LC/MS. In general, these workers found that the use of toluene as a dopant enhanced the sensitivity of APPI toward both high and low proton affinity compounds via either proton transfer or charge exchange reactions.   Further, they reported that use of acetone as a dopant provided analytical advantage only for compounds having a high proton affinity; acetone provided no advantage as a dopant in detecting compounds of low proton affinity. The impact of using a dopant is substantial; e.g., relative to the use of no dopant in APPI, the use of toluene as dopant increased the signal in detecting carbamazepine and acridine by a factor of 100. Further, these workers found that the detection limits for carbamazepine and acridine were eight times lower with APPI than with APCI [63]. The results of a recent study on dopant-assisted APPI using seven naphthalenes and 13 different solvent systems indicated that the ionization efficiency was 1–2 orders of magnitude higher with dopant than without [70]. A method based on APPI with dopant can be used to analyze processed brain tissue for experimental animals for salsolinol and major catecholamines [77].

In a comparative study of APPI, APCI, and ESI in the analysis of lipids such as free fatty acids and their glycerides, APPI was found to be 2–4 times more sensitive than APCI and much more sensitive than ESI without mobile-phase modifiers like ammonium formate [78]. In a comparison of APPI and APCI during the analysis of five pharmaceuticals, APPI produced stronger signals and lower background than APCI [72]. APPI, utilizing toluene as a dopant, provides high ionization efficiency simultaneously to polar and nonpolar compounds that are delivered in a reversed-phase solvent [66]. In an assessment of APPI and APCI for the detection of a group of five neurotransmitters, APPI was found to give better detection limits in both the positive- and negative-ion modes [79]. APPI has been used in the analysis of biological samples from patients with Fabry's disease for globotriaosylceramides, both in the positive- and negative-ion modes [80]. APPI has been used as the basis for analyzing drinking water samples without preliminary purification in an effort to provide high-throughput screening of municipal water supplies [81]. In a feasibility study of analyzing microbial respiratory ubiquinone and menaquinone isoprenologues, the detection limits by APPI were about 30% those by APCI [82]. The effect of a variety of eluent compositions on the ionization efficiency of five flavonoids by ion spray (IS), APCI, and APPI in positive- and negative-ion modes of LC/MS has been reported [83]. APPI MS in the negative-ion mode was studied during the analysis of seven compounds in 17 solvent systems; the results generally showed better detection limits by APPI than by APCI, although APCI gave fewer side-reaction products [84]. In work with electrokinetic chromatography, APPI has been shown to circumvent signal suppression and interferences by the surfactant sodium dodecyl sulfate (SDS) and nonvolatile buffers [85]. Analyses of petroleums by APPI produces radical cations that characterize the core structure of *N*-containing aromatics [86].

A recent report indicates that APPI is useful in applications based on GC/MS [87]. The report covers results of a comparison of APPI, APCI, and EI using either a 9.8-eV or a 10.6-eV lamp, the latter giving more universal ionization; the APPI mass spectra had some similarity to those from EI, which allowed some success with compound identification by comparison with computer libraries [87]. In applications of APPI to peptides, intense fragmentation into b/y- and c-sequence ions was observed, apparently resulting from an electron capture/electron-transfer dissociation-type mechanism following interaction with photoelectrons released during ionization of the dopant [88].

2. Particle Beam Interface

Although no longer commercially available, the particle beam interface is an effective means for connecting the LC with an MS, especially for purposes of obtaining EI mass spectra [89]. Pioneering work on the particle beam (PB) interface was introduced under the name MAGIC [90] (MAGIC was also the copyrighted commercial name given to the particle beam interface by what was then Hewlett-Packard Analytical Instruments, now Agilent Technologies), an acronym for "monodisperse aerosol generation interface coupling" as reviewed by Creaser and Stygal [91] and Cappiello *et al.* [89]. The bulk of the mobile-phase fluid is removed and the analyte concentrated in the residual HPLC solvent principally by aerosol dynamics in the PB interface [92]. As illustrated in Figure 11-11, the major aerodynamic operations are carried out in three distinctly separate, but connected, segments of the overall PB interface. These components of the interface accomplish aerosol formation, desolvation, and subsequent momentum separation of particles and molecules.

The eluent from the LC is forced through a nebulizer (see Figure 11-11) that disperses the liquid phase into a fine mist of droplets. The resulting aerosol passes through a heated desolvation chamber, and finally the previously dissolved analyte

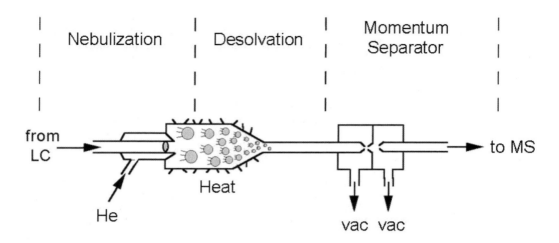

Figure 11-11. *Conceptual illustration of the particle beam interface showing the three major components.*

condenses into solid microscopic particles. The resulting entrained mixture of particles in residual solvent molecules and helium atoms is drawn through a small orifice into a pumped chamber (at approximately 10 Torr) causing a rapid expansion of the gases. The relatively massive solid particles of solute gain substantial momentum during this expansion of gases and continue in a linear beam while the low-mass, relatively low-momentum solvent molecules and helium atoms expand rapidly in the radial orientation. This abrupt expansion of low molecular weight gases allows most of the carrier gas to be drawn off preferentially in the radial direction, enriching the analyte particles as they traverse an alignment of orifices *en route* to the MS. In this way, the particles of condensed analyte are literally shot directly through the skimmer into the next concentration stage to accomplish enrichment of the analyte in the residual solvent gases as the main beam of particles is transported into the ion source of the MS. Operating pressures in the second chamber are on the order of 0.1–1 Torr, and passage of the particles of analyte into a second skimmer for entry into the MS at pressures approaching 10^{-3} Torr permit EI or conventional lower-pressure CI as is carried out in a GC-MS. The so-called "muddy clumps" that enter the ionization chamber of the mass spectrometer impact a heated splatter plate; this results in vaporization of the analyte. Ionization takes place via gas-phase molecules interacting with energetic electrons for EI or with proton-rich reagent ions for CI.

Cappeillo *et al.* have "modernized" the basic PB concept for micro-flow-rate LC-EI MS application to a wide variety of analytes [93]. In this improved interface, a high-velocity gas flows coaxially with the liquid stream to promote pneumatic nebulization to produce a very fine and homogeneous aerosol; the lower amount of solvents minimizes the influence of mobile-phase volatility and increases the rate of desolvation. The new interface design is quite "forgiving" in that no "tuning" of the capillary position is required for efficient operation, a procedure that considerably shortens and simplifies the start-up procedure for the common user [94].

In earlier work, the EI mass spectra of acridines and benzoquinolines [95] and the ECNI mass spectra of retinoic acid [96] have been obtained via use of the PB interface in the LC/MS mode of operation. Unfortunately, there is no longer a particle beam LC/MS system offered commercially. This is partially due to the naivety of many now using LC/MS. The good news is that the LC-MS is a practical and easy-to-use instrument that can be operated by those that have minimal skills and knowledge; the bad news is that those same people are making the decisions as to what instruments to purchase and they make those decisions based on what is popular, rather than on the requirements of the actual job.

3. Electron Ionization and LC/MS

Approaches to using EI with LC have evolved through continuing developments to the particle beam interface, as well as through taking advantage of nanoscale fluid dynamic interfaces that deliver only microliters of liquid per minute to the ion source [12]. Early work on extending EI to LC/MS include the use of eluent-jet formation by means of inductive heating of the micro-LC effluent and momentum separation in a jet separator [97]; a variation of this approach has been described for use under CI MS conditions [98]. Amirav and Granot [99, 100] are investigating a new apparatus for obtaining high-quality library-searchable EI spectra based on the supersonic molecular beam mass spectrometry approach.

IV. Special Operation of LC under MS Conditions

1. Effects of Mobile-Phase Composition

It is important to remember that liquid chromatography/mass spectrometry is as different from either LC or mass spectrometry as LC and mass spectrometry are from one another. Several considerations in HPLC analyses involving chromatographic peak shape, separation of components, interferences with conventional LC detectors such as the diode array, and reproducibility are less important in LC/MS. Several other factors never considered in traditional HPLC are paramount in LC/MS such as signal suppression, mass spectral interferences, adduct formation, and eluent-component volatility, just to name a few [101]. The solvent system in which the analyte is carried into the liquid chromatograph-mass spectrometer (LC-MS) interface can significantly affect the sensitivity and selectivity of the analysis by mass spectrometry. In many cases, the mobile phase used in the HPLC system can be adjusted to meet the needs of mass spectrometry better without undue compromise to the LC separation. In other cases, some compromise of the separation by LC occurs, but the separation and selectivity can be recovered in the processing of the mass spectral data through the use of such data processing "tools" as the mass chromatogram or the uniqueness of the specific m/z values of the ions formed by the analyte.

Components of the sample matrix or the chromatographic mobile phase may have an adverse effect (e.g., signal suppression, adduct formation, spectral interference) on the results of the analysis by mass spectrometry. These adverse effects can occur because of the nature of the interfaces involved in the elimination of the mobile phase. Therefore, the procedures for sample preparation and chromatographic separation must be designed to reduce or eliminate these problems.

A. Signal Suppression

This term is applied to the attenuation of the ion current produced by a given amount (concentration, in the case of ESI) of an analyte due to some chemical and/or physical factor in the sample. (See a practical example in Figure 11-7 of signal suppression as described in the earlier section on APPI as presented in a comparison of ionization techniques.) For APCI, signal suppression can result from ionization of an analyte in the condensed phase, which prevents it from reaching the gas phase as a neutral molecule where it would be ionized in an ion/molecule reaction, the fundamental process in APCI. For ESI, signal suppression can result from ion-pair formation between the analyte and certain substances in the condensed phase such that only neutral species reach the gas phase. Any sample component that suppresses ion formation of the analyte, or that competes in the ionization process, is a cause of signal suppression.

Signal suppression is the most critical of the various interferences with mass spectral formation because its presence and cause are often the most difficult to determine. For both qualitative and quantitative LC/MS, understanding and eliminating signal-suppression effects is essential. In the determination of particular causes of signal suppression for a specific analysis, previous results of the analysis should be studied. Signal suppression of minor constituents in a complex mixture can be caused by the presence of more abundant species as documented in reports on the analyses of protein digests [102]. Signal suppression is as much sample- and matrix-dependent as it is technique-dependent.

It has been demonstrated that even components of the sample that are not represented in the mass spectrum can cause signal suppression [37, 103, 104]. Signal suppression resulting from endogenous substances in biological extracts is often unpredictable and can vary from one sample to the next. A schematic diagram of the plumbing for an apparatus that can be used to characterize these types of signal suppression is shown as an insert in Figure 11-6. The apparatus consists of the components of a typical LC/MS system equipped with a device for post-column addition, which is simply a syringe pump "tee-ed" into the HPLC effluent line for post-column addition of solutions.

A typical experiment designed to assess signal-suppression effects consists of infusing a standard solution of the analyte through the post-column syringe pump connection shown as an insert in Figure 11-6. The mass spectrometer is adjusted to monitor selected ions (for an SIM experiment) or to monitor selected reactions (for a CAD MS/MS experiment; a.k.a. SRM, selected reaction monitoring) that are characteristic for the analyte. While only the mobile phase is flowing from the HPLC column when combined with the constant post-column infusion of the analyte, a constant level of ion current will be detected from the mass spectrometer operated in either the SIM or the SRM mode of operation [37, 103, 104]. Finally, a blank biological sample matrix extract is injected through the HPLC injection system. The blank biological sample is prepared according to the protocol for the analytical samples (except for the addition of standards) and is injected in the same volume and concentration as will be used in the final method. If no signal-suppressing components of the blank biological sample elute from the HPLC, the signal from the mass spectrometer will continue at a constant level. On the other hand, a diminution or disturbance in the otherwise constant signal level from the mass spectrometer will indicate the elution of signal-suppressing components of the blank biological sample from the column. The results of such an experiment are shown in Figure 11-6; in this case, a standard solution of urapidil was infused post-column into the HPLC eluent during analysis of a blank biological sample (injected after preparation by protein precipitation and pH adjustment). The trace in the lower panel in Figure 11-6 shows that ESI signal-suppressing components eluted during the interval of 0.25 to 1.10 minutes in the chromatogram; these results alert the analyst to the fact that erroneous quantitation will occur if urapidil happens to elute during this timeframe. While such experiments do not solve the signal-suppression problem, the results allow the analyst to try to develop an HPLC method in which the analyte and the signal-suppressing components of the biological sample elute from the column at different times, so that deleterious effects of signal suppression for a particular analyte can be avoided or minimized. In the example illustrated in Figure 11-6, it appears that use of APCI would be preferred over ESI for the analysis of this type of biological sample for urapidil [37].

B. Use of Internal Standards in the Face of Signal Suppression

Development of bioanalytical methods must involve consideration of the natural variability in biological matrices. Endogenous compounds in biological materials may have proton affinities that allow them to compete successfully with analyte molecules in electrospray ionization. The variability in matrix composition from one sample to another can result in unacceptable precision in quantitative results for the analyte. Matuszewski *et al.* demonstrated this problem during the development of a quantitative method for finasteride in plasma based on use of an internal standard [105]. Table 11-1 summarizes the results observed for detection of finasteride (I) and its internal standard (II) during analysis of five different human plasma matrices [105]. Varying types and concentrations of endogenous compounds in the plasma compete with the analyte for ion evaporation (i.e., cause a drop in the electrospray signal for the analyte) and thereby cause high CVs (coefficients of variation) in the peak area. However, with quantitation based on an internal standard, this is not normally a problem because the internal standard and the analyte are equally affected by these variables. However, in this case of signal suppression by an endogenous compound, excessive variability is also seen for the ratio of peak areas for the internal standard and analyte; this is possibly caused by the differences in retention times of the analyte and standard and by different degrees of signal suppression for these two compounds. Such imprecision due to variability in signal suppression can compromise the accuracy of the quantitative result.

C. Adjusting the Chromatography in the Face of Signal Suppression during LC/MS

In general, endogenous biological materials remaining in the sample after typical sample preparation (e.g., protein precipitation or solid-phase extraction) are water soluble and will elute early in a reversed-phase separation. However, in cases where some signal-suppressing material coelutes with the analyte under a given elution gradient, further increasing retention of the analyte ("Increased chromatography" in Table 11-1) to move the peaks of interest away from the areas of the chromatogram that show signal suppression may be helpful [104, 105]. In addition to adjusting the chromatography, it is advisable to assess the degree of signal suppression for both the analyte and the internal standard [106].

D. Ion Pairing and Signal Suppression

Ion-pairing agents can improve both chromatographic resolution and retention for basic analytes on reversed-phase columns. This is believed to be accomplished by formation of an acid–base ion pair that behaves chromatographically as a neutral species. However, in some cases this ion pair may not fully dissociate under the very mild ionization conditions found in an electrospray interface, thereby "stealing" some of the analyte ions from the ESI ion beam because they remain tied up as a neutral acid–base pair. An example of unwitting signal suppression with the use of trifluoroacetic acid (TFA) as an ion-pairing agent is illustrated with the results shown in Figure 11-12, which shows three reconstructed total ion current chromatograms resulting from injection of a standard solution of model peptides into an LC-ESI-MS consisting of a 2.1-mm × 150-mm column of 5 micrometer particles of Zorbax^{TM} 300SB-C3 and elution at 0.2 mL min^{-1} by a solvent gradient containing increasing concentrations of TFA [107]. As illustrated in Figure 11-12, the strength of the ESI signal decreases as the concentration of TFA increases in the eluting solvent. Thus, alteration of the chromatography by using ion-pairing agents should be conducted with caution and under method development conditions in

Table 11-1. Variability of ES response for a drug (I) and its internal standard (II) in five different human plasma matrices.

Plasma Number*	Internal Standard (II) peak area mean ($n = 5$)	CV (%) ($n + 5$)	Concentration of I (ng mL^{-1})	Ratio I/II mean ($n = 5$)	CV (%) of of Ratio I/II ($n = 5$)
	Little Chromatography†				
1	262 519	8.8	0.5	0.108	14.6
2	146 886	16.2	1.0	0.235	18.7
3	249 888	11.3	5.0	1.140	29.9
4	271 019	8.0	10.0	2.302	23.6
5	297 677	2.2	50.0	11.14	25.6
			100.0	21.56	25.9
	Increased Chromatography‡				
1	249 740	12.2	0.5	0.78	9.0
2	257 683	11.6	1.0	0.154	9.5
3	246 888	8.0	5.0	0.722	5.9
4	262 492	9.7	10.0	1.439	14.2
5	262 460	9.6	50.0	7.085	12.3
			100.0	13.06	9.1

* pH of plasma adjusted to 9.8 before extraction ("selective" extraction, B).
† Capacity factor, $k = 1.75$, for I.
‡ Capacity factor, $k = 13.25$, for the analyte [104].

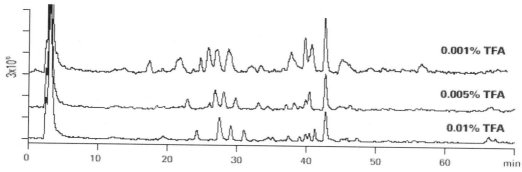

Figure 11-12. *Analysis by ESI LC/MS of a standard solution of model peptides in varying concentrations of TFA.* From Ricker RD, Boyes BE, Nawrocki JP, Pannell LK "Stable Sterically Protected C3 Bonded Phase in LC and LC/MS Applications" 9th ISPPP, Delray Beach, FL, October 1999.

which the impact on the ionization process can be assessed. An example of optimizing LC/MS conditions with the use of 50 mM aqueous triethylammonium bicarbonate as an ion-pairing agent for the analysis of eight 40-mer oligonucleotides on a stationary phase of octadecylated poly(styrene/divinylbenzene) has been described [108].

E.   Influence of the Type and the Nature of LC Buffer

Inappropriate use of a buffer, such as an ammonium ion-based buffer, can cause signal suppression as illustrated by the results seen in Figure 11-13. The signal for 8-hydroxy-2′-deoxyguanosine shows a good response during analysis by ESI LC/MS in a water–acetonitrile mobile phase. It was then decided to try to improve the ESI signal by control of the sample pH using an ammonium acetate buffer; however, as seen in Figure 11-13, this results in severe signal suppression. Whereas ammonium acetate is often cited as a good LC/MS buffer because of its volatility (see Figure 11-16), the presence of the ammonium ion can cause the problem illustrated in Figure 11-13 for analytes of low to moderate proton affinity. In the example illustrated in Figure 11-13, it is possible that the method could have been improved by the use of a simple organic acid for pH control (e.g., acetic or formic acid) or by developing a separation protocol using a higher percentage of organic modifier, if compatible with the stationary phase.

F.   Influence of Solvent Composition on the ESI Signal

The extent to which the ESI signal can be enhanced by increasing the percentage of organic solvent in the mobile phase is shown graphically in Figure 11-14. The data for Figure 11-14 were collected by measuring the ESI ion current obtained from successive injections of a given concentration of protriptyline in an aqueous solution prepared with the indicated increasing concentrations of acetonitrile. It should be noted that the increase in ESI ion current for a given concentration of protriptyline is almost logarithmic as the concentration of acetonitrile is increased from 75 to 100%.

Increasing the organic solvent percentage in the mobile phase is a good tool for use with flow injection and direct infusion (i.e., when no chromatography column is employed). Unfortunately, when a reversed-phase separation of the analytes has been developed and is being used to introduce the sample to the LC/MS, the analyst cannot make dramatic increases to the organic solvent composition without destroying the separation. However, in some cases, selection of a more hydrophobic stationary phase may allow the use of higher percentages of organic solvent in the mobile phase. Consider the example illustrated in Figure 11-15 in which cocaine is analyzed using a

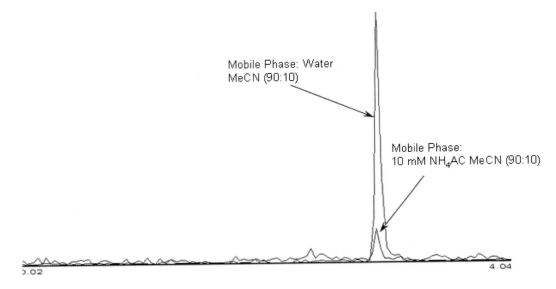

Figure 11-13. *Signal suppression in electrospray of 8-hydroxy-2′-deoxyguanosine as a result of using ammonium acetate buffer in the mobile phase.* Courtesy of Varian, Inc.

standard C_{18} column. In this case, a 12% acetonitrile mobile phase is the strongest solvent system that can be employed to obtain an acceptable retention time; as indicated in Figure 11-15, the ESI response under these conditions is poor (ESI signal = 2000 counts per second). Switching to a column made with a pentafluorophenylpropyl (PFPP) stationary phase (a much more hydrophobic sorbent) allows the use of 90% acetonitrile to obtain the same retention time for cocaine, but with a much improved (18,500 counts per second) ESI signal [109].

It should be noted that for microelectrospray and nanoelectrospray applications, the increase in ESI signal with increased organic percentage is not observed.

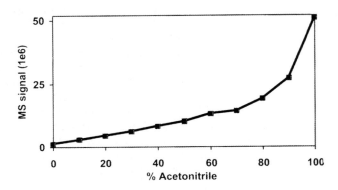

Figure 11-14. *Electrospray signal for protriptyline as a function of the percentage of acetonitrile in the mobile phase during LC/MS.*

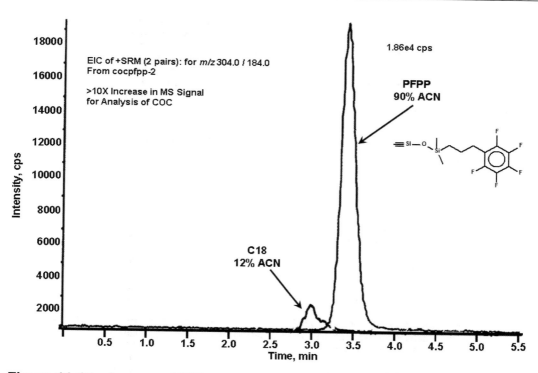

Figure 11-15. *Improved ESI response for cocaine in the 90% ACN mobile phase with PFPP column vs that in 12% ACN with the C$_{18}$ column.*

G. Adduct Formation

This term can have two different meanings. The first meaning pertains only to ESI in the situation where an ion other than a proton forms an adduct with the analyte in solution, resulting in peaks in the mass spectrum at *m/z* values different from those for the protonated molecule (the second meaning is described in the next paragraph). Adduct ions such as sodium, potassium, and ammonium may be readily formed from endogenous substances in the sample. These types of adduct ions also can be formed from compounds in the reagents or the container used for the samples. Glass is a known source of Na and K adduct ions of analytes; no matter what the quality of borosilicate glass (Pyrex®) is, there is still sufficient Na$^+$ and K$^+$ present to form detectable amounts of adduct ions. These types of adduct ions formed from variable amounts of endogenous compounds can result in poor quantitative accuracy. Even in a simple determination of molecular mass of an analyte (qualitative ESI analysis), sodium and potassium adduct ion formation can be problematic, especially if there is a possibility that the analyte will form multiple-charge ions. A very complex mass spectrum results when there are peaks representing protonated, sodiated, potassiated, protonated-potassiated, and sodiated-potassiated molecules. Formation of adduct ions is not always bad. As an example, chloride and bromide adduct ions are often used intentionally in negative-ion ESI of carbohydrates.

The other type of adduct formation is the clustering of molecules of the solvent (such as water, methanol, acetonitrile) with analyte ions. There are even potential problems of the analyte clustering with itself to form dimers and trimers. This type of adduct formation has been observed in all types of ion formation in LC/MS and is not limited to just APCI or ESI. Operating parameters can be adjusted to reduce the problem of adduct formation; examples include the ion-transfer rate between the atmospheric pressure region of the interface and the high-vacuum region of the *m/z* analyzer or by adjusting the temperatures in this region. Sometimes, it may be necessary to modify the solvent system to avoid these adducts, which are often referred to as clusters. These types of adduct ions can be an interference in both quantitative and qualitative analyses. When a suspicious peak is encountered in a mass spectrum, it is often prudent to look at possible adducts that may have been formed based on the composition of the solvent system. Some LC/MS practitioners prepare and keep handy the increments in mass that result from various combinations and numbers of molecules of the potential adducts.

H. Spectral Interference

This term refers to ions that have the same or nearly the same *m/z* value as the analyte. Substances producing ions of the same or similar *m/z* value can generally be separated by the chromatographic or sample preparation processes prior to entry into the mass spectrometer. When chromatographic resolution is not possible, clear distinction can often be made between the components through use of selected reaction monitoring (SRM) with an MS/MS system or the use of accurate mass determination using an *m/z* analyzer with high resolving power.

I. System Compromise

This is a term used to describe mobile-phase components that degrade the performance of the LC/MS system generally through precipitation in the interface. Components of the mobile phase that are likely to precipitate in the interface are generally buffer salts of inorganic acids. As the technique of HPLC developed and the use of UV detectors became more prevalent, there was a tendency to avoid the use of acetate (or salts of other organic acids) as buffering reagents because these compounds have absorbancies in the lower UV range of wavelengths (~240 nm and lower); this practice resulted in the use of potassium phosphate buffers. However, phosphate buffers precipitate in the interface between the LC and the mass spectrometer when the solvent is removed by volatilization. Many LC/MS manufacturers claim to have addressed the problems of phosphate buffer precipitation with the use of elaborate and sometimes bizarre means; however, it is best to avoid use of these solid-forming agents. Often, those beginning the use of LC/MS who are accustomed to HPLC with conventional detectors and who work in a regulated environment such as the pharmaceutical industry will say, "I don't want to change my method by using a different buffer because this will require me to revalidate the method." Changing from the use of a UV detector to a mass spectrometer necessitates a revalidation. The revalidation must be done regardless of whether the buffer system is changed.

Volatile organic solvents used in normal-phase chromatography are generally compatible with all LC-MS interfaces. Being far more volatile than the analytes, these solvents promote rapid desolvation and evacuation from the interface; however, normal-phase chromatography is not compatible with ESI because of ion formation in solution with typical normal-phase solvents (e.g., hexane and methylene chloride).

The common reversed-phase solvents are usable in all LC-MS interfaces. Water, methanol, and acetonitrile all work quite well. It should be noted that the color-coded poly(ether ether ketone) (PEEK) tubing often employed in HPLC laboratories should be avoided because organic solvents have been reported to leach a sufficient amount of the dye to cause interference in the mass spectrum. The relative percentages of aqueous and organic solvents in a reversed-phase mobile phase can affect the signal intensity of the mass spectrometer due to their effect on the performance of the vacuum interfaces. In standard ESI interfaces, a higher organic content in the mobile phase leads to more rapid and complete desolvation, increased ion formation in the interface and, therefore, better signal strength. This is illustrated with the equation:

$$V_{ON} \approx 2 \times 10^5 \, (\gamma \, r_c)^{1/2} \, \ln(4d/r_c) \qquad \text{(Eqn. 11-1)}$$

where γ is the surface tension of the solvent. A higher surface tension requires a higher potential for onset of the ion evaporation process. An experiment to illustrate this was conducted with the Varian 1200 LC/MS system. With all dimensions and other potentials held constant, the V_{ON} settings listed in Table 11-2 were required to induce disruption of the Taylor cone, leading to ion evaporation in various HPLC solvents:

TABLE 11-2. Relative stability of the Taylor cone in various solvents.

Solvent	γ (N/m)	V_{ON} (volts)
CH_3OH	0.0226	2200
CH_3CN	0.030	2500
H_2O	0.073	4000

However, at very low flow rates with ESI interfaces, a greater aqueous percentage improves the signal in many applications. It may be that a high percentage of water prevents the very small droplets formed in this situation from evaporating too quickly and therefore from exposing the analyte to possible thermal degradation in the heated drying gas.

Chemical ionization (CI) efficiency in APCI will be affected by the relative disparity in proton affinities of the analyte and the components of the solvent system. Figure 11-16 *ranks* various solvents according to their relative strength to act as proton donors in positive-ion APCI reactions [1]. Methane is an excellent CI reagent used in GC/MS. If CH_5^+ is mixed with methanol vapor, the CH_5^+ will transfer a proton to methanol to form $CH_3OH_2^+$ – the weaker acid. Similarly, if ammonia is now added to the mix, a proton will transfer from the protonated methanol to ammonia to form the NH_4^+ species. Again, the weaker acid (component with the higher proton affinity) captures the proton to form its conjugate acid. It is clear that it is essential to understand the gas-phase acid–base reactions between molecules of the solvents in the mobile phase and the analyte molecules. Solvents added to the HPLC system to improve separation may greatly alter APCI performance. The important reversed-phase solvents have been highlighted in Figure 11-16. Water, when protonated, is an excellent positive-ion reagent. Also, methanol, when protonated, is a better reagent ion than protonated acetonitrile. Therefore, when performing an analysis by APCI, it may be worthwhile to change the LC method to replace acetonitrile with methanol. Often in conventional HPLC, the only reason for using acetonitrile is to reduce UV background; this problem no longer exists when the method is transferred to LC/MS.

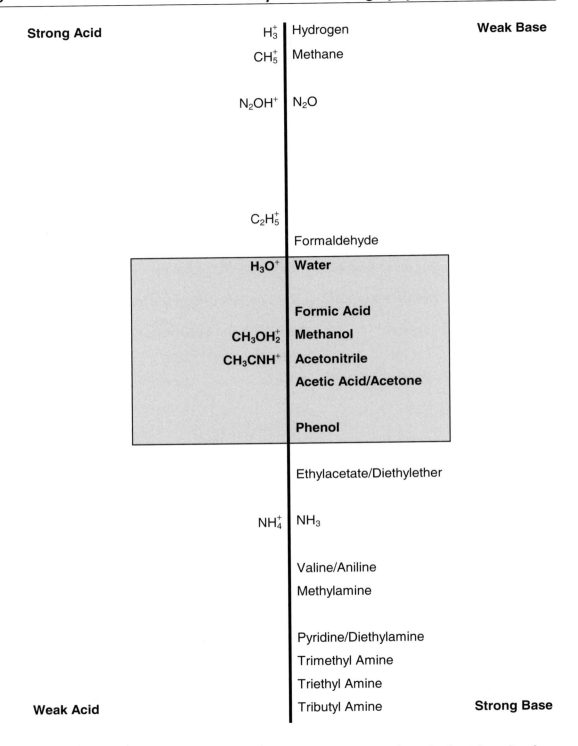

Figure 11-16. *Acid–base scale for positive ions and neutral molecules in the gas phase.*

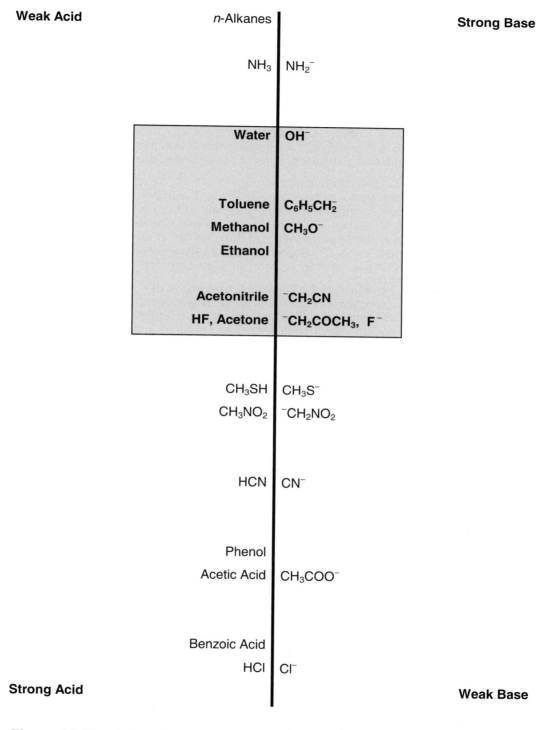

Figure 11-17. *Acid–base scale for negative ions and neutral molecules in the gas phase.*

It is very important to note that proton transfer ion/molecule reactions take place in APCI and ESI interfaces. In ESI, a high cone voltage can result in electrical arcing in the interface; this, in turn, can cause chemical ionization reactions to occur. Even in normal electrospray operation, these reactions will take place and (in positive-ion operation) the species with the highest proton affinity (weakest acid) will become protonated. If, for example, the analyte in ESI is a primary amine and there is a high concentration of tertiary amine present in the sample matrix or mobile phase, then the tertiary amine will be preferentially ionized and very little, if any, signal will be seen for the primary amine. Therefore, another important reason for ESI users to understand the hierarchy in Figure 11-16 is to avoid potential signal-suppression reactions. A scheme for investigating and dealing with signal suppression of this type was presented earlier.

Figure 11-17 is complementary to Figure 11-16 in showing the relative performance of various solvents in negative-ion APCI [1]. It is important to note that water and methanol are excellent negative-ion reagents as well as having the potential to serve as positive-ion reagents as shown in Figure 11-16. In fact, for negative-ion reactions, methanol is a considerably better performer than acetonitrile. (Figures 11-16 and 11-17 are adapted from a figure by A. P. Bruins in "Mass Spectrometry with Ion Sources Operating at Atmospheric Pressure" *Mass Spectrom. Rev.* 1991, *10*(1), 53–77.)

As stated at the beginning of this section, "LC/MS is as different from either liquid chromatography or mass spectrometry as either of these two latter techniques is from the other." Whereas HPLC requires component separation for quantitation, LC/MS does not because of its much higher degree of specificity based on uniqueness of the *m/z* values of ions of individual components. With the development of ESI, LC/MS no longer requires thermal stability and volatility of analytes; HPLC has never had this requirement. HPLC has relied on retention time and a generalized detector response for compound identification. The adsorption wavelength of the UV detector offers some degree of specificity; however, there are many more substances that have the same UV adsorption than have the same exact mass or even the same nominal mass, for that matter. Most single *m/z* analyzer LC/MS systems provide only the molecular mass of the analyte, which sometimes can be sufficiently specific for an unambiguous identification; however, modern instrumentation allows precise control of parameters in the interface that promote nonspecific fragmentation of ions containing the intact analyte molecule, resulting in higher confidence in the identification process. Spectral matching for such results on an instrument-to-instrument basis is now becoming common. LC/MS offers an advantage over HPLC that in addition to the use of chromatographic peak areas, quantitation can be done based on signal strength of an ion of a given *m/z* value even when component separation is not possible. For this reason, most of the analyses by LC/MS do not involve any attempt to separate the analytes. The sample is introduced into the mass spectrometer as a mixture without the need for a chromatographic separation. The additional dimension of *m/z* analysis in MS/MS gives LC/MS/MS even more specificity and better detection limits through improved signal-to-background.

2. Differences in Method Development for ESI vs APCI

Electrospray ionization signals are concentration-dependent and APCI signals are mass-dependent. This concept of concentration dependency on the part of ESI is sometimes difficult to grasp. Most HPLC detectors (including APCI mass spectrometry) are "mass detectors" (i.e., the more material we put through the detector, the greater the detector response). For example, with a UV detector, it is common to increase sensitivity by increasing the injection volume and thereby increasing the mass of analyte introduced. With ESI, a given concentration of analyte in the interface will yield roughly the same response regardless of the volume flow rate. Unfortunately, as illustrated in Scheme 11-7, the increase in ESI response is not linear with analyte concentration.

$$10^{-9} \to 10^{-5} \text{ M:} \qquad \text{Response } \alpha \text{ [ion] (linear)}$$

$$10^{-5} \to 10^{-4} \text{ M:} \qquad \text{Response } \alpha \left[\left(\sqrt{\text{ion}} \right) \right]$$

$$\geq 10^{-4} \text{ M:} \qquad \text{Response } \alpha \text{ [ion]}^{-1}$$

Scheme 11-7

The *m/z*-analyzed ion signal for an ionic analyte increases proportionately to the analyte ion concentration in solution from 10^{-9} M (detection limit) to 10^{-5} M, as illustrated in Scheme 11-7. This linear range lies below the concentration of electrolyte impurities in the solvent (methanol), which is 10^{-5} M. From 10^{-5} to 10^{-4} M, the signal increases with approximately the square root of the concentration and, at higher molalities, the signal "saturates" and begins to decrease. This occurs in both ES and IS (Ion Spray: nebulizer-assisted electrospray). It is concluded that the loss of sensitivity above 10^{-4} M is not only due to an increase of droplet size with concentration in the range but also to a relative decrease in gas-phase ion production from droplets having high electrolyte concentration [110].

The influence of liquid flow rate on relative signal response from API sources is illustrated in Figure 11-18. These data were acquired by monitoring the ion current at *m/z* 167 while making repetitive injections of diphenhydramine at indicated values of flow rate in an experimental combined ESI and APCI source for LC/MS [111]. The authors admit that the APCI source was not optimized in this design (e.g., the interface is operated at only 120 °C); hence, it shows a poorer response than the electrospray

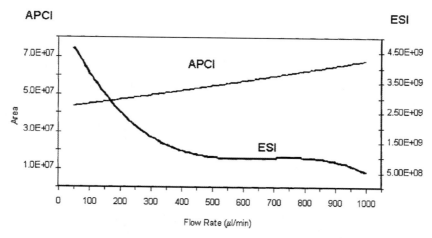

Figure 11-18. Flow-rate effects on chromatographic peak area counts in APCI and ESI [112].

source. The objective of the data in Figure 11-18 is not to compare ESI and APCI, but to show the relative effect of the flow rate on the response from each technique. Note that the electrospray ionization is affected by flow rate, whereas APCI shows relatively little effect. Decreasing the flow rate from 1000 to 100 µL min^{-1} resulted in a near order-of-magnitude increase in electrospray ionization response. This is consistent with the improved sensitivity seen with microelectrospray and nanoelectrospray sources (see an example of this advantage in drug metabolism work [113]). Even with standard electrospray ionization sources, decreasing the flow rate by using a smaller-diameter chromatographic column should show increased response due to this flow-rate effect as well as the effect of increasing "in-peak" analyte concentration observed with the use of smaller column diameters.

V. Applications

The importance of LC/MS in pharmaceutical research has been highlighted in recent publications [114, 115] and on its feasibility for identification of drug metabolites [116]; its importance in biomedical research is highlighted in a book on hyphenated techniques for neuropeptide research [117] and in a book on cancer research [118]. Selective quantitation of HIV protease inhibitors can be achieved by LC-ESI MS/MS [119, 120]. Microsize-exclusion chromatography coupled with capillary LC/MS provides a rapid and simple approach for preliminary screening of active ligands toward a specific target macromolecule in drug discovery work [121]. In other targeted applications, an LC/FTMS has been designed for comparative proteomics [122]. More conventional LC/MS has been used in screening for pharmaceutical discovery compounds [123–125], 2-hydroxyalkyl-cobalamins (vitamin B12 derivatives) [126], opium alkaloids in heroin profiling [35], commercial ceramides in cosmetics for quality control of product formulation [127], thalidomide in human fluids [128], leukotriene B$_4$ [129], oxidized low-density lipoprotein [130], methyldopa [131], fluvastatin [132], coumarin-type anticoagulants [133], quaternary nitrogen muscle relaxants [134], Ritalin [135], cyclosporin [136], ceramides [137], beclomethasone [138], and xenobiotic carboxylic acids [139]. A statistical assessment of quantitation by ESI LC/MS has been reported in the context of analyzing plasma for an anticancer drug [140]. Efaproxiral (RSR13) has been characterized by ESI in LC/MS [141]. Analytical methods for atorvastatin, novobiocin, and roxithromycin using microbore-LC-ESI MS/MS use a positive and negative voltage switching mode [142]; an assay for phencyclidine is based on multiple reaction monitoring in the positive-ion mode [143]. Supercritical fluid chromatography (SFC) is likely to be a powerful and complementary technique to HPLC for the purification of pharmaceutical compounds [52, 144–146].

Metabolomic analysis of plant leaves has been achieved with hydrophobic interaction chromatography-ESI MS [147]; analogous methodology has been used for folates [148]. An LC/MS method allows the quantification of malonyl-CoA (CoA) in physiological tissues [149]; comparable methodology allows quantification of benzo[a]pyrene diol epoxide DNA adducts in biological samples [150]. LC-ESI MS has been used to characterize allergenic proteins from pollen [151] and a family of aldehyde dehydrogenases [152]. A nanospray LC/MS system has been described for proteomic applications [153]; a computer program has been developed to facilitate interpretation of LC/MS results for peptides and proteins [154]. The use of 20 mm i.d. polymeric polystyrene-divinylbenzene monolithic nanocapillary columns for the analysis of tryptic peptides improved detection limits for nano-LC-ESI MS/MS by an order of magnitude [155]. A nano-LC–nano-ESI MS and MS/MS system has been described for on-line analysis of proteolytic digests after SPE with detection limits in the atto-mol range [156]; another application involves limited proteolysis [157]. Using methodology based on LC/MS/MS, it is possible to analyze the relative amounts of amyloid peptides and oxidized metabolites in cerebrospinal fluid from patients with Alzheimer's disease as biomarkers of oxidative stress [158]; on-line immunoaffinity LC/MS has been used to detect amyloid-related polypeptides [159]. A sensitive method for determining amino acids and peptides following derivatization by 2-(9-carbazole)-ethyl chloroformate has been developed, which allows fluorescence detection and analysis by ESI in the context of HPLC [160].

Hydrophilic interaction chromatography on a carbamoyl-derivatized stationary phase in combination with ESI MS has been used to analyze food products for free amino acids and small polar peptides [161]. A capillary ESI LC/MS method has been developed for the analysis of plasma for steroid sulfates [162]. An automated method is available for quantitating a variety of carbamates and phenylureas by LC/MS [163]. The herbicide, atrazine, and metabolites can be detected during the analysis of mouse urine and plasma by LC/MS [164].

A special interface is needed in MS applications of supercritical fluid chromatography (SFC) [51, 52, 165, 166]. Mobile phases in SFC have low viscosities and high diffusion coefficients with respect to those of traditional HPLC. These properties allow the use of higher mobile-phase flow rates and/or longer columns in SFC, resulting in rapid analyses and high-efficiency separations. Most SFC/MS is performed with API sources. Unlike in conventional LC/MS, the interface between the SFC column and the API source must allow control of the downstream (post-column) pressure while also preserving good chromatographic fidelity. The use of packed-column SFC/MS is especially useful in the analysis of enantiomers [167]. SFC/MS has been applied to the analysis of ethoxylated and propoxylated surfactants [168] as well as alkoxylated polymers [169].

An LC/MS/MS assay for a tobacco-specific nitrosamine metabolite has been described [170]. LC/MS/MS has been used in the measurement of a synthetic opiate buprenorphine and its glucuronide [171] and for cocaine metabolites [172] in human subjects. An assay for ethyl glucuronide in hair samples based on ESI MS/MS can monitor and quantify alcohol consumption in human subjects [173]. Quantification of *o*-tyrosine, *o*-nitrotyrosine, and *o,o'*-dityrosine from cat urine samples by LC-ESI MS/MS has been described using an isotope dilution technique in the multiple reaction monitoring mode [174]; similar methodology has been used to analyze *in vitro* samples for testosterone and its metabolites [175]. The hydrolysis products of sulfur mustard compounds have been characterized by LC-ESI MS/MS [176]. Urinary estrogens can be determined by LC-ESI MS/MS [177]. Lactosylceramides derived from porcine blood cells have been characterized by LC-ESI MS/MS [178]. A homogeneous LC radioactivity monitor–ESI MS/MS system has been evaluated for the analysis of drug metabolites [179]. A stopped flow system improves the detection limits for ^{3}H- and ^{14}C-labeled drug metabolites by an order of magnitude during analyses by radioactivity detection LC/MS [180]. The advent of pH-stable columns allows short-lived (<3 sec) arachidonic acid metabolites to be analyzed by negative-ion LC-ESI MS/MS [181]. LC-ESI MS/MS was also used to analyze human atherosclerotic plaques for a variety of eicosanoids [182]; in other work, it was found that several lipoxygenase products of eicosanoic acid are esterified to phospholipids [183, 184]. Detection of DNA adducts was improved by two orders of magnitude by adapting the methodology for capillary LC and micro-ESI [185].

LC-ESI MS has been useful in the analysis of natural waters for naphthenic acids [186], humic acids [187], alkylphenol ethoxylates [188], and alkylbenzene sulfonate surfactants [189]; it also plays an important role in other environmental analyses [190]. LC/MS also played a key role in the identification of platelet-activating factor (PAF) as the inflammatory lipid mediator during oxidative stress in rat liver toxicity [191]; PAF has also been implicated as a pathologic mediator in UV radiation damage to human keratinocytes [192]. LC-ESI was also used to identify a glutathione-containing eicosanoid [193], the metabolites of leukotriene B$_4$ (LTB$_4$) in human hepatocytes [194], and the metabolites of PGE$_2$ in rat hepatocytes [195]; methodology for the metabolism of LTB$_4$ has been reviewed [196]. Certain neutral arachidonic acid derivatives, such as N-arachidonylethanolamide and 2-arachidonylglycerol, have been identified as endogenous ligands for the cannabinoid receptors; these endocannabinoids from various tissues can be monitored by ESI LC/MS as silver cation adducts during studies to elucidate cannabinoid-mediated processes *in vivo* [197]. Monoproanthocyanidins and oligoproanthocyanidins have been detected during the analysis of raw grape products using ESI LC/MS [198]. Eleven flavonoid aglycones have been characterized by LC-ESI with negative-ion detection [199]. Negative ESI detection of high explosives as adducts with chloride, formate, acetate, and nitrate has been reported [200]. Similar methodology has been used to analyze sequence isomers of oligodeoxyribonucleotides [201]; optimization of LC/MS conditions for using ion pairing with an octadecylated poly(styrene/divinylbenzene) stationary phase has been described for the analysis of eight 40-mer oligonucleotides [108]. Mixtures of corticosteroids and of benzodiazepines have been analyzed by electrochemically modulated liquid chromatography with detection by ESI [202].

Quantitative assays based on LC/MS rely heavily on proper calibration as emphasized in a report on the comparison of internal versus external calibration during the validation of analytical methodology for 1-hydroxypyrene in human urine by LC-CAD MS/MS [203]. A variety of phthalates have been quantitatively detected in milk and milk products including infant formulas down to the level of 5–9 mg/kg using LC-ESI-CAD MS/MS [204]; this technology is also useful in drug discovery [205, 206]. The APCI LC/MS method has been developed for screening and quantification of 15 neuroleptic (antipsychotic) drugs [207]. A quantitative LC-APCI MS/MS method for simultaneous determination of multiple illicit drugs, methadone, and their metabolites in oral fluid has been developed and validated [208, 209]. A method for analyzing blood for vitamin K homologs is based on LC-APCI MS/MS [210]. A flow-injection analytical method coupled with APCI LC/MS has been developed for the quantification of polyamines after derivatization with dansyl chloride; the method is reportedly 50-fold more sensitive than the conventional HPLC/fluorometry procedure [211].

With technology related to LC/MS, a homogeneous on-line continuous-flow multiprotein biochemical assay has been demonstrated for the interaction between fluorescein-biotin and streptavidin and for digoxin and anti-digoxigenin using ESI MS/MS with a Q-TOF mass spectrometer [212]. A strategy has been described for the handling of biological media for analysis of intracellular drugs by LC/MS/MS [213].

1. Attention to High Throughput

The demand for rapid throughput has stimulated development of ways to deal with large numbers of samples. To cope with the high demand for protein identification, a fully automated process that significantly reduces analyst intervention has been developed consisting of web-enabled sample submission and registration, automated data processing, data interpretation, and report generation [214]. Another example, well suited for pharmacokinetic studies of fast LC/MS, is based on and can analyze 2000 samples per instrument per day [215] in spite of the use of complex plumbing and valving schemes coupled with efficient timing of injection, run, wash cycles, etc. The use of a 1-min ballistic gradient in LC combined with optimization of an autosampler resulted in an 85-sec analytical cycle for plasma samples based on tandem MS of new chemical entities during drug discovery [216]. Comprehensive analytical systems based on parallel column technology [217–220] make use of a staggered injection schedule with time-shared access to a single electrospray ionization mass spectrometer to analyze only "useful" portions of a given chromatographic separation. A multiplexed ion source reportedly improves throughput by a factor of four over the conventional serial-based use [221]; the use of lithographic procedures to manufacture multiple channels with high geometrical fidelity should improve this approach [222]. The analytical performance of an automated nanospray system based on chip technology interfaced to a triple-quadrupole mass spectrometer as assessed in analyses of a pharmaceutical by infusion showed advantages in terms of throughput and sample consumption (μL min^{-1}) when compared to other methodologies based on LC/MS [223]. A microfluidic device of integrated ESI emitters and monolithic LC columns has been fabricated on a chip made of cycloolefin copolymer, which stabilizes the ESI process as demonstrated in the analysis of a tryptic digest of BSA [224]. An interface consisting of multiple capillaries, each of which has its own ion lens for sprayer control, offers another approach to high-throughput applications [225]. The Aria LX4 staggered parallel HPLC system has been evaluated for application to "good laboratory practice" level for quantitative analysis [226].

Solid-phase extraction plays an important role in high-throughput analytical schemes [227], as exemplified by its use in pharmacokinetic studies of an anticancer drug [228]. A bioanalytical method for morphine and its metabolites in human fluids has been automated for sample transferring, solid-phase extraction, and injection into an LC-MS-MS [229]; other automated systems are also based on 96-well technology for the analysis of plasma samples [40], a study of metabolic stability based on TOFMS [230], or the use of SPE cards to facilitate the analysis of urinary Ritalin [231]. SPE is also the basis for capturing perfluorochemicals from blood for detection by LC/MS [232]. The applicability of liquid-phase microextraction has been optimized for 13 anabolic steroid glucuronides, and compared with conventional liquid–liquid extraction (LLE) and solid-phase extraction (SPE) procedures for processing prior to analysis by ESI-CAD MS/MS [233]. A method for determining residues of 12 sulfonamide antibacterials in cattle and trout muscle tissues is based on LC/MS after sample processing by solid-phase dispersion [234]. A "good laboratory practice" method for isolation and identification of a variety of herbicides in ground water is based on SPE-ESI MS [235]. The use of a diode-array detector has been reported to facilitate the use of libraries of spectra for a range of drugs and toxic substances [236].

Given the high sample load facing the combinatorial chemistry laboratory [237], an eight-channel LC/MS system has been developed to handle up to 240 samples per hour with library identification up to a 70% success rate [218]. Such a complex mixture analysis may benefit from instrumentation combining HPLC with ion mobility spectrometry prior to analysis by mass spectrometry because it offers the analytical advantage of possibly resolving many isomass compounds that have identical retention times [238, 239]; combination with a chirality-based system may also be helpful [240]. A combined ESI-APCI source for use in on-line HPLC applications allows alternate on-line ESI and APCI scans with polarity switching within a single analysis; the new source reduces analysis time by eliminating the need for a source hardware change, source optimization, and repeat analyses [112]. An alternative instrument, based on an array of cylindrical ion traps, has been built with four independent channels and operated using two fully multiplexed channels (sources, ion optics, ion traps, detectors) capable of analyzing different samples simultaneously [241].

References

1. Bruins AP, Mass spectrometry with ion sources operating at atmospheric pressure. *Mass Spectrometry Reviews* **10**(1): 53-77, 1991.
2. Abian J, The coupling of gas and liquid chromatography with MS. *J Mass Spectrom* **34**: 157-168, 1999.
3. Willoughby R, Sheehan E and Mitrovich SA, *Global View of LC/MS*. Global View Publishing, Pittsburgh, PA, 1998.
4. Ettre LS, Pittsburgh Conferences: A Personal Appraisal. *LC/GC* **17**(2): 156-168, 1999.
5. Horning EC, Horning MG, Carroll DI, Dzidic I and Stillwell RN, New picogram detection system based on a mass spectrometer with an external ionization source at atmospheric pressure. *Analytical Chemistry* **45**(6): 936-43, 1973.
6. Carroll DI, Dzidic I, Horning EC and Stillwell RN, Atmospheric-pressure ionization mass spectrometry. *Applied Spectroscopy Reviews* **17**(3): 337-406, 1981.
7. McEwen CN and McKay RG, A Combination Atmospheric Pressure LC/MS:GC/MS Ion Source: Advantages of Dual AP-LC/MS:GC/MS Instrumentation. *Journal of the American Society for Mass Spectrometry* **16**(11): 1730-1738, 2005.
8. Garcia JF and Bacelo D, LC-MS Interfacing Systems. *J. High Res. Chromatogr.* **16**: 633-641, 1993.
9. Arpino P, Baldwin MA and McLafferty FW, Liquid chromatography-mass spectrometry. II. Continuous monitoring. *Biomedical mass spectrometry* **1**(1): 80-2, 1974.
10. Voyksner RD and Keever J, "HPLC-MS". In: *Analysis of Pesticides in Ground and Surface Water II* (Ed. Stan S-V), pp. 109-131, NY, 1995.
11. de Wit JS, Parker CE, Tomer KB and Jorgenson JW, Direct coupling of open-tubular liquid chromatography with mass spectrometry. *Analytical chemistry* **59**(19): 2400-4, 1987.
12. Cappiello A, Famiglini G and Palma P, Electron ionization for LC/MS. *Analytical Chemistry* **75**(23): 496A-503A, 2003.
13. McFadden WH, Schwartz HL and Evans S, Direct Analysis of HPLC Effluents. *J. Chromatogr* **122**: 389-396, 1976.
14. Games DE, McDowall MA, Levsen K, Schafer KH, Dobberstein P and Gower JL, Moving Belt in LC-MS. *Biomed. Mass Spectrom.* **11**: 87-95, 1984.
15. Yergy AL and etal, *Liquid Chromatography/Mass Spectrometry: Techniques and Applications*. Plenum Press, New York, 1990.
16. Blakley CR and Vestal ML, TSP Interface for LC-MS. *Anal. Chem.* **55**: 750-754, 1983.
17. Arpino P, LC-MS. Part II. Techniques and Mechanisms of Thermospray. *Mass Spec. Rev.* **9**: 631-669, 1990.
18. Curcuruto O, Franchi D, Hamdan M, Favretto D and Traldi P, Investigation of some steroids by means of thermospray and tandem mass spectrometry. *Rapid Communications in Mass Spectrometry* **7**(7): 673-5, 1993.
19. Covey TR, Bruins AP and Henion JD, Comparison of thermospray and ion spray mass spectrometry in an atmospheric pressure ion source. *Organic Mass Spectrometry* **23**(3): 178-86, 1988.
20. Blakley CR, Carmody JJ and Vestal ML, A new soft ionization technique for mass spectrometry of complex molecules. *Journal of the American Chemical Society* **102**(18): 5931-3, 1980.
21. Barber M, Bordoli RS, Sedgwick RD and Tyler AN, Fast atom bombardment of solids as an ion source in mass spectrometry. *Nature (London, United Kingdom)* **293**(5830): 270-5, 1981.
22. De Pauw E, Ion promotion in fast-atom bombardment mass spectrometry by charge transfer complexation. *Analytical Chemistry* **55**(13): 2195-6, 1983.
23. Benninghoven A, Organic secondary ion mass spectrometry (SIMS) and its relation to fast atom bombardment (FAB). *International Journal of Mass Spectrometry and Ion Physics* **46**: 459-62, 1983.
24. Benninghoven A, Some aspects of secondary ion mass spectrometry of organic compounds. *International Journal of Mass Spectrometry and Ion Physics* **53**: 85-99, 1983.
25. De Pauw E, Agnello A and Derwa F, Liquid matrices for liquid secondary ion mass spectrometry-fast atom bombardment [LSIMS-FAB]: an update. *Mass Spectrometry Reviews* **10**(4): 283-301, 1991.
26. Caprioli RM and Moore WT, Decreased Ion Suppression by CF-FAB. *Int. J. Mass Spectom.* **86**: 187-199, 1988.
27. Reiser RW and Fogiel AJ, Liquid chromatography/mass spectrometry analyses of small molecules using electrospray and fast-atom bombardment ionization. *Rapid Communications in Mass Spectrometry* **8**(3): 252-7, 1994.
28. Budzikiewicz H, Selected reviews on mass-spectrometric topics. XLIV. Continuous flow FAB mass spectrometry. *Mass Spectrometry Reviews* **11**(2): 153, 1992.
29. Coran SA, Bambagiotti-Alberti M, Giannellini V, Moneti G, Pieraccini G and Raffaelli A, Continuous flow FAB vs ion spray for the simultaneous determination of alkyltrimethylammonium surfactants by MS. *Rapid Commun Mass Spectrom* **12**: 281-284, 1998.

30. Hemling ME, Roberts GD, Johnson W, Carr SA and Covey TR, Analysis of proteins and glycoproteins by on-line HPLC with flow FAB and ESI MS: a comparative evaluation. *Biomed & Environ Mass Spectrom* **19:** 677-91, 1990.

31. Yates JR, McCormack AL, Link AJ, Scheiltz D, Eng J and Hays L, Analysis of Complex Biological Systems using Microcolumn LC-ESI-MS/MS. *journal* **121:** 65R-75R, 1996.

32. Buettering L, Roellgen FW, Wilm MS and Mann M, ESI and Taylor Cone Theory. *Int. J. Mass Spectrom. Ion Proc.* **136:** 139-150, 1989.

33. Kebarle D and Tang L, Solution Ions to Gas-Phase Ions: Mechanism of ESI. *Anal. Chem.* **65:** 972A-986A, 1993.

34. Wilm MS and Mann M, ESI and Taylor Cone Theory. *Int. J. Mass Spectrom. Ion Proc.* **136:** 167-180, 1994.

35. Riet Dams TB, Wolfgang Günther, Willy Lambert, and André De Leenheer*, Sonic Spray Ionization Technology: LC/MS Analysis on a Monolithic Silica Column for Heroin Impurity Profiling. *Anal. Chem.* **74:** 3206 -3212, 2002.

36. Pascoe R, Foley JP and Gusev AI, Reduction in matrix-related signal suppression effects in electrospray ionization mass spectrometry using on-line two-dimensional liquid chromatography. *Anal Chem* **73**(24)**:** 6014-23., 2001.

37. King R, Bonfiglio R, Fernandez-Metzler C, Miller-Stein C and Olah T, Mechanistic investigation of ionization suppression in electrospray ionization. *J Am Soc Mass Spectrom* **11:** 942-950, 2000.

38. Bonfiglio R, King RC, Olah TV and Merkle K, The effects of sample preparation methods on the variability of the electrospray ionization response for model drug compounds. *Rapid Comm. Mass Spectrom.* **13:** 1175-1185, 1999.

39. Gangl ET, Annan M, Spooner N and Vouros P, Reduction of Signal Suppression Effects in ESI-MS Using a Nanosplitting Device. *Anal Chem* **73**(23)**:** 5635-5644, 2001.

40. Berna M, Murphy AT, Wilken B and Ackermann B, Collection, Storage, and Filtration of in Vivo Study Samples by 96-Well Filter Plates for Automated Sample Preparation and LC/MS/MS Analysis. *Anal Chem* **74:** 1197 -1201, 2002.

41. Chang YZ, Chen YR and Her GR, Sheathless capillary electrophoresis/electrospray mass spectrometry using a carbon-coated tapered fused-silica capillary with a beveled edge. *Anal Chem* **73**(21)**:** 5083-7., 2001.

42. Hu S and Dovichi NJ, Capillary electrophoresis for the analysis of biopolymers. *Anal Chem* **74**(12)**:** 2833-50., 2002.

43. Ding J and Vouros P, Advances in CE/MS. *Anal. Chem.* **71**(11)**:** 378A-385A, 1999.

44. Varjo SJ, Jussila M, Palonen S and Riekkola ML, Interface for coupling nonaqueous wide-bore capillary electrophoresis with mass spectrometry. *Electrophoresis* **23**(3)**:** 437-41., 2002.

45. Kim J, Zand R and Lubman DM, Electrophoretic mobility for peptides with post-translational modifications in capillary electrophoresis. *Electrophoresis* **24**(5)**:** 782-93., 2003.

46. Apruzzese WA and Vouros P, Analysis of DNA adducts by capillary methods coupled to mass spectrometry: a perspective. *Journal of Chromatography, A* **794**(1 + 2)**:** 97-108, 1998.

47. Oberacher H, Wellenzohn B and Huber CG, Comparative Sequencing of Nucleic Acids by Liquid Chromatography-Tandem Mass Spectrometry. *Anal. Chem.* **74:** 211 -218, 2002.

48. Ding J and Vouros P, Capillary Electrochromatography and Capillary Electrochromatography-Mass Spectrometry for the Analysis of DNA Adduct Mixtures. *Anal. Chem.* **69**(3)**:** 379-384, 1997.

49. Marzilli LA, Koertje C and Vouros P, Capillary electrophoresis-mass spectrometric analysis of DNA adducts. *Methods Mol. Biol. (Totowa, NJ, U. S.)* **162**(Capillary Electrophoresis of Nucleic Acids, Volume 1)**:** 395-406, 2001.

50. Moini M, Cao P and Bard AJ, Hydroquinone as a buffer additive for suppression of bubbles formed by electrochemical oxidation of the CE buffer at the outlet electrode in CE-ESI/MS. *Anal Chem* **71:** 1658-61, 1999.

51. Pinkston JD and Chester TL, Guidelines for successful SFC/MS. *Analytical Chemistry* **67**(21)**:** 650A-6A, 1995.

52. Pinkston JD, Advantages and drawbacks of popular supercritical fluid chromatography/mass spectrometry interfacing approaches-a user's perspective. *European Journal of Mass Spectrometry* **11**(2)**:** 189-197, 2005.

53. Garcia DM, Huang SK and Stansbury WF, Optimization of APCI for LC-MS. *J. Am. Soc. Mass Spectrom.* **7:** 59-65, 1996.

54. Breithaupt DE, Wirt U and Bamedi A, Differentiation of Lutein Monoester Regioisomers and Diesters from Marigold Flowers (Tagetes erecta L.) and Several Fruits by LC/MS. *J Agric Food Chem* **50:** 66-70, 2002.

55. Byrdwell WC and Neff WE, Parallel ESI/APCI(MS), MS/MS and MS/MS/MS of triacylglycerols and triacylglycerol oxidation products. *Rapid Commun Mass Spectrom* **16:** 300-19, 2002.

56. Kertesz V and Van B, Surface-assisted reduction of aniline oligomers, N-phenyl-1,4-phenylenediimine and thionin in APCI & APPI. *J Am Soc Mass Spectrom* **13:** 109-17, 2002.

57. Van Berkel GJ and Kertesz V, Redox buffering in an electrospray ion source using a copper capillary emitter. *J Mass Spectrom* **36**(10): 1125-32., 2001.
58. Robb DB and Blades MW, Factors Affecting Primary Ionization in Dopant-Assisted APPI for LC/MS. *J Amer Soc Mass Spectrom* **17**: 130-138, 2006.
59. Heller DN, Ruggedness testing of quantitative APPI MS methods: the effect of co-injected matrix on matrix effects. *Rapid Commun Mass Spectrom* **21**: 644-652, 2007.
60. Norlander B, Carlsson B and Bertler A, Sensitive assay of methadone in plasma by using capillary gas chromatography with photoionization detection. *J Chromatogr* **375**(2): 313-9., 1986.
61. Dojahn JG, Wentworth WE and Stearns SD, Characterization of formaldehyde by gas chromatography using multiple pulsed-discharge photoionization detectors and a flame ionization detector. *J Chromatogr Sci* **39**(2): 54-8., 2001.
62. Baim MA, Eatherton RL and Hill HH, Jr., Ion mobility detector for gas chromatography with a direct photoionization source. *Anal Chem* **55**: 1761-1766, 1983.
63. Robb DB, Covey TR and Bruins AP, Atmospheric pressure photoionization: an ionization method for liquid chromatography-mass spectrometry. *Anal Chem* **72**(15): 3653-9., 2000.
64. Hanold KA, Fischer SM, Cormia PH, Miller CE and Syage JA, Atmospheric Pressure Photoionization. 1. General Properties for LC/MS. *Analytical Chemistry* **76**(10): 2842-2851, 2004.
65. Short LC, Hanold KA, Cai S-S and Syage JA, ESI/APPI multimode source for low-flow LC/MS. *Rapid Commun Mass Spectrom* **21**: 1561-1566, 2007.
66. Robb DB and Blades MW, APPI of Both Polar and Nonpolar Compounds in Reversed-Phase LC/MS. *Anal Chem* **78**: 8162-8164, 2006.
67. Syage JA, Hanold KA, Lynn TC, Horner JA and Thakur RA, Atmospheric pressure photoionization. II. Dual source ionization. *Journal of Chromatography, A* **1050**(2): 137-149, 2004.
68. Syage JA, Mechanism of [M + H]+ formation in photoionization mass spectrometry. *Journal of the American Society for Mass Spectrometry* **15**(11): 1521-1533, 2004.
69. Hatano Y, Interaction of photons with molecules--cross-sections for photoabsorption, photoionization, and photodissociation. *Radiat Environ Biophys* **38**(4): 239-47., 1999.
70. Kauppila TJ, Kuuranne T, Meurer EC, Eberlin MN, Kotiaho T and Kostiainen R, APPI MS. Ionization mechanism and the effect of solvent on the ionization of naphthalenes. *Anal Chem* **74**(21): 5470-9., 2002.
71. Short LC, Cai S-S and Syage JA, APPI-MS: Effects of Mobile Phases and VUV Lamps on the Detection of PAH Compounds. *J Amer Soc Mass Spectrom* **18**: 589-599, 2007.
72. Cai S-S, Hanold KA and Syage JA, Comparison of APPI and APCI for Normal-Phase LC/MS Chiral Analysis of Pharmaceuticals. *Anal Chem* **79**: 2491-2498, 2007.
73. Kauppila TJ, Kostiainen R and Bruins AP, Anisole, a new dopant for APPI MS of low proton affinity, low ionization energy compounds. *Rapid Commun Mass Spectrom* **18**: 808-815, 2004.
74. Kauppila TJ, Bruins AP and Kostiainen R, Effect of the Solvent Flow Rate on the Ionization Efficiency in APPI MS. *Journal of the American Society for Mass Spectrometry* **16**: 1399-1407, 2005.
75. Hsieh Y, Merkle K, Wang G, Brisson J-M and Korfmacher WA, HPLC-APPI-MS/MS for Small Molecules in Plasma. *Analytical Chemistry* **75**: 3122-3127, 2003.
76. Wang G, Hsieh Y and Korfmacher WA, Comparison of APCI, ESI, and APPI for the Determination of Cyclosporin A in Rat Plasma. *Anal Chem* **77**: 541-548, 2005.
77. Starkey JA, Mechref Y, Muzikar J, McBride WJ and Novotny MV, Determination of Salsolinol and Related Catecholamines through On-Line Preconcentration and LC/APPI/MS. *Analytical Chemistry* **78**(10): 3342-3347, 2006.
78. Cai S-S and Syage JA, Comparison of APPI, APCI, & ESI MS for Lipids. *Anal Chem* **78**: 1191-1199, 2006.
79. Kauppila TJ, Nikkola T, Ketola RA and Kostiainen R, Atmospheric pressure photoionization-mass spectrometry and atmospheric pressure chemical ionization-mass spectrometry of neurotransmitters. *Journal of Mass Spectrometry* **41**(6): 781-789, 2006.
80. Delobel A, Roy S, Touboul D, Gaudin K, Germain DP, Baillet A, Brion F, Prognon P, Chaminade P and Laprevote O, APPI-porous graphitic carbon LC for globotriaosylceramides in Fabry disease. *J Mass Spectrom* **41**: 50-58, 2006.
81. Calles J, Gottler R, Evans M and Syage J, Early warning surveillance of drinking water by photoionization/mass spectrometry. *Journal - American Water Works Association* **97**(1): 62-73, 2005.
82. Geyer R, Peacock AD, White DC, Lytle C and van Berkel GJ, APCI and APPI for simultaneous MS analysis of microbial respiratory ubiquinones and menaquinones. *J Mass Spectrom* **39**(8): 922-929, 2004.
83. Rauha JP, Vuorela H and Kostiainen R, Effect of eluent on the ionization efficiency of flavonoids by ion spray, APCI, and APPI MS. *J Mass Spectrom* **36**(12): 1269-80., 2001.
84. Kauppila TJ, Kotiaho T, Kostiainen R and Bruins AP, Negative ion-atmospheric pressure photoionization-mass spectrometry. *Journal of the American Society for Mass Spectrometry* **15**(2): 203-211, 2004.

85. Mol R, de Jong GJ and Somsen GW, Atmospheric pressure photoionization for enhanced compatibility in on-line micellar electrokinetic chromatography-mass spectrometry. *Analytical Chemistry* 77(16): 5277-5282, 2005.

86. Purcell JM, Rodgers RP, Hendrickson CL and Marshall AG, Speciation of Nitrogen Containing Aromatics by APPI or ESI FT-ICR MS. *J Amer Soc Mass Spectrom* 18: 1265-1273, 2007.

87. McEwen CN, GC/MS on an LC/MS instrument using atmospheric pressure photoionization. *International Journal of Mass Spectrometry* 259(1-3): 57-64, 2007.

88. Debois D, Giuliani A and Laprevote O, Fragmentation induced in APPI of peptides. *J Mass Spectrom* 41: 1554-1560, 2006.

89. Cappiello A, Famiglini G, Mangani F and Palma P, New trends in the application of electron ionization to liquid chromatography-mass spectrometry interfacing. *Mass Spectrom Rev* 20(2): 88-104., 2001.

90. Willoughby RC and Browner RF, Monodisperse Aerosol Generation Interface for LC-MS. *Anal. Chem.* 56: 2626-2631, 1984.

91. Creaser CS and Stygal JW, Particle Beam LC-MS: Instrumentation and Applications. *Analysis* 118: 1467-1480, 1993.

92. Ligon WV, Jr. and Dorn SB, Particle beam interface for liquid chromatography/mass spectrometry. *Analytical Chemistry* 62(23): 2573-80, 1990.

93. Cappiello A and Famiglini G, Capillary-scale particle-beam LC/MS interface: Can EI sustain the competition? *J Am Soc Mass Spectrom* 9: 993-1001, 1998.

94. Cappiello A, Balogh M, Famiglini G, Mangani F and Palma P, An efficient LC/MSinterface for generation of EI spectra. *Anal Chem* 72: 3841-3846, 2000.

95. Mao J, Pacheco CR, Traficante DD and Rosen W, Acridines and Benzoquinolines by LC-MS with Particle Beam with EI. *J. Liq. Chromatogr.* 18: 903-916, 1995.

96. Lehman PA and Franz TJ, PBI LC-MS for ECNI of Retinoic Acid in Human Plasma. *J. Pharm. Sci.* 85: 287-290, 1996.

97. Kientz C, Huist A, De Jong A and Wils E, Eluent jet interface for combining capillary liquid flows with EI mass spectrometry. *Anal Chem* 68: 675-681, 1996.

98. Dijkstra R, Van Baar B, Kientz C, Niessen W and Brinkman U, An eluent jet interface for CIMS and coupling of microcolumn LC with EIMS. *Rapid Commun Mass Spectrom* 12: 5-10, 1998.

99. Amirav A and Granot O, Liquid chromatography-mass spectrometry with supersonic molecular beams. *J Am Soc Mass Spectrom* 11: 587-591, 2000.

100. Granot O and Amirav A, LC-MS with electron ionization of cold molecules in supersonic molecular beams. *International Journal of Mass Spectrometry* 244(1): 15-28, 2005.

101. Moberg M, Bergquist J and Bylund D, A generic stepwise optimization strategy for liquid chromatography electrospray ionization tandem mass spectrometry methods. *Journal of Mass Spectrometry* 41(10): 1334-1345, 2006.

102. Sun W, Wu S, Wang X, Zheng D and Gao Y, An analysis of protein abundance suppression in data dependent LC/MS/MS with tryptic peptide mixtures of five known proteins. *Eur J Mass Spectrom* 11: 575-580, 2005.

103. Nelson M and Dolan J, Ion Suppression in LC-MS-MS-A Case Study. *LC-GC Europe 2002* 2002(Feb 2-6), 2002.

104. Matuszewski BK, Constanzer ML and Chavez-Eng CM, Strategies for the assessment of matrix effect in quantitative bioanalytical methods based on HPLC-MS/MS. *Analytical Chemistry* 75(13): 3019-3030, 2003.

105. Matuszewski BK, Constanzer ML and Chavez-Eng CM, Matrix Effect in Quantitative LC/MS/MS Analyses of Biological Fluids: Determination of Finasteride in Human Plasma at Pg/mL Concentrations. *Anal Chem* 70: 882-889, 1998.

106. Miller-Stein C, Bonfiglio R, Olah T and King R, Rapid Method Development of Quantitative LC-MS/MS Assays for Drug Discovery. *Am. Pharm. Rev.* 3: 54-61, 2000.

107. Richer R, Boyer B, Nawrocki J and Pannell L, Stable, sterically protected C3 bonded phase in LC and LC/MS applications. In: *9th ISPPP, Delray Beach, FL,1999*.

108. Huber CG and Krajete A, Analysis of Nucleic Acids by Capillary Ion-Pair Reversed-Phase HPLC Coupled to Negative-Ion Electrospray Ionization Mass Spectrometry. *Anal. Chem.* 71: 3730 -3739, 1999.

109. Needham S and Brown P, The Role of the Column for the Analysis of Drugs and Other Components by HPLC/ESI/MS: Part I. *Am. Pharm. Rev.* 4: 45-50, 2000.

110. Ikonomou M, Blades A and Kebarle P, Electrospray-Ion Spray: A Comparison of Mechanisms and Performance. *Anal. Chem.* 63: 1989-1998, 1991.

111. Gallagher R, Balogh M, Davey P, Jackson M, Sinclair I and Southern L, Combined ESI/APCI Source for Use in High Throughput LC MS Applications. *Anal. Chem.* 75: 973-977, 2003.

112. Gallagher RT, Balogh MP, Davey P, Jackson MR, Sinclair I and Southern LJ, Combined ESI-APCI source for use in high-throughput LC-MS applications. *Anal Chem* 75: 973-7., 2003.

113. Andrews CL, Li F, Yang E, Yu C-P and Vouros P, Incorporation of a nanosplitter interface into an LC-MS-RD system to facilitate drug metabolism studies. *Journal of Mass Spectrometry* **41**(1): 43-49, 2006.

114. Mallis LM, Sarkahian AB, Kulishoff JM, Jr. and Watts WL, Jr., Open-access liquid chromatography/mass spectrometry in a drug discovery environment. *J Mass Spectrom* **37**(9): 889-96., 2002.

115. Chen G, Pramanik BN, Liu Y-H and Mirza UA, Applications of LC/MS in structure identifications of small molecules and proteins in drug discovery. *J Mass Spectrom* **42**: 279-287, 2007.

116. Kostiainen R, Kotiaho T, Kuuranne T and Auriola S, Liquid chromatography/atmospheric pressure ionization-mass spectrometry in drug metabolism studies. *J Mass Spectrom* **38**(4): 357-372, 2003.

117. Silberring J and Ekman R, *Mass spectrometry and hyphenated techniques in neuropeptide research.* John Wiley & Sons, Inc.,, New York, NY, 2003.

118. Roboz J, *Mass spectrometry in cancer research.* CRC Press, Boca Raton, FL, USA,, 2002.

119. Crommentuyn KM, Rosing H, Nan-Offeringa LG, Hillebrand MJ, Huitema AD and Beijnen JH, Rapid quantification of HIV protease inhibitors in human plasma by HPLC-ESI-MS/MS. *J Mass Spectrom* **38**: 157-66., 2003.

120. Compain S, Schlemmer D, Levi M, Pruvost A, Goujard C, Grassi J and Benech H, Validation of LC/MS/MS assay for quantitation of nucleoside HIV reverse transcriptase inhibitors in biological matrices. *J Mass Spectrom* **40**: 9-18, 2005.

121. Wabnitz PA and Loo JA, Drug screening of pharmaceutical discovery compounds by micro-size exclusion chromatography/mass spectrometry. *Rapid Commun Mass Spectrom* **16**(2): 85-91., 2002.

122. Masselon C, Pasa-Tolic L, Tolic N, Anderson GA, Bogdanov B, Vilkov AN, Shen Y, Zhao R, Qian W-J, Lipton MS, Camp DG, II and Smith RD, Targeted Comparative Proteomics by LC-FTICR MS. *Anal Chem* **77**: 400-406, 2005.

123. Hsieh Y, Brisson JM, Ng K and Korfmacher WA, Simultaneous determination of drug discovery compounds in monkey plasma by mixed-function column LC/MS/MS. *J Pharm Biomed Anal* **27**: 285-93., 2002.

124. Taylor EW, Jia W, Bush M and Dollinger GD, Accelerating the Drug Optimization Process: Identification, Structure Elucidation, and Quantification of in Vivo Metabolites Using Stable Isotopes with LC/MSn and the Chemiluminescent Nitrogen Detector. *Anal. Chem.* **74**: 3232 -3238, 2002.

125. Hsieh Y, Wang G, Wang Y, Chackalamannil S and Korfmacher WA, Direct plasma analysis of drug compounds by monolithic column LC/MS/MS. *Anal Chem* **75**: 1812-8., 2003.

126. Alsberg T, Minten J, Haglund J and Tornqvist M, Determination of hydroxyalkyl derivatives of cobalamin (vitamin B12) by reversed phase HPLC-ESI/MS/MS & UV diode array detection. *Rapid Commun Mass Spectrom* **15**: 2438-45, 2001.

127. Lee MH, Lee GH and Yoo JS, Analysis of ceramides in cosmetics by reversed-phase LC/ESI-CID. *Rapid Commun Mass Spectrom* **17**: 64-75., 2003.

128. Teo SK, Chandula MS, Harden JL, Stirling DI and Thomas SD, Determination of thalidomide in human plasma and semen by SPE & LC/MS/MS. *J Chromatogr B Biomed Sci Appl* **767**: 145-51., 2002.

129. Profita M, Sala A, Siena L, Henson PM, Murphy RC, Paterno A, Bonanno A, Riccobono L, Mirabella A, Bonsignore G and Vignola AM, Leukotriene B4 production in human mononuclear phagocytes is modulated by interleukin-4-induced 15-lipoxygenase. *J Pharmacol Exp Ther* **300**(3): 868-75., 2002.

130. Harrison KA, Davies SS, Marathe GK, McIntyre T, Prescott S, Reddy KM, Falck JR and Murphy RC, Analysis of oxidized glycerophosphocholine lipids by ESI/MS & microderivatization. *J. Mass Spectrom.* **35**: 224-236, 2000.

131. Oliveira CH, Barrientos-Astigarraga RE, Sucupira M, Graudenz GS, Muscara MN and De NG, Methyldopa in human plasma by HPLC-ESI/MS/MS: bioequivalence study. *J Chromatogr B Biomed Sci Appl* **768**: 341-8., 2002.

132. Nirogi RVS, Kandikere VN, Shrivastava W, Mudigonda K and Datla PV, LC/MS/MS with negative ion detection for the quantification of fluvastatin in human plasma: validation and its application to pharmacokinetic studies. *Rapid Communications in Mass Spectrometry* **20**: 1225-1230, 2006.

133. Kollroser M and Schober C, Determination of Coumarin-type Anticoagulants in Human Plasma by HPLC-Electrospray Ionization Tandem Mass Spectrometry with an Ion Trap Detector. *Clin Chem* **48**(1): 84-91., 2002.

134. Kerskes CH, Lusthof KJ, Zweipfenning PG and Franke JP, The detection and identification of quaternary nitrogen muscle relaxants in biological fluids and tissues by ion-trap LC-ESI-MS. *J Anal Toxicol* **26**(1): 29-34., 2002.

135. Bakhtiar R, Ramos L and Tse FL, Quantification of methylphenidate (Ritalin((R))) in rabbit fetal tissue using a chiral liquid chromatography/tandem mass spectrometry assay. *Rapid Commun Mass Spectrom* **16**(1): 81-3., 2002.

136. Keevil BG, Tierney DP, Cooper DP and Morris MR, Rapid Liquid Chromatography-Tandem Mass Spectrometry Method for Routine Analysis of Cyclosporin A Over an Extended Concentration Range. *Clin Chem* **48**(1): 69-76., 2002.

137. Han X, Characterization and direct quantitation of ceramide molecular species from lipid extracts of biological samples by electrospray ionization tandem mass spectrometry. *Anal Biochem* **302**(2): 199-212., 2002.

138. Guan F, Uboh C, Soma L, Hess A, Luo Y and Tsang DS, LC/MS/MS method for beclomethasone dipropionate and its metabolites in equine plasma and urine. *J Mass Spectrom* **38**: 823-838, 2003.

139. Olsen J, Bjornsdottir I, Tjornelund J and Honore Hansen S, ID of amino acids of human serum albumin in reaction with naproxen acyl coenzyme A thioester by LC with fluorescence & MS detection. *Anal Biochem* **312**: 148-156, 2003.

140. Stokvis E, Rosing H, Lopez-Lazaro L, Rodriguez I, Jimeno JM, Supko JG, Schellens JH and Beijnen JH, Quantitation of anticancer drug Kahalalide F in human plasma by HPLC-ESI/MS/MS. *J Mass Spectrom* **37**: 992-1000., 2002.

141. Thevis M, Krug O and Schaenzer W, Mass spectrometric characterization of efaproxiral (RSR13) and its implementation into doping controls using API LC/MS. *Journal of Mass Spectrometry* **41**(3): 332-338, 2006.

142. Miao XS and Metcalfe CD, Determination of pharmaceuticals in aqueous samples using positive and negative voltage switching microbore LC/ESI/MS/MS. *J Mass Spectrom* **38**: 27-34., 2003.

143. Hendrickson HP, Whaley EC and Owens SM, A validated LC/MS method for the determination of phencyclidine in microliter samples of rat serum. *Journal of Mass Spectrometry* **40**: 19-24, 2005.

144. Wang T, Barber M, Hardt I and Kassel DB, Mass-directed fractionation and isolation of pharmaceutical compounds by packed-column SFC/MS. *Rapid Commun Mass Spectrom* **15**: 2067-75., 2001.

145. Chen J, Hsieh Y, Cook J, Morrison R and Korfmacher WA, Supercrit. fluid chromatography-tandem mass spectrometry for the enantioselective detn. of propranolol and pindolol in mouse blood by serial sampling. *Analytical Chemistry* **78**(4): 1212-1217, 2006.

146. Pinkston JD, Wen D, Morand KL, Tirey DA and Stanton DT, Comparison of LC/MS and SFC/MS for Screening of a Large and Diverse Library of Pharmaceutically Relevant Compounds. *Analytical Chemistry* **78**(21): 7467-7472, 2006.

147. Tolstikov VV and Fiehn O, Analysis of highly polar compounds of plant origin: combination of hydrophilic interaction chromatography and electrospray ion trap mass spectrometry. *Anal Biochem* **301**(2): 298-307., 2002.

148. Garbis SD, Melse-Boonstra A, West CE and van Breemen RB, Determination of folates in human plasma using hydrophilic interaction chromatography-tandem mass spectrometry. *Anal Chem* **73**(22): 5358-64., 2001.

149. Minkler PE, Kerner J, Kasumov T, Parland W and Hoppel CL, Quantification of malonyl-coenzyme A in tissue specimens by LC/MS. *Analytical Biochemistry* **352**: 24-32, 2006.

150. Ruan Q, Kim H-YH, Jiang H, Penning TM, Harvey RG and Blair IA, Quantification of benzo[a]pyrene diol epoxide DNA-adducts by stable isotope dilution LC/MS/MS. *Rapid Commun Mass Spectrom* **20**: 1369-1380, 2006.

151. Barderas R, Villalba M, Lombardero M and Rodriguez R, Identification and Characterization of Che a 1 Allergen from Chenopodium album Pollen. *Int Arch Allergy Immunol* **127**(1): 47-54., 2002.

152. Shen ML, Benson LM, Johnson KL, Lipsky JJ and Naylor S, Effect of enzyme inhibitors on protein quaternary structure determined by on-line size exclusion LC/ESI/MS. *J Am Soc Mass Spectrom* **12**: 97-104., 2001.

153. Shen Y, Zhao R, Berger SJ, Anderson GA, Rodriguez N and Smith RD, High-efficiency nanoscale LC coupled on-line with ESI/MS for proteomics. *Anal Chem* **74**: 4235-49., 2002.

154. Pearcy JO and Lee TD, MoWeD, a computer program to rapidly deconvolute low resolution LC/ESI/MS runs to determine component molecular weights. *J Am Soc Mass Spectrom* **12**: 599-606., 2001.

155. Ivanov AR, Zang L and Karger BL, Low-attomole ESI/MS and MS/MS analysis of protein tryptic digests using 20-mm-i.d. Polystyrene-divinylbenzene monolithic capillary columns. *Anal Chem* **75**: 5306-5316, 2003.

156. Smith R and etal, Ultrasensitive Proteomics Using High-Efficiency On-Line Micro-SPE-NanoLC-NanoESI MS and MS/MS. *Anal. Chem.* **76**: 144 -154, 2004.

157. Stroh JG, Loulakis P, Lanzetti AJ and Xie J, LC-mass spectrometry analysis of N- and C-terminal boundary sequences of polypeptide fragments by limited proteolysis. *J Am Soc Mass Spectrom* **16**(1): 38-45, 2005.

158. Inoue K, Garner C, Ackermann BL, Oe T and Blair IA, LC/MS/MS characterization of oxidized amyloid beta peptides as potential biomarkers of Alzheimer's disease. *Rapid Commun Mass Spectrom* **20**: 911-918, 2005.

159. Sen JW, Bergen HR, 3rd and Heegaard NH, On-line immunoaffinity-liquid chromatography-mass spectrometry for identification of amyloid disease markers in biological fluids. *Anal Chem* **75**(5): 1196-202., 2003.

160. You J, Shan Y, Zhen L, Zhang L and Zhang Y, Determination of peptides and amino acids from wool and beer with sensitive fluorescent reagent 2-(9-carbazole)-ethyl chloroformate by LC/MS. *Anal Biochem* **313**: 17-27, 2003.

161. Schlichtherle-Cerny H, Affolter M and Cerny C, Hydrophilic interaction liquid chromatography coupled to electrospray mass spectrometry of small polar compounds in food analysis. *Anal Chem* **75**(10): 2349-2354, 2003.

162. Liu S, Griffiths WJ and Sjovall J, Capillary liquid chromatography/electrospray mass spectrometry for analysis of steroid sulfates in biological samples. *Anal Chem* **75**(4): 791-7., 2003.

163. Yu K, Krol J, Balogh M and Monks I, A fully automated LC/MS method development and quantification protocol targeting 52 carbamates, thiocarbamates, and phenylureas. *Anal Chem* **75**(16): 4103-12., 2003.

164. Ross MK and Filipov NM, Determination of atrazine and its metabolites in mouse urine and plasma by LC-MS analysis. *Analytical Biochemistry* **351**(2): 161-173, 2006.

165. Pinkston JD, Supercritical fluid chromatography/MS. *Anal. Supercrit. Fluids: Extr. Chromatogr.*: 151-77, 1992.

166. Pinkston JD and Baker TR, Modified ionspray interface for supercritical fluid chromatography/mass spectrometry: interface design and initial results. *Rapid Communications in Mass Spectrometry* **9**(12): 1087-94, 1995.

167. Hoke SH, II, Pinkston JD, Bailey RE, Tanguay SL and Eichhold TH, Packed-column SFC/MS/MS vs LC/MS/MS for determination of (R)- and (S)-Ketoprofen in human plasma by automated 96-well SPE. *Anal Chem* **72**: 4235-4241, 2000.

168. Hoffman BJ, Taylor LT, Rumbelow S, Goff L and Pinkston JD, Determination of alcohol polyether average molar oligomer value/distribution via SFC coupled with UV and MS detection. *Journal of Chromatography, A* **1043**: 285-290, 2004.

169. Pinkston JD, Marapane SB, Jordan GT and Clair BD, Characterization of low molecular weight alkoxylated polymers by SFC/MS and an image analysis based quantitation approach. *J Amer Soc Mass Spectrom* **13**: 1195-1208, 2002.

170. Byrd GD and Ogden MW, Liquid chromatographic/tandem mass spectrometric method for the determination of the tobacco-specific nitrosamine metabolite NNAL in smokers' urine. *J Mass Spectrom* **38**(1): 98-107., 2003.

171. Murphy CM and Huestis MA, LC/ESI/MS/MS for quantification of buprenorphine, norbuprenorphine, buprenorphine-3-b-D-glucuronide and norbuprenorphine-3-b-D-glucuronide in human plasma. *J Mass Spectrom* **40**: 70-74, 2005.

172. Lin SN, Walsh SL, Moody DE and Foltz RL, Detection and time course of cocaine N-oxide and other cocaine metabolites in human plasma by liquid chromatography/tandem mass spectrometry. *Anal Chem* **75**(16): 4335-40., 2003.

173. Morini L, Politi L, Groppi A, Stramesi C and Polettini A, Determination of ethyl glucuronide in hair samples by liquid chromatography/electrospray tandem mass spectrometry. *Journal of Mass Spectrometry* **41**(1): 34-42, 2006.

174. Marvin LF, Delatour T, Tavazzi I, Fay LB, Cupp C and Guy PA, Quantification of o,o'-dityrosine, o-nitrotyrosine, and o-tyrosine in cat urine by LC/ ESI/MS/MS & isotope dilution. *Anal Chem* **75**: 261-7., 2003.

175. Wang G, Hsieh Y, Cui X, Cheng K-C and Korfmacher WA, Ultra-performance LC-MS/MS determination of testosterone and metabolites in in vitro samples. *Rapid Commun Mass Spectrom* **20**: 2215-2221, 2006.

176. Hooijschuur EW, Kientz CE, Hulst AG and Brinkman UA, Hydrolysis products of sulfur mustard by rp-HPLC on-line with sulfur flame detection & ESI/MS of large-volume injections & peak compression. *Anal Chem* **72**: 1199-206., 2000.

177. Xu X, Veenstra TD, Fox SD, Roman JM, Issaq HJ, Falk R, Saavedra JE, Keefer LK and Ziegler RG, Measuring 15 Endogenous Estrogens Simultaneously in Human Urine by HPLC/MS. *Analytical Chemistry* **77**: 6646-6654, 2005.

178. Kaga N, Kazuno S, Taka H, Iwabuchi K and Murayama K, Isolation and characterization of lactosylceramides by LC/ESI/MS. *Anal Biochem* **337**: 316-324, 2005.

179. Egnash LA and Ramanathan R, Heterogeneous and homogeneous radioactivity flow detectors for simultaneous profiling and LC-MS/MS characterization of metabolites. *J Pharm Biomed Anal* **27**: 271-84., 2002.

180. Nassar AE, Bjorge SM and Lee DY, On-line LC-accurate radioisotope counting coupled with a radioactivity detector and MS for metabolite identification in drug discovery and development. *Anal Chem* **75**: 785-90., 2003.

181. Dickinson JS and Murphy RC, Mass spectrometric analysis of leukotriene A4 and other chemically reactive metabolites of arachidonic acid. *J Am Soc Mass Spectrom* **13**(10): 1227-34., 2002.

182. Mallat Z, Nakamura T, Ohan J, Leseche G, Tedgui A, Maclouf J and Murphy RC, Relationship of F2-isoprostanes & hydroxyeicosatetraenoic acids to plaque instability in human carotid atherosclerosis. *J. Clin. Invest.* **103**: 421-427, 1999.

183. Hall LM and Murphy RC, ESI/MS of 5-hydroperoxy and 5-hydroxyeicosatetraenoic acids generated by lipid peroxidation of red blood cell ghost phospholipids. *J. Am. Soc. Mass Spectrom.* **9**: 527-532, 1998.

184. Nakamura T, Bratton DL and Murphy RC, Analysis of epoxyeicosatrienoic and monohydroxyeicosatetraenoic acids esterified to phospholipids in human red blood cells by ESI/MS/MS. *J. Mass Spectrom.* **32**: 888-896, 1997.

185. Gangl ET, Turesky RJ and Vouros P, Detection of in Vivo Formed DNA Adducts at the Part-per-Billion Level by Capillary Liquid Chromatography/Microelectrospray Mass Spectrometry. *Anal Chem* **73**(11): 2397-2404, 2001.

186. Headley JV, Peru KM, McMartin DW and Winkler M, Determination of dissolved naphthenic acids in natural waters by using negative-ion electrospray mass spectrometry. *J AOAC Int* **85**(1): 182-7., 2002.

187. Phillips SL and Olesik SV, Initial Characterization of Humic Acids by LC at the Critical Condition Followed by Size-Exclusion Chromatography and ESI/MS. *Anal Chem* **75**: 5544-5553, 2003.

188. Loyo-Rosales JE, Schmitz-Afonso I, Rice CP and Torrents A, Analysis of Octyl- and Nonylphenol and Their Ethoxylates in Water and Sediments by LC/MS/MS. *Anal Chem* **75**: 4811-4817, 2003.

189. Eichhorn P, Rodrigues SV, Baumann W and Knepper TP, Incomplete degradation of linear alkylbenzene sulfonate surfactants in Brazilian surface waters and pursuit of their polar metabolites in drinking waters. *Sci Total Environ* **284**(1-3): 123-34., 2002.

190. Petrovic M and Barcelo D, Application of liquid chromatography/quadrupole time-of-flight mass spectrometry (LC-QqTOF-MS) in the environmental analysis. *Journal of Mass Spectrometry* **41**(10): 1259-1267, 2006.

191. Marathe GK, Harrison KA, Roberts LJ, II, Morrow JD, Murphy RC, Tjoelker LW, Prescott SM, Zimmerman GA and McIntyre TM, ID of PAF as inflammatory lipid mediator in CCl4-metabolizing rat liver. *J. Lipid Res.* **42**: 587-596, 2001.

192. Barber LA, Spandau DF, Rathman SC, Murphy RC, Johnson CA, Kelley SW, Hurwitz SA and Travers JB, Expression of the platelet-activating factor receptor results in enhanced ultraviolet B radiation-induced apoptosis in a human epidermal cell line. *J. Biol. Chem.* **273**(30): 18891-18897, 1998.

193. Bowers RC, Hevko J, Henson PM and Murphy RC, A novel glutathione containing eicosanoid (FOG7) chemotactic for human granulocytes. *J. Biol. Chem.* **275**(39): 29931-29934, 2000.

194. Wheelan P, Hankin JA, Bilir B, Guenette D and Murphy RC, Metabolic transformations of leukotriene B4 in primary cultures of human hepatocytes. *J. Pharmacol. Exp. Ther.* **288**(1): 326-334, 1999.

195. Hankin JA, Wheelan P and Murphy RC, Identification of novel metabolites of prostaglandin E2 formed by isolated rat hepatocytes. *Arch. Biochem. Biophys.* **340**(2): 317-330, 1997.

196. Murphy RC and Hankin JA, Metabolism of leukotrienes and formation of new leukotriene structures. *Novel Inhib. Leukotrienes*: 63-82, 1999.

197. Kingsley PJ and Marnett LJ, Analysis of endocannabinoids by Ag+ coordination tandem mass spectrometry. *Anal Biochem* **314**(1): 8-15, 2003.

198. Wu Q, Wang M and Simon JE, Determination of Proanthocyanidins in Grape Products by Liquid Chromatography/Mass Spectrometric Detection under Low Collision Energy. *Analytical Chemistry* **75**(10): 2440-2444, 2003.

199. Fabre N, Rustan I, de Hoffmann E and Quetin-Leclercq J, Determination of flavone, flavonol, and flavanone aglycones by negative ion LC/ESI/MS. *J Am Soc Mass Spectrom* **12**: 707-15., 2001.

200. Mathis JA and McCord BR, The analysis of high explosives by LC/ESI/MS: multiplexed detection of negative ion adducts. *Rapid Commun Mass Spectrom* **19**: 99-104, 2005.

201. Rozenski J and McCloskey JA, Characterization of oligonucleotide sequence isomers in mixtures using HPLC/MS. *Nucleosides Nucleotides* **18**(6 & 7): 1539-1540, 1999.

202. Deng H, Van Berkel GJ, Takano H, Gazda D and Porter MD, Electrochemically modulated liquid chromatography coupled on-line with electrospray mass spectrometry. *Anal Chem* **72**(11): 2641-7., 2000.

203. Pigini D, Cialdella AM, Faranda P and Tranfo G, Comparison between external and internal standard calibration in the validation of an analytical method for 1-hydroxypyrene in human urine by high-performance liquid chromatography/tandem mass spectrometry. *Rapid Communications in Mass Spectrometry* **20**(6): 1013-1018, 2006.

204. Soerensen LK, Determination of phthalates in milk and milk products by LC/MS/MS. *Rapid Communications in Mass Spectrometry* **20**(7): 1135-1143, 2006.

205. Xu X, Lan J and Korfmacher WA, Rapid LC/MS/MS method development for drug discovery. *Analytical Chemistry* **77**(19): 389A-394A, 2005.

206. Hsieh Y, Fukuda E, Wingate J and Korfmacher WA, Fast mass spectrometry-based methodologies for pharmaceutical analyses. *Combinatorial Chemistry & High Throughput Screening* **9**(1): 3-8, 2006.

207. Kratzsch C, Peters FT, Kraemer T, Weber AA and Maurer HH, Screening, library-assisted identification and quantification of 15 neuroleptics and 3 metabolites in plasma by LC/APCI/MS. *J Mass Spectrom* **38**: 283-295, 2003.

208. Dams R, Murphy CM, Choo RE, Lambert WE, De Leenheer AP and Huestis MA, LC-APCI-MS/ MS of multiple illicit drugs, methadone, and their metabolites in oral fluid following protein precipitation. *Anal Chem* **75**: 798-804., 2003.

209. Smith WD, LC-ESI-MS/MS draws the line on doping. *Anal Chem* **75**(15): 337A., 2003.

210. Suhara Y, Kamao M, Tsugawa N and Okano T, Method for the Determination of Vitamin K Homologues in Human Plasma by LC/MS/MS. *Anal Chem* **77**: 757-763, 2005.

211. Gaboriau F, Havouis R, Moulinoux J-P and Delcros J-G, Atmospheric pressure chemical ionization-mass spectrometry method to improve the determination of dansylated polyamines. *Anal Biochem* **318**(2): 212-220, 2003.

212. Derks RJE, Hogenboom AC, van der Zwan G and Irth H, On-Line Continuous-Flow, Multi-Protein Biochemical Assays for the Characterization of Bioaffinity Compounds by ESI/QTOFMS. *Anal Chem* **75**: 3376-3384, 2003.

213. Becher F, Pruvost A, Gale J, Couerbe P, Goujard C, Boutet V, Ezan E, Grassi J and Benech H, A strategy for LC/MS/MS assays of intracellular drugs: Application to the validation of the triphosphorylated anabolite of antiretrovirals in peripheral blood mononuclear cells. *J Mass Spectrom* **38**: 879-890, 2003.

214. White WL, Wagner CD, Hall JT, Chaney EE, George B, Hofmann K, Miller LAD and Williams JD, Protein open-access liquid chromatography/mass spectrometry. *Rapid Commun Mass Spectrom* **19**(2): 241-249, 2005.

215. Janiszewski JS, Rogers KJ, Whalen KM, Cole MJ, Liston TE, Duchoslav E and Fouda HG, High-capacity LC/MS system for biosamples generated from plate-based metabolic screening. *Anal Chem* **73**: 1495-501., 2001.

216. Dunn-Meynell KW, Wainhaus S and Korfmacher WA, Optimizing an ultrafast generic high-performance liquid chromatography/tandem mass spectrometry method for faster discovery pharmacokinetic sample throughput. *Rapid Communications in Mass Spectrometry* **19**(20): 2905-2910, 2005.

217. Van Pelt CK, Corso TN, Schultz GA, Lowes S and Henion J, A four-column parallel chromatography system for isocratic or gradient LC/MS analyses. *Anal Chem* **73**(3): 582-8., 2001.

218. Xu R, Nemes C, Jenkins KM, Rourick RA, Kassel DB and Liu CZ, Application of parallel LC/MS for high throughput microsomal stability screening of compound libraries. *J Am Soc Mass Spectrom* **13**: 155-65., 2002.

219. Yang L, Mann TD, Little D, Wu N, Clement RP and Rudewicz PJ, Evaluation of a 4-channel multiplexed ESi/TQMS for the simultaneous validation of LC/MS/MS methods in 4 different preclinical matrixes. *Anal Chem* **73**: 1740-7., 2001.

220. Wang T, Cohen J, Kassel DB and Zeng L, A multiple electrospray interface for parallel mass spectrometric analyses of compound libraries. *Comb Chem High Throughput Screen* **2**(6): 327-34., 1999.

221. Xu R, Wang T, Isbell J, Cai Z, Sykes C, Brailsford A and Kassel DB, High-Throughput Mass-Directed Parallel Purification Incorporating a Multiplexed Single Quadrupole Mass Spectrometer. *Anal. Chem.* **74**: 3055 -3062, 2002.

222. Kameoka J, Orth R, Ilic B, Czaplewski D, Wachs T and Craighead HG, An electrospray ionization source for integration with microfluidics. *Anal Chem* **74**(22): 5897-901., 2002.

223. Corkery LJ, Pang H, Schneider BB, Covey TR and Siu KWM, Automated nanospray using chip-based emitters for the quantitative analysis of pharmaceutical compounds. *J Amer Soc Mass Spectrom* **16**(3): 363-369, 2005.

224. Liu J, Ro K-W, Nayak R and Knapp DR, Monolithic column plastic microfluidic device for peptide analysis using electrospray from a channel opening on the edge of the device. *International Journal of Mass Spectrometry* **259**(1-3): 65-72, 2007.

225. Schneider BB, Douglas DJ and Chen DDY, Multiple sprayer system for high-throughput electrospray ionization mass spectrometry. *Rapid Commun Mass Spectrom* **16**(20): 1982-1990, 2002.

226. King RC, Miller-Stein C, Magiera DJ and Brann J, Description and validation of a staggered parallel HPLC system for good laboratory practice level quantitation by LC/MS. *Rapid Commun Mass Spectrom* **16**: 43-52., 2002.

227. Rossi DT and Zhang N, Automating solid-phase extraction: current aspects and future prospects. *J Chromatogr A* **885**(1-2): 97-113., 2000.

228. Penn LD, Cohen LH, Olson SC and Rossi DT, LC/MS/MS quantitation of an anti-cancer drug in human plasma using a SPE workstation: application to population pharmacokinetics. *J Pharm Biomed Anal* **25**(3-4): 569-76., 2001.

229. Shou WZ, Pelzer M, Addison T, Jiang X and Naidong W, An automatic 96-well solid phase extraction and liquid chromatography-tandem mass spectrometry method for the analysis of morphine, morphine-3-glucuronide and morphine-6-glucuronide in human plasma. *J Pharm Biomed Anal* **27**(1-2): 143-52., 2002.

230. O'Connor D, Mortishire-Smith R, Morrison D, Davies A and Dominguez M, Ultra-performance LC/TOFMS for robust, high-throughput quantitative analysis of an automated metabolic stability assay, with simultaneous determination of metabolic data. *Rapid Communications in Mass Spectrometry* **20**: 851-857, 2005.

231. Wachs T and Henion J, A device for automated direct sampling and quantitation from solid-phase sorbent extraction cards by electrospray tandem mass spectrometry. *Anal Chem* **75**(7): 1769-75., 2003.

232. Kaerrman A, Van Bavel B, Jaernberg U, Hardell L and Lindstroem G, Development of an SPE-HPLC/Single Quadrupole MS Method for Quantification of Perfluorochemicals in Whole Blood. *Anal Chem* **77**: 864-870, 2005.

233. Kuuranne T, Kotiaho T, Pedersen-Bjergaard S, Einar Rasmussen K, Leinonen A, Westwood S and Kostiainen R, Feasibility of a liquid-phase microextraction sample clean-up and liquid chromatographic/mass spectrometric screening method for selected anabolic steroid glucuronides in biological samples. *J Mass Spectrom* **38**(1): 16-26., 2003.

234. Bogialli S, Curini R, Di Corcia A, Nazzari M and Samperi R, A liquid chromatography-mass spectrometry assay for analyzing sulfonamide antibacterials in cattle and fish muscle tissues. *Anal Chem* **75**(8): 1798-804., 2003.

235. Yokley RA, Mayer LC, Huang SB and Vargo JD, Analytical method for the determination of metolachlor, acetochlor, alachlor, dimethenamid, and their corresponding ethanesulfonic and oxanillic acid degradates in water using SPE and LC/ESI-MS/MS. *Anal Chem* **74**(15): 3754-9., 2002.

236. Saint-Marcoux F, Lachatre G and Marquet P, Evaluation of screening procedure by LC/ESI/MS vs GC & HPLC-diode array detection. *J Am Soc Mass Spectrom* **14**: 14-22., 2003.

237. Kassel DB, Combinatonial chemistry & MS in 21st century drug discovery. *Chem Rev* **101**: 255-67., 2001.

238. Matz LM, Dion HM and Hill HH, Jr., Evaluation of capillary liquid chromatography-electrospray ionization ion mobility spectrometry with mass spectrometry detection. *J Chromatogr A* **946**(1-2): 59-68., 2002.

239. Srebalus B, Hilderbrand AE, Valentine SJ and Clemmer DE, Resolving isomeric peptide mixtures: a combined HPLC/ion mobility-TOFMS analysis of a 4000-component combinatorial library. *Anal Chem* **74**(1): 26-36., 2002.

240. Stevens SM, Jr., Prokai-Tatrai K and Prokai L, Screening of Combinatorial Libraries for Substrate Preference by Mass Spectrometry. *Analytical Chemistry* **77**(2): 698-701, 2005.

241. Misharin AS, Laughlin BC, Vilkov A, Takats Z, Ouyang Z and Cooks RG, High-Throughput Mass Spectrometer Using Atmospheric Pressure Ionization and a Cylindrical Ion Trap Array. *Anal Chem* **77**(2): 459-470, 2005.

"Nature uses as little as possible of anything."

~Johannes Kepler

Chapter 12   Analysis of Proteins and Other Biopolymers

Introduction to Mass Spectrometry, 4th Edition: Instrumentation, Applications, and Strategies for Data Interpretation; J.T. Watson and O.D. Sparkman, © 2007, John Wiley & Sons, Ltd

I. Introduction

Biopolymer analysis has been a goal of mass spectrometrists from the early days of organic mass spectrometry. Klaus Biemann, in the early 1970s [1], initiated efforts to characterize proteins after derivatization of proteolytic fragments; his later work with CAD MS/MS of intact peptides led to specific nomenclature for the fragmentation of polypeptides during analysis by mass spectrometry as described in this chapter. The requirement for sample volatility in electron and chemical ionization mass spectrometry precluded direct applications to biopolymers (proteins, oligonucleotides, and carbohydrates), which can now be analyzed readily by MALDI (matrix-assisted laser desorption/ionization) as developed by Michael Karas, Franz Hillenkamp, and coworkers [2] building on laser desorption experiments by Tanaka and coworkers [3, 4], and electrospray ionization (ESI) as developed by John Fenn and coworkers [5]. Developments in field desorption (FD) ionization in the 1960s [6], [252]Californium desorption/ionization in the 1970s [7], and fast atom bombardment (FAB) in the 1980s [8] ushered in ESI and MALDI (see Chapters 9 and 10, respectively) during the 1980/1990s to manifest today's era of mass spectrometry. These latter developments and more specifically protein analyses earned John Fenn and Kochi Tanaka a share of the 2002 Noble Prize in chemistry. MALDI and ESI have played a pivotal role in the development of new types of *m/z* analyzers (mass spectrometers); most of the current literature on the analysis and characterization of biopolymers in mass spectrometry involves these two ionization techniques that only became commercial realities during the past decade. MALDI and ESI have become such a dominant factor in mass spectrometry that a specific chapter is devoted to each of these ionization techniques in this book.

The initially recognized great virtue of electrospray ionization and MALDI was in the capacity to form protonated molecules of large, involatile compounds and to measure the mass to a relatively high accuracy ($\pm 0.01\%$). Because there is very little or no fragmentation of the protonated analytes formed during electrospray and MALDI, these modern instruments are sometimes referred to as "molecular weight machines". Structural information on biopolymers must be obtained by using specialized instrumental techniques (such as CAD, PSD, ECD, etc., as described elsewhere in the book, although representative examples with data interpretation appear in this chapter) or from controlled degradation or **n** of the analyte and subsequent reanalysis for molecular weight determination of the modified analyte. Descriptions of several general procedures for modifying the analyte to obtain structural information are described in the following subsections.

II. Proteins

Proteomics is the identification, study, and assessment of the protein complement of the genome, following the central dogma of molecular biology: DNA $\rightarrow$ RNA $\rightarrow$ proteins. Because proteins are the molecules that carry out operational functions in a physiological cell, study and interpretation of the genome necessarily involves qualitative and quantitative assessment of encoded proteins that are expressed by that cell [9–31]. Results from proteomic studies are useful in drug design and provide a means of recognizing biomarkers for some diseases [32–41]; to this end, efforts have been proposed to define a urinary proteome [42, 43]. Even the mosquito's proteome is becoming defined [44], as well as that of *T. Rex* [45]. Considerable interest is also developing in the area of plant proteomics [46, 47]. In an early effort to establish a phosphoproteome, over 200 phosphopeptides were isolated and identified in an analysis of a whole-cell lysate [48]; a broader net for capturing a phosphoproteome was cast by a method identifying over 2000 phosphopeptides corresponding to nearly 1000 phosphoproteins in a lysate of HeLa cells [49]. Proteomes based on the disulfide-bond structure [50] and on other modifications are being contemplated [51], including ubiquination [52].

In common practice, protein analyses involve separation of complex mixtures by 1D or 2D gel electrophoresis [53–58], physical removal of stained or visualized spots (representing individual proteins or simplified mixtures thereof), often by robotic operation, for proteolytic and/or chemical digestion prior to further analysis [59]. Alternate methodology, sometimes called "shotgun" proteomics [60], employs multidimensional chromatography on the "front end" in methodology called multidimensional protein identification technology (MudPIT) [61–63], in which whole proteomes or fractions are digested and separated using a dual-phase liquid chromatography separation process followed by in-line analysis by ESI MS [64]. MudPIT differs from other approaches in that it relies on high-resolution multidimensional chromatography to resolve all the peptide components of a complex mixture. Therefore, MudPIT has the advantage of complete representation of proteins in the original mixture, but at the cost of high redundancy, which results in high sample complexity and limits throughput. The MudPIT methodology has also been assessed for its capacity to provide quantitative information [65, 66]. Other applications use combinations of SPE and chromatography [67] or conventional LC/MS/MS after simplification of the digest complexity by biotinylation of cysteines [68]. On-line preconcentration of peptides from in-gel digestion using strong cation exchange led to better sequence coverage prior to analysis by ESI LC/MS [69, 70]. In some cases, a chip-based approach using a specially modified sample surface may be appropriate [71]. Novel approaches to sample processing such as "stop and go extraction" will continue to be important in proteomics [72]. Another novel approach takes advantage of the subcellular organization of eukaryotic cells to simplify the mixture of proteins analyzed by LC-ESI-CAD MS/MS [73]. A comparison of two popular mass spectrometry platforms for proteomic analysis was made of the yeast proteome using LC-ESI-CAD MS/MS with linear ion trap (LTQ) and hybrid quadrupole time-of-flight (QqTOF; QSTAR) mass spectrometers [74]. Fundamental aspects of many types of mass spectrometric analyses relating to proteomics are described in the following sections.

Detection of traces of secreted peptides in complex physiological solutions has been demonstrated, based on targeted multistage MS, using a novel MALDI ion trap mass spectrometer [75]; to overcome limitations of current MS methods (limited dynamic range, signal suppression effects, and chemical noise) that impair observation trace constituents, a specific fragmentation signature for the putative secreted peptides is programmed into the multistage MS system. In another application of a MALDI ion trap mass spectrometer, results of a preliminary study indicate that *de novo* sequencing is possible with fragmentation by CAD MS/MS [76]. Selected applications of mass spectrometry to characterize proteins include elucidation of the subunits of human gamma-tubulin complex [77], identification of a 193-kDa SV40 large T antigen-binding protein [78], and characterization of neuropeptides [79]; mass spectrometry is also useful in expediting confirmatory analyses at various stages of protein crystallography [80]. In quite a different application, a simple and rapid means of enzyme kinetic analysis for glutathione *S*-transferase from porcine liver was demonstrated using ESI [81]. A recollection of four decades on bioorganic mass spectrometry by Klaus Biemann presents an interesting historical perspective on its use to characterize proteins [82].

1. Sequencing

A. Nomenclature and Fragmentation in Sequencing of Peptides

The seminal works by Biemann and coworkers [83–85] and others [86, 87] as summarized below provide the basis for elucidating the structure of peptides from mass spectra. This section provides a tutorial on interpretation procedures and strategies that can be used to deduce the sequence [88] of amino acids in a protonated peptide by examining its CAD MS/MS spectrum after ionization by electrospray, for example, or by its PSD spectrum in MALDI TOFMS. Peptides consist of amino acid residues connected through an amide bond. Most proteins are constructed from peptides consisting of 20 common amino acids. These amino acids are shown in the Amino Acid table in the back matter at the end of the book according to their name, abbreviation, symbol, and structure.

1) Nomenclature

As shown in Figure 12-1, the amino acids are connected via the peptide bond (an amide bond) from the carboxyl group of one amino acid to the amino group of the second amino acid through a condensation reaction. The molecular mass of a peptide consists of the sum of the residue masses of all the amino acids in the peptide plus water. Figure 12-1 is a segment of a peptide chain in the region showing a glutamic acid residue of 129 Da connected at its C-terminus to an isoleucine residue of 113 Da to which a serine residue of 87 Da is connected at its C-terminus. Considering the mass of the complete peptide molecule, the second –H at the amino terminus and the –OH attached at the C-terminus constitute the "molecule" of water added to the sum of the residue masses of the amino acids comprising the peptide.

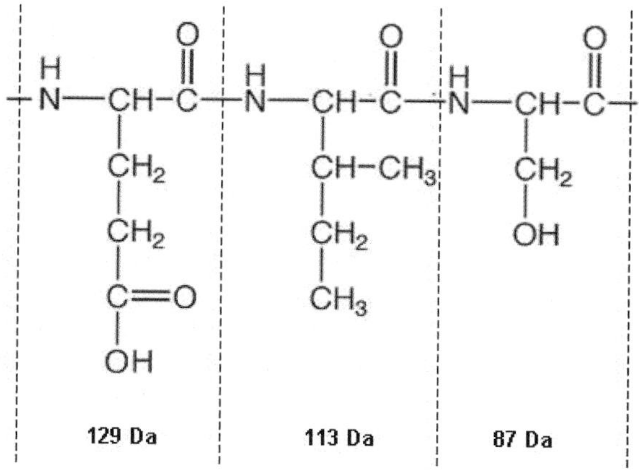

Figure 12-1. *Segment of a peptide showing the residues of glutamic acid, isoleucine, and serine linked by amide bonds, the peptide bonds.*

The residue masses of all the amino acids are also listed in the back matter of the book, which shows the symbol for the amino acid together with its residue mass (ResMass), the masses of the immonium ion (ImmMass), and the side-chain mass (SCMass). Leucine and isoleucine have the same mass, but can usually be distinguished by **w** or **d** ions (ion types are explained below) produced by high-energy CAD [86]; these isomeric residues also can be distinguished by an N-terminal labeling

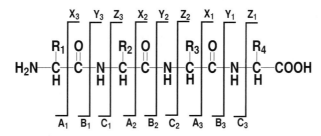

Figure 12-2.   *Roepstorff nomenclature for classifying N-terminal and C-terminal fragment ions of peptides.* *From Roepstorff P; Fohlmann J "Nomenclature for Mass Spectrometry of Peptides" Biomed. Mass Spectrom. 1984, 11, 601, with permission of Wiley Interscience.*

method using a ruthenium(II) complex by the **a** and **d** ions produced during by CAD MS/MS [89]. Lysine and glutamine have the same nominal mass, but their exact masses can be distinguished by the good resolving power and mass accuracy of FTMS or *oa*Q-TOF instrumentation; alternatively, these two residues can be differentiated after acetylation of the lysine residue and a second mass spectral measurement.

The generic structure for the immonium ion is $RCH=NH_2^+$ where R is the side chain on the amino acid residue. The side-chain mass (SCMass) in the back matter of the book is the mass of the R group in this general formula. Imminium ions are formed by fragmentation of $\mathbf{a_2}$ ions [90].

Some of the general fragmentations of a generic protonated peptide are shown in Figure 12-2, which categorizes ions according to the Roepstorff nomenclature [91]. One source of diagnostically useful fragmentation under CAD conditions involves cleavage of the peptide bond; in Figure 12-2, there are three peptide bonds involved in the fragmentation of a tetrapeptide. Fragments retaining the positive charge on the C-terminal component of the peptide are represented by the symbols X, Y, or Z, whereas fragments containing the positive charge on the N-terminal part of the original peptide are indicated with the lettering A, B, and C. Fragmentation at the peptide bond may seem most intuitive, and it is represented by fragments retaining the charge as either a B ion or as a Y ion. Thus, in the tetrapeptide represented in Figure 12-2, there are three possible Y ions resulting from fragmentation of a peptide or amide bond. Charge retention on the C-terminal amino acid residue is required to produce Y_1. Cleavage between the middle two amino acid residues with retention of the charge toward the C-terminus of the peptide would give rise to the Y_2 ion, etc. Similarly, if the charge were retained toward the N-terminus of the peptide upon cleavage of the peptide bond between the third and fourth amino acid residues, the observed ion would be represented by the symbol B_3 and consist of the first three amino acid residues from the N-terminus. Equivalent information is carried by the C and Z ions, which are prominent in mass spectra obtained after ECD or ETD as described in Section 5 later in this chapter.

A more explicit nomenclature was subsequently proposed by Biemann [85], which was acceptable to Roepstorff. The more explicit Biemann nomenclature uses lower case letters to represent structures that more closely relate to the ions being detected; the Biemann nomenclature will be used through the remainder of this section.

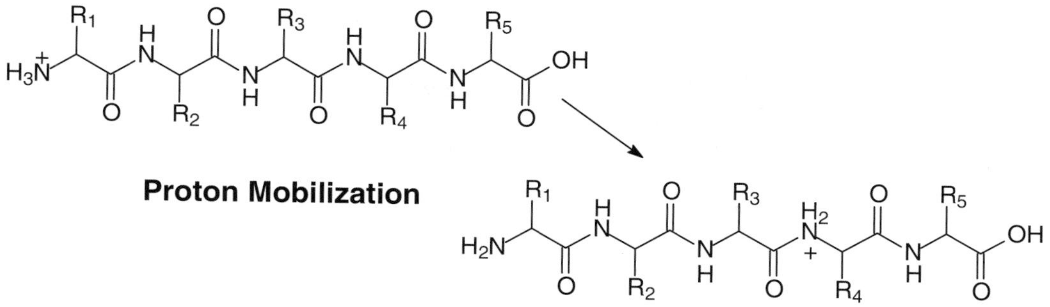

Proton Mobilization

Scheme 12-1

2) Fragmentation

Upon excitation by CAD, the ionizing proton (thought to be located initially at the most basic site within the peptide) becomes "mobile" and samples various amide oxygens and nitrogens as alternative protonation sites [92–96]. In the didactic example in Scheme 12-1, the proton is shown to move from the N-terminus of the protonated peptide to the third amide nitrogen of a generic protonated peptide. Nitrogen-protonated amide bonds are usually cleaved in reactions involving nucleophilic attack by the adjacent N-terminal amide oxygen [94, 97]. (Modeling studies indicate that the other prominent backbone nucleophile, the amino-terminal nitrogen, is usually not close to the sissle amide bond, whereas the attacking adjacent N-terminal amide oxygen is always proximal [93].) An intermediate ion–molecule complex of the protonated peptide, in the form of a protonated heterodimer as represented in the middle of Scheme 12-2, could be the precursor of either the **b** ion or the **y** ion on the b_x–y_z pathway [94, 97]. During dissociation of the protonated heterodimer as illustrated at the bottom of Scheme 12-2, if the monomer in the oxazolone form retains the proton, the ion formed is the **b** ion; if the monomer in the form of a truncated normal protonated peptide retains the proton, a **y** ion is formed [97]. This theoretical prediction is consistent with experimental evidence for the **b** ion being in the form of an oxazolone [98–100] and suggests that the thermochemistry (proton affinities) of peptide fragments plays an important role in determining the abundance of product ions in CAD tandem mass spectra of peptides [101, 102]. Results of other work suggest a diketopiperazine counterpart to the **y** ion [93, 103, 104]. In interpreting experimental data, the primary fragmentation reactions of oligoalanines have been rationalized in terms of the b_x–y_z pathway of amide bond cleavage that results in formation of a proton-bound complex of an oxazolone and a peptide/amino acid; upon decomposition of this complex, the species of higher proton affinity preferentially retains the proton [105]. Fragmentation might then continue via "charge-directed" neighboring-group participation mechanisms involving nucleophilic attack from an adjacent functional group [106]. Ideally, an array of isomeric protonated molecules, with each isomer containing the proton on a different amide, would be created, the decomposition of which would produce an array of **b** or **y** ions as represented by a series of peaks in the mass spectrum, allowing the analyst to "read off" the amino acid sequence.

The simple model represented in Scheme 12-1 probably is not generally applicable because when the protonated peptide consists of more than three residues, the proton is probably "solvated" by or nested among the carbonyl oxygens of the polypeptide backbone. Furthermore, protonation of the carbonyl oxygen of the amide bond is thermodynamically preferred over protonation of the amide nitrogen in peptide ions [107–109]. Alternatively, the proton may be located on the side chain of a basic amino acid such as arginine, histidine, or lysine. Reid and coworkers have characterized "sequence" ions formed by nucleophilic attack at the protonated amide bond by reactive functional groups in side chains [110], as well as "nonsequence" ions produced by the loss of small molecules such as H_2O or NH_3 from the peptide backbone [109, 111].

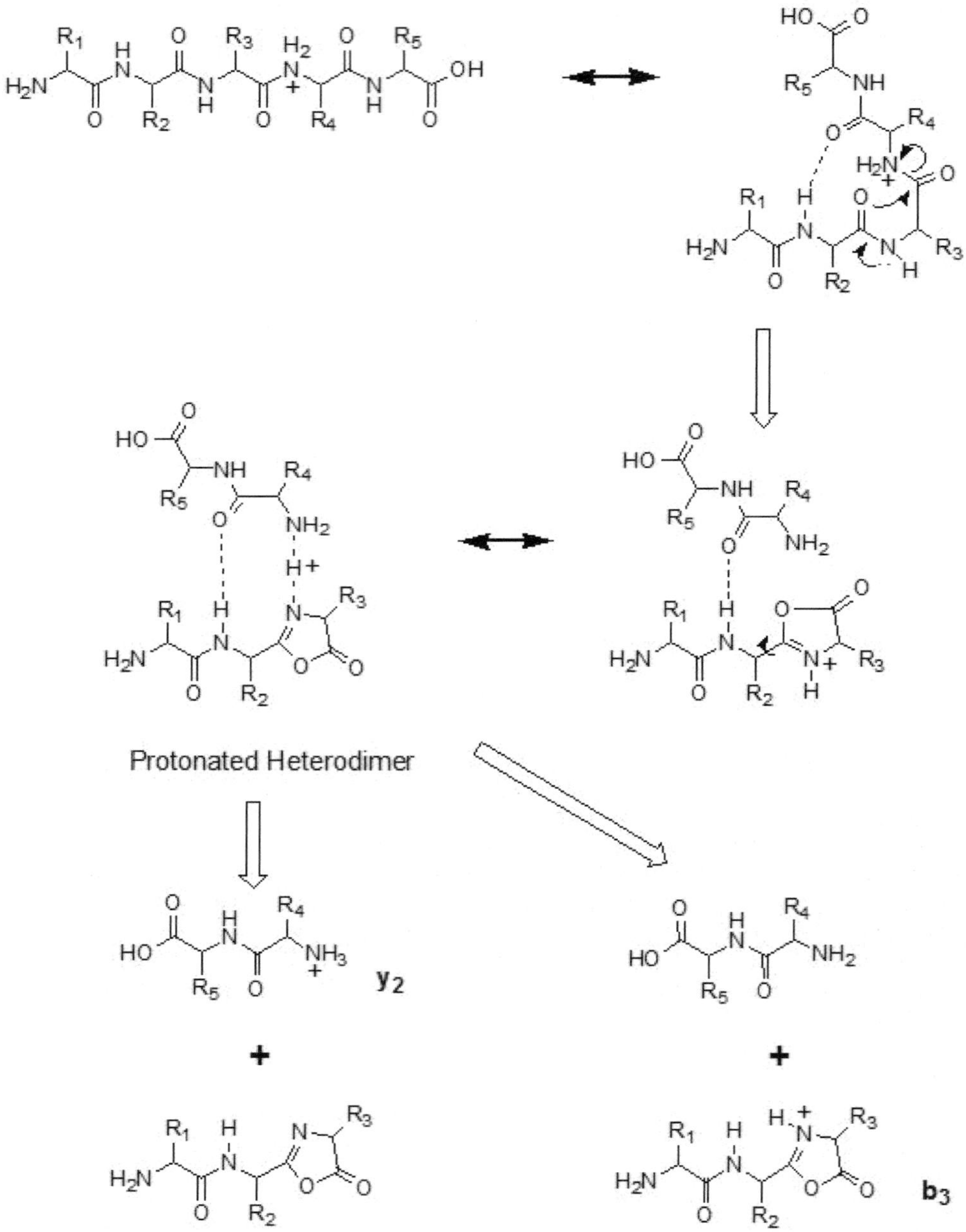

Protonated Heterodimer

y_2

b_3

Scheme 12-2

Unlike the commonly encountered **b**$_n$ ions ($n \geq 2$) shown in Scheme 12-2, which are able to form stable oxazolone structures, the simple **b**$_1$-type acylium product-ion structure formed by cleavage of the N-terminal amide bond within a peptide ion generally decomposes with the loss of CO (28 Da) to yield an immonium ion as illustrated in Scheme 12-3. However, stabilized **b**$_1$ ions may be observed for peptide ions that contain N-terminal lysine, histidine, or methionine residues. These amino acids facilitate cleavage of the adjacent peptide via a neighboring-group pathway involving their nucleophilic side chains, resulting in the formation of alternate cyclic product-ion structures. The **b**$_n$-type ions also may lose CO via an acylium ion intermediate upon further collisional activation to produce **a**-type immonium ions (Scheme 12-3). Immonium ions containing a single amino acid are commonly observed at low m/z values, and may be considered as structurally diagnostic for the presence of certain amino acids within the polypeptide.

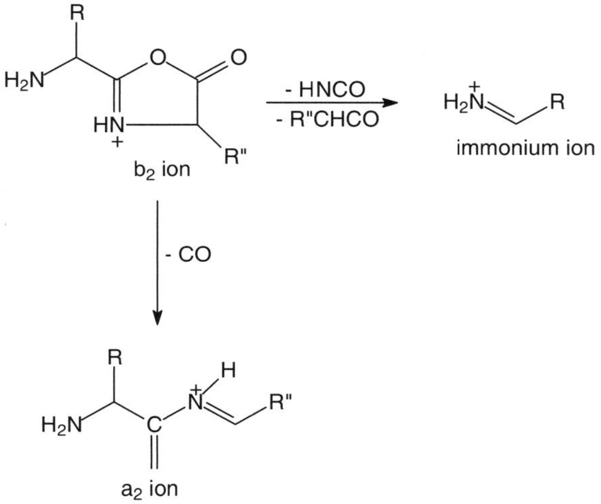

Scheme 12-3

Neighboring-group fragmentation pathways involving peptide backbone nucleophilic attack at a protonated side chain can also result in the formation of nonsequence product ions, such as those corresponding to the loss of small molecules such as NH_3 or H_2O [112], the loss of diagnostic side-chain fragments (e.g., methane sulfonic acid from methionine sulfoxide [113, 114], components of alkylated sulfoxides [115], or phosphoric acid from phosphoserine or phosphothreonine), sequential fragmentations, or intramolecular rearrangement product ions. The neutral loss of H_2O from peptides as induced by the side chain of His is thought to proceed through a retro-Ritter reaction catalyzed by the imidazole nitrogen as proposed in the description of an MS/MS study of the neuropeptide GAHKNYLRFamide [116]. A summary of the types of neighboring-group participation reactions previously observed in peptide ion fragmentation reactions has been reported by O'Hair [106].

Although unusual fragmentation behavior has been noted for Pro-containing peptides [117], a recent report suggests that fragmentation at the Xxx-Pro bond is predictable and that this information may be used to improve the identification of proteins if it is incorporated into peptide sequencing algorithms [118]; information on Xaa-Pro and Asp-Xaa bond cleavages for mobile and nonmobile peptide ion models is also available [119]. A detailed study notes that dominant cleavages adjacent to the acidic residues predominate when the number of ionizing protons equals the number of arginine residues [120].

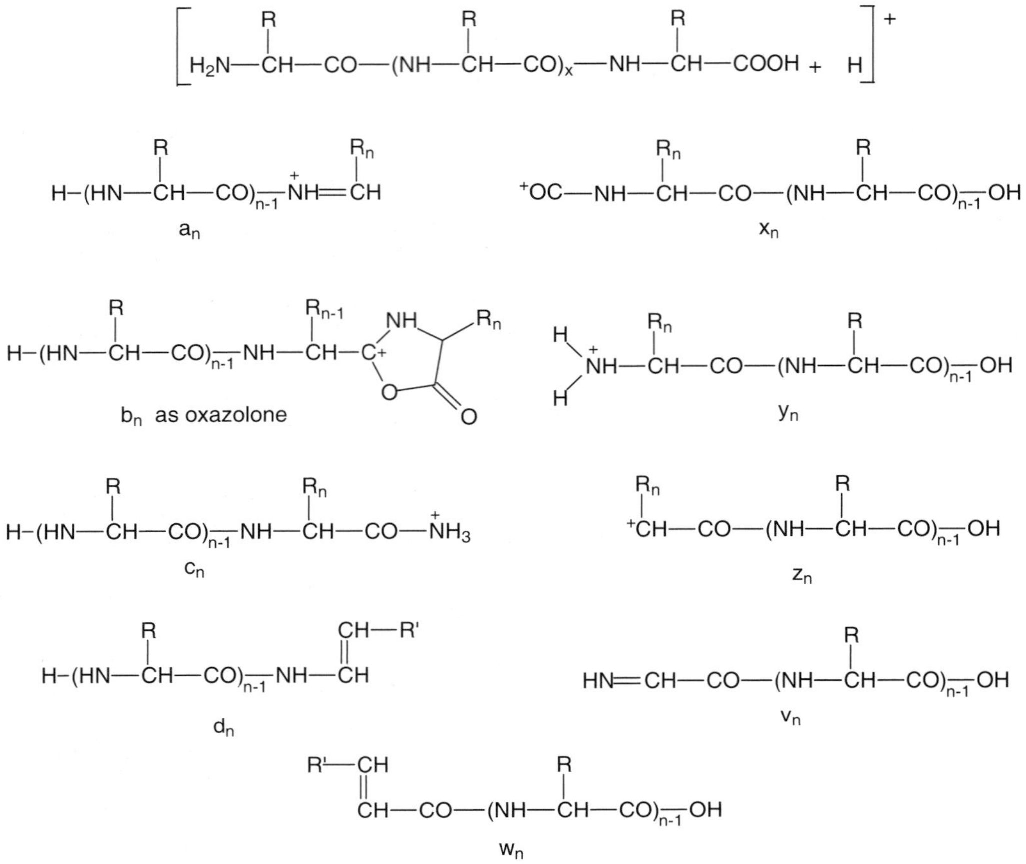

$$\left[H_2N-\underset{\underset{R}{|}}{CH}-CO-(NH-\underset{\underset{R}{|}}{CH}-CO)_x-NH-\underset{\underset{R}{|}}{CH}-COOH + H \right]^+$$

$$H-(HN-\underset{\underset{R}{|}}{CH}-CO)_{n-1}\overset{+}{N}H=\underset{\overset{|}{R_n}}{CH}$$
a_n

$$^+OC-NH-\underset{\overset{|}{R_n}}{CH}-CO-(NH-\underset{\underset{R}{|}}{CH}-CO)_{n-1}-OH$$
x_n

$$H-(HN-\underset{\underset{R}{|}}{CH}-CO)_{n-1}-NH-\underset{\overset{|}{R_{n-1}}}{CH}-\overset{+}{C}$$
b_n as oxazolone

$$\underset{H}{\overset{H}{N}}\overset{+}{N}H-\underset{\overset{|}{R_n}}{CH}-CO-(NH-\underset{\underset{R}{|}}{CH}-CO)_{n-1}-OH$$
y_n

$$H-(HN-\underset{\underset{R}{|}}{CH}-CO)_{n-1}-NH-\underset{\overset{|}{R_n}}{CH}-CO-\overset{+}{N}H_3$$
c_n

$$^+\underset{\overset{|}{R_n}}{CH}-CO-(NH-\underset{\underset{R}{|}}{CH}-CO)_{n-1}-OH$$
z_n

$$H-(HN-\underset{\underset{R}{|}}{CH}-CO)_{n-1}-NH-\underset{\overset{||}{CH-R'}}{CH}$$
d_n

$$HN=CH-CO-(NH-\underset{\underset{R}{|}}{CH}-CO)_{n-1}-OH$$
v_n

$$\underset{\overset{||}{R'-CH}}{CH}-CO-(NH-\underset{\underset{R}{|}}{CH}-CO)_{n-1}-OH$$
w_n

Figure 12-3.   Biemann nomenclature (lower-case letters) for CAD fragment ions of peptides.

Specific structures for the **a**, **b**, **c** and **x**, **y**, **z** ions are shown in Figure 12-3, giving the putative location of the charge on the ion derived from fragmentation of the protonated molecular peptide. Although early on, the b ion was thought to be an acylium ion, recent experimental evidence [98, 121] suggests that it is more likely to be an oxazolone structure as shown in Figure 12-3 and at the bottom of Scheme 12-2. There are some exceptions as, for example, in the case to the VP **b₂** ion, fragmentation data indicate that this species has a diketopiperazine structure rather than the oxazolone structure [122].

The **w** ion shown at the bottom of Figure 12-3 is formed via side-chain cleavage of the **z+1** ion, a protonated species. The **w** ion is critically important in distinguishing the isomass amino acids, leucine and isoleucine [86].

Fragmentation of protonated peptides by CAD and by post-source decay (PSD), a specialized technique associated with MALDI [123] (see Chapter 9), have many similarities in the types of ions formed, but the yield in PSD is lower because of the dependence on metastable decay of the protonated peptide, although this can be altered by choice of the matrix compound or by adjustments in the time delay between ion formation and ion extraction in a TOF instrument. The yield of fragment ions from CAD is much greater because the internal energy of the precursor ions increases efficiently by collisions during the CAD process (see Chapter 3 on MS/MS). Analogous fragment ions of types **a**, **b**, and **y** can be produced by electrosonic spray ionization (ESSI) as the peptide/protein is carried through a heated coiled metal tube at atmospheric pressure [124]; this thermal fragmentation process is gentle in that phosphate groups are not expelled from fragments of phosphopeptides and, also, neutral fragments can be reionized in the corona discharge.

In general, it has been noted that CAD of very low charge states (defined as charge states corresponding to less than or equal to the number of arginine residues present in the protein) results in the loss of structurally uninformative small molecules such as NH_3, H_2O, or both. Dissociation of low- and intermediate-charge states results in preferential cleavage on the C-terminal side of aspartic acid as well as basic residues such as lysine, histidine, and arginine; facile cleavage between adjacent lysine-histidine residues is prevalent in ions of these intermediate-charge states. The proline residue has a characteristic effect on fragmentation of a protonated peptide under metastable or CAD conditions [117]. The results of a CAD MS/MS study of tryptic peptides containing a basic residue remote from the C-terminus (a Lys or Orn residue in this study) indicate facile transfer of one or more residues from the C-terminus of the double-protonated precursor to the epsilon-amino group of Lys (or Orn); unfortunately, no predictive rules for the rearrangements could be established and these reports should serve as a caveate to relying on low-energy CAD of such double-protonated peptides for sequence information [125, 126]. The cleavage of the peptide bond C-terminal to oxidized cysteine is much more facile than that of the peptide bond C-terminal to Asp or Glu [127]. The position of the charge on a gas-phase peptide has been deduced from coupling information from a molecular modeling program with experimental data from a novel technique of determining ion radii by ambient pressure ion mobility spectrometry combined with ESI MS [128]. Dissociation of high-charge states (defined as a number of charges on the peptide exceeding the number of constituent arginine residues) results in preferential cleavage on the N-terminal side of proline residues or at other dominant "nonspecific" fragmentation sites, such as between Leu-Met and Leu-Ser. The fragmentation pathways of a variety of protonated dipeptides and tripeptides containing glutamic acid or glutamine were elucidated using metastable ion studies, energy-resolved mass spectrometry and triple-stage mass spectrometry experiments, and data from H/D exchange [129]. Low-energy CAD mass spectra of the [M − H]⁻ ions of a variety of dipeptides containing glutamic acid indicate that those with the γ linkage can be readily distinguished from those with the α linkage [130]. The CAD mass spectra of deprotonated dipeptides, the [M − H]⁻ ions, containing Met or Cys show the usual backbone fragmentation, allowing the sequence of the peptide to be determined [131, 132]. More fundamental studies of ion chemistry are needed, such as that on the chemically modified side chain of methionine [133], to develop a basis for better identification of peptides and proteins from their fragmentation behavior in mass spectrometry.

The goal of most CAD experiments with a protonated peptide is to obtain a fragmentation pattern from which the analyst can recognize peaks that correspond to a complete series of sequence-related ions. Of course, if a given protonated peptide does not fragment under CAD conditions, no information becomes available for that peptide via the types of fragments described above. Preparation of charged derviatives of such peptides may aid in the sequence analysis [87, 134–137] because the fixed charge on such derivatives often promotes structurally diagnostic fragmentation [138]. As in all charged derivatives of peptides, the charge is an inherent part of the derivative as illustrated in the following structure of the tris-(trimethoxy)-triphenylphosphonium-acetyl (TMPP-Ac) derivative of a generic peptide.

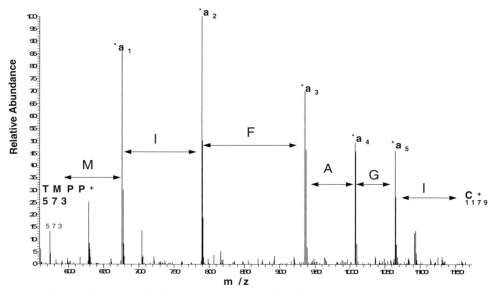

TMPP–Ac–peptide

**Figure 12-4.   *ESI-CAD MS/MS spectrum of TMPP-Ac derivative of Met-Ile-
Phe-Ala-Gly-Ile from a tryptic digest of cytochrome-c.***

As indicated in the above structure of the TMPP-Ac derivative, the positive charge is fixed on the phosphorus atom; therefore, the fragmentation is driven through a remote-site mechanism, as opposed to the mobile-proton mechanism described above, which promotes cleavage at a particular amide nitrogen as illustrated in Scheme 12-1. In some cases, the TMPP-Ac derivative at the N-terminus of a peptide selectively promotes formation of abundant **a**-type ions to give a simplified fragmentation pattern as illustrated in the ESI-CAD MS/MS spectrum in Figure 12-4. The generic structure of the a-type ion is as follows:

$$TMP\overset{+}{P}-Ac-(\,HN-\underset{|}{\overset{R}{CH}}-CO\,)_{n-1}\,N{=}\underset{|}{\overset{R_n}{CH}}$$

In other applications, the TMPP-Ac derivative promotes the **b**-type ion [139]. In any case, the TMPP-Ac derivative facilitates recognition of peaks in the mass spectrum that represent good coverage of N-terminal fragment ions that can be used to determine the sequence of the peptide. The TMPP-Ac derivatization procedure has been successfully applied to the sequence analysis of a complex mixture of tryptic fragments of proteins [134].

B. Strategy for Deducing Amino Acid Sequence via CAD of Peptides

The data interpretation strategy is heavily dependent upon recognition of peaks representing a particular series of ions [140]. Figure 12-5 outlines the strategy for recognizing the start of the **y** series of ions (i.e., the y_{n-1} ion from a peptide of length **n**) in the CAD spectrum of a protonated peptide. It is helpful to remember that a **y** ion is one that is formed by cleavage at a peptide bond with charge retention on the C-terminal component of the original molecule; i.e., a **y** ion will have a mass that consists of all the constituent amino acid residue masses, plus the mass of the OH group (at the C-terminus) and two hydrogens (one to complete the amino group at the N-terminus, the other in the form of a proton to establish a charge as shown on the amino terminus for the generic **y** ion in Figure 12-3). It is also important to realize that formation of a **y** ion

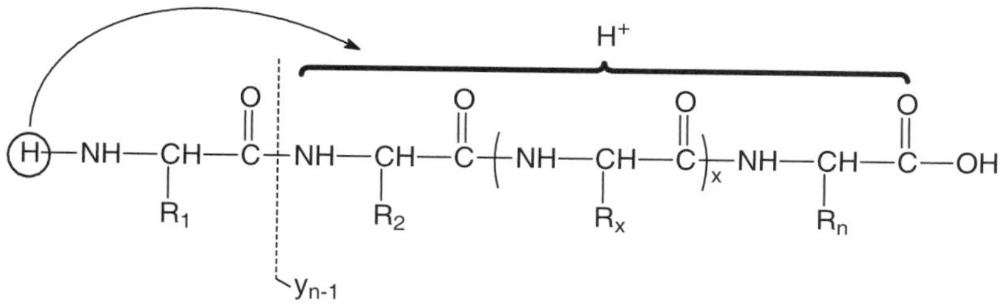

Figure 12-5. *Conceptual movement of a hydrogen and concurrent expulsion of the N-terminal residue to form the y_{n-1} ion from a peptide of length n.*

results in expulsion of residues from the amino terminus of the peptide. These conceptual attributes or features are represented in Figure 12-5; this conceptual representation should help the analyst recognize a candidate peak for the heaviest **y** ion and thereby suggest which amino acid residue is expelled from the amino terminus of the protonated peptide to form the **y** ion. Therefore, the peak representing the y_{n-1} ion will differ from that representing MH$^+$ by a number of mass units equal to the mass of an amino acid residue.

A conceptual model to facilitate recognition of peaks that represent a series of **b** ions is illustrated in Figure 12-6; the actual movement of electrons and hydrogens involved in the formation of the **b** ion is suggested in the pathways illustrated in Scheme 12-2. By definition, a **b** ion is a fragment of the peptide with charge retention on the N-terminal component of the original protonated peptide. Furthermore, because the charge will be retained on the amino terminus of the protonated molecule, this must require that residues from the C-terminus (including the C-terminal hydroxyl group) of the protonated molecule be expelled in the fragmentation process. Therefore, the peak representing the b_{n-1} ion will differ from that representing MH$^+$ by a number of mass units equal to the mass of an amino acid residue *plus* the mass of H_2O.

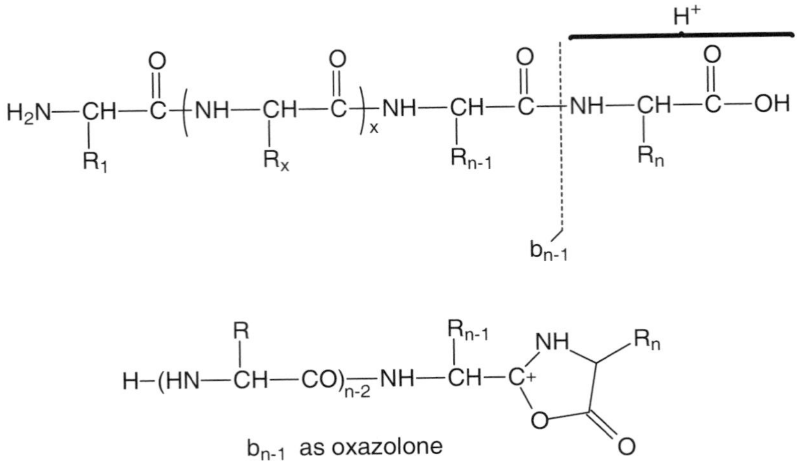

Figure 12-6. *Conceptual expulsion of an amino acid consisting of a residue plus H$_2$O (the proton plus the hydroxyl group) from the C-terminus to form the heaviest b ion.*

1) An Illustrative Example

Consider the high-energy CAD MS/MS spectrum shown in Figure 12-7 as an unknown. (The high-energy CAD process as in a magnetic-sector instrument promotes fragmentation through a somewhat different set of pathways than does low-energy CAD as in an ion trap instrument; however, such differences are not relevant in this illustrative example. Whereas the protonated peptide in this example was generated by FAB, the behavior of a protonated peptide under CAD is independent of the means by which the protonated molecule was formed [141].) It can be assumed at the outset that the major peak at *m/z* 574 represents the protonated molecule of a peptide that is the precursor ion selected for CAD. The mass spectrum shown in Figure 12-7 is a CAD MS/MS spectrum of some of the protonated molecules. The peak at *m/z* 574 in Figure 12-7 represents those protonated molecules that survive the CAD process (i.e., those that did not fragment during exposure to CAD conditions). Those protonated molecules that did fragment during CAD are represented by the peaks at a lower *m/z* value in Figure 12-7. The procedure for analysis or data interpretation consists of examining the mass difference between each of the fragment ion peaks and the peak representing the protonated molecule.

Many of the peaks at high *m/z* values in the mass spectrum in Figure 12-7 do not provide structural information because they represent losses of species lighter than amino acid residues; these peaks can be dispatched as follows. The peak at *m/z* 556 is 18 lower than the peak at *m/z* 574 and probably represents the loss of water from the protonated molecule. This offers no structural information concerning the sequence of amino acids. Consideration of the other peaks down to about *m/z* 500 leads to no particular useful information. The peak at *m/z* 499 is 75 lower than the one representing the protonated molecule; it accounts for the loss of 75 Da, which corresponds to the loss of the side chain of methionine; this loss offers no structural information regarding the sequence, but it is indicative of the presence of a methionine residue within the peptide sequence. (The side-chain masses of the amino acids can be found in the back matter of the book.) The next peak in Figure 12-7 that represents a meaningful loss is that at *m/z* 467 corresponding to the loss of 107 Da from the protonated molecule;

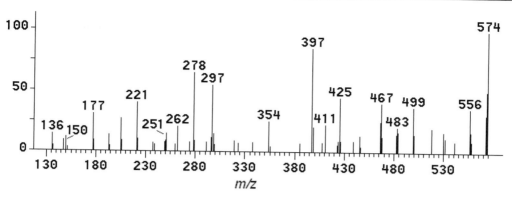

Figure 12-7. *Mass spectrum of an unknown single-protonated peptide of m/z 574 (MH⁺ = 574) obtained by using CAD.*

107 Da corresponds to the side chain of tyrosine; again, this loss offers no structural information regarding the sequence, but it is indicative of the presence of tyrosine in the composition of the peptide.

From the standpoint of sequencing, the goal is to find a peak representing a fragment ion that differs in mass from the protonated molecule by either an amount equal to a residue mass or a residue mass *plus* water. The residue masses of the amino acids can be found in the back matter of the book. As indicated in Figure 12-7, there is a peak at a mass-to-charge value that *differs* from the *m/z* value for the peak representing the protonated molecule by the mass of a residue mass *plus* the mass of water (574 − 425 = 149 = 18 + 131, the residue mass of methionine); this observation indicates that that particular amino acid residue (methionine, in this case) is located at the C-terminus of the original peptide. Thus, observation of the peak at *m/z* 425 suggests that methionine was expelled from the C-terminus to form the heaviest **b** ion (425 Da). With this information, it is possible to begin to sketch out (see Figure 12-8) a sequence of residues with methionine at the C-terminus. Having recognized the beginning of a series of **b** ions, keep this peak at *m/z* 425 as a new reference point from which to find the next peak at a lower *m/z* value that represents the next member of the series of **b** ions.

Having recognized the largest **b** ion in the mass spectrum by its peak at *m/z* 425, the next step is to try to identify a peak that corresponds to the next smaller **b** ion, namely one that has also lost the second amino acid residue from the C-terminus to form the second **b** ion. The second **b** ion will differ in mass from the first **b** ion by *only* the mass of an amino acid residue. The peak at *m/z* 411 is only 14 lower than 425 and, therefore is not a likely candidate to represent the second **b** ion. The peak at *m/z* 397 is 28 lower (i.e., representing an **a** ion) than the peak at *m/z* 425; again, this peak does not represent the second **b** ion as no amino acid residue exists having a mass of only 28 Da. Staying with this strategy, finally, it is the peak at *m/z* 278 that represents a mass difference from 425 that corresponds to *only* the residue mass of an amino acid (i.e., the mass difference between 425 and 278 is 147 Da, which is the residue mass of phenylalanine). This suggests that the second amino acid residue in the sequence from the amino terminus is phenylalanine: –Phe–Met.

Continuing with the **b** series (see the flow chart of sequential losses in Figure 12-8) now represented by the two peaks at *m/z* 425 and 278 in Figure 12-7, the next step involves a search for a peak lower than 278 that corresponds to the loss of another

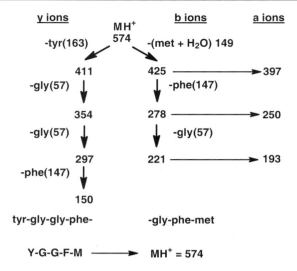

Figure 12-8. Preliminary flow chart of sequential losses from $MH^+ = 574$.

amino acid residue. In this search, the peak at m/z 221 is 57 lower than the one at 278. This mass difference of 57 Da corresponds to the residue mass of glycine: –Gly–Phe–Met. This observation indicates that the third residue from the C-terminus is a glycine residue. Finally, it can be seen that there are no other peaks observed below m/z 221 that correspond to the loss of a residue mass from 221; the search for additional **b** ions comes to a halt.

At this point, a new search of the mass spectrum (Figure 12-7) should focus on peaks that represent **y** ions. Go back to the peak in Figure 12-7 that represents the intact protonated peptide (the peak at m/z 574) and search at lower m/z values for a fragment ion peak that differs from 574 by a value that corresponds exactly to a residue mass. Finding such a peak will indicate the beginning of the **y** series of ions.

In reexamining the mass spectrum in Figure 12-7, it is apparent that the peak at m/z 411 is 163 lower than the peak at m/z 574. The mass of a residue of tyrosine is 163 Da, and observation of a peak at m/z 411 corresponding to the loss of such a residue from the protonated molecule indicates that tyrosine is the amino acid residue located at the N-terminus of the peptide.

Having recognized the start of a series of **y** ions by observing the peak at m/z 411, the search should continue for an m/z value below the new reference point of m/z 411 for a peak corresponding to the second member of the series of **y** ions. With the focus now on m/z 411 as a reference point, observe that the peak at m/z 354 is 57 lower; the residue mass of glycine is 57 Da. Therefore, observation of the peak at m/z 354 indicates that the second amino acid residue from the N-terminus is a glycine residue. Following the flow chart shown in Figure 12-8, the partial sequence at this stage of interpretation appears to be Tyr–Gly–. Using the peak at m/z 354 as a new reference point, the peak at m/z 297 is 57 lower than 354, therefore the peak at m/z 297 suggests the expulsion of another residue of glycine in the formation of this ion. These observations suggest that the third amino acid residue is also a glycine residue in sequence from the N-terminus (see Figure 12-8). Continuing with the peak at m/z 297 as a new reference point, the peak at m/z 150 is 147 lower, suggesting expulsion of a phenylalanine residue, meaning that phenylalanine is the fourth amino acid from the N-terminus, giving the partial sequence: Try–Gly–Gly–Phe–. Because there are no other peaks at lower *m/z* that differ from a new reference point of 150 by a residue mass of an amino acid, recognition of the sequence of **y** ions comes to a halt. Figure 12-8 shows a summary of the losses of mass that correspond to the residues lost to produce the recognized sequence of **y** ions.

Having suggested possibilities for amino acid sequences from either terminus of the peptide, these two suggested partial sequences can be overlaid to compose a complete sequence. For example, the data reviewed above suggest that the sequence from the C-terminus is –Gly–Phe–Met, whereas another series of data suggested that the series of amino acid residues from the N-terminus is: Try–Gly–Gly–Phe–. Each of these two partial sequences shows the C-terminus of a glycine residue connected to the amino terminus of a phenylalanine residue. These two residues might be a redundant observation in the two sequences, and they can be overlaid at that point. This would give an overall sequence starting from the N-terminus of Try–Gly–Gly–Phe–Met.

If the residue masses of the postulated amino acids in the complete putative sequence are summed, 555 Da will be the obtained value. Add 18 Da (1 Da for the hydrogen at the amino terminus and 17 Da for the hydroxyl at the C-terminus) to this sum. Then, add 1 Da for the proton on the protonated molecule. This will give a total of 574 Da for the expected mass of the protonated molecule. This agrees perfectly with the observed peak at *m/z* 574 in the mass spectrum giving credence to the suggested amino acid composition implicit in the sequence of Y–G–G–F–M.

2) Possible Pitfalls in Interpretation

The procedure outlined above, while successful as described, is subject to error if not tested as illustrated in the following scenario (see Figure 12-9). Consider again the peak at *m/z* 574 in the mass spectrum (see Figure 12-7) representing a single-protonated peptide where the peak at *m/z* 425 is the first indication of the loss of an amino acid residue (Met, in this case) *plus* the mass of water from the protonated peptide. This, again, indicates that methionine is the amino acid at the C-terminus of the peptide (see the right side of the flow chart in Figure 12-9).

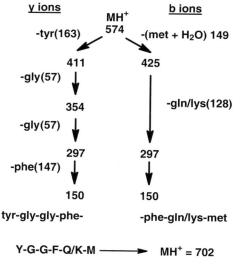

Figure 12-9. *Alternative flow chart for consideration of sequential losses from MH⁺ = 574.*

Considering a new reference point at *m/z* 425, the peak at *m/z* 297 corresponds to the loss of 128, which could correspond to the residue mass of either glutamine or lysine. Therefore, these data suggest that the second amino acid from the C-terminus could be a glutamine or lysine. Taking the new reference point as being 297, the peak at *m/z* 150 again corresponds to the loss of 147, suggesting that phenylalanine is the third amino acid residue from the C-terminus. These data suggest that the sequence from the C-terminus could be as follows: –Phe–Glu/Lys–Met (see Figure 12-9). No further peaks are available to represent other ions of the **b** series.

Returning to the peak at *m/z* 574 for the protonated molecule and searching for losses equal to only residue masses yields the same sequence of information as found previously (compare with Figure 12-8). This suggests that the sequence from the amino terminus is Tyr–Gly–Gly–Phe–. Once again, there are two incomplete suggested series of residues from opposite ends of the peptide molecule that overlap at phenylalanine. Overall, this would suggest a sequence of tyrosine, glycine, glycine, phenylalanine, glutamine or lysine, and methionine. If these six residues were connected as suggested by this interpretation, the computed mass of the protonated molecule would be the sum of the six residues (683 Da) + 18 Da + 1 Da for a total of 702 Da. Because no peak is observed at *m/z* 702 in the mass spectrum, this suggested sequence or amino acid composition must be incorrect.

Many other missteps could have been made during interpretation of the CAD MS/MS spectrum in Figure 12-7. For example, the loss of 75 Da from MH$^+$ as represented by the peak at *m/z* 499 could have been interpreted as the loss of Gly *plus* water (57 + 18 = 75), which would suggest that Gly would be at the C-terminus of the peptide; the correct interpretation of this peak was in its representing the side-chain loss of methionine. In another case, the difference between 278 and 177 is 101, which is the residue mass of Thr; the correct interpretation was in using the difference between 278 and 221 to recognize the residue mass of Gly (57 Da). In yet another case, the difference between 411 and 297 is 114, which could have been interpreted as the residue mass of Asn (114 Da) rather than the mass of two Gly residues. For the record, the mass of methionine sulfoxide is 147 Da; the difference between the peaks at *m/z* 425 and 278 is 147, which was correctly interpreted as the residue mass of Phe. Each of these individual possible misinterpretations would have been recognized eventually as incompatibilities with other findings and "cross-checks" during the interpretation. Whereas it is possible to misinterpret a CAD MS/MS mass spectrum, cross-correlations that are built into algorithms for this kind of data interpretation are responsible for remarkably good success in determining the sequence of an unknown peptide or in confirming the identity of a known peptide.

3) Search for Confirming Ions

A helpful procedure to avoid pitfalls as described above (and summarized in Figure 12-8) would be to try to find other peaks in the mass spectrum that confirm the interpretation of a particular peak as representing a specific type of ion. For example, **b** ions differ in mass from the corresponding **a** ions by the elements of carbon monoxide. Therefore, if a peak at *m/z* 425 is thought to represent a **b** ion, this assertion could be corroborated by observing a peak at *m/z* 397 (i.e., 28 lower than 425) thought to represent an **a** ion. Note, in the mass spectrum in Figure 12-7, that the peak at *m/z* 425 is accompanied by a much larger peak at *m/z* 397, which could represent the **a** ion containing the same residues, thereby giving credence to the assignment of a **b** ion to the peak at *m/z* 425. More important, in this example, is the fact that the peak at *m/z* 278 (which is the second suggested **b** ion in the series as interpreted originally) can be confirmed as a **b** ion by observing a peak at *m/z* 250 for the corresponding **a** ion (see the expanded flow chart in Figure 12-8). Even though the peak at *m/z* 250 is of relatively low intensity, it gives some credence to the suggestion that the peak at *m/z* 278 represents a *bona fide* **b** ion.

Figure 12-9 represents the misinterpretation of the mass spectrum in that the peak at *m/z* 297 was suggested to represent a **b** ion. There is no peak at *m/z* 269 (297 − 28 = 269) in the mass spectrum (Figure 12-7) to corroborate this suggestion; therefore, it is unlikely that the peak at *m/z* 297 represents a **b** ion (the incorrect assumption that led to the misinterpretation). In Figure 12-8, the third **b** ion represented by a peak at *m/z* 221 is corroborated by observing a minor peak at *m/z* 193 for the accompanying **a** ion.

The mass spectrum in Figure 12-10 gives a reasonably comprehensive assignment of ions to the various peaks in the mass spectrum. Those peaks representing the loss of a side chain are identified by the nomenclature (–R with a subscript indicating the amino acid from which the side chain was lost); the side chain itself is represented by R (–R$_F$ represents the loss of the side chain of phenylalanine). The side-chain masses are listed in the back matter of the book. The peak at *m/z* 136 is labeled by Y$_{IM}$ for the immonium ion of tyrosine. The immonium ion (IM) corresponds to $CH_2=NHR^+$ where R is the side chain of the amino acid; masses of the IM are also listed in this table. Peaks at *m/z* 205 and *m/z* 262 represent internal fragments consisting of GF and GGF, respectively; such internal fragments are likely to arise from secondary fragmentation or decomposition of primary product ions from the selected CAD process. For more practice, consider interpreting another unknown CAD MS/MS mass spectrum at the end of this chapter, and comparing the result with the answer provided on the book's Web site.

4) Ladder Sequencing

In principle, the sequence of amino acids from either terminus of the protein could be determined by exposing the protein to an appropriate chemical or enzyme to cleave the amino acid residue at that terminus. Analysis of the protein prior to and after such treatment would give the basis for computing the mass of the expelled amino acid residue, thereby identifying it. Similarly, the protein could be exposed to multiple steps of such degradation with subsequent analysis by MALDI to determine the cumulative mass lost to confirm or identify those residues involved in the terminal sequence of amino acids [142–144].

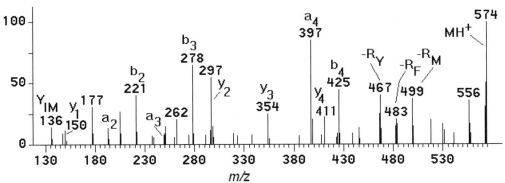

Figure 12-10. Mass spectrum of an unknown peptide obtained using CAD in which most of the peaks have been assigned to represent specific types of peptide fragment ions. The sequence of amino acids in the peptide is Tyr–Gly–Gly–Phe–Met.

A more comprehensive method for amino acid sequencing of peptides by MALDI utilizes multiple steps of partial degradation chemistry prior to analysis of the reaction mixture. This "ladder sequencing" approach [145, 146] generates a set of peptide fragments from step-wise partial degradation of a polypeptide in the presence of a small amount (5%) of phenyl isocyanate (PIC), a terminating reagent that precludes the cleavage step in the Edman cycle for that fraction of the analyte protein. The other 95% of the protein, or peptide, reacts with the usual effective Edman reagent (phenyl isothiocyanate), which allows cleavage at the appropriate step in the cycle. Therefore, a series of successive Edman cycles with the partially poisoned reagent mixture produces a residual series of progressively truncated peptides differing by one amino acid residue. Analysis of the resulting mixture of degradation peptides from the partially poisoned Edman reagent provides a series of peaks representing peptides that were blocked at the amino terminus by reaction with PIC and thus could not undergo further degradation. The difference in mass between the peaks representing these blocked peptides corresponds to the amino acid residues in sequence from the amino terminus of the original peptide or protein.

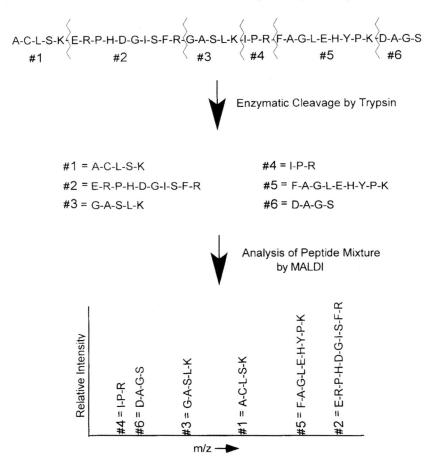

Figure 12-11. Concept of mass mapping peptide fragments resulting from controlled degradation of a protein.

2. Mass Mapping

In many cases, structural information may be deduced from analysis of the peptides formed before and after controlled degradation or modification of the protein of interest in solution. Additionally, direct analysis of the degradation products or shifts in the mass of the analyte following specific chemical modifications of the resultant peptides allow certain conclusions to be drawn about structural features in various portions of the original molecule. Typical chemical modifications include acetylation (mass shift = 42 Da) and methylesterification (mass shift = 14 Da); these chemical procedures are also useful in *de novo* sequencing of an unknown peptide [147].

The procedure for mass mapping parallels that of conventional peptide mapping which employs reversed-phase HPLC with UV detection to identify peptide fragments under controlled protein degradation conditions. In the case of peptide mass mapping, the peptide fragments produced during controlled degradation of the protein are identified according to mass [148, 149] rather than by retention time in chromatography. Often degradation of the targeted protein with a mixture of enzymes, cleanup of the proteolytic peptides via ZipTip, and use of more than one matrix results in the best sequence coverage with analysis by MALDI [150]. On-line techniques for microscale sample processing, such as a miniaturized trypsin membrane reactor housed inside a commonly used capillary fitting, are improving the efficiency of the methodology [151].

A conceptual illustration of mass mapping for confirming the identity of a given protein is provided in Figure 12-11. The amino acid sequence must be known for the protein to be characterized by this technique; the protein is cleaved at selected sites by specific enzymes or chemical reagents. From knowledge of the amino acid sequence in the protein, the masses of the expected peptides can be calculated and then compared to the experimentally determined mass values obtained during mass analysis of the mixture of peptides resulting from enzymatic degradation of the protein.

A. Peptide Mass Fingerprinting

One of the first automated procedures for protein identification is based on the masses of proteolytic fragments as originally reported by Stults and coworkers [152, 153] (for which they received the Distinguished Contribution to Mass Spectrometry Award from ASMS in 2002). An identification of the unknown protein is achieved by matching mass spectral data representing the protonated proteolytic fragments of the unknown with masses calculated according to the "rules" of *in silico* digestion of various entries in a protein database with the same protease. The algorithm is told which enzyme was used experimentally to degrade the unknown protein. The algorithm then "electronically digests" (*in silico*) each sequence entry in a specified protein database, which is simply a list of protein names and the corresponding amino acid sequences. Because the algorithm is given the specificity of the enzyme, the sequence of the proteins, and the masses of the constituent residues, it can calculate the masses of the expected proteolytic fragments for each of the proteins. The automated search consists of a comparison between calculated molecular masses of the proteolytic fragments of each protein in the database with the *m/z* value of peaks for MH$^+$ components in the *real* or experimental digest. Data mining performed on 28,330 unique peptide tandem mass spectra provided the basis for a statistical assessment of gas-phase fragmentation behavior of peptides as a function of sequence, which should facilitate the development of more reliable algorithms for protein identification [154].

B. *De novo* Sequencing

In cases where the protein is not contained in a database, chemical modification (e.g., acetylation, methylesterification, etc.) of the unknown putative protein interleafed with analyses by mass spectrometry can provide useful characteristic information indirectly [147, 155–158]. Formation of a brominated derivative based on sulfhydryl chemistry provides some selectivity based on mass defect that improves protein identification via "shotgun" proteomics provided that adequate mass accuracy is available as with FTMS [159]. Yergy and coworkers use a peptide composition lookup table indexed by residual mass and number of amino acids for *de novo* sequencing of polypeptides up to 1600 Da from MS/MS spectra obtained during metastable decay with TOF/TOF instrumentation [160]. Complementary data from the fragmentation techniques CAD and ECD have been used with a new *de novo* sequencing approach with the same level of efficiency and reliability as conventional database-identification strategies [158].

C. Sequence Tagging

Rather than relying solely on the molecular masses of proteolytic fragments for identification of the parent protein, Mann and Wilm developed a method based on use of a partial sequence of the original protein for identification; in this way, a short sequence, along with fragment ion masses that correspond to the beginning and end of the sequence, constitute a "sequence tag" as a corroborative determinant in the identification [161]. Further improvements in this approach may benefit from a statistical assessment of the distribution of products from the CAD fragmentation of peptides [162]; in-depth studies of the CAD fragmentation process for peptides may help control or minimize intramolecular rearrangements, thereby improving the reliability of peptide identification from the fragmentation pattern [163–165].

A MultiTag method has been reported which assigns statistical significance to matches of multiple error-tolerant sequence tags for a given database entry, thereby providing the possibility of making a sequence-similarity identification using only short (2–4) amino acid residue stretches of peptide sequences [166]. An approach using stable isotope-labeled reagents to modify a protein eventually provides a mass spectrum with an isotope pattern that provides information about the amino acid compositions of trace quantities of a peptide, thereby improving the chances of identifying the parent protein with a minimum of false positives [167]. Better accuracy usually accrues when mass spectral results are coupled with chromatographic retention data [168].

D. Sequest

Yates and coworkers developed and patented an automated routine for calculating the masses of all possible CAD fragments that might be produced by a given peptide sequence and comparing them with experimentally obtained CAD MS/MS data for purposes of determining the sequence of an unknown peptide [169]. Sequest is an example of commercially and publically available software routines to accomplish tasks of sequence determination and/or protein identification; there are other equally good software routines that are associated with various commercial instruments. Database searching has become an essential tool in large-scale proteomics [170–172]. A dot product metric can be used to measure the degree of similarity of tandem mass spectra to provide a means of distinguishing "correct" and "incorrect" identifications using a Sequest database search with a success rate exceeding 90% [173]. A further improvement in the Sequest approach involves a validation routine for identifying phosphopeptides based in part on a machine-learning algorithm (support vector machine) for which tests have shown a 96% accuracy [174].

E. Evaluation of Hits in Automated Searches

Given the ready availability of automated searches, such as that at http://prospector.ucsf.edu, several approaches have been devised to assess the quality of an automated identification procedure [175, 176]. In an attempt to formulate a universal standard for judging the quality of a hit, three simple metrics that describe different aspects of the matching criteria have been suggested: *hit ratio* gives an indication of S/N in the mass spectrum, *mass coverage* measures the extent of the protein sequence matched, and *excess of limit-digested peptides* reflects the completeness of the digestion that precedes the process of peptide mass fingerprinting [177]. The MOWSE (MOlecular Weight SEarch) score is a numerical indicator of a statistical assessment of the quality of a given database match [178–180]; a value of several thousand for the MOWSE score indicates a good match. Additional information on MOWSE can be found at www.matrixscience.com. A database-independent scoring method (S-score), based on the maximum length of the peptide sequence tag detected as combined with complementary fragments simultaneously represented in the available CAD and ECD data, substantially eliminates false positives [181, 182]. Striving to provide a more objective assessment of an identification made by their software package, named ProFound, the Group at Rockefeller University (http://prowl.rockefeller.edu/cgi-bin/ProFound) has incorporated elements of human judgment with approaches based on signal-detection theory into their algorithm [183]. A scoring routine for Sequest has been developed that evaluates cross-correlation values in a way that is independent of peptide size and the database used to perform the search [184, 185]; considerable variance in these scores relates to instrumental effects of spectral variability on the correct identification of peptides [186, 187]. In an effort to reduce the size of the searched library, rules have been developed to identify unlikely missed cleavages and nontryptic proteolysis products, thereby improving accuracy of results and efficiency in data mining [188]. A multinomial algorithm for a spectral profile-based intensity comparison (MASPIC) scorer converts an experimental MS/MS spectrum into an *m/z* profile of probability and then scores peak lists from candidate peptides using a multinomial distribution model incorporating intensity, spectral peak density variations, and *m/z* error distributions as validated with two reference protein standards [189]. Other statistical measures have been used to reduce various protein identification scoring schemes to a common, easily interpretable representation [190–192]; e.g., in one case, an alpha factor is computed that relates to the probability that a given identification has been accomplished by chance [193].

A pattern-recognition algorithm called SALSA (scoring algorithm for spectral analysis) has been developed for the detection and scoring of specific features in MS/MS spectra such as an ion series for specific peptide sequences [194, 195]; other statistical evaluation schemes have also been reported to estimate the accuracy of peptide assignments [119, 196, 197]. However, it must be remembered that protein identification by these automated processes is only possible if the actual protein is correctly listed in the protein database. The average peptide score, which is the ratio of the usual protein score and number of nonredundant peptides has been reported to facilitate initial filtering of database search results in addition to providing a useful measure of confidence for the proteins identified [198]. A statistical assessment of another automated identification process has been reported as coupled with 2D gel patterns for cell lysates from multiple samples of cardiac myocytes with final analysis by MALDI MS [199]; other reports address the statistical performance of automated identification of proteins [191, 200–205]. Complementary data from ECD and CAD increased the confidence level in protein identification by more than one order of magnitude, as demonstrated in the analysis of a cell lysate from *Escherichia coli* [182].

In cases where automated searches fail to identify a protein (not all proteins are listed in databases), at least partial sequence information can be gleaned from CAD during one [206] or multiple stages [207] of MS/MS; in a case involving elucidation of antibodies for which the corresponding DNA sequences were unknown, the cysteine residues were carboxymethylated with a 1:1 mixture of ^{13}C-labeled reagent to facilitate recognition of cysteinyl peptides during analysis of the complex digestion mixture by LC-ESI MS/MS [208]. To meet increasing demands for high-throughput analyses, innovative instrumentation that facilitates automation and reproducible ionization has been reported [209]. A recent report of laser desorption (no matrix) coupled with APCI portends promise for the direct analysis of proteolytic peptides from gels [210].

Protein identification from DNA databases, using mass spectral data, has been demonstrated [169, 211]; cross-correlations between genomic and proteomic analyses are particularly powerful [212, 213]. A variation of this approach has been used to help correlate proteins with regions of a genome that has not yet been annotated; genome fingerprint scanning (GFS) is the name given to this novel approach [214]. Using GFS, an entire sequenced genome is translated *in silico* to its potential proteome without regard for any form of genetic annotation; that potential proteome is then digested *in silico*. Experimentally obtained mass spectral data are then compared with the masses of the *in silico*-produced proteolytic fragments to recognize the region of the genome most likely to be responsible for encoding the proteins found in the sample [214]. These kinds of approaches can be applied to whole-cell lysates after subjecting the digest to chromatofocusing fractionation according to pH [215].

F. Data-Dependent Analysis by Mass Spectrometry

The procedural analysis by mass spectrometry has also been automated considerably. For example, in one commercially available configuration, using the so-called "Triple Play", peaks in the primary mass spectrum (operator-adjustable peak-intensity threshold) are evaluated via a "zoom scan" to determine the charge state of the corresponding ion, which is then subjected to CAD MS/MS.

3. Post-Translational Modifications

The process of post-translational modification of an amino acid residue creates a complex variety of protein isoforms of a given polypeptide chain (gene product) translated from RNA. For example, certain serine and/or threonine residues are susceptible to phosphorylation [216], a process that can regulate the activity of enzymes. Figure 12-12 conceptually illustrates a few of the complex varieties of isoforms that are possible via post-translational modifications (PTMs) of a given polypeptide encoded by RNA for synthesis at the ribosome. Each of the five compounds in Figure 12-12 consists of the same sequence of 19 amino acid residues; therefore, they all represent a single gene product. However, each of the five has been post-translationally modified differently, and each will have a different biological activity; there are many more possible variants of this one gene product than are shown here. Thus, knowing only the sequence of amino acids in a protein is not likely to be sufficient to identify the molecule responsible for a specific process within a cell. It is becoming increasing clear that the composite of PTMs on a protein carries a global message; in part, such a composite picture of PTMs on the massive proteins associated with DNA is being called the "histone code" [217, 218]. As proteomics becomes more sophisticated, it will be necessary to identify all PTMs associated with a given polypeptide in a given state of bioactivity [40, 51]. Simple and common modification of the amino terminus by acetylation (catalyzed by ribosome-associated *N*-acetyltransferase) or by carbamylation (often due to sample processing in high concentrations of urea) can be recognized by net mass shifts of 42 Da and 43 Da, respectively. Studies of gas-phase processes for the dominant fragmentation of methionine sulfoxide-containing peptide ions may help recognize such a PTM [113]. Results of other studies show preferential cleavage of the amide bond at the C-terminal side of the oxidized cysteine residue [219]. The novel discovery of pyrolytic cleavage on the C-terminal side of Asp will surely be useful in peptide mapping [220]; other examples of PTM mapping follow.

Automated procedures for recognizing the presence of and, in some cases, the location of post-translational modifications are beginning to appear in the literature. A new approach in which post-translational modifications are retrieved from a "top-down" database of nearly 50,000 modified histone H4 sequences has been reported [221]. Another approach combines genome fingerprint scanning (GFS) with top-down proteomics in a computer program called PROCLAME to recognize post-translational modifications [222].

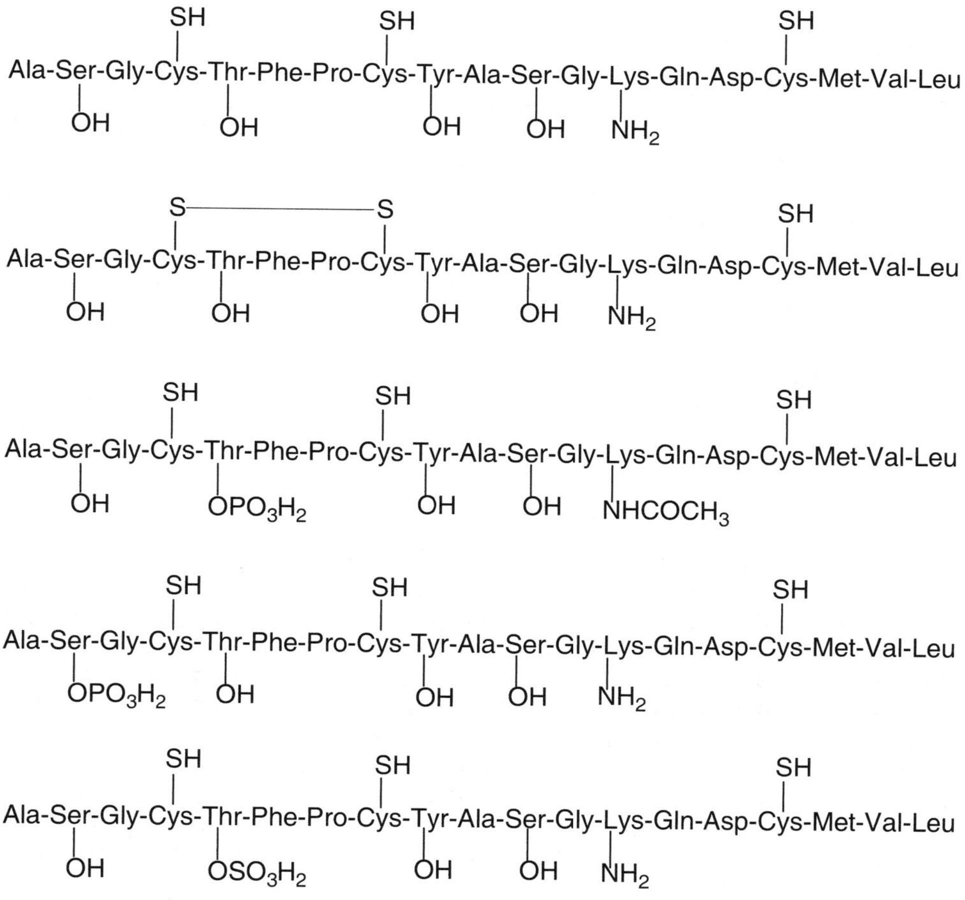

Figure 12-12. *Some examples of post-translational modification to a single gene product.*

In the following sections, interweaved chemical and mass spectrometric techniques for specific post-translational modifications are described in a didactic manner together with representative applications.

A. Recognition of Sites of Protein Phosphorylation

Recognition of the site of phosphorylation [31, 223–229] in a given protein requires much more careful planning than an analysis to determine whether a protein is merely phosphorylated, which is simply indicated by the shift in the molecular weight of the protein by 80N Da where N is the number of phosphate groups. See Scheme 12-4 for a chemical explanation of the 80-Da mass shift upon phosphorylation. As described in several reviews of the methodology [230, 231], the general approach is essentially that used in mass mapping. First, the amino acid sequence of the protein must be known. Second, the masses of the expected proteolytic fragments are computed according to the specificity of cleavage of the degradative enzymes. Third, if no peak is observed in the mass spectrum that corresponds to the expected mass for a given peptide fragment, but a peak is observed that corresponds to a mass shift of N × 80 Da, then that peptide fragment contains N phosphate groups. Fourth, if it is determined that a given peptide fragment contains fewer phosphate groups than there are phosphorylatable amino acids, it will be necessary to design additional experiments to complete the analysis.

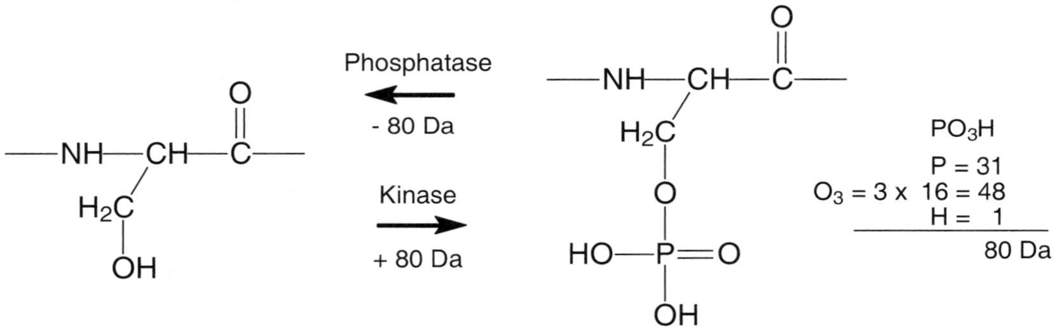

Scheme 12-4

Whereas some mapping experiments can be accomplished by simply determining the mass of proteolytic fragment, additional experiments might involve CAD MS/MS. In some cases, it may be possible to perform further proteolysis to differentiate the various potential phosphorylatable sites in the protein. A handicap to analyses based on CAD MS/MS is the tendency for phosphoserine and phosphothreonine to lose H_3PO_4 (98 Da) under positive-ion conditions and to lose $H_2PO_4^-$ or PO_3^- under negative-ion conditions. In analyses by AP MALDI in an ion trap, protonated peptides containing phosphotyrosine have been reported to undergo an unexpected loss of 98 Da if the peptide also contains lysine or arginine; in sodium-cationized or protonated peptides containing phosphotyrosine, but not a basic residue, no loss of 80 or 98 Da is observed [232]. In other work based on negative-ion ESI CAD MS/MS, high-mass ions resulting from multiple losses of 79 Da have been used as marker ions by which to recognize phosphopeptides with greater reliability than by using a peak at m/z 79, which is more susceptible to interference [233].

1) An Illustrative Example

A realistic example of using MALDI to provide data for recognition of the sites of phosphorylation is illustrated in the context of the analysis of bovine β-casein, which contains several phosphate groups. Following digestion of this 209-residue protein with trypsin, the resulting proteolytic fragments were subjected to coarse separation by HPLC. As the casein was labeled with a radioactive phosphate, two of the peptide fractions, namely fractions 4 and 7, indicated radioactivity and were subjected to analysis by MALDI to yield the data shown in Figure 12-13 [224]. The data in panels A and B of Figure 12-13 show a good match between observed and calculated values for some of the expected fragments from digestion of the β-casein. Treatment of a portion of fraction 4 with alkaline phosphatase and subsequent analysis by MALDI yield the data in panel B, which shows that the peak at m/z 1014 did not shift, whereas that at m/z 2064 shifted to 1984, a shift of 80 m/z units, confirming that this peptide contained a phosphate group at S35 and that serine is the only phosphorylatable site in that particular peptide fragment, as it had previously been determined that phosphorylation does not occur at threonine in bovine β-casein.

Interpretation of the MALDI mass spectrum (panel C in Figure 12-14) obtained from another radioactive fraction (fraction 7) was not as straightforward. At first the origin of the peak at approximately m/z 3150 was not clear as it did not agree with the calculated mass for any expected protonated peptide from the enzymatic digestion. However, upon treatment of an aliquot of fraction 7 with alkaline phosphatase and subsequent analysis by MALDI MS to obtain the spectrum in panel D (lower right-hand panel), a significant peak at m/z 2804.8 appeared. Because losses due to phosphates are quantized by 80 Da, it was assumed that the peak at m/z 2804.8 represented a species that had lost 320 Da (4 × 80 Da) as this was the approximate mass shift from the intense peak observed in the spectrum in panel C of Figure 12-14. By adding 320 mass units to the peak at m/z 2804.8 observed in the spectrum in panel D, it became apparent that the minor peak at m/z 3125.2 in panel C must represent MH$^+$ (320 + 2804.8 = 3124.8), given the experimental error in this relatively early application. Therefore, the major peak at approximately m/z 3150 in the spectrum in panel C represents (M + Na)$^+$, which would be expected to be 22 mass units higher than the peak representing MH$^+$ (3124.8 + 22 = 3146.8), and the subsequent two members of the triplet starting at approximately m/z 3150 represent species in which one and two additional sodiums have replaced hydrogens.

So far, this example has shown some of the complications to data interpretation caused by sodium adducts, but it should also be clear that the peptide represented by the spectra in panels C and D in Figure 12-14 contains four phosphate groups because of the shift of 320 m/z units in the spectrum in panel D. The protonated peptide having a mass of 3124.8 Da contains five serine residues and one threonine residue; because threonine is known not to be phosphorylated in bovine β-casein, four of the five residues of serine are phosphorylated. The question is which four?

In an effort to recognize which four of the five serines were phosphorylated, the peptide indicated in panel C of Figure 12-14 was digested further with endoproteinase Glu-C, an enzyme that cleaves on the carboxyl side of glutamic acid. Subsequent analysis of the peptide mixture resulting from secondary digestion with Glu-C by MALDI is shown in panel E (upper right panel) of Figure 12-14. This MALDI spectrum shows a peak at m/z 1059.8, which corresponds to a mass calculated for the phosphorylated peptide fragment shown in panel E. These data give direct evidence that all four serines in the fragment corresponding to amino acid residues 13–21 are phosphorylated; S_{22} is excluded from this peptide as it was cleaved during digestion with endoproteinase Glu-C.

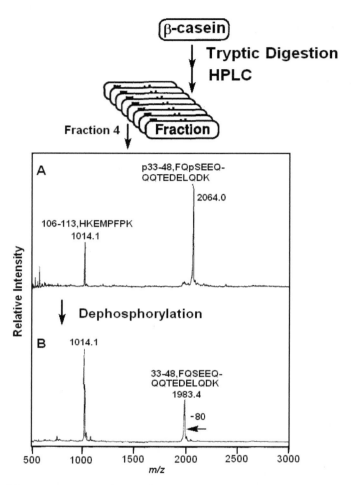

Figure 12-13. Mass spectra of HPLC fraction 4 from bovine β-casein that showed radioactivity indicating phosphorylation. Reprinted from Liao PC, Leykam J, Gage DA, and Allison J "Recognition of phosphorylation sites by MALDI mass-mapping" Anal. Biochem. 1994, 219, 9–20, with permission from Academic Press.

In summary, multiple-enzyme digestions may be needed to pinpoint the location of phosphorylated or unphosphorylated serines or threonines depending on the number of such sites and their relative locations. In an example of using this approach to recognize the site of phosphorylation in an Op-18 phosphorylate protein in leukemia cells, it was necessary to use several enzymes including trypsin, chymotrypsin, endoproteinase Glu-C, and alkaline phosphatase systematically with subsequent analysis of the digest mixtures by MALDI MS. Further, for the approach described in the above example, it is essential that the phosphate group be retained on the residues [234], a feature not always met, which leads to weak signals for the critical ions. A recent study indicates that gas-phase derivatization of phosphorylated peptides by tri-Me borate causes retention of boron at the original site of phosphorylation, thereby marking the residue as originally phosphorylated [235].

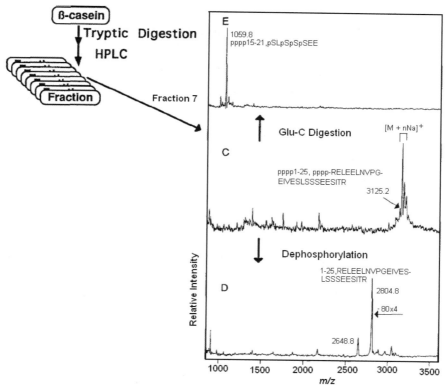

Figure 12-14. Mass spectra of HPLC fraction 7 from bovine β-casein that gave radioactivity indicating phosphorylation; this fraction was digested with Glu-C prior to analysis by MS. Reprinted from Liao PC, Leykam J, Gage DA, and Allison J "Recognition of phosphorylation sites by MALDI mass-mapping" Anal. Biochem. 1994, 219, 9–20, with permission from Academic Press.

An automated approach to the recognition of phosphorylation, termed hypothesis-driven multiple-stage mass spectrometry, is based on the postulate that any or all of the potential sites of phosphorylation in a given protein may be phosphorylated [236]. Using this assumption, the m/z values of all the corresponding single-charge phosphopeptide ions that could, in theory, be produced proteolytically are calculated. The experimental data are tested against these calculated m/z values for the presence of phosphoserine or phosphothreonine residues using tandem mass spectrometry in a vacuum MALDI ion trap where the neutral loss of the elements of H_3PO_4 (98 Da) provides a sensitive assay for the presence of phosphopeptides [236, 237]. The Liao Group has taken the simple, but remarkably effective, approach of splitting a putative phosphoprotein sample mixture and treating one portion with phosphatase before analysis by LC/MS; a computing algorithm is then used to differentiate possible phosphopeptide signals in the data based on the mass shift generated by alkaline phosphatase treatment [238]. Another algorithm for the detection of phosphopeptides interrogates LC/MS/MS data sets from both ECD and CAD for product ions containing phosphorylated serine or threonine generated by either the neutral loss of phosphoric acid or the combined neutral loss of phosphoric acid and water [239]. The phosphorylation sites in histone H1 isoforms have been characterized by proteolytic mapping with FTMS [240].

2) Selective Capture and Detection of Phosphopeptides

Selective capture of the phosphorylated peptides, which are usually the minor constituents of a proteolytic digest, aids and simplifies the analytical procedure. Selective preconcentration of phosphopeptides by on-line immobilized metal affinity chromatography (IMAC) offers analytical advantage [241, 242], including an automated system for sample processing [243]; subsequent separation of the phosphopeptides can be accomplished by capillary electrophoresis with identification by ESI MS/MS [226, 244], although attention must be given to the influence of eluent pH on the recovery of phosphopeptides from IMAC media [245]. The specificity of the method can be optimized for P-His-containing peptides [246], although a poor collection of phosphopeptides containing only one phosphate may occur. In addition, there may be a nonspecific collection of peptides that are rich in Asp and Glu, which tend to have a net negative charge, a problem that can be minimized by esterifying the carboxylic groups [48, 247]. The use of TiO_2 microcolumns reportedly is more effective than IMAC, especially in combination with DHB for analysis by MALDI [248, 249]. Titansphere reportedly achieves 90% efficiency in collecting phosphopeptides from digests [250]. Electrostatic Repulsion-Hydrophilic Interaction Chromatography (ERLIC) efficiently collects, desalts, and isocratically separates phosphopeptides [251]. Metal affinity chromatography (MAC) captures phosphopeptides as ternary complexes with gallium(III) or iron(III) and N_α,N_α-bis(carboxymethyl)lysine (LysNTA); the advantages of MAC-MS/MS over some other methods of phosphopeptide capture and detection are: (1) it uses metal complexes that self-assemble in solution at pH <5, which is favorable for the production of positive ions by ESI; (2) phosphorylation at tyrosine, serine, and threonine can be detected; and (3) the mass spectral peaks for phosphopeptides are encoded with the [69]Ga–[71]Ga isotope pattern, which is especially helpful in analysis of complex mixtures [252]. The use of a metal-ligand exchange column has been used to detect phosphorylated peptides selectively by release of methylcalcein blue during analysis by LC-ESI MS [253]; a diazo-functionalized solid-phase resin has been used to isolate phosphorylated peptides from nonphosphorylated substrates [254]. A 2D chromatographic approach based on strong cation exchange, which does not retain tryptic phosphopeptides in media below pH 2.6, followed efficient capture in the second dimension of reversed-phase chromatography and substantially simplifies the sample with higher phosphopeptide recovery and less nonspecific binding of acidic peptides than some of the commonly used enrichment methods [255]. Also, phosphopeptides can be enriched and fractionated from peptide mixtures based on their differences in isoelectric points after esterification [256]. A promising technique for selective collection of phosphopeptides uses nonretentive solid-phase extraction on highly cross-linked polystyrene-divinylbenzene material (called Strata-X), which hydrophobically binds phosphopeptides regardless of their degee of phosphorylation; this strategy lends itself to selective binding to zirconium-modified MALDI targets for washing prior to direct analysis [257] or to capture by custom-made microcolumns for washing and subsequent off-line analysis by MALDI [258].

Antibodies have been used to enrich proteins phosphorylated on serine/threonine or tyrosine residues by immunoprecipation [259]. Selective capture of tyrosine-phosphorylated peptides by immunoprecipitation with an antiphosphotyrosine antibody and subsequent detection by precursor-ion scanning (detection of the immonium ion of phosphotyrosine), also called phosphotyrosine-specific immonium ion scanning, has been used to analytical advantage [260]. Laser ablation inductively coupled plasma mass spectrometry can screen spots for the presence of phosphorus in SDS-PAGE-separated protein samples for further characterization of phosphoproteins by MALDI FTICRMS as demonstrated in the analysis of protein subunits in yeast mitochondria [261].

The addition of phosphoric acid to 2,5-dihydroxybenzoic acid matrix can facilitate the detection of phosphopeptides [262]. A complementary approach using negative-ion detection for phosphopeptide-specific marker ions at m/z 63 and/or at m/z 79 has been reported [263]. Addition of phosphoric acid (0.1–1.0%) to the sample may substantially improve the detection of phosphopeptides; apparently, phosphoric acid acts as a blocking agent to available silanol groups on both the silica capillary surface and the C18-bonded stationary-phase silica surface during sample processing [264]. The strategy of using the loss of 98 Da to trigger ECD in a data-dependent manner improves the detection and characterization of trace levels of phosphorylated proteins [265].

3) Chemical Modification of Phosphorylation Sites

The phosphate group in proteins and peptides can be easily lost during analysis, as by CAD, which may comprise successful analysis by MS/MS [234, 266]. On the other hand, newer techniques such as electron capture dissociation (ECD) and electron-transfer dissociation (ETD), as described in later sections, can be used directly to recognize the site of phosphorylation [265]. Given the lability of the phosphate group under CAD and some other analytical conditions, trustworthy methods for recognizing the sites of phosphorylation have been developed based on intentional β elimination of the phosphate to produce a dehydroalanine residue from phosphoserine or dehydroaminobutyric acid from threonine [266, 267], thereby marking the residue originally phosphorylated. A variety of subsequent chemical modifications to the acrylate moiety of dehydroalanine has been used to facilitate recognition of this residue by mass spectrometry, which, in turn, distinguishes the site originally occupied by the phosphate group.

A promising approach to mapping sites of phosphorylation in proteins is based on chemical modification of such sites to dehydroalanine, but with further modification to promote proteolytic cleavage at those sites. Specifically, phosphoserine and phosphothreonine residues in the peptides are converted to the protease-sensitive *S*-2-aminoethylcysteine derivatives by β elimination followed by Michael addition of 2-aminoethanethiol [268]. The resultant lysine analogs are then cleaved with *Achromobacter* lysine endopeptidase. The predicted proteolytic fragments can be confirmed by mass spectrometry and N-terminal Edman degradation. When acetylation is carried out as a first step to block extant lysines, direct N-terminal chemical sequencing of the digests yields sequences immediately C-terminal to the phosphorylated residues. Hence, assignment of the sites of modification can be obtained from chemical sequence data or mass spectral data [268]. The same approach has been described by Knight *et al.* [269]. A different chemical modification scheme is based on phosphoramidate chemistry [270, 271].

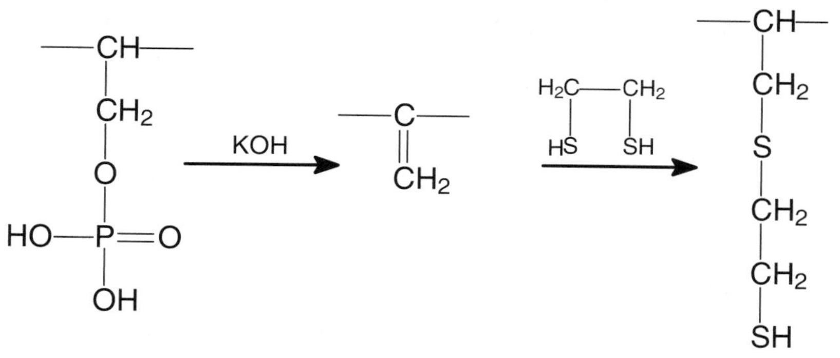

Scheme 12-5

Other methodology based on β elimination of the phosphate moiety converts phosphoserine and phosphothreonine residues to *S*-(2-mercaptoethyl)cysteinyl and β-methyl-*S*-(2-mercaptoethyl)cysteinyl residues, respectively, after β elimination and condensation with 1,2-ethanedithiol [272], leaving a free sulfhydryl group linked to the residue originally phosphorylated as indicated in Scheme 12-5. The newly installed free sulfhydryl (shown at the right in Scheme 12-5) is then conjugated with biotin-HPDP (*N*-[6-(biotinamido-hexyl]-3'-(2'-pyridyldithio)propionamide, product #21341 from Pierce Chemical Co.) to give the generic biotinylated species shown at the left in Scheme 12-6. The biotinylated protein is then digested, usually with trypsin; the tryptic fragments that retain the biotin can be extracted from the complex digestion mixture by reversible affinity attachment of the biotin tag to avidin beads as illustrated conceptually in Scheme 12-6. The biotinylated tryptic fragments are the ones that contain the originally phosphorylated residues; these captured tryptic fragments are released from the affinity column by dithiolthreitol (DTT), which cleaves the disulfide linkage to biotin. The released tryptic fragment is only modestly chemically modified (contains only an ethylenedithiol modification to the dehydroalanine moiety) as shown at the far right in Scheme 12-6; this smaller remnant of the linker is less likely to adversely affect the CAD efficiency or the sequence-specific fragmentation of the tryptic peptide, as might happen if the biotin were still linked to the peptide.

Potential interference from intrinsic sulfhydryls (cysteine residues) is avoided by pretreatment of the protein with performic acid, which converts cysteine to cysteic acid (+48 Da) and methionine to sulfone (+32 Da) or to sulfoxide (+16 Da). This method offers the distinct advantage of mild chemical treatment, including release from the biotin affinity tag by reductive cleavage of the S–S bond in the conjugate side chain, leading to a modest shift in mass (<100 Da) of the peptides, so that they can be readily analyzed by LC/MS/MS for automated identification by database searching. The methodology has been tested sucessfully in the recognition of all known phosphorylation sites in a mixture of α-casein, β-casein, and ovalbumin. With the use of d_0- and d_4-ethane dithiol, this methodology could be readily adapted for quantitative phosphoproteomic applications. An improvement to this methodology has been described in which EDTA is used to minimize a side reaction in which water is eliminated from unmodified serine residues [273].

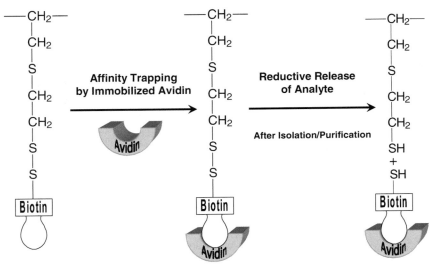

Scheme 12-6

A variation of the above-described theme for recognizing sites of phosphorylation involves chemical modification of the β-elimination product with a 1:1 mixture of d_0- and d_3-methylamine [274]. Following digestion of the now chemically modified protein, the proteolytic fragments containing the originally phosphorylated residues can be recognized during analysis by ESI ion trap analysis with high-resolution zoomscan [275] or MALDI PSD/LIFT [242], allowing easy identification of the presence and determination of the number of such residues in the peptides. Another isotope mass-shift method is based on the same preliminary chemistry, but with the final use of fully deuterated DTT to introduce a 6-Da difference with respect to the unlabeled species [276].

Another method based on β elimination of the original phosphorylated serine residue uses subsequent sulfite addition to transform the dehydroalanine residue to cysteic acid ($-SO_3^-$); in this way, a phosphorylated threonine is converted (in 80% yield) to β-methylcysteic acid [277]. Both cysteic acid and β-methylcysteic acid residues in the sequence were shown to be stable and easily identifiable under general conditions for CAD MS/MS sequencing applicable to common peptides [277]. A related approach using chemical modification by ethanethiol was part of a multitechnique approach to phosphate mapping to characterize the phosphorylation of protein kinase A [278].

These chemical methods for recognizing phosphorylation sites offer great improvement over earlier methods that required retention of the phosphate group during analysis, a condition rarely met with satisfaction [225, 266, 279]. However, screening for tyrosine phosphorylation reportedly can be accomplished by monitoring the phosphotyrosine immonium ion at m/z 216 during ESI MS/MS [280]. A multidimensional LC/MS method has been used to identify 64 unique sites of tyrosine phosphorylation on 32 different proteins from human whole-cell lysates in various states of health and disease [281]. New instrumental methods for detection of phosphopeptides based on the use of ICPMS with selective detection of ^{31}P have the advantage of good quantitative assessment of the distribution of phosphate among several different sites [282]; related methodology using laser ablation of protein spots on a transfer membrane from 1D PAGE allows the quantitative assessment of phosphorus in model proteins such as α-casein [283]. Ion mobility mass spectrometry can be used for rapid screening of phosphopeptides in protein digests because the drift time–m/z relationship for single-charge phosphorylated peptide ions is different from that for nonphosphorylated peptide ions [284].

In an unprecedented application of reasonably conventional mass spectrometric methodology not involving the chemistry described above, over 2000 phosphopeptides corresponding to nearly 1000 phosphoproteins were identified in a whole-cell lysate, which allowed a detailed accounting of known and unknown kinase motifs and substrates in the nucleus of HeLa cells [49]. This heroic effort was achieved by concentrating the nuclear fraction of the whole-cell lysate by strong cation exchange after differential centrifugation, simplifying the complexity of the nuclear fraction by SDS-PAGE, subjecting selected excised bands to in-gel digestion by trypsin, and then analyzing the de-stained extract by LC-ESI MS and CAD MS/MS. Phosphopeptides were recognized by detection of a major mass spectral peak 98 mass units (or multiples thereof) lower in the CAD MS/MS spectrum than that detected in the corresponding MS spectrum; phosphopeptides readily lose the phospho-group during CAD MS/MS (mass of $H_3PO_4 = 98$ Da). The CAD MS/MS spectrum of the ion current corresponding to the protonated dephosphorylated species (MH^+ of 98, etc.) is used to confirm the identity of the phosphopeptide [285]. The site of phosphorylation can be determined if a **b** or **y** ion is detected that corresponds to the dehydroalanine residue in the dephosphorylated peptide during analysis by CAD MS/MS; the route of dehydroalanine formation from a phosphorylated residue is illustrated in Scheme 12-5. Negative-ion CAD of putative phosphopeptide produces characteristic abundance ratios of fragment ions from consecutive Ser/Thr residues allowing sites of phosphorylation to be pinpointed [286]. Quantitative comparisons in the degree of phosphorylation between two different samples of the cells can be determined when cells of one of the two samples are grown in a medium containing stable isotope-labeled amino acids [287, 288].

Phosphorylation site identification has been achieved via ion trap tandem mass spectrometry of a whole protein [289]; analysis of which backbone cleavage products produced by CAD (as identified by ion/ion reactions) contain the phosphate was used to recognize the site of phosphorylation on one of two possible serine residues in bovine α-crystalline A. Development of the ion/ion reaction-based methodology has spawned the technology of electron-transfer dissociation (ETD) as described in a subsequent section; ETD is exquisitely well suited for the determination of sites of phosphorylation [290, 291].

The degree of phosphorylation is also of interest in a biological experiment. In one such experiment, the protocol involved Fischer methylesterification and enzymic (phosphatase) modification steps in concert with d_4-methanol as an accessible isotopic labeling reagent [292]. The use of a phosphoprotein isotope-coded affinity tag (PhIAT), which employs differential isotopic labeling and biotinylation, has been shown to be capable of enriching and identifying mixtures of low-abundance phosphopeptides [293].

B. Recognition of Sites of Sulfation

The fact that $-SO_3$ is isomass with $-OPOOH$ (each has a mass of 80 Da) creates a problem in distinguishing sulfo- from phosphopeptides by the simple mass-shift method; however, because the actual mass difference of these two functionalities is 9.5 mDa, mass spectrometers with good resolving power can distinguish such modified peptides or proteins of reasonable size [294]. Conventional techniques used to capture phosphopeptides can also capture sulfonated peptides [271]; disturbingly, a high fraction of gene products identified as being sulfonated have also been identified as being phosphorylated [295]. Interestingly, a method to distinguish carbohydrates that are sulfated versus those that are phosphorylated has been developed based on the use of an ion-pairing reagent in conjunction with ESI [296]. For peptides/proteins, additional techniques are required to distinguish between these two post-translational modifications.

Under prescribed conditions of in-source dissociation and MS/MS, fragmentation signatures have been described that allow phosphorylation to be distinguished from sulfation [297]. In an MS/MS study of the fragmentation of a set of commercially available sulfo- and phosphopeptide standards by using positive-ion nanoelectrospray and quadrupole time-of-flight mass spectrometry, the sulfo-modification was found to be more labile than phospho-modification. All of the phosphorylated peptides retained their +80 Da modifications under CAD conditions and peptide backbone fragmentation detected by MS/MS allowed for the site-specific identification of the modification. In sharp contrast, sulfated peptides lost SO_3 from the precursor as the collision energy was increased until only the nonsulfated form of the peptide was observed. The number of 80 Da losses indicated the number of sulfated sites. By continuing to increase the collision energy further, it was possible to fragment the nonsulfated peptides and obtain detailed sequence information. However, it was not possible to obtain site-specific information on the location of the sulfate moieties using positive-ion MS/MS as none of the original precursor ions were present at the time of peptide backbone fragmentation. This method was applied to the analysis of recombinant human B-domain deleted factor

VIII (BDDrFVIII), which has six well-documented sulfation sites and several potential phosphorylation sites located in two of the sulfated regions of the protein. Seven peptides with single and multiple +80 Da modifications were isolated and analyzed for their respective PTMs. The fragmentation patterns obtained from the BDDrFVIII peptides were compared with those obtained from the standard peptides. In all cases, the peptides were sulfated; none of the potential phosphorylation sites was found to be occupied. These results are consistent with the literature [297]. Another method enhances the mass spectral signal of mono- and disulfated glycopeptides present in glycoproteins that contain many other nonsulfated glycoforms; the method utilizes the tripeptide Lys–Lys–Lys as an ion-pairing reagent to complex selectively with sulfated species and enhance their ion signal during analysis [298].

C. Recognition of Sites of Glycosylation

Recognition of sites of glycosylation by conventional methods is based on proteolytic mapping, ideally designed so that each glycosylation site is located within a separate peptide [299]. Following isolation of the putative glycopeptides by HPLC and subsequent analysis by ESI-CAD MS/MS, recognition of a particular glycopeptide is achieved by detecting a peak for a characteristic glycan backbone product ion; e.g., at *m/z* 163 (Hex), *m/z* 204 (HexNAc), *m/z* 292 (sialic acid), or *m/z* 366 (HexHexNAc) [279, 300–302]. Occasionally, adequate fragmentation of the glycopeptide backbone can be achieved during CAD producing **b** and **y** ions, the mass shift of which permits the specific site of glycosylation to be recognized. When glycosylation is heterogeneous, however, these approaches can result in spectra that are complex and poorly resolved. Methodology has been developed, based on precursor-ion scanning for ions of high *m/z*, that allows site-specific detection and structural characterization of glycans at high sensitivity and resolution, as demonstrated for the standard glycoprotein, fetuin, and subsequently applied to the analysis of the N-linked glycans attached to the scrapie-associated prion protein [303]. The high resolution furnished by a MALDI-Qq-TOF facilitates structural characterization of site-specific N-glycosylation from limited quantities of material as demonstrated in the analysis of a novel proteinase from tomato, which revealed heterogeneity at different levels, including different glycan side-chain modifications, and heterogeneity of oligosaccharide structures on the same glycosylation site [304]. These techniques have been applied to the characterization of nephrin, a type-1 transmembrane glycoprotein that serves as a principal component of the glomerular filtration barrier [305].

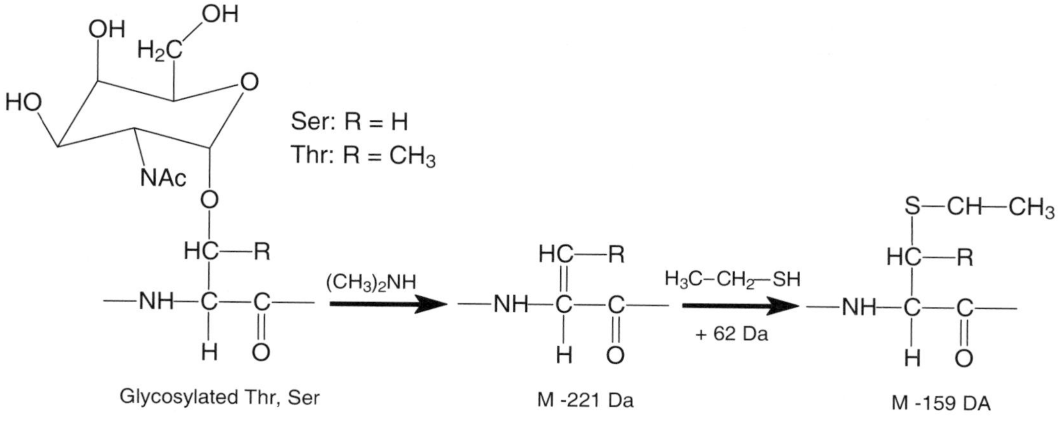

Scheme 12-7

Unfortunately, glycopeptides are not so easily ionized as the corresponding unmodified peptides [299]. Thus, the conventional proteolytic mapping approach is somewhat limited. In cases where the glycopeptide can be ionized, it often undergoes efficient gas-phase deglycosylation during CAD, thereby shedding its site-specific mass shift in the process of forming the product ions [306]. Because of this difficulty, some workers have taken the indirect approach of removing the glycan with peptide N-glycosidase F, sometimes in the presence of ^{18}O-enriched water, with subsequent conversion of the asparagine residue to aspartic acid before analysis by ESI-CAD MS/MS [307, 308].

A novel chemical modification methodology has been introduced for the determination of O-glycosylation sites in mucin-type glycoproteins [139]. The chemical modification involves intentional β elimination of the glycan to form the corresponding dehydroalanine residue at the former site of glycosylation, which subsequently is treated with ethanethiol as illustrated in Scheme 12-7. (Because this reaction will also cause β elimination of O-linked phosphorylation sites, it is necessary to make a preliminary cut of the sample according to glycosylation vs phosphorylation.) Following proteolytic digestion, the glycopeptides are further modified by fixed-charge derivatization of the N-terminus with a phosphonium group to promote **b**-ion formation during CAD MS/MS [87]. The effectiveness of this analytical approach can be appreciated by comparing the mass spectra of the nonglycosylated and formerly glycosylated (now chemically modified by β elimination in the presence of dimethylamine and subsequent condensation with ethanethiol) peptide in Figure 12-15, which shows that the peaks for the b_9 ion (also for b_{10} and higher) shift by 44 *m/z* units, allowing facile recognition of the threonine at position 9 as being the site of glycosylation in this threonine-rich glycopeptide [139]. The mass shift of 44 Da corresponds to the mass difference between a threonine side chain {–CH(CH$_3$)–OH} and a β-elimination/ethanethiol-modified threonine side chain {–CH(CH$_3$)–S–CH$_2$CH$_3$}. Related methodology has been described by Rusnak *et al.* [268].

As described more extensively in a later section on "Top-Down Strategies", Stephenson and McLucky [309] have developed a sequencing approach to analyze the intact protein by ESI-CAD MS/MS in a quadrupole ion trap mass spectrometer. Their method involves selecting a particular high-charge-state precursor ion for fragmentation, often through multiple stages of MS/MS, until sufficient product ions are generated to permit identification and/or characterization of the protein. This promising approach, as described in a later section, takes advantage of the fact that the efficiency of CAD scales directly with the charge state of an ion, offering the possibility of characterizing intact proteins without the need for extensive purification and/or proteolytic digestion prior to analysis [310]. Again, development of the technology of electron-transfer dissociation (ETD) as described in a subsequent section promises to aid efforts to determine sites of glycosylation in peptides and proteins [311, 312]. Related methodology using CAD and ECD can be used to assign the site of the glycan without its release by chemical means [313].

D. Acetylation of Lysine

Post-translational acetylation of lysine in proteins is sometimes difficult to recognize because the acetylated lysine immonium ion of 143.1 Da can be confused with internal fragment ions from some peptides, producing false positive results. A more reliable diagnostic ion of 126.1 Da, generated by secondary loss of NH$_3$ from the acetylated lysine immonium ion, is also approximately 9 times more sensitive to detection than the immonium ion [314]. The utility of this method was demonstrated with acetylated cytochrome c as a model compound [314].

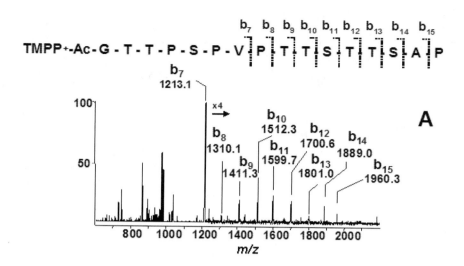

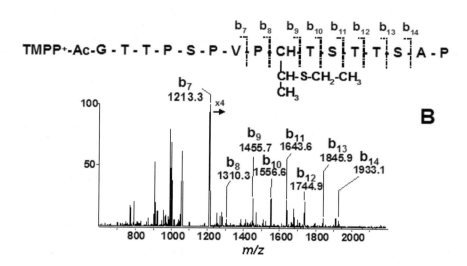

Figure 12-15. *Nanospray-CAD MS/MS spectra of the protonated molecular cations [C+H]²⁺ at m/z 1038 and 1060, corresponding respectively to the Ac-TMPP derivative of (A) deglycosylated GTTPSPVPTTSTTSAP and (B) native GTTPSPVPTGalNAcTSTTSAP after β elimination in the presence of dimethylamine and subsequent condensation with ethanethiol [139].*

E. Cysteine Status in Proteins

1) Are There Any Disulfide Bonds?

Disulfide bonds formed by linking two cysteine residues into a cystine constitute a very important form of post-translational modification. In characterizing a protein of known amino acid sequence and thus a designated number of cysteine residues, it is important to determine whether any of the cysteines exist as cystines. Because the mass difference between a cystine and two cysteines is only 2 Da, chemical modification of the cysteines to exaggerate the mass difference between the reduced and oxidized form of especially large proteins may be desirable [315]. One means of selectively derivatizing sulfhydryl groups in a protein involves using p-hydroxy mercuric benzoate (pHMB), which causes a mass shift of 321 Da for each pHMB added to the protein [316]. Derivatization of any free cysteines in a protein or peptide with 4-dimethylaminophenylazophenyl-4'-maleimide (DABMI) facilitates recognition of the analyte both by UV detection during HPLC and by prompt fragmentation during MALDI [317]. In a related method, a water-soluble thiol-specific fluorogenic reagent, ammonium 7-fluoro-2,1,3-benzoxadiazole-4-sulfonate, is used to facilitate detection of cysteinyl peptides [318]; fluorogenic derivatization that also installs a fixed positive charge facilitates detection by mass spectrometry [319]. The negative-ion mass spectra of cystinyl proteins show a characteristic loss of the elements of H_2S_2, a process diagnostic of the presence of the disulfide moiety [320, 321].

An appreciation for the multiplicity of isomeric structures associated with various redox states of a cysteinyl peptide or protein can be gleaned from Figure 12-16. A peptide or protein containing four cysteines could be completely oxidized to contain two disulfide bonds but, as shown in the bottom row of simple structures in Figure 12-16, there are three different ways to connect two pairs of cysteines (i.e., there are three isomeric disulfide structures for the completely oxidized compound). If the compound is only partially oxidized, there are six isomeric structures containing a single disulfide bond.

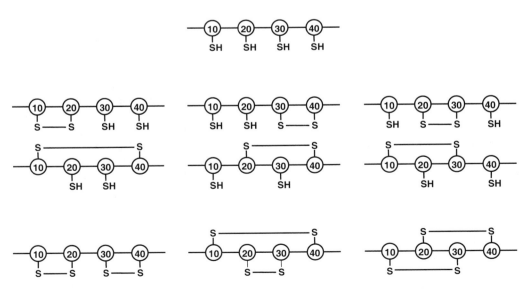

Figure 12-16. Multiplicity of isomeric disulfide structures associated with various redox states of a cysteinyl polypeptide.

In combination with enzymatic digestions and analyses by HPLC, ESI and MALDI can be used to verify the amino acid sequences and identify and locate post-translational modifications involving disulfide-bond formation in proteins [322–324]. Direct analysis of a cystinyl protein rarely allows a determination of the connectivity of disulfide bonds, although some limited success has been reported for two of ten cystines [325]. Collisionally activated multiple-charge anions of cystinyl proteins have been observed to fragment preferentially at the disulfide bonds, which can break at any of three possible locations within and around the sulfur–sulfur bond, giving rise to several products with different partitioning of sulfur atoms [326]; however, the results provide no indication of which cysteines were connected to constitute the original disulfide bonds. Similar fragmentation under MALDI PSD conditions has been reported for positive ions of disulfide-bonded species [327], which have allowed key proteolytic fragments to be identified, but have limited success in disulfide mapping. Technology involving broadband precursor selection has been found to facilitate distinguishing inter- and intramolecular disulfide bonds based on characteristic signature fragmentation following cleavage of cysteine side chains, as demonstrated in partial mapping of disulfide bridges in a 37-kDa protein containing 16 cysteines and complete disulfide mapping of a lysozyme (14.5 kDa) containing 8 cysteines [328].

2) Which Cysteines Are Free?

Methodology based on cyanylation of cysteine and mass mapping allow recognition of the location of free cysteines in a protein [329]. This mapping methodology is unusually direct and simple because the cyanylation reaction is selective for cysteine, and subsequent cyanylation-induced cleavage cuts the peptide backbone only on the N-terminal side of modified cysteines. Therefore, if a protein contains three free cysteines, cyanylation-induced cleavage will generate only four fragments from the entire protein (if additional cysteines are present as disulfides, a reduction step will be necessary to cut any disulfide linkages between fragments prior to analysis after cyanylation). This mapping strategy and methodology has been used to determine sites of cysteine nitration in tyrosine hydroxylase during mechanistic studies of enzyme activity [330].

3) What Is the Linkage of Cysteines in the Disulfide Bonds?

(A) Conventional Proteolytic Mass Mapping of Disulfides

Determination of the connectivity of disulfide bonds in a peptide or protein is essential for complete characterization. Like classical chromatographic approaches to disulfide-bond mapping, those involving mass spectrometry usually involve controlled degradation of the cystinyl protein with a protease [331–333], which is selected to cleave the polypeptide backbone at least once between the cysteines. This ideal situation is nearly achieved in Figure 12-17, which is a conceptual sketch of the principal steps involved in the classical proteolytic approach to disulfide-bond mass mapping.

During the analytical procedure represented in Figure 12-17, the proteolytic enzyme, E, cleaves the polypeptide backbone of the hypothetical cystinyl protein at five sites, including in between all but the second and third cysteines, producing the five proteolytic fragments shown at the upper right. Three of the six expected fragments are detected during analysis of the hypothetical proteolytic digest by MALDI mass spectrometry. The first mass spectrum also shows two peaks at *m/z* values that were not expected, but it can be seen that this is because two of the expected proteolytic fragments are connected by an intermolecular disulfide bond; the other fragment contains an intramolecular disulfide bond and therefore is represented by a peak two mass units lower than the expected mass, assuming the cysteines are free.

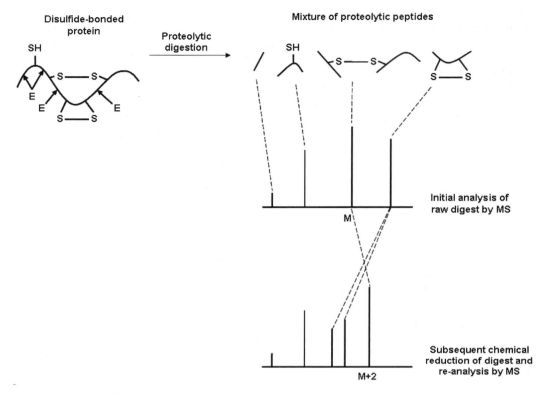

***Figure 12-17. Conceptual representation of requirements and procedural
steps in proteolytic mass mapping of cysteine status in a
protein.***

The second mass spectrum in Figure 12-17 results from reanalysis of the
hypothetical proteolytic digest after it has been treated with a reducing reagent. Now
there are mass spectral peaks for all of the originally expected proteolytic fragments,
assuming that the cysteines are free. Of course, the reason is apparent from the sketch,
which shows reductive cleavage of the intermolecular disulfide bond holding two of the
expected fragments together; the mass spectral peak representing the disulfide-bonded
pair of proteolytic fragments is absent in the second spectrum. The disappearance of the
mass spectral peak for the disulfide-linked fragments, coupled by the appearance of
mass spectral peaks for the individual originally expected fragments, identifies the two
cysteine residues that are connected to one another in the cystinyl protein. The shift of
the mass spectral peak for the proteolytic fragment by two *m/z* units after reduction
confirms that this fragment did contain a disulfide bond, and because there are only two
cysteines, they had to have been connected to one another in the original cystinyl
protein. An algorithm can aid the data interpretation process [334]. Analyses by MALDI
using a reducing matrix such as 1,5-diaminonaphthalene may be helpful in conjunction
with the classical enzymatic approach to disulfide mapping [335].

Ideally, the classical proteolytic method for disulfide mapping relies on at least
one cleavage site between all the cysteines, a constraint that becomes quite serious if
the cysteines lie close to one another in the amino acid sequence. The proteolytic
method is nearly impossible if the cysteines are adjacent in the sequence as no known
protease cleaves at cysteine, although a recent report indicates that chemical

modifications of cysteine (aminoethylation with bromoethylamine or N-(iodoethyl)-trifluoroacetamide, and subsequent guanidination) and lysine (acetylation) prior to tryptic digestion release peptides delineated by cysteine or arginine residues [336]. The proteolytic approach, especially when combined with alkylation of any free cysteines, was used to characterize important classes of cystinyl proteins such as the conotoxins [337]; similar methodology was used to characterize the reversible formation of a selenenylsulfide linkage in mammalian thioredoxin reductase [338]. Characterization of intermolecular disulfides is facilitated by the use of negative-ion ESI-CAD MS/MS for identification of disulfide-linked peptides, as demonstrated in the analysis of a tryptic digest of human 1gG2 antibody to identify seven unique disulfide linkages [339].

However, even in those cases in which the analyst is fortunate enough to find proteolytic sites between all cysteines, the undesirable process of disulfide scrambling is a serious concern because most proteases function optimally above pH 8. Because a sulfhydryl group (RSH) has a pK_a around 10, approximately 1% of it will have dissociated at pH 8 into the thiolate (RS⁻), a reactive nucleophile. As illustrated in Scheme 12-8, the thiolate located at position C in the generic cystinyl protein can attack an extant disulfide bond between cysteines A and B, and cause the formation of a new disulfide bond between cysteines B and C. This reversible process, called disulfide scrambling, leads to artifactual data during attempts to elucidate the disulfide structure of the original cystinyl protein.

In a typical proteolytic digest, the vast majority of fragments provides no useful information about the cysteine status in the original protein. The analyst must find the few cysteinyl and cystinyl fragments, the identification of which can be used to deduce the connectivity of certain cysteines in the original disulfide bonds. A process for enrichment of cysteinyl peptides has been described that is not dependent upon molecular mass, a particularly useful aspect for improving the efficiency of comprehensive proteome-wide analyses [340]; another approach oxidizes the cysteine residue into its highly negative sulfonic form, which facilitates enrichment of such peptides by strong cation chromatography [341]. Borges and Watson [317] described methodology that facilititates isolation and identification of cysteine-containing peptides after reaction with 4-dimethylaminophenylazophenyl-4'-maleimide (DABMI); the DABMI peptides are distinguished by their UV response during separation by HPLC and their MALDI spectra are characterized by "signature sets" of fragment ion peaks. Alternatively, derivatization of the thiol side chain by 1,5-I-AEDANS (5-({2-[(iodoacetyl)amino]ethyl}amino)naphthalene-1-sulfonic acid) provides a means of

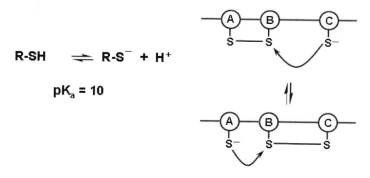

Scheme 12-8

isolating cysteinyl peptides on an *o*-nitrobenzyl-based photocleavable resin more effectively than by IMAC for subsequent analysis by tandem MS [342]. The negative-ion mass spectra of cysteine-containing peptides may also be useful for identifying such components in complex digests [343]. A recent report based on negative-ion detection indicates that electron detachment dissociation (EDD) causes preferential cleavage of S–S and C–S bonds in contrast to previous reports based on positive-ion detection [344]. Results of a recent computational study indicate that high-energy CAD will be required to promote cleavage of S–S and C–S bonds in cystinyl peptides [345]. Results of an experimental study of disulfide-bonded peptides show that complexation with a transition-metal ion promotes structurally diagnostic fragmentation of the analyte [346]. An algorithm can facilitate determination of cystines from the results of proteolytic disulfide mapping [347].

In addition to direct analysis of proteolytic digests by MALDI, post-source decay (PSD) has been reported to generate a series of single-charge fragment ions that, in addition to the peptide sequence ions, provide useful information for assigning a disulfide arrangement in highly bridged disulfide-linked peptides [327, 332]. The technique of partially reducing cystinyl proteins [348] followed by differential alkylation has been successful in some cases [349, 350]. In another fortunate case, LC/MS/MS has been used to provide sufficient fragmentation to assign cysteine connectivity even in the case of adjacent cysteines [351]. Other classical approaches involve chemical degradation with cyanogen bromide [352]. A method for facilitating recognition of disulfide-linked peptides has been reported based on reductive fragmentation of these species following MALDI and detection of the individual peptide fragments by an MS/MS technique [353]; however, it is likely that this approach will not be applicable to cystinyl proteins containing adjacent cysteines, as it indeed was not applicable to the adjacent cysteines in BSA [353]. Methodology for direct evidence of connectivity of cysteines is needed to bolster the common procedure of assigning disulfides by homology [354].

(B) Cyanylation-Based Mass Mapping of Disulfides

A novel chemical approach to disulfide-bond mapping has been developed by Wu and Watson [355, 356] that involves cyanylation(CN)-induced cleavage on the N-terminal side of cyanylated cysteines (see Scheme 12-9) to yield degradation products that can be analyzed by desorption/ionization (DI) for mass mapping of the peptide or protein. By combining this cyanylation approach with the technique of partial reduction (*vide infra*) of proteins containing more than one disulfide bond, it is possible to deduce the connectivity of cysteines involved in a given disulfide bond [355, 357]. Furthermore, the cyanylation procedure is conducted at low pH to avoid disulfide-bond scrambling or exchange [358, 359]; the robust nature of the cyanylation-induced cleavage/mass mapping methodology has been validated by computational analysis [360].

The hypothetical analytical example illustrated in Scheme 12-10 helps provide familarity with the rationale and analytical strategy involved using the CN-induced mass mapping methodology; results of a recent study indicate that methylamine is preferable to ammonia for the cleavage reaction [359]. If the protein consisting of 53 residues, six of which are cysteines, were to be characterized, one possibility would be that the molecule contains one disulfide bond as represented by (A). The fact that the molecule contains only one disulfide bond (or four free cysteines in this case) can be ascertained by cyanylating the compound (B) and determining the shift in its molecular weight. Upon cyanylation, the molecule will increase in mass by 100 Da (the cyano group has a mass of 26 Da, but it replaces a hydrogen to give a net shift in mass of 26 – 1 = 25 Da for each free sulhydryl group); this mass shift of 100 Da indicates that four free cysteines must be present in the protein. The remaining two cysteines can be assumed to be tied up in a disulfide bond.

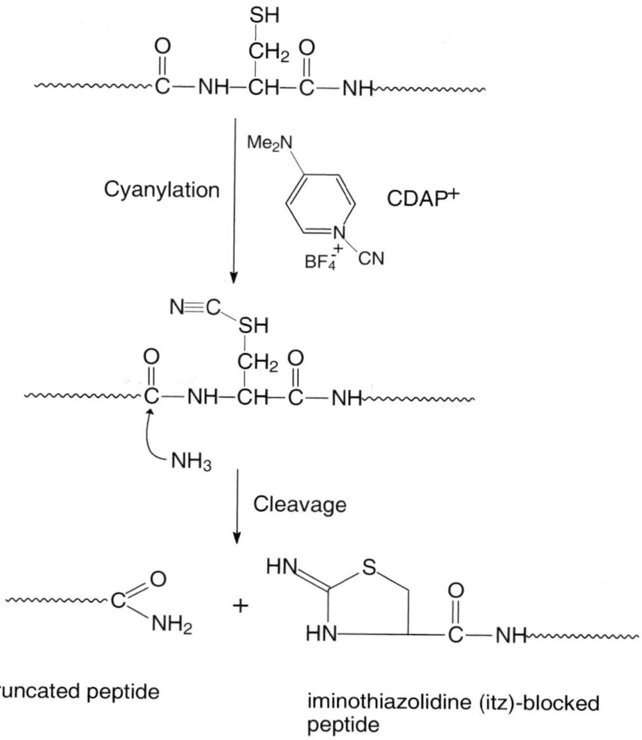

truncated peptide   iminothiazolidine (itz)-blocked peptide

Scheme 12-9

The identity of the two cysteines involved in the disulfide bond of the protein in Scheme 12-10 can be deduced from recognizing the positions of the four free cysteines, which can be determined by mass mapping the CN-induced cleavage products. Exposing the cyanylated protein (B) to nucleophilic attack by ammonia [358] causes cleavage at the N-terminal side of the four modified cysteines giving rise to the five fragments represented at the bottom of the scheme. One of the fragments will be a simple peptide in this case, one consisting of residues 1–5. Each of the other four fragments will have an iminothiazolidine (itz)-modified amino terminus. Agreement between the calculated mass of possible fragments with those experimentally determined to be present in the reaction mixture allows the positions of the four free cysteines in the original protein to be deduced; therefore, in this simple example, the positions of the two cysteines involved in the disulfide bond can also be deduced.

If all of the cysteines in the protein are in the form of cystines (disulfide bonds) as represented by the hypothetical structure at the top of Scheme 12-11, no cyanylation is possible because the reagent, CDAP, only reacts with free sulfhydryl groups. The protein must be reduced to furnish free cysteines, which will react with the cyanylating reagent. However, if the protein is completely reduced, all information relating to the connectivity of cysteines in the cystines will be lost (remember that for a protein containing two disulfide bonds there are two other disulfide structures that are isomeric with the generic structure shown at the top of Scheme 12-11). The best procedure is to reduce the protein partially [348]. Partial reduction is accomplished by providing a stoichiometric excess of reducing agent (preferably a phosphine), but under kinetically limiting conditions of pH and/or temperature. It is necessary to find such reduction conditions (by

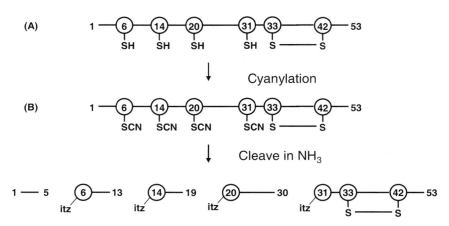

Scheme 12-10

trial and error because each protein is different) from which it can be determined that approximately half of the protein has not been chemically altered (reduced) in any way; under these conditions, most of the other half of the protein molecules will be singly reduced (a small amount will be doubly reduced). The two possible single-reduced isoforms of the protein consisting of two disulfide bonds are shown at the bottom of Scheme 12-11; each single-reduced isoform possesses two free sulfhydryl groups, but each pair originates from reduction of a different disulfide bond in the original cystinyl protein shown at the top.

The chemistry shown in Scheme 12-9 can be applied to each of the single-reduced isoforms shown in Scheme 12-11. Subsequent analysis of the CN-induced cleavage products of each cyanylated single-reduced isoform will provide direct information on the connectivity of the two cysteines involved in the corresponding disulfide bond.

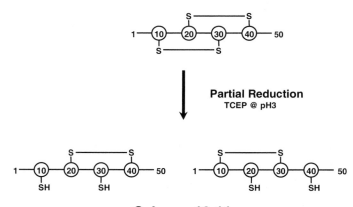

Scheme 12-11

The experimental data that are shown in Figure 12-18 are from analysis of ribonuclease-A (RNase-A), a protein containing four disulfide bonds rather than two as shown in the generic structure at the top of Scheme 12-11, but the strategy is the same; namely, analysis of a single-reduced isoform will provide direct information on the connectivity of two specific cysteines involved in a particular disulfide linkage. The MALDI mass spectrum in Figure 12-18 is of the mixture of CN-induced cleavage products obtained from one of the cyanylated single-reduced isoforms of RNase-A. The relative positions of all eight cysteines in RNase-A are represented just below the mass spectrum. For purposes of expediency, it is assumed that the disulfide bond between Cys 26 and Cys 84 had been reduced to generate this isoform and that the resulting sulhydryls had been cyanylated as shown in Figure 12-18. The mass spectrum is examined for evidence (as described in the next paragraph) that this is the case. After completing the cyanylation of Cys 26 and Cys 84, it is realized that some of the residual disulfide bonds involving cysteines 40, 58, 65, 72, 95, and 110 might still hold some of the CN-induced cleavage products together. Therefore, the cleavage reaction mixture must be treated with a reducing agent (usually a phosphine) to ensure that the anticipated three cleavage fragments (indicated at the lower left in Figure 12-18) will be free [355, 356].

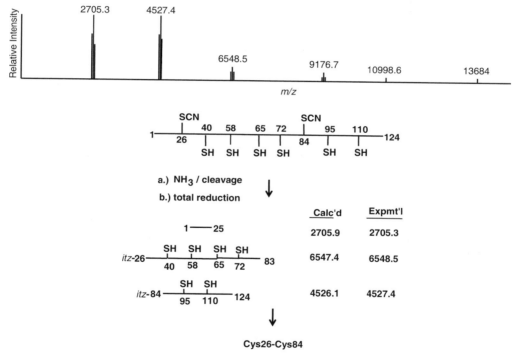

Figure 12-18. Mass mapping the cleavage fragments from a cyanylated single-reduced isoform of RNase-A.

The calculated mass of the expected cleavage fragment corresponding to residues 1–25 is 2705.9 Da; the peak at m/z 2705.3 in Figure 12-18 provides good experimental evidence that this species is present in the cleavage reaction mixture. The calculated mass of the expected CN-induced cleavage fragment consisting of residues 26–83 with residue 26 modified as the iminothiazolidine (itz) derivative is 6547.4 Da; the presence of a peak at m/z 6548.5 provides reasonable evidence that this fragment is present in the reaction mixture. The third expected fragment consisting of itz-84–124 is represented by the peak at *m/z* 4527.4. The peak at *m/z* 9176.7 corresponds to residues 1–83 in which there was cyanylation at residue 26, but no cleavage because of β elimination at that site. The peak at *m/z* 10,998.6 corresponds to itz-26–124 in which there was cyanylation at residue 84, but no cleavage because of β elimination at that site. These latter two peaks corroborate information gleaned from the former three peaks. Taken together, these five peaks provide convincing evidence that Cys 26 and Cys 84 were available for cyanylation and cleavage in this single-reduced isoform of RNase-A, and thus these two cysteine residues must have been joined in a disulfide bond in the original protein molecule [355].

While the mass accuracy and resolution illustrated in Figure 12-18 are not representative of those achievable by modern instrumentation, this example provides a didactic introduction to the partial reduction/cyanylation/cleavage/mass mapping methodology. This novel mass mapping approach to disulfide-bond analysis is applicable to peptides and proteins involving adjacent cysteines [361] in their primary structure, and offers new hope to protein chemists studying tightly knotted proteins that are refractory to conventional methodology [357]. The development and implementation of an algorithm for elucidating a disulfide structure given an input of amino acid sequence and mass spectral data from cyanylation-induced cleavage of the protein obviates the need to isolate single- or double-reduced isoforms of the original cystinyl protein [362]. The utility of the algorithm has been validated in the analysis of a cystinyl protein containing 12 cysteines in the form of six disulfide bonds for which there are 10,395 isomeric disulfide structures; use of the algorithm reduced the number of theoretically possible structures to three, two of which were eliminated by manual interpretation of the data [363]. Variations on the cyanylation theme that also include alkylation have been described [364].

Hydrogen/deuterium exchange (HDX) has been integrated with the cyanylation (CN)-based methodology to determine the conformation of cystinyl proteins and intermediates during refolding. The CN-based methodology can be used to trap, identify, and preserve the disulfide structure of a given cystinyl protein-folding intermediate, whereas the HDX methodology can be used to assess other conformational features of the intermediate [365, 366].

F. Recognition of Ubiquinated Proteins

Ubiquitination plays an important role in the degradation and functional regulation of cellular proteins in organisms ranging from yeasts to mammals. Tryptic digestion of ubiquitinated proteins produces diglycine-branched peptides in which the C-terminal Gly–Gly fragment of ubiquitin is attached to the ε-amino group of a modified lysine residue within the peptide, thereby providing the basis for diagnostic fragmentation. Methods for recognizing ubiquinated proteins as well as determining the site of ubiquination have been described using MALDI [367] or ESI [368]. A MALDI-CAD MS/MS approach involves N-terminal sulfonation of diglycine branched peptides to generate diagnostic fragments that characterize ubiquinated proteins and that can be used to map the site of ubiquination [367]. An ESI-CAD MS/MS approach uses proteolytic digestion by endoproteinase glu-C and by trypsin to characterize ubiquitinated proteins [368]. Chemical derivatization by N-terminal sulfonation of the diglycine-branched tryptic peptides retaining ubiquination modification sites promotes formation of a characteristic fragmentation pattern during CAD MS/MS that can be used to recognize sites of ubiquination in the original protein [369].

G. Other Types of Modifications

Sites of nitration in a protein have been identified by ESI LC/MS/MS [370] and by mass mapping with MALDI [330, 371, 372]. Analyses of the cysteine status in tyrosine hydroxylase shed light on the mechanism of inhibiting the biosynthesis of dopamine by this important enzyme [373]. Lipid modifications on proteins are proving to be of increasing importance in biomedical research; CAD MS/MS techniques can provide marker ions for peptides containing an N-terminally myristoylated glycine, a palmitoylated cysteine, and a farnesylated cysteine [374].

4. Quantitation in Proteomics

Simultaneous quantitative information, including the degree of turnover [375], on a large number of proteins will help make functional proteomics a reality; mass spectrometry is central to realizing this goal [40, 376–378]. For example, while mitochrondria in rat skeletal muscle, heart, and liver tissue may have qualitiative similarities, a quantitative study of 689 proteins extracted from these tissues showed striking differences in abundance [379].

A. ICATs

1) Operating Principles

The isotope-coded affinity tag (ICAT) methodology, as reviewed recently by Tureček [380] and Julka and Regnier [381], provides both qualitative and relative quantitative information about proteins contained in two different samples; e.g., the proteins in hepatic cells isolated from malignant vs normal tissue or in the assessment of enzyme activity [382, 383], a process that has been automated as a "lab on a chip" [384]. The proteins in one sample are "tagged" by a special chemical reagent; the same proteins in a second sample are tagged by a stable isotope-labeled form of the same special chemical reagent. In the original version of ICAT methodology, the special chemical reagent is one that reacts selectively with the sulfhydryl group on cysteine; because cysteine is relatively rare, only a few "tags" are likely to be chemically attached to any given protein. A variant of the original ICAT strategy involves incorporation of a chromaphore to facilitate isolation of low-abundant proteins during purification procedures [385].

The proteins are then digested to produce smaller proteolytic fragments that can be analyzed readily by LC/MS with the possibility of CAD MS/MS. Because the digest is more complex than the original mixture of proteins, biotin is incorporated into the special chemical "tag" that attaches to free cysteines; in this way, the tagged cysteinyl proteolytic peptides can be isolated by affinity chromatographic techniques involving immobilized avidin. Once the tagged proteolytic fragments are free after dissociation from avidin or by chemical cleavage of part of the chemical tag, they are analyzed by LC/MS. The tagged peptides are represented by pairs of mass spectral peaks offset from each other according to the stable isotopes incorporated in the labled form of the chemical tag; relative quantitation of the peptide (and therefore the particular protein in the original sample) is achieved by the ratio of peak intensities. For qualitative purposes, the various proteins are identified by Sequest [169] or other algorithms [21, 386], using inputs of the molecular weights of proteolytic fragments or of CAD data that provide limited sequence information for automated processing. Use of the amine-specific isobaric tagging strategy facilitates quantitative shotgun proteomic analyses of up to four parallel samples based on a reporter ion series by MS/MS [387].

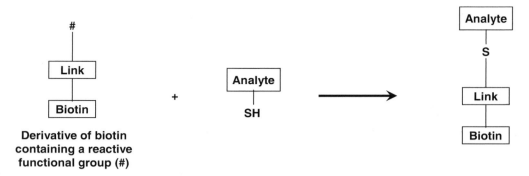

Scheme 12-12

Analyte isolation, recognition, and relative quantitation are achieved with clever chemistry that not only installs a stable isotope-labeled or unlabeled tag on the proteins, but also covalently attaches biotin as an affinity tag to facilitate processing of complex digestion mixtures [388]. Biotin is a widely used affinity "handle" that can be attached to the analyte covalently, in this case using a "linker" containing a reactive group, such as an alkyl halide (represented by #), to react with the sulhydryl group on cysteine as illustrated in Scheme 12-12.

It is the great affinity between biotin and avidin ($K_f = 10^{15}$) that is used to advantage in capturing the biotinylated analyte [389, 390]. Avidin is a protein that in its native state is highly glycosylated, a feature that can cause artifacts or other operational problems; therefore, streptavidin [390], a modified nonglycosylated form of avidin, may be preferred. Avidin (or streptavidin) can be immobilized to some convenient surface (e.g., agarose) and then used to capture the now biotin-tagged analyte (also called biotinylated conjugate), as illustrated in Scheme 12-13.

A key feature of avidin–biotin binding is the fact that this interaction is reversible, so that the captured (biotinylated) analytes can be released from the immobilized avidin (or streptavidin) under mild conditions for subsequent analysis by mass spectrometry. The biotinylated analyte can be released from avidin or streptavidin by exposure to 2 mM biotin in phosphate buffered saline. Harsher conditions have been used (e.g., 8 M guanidine-HCL at pH 1.5) to break the avidin–biotin bond.

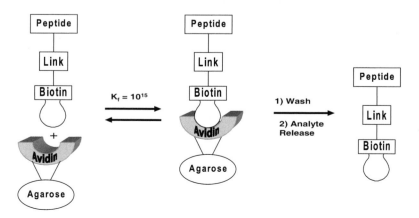

Scheme 12-13

In the original report of the ICAT strategy [391], proteolytic fragments that contained a cysteine residue were treated with a sulfhydryl-reactive linker molecule. Because proteolysis of a complex mixture of proteins significantly increases the complexity of the resulting mixture (possibly by as much as two orders of magnitude), chemical specificity for cysteine reduces the mixture complexity in the following way. Cysteine has a statistical frequency of occurrence of about 1% in the sequence of the "typical" protein; therefore, biotinylation of cysteine-containing proteolytic fragments will offset the increase in the complexity of the mixture due to proteolysis. Further, the affinity capture process also has the advantage that salts, buffers, etc., can easily be removed by washing the agarose–streptavidin surface before releasing the biotinylated analyte to reduce the complexity of the mixture to be analyzed by LC/MS. The linker molecule can be labeled with a given number of stable isotopes (e.g., with eight deuterons) as was the case in the original report by Gygi *et al.* [391]. Because identification of the protein usually will be made by automated "searching" algorithms, it is important that the chemical moiety composing the ICAT does not interfere with fragmentation of the labeled peptide during CAD [392].

It is the quantitative information provided by ICATs [20, 387, 393–397], etc., that can help establish a basis for functional proteomics [398]. For example, the ICAT methodology has been used to study the bacterial pathogen *Pseudomonas aeruginosa* grown under magnesium limitation, an environmental condition previously shown to induce expression of various virulence factors [399]; another ICAT application quantifies chromatin-associated proteins [400]. An ICAT approach has been used to identify oxidant-sensitive cysteine thiols in proteins [401]. Tureček [380] has described applications of ICAT and related affinity methodology in the diagnosis of genetic diseases. However, considerable work remains to be done in the development of ICAT methodology. For example, in ICAT applications to date, of the thousands of proteins encoded by the genome in yeast, fewer than 3% have been detected by this novel approach [380]; therefore, the development of more selective and sensitive methodology for quantitation of the proteome is required. The results of a large-scale assessment of the practical utility of the ICAT methodology in monitoring the proteome of *E. coli* involving the quantitation of more than 24,000 peptides showed good reproducibility (CV = 18%), but a detection bias for acidic proteins (pI < 7) and under-represented small proteins (<10 kDa) [402]. A variation on the ICAT theme is directed toward selected proteins in an application of ion trap selected reaction monitoring (a technique the authors liken to a mass spectrometric version of the Western blot) to maximize sensitivity, enabling analysis of peptides that would otherwise go undetected [403].

Some of the variance in quantitative results is related to the resolution of species containing the labeled vs the unlabeled tags. A study has been made of the structural features responsible for resolution of heavy isotope coded peptides during reversed-phase chromatography [404]; it was concluded that the probability of a deuterium atom interacting with the stationary phase of a reversed-phase column and impacting resolution is greatly diminished by placing it adjacent to a hydrophilic group.

2) Illustrative Example of the ICAT Approach

The following example conceptually describes the relative quantitative assessment of the hypothetical protein "Q" in normal vs abnormal (e.g., malignant) hepatocytes using the ICAT strategy. It is assumed in Scheme 12-14 that the peptide sequence represents one (hypothetical protein Q) of many proteins in a hepatocyte.

Tagging of protein "Q" with unlabled cysteine-reactive biotin

Scheme 12-14

During processing of the normal-cell lysate, protein Q, as well as dozens of other proteins that have a free cysteine, will be tagged (as illustrated in Scheme 12-15) with a d_0-labeled biotin derivative by reacting with a sulfhydryl group (as illustrated earlier in Scheme 12-12). In parallel fashion, the lysate from the malignant cells is treated in a separate container with a d_8-labeled biotin derivative (not illustrated in Scheme 12-15).

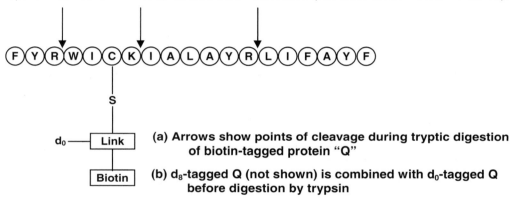

(a) Arrows show points of cleavage during tryptic digestion of biotin-tagged protein "Q"

(b) d_8-tagged Q (not shown) is combined with d_0-tagged Q before digestion by trypsin

Scheme 12-15

The d_0- and d_8-biotin tagged protein samples are combined and digested by trypsin, which cleaves the polypeptide backbone on the C-terminal side of arginine or lysine, as indicated by the three arrows above the sequence in Scheme 12-15, to yield four tryptic fragments. The tryptic peptides of the d_0-labeled protein Q are shown in Scheme 12-16; the d_0-labeled biotin tag remains with the cysteinyl tryptic fragment. The d_8-labeled protein produces the same four tryptic fragments (not shown) except that the d_8-labeled biotin tag remains on the cysteinyl fragment.

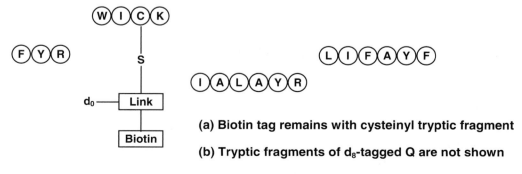

(a) Biotin tag remains with cysteinyl tryptic fragment

(b) Tryptic fragments of d_8-tagged Q are not shown

Scheme 12-16

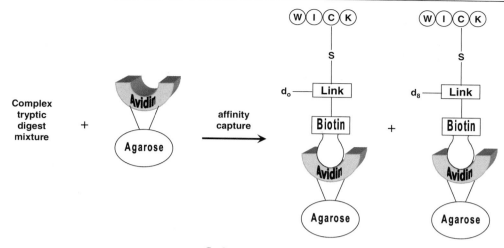

Scheme 12-17

The tryptic digest of a real-cell lysate presents a vastly more complex mixture than is represented by the four tryptic fragments in Scheme 12-16, and use of immobilized avidin reduces the complexity of the sample by capturing only the biotin-tagged tryptic fragments as illustrated in Scheme 12-17. Elution of the captured biotin-tagged tryptic fragments include the d_0- and d_8-labeled species now shown in Scheme 12-17. The ratio of these d_0- and d_8-labeled species can be determined by mass spectrometry as illustrated in Scheme 12-18; these hypothetical results indicate that the concentration of protein Q in the malignant cells (lysate labeled with the d_8-ICAT) is about twice that in the normal cells (lysate labeled with the d_0-ICAT).

Problems associated with poor fragmentation efficiency of the biotin-tagged peptide during CAD MS/MS led to the development of a cleavable isotope-coded affinity tag (cICAT) labeling strategy [405]. The general ICAT strategy is likely to continue to play an important role in striving for quantitative proteomics [20, 394, 395, 398]; an algorithm has been designed to process ICAT-related data [406]. Other quantitative proteomic methodology using affinity capture has been described for assaying enzymes [407]. In some cases, errors in quantitation related to differential elution of isotopically labeled compounds can occur [408]. A novel approach to determining the identity of a protein in the dynamic proteome involves an *in vivo* labeling strategy of replacing methionine with selenomethionine [29].

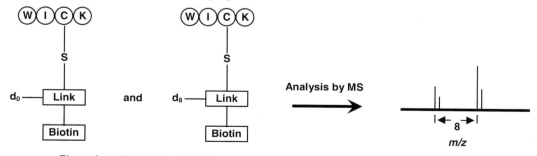

Eluate from immobilized Avidin

Scheme 12-18

The practical utility of acid-cleavable ICATs was recently assessed in a large-scale investigation to determine the degree of reproducibility and depth of proteome coverage by measuring protein changes in *Escherichia coli* treated with triclosan, an inhibitor of fatty acid biosynthesis [402]. The entire ICAT MS experiment was conducted on four independent occasions in which more than 24,000 peptides were quantitated using an ion trap mass spectrometer. The results demonstrated that, quantitatively, the technique provided good reproducibility (the median coefficient of variation of ratios was 18.6%); on average, more than 450 unique proteins were identified per experiment. However, the ICAT method was strongly biased to detect acidic proteins (pl < 7) and under-represented small proteins (<10 kDa); furthermore, the ICAT methodology failed to show clear superiority over 2D electrophoretic methods in monitoring hydrophobic proteins from cell lysates [402].

3) Methodologies Analogous to ICAT

A quantitative technique, related to the ICAT strategy, has been proposed by Chait and coworkers [231, 237] to provide for affinity capture and quantitation of phosphorylated proteins. This method uses base-induced β elimination of phosphate [266] from phosphorylated serine or threonine to produce a reactive acrylate double bond that can be coupled via Michael addition to ethane-1,2-dithiol. The free terminal thiol group is then conjugated by another Michael addition to a biotin-linker-maleinimide reagent, and the resulting succinimide ring is hydrolyzed. The biotinylated proteins are purified by affinity capture release, digested, and the peptide mixture is analyzed by MALDI TOFMS or LC-ESI MS/MS. The presence of the biotinylated tag in the peptide is readily indicated by an abundant fragment at *m/z* 446 originating from the biotinylated side chain. An acid-labile affinity tag has been developed to facilitate removal of the rather heavy biotin tag after isolation of the relevant peptides [409]; in this way, CAD of the still-labeled peptide may be more effective without the substantial mass of the biotin moiety.

Other methodology based on β elimination of the phosphate moiety [266] was reported by Adamczyk, Gebler, and Wu [272], which converts phosphoserine and phosphothreonine residues to *S*-(2-mercaptoethyl)cysteinyl and beta-methyl-*S*-(2-mercaptoethyl)cysteinyl residues, respectively, after β elimination and condensation with 1,2-ethanedithiol. The free sulfhydryl is then conjugated with biotin to permit isolation and purification by affinity capture on avidin beads. With the use of d_0- and d_4-ethane dithiol, this methodology can be adapted for quantitative proteomic applications. Other quantitation methods are also available for phosphoproteins [292, 410].

Absolute quantification has also gathered the attention of many laboratories. A chromophore has been incorporated into the ICAT strategy to allow absolute quantitation based on an absorbance determination in methodology termed visible ICAT (VICAT) [385, 411]. General strategies on isotope coding of peptides have been published by a consortium [412].

B. Alternative Stable Isotope-Based Methodologies

MCAT is the abbreviation for mass-coded abundance tagging, which is based on differential guanidination of C-terminal lysine residues of tryptic peptides [413]. The MCAT approach offers an economical way to quantitatively assess two populations of proteins in a manner somewhat analogous to ICAT, but it provides no affinity tag to facilitate simplification of the digest. On the other hand, an objective of the MCAT approach is to provide more comprehensive analysis than the ICAT method, which is limited to those proteins containing cysteine. In a new approach, termed synthesis/degradation ratio mass spectrometry, stable isotope labeling is employed to calculate a relative synthesis/degradation ratio that reflects the relative rate at which ^{13}C is incorporated into individual proteins, thereby providing a measure of the relative turnover of proteins in cellular systems [375]. Relative quantification of proteomes using accurate mass tags based on ^{18}O-labeled internal standards has been described [414], and the general use of the ^{18}O-labeling technique for determining the ratio of an individual protein's expression level between two samples has been reviewed [415]. Isotope labeling via carbamylation of the primary amines in peptides with ^{13}C-urea is the basis for a comparative proteomic assay as demonstrated with BSA and casein [416]. The effectiveness of an aldehyde-reactive hydrazide-functionalized isotope-coded affinity tag (HICAT) has been demonstrated in the analysis of mitochondrial proteins [417].

SILAC is an acronym representing a relative quantitative technique called "stable isotope labeling with amino acids in cell culture"; it can be used to compare the expression levels of hundreds of proteins in a single experiment [377]. SILAC makes use of ^{12}C- or ^{13}C-labeled amino acids added to the growth media of separately cultured cell lines, giving rise to cells containing either light or heavy proteins, respectively. Upon mixing lysates collected from these two batches of cells, proteins can be identified by the above-described techniques of tandem mass spectrometry. The ratio of the mass spectral peak intensities corresponding to peptides containing the differentially incorporated stable isotopes is the basis for a quantitative comparison of the protein expression between the two samples of cells. In a quantitative study of the proteins in *Arabidopsis thaliana* cell cultures, $^{13}C_6$-arginine proved to be a more suitable probe than $^2H_3$-leucine or $^2H_4$-lysine [418]. SILAC has been used in studies of metastatic processes in prostate cancer [287] and the quantitative assessment of epidermal growth factor-induced changes in nine phosphorylation sites in the extracellular signal-regulated kinase p90 ribosomal S6 kinase-signaling cassette [288]. In a study of the glucose-transporter system, 36 of 603 proteins quantified by SILAC showed an insulin-dependent change of their interaction by more than 1.5-fold in either direction [419]. Dietary administration of stable isotope-labeled amino acids allows turnover information on individual proteins in intact animals [420].

AQUA is an acronym representing a method to achieve absolute quantitation of proteins and their modification states [410, 421]. Peptides are synthesized with incorporated stable isotopes as ideal internal standards to mimic native peptides formed by proteolysis. These synthetic peptides also can be prepared with covalent modifications (e.g., phosphorylation, methylation, acetylation, etc.) that are chemically identical to those with naturally occurring post-translational modifications. Such AQUA internal standard peptides are then used to quantitatively measure the absolute levels of proteins and post-translationally modified proteins after proteolysis by using selected reaction monitoring analysis with tandem mass spectrometry. The AQUA strategy was used to (a) quantify low-abundance yeast proteins involved in gene silencing, (b) quantitatively determine the cell cycle-dependent phosphorylation of Ser-1126 of human separase protein, and (c) identify kinases capable of phosphorylating Ser-1501 of separase in an *in vitro* kinase assay [410]; application of AQUA was also employed in a study of protein ubiquitination [421].

An *in vivo* lysine-specific mass tagging method has been developed that incorporates deuterium-labeled (heavy) lysines into proteins through *in vivo* cell culturing, which provides specific mass tags at the carboxyl termini of proteolytic peptides when cleaved by certain proteases [422]. The mass shift between the unlabeled and the deuterated lysine (d_4-Lys) assigns a mass signature to all lysine-containing peptides in any pool of proteolytic peptides; unlike the MCAT method, which modifies C-terminal lysines and thus misses those internal lysines that result from missed cleavages, the *in vivo* method provides a basis for recognizing all lysines [422, 423]. Although this *in vivo* method provides no affinity tag to facilitate isolation of the labeled peptides, it does provide the possibility for good coverage of quantifying the proteolytic peptides. An alternative approach based on the amidination of lysines increases their basicity and therefore MALDI ionization yields; the amidine labels differ by methylene groups, leading to 14 Da mass differentials [424].

A related strategy for quantitative analysis of proteins focuses on modification of lysine-containing proteolytic fragments, which react with 2-methoxy-4,5-dihydro-1*H*-imidazole to form the corresponding 4,5-dihydro-1*H*-imidazol-2-yl derivative [425]. The reagent can be labeled with four deuterons, thereby allowing the same proteins from two different sources to be compared quantitatively after treating one sample with d_0- and the other with the d_4-2-methoxy-4,5-dihydro-1*H*-imidazole. The peptides containing the chemically modified lysines undergo significantly greater (by approximately an order of magnitude) ionization and fragmentation (to a predominant series of **y**-ions) compared to those with free lysines; this chemical modification thereby compensates for the reported dominance (by approximately an order of magnitude) of arginine-containing peptides or lysine-containing peptides during analysis by MALDI [426].

The use of the endoprotease ^{16}O- to ^{18}O-catalyzed oxygen exchange at the C-terminal carboxylic acid is advantageous because of the specificity assured by the enzymic reaction and the labeling of essentially every protease-derived peptide [427, 428]. Internal standards consisting of ^{18}O-labeled peptides is the basis of another approach to absolute quantification of peptides from in-gel digestion [429].

A differential isotopic labeling technique based on chemical reaction with d_0- vs d_6-acetic anhydride was used to quantitate changes in neuropeptide levels in mouse tissues after exposure to different physiological stress [430]. The role of active site residues in fructose 1,6-bisphosphate aldolase was investigated by "chemical-modification rescue" methodology; ESI FTMS, combined with use of a d_0- and d_4-chemical labeling procedure, allowed precise identification of sites and measurement of degree of protein modification [431]. Another differential isotopic labeling technique involves incorporation of two atoms of ^{18}O into the carboxy terminus of proteolytic fragments to provide quantitative and concurrent comparisons between individual proteins from two entire proteome pools (the second pool is unlabeled) of adenovirus [432]. Proteins of *Saccharomyces cerevisiae* strain S288C grown in either ^{14}N- or ^{15}N-enriched minimal media were mixed, digested proteolytically, and analyzed by LC-ESI MS/MS using a Q-TOF-MS for quantitation and identification by MudPIT [433]. Another approach uses formaldehyde to label globally the N-terminus and ε-amino group of Lys by reductive amination [434]. Peptide dimethylation by isotopically coded formaldehydes gives a more general approach to comparative proteomics than the original cysteine-oriented ICAT method [435].

The identification of intact proteins by determination of molecular mass alone is difficult, if not impossible, for proteins >50 kDa [436]. Partial amino acid composition information on the proteins can be supplied by growing the cells of interest in a medium enriched in a stable isotope-labeled amino acid (e.g., d_{10}-leucine); a comparative sample of cells would be grown in an aliquot of the medium containing d_0-leucine. The shift in mass between the molecular weights of the two variants (labeled and unlabeled) proteins divided by 10 u indicates the number of leucine residues in the protein. This information on the partial amino acid composition combined with a determination of the molecular mass of the protein adds sufficient constraint for identification to be made of the intact protein, provided that microheterogenity due to multiple post-translational modifications is not excessive [436]. A computational method that predicts isotope distributions over a range of enrichments and compares the predicted distributions to experimental peptide isotope distributions obtained by FTICRMS helps avoid systematic errors deriving from large envelopes of isotope peaks for ions composed of thousands of atoms [437].

C. Related Methodologies

A relative quantitative approach, not involving the use of stable isotope tags, normalizes the peak areas of identified peptides from one protein to the total reconstructed peak area of the protein digest, which is further normalized to the peak area of an internal standard protein digest present in the mixture at a constant level [438] The method was shown to be dependable over a concentration range of 10 to 1000 fmol with a standard deviation of 20% in quantitating a variable mixture of five model proteins. A novel MS/MS-based analysis strategy for the accurate quantification of peptides and proteins uses isotopomer labels, referred to as "tandem mass tags" [439].

A nonisotopic method without external standards provides quantitative proteomic and metabolomic information-based LC-ESI MS [440]. The so-called iTRAQ™ approach is based on modification of primary amines with four different chemical tags allowing many parallel samples to be processed; the samples can be distinquished by CAD MS/MS to produce unique product ions at *m/z* 114, 115, 116, and 117 [397, 441].

The concept of "fluorous proteomics" involves tagging of specific peptide subsets in complex biological samples with perfluorinated moieties and subsequently enriching them by solid-phase extraction over a fluorous-functionalized stationary phase [442]. This approach is selective, yet can be readily tailored to enrich different subsets of peptides. The potential of this methodology is demonstrated by the facile enrichment of peptides bearing specific side-chain functionalities or post-translational modifications from tryptic digests of individual proteins as well as whole-cell lysates [442].

Nonquantitative identification of proteins has been achieved by the automated process of mass-analyzing proteolytic fragments of an isolated sample component [21, 386]. Alternate separation processes (e.g., capillary isoelectric focusing (CIEF) [443, 444]) might be used that are more amenable to automation than 2D-PAGE; in addition, more complicated samples might be studied if intact proteins could be identified without the complication of controlled degradation of each. Furthermore, the isoelectric focusing variant of capillary electrophoresis requires a significantly smaller sample size than 2D-PAGE; e.g., in a recent application of capillary isoelectric focusing (CIEF) electrophoresis-electrospray mass spectrometry, nearly 1000 putative proteins (in the range of 2–100 kDa) were revealed during analysis of 0.3 ng of soluble *E. coli* proteins compared to more than 100 µg typically used for 2D-PAGE [444].

An automated label-free method for finding differences in complex mixtures using complete LC/MS data sets selectively finds statistically significant differences in the intensity of sets of peaks, accounting for the variability of measured intensities and the fact that true differences will persist in time [445-447]. The method was used to compare two complex peptide mixtures with known peptide differences to assess the validity of each difference found and so to analyze the method's sensitivity and specificity. The method is more sensitive and gives fewer false positives than subtractive methods that ignore signal variability. Differential mass spectrometry, combined with targeted MS/MS analysis of only components of identified differences, may save both computation time and human effort compared to shotgun proteomics approaches [445-448].

5. "Top-Down" Strategies of Analysis

In recent years, it has been possible to contemplate the analysis of intact proteins, which has been called the "top-down" approach to structural analysis [449–451]. In principle, the difference between the measured mass (molecular weight) of a protein and its calculated monoisotopic mass (based on its DNA-predicted sequence) indicates sequence errors and/or post-translational modifications to the isolated analyte. In the top-down mass spectrometry approach, the protonated analyte is dissociated, and the resulting fragment ion masses are compared with corresponding calculated masses expected from the cDNA-predicted protein sequence in an effort to recognize the locations of any translation errors or post-translational modifications [452, 453].

A. Instrumentation and Fragmentation Requirements

"Top-down" analysis requires the use of a mass spectrometer with sufficient mass range and resolving power to allow for separation of the isotope peaks representing the intact protonated molecule or the product ions formed during MS/MS [450, 452–454]. Therefore, the applications to date have been limited to use of an ICR for FTMS [452, 454] or an ion trap mass spectrometer for the analysis of highly charged species from ESI [309, 310]. A Q-TOF instrument is also suitable for "top-down" analysis for proteins up to 9 kDa [455].

The "top-down" approach to structural analysis also requires some means of causing the high-mass ions to fragment in a way that reveals structural features. The McLafferty Group pioneered the use of electron capture dissociation (ECD) with FTMS for this purpose [452, 456–458], as described in a later section. An algorithm can be used to relate accurate mass data to monoisotopic peak lists and isotope patterns for anticipated fragment ions for a given amino acid sequence [459]. Web-based software is available for use of the top-down strategy of analysis to identify and characterize PTMs of proteins [460]. Kelleher and coworkers are extending the quantitative top-down approach to peptides heavier than 5 kDa as illustrated in applications to yeast proteins [461].

During the same timeframe, the McLuckey Group demonstrated that CAD of protonated proteins in an ion trap can be reasonably effective for ions up to *m/z* 8000 (as with ubiquitin), but more so if the analyte is multiprotonated as with enolase [310, 462]; developments are in progress that will make this process adaptable to commercially available instrumentation [463]. The McLuckey Group also developed methodology for manipulating the charge state of proteins in the gas phase for simplification of the resultant product-ion charge-state distributions [464–467]; the technique is also useful when analyzing multiprotonated proteins exceeding 50 kDa. Dissociation of high-charge states (defined as charge states corresponding to much more than the number of arginine residues present in the protein) results in preferential cleavage on the N-terminal side of proline residues or at other dominant "nonspecific" fragmentation sites, such as between Leu–Met and Leu–Ser. Even though the product-ion spectra may be dominated by ions formed through only a few facile cleavage channels, McLuckey and coworkers showed, under favorable conditions, CAD of a single precursor-ion charge state in the quadrupole ion trap can result in greater than 50% sequence coverage for proteins up to 20 kDa.

The utility of the CAD ion/ion reaction top-down sequencing approach has been demonstrated in the sequence analysis of a variety of proteins up to 20 kDa, including ubiquitin [309, 467]. More recently, they analyzed a protein mixture derived from a whole-cell lysate fraction of *S. cerevisiae* containing roughly 19 proteins by this approach using a quadrupole ion trap tandem mass spectrometer [468]. Collection of the experimental data was facilitated by CAD and ion/ion proton-transfer reactions in multistage mass spectrometry procedures [469]. Ion/ion reactions were used to manipulate charge states of both parent ions and product ions for the purpose of concentrating charge into the parent ion of interest and to reduce the product-ion charge states (preferably to unity) for determination of product-ion mass and abundance [470]. Identification of the protein was achieved by matching the uninterpreted product-ion spectrum against protein sequence databases with varying degrees of annotation and a specialized scoring procedure [468]. Ion/ion reactions between a variety of peptide cations (double- and triple-charge) and SO_2 anions in a 3D quadrupole ion trap have been used to manipulate the charge state by proton transfer as well as to promote fragmentation by electron-transfer dissociation [471, 472]; another application of ion/ion reactions coupled with CAD allows the number of acidic and basic groups at the surface of the gaseous protein to be determined [473].

The McLuckey Group has also shown that multiple-charge protein anions have some advantage in top-down analysis. In subjecting the -8 anion of ribonuclease A to CAD, they found preferential fragmentation at the locus of disulfide bonds [326]. Whereas these fragments could not be used to discern the connectivity of the disulfide bonds, they did allow partitioning of the protein in such a way as to recognize the locus of a glycosylation site as in ribonuclease B.

Other efforts to conduct the top-down approach to protein analysis without the use of expensive FTMS instrumentation have continued in some laboratories. In one case, in-source CAD is used with a Q-TOF instrument to promote protein fragmentation to provide limited sequence information [474]. Results from another research group indicate that heat-assisted (temperatures up to 250 °C) significantly improve the efficiency of CAD for large proteins in the 14–60 kDa range [475].

The following is an example of using CAD with a 3D quadrupole ion trap mass spectromter for a top-down analysis. As stated in the beginning of this section, the top-down approach to protein analysis requires sufficiently good resolving power that the isotope peaks of the protonated molecule can be resolved. This level of performance will allow the charge state to be computed from the spacing of the isotope peaks. The spacing of the isotope peaks at 0.1 *m/z* units indicates a charge state of 10 in the corresponding ion. In this way, a measurement of the *m/z* value of an ion together with an assessment of the charge state allows the analyst to compute the mass of the ion. An example of the special operational protocols for a linear ion trap mass spectrometer for top-down analysis of a 14.6-kDa protein is described here [476].

The first requirement in a top-down analysis is to determine the mass of the intact protein. This can be accomplished by ESI MS, as illustrated by the mass spectrum shown in Figure 12-19, which was acquired at low resolving power at an acquistion rate of 4400 *m/z* units sec^{-1}. A subsequent zoom scan (results not shown) at a resonant ejection scan rate of 1100 *m/z* units sec^{-1} allowed isotope peaks to be resolved sufficiently that the charge state of the protein could be established. As shown in Figure 12-19, the clusters of isotope peaks centered at *m/z* 735.5 and at *m/z* 817.0 represent the protein containing 20 and 18 protons, respectively.

All other discernible clusters of peaks represent protein molecules differing by one in the number of protons attached; e.g., the cluster centered at *m/z* 774.1 indicates 19 protons attached to the molecule while that at *m/z* 1469.6 indicates 10 protons attached to the molecule. These mass spectral results allow the mass of the protein to be experimentally determined as 14,688.3 Da, which is in good agreement with the calculated value of 14,689.8 Da for the known sequence.

The sequence of the protein can be confirmed only if adequate fragmentation of the protonated molecule can be achieved by subjecting it to CAD. The CAD fragmentation pattern is a function not only of energy, but of the charge state of the ion.

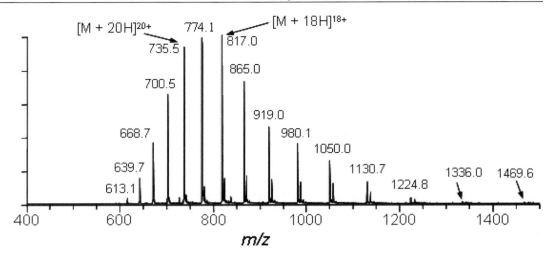

Figure 12-19. ESI mass spectrum of a protein prior to top-down analysis.
Data courtesy of Professor Gavin Reid, Department of Chemistry, Michigan State University, East Lansing, MI, USA.

However, from previous work, it is known that the CAD fragmentation pattern is usually not a continuous function of the charge state, but rather there are abrupt changes in the CAD pattern with the charge state such that there are usually no more than two or three distinct patterns. Therefore, it is necessary to perform a survey scan of the product-ion spectra of the precursor ions of each charge state. The survey scan at 1100 *m/z* units sec^{-1} allows each of the multiple-protonated molecules to be obtained in a few seconds; at this scan rate, the resolving power does not permit unit resolution, but it is adequate for obtaining the general appearance of the fragmentation pattern as described below.

In this example, the CAD behavior of the variously protonated forms of the protein fell into two distinct groups, one group consisted of those molecules containing 18 or fewer protons and the other group consisted of molecules containing 20 or more protons. The CAD fragmentation pattern of the protein molecule containing 19 protons (data not shown) seemed to be a blend of those of the two distinct groups. Because of the high abundance of protein molecules containing 18 and 20 protons (see peaks in Figure 12-19 at *m/z* 817.0 and at *m/z* 735.5, respectively), these species were subjected to CAD MS/MS in separate experiments. The CAD product-ion spectrum of *m/z* 817.0 representing [M + 18H]$^{+18}$ is shown in Figure 12-20, which shows a pattern that is obviously different from that for CAD of *m/z* 735.5 representing [M + 20H]$^{+20}$ shown in Figure 12-21.

Because of the trade-offs in quality of data versus the operating parameters that must be imposed on an ion trap of limited *m/z* range; e.g., *m/z* 2000 (see the effect of the ion trap ion-resonance ejection rate on S/N and resolution of mass spectral data in Figures 2-30 and 2-31 in Chapter 2), the product-ion mass spectra shown in Figures 12-20 and 12-21 were reacquired in the zoom mode at a rate of 1100 *m/z* units sec^{-1} over a period of 15 min [476]. Although the mass spectral data reacquired in the zoom mode have the same general compact appearance as those shown in Figures 12-20 and 12-21, the high-resolution quality is apparent in the expanded format of selected segments of the data as shown in Figure 12-22.

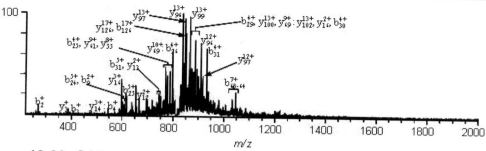

Figure 12-20. CAD product-ion spectrum of m/z 817.0 (from the spectrum in Figure 12-19) representing $[M + 18H]^{+18}$. *Data courtesy of Professor Gavin Reid, Department of Chemistry, Michigan State University, East Lansing, MI.*

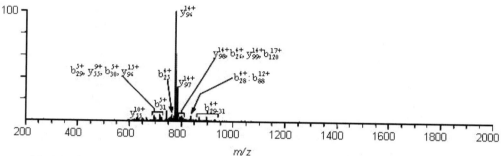

Figure 12-21. CAD product-ion spectrum of m/z 735.5 (from the spectrum in Figure 12-19) representing $[M + 20H]^{+20}$. *Data courtesy of Professor Gavin Reid, Department of Chemistry, Michigan State University, East Lansing, MI.*

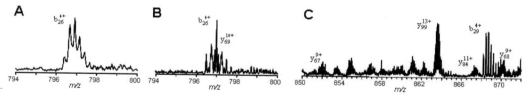

Figure 12-22. Panels A, B, and C show expanded segments of the product-ion spectrum of m/z 817.0 representing $[M + 18H]^{+18}$ at high resolving power allowing for identification of specific ions as described in the text. *Data courtesy of Professor Gavin Reid, Department of Chemistry, Michigan State University, East Lansing, MI.*

The data in panel A of Figure 12-22 are from an expanded segment of the reacquired spectrum (similar to that displayed in Figure 12-20) of product ions of m/z 817.0 representing $[M + 18H]^{+18}$. In the narrow m/z range displayed in panel A of Figure 12-22, one peak was sufficiently resolved for accurate measurement, and its m/z value was sufficiently close to the calculated value for b_{26}^{4+} to qualify as an identification.

The data in panel B of Figure 12-22 are from the same narrow m/z range as displayed in panel A, but were acquired in the ultra-zoom mode at a rate of 28 m/z units sec^{-1} during an acquisition period of 30 minutes. The resolution is now sufficient to

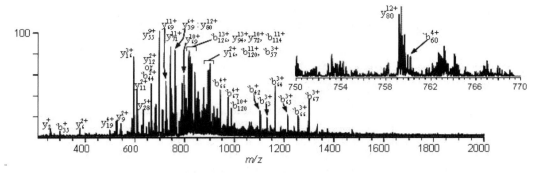

Figure 12-23. CAD MS/MS/MS product-ion scan of m/z 735.5 (from the primary mass spectrum in Figure 12-19). *Data courtesy of Professor Gavin Reid, Department of Chemistry, Michigan State University, East Lansing, MI.*

resolve additional mass spectral peaks, one of which has an experimentally measured value of *m/z* that is sufficiently close to the calculated value for y_{69}^{10+} to constitute an identification.

The data in panel C of Figure 12-22 are from another expanded region of the spectrum similar to that shown in Figure 12-20, but were acquired in the ultra-zoom mode; sufficient mass accuracy was achieved to identify additional ions. Other peaks (data not shown) in the complete scan at this low data acquisition rate (high resolving power) allowed for identification of 36 amide bond cleavages or 28% sequence coverage. In all the MS/MS data available from CAD of representing $[M + 18H]^{+18}$, approximately half of all the amide bonds in the protein were identified [476].

Because the vast majority of the ion current in the CAD MS/MS spectrum shown in Figure 12-21 of *m/z* 735.5 (representing $[M + 20H]^{+20}$ in the primary ESI mass spectrum in Figure 12-19) is due to y_{96}^{14+} (as represented by the predominant peak in Figure 12-21), this primary fragment ion was selected for further CAD and a third stage of analysis by MS (i.e., MS³). Figure 12-23 is the CAD MS/MS product-ion spectrum of y_{96}^{14+} (CAD MS/MS/MS of *m/z* 735.5 from the primary mass spectrum in Figure 12-19) acquired in the zoom mode; these data provided the identification of 29 amide cleavages in the original protein. The inset in Figure 12-23 shows a segment of this CAD MS/MS/MS product-ion spectrum acquired in the ultra-zoom mode; the increased resolution in the data allowed 47 amide bonds to be identified.

Table 12-1 shows the number of amide bonds identified as a function of the zoom and ultra-zoom data acquisition modes. Some amide bonds were identified in all data acquisition modes, but other amide bonds were identified in only one mode. Overall, the CAD MS² and CAD MS³ data provided evidence for 85% sequence coverage in the protein [476]. In some cases, additional amide bonds could be identified from data acquired in the ultra-zoom mode because the lower resonance ejection rate produced better resolution in the mass spectral data. In other cases, amide bonds that were identified in the zoom mode because of adequate resolution and peak intensity could not be identified in the ultra-zoom mode because further reduction in the peak intensity associated with this high-resolution scan mode made it impossible to distinguish the analyte signal from noise. The effect of trade-offs between signal strength and resolution during routine instrument operation is illustrated in Figures 2-30 and 2-31 (Chapter 2).

Table 12-1. Number of amide bonds identified and percent sequence coverage from indicated CAD scan modes.

Mode	MS/MS of [M + 18H]$^{+18}$	MS/MS of [M + 20H]$^{+20}$	MS/MS/MS of [M + 20H]$^{+20}$
Zoom mode	30 amide bonds (24%)	30 amide bonds (24%)	29 amide bonds (23%)
Ultra-Zoom mode	36 amide bonds (28%)	15 amide bonds (12%)	47 amide bonds (37%)

B. Electron Capture Dissociation (ECD)

In pursuing the analysis of macromolecules according to the "top-down" strategy [449, 450], the analyst is likely to find massive ions that are not amenable to CAD, the classical technique for dissociation of even-electron ions from desorption/ionization. In the case of proteins and peptides at least, the developing technique of electron capture dissociation (ECD) shows great promise [325, 477–480]. Most of the applications have been made using FTMS [480–485] although use of ECD with a quadrupole has been reported; the adverse influence of RF voltage on electrons needs to be addressed before ECD can be used with ion traps [486]. ECD has the advantages of cleaving between a high proportion of amino acids, without loss of such post-translational modifications as glycosylation, carboxylation, and phosphorylation [487]. An exception has been reported in the ECD of double-protonated O-sulfated peptides which cleanly expelled SO_3 upon ECD; however, ECD of divalent complexes of these O-sulfated peptides allowed recognition of the sites of sulfation [488].

The mechanism and energetics of bond cleavage following ECD has been well described by Tureček [489, 490]; experiments with labeled peptides show evidence for at least two mechanisms, one slow, the other fast [491]. Early focus was on cleavage of a disulfide bond, a topic recently revisited [492–494]. The single disulfide bond in anthionine bridge-containing antibiotics can be localized by ECD [495]. The tendency of ECD to cause cleavage at disulfide bonds has been used to advantage in the characterization of disulfide cross-linked nanostructures [496].

A fundamental understanding of all possible fragmentation processes following electron capture is necessary if ECD is to succeed in the characterization of unknowns [497, 498]; to this end, the ECD of cyclodepsipeptides were found to yield numerous backbone fragments, but no charge-reduced species, consistent with a radical cascade mechanism [499]. ECD at low temperatures reveals selective dissociations [500]. An interesting ECD study of ubiquitin involving the use of ion mobility spectrometry reveals that the shape of the molecule can have a greater effect on electron capture efficiency than either collisional cross section or charge state alone [501]. ECD has been reported to cleave a cysteine-bonded antibiotic metabolite to a protein [502], and the loss of side chains from model amino acids has been studied [503]. Early work with this promising technique indicates that there may be trade-offs between obtaining simplified ECD spectra and dealing with the relative inefficiency of the ECD process [504]; the issue of low efficiency in ECD may be improved with efforts to increase the total electron current and a larger emitting area [505]. With high ECD rates, the technique can now be used with on-line separation techniques [506]. ECD also results from high-energy collisions between double-protonated peptide ions and Na atoms [507]. A recent report indicates that electron capture also occurs during high-energy collisions between peptide dications and cesium vapor [508]. In ECD, highly positive ions (usually multiprotonated species from ESI) readily interact with thermal electrons to diminish the high charge state, and effectively convert it from an even-electron ion to an odd-electron species as illustrated in Scheme 12-19.

$$MH_2^{2+} + e_{5\,eV}^- \rightarrow MH^{+\cdot} + H^\cdot$$

Scheme 12-19

It has long been realized that from classical mass spectrometry of odd-electron molecular ions formed during electron ionization (EI) that radical sites often promote homolytic cleavage of bonds adjacent to the atom possessing the odd electron. Because the site of electron capture is highly distributed in a protonated protein, extensive cleavage is observed as illustrated in Figure 12-24. Fragment ions associated with side-chain losses are reportedly comparable in abundance to those resulting from backbone cleavage of peptides (up to 14 mers) following ECD and therefore should be taken into account in data interpretation [509]. In advanced studies of protein fragmentation following ECD, considerable loss of CO has been observed from double-charge **b** ions, suggesting the linear open-chain acylium structure as opposed to the protonated oxazolone structure predicted by *ab initio* calculations [510]. In-source decay MALDI, apparently induced by hydrogen atoms generated by a photochemical reaction of the matrix, shares some similarities with ECD; in both reactions, **c** and **z** type ions are formed [511].

The ubiquitous cleavage promoted by ECD has been reported to provide better sequence tags for identifying proteins than can be achieved by CAD, especially for low-level samples [325, 512]. The fragmentation promoted by ECD is complementary to that achieved by infrared multiphoton dissociation (IRMPD) [513]. It has been reported that ECD presents an effective alternative to CAD for fragmenting proline-rich proteins [504]. ECD can be used to distinguish between the presence of aspartyl vs isoaspartyl residues in peptides [514]. For phosphorylated peptides, ECD and CAD spectra give complementary backbone cleavages for identifying modification sites as reported in an ECD study of bovine β-casein, a 24-kDa heterogeneous phosphoprotein [457]. The data-dependent use of ECD following detection of a 98-Da neutral loss during CAD improves sequence coverage during the characterization of phosphoproteins as demonstrated with the analyses of tryptic digests of β-casein and α-casein [265]. In the analysis of sulfated proteins, ECD of diprotonated peptides resulted in the complete loss of SO_3, but ECD of peptide adducts with divalent cations provided information on both sequence and sulfate localization [488]. In another study, ECD was found to promote complementary fragmentation in an N-glycosylated peptide from that observed with infrared multiphoton dissociation on the same instrument [515]. ECD does not perturb noncovalent bonds even though it promotes dissociation of many different covalent bonds, and therefore it has been used to characterize protein-folding intermediates [516, 517]. The distribution of deuterons in a protein exposed to solution-phase HDX can be determined with improved amide resolution following ECD of the polypeptide backbone as compared by the conventional methods of CAD or proteolytic digestion [518].

Intact protein biomarkers from *Bacillus cereus* T spores produce fragmentation-derived sequence tags under ECD [519]; the high-affinity iron-chelating ligands of microorganisms are also well characterized by ECD [520]. Some functional groups such as gamma-carboxyglutamic acid are too sensitive to be analyzed directly by many physical methods, but osteocalcin is amenable to analysis by ECD mass spectrometry [521]. ECD has been used to advantage in the analysis of dendrimers with amide functionalities [522]. ECD has been demonstrated to be an effective fragmentation technique for characterizing the site and structure of the fatty acid modification in ghrelin,

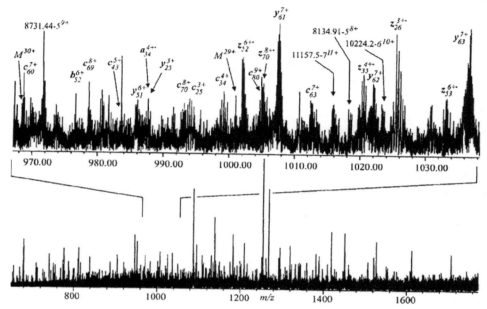

Figure 12-24. A selected segment of the activated ion ECD spectrum of carbonic anhydrase B (29 kDa). *Reprinted from Horn DM, Ge Y, and McLafferty FW, PNAS 2000, 97, 10313, with permission from the National Academy of Science.*

a 28-residue growth-hormone-releasing peptide that has an unusual ester-linked *n*-octanoyl (C8:0) modification at Ser-3; ECD cleaved 21 of 23 possible backbone amine bonds, with the product ions (**c** and **z*** ions) covering a greater amino acid sequence than those obtained by CAD [333]. Consistent with the ECD nonergodic mechanism, the ester-linked octanoyl group is retained on all backbone cleavage product ions, allowing for direct localization of this labile modification; in addition, ECD also induces ester bond cleavage to cause the loss of octanoic acid from protonated ghrelin [333].

"Hot" (11 eV) ECD has been reported to promote secondary fragmentation in z* fragments, following the primary **c**, **z*** cleavage process in multiprotonated polypeptides; such fragmentation has been used to generate **w** ions to distinguish isomeric Ile and Leu residues in tryptic peptides [523]. In one application of top-down analysis, opening the protein by reducing a disulfide bond allowed additional chemistry and cleavage by ECD to facilitate structure elucidation [524].

Electron detachment dissociation (EDD) provides fragmentation similar to ECD for oligodeoxynucleotides, but at enhanced sensitivity, showing promise for characterization of nucleic acid structure [525, 526].

As well as for proteins, ECD has found applications to industrial polymers as described in a report of the analysis of a poly(ethylene/propylene glycol) copolymer mixture consisting of approximately 150 oligomeric formulas. Using CAD to dissociate the copolymer into sequence-specific fragment ions, misleading rearrangements were found, but these were negligible with ECD [527]. A fragmentation study of protonated poly(ethylene glycol) formed by ESI in an FTICR-MS indicated that charge-site cleavage during CAD and radical-site cleavage after ECD led to structurally useful data [528].

The new technology of ECD requires some relatively simple modifications of the MS hardware, but considerable effort to optimize operational conditions to obtain desired results [529].

C. Electron-Transfer Dissociation (ETD)

Electron-transfer dissociation (ETD) appears to be on track to replace electron capture dissociation (described in the next section), in part because a superconducting magnet is not necessary for the instrumentation, which can be simply an ion trap mass spectrometer [472, 530–533]. (The electron capture approach is not practical in quadrupole ion traps because the thermal electrons are quickly excited and ejected by the RF.) In one approach to ETD [472, 531, 532], multiprotonated protein molecules are allowed to react with fluoranthene radical anions. Electron transfer to the multiprotonated protein promotes random dissociation of the N–C$_\alpha$ bonds of the protein backbone. The resultant multiprotonated fragment ions are then deprotonated during a second ion/ion reaction with the carboxylate anion of benzoic acid. The m/z values for the resulting single- and double-charge ions are used to read a sequence of up to 15–40 amino acid residues at the N- and C-termini of the protein. This information and the measured mass of the intact protein are used to search protein or nucleotide databases for possible matches, detect post-translational modifications, and determine possible splice variants [534]. In another application, ETD allows differentiation of isoaspartic acid and aspartic acid residues using the same c + 57 and z – 57 peaks to define both the presence and position of isoaspartic acid in the same way as previously reported using ECD [535].

The use of ETD avoids the problem of the low-mass cutoff when using the ion trap for CAD; i.e., when the ion trap is used for CAD, an RF amplitude is imposed to excite only the precursor ions (not any other ions that may be in the trap). The conditions that make this selective excitation possible also limits the m/z range of product ions that can be trapped during the CAD process down to an m/z value that is one third that of the precursor mass, the so-called low-mass cutoff problem as described in Chapter 2. By using ETD, rather than CAD, to dissociate the precursor ion, the ion trap can be adjusted to trap product ions down to a very low value of m/z; this additional advantage of ETD allows the ion trap to capture low-mass immonium ions that are important for identifying amino acid residues that constitute the protein under investigation. A recent report describes the implementation of ETD on a hybrid linear ion trap–orbitrap to produce accurate mass measurements (2 ppm) at high resolving power (60,000), an advancement that promises to accelerate adoption of top-down protein analyses by research laboratories that are not blessed with super-conducting FTICRMS instrumentation [536].

While highly charged cations are good candidates for ETD, double-protonated molecules of certain compounds like phosphopeptides may be difficult to form. In such cases, the McLuckey Group has found that phosphopeptides readily form negative ions, which can then be converted to double-protonated molecules via an ion/ion reaction multiplied with protonated amino-terminated dendrimers; the indirectly formed double-protonated phosphopeptides can then be subjected to ETD in the usual way to yield characteristic c- and z-type fragment ions by dissociation of the N–C$_\alpha$ bond along the peptide backbone while preserving the labile post-translational modifications [290, 291, 533, 537]. In a developing approach to ETD that circumvents difficulties associated with vaporizing a precursor molecule, ESI is used to produce ions of arenecarboxylic acids that can be dissociated subsequently into ETD reagent ions [538, 539]. The efficiency of dissociation is improved substantially by combining ETD with CAD in a procedure called ETcaD, which reportedly increases dissociation of the protein to permit nearly 90% sequence coverage [540].

D. Applications

The front-end problem of protein solublization in practical top-down analyses can be addressed in some cases by sample processing with an acid-labile detergent [541]. Molecular weight information (LC-ESI MS) for four different full-length membrane proteins up to 61 kDa from three bacterial organisms, two transporters, a channel, and a porin protein have been reported [542]. An analytical procedure for membrane proteins based on isolation by gel electrophoresis, followed by digestion by both trypsin and proteinase K, allows 82–99% sequence coverage based on experience with carnitine palmitoyltransferase-I, long-chain acyl-CoA synthetase, and voltage-dependent anion channel from rat liver mitochondrial outer membranes [543].

6. Noncovalent Interactions

ESI has been used to study a variety of noncovalent interactions [544–552] including those between peptides and polyphenols [553] or polyethers [554], drugs and oligonucleotides [555], proteins and DNA [556], a single-stranded DNA oligonucleotide and various polybasic compounds [557], and receptor and ligands [558–560]; studies of the interaction of peptides with carefully controlled partial pressures of water vapor indicate that the solvation behavior of ions of noncovalent complexes may provide clues to their conformations [561]. A strategy for identifying and characterizing protein interactions among gel-separated proteins and complexes has been developed and tested [562, 563]. A 310-kDa noncovalent homohexamer of the bifunctional enzyme HPr kinase/phosphatase has been characterized by ESI MS [564]. In combination with hydrogen/deuterium exchange, ESI with FTMS has been used to monitor the noncovalent interaction of hyaluronan oligosaccharides with the Link module from human tumor necrosis factor [565]. The noncovalent complexes between amino acid residues and selective ligands, such as sulfonic acid derivatives, have been characterized by ESI and MALDI [566]. A dodecamer of a small heat-shock protein has been reported [567]. As many as 16 proteins have been discovered in the catalytic core of the editosome, a multiprotein complex that catalyzes uridine insertion and deletion RNA editing to produce mature mitochondrial mRNAs in trypanosomes [568]. The bioactivity of sialic acid synthase has been correlated with its pH-dependent quaternary structure as determined during analyses by ESI MS at various cone voltages [569].

An example of mass spectral data for noncovalent interactions between the enzyme citrate synthases (CS) and its allosteric inhibitor, reduced nicotinamide dinucleotide (NADH), is shown in Figure 12-25. Interpretation of preliminary analyses (data not shown) by ESI TOFMS with orthogonal acceleration (*oa*TOF) led to the conclusion that CS is present in solution as a dimer as well as a hexamer [546]. In an effort to determine the specific associations of NADH with CS and to calculate the corresponding dissociation constant K_D, 1 mM of CS was mixed with increasing concentrations (4.5, 9, 18, and 108 mM) of NADH and then analyzed by ESI TOFMS with an *oa*TOF. The mass spectral data in Figure 12-25 are shown in two forms: on the left are the actual mass spectra collected from the *oa*TOF-MS and on the right are the deconvoluted spectra. Deconvolution (also called maximum entropy) calculations use a proprietary algorithm to determinine the molecular mass of the analyte from mass spectral data representing multiple-charge ions. The deconvoluted result is presented as a hypothetical mass spectrum of a single-charge (not protonated) ion that has the peak intensity distribution and resolution that a real spectrum of a single-charge ion would have on the instrument being used if the instrument had sufficient *m/z* range to perform the analysis. As can be seen in Figure 12-25, such hypothetical spectra greatly simplify data interpretation. The numbers over the peaks in the raw mass spectra indicate the charge state of the ion, whereas the numbers over the deconvoluted peaks indicate the

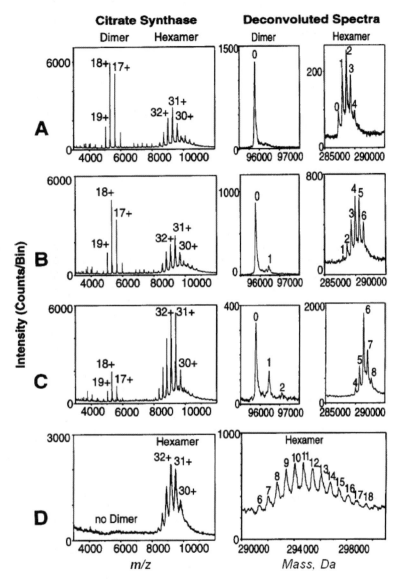

Figure 12-25. *Mass spectra of citrate synthase (CS) obtained in the presence of varying concentrations of NADH. Actual ESI mass spectra are on the left and corresponding deconvoluted (hypothetical single-charge) spectra on the right. Numbers on right-hand spectra indicate the number of NADH molecules bound to one molecule of CS. A: 4.5 mM, B: 9 mM, C: 18 mM, and D: 108 mM.* Reprinted from Krutchinsky AN, Ayed A, Donald LJ, Ens W, Duckworth HW and Standing KG "Studies of noncovalent complexes in an ESI-TOF mass spectrometer" Methods Mol. Biol. 2000, 146, 239–249, with permission from Elsevier.

number of NADH molecules bound to the CS dimer or hexamer. These mass spectra provide a direct indication of the stoichiometry involved in inhibition of CS by NADH: as the ratio of NADH to CS increases, there is a shift from dimeric CS to the hexameric form, and a simultaneous uptake of NADH by the hexamer. The low uptake of NADH by the dimer was shown to be nonspecific, and only uptake by the hexamer has biological significance.

While the focus is usually on hydrogen bonding in noncovalent complexes, there is a growing body of evidence that electrostatic interactions between basic and acidic residues, as well as between basic and aromatic residues, can be critically important [570]. Woods and coworkers have studied the formation and stability of noncovalent complexes linked between basic (K/R-rich) and acidic (D/E-rich or phosphate-rich) segments of peptides and proteins [571]. By using collisional dissociation, they found that salt-bridged complexes were much more stable than hydrogen-bonded dimers. Results of some of these studies led to the design and implementation of an antidote for the adverse effects of dynorphin A, which is released by traumamatic injury to the spinal cord; administration of a Glu/Asp-rich peptide froms a noncovalent complex with the cationic dynorphin A, thereby attenuating its neurotoxicity [572].

ESI mass spectra of nucleosides, recorded in the presence of alkali metals, display alkali metal ion-bound quartets and other clusters that may have implications for understanding noncovalent interactions in DNA and RNA [573]. Insights into mechanistic aspects of metal ion-binding/release by transferrins has been gleaned from studies by ESI [574]. Native hemocyanin (1.3 MDa) was found to consist of an 18-mer noncovalent assembly in equilibrium with dodecamer, hexamer, and monomeric structures [575]. Noncovalent complexes up to 2235 kDa were measured for the associations of 30 subunits of the *Bythograea thermydron* hemocyanins in crab, allowing characterization of various environments in the ocean according to oxygen content, etc. [576]. The subunit structure of the giant hemoglobin from the leech *Nephelopsis oscura* has been examined by ESI MS [577]. The binding specificities of viral fusion peptides for phospholipids has been studied by ESI techniques [578], as have the formation of noncovalent complexes between amyloid-β-peptide and melatonin [579]. Noncovalent complexes between chicken muscle adenylate kinase and two inhibitors, P1,P4-di(adenosine-5')tetraphosphate and P1,P5-di(adenosine-5') pentaphosphate, have been investigated with ESI MS under nondenaturing conditions [580]. The influence of pH and organic composition of aqueous solutions on noncovalent binding has been studied [581]. The use of an in-line desalting device facilitates the removal of salts and detergents that could interfere with the analysis of noncovalent interactions by ESI [582].

Because of the nature of the ESI process, some of the noncovalent complexes observed could be the result of nonspecific interactions, especially those involving small molecules that have a high tendency to form hydrogen bonds [583]. Special precautions and techniques like solution-phase hydrogen/deuterium exchange (HDX) can be used to determine the specificity and to identify the origin (whether in solution or *in vacuo*) of the complexes observed with ESI MS [584]. Protein complexes formed during gas-phase reactions between multiple-charge positive and negative protein ions in a quadrupole ion trap mass spectrometer have also been described [585]. The HDX method was used to investigate noncovalent complexes of some common chemotherapy agents: paclitaxel, doxorubicin, and etoposide by ESI MS [586]. The HDX approach with ESI has been used to evaluate the affinity of cellular retinoic acid binding protein I for its natural ligand all-*trans* retinoic acid (RA), isomers 13-*cis*- and 9-*cis*-RA, and retinol [587]; similar work was done with the retinoic X receptor a [588]. A GlcNAc-6-*O*-sulfotransferase and its complexation with the substrate 3'-phosphoadenosine 5'-phosphosulfate and the inhibitor 3'-phosphoadenosine 5'-phosphate have been studied using ESI FTICRMS [589].

Sites of copper(II) binding in metalloproteins can be marked by metal-catalyzed oxidation reactions to modify the protein by specific oxidation of only the amino acids bound to copper; these modified residues are then identified by MS. Pioneering work in this area was done with small peptides, such as angiotensin I (Agt I) and [Gln11]-amyloid-beta-protein fragment 1–16 (A beta (1–16)), and then with a larger protein, azurin [590]. The noncovalent interaction of beta-amyloid peptide with the natural derived antioxidant oleuropein has been studied by ESI [591]. An LC-nanospray ESI system was used to identify dozens of protein components in a premature mammalian ribosome pull-down from cultured cells as baited with an epitope-tagged protein [592]; a similar system was used to quantitatively assess the noncovalent interaction between RNase A and four oligosaccharide ligands [593]. ESI and size-exclusion chromatography have been used to study protein–protein interactions such as the allosteric inhibition of the lymphocyte function associated antigen-1/intercellullar adhesion molecule [594]. An algorithm facilitates recognition of noncovalent complexes by identifying ion pairs present in a mass spectrum that, when they share a common charge, have an m/z value difference that is an integer fraction of a ligand or binding partner molecular mass [595]. A new mass spectrometry-identifiable cross-linking strategy, using a reversible cross linker, has been developed to study protein–protein interactions [596].

Noncovalent interactions have also been studied with analyses by MALDI [552, 597–599]. An experimental approach called "intensity fading" facilitates the use of MALDI to study some noncovalent interactions; the method shows a reduction in the relative signal intensity for the low-molecular-mass binding partner (e.g., a protease inhibitor) when its target protein (i.e., a protease) is added to the sample as illustrated in experiments with trypsin bound to bovine pancreatic trypsin inhibitor [600]. Given the delicate nature of noncovalent interactions, it is important to use a matrix that does not interfere with the interactions being studied; e.g., in a study of the noncovalent interactions of some opioid peptides [601], it was necessary to use 6-aza-2-thiothymine as a matrix (pH 5.4), as greater acidity disrupted the electrostatic interaction between two adjacent basic residues in one part of the polypeptide with two to five adjacent acidic residues in the other peptide. MALDI has been used to study peptide–DNA complexes [602], as well as interactions between various dyes and peptides or proteins [603]. The noncovalent complex of cytochrome c consisting of four peptide subunits, two hemes, and a variety of lipids and metal ions can be stabilized by additives in the matrix during analysis by MALDI, whereas its dissociation without additives may allow subunit–subunit and subunit–lipid interactions to be monitored [604]. Another application utilized time-resolved limited proteolysis with the high-throughput capability of MALDI TOFMS to determine the binding site in a tetanus toxin C-fragment (51 kDa)–doxorubicin (543 Da) noncovalent complex [605]. The interfaces of noncovalent contact between two proteins can be identified and characterized by recognizing differences in peak intensities in MALDI mass spectra of the digested proteins relative to those of the same mixture treated with a cross-reacting antibody or other binding partner [606]. In this way, the antigenicity of viruses can be characterized by noting that mass spectral peak intensities representing binding domains are selectively suppressed, reflecting the extent of the noncovalent interaction [607].

7. Folding and Unfolding

Protein folding and noncovalent protein interactions have been studied by mass spectrometry in conjunction with pulse labeling techniques [584, 608–613]. Hydrogens located at peptide amide linkages have exchange rates with deuterium that permit the extent of labeling to be measured by MS [614]; use of electrospray is especially useful because a low-pH environment for the samples can be maintained to minimize back-exchange of the deuterium incorporation established during the pulse-labeling experiment [615–617]. Gas-phase hydrogen–deuterium exchange is a useful tool to investigate the gas-phase conformations of proteins when coupled with a mechanistic understanding of exchange [618, 619] and the involvement of Na^+ adduct formation by acidic side chains [620]. An ESI fast-flow technique has been employed to study the H/D exchange reactions of protonated serine with ND_3 and CH_3OD [621, 622]. In another ESI application, HDX kinetics of mildly denatured myoglobin at pD 9.3, in the presence of 27% acetonitrile, were studied with millisecond-time resolution [623]. A stopped-flow ESI MS method has been developed to study fast biochemical reaction kinetics [624]. On-line rapid mixing technology can be combined with ESI to measure enzyme kinetics [625]. An automated system H/D exchange study provides highly reproducible H/D exchange kinetics and facilitates replicate analyses without user intervention [626]. Recognition of EX1 and EX2 kinetic conditions during H/D exchange can be achieved based on the assessement of mass spectral peak widths [617].

In general, the thesis is that when a protein is unfolded, it effectively provides a greater cross section for interaction with a deuterated solvent. Thus, amide hydrogens located in α helixes, β sheets, and the hydrophobic core of a protein (where access to solvent is limited) are replaced by a deuteron slowly. In contrast, amide hydrogens located in peptide bonds between residues on the protein surface are hydrogen-bonded only to water, and they exchange much more rapidly with deuterons available in a pulse of D_2O. Because every residue except proline has a hydrogen in its peptide linkage, the degree of exchange of these amide hydrogens for deuterons provides some insight into their local environment, such as their involvement in intramolecular hydrogen bonding or the exposure of various segments of the polypeptide backbone to the solvent [584, 627, 628]. It is important to minimize the effects of back-exchange as much as possible during the analysis [629]; one way to accomplish this is to miniaturize the analytical system as in nanospray LC/MS to keep analysis times short [630]; however, the use of cryogenic techniques can have a dramatic effect on back-exchange if SFC can be used rather than HPLC [631]. The Woods Group has also contributed improvements to the HDX methodology, striving for near amino acid resolution of the sites of deuterium incorporation by obtaining and analyzing peptic fragments of the targeted protein [632]. The extent of information available from an HDX experiment can be extended by a statistical assessment of the experimentally determined distribution of a heterogeneously deuterated peptide to compute the average deuteration of each component of the mixture [633].

The results of an HDX study of the transmembrane fragment of the M2 protein of influenza A (in the presence of a large molar excess of the detergent Triton X-100) indicate that the accessibility of backbone amide sites to the solvent can be profoundly affected by membrane protein structure and dynamics, as well as the properties of model bilayer systems [634]. This pulse-exchange technique has been used to study the folding of a variety of myoglobins [635] and hemoglobin, obtaining near amino acid resolution through peptic digestion [636]; it has also been used in conformational studies of various kinases [637–641], a variant of cytochrome c2 [642], and enkephalins [643]. The noncovalent interaction of hyaluronan oligosaccharides with the Link module from human tumor necrosis factor has been monitored by HDX using ESI MS [565]. HDX techniques as monitored by MALDI have been used to reveal conformationation changes in peptides and proteins as a function of cysteine redox status [365] and to characterize autonomous folding units for pancreatic tryptsin inhibitor and long-R^3-insulin-like growth factor [366]. Analogous findings have been reported in a study of cytochrome c in which results of HDX revealed increased flexibility in the hinge region that is apparently mediated by oxidation of the methionine sixth heme ligand [642]. A detailed structural study of the solvent accessibility of aprion protein based on HDX used peptic digestion to achieve improved resolution of the loci of deuteriums in the protein [644]. HDX has been used to distinguish chiral forms of a serine octamer [645]. In another report based on H/D exchange methodology, a computer algorithm automatically identifies peptides and their extent of deuterium incorporation from H/D exchange mass spectra of enzymatic digests or fragment ions produced by CAD or ECD [646]. In cases where CAD MS/MS may be involved in HDX studies, it may be advisable to use alkali metal cationization instead of protonation in the ionization process in an effort to minimize scrambling of the deuterium [647]. Severe scrambling of deuterium has been reported in the use of MALDI-CAD MS/MS, which renders such use of high-energy CAD inappropriate in attempts to recognize the location of deuterium at the residue level in HDX studies [648, 649]. With careful instrumental control, the degree of hydrogen/deuterium scrambling can be minimized, if not eliminated [650].

HDX has been monitored by MALDI less frequently than by ESI, but with proper precautions in sample handling MALDI offers convenience to the analytical process. For example, analysis by MALDI offers the possibility of measuring the deuterium incorporation of several peptic fragments simultaneously [651, 652]. In HDX studies achieved using MALDI [653], a 7-kDa subunit was studied in topological association with a 333-kDa yeast protein complex. In all cases, it is important to minimize the effect of back-exchange reactions [654]. A MALDI-based HDX technique, termed SUPREX, has been used to characterize the thermodynamic properties of a series of model protein–peptide complexes [655]. A technique of brief and limited proteolysis with mapping of fragments by MALDI has been used to assess secondary structural features in proteins [656, 657].

While the pulse labeling technique is applicable to observation of folding intermediates of proteins of all types, some folding intermediates of cystinyl proteins contain free cysteines that can be trapped through chemical reactions. Wu and coworkers [658] demonstrated that a simple cyanylation reaction provided a suitable means of trapping folding intermediates of cystinyl proteins. Because the cyanylation reaction is carried out in acid (pH 3) the folding reaction is immediately quenched (acid trapping) and the free sulfhydryls are chemically modified to thiocyanates during the next 15 minutes. The mixture of cyanylated folding intermediates can be analyzed by HPLC to separate and visualize the distribution of species as shown for the refolding intermediates of insulin-like growth factor in Figure 12-26, which is a series of HPLC chromatograms of the refolding mixtures captured at designated times. The HPLC peak for the reduced and denatured species, R, is seen at the far right in the chromatograms in Figure 12-26, and the peak representing the native protein, N, is at the left; the peaks in the middle-left portion of the chromatograms represent various intermediates having different cysteine status [659]. Importantly, once the refolding intermediates of cystinyl proteins are trapped by cyanylation, they are well into the analytical procedure for determination of their disulfide structure or cysteine status, as described in an earlier section [660]. HDX has been integrated with cyanylation (CN)-based methodology to determine the conformation of cystinyl proteins and intermediates during refolding. The CN-based methodology can be used to trap, identify, and preserve the disulfide structure of a given cystinyl protein folding intermediate, while the HDX methodology can be used to assess other conformational features of the intermediate [365].

Other studies of the folding and unfolding of proteins have relied principally on monitoring the different charge-state distributions that distinguish various protein conformations. This approach also allows for monitoring the loss or binding of noncovalent protein ligands during the folding/unfolding process. In one report of this type of methodology, a study of the unfolding mechanism of hMb revealed a short-lived intermediate at acidic pH; no intermediate was observed under basic conditions [661]. The HDX behavior of the low-pH molten globule of human α-lactalbumin, containing all four disulfides, has been examined and compared with that of a single-disulfide variant and of a series of proline variants of α-lactalbumin [662]. HDX methodology can be used to assess the rigidity/flexibility of a protein as in a region near the binding site in cellular retinoic acid binding protein I [588, 663]. An ESI MS method for determining the equilibrium constant and free energy of protein unfolding was used to monitor the denaturation process at different pH values for three metallo-proteins by detecting peaks representing the native holo-protein and the unfolded apo-protein [664]. Intramolecular interactions in folded proteins have been postulated to play a key role in determining the observed charge-state distributions [665].

8. Applications

Some applications have been cited earlier in subsections describing special techniques. One interesting application has used proteomics techniques to identify proteins in old art paintings [666]. A major class of proteins called prions has also been studied by mass spectrometry [667]. The analysis of bacterial outer membrane proteins [668] and the tendency for selenoproteins to lose selenium during sample processing [669] have been described. Serine solutions of alkali metals yield magic-numbered clusters [670]. Accurate mass measurements (few ppm) of peptides up to 3000 Da by FTMS with MALDI have been reported [671]. Special protocols for handling and analyzing individual mammalian cells for peptides by MALDI have resulted in the

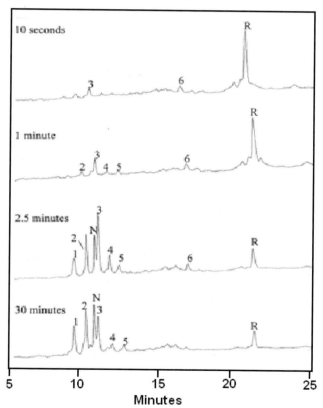

Figure 12-26. HPLC chromatograms obtained at indicated times during the refolding of hEGF as the reduced (R) form converts to the native (N) form via intermediates 1–6. Reprinted from Wu J, Yang Y and Watson JT "Trapping of intermediates during the refolding of hEGF by cyanylation, and analysis by MS" Protein Sci. 1998, 7, 1017–1028, with permission from the Protein Society.

identification of 14 pro-opiomelanocortin prohormone-derived molecules [672]; these protocols permit the classification of mammalian cells by peptide profile to facilitate elucidation of cell-specific prohormone processing and signaling. Highly active cytochrome b(6)f complexes from spinach and the cyanobacterium *Mastigocladus laminosus* have been analyzed by LC/MS/MS [673]. A chemical technique has been developed to identify amino acid residues on the surface of a protein by oxidative labeling as recognized by LC/MS/MS and FTMS [674].

Eukaryotic cells package their DNA with histone proteins to form chromatin that can be regulated to enable transcription, DNA replication, and DNA repair in response to cellular needs and external stimuli. A wealth of information from recent studies of post-translational histone modifications and histone variants have led to an explosion of insights into and more questions about how these processes might be regulated. Work by Donald Hunt and colleagues contributed greatly to understanding of the "histone code" by developing novel methods to study and identify histone modifications in both generic and specialized variant histone proteins [218]. Mass spectrometry has been indispensable in identifying a number of novel histone modification sites that formulate some of the histone code in an intricate system of regulation of genetics [217].

Instrumentation based on ion mobility measurements has been used to assess the time-dependent conformational changes of cytochrome c in the gas phase during residence times ranging from 10 to 200 msec in an ion trap [675]. An ion mobility–H/D exchange ion/molecule reaction chemistry study of bradykinin fragment 1–5 indicates that the protonated molecules exist as three distinct structural forms in the gas phase [676]. Ion mobility/TOFMS techniques have been used to examine distributions of fragment ions generated by CAD in a quadrupole ion trap; the mobility-based separation step prior to m/z analysis reduces spectral congestion and provides information that complements m/z-based assignments of peaks [677–680]. New avenues for enhancing the figures of merit (e.g., sensitivity, limits of detection, dynamic range, and analyte selectivity) and optimizing experimental parameters for combined ion mobility mass spectrometry have been reviewed [681].

III. Oligonucleotides

1. Analytical Considerations

Results of studies of the behavior of nucleosides, nucleotides [682], and oligonucleotides [683] by desorption/ionization laid the basis for sequence and structural analysis of these important biomolecules [684]. Studies on the effects of structural changes in oligonucleotides with respect to their behavior during CAD have been especially important in formulating strategies for data interpretation and sequence determination [685–688]. The high preponderance of phosphate groups in oligonucleotides provides a natural driving force for producing negative ions by deprotonation [689, 690]. Probably because of the low pK_{a1} of the phosphate group, together with the large number of phosphates in oligonucleotides, the charge state of most of the ions is rather high [691]. However, investigations of conditions (i.e., use of solutions of organic ammonium salts) for producing positive ions (polyprotonated species) of oligonucleotides indicate that the positive-ion mode provides some complementary features to results obtained in the negative-ion mode; further, the abundant protonated bases observed in the positive-ion mode offer an advantage in detecting modified nucleotides [692]. Hydrophobic phosphoamidate derivatives of deoxyguanosine adducts, prepared under mild conditions with coupling reagents typically employed in peptide synthesis, gave significantly better detection limits than underivatized counterparts during analysis by ESI LC/MS [693]. A systematic study of ESI tuning parameters revealed that the detectability of multiple-charge biopolymer ions of different molecular size by ESI varies greatly in quadrupole ion trap and triple-quadrupole instruments, depending on whether low-charge-state or high-charge-state tuning is used [694].

MALDI has also been used in the analysis of oligonucleotides, and it holds great promise as a future alternative to conventional sequencing procedures for high-throughput screening for mutations [695]. Distler and Allison have shown that 5-methoxysalicylic acid serves as a useful matrix when little fragmentation of the oligonucleotide is desired; further, they found that the combined use of spermine [696], methylene blue [697], or cobalt hexamine [698] stabilized the duplex and diminished the need for desalting the sample prior to analysis by MALDI. In another work, reduction of cytosine was observed during analysis by MALDI in DHB [699]. In studies of oligonucleotides bearing fixed charge sites at one of the termini, it was discovered that PSD fragment ions always incorporate the charge tag. H/D exchange results showed no evidence for intramolecular migration of protons on the phosphate backbone to initiate the fragmentation event; instead, unique pathways for proton migration are rationalized by a charge-remote fragmentation pathway [700].

Concerning an operational problem in the use of electrospray for the analysis of samples prepared by exonuclease digestion, rapid buildup in the concentration of nucleotides produces interfering cluster ions; Wu and Aboleneen have found that alkaline phosphatase added with snake venom phosphodiesterase to the oligonucleotide solution converts the interfering nucleotides into noninterfering nucleosides and greatly improves the signal-to-background [701]. Preliminary studies with ECD in the analysis of hexanucleotides show that only limited sequence information is available with fragmentation mostly occurring close to the ends of the molecules [702].

2. Sequencing

A. Nomenclature

The conventional manner for presenting the structure of an oligonucleotide is from 5′ to 3′ (i.e., its 5-hydroxy terminus is written on the left and the 3-hydroxyl terminus is written on the right). By analogy to the nomenclature describing fragment ions of proteins (N-terminal fragment ions designated by letters early in the alphabet), ions associated with the 5′ terminus in the mass spectrum of an oligonucleotide are assigned letters early in the alphabet. Those fragment ions associated the 3′-terminus are assigned letters late in the alphabet. The nomenclature represented in Scheme 12-20 is that introduced in the early work by McLuckey *et al.* [682]; the fragmentation processes of nucleic acid components have been studied extensively [685, 686, 692, 703–705].

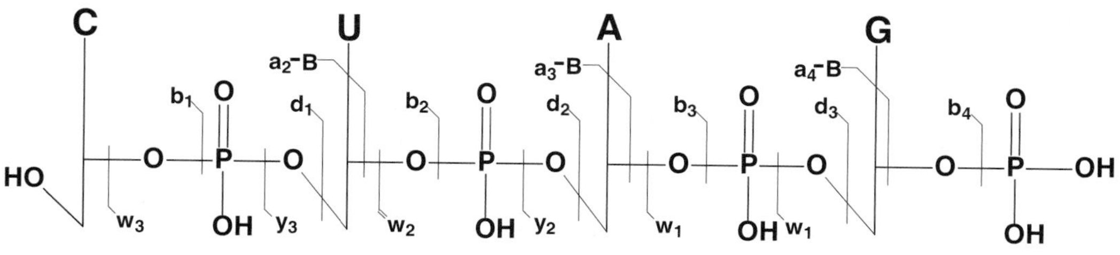

Scheme 12-20

The a_i–B_i ions are important in recognizing the sequence of nucleotides from the 5'-end of the biopolymer; however, some attention to the structural components of this ion is required [686, 692, 703, 704]. The a_i corresponds to the position number of a given sugar (ribose) from the 5'-terminus; B_i corresponds to the base (adenine, cytosine, uracil, etc.), which is lost from that particular ribose. A given a_i–B_i ion represents all the sugar residues from the 5'-terminus up to the designated nucleotide position, but not the base originally connected to that particular residue; e.g., a_3–B represents an ion consisting of the first two complete nucleotides from the 5'-terminus, but only the sugar residue for the third nucleotide, i.e., the base for the third nucleotide is expelled in forming the a_3–B ion. The left panel in Scheme 12-21 shows a general "stick" structure for the a_3–B ion that would be formed from the tetranucleotide shown in Scheme 12-20; adenine (A) has been expelled from the third nucleotide (adenosine) during formation of this ion while the bases cytosine (C) and uracil (U) are retained on the sugars of the first two nucleotides, respectively. The right panel in Scheme 12-21 shows the chemical structure (rather than the stick model) of the a_2–B ion that originates from the tetranucleotide shown in Scheme 12-20. In this a_2–B ion, only the base cytosine is retained (the base uracil has been expelled from the second ribose residue from the 5'-terminus), and the second ribose residue in the original tetranucleotide now appears in the form of a furanyl ring. While the phenomenon of expelling the base from the ith ribose to form the a_i–B_i ion may seem to render such fragment ions useless at first glance, once a series of a_i–B_i ions (or other ions) has been recognized, the serial differences in mass between members of the series provides evidence for a given sequence of nucleotides.

The general stick structures (drawn as neutrals) for other selected CAD fragments from the tetranucleotide CUAG are shown in Scheme 12-22; protons would be removed from various phosphate groups to form the charge state observed for a particular negative ion [686, 692, 703, 704]. The stick structures of the d_2 and b_2 CAD fragments (Scheme 12-22) show that detection of such corresponding ions would provide information to corroborate that gleaned from having detected the a_3–B ion illustrated in Scheme 12-21. Similarly, the stick structures for w_1 and y_1 in Scheme 12-22 illustrate the information provided by detecting such ions for the tetranucleotide shown in Scheme 12-20.

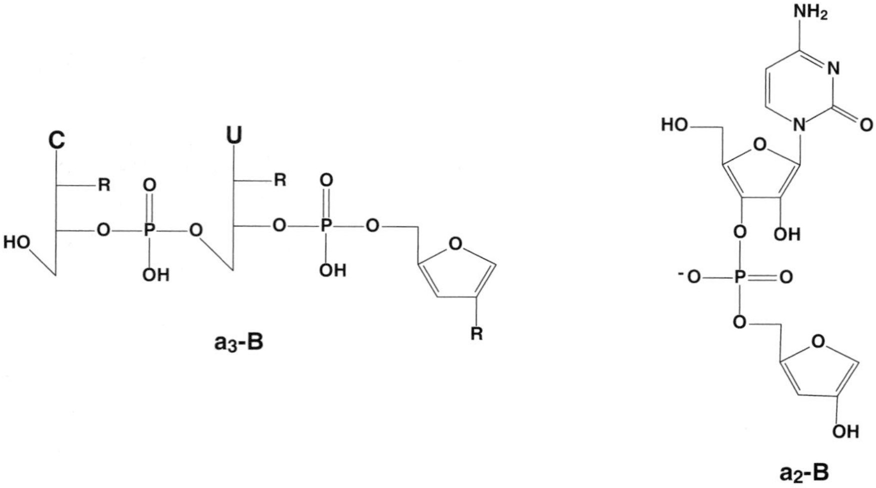

Scheme 12-21

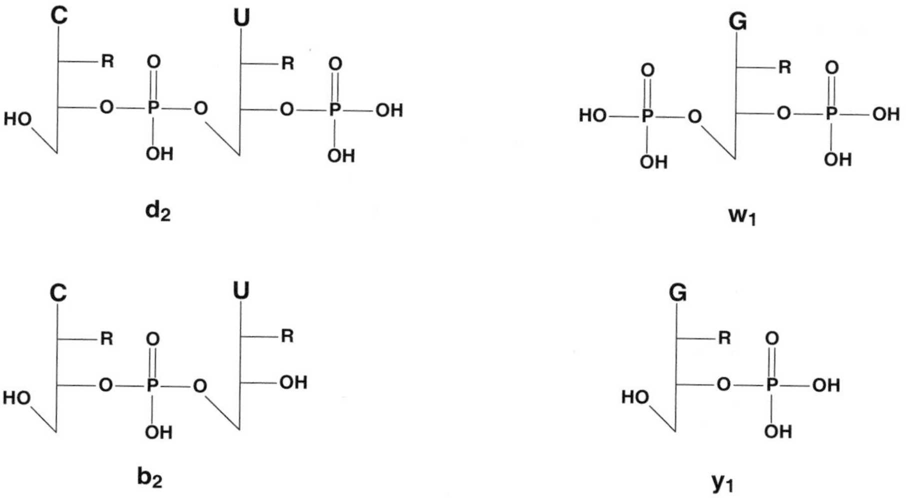

C U G

d₂ w₁

C U G

b₂ y₁

Scheme 12-22

In general, $a_i–B_i$ and w_i ions are the most abundant in the CAD mass spectra of DNA, while y_i ions are the most abundant in the CAD mass spectra of RNA [706, 707]. R = H for DNA, but R = OH for RNA, in the structures in Schemes 12-21 and 12-22. The interpretation usually begins by recognizing a peak for the w_1 ion as the anchor of the 3′ to 5′ ladder. Whereas, there may be peaks for several different w_1 ions; the "correct" one will usually be the most abundant. For example, in the spectrum for the tetranucleotide CUAG, there may be peaks for phospho-U, phospho-A, and phospho-G; however, that for phospho-G is likely to be the most intense because production of the phospho-G species requires only simple cleavage at the 3′-terminus (see Scheme 12-20). The "incorrect" candidates for the w_1 ion (phospho-U and phospho-A) would be less abundant because they would derive from double cuts in the polymer chain, and therefore would be less likely to occur. It is possible to extend the sequence ladder by testing differences in *m/z* values of various peaks against the next four nucleotide mass values, etc. The same strategy is used in interpreting the data for a series of $a_i–B_i$ ions, which will provide sequence information emanating from the 5′-terminus [686, 692, 703, 704].

B. Algorithm for Data Interpretation

Information on the nucleotide sequence is most often represented by fragment ions of a variety of charges; to some extent the more nucleotides there are in a given fragment, the more likely the ion will have more than one charge. For this and other "bookkeeping" reasons, an algorithm greatly facilitates the use of mass spectral data to identify or confirm the sequence of a given oligonucleotide. The SOS algorithm [708] permits user control on a residue-by-residue basis; a particular advantage is the ease with which the analyst can erase and rebuild alternate sequences. Modifications to the parent structure can be defined in any combination of base, sugar, or backbone; sequence ladders can be constructed independently in either the 5′ to 3′ or 3′ to 5′ direction, and graphically compared for homology and overlap.

A different algorithm compares the *m/z* values of peaks in the experimental spectrum to predicted *m/z* values by employing established fragmentation pathways from a known reference sequence; the closeness of matching between the measured spectrum and the predicted set of fragment ions is presented numerically as an index of "fitness" [709]. Another algorithm has been described for *de novo* sequencing of short nucleic acids [710]. Mass spectrometry-based proteomic methodologies can be used to annotate both nucleotide and protein sequence databases [711].

3. Applications

The helix clamp motif of HIV has been characterized using MALDI TOFMS [712]. MALDI has been used to characterize DNA by hybridization with microarrays of gel-immobilized oligonucleotides [713]. MALDI has also been used to analyze DNA methylation carried out by the *Escherichia coli* DNA methyltransferase using oligonucleotide substrates with molecular masses of 5000–10,000 Da per strand; because both unmethylated and methylated DNA are detected, the ratio of methylation can be determined directly and accurately [714]. The use of exonucleases in combination with analyses by MALDI at appropriate times during the course of hydrolysis is the basis for kinetic measurements, which reveal the location of basic sites in an oligonucleotide [715]. Carcinogen-modified oligodeoxynucleotide adducts have been characterized by interweaving partial degradation by an exonuclease with analyses by MALDI PSD [716]. The instrumental parameters that affect detection limits of DNA at high mass have been studied [717]. On-probe sample cleanup of DNA samples has been demonstrated with the use of a polymer-modified cationic surface on a MALDI sample probe that selectively retains nucleic acids, whereas neutral and positive-charge contaminants are washed away with a simple water rinse [718].

Ion-pair reversed-phase HPLC and ESI MS are the most suitable for analysis of nucleic acids [719]; the use of $NH_4(H_2PO_4)$ as an additive improves both the recovery of analyte from chromatography and its detection by mass spectrometry [720]. The complete modification map for SSU rRNA from *Thermus thermophilus* has been characterized by HPLC/ESI MS-based methods [721]. Guidance for the recognition and sequence placement of pseudouridine from LC-ESI MS data has been described for this important post-transcriptional RNA modification [722]. A quantitative method for determination of 5-methyl-2′-deoxycytidine in human DNA based on LC-ESI MS uses stable isotope-labeled internal standards [723]. A systematic CAD MS/MS study of protonated 2′-deoxycytidine, 5-formyl-2′-deoxycytidine, 5-hydroxy-2′-deoxycytidine, 5-hydroxymethyl-2′-deoxycytidine, and their corresponding stable isotope-labeled analogs helps define how chemical modifications of pyrimidine bases are reflected in their product-ion spectra [724]. Complete peptide nucleic acids (PNAs) sequence information has been reported from the unimolecular decomposition of single-charge PNA oligomers in the negative-ion mode using ESI FTICRMS and sustained off-resonance irradiation (SORI) for CAD [725]. In recent studies of noncovalent interactions, ESI MS complexes of double-stranded DNA with cisplatin, daunomycin, and distamycin were obtained that contained only negligible amounts of single-stranded DNA [726]. The influence of the polyphosphate chain on conformational aspects of adenine nucleotide anions has been studied by HDX and analysis by ESI [727]. The results of a comprehensive study of the behavior of bases, nucleotides, and nucleosides under APPI may offer some analytical advantage in method development for nucleic acids [728].

A detection system based on ESI has been developed for single nucleotide polymorphisms (SNPs) requiring no off-line purification of the polymerase chain reaction (PCR) product and achieved by simple addition of reagent solution into a single sample well [729]. SNPs can also be detected by MALDI [730]. MALDI MS has also been used in a DNA mutation detection protocol to identify and characterize a previously unknown change in a given sequence [731]. Methodology for the detection of fetal-specific alleles relating to point mutations as well as SNPs has been described [732]. A model developed in a study of nucleotides by ESI allows the response of a given PCR amplicon to be adjusted in proportion to its hydrophobicity so that the detection limit of a PCR method can be improved, sometimes by an order of magnitude [733]. PCR products, up to 120 mers, can be identified by sub-ppm mass measurements to provide an unambiguous base composition by ESI FTMS, provided that the sample is properly desalted and otherwise purified on an anion-exchange resin [734].

IV. Carbohydrates

1. Analytical Considerations

Although somewhat neglected earlier during the explosive growth of protein characterization by mass spectrometry, carbohydrate structural analysis is now claiming attention [299, 735–738]. The technological status of global and site-specific glycosylation analysis of gel-separated proteins has been critically reviewed [739]. The development of glycomics will probably be much slower than that seen in proteomics because, unlike the rather simple linear primary structure of proteins, the glycans are branched structures.

From an instrumentation point of view, innovations in multistage MS are needed to improve the efficiency of transferring analyte ion current dramatically from one stage of *m/z* analysis to the next. The mind-boggling complexity of a bioactive glycan as it extends outward from a basic core composition of a few saccharides through multiple branches to form a 3D structure consisting of many antennae will require 3 to 5 or more stages of MS for continuity of analysis leading to structure elucidation. With current multistage MS instrumentation, usable signal strength (ion current) beyond MS/MS/MS is simply not available from realistic samples containing glycan structures.

Meanwhile, progress is being made in characterizing model glycan systems by mass spectrometry. Although ESI CAD gives excellent MS/MS spectra, particularly from the MNa^+ ions, the signal falls substantially as the molecular mass of the carbohydrate rises; this problem is reversed by use of a high-pressure (0.1 mbar) MALDI ion source as used with a Q-TOF instrument. Early reports indicate that MS and MS/MS spectra of complex glycans using high-pressure MALDI were less complicated than those recorded with a conventional MALDI-TOF instrument because of the absence of background ions resulting from metastable fragmentations in the flight tube [740]. Transient elevation of the pressure to the 1–10 mbar range during MALDI in an FTMS instrument decreases the metastable fragmentation of gangliosides [741]. The first evidence for pyran cross-ring cleavages that define the inter-residue linkage structure for glycolipids was obtained with a quadrupole ion trap during the analysis of an isomeric disialyl glycosphingolipid extracted from human brain tissue [742]. ESI CAD of protonated disaccharides on a triple quadrupole mass spectrometer produced substantial fragmentation of the glycosidic bond in a systematic study to determine the effect of spatial crowding of incremented alkyl groups in two peralkylated (methyl to pentyl) anomeric pairs (maltose/cellobiose and isomaltose/gentiobiose) [743].

2. Nomenclature

The nomenclature proposed by Domon and Costello [744, 745] for the fragment ions obtained by CAD after desorption/ionization of carbohydrates is illustrated in the generic structure in Scheme 12-23. The exchange of labile hydrogens for sodium or other cations occurs readily. For example, replacement of a hydrogen in a B ion by sodium might be designated as $[B + nNa - nH]^{n+}$ or more simply as $B(nNa)^{n+}$ as recommended by Costello *et al.* [746].

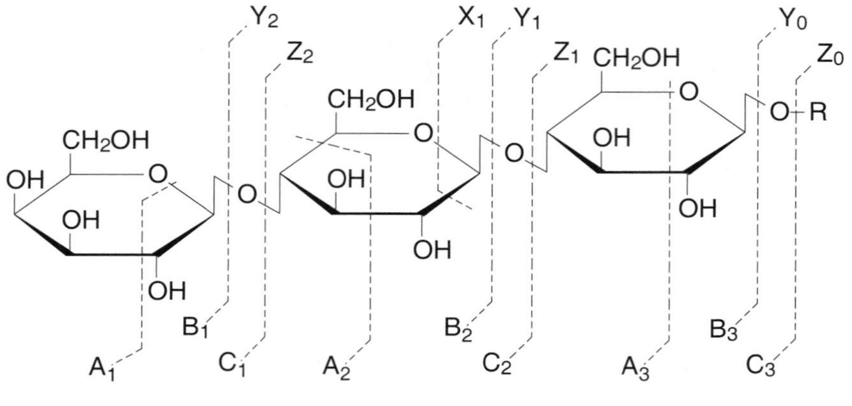

Scheme 12-23

3. Diagnostic Fragmentation

The results of systematic CAD MS/MS studies of model compounds in several laboratories have led to recognition of useful diagnostic fragmentations. However, recent studies indicate that some fragments resulting from expulsion of internal residues from oligosaccharides can give false sequence information [747, 748]. Ionization of oligosaccharides by adduct formation with the divalent cations Mg^{+2}, Ca^{+2}, Mn^{+2}, Co^{+2}, and Cu^{+2} for subsequent CAD has been studied for analytical advantage in producing structurally diagnostic fragmentation [737]; CAD of Ag^+ adducts of neutral high-mannose, hybrid, and complex N-linked glycans is especially efficient in cleaving the glycosidic bonds [749]. CAD MS/MS studies of $[M + NO_3]^-$ and $[M + (NO_3)_2]^{2-}$ of complex N-linked glycans formed during ESI in the presence of ammonium nitrate show predominant C-type glycosidic and cross-ring fragments in contrast to the corresponding spectra of the positive ions that contain mainly B- and Y-type glycosidic fragments [750]. Results of a fragmentation study of chondroitin sulfate oligosaccharides indicate that B- and Y-type ions vs C- and Z-types are produced as a function of the prevalence of competing proton-transfer reactions [751]. Similar negative ions of high-mannose N-linked glycans define the sequence of the constituent monosaccharide residues and are more informative than the corresponding positive-ion spectra regarding the composition of individual antennas not easily obtainable by other means [752].

Because many oligosaccharide antibiotics, such as the everninomicins, do not readily form protonated molecules during ESI, the addition of NaCl can be used to produce sodiated species, which when subjected to CAD MS/MS provide an important series of fragment ions that are specific for sugar sequences and for some sugar-ring openings [753]. A fragmentation study of N-glycosylated peptides by ECD and by infrared multiphoton dissociation (IRMPD) following ESI showed that ECD provided **c** and **z** ions derived from the peptide backbone, except at sites close to the glycosylated asparagine residue, with no observed loss of sugars; IRMPD provided abundant fragment ions, primarily through dissociation at glycosidic linkages from which the monosaccharide composition and the presence of three glycan branch sites could be determined [515]. N-linked oligosaccharides were analyzed successfully from solution by IR in an AP-MALDI experiment [754]. Permethylation of oligosaccharides improved the ESI-CAD MS/MS fragmentation patterns in which the Y- and B-ion series predominated; further evaluation with MS^3 showed that additional information on branching could be obtained from A and X cross-ring fragmentation [755]. Extensive cross-ring fragmentation relating to the linkages of the monosaccharide residues has been observed in the MALDI TOF/TOF MS/MS spectra of oligosaccharides [756, 757]. Fragmentation induced by CAD (TOF/TOF) subsequent to MALDI has been used to facilitate the detailed structural characterization of high-mannose and hybrid-type N-glycans [758]. Analytical strategies involving solid-support assisted sequencing of oligosaccharides may offer advantages [759].

Strategies for data interpretation can be developed based on diagnostic fragmentations observed during CAD MS/MS and other analytical techniques [760]. Reinhold and coworkers have contributed a series of three integrated reports on their congruent strategies for carbohydrate sequencing. The first in their series of seminal reports summarizes the analytical considerations for approaches incorporating the behavior of small oligomer structures during analysis by CAD MS/MS that forms the basis for deducing the structure of glycopolymers based on analysis by mass spectrometry [760]. The second report describes the formation of a library of MS data from known compounds and its use for a "bottom-up" approach to achieve full oligosaccharide and glycan characterization [761]. The second report also describes a set of software tools for storing, organizing, and comparing spectral files, including the identification of isobaric mixtures. These tools provide a facile and objective evaluation of structural details including inter-residue linkage, monomer identification, anomeric configuration, and branching. The third report introduces the operational details of an algorithm defined as the Oligosaccharide Subtree Constraint Algorithm (OSCAR) [762]. This algorithm assimilates analyst-selected MS^n ion fragmentation pathways into oligosaccharide topology (branching and linkage) using what may be considered a "top-down" sequencing strategy. Guided by a series of logical constraints, this *de novo* algorithm provides molecular topology without presumed biosynthetic constraints or external comparisons. In this introductory study, OSCAR is applied to a series of permethylated oligomers and isomeric glycans, and topologies are assigned in a few hundredths of a second.

In numerous situations, it has been found advantageous to derivatize samples by methylation [763, 764] with the products most frequently analyzed as sodium-adducted positive ions [765]. The methylation strategy provides three important advantages for the analysis of carbohydrate samples by MS:

a) It generates a lipophilic product to facilitate organic-phase extraction from complex biological matrices.

b) It is the basis for quantitating modulation of glycan expression by differential CH_3- and CD_3-methylation.

c) It facilitates identification of characteristic structural features that are readily exposed under CAD.

Key structural features in an oligosaccharide include the reducing and nonreducing termini as illustrated in Scheme 12-24. CAD MS/MS of a methylated sample produces a Y-series ion, which defines the reducing terminus, and a B-series ion to define the nonreducing terminus at each glycosidic cleavage point; these fragments would appear as isobars with underivatized samples. Moreover, Y-series fragments are left with an exposed hydroxyl moiety, labeling its point of monomer linkage; branching is indicated by ions corresponding to two such residues as shown in Scheme 12-24 [735, 766].

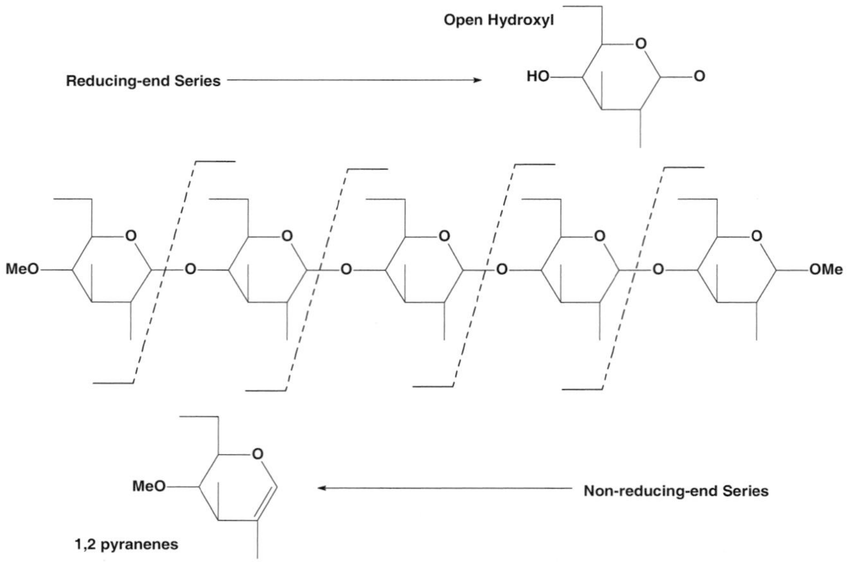

Scheme 12-24

Hydroxyl positions are not discernible in the Y-series of ions in CAD spectra, but are represented in nonreducing fragments as a consequence of cross-ring cleavage [767]. These identifying fragments appear as linkage-specific masses pendant to the B-series and their mass shifts define the linkage point; e.g., 6-O-linkage exhibits B_n+60 and B_n+88 ion fragments; 4-O-linkage, a B_n+88 only fragment; 2-O-linkage, a B_n+74 fragment; and for the 3-O-linkage, a C_n elimination, all as exemplified in Scheme 12-25. These details are most apparent with CAD of small fragments where the energy is dissipated through fewer pathways of dissociation; rarely are such fragments observed from dissociation of high-mass precursors [767, 768].

Determination of the phosphate position in carbohydrates is difficult due to the low-energy loss of the phosphate either as a neutral or as an ion in MS/MS experiments. A possible solution to this problem has been proposed in the use of ion/molecule reactions in which single-charge negative ions from phosphorylated monosaccharides are reacted with trimethyl borate in an FTICR-MS analyzer cell to incorporate boron with the loss of a neutral methanol molecule [769]. The resulting modification stabililizes the phosphorylated monosaccharide under SORI-CAD conditions, allowing generation of ions characteristic of the phosphate linkage; ion structures and dissociation mechanisms explaining these results have been described [769]. Some fragmentation rules for determining the position of phosphorylation in N-glycans were developed based on a study of four kinds of phosphorylated glycans labeled with 2-aminopyridine, purified from yeast mannan and analyzed by MALDI PSD [770]. Sulfated and phosphorylated analogs of a carbohydrate (both of which are 80 Da heavier than the unmodified molecule) can be differentiated by methodology based on MALDI MS [771]; a method based on the use of an ion-pairing reagent in conjunction with ESI can distinguish sulfated from phosphorylated monosaccharides [296, 772].

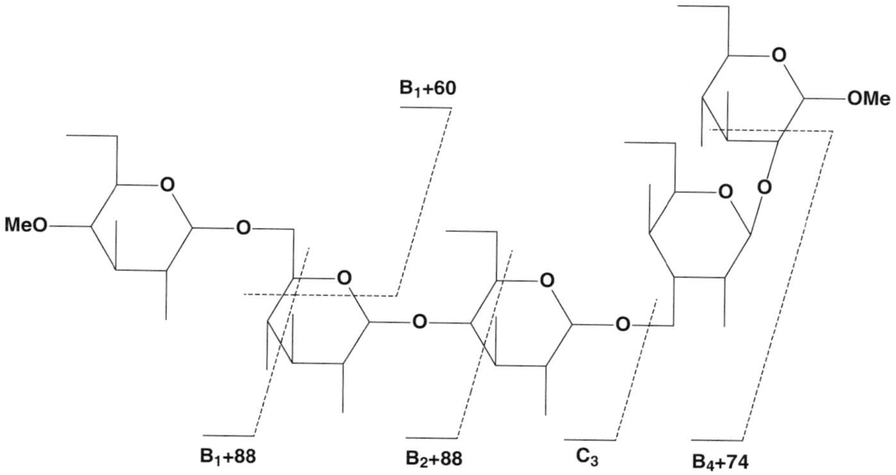

Scheme 12-25

In the first application of electron detachment dissociation (EDD) to carbohydrates, the results show that EDD produces information-rich mass spectra with both cross-ring and glycosidic cleavage product ions as opposed to exclusive glycosidic cleavage when using CAD and IRMPD [773]; in this way, EDD shows great potential as a tool for recognizing sites of sulfation and other modifications in these labile glycosaminoglycan oligosaccharides. EDD also produces diagnostic product ions that can be used to distinguish the epimers iduronic acid and glucuronic acid during the analysis of glycosaminoglycan tetrasaccharides [774].

4. Applications

The initial step in quantitative analysis of O-linked glycans of glycoproteins is to release them in high yield, nonselectively, unmodified, and with a free reducing terminus; it has been demonstrated that O-glycans can be prepared by hydrazinolysis without degradation, thereby making low-microgram levels of naturally occurring glycoproteins amenable to subsequent analysis by MS [775]. Newer methodology using both CAD and ECD can be used to assign the site of the glycan without its release by chemical means [313]. ECD is effective in promoting fragmentation of metal-adducted oligosaccharides to provide saccharide linkage information [776]. The technology of ETD is of great value in analyzing N-glycosylated peptides, in part because loss of the glycan-modified side chain allows deduction and verification of the glycan mass and the nature of the amino acid residue at the glycan attachment site [311, 312]. A high resolving power allows the selective detection of glycan-specific fragment ions minimizing the interference of peptide-derived fragment ions with the same nominal mass [777]. The Q-trap mass spectrometer has been used to characterize protein glycosylation via ESI LC/MS [778]. The microheterogeneity of large sugar chains in glycopeptides can be dissected by applying IRMPD during analysis by FTICRMS [779]. In some cases, it may be possible to analyze by MALDI native glycosphingolipids directly from membranes after transfer from TLC plates [780]. Alkylsulfonates serve as anion dopants to allow neutral and acidic oligosaccharides to be analyzed in the same mixture by MALDI [781]. Partly depolymerized hemicelluloses isolated from wood chips of spruce and aspen [782] and xylans isolated from birch, aspen, spruce, pine, and larch [783] have been characterized by MALDI.

Among the many recent applications of atmospheric MALDI with a quadrupole ion trap, collisional activation of sodiated oligosaccharides, as demonstrated with maltoheptaose, produces primarily B and Y fragments resulting from cleavage of glycosidic bonds; fragments from cross-ring cleavages are also observed following three stages of tandem mass spectrometry, providing additional linkage information for N-linked glycans from chicken egg glycoproteins and a ribonuclease glycan mixture [784]. Using exoglycosidase carbohydrate sequencing with analysis of oligosaccharides by MALDI, several new carbohydrates on apoB100, including truncated complex biantennary N-glycans and hybrid N-glycans, have been identified [785]. The use of capillary electrophoresis combined with ESI is especially useful in the analysis of the oligosaccharides once they have been chemically converted to a charged species for electrophoretic migration [786]. Methodology for confirming the sialic acid linkage in carbohydrates ranging from trisaccharides to biantennary N-linked glycans has been described using MALDI (D-arabinosazone as matrix) and electrospray MS [787]; once cleaved, the sialic acids can be converted to quinoxaline derivatives with 1,2-diamino-4,5-methylenedioxybenzene for analysis by ESI LC/MS [788]. A method based on sequential degradation, *p*-aminobenzoic ethyl ester (ABEE) closed-ring labeling, and negative-ion ESI-CAD MS/MS provides more linkage information for N-linked oligosaccharides than the more popular open-ring reductive amination approach [789]; for sugars containing sialic acid moieties, a protecting group must be used to stabilize sialic acid groups during sequential alkaline degradation, as demonstrated in the analysis of two high-mannose oligosaccharides, M5G2 and M6G2, cleaved from the RNase B and a complex oligosaccharide A2 cleaved from transferrin. A hybrid MALDI Q-*oa*TOF-MS has been used to analyze N-linked oligosaccharides released by N-glycanase F and reductively aminated with anthranilic acid [790]. The N-linked oligosaccharides in hepatocyte growth factor produced in mouse myeloma NS0 cells have been separated and identified according to sequence, branching structure, and linkage using ESI LC/MS with a graphitized carbon column [791]. The analysis of heparin-like glycosaminoglycans is possible by tandem MS once the sulfate groups are deprotonated and stabilized with Ca^{+2}, thereby increasing the relative abundance of backbone cleavage ions and decreasing the abundance of ions produced from SO_3 losses [792, 793]. Other ion-pairing reagents can be used to stabilize the SO_3^- group of the glycopeptide so that useful information can be obtained during the CAD MS/MS experiment [794].

Karas and coworkers [795] have refined methodology for deprotonating underivatized neutral oligosaccharides during analysis by nanoelectrospray ionization and using CAD in a quadrupole ion trap to effect cross-ring cleavages that yield information about linkages of internal monosaccharides; they have applied this methodology to the analysis of human milk for neutral oligosaccharides [795]. Use of a porous graphitic carbon HPLC column allows analysis of underivatized water-soluble oligosaccharides by LC/MS [796]. An in-gel technique for releasing N-linked glycans from glycoproteins for analysis by negative-ion MALDI PSD without significant loss of sulfate groups has been demonstrated in the analysis and characterization of 22 neutral and 15 sulfated N-linked glycans [797]. Cheng and Her have shown that negative-ion ESI-CAD MS/MS of the *p*-aminobenzoic ethyl ester (ABEE) of linear as well as branched oligosaccharides produces linkage-specific fragment ions [798]. The complexity in the structure and composition of N-linked oligosaccharides attached to individual murine immunoglobulin M glycosylation sites has been investigated with ESI LC/MS [799]. Nimtz and coworkers have also pursued negative-ion technology during low-energy ESI-CAD MS/MS with a Q-TOF to produce diagnostic D ions for assignment of 3- and 6-antennae in complex N-glycans carrying three and four antennae in combination with epitope-relevant B- and C-type ions [800]. After validating their new approach through ESI Q-TOF fragmentation of the permethylated analogues in the positive-ion mode, they analyzed products of *in vitro* alpha-1-3-fucosylation of tri- and tetraantennary precursors to test structural requirements for formation of Le(x) vs sLe(x) motifs [800]. On-line TLC with detection by ESI has been described for the analysis of glycolipids [801].

Neutral and acidic glycosphingolipids have been characterized by high-pressure (0.1 mbar) MALDI with a Q-TOF; the technology was applied to the identification of ceramide-trihexosides present in tissues from mice genetically modified to model one of the glycolipid storage diseases (Fabry disease) [802]. The N-glycosylation pattern of polysialylated NCAM from brains of newborn calves was found by methodology using MALDI [803]. Determination of sialylated oligosaccharides is useful in distinguishing chain and blood group types together with branching patterns; the results of one study characterized a novel complex disialylated and monofucosylated tridecasaccharide that is based on the lacto-N-decaose core [804]. Sialylated oligosaccharides have been characterized by IR-MALDI [805]; the influence of sialyation on negative-ion dissociation has also been studied [806]. The results of a study of the N-linked glycosylation (as characterized by exoglycosidase digestions and analysis by MALDI) of recombinant human CD59, expressed in Chinese hamster ovary (CHO) cells with and without a membrane anchor, show that the presence of the anchor increases the extent of glycan processing, possibly as the result of longer exposure to the glycosyltransferases or to a closer proximity of the protein to these enzymes [807]. ESI-CAD MS/MS can be used for the determination of a chondroitin sulfate oligosaccharide sequence with respect to the positions of GalNAc sulfation [808, 809]; this group has also studied the influence of the charge state of chondroitin sulfate oligosaccharides on the product-ion spectra for purposes of determining the sulfate position on N-acetylgalactosamine residues [810]. A study of CAD MS/MS of a variety of keratan sulfates (KS) reveals diagnostic ions that can be used as fingerprint maps to identify unknown KS oligosaccharides as demonstrated in analyses of shark cartilage KS and of two enzyme digests of bovine corneal KS after minimal sample processing [811]. A study of the sulfate position in heparin octasaccharides based on use of FTICRMS revealed that 2-O and N-sulfation of heparin are more important for binding to monocyte chemoattractant protein 1 than 6-O sulfation [812]. A detailed analysis of the N-glycans from the soil nematode *Caenorhabditis elegans* revealed five major classes including high-mannose, mammalian-type complex, hybrid, fuco-pausimannosidic (five mannose residues or fewer substituted with fucose), and phosphocholine oligosaccharides [813]. Sialoglycosphingolipid (gangliosides) membrane components of healthy human volunteers were isolated and analyzed by ESI-CAD MS/MS [814]. ESI-CAD MS/MS has been used to identify a family of complex cell surface glycolipids called the Neisserial lipooligosaccharides [815]. Two unique neutral glycosphingolipids have been isolated from patients with a rare inheritable polyagglutination phenomenon, and identified by ESI-CAD MS/MS [816]. Nano-ESI with a Q-TOF-MS was used for sensitive mapping and sequencing of single molecular species in complex ganglioside mixtures in which fucosylated carbohydrate chains of granulocyte gangliosides carry sLex and VIM-2 epitopes [817]. Accurate mass measurements as from FTMS have been used to facilitate structural analysis of glycated peptides [818]. Identification of disaccharide isomers can be accomplished by high-field asymmetric waveform ion mobility spectrometry [819]. The molecular formulas for the structures and substructures of muraymycin antibiotics were determined by ESI FTMS utilizing capillary-skimmer fragmentation with up to five stages of MS; using the top-down/bottom-up approach, the formulas were determined by summing the elemental formulas of the neutral losses, obtained by measuring the mass differences (<500 Da) between the genetically related sequential parent-ion masses in the MS^n spectra, with the unique elemental formula of the lowest parent-ion mass (<500 Da) [820]. Some isomeric glycosides can be distinguished by ESI-CAD MS/MS [821, 822]; others can be distinguished by photodissociation of lithium-tagged precursors [823].

Exoglycosidase digestion in combination with the catalog library approach (CLA) has been used with MALDI FTMS to obtain the complete structure of oligosaccharides [824]. The sites of advanced glycation end products on human b2-microglobulin have been identified using MALDI TOFMS [825]. The structures of 12 O-linked glycans attached to human neutrophil gelatinase B have been elucidated by LC-ESI-CAD MS/MS [826] in coordinated use of a structural database [827]. A new beta-elimination procedure has been reported for release of O-linked oligosaccharides from glycoproteins; unlike the conventional Carlson degradation, which leads to formation of alditols, this new procedure renders the reducing end intact for analysis by MALDI MS [828]. A strategic method has been developed for the analysis of mucin-type oligosaccharide from the jelly coat of *Xenopus laevis* based on MALDI FTMS using CAD [829]. A bioaffinity probe has been used to analyze the oligosaccharides in the cortical granule lectin released by the egg of the South African toad *Xenopus laevis* upon fertilization [830]. The molecular weight distributions of the phosphatidylglycerols and acyl phosphatidylglycerols, *Trichomonas vaginalis* and *Tritrichomonas foetus*, were determined by negative-ion LSIMS, and the fatty acyl groups within each molecular species were assessed by CAD MS/MS [831]. Phosphatidyl-myo-inositol mannosides have been characterized by multiple-stage ion trap mass spectrometric approaches [832]. The phosphate position in hexose phosphate monosaccharides has been determined in the negative-ion mode using ion/molecule reactions with trimethyl borate as the reagent gas; diagnostic ions were generated enabling differentiation and linkage position determination of the phosphate moiety [833]. The EI fragmentation patterns of monomeric models of O-polysaccharides of *Vibrio cholerae* O:1 have been used to detect these structures in related materials [834].

References

1. Nau H, Kelley J and Biemann K, Determination of the amino acid sequence of the C-terminal cyanogen bromide fragment of actin by GC-MS. *J Am Chem Soc* **95**(21)**:** 7162-4, 1973.
2. Karas M, Bachmann D, Bahr U and Hillenkamp F, Matrix-Assisted Laser Desorption Ionization Mass Spectrometry. *Int. J. Mass Spec. Ion Proc.* **78:** 53-68, 1987.
3. Tanaka K, Waki H, Ido H, Akita S and Yoshida T, Protein and polymer analysis up to 100,000 by laser ionization time-of-flight mass spectrometry. *Rapid Commun Mass Spectrom* **2:** 151-153, 1988.
4. Tanaka K, The Origin of Macromolecule Ionization by Laser Irradiation (Nobel Lecture). *Angew. Chem. Int. Ed.* **42:** 3861-3870, 2003.
5. Fenn JB, Mann M, Meng CK, Wong SK and Whitehouse CM, ESI of Large Biomolecules. *Science* **246:** 64-71, 1989.
6. Lattimer RP and Schulten HR, FI and FDMS: Past, Present and Future. *Anal. Chem.* **61:** 1201A-1215A, 1989.
7. Macfarlane RD and Sundqvist BUR, Cf-252 PDMS. *Mass Spectrom Rev.* **4:** 421-460, 1985.
8. Barber M, Bordoli RS, Sedgwick RD and Tyler AN, F.A.B. *J. Chem. Soc. Chem. Commun.* **1981:** 325-327, 1981.
9. Ahn S, Ramirez J, Grigorean G and Lebrilla CB, Chiral recognition in gas-phase cyclodextrin: amino acid complexes--is the three point interaction still valid in the gas phase? *J Am Soc Mass Spectrom* **12**(3)**:** 278-87., 2001.
10. Figeys D, Proteomics in 2002: a year of technical development and wide-ranging applications. *Analytical Chemistry* **75**(12)**:** 2891-2905, 2003.
11. Aebersold R, A mass spectrometric journey into protein and proteome research. *Journal of the American Society for Mass Spectrometry* **14**(7)**:** 685-695, 2003.
12. Aebersold R and Mann M, Mass spectrometry-based proteomics. *Nature* **422**(6928)**:** 198-207., 2003.
13. Banks RE, Dunn MJ, Hochstrasser DF, Sanchez JC, Blackstock W, Pappin DJ and Selby PJ, Proteomics: new perspectives, new biomedical opportunities. *Lancet* **356**(9243)**:** 1749-56., 2000.
14. Liebler D, *Introduction to Proteomics---Tools for the New Biology.* Humana Press, Totowa, NJ, 2002.
15. Stults JT and Arnott D, Proteomics. *Methods in Enzymology* Vol. 402 *Biological Mass Spectrometry* (Burlingame AL, Ed.), Elsevier: New York, 2005, pp 245-289..
16. Van Oostrum J and Voshol H, The human genome: proteomics. *Am J Psychiatry* **159**(2)**:** 208., 2002.
17. Roberts JK, Proteomics and a future generation of plant molecular biologists. *Plant Mol Biol* **48**(1-2)**:** 143-54., 2002.
18. Rappsilber J and Mann M, What does it mean to identify a protein in proteomics? *Trends Biochem Sci* **27**(2)**:** 74-8., 2002.
19. Rabilloud T, Strub JM, Carte N, Luche S, Van Dorsselaer A, Lunardi J, Giege R and Florentz C, Comparative Proteomics as a New Tool for Exploring Human Mitochondrial tRNA Disorders. *Biochemistry* **41**(1)**:** 144-150., 2002.
20. Peng J and Gygi SP, Proteomics: the move to mixtures. *J Mass Spectrom* **36**(10)**:** 1083-91., 2001.
21. Mann M, Hendrickson RC and Pandey A, Analysis of proteins and proteomes by mass spectrometry. *Annu Rev Biochem* **70:** 437-73., 2001.
22. Graves PR and Haystead TA, Molecular Biologist's Guide to Proteomics. *Microbiol Mol Biol Rev* **66**(1)**:** 39-63., 2002.
23. Field ED, Caldwell JA, Shabanowitz J and Hunt DF, Drugs, bugs and the proteome: sequencing peptides at the attomole level. *Biochemist* **22**(4)**:** 16-19, 2000.
24. Patterson SD, Mass spectrometry and proteomics. *Physiol Genomics* **2**(2)**:** 59-65., 2000.
25. Derrick PJ and Patterson SD, Mass spectrometry of proteomics. *Proteomics* **1**(8)**:** 925-6., 2001.
26. Godovac-Zimmermann J and Brown LR, Perspectives for mass spectrometry and functional proteomics. *Mass Spectrom Rev* **20**(1)**:** 1-57., 2001.
27. Smith RD, Anderson GA, Lipton MS, Masselon C, Pasa-Tolic L, Shen Y and Udseth HR, The use of accurate mass tags for high-throughput microbial proteomics. *Omics* **6**(1)**:** 61-90., 2002.
28. Charlwood J, Hanrahan S, Tyldesley R, Langridge J, Dwek M and Camilleri P, Use of proteomic methodology for the characterization of human milk fat globular membrane proteins. *Anal Biochem* **301**(2)**:** 314-24., 2002.
29. Ogorzalek Loo RR, Loo JA, Du P and Holler T, In vivo labeling: a glimpse of the dynamic proteome and additional constraints for protein identification. *J Am Soc Mass Spectrom* **13**(7)**:** 804-12., 2002.
30. Liebler DC, Proteomic approaches to characterize protein modifications: new tools to study the effects of environmental exposures. *Environ Health Perspect* **110**(Suppl 1)**:** 3-9., 2002.
31. Conrads TP, Issaq HJ and Veenstra TD, New tools for quantitative phosphoproteome analysis. *Biochem Biophys Res Commun* **290**(3)**:** 885-90., 2002.
32. Petricoin EF and Liotta LA, Clinical Applications of Proteomics(). *J Nutr* **133**(7)**:** 2476S-2484S., 2003.
33. Wulfkuhle JD, Liotta LA and Petricoin EF, Proteomic applications for the early detection of cancer. *Nat Rev Cancer* **3**(4)**:** 267-75., 2003.

34. Yoshida M, Loo JA and Lepleya RA, Proteomics as a tool in the pharmaceutical drug design process. *Curr Pharm Des* **7**(4): 291-310., 2001.

35. Figeys D, Proteomics approaches in drug discovery. *Anal Chem* **74**(15): 412A-419A., 2002.

36. Shin BK, Wang H, Yim AM, Le Naour F, Brichory F, Jang JH, Zhao R, Puravs E, Tra J, Michael CW, Misek DE and Hanash SM, Global profiling of the cell surface proteome of cancer cells uncovers an abundance of proteins with chaperone function. *J Biol Chem* **278**(9): 7607-16., 2003.

37. Lubman DM, Kachman MT, Wang H, Gong S, Yan F, Hamler RL, O'Neil KA, Zhu K, Buchanan NS and Barder TJ, Two-dimensional liquid separations-mass mapping of proteins from human cancer cell lysates. *J Chromatogr B Analyt Technol Biomed Life Sci* **782**(1-2): 183-96., 2002.

38. Kachman MT, Wang H, Schwartz DR, Cho KR and Lubman DM, A 2-D liquid separations/mass mapping method for interlysate comparison of ovarian cancers. *Anal Chem* **74**(8): 1779-91., 2002.

39. Petricoin EF and Liotta LA, Mass spectrometry-based diagnostics: the upcoming revolution in disease detection. *Clin Chem* **49**(4): 533-4., 2003.

40. Smith JC, Lambert J-P, Elisma F and Figeys D, Proteomics in 2005/2006: Developments, Applications and Challenges. *Anal Chem* **79**: 4325-4343, 2007.

41. Ackermann BL, Hale JE and Duffin KL, The role of mass spectrometry in biomarker discovery and measurement. *Current Drug Metabolism* **7**(5): 525-539, 2006.

42. Spahr CS, Davis MT, McGinley MD, Robinson JH, Bures EJ, Beierle J, Mort J, Courchesne PL, Chen K, Wahl RC, Yu W, Luethy R and Patterson SD, Towards defining the urinary proteome using LC/MS/MS. I. Profiling an unfractionated tryptic digest. *Proteomics* **1**: 93-107., 2001.

43. Davis MT, Spahr CS, McGinley MD, Robinson JH, Bures EJ, Beierle J, Mort J, Yu W, Luethy R and Patterson SD, Towards defining the urinary proteome using liquid chromatography-tandem mass spectrometry. II. Limitations of complex mixture analyses. *Proteomics* **1**(1): 108-17., 2001.

44. Florens L, Washburn MP, Raine JD, Anthony RM, Grainger M, Haynes JD, Moch JK, Muster N, Sacci JB, Tabb DL, Witney AA, Wolters D, Wu Y, Gardner MJ, Holder AA, Sinden RE, Yates JR and Carucci DJ, A proteomic view of the P. falciparum life cycle. *Nature* **419**(6906): 520-6., 2002.

45. Asara JM, Schweitzer MH, Freimark LM, Phillips M and Cantley LC, Protein Sequences from Mastodon and *Tyrannosaurus Rex* Revealed by Mass Spectrometry. *Science* **316**: 280-285, 2007.

46. Whitelegge JP, Plant proteomics: BLASTing out of a MudPIT. *Proc Natl Acad Sci U S A* **99**(18): 11564-6., 2002.

47. Koller A, Washburn MP, Lange BM, Andon NL, Deciu C, Haynes PA, Hays L, Schieltz D, Ulaszek R, Wei J, Wolters D and Yates JR, 3rd, Proteomic survey of metabolic pathways in rice. *Proc Natl Acad Sci U S A* **99**(18): 11969-74., 2002.

48. Ficarro SB, McCleland ML, Stukenberg PT, Burke DJ, Ross MM, Shabanowitz J, Hunt DF and White FM, Phosphoproteome analysis by mass spectrometry and its application to Saccharomyces cerevisiae. *Nat Biotechnol* **20**(3): 301-5., 2002.

49. Beausoleil SA, Jedrychowski M, Schwartz D, Elias JE, Villen J, Li J, Cohn MA, Cantley LC and Gygi SP, Large-scale characterization of HeLa cell nuclear phosphoproteins. *Proceedings of the National Academy of Sciences of the United States of America* **101**(33): 12130-12135, 2004.

50. Yano H, Kuroda S and Buchanan BB, Disulfide proteome in the analysis of protein function and structure. *Proteomics* **2**(9): 1090-6., 2002.

51. Mann M and Jensen ON, Proteomic analysis of post-translational modifications. *Nat Biotechnol* **21**(3): 255-61., 2003.

52. Kirkpatrick DS, Denison C and Gygi SP, Weighing in on ubiquitin: the expanding role of mass-spectrometry-based proteomics. *Nature Cell Biology* **7**(8): 750-757, 2005.

53. Gevaert K and Vandekerckhove J, Protein ID in proteomics. *Electrophoresis* **21**: 1145-54., 2000.

54. Wilm M, Mass spectrometric analysis of proteins. *Adv Protein Chem* **54**: 1-30., 2000.

55. Jonsson AP, Aissouni Y, Palmberg C, Percipalle P, Nordling E, Daneholt B, Jornvall H and Bergman T, Recovery of gel-separated proteins for in-solution digestion and MS. *Anal Chem* **73**: 5370-7., 2001.

56. Schmidt F, Lueking A, Nordhoff E, Gobom J, Klose J, Seitz H, Egelhofer V, Eickhoff H, Lehrach H and Cahill DJ, Generation of minimal protein identifiers of proteins from two-dimensional gels and recombinant proteins. *Electrophoresis* **23**(4): 621-5., 2002.

57. Nordhoff E, Egelhofer V, Giavalisco P, Eickhoff H, Horn M, Przewieslik T, Theiss D, Schneider U, Lehrach H and Gobom J, Large-gel two-dimensional electrophoresis-MALDI TOF MS: an analytical challenge for studying complex protein mixtures. *Electrophoresis* **22**: 2844-55., 2001.

58. Hanash S, Disease proteomics. *Nature* **422**(6928): 226-32., 2003.

59. Shabanowitz J, Settlage RE, Marto JA, Christian RE, White FM, Russo PS, Martin SE and Hunt DF, Sequencing the primordial soup. *Mass Spectrom. Biol. Med.*: 163-177, 2000.

60. MacCoss MJ, McDonald WH, Saraf A, Sadygov R, Clark JM, Tasto JJ, Gould KL, Wolters D, Washburn M, Weiss A, Clark JI and Yates JR, 3rd, Shotgun identification of protein modifications from protein complexes and lens tissue. *Proc Natl Acad Sci U S A* **99**(12): 7900-5., 2002.

61. Wolters DA, Washburn MP and Yates JR, 3rd, An automated multidimensional protein identification technology for shotgun proteomics. *Anal Chem* **73**(23): 5683-90., 2001.

62. Davis MT, Beierle J, Bures ET, McGinley MD, Mort J, Robinson JH, Spahr CS, Yu W, Luethy R and Patterson SD, Automated LC-LC-MS-MS platform using binary ion-exchange and gradient reversed-phase LC for improved proteomic analyses. *J Chromatogr B Biomed Sci Appl* **752**(2): 281-91., 2001.
63. Kislinger T and Emili A, Multidimensional protein identification technology: current status and future prospects. *Expert Review of Proteomics* **2**: 27-39, 2005.
64. Washburn MP, Wolters D and Yates JR, 3rd, Large-scale analysis of the yeast proteome by multidimensional protein identification technology. *Nat Biotechnol* **19**(3): 242-7., 2001.
65. Washburn MP, Ulaszek RR and Yates JR, III, Reproducibility of quantitative proteomic analyses of complex biological mixtures by multidimensional protein identification technology. *Analytical Chemistry* **75**(19): 5054-5061, 2003.
66. Kislinger T, Gramolini AO, MacLennan DH and Emili A, MudPIT: Technical Overview & Proteomic Investigation of Normal vs Diseased Heart Tissue. *J Am Soc Mass Spectrom* **16**: 1207-1220, 2005.
67. Janini GM, Conrads TP, Veenstra TD and Issaq HJ, Development of a two-dimensional protein-peptide separation protocol for comprehensive proteome measurements. *J Chromatogr B Analyt Technol Biomed Life Sci* **787**(1): 43-51., 2003.
68. Spahr CS, Susin SA, Bures EJ, Robinson JH, Davis MT, McGinley MD, Kroemer G and Patterson SD, Simplification of complex peptide mixtures for proteomic analysis: reversible biotinylation of cysteinyl peptides. *Electrophoresis* **21**(9): 1635-50., 2000.
69. Zhang G, Fan H, Xu C, Bao H and Yang P, On-line preconcentration of in-gel digest by ion-exchange chromatography for protein ID using HPLC-ESI-MS/MS. *Anal. Biochem.* **313**: 327-330, 2003.
70. Le Bihan T, Duewel HS and Figeys D, On-line strong cation exchange m-HPLC-ESI-MS/MS for protein identification and process optimization. *J Am Soc Mass Spectrom* **14**: 719-727, 2003.
71. Nelson RW, Nedelkov D and Tubbs KA, Biosensor chip mass spectrometry: a chip-based proteomics approach. *Electrophoresis* **21**(6): 1155-63., 2000.
72. Rappsilber J, Ishihama Y and Mann M, Stop and go extraction tips for MALDI, nanoelectrospray, and LC/MS sample pretreatment in proteomics. *Anal Chem* **75**: 663-70., 2003.
73. Pflieger D, Le Caer JP, Lemaire C, Bernard BA, Dujardin G and Rossier J, Systematic identification of mitochondrial proteins by LC-MS/MS. *Anal Chem* **74**(10): 2400-6., 2002.
74. Elias JE, Haas W, Faherty BK and Gygi SP, Comparative evaluation of mass spectrometry platforms used in large-scale proteomics investigations. *Nature Methods* **2**(9): 667-675, 2005.
75. Kalkum M, Lyon GJ and Chait BT, Detection of secreted peptides by using hypothesis-driven multistage mass spectrometry. *Proc Natl Acad Sci U S A* **18**: 18, 2003.
76. Zhang W, Krutchinsky AN and Chait BT, \"De novo\" peptide sequencing by MALDI-quadrupole-ion trap mass spectrometry: a preliminary study. *J Am Soc Mass Spectrom* **14**: 1012-1021, 2003.
77. Murphy SM, Preble AM, Patel UK, O'Connell KL, Dias DP, Moritz M, Agard D, Stults JT and Stearns T, GCP5 and GCP6: members of human gamma-tubulin complex. *Mol Biol Cell* **12**: 3340-52., 2001.
78. Tsai SC, Pasumarthi KB, Pajak L, Franklin M, Patton B, Wang H, Henzel WJ, Stults JT and Field LJ, Simian virus 40 large T antigen binds a novel Bcl-2 homology domain 3-containing proapoptosis protein in the cytoplasm. *J Biol Chem* **275**(5): 3239-46., 2000.
79. Silberring J and Ekman R, *Mass Spectrometry and Hyphenated Techniques in Neuropeptide Research.* John Wiley & Sons, NYC, NY, 2003.
80. Cohen SL and Chait BT, Mass spectrometry as a tool for protein crystallography. *Annu Rev Biophys Biomol Struct* **30**: 67-85., 2001.
81. Ge X, Sirich TL, Beyer MK, Desaire H and Leary JA, A strategy for the determination of enzyme kinetics using ESI with an ion trap mass spectrometer. *Anal Chem* **73**: 5078-82., 2001.
82. Biemann K, Four decades of structure determination by mass spectrometry: from alkaloids to heparin. *J Am Soc Mass Spectrom* **13**(11): 1254-72., 2002.
83. Biemann K and Papayannopoulos IA, Sequencing of Proteins. *Acc. Chem. Res.* **27**: 370-378, 1994.
84. Biemann K, Coming of Age of MS in Peptide/Protein Chemistry. *Protein Sci.* **4**: 1920-1927, 1995.
85. Johnson RS, Martin SA, Biemann K, Stults JT and Watson JT, Novel fragmentation process of peptides by CAD-MS/MS: differentiation of leucine and isoleucine. *Anal. Chem.* **59**: 2621-5, 1987.
86. Stults JT and Watson JT, Identification of a new type of fragment ion in the CAD spectra of peptides allows leucine/isoleucine differentiation. *Biomed. Environ. Mass Spectrom.* **14**: 583-6, 1987.
87. Sadagopan N and Watson JT, Mass spectrometric evidence for mechanisms of fragmentation of charge-derivatized peptides. *J Am Soc Mass Spectrom* **12**(4): 399-409., 2001.
88. Stults JT, Peptide sequencing by mass spectrometry. *Methods Biochem Anal* **34**: 145-201., 1990.
89. Ito A, Okamura T, Yamamoto H, Ueyama M, Ake K, Masui R, Kuramitsu S and Tsunasawa S, Distinction of Leu and Ile Using a Ruthenium(II) Complex by MALDI-LIFT-TOF/TOF-MS Analysis. *Analytical Chemistry* **77**(20): 6618-6624, 2005.
90. Harrison AG, Young AB, Schnoelzer M and Paizs B, Formation of iminium ions by fragmentation of a2 ions. *Rapid Commun Mass Spectrom* **18**: 1635-1640, 2004.
91. Roepstorff P and Fohlmann J, MS Peptide Nomenclature. *Biomed. Mass Spectrom.* **11**: 601, 1984.
92. Wysocki V, Tsaprailis G, Smith L and Breci L, Mobile and localized protons: a framework for understanding peptide. *J Mass Spectrom* **35**(12): 1399-406, 2000.

93. Paizs B and Suhai S, Towards understanding the tandem mass spectra of protonated oligopeptides. 1: mechanism of amide bond cleavage. *J Am Soc Mass Spectrom* **15**: 103-113, 2004.

94. Paizs B and Suhai S, Fragmentation of protonated peptides. *Mass Spectrom Rev* **24**: 508-548, 2005.

95. Komaromi I, Somogyi A and Wysocki VH, Proton migration and its effect on the MS fragmentation of N-acetyl OMe proline: MS/MS experiments and ab initio and density functional calculations. *International Journal of Mass Spectrometry* **241**(2-3): 315-323, 2005.

96. Wee S, O'Hair RAJ and McFadyen WD, The role of the position of the basic residue in the generation and fragmentation of peptide radical cations. *Int J Mass Spectrom* **249/250**: 171-183, 2006.

97. Paizs B and Suhai S, Quantum chemical and RRKM modeling of fragmentation pathways of protonated GGG. II. Formation of b2, y1, and y2 ions. *Rapid Commun Mass Spectrom* **16**: 375-389, 2002.

98. Harrison AG, Csizmadia IG and Tang TH, Structure and fragmentation of b2 ions in peptide mass spectra. *J Am Soc Mass Spectrom* **11**(5): 427-36., 2000.

99. Polfer NC, Oomens J, Suhai S and Paizs B, Spectroscopic and Theoretical Evidence for Oxazolone Ring Formation in Collision-Induced Dissociation of Peptides. *Journal of the American Chemical Society* **127**(49): 17154-17155, 2005.

100. Chen X and Turecek F, Simple b Ions Have Cyclic Oxazolone Structures. A Neutralization-Reionization Mass Spectrometric and Computational Study of Oxazolone Radicals. *Journal of the American Society for Mass Spectrometry* **16**(12): 1941-1956, 2005.

101. Bursey MM and Pedersen LG, The barrier to proton transfer from multiply protonated molecules: a quantum mechanical approach. *Organic Mass Spectrometry* **27**(9): 974-5, 1992.

102. Morgan DG and Bursey MM, Linear energy correlation in low-energy MS/MS of protonated tripeptides Xxx-Gly-Gly but failure for Gly-Xxx-Gly. *J Mass Spectrom* **30**: 290-5, 1995.

103. Cordero MM and Wesdemiotis C, Tandem mass spectrometry of peptides: mechanistic aspects and structural information based on neutral losses. II. Tri- and larger peptides. *Organic Mass Spectrometry* **29**(7): 382-90, 1994.

104. Polce MJ, Ren D and Wesdemiotis C, Dissociation of the peptide bond in protonated peptides. *Journal of Mass Spectrometry* **35**(12): 1391-1398, 2000.

105. Harrison AG and Young AB, Fragmentation of protonated oligoalanines: Amide bond cleavage and beyond. *Journal of the American Society for Mass Spectrometry* **15**(12): 1810-1819, 2004.

106. O'Hair RAJ, The role of nucleophile-electrophile interactions in the unimolecular and bimolecular gas-phase ion chemistry of peptides and related systems. *J Mass Spectrom* **35**: 1377-1381, 2000.

107. Zhang K, Zimmerman DM, Chung-Phillips A and Cassady CJ, Experimental & ab initio studies of gas-phase basicities of polyglycines. *J Amer Chem Soc* **115**: 10812-22, 1993.

108. Somogyi A, Wysocki VH and Mayer I, The effect of protonation site on bond strengths in simple peptides: application of Ab initio and modified neglect of differential overlap bond orders and modified neglect of differential overlap energy partitioning. *J Amer Soc Mass Spectrom* **5**: 704-17, 1994.

109. Reid GE, Simpson RJ and O'Hair RAJ, A mass spectrometric and ab initio study of the pathways for dehydration of simple glycine and cysteine-containing peptide [M+H]+ ions. *Journal of the American Society for Mass Spectrometry* **9**(9): 945-956, 1998.

110. Farrugia JM, O'Hair RAJ and Reid GE, Do all b2 ions have oxazolone structures? Multistage mass spectrometry and ab initio studies on protonated N-acyl amino acid methyl ester model systems. *International Journal of Mass Spectrometry* **210/211**(1-3): 71-87, 2001.

111. Reid GE, Simpson RJ and O'Hair RAJ, Gas phase ion chemistry of biomolecules. 17. Probing the fragmentation reactions of protonated glycine oligomers via multistage MS and gas phase ion molecule hydrogen/deuterium exchange. *Int J Mass Spectrom* **190/191**: 209-230, 1999.

112. Reid GE, Simpson RJ and O'Hair RAJ, *J. Am. Soc. Mass Spectrom.* **11**: 1047-1060, 2000.

113. Reid GE, Roberts KD, Kapp EA and Simpson RJ, Statistical and Mechanistic Approaches to Understanding the Gas-Phase Fragmentation Behavior of Methionine Sulfoxide Containing Peptides. *Journal of Proteome Research* **3**(4): 751-759, 2004.

114. Reid GE, Roberts KD, Simpson RJ and O'Hair RAJ, Selective Identification and Quantitative Analysis of Methionine Containing Peptides by Charge Derivatization and Tandem Mass Spectrometry. *Journal of the American Society for Mass Spectrometry* **16**(7): 1131-1150, 2005.

115. Chowdhury SM, Munske GR, Ronald RC and Bruce JE, Evaluation of Low Energy CID and ECD Fragmentation Behavior of Mono-Oxidized Thio-Ether Bonds in Peptides. *J Amer Soc Mass Spectrom* **18**: 493-501, 2007.

116. Fu Q and Li L, Neutral loss of water from the b ions with histidine at the C-terminus and formation of the c ions involving lysine side chains. *Journal of Mass Spectrometry* **41**(12): 1600-1607, 2006.

117. Grewal RN, El Aribi H, Harrison AG, Siu KWM and Hopkinson AC, Fragmentation of Protonated Tripeptides: The Proline Effect Revisited. *Journal of Physical Chemistry B* **108**(15): 4899-4908, 2004.

118. Breci LA, Tabb DL, Yates JR, 3rd and Wysocki VH, Cleavage N-Terminal to Proline: Analysis of a Database of Peptide Tandem Mass Spectra. *Anal Chem* **75**(9): 1963-1971., 2003.

119. Kapp EA, Schutz F, Reid GE, Eddes JS, Moritz RL, O'Hair RA, Speed TP and Simpson RJ, Mining a Tandem Mass Spectrometry Database To Determine the Trends and Global Factors Influencing Peptide Fragmentation. *Anal Chem* **75**(22): 6251-6264., 2003.

120. Tsaprailis G, Nair H, Somogyi A, Wysocki VH, Zhong W, Futrell JH, Summerfield SG and Gaskell SJ, Influence of Secondary Structure on the Fragmentation of Protonated Peptides. *Journal of the American Chemical Society* **121**(22): 5142-5154, 1999.

121. Paizs B, Suhai S and Harrison AG, Experimental and theoretical investigation of the main fragmentation pathways of protonated H-Gly-Gly-Sar-OH and H-Gly-Sar-Sar-OH. *Journal of the American Society for Mass Spectrometry* **14**(12): 1454-1469, 2003.

122. Smith LL, Herrmann KA and Wysocki VH, Investigation of Gas Phase Ion Structure for Proline-Containing b2 Ion. *Journal of the American Society for Mass Spectrometry* **17**(1): 20-28, 2006.

123. Fournier I, Chaurand P, Bolbach G, Lutzenkirchen F, Spengler B and Tabet JC, Sequencing of a branched peptide using MALDI-TOF-MS. *J Mass Spectrom* **35**: 1425-33., 2000.

124. Chen H, Eberlin LS and Cooks RG, Neutral Fragment Mass Spectra via Ambient Thermal Dissociation of Peptide and Protein Ions. *J Amer Chem Soc* **129**(18): 5880-5886, 2007.

125. Tang X and Boyd R, Rearrangements of doubly charged acylium ions from lysyl and ornithyl peptides. *Rapid Commun Mass Spectrom* **8**(9): 678-86, 1994.

126. Tang X, Thibault P and Boyd R, Fragmentation reactions of multiply-protonated peptides and implications for sequencing by low-energy CAD-MS/MS. *Anal Chem* **65**: 2824-34, 1993.

127. Men L and Wang Y, Fragmentation of protonated peptides containing aspartic, glutamic, cysteine sulfinic acid, and cysteine sulfonic acid. *Rapid Commun Mass Spectrom* **19**: 23-30, 2005.

128. Hill HH, Hill CH, Asbury GR, Wu C, Matz LM and Ichiye T, Charge location on gas phase peptides. *International Journal of Mass Spectrometry* **219**(1): 23-37, 2002.

129. Harrison AG, Fragmentation reactions of protonated peptides containing glutamine or glutamic acid. *J Mass Spectrom* **38**(2): 174-87., 2003.

130. Harrison AG, Characterization of a- and g-glutamyl dipeptides by negative ion collision-induced dissociation. *Journal of Mass Spectrometry* **39**(2): 136-144, 2004.

131. Waugh RJ, Bowie JH and Gross ML, Collision-induced dissociations of deprotonated peptides. Dipeptides containing methionine or cysteine. *Rapid Commun Mass Spectrom* **7**: 623-5, 1993.

132. Bradford AM, Waugh RJ and Bowie JH, Characterization of underivatized tetrapeptides by negative-ion FAB MS. *Rapid Communications in Mass Spectrometry* **9**: 1082-5, 1995.

133. Amunugama M, Roberts KD and Reid GE, Mechanisms for the Selective Gas-Phase Fragmentation Reactions of Methionine Side Chain Fixed Charge Sulfonium Ion Containing Peptides. *Journal of the American Society for Mass Spectrometry* **17**(12): 1631-1642, 2006.

134. Shen TL, Huang ZH, Laivenieks M, Zeikus JG, Gage DA and Allison J, Evaluation of charge derivatization of a proteolytic protein digest for improved mass spectrometric analysis: de novo sequencing by MALDI-PSD-MS. *J Mass Spectrom* **34**: 1154-65., 1999.

135. Keough T, Youngquist RS and Lacey MP, Sulfonic acid derivatives for peptide sequencing by MALDI MS. *Anal Chem* **75**(7): 156A-165A., 2003.

136. Roth KD, Huang ZH, Sadagopan N and Watson JT, Charge derivatization of peptides for analysis by mass spectrometry. *Mass spectrometry reviews* **17**(4): 255-74, 1998.

137. Sadagopan N and Watson JT, Investigation of the tris(trimethoxyphenyl)phosphonium acetyl charged derivatives of peptides by ESI-MS/MS. *Journal of the American Society for Mass Spectrometry* **11**: 107-119, 2000.

138. Shen TL and Allison J, Interpretation of MALDi-PSD spectra of charge-derivatized peptides: some examples of tris[(2,4,6-trimethoxyphenyl) phosphonium]-tagged proteolytic digestion products of phosphoenolpyruvate carboxykinase. *J Am Soc Mass Spectrom* **11**: 145-52., 2000.

139. Czeszak X, Ricart G, Tetaert D, Michalski JC and Lemoine J, Identification of substituted sites on MUC5AC mucin motif peptides after enzymatic O-glycosylation combining beta-elimination and fixed-charge derivatization. *Rapid Commun Mass Spectrom* **16**(1): 27-34., 2002.

140. Renner D and Spiteller G, Mechanism of Fragmentation of [MH]+ Ions from Peptides by LSIMS. *Biomed. Environ. Mass Spectrom.* **13**: 405-410, 1986.

141. Pittenauer E, Zehl M, Belgacem O, Raptakis E, Mistrik R and Allmaier G, Comparison of CID spectra of singly charged polypeptide antibiotic precursor ions obtained by positive-ion vacuum MALDI IT/RTOF and TOF/RTOF, AP-MALDI-IT and ESI-IT mass spectrometry. *Journal of Mass Spectrometry* **41**(4): 421-447, 2006.

142. Schar M, Bornsen KO and Gassmann E, Fast Protein Sequencing by MALDI-MS. *Rapid Commun. Mass Spectrom.* **5**: 319-326, 1991.

143. Knierman MD, Coligan JE and Parker KC, Peptide Fingerprints by Partial Acid Hydrolysis and Analysis by MALDI-MS. *Rapid Commun. Mass Spectrom.* **8**: 1007-1010, 1994.

144. Kratzer R, Eckerskorn C, Karas M and Lottspeich F, Suppression effects in enzymic peptide ladder sequencing using UV-MALDI-MS. *Electrophoresis* **19**: 1910-1919, 1998.

145. Chait BT, Wang R, Beavis RC and Kent SBH, Protein ladder sequencing. *Science* **262**: 89-92, 1993.

146. Bartlett-Jones M, Jeffery WA, Hansen HF and Pappin DJC, Peptide Ladder-Sequencing Using Volatile Reagents and MALDI. . *Rapid Commun. Mass Spectrom.* **8**: 737-742, 1994.

147. Hunt DF, Yates JR, III, Shabanowitz J, Winston S and Hauer CR, Protein sequencing by tandem mass spectrometry. *Proc Natl Acad Sci USA* **83**: 6233-7, 1986.

148. Stults JT, Tryptic Mapping by MALDI-MS. *Anal. Chem.* **65:** 1709-1716, 1993.
149. Billeci TM and Stults JT, Tryptic mapping of recombinant proteins by matrix-assisted laser desorption/ionization mass spectrometry. *Anal Chem* **65**(13)**:** 1709-16., 1993.
150. Wa C, Cerny R and Hage DS, Obtaining high sequence coverage in matrix-assisted laser desorption time-of-flight mass spectrometry for studies of protein modification: Analysis of human serum albumin as a model. *Analytical Biochemistry* **349**(2)**:** 229-241, 2006.
151. Cooper JW, Chen J, Li Y and Lee CS, Membrane-based nanoscale proteolytic reactor enabling protein digestion, peptide separation, and protein ID by MS. *Anal Chem* **75:** 1067-74., 2003.
152. Henzel WJ, Billeci TM, Stults JT, Wong SC, Grimley C and Watanabe C, Identifying proteins from two-dimensional gels by molecular mass searching of peptide fragments in protein sequence databases. *Proc Natl Acad Sci U S A* **90**(11)**:** 5011-5., 1993.
153. Henzel WJ, Watanabe C and Stults JT, Protein identification: the origins of peptide mass fingerprinting. *Journal of the American Society for Mass Spectrometry* **14**(9)**:** 931-942, 2003.
154. Huang Y, Triscari JM, Tseng GC, Pasa-Tolic L, Lipton MS, Smith RD and Wysocki VH, Statistical Characterization of the Charge State and Residue Dependence of Low-Energy CID Peptide Dissociation Patterns. *Analytical Chemistry* **77**(18)**:** 5800-5813, 2005.
155. Hunt DF, Zhu NZ and Shabanowitz J, Oligopeptide sequence analysis by CAD of multiply charged ions. *Rapid Communications in Mass Spectrometry* **3**(4)**:** 122-4, 1989.
156. Bieber AL, Becker RR, McParland R, Hunt DF, Shabanowitz J, Yates JR, III, Martino PA and Johnson GR, Sequence of acidic subunit of Mojave toxin by Edman degradation and MS. *Biochimica et Biophysica Acta, Protein Structure and Molecular Enzymology* **1037:** 413-21, 1990.
157. Hunt DF, Shabanowitz J, Martino PA, McCormack AL, Michel H, Alexander JE and Sherman N, Protein sequence analysis. New methods and instrumentation. *Nippon Iyo Masu Supekutoru Gakkai Koenshu* **15:** 47-60, 1990.
158. Savitski MM, Nielsen ML, Kjeldsen F and Zubarev RA, Proteomics-Grade de Novo Sequencing Approach. *Journal of Proteome Research* **4**(6)**:** 2348-2354, 2005.
159. Hernandez H, Niehauser S, Boltz SA, Gawandi V, Phillips RS and Amster IJ, Mass Defect Labeling of Cysteine for Improving Peptide Assignment in Shotgun Proteomic Analyses. *Anal Chem* **78**(10)**:** 3417-3423, 2006.
160. Olson MT, Epstein JA and Yergey AL, De Novo Peptide Sequencing Using Exhaustive Enumeration of Peptide Composition. *Journal of the American Society for Mass Spectrometry* **17**(8)**:** 1041-1049, 2006.
161. Mann M and Wilm M, Error Tolerant Identification of Peptides in Sequence Databases by Peptide Sequence Tags. *Analytical Chemistry* **66:** 4390-4399, 1994.
162. Tabb DL, Smith LL, Breci LA, Wysocki VH, Lin D and Yates JR, 3rd, Statistical characterization of ion trap tandem mass spectra from doubly charged tryptic peptides. *Anal Chem* **75**(5)**:** 1155-63., 2003.
163. Yague J, Paradela A, Ramos M, Ogueta S, Marina A, Barahona F, Lopez de Castro JA and Vazquez J, Peptide rearrangement during quadrupole ion trap fragmentation: added complexity to MS/MS spectra. *Anal Chem* **75**(6)**:** 1524-35., 2003.
164. Witt M, Fuchser J and Baykut G, FT-ICR-MS with nanolc/micro ESI and MALDI: analytical performance in peptide mass fingerprinting. *J Am Soc Mass Spectrom* **14:** 553-561, 2003.
165. Hernandez P, Müller M and Appel RD, Automated protein identification by MS/MS: Issues and strategies. *Mass Spec Rev* **25:** IN PRESS, 2006.
166. Sunyaev S, Liska AJ, Golod A and Shevchenko A, MultiTag: multiple error-tolerant sequence tag search for the sequence-similarity ID of proteins by MS. *Anal Chem* **75:** 1307-15., 2003.
167. Pan S, Gu S, Bradbury EM and Chen X, Single peptide-based protein ID in human proteome by MALDI-TOF MS coupled with amino acids coded mass tagging. *Anal Chem* **75:** 1316-24., 2003.
168. Reid GE, Characterization of proteins by MS. *Purifying Proteins for Proteomics*: 489-516, 2004.
169. Yates JR, 3rd, Eng JK and McCormack AL, Mining genomes: correlating MS/MS spectra of modified/unmodified peptides to sequences in nucleotide databases. *Anal Chem* **67:** 3202-10., 1995.
170. Sadygov Rovshan G, Cociorva D and Yates John R, 3rd, Large-scale database searching by MS/MS: looking up the answer in the back of the book. *Nat Methods* **1:** 195-202, 2004.
171. Cantin Greg T and Yates John R, 3rd, Strategies for shotgun identification of post-translational modifications by mass spectrometry. *Journal of chromatography. A* **1053**(1-2)**:** 7-14, 2004.
172. Kapp EA, Schutz F, Connolly LM, Chakel JA, Meza JE, Miller CA, Fenyo D, Eng JK, Adkins JN, Omenn GS and Simpson RJ, Evaluation, comparison, and accurate benchmarking of public MS/MS search algorithms: Sensitivity and specificity analysis. *Proteomics* **5:** 3475-3490, 2005.
173. Frewen BE, Merrihew GE, Wu CC, Noble WS and MacCoss MJ, Analysis of Peptide MS/MS Spectra from Large-Scale Proteomics Experiments Using Spectrum Libraries. *Analytical Chemistry* **78**(16)**:** 5678-5684, 2006.
174. Lu B, Ruse C, Xu T, Park SK and Yates J, III, Automatic Validation of Phosphopeptide Identifications from Tandem Mass Spectra. *Anal Chem* **79:** 1301-1310, 2007.
175. Hilario M, Kalousis A, Pellegrini C and Muller M, Processing and classification of protein mass spectra. *Mass spectrometry reviews* **25**(3)**:** 409-49, 2006.

176. Zimmer JSD, Monroe ME, Qian W-J and Smith RD, Advances in proteomics data analysis and display using an accurate mass and time tag approach. *Mass spectrometry reviews* **25**(3): 450-82, 2006.

177. Stead DA, Preece A and Brown AJP, Universal metrics for quality assessment of protein identifications by mass spectrometry. *Molecular and Cellular Proteomics* **5**(7): 1205-1211, 2006.

178. Pappin DJ, Peptide mass fingerprinting by MALDI-TOF MS. *Methods Mol Biol* **64**: 165-73., 1997.

179. Pappin DJ, Peptide mass fingerprinting by MALDI-TOF MS. *Methods Mol Biol* **211**: 211-9., 2003.

180. Perkins DN, Pappin DJ, Creasy DM and Cottrell JS, Probability-based protein identification by searching sequence databases using MS data. *Electrophoresis* **20**(18): 3551-67., 1999.

181. Savitski MM, Nielsen ML and Zubarev RA, New data base-independent, sequence tag-based scoring of peptide MS/MS data validates MOWSE scores, recovers below threshold data, singles out modified peptides, and assesses the quality of MS/MS techniques. *Molecular and Cellular Proteomics* **4**(8): 1180-1188, 2005.

182. Nielsen ML, Savitski MM and Zubarev RA, Improving protein identification using complementary fragmentation techniques in Fourier transform mass spectrometry. *Molecular and Cellular Proteomics* **4**(6): 835-845, 2005.

183. Zhang W and Chait BT, ProFound: An Expert System for Protein Identification Using MS Peptide Mapping Information. *Anal. Chem.* **72**: 2482-2489, 2000.

184. MacCoss MJ, Wu CC and Yates JR, 3rd, Probability-based validation of protein identifications using a modified SEQUEST algorithm. *Anal Chem* **74**(21): 5593-9., 2002.

185. Gao J, Friedrichs MS, Dongre AR and Opiteck GJ, Guidelines for the Routine Application of the Peptide Hits Technique. *J Am Soc Mass Spectrom* **16**: 1231-1238, 2005.

186. Venable JD and Yates JR, III, Impact of Ion Trap Tandem Mass Spectra Variability on the Identification of Peptides. *Analytical Chemistry* **76**(10): 2928-2937, 2004.

187. Wenner BR and Lynn BC, Factors that affect ion trap data-dependent MS/MS in proteomics. *Journal of the American Society for Mass Spectrometry* **15**(2): 150-157, 2004.

188. Yen C-Y, Russell S, Mendoza AM, Meyer-Arendt K, Sun S, Cios KJ, Ahn NG and Resing KA, Improving Sensitivity in Shotgun Proteomics by a Peptide-Centric Database with Reduced Complexity: Protease Cleavage and SCX Elution Rules from Data Mining of MS/MS Spectra. *Anal Chem* **78**: 1071-1084, 2006.

189. Narasimhan C, Tabb DL, VerBerkmoes NC, Thompson MR, Hettich RL and Uberbacher EC, MASPIC: Intensity-Based Tandem Mass Spectrometry Scoring Scheme That Improves Peptide Identification at High Confidence. *Analytical Chemistry* **77**(23): 7581-7593, 2005.

190. Fenyo D and Beavis RC, A method for assessing the statistical significance of mass spectrometry-based protein identifications using general scoring schemes. *Anal Chem* **75**(4): 768-74., 2003.

191. Gibbons FD, Elias JE, Gygi SP and Roth FP, SILVER helps assign peptides to MS/MS spectra using intensity-based scoring. *J Am Soc Mass Spectrom* **15**: 910-912, 2004.

192. Sadygov R, Wohlschlegel J, Park SK, Xu T and Yates JR, Central Limit Theorem as an Approximation for Intensity-Based Scoring Function. *Anal Chem* **78**: 89-95, 2006.

193. English RD, Warscheid B, Fenselau C and Cotter RJ, Bacillus spore identification via proteolytic peptide mapping with a miniaturized MALDI TOF MS. *Analytical Chemistry* **75**: 6886-6893, 2003.

194. Hansen BT, Jones JA, Mason DE and Liebler DC, SALSA: pattern recognition algorithm for electrophile-adducted peptides by automated evaluation of LC-CID-MS/MS spectra. *Anal Chem* **73**: 1676-83., 2001.

195. Liebler DC, Hansen BT, Davey SW, Tiscareno L and Mason DE, Peptide sequence motif analysis of tandem MS data with the SALSA algorithm. *Anal Chem* **74**(1): 203-10., 2002.

196. Eddes JS, Kapp EA, Frecklington DF, Connolly LM, Layton MJ, Moritz RL and Simpson RJ, CHOMPER: a bioinformatic tool for rapid validation of tandem mass spectrometry search results associated with high-throughput proteomic strategies. *Proteomics* **2**(9): 1097-103., 2002.

197. Keller A, Nesvizhskii AI, Kolker E and Aebersold R, Empirical statistical model to estimate the accuracy of peptide ID made by MS/MS and database search. *Anal Chem* **74**: 5383-92., 2002.

198. Chepanoske CL, Richardson BE, von Rechenberg M and Peltier JM, Average peptide score: A useful parameter for ID of proteins derived from database searches of LC/MS/MS data. *Rapid Commun Mass Spectrom* **19**: 9-14, 2005.

199. Arnott D, O'Connell KL, King KL and Stults JT, An integrated approach to proteome analysis: identification of proteins associated with cardiac hypertrophy. *Anal Biochem* **258**(1): 1-18., 1998.

200. Eriksson J, Chait BT and Fenyo D, A statistical basis for testing the significance of mass spectrometric protein identification results. *Anal Chem* **72**(5): 999-1005., 2000.

201. Cargile BJ, Bundy JL and Stephenson JL, Jr., Potential for False Positive Identifications from Large Databases through MS/MS. *Journal of Proteome Research* **3**: 1082-1085, 2004.

202. Eriksson J and Fenyoe D, The Statistical Significance of Protein Identification Results as a Function of the Number of Protein Sequences Searched. *Journal of Proteome Research* **3**(5): 979-982, 2004.

203. Geer LY, Markey SP, Kowalak JA, Wagner L, Xu M, Maynard DM, Yang X, Shi W and Bryant SH, Open Mass Spectrometry Search Algorithm. *Journal of Proteome Research* **3**(5): 958-964, 2004.

204. Rejtar T, Chen H-S, Andreev V, Moskovets E and Karger BL, Increased identification of peptides by enhanced data processing of high-resolution MALDI TOF/TOF mass spectra prior to database searching. *Analytical Chemistry* **76**(20): 6017-6028, 2004.

205. Monigatti F and Berndt P, Algorithm for accurate similarity measurements of peptide mass fingerprints and its application. *J Am Soc Mass Spectrom* **16:** 13-21, 2005.

206. Shevchenko A, Chernushevic I, Wilm M and Mann M, "De novo" sequencing of peptides recovered from in-gel digested proteins by nano ESI-MS/MS. *Mol Biotechnol* **20:** 107-18., 2002.

207. Arnott D, Henzel WJ and Stults JT, Rapid identification of comigrating gel-isolated proteins by ion trap-mass spectrometry. *Electrophoresis* **19**(6): 968-80., 1998.

208. Adamczyk M, Gebler JC, Wu J and Yu Z, Complete sequencing of anti-vancomycin fab fragment by LC-ESI ion trap MS with a combination of database searching and manual interpretation of the MS/MS spectra. *J Immunol Methods* **260:** 235-49., 2002.

209. Krutchinsky AN, Kalkum M and Chait BT, Automatic identification of proteins with a MALDI-quadrupole ion trap mass spectrometer. *Anal Chem* **73**(21): 5066-77., 2001.

210. Coon JJ, Steele HA, Laipis PJ and Harrison WW, Laser desorption-APCI: a novel ion source for the direct coupling of PAGE to MS. *J Mass Spectrom* **37:** 1163-7., 2002.

211. Shevchenko A, Jensen ON, Podtelejnikov AV, Sagliocco F, Wilm M, Vorm O, Mortensen P, Boucherie H and Mann M, Linking genome and proteome by mass spectrometry: large-scale identification of yeast proteins from 2D gels. *Proc Natl Acad Sci U S A* **93:** 14440-5., 1996.

212. Ideker T, Thorsson V, Ranish JA, Christmas R, Buhler J, Eng JK, Bumgarner R, Goodlett DR, Aebersold R and Hood L, Integrated genomic and proteomic analyses of a systematically perturbed metabolic network. *Science* **292**(5518): 929-34., 2001.

213. Holmes MR, Ramkissoon KR and Giddings MC, Proteomics and protein identification. *Bioinformatics (3rd Edition):* 445-472, 2005.

214. Wisz MS, Suarez MK, Holmes MR and Giddings MC, GFSWeb: A web tool for genome-based identification of proteins from MS samples. *Journal of Proteome Research* **3:** 1292-1295, 2004.

215. Yoo C, Patwa TH, Kreunin P, Miller FR, Huber CG, Nesvizhskii AI and Lubman DM, Comprehensive analysis of proteins of pH fractionated samples using monolithic LC/MS/MS, intact MW measurement and MALDI-QIT-TOF MS. *J Mass Spectrom* **42:** 312-334, 2007.

216. Mann M, Ong SE, Gronborg M, Steen H, Jensen ON and Pandey A, Analysis of protein phosphorylation using mass spectrometry: deciphering the phosphoproteome. *Trends Biotechnol* **20**(6): 261-8., 2002.

217. Ueberheide BM and Mollah S, Deciphering the histone code using mass spectrometry. *International Journal of Mass Spectrometry* **259**(1-3): 46-56, 2007.

218. Taverna SD, Allis CD and Hake SB, \"Hunt\"-ing for post-translational modifications that underlie the histone code. *International Journal of Mass Spectrometry* **259**(1-3): 40-45, 2007.

219. Wang Y, Vivekananda S, Men L and Zhang Q, Fragmentation of protonated ions of peptides containing cysteine, cysteine sulfinic acid, and cysteine sulfonic acid. *Journal of the American Society for Mass Spectrometry* **15**(5): 697-702, 2004.

220. Zhang S and Basile F, Site-Specific Pyrolysis-Induced Cleavage at Aspartic Acid Residue in Peptides and Proteins. *J Proteome Research* **6**(5): 1700-1704, 2007.

221. Pesavento JJ, Kim Y-B, Taylor GK and Kelleher NL, Shotgun Annotation of Histone Modifications: A New Approach for Streamlined Characterization of Proteins by Top Down Mass Spectrometry. *Journal of the American Chemical Society* **126**(11): 3386-3387, 2004.

222. Holmes MR and Giddings MC, Prediction of posttranslational modifications using intact-protein mass spectrometric data. *Analytical Chemistry* **76**(2): 276-282, 2004.

223. Zhang W, Czernik AJ, Yungwirth T, Aebersold R and Chait BT, Recognition of Phosphorylation Sites in Synapsin I by 2D-PAGE and MALDI Peptide Mapping. *Protein Sci* **3:** 677-686, 1994.

224. Liao PC, Leykam J, Gage DA and Allison J, Recognition of Phosphorylation Sites by MALDI Mass-Mapping. *Anal. Biochem.* **219:** 9-20, 1994.

225. Shou W, Verma R, Annan RS, Huddleston MJ, Chen SL, Carr SA and Deshaies RJ, Mapping phosphorylation sites in proteins by mass spectrometry. *Methods Enzymol* **351:** 279-96., 2002.

226. Cao P and Stults JT, Mapping the phosphorylation sites of proteins using on-line IMAC/capillary electrophoresis/ESI-MS/MS. *Rapid Commun Mass Spectrom* **14:** 1600-6., 2000.

227. Zappacosta F, Huddleston MJ, Karcher RL, Gelfand VI, Carr SA and Annan RS, Improved Sensitivity for Phosphopeptide Mapping Using Capillary Column HPLC and Microionspray MS: Comparative Site Mapping from Gel-Derived Proteins. *Anal. Chem.* **74:** 3221 -3231, 2002.

228. Garcia BA, Shabanowitz J and Hunt DF, Analysis of protein phosphorylation by mass spectrometry. *Methods (San Diego, CA, United States)* **35**(3): 256-264, 2005.

229. Salih E, Phosphoproteomics by mass spectrometry and classical protein chemistry approaches. *Mass Spectrometry Reviews* **24**(6): 828-846, 2005.

230. McLachlin DT and Chait BT, Analysis of phosphorylated proteins and peptides by mass spectrometry. *Curr Opin Chem Biol* **5**(5): 591-602., 2001.

231. Oda Y, Nagasu T and Chait BT, Enrichment analysis of phosphorylated proteins as a tool for probing the phosphoproteome. *Nat Biotechnol* **19**(4): 379-82., 2001.

232. Moyer SC, VonSeggern CE and Cotter RJ, Fragmentation of cationized phosphotyrosine containing peptides by atmospheric pressure MALDI/Ion trap mass spectrometry. *Journal of the American Society for Mass Spectrometry* **14**(6): 581-592, 2003.

233. Edelson-Averbukh M, Pipkorn R and Lehmann WD, Phosphate Group-Driven Fragmentation of Multiply Charged Phosphopeptide Anions. Improved Recognition of Peptides Phosphorylated at Serine, Threonine, or Tyrosine by Negative Ion ESI MS/MS. *Anal Chem* **78:** 1249-1256, 2006.

234. Sadagopan N, Malone M and Watson JT, Effect of charge derivatization in the determination of phosphorylation sites in peptides by ESI-CAD-MS/MS. *J Mass Spectrom* **34:** 1279-82., 1999.

235. Gronert S, Huang R and Li KH, Gas phase derivatization in peptide analysis I: the utility of trimethyl borate in identifying phosphorylation sites. *Int J Mass Spectrom* **231:** 179-187, 2004.

236. Chang EJ, Archambault V, McLachlin DT, Krutchinsky AN and Chait BT, ID of protein phosphorylation by hypothesis-driven multiple-stage MS. *Analytical Chemistry* **76:** 4472-4483, 2004.

237. Jin M, Bateup H, Padovan JC, Greengard P, Nairn AC and Chait BT, Quantitative Analysis of Protein Phosphorylation in Mouse Brain by Hypothesis-Driven Multistage Mass Spectrometry. *Analytical Chemistry* **77**(24): 7845-7851, 2005.

238. Wu H-Y, Tseng VS-M and Liao P-C, Mining Phosphopeptide Signals in LC/MS Data for Protein Phosphorylation Analysis. *J. Proteome Res* (in press), 2007.

239. Koecher T, Savitski MM, Nielsen ML and Zubarev RA, PhosTShunter: A Fast and Reliable Tool to Detect Phosphorylated Peptides in Liquid Chromatography Fourier Transform Tandem Mass Spectrometry Data Sets. *Journal of Proteome Research* **5**(3): 659-668, 2006.

240. Garcia BA, Busby SA, Barber CM, Shabanowitz J, Allis CD and Hunt DF, Characterization of phosphorylation sites on histone H1 isoforms by tandem mass spectrometry. *Journal of Proteome Research* **3**(6): 1219-1227, 2004.

241. Black TM, Andrews CL, Kilili G, Ivan M, Tsichlis PN and Vouros P, Characterization of Phosphorylation Sites on Tpl2 Using IMAC Enrichment and a Linear Ion Trap MS. *J Proteome Research* **6:** 2269-2276, 2007.

242. Thompson AJ, Hart SR, Franz C, Barnouin K, Ridley A and Cramer R, Characterization of protein phosphorylation by mass spectrometry using immobilized metal ion affinity chromatography with on-resin b-elimination and Michael addition. *Analytical Chemistry* **75**(13): 3232-3243, 2003.

243. Ficarro SB, Salomon AR, Brill LM, Mason DE, Stettler-Gill M, Brock A and Peters EC, Automated IMAC/nano-LC-ESI-MS platform for profiling protein phosphorylation sites. *Rapid Commun Mass Spectrom* **19:** 57-71, 2005.

244. Kocher T, Allmaier G and Wilm M, Nano ESI-based detection and sequencing of substoichiometric amounts of phosphopeptides in complex mixtures. *J Mass Spectrom* **38:** 131-7., 2003.

245. Hart SR, Waterfield MD, Burlingame AL and Cramer R, Factors governing the solubilization of phosphopeptides retained on ferric NTA IMAC beads and their analysis by MALDI TOFMS. *J Am Soc Mass Spectrom* **13**(9): 1042-51., 2002.

246. Napper S, Kindrachuk J, Olson DJ, Ambrose SJ, Dereniwsky C and Ross AR, Selective extraction and characterization of a histidine-phosphorylated peptide using immobilized copper(II) ion affinity chromatography and MALDI-TOF-MS. *Anal Chem* **75**(7): 1741-7., 2003.

247. Martin SE, Shabanowitz J, Hunt DF and Marto JA, Subfemtomole MS and MS/MS peptide sequence analysis using nano-HPLC micro-ESI FT-ICR-MS. *Anal Chem* **72**(18): 4266-74., 2000.

248. Larsen MR, Thingholm TE, Jensen ON, Roepstorff P and Jorgensen TJD, Enrichment of phosphorylated peptides using titanium dioxide microcolumns. *Molecular and Cellular Proteomics* **4:** 873-886, 2005.

249. Klemm C, Otto S, Wolf C, Haseloff RF, Beyermann M and Krause E, Evaluation of the titanium dioxide approach for MS analysis of phosphopeptides. *Journal of Mass Spectrometry* **41**(12): 1623-1632, 2006.

250. Schlosser A, Vanselow JT and Kramer A, Mapping Phosphorylation Sites by Multi-Protease Approach with Phosphopeptide Enrichment and NanoLC-MS/MS. *Anal Chem* **77:** 5243-5250, 2005.

251. Alpert AJ, Gygi SP and Shukla AK, Desalting Phosphopeptides by Solid-Phase Extraction. In: *55th Annual Conf. of Amer Soc for Mass Spectrom, Indianapolis,2007*, pp. MP 438.

252. Blacken GR, Gelb MH and Turecek F, Metal Affinity Capture MS/MS for Selective Detection of Phosphopeptides. *Anal Chem* **78:** 6065-6073, 2006.

253. Krabbe JG, Lingeman H, Niessen WMA and Irth H, Ligand-Exchange Detection of Phosphorylated Peptides by LC-ESI-MS. *Analytical Chemistry* **75:** 6853-6860, 2003.

254. Lansdell TA and Tepe JJ, Isolation of phosphopeptides using solid phase enrichment. *Tetrahedron Letters* **45**(1): 91-93, 2004.

255. Lim KB and Kassel DB, Phosphopeptides enrichment using on-line two-dimensional strong cation exchange followed by reversed-phase liquid chromatography/mass spectrometry. *Analytical Biochemistry* **354**(2): 213-219, 2006.

256. Xu C-F, Wang H, Li D, Kong X-P and Neubert TA, Selective Enrichment and Fractionation of Phosphopeptides from Peptide Mixtures by Isoelectric Focusing after Methyl Esterification. *Anal Chem* **79:** 2007-2014, 2007.

257. Blacken GR, Volny M, Vaisar T, Sadilek M and Turecek F, In Situ Enrichment of Phosphopeptides on MALDI Plates Functionalized by Reactive Landing of Zirconium(IV)-n-Propoxide Ions. *Anal Chem* **79**(in press), 2007.

258. Kapkova P, Lattova E and Perreault H, Nonretentive Solid-Phase Extraction of Phosphorylated Peptides for analysis by MALDI MS. *Anal Chem* **78:** 7027-7033, 2006.

259. Gronborg M, Kristiansen TZ, Stensballe A, Andersen JS, Ohara O, Mann M, Jensen ON and Pandey A, MS-based proteomic approach to ID of serine/threonine-phosphorylated proteins by enrichment with phospho-specific antibodies: ID of a novel protein, Frigg, as a protein kinase A substrate. *Mol Cell Proteomics* **1**: 517-27., 2002.

260. Steen H, Fernandez M, Ghaffari S, Pandey A and Mann M, Phosphotyrosine Mapping in Bcr/Abl Oncoprotein Using Specific Immonium Ion Scanning. *Mol Cell Proteomics* **2**: 138-45., 2003.

261. Krause-Buchholz U, Becker JS, Zoriy M, Pickhardt C, Przybylski M, Roedel G and Becker JS, Phosphorylated subunits in yeast mitochondria by LA-ICP-MS and MALDI-FTICR-MS after SDS-PAGE. *Int J Mass Spectrom* **248**: 56-60, 2006.

262. Kjellstroem S and Jensen ON, Phosphoric Acid as a Matrix Additive for MALDI MS Analysis of Phosphopeptides and Phosphoproteins. *Analytical Chemistry* **76**(17): 5109-5117, 2004.

263. Annan RS, Huddleston MJ, Verma R, Deshaies RJ and Carr SA, A multidimensional electrospray MS-based approach to phosphopeptide mapping. *Anal Chem* **73**(3): 393-404., 2001.

264. Kim J, Camp DG, II and Smith RD, Improved detection of multi-phosphorylated peptides in the presence of phosphoric acid in LC/MS. *J Mass Spectrom* **39**: 208-215, 2004.

265. Sweet SMM, Creese AJ and Cooper HJ, Strategy for ID of Sites of Phosphorylation in Proteins: Neutral Loss Triggered ECD. *Anal Chem* **78**: 7563-7569, 2006.

266. Resing KA and Ahn NG, Protein phosphorylation analysis by electrospray ionization-mass spectrometry. *Methods Enzymol* **283**: 29-44., 1997.

267. Molloy MP and Andrews PC, Phosphopeptide derivatization signatures to identify serine and threonine phosphorylated peptides by mass spectrometry. *Anal Chem* **73**(22): 5387-94., 2001.

268. Rusnak F, Zhou J and Hathaway GM, Identification of Phosphorylated and Glycosylated Sites in Peptides by Chemically Targeted Proteolysis. *J Biomol Tech* **13**: 228-237, 2002.

269. Knight ZA, Schilling B, Row RH, Kenski DM, Gibson BW and Shokat KM, Phosphospecific proteolysis for mapping sites of protein phosphorylation. *Nat Biotechnol* **21**(9): 1047-54., 2003.

270. Bodenmiller B, Mueller Lukas N, Pedrioli Patrick GA, Pflieger D, Junger Martin A, Eng Jimmy K, Aebersold R and Tao WA, An integrated chemical, mass spectrometric and computational strategy for (quantitative) phosphoproteomics: application to Drosophila melanogaster Kc167 cells. *Mol Biosyst* **3**(4): 275-86, 2007.

271. Bodenmiller B, Mueller Lukas N, Mueller M, Domon B and Aebersold R, Reproducible isolation of distinct, overlapping segments of the phosphoproteome. *Nat Methods* **4**(3): 231-7, 2007.

272. Adamczyk M, Gebler JC and Wu J, Selective analysis of phosphopeptides within a protein mixture by chemical modification, reversible biotinylation and mass spectrometry. *Rapid Commun Mass Spectrom* **15**(16): 1481-8., 2001.

273. McLachlin DT and Chait BT, Improved b-Elimination-Based Affinity Purification Strategy for Enrichment of Phosphopeptides. *Analytical Chemistry* **75**(24): 6826-6836, 2003.

274. Adamczyk M, Gebler JC and Wu J, Identification of phosphopeptides by chemical modification with an isotopic tag and ion trap MS. *Rapid Commun Mass Spectrom* **16**(10): 999-1001., 2002.

275. Adamczyk M, Gebler JC and Wu J, A simple method to identify cysteine residues by isotopic labeling and ion trap mass spectrometry. *Rapid Commun Mass Spectrom* **13**(18): 1813-7., 1999.

276. Amoresano A, Marino G, Cirulli C and Quemeneur E, Mapping phosphorylation sites: A new strategy based on the use of isotopically-labelled dithiothreitol and mass spectrometry. *European Journal of Mass Spectrometry* **10**(3): 401-412, 2004.

277. Li W, Boykins RA, Backlund PS, Wang G and Chen HC, Identification of phosphoserine and phosphothreonine as cysteic acid and beta-methylcysteic acid residues in peptides by tandem mass spectrometric sequencing. *Anal Chem* **74**(22): 5701-10., 2002.

278. Shen J, Smith RA, Stoll VS, Edalji R, Jakob C, Walter K, Gramling E, Dorwin S, Bartley D, Gunasekera A, Yang J, Holzman T and Johnson RW, ID of protein kinase A phosphorylation: multi-technique approach to phosphate mapping. *Anal. Biochem.* **324**: 204-218, 2004.

279. Annan RSC, S. A., The essential role of mass spectrometry in characterizing protein structure: mapping posttranslational modifications. *J. Protein Chem.* **16**: 391-402, 1997.

280. Salek M, Alonso A, Pipkorn R and Lehmann WD, Analysis of protein tyrosine phosphorylation by nano ESI-MS/MS and tyrosine-targeted product ion scanning. *Anal Chem* **75**: 2724-2729, 2003.

281. Salomon AR, Ficarro SB, Brill LM, Brinker A, Phung QT, Ericson C, Sauer K, Brock A, Horn DM, Schultz PG and Peters EC, Profiling of tyrosine phosphorylation pathways in human cells using mass spectrometry. *Proc Natl Acad Sci U S A* **100**(2): 443-8., 2003.

282. Wind M, Kelm O, Nigg EA and Lehmann WD, Identification of phosphorylation sites in the polo-like kinases Plx1 and Plk1 by a novel strategy based on element and electrospray high resolution mass spectrometry. *Proteomics* **2**(11): 1516-23., 2002.

283. Wind M, Feldmann I, Jakubowski N and Lehmann WD, Spotting and quantification of phosphoproteins purified by gel electrophoresis and laser ablation-element mass spectrometry with phosphorus-31 detection. *Electrophoresis* **24**(7-8): 1276-80., 2003.

284. Ruotolo BT, Gillig KJ, Woods AS, Egan TF, Ugarov MV, Schultz JA and Russell DH, Phosphorylated peptides by ion mobility-MS. *Anal Chem* **76**: 6727-6733, 2004.

285. Schwartz D and Gygi SP, An iterative statistical approach to the identification of protein phosphorylation motifs from large-scale data sets. *Nature Biotechnology* **23**(11): 1391-1398, 2005.

286. Edelson-Averbukh M, Pipkorn R and Lehmann WD, Analysis of Protein Phosphorylation in the Regions of Consecutive Serine/Threonine Residues by Negative Ion ESI-CAD. Approach to Pinpointing of Phosphorylation Sites. *Anal Chem* **79**: 3476-3486, 2007.

287. Everley PA, Krijgsveld J, Zetter BR and Gygi SP, Quantitative cancer proteomics: Stable isotope labeling with amino acids in cell culture (SILAC) as a tool for prostate cancer research. *Molecular and Cellular Proteomics* **3**(7): 729-735, 2004.

288. Ballif BA, Roux PP, Gerber SA, MacKeigan JP, Blenis J and Gygi SP, Quantitative phosphorylation profiling of the ERK/p90 ribosomal S6 kinase-signaling cassette and its targets, the tuberous sclerosis tumor suppressors. *Proc Natl Acad Sci USA* **102**: 667-672, 2005.

289. Hogan JM, Pitteri SJ and McLuckey SA, Phosphorylation site identification via ion trap MS/MS of whole protein and peptide ions: Bovine a-crystallin A chain. *Anal Chem* **75**: 6509-6516, 2003.

290. Zarling AL, Polefrone JM, Evans AM, Mikesh LM, Shabanowitz J, Lewis S, Engelhard VH and Hunt DF, Identification of class I MHC-associated phosphopeptides as targets for cancer immunotherapy. *Proc Nat Acad Sci USA* **103**: 14889-14894, 2006.

291. Chi A, Huttenhower C, Geer LY, Coon JJ, Syka JEP, Bai DL, Shabanowitz J, Burke DJ, Troyanskaya OG and Hunt DF, Analysis of phosphorylation sites on proteins from saccharomyces cerevisiae by ETD MS. *Proc Nat Acad Sci USA* **104**: 2193-2198, 2007.

292. Hegeman AD, Harms AC, Sussman MR, Bunner AE and Harper JF, An isotope labeling strategy for quantifying the degree of phosphorylation at multiple sites in proteins. *Journal of the American Society for Mass Spectrometry* **15**(5): 647-653, 2004.

293. Goshe MB, Veenstra TD, Panisko EA, Conrads TP, Angell NH and Smith RD, Phosphoprotein isotope-coded affinity tags: Application to the enrichment and identification of low-abundance phosphoproteins. *Analytical Chemistry* **74**(3): 607-616, 2002.

294. Bossio RE and Marshall AG, Baseline resolution of isobaric phosphorylated and sulfated peptides and nucleotides by electrospray ionization FTICR ms: another step toward mass spectrometry-based proteomics. *Anal Chem* **74**(7): 1674-9., 2002.

295. Gerrits B, Bodenmiller B, Panse C, Barkow S, Aebersold R and Schlapbach R, On the Co-Purification of Sulfonated Peptides from Phosphopeptide-Enriched Samples. In: *55th Annual Conf of Amer Soc Mass Spectrom, Indianapolis,2007*, pp. MP 387.

296. Zhang Y, Go EP, Jiang H and Desaire H, A Novel Mass Spectrometric Method to Distinguish Isobaric Monosaccharides that are Phosphorylated or Sulfated Using Ion-Pairing Reagents. *Journal of the American Society for Mass Spectrometry* **16**(11): 1827-1839, 2005.

297. Nemeth-Cawley JF, Karnik S, Rouse JC and Vath JE, Analysis of sulfated peptides using positive electrospray ionization tandem mass spectrometry. *J Mass Spectrom* **36**(12): 1301-11., 2001.

298. Jiang H, Irungu J and Desaire H, Enhanced detection of sulfated glycosylation sites in glycoproteins. *J Amer Soc Mass Spectrom* **16**(3): 340-348, 2005.

299. Morris HR, Chalabi S, Panico M, Sutton-Smith M, Clark GF, Goldberg D and Dell A, Glycoproteomics: Past, present and future. *International Journal of Mass Spectrometry* **259**(1-3): 16-31, 2007.

300. Harvey DJ, ID of protein-bound carbohydrates by mass spectrometry. *Proteomics* **1**: 311-328, 2001.

301. An HJ, Peavy TR, Hedrick JL and Lebrilla CB, Determination of N-glycosylation sites and site heterogeneity in glycoproteins. *Analytical Chemistry* **75**(20): 5628-5637, 2003.

302. Peterman SM and Mulholland JJ, A Novel Approach for Identification and Characterization of Glycoproteins Using a Hybrid Linear Ion Trap/FT-ICR Mass Spectrometer. *Journal of the American Society for Mass Spectrometry* **17**(2): 168-179, 2006.

303. Ritchie MA, Gill AC, Deery MJ and Lilley K, Precursor ion scanning for structural characterization of heterogeneous glycopeptide mixtures. *J Am Soc Mass Spectrom* **13**: 1065-77., 2002.

304. Bykova NV, Rampitsch C, Krokhin O, Standing KG and Ens W, Determination and Characterization of Site-Specific N-Glycosylation Using MALDI-Qq-TOF Tandem Mass Spectrometry: Case Study with a Plant Protease. *Analytical Chemistry* **78**(4): 1093-1103, 2006.

305. Khoshnoodi J, Hill S, Tryggvason K, Hudson B and Friedman DB, Identification of N-linked glycosylation sites in human nephrin using MS. *J Mass Spectrom* **42**: 370-379, 2007.

306. Huddleston MJ, Bean MF and Carr SA, CAD of glycopeptides by ESI LC/MS and LC/MS/MS: selective method for detection of glycopeptides in protein digests. *Anal. Chem.* **65**: 877-884, 1993.

307. Nemeth JFH, G. P.; Marnett, L. J.; Capriolli, R. M., Characterization of the Glycosylation Sites in Cyclooxygenase-2 Using Mass Spectrometry. *Biochemistry* **40**: 3109-3116, 2001.

308. Angel PM, Lim J-M, Wells L, Bergmann C and Orlando R, A potential pitfall in 18O-based N-linked glycosylation site mapping. *Rapid Commun Mass Spectrom* **21**: 674-682, 2007.

309. Stephenson JL, Jr. and McLuckey SA, Simplification of Product Ion Spectra Derived from Multiply Charged Parent Ions via Ion/Ion Chemistry. *Anal. Chem.* **70**: 3533-3544, 1998.

310. Reid GE, Stephenson JL, Jr. and McLuckey SA, Tandem Mass Spectrometry of Ribonuclease A and B: N-Linked Glycosylation Site Analysis of Whole Protein Ions. *Anal. Chem.* **74**: 577 -583, 2002.

311. Catalina MI, Koeleman CAM, Deelder AM and Wuhrer M, ETD of N-glycopeptides: loss of the entire N-glycosylated asparagine side chain. *Rapid Communications in Mass Spectrometry* **21**: 1053-1061, 2007.

312. Zhang Q, Frolov A, Tang N, Hoffmann R, van de Goor T, Metz TO and Smith RD, Application of ETD MS in analyses of non-enzymatically glycated peptides. *Rapid Commun Mass Spectrom* **21**: 661-666, 2007.

313. Deguchi K, Ito H, Baba T, Hirabayashi A, Nakagawa H, Fumoto M, Hinou H and Nishimura S-I, Structural analysis of O-glycopeptides employing negative- and positive-ion multi-stage MS obtained by CAD and ECD in linear ion trap TOF MS. *Rapid Commun Mass Spectrom* **21**: 691-698, 2007.

314. Kim JY, Kim KW, Kwon HJ, Lee DW and Yoo JS, Probing lysine acetylation with a modification-specific marker ion using LC-ESI-CAD-MS/MS. *Anal Chem* **74**: 5443-9., 2002.

315. Solouki T, Emmett MR, Guan S and Marshall AG, Detection, number, and sequence location of sulfur-containing amino acids and disulfide bridges in peptides by ultrahigh-resolution MALDI FTICR mass spectrometry. *Anal Chem* **69**(6): 1163-8., 1997.

316. Zaluzec EJ, Gage DA and Watson JT, Cysteine and Cystine status by Organomercurial Derivatization and MALDI-MS. *J. Am. Soc. Mass Spectrom.* **5**: 359-366, 1994.

317. Borges CR and Watson JT, Recognition of cysteine-containing peptides through prompt fragmentation of the 4-dimethylaminophenylazophenyl-4'-maleimide derivative during analysis by MALDI-MS. *Protein Sci* **12**(7): 1567-72., 2003.

318. Toriumi C and Imai K, An Identification Method for Altered Proteins in Tissues Utilizing Fluorescence Derivatization, Liquid Chromatography, Tandem Mass Spectrometry, and a Database-Searching Algorithm. *Analytical Chemistry* **75**(15): 3725-3730, 2003.

319. Masuda M, Toriumi C, Santa T and Imai K, Fluorogenic derivatization reagents suitable for isolation and identification of cysteine-containing proteins utilizing high-performance liquid chromatography-tandem mass spectrometry. *Analytical Chemistry* **76**(3): 728-735, 2004.

320. Bilusich D, Maselli VM, Brinkworth CS, Samguina T, Lebedev AT and Bowie JH, ID of intramolecular disulfide links in peptides using negative ion ESI of underivatised peptides. A joint experimental and theoretical study. *Rapid Commun Mass Spectrom* **19**: 3063-3074, 2005.

321. Bilusich D and Bowie JH, Identification of intermolecular disulfide linkages in underivatised peptides using negative ion ESI MS. A joint experimental and theoretical study. *Rapid Commun Mass Spectrom* **21**: 619-628, 2007.

322. Hodder AN, Cewther PE, Matthew MLSM, Reid GE, Moritz RL, Simpson RJ and Anders RF, The disulfide bond structure of Plasmodium apical membrane antigen-1. *Journal of Biological Chemistry* **271**(46): 29446-29452, 1996.

323. Cole AR, Hall NE, Treutlein HR, Eddes JS, Reid GE, Moritz RL and Simpson RJ, Disulfide bond structure and N-glycosylation sites of the extracellular domain of the human interleukin-6 receptor. *Journal of biological chemistry* **274**(11): 7207-15, 1999.

324. Lim A, Wally J, Walsh MT, Skinner M and Costello CE, ID and location of a cysteinyl PTM in an amyloidogenic kappa1 light chain protein by ESI and MALDI. *Anal Biochem* **295**: 45-56., 2001.

325. Ge Y, Lawhorn BG, ElNaggar M, Strauss E, Park JH, Begley TP and McLafferty FW, Top down characterization of larger proteins (45 kDa) by ECD MS. *J Am Chem Soc* **124**: 672-8., 2002.

326. Chrisman PA and McLuckey SA, Dissociations of disulfide-linked gaseous polypeptide/protein anions: ion chemistry with implications for protein ID. *J Proteome Res* **1**: 549-57., 2002.

327. Jones MD, Patterson SD and Lu HS, Determination of disulfide bonds in highly bridged disulfide-linked peptides by MALDI MS with PSD. *Anal Chem* **70**: 136-43., 1998.

328. Zhang M and Kaltashov IA, Mapping of Protein Disulfide Bonds Using Negative Ion Fragmentation with a Broadband Precursor Selection. *Analytical Chemistry* **78**(14): 4820-4829, 2006.

329. Wu J, Gage DA and Watson JT, A strategy to locate cysteine residues in proteins by specific chemical cleavage followed by MALDI-TOF-MS. *Anal. Biochem.* **235**: 161-74, 1996.

330. Kuhn DM, Sadidi M, Liu X, Kreipke C, Geddes T, Borges C and Watson JT, Peroxynitrite-induced nitration of tyrosine hydroxylase: identification of tyrosines 423, 428, and 432 as sites of modification by MALDI and tyrosine-scanning mutagenesis. *J Biol Chem* **277**: 14336-42., 2002.

331. Sun Y, Bauer MD, Keough TW and Lacey MP, Disulfide bond location in proteins. *Methods Mol Biol* **61**: 185-210., 1996.

332. Merewether LA, Le J, Jones MD, Lee R, Shimamoto G and Lu HS, Development of disulfide peptide mapping and determination of disulfide structure of recombinant human osteoprotegerin chimera produced in Escherichia coli. *Arch Biochem Biophys* **375**(1): 101-10., 2000.

333. Guan Z, Identification and localization of the fatty acid modification in ghrelin by electron capture dissociation. *J Am Soc Mass Spectrom* **13**(12): 1443-7., 2002.

334. Wefing S, Schnaible V and Hoffmann D, SearchXLinks. A Program for the Identification of Disulfide Bonds in Proteins from Mass Spectra. *Analytical Chemistry* **78**(4): 1235-1241, 2006.

335. Fukuyama Y, Iwamoto S and Tanaka K, Rapid sequencing and disulfide mapping of peptides containing disulfide bonds by using 1,5-diaminonaphthalene as a reductive matrix. *Journal of Mass Spectrometry* **41**(2): 191-201, 2006.

336. Thevis M, Ogorzalek Loo RR and Loo JA, In-gel derivatization of proteins for cysteine-specific cleavages and their analysis by MS: Studying noncovalent protein complexes by ESI MS. *J Proteome Res* **2**: 163-72., 2003.

337. Jakubowski JA and Sweedler JV, Sequencing and mass profiling highly modified conotoxins using global reduction/alkylation followed by MS. *Analytical Chemistry* **76**: 6541-6547, 2004.

338. Ma S, Hill KE, Burk RF and Caprioli RM, Mass spectrometric determination of selenenylsulfide linkages in rat selenoprotein P. *Journal of Mass Spectrometry* **40**(3): 400-404, 2005.

339. Chelius D, Huff Wimer ME and Bondarenko PV, rp LC In-Line with Negative ESI for ID of Disulfide-Linkages of an Immunoglobulin Gamma Antibody. *J Amer Soc Mass Spectrom* **17**: 1590-1598, 2006.

340. Liu T, Qian W-J, Chen W-NU, Jacobs JM, Moore RJ, Anderson DJ, Gritsenko MA, Monroe ME, Thrall BD, Camp DG, II and Smith RD, Improved proteome coverage by using high efficiency cysteinyl peptide enrichment: Human mammary epithelial cell proteome. *Proteomics* **5**: 1263-1273, 2005.

341. Dai J, Wang J, Zhang Y, Lu Z, Yang B, Li X, Cai Y and Qian X, Enrichment and Identification of Cysteine-Containing Peptides from Tryptic Digests of Performic Oxidized Proteins by Strong Cation Exchange LC and MALDI-TOF/TOF MS. *Analytical Chemistry* **77**(23): 7594-7604, 2005.

342. Clements A, Johnston MV, Larsen BS and McEwen CN, Fluorescence-Based Peptide Labeling and Fractionation Strategies for Analysis of Cysteine-Containing Peptides. *Analytical Chemistry* **77**(14): 4495-4502, 2005.

343. Bilusich D, Brinkworth CS and Bowie JH, Negative ion mass spectra of Cys-containing peptides. The characteristic Cys g backbone cleavage: A joint experimental and theoretical study. *Rapid Communications in Mass Spectrometry* **18**(5): 544-552, 2004.

344. Kalli A and Hakansson K, Preferential cleavage of SS and CS bonds in ETD and infrared multiphoton dissociation of disulfide-linked peptide anions. *Int J Mass Spectrom* **263**: 71-81, 2007.

345. Lioe H and O'Hair RAJ, A Novel Salt Bridge Mechanism Highlights the Need for Nonmobile Proton Conditions to Promote Disulfide Bond Cleavage in Protonated Peptides Under Low-Energy Collisional Activation. *J Amer Soc Mass Spectrom* **18**: 1109-1123, 2007.

346. Mihalca R, van der Burgt YEM, Heck AJR and Heeren RMA, Disulfide bond cleavages observed in SORI-CID of three nonapeptides complexed with divalent transition-metal cations. *J Mass Spectrom* **42**: 450-458, 2007.

347. Craig R, Krokhin O, Wilkins J and Beavis RC, Implementation of an algorithm for modeling disulfide bond patterns using mass spectrometry. *Journal of Proteome Research* **2**(6): 657-661, 2003.

348. Gray WR, Disulfide structures of highly bridged peptides: a new strategy for analysis. *Protein Sci* **2**(10): 1732-48., 1993.

349. van den Hooven HW, van den Burg HA, Vossen P, Boeren S, de Wit PJ and Vervoort J, Disulfide bond structure of the AVR9 elicitor of the fungal tomato pathogen Cladosporium fulvum: evidence for a cystine knot. *Biochemistry* **40**(12): 3458-66., 2001.

350. Nair SS, Nilsson CL, Emmett MR, Schaub TM, Gowd KH, Thakur SS, Krishnan KS, Balaram P and Marshall AG, De Novo Sequencing and Disulfide Mapping of a Bromotryptophan-Containing Conotoxin by Fourier Transform Ion Cyclotron Resonance Mass Spectrometry. *Analytical Chemistry* **78**(23): 8082-8088, 2006.

351. Bauer M, Sun Y, Degenhardt C and Kozikowski B, Assignment of all four disulfide bridges in echistatin. *J Protein Chem* **12**(6): 759-64., 1993.

352. Sturrock ED, Yu XC, Wu Z, Biemann K and Riordan JF, Assignment of Free and Disulfide-Bonded Cysteine Residues in Testis Angiotensin-Converting Enzyme: Functional Implications. *Biochemistry* **35**(29): 9560-9566, 1996.

353. Schnaible V, Wefing S, Resemann A, Suckau D, Bucker A, Wolf-Kummeth S and Hoffmann D, Screening for disulfide bonds in proteins by MALDI in-source decay and LIFT-TOF/TOF-MS. *Anal Chem* **74**(19): 4980-8., 2002.

354. Krokhin OV, Cheng K, Sousa SL, Ens W, Standing KG and Wilkins JA, MS-based mapping of the disulfide bonding patterns of integrin alpha chains. *Biochemistry* **42**(44): 12950-9., 2003.

355. Wu J and Watson JT, A novel methodology for assignment of disulfide bond pairings in proteins. *Protein Sci.* **6**: 391-398, 1997.

356. Wu J and Watson JT, Assignment of disulfide bonds in proteins by chemical cleavage and peptide mapping by mass spectrometry. In: *Methods Mol Biol*, Vol. 194, pp. 1-22., 2002.

357. Qi J, Wu J, Somkuti GA and Watson JT, Disulfide Structure of Sillucin, a Knotted, Cysteine-Rich Peptide, by Cyanylation/Cleavage Mass Mapping. *Biochemistry* **40**(15): 4531-4538, 2001.

358. Wu J and Watson JT, Optimization of the cleavage reaction for cyanylated cysteinyl proteins for efficient and simplified mass mapping. *Anal. Biochem.* **258**(2): 268-276, 1998.

359. Gallegos-Perez J-L, Rangel-Ordonez L, Bowman SR, Ngowe CO and Watson JT, Study of primary amines for nucleophilic cleavage of cyanylated cystinyl proteins in disulfide mass mapping methodology. *Analytical Biochemistry* **346**(2): 311-319, 2005.

360. Wu W, Huang W, Qi J, Chou Y-T, Torng E and Watson JT, 'Signature sets', minimal fragment sets for identifying protein disulfide structures with cyanylation-based mass mapping methodology. *Journal of Proteome Research* **3**(4): 770-777, 2004.

361. Yang Y, Wu J and Watson JT, Disulfide Mass Mapping in Proteins Containing Adjacent Cysteines Is Possible with Cyanylation/Cleavage Methodology. *J. Am. Chem. Soc.* **120**(23): 5834-5835, 1998.
362. Qi J, Wu W, Borges CR, Hang D, Rupp M, Torng E and Watson JT, Automated data interpretation based on the concept of \"negative signature mass\" for mass-mapping disulfide structures of cystinyl proteins. *Journal of the American Society for Mass Spectrometry* **14**(9): 1032-1038, 2003.
363. Borges CR, Qi J, Wu W, Torng E, Hinck AP and Watson JT, Algorithm-assisted elucidation of disulfide structure: application of the negative signature mass algorithm to mass-mapping the disulfide structure of the 12-cysteine transforming growth factor beta type II receptor extracellular domain. *Anal Biochem* **329**(1): 91-103., 2004.
364. Schnaible V, Wefing S, Bucker A, Wolf-Kummeth S and Hoffmann D, Partial reduction and two-step modification of proteins for identification of disulfide bonds. *Anal Chem* **74**(10): 2386-93., 2002.
365. Li X, Chou Y-T, Husain R and Watson JT, Integration of hydrogen/deuterium exchange and cyanylation-based methodology for conformational studies of cystinyl proteins. *Analytical Biochemistry* **331**(1): 130-137, 2004.
366. Li X, Hood RJ, Wedemeyer WJ and Watson JT, Characterization of peptide folding nuclei by hydrogen/deuterium exchange-mass spectrometry. *Protein Science* **14**(7): 1922-1928, 2005.
367. Wang D and Cotter RJ, Approach for Determining Protein Ubiquitination Sites by MALDI-TOF Mass Spectrometry. *Analytical Chemistry* **77**(5): 1458-1466, 2005.
368. Warren MRE, Parker CE, Mocanu V, Klapper D and Borchers CH, ESI MS/MS of model peptides reveals diagnostic fragment ions for protein ubiquitination. *Rapid Commun Mass Spectrom* **19**: 429-437, 2005.
369. Wang D, Kalume D, Pickart C, Pandey A and Cotter RJ, ID of Protein Ubiquitylation by ESI-MS/MS of Sulfonated Tryptic Peptides. *Anal Chem* **78**: 3681-3687, 2006.
370. Willard BB, Ruse CI, Keightley JA, Bond M and Kinter M, Site-Specific Quantitation of Protein Nitration Using LC/MS/MS. *Analytical Chemistry* **75**: 2370-2376, 2003.
371. Borges CR, Kuhn DM and Watson JT, Mass mapping sites of nitration in tyrosine hydroxylase: random vs selective nitration of three tyrosine residues. *Chem Res Toxicol* **16**(4): 536-40., 2003.
372. Zhan X and Desiderio DM, Linear ion-trap mass spectrometric characterization of human pituitary nitrotyrosine-containing proteins. *International Journal of Mass Spectrometry* **259**(1-3): 96-104, 2007.
373. Borges CR, Geddes T, Watson JT and Kuhn DM, Dopamine biosynthesis is regulated by S-glutathionylation. Potential mechanism of tyrosine hydroxylast inhibition during oxidative stress. *J Biol Chem* **277**(50): 48295-302., 2002.
374. Hoffman MD and Kast J, Mass spectrometric characterization of lipid-modified peptides for the analysis of acylated proteins. *Journal of Mass Spectrometry* **41**(2): 229-241, 2006.
375. Cargile BJ, Bundy JL, Grunden AM and Stephenson JL, Synthesis/Degradation Ratio MS for Measuring Relative Dynamic Protein Turnover. *Analytical Chemistry* **76**: 86-97, 2004.
376. Hamdan M and Righetti PG, Modern strategies for protein quantification in proteome analysis: advantages and limitations. *Mass Spectrom Rev* **21**(4): 287-302., 2002.
377. Ong SE, Foster LJ and Mann M, Mass spectrometric-based approaches in quantitative proteomics. *Methods* **29**(2): 124-30., 2003.
378. Ong SE and Mann M, Mass spectrometry-based proteomics turns quantitative. *Nat Chem Biol* **1**(5): 252-62, 2005.
379. Forner F, Foster LJ, Campanaro S, Valle G and Mann M, Quantitative proteomic comparison of rat mitochondria from muscle, heart and liver. *Mol Cell Proteomics*, 2006.
380. Turecek F, Mass spectrometry in coupling with affinity capture-release and isotope-coded affinity tags for quantitative protein analysis. *J Mass Spectrom* **37**(1): 1-14., 2002.
381. Julka S and Regnier F, Quantification in proteomics through stable isotope coding: A review. *Journal of Proteome Research* **3**(3): 350-363, 2004.
382. Gerber SA, Scott CR, Turecek F and Gelb MH, Direct profiling of multiple enzyme activities in human cell lysates by affinity chromatography/electrospray ionization mass spectrometry: application to clinical enzymology. *Anal Chem* **73**(8): 1651-7., 2001.
383. Zhou X, Turecek F, Scott CR and Gelb MH, Quantification of cellular acid sphingomyelinase and galactocerebroside beta-galactosidase activities by ESI MS. *Clin Chem* **47**: 874-81., 2001.
384. Ogata Y, Scampavia L, Ruzicka J, Scott CR, Gelb MH and Turecek F, Automated affinity capture-release of biotin-containing conjugates using a lab-on-valve apparatus coupled to UV/visible and electrospray ionization mass spectrometry. *Anal Chem* **74**(18): 4702-8., 2002.
385. Lu Y, Bottari P, Aebersold R, Turecek F and Gelb MH, Absolute quantification of specific proteins in complex mixtures using visible isotope-coded affinity tags. *Methods in Molecular Biology* **359**(Quantitative Proteomics by Mass Spectrometry): 159-176, 2007.
386. Lahm HW and Langen H, Mass spectrometry: a tool for the identification of proteins separated by gels. *Electrophoresis* **21**(11): 2105-14., 2000.
387. Aggarwal K, Choe LH and Lee KH, Quantitative analysis of protein expression using amine-specific isobaric tags in Escherichia coli cells expressing rhsA elements. *Proteomics* **5**: 2297-2308, 2005.

388. Gerber SA, Turecek F and Gelb MH, Design and synthesis of substrate and internal standard conjugates for profiling enzyme activity in the Sanfilippo syndrome by affinity chromatography/electrospray ionization mass spectrometry. *Bioconjug Chem* **12**(4): 603-15., 2001.

389. Bayer EA and Wilchek M, Application of avidin-biotin technology to affinity-based separations. *J Chromatogr* **510**: 3-11., 1990.

390. Gitlin G, Bayer EA and Wilchek M, Studies on the biotin-binding sites of avidin and streptavidin. Tyrosine residues are involved in the binding site. *Biochem J* **269**(2): 527-30., 1990.

391. Gygi SP, Rist B, Gerber SA, Turecek F, Gelb MH and Aebersold R, Quantitative analysis of complex protein mixtures using isotope-coded affinity tags. *Nat Biotechnol* **17**(10): 994-9., 1999.

392. Borisov OV, Goshe MB, Conrads TP, Rakov VS, Veenstra TD and Smith RD, Low-Energy CAD of Cysteinyl-Modified Peptides. *Anal Chem* **74**: 2284 -2292, 2002.

393. Aebersold R, Rist B and Gygi SP, Quantitative proteome analysis: methods and applications. *Ann N Y Acad Sci* **919**: 33-47., 2000.

394. Griffin TJ, Han DK, Gygi SP, Rist B, Lee H, Aebersold R and Parker KC, Toward a high-throughput approach to quantitative proteomic analysis: expression-dependent protein identification by mass spectrometry. *J Am Soc Mass Spectrom* **12**(12): 1238-46., 2001.

395. Griffin TJ, Gygi SP, Rist B, Aebersold R, Loboda A, Jilkine A, Ens W and Standing KG, Quantitative proteomic analysis using a MALDI quadrupole TOF MS. *Anal Chem* **73**: 978-86., 2001.

396. Gygi SP, Rist B and Aebersold R, Measuring gene expression by quantitative proteome analysis. *Curr Opin Biotechnol* **11**(4): 396-401., 2000.

397. Aggarwal K, Choe LH and Lee KH, Shotgun proteomics using the iTRAQ isobaric tags. *Briefings in Functional Genomics & Proteomics* **5**(2): 112-120, 2006.

398. Andersen JS and Mann M, Functional genomics by MS. *FEBS Lett* **480**(1): 25-31., 2000.

399. Guina T, Wu M, Miller SI, Purvine SO, Yi EC, Eng J, Goodlett DR, Aebersold R, Ernst RK and Lee KA, Proteomic analysis of Pseudomonas aeruginosa grown under magnesium limitation. *Journal of the American Society for Mass Spectrometry* **14**(7): 742-751, 2003.

400. Shiio Y, Eisenman RN, Yi EC, Donohoe S, Goodlett DR and Aebersold R, Quantitative proteomic analysis of chromatin-associated factors. *J Am Soc Mass Spectrom* **14**: 696-703, 2003.

401. Sethuraman M, McComb ME, Huang H, Huang S, Heibeck T, Costello CE and Cohen RA, ICAT Approach to Redox Proteomics: Identification and Quantitation of Oxidant-Sensitive Cysteine Thiols in Complex Protein Mixtures. *Journal of Proteome Research* **3**(6): 1228-1233, 2004.

402. Molly MP, Donohoe S, Brzezinski EE, Kilby GW, Stevenson TI, Baker JD, Goodlett DR and Gage DA, Large-scale evaluation of quantitative reproducibility and proteome coverage using acid cleavable ICAT mass spectrometry for proteomic profiling. *Proteomics* **5**(5): 1204-1208, 2005.

403. Arnott D, Kishiyama A, Luis EA, Ludlum SG, Marsters JC, Jr. and Stults JT, Selective detection of membrane proteins without antibodies: a mass spectrometric version of the Western blot. *Mol Cell Proteomics* **1**(2): 148-56., 2002.

404. Zhang R, Sioma CS, Thompson RA, Xiong L and Regnier FE, Controlling deuterium isotope effects in comparative proteomics. *Anal Chem* **74**(15): 3662-9., 2002.

405. Li J, Steen H and Gygi SP, Protein profiling with cleavable ICAT reagents. The yeast salinity stress response. *Molecular and Cellular Proteomics* **2**: 1198-1204, 2003.

406. Griffin TJ, Lock CM, Li XJ, Patel A, Chervetsova I, Lee H, Wright ME, Ranish JA, Chen SS and Aebersold R, Abundance ratio-dependent proteomic analysis by MS. *Anal Chem* **75**: 867-74., 2003.

407. Li Y, Ogata Y, Freeze HH, Scott CR, Turecek F and Gelb MH, Affinity capture and elution/electrospray ionization mass spectrometry assay of phosphomannomutase and phosphomannose isomerase for the multiplex analysis of congenital disorders of glycosylation types Ia and Ib. *Anal Chem* **75**(1): 42-8., 2003.

408. Zhang R, Sioma CS, Wang S and Regnier FE, Fractionation of isotopically labeled peptides in quantitative proteomics. *Anal Chem* **73**: 5142-9., 2001.

409. Qiu Y, Sousa EA, Hewick RM and Wang JH, Acid-labile isotope-coded extractants: a class of reagents for quantitative MS of protein mixtures. *Anal Chem* **74**: 4969-79., 2002.

410. Gerber SA, Rush J, Stemman O, Kirschner MW and Gygi SP, Absolute quantification of proteins and phosphoproteins from cell lysates by tandem MS. *Proc Natl Acad Sci U S A* **100**(12): 6940-5., 2003.

411. Lu Y, Bottari P, Turecek F, Aebersold R and Gelb MH, Absolute quantification of specific proteins in complex mixtures using visible ICATs. *Analytical Chemistry* **76**: 4104-4111, 2004.

412. Goodlett DR, Purvine S, Yi E, Eng JK, Aebersold R, Watts JD, von Haller P, Newitt R, Gygi SP, Mylchreest I and Hemenway T, Differential isotopic labelling of peptides and other tricks with isotope distribution encoded tags. *Advances in Mass Spectrometry* **15**: 365-367, 2001.

413. Cagney G and Emili A, De novo peptide sequencing and quantitative profiling of complex protein mixtures using mass-coded abundance tagging. *Nat Biotechnol* **20**(2): 163-70., 2002.

414. Johnson KL and Muddiman DC, A method for calculating 16O/18O peptide ion ratios for the relative quantification of proteomes. *J Am Soc Mass Spectrom* **15**: 437-445, 2004.

415. Miyagi M and Rao KCS, Proteolytic 18O-labeling strategies for quantitative proteomics. *Mass Spectrometry Reviews* **26**: 121-136, 2007.

416. Angel PM and Orlando R, Quantitative carbamylation as a stable isotopic labeling method for comparative proteomics. *Rapid Commun Mass Spectrom* 21: 1623-1634, 2007.
417. Han B, Stevens JF and Maier CS, Design, Synthesis, and Application of a Hydrazide-Functionalized Isotope-Coded Affinity Tag for the Quantification of Oxylipid-Protein Conjugates. *Anal Chem* 79: 3342-3354, 2007.
418. Gruhler A, Schulze WX, Matthiesen R, Mann M and Jensen ON, Stable isotope labeling of Arabidopsis thaliana cells and quantitative proteomics by mass spectrometry. *Mol Cell Proteomics* 4(11): 1697-709, 2005.
419. Foster LJ, Rudich A, Talior I, Patel N, Huang X, Furtado LM, Bilan PJ, Mann M and Klip A, Insulin-dependent interactions of proteins with GLUT4 revealed through stable isotope labeling by amino acids in cell culture (SILAC). *J Proteome Res* 5(1): 64-75, 2006.
420. Doherty MK, Whitehead C, McCormack H, Gaskell SJ and Beynon RJ, Proteome dynamics in complex organisms: stable isotopes to monitor protein turnover rates. *Proteomics* 5: 522-533, 2005.
421. Kirkpatrick DS, Gerber SA and Gygi SP, The absolute quantification strategy: a general procedure for the quantification of proteins and post-translational modifications. *Methods (San Diego, CA, United States)* 35(3): 265-273, 2005.
422. Gu S, Pan S, Bradbury EM and Chen X, Peptide sequencing and protein quantification in the human proteome through in vivo lysine-specific mass tagging. *J Am Soc Mass Spectrom* 14: 1-7., 2003.
423. Gu S, Pan S, Bradbury EM and Chen X, Use of deuterium-labeled lysine for efficient protein identification and peptide de novo sequencing. *Anal Chem* 74(22): 5774-85., 2002.
424. Beardsley RL and Reilly JP, Quantitation using enhanced signal tags: a technique for comparative proteomics. *J Proteome Res* 2(1): 15-21., 2003.
425. Peters EC, Horn DM, Tully DC and Brock A, A novel multifunctional labeling reagent for enhanced protein characterization with MS. *Rapid Commun Mass Spectrom* 15: 2387-92, 2001.
426. Krause E, Wenschuh H and Jungblut PR, The dominance of arginine-containing peptides in MALDI-derived tryptic mass fingerprints of proteins. *Anal Chem* 71(19): 4160-5., 1999.
427. Heller M, Mattou H, Menzel C and Yao X, Trypsin catalyzed 16O-to-18O exchange for comparative proteomics: tandem mass spectrometry comparison using MALDI-TOF, ESI-QTOF, and ESI-ion trap mass spectrometers. *Journal of the American Society for Mass Spectrometry* 14(7): 704-718, 2003.
428. Sevinsky JR, Brown KJ, Cargile BJ, Bundy JL and Stephenson JL, Jr., Minimizing Back Exchange in 18O/16O Quantitative Proteomics Experiments by Incorporation of Immobilized Trypsin into the Initial Digestion Step. *Anal Chem* 79: 2158-2162, 2007.
429. Havlis J and Shevchenko A, Absolute quantification of proteins in solutions and in polyacrylamide gels by mass spectrometry. *Analytical Chemistry* 76(11): 3029-3036, 2004.
430. Che F-Y and Fricker LD, Quantitation of Neuropeptides in Cpefat/Cpefat Mice Using Differential Isotopic Tags and Mass Spectrometry. *Anal. Chem.* 74: 3190 -3198, 2002.
431. Hopkins CE, O'Connor P B, Allen KN, Costello CE and Tolan DR, Chemical-modification rescue assessed by mass spectrometry demonstrates that gamma-thia-lysine yields the same activity as lysine in aldolase. *Protein Sci* 11(7): 1591-1599., 2002.
432. Yao X, Freas A, Ramirez J, Demirev PA and Fenselau C, Proteolytic 18O labeling for comparative proteomics: model studies with two serotypes of adenovirus. *Anal Chem* 73(13): 2836-42., 2001.
433. Washburn MP, Ulaszek R, Deciu C, Schieltz DM and Yates JR, 3rd, Analysis of quantitative proteomic data generated via MudPIT. *Anal Chem* 74: 1650-7., 2002.
434. Hsu J-L, Huang S-Y, Chow N-H and Chen S-H, Stable-Isotope Dimethyl Labeling for Quantitative Proteomics. *Analytical Chemistry* 75(24): 6843-6852, 2003.
435. Melanson JE, Avery SL and Pinto DM, High-coverage quantitative proteomics using amine-specific isotopic labeling. *Proteomics* 6(16): 4466-4474, 2006.
436. Martinovic S, Veenstra TD, Anderson GA, Pasa-Tolic L and Smith RD, Selective incorporation of isotopically labeled amino acids for identification of intact proteins on a proteome-wide level. *J Mass Spectrom* 37(1): 99-107., 2002.
437. MacCoss MJ, Wu CC, Matthews DE and Yates JR, III, Measurement of the Isotope Enrichment of Stable Isotope-Labeled Proteins Using High-Resolution Mass Spectra of Peptides. *Analytical Chemistry* 77(23): 7646-7653, 2005.
438. Bondarenko PV, Chelius D and Shaler TA, Identification and relative quantitation of protein mixtures by enzymatic digestion and LC/MS/MS. *Anal Chem* 74: 4741-9., 2002.
439. Thompson A, Schafer J, Kuhn K, Kienle S, Schwarz J, Schmidt G, Neumann T and Hamon C, Tandem mass tags: a novel quantification strategy for comparative analysis of complex protein mixtures by MS/MS. *Anal Chem* 75(8): 1895-904., 2003.
440. Wang W, Zhou H, Lin H, Roy S, Shaler TA, Hill LR, Norton S, Kumar P, Anderle M and Becker CH, Quantification of Proteins and Metabolites by Mass Spectrometry without Isotopic Labeling or Spiked Standards. *Analytical Chemistry* 75(18): 4818-4826, 2003.
441. DeSouza L, Diehl G, Rodrigues MJ, Guo J, Romaschin AD, Colgan TJ and Siu KWM, Search for cancer markers from endometrial tissues using differentially labeled tags iTRAQ and cICAT with multidimensional LC/MS/MS. *J Proteome Res* 4: 377-386, 2005.

442. Brittain SM, Ficarro SB, Brock A and Peters EC, Enrichment and analysis of peptide subsets using fluorous affinity tags and mass spectrometry. *Nature Biotechnology* **23**(4): 463-468, 2005.

443. Tang W, Harrata AK and Lee CS, Two-dimensional analysis of recombinant E. coli proteins using capillary isoelectric focusing ESI MS. *Anal Chem* **69**(16): 3177-82., 1997.

444. Jensen PK, Pasa-Tolic L, Anderson GA, Horner JA, Lipton MS, Bruce JE and Smith RD, Probing proteomes using capillary isoelectric focusing-ESI-FT-ICR-MS. *Anal Chem* **71**: 2076-84., 1999.

445. Higgs RE, Knierman MD, Gelfanova V, Butler JP and Hale JE, Comprehensive Label-Free Method for the Relative Quantification of Proteins from Biological Samples. *Journal of Proteome Research* **4**(4): 1442-1450, 2005.

446. Higgs RE, Knierman MD, Freeman AB, Gelbert LM, Patil ST and Hale JE, Estimating the Statistical Significance of Peptide Identifications from Shotgun Proteomics Experiments. *Journal of Proteome Research* **6**(5): 1758-1767, 2007.

447. Wiener MC, Sachs JR, Deyanova EG and Yates NA, Differential Mass Spectrometry: A Label-Free LC-MS Method for Finding Significant Differences in Complex Peptide and Protein Mixtures. *Analytical Chemistry* **76**(20): 6085-6096, 2004.

448. Meng F, Wiener MC, Sachs JR, Burns C, Verma P, Paweletz CP, Mazur MT, Deyanova EG, Yates NA and Hendrickson RC, Quantitative Analysis of Complex Peptide Mixtures Using FTMS and Differential Mass Spectrometry. *Journal of the American Society for Mass Spectrometry* **18**(2): 226-233, 2007.

449. Kelleher NL, Top-down proteomics. *Analytical Chemistry* **76**(11): 196A-203A, 2004.

450. Meng F, Forbes AJ, Miller LM and Kelleher NL, Detection and localization of protein modifications by high resolution tandem mass spectrometry. *Mass Spectrometry Reviews* **24**(2): 126-134, 2005.

451. Bogdanov B and Smith RD, Proteomics by FTICR mass spectrometry: Top down and bottom up. *Mass Spectrometry Reviews* **24**(2): 168-200, 2005.

452. Sze SK, Ge Y, Oh H and McLafferty FW, Top-down mass spectrometry of a 29-kDa protein for characterization of any posttranslational modification to within one residue. *Proc Natl Acad Sci U S A* **99**(4): 1774-9., 2002.

453. Stephenson JL, McLuckey SA, Reid GE, Wells JM and Bundy JL, Ion/ion chemistry as a top-down approach for protein analysis. *Curr Opin Biotechnol* **13**(1): 57-64., 2002.

454. Fridriksson EK, Beavil A, Holowka D, Gould HJ, Baird B and McLafferty FW, Heterogeneous Glycosylation of Immunoglobulin E Constructs Characterized by Top-Down High-Resolution 2-D Mass Spectrometry. *Biochemistry* **39**: 3369-3376, 2000.

455. Moehring T, Kellmann M, Juergens M and Schrader M, Top-down identification of endogenous peptides up to 9 kDa in cerebrospinal fluid and brain tissue by nanoelectrospray quadrupole time-of-flight tandem mass spectrometry. *Journal of Mass Spectrometry* **40**(2): 214-226, 2005.

456. McLafferty FW, Horn DM, Breuker K, Ge Y, Lewis MA, Cerda B, Zubarev RA and Carpenter BK, ECD of gaseous multiply charged ions by FT-ICR. *J Am Soc Mass Spectrom* **12**: 245-9., 2001.

457. Shi SD, Hemling ME, Carr SA, Horn DM, Lindh I and McLafferty FW, Phosphopeptide/phosphoprotein mapping by ECD-MS. *Anal Chem* **73**: 19-22., 2001.

458. Sze SK, Ge Y, Oh H and McLafferty FW, Plasma electron capture dissociation for the characterization of large proteins by top down MS. *Anal Chem* **75**: 1599-603., 2003.

459. Kaur P and O'Connor PB, Algorithms for Automatic Interpretation of High Resolution Mass Spectra. *Journal of the American Society for Mass Spectrometry* **17**(3): 459-468, 2006.

460. Taylor GK, Kim Y-B, Forbes AJ, Meng F, McCarthy R and Kelleher NL, Web and database software for identification of intact proteins using \"top down\" MS. *Analytical Chemistry* **75**: 4081-4086, 2003.

461. Du Y, Parks BA, Sohn S, Kwast KE and Kelleher NL, Top-Down Approaches for Measuring Expression Ratios of Intact Yeast Proteins Using Fourier Transform Mass Spectrometry. *Analytical Chemistry* **78**(3): 686-694, 2006.

462. Reid GE and McLuckey SA, 'Top down' protein characterization via tandem mass spectrometry. *J Mass Spectrom* **37**(7): 663-75., 2002.

463. Chrisman PA, Pitteri SJ and McLuckey SA, Parallel Ion Parking of Protein Mixtures. *Anal Chem* **78**: 310 - 316, 2006.

464. He M and McLuckey SA, Two ion/ion charge inversion steps to form a doubly protonated Peptide from a singly protonated Peptide in the gas phase. *J Am Chem Soc* **125**(26): 7756-7., 2003.

465. Stephenson JL, Jr.; McLuckey, S. A., Charge manipulation for improved mass determination of high-mass species and mixture components by ESI. *J. Mass Spectrom.* **33**: 664-672, 1998.

466. Xia Y, Wu J, McLuckey SA, Londry FA and Hager JW, Mutual storage mode ion/ion reactions in a hybrid linear ion trap. *Journal of the American Society for Mass Spectrometry* **16**(1): 71-81, 2005.

467. Xia Y, Liang X and McLuckey SA, Ion Trap versus Low-Energy Beam-Type Collision-Induced Dissociation of Protonated Ubiquitin Ions. *Analytical Chemistry* **78**(4): 1218-1227, 2006.

468. Amunugama R, Hogan JM, Newton KA and McLuckey SA, Whole Protein Dissociation in a Quadrupole Ion Trap: ID of an a Priori Unknown Modified Protein. *Anal Chem* **76**: 720-727, 2004.

469. Liang X, Xia Y and McLuckey SA, Alternately Pulsed ESI/APCI for Ion/Ion Reactions in an Electrodynamic Ion Trap. *Anal Chem* **78**: 3208-3212, 2006.

470. He M, Emory JF and McLuckey SA, Reagent Anions for Charge Inversion of Polypeptide/Protein Cations in the Gas Phase. *Anal Chem* **77:** 3173-3182, 2005.
471. Pitteri SJ, Chrisman PA, Hogan JM and McLuckey SA, Electron Transfer Ion/Ion Reactions in a Three-Dimensional Quadrupole Ion Trap: Reactions of Doubly and Triply Protonated Peptides with SO2.bul. *Analytical Chemistry* **77**(6)**:** 1831-1839, 2005.
472. Syka JEP, Coon JJ, Schroeder MJ, Shabanowitz J and Hunt DF, Peptide and protein sequence analysis by ETD-MS. *Proc Natl Acad Sci USA* **101:** 9528-9533, 2004.
473. Verkerk UH and Kebarle P, Ion-Ion and Ion-Molecule Reactions at the Surface of Proteins Produced by Nanospray. the Number of Acidic Residues and Control of the Number of Ionized Acidic and Basic Residues. *J Am Soc Mass Spectrom* **16:** 1325-1341, 2005.
474. Johnson RW, Ahmed TF, Miesbauer LJ, Edalji R, Smith R, Harlan J, Dorwin S, Walter K and Holzman T, Protein fragmentation via LC-ESI-in source-CAD-QTOF-MS/MS: use of limited sequence information in structural characterization. *Anal. Biochem.* **341:** 22-32, 2005.
475. Yamada N, Suzuki E-i and Hirayama K, Novel dissociation methods for intact protein: Heat-assisted CAD & IR-multiphoton dissociation using ESI-FTICR-MS/MS. *Anal. Biochem.* **348:** 139-147, 2006.
476. Scherperel G, Yan H, Wang Y and Reid GE, 'Top-down' characterization of site-directed mutagenesis products of Staphylococcus aureus dihydroneopterin aldolase by multistage tandem mass spectrometry in a linear quadrupole ion trap. *Analyst (Cambridge, United Kingdom)* **131**(2)**:** 291-302, 2006.
477. Zubarev RA, Reactions of polypeptide ions with electrons in the gas phase. *Mass Spectrom Rev* **22**(1)**:** 57-77., 2003.
478. Ge Y, ElNaggar M, Sze SK, Oh HB, Begley TP, McLafferty FW, Boshoff H and Barry CE, Top down characterization of secreted proteins from Mycobacterium tuberculosis by ECD MS. *J Am Soc Mass Spectrom* **14:** 253-261, 2003.
479. Kjeldsen F, Haselmann KF, Budnik BA, Sorensen ES and Zubarev RA, Complete characterization of posttranslational modification sites in the bovine milk protein PP3 by tandem mass spectrometry with electron capture dissociation as the last stage. *Analytical Chemistry* **75**(10)**:** 2355-2361, 2003.
480. Cooper HJ, Håkansson K and Marshall AG, The role of electron capture dissociation in biomolecular analysis. *Mass Spec Rev* **24:** 201-222, 2005.
481. Davidson W and Frego L, Micro-HPLC/FTMS with ECD of protein enzymatic digests. *Rapid Commun Mass Spectrom* **16:** 993-8., 2002.
482. Haselmann KF, Budnik BA, Olsen JV, Nielsen ML, Reis CA, Clausen H, Johnsen AH and Zubarev RA, Advantages of external accumulation for ECD in FTMS. *Anal Chem* **73:** 2998-3005., 2001.
483. Palmblad M, Tsybin YO, Ramstrom M, Bergquist J and Hakansson P, Liquid chromatography and ECDin FT-ICR-MS. *Rapid Commun Mass Spectrom* **16:** 988-92., 2002.
484. Polfer NC, Haselmann KF, Zubarev RA and Langridge-Smith PR, Electron capture dissociation of polypeptides using a 3 Tesla FT-ICR-MS. *Rapid Commun Mass Spectrom* **16:** 936-43., 2002.
485. McFarland MA, Chalmers MJ, Quinn JP, Hendrickson CL and Marshall AG, Evaluation and Optimization of ECD Efficiency in FT-ICR MS. *J Amer Soc Mass Spectrom* **16:** 1060-1066, 2005.
486. Silivra OA, Kjeldsen F, Ivonin IA and Zubarev RA, Electron capture dissociation of polypeptides in a 3D quadrupole ion trap: Implementation and first results. *J Am Soc Mass Spectrom* **16:** 22-27, 2005.
487. Adams CM, Kjeldsen F, Zubarev RA, Budnik BA and Haselmann KF, ECD distinguishes a single-amino acid and probes the tertiary structure. *J Am Soc Mass Spectrom* **15:** 1087-1098, 2004.
488. Liu H and Hkansson K, Electron Capture Dissociation of Tyrosine O-Sulfated Peptides Complexed with Divalent Metal Cations. *Analytical Chemistry* **78**(21)**:** 7570-7576, 2006.
489. Turecek F, N-C(alpha) Bond Dissociation Energies and Kinetics in Amide and Peptide Radicals. Is the Dissociation a Non-ergodic Process? *J Am Chem Soc* **125**(19)**:** 5954-63., 2003.
490. Turecek F and Syrstad EA, Mechanism and energetics of intramolecular hydrogen transfer in amide and Peptide radicals and cation-radicals. *J Am Chem Soc* **125**(11)**:** 3353-69., 2003.
491. O'Connor PB, Lin C, Cournoyer JJ, Pittman JL, Belyayev M and Budnik BA, Long-Lived Electron Capture Dissociation Product Ions Experience Radical Migration via Hydrogen Abstraction. *Journal of the American Society for Mass Spectrometry* **17**(4)**:** 576-585, 2006.
492. Uggerud E, Electron capture dissociation of the disulfide bond-a quantum chemical model study. *International Journal of Mass Spectrometry* **234**(1-3)**:** 45-50, 2004.
493. Uggerud E, Update on ECD of disulfide bond. *Int J Mass Spectrom* **235:** 279, 2004.
494. Chalkley RJ, Brinkworth CS and Burlingame AL, Side-Chain Fragmentation of Alkylated Cysteine Residues in Electron Capture Dissociation Mass Spectrometry. *Journal of the American Society for Mass Spectrometry* **17**(9)**:** 1271-1274, 2006.
495. Kleinnijenhuis AJ, Duursma MC, Breukink E, Heeren RMA and Heck AJR, Localization of Intramolecular Monosulfide Bridges in Lantibiotics Determined with Electron Capture Induced Dissociation. *Analytical Chemistry* **75**(13)**:** 3219-3225, 2003.
496. Mirgorodskaya OA, Haselmann KF, Kjeldsen F, Zubarev RA and Roepstorff P, Standard-module approach to disulfide-linked polypeptide nanostructures. I. Methodological prerequisites and MS characterization of the test two-loop structure. *Eur J Mass Spectrom (Chichester, Eng)* **9:** 139-48., 2003.

497. Syrstad EA and Turecek F, Toward a general mechanism of electron capture dissociation. *Journal of the American Society for Mass Spectrometry* **16**(2): 208-224, 2005.

498. Patriksson A, Adams C, Kjeldsen F, Raber J, van der Spoel D and Zubarev RA, Prediction of NCa bond cleavage frequencies in electron capture dissociation of Trp-cage dications by force-field molecular dynamics simulations. *International Journal of Mass Spectrometry* **248**(3): 124-135, 2006.

499. Cooper HJ, Hudgins RR and Marshall AG, ECD FT-ICR-MS of cyclodepsipeptides, branched peptides, and .vepsiln.-peptides. *Int J Mass Spectrom* **234**: 23-35, 2004.

500. Mihalca R, Kleinnijenhuis AJ, McDonnell LA, Heck AJR and Heeren RMA, ECD at low temperatures reveals selective dissociations. *J Am Soc Mass Spectrom* **15**: 1869-1873, 2004.

501. Robinson EW, Leib RD and Williams ER, The Role of Conformation on ECD of Ubiquitin. *J Amer Soc Mass Spectrom* **17**: 1469-1479, 2006.

502. Fagerquist CK, Hudgins RR, Emmett MR, Hakansson K and Marshall AG, An antibiotic linked to peptides and proteins is released by ECD FT-ICR-MS. *J Am Soc Mass Spectrom* **14**: 302-310, 2003.

503. Fung YME and Chan TWD, Experimental and Theoretical Investigations of the Loss of Amino Acid Side Chains in ECD of Model Peptides. *J Amer Soc Mass Spectrom* **16**: 1523-1535, 2005.

504. Leymarie N, Berg EA, McComb ME, O'Connor PB, Grogan J, Oppenheim FG and Costello CE, Tandem mass spectrometry for structural characterization of proline-rich proteins: application to salivary PRP-3. *Anal Chem* **74**(16): 4124-32., 2002.

505. Tsybin YO, Hakansson P, Budnik BA, Haselmann KF, Kjeldsen F, Gorshkov M and Zubarev RA, Improved low-energy electron injection systems for high rate ECD in FT-ICR-MS. *Rapid Commun Mass Spectrom* **15**: 1849-54., 2001.

506. Tsybin YO, Ramstroem M, Witt M, Baykut G and Hakansson P, Peptide and protein characterization by high-rate ECD-FT-ICR-MS. *Journal of Mass Spectrometry* **39**: 719-729, 2004.

507. Hvelplund P, Liu B, Nielsen SB and Tomita S, Electron capture induced dissociation of peptide dications. *International Journal of Mass Spectrometry* **225**(1): 83-87, 2003.

508. Hvelplund P, Liu B, Nielsen SB, Panja S, Poully J-C and Stochkel K, Electron capture induced dissociation of peptide ions: Identification of neutral fragments from secondary collisions with cesium vapor. *Int J Mass Spectrom* **263**: 66-70, 2007.

509. Cooper HJ, Hudgins RR, Hakansson K and Marshall AG, Characterization of amino acid side chain losses in electron capture dissociation. *J Am Soc Mass Spectrom* **13**(3): 241-9., 2002.

510. Haselmann KF, Budnik BA and Zubarev RA, ECD of b (2+) peptide fragments reveals the presence of the acylium ion structure. *Rapid Commun Mass Spectrom* **14**: 2242-6., 2000.

511. Koecher T, Engstroem A and Zubarev RA, Fragmentation of Peptides in MALDI In-Source Decay Mediated by Hydrogen Radicals. *Analytical Chemistry* **77**(1): 172-177, 2005.

512. Hakansson K, Emmett MR, Hendrickson CL and Marshall AG, High-sensitivity ECD tandem FTICR mass spectrometry of microelectrosprayed peptides. *Anal Chem* **73**: 3605-10., 2001.

513. Tsybin YO, Witt M, Baykut G, Kjeldsen F and Hakansson P, Combined IR multiphoton dissociation and ECD in FT-ICR-MS. *Rapid Commun Mass Spectrom* **17**: 1759-68., 2003.

514. Cournoyer JJ, Lin C and O'Connor PB, Detecting Deamidation Products in Proteins by Electron Capture Dissociation. *Analytical Chemistry* **78**(4): 1264-1271, 2006.

515. Hakansson K, Cooper HJ, Emmett MR, Costello CE, Marshall AG and Nilsson CL, Electron capture dissociation and infrared multiphoton dissociation MS/MS of an N-glycosylated tryptic peptic to yield complementary sequence information. *Anal Chem* **73**(18): 4530-6., 2001.

516. Horn DM, Breuker K, Frank AJ and McLafferty FW, Kinetic intermediates in the folding of gaseous protein ions characterized by ECD MS. *J Am Chem Soc* **123**: 9792-9., 2001.

517. Breuker K, Oh H, Horn DM, Cerda BA and McLafferty FW, Detailed unfolding and folding of gaseous ubiquitin ions characterized by ECD. *J Am Chem Soc* **124**: 6407-20., 2002.

518. Charlebois JP, Patrie SM and Kelleher NL, Electron Capture Dissociation and 13C,15N Depletion for Deuterium Localization in Intact Proteins after Solution-Phase Exchange. *Analytical Chemistry* **75**(13): 3263-3266, 2003.

519. Demirev PA, Ramirez J and Fenselau C, Tandem mass spectrometry of intact proteins for characterization of biomarkers from Bacillus cereus T spores. *Anal Chem* **73**(23): 5725-31., 2001.

520. Liu H, Hakansson K, Lee JY and Sherman DH, CAD, IR Multiphoton Dissociation, and ECD of the Bacillus anthracis Siderophore Petrobactin and Its Metal Ion Complexes. *J Amer Soc Mass Spectrom* **18**: 842-849, 2007.

521. Niiranen H, Budnik BA, Zubarev RA, Auriola S and Lapinjoki S, HPLC/MS and ECD-MS/MS of osteocalcin. Determination of gamma-carboxyglutamic residues. *J Chromatogr A* **962**: 95-103., 2002.

522. Lee S, Han SY, Lee TG, Chung G, Lee D and Oh HB, Observation of Pronounced b.bul.,y Cleavages in the Electron Capture Dissociation Mass Spectrometry of Polyamidoamine (PAMAM) Dendrimer Ions with Amide Functionalities. *Journal of the American Society for Mass Spectrometry* **17**(4): 536-543, 2006.

523. Kjeldsen F, Haselmann KF, Sorensen ES and Zubarev RA, Distinguishing of Ile/Leu amino acid residues in the PP3 protein by (hot) ECD in FT-ICR-MS. *Anal Chem* **75**: 1267-74., 2003.

524. Thevis M, Loo RRO and Loo JA, MS characterization of transferrins and their fragments derived by reduction of disulfide bonds. *J Am Soc Mass Spectrom* **14**: 635-647, 2003.

525. Yang J, Mo J, Adamson JT and Haakansson K, Characterization of Oligodeoxynucleotides by Electron Detachment Dissociation FT-ICR-MS. *Analytical Chemistry* **77:** 1876-1882, 2005.

526. Yang J and Hakansson K, Fragmentation of Oligoribonucleotides from Gas-Phase Ion-Electron Reactions. *Journal of the American Society for Mass Spectrometry* **17**(10): 1369-1375, 2006.

527. Cerda BA, Horn DM, Breuker K and McLafferty FW, Sequencing of specific copolymer oligomers by electron-capture-dissociation mass spectrometry. *J Am Chem Soc* **124**(31): 9287-91., 2002.

528. Cerda BA, Breuker K, Horn DM and McLafferty FW, Charge/radical site initiation versus coulombic repulsion for cleavage of multiply charged ions. Charge solvation in poly(alkene glycol) ions. *J Am Soc Mass Spectrom* **12**(5): 565-70., 2001.

529. Chan TW and Ip WH, Optimization of experimental parameters for ECDof peptides in a Fourier transform mass spectrometer. *J Am Soc Mass Spectrom* **13:** 1396-406., 2002.

530. Coon JJ, Syka JEP, Shabanowitz J and Hunt DF, Tandem mass spectrometry for peptide and protein sequence analysis. *BioTechniques* **38**(4): 519,521,523, 2005.

531. Coon JJ, Shabanowitz J, Hunt DF and Syka JEP, Electron Transfer Dissociation of Peptide Anions. *Journal of the American Society for Mass Spectrometry* **16**(6): 880-882, 2005.

532. Chi A, Bai DL, Geer LY, Shabanowitz J and Hunt DF, Analysis of intact proteins on a chromatographic time scale by ETD-MS/MS. *Int J Mass Spectrom* **259:** 197-203, 2007.

533. Mikesh LM, Ueberheide B, Chi A, Coon JJ, Syka JEP, Shabanowitz J and Hunt DF, The utility of ETD mass spectrometry in proteomic analysis. *Biochimica et Biophysica Acta, Proteins and Proteomics* **1764**(12): 1811-1822, 2006.

534. Coon JJ, Ueberheide B, Syka JEP, Dryhurst DD, Ausio J, Shabanowitz J and Hunt DF, Protein identification using sequential ion/ion reactions and tandem mass spectrometry. *Proceedings of the National Academy of Sciences of the United States of America* **102**(27): 9463-9468, 2005.

535. O'Connor PB, Cournoyer JJ, Pitteri SJ, Chrisman PA and McLuckey SA, Differentiation of Aspartic and Isoaspartic Acids Using ETD. *Journal of the American Society for Mass Spectrometry* **17**(1): 15-19, 2006.

536. McAlister GC, Phanstiel D, Good DM, Berggren WT and Coon JJ, Implementation of ETD on a Hybrid Linear Ion Trap-Orbitrap MS. *Anal Chem* **79:** 3525-3534, 2007.

537. Gunawardena HP, Emory JF and McLuckey SA, Phosphopeptide Anion Characterization via Sequential Charge Inversion and ETD. *Anal Chem* **78:** 3788-3793, 2006.

538. Liang X and McLuckey SA, Transmission Mode Ion/Ion Proton Transfer Reactions in a Linear Ion Trap. *J Amer Soc Mass Spectrom* **18:** 882-890, 2007.

539. Huang T-Y, Emory JF, O'Hair RAJ and McLuckey SA, Electron-Transfer Reagent Anion Formation via ESI and CAD. *Anal Chem* **78:** 7387-7391, 2006.

540. Swaney DL, McAlister GC, Wirtala M, Schwartz JC, Syka JEP and Coon JJ, Supplemental Activation Method for High-Efficiency ETD of Doubly Protonated Peptide Precursors. *Anal Chem* **79:** 477-485, 2007.

541. Meng F, Cargile BJ, Patrie SM, Johnson JR, McLoughlin SM and Kelleher NL, Processing Complex Mixtures of Intact Proteins for Direct Analysis by MS. *Anal. Chem.* **74:** 2923 -2929, 2002.

542. le Coutre J, Whitelegge JP, Gross A, Turk E, Wright EM, Kaback HR and Faull KF, Proteomics on full-length membrane proteins using mass spectrometry. *Biochemistry* **39**(15): 4237-42., 2000.

543. Distler AM, Kerner J, Peterman SM and Hoppel CL, A targeted proteomic approach for the analysis of rat liver mitochondrial outer membrane proteins with extensive sequence coverage. *Analytical Biochemistry* **356**(1): 18-29, 2006.

544. Loo JA, Studying noncovalent protein complexes by ESI-MS: The tools of proteomics. *Mass Spectrom Rev* **16:** 1-23., 1997.

545. Gabelica V, Galic N, Rosu F, Houssier C and De Pauw E, Influence of response factors on association constants of non-covalent complexes by ESI MS. *J Mass Spectrom* **38:** 491-501, 2003.

546. Krutchinsky AN, Ayed A, Donald LJ, Ens W, Duckworth HW and Standing KG, Studies of noncovalent complexes in an ESI-TOF mass spectrometer. *Methods Mol Biol* **146:** 239-49., 2000.

547. Loo JA, The tools of proteomics. *Adv Protein Chem* **65:** 25-56., 2003.

548. Chen X-L, Qu L-B, Zhang T, Liu H-X, Yu F, Yu Y, Liao X and Zhao Y-F, The Nature of Phosphorylated Chrysin-Protein Interactions Involved in Noncovalent Complex Formation by Electrospray Ionization Mass Spectroscopy. *Analytical Chemistry* **76**(1): 211-217, 2004.

549. Heck A and Van Den Heuvel R, Investigation of intact protein complexes by mass spectrometry. *Mass Spectrom Rev* **23**(5): 368-89, 2004.

550. McCammon MG and Robinson CV, Me, my cell, and I: The role of the collision cell in the tandem mass spectrometry of macromolecules. *BioTechniques* **39**(4): 447, 449, 459, 453, 2005.

551. Gabelica V and De Pauw E, Internal energy and fragmentation of ions produced in ESI. *Mass Spec Rev* **24:** 566-587, 2005.

552. Borch J, Jorgensen TJD and Roepstorff P, MS analysis of protein interactions. *Current Opinion in Chemical Biology* **9:** 509-516, 2005.

553. Sarni-Manchado P and Cheynier V, Study of non-covalent complexation between catechin derivatives and peptides by electrospray ionization mass spectrometry. *J Mass Spectrom* **37**(6): 609-16., 2002.

554. Colgrave ML, Bramwell CJ and Creaser CS, Nanoelectrospray ion mobility spectrometry and ion trap mass spectrometry studies of the non-covalent complexes of amino acids and peptides with polyethers. *International Journal of Mass Spectrometry* **229**(3): 209-216, 2003.

555. Gabelica V, De Pauw E and Rosu F, Interaction between antitumor drugs and a double-stranded oligonucleotide studied by ESI-MS. *J Mass Spectrom* **34**: 1328-37., 1999.

556. Gabelica V, Vreuls C, Filee P, Duval V, Joris B and De Pauw E, Advantages and drawbacks of nanospray for studying noncovalent protein-DNA complexes by mass spectrometry. *Rapid Communications in Mass Spectrometry* **16**(18): 1723-1728, 2002.

557. Terrier P, Tortajada J and Buchmann W, A Study of Noncovalent Complexes Involving Single-Stranded DNA and Polybasic Compounds Using Nanospray MS. *J Amer Soc Mass Spectrom* **18**: 346-358, 2007.

558. van der Kerk-van Hoof A and Heck AJ, Covalent and non-covalent dissociations of gas-phase complexes of avoparcin and bacterial receptor mimicking precursor peptides studied by collisionally activated decomposition mass spectrometry. *J Mass Spectrom* **34**(8): 813-9., 1999.

559. Powell KD, Ghaemmaghami S, Wang MZ, Ma L, Oas TG and Fitzgerald MC, A general mass spectrometry-based assay for the quantitation of protein-ligand binding interactions in solution. *J Am Chem Soc* **124**(35): 10256-7., 2002.

560. Weinglass AB, Whitelegge JP, Hu Y, Verner GE, Faull KF and Kaback HR, Elucidation of substrate binding interactions in a membrane transport protein by MS. *Embo J* **22**: 1467-77., 2003.

561. Zhan D and Fenn JB, Gas phase hydration of electrospray ions from small peptides. *International Journal of Mass Spectrometry* **219**(1): 1-10, 2002.

562. Mackun K and Downard KM, Strategy for identifying protein-protein interactions of gel-separated proteins and complexes by mass spectrometry. *Analytical Biochemistry* **318**(1): 60-70, 2003.

563. McCammon MG and Robinson CV, Structural change in response to ligand binding. *Current Opinion in Chemical Biology* **8**(1): 60-65, 2004.

564. Sanglier S, Ramstrom H, Haiech J, Leize E and Van Dorsselaer A, ESI MS revealS 310 kDa noncovalent hexamer of HPr kinase/phosphatase from B. subtilis. *Int J Mass Spectrom* **219**: 681-696, 2002.

565. Seyfried NT, Atwood JA, III, Yongye A, Almond A, Day AJ, Orlando R and Woods RJ, FT-MS to monitor hyaluronan-protein interactions: use of hydrogen/deuterium amide exchange. *Rapid Commun Mass Spectrom* **21**: 121-131, 2007.

566. Friess SD, Daniel JM, Hartmann R and Zenobi R, Mass spectrometric noncovalent probing of amino acids in peptides and proteins. *International Journal of Mass Spectrometry* **219**(1): 269-281, 2002.

567. Kennaway CK, Benesch JLP, Gohlke U, Wang L, Robinson CV, Orlova EV, Saibi HR and Keep NH, Dodecameric Structure of the Small Heat Shock Protein Acr1 from Mycobacterium tuberculosis. *Journal of Biological Chemistry* **280**(39): 33419-33425, 2005.

568. Panigrahi AK, Allen TE, Stuart K, Haynes PA and Gygi SP, MS of editosome and multiprotein complexes in T. brucei. *J Am Soc Mass Spectrom* **14**: 728-735, 2003.

569. Huang H-H, Liao H-K, Chen Y-J, Hwang T-S, Lin Y-H and Lin C-H, Structural characterization of sialic acid synthase by Electrospray Mass Spectrometry--a tetrameric enzyme composed of dimeric dimers. *J Amer Soc Mass Spectrom* **16**(3): 324-332, 2005.

570. Beene DL, Brandt GS, Zhong W, Zacharias NM, Lester HA and Dougherty DA, Cation-pi interactions in ligand recognition by serotonergic (5-HT3A) and nicotinic acetylcholine receptors: the anomalous binding properties of nicotine. *Biochemistry* **41**(32): 10262-10269, 2002.

571. Jackson SN, Wang H-YJ, Yergey A and Woods AS, Phosphate Stabilization of Intermolecular Interactions. *Journal of Proteome Research* **5**(1): 122-126, 2006.

572. Woods AS, Kaminski R, Oz M, Wang Y, Hauser K, Goody R, Wang H-YJ, Jackson SN, Zeitz P, Zeitz KP, Zolkowska D, Schepers R, Nold M, Danielson J, Graeslund A, Vukojevic V, Bakalkin G, Basbaum A and Shippenberg T, Decoy Peptides that Bind Dynorphin Noncovalently Prevent NMDA Receptor-Mediated Neurotoxicity. *Journal of Proteome Research*: ACS ASAP.

573. Aggerholm T, Nanita SC, Koch KJ and Cooks RG, Clustering of nucleosides in the presence of alkali metals: Biologically relevant quartets of guanosine, deoxyguanosine and uridine observed by ESI-MS/MS. *J Mass Spectrom* **38**(1): 87-97., 2003.

574. Gumerov DR and Kaltashov IA, Dynamics of iron release from transferrin N-lobe studied by electrospray ionization mass spectrometry. *Anal Chem* **73**(11): 2565-70., 2001.

575. Zal F, Chausson F, Leize E, Van Dorsselaer A, Lallier FH and Green BN, QTOFMS of Native Hemocyanin of Deep-Sea Crab B. thermydron. *Biomacromolecules* **3**: 229-231., 2002.

576. Sanglier S, Leize E, Van Dorsselaer A and Zal F, Comparative ESI-MS study of .apprx.2.2 MDa native hemocyanins from deep-sea and shore crabs: from protein oligomeric state to biotope. *Journal of the American Society for Mass Spectrometry* **14**(5): 419-429, 2003.

577. Green BN and Vinogradov SN, ESI MS study of subunit structure of the giant hemoglobin from the leech N. oscura. *J Am Soc Mass Spectrom* **15**: 22-27, 2004.

578. Li Y, Heitz F, Le Grimellec C and Cole RB, Fusion Peptide-Phospholipid Noncovalent Interactions As Observed by Nanoelectrospray FTICR-MS. *Analytical Chemistry* **77**(6): 1556-1565, 2005.

579. Bazoti FN, Tsarbopoulos A, Markides KE and Bergquist J, Study of non-covalent interaction of amyloid-b-peptide and melatonin by ESI MS. *J Mass Spectrom* **40**: 182-192, 2005.

580. Daniel JuM, McCombie G, Wendt S and Zenobi R, MS determination of association constants of adenylate kinase with two noncovalent inhibitors. *J Am Soc Mass Spectrom* **14**: 442-448, 2003.

581. Clark SM and Konermann L, Diffusion measurements by ESI MS for studying solution-phase noncovalent interactions. *J Am Soc Mass Spectrom* **14**: 430-441, 2003.

582. Cavanagh J, Benson LM, Thompson R and Naylor S, In-Line Desalting Mass Spectrometry for the Study of Noncovalent Biological Complexes. *Analytical Chemistry* **75**(14): 3281-3286, 2003.

583. Sobott F and Robinson CV, ESI-MS/MS of biomolecules in the noncovalent GroEL chaperonin assembly. *Int J Mass Spectrom* **236**: 25-32, 2004.

584. Englander SW, Hydrogen Exchange and Mass Spectrometry: A Historical Perspective. *Journal of the American Society for Mass Spectrometry* **17**(11): 1481-1489, 2006.

585. Wells JM, Chrisman PA and McLuckey SA, Formation and characterization of protein-protein complexes in vacuo. *J Am Chem Soc* **125**(24): 7238-49., 2003.

586. Lorenz SA, Maziarz EP, 3rd and Wood TD, Using solution phase hydrogen/deuterium (H/D) exchange to determine the origin of non-covalent complexes observed by electrospray ionization mass spectrometry: in solution or in vacuo? *J Am Soc Mass Spectrom* **12**(7): 795-804., 2001.

587. Xiao H, Kaltashov IA and Eyles SJ, Indirect assessment of small hydrophobic ligand binding to a model protein using a combination of ESI MS and HDX/ESI MS. *J Am Soc Mass Spectrom* **14**: 506-515, 2003.

588. Yan X, Deinzer ML, Schimerlik MI, Broderick D, Leid ME and Dawson MI, Investigation of Ligand Interactions with Human RXRa by Hydrogen/Deuterium Exchange and Mass Spectrometry. *Journal of the American Society for Mass Spectrometry* **17**(11): 1510-1517, 2006.

589. Yu Y, Kirkup CE, Pi N and Leary JA, Noncovalent protein-ligand complexes with enzyme intermediates of GlcNAc-6-O-sulfotransferase by ESI FT-ICR MS. *J Am Soc Mass Spectrom* **15**: 1400-1407, 2004.

590. Lim J and Vachet RW, Method based on metal-catalyzed oxidation reactions and MS to determine the metal binding sites in copper metalloproteins. *Anal Chem* **75**: 1164-72., 2003.

591. Bazoti FN, Bergquist J, Markides KE and Tsarbopoulos A, Noncovalent Interaction Between Amyloid-b-Peptide (1-40) and Oleuropein by ESI MS. *J Amer Soc Mass Spectrom* **17**: 568-575, 2006.

592. Natsume T, Yamauchi Y, Nakayama H, Shinkawa T, Yanagida M, Takahashi N and Isobe T, A direct nanoflow LC-MS/MS system for interaction proteomics. *Anal Chem* **74**: 4725-33., 2002.

593. Zhang S, Van Pelt CK and Wilson DB, Quantitative Determination of Noncovalent Binding Interactions Using Automated Nano-ESI MS. *Anal Chem* **75**: 3010-3018, 2003.

594. Davidson W, Hopkins JL, Jeanfavre DD, Barney KL, Kelly TA and Grygon CA, Characterization of the allosteric inhibition of a protein-protein interaction by MS. *J Am Soc Mass Spectrom* **14**: 8-13., 2003.

595. Wong JWH and Downard KM, COMPLX: A computer algorithm for the detection of protein-ligand and other macromolecular complexes in mass spectra. *J Mass Spectrom* **38**: 573-581, 2003.

596. Tang X, Munske GR, Siems WF and Bruce JE, Mass Spectrometry Identifiable Cross-Linking Strategy for Studying Protein-Protein Interactions. *Analytical Chemistry* **77**(1): 311-318, 2005.

597. Farmer TB and Caprioli RM, Determination of protein-protein interactions by matrix-assisted laser desorption/ionization mass spectrometry. *J Mass Spectrom* **33**(8): 697-704., 1998.

598. Marx MK, Mayer-Posner F, Soulimane T and Buse G, MALDI MS thiol-group determination of isoforms of bovine cytochrome c oxidase, a hydrophobic multisubunit membrane protein. *Anal Biochem* **256**: 192-9., 1998.

599. Villanueva J, Yanes O, Querol E, Serrano L and Aviles FX, ID of Protein Ligands in Complex Biological Samples Using Intensity-Fading MALDI-TOF MS. *Anal Chem* **75**: 3385-3395, 2003.

600. Yanes O, Aviles FX, Roepstorff P and Jorgensen TJD, Exploring the \"Intensity Fading\" Phenomenon in the Study of Noncovalent Interactions by MALDI-TOF MS. *J Amer Soc Mass Spectrom* **18**: 359-367, 2007.

601. Woods AS and Huestis MA, A study of peptide--peptide interaction by matrix-assisted laser desorption/ionization. *J Am Soc Mass Spectrom* **12**(1): 88-96., 2001.

602. Luo S-Z, Li Y-M, Qiang W, Zhao Y-F, Abe H, Nemoto T, Qin X-R and Nakanishi H, Noncovalent interaction of peptide with DNA by MALDI-TOF. *J Am Soc Mass Spectrom* **15**: 28-31, 2004.

603. Salih B and Zenobi R, MALDI Mass Spectrometry of Dye-Peptide and Dye-Protein Complexes. *Anal. Chem.* **70**(8): 1536-1543, 1998.

604. Distler AM, Allison J, Hiser C, Qin L, Hilmi Y and Ferguson-Miller S, MS detection of protein, lipid and heme of cytochrome c oxidase in R. sphaeroides& stabilization of non-covalent complexes. *Eur J Mass Spectrom* **10**: 295-308, 2004.

605. Shields SJ, Oyeyemi O, Lightstone FC and Balhorn R, MS and non-covalent protein-ligand complexes: binding sites and tertiary structure. *J Am Soc Mass Spectrom* **14**: 460-470, 2003.

606. Kiselar JG and Downard KM, Direct identification of protein epitopes by MS without immobilization of antibody and isolation of antibody-peptide complexes. *Anal. Chem.* **17**: 1792-1801, 1999.

607. Morrissey B and Downard KM, A Proteomics Approach to Survey the Antigenicity of the Influenza Virus by MS. *Proteomics* **6**: 2034-2041, 2006.

608. Konermann L and Simmons DA, Protein-folding kinetics and mechanisms studied by pulse-labeling and mass spectrometry. *Mass Spectrom Rev* **22**(1): 1-26., 2003.

609. Robinson CV, Protein folding monitored by mass spectrometry. *Frontiers in Molecular Biology* **32**(Mechanisms of Protein Folding (2nd Edition)): 105-117, 2000.

610. Nettleton EJ and Robinson CV, Probing conformations of amyloidogenic proteins by hydrogen exchange and MS. *Methods in Enzymology* **309:** 633-646, 1999.

611. Chung EW, Nettleton EJ, Morgan CJ, Grob M, Miranker A, Radford SE, Dobson CM and Robinson CV, Hydrogen exchange properties of proteins in native and denatured states monitored by mass spectrometry and NMR. *Protein Science* **6**(6): 1316-1324, 1997.

612. Robinson CV, Gross M, Miranker A, Radford SE, Aplin RT and Dobson CM, Protein structure and folding by mass spectrometry. *Perspectives on Protein Engineering & Complementary Technologies, Collected Papers, International Symposium, 3rd, Oxford, Sept. 13-17, 1994:* 57-59, 1995.

613. Raza AS and Smith DL, Optimization of conditions for protein unfolding by HDX MS. *Eur J Mass Spectrom* **10:** 289-294, 2004.

614. Wales TE and Engen JR, Hydrogen exchange mass spectrometry for the analysis of protein dynamics. *Mass Spec Rev* **25:** 158-170, 2006.

615. Engen JR and Smith DL, Investigating protein structure and dynamics by hydrogen exchange MS. *Anal Chem* **73**(9): 256A-265A., 2001.

616. Kaltashov IA and Eyles SJ, Crossing the phase boundary to study protein dynamics and function: combination of amide hydrogen exchange in solution and ion fragmentation in the gas phase. *J Mass Spectrom* **37**(6): 557-65., 2002.

617. Weis DD, Wales TE, Engen JR, Hotchko M and Ten Eyck LF, Identification and Characterization of EX1 Kinetics in H/D Exchange Mass Spectrometry by Peak Width Analysis. *Journal of the American Society for Mass Spectrometry* **17**(11): 1498-1509, 2006.

618. Evans SE, Lueck N and Marzluff EM, Gas phase hydrogen/deuterium exchange of proteins in an ion trap mass spectrometer. *International Journal of Mass Spectrometry* **222**(1-3): 175-187, 2003.

619. Mao D and Douglas DJ, H/D exchange of gas phase bradykinin ions in a linear quadrupole ion trap. *Journal of the American Society for Mass Spectrometry* **14**(2): 85-94, 2003.

620. Jurchen JC, Cooper RE and Williams ER, Role of acidic residues and Na adductS on the gas-phase H/DX of peptides and peptide dimers. *J Am Soc Mass Spectrom* **14:** 1477-1487, 2003.

621. Ustyuzhanin P, Ustyuzhanin J and Lifshitz C, An ESI-flow tube study of H/D exchange in protonated serine. *International Journal of Mass Spectrometry* **223-224:** 491-498, 2003.

622. Geller O and Lifshitz C, An ESI-flow tube study of H/D exchange in the protonated serine dimer and protonated serine dipeptide. *International Journal of Mass Spectrometry* **227**(1): 77-85, 2003.

623. Simmons DA, Dunn SD and Konermann L, Conformational dynamics of partially denatured myoglobin by time-resolved ESI MS online hydrogen-deuterium exchange. *Biochemistry* **42:** 5896-905., 2003.

624. Kolakowski BM and Konermann L, From small-molecule reactions to protein folding: studying biochemical kinetics by stopped-flow ESI MS. *Anal Biochem* **292:** 107-14., 2001.

625. Konermann L and Douglas DJ, Pre-steady-state kinetics of enzymatic reactions studied by ESI MS with on-line rapid-mixing techniques. *Methods Enzymol* **354:** 50-64., 2002.

626. Chalmers MJ, Busby SA, Pascal BD, He Y, Hendrickson CL, Marshall AG and Griffin PR, Probing Protein Ligand Interactions by Automated Hydrogen/Deuterium Exchange Mass Spectrometry. *Analytical Chemistry* **78**(4): 1005-1014, 2006.

627. Englander SW and Kallenbach NR, Hydrogen exchange and structural dynamics of proteins and nucleic acids. *Q Rev Biophys* **16**(4): 521-655., 1983.

628. Hossain BM and Konermann L, Pulsed H/D Exchange MS/MS for Studying the Relationship between Noncovalent Protein Complexes in Solution and in the Gas Phase after ESI. *Anal Chem* **78:** 1613-1619, 2006.

629. Wu Y, Kaveti S and Engen JR, Extensive Deuterium Back-Exchange in Certain Immobilized Pepsin Columns Used for H/D Exchange Mass Spectrometry. *Analytical Chemistry* **78**(5): 1719-1723, 2006.

630. Wang L and Smith DL, Downsizing improves sensitivity 100-fold for hydrogen exchange-mass spectrometry. *Analytical Biochemistry* **314**(1): 46-53, 2003.

631. Emmett MR, Kazazic S, Marshall AG, Chen W, Shi SDH, Bolanos B and Greig MJ, SFC Reduction of H/D Back Exchange in Solution-Phase by MS. *Anal Chem* **78:** 7058-7060, 2006.

632. Woods VL, Jr. and Hamuro Y, Amide deuterium exchange-MS of protein binding site structure and dynamics: utility in pharmaceutical design. *J Cell Biochem Suppl* **Suppl**(37): 89-98., 2001.

633. Chik JK, Vande Graaf JL and Schriemer DC, Quantitating the Statistical Distribution of Deuterium Incorporation To Extend the Utility of HDX MS Data. *Anal Chem* **78:** 207 - 214, 2006.

634. Hansen RK, Broadhurst RW, Skelton PC and Arkin IT, Hydrogen/deuterium exchange of hydrophobic peptides in model membranes by ESI MS. *J Am Soc Mass Spectrom* **13:** 1376-87., 2002.

635. Simmons DA and Konermann L, Transient Protein Folding Intermediates IN Myoglobin Reconstitution by Time-Resolved ESI MS Isotopic Pulse Labeling. *Biochemistry* **41:** 1906-14., 2002.

636. Englander JJ, Del Mar C, Li W, Englander SW, Kim JS, Stranz DD, Hamuro Y and Woods VL, Jr., Protein structure change studied by hydrogen-deuterium exchange, functional labeling, and mass spectrometry. *Proc Natl Acad Sci U S A* **100**(12): 7057-62., 2003.

637. Hamuro Y, Burns L, Canaves J, Hoffman R, Taylor S and Woods V, Domain organization of D-AKAP2 revealed by enhanced deuterium exchange-MS. *J Mol Biol* **321:** 703-14., 2002.

638. Hamuro Y, Wong L, Shaffer J, Kim JS, Stranz DD, Jennings PA, Woods VL, Jr. and Adams JA, Phosphorylation driven motions in the COOH-terminal Src kinase, CSK, revealed through enhanced hydrogen-deuterium exchange and mass spectrometry (DXMS). *J Mol Biol* **323**(5): 871-81., 2002.

639. Hoofnagle AN, Resing KA, Goldsmith EJ and Ahn NG, Changes in protein conformational mobility upon activation of extracellular regulated protein kinase-2 as detected by hydrogen exchange. *Proc Natl Acad Sci U S A* **98**(3): 956-61., 2001.

640. Resing KA, Hoofnagle AN and Ahn NG, Modeling deuterium exchange behavior of ERK2 using pepsin mapping to probe secondary structure. *J Am Soc Mass Spectrom* **10**(8): 685-702., 1999.

641. Lewis TS, Hunt JB, Aveline LD, Jonscher KR, Louie DF, Yeh JM, Nahreini TS, Resing KA and Ahn NG, Identification of novel MAP kinase pathway signaling targets by functional proteomics and mass spectrometry. *Mol Cell* **6**(6): 1343-54., 2000.

642. Cheng G, Wysocki VH and Cusanovich MA, Local Stability of Rhodobacter capsulatus Cytochrome c2 Probed by Solution Phase Hydrogen/Deuterium Exchange and Mass Spectrometry. *Journal of the American Society for Mass Spectrometry* **17**(11): 1518-1525, 2006.

643. Cai X and Dass C, Structural characterization of methionine and leucine enkephalins by hydrogen/deuterium exchange & ESI MS. *Rapid Commun Mass Spectrom* **19**: 1-8, 2005.

644. Nazabal A, Bonneu M, Saupe SJ and Schmitter J-M, High-resolution HDX of HET-s218-295 prion protein. *J Mass Spectrom* **40**: 580-590, 2005.

645. Takats Z, Nanita SC, Schlosser G, Vekey K and Cooks RG, Atmospheric Pressure Gas-Phase H/D Exchange of Serine Octamers. *Analytical Chemistry* **75**(22): 6147-6154, 2003.

646. Palmblad M, Buijs J and Hakansson P, Hydrogen/deuterium exchange MS of peptides/proteins using calculations of isotopic distributions. *J Am Soc Mass Spectrom* **12**: 1153-62., 2001.

647. Demmers JA, Rijkers DT, Haverkamp J, Killian JA and Heck AJ, Factors affecting gas-phase deuterium scrambling in peptide ions and their implications for protein structure determination. *J Am Chem Soc* **124**(37): 11191-8., 2002.

648. Joergensen TJD, Bache N, Roepstorff P, Gaardsvoll H and Ploug M, CAD in MALDI TOF/TOF induces intramolecular migration of amide hydrogens in protonated peptides. *Molecular and Cellular Proteomics* **4**: 1910-1919, 2005.

649. Jorgensen TJD, Grdsvoll H, Ploug M and Roepstorff P, Intramolecular Migration of Amide Hydrogens in Protonated Peptides upon CAD. *J Amer Chem Soc* **127**: 2785-2793, 2005.

650. Hoerner JK, Xiao H, Dobo A and Kaltashov IA, Is There Hydrogen Scrambling in the Gas Phase? Energetic and Structural Determinants of Proton Mobility within Protein Ions. *Journal of the American Chemical Society* **126**(24): 7709-7717, 2004.

651. Mandell JG, Falick AM and Komives EA, Measurement of amide hydrogen exchange by MALDI-TOF mass spectrometry. *Anal Chem* **70**(19): 3987-95., 1998.

652. Mandell JG, Falick AM and Komives EA, Identification of protein-protein interfaces by decreased amide proton solvent accessibility. *Proc Natl Acad Sci U S A* **95**(25): 14705-10., 1998.

653. Nazabal A, Laguerre M, Schmitter J-M, Vaillier J, Chaignepain St and Velours J, Hydrogen/deuterium exchange on yeast ATPase supramolecular protein complex analyzed at high sensitivity by MALDI mass spectrometry. *Journal of the American Society for Mass Spectrometry* **14**(5): 471-481, 2003.

654. Kipping M and Schierhorn A, Improving hydrogen/deuterium exchange mass spectrometry by reduction of the back-exchange effect. *Journal of Mass Spectrometry* **38**(3): 271-276, 2003.

655. Powell KD and Fitzgerald MC, Accuracy and precision of h/d exchange-MS-based measurement of thermodynamic properties of protein-Peptide complexes. *Biochemistry* **42**: 4962-70., 2003.

656. Villanueva J, Villegas V, Querol E, Aviles FX and Serrano L, Protein secondary structure and stability determined by combining exoproteolysis & MALDI-TOF-MS. *J Mass Spectrom* **37**: 974-84., 2002.

657. Yang HH, Li XC, Amft M and Grotemeyer J, Protein conformational changes determined by matrix-assisted laser desorption mass spectrometry. *Anal Biochem* **258**(1): 118-26., 1998.

658. Wu J, Yang Y and Watson JT, Trapping of intermediates during the refolding of hEGF by cyanylation, and analysis by MS. *Protein Sci.* **7**: 1017-1028, 1998.

659. Yang Y, Wu J and Watson JT, Probing the folding pathways of long R3 insulin-like growth factor-I (LR3IGF-I) and IGF-I via capture and identification of disulfide intermediates by cyanylation methodology and mass spectrometry. *J. Biol. Chem.* **274**(53): 37598-37604, 1999.

660. Watson JT, Yang Y and Wu J, Capture and identification of folding intermediates of cystinyl proteins by cyanylation and mass spectrometry. *J Mol Graph Model* **19**(1): 119-28., 2001.

661. Sogbein OO, Simmons DA and Konermann L, Effects of pH on reaction mechanism of myoglobin unfolding studied by time-resolved ESI MS. *J Am Soc Mass Spectrom* **11**: 312-9., 2000.

662. Last AM, Schulman BA, Robinson CV and Redfield C, Probing Subtle Differences in the Hydrogen Exchange Behavior of Variants of the Human a-Lactalbumin Molten Globule Using Mass Spectrometry. *Journal of Molecular Biology* **311**(4): 909-919, 2001.

663. Xiao H and Kaltashov IA, Transient Structural Disorder as Facilitator of Protein-Ligand Binding: Native HDX Study of Cellular Retinoic Acid Binding Protein I. *J Am Soc Mass Spectrom* **16**: 869-879, 2005.

664. Cunsolo V, Foti S, La Rosa C, Saletti R, Canters GW and Verbeet MP, Monitoring of unfolding of metalloproteins by ESI MS. *J Mass Spectrom* **38**: 502-509, 2003.

665. Grandori R, Origin of the conformation dependence of protein charge-state distributions in electrospray ionization mass spectrometry. *J Mass Spectrom* **38**(1): 11-5., 2003.

666. Tokarski C, Martin E, Rolando C and Cren-Olive C, Identification of Proteins in Renaissance Paintings by Proteomics. *Analytical Chemistry* **78**(5): 1494-1502, 2006.

667. Baldwin MA, Mass spectrometric analysis of prion proteins. *Adv Protein Chem* **57**: 29-54., 2001.

668. Molloy MP, Phadke ND, Maddock JR and Andrews PC, Two-dimensional electrophoresis and peptide mass fingerprinting of bacterial outer membrane proteins. *Electrophoresis* **22**(9): 1686-96., 2001.

669. Ma S, Caprioli RM, Hill KE and Burk RF, Loss of selenium from selenoproteins: conversion of selenocysteine to dehydroalanine in vitro. *J Am Soc Mass Spectrom* **14**: 593-600, 2003.

670. Nanita SC, Sokol E and Cooks RG, Alkali Metal-Cationized Serine Clusters Studied by Sonic Spray Ionization MS/MS. *J Amer Soc Mass Spectrom* **18**: 856-868, 2007.

671. Chan TW, Duan L and Sze TP, Accurate mass measurements for peptide and protein mixtures by using MALDI-FT-MS. *Anal Chem* **74**: 5282-9., 2002.

672. Rubakhin SS, Churchill JD, Greenough WT and Sweedler JV, Profiling Signaling Peptides in Single Mammalian Cells Using Mass Spectrometry. *Analytical Chemistry* **78**(20): 7267-7272, 2006.

673. Whitelegge JP, Zhang H, Aguilera R, Taylor RM and Cramer WA, Full Subunit Coverage LC ESI MS of an Oligomeric Membrane Protein: Cytochrome b(6)f Complex From Spinach and the Cyanobacterium Mastigocladus Laminosus. *Mol Cell Proteomics* **1**: 816-27., 2002.

674. Sharp JS, Becker JM and Hettich RL, Protein surface mapping by chemical oxidation: Structural analysis by mass spectrometry. *Analytical Biochemistry* **313**(2): 216-225, 2003.

675. Badman ER, Hoaglund-Hyzer CS and Clemmer DE, Monitoring structural changes of proteins in ion trap for 10-200 ms: unfolding transitions in cytochrome c. *Anal Chem* **73**: 6000-7., 2001.

676. Sawyer HA, Marini JT, Stone EG, Ruotolo BT, Gillig KJ and Russell DH, Structure of Gas-Phase Bradykinin Fragment 1-5 (RPPGF) Ions: An Ion Mobility Spectrometry and H/D Exchange Ion-Molecule Reaction Chemistry Study. *J Am Soc Mass Spectrom* **16**: 893-905, 2005.

677. Badman ER, Myung S and Clemmer DE, Gas-phase separations of protein and peptide ion fragments generated by CAD in an ion trap. *Anal Chem* **74**: 4889-94., 2002.

678. Jarrold MF, Peptides and proteins in the vapor phase. *Annu Rev Phys Chem* **51**: 179-207., 2000.

679. Koeniger SL and Clemmer DE, Resolution and Structural Transitions of Elongated States of Ubiquitin. *J Amer Soc Mass Spectrom* **18**: 322-331, 2007.

680. Badman ER, Myung S and Clemmer DE, Evidence for Unfolding and Refolding of Gas-Phase Cytochrome c Ions in a Paul Trap. *Journal of the American Society for Mass Spectrometry* **16**(9): 1493-1497, 2005.

681. McLean JA, Ruotolo BT, Gillig KJ and Russell DH, Ion mobility-mass spectrometry: a new paradigm for proteomics. *International Journal of Mass Spectrometry* **240**(3): 301-315, 2005.

682. McLuckey SA, Van Berkel GJ and Glish GL, Tandem mass spectrometry of small, multiply charged oligonucleotides. *J. Am. Soc. Mass Spectrom.* **3**: 60-70, 1992.

683. McLuckey SA and Stephenson JL, Jr., Ion/ion chemistry of high-mass multiply charged ions. *Mass Spectrom Rev* **17**(6): 369-407., 1998.

684. Wu J and McLuckey SA, Gas-phase fragmentation of oligonucleotide ions. *International Journal of Mass Spectrometry* **237**(2-3): 197-241, 2004.

685. Bartlett MG, McCloskey JA, Manalili S and Griffey RH, The effect of backbone charge on the collision-induced dissociation of oligonucleotides. *J Mass Spectrom* **31**(11): 1277-83., 1996.

686. Ni J, Pomerantz C, Rozenski J, Zhang Y and McCloskey JA, Interpretation of oligonucleotide mass spectra for determination of sequence using ESI-MS/MS. *Anal Chem* **68**: 1989-99., 1996.

687. Tost J and Gut IG, DNA analysis by mass spectrometry - past, present and future. *Journal of Mass Spectrometry* **41**(8): 981-995, 2006.

688. Rozenski J and McCloskey JA, Determination of Nearest Neighbors in Nucleic Acids by Mass Spectrometry. *Anal. Chem.* **71**(7): 1454-1459, 1999.

689. Wu KJ, Steding A and Becker CH, MALDI-TOF-MS of oligonucleotides using 3-hydroxypicolinic acid as an UV-sensitive matrix. *Rapid Commun Mass Spectrom* **7**: 142-6., 1993.

690. Huber CG and Buchmeiser MR, On-line cation exchange for suppression of adduct formation in negative-ion electrospray mass spectrometry of nucleic acids. *Anal Chem* **70**(24): 5288-95., 1998.

691. Scalf M, Westphall MS, Krause J, Kaufman SL and Smith LM, Controlling charge states of large ions. *Science* **283**(5399): 194-7., 1999.

692. Ni J, Mathews MA and McCloskey JA, Collision-induced dissociation of polyprotonated oligonucleotides produced by ESI. *Rapid Commun Mass Spectrom* **11**: 535-40., 1997.

693. Flarakos J, Xiong W, Glick J and Vouros P, A Deoxynucleotide Derivatization Methodology for Improving LC-ESI-MS Detection. *Analytical Chemistry* **77**(8): 2373-2380, 2005.

694. Oberacher H, Walcher W and Huber CG, Effect of instrument tuning on the detectability of biopolymers in electrospray ionization mass spectrometry. *J Mass Spectrom* **38**(1): 108-16., 2003.

695. Kirpekar F, Nordhoff E, Larsen LK, Kristiansen K, Roepstorff P and Hillenkamp F, DNA sequence analysis by MALDI mass spectrometry. *Nucleic Acids Res* **26**(11): 2554-9., 1998.

696. Distler AM and Allison J, 5-Methoxysalicylic acid and spermine: a new matrix for MALDI-MS of oligonucleotides. *J Am Soc Mass Spectrom* **12**: 456-62., 2001.

697. Distler AM and Allison J, Additives for the stabilization of double-stranded DNA in UV-MALDI MS. *J Am Soc Mass Spectrom* **13**(9): 1129-37., 2002.

698. Distler AM and Allison J, Stabilization of duplex oligonucleotides for MALDI-MS using the crystallographic condensing agent cobalt(III) hexammine. *Analytical Biochemistry* **319**: 332-334, 2003.

699. Koomen JM and Russell DH, UV/MALDI-MS of 2,5-dihydroxybenzoic acid-induced reductive hydrogenation of oligonucleotides on cytosine residues. *J Mass Spectrom* **35**: 1025-34., 2000.

700. Chou CW, Limbach PA and Cole RB, Fragmentation pathway studies of oligonucleotides in MALDI-MS by charge tagging and H/D exchange. *J Am Soc Mass Spectrom* **13**: 1407-17., 2002.

701. Wu H and Aboleneen H, Improved oligonucleotide sequencing by alkaline phosphatase and exonuclease digestions with mass spectrometry. *Anal Biochem* **290**(2): 347-52., 2001.

702. Hakansson K, Hudgins RR, Marshall AG and O'Hair RA, ECD and infrared multiphoton dissociation of oligodeoxynucleotide dications. *J Am Soc Mass Spectrom* **14**(1): 23-41., 2003.

703. Crain PF and McCloskey JA, Applications of mass spectrometry to the characterization of oligonucleotides and nucleic acids. *Curr. Opin. Biotechnol.* **9**(1): 25-34, 1998.

704. McCloskey JA, Mass Spectrometry of Biological Materials. Second Edition, Revised and Expanded. Edited by B. S. Larsen and C. N. McEwen. *J. Med. Chem.* **41**(26): 5334-5335, 1998.

705. Andersen TE, Kirpekar F and Haselmann KF, RNA Fragmentation in MALDI MS with H/D-Exchange. *J Amer Soc Mass Spectrom* **17**: 1353-1368, 2006.

706. McCloskey JA, personal communication. 2002.

707. Kirpekar F and Krogh TN, RNA fragmentation studied in a MALDI-tandem quadrupole/orthogonal time-of-flight mass spectrometer. *Rapid Commun Mass Spectrom* **15**: 8-14., 2001.

708. Rozenski J and McCloskey JA, SOS: a simple interactive program for ab initio oligonucleotide sequencing by mass spectrometry. *J Am Soc Mass Spectrom* **13**(3): 200-3., 2002.

709. Oberacher H, Wellenzohn B and Huber CG, Comparative sequencing of nucleic acids by liquid chromatography-tandem mass spectrometry. *Anal Chem* **74**(1): 211-8., 2002.

710. Oberacher H, Mayr BM and Huber CG, Automated de novo sequencing of nucleic acids by liquid chromatography-tandem mass spectrometry. *J Am Soc Mass Spectrom* **15**: 32-42, 2004.

711. Mann M and Pandey A, Use of mass spectrometry-derived data to annotate nucleotide and protein sequence databases. *Trends Biochem Sci* **26**(1): 54-61., 2001.

712. Lin S, Long S, Ramirez SM, Cotter RJ and Woods AS, Helix clamp motif of HIV-1 reverse transcriptase by MALDI MS & surface plasmon resonance. *Anal Chem* **72**: 2635-40., 2000.

713. Stomakhin AA, Vasiliskov VA, Timofeev E, Schulga D, Cotter RJ and Mirzabekov AD, DNA sequence analysis by hybridization with oligonucleotide microchips: MALDI mass spectrometry identification of 5mers contiguously stacked to microchip oligonucleotides. *Nucleic Acids Res* **28**(5): 1193-8., 2000.

714. Humeny A, Beck C, Becker C-M and Jeltsch A, Detection and analysis of enzymatic DNA methylation of oligonucleotides by MALDI-TOF-MS. *Anal. Biochem.* **313**: 160-166, 2003.

715. Zhang LK, Rempel D and Gross ML, MALDI-MS in locating abasic sites and determining the rates of enzymatic hydrolysis of oligodeoxynucleotides. *Anal Chem* **73**: 3263-73., 2001.

716. Brown K, Harvey CA, Turteltaub KW and Shields SJ, Structural characterization of carcinogen-modified oligodeoxynucleotide adducts by MALDI-MS. *J Mass Spectrom* **38**: 68-79., 2003.

717. Chen X, Westphall MS and Smith LM, MS of DNA mixtures: Instrumental effects responsible for decreased sensitivity with increasing mass. *Analytical Chemistry* **75**: 5944-5952, 2003.

718. Xu Y, Bruening M and Watson J, Use of polymer-modified MALDI-MS probes to improve analyses of protein. *Anal Chem* **76**(11): 3106-11, 2004.

719. Huber CG and Oberacher H, Analysis of nucleic acids by on-line liquid chromatography-mass spectrometry. *Mass Spectrom Rev* **20**(5): 310-43., 2001.

720. Tuytten R, Lemiere F, Dongen WV, Esmans EL and Slegers H, Short capillary ion-pair HPLC-ESI-MS/MS for the simultaneous analysis of nucleoside mono-, di- and triphosphates. *Rapid Commun Mass Spectrom* **16**: 1205-15., 2002.

721. Guymon R, Pomerantz SC, Crain PF and McCloskey JA, Influence of Phylogeny on Posttranscriptional Modification of rRNA in Thermophilic Prokaryotes: The Complete Modification Map of 16S rRNA of Thermus thermophilus. *Biochemistry* **45**(15): 4888-4899, 2006.

722. Pomerantz SC and McCloskey JA, Detection of the common RNA nucleoside pseudouridine in mixtures of oligonucleotides by mass spectrometry. *Analytical Chemistry* **77**(15): 4687-4697, 2005.

723. Friso S, Choi SW, Dolnikowski GG and Selhub J, A method to assess genomic DNA methylation using HPLC-ESI-MS. *Anal Chem* **74**: 4526-31., 2002.

724. Cao H and Wang Y, CAD of Protonated 2'-Deoxycytidine, 2'-Deoxyuridine, and their Oxidatively Damaged Derivatives. *J Amer Soc Mass Spectrom* **17**: 1335-1341, 2006.

725. Flora JW and Muddiman DC, Complete sequencing of mono-deprotonated peptide nucleic acids by sustained off-resonance irradiation CAD. *J Am Soc Mass Spectrom* **12**: 805-9., 2001.

726. Gupta R, Kapur A, Beck JL and Sheil MM, Positive ion electrospray ionization mass spectrometry of double- stranded DNA/drug complexes. *Rapid Commun Mass Spectrom* **15**(24): 2472-80, 2001.

727. Crestoni ME and Fornarini S, Gas-phase hydrogen/deuterium exchange of adenine nucleotides. *Journal of Mass Spectrometry* **38**(8): 854-861, 2003.

728. Bagag A, Giuliani A and Laprevote O, Atmospheric pressure photoionization mass spectrometry of nucleic bases, ribonucleosides and ribonucleotides. *International Journal of Mass Spectrometry* 264(1): 1-9, 2007.

729. Zhang S, Van Pelt CK, Huang X and Schultz GA, Detection of single nucleotide polymorphisms using ESI MS: validation of assay & quantitative pooling studies. *J Mass Spectrom* 37: 1039-50., 2002.

730. Lau CC, Yue PYK, Chui SH, Chui AKK, Yam WC and Wong RNS, Detection of single nucleotide polymorphisms in hepatitis B virus precore/basal core promoter region by MALDI TOF MS. *Anal Biochem* 366(1): 93-95, 2007.

731. Elso C, Toohey B, Reid GE, Poetter K, Simpson RJ and Foote SJ, Mutation detection using MS separation of tiny oligonucleotide fragments. *Genome Research* 12: 1428-1433, 2002.

732. Ding C, Chiu RW, Lau TK, Leung TN, Chan LC, Chan AY, Charoenkwan P, Ng IS, Law HY, Ma ES, Xu X, Wanapirak C, Sanguansermsri T, Liao C, Ai MA, Chui DH, Cantor CR and Lo YM, MS analysis of single-nucleotide differences in circulating nucleic acids. *Proc Natl Acad Sci USA* 101: 10762-7, 2004.

733. Null AP, Nepomuceno AI and Muddiman DC, Implications of hydrophobicity and free energy of solvation for characterization of nucleic acids by ESI MS. *Anal Chem* 75: 1331-9., 2003.

734. Jiang Y and Hofstadler SA, A highly efficient and automated method of purifying and desalting PCR products for analysis by ESI MS. *Analytical Biochemistry* 316: 50-57, 2003.

735. Reinhold VN, Reinhold BB and Costello CE, Carbohydrate molecular weight profiling, sequence, linkage, and branching data: ES-MS and CID. *Anal Chem* 67(11): 1772-84., 1995.

736. Harvey DJ, Matrix-assisted laser desorption/ionization mass spectrometry of carbohydrates. *Mass Spectrom Rev* 18(6): 349-450., 1999.

737. Harvey DJ, Ionization and collision-induced fragmentation of N-linked and related carbohydrates using divalent cations. *J Am Soc Mass Spectrom* 12(8): 926-37., 2001.

738. Hanneman AJ and Reinhold VN, Oligosaccharide analysis by mass spectrometry. *Encyclopedia of Biological Chemistry* 3: 155-160, 2004.

739. Kuster B, Krogh TN, Mortz E and Harvey DJ, Glycosylation analysis of gel-separated proteins. *Proteomics* 1(2): 350-61., 2001.

740. Harvey DJ, Bateman RH, Bordoli RS and Tyldesley R, Ionisation and fragmentation of complex glycans by MALDI-QTOF-MS. *Rapid Commun Mass Spectrom* 14: 2135-42., 2000.

741. O'Connor PB, Mirgorodskaya E and Costello CE, High pressure MALDI-FtMS for minimization of ganglioside fragmentation. *J Am Soc Mass Spectrom* 13: 402-7., 2002.

742. Reinhold VN and Sheeley DM, Detailed characterization of carbohydrate linkage and sequence in an ion trap mass spectrometer: glycosphingolipids. *Anal Biochem* 259(1): 28-33., 1998.

743. Mendonca S, Cole RB, Zhu J, Cai Y, French AD, Johnson GP and Laine RA, Alkyl derivatives enhance CAD of glycosidic bond in MS of disaccharides. *J Am Soc Mass Spectrom* 14: 63-78, 2003.

744. Domon B and Costello CE, Structure elucidation of glycosphingolipids and gangliosides using high-performance tandem mass spectrometry. *Biochemistry* 27(5): 1534-43., 1988.

745. Costello CE and Vath JE, Tandem mass spectrometry of glycolipids. In: *'Mass Spectrometry' in Methods Enzymol*, Vol. 193 (Ed. McCloskey JA), pp. 738-68. Academic Press, 1990.

746. Costello CE, Juhasz P and Perreault H, New mass spectral approaches to ganglioside structure determinations. *Prog Brain Res* 101: 45-61., 1994.

747. Ma YL, Vedernikova I, Van den Heuvel H and Claeys M, Internal glucose residue loss in protonated O-diglycosyl flavonoids in low-energy CAD. *J Am Soc Mass Spectrom* 11: 136-44., 2000.

748. Harvey DJ, Mattu TS, Wormald MR, Royle L, Dwek RA and Rudd PM, "Internal residue loss"occurs in carbohydrates derivatized at the reducing terminus. *Anal Chem* 74: 734-40., 2002.

749. Harvey DJ, Ionization and fragmentation of N-linked glycans as silver adducts by electrospray mass spectrometry. *Rapid Communications in Mass Spectrometry* 19(4): 484-492, 2005.

750. Harvey DJ, Fragmentation of Negative Ions from Carbohydrates: Part 3. Fragmentation of Hybrid and Complex N-Linked Glycans. *Journal of the American Society for Mass Spectrometry* 16(5): 647-659, 2005.

751. Zaia J, Miller MJC, Seymour JL and Costello CE, The Role of Mobile Protons in Negative Ion CID of Oligosaccharides. *J Amer Soc Mass Spectrom* 18: 952-960, 2007.

752. Harvey DJ, Fragmentation of Negative Ions from Carbohydrates: Part 2. Fragmentation of High-Mannose N-Linked Glycans. *Journal of the American Society for Mass Spectrometry* 16(5): 631-646, 2005.

753. Chen G, Pramanik BN, Bartner PL, Saksena AK and Gross ML, Multiple-stage mass spectrometric analysis of complex oligosaccharide antibiotics (everninomicins) in a quadrupole ion trap. *J Am Soc Mass Spectrom* 13(11): 1313-21., 2002.

754. Tan PV, Taranenko NI, Laiko VV, Yakshin MA, Prasad CR and Doroshenko VM, MS of N-linked oligosaccharides using AP IR laser ionization from solution. *J Mass Spectrom* 39: 913-921, 2004.

755. Delaney J and Vouros P, Liquid chromatography ion trap MS analysis of oligosaccharides using permethylated derivatives. *Rapid Commun. Mass Spectrom.* 15: 325-334, 2001.

756. Mechref Y, Novotny MV and Krishnan C, Structural characterization of oligosaccharides using MALDI-TOF/TOF tandem mass spectrometry. *Analytical Chemistry* 75(18): 4895-4903, 2003.

757. Stephens E, Sugars J, Maslen SL, Williams DH, Packman LC and Ellar DJ, The N-linked oligosaccharides of aminopeptidase N from Manduca sexta. Site localization and identification of novel N-glycan structures. *European Journal of Biochemistry* 271(21): 4241-4258, 2004.

758. Stephens E, Maslen SL, Green LG and Williams DH, Fragmentation Characteristics of Neutral N-Linked Glycans by MALDI-TOF/TOF MS/MS. *Anal Chem* **76**: 2343-2354, 2004.

759. Guillaumie F, Justesen SFL, Mutenda KE, Roepstorff P, Jensen KJ and Thomas ORT, Fractionation, solid-phase immobilization and chemical degradation of long pectin oligogalacturonides. Initial steps towards sequencing of oligosaccharides. *Carbohydrate Research* **341**(1): 118-129, 2005.

760. Ashline D, Singh S, Hanneman A and Reinhold V, Congruent Strategies for Carbohydrate Sequencing. 1. Mining Structural Details by MSn. *Analytical Chemistry* **77**(19): 6250-6262, 2005.

761. Zhang H, Singh S and Reinhold VN, Congruent Strategies for Carbohydrate Sequencing. 2. FragLib: An MSn Spectral Library. *Analytical Chemistry* **77**(19): 6263-6270, 2005.

762. Lapadula AJ, Hatcher PJ, Hanneman AJ, Ashline DJ, Zhang H and Reinhold VN, Congruent Strategies for Carbohydrate Sequencing. 3. OSCAR: An Algorithm for Assigning Oligosaccharide Topology from MSn Data. *Analytical Chemistry* **77**(19): 6271-6279, 2005.

763. Carr SA and Reinhold VN, Structural characterization of glycosphingolipids by direct chemical ionization mass spectrometry. *Biomed Mass Spectrom* **11**(12): 633-42., 1984.

764. Reinhold B, Chan S-Y, Chan S and Reinhold V, Profiling Glycosphingolipid Structural Detail: Periodate Oxidation, ESI-CAD-MS/MS. *Org Mass Spectrom* **29**: 736-746, 1994.

765. Linsley KB, Chan SY, Chan S, Reinhold BB, Lisi PJ and Reinhold VN, Applications of electrospray mass spectrometry to erythropoietin N- and O-linked glycans. *Anal Biochem* **219**(2): 207-17., 1994.

766. Reinhold BB, Chan SY, Reuber TL, Marra A, Walker GC and Reinhold VN, Detailed structural characterization of succinoglycan, the major exopolysaccharide of Rhizobium meliloti Rm1021. *J Bacteriol* **176**(7): 1997-2002., 1994.

767. Sheeley DM and Reinhold VN, Structural characterization of carbohydrate sequence, linkage, and branching in a quadrupole Ion trap mass spectrometer: neutral oligosaccharides and N-linked glycans. *Anal Chem* **70**(14): 3053-9., 1998.

768. Reinhold BB, Hauer CR, Plummer TH and Reinhold VN, Detailed structural analysis of a novel, specific O-linked glycan from F. meningosepticum. *J Biol Chem* **270**: 13197-203., 1995.

769. Leavell MD, Kruppa GH and Leary JA, Determination of phosphate position in hexose monosaccharides using an FTICR mass spectrometer: ion/molecule reactions, labeling studies, and dissociation mechanisms. *International Journal of Mass Spectrometry* **222**(1-3): 135-153, 2003.

770. Takashiba M, Chiba Y and Jigami Y, ID of Phosphorylation Sites in N-Linked Glycans by MALDI TOFMS. *Anal Chem* **78**: 5208-5213, 2006.

771. Harvey DJ and Bousfield GR, Differentiation between sulphated and phosphated carbohydrates in MALDI MS. *Rapid Commun Mass Spectrom* **19**: 287-288, 2005.

772. Zhang Y, Jiang H, Go EP and Desaire H, Distinguishing Phosphorylation and Sulfation in Carbohydrates and Glycoproteins by Ion-Pairing & MS. *J Amer Soc Mass Spectrom* **17**: 1282-1288, 2006.

773. Wolff JJ, Amster IJ, Chi L and Linhardt RJ, Electron Detachment Dissociation of Glycosaminoglycan Tetrasaccharides. *J Amer Soc Mass Spectrom* **18**: 234-244, 2007.

774. Wolff JJ, Chi L, Linhardt RJ and Amster IJ, Distinguishing Glucuronic from Iduronic Acid in Glycosaminoglycan Tetrasaccharides by Using EDD. *Anal Chem* **79**(5): 2015-2022, 2007.

775. Merry AH, Neville DC, Royle L, Matthews B, Harvey DJ, Dwek RA and Rudd PM, Intact 2-aminobenzamide-labeled o-glycans from glycoproteins by hydrazinolysis. *Anal Biochem* **304**: 91-9., 2002.

776. Adamson JT and Hkansson K, ECD of Oligosaccharides Ionized with Alkali, Alkaline Earth, and Transition Metals. *Anal Chem* **79**: 2901-2910, 2007.

777. Jebanathirajah J, Steen H and Roepstorff P, Using optimized collision energies and high resolution, high accuracy fragment ion selection to improve glycopeptide detection by precursor ion scanning. *Journal of the American Society for Mass Spectrometry* **14**(7): 777-784, 2003.

778. Sandra K, Devreese B, Van Beeumen J, Stals I and Claeyssens M, Q-Trap MS, a novel tool in the study of protein glycosylation. *J Am Soc Mass Spectrom* **15**: 413-423, 2004.

779. Bindila L, Steiner K, Schaeffer C, Messner P, Mormann M and Peter-Katalinic J, Sequencing of O-Glycopeptides Derived from an S-Layer Glycoprotein of Geobacillus stearothermophilus NRS 2004/3a Containing up to 51 Monosaccharide Residues at a Single Glycosylation Site by FT-ICR IR Multiphoton Dissociation MS. *Anal Chem* **79**: 3271-3279, 2007.

780. Guittard J, Hronowski XL and Costello CE, MALDI MS of glycosphingolipids on TLC plates and transfer membranes. *Rapid Commun Mass Spectrom* **13**: 1838-49., 1999.

781. Wong AW, Wang H and Lebrilla CB, Selection of anionic dopant for quantifying desialylation reactions with MALDI-FTMS. *Anal Chem* **72**(7): 1419-25., 2000.

782. Jacobs A, Lundqvist J, Stalbrand H, Tjerneld F and Dahlman O, Water-soluble hemicelluloses from spruce and aspen employing SEC/MALDI MS. *Carbohydr Res* **337**: 711-7., 2002.

783. Jacobs A, Larsson PT and Dahlman O, Distribution of uronic acids in xylans from various species of soft- and hardwood as determined by MALDI MS. *Biomacromolecules* **2**: 979-90., 2001.

784. Creaser CS, Reynolds JC and Harvey DJ, Analysis of oligosaccharides by AP MALDI quadrupole ion trap mass spectrometry. *Rapid Commun Mass Spectrom* **16**: 176-84., 2002.

785. Garner B, Harvey DJ, Royle L, Frischmann M, Nigon F, Chapman MJ and Rudd PM, Human apolipoprotein B100 oligosaccharides in LDL of normal and hyperlipidemic plasma: deficiency of alpha-N-acetylneuraminyllactosyl-ceramide. *Glycobiol* **11**: 791-802., 2001.

786. Gennaro LA, Delaney J, Vouros P, Harvey DJ and Domon B, Capillary electrophoresis/electrospray ion trap mass spectrometry for the analysis of negatively charged derivatized and underivatized glycans. *Rapid Commun Mass Spectrom* **16**(3): 192-200., 2002.

787. Wheeler SF and Harvey DJ, Negative ion MS of sialylated carbohydrates: discrimination of N-acetylneuraminic acid linkages by MALDI-TOF and ESI-TOF MS. *Anal Chem* **72**: 5027-39., 2000.

788. Morimoto N, Nakano M, Kinoshita M, Kawabata A, Morita M, Oda Y, Kuroda R and Kakehi K, Specific distribution of sialic acids in animal tissues as examined by LC-ESI-MS after derivatization with 1,2-diamino-4,5-methylenedioxybenzene. *Anal Chem* **73**(22): 5422-8., 2001.

789. Cheng H-L, Pai P-J and Her G-R, Linkage and Branch Determination of N-linked Oligosaccharides Using Sequential Degradation/Closed-Ring Chromophore Labeling/Negative Ion Trap Mass Spectrometry. *J Amer Soc Mass Spectrom* **18**: 248-259, 2007.

790. Qian J, Liu T, Yang L, Daus A, Crowley R and Zhou Q, Structural characterization of N-linked oligosaccharides on monoclonal antibody cetuximab by MALDI Q-oaTOFMS and sequential enzymatic digestion. *Anal Biochem* **364**: 8-18, 2007.

791. Kawasaki N, Itoh S, Ohta M and Hayakawa T, Microanalysis of N-linked oligosaccharides in glycoprotein by capillary LC/MS & LC/MS/MS. *Anal. Biochem.* **316**: 15-22, 2003.

792. Zaia J and Costello CE, Tandem Mass Spectrometry of Sulfated Heparin-Like Glycosaminoglycan Oligosaccharides. *Analytical Chemistry* **75**(10): 2445-2455, 2003.

793. Henriksen J, Roepstorff P and Ringborg LH, Ion-pairing reversed-phased LC/MS of heparin. *Carbohydrate Research* **341**: 382-387, 2006.

794. Irungu J, Dalpathado DS, Go EP, Jiang H, Ha H-V, Bousfield GR and Desaire H, Method for Characterizing Sulfated Glycoproteins in a Glycosylation Site-Specific Fashion, Using Ion Pairing and Tandem Mass Spectrometry. *Analytical Chemistry* **78**(4): 1181-1190, 2006.

795. Pfenninger A, Karas M, Finke B and Stahl B, Structural analysis of underivatized neutral human milk oligosaccharides in the negative ion mode by nano-electrospray MS(n) (part 1: methodology). *J Am Soc Mass Spectrom* **13**(11): 1331-40., 2002.

796. Robinson S, Bergstroem E, Seymour M and Thomas-Oates J, Screening of Underivatized Oligosaccharides Extracted from the Stems of Triticum aestivum Using Porous Graphitized Carbon LC/MS. *Anal Chem* **79**: 2437-2445, 2007.

797. Wheeler SF and Harvey DJ, Extension of the in-gel release method for structural analysis of neutral and sialylated N-linked glycans to the analysis of sulfated glycans: application to the glycans from bovine thyroid-stimulating hormone. *Anal Biochem* **296**(1): 92-100., 2001.

798. Cheng HL and Her GR, Determination of linkages of linear and branched oligosaccharides using closed-ring chromophore labeling and negative ion trap mass spectrometry. *J Am Soc Mass Spectrom* **13**(11): 1322-30., 2002.

799. Wang F, Nakouzi A, Hogue Angeletti R and Casadevall A, Site-specific characterization of the N-linked oligosaccharides of a murine immunoglobulin M by ESI LC/MS. *Anal Biochem* **314**: 266-280, 2003.

800. Sagi D, Peter-Katalinic J, Conradt HS and Nimtz M, Sequencing of tri- and tetraantennary N-glycans contg sialic acid by negative ESI QTOF MS/MS. *J Am Soc Mass Spectrom* **13**: 1138-48., 2002.

801. Chai W, Leteux C, Lawson AM and Stoll MS, On-line overpressure TLC separation and ESI MS of glycolipids. *Anal Chem* **75**: 118-25., 2003.

802. Hunnam V, Harvey DJ, Priestman DA, Bateman RH, Bordoli RS and Tyldesley R, Ionization and fragmentation of neutral and acidic glycosphingolipids with a Q-TOF mass spectrometer fitted with a MALDI ion source. *J Am Soc Mass Spectrom* **12**(11): 1220-5., 2001.

803. von Der Ohe M, Wheeler SF, Wuhrer M, Harvey DJ, Liedtke S, Muhlenhoff M, Gerardy-Schahn R, Geyer H, Dwek RA, Geyer R, Wing DR and Schachner M, Localization and characterization of polysialic acid-containing N-linked glycans from bovine NCAM. *Glycobiology* **12**(1): 47-63., 2002.

804. Chai W, Piskarev VE, Mulloy B, Liu Y, Evans PG, Osborn HMI and Lawson AM, Analysis of Chain and Blood Group Type and Branching Pattern of Sialylated Oligosaccharides by Negative Ion Electrospray Tandem Mass Spectrometry. *Analytical Chemistry* **78**(5): 1581-1592, 2006.

805. Von Seggern CE, Moyer SC and Cotter RJ, Liquid IR APMALDI-ion trap-MS of sialylated carbohydrates. *Analytical Chemistry* **75**: 3212-3218, 2003.

806. Seymour JL, Costello CE and Zaia J, The Influence of Sialylation on Glycan Negative Ion Dissociation and Energetics. *Journal of the American Society for Mass Spectrometry* **17**(9): 1324, 2006.

807. Wheeler SF, Rudd PM, Davis SJ, Dwek RA and Harvey DJ, Comparison of the N-linked glycans from soluble and GPI-anchored CD59 expressed in CHO cells. *Glycobiology* **12**(4): 261-71., 2002.

808. Zaia J, McClellan JE and Costello CE, Tandem mass spectrometric determination of the 4S/6S sulfation sequence in chondroitin sulfate oligosaccharides. *Anal Chem* **73**(24): 6030-9., 2001.

809. Zaia J and Costello CE, Compositional analysis of glycosaminoglycans by electrospray mass spectrometry. *Anal Chem* **73**(2): 233-9., 2001.

810. McClellan JE, Costello CE, O'Connor PB and Zaia J, Influence of charge state on product ion mass spectra and the determination of 4S/6S sulfation sequence of chondroitin sulfate oligosaccharides. *Anal Chem* **74**(15): 3760-71., 2002.

811. Zhang Y, Kariya Y, Conrad AH, Tasheva ES and Conrad GW, Keratan Sulfate Oligosaccharides by ESI-MS/MS. *Analytical Chemistry* **77**: 902-910, 2005.

812. Sweeney MD, Yu Y and Leary JA, Effects of Sulfate Position on Heparin Octasaccharide Binding to CCL2 Examined by Tandem Mass Spectrometry. *J Amer Soc Mass Spectrom* **17**: 1114-1119, 2006.

813. Cipollo JF, Costello CE and Hirschberg CB, The Fine Structure of Caenorhabditis elegans N-Glycans. *J Biol Chem* **277**(51): 49143-57., 2002.

814. Yohe HC, Wallace PK, Berenson CS, Ye S, Reinhold BB and Reinhold VN, Gangliosides of human blood monocytes/macrophages: absence of ganglio structures. *Glycobiol* **11**: 831-41., 2001.

815. Tong Y, Reinhold V, Reinhold B, Brandt B and Stein DC, Structural and immunochemical characterization of the lipooligosaccharides of N. subflava 44. *J Bacteriol* **183**: 942-50, 2001.

816. Duk M, Reinhold BB, Reinhold VN, Kusnierz-Alejska G and Lisowska E, Structure of a neutral glycosphingolipid recognized by human antibodies in polyagglutinable erythrocytes from the rare NOR phenotype. *J Biol Chem* **276**(44): 40574-82., 2001.

817. Metelmann W, Peter-Katalinic J and Muthing J, Gangliosides from human granulocytes: a nano-ESI QTOF ms fucosylation study. *J Am Soc Mass Spectrom* **12**: 964-73., 2001.

818. Marotta E, Lapolla A, Fedele D, Senesi A, Reitano R, Witt M, Seraglia R and Traldi P, Accurate mass in FTMS of advanced glycation end peptides. *J Mass Spectrom* **38**: 196-205, 1 Plate, 2003.

819. Gabryelski W and Froese KL, Rapid and sensitive differentiation of anomers, linkage, and position isomers of disaccharides FAIMS. *J Am Soc Mass Spectrom* **14**: 265-277, 2003.

820. McDonald LA, Barbieri LR, Carter GT, Kruppa G, Feng X, Lotvin JA and Siegel MM, FTMS of Natural Products: Muraymycin Antibiotics Using ESI Multi-CHEF SORI-CID FTMSn, the Top-Down/Bottom-Up Approach. *Anal Chem* **75**: 2730-2739, 2003.

821. March RE, Lewars EG, Stadey CJ, Miao X-S, Zhao X and Metcalfe CD, Flavonoid glycosides by ESI-MS/MS. *Int J Mass Spectrom* **248**: 61-85, 2006.

822. Reddy PN, Ramesh V, Srinivas R, Sharma GVM, Nagendar P and Subash V, Differentiation of some positional and diastereomeric isomers of Boc-carbo-b3 dipeptides containing galactose, xylose and mannose sugars by ESI MS/MS. *Int J Mass Spectrom* **248**: 115-123, 2006.

823. Polfer NC, Valle JJ, Moore DT, Oomens J, Eyler JR and Bendiak B, Differentiation of Isomers by Wavelength-Tunable Infrared Multiple-Photon Dissociation-Mass Spectrometry: Application to Glucose-Containing Disaccharides. *Analytical Chemistry* **78**(3): 670-679, 2006.

824. Xie Y, Tseng K, Lebrilla CB and Hedrick JL, Exoglycosidase digestion in structural elucidation of neutral O-linked oligosaccharides. *J Am Soc Mass Spectrom* **12**: 877-84., 2001.

825. Cocklin RR, Zhang Y, O'Neill KD, Chen NX, Moe SM, Bidasee KR and Wang M, Advanced glycation end products of human b2-microglobulin using MALDI-TOF-MS. *Anal Biochem* **314**: 322-325, 2003.

826. Mattu TS, Royle L, Langridge J, Wormald MR, Van den Steen PE, Van Damme J, Opdenakker G, Harvey DJ, Dwek RA and Rudd PM, O-glycans human neutrophil gelatinase B by LC-MS/MS: implications for domain organization in enzyme. *Biochem* **39**: 15695-704., 2000.

827. Royle L, Mattu TS, Hart E, Langridge JI, Merry AH, Murphy N, Harvey DJ, Dwek RA and Rudd PM, Structural database sequencing of o-glycans in glycoproteins. *Anal Biochem* **304**: 70-90., 2002.

828. Huang Y, Mechref Y and Novotny MV, Nonreductive release of O-linked glycans for MALDI MS and capillary electrophoresis. *Anal Chem* **73**: 6063-9., 2001.

829. Tseng K, Xie Y, Seeley J, Hedrick JL and Lebrilla CB, Neutral and anionic O-linked oligosaccharides in the egg jelly coat of Xenopus laevis by FTMS. *Glycoconj J* **18**: 309-20., 2001.

830. Tseng K, Wang H, Lebrilla CB, Bonnell B and Hedrick J, ID of lectin-binding oligosaccharides by bioaffinity MALDI-TOF-MS. *Anal Chem* **73**: 3556-61., 2001.

831. Costello CE, Beach DH and Singh BN, Acidic glycerol lipids of Trichomonas vaginalis and Tritrichomonas foetus. *Biol Chem* **382**(2): 275-82., 2001.

832. Hsu F-F, Turk J, Owens RM, Rhoades ER and Russell DG, Structural Characterization of Phosphatidyl-myo-Inositol Mannosides from Mycobacterium bovis Bacillus Calmette Guerin by Multiple-Stage Quadrupole Ion-Trap ESI MS II. Monoacyl- and Diacyl-PIMs. *J Amer Soc Mass Spectrom* **18**: 479-492, 2007.

833. Leavell MD, Kruppa GH and Leary JA, Analysis of phosphate position in hexose monosaccharides by ion-molecule reactions & SORI-CID FT-ICR MS. *Anal Chem* **74**: 2608-11., 2002.

834. Kovacik V, Paetoprsty V, Oksman P, Mistrik R and Kovac P, EI MS of monomeric models of O-polysaccharides of Vibrio cholerae O:1, serotypes Ogawa. *J Mass Spectrom* **38**: 924-930, 2003.

INDEX

Page numbers in *italics* refer to figures. Page numbers in **bold** refer to tables.

*Introduction to Mass Spectrometry, 4th Edition: Instrumentation, Applications, and Strategies for Data
Interpretation*; J.T. Watson and O.D. Sparkman, © 2007, John Wiley & Sons, Ltd

Abbreviations

μ	micro
3D QIT	original three-dimensional QIT
ADC	analog-to-digital converter
Ac	acetate
AGC	automatic gain control
AE	appearance energy; formerly appearance potential (AP)
amu	obsolete symbol for unit of mass
AMDIS	Automated Mass spectral Deconvolution and Identification System
AP	appearance potential
APCI	atmospheric pressure chemical ionization
API	atmospheric pressure ionization
AP MALDI	atmospheric pressure MALDI
atm	atmosphere (760 Torr)
APPI	atmospheric pressure photoionization
ASAP	atmospheric-pressure solids analysis probe
B	magnetic sector
BIRD	blackbody infrared radiative dissociation
BFB	bromofluorobenzene
CA	collisional activation
CAD	collisionally activated dissociation
cal	calorie
CE	capillary electrophoresis
CEMA	channel electron multiplier array
CEZ	capillary zone electrophoresis
CHCA	cyano-4-hydroxycinnamiic acid (a MALDI matrix)

CI	chemical ionization
CID	collision-induced dissociation
CREMS	charge-reduction electrospray mass spectrometry
CRIMS	chemical reaction interface mass spectrometry
Da	dalton (unit of mass on the atomic scale)
DAC	digital-to-analog converter
DAPCI	direct atmospheric pressure chemical ionization
DART	direct analysis in real time
DB	database
DCI	desorption (direct) chemical ionization
DE	delayed extraction
DESI	desorption electrospray ionization
DFTPP	decafluorotriphenylphosphine
DI	desorption/ionization
DIOS	desorption ionization from silicon
DIP	direct insertion probe
DMSC	dimethyldichlorosilane
e^-	electron
E	electric field
EA	electron affinity
ECD	electron capture dissociation
ECNI	electron capture negative ionization
EDD	electron detachment dissociation
eV	electron volt
EE	even electron

EI	electron ionization		IMS	ion mobility spectrometry
EIC	extracted ion current		IP	ion potential
ESI	electrospray ionization		IRMPD	infrared multiphoton dissociation
ETcaD	electron transfer/collisionally activated dissociation		ITMS	ion trap mass spectrometer
ETD	electron-transfer dissociation		J	joule
EM	electron multiplier		KE	kinetic energy
ESA	electrostatic analyzer		LID	laser-induced dissociation
FAB	fast atom bombardment		LIFT	NOT an abbreviation (see end of this list)
FAIMS	field-asymmetric waveform ion mobility spectrometry		L	liter
FFR	field-free region		LIT	linear quadrupole ion trap
FI	field ionization		LC-MS	liquid chromatograph-mass spectrometer
FD	field desorption		LC/MS	liquid chromatography/mass spectrometry
FT	Fourier transform			
FTMS	Fourier transform mass spectrometry (usually ICR)		LOD	limit of detection
FWHM	full width half maximum		LOQ	limit of quantitation
GC-MS	gas chromatograph-mass spectrometer		$M^{+\bullet}$ or $M^{-\bullet}$	molecular ion
GC/MS	gas chromatography/mass spectrometry		MH^+	protonated molecule
			$[M - H]^-$	deprotonated molecule
H	magnetic field		MALDI	matrix-assisted laser desorption/ionization
H/D	hydrogen/deuterium		MCAT	mass-coded abundance tagging
HDX	hydrogen/deuterium exchange		MELDI	material-enhanced laser desorption/ionization
HPA	3-hydroxypicoloc acid (a MALDI matrix)		MS/MS	tandem mass spectrometry
HT	Hadamard transform		MS^n	multiple stages of MS/MS
Hz	hertz (cycles sec^{-1})		MIMS	membrane introduction mass spectrometry
IC	ion current		mmu	millimass unit 0.001 u a millidalton
ICAT	isotope-coded affinity tag			
ICP	inductively coupled plasma		MOWSE	molecular weight search
ICR	ion cyclotron resonance		MPD	multiphoton dissociation
IMAC	immobilized metal affinity chromatography		MPI	multiphoton ionization
			m/z	mass-to-charge ratio

MRM	multiple reaction monitoring (synonym for SRM)		R	resolving power
			RA	relative abundance
NBS	National Bureau of Standards		RC	reconstructed
NCI	negative-ion CI; *not* the same as ECNI		rdbe	rings-plus-double-bond equivalents
NIST	National Institute of Standards and Technology (formerly NBS)		*re*TOF	reflectron TOF
			*re*ISD	in-source decay in reflectron mode
OE	odd electron			
*oa*TOF	orthogonal injection TOF		REMPI	resonance-enhanced multiphoton ionization
Pa	pascal, SI unit of pressure		RF	radio frequency
PA	proton affinity		RGA	residual gas analyzer
PAD	post-acceleration detector		RI	relative intensity
PBM	probability-based matching		RIC	reconstructed ion current
PCI	positive-ion CI		ROOMS	regular old ordinary mass spectrometry
PEG	polyethyleneglycol			
PFK	perfluorokerosene		RTIC	reconstructed total ion current
PFTBA	perfluorotributylamine		SA	sinapinic acid (a MALDI matrix)
PMF	peptide mass figure printing		S/B	signal-to-background ratio
PSD	post-source decay		SELDI	surface-enhanced laser desorption/ionization
PPG	polypropyleneglycol			
PTM	post-translational modification		sec	second
PTR	proton-transfer reaction		SFC	supercritical fluid chromatography
PyMS	pyrolysis mass spectrometry			
q	collision cell in a *tandem-in-space* instrument		SFE	supercritical fluid extraction
			SID	surface-induced dissociation
Q	a quadrupole *m/z* analyzer in a *tandem-in-space* instrument		SIFT	selected ion flow tube
			SIM	selected ion monitoring
QET	quasi-equilibrium theory		S/N	signal-to-noise ratio
QIT	quadrupole ion trap		SPE	solid-phase extraction
QMF	quadrupole mass filter		SPME	solid-phase microextraction
QMS	transmission quadrupole mass spectrometer		SRM	selected reaction monitoring
Q-TOF	*tandem-in-space* instrument with a Q followed by a TOF		STIRS	self-training interpretive and retrieval system
r	radius of electric or magnetic sector		SWIFT	stored-waveform inverse Fourier transform

T	tesla	TSD	time-slice detection
TAD	time-array detection	TSP	thermospray
TG	Thermogravimetric	u	unified atomic mass unit. SI symbol = 1 / 12 mass of ^{12}C
THAP	2,4,6-trihydroxyacetophenone (a MALDI matrix)		
		vMALDI	vacuum MALDI
TIC	total ion current	V	volt
TMS	trimethylsilyl	VOC	volatile organic compound
TOF	time-of-flight	z	number of charges on an ion
Torr	pressure unit equaling 133 Pa		
TQMS	triple-quadrupole mass spectrometer		

LIFT™ is not an abbreviation; it is a proprietary trademarked term of Bruker Daltonics used with their TOF-TOF mass spectrometer to describe the process of elevating the potential of the collision cell above that of the ion source. Unfortunately, this neologism gets incorporated into journal-article titles with no explanation. In articles where the term MALDI LIFT TOF/TOF mass spectrometry is used, the abbreviations TOF and MALDI are defined, but there is no explanation for LIFT. The impression is that LIFT must be an abbreviation like MALDI or TOF but is so common that it does not require a definition. This poor journalism confuses the reader. Journal editors need to be more circumspect when it comes to such proprietary terms and require authors to provide an unambiguous explanation of such terminology.

Symbols for LC/MS

Pictograms–Scan Mode

MS Scan Modes

+ ESI	——— ———	Positive ESI, using a Q or a QqQ instrument; <u>regular</u> <u>scan</u> <u>mode</u>.
+ APCI	——— 762	Positive APCI, using a Q or a QqQ instrument; <u>SIM</u> of *m/z* 762.
EI	——— ———	Particle beam or moving belt with an EI source, using a Q or a QqQ instrument; <u>regular</u> <u>scan</u> <u>mode</u>.
+ ESI	⌐⌐	Positive ESI, using a Q or a QqQ instrument; <u>SIM</u> of more than one *m/z* value.
+ FAB	⬠ ≋	Positive CF-FAB, using a BE sector instrument; <u>regular</u> <u>scan</u> <u>mode</u>.
⌐⌐ 28 + CI	⬠ ≋	Particle beam or moving belt with a positive CI source, using a BE sector instrument, electric sector (E) <u>SIM</u>, and magnetic sector (B) set at *m/z* 281 at full acceleration voltage.
+ ESI ✳ 40	——— ———	Positive ESI, using a Q or a QqQ instrument; <u>skimmerCID</u> (sCID or in-source CAD) at 40 V offset and <u>regular</u> <u>scan</u> <u>mode</u>.
–/+ –/+ ESI	⌐⌐	Dual-polarity ESI-programmed sCID, using a Q or a QqQ instrument; fragment ion detection in <u>SIM</u> mode in low-mass region; positive ESI for molecular ion detection in high-mass region, using <u>regular</u> <u>scan</u> <u>function</u>. Polarity switch occurs when sCID offset is switched off.
–/+/+ –/+ ESI	——— ———	Dual-polarity ESI-programmed sCID, using a Q or a QqQ instrument; negative and positive polarity for fragment ion detection in <u>regular</u> <u>scan</u> <u>mode</u>; positive polarity for molecular ion detection in high-mass region, using <u>regular</u> <u>scan</u> <u>mode</u>. Polarity switch occurs when sCID offset is switched off.

Adapted with permission from Lehmann WD "Pictograms for Experimental Parameters in Mass Spectrometry" J. Am. Soc. Mass Spectrom. 1997, 8, 756–759.

Pictograms–MS/MS Modes

MS Scan Modes

+ ESI ___ 760 ✳ Ar 25 ___ Positive ESI, using a QqQ instrument; <u>product-ion</u> <u>scan</u> of *m/z* 760 (MS^2). Argon (Ar) as collision gas at 25 V offset.

+ ESI ___ ✳ N$_2$ 25 ___ 184 Positive ESI, using a QqQ instrument; <u>precursor-ion</u> <u>scan</u> of *m/z* 184. Nitrogen (N$_2$) as collision gas at 25 V offset.

+ APCI ___ 98 ✳ Ar 20 ___ Positive APCI, using a QqQ instrument; <u>neutral-loss</u> <u>scan</u> for loss of mass 98. Argon (Ar) as collision gas at 20 V offset.

+ APCI ___ 898 ✳ Ar 20 ___ 800 Positive APCI, using a QqQ instrument; <u>selected</u> <u>reaction</u> <u>monitoring</u> for the fragmentation *m/z* 898 to 800 (SRM). Argon (Ar) as collision gas at 20 V offset.

+ ESI ✳ 50 ___ ✳ Ar 30 ___ 211 Positive ESI, using a QqQ instrument; <u>sCID</u> (50 V offset) plus <u>precursor-ion</u> <u>scan</u> for *m/z* 211. Argon (Ar) as collision gas at 30 V offset.

+ ESI ___ 465 ✳ Ar 30 ⎣V⎦ Negative ESI, using a QqQ-TOF (orthogonal extraction) instrument; <u>product-ion</u> <u>scan</u> for *m/z* 465 (MS^2). Argon (Ar) as collision gas at 25 V offset.

+ ESI ✳ 1000 902 ● — ● — ○ Positive ESI, using an IT instrument; <u>product-ion</u> <u>scan</u> of *m/z* 902, formed by CID of *m/z* 1000 (MS^3).

Adapted with permission from Lehmann WD "Pictograms for Experimental Parameters in Mass Spectrometry" J. Am. Soc. Mass Spectrom. 1997, 8, 756–759.

Peak Patterns Representing Ions with Atoms of Cl/Br

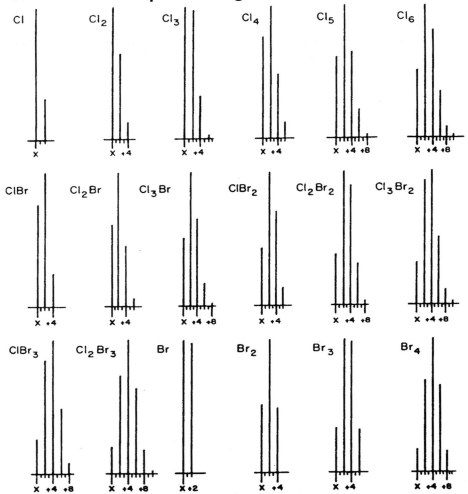

Graphical representation of relative isotope peak intensities for any given ion containing the indicated number of chlorine and/or bromine atoms. Numeric values are on the following page.

Steps to Determine Elemental Composition Based on Isotope Peak Ratios

1. Determine the nominal *m/z* value peak (peak at lowest *m/z* value, above which other peaks can be attributed to isotopic multiplicity or background).
2. Assign the X+2 elements, except oxygen.
3. Assign the X+1 elements (remember to normalize X+1 to X, if necessary).
4. Balance the mass.
5. Assign the atoms of oxygen.
6. Balance the mass.
7. Assign the X elements.
8. From the elemental composition, determine the number of rings plus double bonds.
9. Propose a possible structure.
10. Does it make sense?

Chlorine and Bromine Isotopic Abundance Ratios

Atoms of ClBr	X	X+2	X+4	X+6	X+8	X+10
Cl	100	32.5				
Cl_2	100	65.0	10.6			
Cl_3	100	97.5	31.7	3.4		
Cl_4	76.9	100	48.7	0.5	0.9	
Cl_5	61.5	100	65.0	21.1	3.4	0.2
Cl_6	51.2	100	81.2	35.2	8.5	1.1
ClBr	76.6	100	24.4			
Cl_2Br	61.4	100	45.6	6.6		
Cl_3Br	51.2	100	65.0	17.6	1.7	
$ClBr_2$	43.8	100	69.9	13.7		
Cl_2Br_2	38.3	100	89.7	31.9	3.9	
Cl_3Br_2	31.3	92.0	100	49.9	11.6	1.0
$ClBr_3$	26.1	85.1	100	48.9	8.0	
Cl_2Br_3	20.4	73.3	100	63.8	18.7	2.0
Br	100	98.0				
Br_2	51.0	100	49.0			
Br_3	34.0	100	98.0	32.0		
Br_4	17.4	68.0	100	65.3	16.0	

Common Neutral Losses

M – 1	loss of hydrogen radical	M – $^\bullet$H
M – 15	loss of methyl radical	M – $^\bullet CH_3$
M – 29	loss of ethyl radical	M – $^\bullet CH_2CH_3$
M – 31	loss of methoxyl radical	M – $^\bullet OCH_3$
M – 43	loss of propyl	M – $^\bullet CH_2CH_2CH_3$
M – 45	loss of ethoxyl	M – $^\bullet OCH_2CH_3$
M – 57	loss of butyl radical	M – $^\bullet CH_2CH_2CH_2CH_3$
M – 2	loss of hydrogen	M – H_2
M – 18	loss of water	M – H_2O
M – 28	loss of CO or ethylene	M – CO or M – C_2H_4
M – 32	loss of methanol	M – CH_3OH
M – 44	loss of CO_2	M – CO_2
M – 60	loss of acetic acid	M – CH_3CO_2H
M – 90	loss of silanol: $HO–Si(CH_3)_3$	M – $HO–Si–(CH_3)_3$

Amino Acids with Apolar Side Chains

Name And Res Comp	Abb	Res Nom	Residue Monoiso	Residue Ave.	Immo Mass	SC Mass	Structure
Glycine C_2H_3NO	gly G	57	57.02146	57.0520	30	–	
Alanine C_3H_5NO	ala A	71	71.03711	71.0788	44	15	
Valine C_5H_9NO	val V	99	99.06841	99.1326	72	43	
Leucine $C_6H_{11}NO$	leu L	113	113.08406	113.1595	86	57	
Isoleucine $C_6H_{11}NO$	ile I	113	113.08406	111.1595	86	57	
Proline C_5H_7NO	pro P	97	97.05276	97.1167	70	–	
Phenyl-alanine C_9H_9NO	phe F	147	147.06841	147.1766	120	91	
Tryptophan $C_{11}H_{10}N_2O$	trp W	186	186.07931	186.2133	159	130	
Methionine C_5H_9NOS	met M	131	131.04049	131.1986	104	75	

Amino Acids with Uncharged Polar Side Chains

Name And Res Comp	Abb	Res Nom	Residue Monoiso	Residue Ave.	Immo Mass	SC Mass	Structure
Serine $C_3H_5NO_2$	ser S	87	87.03203	87.0782	60	31	
Threonine $C_4H_7NO_2$	thr T	101	101.04768	101.1051	74	45	
Cysteine C_3H_5NOS	cyc C	103	103.00919	103.1448	76	47	
Tyrosine $C_9H_9NO_2$	tyr Y	163	163.06333	163.1760	136	107	
Asparagine $C_4H_6N_2O_2$	asn N	114	114.04293	114.1039	87	58	
Glutamine $C_5H_8N_2O_2$	gln Q	128	128.05856	128.1308	101	72	

Amino Acids with Charged Polar Side Chains

Name And Res Comp	Abb	Res Nom	Residue Monoiso	Residue Ave.	Immo Mass	SC Mass	Structure
Aspartic acid $C_4H_5NO_3$	asp D	115	115.02694	115.0886	88	59	
Glutamic acid $C_5H_7NO_3$	glu E	129	129.04259	129.1155	102	73	
Lysine $C_6H_{12}N_2O$	lys K	128	128.09496	128.1742	101	72	
Arginine $C_6H_{12}N_4O$	arg R	156	156.10111	156.1876	129	100	
Histidine $C_6H_7N_3O$	his H	137	137.05891	137.1412	110	81	

Proof of a Molecular Ion Peak – $M^{+\bullet}$

1. If a compound is known, the molecular ion has a mass-to-charge ratio (*m/z*) value equal to the sum of the atomic masses of the most abundant isotope of each element that comprises the molecule (assuming the ion is a single-charge ion).

2. The nominal mass of a compound, or the *m/z* value for the molecular ion, is an even number for any compound containing only C, H, O, S, Si, P, and the halogens.

    Fragment ions, derived via homolytic, heterolytic, or sigma-bond cleavage from these molecular ions (even *m/z*) have an odd *m/z* value and an even number of electrons.

    Fragment ions derived from these molecular ions (even *m/z*) via expulsion of neutral components (e.g., H_2O, CO, ethylene, etc.) have an even *m/z* value and an odd number of electrons.

3. **Nitrogen Rule**: A compound containing an odd number of nitrogen atoms—in addition to C, H, O, S, Si, P, and the halogens—has an odd nominal mass.

    Molecular ions of these compounds fragment via homolytic, heterolytic, or sigma-bond cleavage to produce ions of an even *m/z* value unless the nitrogen atom is lost with the neutral radical.

    An even number of nitrogen atoms in a compound results in an even nominal mass.

4. The molecular ion peak must be the highest *m/z* value of any significant (nonisotope or nonbackground) peak in the spectrum. Corollary: The highest *m/z* value peak observed in the mass spectrum need not represent a molecular ion.

5. The peak at the next lowest *m/z* value in the mass spectrum must not correspond to the loss of an impossible or improbable combination of atoms.

6. No fragment ion may contain a larger number of atoms of any particular element than the molecular ion.

Courses on the interpretation of mass spectra and techniques of mass spectrometry are offered by LC Resources (http://www.LCResources.com).

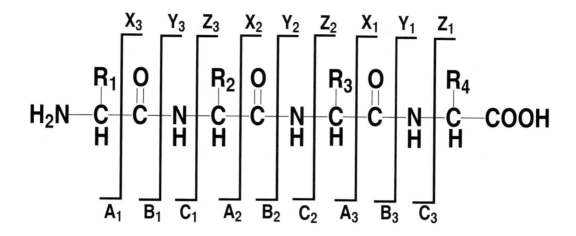

The general fragmentation of a generic-protonated peptide according to the nomenclature of Peter Roepstorff

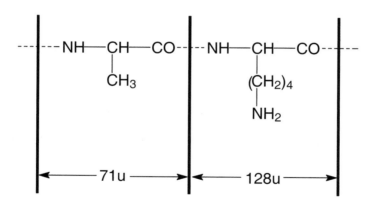

Amino acid residue masses

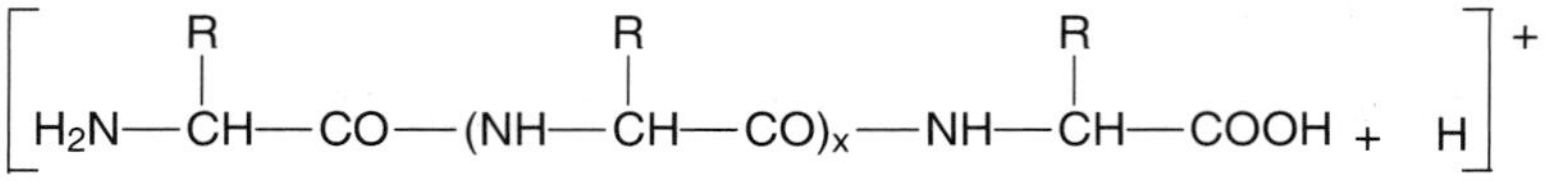

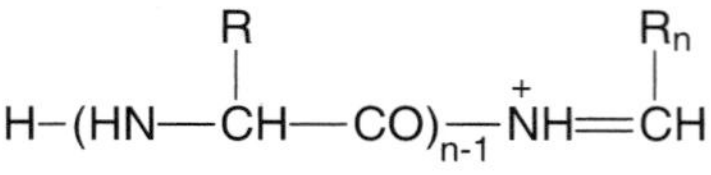

$$\left[H_2N-\underset{\underset{R}{|}}{CH}-CO-(NH-\underset{\underset{R}{|}}{CH}-CO)_x-NH-\underset{\underset{R}{|}}{CH}-COOH \ + \ H \right]^+$$

$$H-(HN-\underset{\underset{R}{|}}{CH}-CO)_{n-1}\overset{+}{N}H=\underset{\underset{R_n}{|}}{CH}$$

a_n

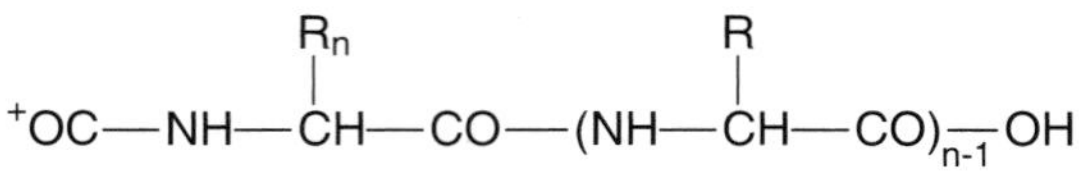

$$\overset{+}{O}C-NH-\underset{\underset{R_n}{|}}{CH}-CO-(NH-\underset{\underset{R}{|}}{CH}-CO)_{n-1}-OH$$

x_n

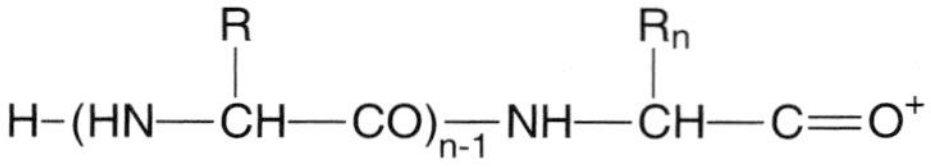

$$H-(HN-\underset{\underset{R}{|}}{CH}-CO)_{n-1}-NH-\underset{\underset{R_n}{|}}{CH}-C=O^+$$

b_n

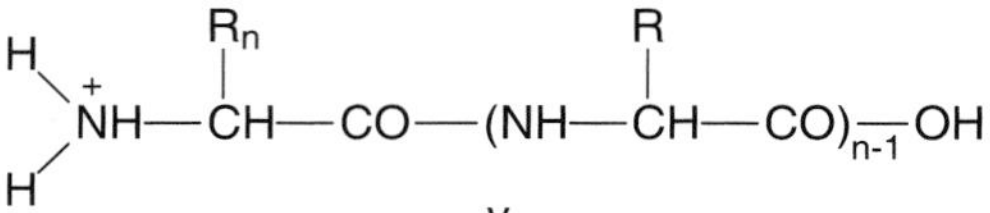

$$\underset{\underset{H}{|}}{\overset{\underset{H}{|}}{N}}\overset{+}{H}-\underset{\underset{R_n}{|}}{CH}-CO-(NH-\underset{\underset{R}{|}}{CH}-CO)_{n-1}-OH$$

y_n

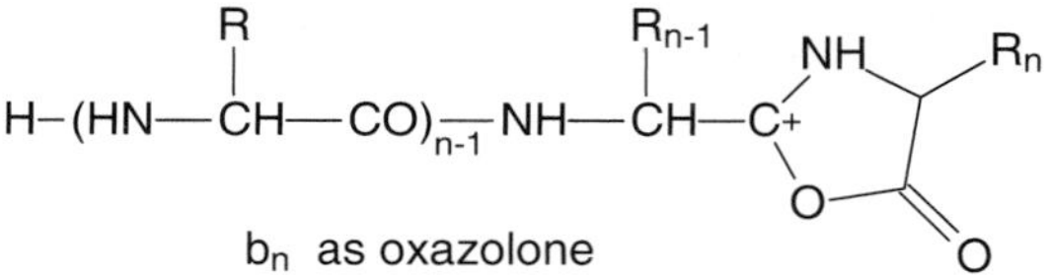

$$H-(HN-\underset{\underset{R}{|}}{CH}-CO)_{n-1}-NH-\underset{\underset{R_{n-1}}{|}}{CH}-\underset{\underset{O}{|}}{\overset{+}{C}}\cdots$$

b_n as oxazolone

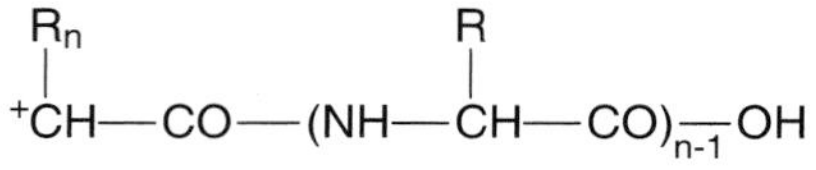

$$\overset{+}{C}H-CO-(NH-\underset{\underset{R}{|}}{CH}-CO)_{n-1}-OH \quad \scriptstyle R_n$$

z_n

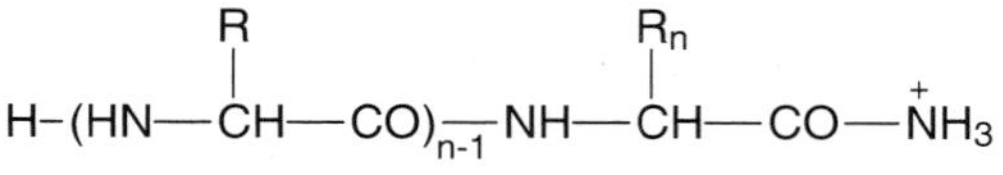

$$H-(HN-\underset{\underset{R}{|}}{CH}-CO)_{n-1}-NH-\underset{\underset{R_n}{|}}{CH}-CO-\overset{+}{N}H_3$$

c_n

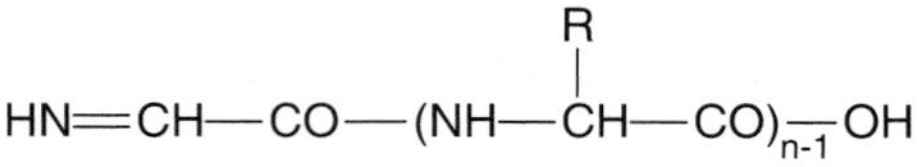

$$HN=CH-CO-(NH-\underset{\underset{R}{|}}{CH}-CO)_{n-1}-OH$$

v_n

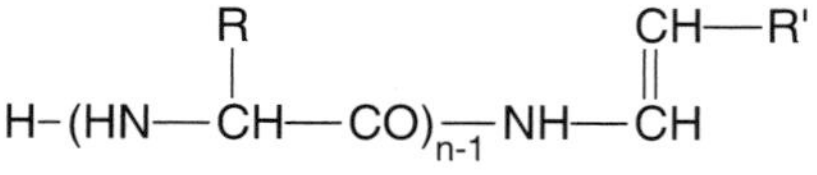

$$H-(HN-\underset{\underset{R}{|}}{CH}-CO)_{n-1}-NH-\underset{\overset{\|}{CH-R'}}{CH}$$

d_n

$$\underset{\overset{\|}{CH-CO}}{R'-CH}-(NH-\underset{\underset{R}{|}}{CH}-CO)_{n-1}-OH$$

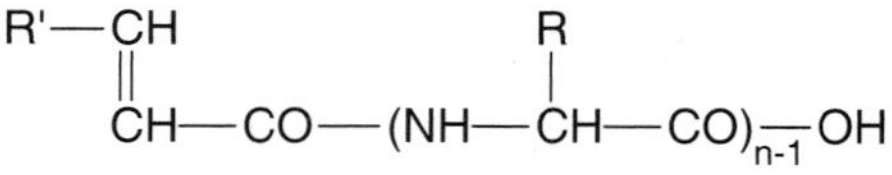

w_n

***Specific structures for the a, b, c and x, y, z ions per the
Klaus Biemann nomenclature***

Peptide Fragmentation

To recognize the start of a **y** series:

    (a) It will be a high-mass ion.

    (b) It will correspond to the loss of a residue mass from MH^+.

    (c) Note that the **y** ion accumulates two hydrogens:
        (1) one is from the original protonation;
        (2) the other is abstracted from the N-terminal portion as it is expelled.

To recognize the start of a **b** series:

    (a) It will be a high-mass ion.

    (b) It will correspond to the loss of a residue plus H_2O from MH^+.

b_n as oxazolone

Practical Considerations for LC/MS

Common Mobile-Phase Clusters in Positive-Ion Mode

Acetonitrile		Methanol		Water	
42	$(CH_3CN)H^+$	33	$(CH_3OH)H^+$	19	$(H_2O)H^+$
83	$(CH_3CN)_2H^+$	65	$(CH_3OH)_2H^+$	37	$(H_2O)_2H^+$
124	$(CH_3CN)_3H^+$	97	$(CH_3OH)_3H^+$	55	$(H_2O)_3H^+$
165	$(CH_3CN)_4H^+$	129	$(CH_3OH)_4H^+$	73	$(H_2O)_4H^+$
206	$(CH_3CN)_5H^+$	161	$(CH_3OH)_5H^+$	91	$(H_2O)_5H^+$
247	$(CH_3CN)_6H^+$			199	$(H_2O)_{11}H^+$
288	$(CH_3CN)_7H^+$			379	$(H_2O)_{21}H^+$
				505	$(H_2O)_{28}H^+$

Common Adducts

Positive-Ion Mode

M + 23 (Na^+)
M + 32 (MeOH)
M + 39 (K^+)
M + 41 (CH_3CN)
m/z 159 (TFA + Na^+)
m/z 242 (tetrabutyl ammonium)
m/z 391 (DOP)

Negative-Ion Mode

M + 45 (Formate)
M + 59 (Acetate)
M + 58 (NaCl salt)
M + 78 (DMSO)
M + 113 (TFA)

Common Artifacts in Negative-Ion Mode

26   CN^{1-} from acetonitrile
35   Cl^{1-} from inorganic or organic chlorides
59   Acetate^{1-}
79   Phosphite PO_3^- (several sources: phosphoric acid, oligonucleotides)
80   Sulfite SO_3^{-2}
96   SO_4^{2-} adduct (proteins and peptides)
97   HSO_4^- and $H_2PO_4^-$
113   TFA^{1-}

Common Artifacts and Adducts in Positive-Ion Mode

28   Series of peaks *m/z* 300 to 600 separated by 28 *m/z* units – triglycerides from fingerprints or contamination

41   Acetonitrile adduct

44   Series of peaks separated by 44 *m/z* units – ethylene oxide

50   Series of peaks separated by 50 *m/z* units – CF_2 fluorinated surfactants

58   Series of peaks separated by 58 *m/z* – propylene oxide, or NaCl adduct if 58/60

61.5   Series of peaks separated by 61.5 *m/z* – copper adducts

63       $H+H_2CO_3$

64       $ACN+Na^+$

71/72 THF

74   DMF + H or diethyl amine + H^+

79       $DMSO + H^+$

83 2 $ACN + H^+$

88   Formic acid + acetonitrile + H^+

101, 138, 183 $MeOH + H_2O$ clusters

102   Triethylamine + H^+

102   ACN + HOAc

105   2× acetonitrile + Na^+

146   3× acetonitrile + Na^+

149   Fragment from dioctyl phthalate

158   Amino sugar; common from antibiotic analyses

159     $NaTFA + Na^+$

163   Nicotine

169, 165, 195   (possibly 133 and 135) – dimethyl phthalate

181   BHA (butylated hydroxyanisole – food additive, preservative)

186   Tributylamine + H^+ (ion-pair reagents)

211, 227, 241, Detergents from glassware
253, 269, 281

219   Tri-*tert*-butylphosphine oxide from peptide synthesis

221   BHT (butylated hydroxytoluene – food additive, preservative)

241, 253, 255, Fatty acids and/or soap
269, 281

279   Dibutyl phthalate

281/282 Oleic acid soap or oleamide (Na^+, K^+ or NH_4^+ from mold release agents used in plastic production)

317, 361, 405   Triton detergent

362   Dioctyl diphenyl phthalate

388, 437, 444, Lubricants from HPLC components
463

391   Dioctyl phthalate + H^+ (common from contaminated solvents and from plastic tubing)

413   Dioctyl phthalate + Na^+

427   Dioctyl sebacate

447   Diisodecyl phthalate + H^+

481, 525, 569   PEG as sodium adducts

503, 547, 591   Protonated PEG

563   Oleic acid soap

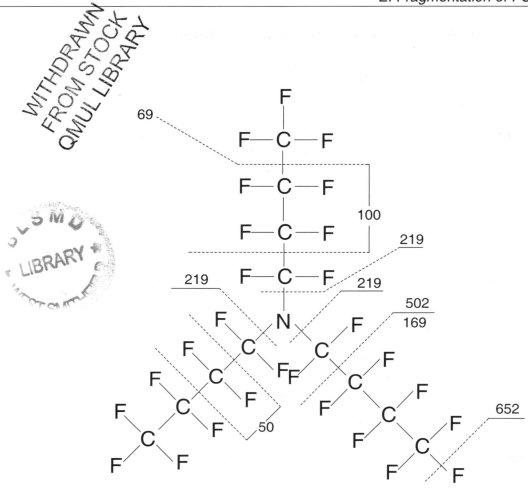

The molecular mass of perfluorotributylamine (PFTBA, a.k.a. FC-43), used to calibrate the *m/z* scale of transmission quadrupole and quadrupole ion trap mass spectrometers operated in the electron ionization mode, is 671. The following is an explanation of the origin of some of the peaks observed in its EI mass spectrum:

$$
\begin{array}{l}
671 \\
\underline{-207} \quad (3 \times 69) \\
464 \\
\\
\underline{-50} \\
414
\end{array}
\qquad
\begin{array}{l}
671 \\
\underline{-507} \quad (3 \times 169) \\
164 \\
\\
\underline{+100} \\
264
\end{array}
\qquad
\begin{array}{l}
671 \\
\underline{-57} \quad (3 \times 19) \\
614 \\
\\
169 \\
\underline{-38} \quad (2 \times 19) \\
131
\end{array}
$$